W9-AOZ-139

FUSION TECHNOLOGY 1988

FUSION TECHNOLOGY 1988

Proceedings of the 15th Symposium on Fusion Technology,
Utrecht, The Netherlands, 19–23 September 1988

VOLUME 2

edited by:

A.M. VAN INGEN
FOM–EUR, Nieuwegein

A. NIJSEN–VIS
FOM–EUR, Nieuwegein

H.T. KLIPPEL
ECN, Petten

1989

NORTH-HOLLAND
AMSTERDAM • OXFORD • NEW YORK • TOKYO

ISBN: 0 444 87369 4

Publishers:

ELSEVIER SCIENCE PUBLISHERS B.V.
P.O. Box 1991
1000 BZ Amsterdam
The Netherlands

Sole distributors for the U.S.A. and Canada:

ELSEVIER SCIENCE PUBLISHING COMPANY, INC.
655 Avenue of the Americas
New York, N.Y. 10010
U.S.A.

Publication No. EUR 12184 EN of the
Commission of the European Communities,
CEC – DG XII-Fusion Programme,
Brussels, Belgium

PRINTED IN THE NETHERLANDS

TABLE OF CONTENTS

VOLUME 1

CONTRIBUTED PAPERS

Section I: Experimental Systems

Section III: Fuel Cycle Engineering

Section IV: First-Wall Engineering and Vacuum Technology

VOLUME 2

Section V: Materials for Fusion Devices

Section VI: Blanket Technology

Section VIII: Remote Operations and Maintenance

Section XII: Reactor Studies

FUSION TECHNOLOGY 1988
A.M. Van Ingen, A. Nijsen-Vis, H.T. Klippel (editors)

LIFETIME PREDICTIONS FOR CERAMIC WINDOWS IN FUSION REACTORS

Theo FETT, Dietrich MUNZ

Association KfK-EURATOM,
Kernforschungszentrum Karlsruhe GmbH, Postfach 3640, 7500 Karlsruhe 1,
W-Germany, Institut für Material- und Festkörperforschung IV

1. INTRODUCTION

The electron cyclotron resonance heating is considered as an external plasma heat source for fusion reactor systems. The window which separates the reactor vacuum from the electron tube vacuum, is heated up due to dielectric losses. The temperature gradients in the window cause thermal stresses and, especially in case of ceramic windows, the tensile stresses are of high importance in lifetime predictions. These stresses are the basis of lifetime evaluations for different candidate materials. The power distribution in the waveguide depends on the mode of the electrical field. Ferber et al [1] considered the transverse electric mode TE_{02}.

The KfK gyrotron with a cylindrical waveguide which works in the TE_{03}-mode. Its power distribution (fig.1) is given by

$$p = p_0 J_1^2(\lambda_{14}\frac{\rho}{R}) = p_0 \bullet f(\rho/R) \qquad (1)$$

where ρ is the radial coordinate and J_1 denotes the first order Bessel function and λ_{14} is its fourth zero (λ_{14}=10.17). The total electric power crossing the waveguide with radius R then becomes

$$W = 2\pi \int_0^R p_0 J_1^2(\lambda_{14}\frac{\rho}{R})\, \rho\, d\rho = \pi\, p_0 R^2 J_2^2(\lambda_{14}) \qquad (2)$$

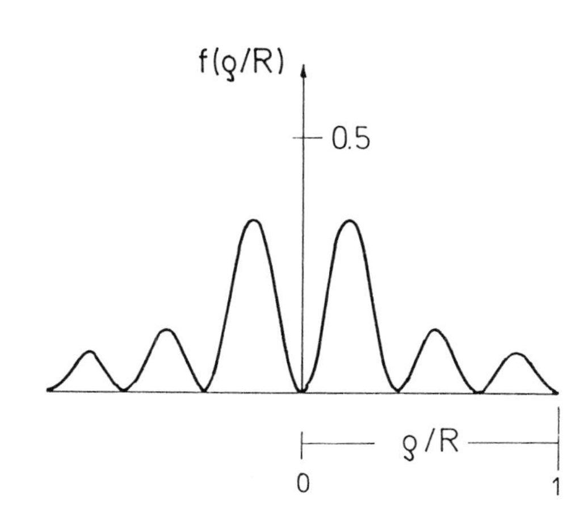

FIGURE 1
Radial heat source distribution

Due to the dielectric losses in the ceramic material, heat is generated in the window. The volumetric heat generation rate is given by

$$\dot{q} = \dot{q}_0 \bullet f(\rho/R) \quad , \ \dot{q}_0 = 2\pi\, \nu \sqrt{\varepsilon}\, \tan\delta\; p_0/c \qquad (3)$$

where ν is the frequency, ε the dielectric constant, $\tan\delta$ the loss tangent and c the electromagnetic wave velocity. ε and $\tan\delta$ are dependent on the frequency. The dielectric losses per unit of volume are distributed with respect to ρ in the same way as the dielectric power described by eq.(1). Neglecting interference effects, one can assume the heat sources to be constant with respect to the z-axis.

2. TEMPERATURES IN THE HF-WINDOW

The stationary thermal problem is completely described by the heat conduction equation

$$\frac{\partial^2 T}{\partial \rho^2} + \frac{1}{\rho}\frac{\partial T}{\partial \rho} + \frac{\partial^2 T}{\partial z^2} = -\frac{1}{\lambda}\dot{q}(\rho) \tag{4}$$

and by the thermal boundary conditions. These conditions as well as the geometric data are illustrated in fig.2 , i.e.,

- liquid cooling on one of the plane faces with the heat transfer coefficient h_1;
- no heat exchange on the vacuum side;
- cooling by air convection at the convex surface with the heat transfer coefficient h_2

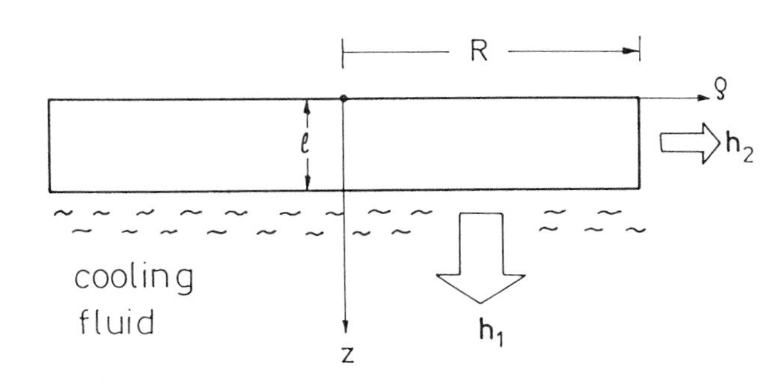

FIGURE 2
Geometrical quantities and cooling conditions of the HF-window

2.1 Solution of the differential equation (4)

A particular solution of the inhomogeneous differential equation (4) is found to be

$$\begin{aligned}\frac{2\lambda}{\dot{q}_0}T_i &= \frac{R^2}{2}[J_0^2(\beta R)-J_0(\beta R)J_1(\beta R)] \\ &- \frac{\rho^2}{2}[2J_1^2(\beta\rho)+J_0^2(\beta\rho)-J_0(\beta\rho)J_2(\beta\rho)] \\ &+ \frac{1}{\beta^2}[J_0^2(\beta R)-J_0^2(\beta\rho)] + \frac{R\lambda}{h_2}[J_1^2(\beta R)-J_0(\beta R)J_2(\beta R)]\end{aligned} \tag{5}$$

where $\beta = \lambda_{14}/R$

The homogeneous part of eq.(4) can be solved by separation of the variables which yields

$$T = \sum_{\nu=0}^{\infty} B_\nu\left[1-h_1\frac{\cosh(\alpha z)}{\lambda\alpha_\nu \sinh(\alpha_\nu \ell)+h_1\cosh(\alpha_\nu \ell)}\right]J_0(\alpha_\nu \rho) \tag{6}$$

where the values α_ν are the roots of

$$\lambda\alpha_\nu J_1(\alpha_\nu R) - h_2 J_0(\alpha_\nu R) = 0 \tag{7}$$

and the coefficients B_ν are

$$B_\nu = \frac{2}{R^2}[J_0^2(\alpha_\nu R)+J_1^2(\alpha_\nu R)]^{-1}\int_0^R T_i(\rho)J_0(\alpha_\nu \rho)\rho\, d\rho \tag{8}$$

3. ESTIMATION OF THERMAL STRESSES

Due to the very strongly inhomogeneous temperature distributions thermal stresses are generated. Especially radial and tangential stresses are of interest because a thin disc is characterized by $\sigma_z=0$. In a first rough estimation, the temperature changes in axial direction shall be neglected. In this approximation, the radial and tangential stresses σ_ρ and σ_ϕ are given by

$$\sigma_\rho = \alpha E \sum_{\nu=0}^{\infty} C_\nu\left[\frac{1}{\alpha_\nu R}J_1(\alpha_\nu R) - \frac{1}{\alpha_\nu \rho}J_1(\alpha_\nu \rho)\right] \tag{9}$$

$$\sigma_\phi = \alpha E \sum_{\nu=0}^{\infty} C_\nu\left[-J_0(\alpha_\nu \rho) + \frac{1}{\alpha_\nu R}J_1(\alpha_\nu R) + \frac{1}{\alpha_\nu \rho}J_1(\alpha_\nu \rho)\right] \tag{10}$$

with

$$C_\nu = B_\nu\left[1 - \frac{h_1}{\alpha_\nu \ell}\frac{\sinh(\alpha_\nu \ell)}{\lambda\alpha_\nu \sinh(\alpha_\nu \ell)+h_1\cosh(\alpha_\nu \ell)}\right] \tag{11}$$

where α is the coefficient of thermal expansion, and E denotes the Young's modulus. Due to the deviations between the local temperatures T and the mean temperatures, additional stress components result. Their values can be calculated using two special limit cases of the mechanical boundary conditions, the plate with totally prevented bending and the free bending plate, both treated by Munz and Fett[2,3]. Figure 3 represents the temperature distribution after the stationary state has been reached.

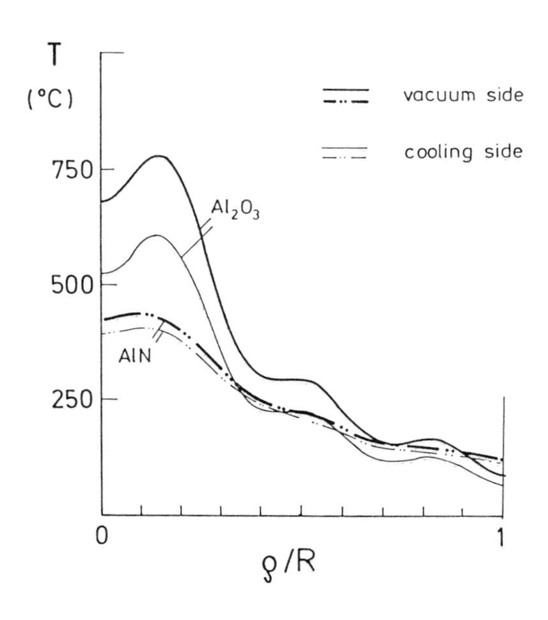

FIGURE 3
Temperature distribution in the HF-window

The calculations are performed for the geometrical data given in table 1 and the material data represented in table 2. The related stresses are plotted in fig. 4.

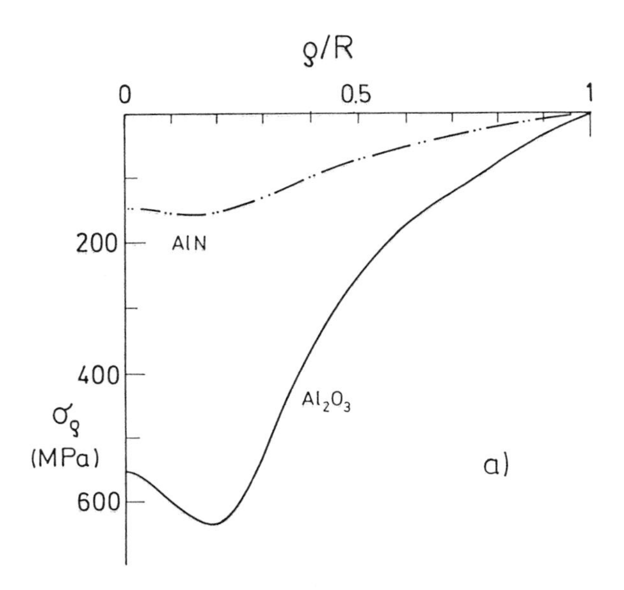

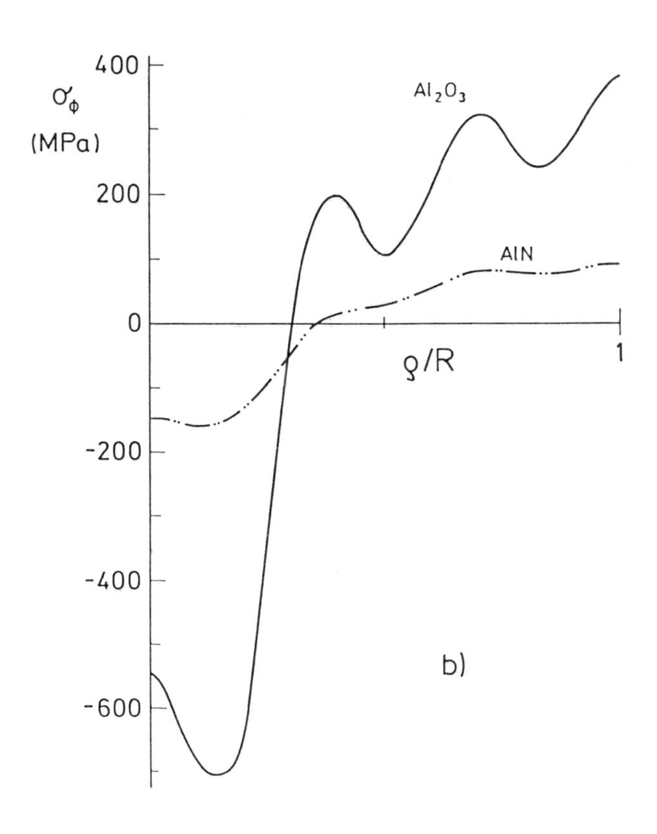

FIGURE 4
Thermal stresses (averaged over the disc thickness) a) tangential stresses b) radial stresses

The radial stress component is a compressive stress acting throughout the disc. The tangential stress is also compression in the center but it reaches high tensile stresses in the outer disc parts.

4. SPONTANEOUS FAILURE

A first possibility of failure is given when the thermal stresses reach the tensile strength. The strength, usually measured in a bending test, can be described by a Weibull distribution with the cumulative density function given by

$$F = 1 - \exp[-(\sigma_c/\sigma_0)^m] \qquad (12)$$

where m and σ_0 are the parameters of the Weibull distribution. The parameter σ_0 is dependent on the size and stress distribution of the entire component. Equation (12) has to be modified to take into account the size effect. For surface flaws eq.(12) has to be replaced by

$$F = 1 - \exp[-\frac{S_{eff}}{S_0}(\sigma^*/\sigma_s)^m] \tag{13}$$

where S_0 is the unit area and σ_s is a size independent parameter. S_{eff} is an effective area given by

$$S_{eff} = \int (\sigma/\sigma^*)^m \, dS \tag{14}$$

where σ^* is a reference stress in the component - e.g. the maximum stress. The failure stress of a component for a given failure probability F is obtained from eq.(13) as

$$\sigma^* = \sigma_0[S_{eff,sp}/S_{eff,c}]^{1/m}[\ln \frac{1}{1-F}]^{1/m} \tag{15}$$

where σ_0 is the Weibull-parameter measured in the bending test and $S_{eff,sp}$ and $S_{eff,c}$ are the effective areas of the specimens and the component, respectively. This equation provides a first possibility of deciding whether finite lifetimes can be expected or failure by spontaneous fracture may occur.

5. LIFETIME ESTIMATION

Even if the strength is not exceeded the stresses in the ceramic window may cause failure by subcritical extension of small flaws or cracks after some time of operation. The crack growth rate is a function of the stress intensity factor K_I which usually can be described by a power law:

$$v = \frac{da}{dt} = A\,K_I^n = A^*(K_I/K_{Ic})^n \tag{16}$$

A and n are parameters dependent on the material and the environment. The stress intensity factor K_I is related to the stress σ and the crack size a by

$$K_I = \sigma\sqrt{a}\,Y \tag{17}$$

and $Y \approx 1.28$ for the small cracks in ceramic materials. The lifetime t_f of a component loaded by the stress σ is

$$t_f = B\sigma_c^{n-2}\,\sigma^n \tag{18}$$

with

$$B = \frac{2}{A^*(n-2)Y^2}\,K_{Ic}^2$$

where K_{Ic} is the fracture toughness. The scatter in the lifetime can be related to the scatter in strength. For a given failure probability the time to failure is

$$t_f = [\ln \frac{1}{1-F}]^{(n-2)/m}\,B\sigma_0^{n-2}\sigma^{-n} \cdot \tag{19}$$

$$\cdot [S_{eff,sp}/S_{eff,c}]^{(n-2)/m}$$

The lifetime includes the frequency of the microwaves and the electric power in the form $t_f \sim (p_0\nu)^{-n}$. It is important that the physical parameters and the microwave data enter the lifetime relation with the power of the crack growth exponent n .

6. EXPERIMENTAL RESULTS

For determination of v-K curves the authors developed a procedure based on lifetime measurements which allows to measure crack growth rates less than 10^{-12} m/s [4,5]. The evaluating equation requires the knowledge of the inert strength σ_c measured in bending tests at high loading rates as well as the lifetimes t_f in a static bending test to be known.

The Weibull parameters of the inert strength were found to be

Al_2O_3 : σ_0= 420 MPa m= 10.1

AlN : σ_0= 310 MPa m= 15.3

With these m-values the effective surfaces are

$S_{eff,c}(Al_2O_3) = 6.7 cm^2$

$S_{eff,c}(AlN) = 5.5\ cm^2$

First measurements of subcritical crack growth were performed on Al_2O_3 and AℓN at 60^oC in fluor-carbon FC43 as the environment. The resulting v-K-curves are represented in fig. 5.

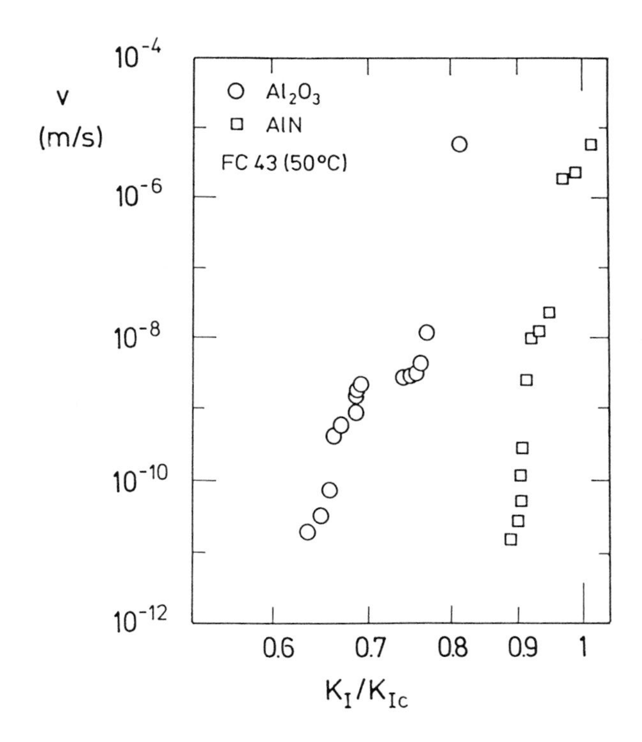

FIGURE 5
v-K-curves for Al_2O_3 and AlN

From these curves the subcritical crack growth parameters were determined

Al_2O_3: n = 33 $A^* = 1.5 \cdot 10^{-4}$ m/s

AlN : n = 135 $A^* = 1.5 \cdot 10^{-4}$ m/s

Inserting the data for the two materials into eq.(13) gives the probability of spontaneous failure

F= 0.975 for Al_2O_3 and F= $5 \cdot 10^{-8}$ for AlN

The lifetime for AlN is approximatively infinite. For a survival probability of 99.9% the lifetime is >10^{36} s.

8. SUMMARY AND CONCLUSION

The temperature and the stress distribution in a gyrotron window have been calculated. Due to the radial distribution of the electric power, temperature gradients develop in the radial direction which lead to tensile tangential stresses. Two candidate materials, Al_2O_3 and AlN, were considerd.

For a microwave energy of 200 kW:

- The Al_2O_3 window will fail by spontaneous extension of cracks during the first cycle.
- AlN will not fail by spontaneous extension of flaws; due to the high n-value there is only low probability for delayed failure by subcritical crack growth.
- There may be problems for AlN resulting from cyclic fatigue.
- The temperatures for both materials are too high for the coolant medium FC-43

ACKNOWLEDGEMENTS

A part of this work has been performed in the framework of the Nuclear Fusion Project of the Kernforschungszentrum Karlsruhe and is supported by the European Communities within the European Technology Program.

REFERENCES

1. M.K. Ferber, H.D. Kimrey, P.F. Becher, Journ. Mat. Sci. 19(1984),3767-3777.
2. D. Munz , ICFRM-3, to be published in conference volume
3. T. Fett, D. Munz, in preparation.
4. T. Fett, D. Munz, Comm. Amer. Ceram. Soc. 68(1985),C213-215.
5. T. Fett, K. Keller, D. Munz, Int. J. Fract. 36(1988),3-14.
6. R. Heidinger, W. Dienst, personal communication

TABLE 1 Geometric and power data

R	3.5 cm
ℓ	0.3 cm
W	200 kW
ν	150 GHz
h_1	5kW/m²/K
h_2	0.5kW/m²/K

TABLE 2 Physical properties[6]

	Al_2O_3	AlN
ε_r	10.0	8.27
λ	0.2 W/cm/K	1 W/cm/K
E	370 GPa	310 GPa
α	$7 \cdot 10^{-6}$/K	$4.4 \cdot 10^{-6}$/K

It is assumed $\tan\delta = 5 \cdot 10^{-4}$ for both materials at 140GHz

TABLE 3 Strength and crack growth data

	Al_2O_3	AlN
m	10.1	15.3
σ_0	420 MPa	310 MPA
n	33	135
A^*	$1.5 \cdot 10^{-4}$ m/s	$1.5 \cdot 10^{-4}$ m/s
K_{Ic}	4.5 MPa$\sqrt{m}$	3.2 MPa$\sqrt{m}$

FUSION TECHNOLOGY 1988
A.M. Van Ingen, A. Nijsen-Vis, H.T. Klippel (editors)

NEUTRON IRRADIATION CREEP EXPERIMENTS ON AUSTENITIC STAINLESS STEEL ALLOYS

H. Hausen*, W. Schüle** and M.R. Cundy*

*High Flux Reactor Division, Joint Research Centre, Petten, The Netherlands.
**Materials Science Division, Joint Research Centre, Ispra (VA), Italy.

Results of measurements of the neutron induced creep elongation on AMCR-steels (Mn-base), on 316 CE-reference steels, and on US 316 and US PCA steels are reported. It was found that the stationary creep rate is not very sensitive to variations of the irradiation temperature between 300 and 420°C and that the stress-exponent of plastically deformed and of annealed materials is $n \approx 1$ and $n \approx 1.59$, respectively. A small primary creep stage is found in annealed materials. Deformed materials show a negative creep elongation at the beginning of the irradiation, which increases for decreasing stresses and decreases for increasing irradiation temperatures.

1. INTRODUCTION

Austenitic stainless steels have been considered as candidate structural materials for fusion reactors. In the Joint Research Centre Ispra manganese containing steels are developed because the helium production rate of these alloys is smaller, the corrosion resistance against lithium is better, and the neutron avtivation is lower compared to nickel-based austenitic stainless steel alloys. In order to study the creep behaviour of these materials two irradiation creep facilities were developed for the HFR at Petten /1/.

In the creep facilities "TRIESTE" (Project E167) a series of tests are carried out on 294 samples. The elongations are measured at intervals of irradiation in hot cells. In a single irradiation rig, 49 specimens can be irradiated simultaneously in uniaxial tension. In the irradiation device"CRISP" (Project E157) strain registrations are performed in situ. Three specimens, each with independent control of stress and temperature are irradiated simultaneously in uniaxial tension. The results obtained from these experiments will be complementary to those from the TRIESTE series. In both rigs irradiations can be performed in the temperature range between 200 and 600°C for uniaxial stresses between 25 and 300 MPa. The irradiations were performed with a neutron flux density = ø 2 x 10^14 n. cm^-2 .s^-1 (E>0.1 MeV), which corresponds to a displacement rate of K = 1,7 x 10^-7 dpa .s^-1 according to Hip-Teddi code /2/.

In the present paper further results obtained in the past two years are reported /3,4/.

2. EXPERIMENTS

2.1. Irradiation facility TRIESTE

The irradiation facility TRIESTE is designed for irradiations of a large number of samples under various combinations of temperature and stress. A single sample column comprises seven cylindrical dumbbell-shaped tensile samples and associated reference shells, all contained in a supporting tube. The specimens have a gauge length of 5 cm and diameters varying between 1.5 and 3 mm. Seven independent sample columns are inserted into separate channels of the TRIESTE carrier. The columns can be loaded independently by a self-contained spring washer system at the head of the stems. The stresses in the tensile specimens during irradiations were set to 25, 50, 75, 100, 130, 200 and 300 MPa. The coresponding tensile loads on the sample columns are adjusted to an accuracy of ± 2 %. The strain measurements on the individual tensile samples are performed through slot-shaped windows in the supporting tubes by a photoelectric incremental linear measuring system. Dimensional changes of ± 1.5 microns can be detected, which corresponds to a measurement accuracy of about ± 2 %. Irradiation temperatures are maintained by controlling gas mixtures in a stepped gap between the multiple channel specimen carrier and the outer reloadable thimble. The tests are performed at temperatures in the range between 300 and 420°C, with a detection precision of ± 2.5 %. Neutron metrology is mainly performed by evaluating activation detectors which are installed internally in each of the unstressed reference half-shells.The fluence values are determined with an accuracy of ± 10 %.

The entire experimental TRIESTE programme comprises six irradiation facilities where each facility is irradiated for several steps and dimensional measurements on the individual tensile samples are performed in hot cells between the irradiation steps /5,6/. During the whole irradiation campaign a total of 294 tensile samples and a large number of reference half-shell pairs are tested. Irradiation samples and half-shell pairs are manufactured from ten different materials. The chemical composition of these materials is given in Table I. Two cold work levels (10 and 20 %), three separate aging conditions as well as the not pre-treated condition were employed.

2.2. Irradiation facility CRISP

In the device CRISP the irradiation creep elongation of three specimens in three different rigs can be measured simultaneously. All three rigs, combined in one irradiation facility, are independent with respect to the irradiation temperature and the applied stresses which can be varied between 300 and 600 °C, and between 25 and 300 MPa, respectively. The experimental programme comprises three irradiation thimbles with a total of nine individual creep rigs /6/.

In each rig a single cylindrical dumb-bell-shaped sample, which is submerged in NaK, is stressed in tension by a bellows system. Three identical tensile specimens were irradiated separately in the first prototype experiment for tensile stresses of 50, 90 and 130 MPa. The stresses were adjusted with an accuracy of ± 2 %. Strain measurements are taken semi-continuously by comparing the sample length with the length of an unstressed reference piece of the same material. Length changes are detected by a displacement transducer remote from the reactor core. Dimensional changes can be detected with an accuracy of ± 2 microns, i.e. with 0.5 %. The temperatures are determined with a precision of ± 2.5 %. In order to measure thermal creep strains the irradiation temperatures are maintained during reactor shut-down periods by in built electrical heaters.

Designation	C	Mn	Ni	Cr	Si	Mo	Ti	N	S	P
AMCR-0033 Creusot-Loire (France)	0.105	17.50	<0.10	10.12	0.555	<0.06		0.19	0.008	0.016
AMCR-0034 Creusot-Loire (France)	0.100	17.69	0.15	10.11	0.64	1.52		0.16	0.008	0.025
AMCR-0035 Creusot-Loire (France)	0.029	19.88	0.265	14.09	0.63	<0.06		0.048	0.006	0.018
AISI 316 L Creusot-Loire (France)	0.024	1.81	12.32	17.44	0.46	2.5		0.06	0.002	0.027
7758 Vakuum Schmelze (Germany)	0.062	28.6		10.0	0.87		0.87			
7761 Vakuum Schmelze (Germany	0.11	29.4		10.2	1.01					
7763 Vakuum Schmelze (Germany)	0.10	19.4		10.2	0.94		0.85			
AISI 316 ORNL-stockpile SS 316	0.06	2.0	10-14	16-18	1.0	2.3	0.2-0.4			
PCA ORNL-stockpile Path A PCA	0.06	1.5-2.25	15-17	13-15	0.4-0.6	1.8-2.2				

TABLE I
Composition of steels (wt%) tested in TRIESTE and CRISP

3. RESULTS AND DISCUSSION

3.1. Annealed Materials

3.1.1. AMCR 0033

An example of the primary creep characteristics is shown in Fig. 1, in which the creep elongation is plotted versus the number of displaced atoms. The irradiation temperature was 380 °C and the applied stress 50 MPa. The magnitude of the primary creep stage up to about 0.6 dpa does not show uniformity i.e. does not increase with increasing applied stress. In some cases (not shown here) the pretreatment of the materials - as received, 1 h at 400°C, and 1 h at 600°C - does not result in a different primary creep behaviour in contrast to the example shown in Fig.1, according to which the magnitude of the primary creep stage is larger for the as received material in comparison to the heat treated specimens. For an applied stress of 50 MPa no primary creep stage is found. It is assumed that the annealed specimens were very slightly plastically deformed during mounting in the irradiation rig. As we will show later even a negative creep elongation is found at the beginning of an irradiation if the specimens are deformed prior irradiation. This means in summary that the primary creep feature always present in annealed materials can be suppressed through slight plastic deformation and that the magnitude of the primary creep stage is very sensitive to slight plastic deformation prior irradiation.

The yield stress of annealed AMCR 0033 is in the vicinity of 130 MPa. If higher stresses are applied to this material during irradiation a large primary creep stage is observed up to 0.3 dpa.

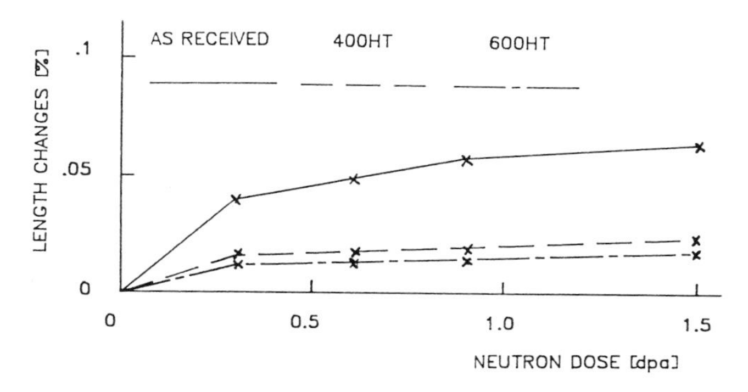

FIGURE 1
Length changes of AMCR 33 (annealed and as received), tensile stresses 50 MPa, irradiation temperature 380°C.

In Fig. 2 the secondary creep rate is plotted versus the applied stress and a straight line results, from the slope of which a stress exponent of n =1.59 ±0.09 is obtained for 2.1 dpa regardless of the magnitude of the primary creep elongation.

First results of in situ elongation measurements were obtained during a test of the CRISP facility. Three specimens were irradiated at 400 °C up to a neutron dose of 1.5 dpa and then the temperature was decreased to 270 °C for further 0.5 dpa irradiation. After a dose of 2 dpa was achieved the functioning of the CRISP facility was impaired and the experiment was stopped. The essential result obtained is seen in the fact that also a negative creep elongation is found as with the TRIESTE facility and that in situ-results correspond to the results measuring the creep elongation out of pile.

3.1.2. AMCR 0034

The results obtained for alloy AMCR 0034 are limited due to technical problems with the irradiation rigs.

The primary creep elongations observed for alloy AMCR 0034 are very similar in magnitude with respect to those found for alloy AMCR 0033. Also the sensitivity of the magnitude of primary creep with respect to previous plastic deformation is noted.

The secondary creep rates plotted versus the applied stress for 2.1 dpa in Fig. 2 agree with the data obtained for alloy AMCR 0033.

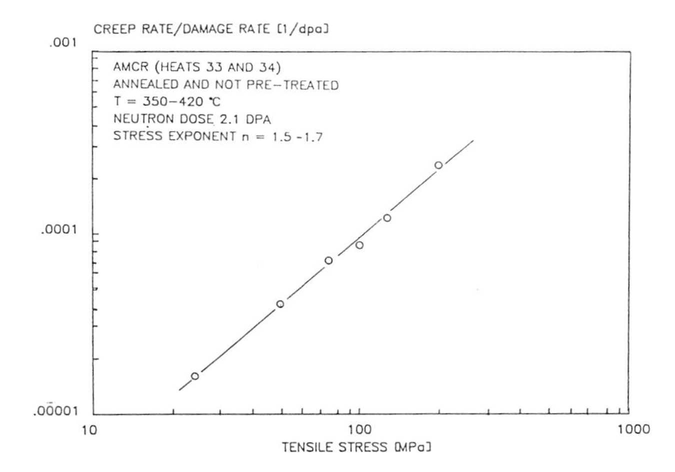

FIGURE 2
Secondary creep rate as a function of the applied stress for AMCR 33 and 34

3.1.3. AMCR 0035

The results obtained for alloy AMCR 0035 are again limited due to technical difficulties.The results obtained so far are, within the error limits, very similar to those obtained for alloys AMCR 0033 and AMCR 0034. The yield stress of alloy AMCR 0035 for the as received state, for the annealed state (400 or 600 °C) appears to be slightly below 130 MPa and consequently slightly lower than the yield stresses of the two other alloys.

3.1.4. Alloys 7758, 7761 and 7763, alloy 316 CE-reference steel and the US steels 316 and PCA

The alloys 7758, 7761 and 7763 were irradiated at 325 °C for 50 MPa up to a neutron dose of 0.9 dpa. It is interesting to note that alloy 7763 shows a much larger primary irradiation creep in the as received state than in the annealed state (1 h at 700 °C). The primary irradiation creep for all three alloys is again, as for the AMCR-alloys, very small. Alloy 7758 shows even a small negative creep elongation for 50 MPa and for an irradiation temperature of 325 °C indicating that the annealed specimen was again slightly deformed during mounting in the irradiation rig. For a higher stress (100 MPa) and for the same irradiation temperature a small primary creep elongation is found comparable to those observed for alloys of the AMCR-type.

The annealed 316 reference steel (1 h at 700 °C) shows no creep elongation up to 0.9 dpa for an irradiation temperature of 320 °C and 75 MPa. For 50 MPa and an irradiation temperature of 325 °C the creep elongation of the 316 reference steel is even slightly negative.

Annealed US 316 stainless steel and US PCA alloys show a very small primary irradiation creep elongation very similar to that of 20 % cold-worked US 316 stainless steel alloys for an irradiation temperature of 360 °C (Fig. 3).

One can say that the primary creep elongations obtained for the nickel containing stainless steel alloys are very similar to those of the AMCR-type alloys.

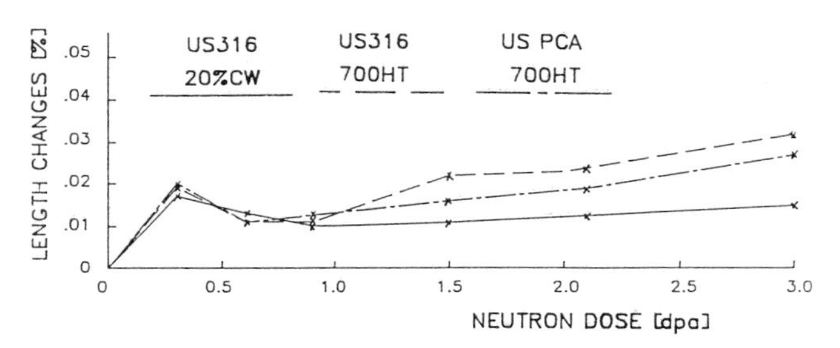

FIGURE 3
Length changes of US 316 (annealed and 20% cold-worked) and US PCA (annealed), tensile stresses 200 MPa, irradiation temperature 360°C.

3.2. Plastically Deformed Materials

3.2.1. AMCR-alloys

AMCR alloys, which have been 20 % cold-worked prior irradiation show initially a negative creep elongation for all stresses applied (Fig. 4). The negative creep elongation feature decreases with increasing irradiation temperature for irradiation temperatures between 320 °C and 420 °C and for stresses of 100 MPa. The negative creep feature is more pronounced for 75 MPa than for 100 MPa for an irradiation temperature of 320 °C. For stresses of 200 or 300 MPa only a positive creep elongation is found with a pronounced primary creep feature even for irradiation temperatures as low as 360 °C.

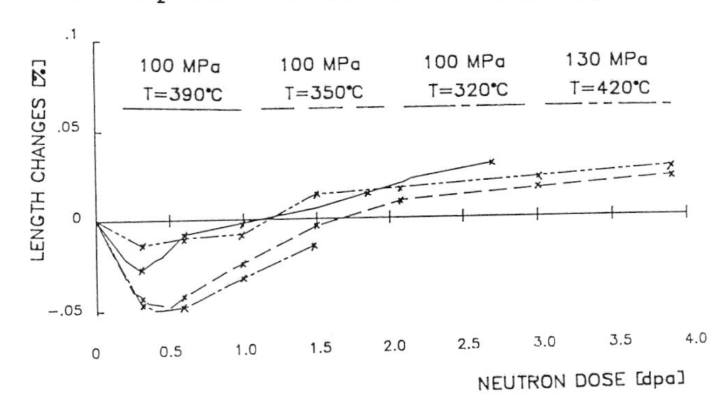

FIGURE 4
Length changes of AMCR 33 (20% CW), tensile stresses 100 and 130 MPa, irradiation temperatures 320-420°C.

The secondary creep rate is plotted versus the applied stress and a straight line results (Fig. 5), from which the stress-exponent n ≈ 1 is derived.

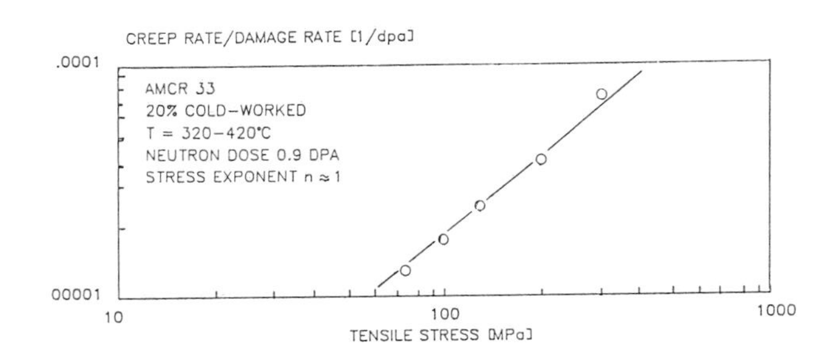

FIGURE 5
Secondary creep rate as a function of the applied stress for 20% CW AMCR 33

At an irradiation temperature of 420 °C the creep characteristics of 20 % cold-worked AMCR 0033 alloys and of annealed materials are very similar for irradiation doses larger than about 1.5 dpa and for stresses of 130 MPa.

3.2.2. US 316- and US PCA-alloys

Irradiation creep rates of 20 % cold-worked US 316 and for US PCA steels are not known from the literature for doses below about 5 dpa. Both American alloys show primary irradiation creep, however, not as large as the AMCR alloys, for 130, 200 and 300 MPa. The irradiation creep exponents for the steady state creep of both types of deformed materials is $n \approx 1$ (Fig. 5).

For cold-worked material irradiated at 325 °C and 100 MPa the American steels, US 316 and US PCA, and the CE 316 reference steel also show a negative creep feature similar to that observed for the AMCR alloys (Fig. 6).

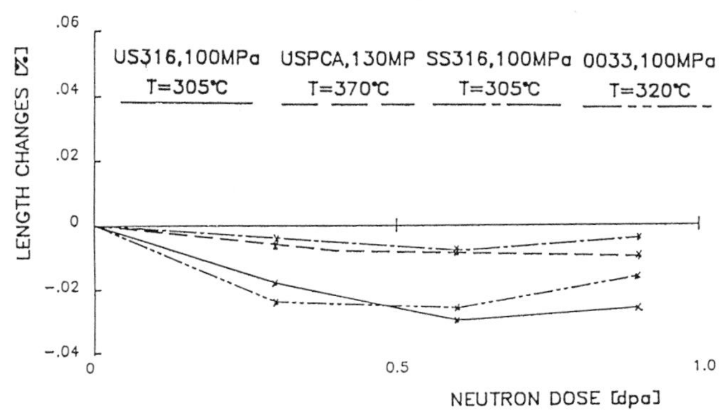

FIGURE 6
Length changes of 20% cold-worked US 316, US PCA, 316 CE-reference and AMCR 33, tensile stresses 100 and 130 MPa, irradiation temperatures 305-370 °C.

In the following figure (Fig. 7) the creep rates are plotted for 20 % cold-worked US 316, US PCA and AMCR 0033 at 130 MPa and an irradiation temperature of about 370 °C. It can be seen that up to 4 dpa US 316 and AMCR 0033 behave very similary but that the creep elongation of US PCA is a factor five smaller than the other two alloys.

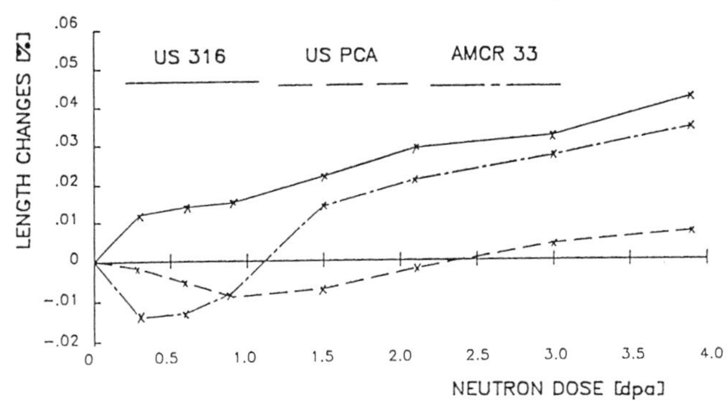

FIGURE 7
Length changes of 20% cold-worked US 316, US PCA and AMCR 33, tensile stress 130 MPa, irradiation temperature 370 °C.

4. CONCLUSIONS

The creep elongations measured for all alloys were smaller than predicted from literature. Primary creep elongations were found in all annealed materials i.e. in AMCR-type and in 316-type alloys, and also in 20 % cold-worked materials for stresses larger than 200 MPa.

A negative creep elongation is found for previously cold-worked material for stresses equal or below about 100 MPa. An increase of the negative creep elongation is found for decreasing irradiation temperatures and decreasing applied stresses. This negative creep feature is attributed to the formation of radiation induced precipitates in both AMCR-type and 316-type alloys /7/.

The stress exponent of secondary irradiation creep for annealed and for cold-worked materials is $n \approx 1.59$ and $n \approx 1.0$, respectively. The irradiation creep rate increases slightly at higher irradiation temperatures.

REFERENCES

1. H. Röttger, A. Tas, P. Von der Hardt, W.P. Voorbraak, Characteristics of Facilities and Standard Irradiation Devices. EUR 5700 EN (1986/1987) High Flux Materials Testing Reactor, HFR Petten.
2. A. Tas, G.J.A.Teunissen, Users Report of HIP-TEDDI, EUR 6001 EN (1978).
3. H. Hausen, R. Lölgen, R. Scholz and W. Schüle, Proc. of the 14th Symp. on Fusion Technology (SOFT), Avignon, Sept. 8-12 (1986), Vol. 2 p. 1095.
4. R. Conrad, M. Cundy, L. Debarberis, H. Hausen, H. Scheurer, C. Sciolla, G. Tsotridis, Proc. of the Int. Symp. on the Utilization of Multi-Purpose Research Reactors, Grenoble, October 19-23 (1987).
5. R. Lölgen, M. Cundy, W. Schüle, Transactions of the 4th Conf.on Structural Mechanics in Reactor Technology, San Francisco, August 15-19 (1977), Vol.L paper L8/11.
6. R. Lölgen, H. Hausen, W. Schüle, Proc. of an Int. Conf. on Fast, Thermal and Fusion Reactor Experiments, Salt Lake City, April 11-15 (1982), Recent Developments in In-Pile Creep and Fatigue Facilities for the JRC Fusion Progr.
7. W. Schüle, E. Lang and A. Panzarasa, EUR 11756 EN (1988).

FUSION TECHNOLOGY 1988
A.M. Van Ingen, A. Nijsen-Vis, H.T. Klippel (editors)

METAL-OXIDE INSULATING MATERIALS FOR AN 18T RESISTIVE INSERT COIL FOR A TANDEM MIRROR FUSION REACTOR

K.L. AGARWAL AND J.F. PARMER

General Dynamics Space Systems Division
P.O. Box 85990, San Diego, CA 92138, USA

18- to 30-tesla solenoids are used in the end cell of a tandem mirror reactor to prevent plasma from leaking out of the central reaction chamber. Superconductors can carry current only up to 16 tesla, so the solenoid would be a hybrid that is part superconducting and part normal conducting. The normal conducting part is a water-cooled coil insert that is subjected to intense neutron irradiation. Organic electrical insulators would break down under these radiation conditions, losing both mechanical and dielectric strength. In order to alleviate this problem, we developed and tested three possibilities for an electrical insulation for a double pancake magnet design configuration. We (1) tested thin anodized strips of four metal alloys under simulated insulator load and slippage conditions, (2) co-wound and tested strips of anodized aluminum and titanium with the copper conductor, and (3) ion-vapor deposited and anodized a coating of aluminum on titanium, then tested this composite. This paper records the results, conclusions, and recommendations of our efforts.

1. INTRODUCTION

Mirror fusion reactors require hybrid solenoid magnets that generate fields of from 18 to 30 tesla. Because of their intense magnetic fields and severe neutron irradiation environment, the materials used for normal conducting inserts are expected to satisfy a diverse range of long-term material properties requirements. Organic composite insulators, such as glass fabric epoxy/polyimide composites, will degrade in mechanical strength and electrical insulation properties in a short period of time. Ceramic insulators, like aluminum oxide (Al_2O_3) or spinel ($MgAl_2O_4$), are more radiation-resistant than glass fabric epoxy/polyimide insulators and may perform better over extended periods. But these ceramic insulators have serious fabrication problems. They are brittle and have very little ductility and tensile strength. Metal-oxide insulators, or metals that form adherent, insulating, oxide layers (notably aluminum, titanium, and tantalum), offer attractive alternatives. These insulators are expected to be ductile as a metal and would have the superior radiation-resistant insulation properties of a ceramic. Such a radiation-hardened ductile insulator could perform as long as the rest of the magnet components in a radiation environment.

2. MATERIALS AND PROCESSES

2.1 Base materials

For a preliminary study to determine suitability of anodized strips of the metals aluminum, titanium, and tantalum, we selected the following alloys for anodizing and further testing:

- Aluminum 6061-0
- Titanium-6Al-4V
- Titanium-15V-3Al-3Cr-3Sn
- Tantalum-8W-2Hf

These alloys were readily available. They were mechanically stronger than the parent metals, and their anodizing characteristics were mostly known. Strips of each were anodized and tested for electrical insulation properties.

2.2 Reasons for anodizing

There are many reasons for anodizing metallic surfaces. Anodizing will:

- Increase corrosion resistance
- Increase paint adhesion and improve bonding
- Permit subsequent plating
- Improve decorative appearance
- Permit application of photographic and lithographic emulsions
- Increase emissivity
- Increase abrasion resistance
- Detect surface flaws
- Prevent galling

- Provide electrical insulation

In this application, the last reason, that is, to provide electrical insulation, was of most interest to us.

2.3 Criteria for adherent protective coating

To be a protective coating, an oxide coating should fulfill many requirements. The Pilling-Bedworth ratio, i.e., the volume of oxide formed from a unit volume of metal, should be close to 1. A volume ratio less than 1 produces insufficient oxide to cover metal surfaces completely. A ratio much greater than 1 tends to introduce large compressive stresses in the oxide coating and causes cracking. To be adherent, the coefficient of thermal expansion of the coating should be close to that of the metal. The coating should have good plasticity to resist fracture. Aluminum has one of the most favorable oxide-metal volume ratios (1.28). Titanium and tantalum are known to form spontaneous adherent coatings. Their oxide-metal volume ratios are 1.95 and 2.33, respectively.

2.4 Anodizing process details

The process cycle used for in-house anodizing of metal strips was as follows:

- Cleaned with methyl-ethyl ketone
- Cleaned in alkaline bath @ 22°C for 15 minutes
- Rinsed in hot tap water
- Rinsed in deionized water
- Etched in $HF\text{-}HNO_3$ mixture @ room temperature
- Rinsed in cold water
- Anodized
- Rinsed in hot tap water for 15 minutes
- Rinsed in deionized water for 1 minute
- Oven dried at 70°C for 30 minutes

Process details for samples anodized by vendors, using their proprietary processes, are not available.

2.5 Anodizing bath formulations and operating parameters

Both alkaline and acid baths have been used for anodizing titanium. We chose to experiment with alkaline baths because relatively thick and adherent proprietary coatings had been produced in alkaline baths. We tried three formulations:

- Sodium hydroxide (1–4 molar solutions @ 20–50°C)
- Sodium hydroxide (2 molar) + hydrogen peroxide (0.5 molar) @ 20–60°C
- Sodium hydroxide (2–6 molar) + sodium or potassium silicate (0.2–1.0 molar) @ 22°C

In titanium samples, thick (1 mil), insulating coatings were produced by the third formulation at anodizing current density of 1.5 A/dm^2 for 24 hours.

Aluminum alloys were anodized by vendors using sulphuric acid baths. Aluminum was deposited on titanium alloy by a proprietary ion-vapor deposition process. The aluminum coating was subsequently anodized by a vendor.

3. ELECTRICAL PROPERTIES: TEST RESULTS

3.1 Electrical resistance under compressive loading

The experimental setup used for testing electrical insulation properties of anodized metallic strips is shown in Figure 1. A stack of four anodized strips (1-inch width × 4-inch length) is placed between two forms. The forms are made of aluminum, and the upper form is attached to the loading arm of the MTS-810 fatigue-testing machine. The forms have a curvature with a 10-inch radius. This curvature permitted a slippage of 3.5 mils between the adjacent strips on loading. The calculated slippage between the adjacent strips for an 18-tesla magnet is expected to be less than 1 mil. Thus, experimental arrangement allows far more slippage than would be experienced by the strips in the actual application.

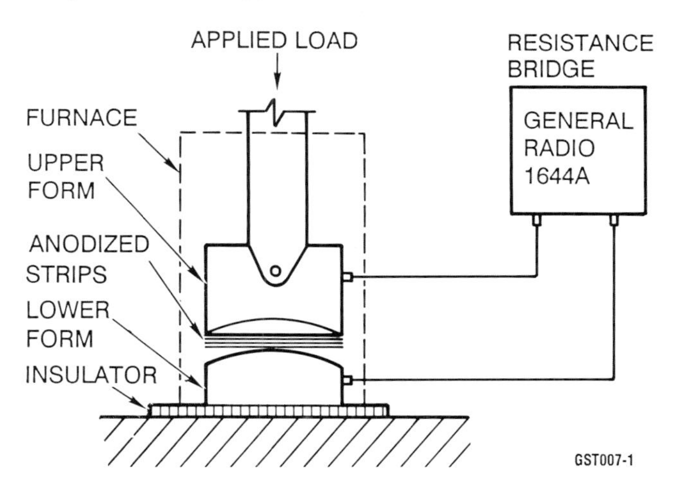

Figure 1. Experimental setup for resistance measurement of anodized metallic strips under compressive loading.

The anodized strips and the forms are surrounded by a temperature-controlled furnace. Electrical resistance was measured using a General Radio Model 1644A resistance bridge. Resistance measurements were made at ambient temperature and at 150°C. The maximum stress applied was 2,500 pounds per square inch. Measurements were limited to 1000 cycles.

Results of measurements on Ti-6Al-4V and Al 6061-0 alloy anodized strips are summarized in Table 1. It can be seen from the table that ion-vapor deposited and then anodized aluminum/titanium composite strips showed erratic electrical resistance at room temperature, and the strips showed electrical short at 150°C testing. Al 6061-0 strips anodized at General Dynamics and also by a vendor performed satisfactorily electrically, but were mechanically deformed.

The titanium alloy strips performed satisfactorily both electrically and mechanically.

3.2 Insulation properties of wound coil

A demo coil was designed to test insulation properties of anodized aluminum and titanium alloy strips. Two 12-foot lengths of copper bus bars were used as conductor. The bus bars were annealed for easy winding. A stack of four anodized alloy strips was placed between the copper bus bars, and the whole assembly was wound into a coil.

Two methods were used to test the demo coil electrical insulation: (1) a Hypot tester and (2) an ammeter, voltmeter, and power supply. The test coil wound with anodized titanium strips was found shorted, so no further tests were conducted on it. The coil wound with anodized Aluminum 1100-0 strips was tested by both methods, and the results are summarized in Tables 2 and 3. The anodized aluminum strips proved to have insulating properties.

4. DISCUSSION

4.1 Current leakage

The use of electrical insulation made from a metallic base material raises the question of current leakage. If the insulating metal-oxide layer should wear, how high should the parent material resistivity be so that the bypass current is tolerable? The answer is obtained by comparing the resistivity of a single turn of the coil to the resistivity of an electrical short through the insulation. The average conductor for the 18-tesla magnet has a resistance of 0.276 milliohms at 160°C. If the conductor carries approximately 17,000 amperes, the voltage drop will be 4.69 volts. Each conductor makes four turns spiralling in and out of the winding; therefore, the average voltage drop per turn is 1.17 volts. Because the coil is wound two-in-hand, the average turn-to-turn drop will be 1.17/2 = 0.585 volts. If we arbitrarily select the allowable leakage current to be less than 1.0 per-

Table 1. Electrical resistance of anodized metallic strips under compressive loading: four metallic strips; load 1,250 psi.

Material	Test Temp (°C)	Electrical resistance (ohms) under cyclic load: Initial	200 cycles	400 cycles	600 cycles	800 cycles	1,000 cycles	Comments
Al 6061-0	22	1.23×10^{8}	3.0×10^{7}	2.2×10^{7}	2.0×10^{7}	1.4×10^{7}	1.5×10^{7}	Strips deformed
Al 6061-0	148		2.4×10^{6}	2.0×10^{6}	1.9×10^{6}	1.7×10^{6}	2.1×10^{6}	Strips deformed
Ti-6Al-4V	22	6.5×10^{7}	–	–	–	–	2.4×10^{5}	
Ti-6Al-4V (Ion-vapor deposited aluminum)	22	4.5×10^{6}	12.0×10^{6}	14.4×10^{6}	1.8×10^{6}	10.6×10^{6}	9.4×10^{6}	Erratic behavior, galling
	148	–	–	–	–	–	–	Shorted

Note: Cyclic loading tests were performed two weeks after initial testing. The samples may have absorbed moisture from the environment.

GST007-2

Table 2. Results of insulation measurements using HP 6291 power supply and Fluke 8050A DMM.

Insulation: Anodized Aluminum

Voltage (volts)	Current (microamps)	Resistance (megaohms)
10.0	0.07	143
20.0	0.21	95
30.0	0.42	71.5
40.0	0.75	53.5
50.0	1.21	41.3
60.0	1.86	32.3
70.0	2.67	26.2
80.0	3.58	22.3
85.6	4.08	21.0

GST007-3

Table 3. Results of insulation measurements using Hypot tester.

Conductor: Anodized Aluminum

Voltage (volts)	Current (microamps)	Resistance (megaohms)	Remarks
100	3.0	33.3	
150	7.5	20.0	
200	13	15.4	
250	25	10.0	
300	36	8.3	
350	54	6.5	
400	73	5.5	
450	95	4.7	
500	126	4.0	
550	156	3.5	
600	200	3.0	
650	270	2.4	
700	340	2.1	
750	450	1.7	Sparks
800	610	1.3	Sparks
750	450	1.7	Sparks
700	330	2.1	
650	250	2.6	
600	178	3.4	
550	136	4.0	
500	103	4.9	
450	77	5.8	
400	56	7.1	
350	39	9.0	
300	26	11.5	
250	14.7	17.0	
200	9.3	21.5	
150	4.1	36.6	
100	1.8	55.5	

GST007-4

cent or 170 amperes, the resistance of the insulation would have to be 0.59/170 = 3.44 milliohms minimum. The results of resistivity tests on worn anodized titanium strips indicate that a four-layer stack of these strips would yield a resistance of 0.325 megaohms, which is many orders of magnitude higher than needed.

4.2 Anodized aluminum versus anodized titanium

That the anodized titanium strips did not perform according to expectations came as a surprise. Both aluminum and titanium form a spontaneous oxide coating when exposed to air. This oxide coating is protective and adheres to the base metal. The titanium oxide coating had microcracks, but the aluminum oxide coating did not. This may be explained on the basis of a favorable Pilling-Bedworth ratio for aluminum, which is 1.28, versus a ratio of 1.95 for titanium.

A coil wound with titanium strips did not perform electrically, though it worked well when the strips were stacked manually. This may be due in part to a lower value of hardness for the titanium oxide coating, which might have been scratched and therefore unable to stand the stresses of the winding process.

It appears that the aluminum oxide coating is superior to the titanium oxide coating in three respects: it is more adherent, harder, and can stand the stresses of the co-winding process.

5. CONCLUSIONS

Two conclusions can be drawn from this study:

- Both anodized aluminum and anodized titanium alloy strips can be used as ductile insulation materials in a pancake design coil. A stack of anodized strips can be inserted into the conductor-to-conductor gap cut by electrical discharge machining or abrasive water jet cutting.
- Only anodized aluminum strips can be successfully co-wound with a conductor.

The effect of neutron irradiation on metal-oxide interfaces is not very well understood. More experimental work is needed in this area.

REFERENCES

1. H. Geduld, Anodizing Titanium, Metal Finishing, 65 (1967) 62.
2. A. Mahoon and R.P.J. Kohler, Anodizing Titanium, U.S. Patent No. 4,394,224, July 19, 1983.

FUSION TECHNOLOGY 1988
A.M. Van Ingen, A. Nijsen-Vis, H.T. Klippel (editors)
Elsevier Science Publishers B.V., 1989

DIELECTRIC MEASUREMENTS BY THE T.S.D. TECHNIQUE IN INSULATOR MATERIALS

Rafael VILA, Angel IBARRA, Miguel JIMENEZ DE CASTRO

Asociación EURATOM/CIEMAT (Fusión), Av. Complutense 22,
28040 Madrid, Spain

Thermally stimulated depolarization measurements in some candidate insulator materials for fusion applications are presented in this work. A number of dipolar reorientation processes has been detected and their contribution to the dielectric loss in a wide frequency range has been calculated. This shows that this type of measurements is interesting to make a preliminary screening of insulator materials related to their dielectric properties.

1. INTRODUCTION

Among other applications, ceramic insulator materials are needed for the r.f. heating systems in a fusion device. Measurements of dielectric losses in a wide frequency range are thus needed. It is known that r.f. induced dipole oscillations in insulators may lead, together with other enegy absorption processes, to dielectric losses at the working temperature of a fusion device in the frequency range of interest for plasma heating methods. Since its introduction by Bucci et al.[1] in 1964, the thermally stimulated depolarization (TSD) technique has been a powerful tool for the study of dipolar defects in dielectric materials.

The different steps in a TSD measurement are: i) At a temperature T_p, at which the dipolar defect under study has a short relaxation time $\tau(T_p)$, an electric field E_p is applied to the sample in order to orient dipoles. The polarization time must be much longer than $\tau(T_p)$ to achieve a saturing polarization. ii) The sample is then cooled down with the field on to a temperature T_o, at which $\tau(T_o)$ is very large. The crystal polarization is so "frozen in". The field is then removed and the two sample sides are connected to an electrometer. iii) The sample is kept at T_o until the fast polarization processes decay. It is then heated at a linear rate. As the sample temperature increases, the dipole relaxation time becomes shorter and shorter, thus dipoles will be again disoriented. Dipole disorientation leads to a measurable electric current with a peak shaped dependence on temperature.

The TSD technique has been mainly applied to point defect relaxations but, in principle, it can also be used to study any type of dielectric relaxation processes, as Maxwell-Wagner effects or space-charge polarization.

This paper presents a summary of TSD measurements in single and polycrystalline MgO and Al_2O_3 as well as in polycrystalline $MgAl_2O_4$. Loss tangent data at several temperatures calculated from the measured TSD parameters are also shown.

2. EXPERIMENTAL

MgO and Al_2O_3 single crystalline samples of about 10x10x1 mm^3 were cut and polished from blocks purchased from W.C. Spicer and Union Carbide respectively. Disc shaped ceramic samples, 10 mm in diameter and 1 mm thick, of MgO and Al_2O_3 were obtained from Friedrichsfeld (GFR) (MG23 and AL23 types respectively). $MgAl_2O_4$ ceramic samples from Raytheon (USA) were kindly supplied by Kernforschungszentrum Karlsruhe, IMF-I (GFR).

TSD measurements between 10 and 300 K, at a heating rate of 0.13 Ks^{-1}, have been performed in a Displex CSW-202 closed-cycle refrigeration system. Electric currents have been measured with a Keythley 617 electrometer. A Fluke 415B high voltage power supply has been used for sample polarization up to an electric field value of 30000 Vcm^{-1}.

3. RESULTS AND DISCUSSION

Figure 1 shows TSD spectra in single crystal and polycrystalline MgO. The polarization temperature, T_p, for each case is 295 and 180 K respectively.

As it can be seen, the single crystalline material only shows a TSD peak at about 280 K, while a high peak at 165 K, as well as the peak at 280 K when the sample has been polarized

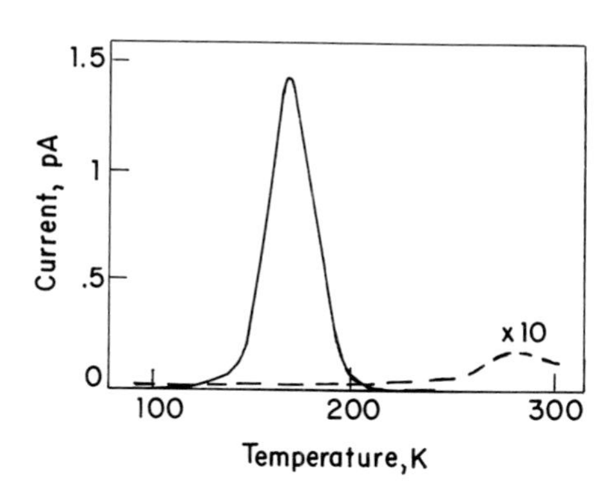

FIGURE 1
TSD spectra of single crystal (---) and polycrystalline(——) MgO for an applied field of 7500 Vcm^{-1}.

above 270 K, are observed in the MG23-type ceramic.

A linear relationship between the peak height and the polarizing field, E_p, at least for E_p values ranging between -30000 and +30000 $V.cm^{-1}$, has been found to occur for all these peaks. In spite of the fact that the 165 K peak position does not depend on the polarization temperature, its shape does not follow the usual equation governing TSD processes related to dipoles[1], perhaps because it is actually composed by two strongly overlapped peaks. For this reason it has not yet been analyzed in detail. The position in temperature of the peak at about 280 K in both types of materials depends on T_p, this might indicate that it is related to space charge release.

A peak above 300 K is seen in the TSD spectrum of both single crystal and AL23-type ceramic Al_2O_3, when polarizing above 270K

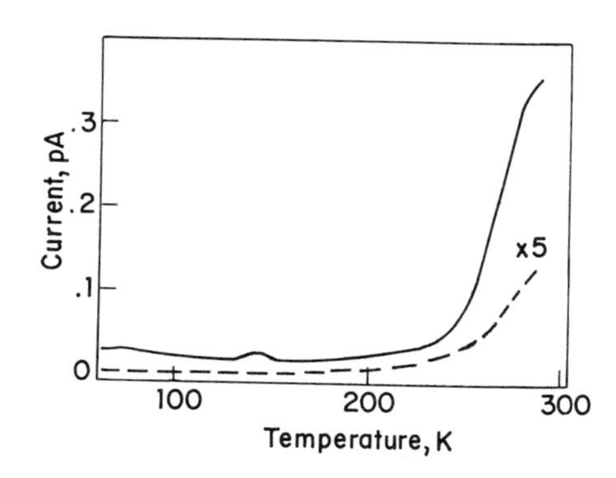

FIGURE 2
TSD spectra of single crystal (---) and polycristalline (——) Al_2O_3 for an applied field of 7500 Vcm^{-1}.

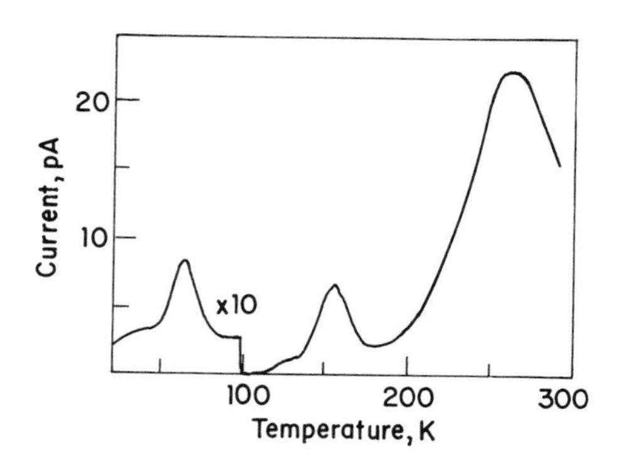

FIGURE 3
TSD spectrum of $MgAl_2O_4$ polarized at 250K with an electric field of 20000 Vcm^{-1}.

(figure 2). A very low intensity peak at 142K is also obtained in the ceramic material.

The peak heights depend linearly on E_p, but the higher temperature peak position depends on T_p, thus indicating that it might be again due to space charge polarization processes.

From these results it may be concluded that dielectric losses due to dipolar relaxations in our as received MgO and Al_2O_3 samples are not important for fusion applications. This agrees with the low value (around $5x10^{-4}$) for the loss tangent in Al_2O_3 due to dipolar processes, which can be estimated from the values of the real part of the dielectric constant at low and high frequency, taken from the literature[2,3].

The TSD spectrum of polycrystalline $MgAl_2O_4$ is plotted in figure 3.

Peaks at about 65(I), 120(II), 155(III) and 275(IV) K are clearly observed. A linear dependence of peak intensities on E_p has been found to occur for peaks I to III. The temperature at the maximum of these three peaks does not vary with T_p. Besides this, it is unlikely to observe spatial charge processes at those low temperatures; so these TSD peaks may be ascribed to dipolar relaxation processes. The behaviour of peak IV is different from the others. It is very wide and the temperature and the current intensity at its maximum strongly depend on T_p. This indicates that it might be due to a spatial charge process.

All peaks appearing in the TSD spectrum of $MgAl_2O_4$ have been isolated by either using a thermal cleaning method or by polarizing at an appropriate temperature to avoid peaks other than that under study to be obtained. The peak shapes of dipolar processes I to III have been fitted by a recently proposed method[4] to the TSD equations[1], and values for activation energies and relaxation times have been obtained. From other experimental results it has been proposed that dipolar defects related to OH^- ions, to the non-stoichiometry of this material and to the

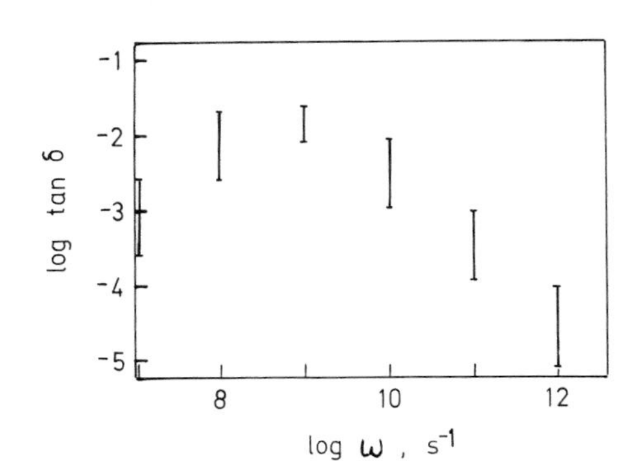

FIGURE 4
Contribution of dipolar relaxation processes II and III of $MgAl_2O_4$ to the loss tangent at a temperature of 300K.

presence of some type or impurity might be then origin of these three TSD peaks[5].

Assuming the Debye model[6] to be applicable, the contribution of each dipolar relaxation process to the real and imaginary parts of the dielectric constant can be calculated as a function of frequency. Since they are also a function of the dipole relaxation time and of the difference between the static and the infinite frequency dielectric constants, which in turn depend on the temperature[1,7] at which the dielectric constant values are to be obtained, those values are temperature dependent. As an example, the contribution of processes II and III of $MgAl_2O_4$ to the loss tangent is shown in figure 4 for an insulator working temperature of 800K. The contribution of process I is much less because of the low intensity of peak I.

4. CONCLUSIONS

This work shows that dipolar relaxation processes in insulators can be easily detected by the TSD technique. Their contribution to loss tangent values can be assessed from the TSD parameters. The obtained results in our unirradiated samples allow us to conclude that there are, in $MgAl_2O_4$ only, such type of processes whose contribution can be important in the frequency range of interest for fusion applications.

However, it is important to point out that, usually, the lower the temperature at the maximum of a TSD peak the higher the frequency of its associated resonance peak. On the other hand, irradiations in which displacements occur can be very effective in producing additional dipolar complexes, as divacancies, interstitial-vacancy and impurity-vacancy dipoles, the lattest as a result of transmutation products in neutron irradiated materials. In this sense, preliminary results on neutron-irradiated Al_2O_3 single crystals and polycrystalline samples show new TSD peaks which are likely due to dipolar defects. Their calculated contribution to loss tangent values seems to be in good agreement with previous dielectric loss measurements. Further work is, however, needed. Ionizing radiation, which is expected only to induce electric charge rearrangement processes, do not produce significant changes on the TSD spectra.

Therefore, this experimental technique is an easy method to be used for a preliminary screening of insulator materials related to their dielectric properties.

REFERENCES

1. C. Bucci, R. Fieschi and G. Guidi, Phys. Rev. 148 (1966) 816.

2. R. Selby, J. Fontanella and C. Andeen, J. Phys. Chem. Solids 41 (1980) 69.

3. M.N. Afsar and K.J. Button, in: Infrared and Millimeter Waves, ed. K.J. Button (Academic Press, Orlando, Florida, 1984) p.27.

4. R. Vila, A. Ibarra and M. Jiménez de Castro, Phys. Status Solidi (a) 105 (1988) 601.

5. A. Ibarra, R. Vila and M. Jiménez de Castro, Int. Conf. on Defects in Insulating Crystals, Parma (Italy, 1988).

6. P. Debye, Polar Molecules (Chem. Catalog Co., New York, 1929).

7. H. Frölich, Theory of Dielectrics (Oxford Press, London, 1950).

FUSION TECHNOLOGY 1988
A.M. Van Ingen, A. Nijsen-Vis, H.T. Klippel (editors)

STRUCTURAL RESPONSE OF TUNGSTEN AND MOLYBDENUM TO HIGH HEAT FLUXES

H. Bolt[1], K. Kiuchi[2], M. Araki[3], M. Seki[3]

1: KFA Jülich GmbH, Institute for Reactor Materials, KFA-EURATOM Association, D-5170 Jülich, FRG
2: Japan Atomic Energy Research Institute, Dept. of Fuels and Materials Research, Tokai, Japan
3: Japan Atomic Energy Research Institute, Dept. of Thermonuclear Fusion Research, Naka, Japan

Refractory metals like tungsten and molybdenum are regarded as candidate materials for divertor plate applications in fusion devices of the next generation. During operation these materials would supposedly be subjected to severe thermal shocks in the course of plasma disruptions. In order to determine the cracking and melting behaviour of tungsten and molybdenum high heat flux experiments were carried out by use of the JT-60 Neutral Beam Injection Test Stand. Material samples were subjected to hydrogen beam pulses with power densities of 82 and 200 MW/m^2 and pulse lengths in the range of 45 to 200 ms.
After the experiments the tungsten samples showed crack formation already at the lowest heat loads but the material has a very high threshold for melting which was reached at a pulse of 200 MW/m^2 for 85 ms. Molybdenum did not show obvious crack formation but has a considerably lower melting threshold than tungsten.

1. INTRODUCTION

Refractory metals like tungsten and molybdenum are regarded as potential materials for divertor targets in tokamak fusion devices[1]. Their main advantages for such applications are low sputtering yields in the low edge plasma temperature regime and high temperature resistance. In this second respect the very high melting points of the materials (2620°C for molybdenum, 3410°C for tungsten), good thermal conductivity (88 W/mK for molybdenum and 105 W/mK for tungsten at 1500°C) and low coefficients of thermal expansion (6.5 x 10^{-6}/°C for molybdenum and 5.3 x 10^{-6}/°C for tungsten) have to be mentioned[2]. However with respect to the thermal shock resistance of these materials under high surface heat loads only insufficient data are available. Especially this aspect is of importance for applications of these materials since very serious thermal shocks are anticipated to occur on divertor targets during disruptions and run-away electron events[3]. Here heat loads of several thousand MW/m^2 which are deposited in the range of milliseconds may seriously affect the integrity of plasma-facing materials.

Regarding the thermal shock behaviour of tungsten under surface heat loads studies indicate a very high susceptibility to crack formation under thermal shocks[4,5,6]. Molybdenum has a significantly lower melting threshold than tungsten but is more resistent against cracking[4,6]. Most of the previous studies were performed using electron beam facilities with irradiated areas of about 1 cm^2. It is assumed that the high heat flux behaviour of these materials differs in the case of heat loads on large areas as they are relevant to fusion issues. For this reason a series of thermal shock tests was performed by use of a hydrogen beam with a very broad distribution of the power density.

2. EXPERIMENTAL PROCEDURE

For the high heat flux experiments rectangular molybdenum samples (TZM, Metallwerke Plansee, Austria) of 25 mm x 25 mm with 10 mm thickness and tungsten plates (Metallwerke Plansee) of 40 mm diameter and 10 mm thickness were used. In the case of the tungsten samples the grain orientation was perpendicular to the heated surface.

The high heat flux experiments were performed using the Neutral Beam Injection Test Stand to JT-60 of JAERI, Naka[7]. Figure 1 shows a schematic of the test stand and the experimental set-up. Material samples were inserted in the beam line of the hydrogen beam which was extracted from the ion source of the test stand and had currents up to 44 A at acceleration voltages of up to 75 kV. The samples were driven into the main tank from a separate vacuum chamber via a lock system. The sample holder on which the samples were placed loosely was actively cooled. Since the beam hit the sample surfaces in vertical direction, the formation of melt layers was not affected by gravity.

Beam calibration was performed with a calorimeter consisting of an array of 9 thermocouples which was inserted in the beam line at the sample position. From the temperature rise of the thermocouples the deposited energy at discrete locations during the beam pulses could be determined. The beam profile of the power density was nearly circular and gaussian shaped with a flat top region covering the sample area with a power density variation of less than 10 %. After the thermal shock tests the samples were investigated by optical microscopy, SEM, and metallography.

3. EXPERIMENTAL RESULTS

The parameters of the experiments performed on tungsten and molybdenum are shown in Tab. 1.

3.1 SINGLE PULSE TESTS

Tungsten:

Table 1 shows the experimental results on tungsten. Already the lowest heat load (82 MW/m^2, 50 ms) resulted in severe cracking of the surface in a net-like pattern which is homogeneously distributed over the heated surface (fig. 2). With increasing pulse duration the crack opening decreases which indicates that crack formation is stronger at shorter pulse durations even at the same power density.

By SEM cracks were not found on the surfaces of melted samples (200 MW/m^2, 85 ms/100 ms). However metallographic cross sections releaved fine intergranular cracks which reach into depths much larger than the thickness of the melted layer (fig. 3). The surfaces of samples which underwent melting and resolidification remained very smooth with the resolidified material showing epitaxial grain growth.

Molybdenum:

No cracks were observed on any of the molybdenum samples after the thermal shock tests (tab. 1). In contrast to tungsten the melted molybdenum samples show very rough surfaces which is attributed to melt layer motion due to high surface tension at the solid-liquid interface (fig. 4).

3.2 MULTIPLE SHOT TESTS

Tungsten samples subjected to five subsequent pulses of 82 MW/m^2 and 50 ms/100 ms show an increase in the crack opening on the heat loaded surfaces compared to the samples loaded with single pulses. However a significant increase of the penetration depth of these cracks into the material was not observed (fig. 5).

Molybdenum did not show any difference between single and multiple loaded samples.

4. DISCUSSION

The crack behaviour of tungsten seems to depend mainly on the ductile-brittle transition (DBT) behaviour of this material and on the temperature/stress field which is built up in the sample during the experiment. Short pulses allow the material to stay below the DBT temperature in a region very close to the heated surface. This temperature is 100 to 400°C for tungsten[2]. Additionally a very steep temperature and stress gradient occurs. Both effects result in the pronounced cracking of this material already under low and short heat loads. The crack behaviour slightly improves with increasing duration of the heat load. In such a case the heat is allowed to be transferred into greater depths of the material and a thicker layer can reach a ductile state above the DBT.

On molybdenum (TZM) no cracks were observed. The DBT of this material occurs at about room temperature[2]. The melting threshold as product of the incident power density and the square root of the pulse duration[8] is much lower for molybdenum than for tungsten. For molybdenum the experimentally determined threshold is about 37 $MW/m^2 \sqrt{s}$, for tungsten 55 $MW/m^2 \sqrt{s}$. The data for molybdenum are about 25 % lower than previously published ones which may be due to the different calibration of the facilities used[8].

5. CONCLUSIONS

High heat flux experiments were carried out on tungsten and molybdenum by use of a hydrogen beam. Power densities and pulse durations were in the range of 82 to 200 MW/m^2 and 50 to 200 ms.

The results obtained on tungsten can be summarized as follows:

- The material shows intergranular crack formation already under the lowest heat flux, 82 MW/m^2, 50 ms.
- Longer pulse durations at the same power density slightly reduce cracking.
- Cracks also form on resolidified surfaces and reach deep into the unmelted bulk material.
- The melt threshold is about 55 $MW/m^2 \sqrt{s}$.
- Resolidified surfaces are very smooth.

The results obtained on molybdenum (TZM) are:

- No cracks were found on the tested samples.
- The melt threshold is about 37 $MW/m^2 \sqrt{s}$.
- Resolidified surfaces show a very rough structure.

REFERENCES

1. International Tokamak Reactor, Phase Two A, Part I (IAEA, Vienna, 1983)
2. Metallwerke Plansee, Austria, product information on TZM and tungsten (1985, 1986)
3. G. Vieider, A. Cardella, M. Chazalon, et al. in: Proc. 14 Symp. on Fusion Technology, Avignon (1986) 573
4. A. Tobin, J. Nucl. Mater. 85/86 (1979) 197
5. M. Shibui, J. Ohmori, S. Itoh, et al. in: Proc. 11 Symp. on Fusion Engineering, Austin, Texas (1986) 877

6. H. Bolt, H. Hoven, E. Kny, K. Koizlik, J. Linke, H. Nickel, E. Wallura, Berichte der Kernforschungsanlage Juelich, Juel-2086, Juelich (1986).

7. H. Horiike, M. Akiba, M. Araki et al., Rev. Sci. Instrum. 55 (3) (1984) 332.

8. H. Bolt, A. Miyahara, T. Kuroda, Y. Oka, Institute of Plasma Physics, Nagoya University, IPPJ-AM 53 (1987)

material	power density (MW/m²)	pulse duration (ms)	single pulse	multiple pulse	no damage	cracks	melting
tungsten	82	50	x			x	
	82	50		x		x	
	82	100	x			x	
	82	100		x		x	
	82	200	x			x	
	200	45	x			x	
	200	85	x			x	x
	200	100	x			x	x
molybd.	82	50		x	x		
	82	100	x		x		
	82	100		x	x		
	82	200	x				x
	200	45	x		x		
	200	85	x				x
	200	100	x				x

remark: multiple shot test: 5 subsequent pulses with about 180 s between pulses

TABLE 1
Experimental parameters and results

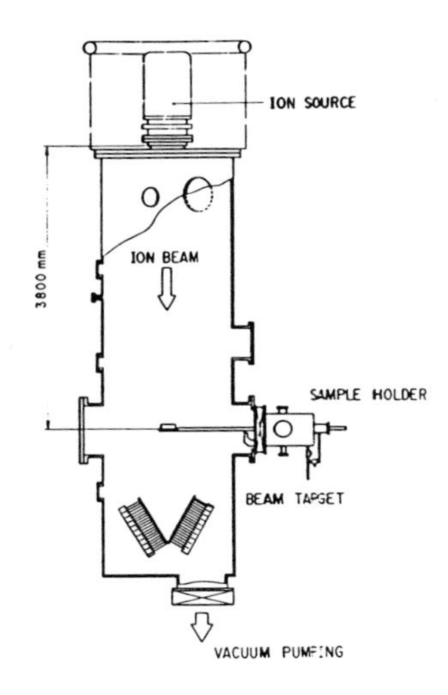

FIGURE 1
Schematic of the Neutral Beam Injection Test Stand to JT-60 used for the high heat flux experiments

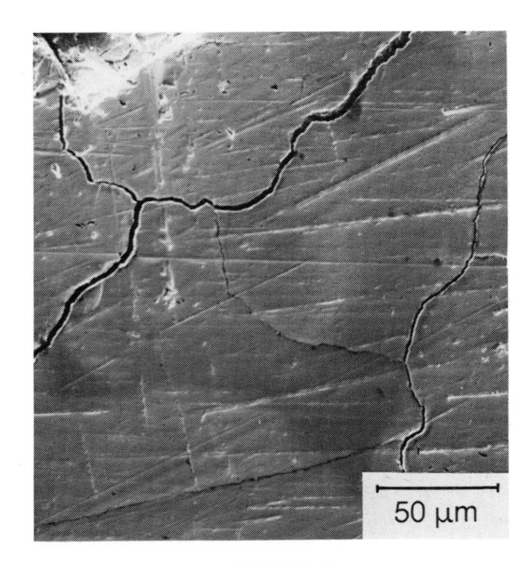

FIGURE 2
Cracks on a tungsten surface after a heat load of 82 MW/m^2 for 50 ms, SEM

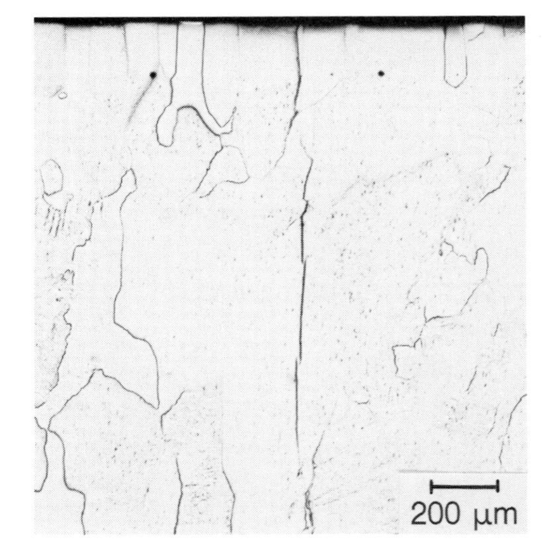

FIGURE 3
Resolidified surface and cracks on tungsten after a heat load of 200 MW/m^2 and 85 ms, metallographic cross section

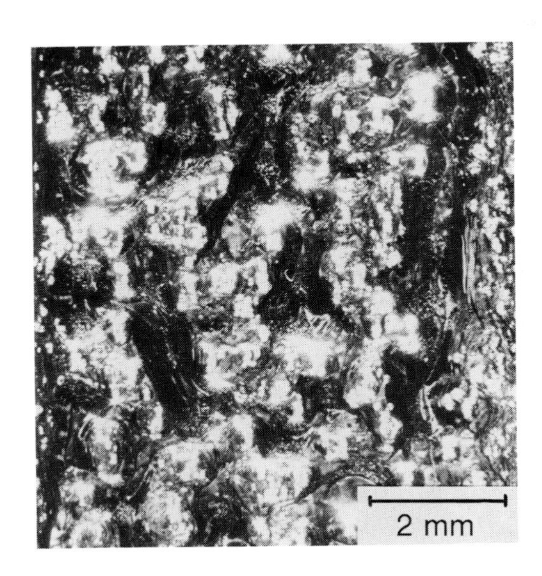

FIGURE 4
Structure of a resolidified molybdenum surface after a heat load of 200 MW/m^2, 100 ms

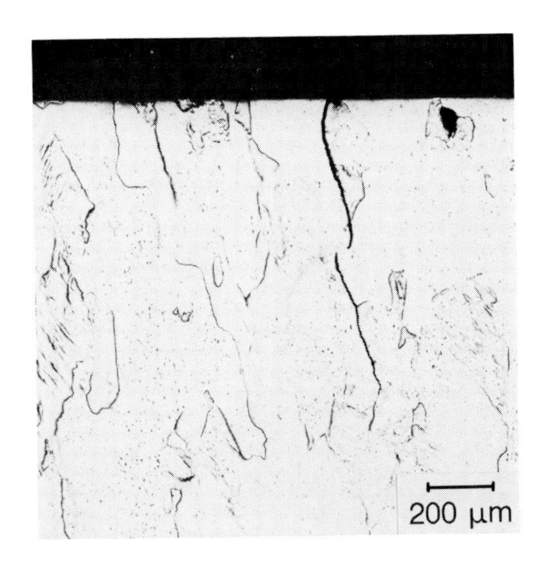

FIGURE 5
Cracks on a tungsten surface after 5 subsequent pulses with a heat load of 82 MW/m^2, 100 ms; duration between the pulses 180 s

FUSION TECHNOLOGY 1988
A.M. Van Ingen, A. Nijsen-Vis, H.T. Klippel (editors)
Elsevier Science Publishers B.V., 1989

TRAPPING AND DETRAPPING OF DEUTERIUM AND HYDROGEN IN NEUTRON IRRADIATED MOLYBDENUM.

D.T.BRITTON, K.R.BIJKERK, A.VAN VEEN, J.R.HERINGA, H.A.FILIUS, J.DE VRIES, J.H.EVANS[1] and W.H. SEGETH[2]

Interfaculty Reactor Institute, Delft University of Technology, NL-2629JB Delft, The Netherlands; [1] Harwell Laboratory, Oxon OX11 ORA, UK.; [2] Lab. voor Alg. Natuurkunde, Rijksuniversiteit Groningen, NL-9718CM Groningen

Hydrogen desorption spectroscopy was applied to molybdenum samples which contained neutron irradiation produced voids. Hydrogen was introduced into the samples by low energy (1 keV) H_2^+ or D_2^+ ion irradiation: ion fluences and injection temperatures were varied from 10^{16} to 10^{20} cm^{-2} and from 300 to 625 K respectively. During the ion irradiation the voids become highly pressurized, but the pressure dropped considerably when the irradiation was stopped and the sample was cooled to RT. The presence of surface impurities causes a drastical enhancement of hydrogen trapping in the voids.

1. INTRODUCTION

During the past years a considerable effort has been put in modelling the behaviour of hydrogen and hydrogen isotopes in first wall components that are exposed to the plasma in a fusion device [1,2,3]. Important model parameters are diffusion constants, surface recombination rates and the trapping or detrapping rates at defects.

Experimental data on the interaction of hydrogen with vacancies and helium bubbles in molybdenum obtained by Nuclear Reaction Analysis (NRA) methods are reported by Meyers et al[4]. Positron lifetime spectroscopy has been employed by Nielsen et al[5] to study the interaction of implanted or electrochemically dissolved hydrogen with vacancies and small voids. Tanabe et al[6] have reported on effects caused by radiation damage on deuterium permeation and reemission when 20 keV deuterium ions are implanted. The first desorption measurements indicating binding of deuterium to vacancy clusters were reported by Erents and McCracken[7]. In this article results will be presented of hydrogen and deuterium desorption spectrometry on neutron irradiated molybdenum with well characterised void population, uniformly distributed throughout the sample. Additional measurements are performed with NRA and Positron Annihilation techniques. Low energy hydrogen injection at varying temperatures is used to simulate conditions as present in a plasma fusion device.

2. EXPERIMENTAL

The molybdenum samples used for the desorption study are 3 mm diameter Mo discs of 1 mm thickness. Neutron irradiation to a fluence of $8x10^{22}$ cm^{-2} and subsequent annealing to 650 ^{o}C has led to a defect structure[8] which is dominated by the presence of 1.2 appm cavities with an average diameter of 5.4 nm. The grainsize is larger than 1 μm. After irradiation the estimated concentration of metallic transmutation products (Zr,Tc,Nb) was 100 appm. These products may have been segregated at the inner surfaces of the cavities and the outer surface of the samples.

Hydrogen and deuterium ion irradiation was performed by negatively biasing the sample with respect to a low pressure hydrogen/deuterium plasma (10^{-3} Pa). The plasma was sustained by electrons from two filaments and was constricted to a ribbon like shape by a magnetic field. At the standard bias voltage of 1 kV, the distance between sample and plasma ribbon was ~ 2 mm and the ion current density amounted to ~0.5 mA/cm^2. The sample was heated by the ion-current (for low injection temperatures currents had to be reduced) or by radiative heating by a filament at the rear-side of the target. After completion of the hydrogen or deuterium injection the sample was heated by electron bombardment to a temperature of 1000 K with a constant heating rate of 40 K/s. The hydrogen or deuterium release rate was monitored with a calibrated quadrupole mass-spectrometer. Most of the desorption measurements were performed for deuterium because of a better signal to background ratio.

In a few cases hydrogen and deuterium irradiated samples have been analysed with a positron lifetime spectrometer (resolution 150 ps)[9]. NRA has been used to measure the absolute concentration of deuterium up to a depth of 200 nm below the sample surface (via the $^3He(d,^4He)p$ reaction).

3. RESULTS

The measured desorption spectra contained usually one dominant desorption peak at temperatures varying from 400 to 800 K depending on the temperature of injection and on the hydrogen or deuterium fluence applied. The peak width (FWHM) amounted to 25% of the peak temperature. The peak shape, a relatively steep rise of the release rate at the low temperature side and a slower decrease at the high temperature side, is typical for a diffusional release mechanism of the gas.

The spectra shown in fig.1 obtained for varying doses are examples of this type of release. With increasing fluence the peak position shifts to a higher temperature indicating an increase of the average depth of the trapped deuterium. At injection temperatures starting from 325 K it was found that the trapped amount

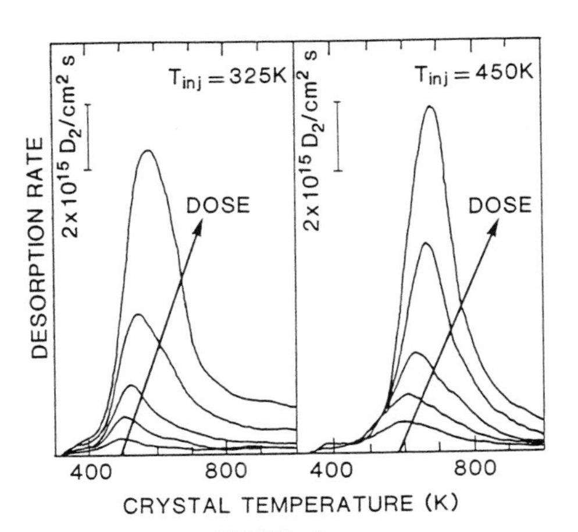

FIGURE 1
Deuterium desorption spectra obtained for the MoH1 sample. Doses for injection temperature of 325 K: 0.13, 0.40, 1.2, 4.0, 12.0×10^{17} cm^{-2}. For 450 K: 0.4, 1.2, 4.0, 12.0, 20.0×10^{17} cm^{-2}. The dose rate was 10^{15} $cm^{-2}s^{-1}$.

of deuterium increased with the square root of the dose. In fig.2 this is demonstrated for deuterium injection in one of the (as-received) samples (MOH1) at 325 and 450 K. An interesting observation was that another sample (MOH2) which was electrochemically polished before insertion in the apparatus showed a similar dose dependence but at a factor 4 lower absolute level. In-situ sputter cleaning of

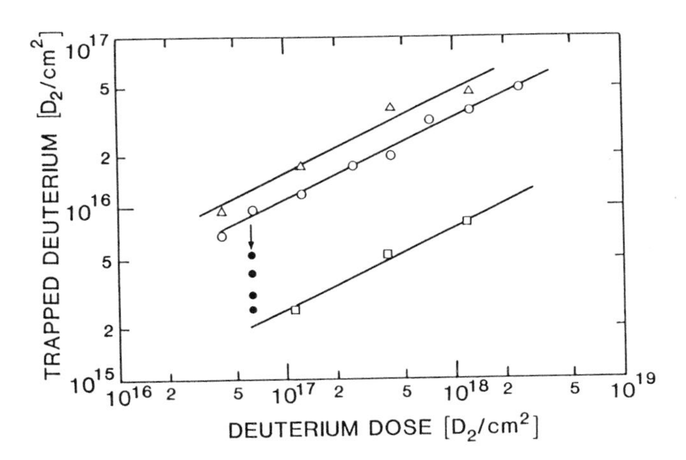

FIGURE 2
Trapped amounts of deuterium vs dose for the as-received MOH1 sample at T_{inj} = 325 K (Δ) and T_{inj} = 450 K (o); The arrow indicates the decrease (●) observed during stepwise sputter-cleaning of MOH1. Trapped amounts in the electro-polished MOH2 sample are also plotted (□).

MOH1 by 250 eV Ar-ions (up to $2x10^{18}$ Ar^+ cm^{-2}) reduced the amount of trapped deuterium also in this sample. Apparently surface conditions played an important role.

Fig.3 shows results obtained when varying the injection temperature. The peak temperature increases with injection temperature and the amounts released decrease.

When the dose rate was increased from the regular $2x10^{15}$ D_2^+/cm^2s to twice this value, the amounts released were about the same but the release took place at 20 K lower temperature (for T_{inj}=450 K and dose 1.25x $10^{18}cm^{-2}$).

The exchange of hydrogen isotopes during hydrogen injection at 325 K is demonstrated in fig.4. Deuterium desorption spectra are shown for equal total doses of hydrogen and deuterium, but a different sequence of implantation: deuterium only (D/D), hydrogen followed by deuterium (H/D) and deuterium followed by hydrogen (D/H). If no isotope exchange had taken place during the injection then in both the (H/D) and (D/H) case the released amount of deuterium would have been equal, i.e. 50% of the release in the (D/D) case. However 75% is observed in the (H/D) and 25% in the (D/H) experiment implying that the latter arriving isotope is very effective in replacing

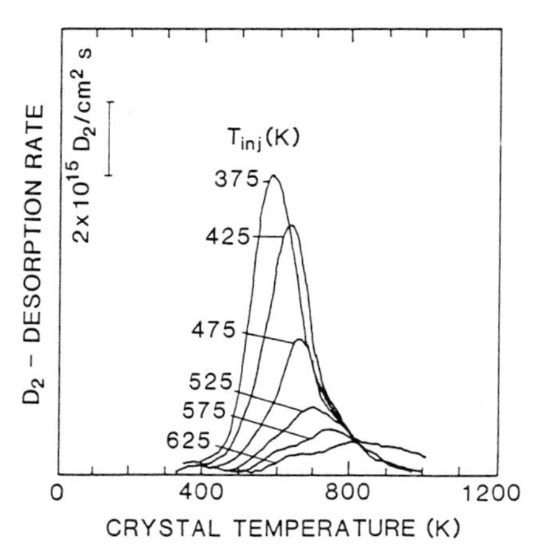

FIGURE 3
Deuterium desorption spectra for MOH1 (as-received) irradiated at the indicated temperatures.

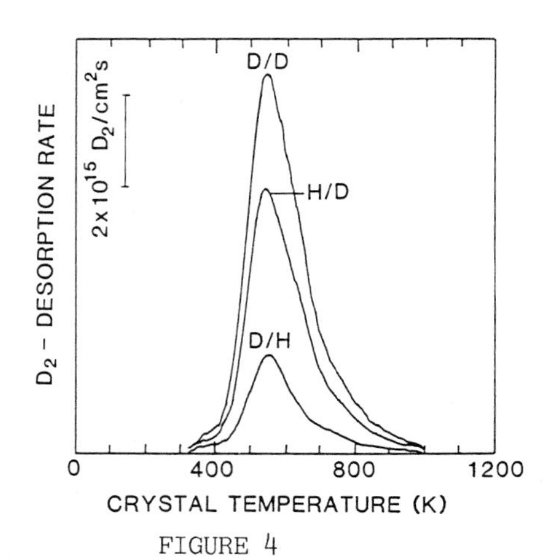

FIGURE 4
Deuterium desorption spectra obtained for MOH1 irradiated with a total dose of $4x10^{17}/cm^2$ hydrogen isotopes: D/D, deuterium only; H/D, 50% H_2 followed by 50 % D_2; D/H, 50% D_2+50% H_2.

earlier trapped isotopes.

Exposure of the sample to the plasma without applying a bias voltage resulted in desorption spectra similar to those obtained with 1 keV D_2^+ irradiation (equal time of exposure). In this case only deuterium particles (neutral deuterium atoms, molecules and ions) with very low energy (<2eV) were hitting the sample. Release temperatures were lower by 50K and released amounts were lower by one factor.

Positron lifetime measurements on the samples irradiated with 10^{20} H_2^+- ions /cm^2 at T_{inj}= 380 K revealed no reduction of the positron lifetime in the voids (470 ps). This indicates that no large amount of hydrogen was present in the voids. Reduction of the positron lifetime was observed only for a neutron irradiated sample ($4x10^{19}$ n/cm^2; annealed at 1100 K) which contained smaller voids. The lifetime reduced from 390 ps to 360 ps.

Deuterium depth profiling of the MOH1 sample by the NRA method indicated that a uniform concentration of deuterium was present at least to a depth of 200 nm (for larger depths the method looses sensitivity). The concentration amounted to $(2\pm0.2)x10^{-3}$ atomic fraction. Assuming that all deuterium was trapped in voids (1.2 appm) the concentration of deuterium atoms in the voids was 1600± 200.

4. DISCUSSION

In the description of the results the behaviour of the hydrogen in the three subsequent stages of the experiment should be taken into account: the implantation at elevated temperature, cooling to RT, and the desorption.

During the implantation a steady concentration C_o of interstitial H or D will be present below the sample surface. The magnitude of C_o can be calculated according to models earlier developed for ion-driven permeation[3]. The dose rate J, the implantation depth L_H, the hydrogen diffusivity D and the surface recombination rate k_r form the relevant parameters.

The amount trapped is found by solving the time dependent diffusion equation for interstitial H or D (concentration C_H) with a sink term equal to the trapping rate: $- Z\ C_H(x,t)\ C_V$, and a source term equal to the detrapping rate: $Z\ C_{H,eq}(N_H,x,t)\ C_V$, where $C_{H,eq}$ is the H or D concentration in thermal equilibrium with the hydrogen in the void, $Z = 4\pi r_V D / N_o$ with r_V the void radius, N_H the number of H or D per void, N_o the atomic density and C_V the void concentration. For voids it can be expected that hydrogen is chemisorbed at the inner surface of the void and that molecular gas will be present in the free volume of the void. The concentrations of chemisorbed, molecular and interstitially dissolved hydrogen will be in thermal equilibrium, i.e. the chemical potentials will be equal $\mu_H=1/2\ \mu_{H2}$.

In fig.5 $C_{H,eq}$ is plotted vs N_H for different temperatures where we assume that only molecular hydrogen exists in the voids, i.e. $N_H=1/2\ N_{H2}$. The values are calculated with the chemical potential derived from the equation of state, valid to high pressures, for molecular deuterium proposed by Mills et al[10]. For low values of N_H the hydrogen concentration is proportional to $N_H^{1/2}$ (ideal gas law) but for $N_H>1000$ the concentration increases much faster with N_H due to high pressure effects. During the implantation the concentration profile of the hydrogen will be such that a local equilibrium exists between hydrogen inside and outside the voids; in other words the voids must be saturated before hydrogen can reach larger depths. Assuming that the concentration in the zone with saturated voids is of the order of the concentration C_o an estimate can be made of the number N_H of hydrogen in the voids. In fig.5 the value of C_o is indicated for different temperatures, assuming that the dose rate J amounted to 2×10^{15} $cm^{-2}s^{-1}$. The corresponding values of N_H vary from 2000 for T=450 K to >10000 for T= 250 K; the pressures vary from 0.1 GPa to >5 GPa.

The depth integrated amount of trapped hydrogen after the implantation of a dose η J t (η = the ion penetration probability ≈ 0.5 at 1 keV) can be approximated by[11]:

$$n_{trapped} = (\eta\ J\ t\ L_H\ C_V\ N_H\ N_o)^{1/2},$$

and the thickness of the layer containing the saturated voids by:

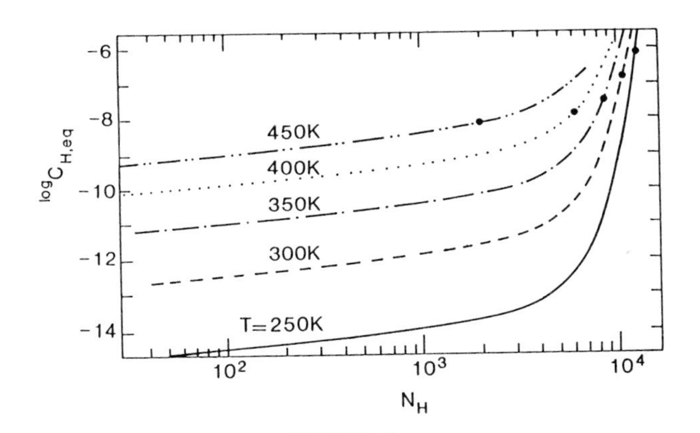

FIGURE 5

The calculated equilibrium concentration $C_{H,eq}$ (see text) vs the number of deuterium in voids (r_V = 2.7 nm) at the indicated temperatures. Values of C_o the concentration during implantation with a dose rate $J=2\times10^{15}$ D_2^+/cm^2s are indicated on the curves(●).Data used in the calculation: $D= 10^{-3}\exp(-0.2(eV)/kT)$ cm^2/s and heat of solution E_s= 0.54 eV.

$x_s = 2\, n_{trapped} / (C_V\, N_H\, N_o)$.
With C_o taken from fig.5 for T= 450 K we calculated $n_{trapped} = 3.7 \times 10^{15}$ D_2/cm^2 with $x_s = 400$ nm (J t $=10^{17}$ D_2^+/cm^2 and $L_H = 20$ nm) which is only 50% higher than the experimental value found for the cleaned samples (fig.2). The amounts measured for the sample with an impurity covered surface are much higher which might be explained by an enhancement of C_o due to a strong reduction of the surface recombination rate[12]. In all cases the predicted dose dependence is found.

When the implantation is stopped the voids are quickly loosing the hydrogen in excess of $N_H \approx 4000$ (beyond this level hydrogen detrapping rates are excessively high). A part of this hydrogen becomes retrapped and the other part is released via the surface. Upon cooling to RT the release will proceed at a slower rate.
The NRA-result obtained for a sample 3 weeks after it had been implanted could well be explained by a continuous room temperature release bringing N_H down to 1600. Part of this (at maximum 930 D-atoms at full surface coverage) could be ascribed to chemisorbed D. Reduction of the positron lifetime is expected[13] when the pressure in the cavities is >5 GPa;in our case corresponding with $N_H > 10^4$ which probably is reached only during implantation.

The observed exchange of isotopes during the implantation supports the assumption of a local (dynamical) equilibrium at the hydrogen filled voids.

We have attempted to describe the desorption spectra by a single-activated diffusion process. Though the trends found in the experimental work could be reproduced, the peak from the experiments is wider than from the calculations. It is clear that trapping and detrapping rates must be included which depend on N_H. Also the release of chemisorbed gas must be taken into account. The contribution to the release spectrum is probably merging with the high temperature tail of the molecular gas release peak. Work on a full analysis of the desorption spectra is in progress.

ACKNOWLEDGEMENT

This work forms part of the European Fusion Technology program (Task Mat. 20.2).

REFERENCES

1. B.L.Doyle,J.Nucl.Mater. 111/112 (1982) 628.

2. F. Waelbroeck, P. Wienhold, and J. Winter, J. Nucl. Mater. 111/112 (1982) 185.

3. M.A. Pick and K. Sonnenberg , J. Nucl. Mat. 131 (1985) 208.

4. S.M. Meyers, D.M. Follstaedt, and F. Besenbacher, in: Surface Alloying by Ion, Electron, and Laser Beams, eds.,L.E. Rehn, S.T. Picreaux and H. Wiedersich (ASM, Ohio,1987) pp. 223-269.

5. B. Nielsen, H.E. Hansen, H.K. Nielsen, M.D. Bentzon and K. Petersen, in: Positron Annihilation, eds., P.C. Jain, R.M. Singru and K.P. Copinathan (World Scientific Publ. Co.,Singapore, 1985) pp. 497-499.

6. T. Tanabe, N. Saito, Y. Etoh and S. Imoto, J. Nucl. Mat. 103 & 104 (1981) 483.

7. K. Erents and G. McCracken, Brit. J. Appl. Phys. Ser. 2, 2 (1969) 1397.

8. J.H. Evans, J. Nucl. Mat. 88 (1980) 31.

9. J. de Vries and F.E.T. Kelling, IEEE Trans. on Nucl. Science 35 (1988) 392.

10. R.L. Mills, D.H. Liebenberg and J.C. Bronson, J. Chem. Phys. 68 (1978) 2663.

11. A. van Veen, H. A. Filius, J. de Vries, K.R. Bijkerk, G.J. Rozing and D. Segers, J. Nucl. Mat. 155/157 (1988) 1113.

12. M.A. Pick, J.Nucl.Mat. 145/147 (1987) 297.

13. H.E. Hansen, H. Rajainmaki, R.Talja, M.D. Bentzon, R.M. Nieminen and K.Petersen, J. Phys. F: Met. Phys. 15 (1985) 1.

FUSION TECHNOLOGY 1988
A.M. Van Ingen, A. Nijsen-Vis, H.T. Klippel (editors)
Elsevier Science Publishers B.V., 1989

TYPE 316L SOLUTION-ANNEALED AUSTENITIC STEEL AS STRUCTURAL MATERIAL FOR THE BASIC NET MACHINE

J.L. BOUTARD AND J. NIHOUL

The NET Team, c/o Max-Planck Institut fuer Plasmaphysik, Bolzmannstrasse 2,
D-8046 Garching bei Muenchen, W. Germany

Type 316L solution-annealed austenitic steel has been selected for use as structural material for the basic first wall and shielding blankets of NET.The constitution and structure of the CEC 316 reference steel are reviewed. The effects of irradiation on mechanical properties and dimensional stability are discussed in terms of the operating conditions of the basic components of NET. Finally, areas where additional data are required are indicated.

1. INTRODUCTION

The structural materials for the first wall and blankets of NET were originally selected to meet the requirements of the two main types of preliminary design concepts, namely the water-cooled liquid breeder and the He-cooled solid breeder[1-3] and to give satisfactory operation at temperatures in the range 250 to 500°C which would be induced in the first wall. Based on the experience gained in the development and/or use of structural steels in fast breeder reactors, austenitic 316L type and 10CrMoVNb martensitic steels have been selected as potential structural materials[4]. As part of the coordinated European Fusion Technology Programme initiated in 1982 and extended for the period 85-89, studies have been launched to acquire the additional NET relevant data which are needed before the choice of the optimised material can be made for the detailed design phase of NET, scheduled to start at the beginning of 1990[5].

Since these initial choices were made, the design studies have proceeded within the NET Team and have progressively favoured a first wall concept with non-boiling, low pressure cooling water because it allows low temperature and primary stress operating conditions[6]. This concept is designed to give good reliability and availability under the highly uncertain loading conditions and has therefore recently been chosen as the reference design for the basic first wall and shielding blankets in the Physics and Technology Phases of NET[7].

The martensitic steels exhibit high embrittlement under neutron irradiation at low temperature[8] which make them unsuitable for the basic first wall and blankets of NET[6]. However they are likely to have in service properties such as good dimensional stability and resistance to thermal fatigue and He-embrittlement which make them reactor relevant. Therefore they are foreseen as structural material for the Tritium breeding blankets to be tested in the Technology phase of NET, especially for the LiPb type because of good compatibility with this liquid metal[9].

The objectives of this paper are to review the data now available on 316 type stainless steels from both the open literature and the European Fusion Technology Programme and reassess the use of the solution-annealed CEC reference 316L steel as structural material for the basic first wall and blanket of NET. However, a complete review of the behaviour of 316 type steels is beyond the scope of this paper. Consequently, emphasis will be given to addressing the effects of irradiation on the mechanical properties and dimensional stability.

2. BASIC FIRST WALL OPERATING CONDITIONS

The reference parameters of the basic first wall of NET have not been totally finalised yet. However the following specifications are being assumed for design and material development of this component[7]:

TABLE I:

Phase	Physics	Technology
Average neutron flux (MW/m^2)	1	1
Average heat flux (MW/m^2)	0.15	0.15
Peak heat flux (MW/m^2)	<1	<1
Number of pulses (10^4)	1	8
Pulse duration (s)	100	300
Integrated time of burn (h)	300	7000
Neutron fluence ($MWyr/m^2$)	0.03	0.8
Neutron dose (dpa)	0.3	9
He production in 316-Type (appm)	4	118
H production in 316-Type (appm)	15	455
Coolant temperature (°C)	60/100	60/100

The basic first wall is assumed to be fully protected by conductively or radiatively cooled graphite tiles. The minimum operating temperature is that of the cooling water i.e. 60/100°C. The maximum temperature is strongly design dependent and thus rather uncertain. Thermomechanical studies performed with conductive tiles show that ~425°C is a reasonable value for a peak heat flux of 0.5 MW/m^2 [10]. However due to numerous uncertainties including that of the peak heat flux, some hot spots with temperatures of $\gtrsim$500°C are not to be excluded. Therefore the bulk of the paper will be devoted to irradiation effects below ~400 to 450°C but typical high temperature features such as He-embrittlement will also be addressed.

The stress and strain levels in the first wall are also very dependent on the design. However the plasma facing side is the most fatigue loaded and has to sustain total strain range up to ~0.4% due to the pulsed operating regime of NET[6]. This side is in compression during the burn phase and tension during the dwell phase. Creep during burn will relax the stress and increase the tensile stress during off-burn as well as the mean stress of the total cycle[5].

Irradiation conditions such as H and He to dpa ratio are typical of the fusion 14MeV neutrons. In the absence of a source of such neutrons which is capable of irradiating large specimens at a representative dose rate, data on the effects of irradiation on the dimensional stability and mechanical behaviour of structural materials have to be simulated by mixed spectrum, thermal and fast reactor irradiations, supplemented as appropriate by charged particles bombardments in high voltage electron microscopes, accelerators and cyclotrons.

3. STRUCTURAL MATERIAL: CONSTITUTION, STRUCTURE

The chemical composition of the solution-annealed CEC 316L reference steel is within the commercial specification of 316L-167-SPH used as the structural material for the primary vessel of SUPERPHENIX. The composition of the reference heat studied in the Fusion Technology Programme is given in Table II (weight %):

TABLE II:

C	Cr	Ni	Mn	Mo	Si	N	B
0.024	17.5	12.3	1.82	2.30	0.45	0.060	0.0008

The solution-annealed condition associated with a low carbon content results in excellent micro-structural stability during fabrication processes such as welding or brazing. In particular, the risk of sensitization and subsequent intergranular corrosion is greatly reduced[11].

The controlled nitrogen content gives allowable stresses comparable with those of heat resistant grades such as 316H or 304H (0.04% <C<0.1%) specified in ASME standards[12]. This is shown in Fig.1 where the allowable stresses of the low temperature ASME grades 304L and 316L (<0.03%) are also given for comparison: they are 20 % below those of the heat resistant grades.

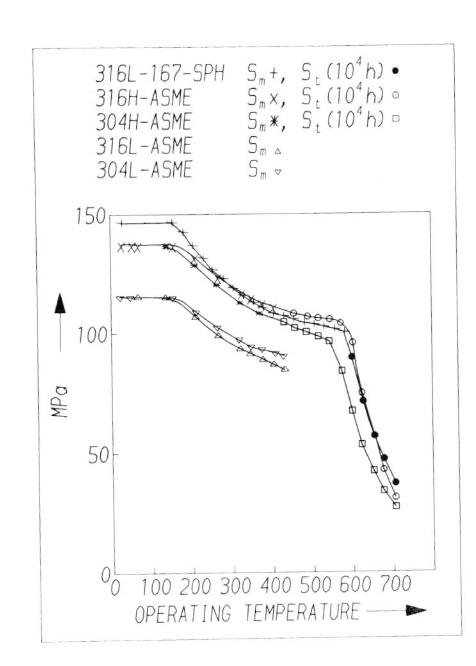

FIGURE 1
Allowable stress versus temperature for various grades of solution-annealed 304 and 316 type austenitic steels

The structure following rapid cooling from the solution-annealing heat treatment is mainly austenitic with ~0.5 to 1% by volume of . ferrite measured by ferrometer[13]. This structure is in good agreement with predictions by Schaeffler-type diagrams[5]. It results in a nearly paramagnetic material and limits the risk of hot cracking during welding[14]. The latter point could be balanced by the embrittlement due to the ($\alpha + \alpha'$) decomposition of the δ ferrite under irradiation at 300 to 500°C. However due to the small volume fraction of the δ ferrite the overall loss of ductility and fracture toughness of the steel is expected to be limited[15].

4. IRRADIATION HARDENING, EMBRITTLEMENT

The effect of neutron irradiation on the tensile properties of the CEC reference steel has been largely addressed in the Fusion Technology Programme for temperatures and doses ranging from 250 to 550°C and 3 to 10 dpa[16-20]. Similarly, tensile data on solution-annealed 316 type steels irradiated at 50 to 100, 250 and 400 to 450°C are rather abundant in the open literature[21-24]. The irradiations having been performed in different reactors with results reported versus dose or fluence, the main characteristics concerning fluence per dpa and He to dpa ratio are gathered in Table III.

The H/dpa ratio in 316-type steels is also strongly dependent on the reactor: ~10 for mixed neutron spectrum reactors such as BR2[31], HFR or R2 but only ~1 for fast breeder reactors.

TABLE III:

Reactors	BR2 (BELGIUM)	HFR (NL)	R2 (SWEDEN)	HFIR (USA)	RAPSODE PHENIX (FRANCE)
Fluence per dpa					
E > 0.1 MeV	~1.5x10^{21} [25]	~BR2	~BR2	~1.2x10^{21} [29]	~1.7x10^{21} [30]
E > 1 MeV	~0.7x10^{21} [25]	~BR2	~BR2		
(n/cm^2)					
appm He/dpa (316-type)	~10 [26]	~10 [27]	~5 (at 3dpa)[28] ~10(at 10 dpa)[28]	~25 to 55[29]	~1 (Fast Reactors)

The effect of irradiation at temperatures below 400-450°C will be discussed first.

4.1. Effect of spectrum and/or chemical composition (fig. 2)

Irradiations in HFR and HFIR at 250°C up to 10 dpa showed that, the strengthening of solution-annealed 316 steels is not significantly dependent on the chemical composition or neutron spectrum[24,29]. Fig.2 shows that this conclusion has to be extended to the CEC reference 316L heat irradiated in R2 up to 5 dpa[19]. In addition, the CEC reference heat irradiated at 430°C up to ~10 dpa in HFR[16], BR2[18] or ~20 dpa in Phenix[20] has the same behaviour as medium carbon 316 type steels in Phenix and Rapsodie[23].

From the results shown in figure 2 after irradiation at 50 to 100°C in HFR[21] and HFIR[22] it is difficult to draw any conclusion, because the doses are too varied. However, yield stress increases of solution-annealed 316 steel irradiated at 90°C with 14 MeV or mixed spectrum neutrons up to low doses ($\lesssim 10^{-2}$dpa) correlate quite well if dpa are used[32], indicating no strong dependence on neutron spectrum.

4.2. Strength saturation behaviour (fig. 2)

Solution-annealed 316-type steels exhibit saturation of the yield strength which depends markedly on the irradiation temperature. Irradiated at ~425°C in the Rapsodie and Phenix fast breeder reactors they show saturation of the yield stress at 3 to 4 x 10^{22} n/cm^2 (E>0.1 MeV) i.e. 15 to 20 dpa[23]. On irradiation in a mixed spectrum the saturation fluences are either smaller than 0.2 x 10^{20} n/cm^2 (E>1 MeV) i.e. $\lesssim$ 0.1 dpa for irradiation at 50 to 100°C[21] or $\lesssim$ 5pda for irradiated 250°C[19,29]. In all cases the saturated yield stresses are high compared to the initial ones and do not seem to depend significantly on the irradiation temperature.

Consequently, locations of the first wall of NET at temperatures below $\lesssim$ 100°C should have their strength saturated before the end of the Physics Phase whereas locations at 400-450°C should not have reached strength saturation even at the end of the Technology Phase.

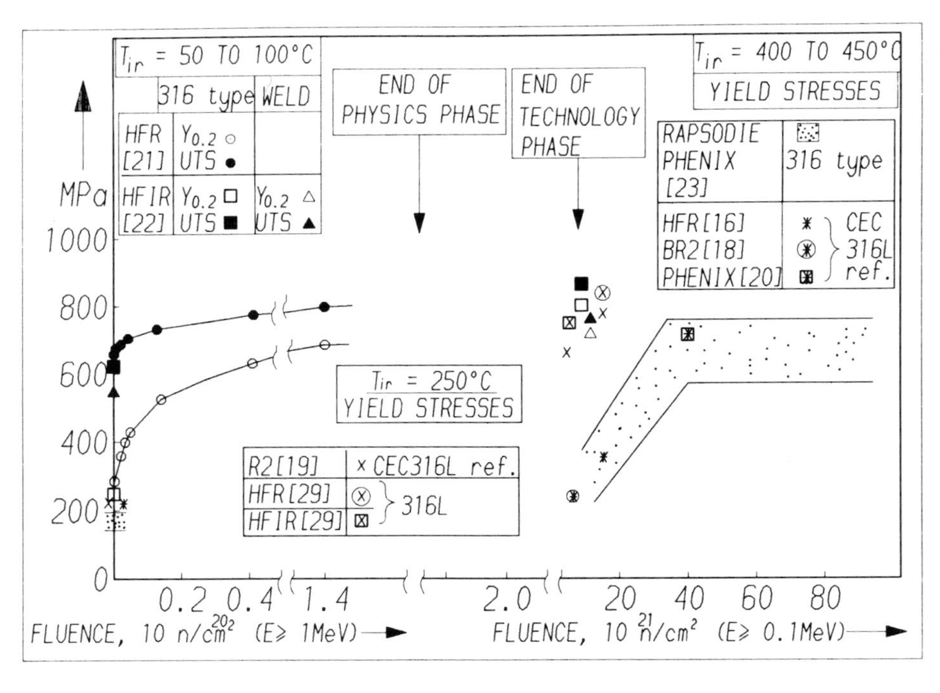

FIGURE 2
Yield and ultimate tensile strength versus fluence for solution-annealed 316 type steel irradiated between ~50 and ~400°C in mixed and fast neutron spectra

4.3. Dependence of ductility on temperature and initial metallurgical condition

The strengthening of the material is accompanied by a corresponding loss of ductility. However the CEC reference 316L retains acceptable uniform elongation of ~ 25% and ~2% after irradiation at 430°C up to 5 and 20 dpa respectively[16,20]. For the latter irradiation conditions electron beam welds exhibit uniform elongation of ~2%, which is comparable to the parent material[20].

At lower temperature the post irradiation strain behaviour seems to be significantly more dependent on the initial metallurgical condition. After irradiation up to 5 and 10 dpa at 250°C in HFIR[24], solution-annealed 316L steel retains uniform elongation ≳5% whereas softening appears during tensile testing for plastic strains ≲1% in the case of 20% cold-worked 316 steel.

After irradiation at 50-100°C the difference in ductility between solution-annealed and cold-worked steels is even higher. Solution-annealed 316 type steels retain uniform elongation higher than 20 % after irradiation up to ~2 dpa in HFR[21] or 10 dpa in HFIR[22]. In this latter irradiation, 20% cold-worked 316 steel specimens were also present and exhibited plastic instability with subsequent softening after plastic strain ~0.5%, the total elongation being ~10%. In the same condition weldments of a 16-8-2 type steel showed strain behaviour intermediate between those of solution-annealed and cold-worked parent materials[22].
This difference in ductility between solution-annealed and cold-worked steel is in a good agreement with microstructural examination. In addition to Franck loops and black dots created by irradiation at ~100°C in both metallurgical conditions, the initial dislocation network of the cold-worked steel is modified by irradiation but still present[22]. This result in a significantly higher yield stress and limited strain-hardening.

Finally it should be noted that the measured ductility after irradiation at ~50 to 100°C could not be totally conservative. In fact hydrogen embrittlement could have been underestimated due to too low a H-production, since the appm H/dpa ratio was ~ 10 instead of 50 in fusion environment (see § 9).

4.4. Tensile properties after irradiation in the range 550 to 650°C

The major feature is that of He-embrittlement. Specimens of the CEC 316L reference heat irradiated in HFR up to ~5 dpa exhibited intergranular failure when tensile tested at the irradiation temperature of 625°C under low strain rate ($10^{-6}s^{-1}$)[16].

4.5. Main conclusions

The relative independence of strengthening on neutron spectrum, He/dpa and chemical composition gives some confidence in the applicability of these results to the NET first wall and shielding blanket. Compared with the cold-worked material, 316-type steels in the solution-annealed condition give a better compromise between ductility and strength when irradiated at low temperature.

5. LOW CYCLE FATIGUE (LCF)

The first wall will be subjected to biaxial thermal fatigue as a result of the cyclic strains produced by the temperature changes during the burn and off-burn periods of the plasma. The design studies have amply demonstrated that this loading will limit the lifetime of the first wall of NET[6]. The acquisition of biaxial thermal fatigue data is difficult. Therefore most of the data on LCF of the CEC 316L reference steel were obtained under classical uniaxial strain-controlled isothermal conditions[16,18,34], bearing in mind that experimentally based procedures to predict biaxial thermal fatigue lives of 316 type steels using the classical LCF data do exist[33].

5.1. Non-irradiated material

The fatigue life consists of the initiation and growth of the crack damage. Analyses of striated fracture surfaces show that for 10^4 and 10^5 cycles to failure, the initiation phase controls respectively ~50 and ~90% of the total life at 425°C[35]. This is in good agreement with the large decrease in fatigue life resulting from 'surface' micro-cracks induced by simulated disruption damage[35] and justifies the use of extensive first wall protection in NET to take advantage of this long initiation phase. The fatigue life therefore, should not be reduced despite the occurrence of disruptions.

The effect of hold time on the fatigue life has been investigated by changing the frequency of the strain cycles from 0.1 to ~0.001 Hz. Either insignificant[18] or small[16] effects are reported. In the latter case the fatigue life is decreased by a factor ~2 when the frequency is reduced by two orders of magnitude. Such an effect is surprising at temperatures in the range of 325 to 425°C because both thermal creep and strain rate effects on tensile properties are negligible. However it could reveal some sensitivity of the fatigue properties to the final heat treatment: the CEC reference steel having been given additional swaging and solution-annealing before irradiation in one experiment[18] but irradiated in the as-received condition in the other[16].

At higher temperatures thermal creep during hold times is known to cause a change in the fracture mode from trans- to intergranular, resulting in a significant reduction of the number of cycles to failure[36].

Finally, as expected, the solution-annealed 316L reference steel cyclicly hardens as shown in figure 3b.

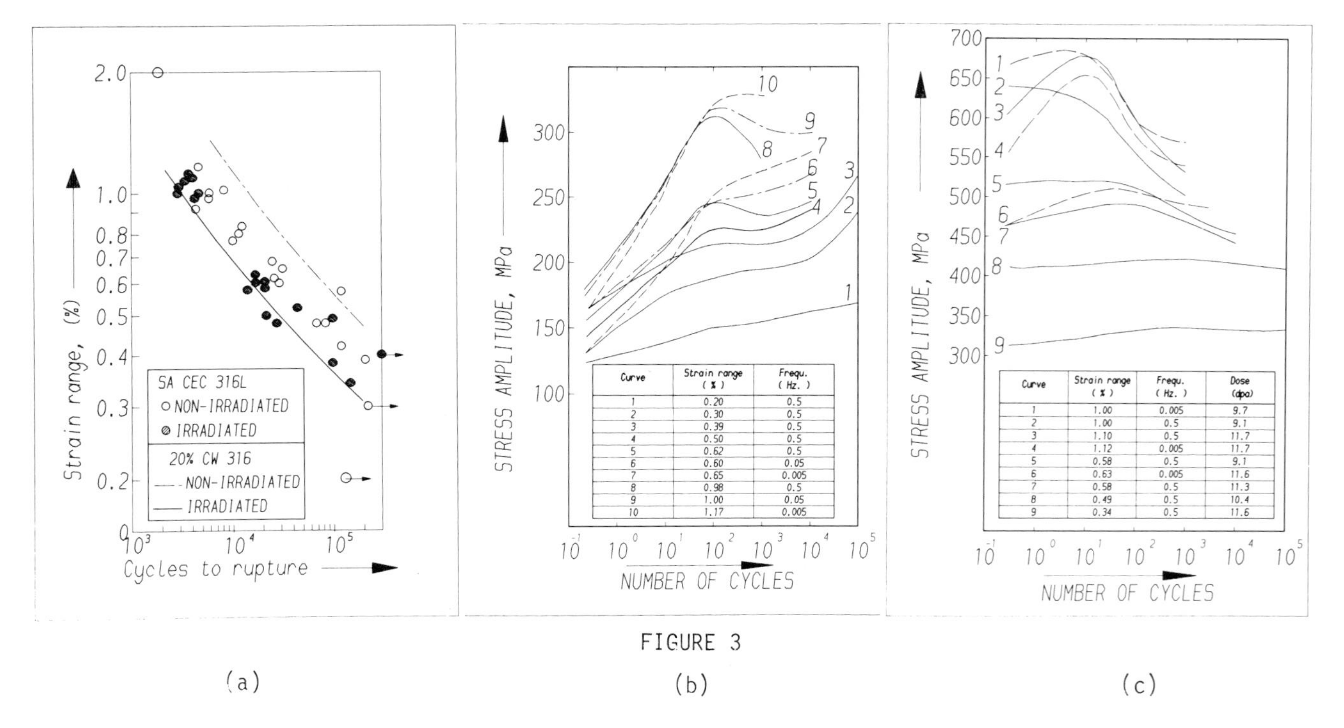

Curve	Strain range (%)	Frequ. (Hz.)
1	0.20	0.5
2	0.30	0.5
3	0.39	0.5
4	0.50	0.5
5	0.62	0.5
6	0.60	0.05
7	0.65	0.005
8	0.98	0.5
9	1.00	0.05
10	1.17	0.005

Curve	Strain range (%)	Frequ. (Hz.)	Dose (dpa)
1	1.00	0.005	9.7
2	1.00	0.5	9.1
3	1.10	0.5	11.7
4	1.12	0.005	11.7
5	0.58	0.5	9.1
6	0.63	0.005	11.6
7	0.58	0.5	11.3
8	0.49	0.5	10.4
9	0.34	0.5	11.6

FIGURE 3

(a) (b) (c)

Fatigue behaviour of the solution-annealed CEC reference 316 L: a) fatigue life; b) cyclic hardening (non-irradiated); c) cyclic softening (irradiated)

5.2. Irradiated material

The effect of irradiation on the LCF behaviour of the CEC 316L reference heat has been assessed at the irradiation temperature of (i) 430°C up to 5 and 10 dpa in HFR[34b] and BR2[18] respectively and (ii) 525°C up to 5 dpa in HFR[34a].

At 430°C irradiation decreases the fatigue life of the CEC 316 reference steel by a factor of $\lesssim 2$ as shown in figure 3-a, where results on 20% cold-worked 316 type steel irradiated also at 430°C and up to 10 dpa but in HFIR[37] are given for comparison. The better initial resistance to fatigue of the 20% cold-worked 316 steel is smoothed out by irradiation.

Similar to the non-irradiated material insignificant[18] or small[34b] effects of the frequency are reported after irradiation at 430°C. At 525°C a frequency effect is reported[34a] and attributed to the time dependent intergranular He-embrittlement fracture mechanisms which will be discussed in § 7.2.

In contrast to the non-irradiated material, the irradiated specimens exhibit cyclic softening after 10 or 10^2 cycles when tested under a total strain range $\gtrsim 0.6\%$ as shown in fig. 3-c[17]. Below this strain range significant cyclic hardening or softening is not detected.

5.3. Main issues

In the above LCF testing, a previously irradiated microstructure is submitted in hot cell to cyclic plasticity. In an actual first wall, the cyclic hardening, occurring within the first 10^2 cycles, will be achieved in a structural material which basically retains its solution-annealed initial condition. It is this cyclicly hardened microstructure which will be submitted to both irradiation and cyclic plasticity. The procedure of post-irradiation testing is thus a very simplified one and its relevance to NET first wall behaviour is being tested by additional in-pile and in-beam fatigue experiments.

At low temperatures, typically below 200°C, 300 series steels exhibit cold creep that can induce significant plastic strains especially when relaxation occurs at stresses close to the yield stress[39,40]. Due to the low dose of the Physics Phase the irradiation creep strains will be negligible (see § 7) and consequently the cold creep will control the stress relaxation during the burn. Therefore its effect on fatigue life and cyclic plasticity, which is poorly documented, is to be assessed.

Finally, the endurance limit and the margins taken to derive the design fatigue curve, a factor of 1/20 or 1/2 of the experimental number of cycles to failure or strain range respectively, have to be determined or reassessed for irradiated material. In the process, the understanding of the effect of irradiation on the crack initiation and growth phases should provide useful guidelines.

6. FRACTURE TOUGHNESS

After irradiation at ~250 and ~400°C up to ~10, ~15 and ~50 dpa in fast breeder reactors such as DFR[41] or EBRII[42,43] or in test reactors[43], stainless steels of 300 series show a decrease in fracture toughness down to ~100 MPa $\sqrt{m}$ and ~70 MPa$\sqrt{m}$ for solution-annealed and welded conditions respectively. The fracture toughness of solution-annealed 348-type steel obtained after irradiation in Advanced Test Reactor (ATR) at ~400°C up to ~50 dpa with a final He content of ~250 appm was 30% lower than that measured after irradiation in EBRII in the same condition of dose and temperature but with a final He content of ~25 appm[43] implying a possible effect of He-content on fracture toughness.

After irradiation at temperatures down to ~ 50°C, no data seem to be available. In addition, the good tensile ductility after irradiation at these temperatures is obtained by extensive shearing[21], so it would be misleading to use this as an index of excellent fracture toughness.

Consequently a significant but acceptable reduction of fracture toughness of the CEC reference steel is expected and needs to be assessed in all the operating temperature ranges of basic components for solution-annealed and welded conditions.

7. CREEP AND RUPTURE

Austenitic steels do not deform by thermal creep processes at temperatures <480°C; at these temperatures viscoplastic deformation will occur by irradiation creep.

7.1. Irradiation Creep

It is desirable that the irradiation creep rate should be low enough to restrict the degree of deformation under the primary stresses. The irradiation creep relaxes the secondary stresses due to thermal or swelling gradients. It is thus benefitial. However for any location in compression during burn, such as the plasma facing surface of the first wall, the relaxation of the secondary stresses will increase the mean stress of the cycle and result in a decrease of the fatigue life[5]. Therefore it is also desirable to limit the irradiation creep rate due to the secondary stresses when fatigue controls the lifetime.

The existing evidence suggests that such relaxation controlled by irradiation creep is not accompanied by any cracking and ductility exhaustion phenomena. Therefore the interaction between irradiation creep and fatigue is only expected to decrease the life-time slightly.

Light-ion irradiations have confirmed the results of the open literature on the steady state irradiation creep rate obtained mostly under fast neutron irradiation up to a few 10 dpa [see for example[44]]. Down to 200°C steady state creep rate is (i) weakly dependent on temperature (ii) linear with dose rate[45] and stress with quadratic stress behaviour above ~150 MPa[46] and (iii) rather insensitive to parameters such as cold-working, He[47], C or N contents[48].

In fact, the irradiation regime of NET will be pulsed implying at the same time a pulsing of the neutron flux, temperature and stress. Considering the dynamic preference of the faster interstitials to migrate to dislocations during the transient loading conditions of start up and shutdown of the plasma, theoretical calculations predicted enhancement of the creep rate[49,50]. However these calculations over-estimate the phenomenon due to too high a migration energy of vacancies and the maintenance of high temperature during shutdown[51]. Cycling the flux of light ions with and without temperature changes does not significantly enhance the irradiation creep rate of 316-type steels[52]. Stress changes cause transient strains which completely recovered upon returning to the original stress level[46]. Consequently, stationary creep rates apply to the pulsed conditions of NET so that negligible stress relaxation (~10%) is anticipated in the Physics Phase. Conversely, the irradiation creep will make a major contribution to stress relaxation in the Technology Phase.

7.2. Thermal creep

The rupture life and ductility of 316 steel above 500°C are significantly reduced during or following irradiation, mainly due to He-embrittlement and premature intergranular fracture. This effect produces strong creep fatigue interaction and dominates any mean stress effect on the fatigue life when thermal creep is present.

This He-embrittlement occurs even at very low He-content. CEC 316L reference steel containing 10 appm He produced by ^{10}B (n,α) Li under thermal neutron irradiation at 50°C, exhibits only 10% of the initial rupture life when creep tested at 625°C under 190 MPa[53]. No bubbles were observed. The rupture is due to wedge-type cracks. It is suggested that they are promoted by a He-induced decrease in surface energy of the grain boundaries[54].

At higher He content, from 10 to 1000 appm resulting from cyclotron irradiation, He-bubbles are present throughout the whole material. Their nucleation and growth have been extensively studied[55]. The onset of unstable growth of the grain-boundary bubbles controls the creep rupture time: 316 samples creep tested after He implantation resulting in bubbles with the critical radius fail instantaneously[38].

Finally it is to be noted that since the plasma facing side of the first wall will be in compression during the burn, the mechanism of critical growth of bubbles should not occur. However wedge crack initiation due to the weakening of grain-boundaries occurring under compression as well as tension should be effective in damaging the hottest parts of the first wall during the burn.

8. VOID SWELLING

Irradiation-induced void swelling could have several consequences. The whole of the first wall and blanket structure may dilate as a result of uniform swelling, or distort due to non-uniform swelling induced by temperature and neutron flux spatial variations. The dilation should not produce failure but might impair the operation of the component. The distortion is generally not free and will induce internal stresses and possibly additional damage. The voids can also act as nuclei for dimples and impair tensile ductility and ultimate tensile strength[56].

The void swelling of solution-annealed 316L austenitic steel irradiated in Phenix in the range from 400 to 435°C is only ~0.5% by volume at 10 dpa but becomes as high as ~4% at 35 dpa[5]. In the range 200 to 300°C, the recombination-controlled low temperature range is approached. A low swelling rate is expected at these temperatures and even no swelling would be anticipated at lower temperatures.

The effects on the void swelling of He and He plus H concentration higher than those obtained in fast reactors have been studied by charged particles to doses of <200 dpa[57,58]: increased concentrations of H and He appear to increase the void swelling of 316 type and austenitic steels.

9. HYDROGEN EMBRITTLEMENT

The principal sources of H isotopes in protected first wall and shielding blanket structural materials are: (i) hydrogen formed in the steel by (n,p) reactions, (ii) hydrogen produced by corrosion at the steel - coolant water interface (iii) hydrogen added to the water to inhibit radiolytic decom-position.

The solubilities, diffusivities and permeabilities of H isotopes are reasonably well established in non-irradiated 316 type steel[5]. At low temperature ~100°C the diffusion mean path $\sqrt{2Dt}$ during a lifetime of ten years is $\lesssim$ 1cm i.e. comparable to or smaller than the thickness of the plates involved in the first wall and shielding blankets. In addition, there is some evidence that irradiation-induced point defects and/or transmutation products such as He act as traps for hydrogen and thereby decrease the diffusivity and permeability[59]. Additional investigations are required to confirm and further quantify these effects, especially at low temperatures.

The austenite in Type 316 steels containing 13-14% Ni is stable towards quenching, deformation and hydrogen charging induced transformation to ε and α' martensites, which promote hydrogen cracking and loss of ductility in the more unstable (low Ni) austenitic steels. However the low temperature and possible He trapping effect could result in the trapping of most of the H produced in the alloy and have a higher impact on mechanical properties and fracture behaviour than observed on material irradiated in mixed spectrum reactors where the H to dpa ratio is only ~10 appm/dpa.

10. CONCLUSION

The solution-annealed CEC 316L reference steel seems to retain a reasonable compromise between strength, ductility and dimensional stability in tests simulating the operating conditions of the basic first wall and shielding blankets of NET. However, some points need additional clarification:

(i) Tensile strength and ductility of materials irradiated in the temperature range $\lesssim$250°C are more dependent on the initial metallurgical condition than at higher temperatures ~400°C. Consequently special care needs to be taken in evaluating tensile properties of welds that are generally less ductile than parent material.

(ii) The low cycle fatigue behaviour at temperatures $\lesssim$250°C will be controlled by cold creep mechanisms, at least during the Physics Phase. Effect of cold creep relaxation during hold time on fatigue life, and cyclic plasticity is poorly documented and clearly needs further assessment.

(iii) The fracture toughness of parent materials and welds is significantly reduced on irradiation below ~400°C, and has to be evaluated.

(iv) The behaviour of hydrogen isotopes in irradiated microstructures is poorly documented and hydrogen embrittlement due to irradiation at low temperatures could be underestimated due to too low a H/dpa ratio in neutron mixed spectrum irradiation.

REFERENCES

1. A. Cardella, F. Farfaletti-Casali, M. Biggio and A. Inzaghi, Proc. Thirtheenth Symposium on Fusion Technology, Varese (Italy), Sept. 1984, 1, p.297.
2. G. Vieider, W. Daenner and B. Haferkamp, Ibidem, 2, p.1363.
3. M. Biggio, G. Casini, F. Farfaletti-Casali, C. Ponti, M. Rieger and P. Li Bassi, Ibidem, 2, p.1355.
4. D.R. Harries, J.M. Dupouy and C.H. Wu, J.Nucl. Mater., 133-134 (1985) 25.
5. D.R. Harries, Rad. Effects, 101 (1986) 3.
6. M. Chazalon, J.L. Boutard, M.I. Budd, A. Cardella, W. Daenner, P. Dinner, D.S. Evans, M. Iseli, B. Libin, F. Moons, J. Nihoul, M.A. Vassiliadis, G. Vieider, C.H. Wu and E. Zolti, Fusion Technology, 14, No.1 (1988) 82.
7. G. Vieider, A. Cardella, M. Chazalon, F. Engelmann, H. Gorenflo, B. Libin, B. Pavan, J. Raeder, E. Theisen and C.H. Wu, Proc. of the Int. Symposium on Fusion Nuclear Technology, Tokyo (Japan), April 1988, in print.
8. K. Ehrlich and K. Anderko, Proc. of Int. Symp. on Fast Breeder Reactors: Experience and Future Trends, Lyon (France) July 1985, p. 231.
9. T. Flament, P. Fauvet, B. Hocde and J. Sannier, Corrosion of Martensitic Steels in Flowing 17Li83Pb Alloy, this volume.
10. E. Zolti, NET Internal Note NET/IN/88-01.
11. D. Peckner and I.M. Bernstein, Handbook of Stainless Steels, MacGraw Hill Book Company, Chapter 4, p.45.
12. ASME-SA240. Specification for Heat Resisting Chromium and Chromium Nickel Stainless Steel Plates, Sheets and Strips for Pressure Vessels, p.261.
13. Compte-Rendu de Fabrication - Creusot-Loire (1981).
14. Metals Handbook, 9th Ed., Vol. 3: Properties and Selection: Stainless Steels, Tool Materials and Special Purpose Materials, ASME 1980, p.51.
15. A.A. Tavassoli, J. Nucl. Mater. Proceedings of ICFRM3, Karlsruhe (Germany), Oct. 1987, in print.
16. B. Van der Schaaf, Proc. Fourteenth Symposium on Fusion Technology, Avignon (France), 1986, Pergamon Press, p. 993.
17. W. Vandermeulen, Ibidem, p.1025.

18. W. Vandermeulen, W. Hendrix, V. Massault and J. Van de Velde, Report FT/MOL87-08, SCK/CEN Mol (Belgium), J. Nucl. Mater. Proceedings of ICFRM3, Karlsruhe (Germany), Oct. 1987, in print.

19. U. Bergenlid, Studsvik (Sweden), private communication (21.5.88).

20. A. Tavassoli, R. Schauff, J. Menard and R. Cauvin, CEN-Saclay (France), Note technique no.88-1653.

21. H.R. Higgy and F.H. Hammad, J. Nucl. Mater. 55 (1975) 177.

22. F.W. Wiffen and P.J. Maziasz, J. Nucl. Mater. 103-104 (1981) 821.

23. J.M. Dupouy, J. Erler and R. Huillery, Proc. of Int. Conf. on Radiation Effects in Breeder Reactor Structural Materials, Scottsdale, Arizona (USA), June 1977, Ed. M.L. Bleiberg and J.W. Bennett, p. 83.

24. K. Tangri and P. Schiller, J. Nucl. Mater., Proceedings of ICFRM3, Karlsruhe (Germany), Oct. 1987, in print.

25. Ch. de Raedt, Note 205/86-02, Reactor Physics - SCK/CEN Mol (Belgium).

26. Ch. de Raedt, Note 205/86-04, Reactor Physics - SCK/CEN Mol (Belgium).

27. Fusion Technology Programme - ECN 157, compiled by J.D. Elen - July 1984, ECN Petten, (The Netherlands).

28. K. Pettersson, NS 87/136, Studsvik (Sweden).

29. B. Van der Schaaf, M. Grossbeck and H. Scheurer, EUR 10659 EN, Commission of the European Communities, Luxembourg.

30. A. Boltax, A. Bianchera and P.J. Levine, Proc. of Int. Conf. on Irradiation Behaviour of Metallic Materials for Fast Reactor Core Components, Ajaccio, Corsica (France), 1979, Ed. J.P. Poirier and J.M. Dupouy, p. 331.

31. Ch. de Raedt, SCK/CEN Mol (Belgium), private communication.

32. H.L. Heinish, S.D. Atken and C. Martinez, J. Nucl. Mater. 141-143 (1986) 807.

33. D.J. Marsh, Fatigue of Engineering Materials and Structures 4 (1981) 179.

34a. M.I. de Vries and B. Van der Schaaf, this volume.

34b. B. Van der Schaaf, this volume.

35. Fusion Technology Programme, ECN-185. Compiled by J.D. Elen, p.44.

36. P. Dabo, C. Levaillant, A. Pineau, I. Grattier and M. Mottot, Fatigue a haute temperature, Journee Internationale de Printemps, Paris, Juin 1986, Societe Francaise de Metallurgie, p.40.

37. M.L. Grossbeck and K.C. Liu, Nucl. Techn., 58 (1982) 538.

38. I.S. Batra, H. Ullmaier and K. Sonnenberg, J. Nucl. Mater., 116 (1983) 136.

39. A.P.L. Turner and D.G. Martin, Metall. Trans. 11a (1980) 475.

40. D.S. Wood, A.B. Baldwin, T.J. Sarbutts and K. Williamson, Proc. of Int. Conf. on Materials for Nuclear Reactor Core Application, Bristol (U.K.), Oct. 1987, BNES, p.37.

41. E.A. Little, J. Nucl. Mater. 139 (1986) 261.

42. D.J. Mitchel and R.A. Gray, J. Nucl. Mater., 148 (1987) 194.

43. F.M. Haggag, E.G and G. Idaho Inc., Report EGG-M-14984/DE008440.

44. D. Mosedale, D.R. Harries, J.A. Hudson, G.W. Lewthwaite and J.R. McElroy, Proc. of Int. Conf. on Radiation Effects in Breeding Reactor Structural Materials, Scottsdale, Arizona (USA), June 1977, Ed. M.L. Bleiberg and J.W. Bennett, p. 209.

45. P. Jung, Proc. of Int. Conf. on Dimensional Stability and Mechanical Behaviour of Irradiated Metals and Alloys, Brighton (UK), April 1983, BNES, p.133.

46. P. Jung, J. Nucl. Mater. 113 (1983) 133.

47. R. Scholtz, JRC-Ispra, private communication (July 1987).

48. P. Jung and H. Klein, KfK Juelich, in print.

49. H. Gurol and N.M. Ghoniem, Rad. Effects 52 (1980) 103.

50. H. Gurol, N.M. Ghoniem and W.G. Wolfer, J. Nucl. Mater. 99 (1981) 1.

51. P. Jung and H. Ullmaier, Rad. Effects 103 1987) 21.

52. P. Jung, J. Nucl. Mater. 113 (1983) 163.

53. Fusion Technology Programme, ECN-177. Compiled by J.D. Elen, p.41.

54. B. Van der Schaaf and P. Marshall, Proc. of the Int. Conf. on Dimensional Stability and Mechanical Properties of Irradiated Metals and Alloys, BNES, Brighton (UK), April 1983, p.141.

55. H. Ullmaier, Rad. Effects 101 (1986) 147.

56. J.M. Dupouy, J.P. Sagot and J.L. Boutard, Proc. Int. Conf. on Dimensional Stability and Mechanical Behaviour of Irradiated Metals and Alloys, Brighton (UK), April 1983, BNES, Paper 30, Vol I, p.155.

57. D. Gilbon and C. Rivera, Proc. Thirteenth Symposium on Fusion Technology, Varese (Italy), September 1984, Vol. 2, p.1069.

58. D.J. Mazey et al., AERE-R 11710.

59. K. Fukushima and K. Ebisawa, J. Nucl. Mater. 127 (1985) 109.

FUSION TECHNOLOGY 1988
A.M. Van Ingen, A. Nijsen-Vis, H.T. Klippel (editors)
Elsevier Science Publishers B.V., 1989

DESIGN FOR A FUSION MATERIALS IRRADIATION FACILITY

Carl E. WALTER and F. H. COENSGEN

Lawrence Livermore National Laboratory, P. O. Box 808, Livermore, CA, U.S.A.

A fusion materials irradiation facility is required for the timely and cost-effective development of economical fusion power. Our conceptual machine provides sufficient neutron fluence for accelerated lifetime material tests in a time span of 1-2 y while producing less than 1 MW of fusion power. Neutral deuterium beams at 150 keV are injected into the center of a high-density warm tritium plasma housed in a 12-m-long cylindrical vessel. Superconducting magnets hold the plasma, which transfers the power to each end of the solenoid. The stainless steel end sections absorb the beam power and are externally cooled by high-pressure water to maintain the plasma-side wall temperature below 740 K. A service loop separates tritium from deuterium in the plasma effluent. Tritium is reinjected at each end.

1. INTRODUCTION

In addition to development of an adequate means of containing a reacting deuterium-tritium (D-T) plasma at temperatures as high as 10^8 K, the successful entry of fusion as a commercial source of electric power requires development of new, structurally sound, low-activation materials. For fusion to be an environmentally benign source of power, these new fusion reactor materials should not activate into the high-level waste category at end of life.

Fortunately, it has been possible to commence fundamental materials development efforts with fission irradiation facilities such as the Fast Flux Test Reactor at Richland, WA. Useful data have also been obtained from low-flux 14-MeV neutron sources such as the recently shut down Rotating Target Neutron Source II.[1] However, final selection, acquisition of engineering data, and qualification of fusion reactor materials will require life tests in a facility that provides the appropriate D-T neutron energy spectrum at an accelerated rate.

The appropriate spectrum is best provided by D-T fusion. Although fusion neutrons are initially monoenergetic (14.1 MeV), collisions with the structure surrounding the fusion reaction and the materials being irradiated cause the neutron energy to decrease and additional neutrons (at a lower energy) to be produced so that a spectrum results which peaks and cuts off at 14.1 MeV. An approximation to the correct spectrum is produced by the nuclear stripping reaction of an energetic D^+ beam on solid or liquid lithium. However, for the neutron energy distribution to peak at 14.1 MeV, the D^+ ion must have an energy of 33.8 MeV.[2] As a result, an appreciable number of neutrons have energies between 14.1 and 33.8 MeV, which may be sufficient to exceed the threshold for material damage reactions, but which would not occur in a fusion reactor. Consequently, material damage or activation would be over-predicted, and suitable materials could be excluded from consideration.

Considerable effort has been expended on developing the D^+/Li neutron source, but construction of a suitable irradiation facility in the U. S. was cancelled. Since then, several design concepts have been proposed for D-T fusion neutron sources.

We believe that our design concept of a Beam Plasma Neutron Source, discussed below, satisfies the requirements of a program for materials development for an irradiation facility in a cost-effective manner. In addition, its construction and operation would contribute to the technology relevant to fusion systems.

2. CONCEPT DESCRIPTION

Our concept[3] is based on the successful 2XIIB experiment.[4,5] Here we briefly describe the concept in terms of a baseline design and an option that indicates the trade-off between costs and capability.

The concept is based on well-understood plasma containment principles, neutral beam injection techniques, and methods of transferring energy from the plasma to a cooled wall. Neutral deuterium beams at 150 keV with a combined power of 60 MW are injected into a warm tritium plasma with a density of 3.2 x 10^{21} m^{-3} at the center of the machine. Less than 1 MW of fusion power is produced. Deuterium is ionized in the plasma, and the D^+ ions transfer their energy to the warm plasma primarily by electron drag. A two-component plasma results with an average D^+ energy of 50 keV and a T^+ temperature of 0.22 KeV. Our baseline design uses a quadrupole magnet at the reaction zone to ensure magneto-hydrodynamic (MHD) stability. (Recent calculations indicate that this magnet may be unnecessary.) On either side, superconducting solenoid magnets guide the plasma and the power toward each end. Outside the solenoid region, the field lines diverge and the power density decreases to a manageable level so that heat can be transferred to the walls of externally cooled end cells.

2.1. General arrangement

An isometric cutaway view of the LLNL Beam Plasma Neutron Source is shown in Fig. 1. The machine would be located in a pit that would be covered during operation by removable concrete shield sections. This feature permits manned access at all times to the irradiation building, which would also house the ancillary systems and the operation and maintenance areas. The shield sections could be removed to permit easy installation and removal of equipment from the pit.

The machine consists of six major assemblies. Since the machine is symmetrical about its orthogonal centerlines, each assembly appears twice except for the central core assembly. These assemblies are briefly described below.

Core Assembly - The core assembly consists of a stainless steel vacuum vessel, a water-cooled copper quadrupole magnet, and a support structure. The vessel is suitably stiffened to restrain the magnet loads. Six major flanged openings are placed on the three orthogonal axes of the vessel. Plasma tubes attach to openings at each end of the plasma axis (z axis). On the horizontal axis (x axis) two openings 180° apart are used for connecting the neutral beam assemblies. Two openings on the vertical axis (y axis) allow for attachment of upper and lower specimen holders. All mating connections are made with split clamp rings suitable for simple (remote) pneumatic actuation. An integral structure on the machine pit floor supports the core assembly, the central ends of the neutral beam assemblies, the plasma tubes, and the specimen holders. The support load is estimated to be 20 Mg.

Plasma Tube - The plasma tube encloses the plasma column, which equilibrates the D^+ energy, by means of electron drag, with the warm tritium plasma. The stainless steel plasma tube consists of a 5-m-long cylindrical section of 0.4 m diameter. It is supported at one end by the mating core assembly and at the other end by the end cell.

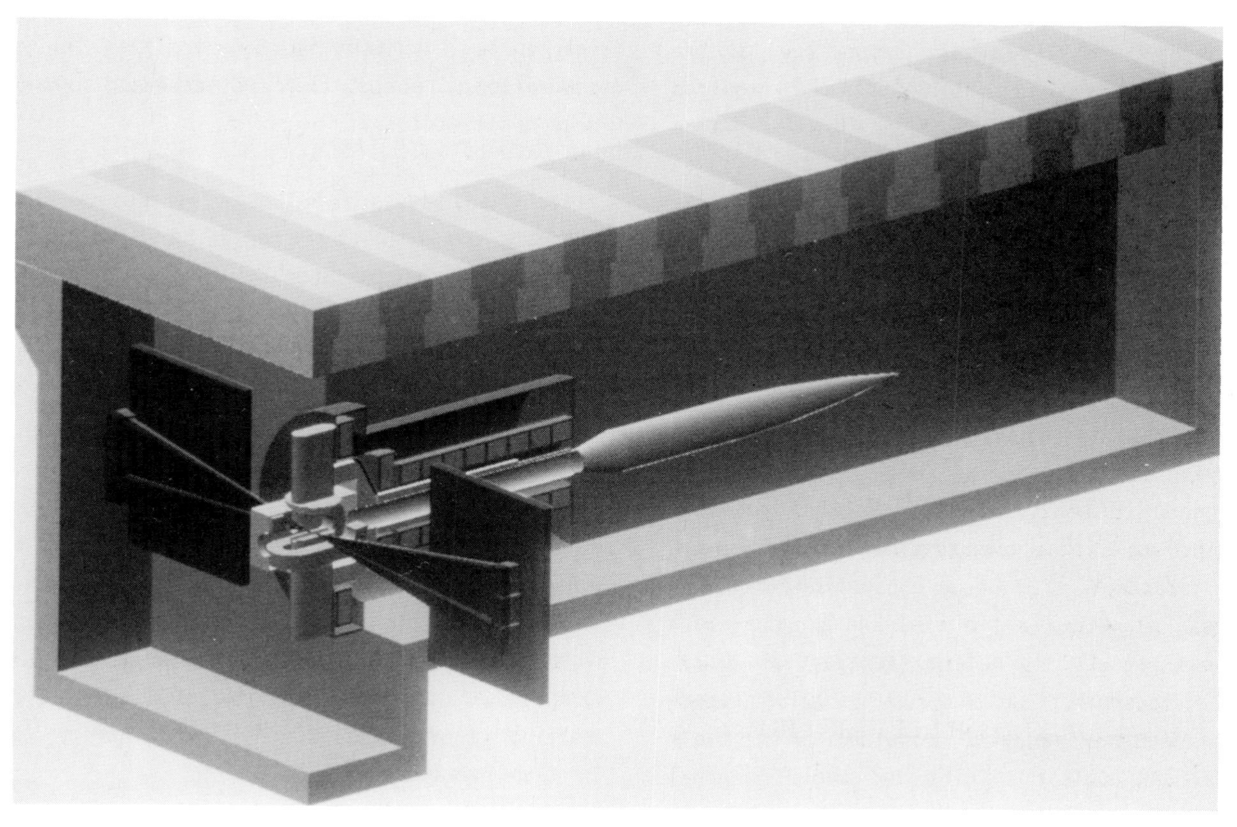

Figure 1
Isometric cutaway view of one-half of the Beam Plasma Neutron Source irradiation facility

End Cell - The function of the end cell is to remove energy from the plasma. The end cell is a stainless steel vessel that is externally cooled by high-pressure water. The vessel is a tubular weldment consisting of an uncooled conical transition section that mates with the plasma tube and increases in diameter to 0.75 m. No heat is transferred from the plasma to the end cell in this section. The transition section is followed by a cooled cylindrical section and three cooled conical sections, as seen in Fig. 1. The overall length of the end cell is 7.5 m. The vessel diameters, cone angles, and lengths are chosen to limit the maximum heat flux at the wall to 3 MW/m^2. In the solenoid region, and in the near portion of the end cell, energy flows primarily in the direction of the field lines, which diverge as they leave the influence of the solenoid. In the solenoid region the thermal conductivity of the plasma along the field lines is much greater than its value normal to the field lines. As the field strength decreases (and the field lines diverge away from the solenoid), the normal thermal conductivity increases until thermal diffusion becomes isotropic in the plasma. We have neglected (conservatively) this effect in sizing the end cell and have maintained a simple axisymmetric shape.

The walls of the end cell have numerous circumferential (square cross section) flow passages. Horizontal manifolds 180° apart along the end cell provide for supply and return coolant. Thus, the cooling water heated path length is half the vessel circumference. Water pressure is maintained sufficiently high (1.6 MPa) that film boiling will not occur.

Although the maximum wall temperature at the plasma is expected to be 740 K, the required water pressure corresponds to the water-side wall temperature of 475 K (saturation pressure: 1.6 MPa). The large temperature drop results from the choice of stainless steel for the end cell and the high heat flux. The average exit water temperature is 312 K for an inlet temperature of 300 K. Thus, the outer wall of the end cell, which is the principal structural member, will operate only slightly above room temperature.

Solenoid Assembly - The solenoid assembly contains one large, one intermediate size, and eight standard size superconducting magnets. The mean diameter, radial thickness, and width of these magnets (in meters) are 3, 0.56, 0.28; 1.46, 0.56, 0.28; and 1.2, 0.4, 0.3, respectively. These magnets, together with their thermal and nuclear shields and their internal support structure, are contained in one large annular vacuum vessel. Its inner diameter accommodates the plasma tube. The combined magnetic field of the superconducting magnets and the quadrupole magnet in the core assembly is 4 T at the center of the machine. The field increases to 12 T over most of the solenoid region. The annular vessel is supported in a structural cradle that can be moved along the z axis of the machine for maintenance. The weight of each solenoid assembly is about 200 Mg.

Neutral Beam Assembly - Each neutral beam assembly has six 150-keV D° beams, which provide 30 MW of power. The beams share a common vacuum enclosure. One neutral beam assembly is located on each side of the machine along the x axis.

The neutral beams are based on an extension of the technology developed for use on the Tokamak Fusion Test Reactor and Doublet III-D experiments.[6] To reduce power consumption, the energy in ions not neutralized in the neutralizer would be recirculated in a regenerative circuit.[7]

Specimen Holder - The specimen holder provides the necessary holding fixtures and the correct thermal, atmospheric, and mechanical stress environment for the specimen being irradiated. To date, we have considered only the external configuration of the specimen holder. The specimen holder extends into the upper and lower (y axis) openings in the source assembly vessel.

2.2 Ancillary equipment

Four principal types of ancillary systems are required for provision of vacuum, cryogenic cooling, heat removal, and tritium management. These systems require conventional design and operational practices. Care is taken in the tritium management system to minimize the tritium inventory.

2.3 Controls and data acquisition

Machine controls and data acquisition are accomplished in our conceptual design with a commercial process control system.

2.4 Operating characteristics

The major operating characteristics of our Beam Plasma Neutron Source concept are given in Table 1. A key parameter determining accumulation of fluence on material specimens is machine availability. We project availability to approach 90% by providing design features and operating procedures that facilitate component replacement.

Using reflectors to increase the neutron flux at the specimen location has been studied.[8] Substantial enhancements in flux for neutrons having energies greater than 0.1 MeV can be obtained, as shown in Fig. 2. It is apparent from Table 1 that sufficient neutron fluence is provided over a reasonable volume for accelerated lifetime material tests in a time span of 1-2 y.

Table 1. Major operating characteristics of the baseline design.

Beam kinetic energy, keV	150
Total beam power, MW	60
Quadrupole power, MW	6.8
Fusion power, MW	1
Neutron output, n/s	3.6×10^{17}
Unperturbed neutron flux at plasma surface, $n/m^2.s$	3.2×10^{18}
Unperturbed neutron power density at plasma surface, MW/m^2	7.2
High flux volume, 10^{-3} m^3	
for loading > 3 MW/m^2	5.8
for loading > 5 MW/m^2	1.8
for loading > 7 MW/m^2	0.2
Annual tritium use, kg	0.05
Central magnetic field, T	4
End magnetic field, T	12
Hot ion number density, m^{-3}	8×10^{20}
Electron number density, m^{-3}	3.2×10^{21}
Plasma/magnetic pressure ratio (ß)	1

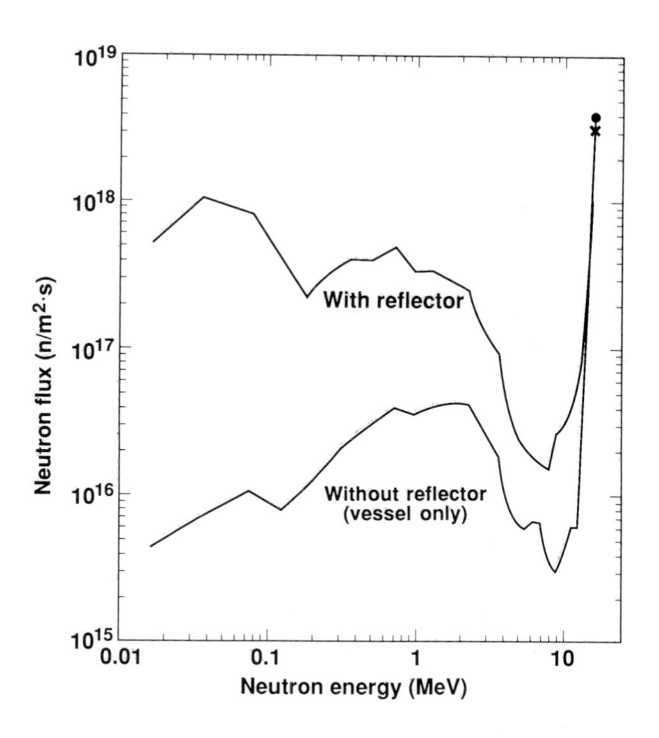

Figure 2
Example of enhancement in neutron flux with a 0.30-m thick aluminum reflector. In both cases the reflector is outside an 8-mm thick aluminum vessel. (From T. Kawabe, University of Tsukuba and H. Sagawa, JAERI[8])

3. TECHNOLOGY BASE

The technology necessary to construct a Beam Plasma Neutron Source is essentially on hand. Experimental verification is desirable in three areas: (1) MHD stability of high-density plasma in this application, (2) neutral beam injector qualification, and (3) plasma heat transfer.

Both MHD stability and plasma microstability issues were largely resolved in the 2XIIB experiment.[4,5] However, further examination of these issues at the higher plasma density needed for this application is desirable and could be accomplished with a subscale experimental investigation by modifying equipment available at our laboratory.

Over $58M have been spent in the U. S. alone over a ten-year period to develop positive-ion, neutral beam injectors.[6] Much has been learned about integrating their components into working systems. New types of components do not appear to be required, and a short development effort should provide the necessary information needed to build continuously operating, long-lived, regenerative neutral beam assemblies.

The mechanism by which heat is transferred between the high-density energetic plasma through a thin layer of neutral gas to a solid wall (the end cell) would be investigated in the experimental equipment mentioned above. Some experimental measurements of this mechanism have already been made.[9]

4. COST ESTIMATE

To assess the practicality of the beam plasma neutron source we estimated the capital and operating costs for our baseline design and for a lower beam power (smaller irradiation volume) option to determine the sensitivity of these costs to beam power. We found that both capital and operating costs are roughly proportional to the beam power. Table 2 lists the operating and capital costs of these designs. A breakdown of the capital costs shown in Fig. 2 shows that the neutral beam assembly clearly dominates the cost in the baseline design and represents 33% of the total cost for the lower power, 13-MW option. Innovatively designing the neutral beam components to lower their cost would have a strong effect on reducing the total cost of the machine.

The major component of the operating cost of the irradiation facility is power.

5. DESIGN OPTIONS AND PROGRAM PROJECTIONS

Recently we extended our study to consider a lower cost option, which would provide smaller

Table 2. Capital and operating costs.

	Baseline (60-MW) $M	%	Low power (13-MW) $M	%
CAPITAL COST				
Neutral beam assemblies	123.0	62	26.7	33
Basic machine	19.5	10	17.8	22
Ancillary equipment	24.1	12	14.5	18
Control and instr.	13.6	7	12.5	15
Facilities	6.5	3	5.1	6
Integration, proj. mgmt.	13.1	7	5.4	7
TOTAL CAPITAL COST	199.8	100	81.9	100
ANNUAL OPERATING COST	55.0	---	17.0	---

irradiation volumes. In the lower cost option, the hot plasma size (at the center of the machine) would be reduced by increasing the central magnetic field from 4 to 8 T. This would reduce the required neutral beam current and significantly reduce the major cost item, the neutral beam assemblies, as seen in Table 2.

The ratio of plasma pressure to magnetic pressure for these cases is reduced from about 1 to about 0.15. It may be possible to assure MHD stability by "line tying" and thus eliminating the quadrupole magnet. Elimination of the quadrupole would further reduce cost and would provide greater flexibility in the design of the specimen holder.

Comparison of the 13-MW option with a system using an accelerator and the D^+/Li stripping reaction indicates that about twice the volume (at the same flux) can be provided at half the cost with a Beam Plasma Neutron Source. On the other hand, if larger irradiation volumes than those provided by our baseline design are desired, our design could be scaled and the cost would rise more slowly than the irradiation volume.

ACKNOWLEDGMENTS

This work was performed under the auspices of the U.S. Department of Energy by Lawrence Livermore National Laboratory under Contract W-7405-Eng-48. Many of our colleagues at LLNL contributed to this design concept and provided invaluable assistance and encouragement. We single out for special mention the work done by T. Casper, C. Damm, A. Futch, A. Molvik, W. Neef, and W. Sterbenz.

REFERENCES

1. The Rotating Target Neutron Source, in: Energy and Technology Review, Lawrence Livermore National Laboratory, Livermore, CA, UCRL-52000-88-3 (1988) p.1.

2. E. K. Opperman, FMIT Experimental Capabilities, Hanford Engineering Development Laboratory, Richland, WA, TC-1633 (1980).

3. F. H. Coensgen, et al., A D-T Neutron Source for Fusion Materials Testing, Lawrence Livermore National Laboratory, Livermore, CA, UCRL-97280, Rev. 1 (1987).

4. W. C. Turner, et al., Nucl. Fusion 19 (1979) 1011.

5. D. L. Correll, et al., Nucl. Fusion 20 (1980) 655.

6. W. B. Lindquist and S. H. Staten, Evolution of the Common Long Pulse Neutral Beam Source Program, Lawrence Livermore National Laboratory, Livermore, CA, UCRL-97484 (1987).

7. M. Fumelli, F. Jequier, and J. Pamela, First Experimental Results of Energy Recovery on the Tore Supra Neutral Beam Injector Prototype, presented at the 15th European Conference on Controlled Fusion and Plasma Heating, Dubrovnic, Yugoslavia, May, 1988.

8. T. Kawabe, University of Tsukuba, Ibarki, Japan, personal communication, July 1988.

9. W. L. Hsu, M. Yamada, and F. H. Tenney, J. Nucl. Mater. 111 & 112 (1982) 311.

FUSION TECHNOLOGY 1988
A.M. Van Ingen, A. Nijsen-Vis, H.T. Klippel (editors)

FATIGUE OF NEUTRON IRRADIATED AUSTENITIC STEEL BELOW THE CREEP RANGE

B. van der Schaaf and M.I. de Vries

Netherlands Energy Research Foundation, ECN, Westerduinweg 3, 1755 LE Petten, The Netherlands

In the present design of the Next European Torus fatigue with strain amplitudes in stainless steel as high as 0.3% are anticipated in the first wall operating at temperatures up to 700 K. Fatigue and tensile test results of the European type 316 reference heat irradiated up to 5 dpa and upto 40 appm helium are presented. In the test frequency range from 0.5 mHz to 0.5 Hz it was observed that the fatigue endurance is dependent on the frequency, though the test temperatures were in the range of 600-800 K, thus below the creep range. At normal fatigue strain rates of 10^{-3} s^{-1} the number of cycles to failure is hardly affected by radiation which can be explained on the basis of tensile test results.

1. INTRODUCTION

Austenitic stainless steel type 316 is widely applied in core components of nuclear fission reactors such as the gas cooled reactor and the sodium cooled fast reactor. In these applications the steel is subjected to alternating mechanical loads under neutron irradiation at elevated temperatures. These operating conditions are very demanding and therefore a considerable research and development effort has been devoted to ensure safe and economic service of this type of alloy[1]. Most of the fatigue test results for those studies are aimed at the creep range because of the operating temperatures of these reactors.

The design of the first wall structures of the Next European Torus aims at operating temperatures just under the creep range of type 316 stainless steel[2] thus below 800 K. In this temperature domain there exists a limited set of fatigue data on type 316 irradiated up to 10 dpa. In the frame of the European Fusion Technology Programme a coordinated effort to provide fatigue data in this temperature regime is underway. Several European laboratories irradiate and test the European reference heat of type 316 mechanically to provide the NET design team with appropriate design data.

The ECN contribution consists of irradiating type 316 up to 5 and 10 dpa in the HFR, Petten, at temperatures in the range of 500 to 800 K in this programme. In this paper the type 316 post-irradiation fatigue results of 5 dpa irradiations outside the creep range are presented. The observations are compared with results of other austenitic steels both in irradiated and thermal control conditions. Especially the strain rate effect is addressed since this parameter is mostly variable in the loading patterns of primary fusion reactor structures. This paper is one step in an ongoing larger programme. More experimental material will become available in the near future.

2. EXPERIMENTAL PROCEDURE

2.1. Material and specimens

The material used in this investigation is the type 316 European reference heat, which has an 0.06% N content to retain sufficient strength despite its low 0.02% C content. The fatigue and tensile specimens have been taken perpendicular to the rolling direction of the 30 mm thick plate. The material is in solution annealed condition, with a few small carbide particles only in the matrix. The grain diameter of the material is between 50 and 100 μm.

The tensile specimens have a gauge length and diameter of resp. 40 and 8 mm. The low cycle fatigue specimens are thread ended hour glass type specimens with a minimum diameter of 8.8 mm. The fatigue specimens are subjected to axial control over their 21 mm gauge length. More details are given elsewhere[3].

2.2. Irradiations

The tensile and fatigue specimens have been irradiated in the HFR, Petten in a sodium filled rig of the TRIO type. During irradiation the temperatures are controlled, adjusting the gas mixture in the gap surrounding the sodium filled canisters. The target irradiation temperatures are 600, 700 and 800 K ± 20 K for 90% of the irradiation time. These temperatures are measured for each specimen.

Fast and thermal fluences were measured with two different detector sets. The analyses indicate that the radiation damage in the specimens is in the range from 3 to 5 dpa, whereas the helium content is in the range from 30 - 40 appm. The latter figures have been verified with helium measurements in irradiated samples by Rockwell Int. Cy. The displacement damage has been observed in Transmission Electron Microscopy, TEM, but that does not produce a quantitative indication.

2.3. Mechanical testing

The post-irradiation tensile and low cycle fatigue tests are conducted on servo-hydraulic and electro-mechanical testing machines shielded with 250 mm lead bricks. The two zone resistance furnaces used have a temperature gradient of 2.5 K maximally and a temperature stability of about 2 K. Independent computers control each test machine and perform the data acquisition. The thermal control tests are performed on similar unshielded machines.

3. RESULTS

3.1. Tensile properties

Tensile tests have been performed at 500, 600, 700 and 800 K up to 5 dpa. In two sets of tests the test and irradiation temperature are similar. The strain rate has been varied from 10^{-6} s^{-1} to 10^{-2} s^{-1}. For each irradiated condition thermal control tests have been conducted.

In the temperature range investigated strength nor ductility values were dependent on strain rate. With increasing temperature both strength and ductility decrease gradually in both irradiated and reference condition.

The main effects of radiation on strength are:

- An increase in 0.2% YS of about 100 MPa at test temperatures of 500 to 700 K, at 800 K the effect is negligible.
- An increase in UTS of about 70 MPa in the range from 500 to 700 K, at 800 K the effect is negligible.

In the temperature range of 500 to 800 K the uniform elongation decreases about 25%, whereas the total elongation is reduced by about 10% strain. The major effect of the 5 dpa, 40 appm He radiation is thus a slight embrittlement together with a relatively comparable strength increase. More details of radiation effects on tensile results are given in [4].

3.2. Low cycle fatigue properties

Low cycle fatigue specimens have been irradiated and tested at 600, 700 and 800 K. Due to an accidental interchange of specimens in three pairs of tests the test and irradiation temperatures are 600 and 700 K respectively. The total cyclic strain ranges were chosen in the interval from 0.6 to 1.2%. The test frequencies were selected in the range from 0.0005 to 0.05 Hz corresponding to strain rates from 10^{-5} to 10^{-3} s^{-1}.

In figure 1 the cyclic hardening curves of irradiated steel tested at 600, 700 and 800 K are shown. Only the steel tested at 700 K at a 0.05 Hz frequency shows pronounced softening. In the other conditions tested at 600 and 700 K

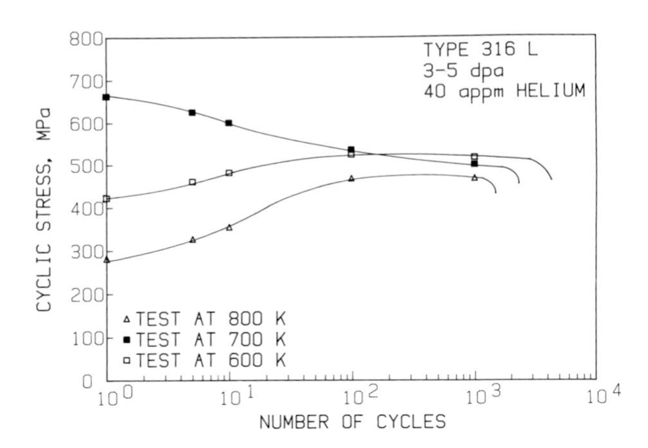

FIGURE 1
Effect of temperature on cyclic strain hardening of irradiated type 316 steel.

a slight hardening followed by softening is observed. Testing at 800 K results in a pronounced cyclic strain hardening. The latter behaviour is also observed in the reference condition at all temperatures.

Figure 2 shows the total strain range plotted versus the observed number of cycles to failure for the limited number of tests conducted with a strain rate of 10^{-3} s^{-1}. The increase in temperature reduces the number of cycles to failure in both irradiated and reference condition, but the irradiation effect on fatigue life is limited.

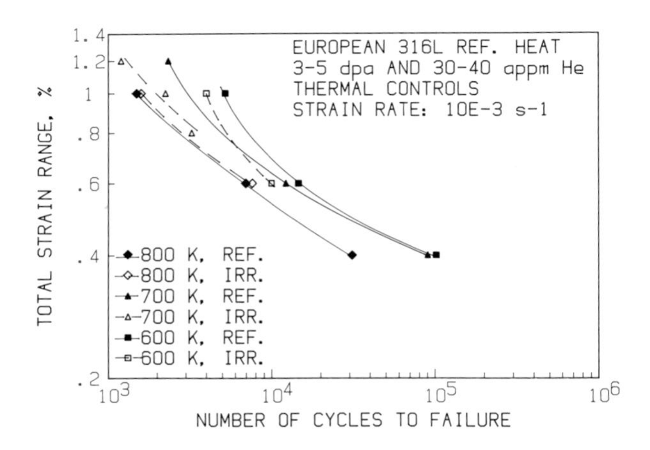

FIGURE 2
Effect of radiation, strain range and temperature on the number of cycles to failure.

The effect of cyclic strain rate and irradiation on the fatigue life is shown in figure 3. The effect of decreasing frequency is to reduce the number of cycles to failure to a certain extent. This effect is retained after radiation.

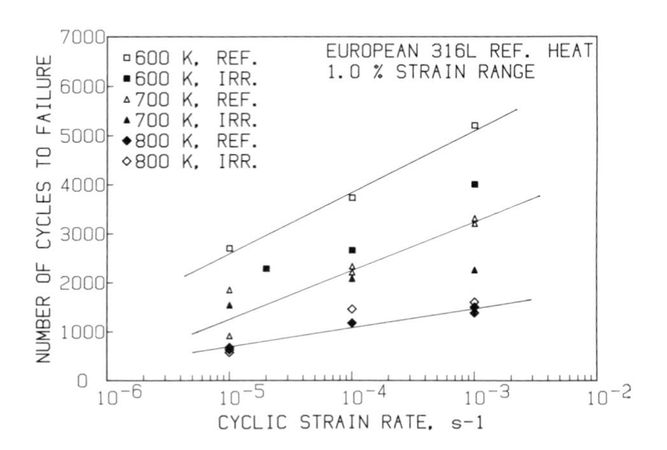

FIGURE 3
The effect of cyclic strain rate, temperature and radiation on the number of cycles to failure.

4. DISCUSSION

4.1. Effect of strain rate and cyclic strain range

The effect of strain rate on type 316 fatigue endurance reported here, has also been observed for other type 300 alloys in the temperature range below the creep range. In figure 4 the results of unirradiated type 316 are plotted together with data of Kanazawa[5] for type 310 tested under similar conditions. The effects of strain rate on number of cycles to failure are typical for type 300 austenitic steel: the fatigue life decreases with decreasing strain rate. Kanazawa has observed transgranular fatigue failure only up to 800 K. This is also the case for our type 316 thermal control specimens as well as the irradiated steel. This means that the empirical frequency modified Manson-Coffin model is applicable for both reference and irradiated material.

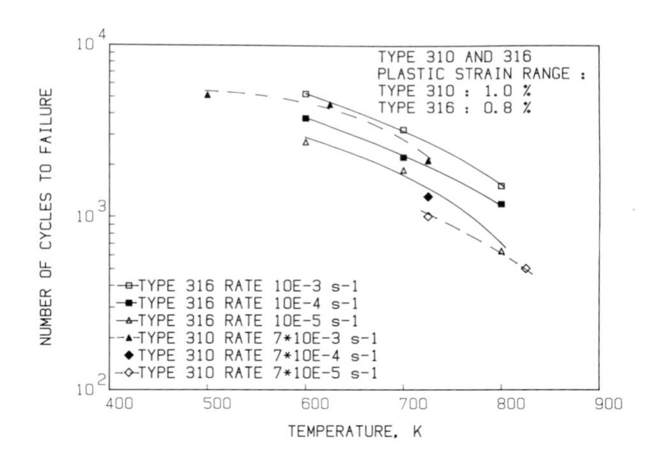

FIGURE 4
Comparison of fatigue endurances as dependent on strain rate of type 310 (reported by Kanazawa[5]) and type 316 in unirradiated condition.

The cyclic hardening behaviour observed by Kanazawa at different strain rates is similar to that for the European reference heat in reference condition. For temperatures up to 800 K oxidation plays no role, so secondary cyclic hardening as observed by Driver[5] under vacuum is observed at low total strain ranges. Due to the interchange of specimens the tests conducted at 700 K show continuous softening, because the material has been irradiation hardened at 600 K, as is indicated by the tensile tests. The observed reduction in fatigue endurance with decreasing strain rate cannot be explained by oxidation: Driver[6] found similar results with tests in vacuum. Creep must also be excluded: there is no intergranular failure and the dislocation structures observed in transmission electron microscopy of fatigue tested specimens in and below the creep range differ considerably. At temperatures in excess of 800 K dislocations form cells according to Challenger et al[7] whereas below that temperature planar dislocation arrays are observed. Since there is no effect of strain rate on total tensile elongation exhaustion of ductility cannot explain the strain rate effect on fatigue endurance.

4.2. Effects of temperature and neutron radiation

Up to 600 K the neutron radiation produces interstitials, vacancies and dislocation loops. In the temperature range of 600 K to 800 K (800 K forms the lower level of the type 300 thermal creep range) small voids are formed (tens of Å's in diameter) and dislocation networks start to develop, which both can be resolved with a transmission electron microscope.
The present 5 dpa irradiation also results in 30-40 appm He content due to thermal neutron reactions with boron and nickel, contributing 25 and 5-15 appm respectively. Helium bubbles have not been observed in the European reference heat as irradiated, but after heat treatments at temperatures over 800 K the bubbles could be resolved. In the post-irradiation mechanical test material bubbles must be small in diameter or non-existent.
The main effects of the lattice defects and clusters of helium atoms are an increase in tensile strength and a reduction in ductility. This is in line with the present tensile test results. The effect of radiation on low cycle fatigue can be estimated by the relation:

$$\Delta\varepsilon_t = \Phi_u \, A \, N_f^a + \Phi_t \, B \, N_f^b \qquad (1)$$

where $\Delta\varepsilon_t$ = total strain range
Φ_u = constant for radiation effect on tensile strength
Φ_t = constant for radiation effect on tensile ductility
N_f = number of cycles to failure
A, a, B, b = constants

Michel and Korth[8] and De Vries[9] have shown in their studies that this is a practical relation for type 300 steels in irradiated condition. In the present study the effects of radiation on tensile properties counter balance each other so the limited effect of 5 dpa radiation on fatigue endurance at 0.05 Hz frequency is in line with the test results at 700 and 800 K.

The frequency effect on thermal control and irradiated type 316 steel has also been observed below the creep range for DIN 1.4948 steel, similar to type 304 steel. In figure 5 the normalized fatigue life of this steel has been plotted versus cyclic strain rate at 700 K for two radiation levels. In this case the strongest effect is observed at a dose of 0.5 dpa. At 5 dpa radiation damage the effect is reduced and similar to the reductions in cyclic frequency observed for type 316 with decreasing strain rate. The strain rate effect below the creep range seems to be influenced by the radiation level for some heats, but more experimental work will have to be available to confirm the phenomenon and to explain it.

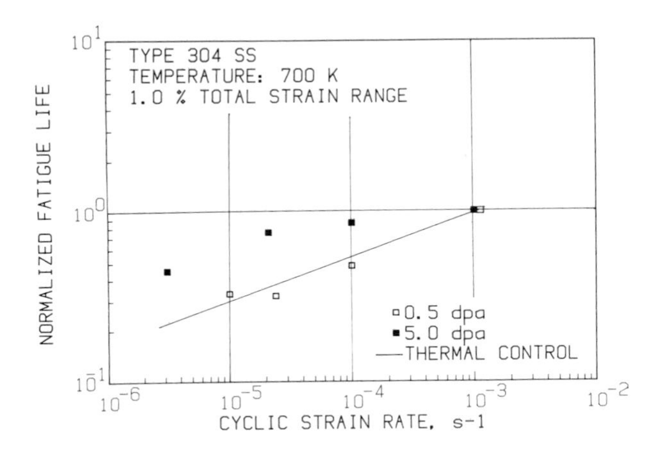

FIGURE 5
Effect of radiation damage level and cyclic strain rate on type 304 stainless steel in irradiated condition.

Though the tensile properties show no dependence on rate below the creep range low cycle fatigue properties of irradiated steel do. In the NET first wall design the effect must be taken into account.

5. CONCLUSION

The European reference heat type 316 L irradiated to 5 dpa and 40 appm helium shows a limited effect of radiation on tensile strength and ductility which is reflected in the low cycle fatigue properties in the temperature range from 600 to 800 K at moderate test frequencies.
The fatigue endurance is dependent on strain rate, whereas the tensile properties are not in the particular temperature range. An explanation for this effect cannot be given yet. Additional experimental results will have to support the development of a model. The effect must be taken into account when designing first walls of type 300 alloys operating below the creep range.

REFERENCES

1. D.R. Harries, J. Nucl. Mat. 82 (1979) 2.
2. D.R. Harries and E. Zolti, Nucl. Eng. and Design Fusion 3 (1986) 331.
3. B. van der Schaaf, 13th Symp. on Fusion Techn., Varese (Pergamon Press, Oxford 1984), Vol. 2, 1045.
4. B. van der Schaaf, 14th Symp. on Fusion Techn., Avignon (Pergamon Press, Oxford 1986), Vol. 2, 993.
5. K. Kanazawa, K. Yamaguchi, S. Nishijima, Symp. on Low Cycle Fatigue, Bolton Landing, ASTM STP 942 (ASTM, Philadelphia, 1987) 519.
6. J.H. Driver et al., Symp. on Low Cycle Fatigue, Bolton Landing, ASTM STP 942 (ASTM, Philadelphia, 1987) 438.
7. K.D. Challenger, J. Moteff., Symp. on Fatigue at Elevated Temperatures, Storrs, ASTM STP 520 (ASTM Philadelphia, 1973) 68.
8. D.J. Michel and G.E. Korth, Int. Conf. on Rad. Effects in Breeder Reactor Structural Materials, Scottsdale (Met. Soc. of AIME, New York, 1977) 117.
9. M.I. de Vries, 11th Int. Symp. on Effects of Radiation on Materials, Scottsdale, ASTM STP 782 (ASTM, Philadelphia, 1982) 665.

FUSION TECHNOLOGY 1988
A.M. Van Ingen, A. Nijsen-Vis, H.T. Klippel (editors)

CYCLE-PERIOD DEPENDENT FATIGUE OF NEUTRON IRRADIATED AUSTENITIC STEELS AT ELEVATED TEMPERATURES

M.I. de Vries and B. van der Schaaf

Netherlands Energy Research Foundation, ECN, Westerduinweg 3, 1755 LE Petten, The Netherlands

Fatigue resistance is a major concern in the design studies of austenitic stainless steel first wall structures. Although the operating temperatures of the first wall of the Next European Torus (NET) are envisaged to be less than 700 K this temperature might be exceeded locally. In this respect the results from post-irradiation fatigue tests at elevated temperatures are of interest for the analyses of the fatigue damage at such locations.

Cylindrical specimens and compact-tension ones of mill-annealed Type 316 and Type 304 stainless steel have been neutron irradiated from less than 0.5 dpa up to 5 dpa at elevated temperatures. Fatigue tests were subsequently performed at 600 to 1000 K.

Irradiations at the low temperatures of the range caused significant irradiation hardening from displacement damage. After 4 dpa at 700 K the 0.2 yield stress of Type 316 and Type 304 increased by a factor of 2 and 4.5 respectively.

Transmission electron microscopy (TEM) revealed displacement damage in the form of loop-type defects. The large difference between Type 316 and Type 304 is attributed to the swelling of the latter. TEM of Type 316 did not reveal helium bubbles or voids whereas Type 304 showed a high density of cavities with diameters ranging up to 20 nm. The cavities were mainly voids as was demonstrated by the disappearance during annealing at 1000 K. The incubation dose level appeared to be less than 3 dpa for Type 304 and more than 5 dpa for Type 316.

Cyclic straining caused softening of the irradiation-hardened materials whereas the unirradiated materials hardened. However, at half-life ($N_f/2$) the cyclic stress values varied less than about 20 percentage points.

Low cycle fatigue tests and (linear elastic) fatigue crack growth tests showed degradation of the fatigue resistance with increasing temperature and decreasing loading frequency over the entire temperature range. The degradation was enhanced by irradiation.

Under non-creep-contributing conditions, cycle-periods less than about 0.5 minutes at temperatures up to 900 K, the irradiation effect was rather moderate. The fatigue (irradiation) reduction factor for the conventional time-independent fatigue did not exceed the value of 1.5.

The fatigue resistance was much more detrimentally affected under cyclic loading with extended cycle-periods in the creep-range. Reduction factors ranged up to values of about 10 for cycle-periods up to about 30 minutes at temperatures up to 900 K. The reduction of the fatigue resistance was nearly the same for 0.5 and 5 dpa irradiated materials. The strong irradiation effect is attributed to the time-dependent intergranular helium embrittlement failure mechanism.

FUSION TECHNOLOGY 1988
A.M. Van Ingen, A. Nijsen-Vis, H.T. Klippel (editors)

MEASUREMENT OF RADIOACTIVITIES INDUCED IN Ti, Zr, Nb AND Mo FOR FUSION REACTOR MATERIALS

Takashi NAKAMURA

Cyclotron and Radioisotope Center, Tohoku University, Aoba, Aramaki, Sendai 980, Japan

Tohru OHKUBO and Yoshitomo UWAMINO

Institute for Nuclear Study, University of Tokyo, Midori-cho 3-2-1, Tanashi, Tokyo 188, Japan

The induced activities of Ti, Zr, Nb and Mo to be used as future fusion reactor materials were measured by irradiating semi-monoenergetic neutrons from the p-Be reaction and quasi 15-MeV neutrons from the d-Be reaction. The activation rates are tabulated for six neutron fields, 33-MeV d-Be netrons and 20, 25, 30, 35 and 40-MeV p-Be neutrons. The decay curves after irradiation are also shown for Ti, Zr, Nb and Mo and the induced activities increased in the order of Mo, Ti, Zr and Nb. The measured activities were used as the data of integral tests of neutron activation cross sections above 15 MeV.

1. INTRODUCTION

It is quite evident that the structural components of the first-wall/blanket portions of a deuterium-tritium (D-T) fusion reactor will be exposed to very intense high-energy neutron fluxes. The neutrons can interact with the various isotopes in the materials to cause transmutations which in turn can alter the mechanical and physical properties of a metal. Quite often the new isotopes are radioactive and the increase of residual activities after long operation might arise serious problems for workers in the fusion reactor facilities.

High-energy accelerator-based neutron sources are now being routinely used to investigate damage and end-of-life test in fusion reactor materials, since there are no intense 14-MeV d-T neutron sources presently available. These sources produce a broad neutron spectrum centered around 14 MeV by stopping a deuteron beam having energy of 30 to 40 MeV in a thick beryllium or lithium target. Unfortunately, neutron activation cross sections are very poorly known above 14 MeV, even though the ENDF/B-IV files extend to 20 MeV.

We have developed the quasi-monoenergetic p-Be neutron beams of energy from 15 MeV to 35 MeV by bombarding 20 to 40 MeV protons on a beryllium target,[1] and the activation cross section measurements were performed by using these neutron beams.[2] The object of this paper is to investigate the radioactivity induced in titanium, zirconium, niobium and molybdenum metals by irradiating these quasi-monoenergetic p-Be neutrons and the quasi 15-MeV d-Be neutrons. Molybdenum is contained in the stainless steel, titanium is to be used as V-20%Ti alloy or TiC, and niobium and zirconium as Nb-1%Zr alloy, which are current candidates for the first-wall and blanket structure of controlled thermonuclear reactor. The measured activities were used for the integral tests of neutron cross sections above 15 MeV, which are very poorly known.

2. EXPERIMENTAL PROCEDURE

The quasi-monoenergetic neutrons were produced from the beryllium target of 1 mm or 2 mm in thickness bombarded by 20, 25, 30, 35 and 40 MeV protons extracted from the cyclotron of the Institute for Nuclear Study (INS), University of Tokyo. The residual energy of protons transmitted through the beryllium target was absorbed in the cooling water just behind the target. The quasi 15-MeV neutrons were also produced from the 5-mm thick beryllium target where injected 33-MeV deuteron stops fully.

The beryllium target assembly is shown in Fig. 1. Measurements of the neutron energy spectra in the forward direction to the incident particle beam were made by using the NE-213 organic liquid scintillator coupled with the spectrum unfolding technique. The results for p-Be neutrons are shown in Fig.2.

The proton and deuteron beam intensities (about 5 μA) were monitored by integrating the current on the beryllium target assembly. The irradiated samples for activation measurements were placed 20 cm behind the target and the neutron fluxes on the samples were measured during irradiation with the aluminum and gold foils directly attached on the sample. The

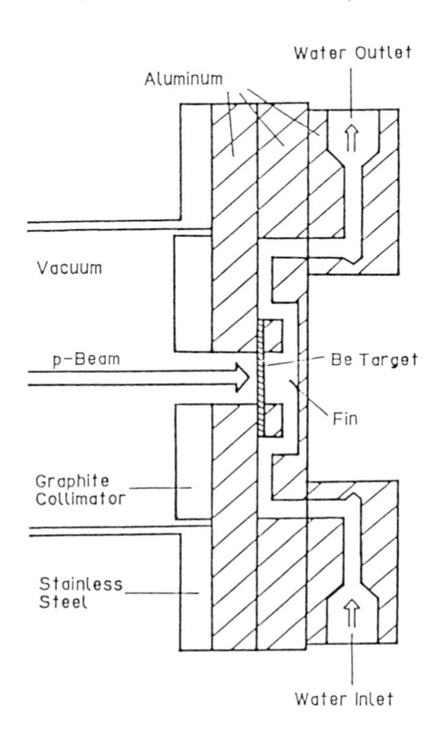

FIGURE 1
Cross sectional view of beryllium target assembly

neutron fluxes were estimated from the ^{27}Al (n,α)^{24}Na and ^{197}Au(n,2n)^{196}Au reaction rates.

The gamma-ray activities of the irradiated samples were measured by a high purity Ge detector which had been carefully calibrated in advance. The gamma-ray measurements were repeated several times after irradiation to obtain decay curves. The gamma-ray spectra were analyzed by the KEI11F computer code[3]. The radionuclides were identified from the peak energy in the gamma-ray spectra and the half-life of the decay curve. The coincidence summing effects coming from cascading photons were corrected with the SUMECC code prepared by Torii et al.[4] For some gamma-ray peaks, there are two kinds of mixed contributions; one is a mixture from parent and daughter nuclides and another a mixture of two photons of very close energies emitted from different nuclides. Since existence of both mixtures appears on a decay curve with bending, they could be separated and nuclides were identified by analyzing the decay curves of the peak counts with the DECFIT code prepared by Torii et al.[4]

The saturated activities corrected for isotopic abundance and decay during irradiation were normalized to the total neutron flux above 4 MeV at the irradiation point, 20 cm behind the target. Here in this study, these normalized saturated activities are specified to be activation rates.

3. RESULTS AND DISCUSSION

The activation rates of radionuclides thus obtained are exemplified in Fig. 3 for Nb as a function of the projectile and its accelerating energy. The abscissa is approximated to correspond to the neutron energy. The 33-MeV d-Be neutrons have a different energy spectrum from that of the p-Be neutrons. However, its average energy is about 14.7 MeV and close to

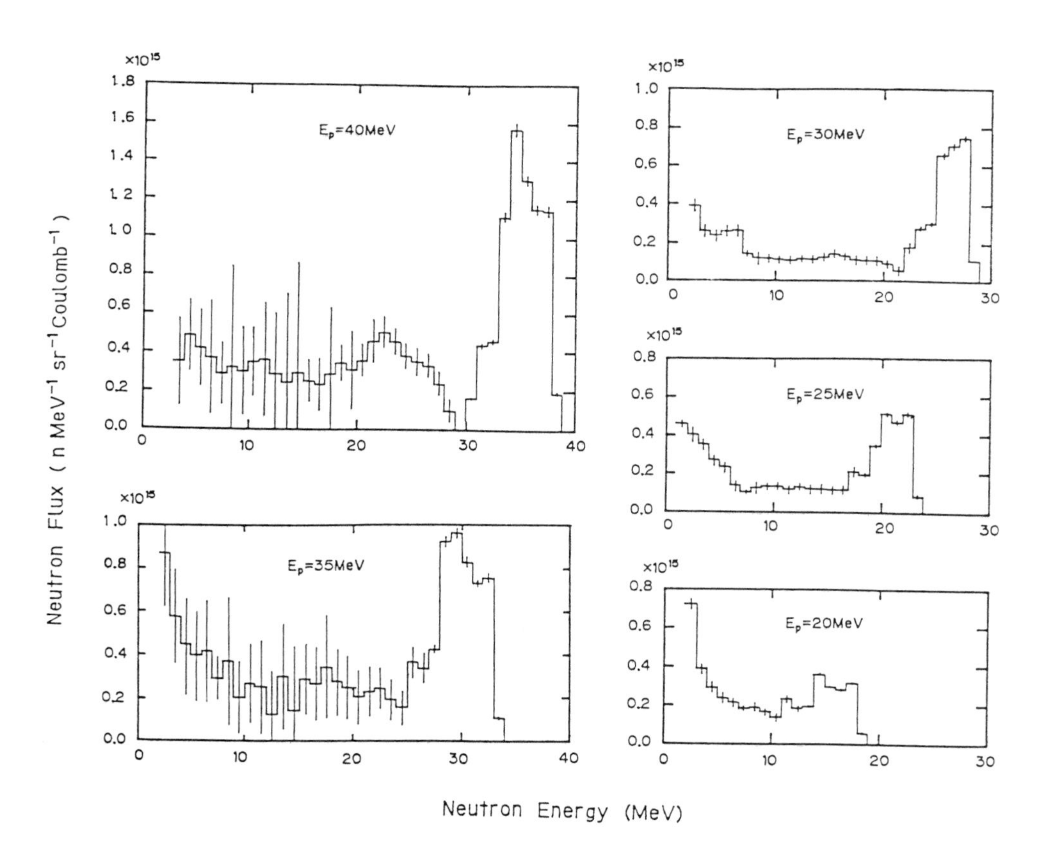

FIGURE 2
Neutron energy spectra emitted at 0 degree from a beryllium target for proton energies of 40, 35, 30, 25 and 20 MeV

the quasi-monoenergetic peak energy of 20 MeV p-Be neutrons.

In Fig. 3, a general tendency can be seen that the activation rates produced by low threshold energy reactions such as (n,γ), (n,p), (n,α) and (n, 2n) increase for lower energy neutrons, while the activation rates for high threshold energy reactions like (n,3n), (n, 4n) and (n,Xnp) increase steeply with the neutron energy, and the higher threshold energy, the more steeply. For the (n,γ) reaction products, such as ^{93}Nb (n,γ)^{94m}Nb, the activation rates are comparatively low for the 33-MeV d-Be neutron irradiation. This indicates that the d-Be neutron field has relatively small thermal neutron components than the p-Be neutron field, because the (n,γ) reactions are mainly caused by thermal neutrons. Since the natural niobium metal consists of a single stable isotope of ^{93}Nb, the reaction channels for each radionuclide can be clearly identified and are shown in Fig. 3.

The 40-MeV p-Be neutron irradiation produced the widest variety of radioactive products. For 1 hour irradiation, ^{46}Sc (a half-life $T_{1/2}$ of 83.8 day) in Ti, ^{88}Y (106.6 day), ^{88}Zr (83.4 day) and ^{95}Zr (64.0 day) in Zr, ^{91}Y (58.5 day) and ^{88}Y (106.6 day) in Nb, ^{88}Zr (83.4 day) and ^{95}Nb (35.0 day) in Mo remain even beyond 10^7 sec (about 4 months). Our results give the fact that the total induced activity per gramme increases in the order of Mo, Ti, Zr and Nb, and that the total activities of Nb and Ti are about three and two times of that of Mo, respectively.

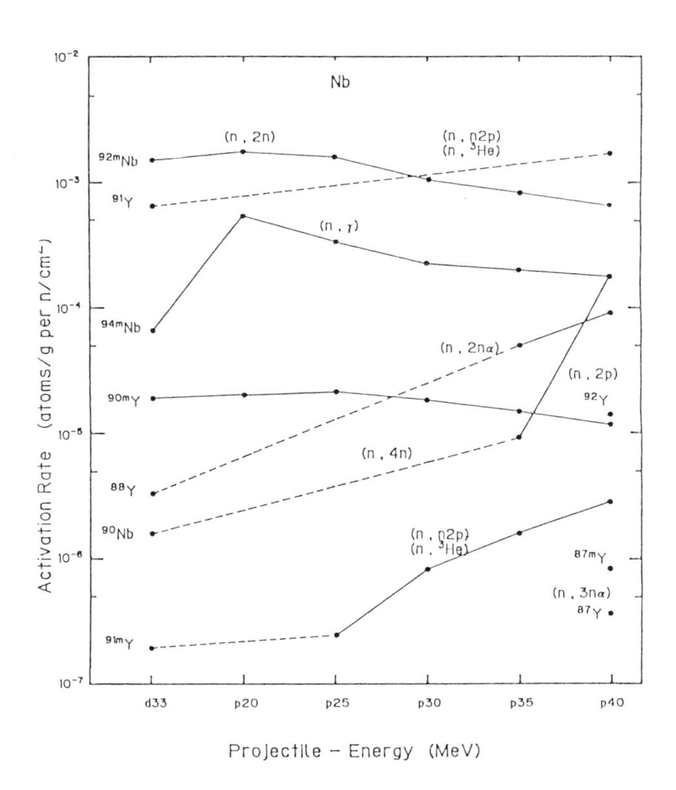

FIGURE 3
Dependence of activation rates of Nb on the projectile energy of d-Be and p-Be neutron irradiation fields

The measured activation rates were compared with calculations as an integral test of reaction cross sections for neutron energy region higher than 14 MeV. Some reaction cross sections are given in Ref. (5) up to 44 MeV by extrapolating the calculated and the experimental data. Among radionuclides produced in Ti, Zr, Nb and Mo, the five reaction cross sections of $^{93}Nb(n,2n)^{92m}Nb$, $Ti(n,Xnp)^{48}Sc$, $Ti(n,Xnp)^{47}Sc$, $Ti(n,Xnp)^{46}Sc$ and $Zr(n,Xn)^{89}Zr^{g+m}$ are given in Ref. (5). The activation rates of these reaction products, A_{cal}, can be calculated as follows,

$$A_{cal} = \int_{E_{th}}^{E_{max}} \sigma(E)\ \Phi(E)\ dE, \qquad (1)$$

where Emax : maximum neutron energy,
Eth : threshold energy,
$\sigma(E)$: reaction cross section,
$\phi(E)$: neutron flux.

The ratio of experimental activation rates to the calculations, $R=A_{exp}/A_{cal}$, is shown in Figs. 4 and 5 and can be used as an indicator of the integral check of the reaction cross section data, $\sigma(E)$, given in Ref. (5):

1) $^{93}Nb(n,2n)$ reaction

The cross section data of Ref. (5) show very good agreement with the present experimental results for 20- and 25-MeV p-Be neutrons (12 - 23 MeV neutrons), but it gives about 30% and 15% underestimation for 30-MeV p-Be neutrons (23 - 28 MeV neutrons) and for 35-MeV p-Be neutrons (28 - 33 MeV neutrons), respectively, compared with our experimental results.

2) $Zr(n,Xn)^{89}Zr^{g+m}$ reaction

The cross section data give about 20% underestimation for 12 - 18 MeV neutrons and about 15% overestimation for 33 - 38 MeV neutrons, and good agreement in between them, compared with our results.

3) $Ti(n,Xnp)^{46}Sc$ reaction.

Agreement is good for neutrons of energies lower than 18 MeV and higher than 33 MeV, but the cross section data give about 25 to 40% overestimation for energy region of 18 to 33 MeV.

4) $Ti(n,Xnp)^{47}Sc$ reaction

Agreement is very good in the energy region below 33 MeV, but a reference data gives about 50% overestimation for neutrons above 33 MeV.

5) $Ti(n,Xnp)^{48}Sc$ reaction

On the contrary, the cross section data has about 15% underestimation for 12 to 28 MeV neutrons and about 15% overestimation for 33 to 38 MeV neutrons.

4. CONCLUDING REMARKS

Generally speaking, it can be said from the comparison with our experimental data that these five reaction cross section data given

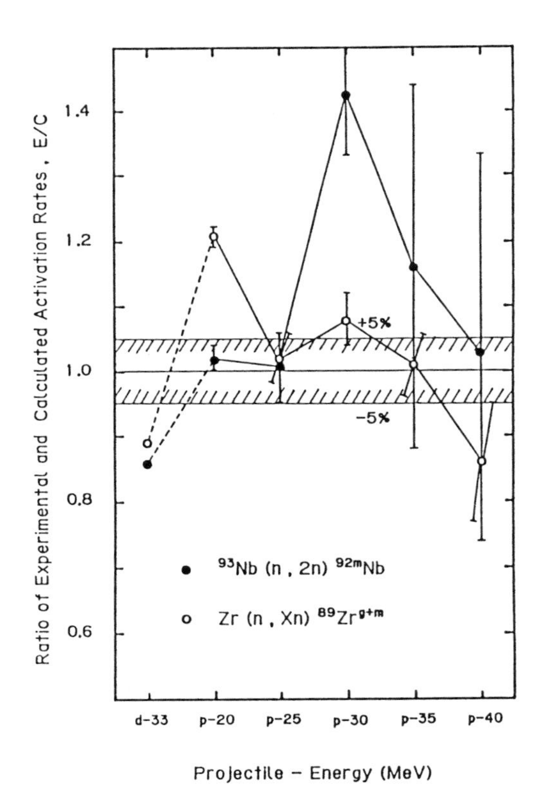

FIGURE 4
Ratio of measured and calculated activation rates of $^{93}Nb(n,2n)^{92m}Nb$ and $Zr(n,Xn)^{89}Zr^{g+m}$ reactions

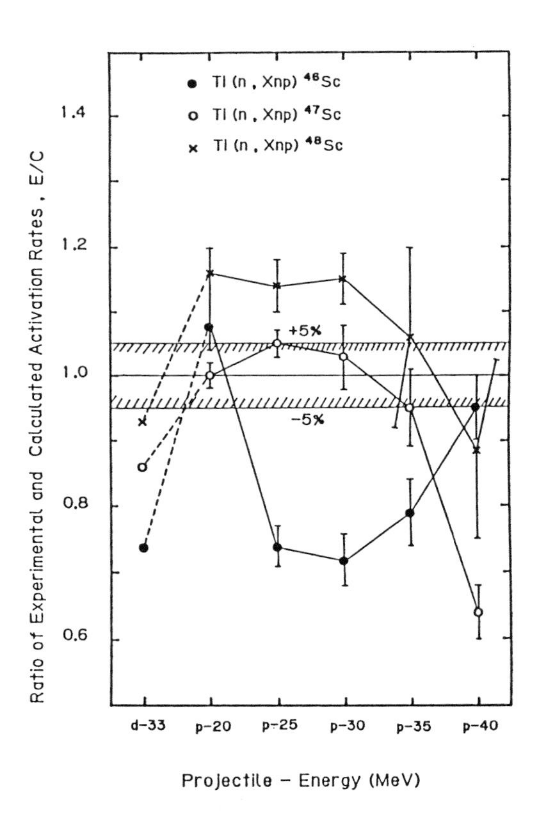

FIGURE 5
Ratio of measured and calculated activation rates of $Ti(n,Xnp)^{48}Sc$, ^{47}Sc, ^{46}Sc reactions

in Ref. (5) are reasonable for neutron energy below about 30 MeV and they give overestimated values gradually with increasing neutron energy beyond 30 MeV. The present results will be useful as the basic data for estimating the induced activities of these materials which are intended to be used in fusion reactors.

ACKNOWLEDGMENTS

The authors wish to thank the machine group of the Cyclotron at INS for operating the cyclotron during experiment. This work is financially supported by the Grant-in-Aid for Scientific Research of Japanese Ministry of Education.

REFERENCES

1. Y. Uwamino, T. Ohkubo, A. Torii and T. Nakamura, Nucl. Instrum. Methods, in press.

2. A. Torii, Y. Uwamino, T. Ohkubo and T. Nakamura, Activation Experiment by Neutrons above 15 MeV, Genshikaku Kenkyu 32 (1987) 89-103, in Japanese.

3. K. Komura, A Computer Program for the Analysis of Gamma-Ray Spectrum, Institute for Nuclear Study, University of Tokyo, INS-TCH-9 (June 1974).

4. A. Torii, Y. Uwamino and T. Nakamura, Computer Codes for Activation Data Analysis; Coincidence Summing Correction Code, SUMECC and Decay Correction Code, DECFIT, Institute for Nuclear Study, University of Tokyo, INS-T-468 (March 1987).

5. L.R. Greenwood, Extrapolated Neutron Activation Cross Sections for Dosimetry to 44 MeV, Argonne National Laboratory, ANL/FPP/TM-115 (September 1978).

FUSION TECHNOLOGY 1988
A.M. Van Ingen, A. Nijsen-Vis, H.T. Klippel (editors)
Elsevier Science Publishers B.V., 1989

EVALUATION OF $^{27}AL(N,2N)^{26}AL(T_{1/2}=7.16x10^5YR)$ CROSS SECTION AND ESTIMATION OF RESIDUAL ACTIVITY OF ALUMINUM EXPOSED TO THE FUSION NEUTRONS

Shin IWASAKI, Naoteru ODANO, Johannes R. DUMAIS, Satoru TANAKA, Kazusuke SUGIYAMA

Department of Nuclear Engineering, Tohoku University, Aramaki-Aza-Aoba, Sendai 980, Japan

In order to estimate the residual activity of aluminum exposed to the fusion neutrons accurately, the (n,2n) reaction cross section around 14MeV was evaluated by using a multistep Hauser-Fechbach reaction model code with parameters based on the recently obtained (n,2n) cross section data base. The obtained result was less cross section than that of the ENDF/B-IV by about factor of two.
Estimated residual activity of the ^{26}Al due to the (n,2n) reaction is 3-4 times smaller than conventional neutron transport calculation using multigroup constant which has crude mesh structure around 14MeV and not take into considering the fine spectrum of the fusion neutrons. For the plasma diagnostics, the aluminum activation technique would not be so attractive method.

1. INTRODUCTION

Aluminum has been considered to be a potential candidate of the low activation materials[1], because major activities of aluminum are of short life. One exception is due to the activity of ^{26}Al which has very long life of $7.2x10^5$yr. Recently, Fetter et al.[2] pointed out that aluminum is one of the materials which should be limited to concentration of 0.1 to 10 %, if the specific activity limit for shallow land waste disposal of long lived radionuclide according to the present U.S. regulation, because of its high energy gamma rays.

It is important to note, however, that above discussions were made using the ENDF/B-V data library, in which the $^{27}Al(n,2n)$ cross section was estimated only by simple model because no available data existed at that time, and therefore very uncertain. In the present work, the reliable and consistent cross sections have been estimated using modern nuclear reaction model from threshold to 20MeV on the basis of the recent experimental data.

Activation rate of aluminum exposed to the fusion neutrons has been estimated using present cross section and exact neutron spectra from the fusion plasma.

The feasibility of the aluminum activation technique to the plasma diagnostics was discussed using the calculated activation rate.

2. EXPERIMENTAL DATA BASE

Because, residual activity of $^{27}Al(n,2n)$ ^{26}Alg.s. reaction has a very long life, and cross section is very small just above 14MeV, it has been impossible to measure the cross section using conventional 14MeV sources before 1984.

Recently, several high intence 14MeV-neutron sources have been developed in the world especially for the material irradiation purpose and neutronics. First, Smither and Greenwood[3] measured the cross section of this reaction using RTNS-II of LLNL with accelerator mass spectrometry technique. Second, Sasao et al.[4] measured the cross section using an intence neutron source OKTAVIAN of Osaka Univ. with activation technique. Last, Iwasaki et al.[5] have measured the cross section from 14.1 to 14.7MeV using the RTNS-II source with activation technique. The Iwasaki et al's data were obtained using one of the chances of the world most intence and longest 14-MeV neutron irradiation condition in which the achieved average fluence on their aluminum samples was

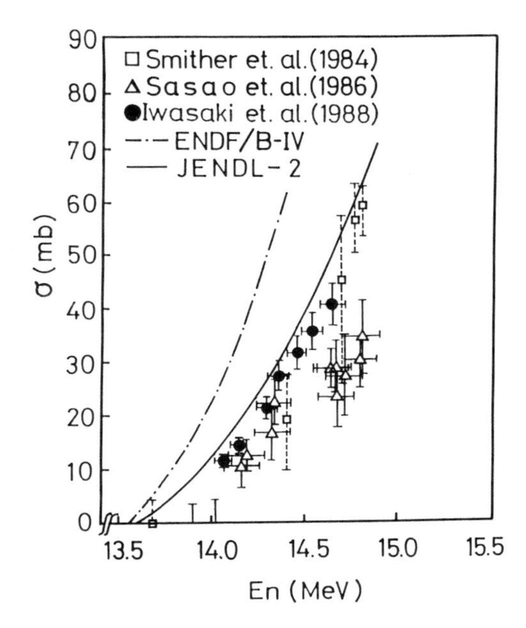

Figure 1
Comparison of experimental data for ^{27}Al (n,2n)^{26}Alg.s. (T1/2=7.16x10^5yr) reaction cross sections by Smither and Greenwood[3], Sasao et al[4]. and Iwasaki et al[5]. with those of evaluated nuclear data libraries.

up to 5x10^{16}(n/cm^2) in the period of 6 months. This condition enabled them to get good counting statistics[5] in relatively short time.

In Fig. 1, three experimental data above mentioned were compared with each other and two evaluated nuclear data libraries. The three experimental data show roughly the same trend. In detail, the Smither's data show similar with that of the Iwasaki et al, except for at 14.3MeV. On the other hand, the Sasao et al's data are different with that of the Iwasaki et al. except for the energy region of about 14.3MeV and are scattered.

Among three experimental data, the Iwasaki et al's data was selected as main data base from the following reasons: that data had smallest error band and was smooth possibly because of its high counting statistics; above 14.5 MeV both Smither's and Iwasaki's data were not inconsistent, and below 14.2 MeV the Iwasaki's data was more reasonable than the trends of the Smither's and Sasao's data because the threshold energy of the reaction was considered to be nearly 13.5MeV and not around 14MeV[3] according to a preliminary theoretical calculation.

3. EVALUATION OF THE CROSS SECTION

As seen in the previous section, available cross section data are only above 14MeV. Concerning the activation rate estimation, it is necessary to extrapolate accurately from the threshold to at least 16MeV, with the aid of suitable nuclear theory because the fusion neutron spectrum remarkably changes with plasma temperature.

3.1 CALCULATIONAL MODEL

In this study the complete multistep Hauser-Feshbach model with the precompound reaction process effect was adopted. In this model, the effects of competing compound and precompound processes and of angular momentum are automatically introduced and consistent cross section could be estimated. Among several multistep Hauser-Feshbach model codes, the GNASH code[6] was chosen because this code was widely used for the evaluation of the JENDL-3 in Japan and provided successful results. Further, in the code the direct reaction process by separate reaction model code can be taken into account.

Recently, Yamamuro developed a new combined system SINCROS-I[7,8] of the ELIESE-3[9], GNASH and DWUCK4[10] of which many input parameters were determined by himself and given in default mode. In the present study, this system has basically been used, however, important parameters were further searched so that the available cross section curves not only of the (n,2n)g.s. and (n,2n)isomer but also of the (n,p) and (n,α) reactions, including secondary particle emission cross section should be reproduced as accurately as possible.

3.2 INPUT PARAMETERS

In the fitting process, we have taken into account the recent works on aluminum, by Kitazawa and Harima[11], and Yamamuro[8], respectively. Direct inelastic cross section to the lower excited states were taken to be the same values with Yamamuro's calculation[12].

Finally, as the optical potential parameter sets for neutron and charged particles, the default potential ones were chosen in the SINCROS-I. Some different parameters sets were tested, but degree of fitting was not improved within the framework of the present calculational model.

Level density parameters for the compound nucleus and final nuclei for (n,p) and (n,α), and some F2 and F3 values were slightly varied from the default values by Yamamuro, where the F2 and F3 were the adjustable parameters[8] for precompound reaction and direct reaction (pick up), respectively.

3.3 CALCULATIONAL RESULTS

Figure 2, 3 and 4 shows the present calculated results with the adopted experimental data or evaluated nuclear data libraries in the energy range between 7 to 20MeV. The (n,2n) cross section curves around the threshold energy are also shown in figure 5.

As seen in figure 2, complete fitting has not been attained in particular for the standard reaction (n,α) above 17 MeV and for (n,p) reaction above 14 MeV due to insufficient knowledge of the various input parameters. In the latter case, however, the cross section is still uncertain since the recent experimental data are less than that of the ENDF/B-V and closer to the present calculation as shown in the figure.

Present (n,2n) cross section shows very good agreement with adopted data base by Iwasaki et al. Throughout three figures, rather good agreement was obtained.

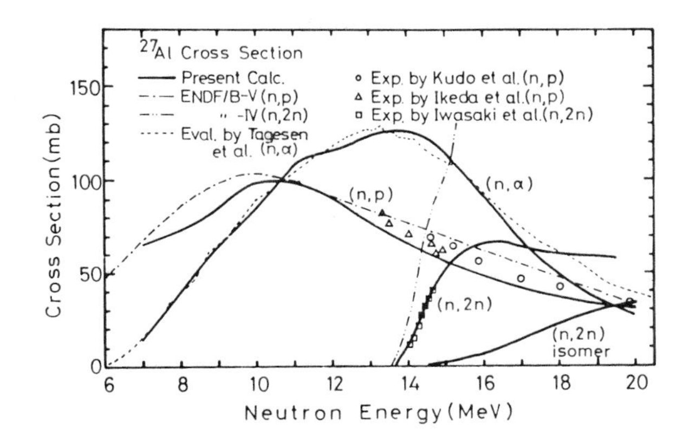

Figure 2
Present calculational results compared with some recent experimental data for (n,p) by Ikeda et al.[13] and Kudo et al.[14] and evaluated nuclear data for (n,α) by Tagesen et al.[15], for (n,p) of ENDF/B-V, and for (n,2n) reactions by the ENDF/B-IV and Iwasaki et al[5].

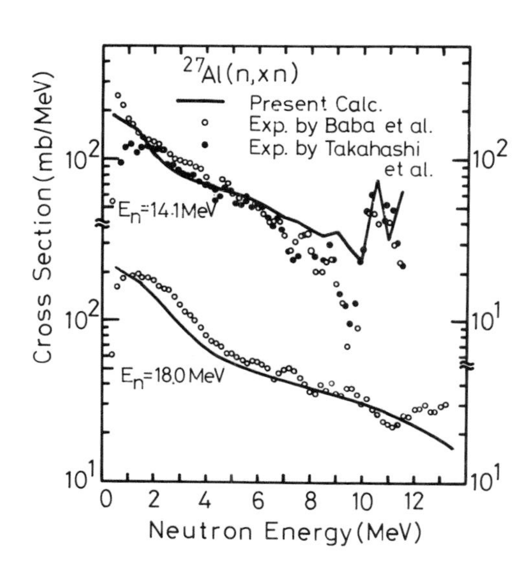

Figure 3
Present calculational results for neutron emission cross section compared with some recent experimental data at 14.1 by Takahashi et al[16], and at 14.1 and 18MeV by Baba et al[17].

4. ESTIMATION OF LONG TERM ACTIVITY OF ALUMINUM

4.1 NEUTRON SPECTRUM

In the present study, the neutron spectrum of a fusion device was assumed to be determined

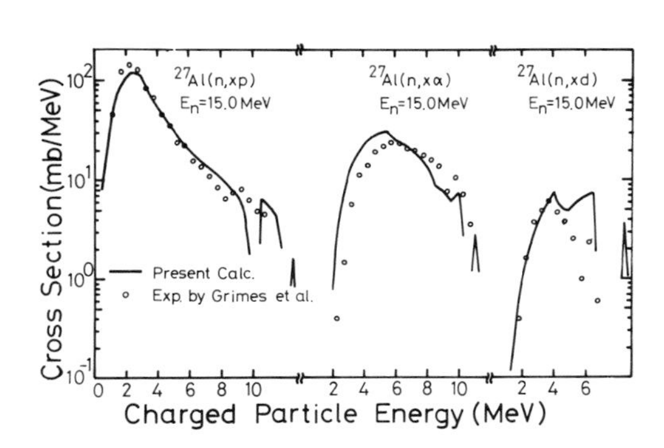

Figure 4
Present calculational results compared with some recent experimental data by Grimes et al.[18] for charged particle emission cross section.

only by the plasma temperature. In this case, the outgoing neutron spectrum can be calculated from the plasma spectrum, reaction cross section and kinematics. Recently, Talley et al.[19] formulated an exact model and gave numerical results. Some examples are shown in figure 5.

4.2 ACTIVATION RATE CALCULATION

The activation rate of the aluminum by (n,2n)g.s. reaction was calculated as a function of the plasma temperature. In figure 6, the calculated result is compared with that using a usual multigroup constant library, for example, CROSSLIB for THIDA code[13], which has broad group structure in the vicinity of 14MeV and does not take into account the fusion neutron spectrum as shown in figure 5. The latter case is 3 to 4 times larger than that of the present calculation.

Dependency of the activation rate on the plasma temperature is rather weak than expected as shown in figure 6.

5. CONCLUSION

The ^{27}Al(n,2n) cross sections have been evaluated base on the multistep Hauser-Feshbach reaction model in the energy range between 7 and

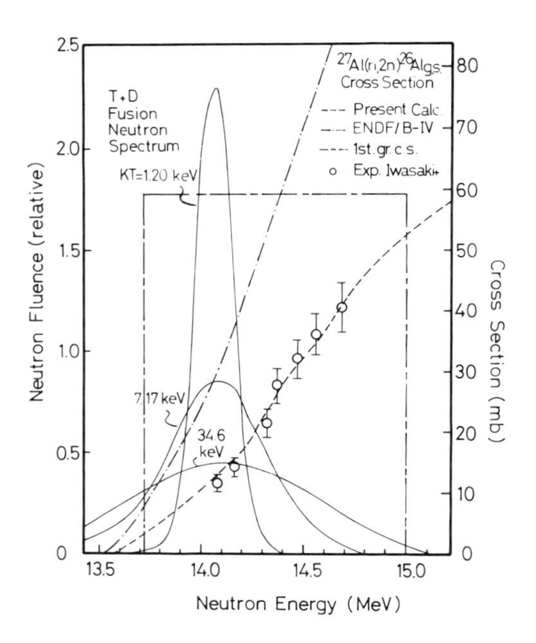

Figure 5
Comparison of the present calculated cross section curve for the ^{27}Al(n,2n)^{26}Al reaction and the source neutron flux spectrum of the exact one from fusion plasma given by Talley et al[19]. and the 1st group cross section and flux in a neutron transport calculation.

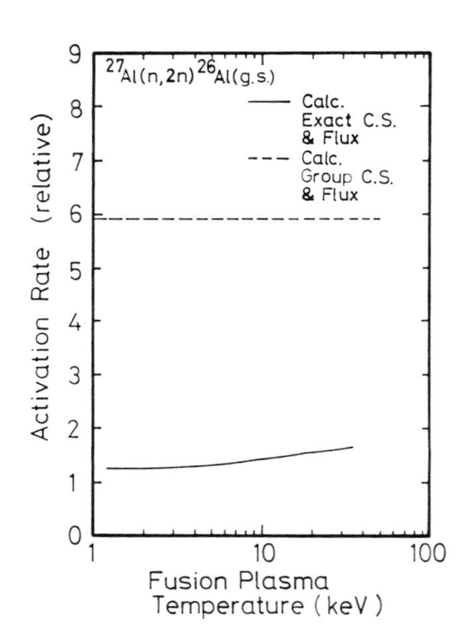

Figure 6
Comparison of activation rate of aluminum exposed to the fusion neutrons; exact neutron spectrum case, and group cross section and flux case.

20MeV. Model parameters were adjusted to fit the calculational result to the recent cross section data set by S. Iwasaki et al for (n,2n) reaction, together with other reaction cross sections, particularly for the (n,α) and (n,p) reactions. The over all agreement was rather good. The cross section curve was lower by factor of 2 than that of the ENDF/B-IV.

The estimation of the activation rate of the aluminum due to the fusion neutrons have been calculated using the present result were 3-4 times smaller than that of the calculation with a group constant and flux.

The obtained results showed rather flat function of the plasma temperature, and the aluminum activation technique is not so attractive as a method of plasma diagnostics.

ACKNOWLEDGEMENT

The authors would like to express their appreciations to Prof. N. Yamamuro for his kind offering the parameter values and calculated results for aluminum, and valuable suggestions. Drs. Nakagawa and Narita of the NDC-JAERI are acknowledged for their great help to transfer nuclear codes and data to Tohoku Univ. Comp. Center. We wish to thank Dr. Talley of LANL for his kind sending the numerical data.

REFERENCES

1. E.T.Cheng, Nuclear Technology/Fusion 4 (1983) 545.

2. S. Fetter, E.T. Cheng, and F.M. Mann, Fusion Engineering and Design, 6 (1988) 123.

3. R.K. Smither and L.R. Greenwood, Jour. Nucl. Mat., 122&123 (1984) 1071.

4. M. Sasao, T. Hayashi, K. Taniguchi, A. Takahashi, and T. Iida, Phys. Rev., C35, (1987) 2327.

5. S. Iwasaki, J.R. Dumais, and K. Sugiyama, Proc. Internat. Conf. Nucl. Data. Sci. Tech., May 30-June 3, Mito, Japan, in press, (1988).

6. P.G. Young, and E.D. Arthur, "GNASH: A Preequilibrium, Statistical Nuclear-Model Code for Calculation of Cross Sections and Emission Spectra",LA-6942 (1977).

7. N. Yamamuro, Proc. Internat. Conf. Nucl. Data. Sci. Tech., May 30-June 3, Mito, Japan, in press, (1988).

8. N. Yamamuro,in Japanese, JAERI-M 88-140 (1988).

9. S. Igarashi,"Program ELIESE-3: Program for Calculation of the Nuclear Cross Sections by using Local and Non-Local Optical Models and Statistical Model", JAERI 1224 (1972).

10. P.W. Kunz,"Distorted Wave Code DWUCK4", University of Cololado (1974).

11. H. Kitazawa and Y. Harima, Proc. Internat. Conf. Nucl. Data. Sci. Tech., May 30-June 3, Mito, Japan, in press, (1988).

12. N. Yamamuro, Private Communication (1988).

13. Y. Ikeda, C. Konno, K. Oishi, T. Nakamura, H. Miyade, K. Kawade, H. Yamamoto, and T. Katoh,"Activation Cross Section Measurements for Fusion Reactor Structural Materials at Neutron Energy from 13.3 to 15.0 MeV Using FNS Facility, JAERI 1312 (1988).

14. K. Kudo, T. Kinoshita, Y. Hino, A. Fukuda Y. Kawade, K. Takeuchi, and A. Iwahara, Proc. Int. Conf. Nucl. Data. Sci. Tech., May 30-June 3, Mito, Japan, in press, (1988).

15. S. Tagesen and H. Vonach, INDC/NEANDC Nuclear Standard File 1980 Version, INDC36/LN (1981).

16. A. Takahashi, E. Ichimura, Y. Sasaki and H. Sugimoto,"Double and Single Differential Neutron Emission Cross Sections At 14.1 MeV", Vol.1, OKTAVIAN Report A-87-03, Osaka Univ. (1987).

17. M. Baba, M. Ishikawa, T. Kikuchi, N. Yabuta, H. Wakabayashi, and N. Hirakawa, Proc. Int. Conf. Nucl. Data. Sci. Tech., May 30-June 3, Mito, Japan, in press, (1988).

18. S.M. Grimes, R.C. Haight, and J.D. Anderson, Nucl. Sci. Eng., 62 (1977) 187.

19. T.L. Talley, and G. M. Hale, Proc. Int. Conf. Nucl. Data. Sci. Tech., May 30-June 3, Mito, Japan, in press, (1988)

20. Y. Seki, H. Iida, H. Kawasaki, and K. Yamada,"THIDA-2: An Advanced Code System for Calculation of Transmutation, Activation, Decay Heat and Dose Rate, JAERI 1301 (1986).

FUSION TECHNOLOGY 1988
A.M. Van Ingen, A. Nijsen-Vis, H.T. Klippel (editors)

DEVELOPMENT OF LOW ACTIVATION VANADIUM ALLOYS FOR FUSION REACTOR APPLICATION

P. Hokkeling* and W. van Witzenburg**

* Philips Research Laboratories, Eindhoven, The Netherlands
** Netherlands Energy Research Foundation, ECN, Petten, The Netherlands

ABSTRACT
Results are reported concerning the first phase of a comprehensive vanadium-base alloy development programme. This phase implies systematic investigation of the effect of alloying additions, with allowable quantities as recently calculated for recycling and shallow land burial, on strength and ductility at room temperature. So far, about forty alloys were investigated. Preliminary high temperature tensile properties data of some promising alloys are presented as well.

1. INTRODUCTION

Development of low activation materials is increasingly considered as being of essential importance for fusion reactor application. Besides steel, which has limited perspectives as low activation material by adjustment of conventional compositions, vanadium-base alloys offer a potential alternative. Even though vanadium alloys investigated so far may qualify as low activation materials, depending on which criteria and nuclear activation data are used, they may not represent optimum compositions from the activation point of view, nor from that of other properties. Some of these properties, for example strength, ductility, compatibility with coolants and high temperature radiation embrittlement, are strongly influenced by the choice of alloying additions[1, 2].

Addition of substitutional elements in solid solution is an established and effective way of improving most of these properties of metals, including vanadium. Precipitations and metastable phases could be less reliable as means of improving material properties, because prolonged operation at high temperatures, as is envisaged for the first wall structure of fusion reactors, may cause modification of the microstructure, removing its beneficial aspects. Maintenance of the critical size and spacing of precipitates, for example, is questionable at the least. Strength of vanadium is also strongly influenced by interstitial elements in solid solution, notably O, N and C. The concentration of these elements, however, is difficult to control. During production of the alloys and also during the operational period, significant quantities of the elements could be absorbed from, or in the case of oxygen leached out by, liquid lithium. For these various reasons the emphasis in this programme so far has been on alloying with substitutional elements in solid solution. Most of the elements used have formed part of previous V-alloy development programmes: Ti, Cr, W, Fe, Si and Y, e.g. [3-8]. Some elements that have been shown to enhance the mechanical properties of vanadium, in particular Nb[9], were not considered here because of their high activation character.

2. SELECTION OF ALLOYING ELEMENTS

2.1. Low activation requirements

Recent results of calculations by Jarvis[10], Ponti[11, 12] and Fetter et al.[13] concerning concentration limits of a number of elements in the first wall structure, are tabulated in table 1.

Significant discrepancies exist for some elements, e.g. Si, Ti and W. There are various reasons for these discrepancies:

Table 1. Concentration limits to satisfy low activation requirements for a number of elements in the first wall structure. Calculated data from different sources and for different criteria.

Authors	Jarvis 1983[1]	Ponti 1986[2]	Ponti 1986[2]	Fetter et al. 1988[3]	Ponti 1988[4]	Ponti 1988[4]
Meth. of disposal	recycl.	recycl.	SLB	SLB	SLB US-limit	SLB UK-limit
Int. wall load (MWa/m²)	20	10	10	20	12.5	12.5
C	NR	NR	NR	NR		
N	>10%	NR	0.1%	4-50%	0.1%	5 m
O	>10%	NR	NR	NR		
Si	10-50%	20%	NR	2.5%	30%	NR
Ti	0.1-1%	5-10%	NR	NR	NR	10%
V	NR	NR	NR	NR	NR	NR
Cr	NR	NR	NR	NR	NR	NR
Mn	NR	NR	NR	NR	NR	NR
Fe	NR	NR	NR	NR	NR	20%
Ge	NR	NR	NR	NR		
Y	NR	NR	NR	NR		
Nb	1-10 m	1 m	10 m	0.4-9 m	5 m	10 m
Mo	10-1000m	50 m	500 m	0.7-6 m	100 m	10 m
Rh	NR	NR	NR	0.06-0.7%		
Ta	NR	50 m	NR	6%	NR	1%
W	NR	0.5%	5-10%	0.1-7%	2%	0.8%
Au	NR	50%	NR	0.5-1%		
Zr	10-1000m	1-5%	50%	4%	7%	0.1%
Ni	0.1-1%	0.1%	10%	10%	10%	5 m
Al	10-1000m	100-500m	0.5%	0.09%	0.2%	5%

NR = no restriction
m = parts per million
SLB = shallow land burial

[1] Ref. 10
[2] Ref. 11
[3] Ref. 13
[4] Ref. 12

1. The activation cross-section data-base used in the different calculations has not been the same. The most up-to-date data, contained in the REAC-ECN-3 library[14], has been used in the second set of calculations by Ponti[12], assuming 2.5 years of operation at 5 MW/m²: 12.5 MWa/m².
2. No uniform criteria exist for classification of radioactive materials, neither for recycling of waste products, nor for shallow land burial. This is illustrated by the results of Ponti[12], starting from the same data-base for two different shallow land burial criteria, one derived from US-codes and the other from UK-codes.
3. The neutron spectrum at the first wall depends on the blanket design and the constituting materials. The results are therefore, strictly speaking, applicable only for that particular design.
4. For multi-step reactions the induced activity caused by nuclides, which are themselves produced by transmutation, will depend on the neutron flux, not merely on the integral wall loading.

There is, however, agreement on a number of elements. V, Cr and Mn can be used without restriction, while Fe is acceptable to a considerable concentration. For elements which are restricted our approach was to adhere to the lower concentration limits in table 1. For two or more restricted elements, the sum of their relative concentration limits (tabulated concentration limits divided by their respective weight fractions) has to be at least unity.

2.2. Additional considerations

Summary for elements of interest as indicated by previous investigators:

Titanium has been used extensively in the past to improve mechanical properties of vanadium. It is a solid solution strengthener, but its effect at lower concentrations appears to depend strongly on the impurity content of a particular alloy. Ti, with unlimited solubility in V, interacts strongly with interstitial impurities (O, N, C) which strengthen vanadium considerably, but also reduce its ductility. Addition of 3% Ti reduces the solubility of oxygen in vanadium from about 2% at room temperature to 400 ppm. The result is increased ductility and workability. For some properties, for example creep, the Ti-content should not exceed 3% for maximum benefit[9]. An additional positive aspect of Ti includes decreasing irradiation induced swelling. A negative side of Ti is related to the tritium inventory, because Ti increases the solubility of hydrogen isotopes.

The effect of chromium, also a solid solution strengthener and with large affinity for interstitial impurities, is more ambiguous. Cr also has no solubility restrictions in V. Negative aspects of Cr are linked to embrittlement of different origins: radiation embrittlement[15], hydrogen embrittlement[8] and embrittlement caused by exposure to lithium[16]. Improved oxidation resistance due to Cr will be of no value when in contact with liquid lithium, because Li_2O is much more stable than Cr_2O_3. In contrast to Ti, Cr decreases the solubility of hydrogen isotopes in V. The overall impression is that Cr should be avoided, if possible.

Silicon has also been used in V-alloys to increase its strength[3, 4, 17]. Its solubility in V is limited to ~ 2% and is reduced even further by addition of Ti. The ternary phase diagram of V-Ti-Si at room temperature shows that with 3% Ti the Si-content should not exceed ~ 1% to avoid formation of V_3Si. Si has a tendency to segregate to free surfaces, which may lead to embrittlement.

A potent strengthener, is also tungsten[5], although its content must be severely restricted (< 1%) because of its activation characteristics.

Then there are the elements Mn and Ge, which have been encountered in the literature only in relation to phase diagrams. Mn has unrestricted solubility in V. A complication with Mn is its high vapour pressure at the melting temperature of vanadium, which makes it difficult to dose in the desired quantity. The solubility of Ge in V is similar to that of Si. Ge only dissolves up to ~ 2,5%. At higher concentrations V_3Ge is formed. Restrictions are likely to be even more severe when additional elements are present. The tendency to segregate to free surfaces is even stronger than it is for Si.

Rhodium with solubility up to 20% and gold with unrestricted solubility have been included in the programme for curiosity reasons. Their cost obviously would preclude large scale application in fusion reactors.

Iron, also mentioned in the literature as strenghtener of V-alloys[3, 6], could be a very interesting element. For low activation requirements as much as 20% is permitted, at least according to the latest figures. The binary phase diagram V-Fe also does not show any restrictions. Even the ternary system V-Ti-Fe indicates that Fe is soluble up to 10% at ~ 3% Ti. The effect of Fe on various properties has yet to be established.

Finally, yttrium and other rare earth metals appear to be interesting to control interstitial impurity content.

3. PREPARATION OF ALLOYS AND SPECIMENS

Ingots (10-30 grams), made by argon-arc melting, were inserted in close fitting steel jackets and flattened under a load of 100 tons at room temperature. Hot-rolling, which is a tedious procedure because of the necessary sheathing, could be avoided. These products were cold-rolled, without intermediate annealing, to plate of 0.5 mm thickness, with the exception of V-25% Cr which crumbled during cold-rolling. Tensile specimens (gauge width 3 mm and length 12 mm) were produced by laser cutting from the plates following annealing at 1100°C in vacuum (10^{-4} Pa). The flat surfaces of the specimens were not further treated before tensile testing. Grain sizes of the alloys in annealed condition ranged from 20-40 μm. The composition of all alloys produced so far are listed in table 3. Numbered alloys are alloys with more than one additional element.

TABLE 2. Specification of V:

Series	Source	Nom. purity	Interst. imp. cont. (ppm) O	N	C
I	MRC-VP	99.8%	1400	35	15
II	MRC-VP	99.8%	400	35	15
III	Highw. Int.	99.8%	370	58	40
IV	" "	"	"	"	"

4. TEST RESULTS

4.1. Mechanical properties at room temperature

Initial evaluation of the alloys was done by hardness and tensile properties measurements at room temperature. Results are listed in table 3 and plotted as a function of volume percentage of added elements per alloy in figure 1. The hardness data, determined with a 10 N load, represent averages of 5 measurements, with a typical standard deviation of ~ 4%. Tensile tests were performed in duplo at a strain rate of 10^{-3} s^{-1}. Measured values generally are within 10% of the average values indicated.

Some alloys have considerable strength. For example, alloy 29 has a UTS value of over 800 MPa while still showing a total elongation of about 20% (for comparison: the UTS of stainless steel AISI 316 in solution annealed condition is ~ 575 MPa at 24°C). The total elongation of our alloys, incidently, is significantly lower than those reported recently by Loomis et al.[8] for various V-alloys. For V - 15 Cr - 5 Ti (alloy nr. 2) ~ 20% was obtained, while Loomis et al. found for the same alloy more than 30%. The difference may be due to our specimen surfaces not having received a polishing treatment, in contrast to those of Loomis et al.

The hardness and UTS (fig. 1 A and B), as

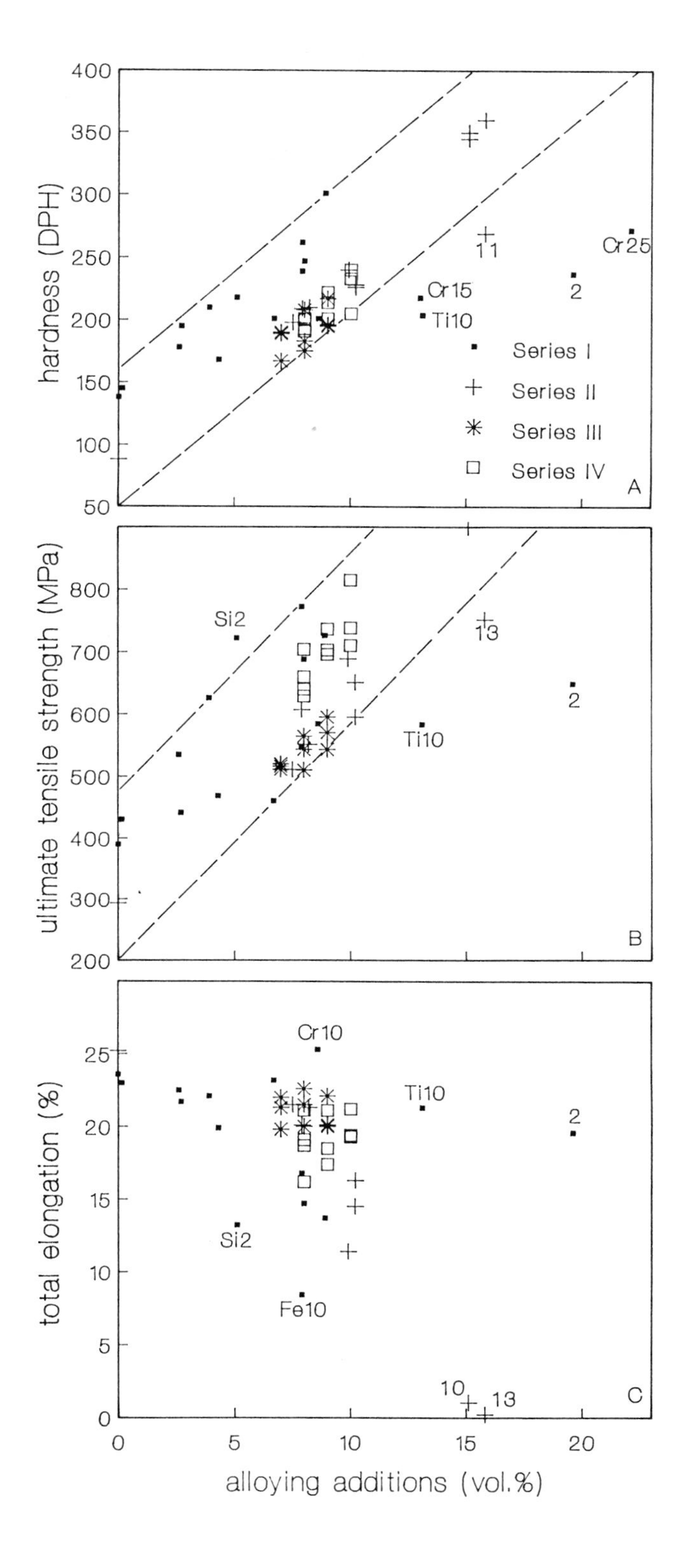

FIGURE 1
Effect of alloying additions on the tensile properties at room temperature.

Table 3. Alloy compositions and hardness and tensile properties at room temperature.

Series	Nr.	V	Si	Ti	Cr	Mn	Fe	Ge	Y	Rh	W	Au	total added elements		Room temperature properties: Hard-ness	yield strength	UTS	Total elong.
		W. %											at.%	vol.%	(10N)	(MPa)	(MPa)	(%)
I	V (1)	99.9													138	341	390	23.6
	W 0.5	99.5									0.5		0.14	0.16	145	380	430	23.0
	Cr 5	95.0			5.0								4.9	4.3	168	398	468	19.9
	Cr 10	90.0			10.0								9.8	8.6	201	517	584	25.3
	Cr 15	85.0			15.0								14.7	13.0	218	-	-	-
	Cr 25	75.0			25.0								24.6	22.1	272	-	-	-
	Ti 2	98.0		2.0									2.1	2.7	195	376	441	21.7
	Ti 5	95.0		5.0									5.3	6.7	201	363	460	23.2
	Ti 10	90.0		10.0									10.6	13.1	204	496	583	21.3
	Fe 5	95.0					5.0						4.6	3.9	210	545	626	22.1
	Fe 10	90.0					10.0						9.2	7.9	262	703	772	8.4
	Si 1	99.0	1.0										1.8	2.6	178	466	535	22.5
	Si 2	98.0	2.0										3.6	5.1	218	628	722	13.2
	Rh 15	85.0								15.0			8.0	8.0	247	655	688	14.7
	1	95.0	2.0	3.0									6.7	8.9	301	620	726	13.7
	2	80.0		5.0	15.0								20.0	19.6	237	545	649	19.6
	3	92.0		3.0			5.0						7.8	7.9	239	440	548	16.8
II	V(2)	99.9													88	240	293	25.2
	4	92.0		3.0		5.0							7.8	8.2	210	442	551	21.3
	5	92.0		3.0			5.0						7.8	7.9	209	485	607	20.1
	6	89.5		3.0			7.5						10.1	9.9	240	558	689	11.4
	7	94.0		3.5				2.5					5.5	7.5	198	426	511	21.5
	8	85.5		3.0						11.5			9.4	10.2	228	548	651	14.5
	9	82.0		3.5								14.5	8.4	10.2	226	500	595	16.3
	10	87.0	2.0	3.0			8.0						14.0	15.1	345	800	900	1.0
	11	86.5	2.0	3.0			8.0		0.5				14.3	15.8	269	-	-	-
	12	87.0	2.0	3.0			8.0						14.0	15.1	350	-	-	-
	13	86.5	2.0	3.0			8.0		0.5				14.3	15.8	360	-	751	0.2
III	14	93.1		3.0			3.2		0.2		0.5		6.4	7.0	167	394	516	21.3
	15	91.8		3.0			4.5		0.2		0.5		7.6	8.0	175	412	543	20.0
	16	90.5		3.0			5.8		0.2		0.5		8.8	9.0	196	450	570	20.0
	17	95.5	1.0	3.0			0.3		0.2				5.3	7.0	189	421	511	19.8
	18	94.2	1.0	3.0			1.6		0.2				6.5	8.0	208	469	564	21.5
	19	92.9	1.0	3.0			2.9		0.2				7.7	9.0	217	489	595	22.1
	20	95.2	1.0	3.0		0.07			0.2		0.5		5.3	7.0	190	424	520	22.0
	21	94.0	1.0	3.0		1.3			0.2		0.5		6.4	8.0	183	422	510	22.6
	22	93.8	1.0	3.0		2.5			0.2		0.5		7.6	9.0	195	453	543	20.1
IV	23	92.1		3.0			4.7		0.2				7.6	8.0	200	528	640	16.2
	24	91.8		3.0			4.5		0.2		0.5		7.6	8.0	193	493	629	18.7
	25	90.5		3.0			5.8		0.2		0.5		8.8	9.0	214	578	703	21.1
	26	89.3		3.0			7.0		0.2		0.5		9.9	10.0	240	596	738	19.3
	27	94.3	1.0	3.0			1.4		0.3				6.4	8.0	191	605	704	19.1
	28	93.0	1.0	3.0			2.7		0.3				7.6	9.0	222	638	736	18.5
	29	91.7	1.0	3.0			4.0		0.3				8.8	10.0	233	698	815	19.4
	30	94.1	1.0	3.0		1.1			0.3		0.5		6.3	8.0	201	575	660	21.1
	31	92.8	1.0	3.0		2.4			0.3		0.5		7.5	9.0	201	616	696	17.4
	32	91.6	1.0	3.0		3.6			0.3		0.5		8.6	10.0	205	600	710	21.2

well as the yield strength, clearly increase with the quantity of added elements. In fact, it does not appear to make much difference which elements are added, with the exception of Ti and Cr, which show a weaker influence. This general behaviour is more obvious when properties are plotted versus volume percentage of added elements, instead of weight or atomic percentage as is usually done. The alloys of series I, which contain more oxygen than the alloys of the other series (table 2), generally are higher up in the band of data points. The ductility shows a similar common behaviour, although not as clear. The total elongation (fig. 1C) generally decreases with increasing solute content, with the series I alloys having lowest elongations, except for Ti and Cr containing alloys.

The results of series I and II alloys suggested the following tentative guiding principles:

1. For a total elongation of ~ 20% at room temperature a V-alloy generally should contain no more than 10 vol. % of alloying additions.
2. In order to have their full strengthening effect, the added elements should remain in solution. A room temperature UTS of 600 - 800 MPa should then be possible.
3. The vanadium used as well as the alloying additions should be of low O and N content.

Application of these principles led to the compositions of series III and IV alloys, which more or less appeared to behave accordingly.

Further analysis of the data shows deviations from the general behaviour, however. The alloy V - 15 Cr - 5 Ti, for example, has a relatively low hardness and UTS and high ductility for its alloy content. This, by the way, is in line with the hardness and tensile properties of the binary Ti and Cr containing V-alloys mentioned before. The behaviour of Cr and Ti containing alloys may be complicated by interaction of these elements with the interstitial impurities O, N and C.

There are also some as yet unexplained tendencies. For instance, the series IV alloys are relatively strong in UTS and yield strength, while the hardness and elongation seem to fit the general behaviour.

4.2. Tensile properties at elevated temperatures

The tensile properties of the alloys 23-32 were measured also at 500, 600 and 700°C in vacuum (10^{-4} Pa). The temperature dependence of the yield strength, UTS and total elongation is shown in figure 2 for the alloy 29, which is

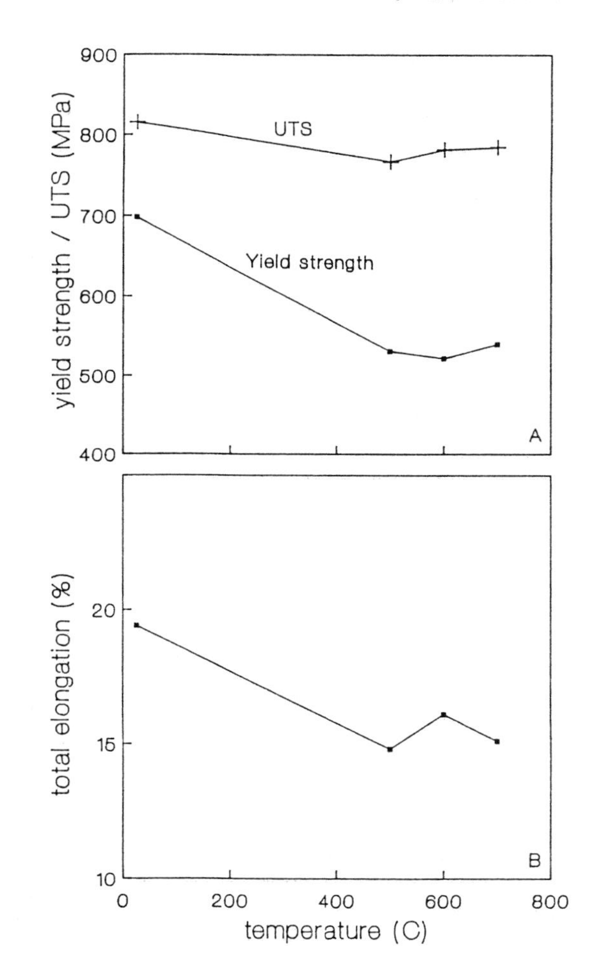

FIGURE 2
Tensile properties of the alloy V-1Si-3Ti-4Fe-0.3Y as function of test temperature.

the strongest alloy. The tensile properties of the other alloys have similar temperature dependence. The UTS between 500 and 700°C is hardly different from the room temperature value, while yield strength as well as total elongation remain at levels, somewhat below their respective values at room temperature. All stress-strain curves of tests at 500 and 600°C show serrated yielding.
The strength remaining at a high level up to high temperatures, also observed recently by Loomis et al.[8] for various vanadium alloys, may be explained by dissociation of vanadium oxides (V_4O, V_9O) at higher temperatures, as indicated by the V-O phase diagram. The resulting increase in concentration of dissolved oxygen could then counteract the normally weakening trend with increasing temperature.

5. CONCLUSION

A number of ductile low activation vanadium-base alloys were produced having considerable strength at room temperature (UTS ~ 700 MPa). The strength of these alloys is maintained to at least 700°C, while the total elongation is not much lower at that temperature than at room temperature. These alloys should not be considered as optimum results. Further testing obviously is required and will include impact properties, long duration mechanical behaviour at elevated temperatures, radiation effects and corrosion in flowing lithium. Results of these various tests are likely to lead to readjustment of the alloy compositions.

REFERENCES

1. D.L. Harrod and R.E. Gold, Int. metals Reviews 4 (1980) 163.
2. D.R. Diercks and B.A. Loomis, J. Nucl. Mater. 141-143 (1986) 1117.
3. W. Rostoker, A.S. Yamamoto and R.E. Riley, Trans. ASM 48 (1956) 560.
4. B.R. Rajala and R.J. van Thyne, J. less-Common Met. 3 (1961) 489.
5. D.R. Mathews and H.G. Iverson, U.S. Bureau of Mines Report 6929 (1966).
6. H. Böhm, H.U. Borgstedt, M. Rühle and P. Wincierz, Proc. 6. Plansee Seminar (1968) 256.
7. G.H. Keith and J.S. Winston, U.S. Bureau of Mines Report 7393 (1970).
8. B.A. Loomis, R.H. Lee, D.L. Smith and J.R. Peterson, Proc. ICFRM-3, Karlsruhe (1987).
9. H. Böhm and M. Schirra, KfK-report 774 (1968).
10. N. Jarviș, AERE-report R 10860 (1983).
11. C. Ponti, Fusion Technology 13 (1988) 157.
12. C. Ponti, Proc. ISFNT, Tokyo (1988).
13. S. Fetter, E.T. Cheng and F.M. Mann, Fusion Engineering and Design 6 (1988) 123.
14. J. Kopecky and H. Gruppelaar, Proc. 15e SOFT, Utrecht (1988).
15. D. Braski, J. Nucl. Mater. 141-143 (1986) 1125.
16. O.K. Chopra and D.L. Smith, Proc. ICFRM-3, Karlsruhe (1987).
17. M. Schirra, Metall 33 (1979) 455.

FUSION TECHNOLOGY 1988
A.M. Van Ingen, A. Nijsen-Vis, H.T. Klippel (editors)

ACTIVATION OF MATERIALS FOR THE FUSION REACTOR FIRST WALL

A. KHURSHEED*, G.J. BUTTERWORTH+, A.J.H. GODDARD* and J.A. MASON*

* Centre for Fusion Studies, Imperial College, London SW7 2BX, U.K.
+ Culham Laboratory, Abingdon, Oxon., OX14 3DB, U.K.

(Euratom/UKAEA Fusion Association)

Activation calculations, taking irradiation conditions derived from a first wall position in the DEMO design, have been performed for all stable elements with Z numbers between 3 and 83, with the aid of a recently upgraded database. Activity and surface gamma doserate are used as indices to compare the activation properties of elements. Comparisons with previous results have been made, and improvements to the database discussed. The activation properties of several candidate first wall compositions, including 2 vanadium alloys, have been considered.

1. INTRODUCTION

An intrinsic feature of D-T fusion reactors will be activation of structural components adjacent to the plasma chamber. The reactor first wall, as the part of the structure closest to the plasma and also the component requiring the most frequent replacement, will contain the major part of a fusion reactor's radioactive inventory.

Previous work has shown that the activation properties of elements vary greatly, and there is the potential, through the careful selection of composition, to achieve considerable reductions in first wall and blanket activation. Relative merits have been assigned to individual elements, usually on the basis of activity, surface gamma doserate or waste disposal criteria. Existing results are far from conclusive, however, on account of uncertainties in the activation database and the gaps in dosimetric assessments. In this paper, the first of these problems has been addressed. Activation of pure elements and several candidate first wall materials has been calculated with the updated UK fusion activation database UKACT1 and the recently developed inventory code FISPACT.

2. ACTIVATION CALCULATIONS

2.1 Calculational method

The method for fusion activation calculations is well established : the set of inventory equations is solved by a standard code, employing a constant neutron flux obtained from a preceding neutron transport calculation. The inventory code in this case, FISPACT[2], was derived from FISPIN6[10] and provides, in addition to potential ingestion and inhalation doses, a sensitivity analysis of nuclide concentrations with respect to library decay and cross-section data. The neutron energy spectrum was calculated for the first wall of the solid-breeder Culham DEMO reactor design[11], and is identical to that used earlier by Giancarli[3].

The UKACT1 database represents a major improvement on the previous UK fusion activation database UKCTR-III, which was patchy in its coverage of elements and contained data with large uncertainties. It is based on the REAC library[4] and the current version contains data for 625 target nuclei and 8719 reactions. As with UKCTR-III, the bulk of the cross-section data has been obtained from nuclear models, but improvements have been made in the following areas ;

a) In the treatment of short-lived nuclei, an effort has been made to include all reactions on nuclides with half-lives of a day or more. It can be shown, however, that in certain

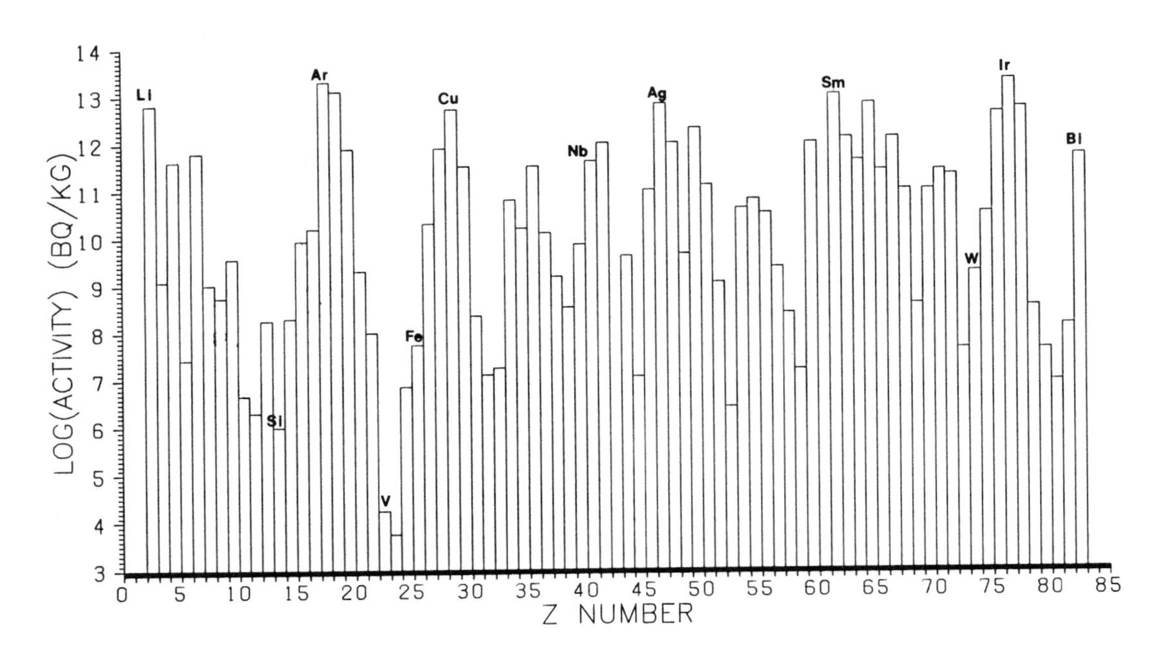

Figure 2.1 : Activities of Pure Elements after 100 years

conditions nuclides with half-lives of less than a day can be significant if they are a small number of reactions removed from a long-lived product[5]. Examples which have been included are the reactions Si^{31}(n,g), Pb^{209}(n,g) and Ti^{44}(n,2n).

b) Reactions leading to the first or second isomeric states have been considered.

c) A more uniform coverage over the entire periodic table has been achieved.

d) Gaps in the cross-section data, notably at large Z numbers, have been filled.

e) Additional reaction types, such as (n,2p) and (n,3n) have been included.

f) Improved systematics have been applied to existing cross-sections[6].

The irradiation was considered to be a continuous 2.5 year burn at a wall-loading of $5MW/m^2$, corresponding to a cumulative loading of $12.5MWy/m^2$, followed by a cooling period of 100 years. Surface gamma doserate, including a correction for bremsstrahlung, was obtained in the manner suggested by Jarvis[7]. In order to treat self-absorption in this calculation, it was assumed that each element or composition had the mass-attenuation coefficients of iron.

Where the concentrations of important inventory nuclides have changed significantly from previous calculations, the relevant cross-sections have been identified by means of sensitivity coefficients[5].

2.2 Pure Element Calculations

The overall pattern of the activation may be discerned from figures 2.1 and 2.2. Peaks of high activation, associated with long-lived nuclei, occur over the whole range of atomic numbers, interspersed with low activation windows which may be exploited for alloy development. Of particular relevance to current low-activation materials development programmes are those found about Si(Z=14), Cr(Z=24) and W(Z=74). Other windows occur around As(Z=33), I(Z=53), Tl(Z=81) and elsewhere, but the properties of these elements as alloying additions appear to be less favourable.

There are significant differences between the figures presented here and those given by calculations performed by Jarvis[1] and Giancarli[3] with versions of the UKCTR-III library and the ORIGEN inventory code, and with the calculations performed by Ponti[8].

Concentrating on elements which are of interest to current low-activation compositions, it is noted that in comparison with Giancarli's figures, data for the common elements Fe, Co, Ni, Cu, Zn, Nb, Pb and Bi have not changed by a great deal - less than a factor of 2 in most cases, owing to the more certain cross-section data available for these elements.

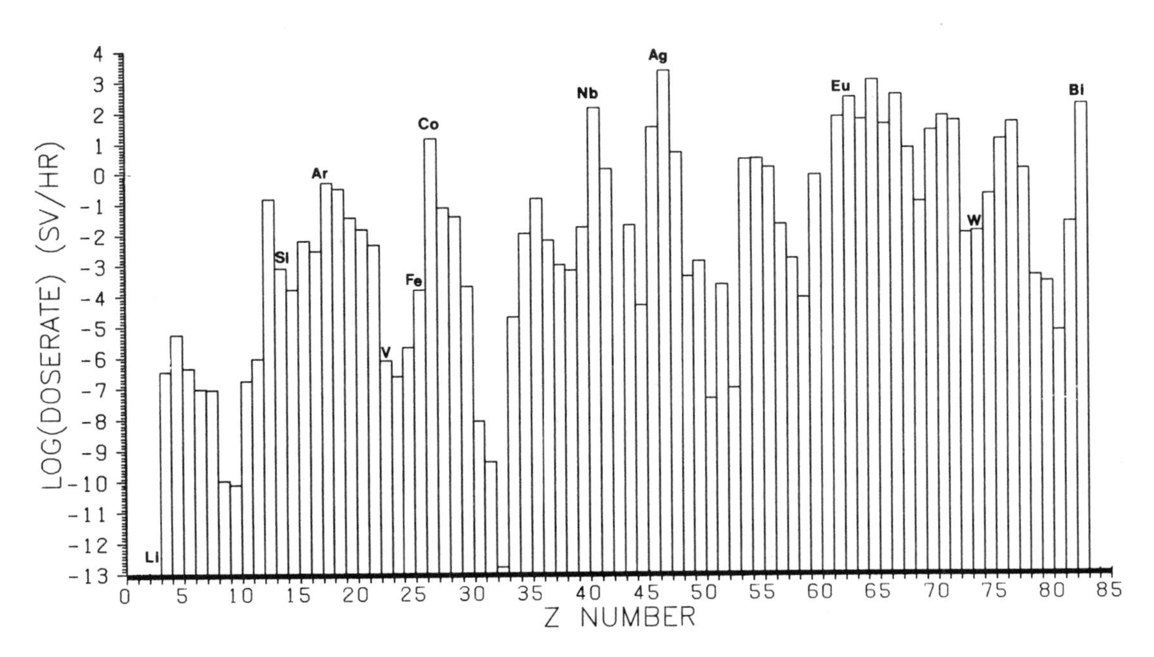

Figure 2.1 : Gamma Doserates (Sv/hr) from Pure Elements

Molybdenum

A significant change is noted for Mo, for which surface doserate has increased by about a factor of 7. The principal gamma-emitting nuclide in the Mo inventory is Nb^{94}, produced by the Mo^{94}(n,p) (~ 47%) and Mo^{95}(n,np) (~ 51%) channels. Recent determinations of the cross-sections[12] of these reactions at 14MeV are in general agreement with the UKACT1 values, allowing some confidence to be attached to the increased figure for gamma doserate.

Silicon and aluminium

The gamma doserates from Si and Al have increased by a factor of 12 and decreased by a factor of 4, respectively. The principal long-lived gamma emitter in both inventories is Al^{26}. The reaction channel Si^{28}(n,np)Al^{27}(n,2n)Al^{26} accounts for 97% of the Al26 production in silicon and all of the production in aluminium. Recent experimental determinations[12] have reduced the cross-section for the (n,2n) reaction, but this is more than compensated by an increase in the (n,np) cross-section. The cross-sections for both of these reactions may be subject to change.

Tungsten and Tantalum

The doserates from Ta and W have risen, the latter because of the inclusion of successive capture reactions to Ir^{192}n, the former because of the inclusion of reactions to Hf^{178}n. Since the treatment of isomeric branching ratios is approximate (adopting a value of 0.5 or 0.01), there are large uncertainties in the results for these two elements, and it is quite possible that their activation properties will differ from present predictions as data from new determinations become available. A sensitivity analysis suggests that the critical reactions in question are $Ir^{191}(n,\gamma)Ir^{192n}$ (~ 70%) in W, and Hf^{178}(n,n')Hf^{178n} (~ 48%) and Hf^{179}(n,2n)Hf^{178n} (~ 28%) in Ta. It is noted that in the case of W, if the concentration of Ir^{192n} is substantially reduced, then the dominant long-lived gamma-emitter becomes Re^{186m}, produced via Re^{187}(n,2n)Re^{186m} (~ 72%) and $Re^{185}(n,\gamma)Re^{186m}$ (~ 25%).

Titanium

The doserate from titanium, an important constituent in vanadium alloys, has been decreased by roughly a factor of 2, due mostly to a reduction in the cross-section for the $Ca^{45}(n,\alpha)Ar^{42}$ reaction from new systematics. The reasons for this, and other aspects of the activation of titanium, have been discussed elsewhere[9].

On the whole, the level of activation is higher than that given by Giancarli, owing to the inclusion in the UKACT1 library of new reaction pathways to long-lived products - especially capture reactions on heavy nuclei, some of which result in high burn-up rates (e.g. in excess of 90% in Eu). Where activation levels have been reduced, the principal cause is the renormalisation of cross-sections according to improved systematics.

In comparison with Ponti's figures, it is observed that

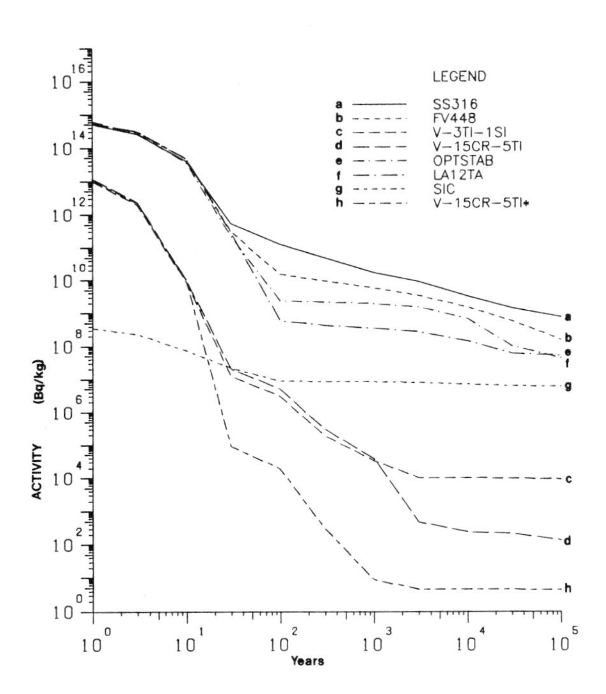

Figure 2.3 : Activity from First Wall Compositions

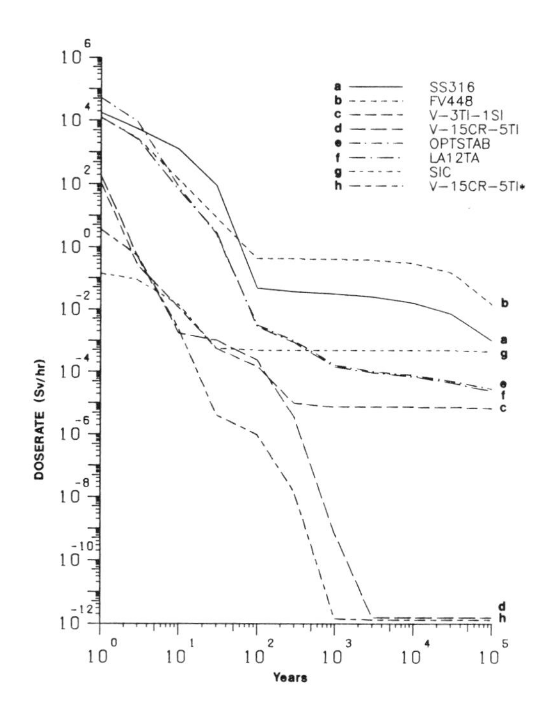

Figure 2.4 : Doserate from First Wall Compositions (Sv/hr)

there are a large number of disparities, even for elements such as Fe and Ti, for which sound cross-section data exist. This fact points to the calculation rather than data as the source of the discrepancies. A possible source of the disagreements may be the neutron flux used by Ponti (calculated for the first wall of the NET design and a $1MW/m^2$ wall-loading). It is relevant to note here that for stages of an activation chain involving decaying target nuclei, the wall-loading is a unique combination of flux and irradiation time, and the activation corresponding to a higher wall-loading cannot be obtained by simple scaling.

It is interesting to note,too, that a study of the reaction rates of capture and threshold reactions in DEMO and NET first walls[5], shows that though the neutron energy spectrum in the latter is highly moderated by the presence of water in the blanket, the reaction rates are similar.

2.3 Candidate First Wall Compositions

The specific activities and doserates resulting from candidate first wall compositions are shown in figures 2.1 and 2.2. It must be kept in mind that though impurity levels have been estimated for the four steels, no impurities havebeen included for vanadium alloys and SiC.

If the steels are taken first, it is seen, as in the previous study[3], that the gains offered in the immediate post-irradiation period by low-activation steels are small, but they increase after about 30 years, such that after 1000 years they give doserates which are approximately 2 and 3 orders of magnitude better than SS316 and FV448 respectively. The pure vanadium alloys give substantially lower doserates than the standard steels at all times, particularly at long times, when the advantage is very substantial indeed. It is observed that the V-Ti-Si alloy becomes less attractive than the V-Cr-Ti alloy after about 100 years, due to Al^{26} produced from the Si component. An isotopically tailored version of the V-Cr-Ti alloy, in which the Ti component is present only as Ti^{50}, shows some advantage over the 10 to 1000 year time interval in which the dominant radionuclide is Ar^{42} ($t_{1/2}$ = 33yrs).

With regard to SiC, the obvious features of the doserate curve are unaltered : a rapid fall in doserate following removal from the reactor, followed by a high residual level, resulting from the long-lived species Al^{26}. However, the altered doserate for Si discussed above increases the level of this tail by about an order of magnitude, and the short-term doserate is higher too, presumably from an increase in the Na^{22} concentration. The prospects for SiC suffer further when activation of the C component is examined : though the gamma doserate would be low, the potential ingestion dose from C^{14}, produced by $C^{13}(n,\gamma)$, would be high in some hazard scenarios, given the selective absorption of C by the biosphere.

It must be stressed again that impurities have not been considered for the vanadium alloys and SiC. Preliminary determinations suggest that Nb, Mo and Ag would be present in vanadium. Ideally, it would be necessary to keep the levels of these elements down to about 0.005ppm, 0.5ppm and 0.0004ppm respectively, if the low intrinsic doserate of the vanadium host material is not to be compromised. It should be added, however, that since the limiting factor in vanadium alloys may be the long-term dose from the activation products of the titanium component, namely $Ar^{42}(K^{42})$, some relaxation of the allowable impurity levels may be permitted.

3. CONCLUSIONS

Improvement of the activation database has resulted in significant changes to activation inventories. The database is far from complete however, and it is to be expected that further improvements, particularly to data for minority elements, will have a further effect on inventories. Of particular value to the development of vanadium alloys would be a more accurate determination of the $Ca^{45}(n,\alpha)$ cross-section. With respect to the use of W and Ta in LA materials, determinations of branching ratios of reactions to $Ir^{192}n$, Re^{186m} and $Hf^{178}n$ are required. At high atomic numbers, the unexpected dominance of capture reactions, giving rise to high burn-up of source nuclei and long activation chains, demands that cross-section determinations for these reactions receive more attention than hitherto.

The updated database has not altered conclusions with respect to the activation properties of low-activation steels significantly : in comparison with commercially available standards, substantial reductions in surface gamma doserate are achievable at cooling times of greater than 30 years. The gamma dose rate is likely to be dominated by unavoidable impurities and minority constituents such as tungsten and tantalum, the activation properties of which are still subject to large uncertainties. The potentially greater reductions offered by vanadium alloys are confirmed and have improved marginally relative to earlier predictions, but again much will depend on the concentration levels of impurities. The prospects for SiC are slightly worse than before, the present predictions indicating a significantly higher long-term dose rate as a result of the increased Al^{26} inventory.

REFERENCES

1. O.N. Jarvis, "Low-Activity Material : Reuse and Disposal" (AERE-R-10860, Harwell, 1983).
2. R.A. Forrest, D.A. Endacott and A. Khursheed, "Fispact - User Manual" (AERE-M-3655, Harwell, April 1988).
3. L. Giancarli, "On the Behaviour of First Wall Fusion Structural Materials" (CLM-R275, Culham, February 1987).
4. R.A. Forrest, "Status of the UK Activation Cross-Section Library for Fusion" (IAEA Advisory Group Meeting on Nuclear Data, Garching, 1986).
5. A. Khursheed, PhD. Thesis (Imperial College, London, in preparation).
6. R.A. Forrest, "Systematics of Neutron-Induced Threshold Reactions with Charged Particle Products at about 14.5 MeV" (AERE-R-12419, Harwell, December 1986).
7. O.N. Jarvis, "Selection of Low-Activity Elements for Fusion Reactor Materials" (AERE-R-10496, Harwell, 1982).
8. C. Ponti, "Low-Activation Elements for Fusion Reactor Materials" (IAEA - Technical Committee on Fusion Safety, Culham, November 1986).
9. G.J. Butterworth, R.A. Forrest and A. Khursheed, "On the Use of Titanium in Low-Activation Alloys" (Culham, in preparation).
10. R.F. Burstall, "Fispin- a Computer Code for Nuclide Inventory Calculations" (ND-R-328(R), 1979).
11. P. Reynolds *et al*, "A Demo Tokamak Reactor - Aspects of the Conceptual Design" (CLM-R 254, Culham, 1985). Technology, Tokyo, April 1988).
12. L.R. Greenwood and D.L. Bowers, "Production of Long-Lived Activity in Fusion Materials" (Third International Conference on Reactor Materials, Karlsruhe, October 1987).

FUSION TECHNOLOGY 1988
A.M. Van Ingen, A. Nijsen-Vis, H.T. Klippel (editors)

RELEASE OF MANGANESE RADIOISOTOPES FROM FUSION REACTOR STEELS

C. PONTI, E. RUEDL and G. CASINI

Commission of the European Communities, Joint Research Centre, Ispra Establishment
21020 Ispra (Va) - Italy

Mn-Cr steels produce a small amount of long-lived radioactivity and offer better prospects for decommissioning and waste management. However, the high Mn content is responsible for the high inventory of ^{54}Mn and ^{56}Mn, and since the vapour pressure of Mn is relatively high, in accident conditions these isotopes could be released, possibly with severe environmental effects. The mobility of Mn atoms in accident situations has been investigated experimentally and the results show that the increased amount of Mn in these steels does not increase the potential hazard in the case of an accident, and that the release of Mn isotopes is much reduced by the formation of oxide layers on the surface.

1. INTRODUCTION

The Joint Research Centre has been working for many years on R&D of structural materials for fusion reactors. Particular attention has been paid to austenitic steels with low Ni and high Mn content[1,2]. The main reason for this choice is the favourable activation behaviour of this type of material, particularly for long-lived radioactivity, so that it offers better prospects for plant decommissioning and waste management.

Results of previous studies[3-6] show that, of the main constituents of the steels, Ni is the most troublesome, while Cr and Mn are the safest. Of the minor constituents, Mo and Nb should be avoided, and W and V should be used instead.

When attention moves towards medium and short term aspects of the activation, however, one observes that Mn is responsible for a considerable fraction of the induced radioactivity for decay times up to a few years. Furthermore, it is considered to be one of the most volatile of the steel constituents. An accident situation that would increase the temperature of the metal to $\gtrsim$ 1000 K could, in principle, produce the release of considerable amounts of radioactive Mn.

This paper describes the results of the study of these problems. The second section deals with radioactivity aspects related to the high Mn content, and the third section with the behaviour in accident situations.

2. ACTIVATION OF MANGANESE

Three radioactive isotopes of Mn can be present in non-negligible amounts in activated steels. They are shown in Table 1, together with their halflife T, the energy G of the gamma radiation associated with the decay, the total energy Q emitted in the decay (available as decay heat), and the main nuclear reactions that produce them.

The long-lived ^{53}Mn is not produced by Mn but by Fe isotopes, and its amount is independent of the Mn content of the steel. ^{56}Mn will completely decay within a few days from shut-down; ^{54}Mn needs about 20 years to become negligible. A hypothetical first wall in pure Mn, without impurities, could be reworked and recycled 25 years after shut-down, or disposed of in a near surface repository.

^{56}Mn and ^{54}Mn are produced in large quantities in steels: they pose problems because of their contribution to the decay heat, to the gamma-ray dose, and finally in their possible

TABLE 1
Radioactive isotopes of manganese

Isotope	T	G (MeV)	Q (MeV)	Reactions
^{56}Mn	2.6 h	1.68	2.30	$^{55}Mn(n,\gamma)$, $^{56}Fe(n,p)$
^{54}Mn	312 d	0.84	0.85	$^{55}Mn(n,2n)$, $^{54}Fe(n,p)$
^{53}Mn	$3.7\cdot10^{6}$ a	-	0.005	$^{54}Fe(n,2n)$, $^{54}Fe(n,d)$

release in accident situations. These aspects will be considered separately.

2.1 Decay heat

The activation products of Mn make an important contribution to the decay heat of the irradiated steel. This contribution increases with the Mn content and with the softening of the neutron spectrum. Table 2 shows the decay heat in a first wall built of the Mn-Cr steel AMCR 0033, and, for comparison of AISI 316 and the ferritic steel MANET-2, after one year of irradiation at 1 MW/m^2, with two neutron spectra representing the case of a softer (S) and harder (H) spectrum, respectively. The table gives the decay heat vs cooling time and the total decay heat released within the first ten days of decay. During the first 12 h the after-heat is higher in AMCR 0033 than in AISI 316: this is due to ^{56}Mn which is the main contributor during this time. ^{56}Mn is the product of a (n,γ) reaction, and is more abundant in a soft neutron spectrum. After 12 h the decay heat becomes similar in the two steels; the greater amount of ^{54}Mn present in AMCR is balanced by the nuclides produced from Ni in AISI 316. The total heat released within the first 10 days in the two steels is about the same.

Remarkably smaller values are found for MANET-2 where the total decay heat is about a factor 3 smaller than that of the two previous steels. The content of both Mn and Ni is lowest in the ferritic steel.

2.2 Dose rate

In the three steels mentioned the gamma-ray sources produced by the neutron activation are similar and very intense. No contact maintenance could be allowed for components within the outer shield. The higher or lower Mn content will not change the picture.

TABLE 2
Decay heat (mW/cm^3) vs cooling time and total decay heat released within ten days (KJ/cm^3) in AMCR 0033, AISI 316 and MANET-2, in a hard spectrum (H) and in a soft one (S).

Cooling time	AMCR 0033 H	AMCR 0033 S	AISI 316 H	AISI 316 S	MANET-2 H	MANET-2 S
0	770	1440	348	449	294	359
2 h	447	835	207	273	163	200
3 h	360	657	179	234	130	159
6 h	203	334	128	164	71	86
12 h	101	128	94	118	33	39
1 d	76	78	85	105	23	27
3 d	75	76	81	101	22	25
10 d	74	74	75	93	20	24
0-10 d	73	82	71	89	22	26

2.3 Potential hazard

In off-normal conditions with increased temperature, with or without oxidizing atmosphere, radioactive material could be released either through the transport from the bulk to the surface, or by spalling of oxidized layers. The release to the environment of one kg of steel, irradiated for one year at 1 MW/m^2, produces a dose commitment a few km from the plant[7] in the range from 1 to 5 mSv; the difference in the steel composition has little effect. In the case of AMCR, however, about 80% of this dose is due to Mn isotopes (mainly ^{54}Mn).

In the presence of oxidizing gas, a surface layer containing mainly MnO could be produced. This layer is not strongly bound and its spalling could occur. Assuming that the amount of

MnO corresponding to a thickness of 1 μm, over a surface of 1 m^2, is released as particulate (aerosol) to the environment, the dose to an individual 3 km from the plant, in adverse weather conditions, will be about 0.1 mSv. It is hence important to understand the behaviour of Mn atoms in off-normal situations, and the possible factors that could favour their mobility.

3. EXPERIMENTAL DETERMINATION OF Mn MOBILITY IN Mn-Cr STEELS AT HIGH TEMPERATURE IN AIR

No published data are available on the volatilization of Mn from Mn-Cr steels. Some exploratory tests were, therefore, carried out in collaboration with researchers of the Materials Technology Group of INEL[8]. As material was chosen the steel AMCR 0033, supplied by Creusot-Loire, for which results of activation calculations are given in section 2. This material was extensively studied in the past at the Materials Science Division of the Ispra Centre and its mechanical properties, welding behaviour, liquid metal corrosion, phase stability and radiation damage behaviour have been determined. The high temperature tests of the present work were carried out at INEL, while the microstructural analyses of the resulting oxide scales were performed at Ispra.

3.1 Experimental details

The composition of AMCR 0033 is as follows: Mn: 17.5%, Cr: 10.41%, Ni: 0.09%, Mo: 0.01%, Cu: 0.024%, Si: 0.53%, C: 0.11%, N: 0.19%, P: 0.016%, As,Zr,Ti,Nb,Ta,W,S: < 0.005% (wt%), Fe: balance. Samples of this material, 2.5 mm thick, were mechanically polished, ultrasonically cleaned and degreased. The samples were suspended in a SiO_2 tube, fitted to an air-cooled plug filled with silica wool, and were exposed to an air-stream at temperatures from 1073 to 1473 K, the details being given in Table 3. For the analysis of the material volatilized from the specimens, the silica wool was dissolved. The liquid was processed and examined by inductively coupled plasma spectroscopy (ICP). The weight change of the samples was also measured and the resulting oxide scales studied by X-ray diffraction. Furthermore, cross sections of the specimens were prepared by mechanical polishing; they were examined in the optical microscope and analyzed in the scanning electron microscope (SEM).

3.2 Structural and chemical examination of the resultant oxide scales and the contacting matrix

The study of the cross section of the samples indicates that presumably three types of oxidation occur (Figures 1-3).

Type one is characterized by the formation of a fairly protective, thin oxide layer which is strongly adherent to the substrate (Figure 1).

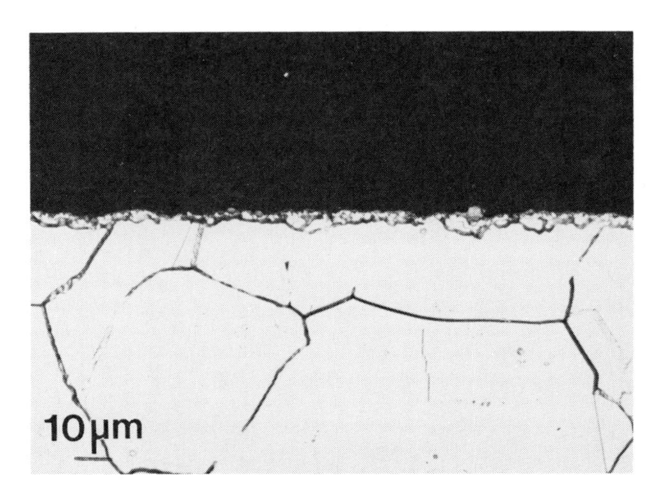

FIGURE 1
Cross section of AMCR 0033 sample exposed to air at 1073 K for 5 h. Formation of protective oxide scale of < 1 μm thickness. Contacting matrix transformed to ferrite to ∿ 2 μm depth. Etched.

According to the X-ray diffraction and X-ray microanalyses, as reported in Table 4, the scale corresponding to type I oxidation consists mainly of Mn_2O_3 with discrete nodules of spinel underneath. Internal oxidation does not occur, and the contacting matrix is transformed to ferrite. The samples exhibiting this type of oxidation are characterized by a weight gain.

TABLE 3
Experimental conditions for oxidation/volatilization tests on AMCR 0033

Test Nº	Temperature (K)	Time (h)	Ramp atmosphere (2ℓ/min flow)	Test atmosphere (2ℓ/min flow)	Surface area (cm^2)
Mn-1	1073	5	argon	dry air	15.02
Mn-2	1273	1	argon	dry air	15.08
Mn-3	1273	5	argon	dry air	15.02
Mn-4	1273	10	argon	dry air	14.99
Mn-5	1273	20	argon	dry air	15.10
Mn-6	1473	5	argon	dry air	15.10

TABLE 4
Type and sequence of oxides in scales formed on AMCR 0033 as determined by X-ray diffraction, X-ray microanalysis and optical microscopy

Test Nº *	Outer layer	Intermediate layers	Inner layer	Internal oxidation	Ferrite layer underneath	Type of oxidation
Mn-1	Mn_2O_3	-	Spinel in the form of nodules	-	yes	I
Mn-2	Mn_2O_3	MnO, Fe-oxide	spinel	yes	yes	II
Mn-3	Mn_2O_3	MnO, Fe-oxide	spinel	yes	yes	II
Mn-4	Mn_2O_3	MnO, Fe-oxide	spinel	yes	yes	II
Mn-5	Mn_2O_3	-	Spinel in the form of nodules	-	yes	I
Mn-6		spinel		-	(not continuous)	III

*see Table 3

The second type of oxidation (Figures 2a,b) leads to a thick, porous oxide scale which shows a tendency to spalling. This type of scale consists of various oxides (see Table 4) including mainly MnO at an intermediate level. The oxidation of type II leads to internal oxidation. In contact with the scale, a layer of ferrite of irregular thickness is observed (Figure 2b). These samples undergo a weight loss.

The third type of oxidation (Figure 3), observed at the highest test temperature, leads to the formation of a badly adherent spinel layer and to a thin, discontinuous ferrite layer underneath. No internal oxidation occurs. The different types of oxidation, observed for air-exposed AMCR 0033, can be understood, at least for the test temperatures of 1073 and 1273 K, by considering a recent oxidation study of an AMCR steel of similar composition[9] and studies of another Mn-Cr steel (Nitronic 32) of higher Cr-content[10,11]. According to these results, the protective oxidation (type I in the present work) is related to the presence of a worked layer near the surface which gives rise to an enhanced diffusion of Mn and its preferential oxidation to Mn_2O_3. At this stage no internal oxidation can occur. This oxide is only metastable, however, and its protective action breaks down when the worked layer recrystallizes. Its recrystallization is difficult to predict, however, since it depends on various factors such as the temperature as well as the amount of deformation and the thickness of the worked layer.

As a consequence of the recrystallization of the matrix to some depth, the oxidation of type I changes to type II. In the case of a lower Cr-content such as in AMCR 0033, the rapidly

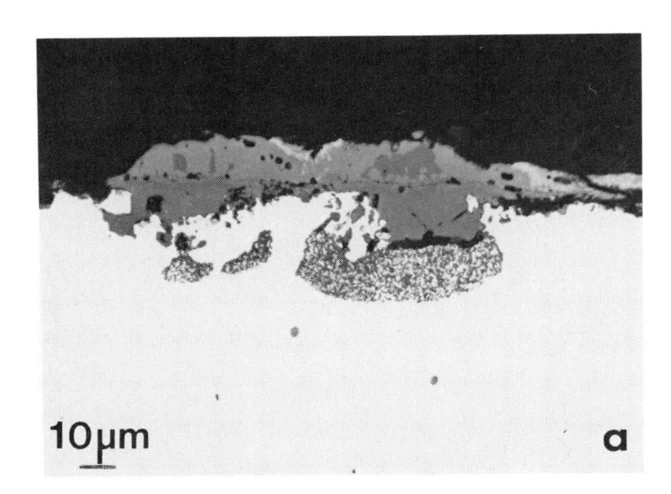

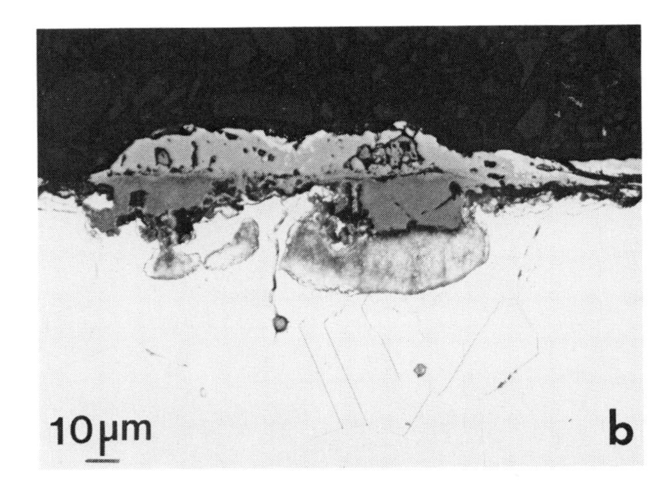

FIGURE 2
Cross section of AMCR 0033 sample exposed to air at 1273 K for 5 h. (a) Unetched, indicating internal oxidation; and (b) etched to show the formation of irregular ferrite layer underneath the scale.

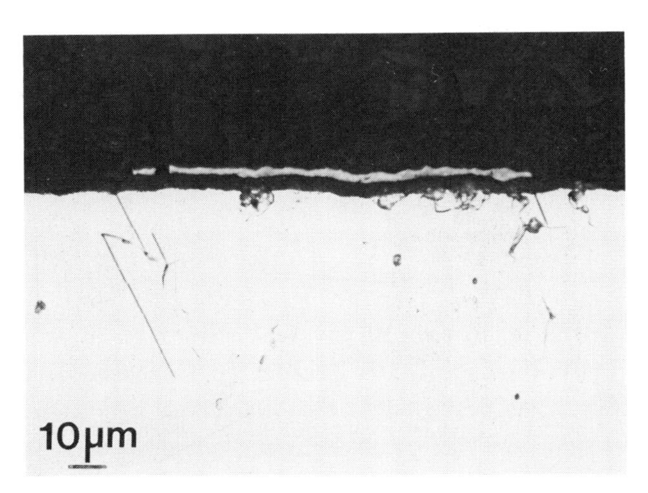

FIGURE 3
Cross section of AMCR 0033 sample exposed to air at 1473 K for 5 h. Formation of spinel layer showing a tendency to spalling. Slightly etched.

growing MnO (manganowustite) is formed. The persistance of type I oxidation observed at 1273 K after 20 h (test Mn-5) is, therefore, surprising and seems to indicate that further, as yet unexplored factors, may also influence the oxidation behaviour of Mn-Cr austenitic steels.

More tests are required to interpret the oxidation of type III observed at 1473 K, a temperature not examined in previous work. Since for this case no internal oxidation was observed, it is assumed that the spinel layer observed represents the complete oxide formed and is not the rest of a thick layer, partly removed by spalling. Type III oxidation is presumably associated with losses of metallic species as well as of non-metallic species which escaped detection as they were not condensed by air-cooling, and considerable compositional changes of the matrix.

3.3 Analysis of volatilized species

Although the number of tests was limited and nowhere near enough for a detailed understanding of the volatilization phenomena related to AMCR 0033, the data obtained and summarized in Table 5 allow some important conclusions to be drawn. It should be pointed out, however, that the release data of this table do not represent the complete analysis of all the species volatilized from AMCR 0033 since the elements present in this material below 0.005 wt% could not be analyzed by ICP due to lack of resolution. Moreover, the detection of Si was also not possible since silica wool was used for collection. Finally, the possible loss of non-metallic species including C, N, P

TABLE 5
Total weight change due to oxidation and volatilization and weight losses due to release of the elements Mn, Fe, Cr, Cu, Ni and Mo from AMCR 0033

Test Nº *	Weight change (mg/cm^2)	Weight of volatilized elements (mg/cm^2)						
		Mn	Fe	Cr	Ni	Cu	Mo	Total
Mn-1(1)	+ 1.5	1.3×10^{-4}	5.0×10^{-4}	-	2.0×10^{-4}	5.0×10^{-4}	-	1.3×10^{-3}
Mn-2(2)	- 2.4	2.0×10^{-2}	5.0×10^{-2}	5.0×10^{-3}	4.6×10^{-4}	1.0×10^{-3}	2.0×10^{-4}	7.6×10^{-2}
Mn-3(2)	- 3.2	1.6×10^{-2}	3.7×10^{-2}	1.6×10^{-3}	4.0×10^{-4}	2.4×10^{-3}	3.5×10^{-4}	5.8×10^{-2}
Mn-4(2)	- 4.8	4.0×10^{-2}	9.5×10^{-2}	6.0×10^{-3}	5.0×10^{-4}	6.7×10^{-4}	4.0×10^{-4}	1.4×10^{-1}
Mn-5(1)	+ 0.4	1.5×10^{-4}	2.3×10^{-4}	-	6.6×10^{-4}	7.2×10^{-4}	3.2×10^{-4}	6.0×10^{-3}
Mn-6(3)	- 1.5	6.0×10^{-2}	3.3×10^{-2}	4.4×10^{-3}	4.0×10^{-4}	4.6×10^{-4}	3.5×10^{-4}	1.4×10^{-1}

*see Table 3. (1) oxidation type I; (2) oxidation type II; (3) oxidation type III

or S, not condensing on the cooled plug, could also not be studied. Nevertheless, the data of Table 5 are useful for information about the volatilization behaviour of the major constituents of AMCR 0033 and particularly of Mn, which is the aim of the present work.

A check of the data of Table 5 shows that the observed weight changes depend on the type of oxidation discussed in section 3.2. The change is positive in the case of oxidation type I, the weight gain due to oxidation being predominant. The weight change instead is negative for type II and III because in these cases the weight losses due to spalling and volatilization predominate. As a relevant point, a dependence on the type of oxidation is also observed for the volatilization of the elements Mn, Cr and Fe while the losses observed for the minor constituents Ni, Cu and Mo seem to be independent of the type of oxidation. As an important conclusion, it can be deduced from Table 5 that the amount of Mn mobilized from air-exposed AMCR 0033 is not relevant, particularly when this steel undergoes type I oxidation. The greatest release at 1073 and 1273 K is not that of Mn, which is always smaller than that of Fe. Cr was not volatilized at all in the case of type I oxidation. Furthermore, we observe that the loss of Mn (and probably also of Fe and Cr) observed for air-exposed AMCR, is not related to the vapour pressure of the pure metal as reported in ref.12. The processes leading to the release of Mn are presumably due to diffusion processes through oxide scales and to the volatilization and decomposition of oxides. We also observe that the sum of the weights of the released elements is always much smaller than the measured weight change of the sample. Spalling and volatilization of non-metallic elements (not condensed on the air-cooled plug) are probably responsible for the difference[13,14].

4. COMPARISON WITH VOLATILIZATION DATA OBTAINED ON A Mn-Cr STEEL AT 1123 K IN A VACUUM OF 1.3×10^{-6} Pa

It is of interest to compare the present results with a release study currently carried out at MPI-Garching on samples of an AMCR steel of similar composition (Mn: 17.7 wt%, Cr: 9.53%, Ni: 0.99%, C: 0.27%, N: 0.03%, Si: 0.53%), the results of which will be reported in detail elsewhere[15]. The samples were tested in a vacuum of 1.3×10^{-6} Pa at 1123 K for 24 h. Before the test, the worked layer was removed from the samples by annealing at 1073 K and subsequent electropolishing, in order to avoid a possible enhanced diffusion of Mn. According to electron spectroscopy (ESCA), the samples treated in this way were covered by a ∿ 500 nm thick oxide film. After the test, the material volatilized and condensed on Si-catchers, was analyzed by Rutherford backscattering. The results of this experiment confirm that the amount of Mn released is much smaller than was expected on the basis of vapour pressure data. It appears, therefore, that the volatilization of Mn, even during a high vacuum treatment, seems to be controlled by an oxide film whose thickness and composition is being currently determined and will be reported in ref.15.

ACKNOWLEDGEMENTS

The authors are indepted to Mrs. R. Van Heusden for the execution of activation calculations with the ANITA code, Mr. M. Mariotto (Optical Metallography), Mr. N. Toussaint (X-ray diffraction) and Dr. A. Manara (ESCA analysis).

REFERENCES

1. P. Fenici, D. Boerman, V. Coen, E. Lang, C. Ponti and W. Schüle, Nucl. Eng. and Design/Fusion, 1 (1984) 167.

2. G. Piatti, D. Boerman and J. Heritier, Proc. 15th SOFT Conf., Utrecht, September 1988, The Netherlands.

3. C. Ponti, Fusion Technol. 13 (1988) 157.

4. S. Fetter, E. Cheng and F. Mann, Fusion Eng. and Design 6 (1988) 123.

5. L. Giancarli and G.J. Butterworth, Proc. 14th SOFT Conf., Avignon, France, 1986, Vol.2, p.1343.

6. C. Ponti, ISFNT, Tokyo, April 1988, to be published in Fusion Eng. and Design.

7. O. Edlund, to be published.

8. S.J. Piet and R.M. Neilson, Work performed under contract with JRC-Ispra.

9. D.L. Douglass, F. Gesmundo and C. de Asmundis, Oxid. Met. 25 (1986) 235.

10. F. Gesmundo, C. de Asmundis, G. Battilana and E. Ruedl, Werkst. Korros. 38 (1987) 367.

11. D.L. Douglass and F. Rizzo-Assuncao, Oxid. Met. 29 (1988) 271.

12. F.A. Garner and H.R. Brager, Proc. 13th ASTM Symposium on Effects of Radiation on Materials, Seattle, Wa, June 1986.

13. I. Maya, F. Montgomery, P. Trester, R. Burnette, W. Johnson and K. Schultz, ICFRM-1, Tokyo 1984; J. Nucl. Mater. 133 & 134 (1985) 912.

14. S.J. Piet, H.G. Kraus, R.M. Neilson and J.L. Jones, ICFRM-2, Chicago 1986; J. Nucl. Mater. 141-143 (1986) 24.

15. J. Bodansky, E. Ruedl and A. Martinelli, to be published.

FUSION TECHNOLOGY 1988
A.M. Van Ingen, A. Nijsen-Vis, H.T. Klippel (editors)

DEVELOPMENT OF LOW ACTIVATION Cr-Mn AUSTENITIC STEELS FOR FUSION REACTOR APPLICATIONS

G. PIATTI, D. BOERMAN
Commission of the European Communities, Joint Research Centre - Ispra Establishment
Materials Science Division, 21020 Ispra (Va) - Italy

and

J. HERITIER
Unirec, Centre Commun de Recherche du Groupe Usinor-Sacilor
42702 Firminy - France

With a view to the disposal by shallow land burial of the radioactive wastes of the future Tokamak reactors after an appropriate decay time, to be achieved by limiting the long-lived radioactivity of the first wall and of the breeding blanket through the choice of appropriate materials, the authors have identified five new low-activation high-Mn stainless steel compositions on the basis of recent activation calculations and alloy design considerations. The results of the first stage of the development of these possible steels concerning: fabrication, constitution, phase stability and hardness, show that these compositions can be considered as potential low-activation substitute materials for Type 316 austenitic stainless steel.

1. INTRODUCTION

The purpose of the development of low-activation materials is to reduce the radioactive wastes of a Tokamak reactor by limiting the long-lived radioactivity after exposure to high-energy neutrons of the material, so that the material itself can be recycled or disposed of by shallow land burial (SLB) after an appropriate decay time[1]. Very recent activation calculations of the acceptability of each natural element in fusion reactor materials showed that there is a strong incentive in the case of stainless steels to optimize the composition, in order to minimize long-lived radioactive nuclei and to satisfy the SLB conditions, while the more ambitious goal of developing steels for recycling does not appear feasible at the moment[2,3]. According to this phylosophy the authors developed five low-activation stainless steels as possible structural materials in a fusion reactor by:

a) An alloy design based on the decrease of the content of Ni, N, Mo and Nb below values acceptable for SLB (Ni<2%, N<500, Mo<500 and Nb < 10 ppm) and on replacing them with Mn, C, W and V, respectively. These elements display similarities to the replaced elements when they are used as alloying elements in steel but have lower long-lived radioactivity.
b) An alloy fabrication on a pilot plant scale in order to assess the feasibility of the industrial production method of proposed steels characterized according to SLB phylosophy by low concentrations of some undesirable elements (Cu < 1000, Al < 500, Co < 1000 ppm) and by very low contents of some unacceptable elements (Nb < 10, Ag < 2, Bi < 1 ppm). This feasibility should be evaluated by chemical analysis and non-destructive testing interpreted as effective production control, as in the case of high quality fabrication or in the nuclear industry.
c) Microstructural and mechanical evaluation of an alloy in the temperature range 20-450°C having as reference properties those corresponding to the AMCR-0033, a high-Mn austenitic stainless steel previously developed[4] under an Ispra-Creusot Loire colla-

boration and the conventional Type 316L stainless steel[4], which is the European austenitic stainless steel fusion reference composition and which, at present, is also the first choice for NET[5].

The present paper describes the first stage of the development which concerned fabrication, constitution, phase stability and Vickers hardness (indentation load 10 kg) of the above low-activation alloys.

2. ALLOY DESIGN CONSIDERATIONS

The proposed compositions are given in Table 1. The alloys were designed on the basis of the available Fe-Cr-Mn phase diagrams[6] and the statement, widely recognized in recent literature reviews[7,8], that Mn is not as strong a γ-austenite stabilizer as Ni. It was then necessary to accept either some Ni (alloys IF-A and IF-C) limited, however, to 2% by reduced activation considerations[2,3], or to increase C (alloys IF-B and IF-D), which is a very effective γ-stabilizer, to a high level (0.3%) in order to obtain fully austenitic structures. On the other hand, the composition of the fifth alloy was balanced to obtain intentionally for purposes of comparison a duplex γ-austenite + δ-ferrite microstructure, with a high content of Cr (17%), Ni as trace constituent ($<$ 0.25%) and a moderate level of C (0.1%).

It should be observed that all the alloys were designed to be solution annealed, followed, however, in the case of the alloys containing a high C percent, IF-B and IF-D, by an ageing treatment in order to form fine precipitates within the grains. In fact, in Cr-Mn-C stainless steels, the susceptibility to grain boundary reactions, typical of conventional high C austenitic stainless steels, is avoided by a suitable ageing treatment, as was previously shown[9]. In IF-B and IF-D alloys the precipitates could be in the form of $M_{23}C_6$ carbides or of V and W complex carbides depending on temperature.

Moreover, any V and W in the five alloys proposed should also contribute to hardening by a solution strengthening mechanism according to recent results on design and developments of some new stainless steels for high temperature use[10].

In addition, the authors have reported in Table 1 the chemical composition of two austenitic stainless steels studied in Ispra as part of the thermonuclear fusion program mentioned previously: AMCR-0033 and AISI 316L[4].

3. ALLOY FABRICATION

The five alloys were fabricated on a pilot scale according to the following steps:

a) selected constituent elements were used as ferro-alloys and came from high purity minerals. The Fe element came from the Firminy steel-works as a vacuum cast steel with very low non-metallic inclusions and gas content.

b) A 500 kg high-frequency induction furnace and a bottom pouring ladle were used to cast a single ingot of about 300 kg for each alloy. The linings of furnace and ladle were completely renewed for this task and argon gas was used as protection.

c) Every ingot (with a thickness of 450 mm) was fully X-ray checked with a special high-voltage accelerator. In this way the top and bottom cutting was made more precise and it was also possible to crop,off the centre line, the axial shrinkage and the porosity formed in some ingots.

d) All blocks were forged and rolled at temperatures in the range 1000-1150°C to obtain bars with Ø 19.5 mm. They were then solution heat-treated at 1100-1150°C and water quenched.

e) All bars were cut and numbered in such a way that the top-centre and bottom of every ingot could be identified. Then the bars were turned and ground to Ø 18.0 mm and checked by ultrasonic non-destructive testing. No

TABLE 1 - Designed composition and chemical analysis of the low-activation alloys

No.	Alloy designation	Composition	Cr	Mn	Ni	Mo	C (wt%)	N	Si	V	W	Heat treatment
1	IF-A	Nominal	12.5	10.5	2.0	<0.05	0.10	<0.05	0.5	0.75	1.4	solution
		Effective	13.57	11.34	2.04	0.031	0.10	0.047	0.20	0.63	1.42	annealed
2	IF-B	Nominal	12.5	10.5	<0.25	<0.05	0.30	<0.05	0.5	0.75	1.4	solution annealed or
		Effective	12.37	10.62	0.23	0.023	0.31	0.036	0.17	0.64	1.38	sol. ann. + ageing
3	IF-C	Nominal	12.5	17.5	2.0	<0.05	0.10	<0.05	0.5	<0.1	2.0	solution
		Effective	13.14	18.00	2.14	0.037	0.10	0.042	0.20	0.021	1.92	annealed
4	IF-D	Nominal	10.0	17.5	<0.25	<0.05	0.30	<0.05	0.5	<0.1	2.0	solution annealed or
		Effective	10.24	16.92	0.13	0.026	0.26	0.080	0.50	0.032	2.04	sol. ann. + ageing
5	IF-E	Nominal	17.0	10.5	2.0	<0.05	0.10	<0.05	0.5	0.75	2.0	solution
		Effective	17.86	11.00	2.08	0.041	0.08	0.054	0.30	0.74	2.02	annealed
6	AMCR-0033*	Effective	10.1	17.5	<0.10	<0.06	0.10	0.19	0.55	<0.02	N.D.	solution annealed
7	AISI 316L**	Effective	17.4	1.8	12.3	2.5	0.024	0.06	0.46	N.D.	N.D.	solution annealed

No.	Alloy designation	Composition	S	P	Cu	Al	Nb (ppm)	Ta	Pb	Co	B	Bi	Ag	Ti
1	IF-A	Nominal	<100	<200	<1000	<500	<50	<50	<10	<1000	<50	<1	<2	<100
		Effective	70	130	370	<30	<50	<50	2	220	<3	<1	<1	<10
2	IF-B	Nominal	<100	<200	<1000	<500	<50	<50	<10	<1000	<50	<1	<2	<100
		Effective	70	140	290	<30	<50	<50	1	200	<3	<1	<1	<10
3	IF-C	Nominal	<100	<200	<1000	<500	<50	<50	<10	<1000	<50	<1	<2	<100
		Effective	50	130	360	<30	<50	<50	2	210	<3	<1	<1	<10
4	IF-D	Nominal	<100	<200	<1000	<500	<50	<50	<10	<1000	<50	<1	<2	<100
		Effective	30	80	240	45	<50	<50	1	200	<3	<0.5	<1	<20
5	IF-E	Nominal	<100	<200	<1000	<500	<50	<50	<10	<1000	<50	<1	<2	<100
		Effective	70	140	370	40	<50	<50	1.5	220	<3	<1	<1	<10
6	AMCR-0033	Effective	80	160	<600	<500	N.D.	N.D.	1	N.D.	25	N.D.	N.D.	N.D.
7	AISI 316L	Effective	20	270	2000	N.D.	100	N.D.	N.D.	1700	80	N.D.	N.D.	N.D.

* Trademark of Creusot Loire (France)

**European 316L fusion reference heat[4,5].

defects were found.

The chemical analysis of the bars is given in Table 1 in addition to the design compositions discussed above. There is good agreement between the nominal and the experimental compositional values.

4. PREDICTION OF METALLURGICAL CHARACTERISTICS

The performance of the proposed new low-activity alloys can first be evaluated by the prediction of some important metallurgical characteristics with the appropriate use of some empirical relationships, reported in the literature[7,8,12-14] and of chemical composition data from Table 1.

4.1 Constitution

The constitution of the austenitic stainless steels at room temperature after a rapid cooling from the solution heat treatment temperature can be predicted according to a standard Schaeffler constitution diagram[12]. Several Cr and Ni equivalent formulations have been suggested: one such being applicable to Cr-Mn stainless steel according to Climax Molybdenum Company researchers as quoted in refs.[7,13]:

$$\text{Cr equivalent (wt\%)} = \%\text{Cr}+2(\%\text{Si})+1.5(\%\text{Mo})+5(\%\text{V})+5.5(\%\text{Al})+1.75(\%\text{Nb})+1.5(\%\text{Ti})+0.75(\%\text{W}) \quad (1)$$

$$\text{Ni equivalent (wt\%)} = \%\text{Ni}+\%\text{Co}+0.5(\%\text{Mn})+30(\%\text{C})+0.3(\%\text{Cu})+25(\%\text{N}) \quad (2)$$

The calculated values for the present alloys located on the Schaeffler diagram of Figure 1 show that alloy IF-A is expected to be austenitic with a maximum 1% δ-ferrite content whereas alloys IF-B, IF-C and IF-D are predicted to be fully austenitic and the fifth alloy, IF-E, is austenitic plus ferrite (about 10% δ-ferrite), as expected. The AMCR-0033, chosen for comparison, is located in the austenitic region and the previous experimental results[4] agree with

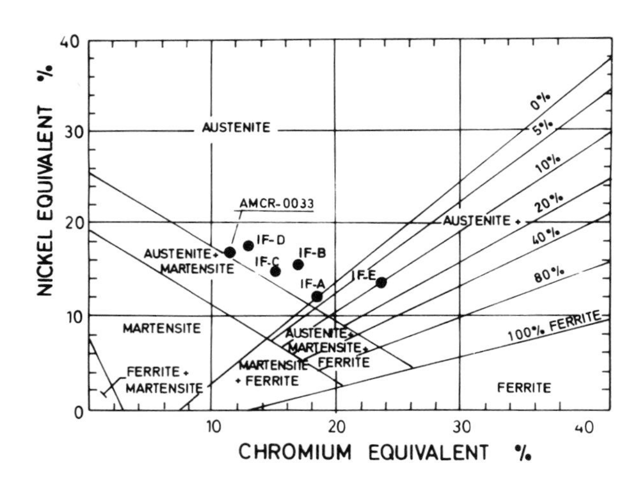

FIGURE 1
Prediction of phase constitution of the low-activation alloys and of the AMCR-0033 alloy from a Schaeffler constitutional diagram and from Cr and Ni equivalents cited by Harries[7].

this prediction, confirming the validity of Eqs.(1) and (2). The fully austenitic structures of the alloy IF-B and IF-D are principally due to high C content whereas that of the alloy IF-C is due to a relatively high level of Mn and Ni elements (Mn+Ni = 19.5%) together.

4.2 Transformations

Transformations of the para-magnetic (fcc) γ-austenite to the para-magnetic (hcp) ε-martensite and to the ferro-magnetic (bcc) α'-martensite can be induced in austenitic steels by quenching to below temperature M_s or straining below temperature M_d, where M_s is the temperature corresponding to spontaneous formation of martensite by quenching and M_d is the temperature at which 50% martensite is formed during the straining to a 0.30 true strain. The tendency to form martensite can be calculated from the relationships[8,14]:

$$M_s = 502-810(\%\text{C})-1230(\%\text{N})-13(\%\text{Mo})-30(\%\text{Ni})-12(\%\text{Cr})-54(\%\text{Cu})-6(\%\text{Mo}) \quad (3)$$

$$M_d \;(30/50) = 497-462\%(\text{C+N})+9.2(\%\text{Si})-8.1(\%\text{Mn})-13.7(\%\text{Cr})-20(\%\text{Ni})-18.5(\%\text{Mo}) \quad (4)$$

It should be observed, however, that Eq.(4) is

based on data for α'rather than ε martensite[14] and care must be taken in interpreting the results. The γ→ε transformation during straining is an important feature of the high-Mn stainless steel because, as was previously shown in the case of AMCR-0033[11], the work hardening behaviour increases relative to flow-stress and high uniform elongation values can be obtained. The M_s and M_d values obtained by applying Eqs.(3) and (4) to the present alloys are listed in Table 2. IF-A alloy is less stable with respect to the spontaneous martensitic transformation whereas IF-E, IF-B and IF-C are, in turn, less stable than IF-D. It should be observed that the latter low-activation composition presents M_s values similar to those of the AMCR-0033 alloy. Concerning M_d values, these are included in the 50-110°C temperature range as well as AMCR-0033 alloy. This means that the tensile behaviour at room temperature of these low-activation alloys could be characterized by a partial transformation of γ-austenite to ε-martensite and/or α'-martensite as shown for AMCR-0033[11].

4.3 Stacking fault energy

The formation of ε-martensite is favoured by the low stacking fault energy (SFE) which can be predicted by the empirical equation[8]:

TABLE 2 - Predicted transformation temperatures (M_s and M_d), stacking fault energy (SFE), 0.2% yield strength ($R_{0.2}$) and ultimate tensile strength (R_m) of the low-activation alloys

Alloy	M_s (°C)	M_d(30/50) (°C)	SFE (mJ/m^{-2})	$R_{0.2}$ (MPa)	R_m (MPa)
IF-A	- 11.7	107.37	37.7	243.6	627.1
IF-B	- 89.4	75.07	124.4	301.8	716.5
IF-C	- 94.2	57.9	33.7	222.3	607.8
IF-D	-158.2	52.7	95.0	293.4	734.4
IF-E	- 58.4	52.5	29.5	289.9	672.1
AMCR-0033	-170.4	74.73	15	277.1 360*	732.4 840*

*Experimental values

$$SFE\ (mJm^{-2}) = 25.7+2(\%Ni)+410(\%C)-0.9(\%Cr)-77(\%N)-13(\%Si)-1.2(\%Mn) \quad (5)$$

Table 2 reports the SFE values calculated for the present alloys. In all the alloys the SFE is higher than the corresponding value for AMCR-0033 alloy, 14.4 mJm^{-2}. Moreover, remembering that according to Harries[7] ε-martensite forms on deformation at room temperature if the SFE ≦ 30 mJm^{-2}, it is evident that during deformation at room temperature α'-martensite has a higher formation probability than ε-martensite.

4.4 Strength

Finally, the strengths of the designed alloys were predicted using the following equations which relate the 0.2% yield strength ($R_{0.2}$) and the ultimate tensile strength (R_m) to the chemical composition and the microstructure[14]:

$$R_{0.2}\ (MPa) = 68+354(\%C)+20(\%Si)+4(\%Cr)+14(\%Mo)+18(\%V)+4(\%W)+26(\%Si)+12(\%Al)+493(\%N)+7.1d^{-\frac{1}{2}} \quad (6)$$

$$R_m\ (MPa) = 447+539(\%C)+847(\%N)+37(\%Si)+2(\%Ni)+18(\%Mo)+77(\%Nb)+46(\%Ti)+18(\%Al)+12.7d^{-\frac{1}{2}} \quad (7)$$

where d is the mean austenitic grain size in mm. Calculation results obtained by applying Eqs.(6) and (7) and taking into account a grain size of ASTM 5 for the present alloys are reported in Table 2. Caution is necessary in the use of the calculated data because the effect of δ-ferrite is excluded from the equations. The alloys show high strength similar to that of AMCR-0033. However, comparison in the case of the latter alloy between calculated and experimental values seems to indicate that Eqs.(6) and (7) underestimate the true strength.

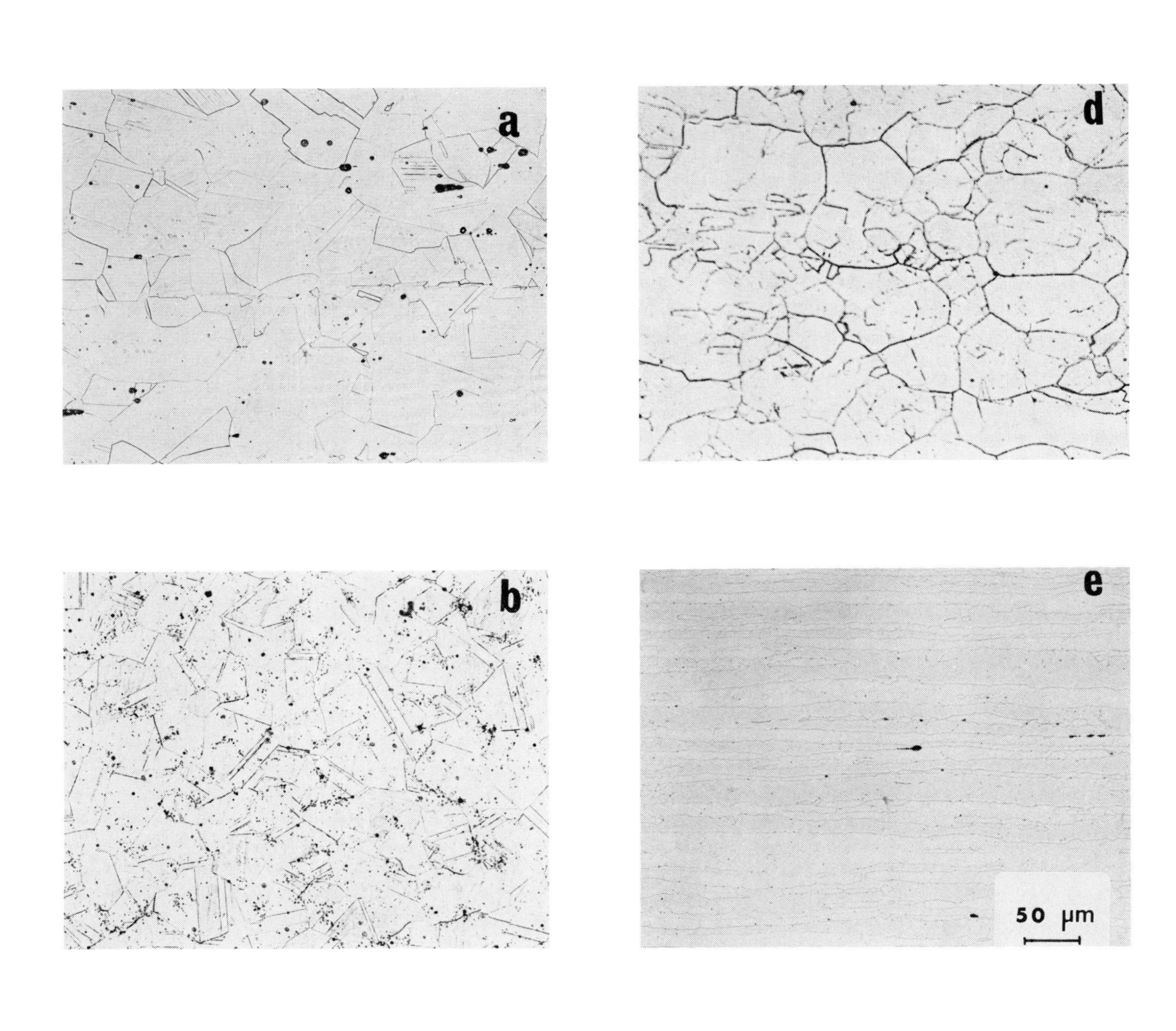

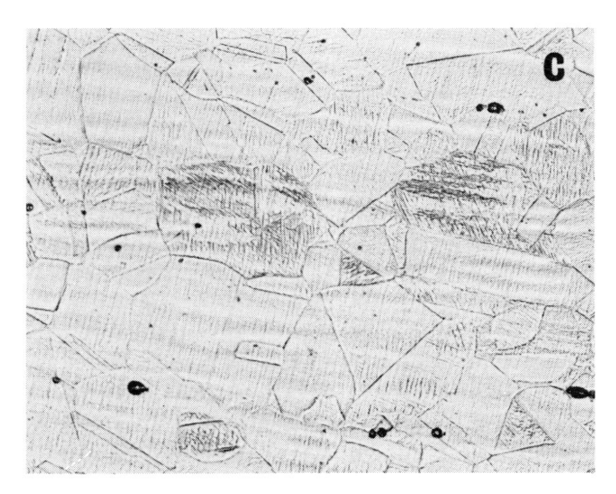

FIGURE 2
Microstructure of the low-activation alloys: (a) IF-A, (b) IF-B, (c) IF-C, (d) IF-D and (e) IF-E (longitudinal sections of the bar except in the case of the microstructure of the alloy B which corresponds to a transverse section).

5. EXPERIMENTAL OBSERVATIONS

5.1 Solution annealed structures

Examination of the microstructures of the low-activation composition (see micrographs of Figure 2) revealed some δ-ferrite precipitates (evaluated by magnetic measurements of the order of 1%) in elongated form within the austenite matrix in alloy IF-A (Figure 2a), a fully austenitic structure in alloys IF-B (Figure 2b), IF-C (Figure 2c) and IF-D (Figure 2d), and a duplex structure of elongated grains of ferrite and austenite in the IF-E alloy (Figure 2e). There is then good agreement between the above experimental observations and the constitution prediction previously shown in Figure 1 for all the low compositions, even if the ferrite amount seen on Figure 2e (evaluated by magnetic measurement of the order of 20%) is 10% higher than predicted in Figure 1. Moreover, alloys IF-B and IF-D are characterized by carbide precipitations at the grain boundaries, as expected and also inside the grains (alloy IF-B). Strength was evaluated by hardness testing and the results are presented in Table 3. The data show

TABLE 3 - Vickers hardness values (indentation load 10 kg) for the low-activation, AMCR-0033 and AISI 316L alloys (solution annealing conditions)

IF-A	IF-B	IF-C	IF-D	IF-E	AMCR-0033	AISI 316
199	250	160	235	223	230	160

the highest Vickers values, 250 and 235 for alloys IF-B and IF-D, respectively, due to the high C content (0.3%), intermediate values 233 and 199 for alloys IF-E and IF-A, respectively, and the lowest value 160 for alloy IF-C. This value scale corresponds to that of predicted yields strength and ultimate tensile strength (Table 2). All the alloys show hardness characteristics similar to those of AMCR-0033 and 316L stainless steels.

5.2 Aged structures

According to hardness values reported in Table 3 and concerning precipitation heat treatment performed on alloys IF-B and IF-D (in other words the alloys with high C percent) (see Figure 3), it is evident that the maximum hardness increases are reported for both alloys after an ageing of about 1 h. The temperature of 900°C was chosen on the basis of preliminary tests at different temperatures. The 310 and 265 HV maximum values for the IF-B and IF-D alloys, respectively, confirm the occurrence of a precipitation hardening mechanism.

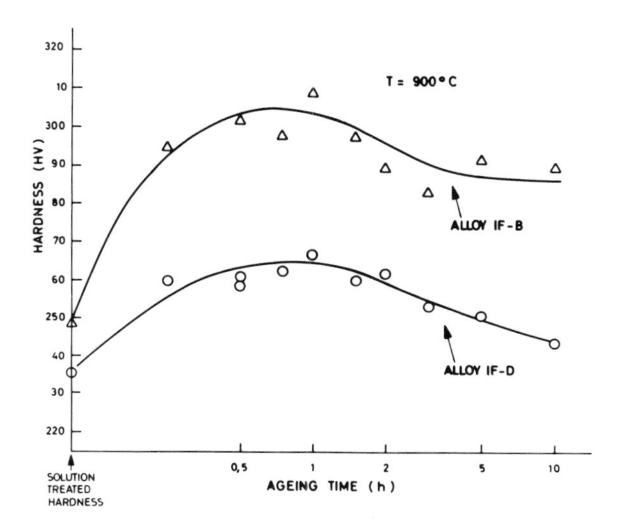

FIGURE 3
Vickers hardness as a function of ageing time at 900°C for the alloys IF-B and IF-D.

6. CONCLUSIONS

The results of the first stage of the development of low-activation stainless steels for fusion purposes based on Fe-Cr-Mn system can be summarized as follows:

1) Five new low-activation compositions have been identified through design considerations based on the one hand on element acceptability according to a shallow land burial phylosophy of the radioactive waste problem, and on the other hand on metallurgical rules.

2) The proposed alloys can be successfully fabricated industrially on a pilot scale achieving very low amounts of undesirable constituents such as Nb or unacceptable impurities such as Ag and Bi.
3) Microstructural examination revealed that four alloys were austenitic and one duplex (austenitic plus ferritic) in agreement with predictions made using a Schaeffler constitution diagram but using particular Cr and Ni equivalents.
4) The alloys show hardness characteristics similar to those of high-Mn stainless steel such as AMCR-0033 alloy and of type 316L.

ACKNOWLEDGEMENTS

The authors are grateful to Dr. P. Schiller, Dr. E. Ruedl, Dr. C. Ponti and Dr.Ing. R. Ponte of the JRC-Ispra Establishment for many helpful discussions and to S. Colpo and H.A. Weir (Materials Science Division, JRC-Ispra Establishment) for assistance in the experimental work. Thanks are also due to the research team of the Unirec Laboratories who contributed to this work.

REFERENCES

1. R.W. Conn, E.E. Bloom, J.W. Davis, R.E. Gold, R. Little, K.R. Schultz, D.L. Smith and F.W. Wiffen, Nucl. Technol./Fusion 5 (1984) 291.
2. C. Ponti, Fusion Technol. 13 (1988) 157.
3. C. Ponti, paper presented to the Int. Symp. on Fusion Nuclear Technology (Tokyo, April 1988) to be published in Fusion Engineering and Design.
4. G. Piatti and P. Schiller, J. Nucl. Mater. 141-143 (1986) 417.
5. D.R. Harries, J.M. Dupouy and C.H. Wu, J. Nucl. Mater. 133-134 (1985) 25.
6. G.V. Rivlin and G.V. Raynor, Intern. Met. Rev. 28 (1983) 23.
7. D.H. Harries, Proc. Varese (Italy) Conf. on Mechanical Properties and Nuclear Applications of Stainless Steels (The Metals Soc., London, 1982) Book 280, p.1.
8. F.B. Pickering, Proc. Göteborg (Sweden) Conf.on Stainless Steels 84 (The Institute of Metals, London, 1985) Book 320, p.2.
9. C.M. Hsiao and E.J. Dulis, Trans. ASM 50 (1957) 773.
10. R.P.H. Fleming, Proc. Varese (Italy) Conf. on Mechanical Properties and Nuclear Applications of Stainless Steels (The Metals Soc., London, 1982) Book 280, p.15.
11. G. Piatti, S. Matteazzi and G. Petrone, Nucl. Eng. Design/Fusion 2 (1985) 391.
12. H. Schneider, Foundry Trade J. 108 (1962) 562.
13. A.H. Bott, F.B. Pickering and G.J. Butterworth, J. Nucl. Mater. 141-143 (1986) 1097.
14. F.B. Pickering, Physical Metallurgy and Design of Steels (Applied Science Publisher, London, 1978).

FUSION TECHNOLOGY 1988
A.M. Van Ingen, A. Nijsen-Vis, H.T. Klippel (editors)
Elsevier Science Publishers B.V., 1989

A STUDY ON THE PROSPECTS FOR DEVELOPMENT OF A NEW LOW ACTIVITY AUSTENITIC STAINLESS STEEL FOR FUSION APPLICATIONS

Massimo ZUCCHETTI, Manuela ZUBLENA

Dipartimento di Energetica, Politecnico di Torino, Corso Duca degli Abruzzi 24, 10129, Torino, Italia.

The properties of three austenitic stainless steels (AISI316L, AMCR0033 and VA64) are examined from a fusion point of view, and a new composition for a low-activity austenitic stainless steel is suggested.

Some important properties of the three alloys are compared; some outstanding properties of the age-hardenable high-Mn stainless steel VA64 are stressed. Induced radioactivity in the three alloys for various times after shutdown is calculated; as far as long-term radioactivity is concerned, VA64 suffers its relatively high Mo and Nb content.

Elemental substitutions on VA64 are examined, and a new composition, giving a low-activity version of VA64, is suggested. Effects of elemental substitution on the properties of VA64 are predicted; the new alloy is a possible good candidate first wall material for fusion reactors.

1. INTRODUCTION

Radioactivity in fusion reactors can be effectively controlled through materials selection. This general rule [1] primarily applies to first wall and blanket structures selection. Since it has frequently been pointed out that these are the source of most part of radioactive wastes from fusion machines [2,3], several materials have been envisaged for these structures, including stainless steels, refractory metals, vanadium based alloys and ceramics [4,5]. Of these, austenitic stainless steels seem to be the material of choice for near-term fusion machines [6].

Ferritic steels have some attractive properties, mainly greater radiation damage resistance and a better compatibility with liquid breeders [5,7,8]. Austenitic steels, however, display better formability, weldability and ferromagnetic properties [5,9]. In view of the expected frequency and scale of the replacement of the first wall and blanket in near-term fusion reactors, formability and weldability of austenitic steels could be major factors in favour of their selection.

2. MATERIALS AND THEIR PROPERTIES

The compositions of the three alloys investigated in this study are listed in tab.1.

AMCR0033 is a Cr-Mn (Ni-free) austenitic steel of interest for structural use in fusion

TABLE 1
Composition of some austenitic stainless steels

	AMCR033	AISI316L	VA64	VA64 low-act
Cr	10.12	17.44	20.76	20.25
Ni	0.10	12.32	0.25	0.24
Mn	17.50	1.81	10.59	10.90
Si	0.55	0.46	0.12	0.12
S	0.008	0.002	0.006	0.006
P	0.016	0.027	0.024	0.024
Mo	0.06	2.50	1.04	0.005
V	0.02	--	1.0	0.98
N	0.19	0.06	0.50	0.10
C	0.10	0.024	0.62	0.78
Nb	0.00002	0.01	1.20	0.00002
Co	--	0.17	0.03	0.03
Al	0.05	0.001	0.001	0.001
Ta	--	--	--	2.28
W	--	--	--	1.94
Cu	0.06	0.2	--	--
Fe	bal.	bal.	bal.	bal.

Others: Sn 0.005 - Ba 0.0002 - Sm 0.00005
Bi 0.0005 - Tb 0.00005 - Eu 0.00002
Ir 0.00001 - Ag 0.00005

reactors due to its low activation levels. VA64 is a non-fusion designed age-hardenable high-Mn stainless steel. AISI316L, chosen for the sake of comparison, is the first choice for NET's first wall [10].

In this section, some properties of the three alloys determin ed by one of us (M.Zublena) in [11] will be compared. Reference can be made to [11-15] for experimental details.

Thermal conductivity and thermal dilatation coefficient values vs. temperature are shown in fig.1; in the interval of interest ([100,400 °C]), VA64 has in both cases the best results.

In general, disruption damage is less for AISI316L than the other two alloys; for example, fig.2 shows the per cent loss of weight in the three alloys, in function of the shot time of the electron gun used to simulate disruption (for details, see [11,12]).

The ultimate tensile strength of VA64 is far better than AMCR0033 and AISI316L, especially at high temperatures (see fig.3); this is mainly due to the presence of carbides and nitrides, precipitated during the age-hardening treatment.

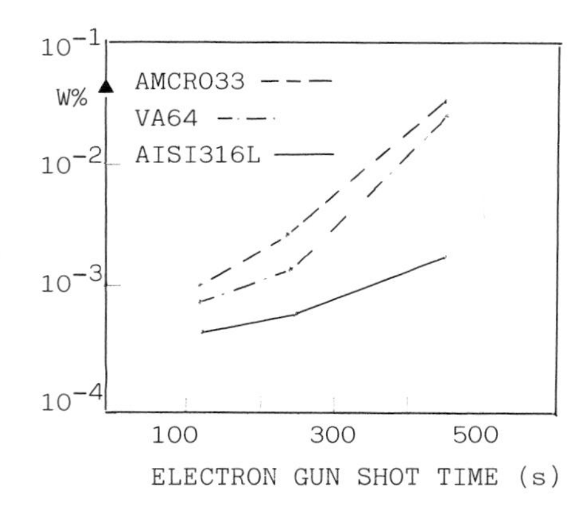

FIGURE 2
Per cent loss of weight (W%) in three alloys, in function of the shot time of the electron gun used to simulate disruption.

Thermal stress factor vs. temperature is shown in fig.4; it represents the maximum allowable superficial thermal flux (in MW/m^2) on a plane 1 mm. deep. The results of VA64 are much better than the other two alloys, thanks to a good combination of high-temperature tensile strength and thermal dilatation coefficient.

The ASME standards for pressure vessels [16]

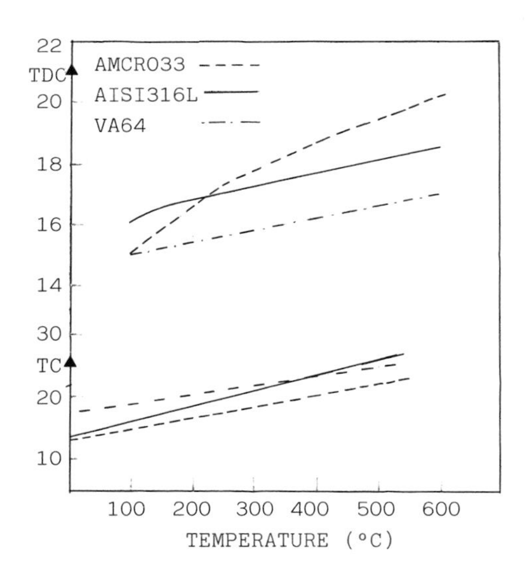

FIGURE 1
Thermal dilatation coefficient (TDC - $10^{-6}K^{-1}$) and thermal conductivity (TC - W/m °C) vs. temperature for three alloys.

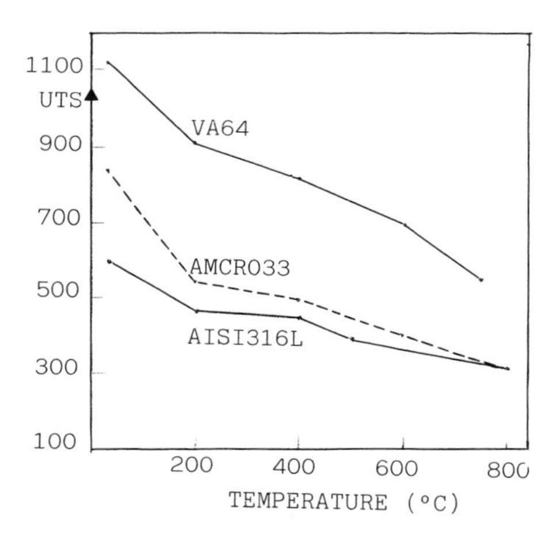

FIGURE 3
Ultimate tensile strength (UTS - MPa) versus temperature for three alloys.

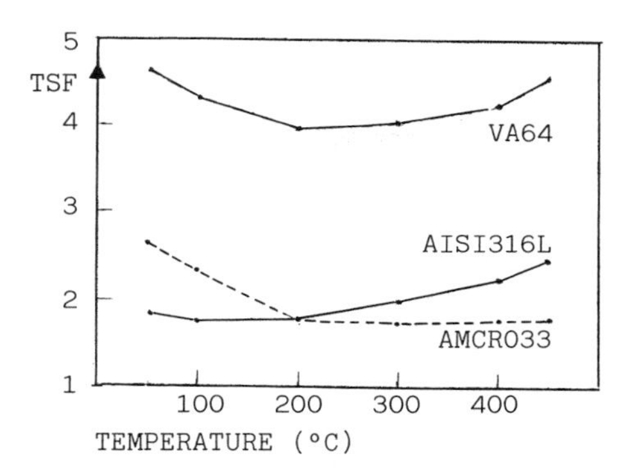

FIGURE 4
Thermal stress factor (TSF - W/m) vs. temperature for three alloys.

TABLE 2
Induced radioactivity levels (Bq/cm^3) in some stainless steels at various times after shutdown. (1 Ci = 3.7 10^{10} Bq)

Time	AMCR033	AISI316L	VA64	VA64 l.a.
zero	6.80+12	3.96+12	5.46+12	8.50+12
1 day	1.16+12	1.62+12	1.37+12	2.34+12
1 month	9.30+11	1.05+12	9.22+11	1.38+12
1 year	5.01+11	4.28+11	4.10+11	5.25+11
10 y	3.44+10	3.84+10	3.17+10	3.10+10
10^2 y	1.28+07	3.24+08	4.45+07	8.96+06
10^3 y	5.10+06	2.50+07	2.26+07	2.60+06
10^4 y	1.67+06	6.97+06	7.26+06	9.69+05

Read: 6.80+12 = 6.80 10^{12} ; l.a.=low activity.

can be used to calculate primary and secondary maximum allowable stresses, with suitable modifications to take the high radiation levels on a first wall into account; calculated primary maximum allowable stress values are listed below:

AMCR0033 : 159.5 MPa

AISI316L : 143.0 MPa

VA64 : 270.2 MPa

VA64 displays by far the best results, due to its good high temperature behaviour.

In conclusion, VA64 is an austenitic stainless steel with very attractive properties, especially mechanical properties and resistance to thermal stresses.

3. ACTIVATION ANALYSIS RESULTS

In this section, induced radioactivity levels in AISI316L, AMCR0033 and VA64 have been calculated for several times after shutdown, using the ANITA code [17], with NET's first wall input parameters (Li-Pb water-cooled blanket [10]). A pulsed operating time of 8.5 years has been considered, with 0.95 full power years at a wall loading of 2 MW/SQM.

The calculated total induced radioactivity levels for the three alloys are shown in tab.2. Both AMCR0033 and VA64 have higher shutdown radioactivity levels than AISI316L, mainly due to their high content of Mn.

A word should be added about compositions in tab.1; since long-term radioactivity levels may be substantially affected by impurity elements, some of them were included in the compositions, according to present estimates of the impurity levels in commercial production of low activation steels [18].

Table 2 also shows that AMCR0033 has low long-term activity levels, thanks to its low content of four of the "proscribed elements" for long-term activity: Ni, Mo, Nb and N. When exposed to a fusion reactor neutronic flux, all these elements start radioactive decay chains containing long-lived radioactive nuclides. The poor performance of VA64 is equally the result of its high content of Mo, Nb and N.

Some other versions of AMCR0033 with lower N content (less than 0.1%) have even lower long term radioactivity levels.

The per cent contribution of some important elements as precursors in the generation of total induced radioactivity levels in VA64 for T = 100,1000,10000 years is shown in tab.3; it is clear that a replacement of Nb, Mo and N with other elements generating short-lived activation products would strongly reduce induced long-term radioactivity levels.

TABLE 3
Per cent contribution of some alloying elements as precursors in the generation of induced radioactivity levels in VA64 at various times after shutdown.

Element	$T=10^2$y	$T=10^3$y	$T=10^4$y
Ni	12.27%	0.47%	1.24%
Mo	29.00%	32.05%	10.27%
Nb	24.61%	14.66%	32.65%
N	32.90%	52.59%	55.12%
Others	1.22%	0.23%	0.72%
Total	100.00	100.00	100.00

4. ELEMENTAL SUBSTITUTION ON VA64 ALLOY

Too high long-term radioactivity levels of VA64, deriving from its original non fusion-oriented composition design, can be strongly reduced removing niobium, molibdenum and nitrogen. These are also important alloying elements; our task is therefore to find good substitutes, trying to reduce the alloy's long-term activity levels without worsening its other properties.

From studies on the development of low-activation austenitic steels [9,18], the following elemental substitutions can be suggested:

- replace Nb with Ta (or Hf)
- replace Mo with W (or V)

Tantalum, in fact, is a strong carbide-forming element, whose alloying effects are similar to Nb. Tungsten is a ferrite stabilising element and a solid solution strengthener at concentrations up to 1%; its effects as an alloying element are well known.

Nitrogen can be partially replaced by increasing the carbon content of the alloy, since the two elements have similar alloying effects [19]. Possible effects on weldability and lithium corrosion resistance preclude a total substitution of nitrogen with carbon only.

We therefore propose an elemental composition (tab.1) for a new low-activity austenitic stainless steel, that should preserve the good properties of VA64 alloy, from which its composition is derived. Nb and Mo have been removed and N content has been reduced.

5. PREDICTED PROPERTIES OF THE NEW ALLOY

The induced radioactivity levels for the new alloy calculated under the same conditions listed in section 3 are shown in tab.2.

It can be seen that they are distinctly better than those for old VA64 AISI316L and AMCR0033, as far as long-term activity is concerned.

The replacement of one element by another could give rise to changes in the properties of the alloy.

The first consideration is to ensure that the alloy would be fully austenitic after solution treatment at high temperature; the presence of delta ferrite must be avoided, since it causes impaired hot working characteristics, loss of strength and a heterogeneous distribution of the alloying elements.

The constitution of stainless steels at room temperature after cooling from high temperatures can be predicted by means of a Schaeffler diagram [20], and we have found that also the new alloy obviously falls in the fully austenitic zone.

TABLE 4
Yield and tensile strength values for four stainless steels (MPa).

Alloy	Yield strength		Tensile strength	
	Pred.	Meas.	Pred.	Meas.
AMCR033	278.4	391.0	734.6	841.0
AISI316L	248.1	291.0	648.4	596.0
VA64	680.0	766.0	1370.9	1123.0
VA64 l.a.	530.9	--	1217.5	--

A rough estimate of 0.2% proof stress and tensile strength values can be obtained with the formulae in [20]; estimates for the four alloys are given in tab.4, together with available test values from [11]. It can be seen that the calculated values overestimate tensile strength and underestimate the 0.2% proof stress; for the new alloy there are slight decreases in both values, but they still remain appreciably higher than those of AISI316L and AMCR0033.

The effect of some elemental substitutions on alloy's behaviour under irradiation, especially on swelling, could be another critical problem, since the swelling values of austenitic steels are already higher than those of ferritic steels, under the same irradiation and temperature conditions [5]; swelling must be determined experimentally, but from our predictions based on He production, it can be expected the new alloy not to have very different behaviour from that of VA64.

Difficulties in fabrication will be certainly created by tantalum, due to its high cost and the difficulty of obtaining a steelmaking grade of the metal with a sufficiently low niobium content; impurities and their activation levels are one of the central problems in low-activity materials design [21] and niobium is a troublesome element: present estimates limit its concentration under 1 ppm, so as to remain under the permitted levels of induced radioactivity in the alloy [21].

6. SUMMARY AND CONCLUSIONS

The age-hardenable high-Mn stainless steel VA64 is an alloy with some outstanding properties, compared with other austenitic stainless steels, especially thermal stresses resistance and tensile properties (section 2).

Since VA64 is a non fusion-designed alloy, some of its elements (Mo,Nb,N) cause over high induced long-term radioactivity levels, when the steel is used as first-wall material for fusion reactors (section 3).

Some elemental substitutions in the alloy's composition have originated a new low-activation austenitic stainless steel; properties of the new alloy will be comparable with VA64, due to the particular attention devoted to research in this field (section 4).

Induced radioactivity levels for the new alloy are strongly reduced. Its austenitic constitution has also been tested; possible problems in strength, swelling and fabrication have been discussed (section 5).

In conclusion, we feel that our results justify further investigation of this low-activity stainless steel.

The fabrication and testing of specimen (a special precipitation hardening treatment will be necessary) may be expected to corroborate its attractive properties, with further slight adjustments of its composition. The new alloy could result as a possible good candidate first wall and blanket material for fusion reactors.

AKNOWLEDGEMENTS

The authors wish to thank Prof. G.Piatti and Dr. C.Ponti of the JRC-Ispra (VA) and Prof. G.Del Tin and Prof.P.Ravetto of the Energy Department - Politechnic of Turin, for their help during the drafting of this work.

REFERENCES

1. R.W.Conn et al., Nucl. Techn./Fus. 5 (1984) 291.

2. G.Casini, C.Ponti, P.Rocco, Environmental Aspects of Fusion Reactors (EUR-10728, European Atomic Energy Community ,1986).

3. C.Ponti, Fus. Techn. 13 (1988) 157.

4. R.E.Gold et al., Nucl. Techn./Fus. 1 (1981) 169.

5. D.L.Smith et al., Fus. Techn. 8 (1985) 10.

6. P.Schiller, J.Nihoul, Projected Materials Requirements for the Next European Torus and for Long-term Tokamak-Based Demo Fusion Devices, in: Proc. 3rd Int. Conf. Fus. React. Mat. (Karlsruhe, 1987), in print.

7. S.Majumdar, Fus. Techn. 8 (1985) 1944.

8. O.K.Chopra, Fus. Techn. 8 (1985) 1956.

9. A.H.Bott et al., The development of Austenitic Stainless Steels as Low-Activity Structural Materials for Fusion Reactors (CLM-R255 Culham Report 1985).

10. THE NET TEAM, NET Status Report (EU-FU/XII-80/86/51 1985).

11. M.Zublena, Thesis, Politecnico di Torino, 1986.

12. G.Piatti, M.Zublena, G.Del Tin, Mechanical Effects of Simulated Plasma Disruptions in Cr-Ni and Cr-Mn Stainless Steels in: Proc. 3rd Int. Conf. Fus. React. Mat. (Karlsruhe, 1987), in print.

13. G.Piatti, P.Schiller, Journ. Nucl. Mat. 141-143 (1986) 417.

14. G.Piatti, M.Zublena et al., Thermal and Mechanical Properties of Cr-Mn Austenitic Steels, in: JRC Programme Progress Report Cat.1.6,Nr.4311 (JRC Ispra, 1986) pp. 43-45.

15. P.Fenici et al., Nucl. Eng. and Des./Fus. 1

16. ASME Standards, Section VIII part 1: Pressure Vessels (New York, 1980).

17. C.Ponti, Calculation of Radioactive Decay Chains (EUR9389 Ispra, 1984).

18. L.Giancarli, J.Butterworth, The implications of gamma dose rate limits for the recycling of fusion reactor first wall structural material, in: Proc. 14th Symp. on Fus. Techn. (Avignon, 1986) pp. 1343-1348.

19. A.H.Bott et al., An Investigation of the Constitution, Structure and Properties of Some Low Activation Austenitic Stainless Steels Proposed for Fusion (CLM-R265 Culham Report, 1986).

20. E.Ruedl et al., On the Role of Carbon and Nitrogen in Austenitic Cr-Mn Steels for Fusion Reactor Structural Applications,in: Proc. 13th Symp. on Fus. Techn. (Varese, 1984) pp.1029-1036.

21. F.B.Pickering, Physical Metallurgy and the Design of Steels (Applied Science Publishers, 1978).

22. L.Giancarli, Journ. Nucl. Mat. 139 (1986) 1.

FUSION TECHNOLOGY 1988
A.M. Van Ingen, A. Nijsen-Vis, H.T. Klippel (editors)

EUROPEAN TRANSPORT AND ACTIVATION LIBRARIES FOR FUSION REACTOR TECHNOLOGY

J. Kopecky and H. Gruppelaar

Netherlands Energy Research Foundation ECN P.O. Box 1, 1755 ZG Petten, The Netherlands

The European Fusion File (EFF) is designed for neutron and photon transport calculations in fusion-reactor technology. A separate file (EAF) contains cross-sections for activation and transmutation reactions. A detailed survey of the above-mentioned datafiles is presented, together with plans for further updates to satisfy the needs of fusion-reactor calculations.

1. INTRODUCTION

The EFF project is part of the European Fusion Technology Programme of the European Community (EC). The following laboratories are contractors: CEA (Saclay), ECN (Petten), ENEA (Bologna), KfK (Karlsruhe) and the University of Birmingham . Moreover, JRC (Ispra) and CBNM (Geel) are involved as EC institutes. The project is guided by the NET team (Next European Torus) at Garching and by EC, Brussels. The management and maintanance is performed at ECN (Petten). Other European laboratories are also involved: IRK (Vienna), AERE (Harwell), Culham (JET), IKE (Stuttgart), KfA (Jülich) and EIR (Würenlingen). Technical support is received from the NEA Data Bank (NDB) at Saclay.

The main objective of the EFF project[1] is to supply the NET team with reliable evaluated data for neutron and photon transport calculations. There is a related project with the name EAF (European Activation File) to achieve activation cross-sections[2].

2. THE EFF-1 FILE

2.1. Contents

The materials included in the EFF-1 file are: H, D, T, ^{6}Li, ^{7}Li, ^{9}Be, ^{10}B, ^{11}B, C, O, Al, Si, Ti, V, Cr, Mn, Fe, Ni, Cu, Zr, Nb, Mo, Ba, W, Pb and Bi. Most of the evaluations are identical to those of JEF-1, except that gas production cross-sections were explicitly added. For the materials ^{6}Li, ^{7}Li, ^{9}Be, Al, Si and Pb different evaluations were adopted. Covariance information is available for most of the light materials on EFF-1, whereas this information is given in a multigroup library for the important shielding materials.

The format of EFF-1 is ENDF-V with some extensions: the Reich-Moore description in the resonance range, the energy-angle distribution file MF6 of ENDF-VI and the addition of data type MT=10 (continuum particle emission). For most of the light materials the pseudo level representation has been used to store the energy-angle distributions; a deviation from the official ENDF-V format was adopted for Be by storing the (n,2n) reaction as pseudo inelastic scattering, followed by break-up neutron emission. In addition to the basic EFF-1 file a point-wise given version of the file has been generated.

For the tritium production cross-sections ^{6}Li(n,t) and ^{7}Li(n,n't) the ENDF/B-V were adopted. The ^{7}Li(n,n't) cross-section is quite accurate at 14.1 MeV: 303 ± 10 mb. A very recent evaluation[5], including the most recent experimental data yielded almost the same value (with smaller uncertainty), confirming the EFF-1 value, cf. Sect. 3.

In view of their importance for ceramic breeder materials revisions of the ENDF/B-IV evaluations for Al and Si have been performed at ENEA-Bologna. These revisions include a better description of the resolved resonance range and updates of the charged-particle emission cross-sections.

For the Be neutron multiplier the recent evaluation of LASL[4] using the pseudo-level representation for (n,2n) has been used. This guarantees correct energy-angle coupling if the data are processed with the NJOY code[5].

For the other important neutron multiplier, lead, a revision of the ENDF/B-IV evaluation was made at ECN, Petten. The modifications concern the continuum part of the inelastic scattering, the (n,2n) reaction and the (n,3n) reaction. The energy and angle integrated sum of these reactions (MT=10) was not altered. For these continuum emission reactions the energy-angle distributions were re-evaluated and were stored in file MF6 of the ENDF-VI format, using a rather fine mesh in E'. These distributions are in agreement with recent double-differential measurements at 14 to 15 MeV. The absolute value of the (n,2n) cross-section is about 2100 mb at 14.1 MeV, which is still in agreement with the existing experimental data. Very recent studies indicate a 50 to 100 mb higher value, see Sect. 3, but confirm the shape of the neutron emission spectrum.

For the major structural materials the JEF-1 evaluations have been adopted, which have excellent low-energy cross-sections, based upon recent Reich-Moore analyses. At high energies revisions are needed, cf. Sect. 3.

2.2 Processing and multi-group libraries

The main tool for calculating multi-group constants from the EFF-1 file is the NJOY[5] code or its French version THEMIS. However, this code should at present be supplemented with the code GROUPXS[6] for the processing of continuum reactions stored in the MF6 format. An important part of the code concerns the c.m. to lab. conversion. It is interesting to note[1] that even for a heavy nucleus like lead this conversion has an important effect on the first-order Legendre coefficient of the neutron emission cross-section at 14 MeV at emission energies below 5 MeV. In view of the importance of the P1 term at emission energies from 1 to 5 MeV (cf. Sect. 2.3) this is an interesting result.

With the above-mentioned tools the GEFF-1 multi-group transport library was constructed (at various European laboratories and NDB). This library is based upon EFF-1 and consists of 175 neutron groups and 42 γ-groups. The group structure is VITAMIN-J, which contains all group boundaries of the VITAMIN-C and -E libraries developed at Oak Ridge. A larger set of data, in the same group structure (MATXS format), has been made at EIR, Würenlingen[7] by combining the EFF-1 data file with the more extensive JEF-1 data file and by increasing the number of temperatures and σ_p values. A covariance file is available from the NDB for the shielding materials taken from JEF-1. For the light materials including Li and Be uncertainly information is given on EFF-1. Efforts to create a Monte Carlo library for use in the MCNP code are in progress at various laboratories.

2.3. Benchmarking

In the framework of the benchmarking of the JEF-1 library the major shielding materials, also relevant to the NET blanket have been tested. The results were quite satisfactory[8].

In order to test the lead data an international benchmark exercize was defined[9] in which the neutron leakage spectrum of lead shells surrounding a 14 MeV neutron source has to be calculated. The EFF-1 lead file was released and distributed in different versions to allow a study of the effect of anisotropy. The results will be discussed at a meeting at the IAEA in Vienna, in early 1989.

Fischer et al.[10] already recently performed some calculations on the Dresden[11] and Osaka[12] integral lead shell experiments. The results are that the EFF-1 data file predicts the shape of the neutron leakage spectra quite well, although there is still a discrepancy with respect to the absolute magnitude, indicating that the (n,2n) cross-section needs to be increased by about 10% to 13%, respectively. In view of the fact that there are also indications from differential measurements that this cross-section needs to be increased (by about 5%) this is an encouraging result. The correct shape of the leakage spectrum is partly due to the anisotropy effect (mainly the P1-term) of continuum neutron emission included in EFF-1. This effect is still important at emission energies below 5 MeV.

There are also some calculations performed by Pelloni and Cheng[13] for blankets including regions with $Li_{17}Pb_{83}$ (30% ^{6}Li enrichment) of different thicknesses. The calculations were performed with ENDF/B-V.2 and EFF-1 libraries in which there are no differences in the cross-sections for ^{6}Li and ^{7}Li. The results indicate differences in the tritium breeding rate (main component due to ^{6}Li), apparently caused by the different energy and angle distributions of lead in the two evaluations (the energy-angle integrated cross-sections are not much different, except perhaps near the threshold). Some of the results, which still need further interpretation, are given in Table 1.

Table 1. Calculated reaction rates[13]

	20 cm		60 cm	
Reaction	EFF-1	ENDF/B-V	EFF-1	ENDF/B-V
$^{6}Li(n,t)$	0.718	0.769	1.225	1.282
$^{7}Li(n,n't)$	0.014	0.015	0.018	0.018
TBR	0.732	0.784	1.243	1.300
Pb(n,2n)	0.585	0.625	0.714	0.714
Pb(n,γ)	0.027	0.032	0.059	0.064

For thick breeding zones the (n,2n) reaction rates are equal because the cross-sections for (n,2n) are almost the same. Still there are differences in the breeding rate, due to the different energy-angle distributions. If this explanation is correct, it demonstrates the importance of a very accurate treatment of double-differential cross-sections. Further benchmarking of the EFF-1 file with Li, Be or LiPb shells is recommended.

3. PLANS FOR EFF-2

The EFF-1 file already means a large improvement as compared to the until recently adopted data files based upon ENDF/B-IV. However, further updating is necessary, in particular with respect to the double-differential neutron emission cross-sections in the continuum and consistent differential photon-production cross-sections. Also more work is demanded to create reliable covariance files.

The contents of EFF-2 will be the same as that of EFF-1, with the following extensions requested by the NET team: He, Mg, S, Ca, Co, Sn, Ta and In. The EFF-2 file will be supplementary to the JEF-2 file, with emphasis on evaluations for ^{7}Li, ^{9}Be, Al, Si, Fe, Cr, Ni and Pb. The format will be ENDF-VI with some restrictions on the use of the rather extensive possiblitites. The evaluations for the ENDF/B-VI standard reactions: H(n,n), $^{6}Li(n,t)$ and $^{10}B(n,\alpha)$ will be inserted in EFF-2.

For ^{7}Li a completely new evaluation is made in a cooperative effort of Birmingham University, CBNM-Geel, ECN-Petten and Los Alamos. The energy-angle integrated data come from the very recent re-evaluation by Young[3] based upon a least-squares analysis including the newest data. The results are not much different from those of EFF-1. However, the uncertainty has been significantly reduced for the important tritium production reaction $^{7}Li(n,n't)$. The main improvement of the ^{7}Li

evaluation should come from revised double-differential continuum neutron emission cross-sections[14] of which the main component is the $(n,n_c)t\alpha$ reaction.

For ^9Be the plans are very similar to those for ^7Li. Here the important quantity is (n,2n) cross-section and its coupled energy angle distributions.

For Al the new evaluation of ENEA-Bologna will be used, combined with the low-energy range of EFF-1. For Si the EFF-1 evaluation will be compared with other evaluations.

For the structural materials Fe, Cr and Ni the revisions are made in a joint JEF-2 and EFF-2 effort. The new evaluations will be made for the isotopes rather than for the elements. The resolved-resonance ranges, extending to energies between 0.5 and 1.0 MeV have already been re-evaluated. At higher energies the revisions fit into the EFF-2 programme. For Fe there is a cooperation between KfK-Karlsruhe and ENEA-Bologna; for Cr progress is made at Bologna and for Ni there is a cooperation between ECN-Petten and M. Uhl from IRK-Vienna.

For Pb some modifications will be applied to the present EFF-1 evaluation. The main improvement is an increase of the (n,2n) cross-section at 14.1 MeV by about 5%, based upon a careful re-evaluation of Vonach[15]: $\sigma(n,2n) = 2193 \pm 71$ mb taking into account the accurate value of the non-elastic cross-section. The shape of the EFF-1 neutron emission cross-section spectrum at about 14.5 MeV is quite close to the recommended curve of Vonach[15]. Still, some slight improvements are possible, cf. Ref. 1.

4. EUROPEAN ACTIVATION FILE

The EFF-1 and -2 files are useful for neutron and photon transport calculations. However, for the calculation of activation and transmutation the number of materials in EFF-1,2 is much too low. Therefore a separate file, if possible consistent with EFF-2, is needed. The recent interest in low-activation materials in view of recycling or simple waste disposal (like shallow land burial) has accelerated the project to achieve such a file. The requirements of the European Activation File are: a complete data base for all stable isotopes ($A \leq 210$) and isotopes (including isomers) with half lifes longer than about 1 day, including uncertainty estimates. For the most important activation reactions detailed evaluations are necessary.

Important steps in the direction of a European Activation File have been made at ECN-Petten and AERE-Harwell. The results were presented in two contributions to the Mito conference[2,16]. The two laboratories have made libraries that are based upon the REAC-2 data file of Mann et al.[17], with extensions and important renormalizations at 14.5 MeV to experimental data or results from systematics. In particular with respect to the systematics of cross-sections[18] and of isomer ratios[19] important progress has been made. As a by-product of the systematics uncertainty estimates are available for all reactions that have been normalized to the systematics. The REAC-ECN-3 data file[2] contains about 8500 reactions with cross-sections between 0 and 20 MeV. There are 51 reactions that were completely re-evaluated to obtain reliable predictions of long-lived products, including uncertainty information. The current UK library is based upon REAC-ECN-2 with extensions and improvements. It also contains a consistent decay library. Because of intense cooperation the REAC-ECN-3 and the current UK activation library are very similar. Practice is that revisions are exchanged.
The present work on the REAC-ECN-3 file was supported by the JRC Ispra and the file was applied in their activation studies[20]. The data file has been processed into a multigroup library of 100 groups (GAM-structure) and 175

groups (VITAMIN-J structure, consistent with GEFF-1).

Recently the first version of the European Activation File (EAF-1), using as a starting point the current REAC-ECN-3 library, has been assembled[21]. It has been extended by the addition of reactions on 142 targets (many of them isomers) and improved by applying proper effective thresholds for isomeric states. It is at present probably the best data base for activation calculations of the first wall and inner blanket, since all 14.5 MeV data have been checked rather carefully.

The main emphasis of the next version (EAF-2) will be in the completeness and quality of the (n,γ) and (n,n'γ) data. These reactions are important at relatively low energies and therefore also the thermal and resonance ranges will be reconsidered. This is of importance in "thermal" blankets and the NET water-cooled blanket. Also the existing information on uncertainties will be included in a more systematic way. Thus, EAF-2 will become a general activation file, applicable in fission and fusion reactor technology. It is aimed to include evaluations of various European groups (e.g. AERE-Harwell, IRK-Vienna and ECN-Petten). A further development is an agreed tripartite cooperation with the U.S. (Dr. F.M. Mann, Richland, U.S.) to create a joint European-U.S. activation file.

5. CONCLUSION

The status of the EFF- and EAF-projects has been discussed. The EFF-1 library has been checked with integral experiments and benchmark calculations. Multigroup transport libraries based upon EFF-1/JEF-1 and EAF-1 are being introduced for routine calculations of fusion reactor systems in Europe. Their performance is expected to be much better than the previously used libraries based upon ENDF/B-IV. Meanwhile, work for EFF-2/JEF-2 and EAF-2 is in good progress. Further international cooperation is required to meet the increasing demands for high-quality data files supplied with covariance information.

REFERENCES

1. H. Gruppelaar, Int. Conf. on Nuclear Data for Science and Technology, Mito, Japan, May 30 - June 3, 1988, paper ID04.
2. J. Kopecky and H. Gruppelaar, ibid, paper CB03.
3. P.G. Young, ibid, paper DD25.
4. P.G. Young and L. Stewart, LA-7932-MS (1979).
5. R.E. Macfarlane, D. Muir and R.M. Boicourt, LA-9303-M (1982).
6. H. Gruppelaar, J.M. Akkermans and D. Nierop, ECN-182 (1986). GROUPXS code available from NEA Data Bank and RSIC.
7. P. Vontobel and S. Pelloni, EIR-535 (1987).
8. M. Salvatores et al., Int. Conf. Radiation Shielding, Bournemouth, Sept. 1988.
9. E.T. Cheng, Specifications for the IAEA lead benchmark problem, 1988.
10. U. Fischer et al. Int. Symp. on Fusion Technology, Tokyo, April 1988 and priv. comm. 1988.
11. T. Elfruth et al., Kerntechnik 49, 121 (1987).
12. A. Takahashi et al., Proc. Int. Conf. Nuclear Data for Basic and Applied Science, Santa Fe, 1985, p. 59.
13. S. Pelloni and E.T. Cheng, Int. Conf. on Nuclear Data for Science and Technology, Mito, Japan, May 30 - June 3, 1988, paper CB09.
14. T.D. Beynon and A.J. Oastler, Ann. Nucl. En. 6, 437 (1979) and private comminication (1988).
15. H. Vonach, NEANDC Spec. Mtg. on Prequilibrium Nuclear Reactions, Semmering, Febr. 1988 and A. Pavlik and H. Vonach, extended IRK-report (in press).
16. R.A. Forrest et al., Int. Conf. on Nuclear Data for Science and Technology, Mito, Japan, May 30 - June 3, 1988, paper CH06.
17. F.M. Mann et al., Proc. Conf. Nuclear Data for Basic and Applied Science, Santa Fe, May 1985, p. 207.
18. R.A. Forrest, AERE-R-12419 (1986).
19. J. Kopecky and H. Gruppelaar, ECN-200, (1987).
20. C. Ponti, Fusion Technology 13 (1988) 157; C. Ponti, Proc. Int. Symp. on Fusion Nuclear Technology, Tokyo, Japan, April 1988.
21. J. Kopecky and H. Gruppelaar, ECN report, to be published.

FUSION TECHNOLOGY 1988
A.M. Van Ingen, A. Nijsen-Vis, H.T. Klippel (editors)

TRANSMUTATION IN MARTENSITIC REDUCED ACTIVATION ALLOYS BASED ON DIN 1.4914 MARTENSITIC STEEL

Javier SANZ and José M. PERLADO

Instituto de Fusion Nuclear (DENIM). Universidad Politécnica de Madrid,
José Gutiérrez Abascal, 2. 28006 Madrid, Spain.

The objective of the present paper is to calculate irradiation damage behaviour, in terms of displacement and transmutant production in inertial and magnetic fusion reactors. For purposes of planning irradiation experiments in fission reactors, calculations has been also carried out for the peripheral target position (PTP) in the High Flux Isotope Reactor (HFIR). The material included in the calculations is the reduced activation martensitic steel, B-TAHF, based on the specification of 1.4914 steel. Results of this work would be useful to gain insight into the behaviour in different reactor confinement fusion concepts and to guide and assess efforts in fission-fusion correlations.

1. INTRODUCTION

To develop and to asses the suitability of a material for use as the first structural wall in a fusion reactor, it is necessary to know the displacement damage and the transmutation behaviour of the material.

It is believed that in neutron environments the change in material behaviour will be very sensitive to displacement damage, and that for many purposes, it can be well characterized in terms of rate and total production of atoms displaced from their sites.

Transmutations by nuclear reactions has been shown to greatly affect the properties of reactor materials. Gaseous transmutation products are particularly important[1], but the importance of solid transmutants is also recognized as significant[2,3,4]. Hence, complete information about transmutations is needed for reactor design.

On the other hand, because no facility exists which has a neutron environment similar to that expected in a fusion reactor, existing fission reactors are being used to test the behaviour of materials. It is worth to point out that at the present time, there is a great interest for conceptual design studies of high-energy intense neutron sources for material testing.

A code system called CIBELES[5] has been developed to (a) calculate transmutation and atom displacement production, and (b) analyze the composition change of a particular nuclide or element from others initially present in the alloy, at fusion reactors and several neutron experimental facilities.

This paper has as a major goal to determine the neutron-induced composition change of B-TAHF alloy as a function of the number of displacements produced (also as a function of neutron fluence and time) in both magnetic confinement fusion (MCF) and inertial confinement fusion (ICF) reactors. As currently, high exposure data can be obtained only in fission reactors, calculations of damage has been also carried out in PTP positions of the HFIR reactor where most of the high fluence irradiations, for fusion reactor material aplications has been performed.

The alloy B-TAHF was designed by Kernforschungszentrum Karlsruhe (KfK), it has been provided by German and Spanish manufactures and it is explored under the KfK-

CIEMAT-EEC program on "development of low activation structural steels based on the specification of the martensitic steel 1.4914". Begin of irradiations in fission reactors was planed for 1989.

Damage results from fusion and HFIR reactors will be compared searching for differences on transmutant and displacement production, which would be useful to guide and validate work for fission-fusion correlations. Differences in ICF and MCF reactors will be remarked.

2. CALCULATIONAL METHODOLOGY

2.1. Displacement damage

For the calculation of displacement damage in alloys, it is assumed that the appropriate displacement can be determined by simply adding up the individual contributions from each element weighted by the appropriate atomic fraction. Compound dpa effects are quite small for structural materials such as stainless steel[6].

Calculations of spectral-averaged displacement cross sections for the constituent elements of the martensitic steel is made as follows.

For major elements, in a first step, the displacement damage function is obtained by damage numerical simulation and then it is used for calculations of cross sections. MARLOWE[7] code has been used to obtain the damage function giving the number of Frenkel Pairs, vacancy-interstitial pairs, which are separated by more than one lattice constant (distant FP's). For iron, MARLOWE results from Diercks[8] calculations has been used. The corresponding displacement cross sections are subsequently obtained considering recoil atom energy distributions from calculations using the NJOY[9] nuclear data processing code and ENDF/B-IV nuclear data.

For minor elements, damage energy cross sections were calculated using the NJOY code and ENDF/B-IV library. These were converted to displacement cross sections multiplying by a factor of 0.8/2 E_d where E_d is the "effective" displacement thereshold energy. The adopted values for E_d have been those used in the SPECTER[10] code. Most of those values are in good agreement with our MARLOWE calculations.

2.2. Transmutations

The calculational procedure for the transmutation problem is based on the concept of "generating function[5]. The transmutation of nuclides induced by neutronic reactions and radioactive decay can be described by a system of ordinary linear differential equations, so we may write the solution of this system as:

$$n_i(t) = \sum_l p_l^i r_l \quad , \tag{1}$$

where n_i is the number of atoms produced of element i per atom of alloy, and t denotes time. Hence, p^i_l are the element i generating functions giving the cumulative number of atoms of element i produced per initial atom of element l at time t[11,12,13]. The symbol r_l is the atom fraction of element l in the alloy. Eq. (1) can be also rewritten as:

$$n_i(t) = \sum_l \sum_j p_{lj}^i q_{lj} r_l \quad . \tag{2}$$

where p^i_{lj} are the element i generating functions from the isotope j of the element l, giving the number of atoms produced per initial atom of isotope j. The symbol q_{lj} is the initial fraction of atoms of isotopes j of element l to all atoms of element l in the alloy.

Four different modules of CIBELES system are involved in the transmutation problem. The system includes powerful computing methods for analysing results, and uses the numerical calculation techniques of the ORIGEN[14] code. It has associated a transmutation modified ACFA[15] code; a one-energy group cross section library, that is presently produced from the neutron cross section library UKCTRIIIA[16]; a plotting code; and several driver and functional codes.

Two major functions are performed by the system: calculation and analysis. In the calculation, elements with major transmutation rates are identified and only temporal evolution of the elements (initially present or produced in the alloy) specified by the user are given. In the analysis, it is carried out the evaluation of sources (elements and isotopes) and sequences for the production of a particular nuclide or element from others initially present. Displacement damage production from reactions that are not the first ones in a transmutation sequence can be easily calculated by CIBELES, when energy of the reaction products is given. In this paper displacements from $^{59}Ni(n,\alpha)^{56}Fe$ reaction are assessed for natural nickel, and the two alloys.

3. NEUTRON DAMAGE CALCULATIONS

3.1. Material and irradiation characteristics

Calculations have been performed for B-TAHF steel, designed in the KfK Center. The nominal composition (in wt%) of that alloy used in the calculations is as follows:

B-TAHF: 0.008 B, 0.16 C, 0.005 N, 0.05 Si, 0.005 P, 0.005 S, 0.01 Ti, 0.5 V, 9.5 Cr, 1.0 Mn, 87.397 Fe, 0.01 Ni, 0.08 Cu, 0.01 Nb, 0.01 Mo, 0.25 Hf, 0.25 Ta, 0.75 W.

The alloy is assumed to be irradiated by three quite different neutron environments:

i) Magnetic confinement fusion reactor. For calculations the adopted reactor model has been that of the Culham Conceptual Tokamak Reactor Mark IIA (CCTRII). The neutron average energy in the first wall(FW) is 4.11 MeV. The martensitic steel for the first wall is assumed to be subjected to a neutron power loading of 5 $MW.m^{-2}$ (resulting in a neutron flux in the FW of 1.85×10^{15} $cm^{-2}s^{-1}$) and to an irradiation exposure time of thirty years, which corresponds to a notional long term commercial reactor.

ii) Inertial confinement fusion reactor. Results presented here are based on: (a) the chamber concept of HIBALL-II (a Heavy Ion Beam Driven Fusion Reactor), (b) a DT yield of 400MJ and a repetition rate of 5Hz yielding 7.1×10^{20} fusion neutrons per second, and (c) a $\langle\rho R\rangle_b$ (burnup-averaged ρR)[17,18] target parameter of 4 $g.cm^{-2}$. The resulting neutron flux at the first structural wall was 9.96×10^{14} $cm^{-2}.s^{-1}$ and the average energy was 0.15 MeV. An irradiation exposure time of thirty years was assumed, corresponding to a notional reactor lifetime.

iii) High-Flux Isotope Reactor (HFIR). This mixed-spectrum reactor presents in PTP positions an average energy of 0.41 MeV and a total flux about 5.1×10^{15} $cm^{-2}s^{-1}$. It has been assumed an irradiation time of fourteen years, though too long, it is the neccesary to obtain a dose approaching to 400 dpa, which at least must be withstanded by the first wall of a magnetic fusion power reactor for economic feasibility.

3.2. Displacement Damage

Spectral-average displacement cross sections of B-TAHF for CCTR-II, HIBALL-II and HFIR reactors are 196, 119 and 682 barns, respectively. The displacement rate in MCF, ICF and HFIR reactors are 31.5, 3.75 and 40 dpa's/year, resulting in an accumulated damage at the end of irradiation of 441, 112.5 and 1200 dpa's, respectively. These results are nearly similar than those of 1.4914 steel[19].

The additional contribution to the displacement rate from the $^{59}Ni(n,\alpha)^{56}Fe$ reaction has been calculated by CIBELES.

Results at the end of the irradiation exposure periods, above postulated, of MCF, ICF, and HFIR reactors show: (a) for natural nickel, a displacement production of 148, 4.66 and 1 dpa, respectively, and (b) for the B-TAHF alloy, a production of 0.0139 dpa's in HFIR and no contribution in both fusion reactors.

3.3. Transmutation products

Figure 1, 2 and 3 show transmutation products generated in B-TAHF when exposed to neutron environments of the MCF, HFIR and ICF reactors, respectively.

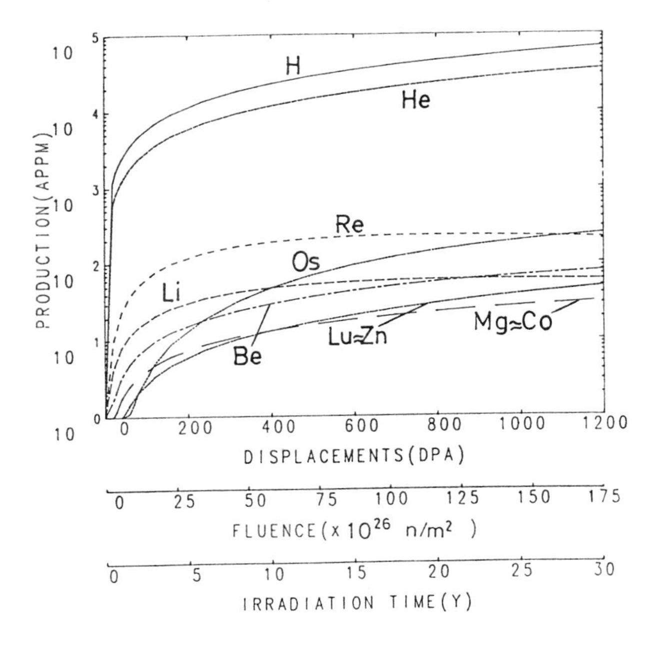

FIGURE 1

Principal elements produced in B-TAHF for the magnetic fusion reactor concept.

Figure 4 (part a and b), figure 5 (part a and b) and figure 6 show the relative change in the population (see eq.3) of the initial constituents of B-TAHF steel, for the MCF, HFIR and ICF reactors, respectively, as a function of dpa's, neutron fluence, and time.

Relative change (%) =

$$= \frac{\textbf{Population at time t} - \textbf{Initial population}}{\textbf{Initial population}} \times 100 \;. \quad (3)$$

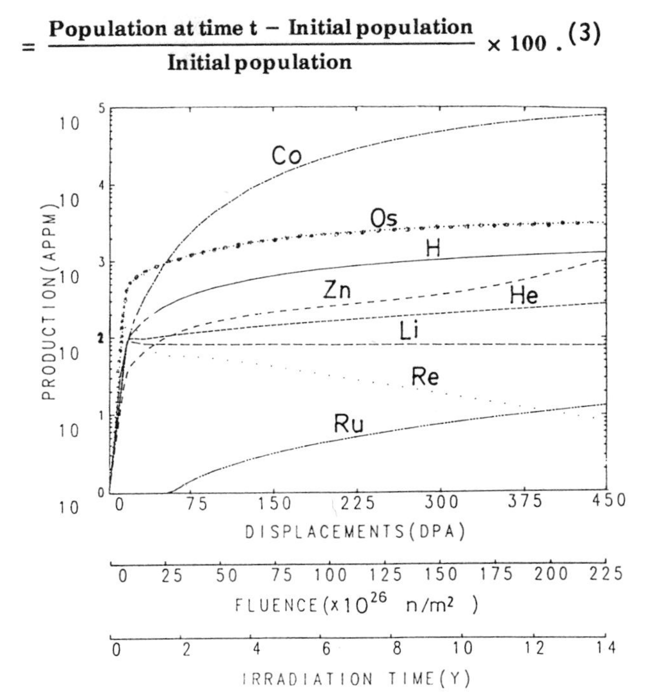

FIGURE 2

Principal elements produced in B-TAHF for the HFIR reactor.

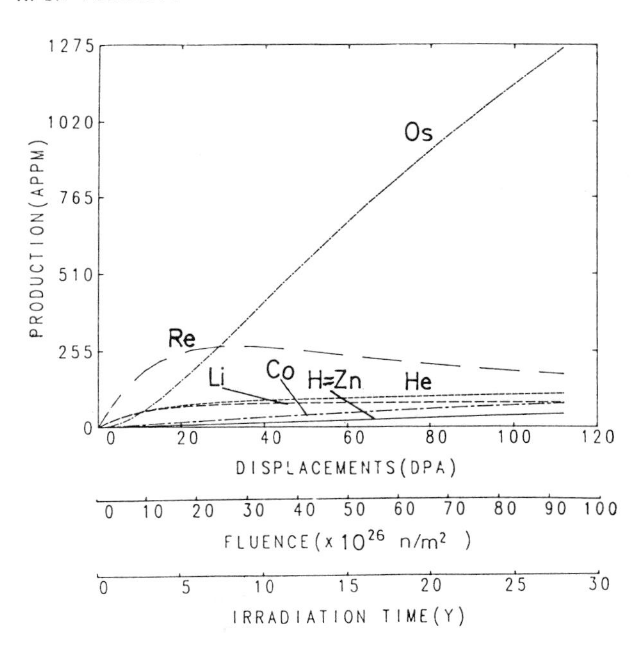

FIGURE 3

Principal elements produced in B-TAHF for the inertial confinement fusion reactor.

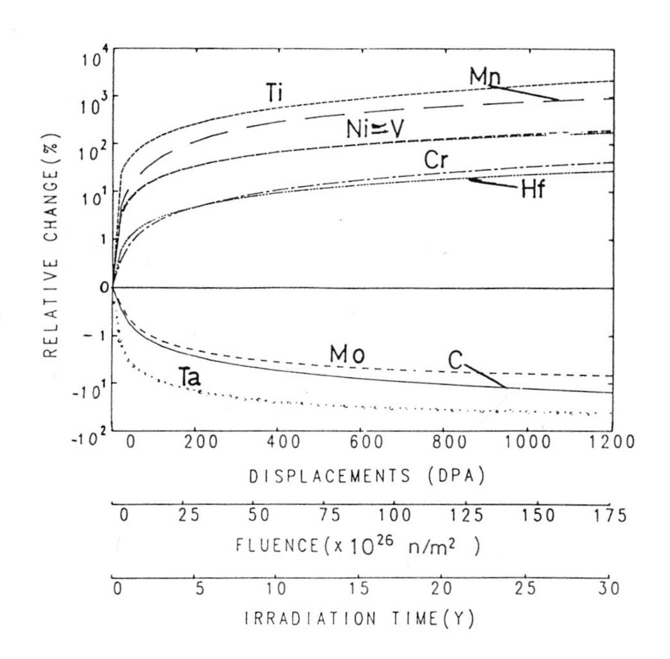

FIGURE 4(a)

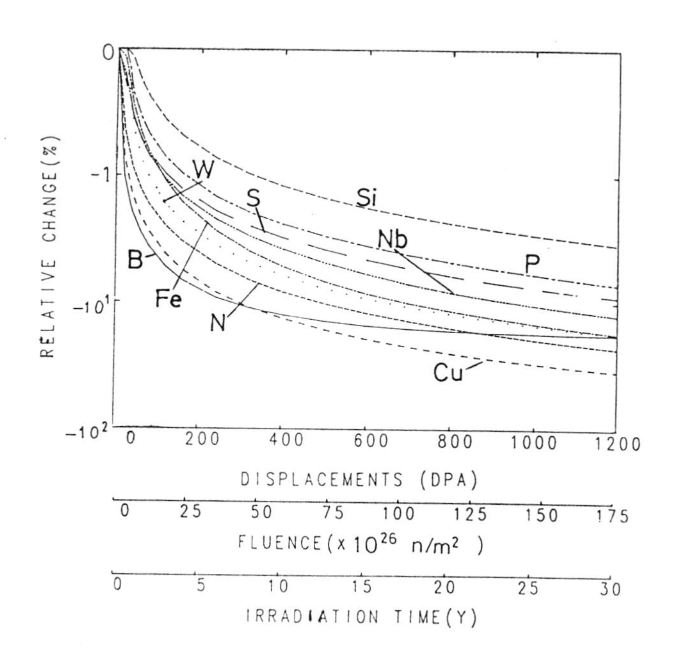

FIGURE 4(b)

Atomic relative changes (%) in the population of constituents of B-TAHF for the magnetic fusion reactor concept. Values between + 0.1% and -0.1% are set to zero.

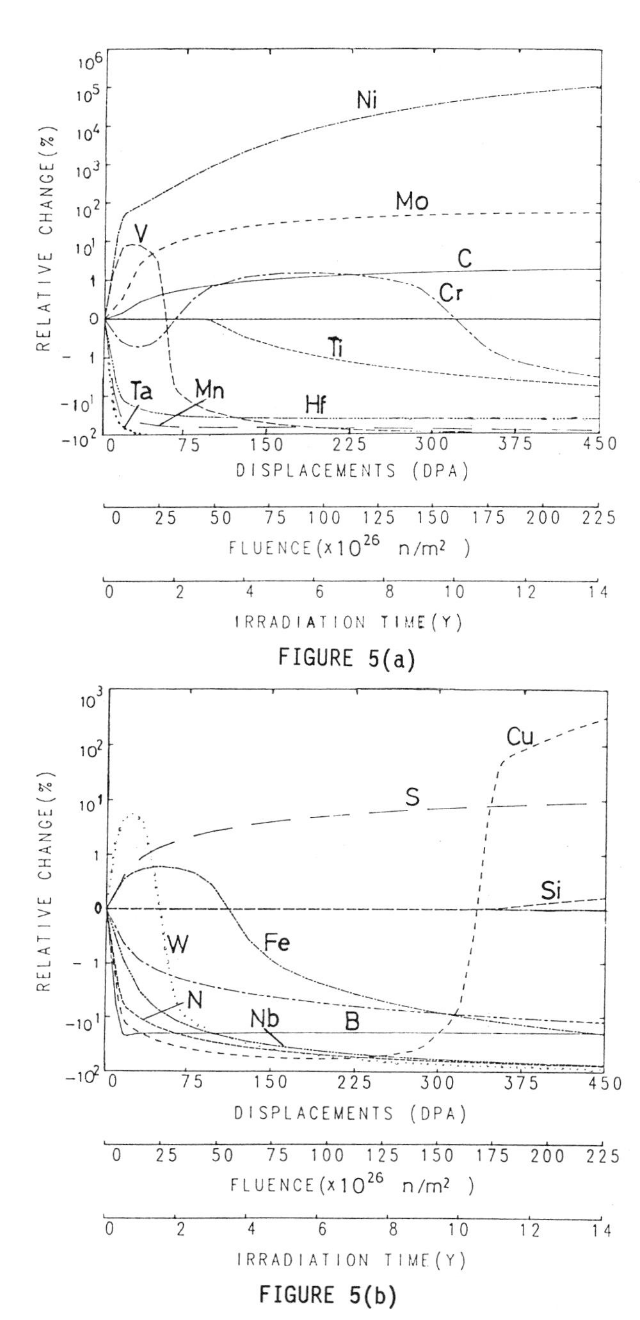

FIGURE 5(b)

Atomic relative changes (%) in the population of constituents of B-TAHF for the HFIR reactor. Values between + 0.1% and -0.1% are set to zero.

It is significant to note that in the ICF reactor, changes occurring in C, N, Si, P, S, Ti, Cr and Fe elements are less than 1% (in atoms) and less than 2% for vanadium.

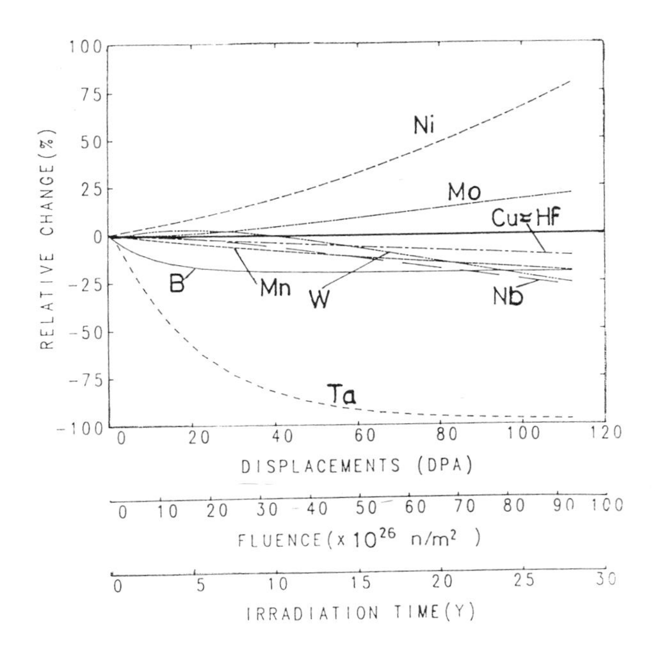

Figure 6
Atomic relative changes (%) in the population of constituents of B-TAHF for the ICF reactor.

4. CONCLUSIONS

The displacement production rate in the FW of HIBALL-II reactor is approximately an order of magnitude lower than that of both HFIR, and MCF reactors. The MCF reactor type CCTR-II has a dpa rate slightly higher than that of HFIR.

For the alloy considered the displacement production in the three environments caused by $^{59}Ni(n,\alpha)^{56}Fe$ reactions is negligible. For HFIR, this is due to the low Ni content of the martensitic steel. Nevertheless, the HFIR spectral average displacement from Ni by the two-step reaction $^{58}Ni(n,\gamma)^{59}Ni(n,\alpha)^{56}Fe$ is important.

There are significant changes in the composition of the steel irradiated in both the MCF, and the HFIR reactors. At the first estructural wall of the ICF reactor, HIBALL-II, changes are much less pronounced than in the other two cases. In this case significant changes occur only in Ta, Ni, Os and Re elements.

The evolution of the elements initially present in B-TAHF is strongly dependent on the neutron spectrum. Differences in neutron spectra have also an important effect on the generation of new elements. There are remarkable differences in the amounts, the rate of production, and the variety of solid and gas transmutants.

In martensitic steels, the helium production is an increasing function of average spectrum energy, mainly because of the iron effect. In the CCTRII spectrum the generation of helium is linear with fluence, while in reactors containing appreciable thermal neutron fluxes nonlinear increases in helium production take place due to multi-step reactions and burnup of nickel, boron and a few other elements.

In high flux fission reactors it appears to be possible to obtain displacement damage and transmutation levels very similar to those of inertial fusion reactors, within a reasonable irradiation time. Therefore, we may conclude that HFIR provides a means of testing several properties of developmental alloys in damage regimes of engineering interest for fusion inertial reactor applications. On the other hand, it is very unlikely that reliable engineering data for MCF reactors can be obtained in a reasonable time in HFIR.

ACKNOWLEDGEMENTS

We are greatly indebted to the personnel of the Institute of Nuclear Fusion (DENIM) for their help, and to Prof. G. Velarde for encouragement in the preparation of this article.

The calculational models used in this work have been developed under a research project supported by PIE-UNESA. The applications here presented are included in the cooperation agreement between DENIM and the Spanish Nuclear Research Center (CIEMAT), as a part of the general KfK/CIEMAT/EEC research project, entitle "Development of Low Activation Ferritic-Martensitic Steels.

REFERENCES

1. H. Ullmaier, Nucl. Fusion 24(8) (1984) 1039.

2. G.J. Butterworth, J. Nucl. Mater. 135 (1985) 160.

3. J.F. Bates, F.A. Garner and F.M. Mann, J. Nucl. Mater. 103 & 104 (1981) 999.

4. F.A. Garner, H.R. Brager, J. Nucl. Mater. "in print". Presented at III ICFRM, 1987, Karlsruhe.

5. J. Sanz, CIBELES system: A computer code package for calculation and analysis of transmutant products. Preliminary user's manual. DENIM 134 (1987).

6. L.R. Greenwood, "Radiation Damage Calculations for Compound Materials". Presented at 14th International Symposium on Effects of Radiation on Materials, 1988. Andover, MA.

7. M.T. Robinson, I.M. Torrens, Phys. Rev. B9 (1974) 5008.

8. R. Dierckx, J. Nucl. Mat. 144 (1987) 214.

9. R.E. MacFarlane, R.J. Barrett, D.W. Muirand, R.B. Boccourt. The NJOY Nuclear Data Processing System. Vol. I. Users Manual. LA-7584-M, (1978).

10. L.R. Greenwood and R.K. Smither, SPECTER: Neutron damage calculations for materials irradiations ANL/FPP/TM-197 (1985).

11. L.K. Mansur, A.F. Rowcliffe, M.L. Grossbeck and R.E. Stoller, J. Nucl. Mater. 139 (1986) 228.

12. J. Sanz, J.M. Santolaya, J.M. Perlado, in: Proc. 14th Symp. on Fusion Technology, Avignon, 1986 (Pergamon Press, 1987) p. 1315.

13. J. Sanz, R. de la Fuente, J.M. Perlado, J. Nucl. Mater. "in print". Presented at III ICFRM 1987. Karlsruhe.

14. A.G. Croff, a user's manual for the ORIGEN2 computer code, ORNL/TM-7175 (1980).

15. H. Brockmann, U. Ohlig, ACFA: A general purpose activation code, JÜLICH 1986 (1983).

16. O.N. Jarvis, Description of the transmutation and activation data library UKCTRIIIA, AERE-R-9601 (1980).

17. G. Velarde et al., Fus. Tech. 8 (1985) 1850.

18. J. Sanz, P. Hernan, J.M. Perlado, "Coupled Analysis of the Set Target-Blanket" 17th ECLIM, Rome (1985), DENIM 071.

19. J. Sanz and J.M. Perlado. "Transmutation in martensitic reduced activation alloys based on DIN 1.4914 martensitic steel". Poster presented at the 15th SOFT, Utrecht, 1988. DENIM 185 (1988).

FUSION TECHNOLOGY 1988
A.M. Van Ingen, A. Nijsen-Vis, H.T. Klippel (editors)

PREPARATION, CHARACTERIZATION AND PROPERTIES OF Li_2ZrO_3 FABRICATED FROM Li_2CO_3 AND ZrO_2

A. Jean FLIPOT, Etienne BRAUNS, Paul DIELS

Centre d'Etude de l'Energie Nucléaire, Studiecentrum voor Kernenergie (CEN/SCK), Ceramic Research Department, Boeretang, 200, B-2400 Mol, Belgium

The preparation of lithium metazirconate by calcining at low temperature an intimate mixture of lithium carbonate and zirconia is discussed. Such a powder can be cold pelletized and sintered to densities up to 92% T.D. The pellets are contamination free, fine-grained and have a completely open porosity. Some properties are reported.

1. INTRODUCTION

Lithium metazirconate is a promising ceramic candidate for breeding tritium in a fusion reactor as it seems to fulfill the following requirements according to the first irradiation experiments : low tritium and helium retention [1, 2, 3, 4], good dimensional stability[1, 2, 3] and good compatibility with steels[5].

In addition to the study of lithium metasilicate, which is in a rather advanced stage already, CEN/SCK started in 1986 a development programme on lithium metazirconate in view of preparing well characterized samples for irradiation in the HFR and Osiris reactors. Both compounds are investigated under the auspices of the European Community (contract No. 200/85-FUA-B) as part of a coordinated programme on ceramic breeding materials for fusion reactors and with the collaboration of FBFC International (Dessel - Belgium).

In the paper a description is given of the preparation method for lithium metazirconate and the resulting pellet characteristics. Some physical property data are also presented.

2. FABRICATION ROUTE

As for the earlier fabrication of lithium metasilicate, the process is based on commercially available raw materials and equipment currently used in the ceramic industry in order to facilitate future industrialization[6,10]. Lithium metazirconate is synthesized from lithium carbonate and zirconia. Subsequently the powders are cold pelletized and sintered.

2.1. Calcination

Complete reaction can be achieved below 700°C with stoichiometric and understoichiometric (some zirconia in excess) mixtures, thus avoiding to pass through a liquid phase. The reaction usually starts around 500°C and completes between 625°C and 675°C, though

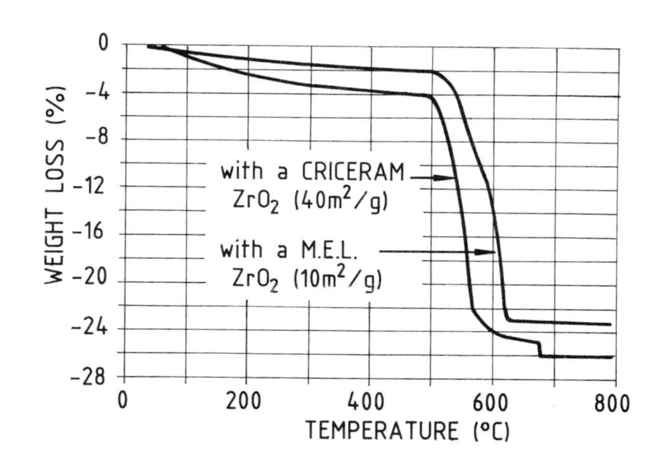

FIGURE 1
Effect of the zirconia powder quality on the reaction rate between Li_2CO_3 and ZrO_2 at various temperatures

depending on the characteristics of the zirconia powders. This is illustrated by the thermo- gravimetric curves of Fig. 1 obtained when using either a M.E.L. zirconia (Magnesium Elektron Limited) with a specific surface area of 10 m^2/g (upper curve) or a high specific surface (40 m^2/g) zirconia powder (lower curve), manufactured by CRICERAM (Jarrie-France). In each case, the maximum weight loss corresponds to complete reaction, the difference between both experiments resulting from the water content in the respective zirconia powders. Though the mixture containing a M.E.L. zirconia reacts completely at low temperature and gives a calcined lithium metazirconate with a specific surface area of 8 m^2/g, this can only be sintered up to maximum densities below 75 % T.D. Other lithium metazirconate qualities, synthesized from other zirconia powders with a similar specific surface area, reach, on the contrary, sintered densities ranging from 85 % to 90 % T.D. This demonstrates that some general characteristics of the raw materials such as morphology can sometimes be more important than specific surface area. A survey of the results obtained when using various zirconia powders from different manufacturers shows that the CRICERAM powders are to be preferred for their physical and chemical (hafnium-free) characteristics.
The reaction was examined in detail and found to be first order with an activation energy of 276 KJ/mole.

Unlike lithium metasilicate[6], sinterable lithium metazirconate can indifferently be synthesized under dry or wet air, from powders or compacts. When examining a calcined sample by X-ray diffraction, one observes next to lithium metazirconate and possible traces of unreacted raw materials, a compound, which has not been identified yet, called hereafter Y-phase. The quantity of Y-phase is less than that of the X-phase detected when synthesizing lithium metasilicate. Its rôle is not clear. Though theoretically unstable below 1100°C[8], the tetragonal lattice represents more than 90 % of the lithium metazirconate formed around 650°C. According to the diffraction intensities, the Y-phase in this case is about 15-20 % of the total mass. These values are valid whatever the zirconia crystal lattice may be (monoclinic or tetragonal). At higher temperature, both tetragonal lithium metazirconate and Y-phase disappear progressively to form monoclinic metazirconate. Pellets treated above 900°C do not contain any Y-phase and an essentially monoclinic lattice ($\geq$ 90 %) is formed.

2.2. Sintering

Though dilatometric experiments carried out on green pellets already show some shrinkage between 800 and 850°C, effective sintering of the metazirconate only begins above 950°C. Densities as high as 85-92 % T.D. can be reached by sintering pellets at about 1000°C. Equilibrium proceeds slowly and more than 20 hours at maximum temperature are needed to reach the maximum attainable densification.

3. PELLET CHARACTERISTICS

3.1. Chemical purity

All pellets sintered above 975°C are fluorine and chlorine free. Lithium metazirconate reacts easily with carbon dioxide and water. When pellets are exposed to ambient air, even for a short time, carbon contamination can reach 900 to 1000 ppm and also the water content increases rapidly. A water content of 200 ppm may be analysed in 90 % T.D. pellets and values as high as 700 ppm in 75 % T.D. pellets. The pellets have therefore to be dried after centerless grinding at the air and stored under a controlled atmosphere. Wet milling causes an aluminium contamination of about 600 ppm resulting from the wear of the alumina milling tools. Though this contamination is not embarassing because aluminium only activates slightly under irradiation, it can be reduced by using zirconia balls.

3.2. Grain size

The average grain size ranges from 1 to 3 microns in medium- and high-density pellets, depending on the process parameters used.

3.3. Open porosity

The open porosity of numerous pellets, sintered at different temperatures, has been measured by mercury porosimetry. The results are summarized in Fig. 2. The graph shows that, within the limits of the precision of the measurements and the reproducibility of the pellet characteristics, all pores can be considered as open in a density range from 70 to 93 % T.D. No effect of the calcination or sintering temperature could be evidenced. Since open porosity could favour tritium release during irradiation, this result is

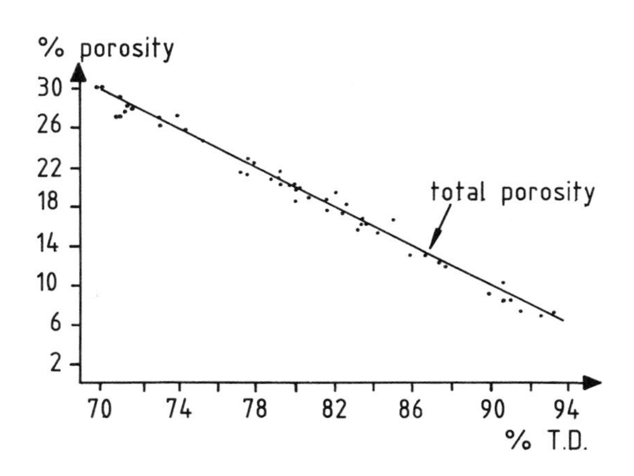

FIGURE 2
Open porosity of Li_2ZrO_3 pellets with a density up to 92% T.D.

even better than that obtained on lithium metasilicate pellets[9] where an increasing amount of close pores was detected above 76-77 % T.D. The open pore distribution is unimodal in low-density pellets and bimodal in medium- or high-density pellets. Fig. 3 shows how the percentage of the most frequent pores changes as a function of the pellet density. The fraction of small pores is hardly affected by density below 82% T.D. Above 82% T.D., the fraction of small pores increases linearly with density, the quantity of large pores decreasing accordingly. Both curves cross each other at about 86% T.D. Complete pore spectra are illustrated in Fig. 4 for a 86% T.D. pellet and

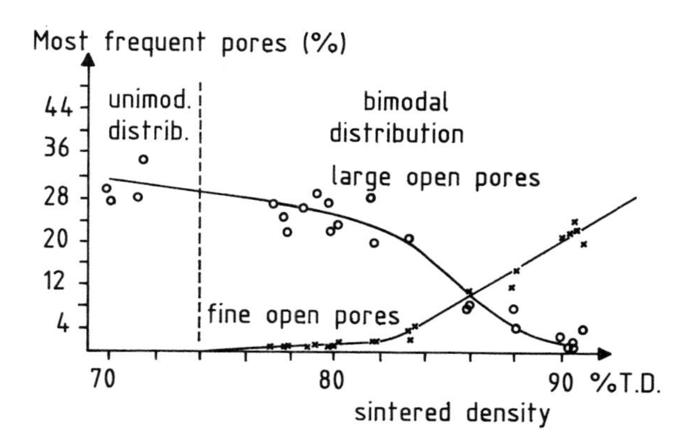

FIGURE 3
Percentage of the most frequent pores as a function of the pellet density

in Fig. 5 for a 80% T.D. pellet. These spectra let anticipate a much lower tritium release in 86% T.D. pellets than in 80% T.D. pellets because of the cumulative effect of decreasing total open pore volume and mean pore size.

The most frequent large pores have a diameter between 0.5 and 0.6 µm, while the diameter of the most frequent fine pores is about fifteen times smaller. These typical data are valid in a broad range of densities but can be affected by the characteristics of the raw materials and the process parameters.

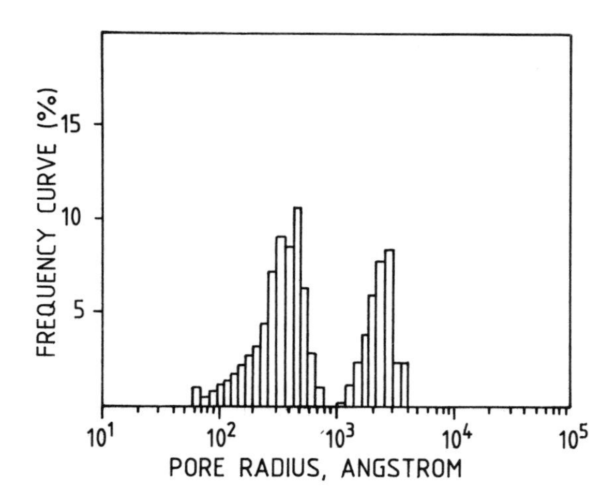

FIGURE 4
Pore size distribution in a 86% T.D. pellet

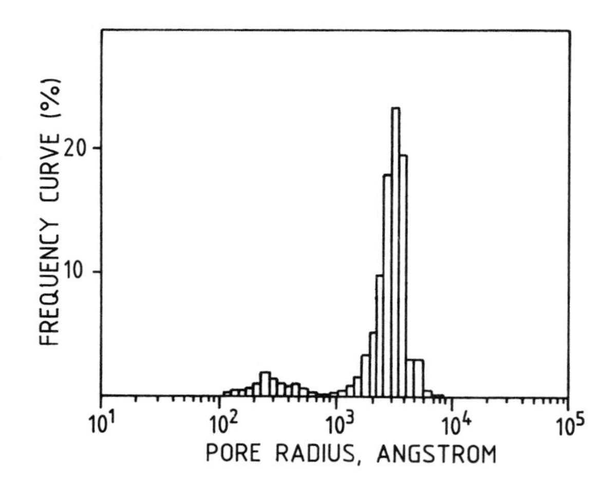

FIGURE 5
Pore size distribution in a 80% T.D. pellet

4. PROPERTIES

4.1. Linear thermal expansion

Linear thermal expansion of lithium metazirconate has been determined by dilatometry. Three experiments have been performed on sintered pellets with a density of about 90 % T.D. They have been stabilized by an additional treatment at 1000°C in the dilatometer just before measuring the thermal expansion between room temperature and 950°C. The mean result is plotted in Fig. 6 together with the results of former determinations on lithium metasilicate. The relatively low thermal expansion coefficient of lithium metazirconate is a favourable characteristic with respect to thermal shock resistance.

4.2. Specific heat

Specific heat was measured using a differential scanning calorimeter. The results plotted in Fig. 6 are in good agreement with the results reported by Kennedy[10]. The specific heat varies little with temperature, is lower than that of $LiAlO_2$ and Li_2SiO_3 and even 2-3 times lower than for lithium oxide.

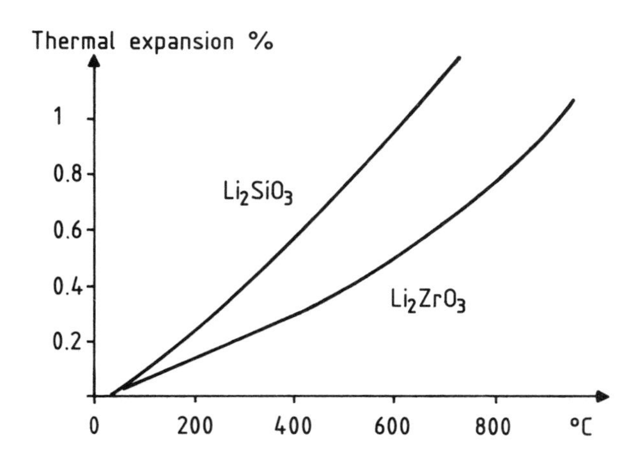

FIGURE 6
Thermal expansion of Li_2SiO_3 and Li_2ZrO_3

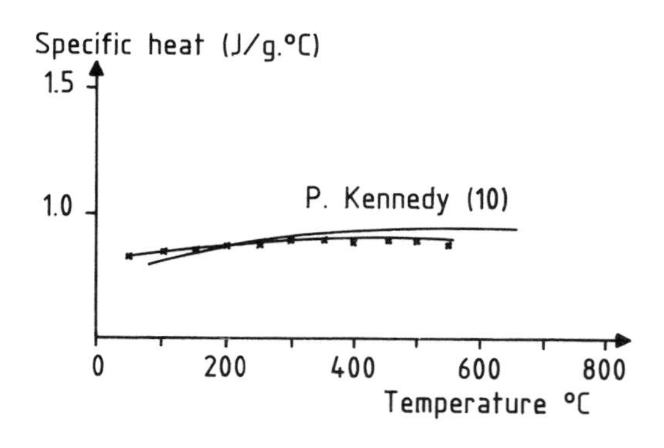

FIGURE 7
Specific heat of lithium metazirconate[10, 11]

4.3. Thermal conductivity

The thermal conductivity data shown in Fig.8 are derived from thermal diffusivity measurements made using a modulated electron beam thermal diffusivity apparatus. The thermal conductivity of 87 % dense zirconate is much lower than that of lithium oxide and in good agreement with the results of Hollenberg and Kennedy[10].

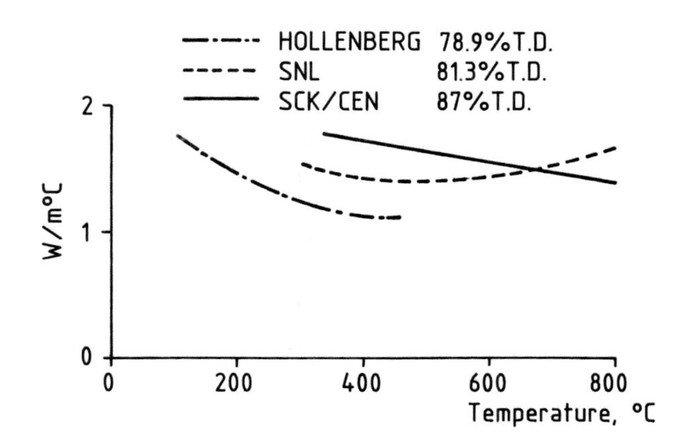

FIGURE 8
Thermal conductivity of lithium metazirconate[10, 12]

5. CONCLUSIONS

Lithium metazirconate pellets can be synthesized from lithium carbonate and zirconia below 680°C, thus avoiding formation of a liquid phase. The powder can be used for obtaining medium- and high-density pellets by cold pelletizing and sintering. The pellets are pure, have a grain size of 1-2 µm and an essentially open porosity over a wide density range (73-93 % T.D.). Pore distribution depends on density. The fraction of fine pores increasing linearly with density above 82 % T.D., it is anticipated that tritium release is much easier in 80 % dense pellets than in 86 % ones.

Linear thermal expansion, specific heat and thermal conductivity have been determined. The results are in good agreement with the results published in the literature.

REFERENCES

1. G.W. Hollenberg and D.L. Baldwin, J. Nucl. Mater., 133 and 134 (1985), 242

2. G.W. Hollenberg et al., J. Nucl. Mater., 141-143 (1986), 271

3. H. Kwast et al., J. Nucl. Mater., 154-156 (1988), in print

4. C.E. Johnson et al., J. Nucl. Mater., 154-156 (1988), in print

5. P. Hofmann and W. Dienst, J. Nucl. Mater., 141-143 (1986), 289

6. A.J. Flipot et al., J. Nucl. Mater., 133 and 134 (1985), 226

7. A.J. Flipot et al., European Patent published under No. 0175423 B1

8. D.J. Suiter, Report MDS E2677 UC 20 - June 1983

9. A.J. Flipot et al., Presented at the International Symposium on Fusion Nuclear Technology, Tokyo, 10-15 April, 1988, Fusion Engineering and Design, in print

10. P. Kennedy, The preparation, characteris-ation and properties of lithium oxyde and lithium meta-zirconate specimens irradiated in HFR Petten in second and third EXOTIC experiments, in : Proceedings of the 14th Symposium on Fusion Technology, eds Pergamon Press (1986) pp. 1013-1018

11. W. Timmermans - SCK/CEN - personal communication

12. R. De Coninck - SCK/CEN - personal commmunication

FUSION TECHNOLOGY 1988
A.M. Van Ingen, A. Nijsen-Vis, H.T. Klippel (editors)

IN-PILE BEHAVIOUR OF LITHIUM METASILICATE PELLETS IRRADIATED IN THE EXOTIC EXPERIMENTS 1, 2 AND 3.

Paul DIELS*, A. Jean FLIPOT*, Leonard SANNEN*, Gilbert VERSTAPPEN*, Rainer CONRAD** and Henk KWAST***

* Studiecentrum voor Kernenergie (CEN/SCK), Ceramic Research Department, Boeretang, 200, B-2400 Mol, Belgium
** C.E.C.-J.R.C. Petten Establishment, P.O. Box 2, ZG 1755 Petten, The Netherlands
*** E.C.N., P.O. Box 1, 1755 ZG Petten, The Netherlands

The irradiation experiments EXOTIC-1, 2 and 3 are part of the European Fusion Technology Programme on Blanket Technology and are carried out as a joint project by SL Springfields, SCK/CEN Mol and ECN Petten in collaboration with JRC Petten. In the present paper, the characteristics of the as-fabricated and of the irradiated pellets of the EXOTIC-3 experiment are described and compared to the results obtained in the two preceding experiments, with particular attention for the integrity of the irradiated pellets, their dimensional stability, vitrification effects and tritium retention.

1. INTRODUCTION

The experimental programme, EXOTIC[1], aims to investigate the tritium release of various candidate solid breeder materials, and their in-pile behaviour. The EXOTIC-1, 2 and 3 experiments, including the in-situ release measurements have been conducted in the High Flux Reactor (HFR) at Petten[2,3,4]. The fabrication route and the characteristics of the SCK/CEN samples have been given in detail in several papers[5,6,7].

2. EXPERIMENTAL

The SCK/CEN contribution to the EXOTIC-1 and 2 experiments have been communicated at the ICFRM-3 Conference held in Karlsruhe in October 1987[8]. The SCK/CEN contribution to the EXOTIC-3 experiment involves 4 vented capsules (He-0.1 % H_2) irradiated during 75 FPD in the HFR reactor in Petten. Two capsules irradiated at 600°C and 620°C, contain low-density (76 % T.D.) hyperstoichiometric pellets, the other two low-density (75 % T.D.) or high-density (87 % T.D.) hypostoichiometric pellets, irradiated at respectively 630°C and 600°C. Each capsule was loaded with a different lithium metasilicate pellet batch prepared from depleted lithium carbonate (^{6}Li = 0.55 at. %).

The low-density pellets exhibit a nearly completely open porosity and a submicron grain size. The high-density pellets have a 1.7 µm grain size and show an open porosity fraction of at least 70 %. The main characteristics of the as-fabricated pellets and the irradiation conditions are summarized in Table 1.

3. RESULTS

3.1. Pellet integrity

The EXOTIC-3 pellets show a marked tendency to fracture during irradiation : nearly 100 % of the low-density pellets and 60 % of the high-density pellets were found to be broken during unloading of the capsules. This phenomenon is believed to result from a combined effect : the EXOTIC-3 experiment has been subjected to rather severe thermal cycling during irradiation and the low-density pellets were sintered at too low a temperature for obtaining sufficient mechanical strength. It is therefore indicated to apply a fabrication process involving higher sintering temperatures[7]. It is interesting to note that pellets without thermocouple holes all break in two equal pieces whereas the pellets with ther-

TABLE 1
EXOTIC-3 : Material and irradiation data

Caps. No.	Batch No.	Composition	Density % T.D.	Grain size µm	Open porosity		Irrad. Temp. °C
					Fract.(%)	Mean Ø (µm)	
09.1	5	M + 0.2 % O + 0.2 % C	76.3	0.23	23.2	0.126	600
10.1	6	M + 0.2 % C	76.7	0.27	22.7	0.111	620
09.2	7	M + 5.6 % D	75.5	0.38	24.1	0.122	630
10.2	8	M + 6.6 % D	87.4	1.7	>8.8	<0.014	600

Irradiation time : 75 FPD; ^{6}Li : 0.55 at.%
M = lithium metasilicate; D = lithium disilicate
O = lithium orthosilicate; C = lithium carbonate

mocouple holes all break along these holes.

3.2. Dimensional stability

The low-density EXOTIC-3 pellets (capsules 09.1, 10.1 and 09.2) show only negligible swelling : after irradiation a decrease in density of only 0.16 % is measured which is in good agreement with the results of the EXOTIC-1 experiment[8], when taking into account the respective densities and irradiation temperatures. However, the high-density EXOTIC-3 pellets (capsule 10.2) exhibit a 2.0 % decrease in density which does not confirm the results of the EXOTIC-2 experiment[8]. The EXOTIC-2 pellets (95 % T.D.- 96 % T.D.) irradiated at 575°C and 630°C indeed revealed only limited swelling, the corresponding decrease in density being less than 0.4 %. The supposition made for EXOTIC-2 that the accommodation of tritium and helium might be responsible for swelling appears not to hold for the EXOTIC-3 experiment. However the EXOTIC-3 high-density pellets, unlike the EXOTIC-2 pellets, are hypostoichiometric, present an extremely fine porosity and have been irradiated 1.5 times as long.

3.3. Structural characteristics

The irradiated pellets, like the as-fabricated ones, were characterized by scanning electron microscopy (SEM), mercury porosimetry and X-ray diffraction analysis (XRD).

3.3.1. Grain size

Examination of fracture surfaces by SEM reveals the grain size and shape of all EXOTIC-3 pellets not to be affected under irradiation, confirming the results of the EXOTIC-1 and -2 experiments[8].

3.3.2. Porosity

The open porosity, as well as the mean open pore diameter of both the low-density and high-density EXOTIC-3 pellets are unaffected by the irradiation. Again, the hypostoichiometric EXOTIC-3 high-density pellets behave differently from the hyperstoichiometric high-density EXOTIC-2 pellets which indeed show a marked increase of the mean open pore diameter after irradiation.

3.3.3. Crystallographic structure

The XRD results obtained from EXOTIC-3 batch 5 (capsule 09.1) suggest some in-pile instability of orthosilicate : this compound is indeed not detected in the irradiated pellets. This confirms the results of the EXOTIC-2 experiment. The irradiated hyperstoichiometric pellets (batches 5 and 6) still contain some free lithium carbonate. However, in contrast with the as-fabricated pellets, the carbonate disappears after thermal treatment at 720°C under Ar atmosphere, indicating that the compound probably forms during the prolonged storage of the irradiated pellets under ambient air.

A slight decrease of the crystallinity of the metasilicate is observed for all 4 batches. A decrease of 22 % and 10 % is found after irradiation at temperatures of 600°C and 630°C respectively. This compound also shows some tendency to line broadening. These effects vanish after a 46 h thermal treatment at 800°C under argon atmosphere even when, as observed for EXOTIC-1 capsule 02.1 (505°C)[8], extensive vitrification of metasilicate had taken place.

On the other hand, as shown in Table 2, the intensities of the diffraction lines of the disilicate secondary phase (batches 7 and 8)

TABLE 2
Vitrification of disilicate

Batch No.	Treatment	$I_D(111)$ 10^3c/s
7	As-fabricated	1.10
	Irradiated at 630°C	0.22
	Irradiated at 630°C + 46 h 800°C-Ar	0.67
8	As-fabricated	1.18
	Irradiated at 600°C	0.10
	Irradiated at 600°C + 46 h 800°C-Ar	0.64

drops drastically, the effect being more pronounced after irradiation at 600°C. It is clear that this phenomenon results from vitrification and not from decomposition since appreciable restoration of the line intensity is obtained after a thermal treatment at 800°C. Since no vitrification of disilicate was observed for the hypostoichiometric EXOTIC-1 pellets (batch 3)[8] irradiated at 395°C and 505°C it can be stated that, in comparison with metasilicate, vitrification of disilicate occurs at higher temperature. Indeed, severe vitrification of metasilicate occurs during irradiation at 505°C[8].

3.4. Linear thermal expansion

The linear thermal expansion has been measured on both the as-fabricated and irradiated EXOTIC-3 pellets.

All batches exhibit a nearly identical linear thermal expansion, indicating that the presence of traces of orthosilicate or some disilicate have no influence upon this characteristic. It also shows that the linear thermal expansion coefficient is not affected by irradiation : the measurements on the as-fabricated and irradiated pellets give identical results.

3.5. Retained tritium

A new equipment for the determination of the total residual tritium content has been built and tested by the LHMA department at SCK/CEN. Tritium is released by heating the irradiated samples for 2 hours at 950°C to 1000°C and transported to a hot CuO bed by a humidified constant $Ar-H_2$ (90/10) gas stream of about 5 $l.h^{-1}$. The oxidized hydrogen and tritium are captured in a liquid nitrogen trap. The experimental set-up and working procedure allows to control the yield of both the oxidation and cold trapping processes and to determine quantitatively the amount of tritium which could have been retained in the equipment. The performance of the procedure is experimentally verified by measuring at regular intervals a gaseous standard mixture of tritium and hydrogen. The precision of the method was found to be 3 %. Complete extraction of tritium from the samples is checked by partial dissolution[8] of the analysed sample and measuring the tritium activity of the resulting solution. The yield of tritium outgassing by the hot extraction method is found to be better than 98 %. Tritium was measured by liquid scintillation counting.

The results of the determinations on the EXOTIC-3 as well as on some EXOTIC-1 and -2 pellets are summarized in Table III. The measurements have been corrected for decay to end of irradiation.

It appears that earlier determinations[8] performed on EXOTIC-1 and 2 pellets using high-vacuum extraction (HVE) and partial dissolution yielded considerably lower retained tritium values in comparison with the hot extraction method (HE), probably resulting from incomplete extraction and/or tritium losses. Consequently only the results obtained from the HE method will be used for further interpretation.

Mutual comparison of the EXOTIC-1 and -2 capsules shows tritium retention to be dependent upon the irradiation temperature suggesting tritium release to be governed by a diffusion-controlled mechanism. Tritium retention in the EXOTIC-2 pellets (95 % T.D.-

TABLE 3
EXOTIC-1, 2 and 3 : Results of tritium retention as obtained from hot extraction

Experiment	Capsule No.	Irrad. temp. °C	Batch No.	Density % T.D.	Grain size µm	Open porosity		Retained* tritium %
						Fraction %	Mean Ø µm	
EXOTIC-1	01.1	395	3	81.3	2	17.3	0.34	17.2
			4	81.8	5	15.2	0.13	13.6**
	02.1	505	2	80.8	0.7	15.3	0.30	14.5
			3	81.3	2	17.3	0.34	13.4
			4	81.8	5	15.2	0.13	10.7
EXOTIC-2	06.1	425	2c	96.0	11	3.4	0.07	76.7
	05.2	575	2c	96.0	11	3.4	0.07	40.0
	05.1	630	1c	95.2	13	4.1	0.16	31.6
EXOTIC-3	09.1	600	5	76.3	0.23	23.2	0.126	0.17
	10.1	620	6	76.7	0.27	22.7	0.111	0.12
	09.2	630	7	75.5	0.38	24.1	0.122	0.18
	10.2	600	8	87.4	1.7	>8.8	<0.014	0.72

* Decay corrected

** Measured by E.C.N. : hot extraction 800°C He-0.1 % H_2

96 % T.D.) is exceptionally high : even when irradiated at 630°C the retained tritium still amounts to 31.6 %. On the other hand, tritium retention does not seem to be significantly influenced by the grain size or the open pore diameter.

The EXOTIC-3 experiment exhibits a nearly complete tritium release for both the low-density and high-density (87.4 % T.D.) pellets, irrespective of differences in grain size, open pore diameter and composition.

With the exception of the EXOTIC-2 experiment the results appear to be in good agreement with those obtained by E.C.N.[4].

4. CONCLUSIONS

The fabrication of low-density lithium metasilicate pellets with convenient mechanical strength probably requires the application of an appropriate fabrication process[7].

Swelling might be governed by both the duration and temperature of the irradiation as well as by several material characteristics such as pellet density composition and open pore diameter.

Grain growth or changes in grain shape were not observed after irradiation. The open porosity also remained unchanged. On the other hand a rather marked increase of the open pore diameter was noticed for all EXOTIC-2 pellets (95 % T.D.- 96 % T.D.) and for batch 4 of the EXOTIC-1 experiment.

XRD analyses suggest a certain in-pile instability of orthosilicate.

Metasilicate and disilicate both vitrify at rather well-defined temperatures, vitrification of disilicate occurring at higher temperature. The question arises whether or not this phenomenon also occurs when these compounds are formed in the reactor as a result of prolonged orthosilicate irradiation. The restoration of the crystalline character by thermal treatment appears to be more difficult in the case of disilicate.

The irradiation temperature and pellet density are undoubtedly the predominant parameters governing tritium release. Material characteristics such as grain size, open porosity and composition are apparently of minor importance. High-density metasilicate pellets (95 % T.D. - 96 % T.D.) are not convenient as breeder material. Indeed, when irradiated at 630°C retained tritium still amounts to 31.6 %.

REFERENCES

1. H. Kwast, R. Conrad and J.D. Elen, J. Nucl. Mater., 133-134 (1985), 246

2. H. Kwast, R. Conrad et al., J. Nucl. Mater., 141-143 (1986), 300

3. H. Kwast, R. Conrad et al., J. Nucl. Mater., 154-156 (1988), in print

4. H. Kwast, A. Kout et al., International Symposium on Fusion Nuclear Technology, Tokyo, 10-15 April 1988. Fusion Engineering and Design, in print

5. A.J. Flipot P. Diels and R. Lecocq, J. Nucl. Mater., 133 and 134 (1985), 226

6. A.J. Flipot and P. Diels, J. Nucl. Mater., 141-143 (1986), 339

7. A.J. Flipot et al., Presented at the International Symposium on Fusion Nuclear Technology, Tokyo, 10-15 April, 1988, Fusion Engineering and Design, in print

8. P. Diels, A.J. Flipot et al., J. Nucl. Mater., 154-156 (1988), in print

FUSION TECHNOLOGY 1988
A.M. Van Ingen, A. Nijsen-Vis, H.T. Klippel (editors)

IN-SITU TRITIUM RELEASE FROM VARIOUS LITHIUM ZIRCONATES

H. Kwast(*), R. Conrad(**), L. Debarberis(**), P. Kennedy(***), A.J. Flipot(****) and J.D. Elen(*)

(*) Netherlands Energy Research Foundation, ECN, P.O. Box 1, 1755 ZG Petten, The Netherlands
(**) JRC-Petten, P.O. Box 2, 1755 ZG Petten, The Netherlands.
(***) Northern Research Laboratories, Springfields, U.K.
(****) SCK/CEN-Mol, Boeretang, Mol, Belgium.

In-situ tritium release experiments have been performed on samples of Li_2ZrO_3, $Li_6Zr_2O_7$ and Li_8ZrO_6 contained in vented capsules. The irradiation was carried out in the High Flux Reactor (HFR) at Petten during 97 full power days. The capsules were purged with helium +0.1 vol % H_2. During the irradiation several temperature transients were performed and the corresponding tritium release measured. Tritium residence times were determined using the diffusion and first order desorption model. A tritium residence time of 1 day was achieved at temperatures of approximately 560°C for Li_8ZrO_6, 400°C for Li_2ZrO_3 (85% TD), 350°C for Li_2ZrO_3 (80% TD) and << 350°C for $Li_6Zr_2O_7$. The grain size and OPV, which differed for the various materials, have a significant influence on the tritium residence time.

1. INTRODUCTION

Lithium-based zirconates are receiving increasing interest as tritium breeding material for fusion reactors, especially Li_2ZrO_3 and Li_8ZrO_6[1-9]. Although, irradiation data of these materials are limited it has been shown that the tritium diffusivity is larger than that of Li_2SiO_3 and $LiAlO_2$[6,8]. In particular, the Li_2ZrO_3 is thermally stable[10] whereas the swelling, the linear expansion and the tritium and helium retention is low [1,2,10]. Moreover, the chemical interaction of Li_2ZrO_3 with stainless steel between 500 and 800°C is weaker compared to Li_2O, Li_4SiO_4, Li_2SiO_3 and $LiAlO_2$[3]. Information on preparation and thermochemical properties are given in [9,11-16]. Because the lithium-zirconates appear to be attractive alternatives for Li_2O and $LiAlO_2$, various zirconates have been included in the experimental programme, EXOTIC, currently being carried out in the High Flux Reactor (HFR) at Petten. This programme includes in-situ tritium release experiments and annealing experiments after irradiation. Some preliminary results of Li_2ZrO_3 irradiated in EXOTIC-3 are given in[7]. In this paper results on in-situ tritium release, in particular tritium residence times, are given for Li_2ZrO_3, $Li_6Zr_2O_7$ and Li_8ZrO_6, irradiated in EXOTIC-4.

2. EXPERIMENTAL

2.1. Materials

The samples of EXOTIC-4 were supplied by SCK/CEN-Mol (Li_2ZrO_3) and NRL-Springfields (Li_2ZrO_3, $Li_6Zr_2O_7$ and Li_8ZrO_6). The SCK/CEN samples are annular pellets with outer and inner diameter of 14 mm and 4 mm, respectively. The Springfields samples have six different shapes, modelled as closely as possible to the shape required for the various post-irradiation measurements.

A systematic investigation has been carried out to fabricate various lithium-zirconates. From this it was found that Li_2ZrO_3, Li_8ZrO_6 and $Li_6Zr_2O_7$ are the stable phases and that the existance of Li_4ZrO_4 is doubtful. Sofar, the existance of $Li_6Zr_2O_7$ has not been recognized. The investigations, however, indicated that $Li_6Zr_2O_7$ was chemically stable by heating at temperatures in the range 600-1100 °C. Most likely the X-ray diffraction pattern given by Scholder[12] and Neubert[13] for Li_4ZrO_4 were based on incorrect data. In fact the JCPDS has been changed in 1984 into JCPDS 34-312 which corresponds to $Li_6Zr_2O_7$. Consequently, if the ortho-lithium zirconate exists, it is metastable.

Sofar, there are no material properties reported in the literature on $Li_6Zr_2O_7$. Some thermal, mechanical and gas flow properties have been measured at Springfields. However, some properties in the literature quoted for Li_4ZrO_4 could be valid for $Li_6Zr_2O_7$. Characteristics and material data of EXOTIC-4 samples are given in table 1. The microstructure of the materials can be observed in fig. 1.

2.2. Irradiation device

The irradiation was performed in a four channel rig containing eight independently controlled capsules, like EXOTIC experiments -1, -2 and -3. All eight capsules, of which five were loaded with the various Li-zirconates, were continuously purged. In-situ tritium release rates were measured by eight ionization chambers, one for each capsule. A detailed description of the irradiation facility is given in[18].

Table 1. Characteristics of EXOTIC-4 samples.

Material	Li_2ZrO_3	Li_2ZrO_3	Li_8ZrO_6	$Li_6Zr_2O_7$
Supplier	SCK/CEN	NRL	NRL	NRL
Capsule no.	14.1/14.2	16.1	15.1	15.2
Li-6 content, (%)	0.54	0.60	0.60	0.60
Li-density, (g/cm^3)	0.32	0.30	0.57	0.35
Density, (%TD)	85.5	79	82	80
(g/cm^3)	3.55	3.29	2.48	2.86
Open porosity, (%)	14.5	21	1	17
Pore size (μm)	0.14	0.3	*	0.3
Permeab.coeff., Bo, (cm^2*E-12)		1.1	*	0.4
Grain size, (μm)	1.0	0.4-1.0	0.5-1.0	0.3-0.5
Weight, (g)	59.4/58.9	34.6	25.7	29.9

* Unable to measure due to very small OPV.

2.3. Irradiation conditions

EXOTIC-4 was performed in a peripheral in-core position of the HFR during four reactor cycles, i.e. approximately 97 full power days. In the first cycle the temperatures have been kept at nominal values, being 400 °C for capsule 14.1 and 600 °C for the other capsules, in order to achieve steady state conditions. Temperature transients have been carried out in the second, third and fourth cycle. The nuclear conditions remained constant throughout the irradiation. The capsules were purged with dry helium, doped with 0.1 vol % hydrogen at a flow rate of 100 cm^3/min. The relevant process parameters were scanned every two minutes by a computerized data-logger. The irradiation conditions are given in table 2.

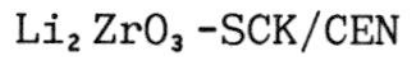

Li_2ZrO_3-SCK/CEN Li_2ZrO_3-NRL

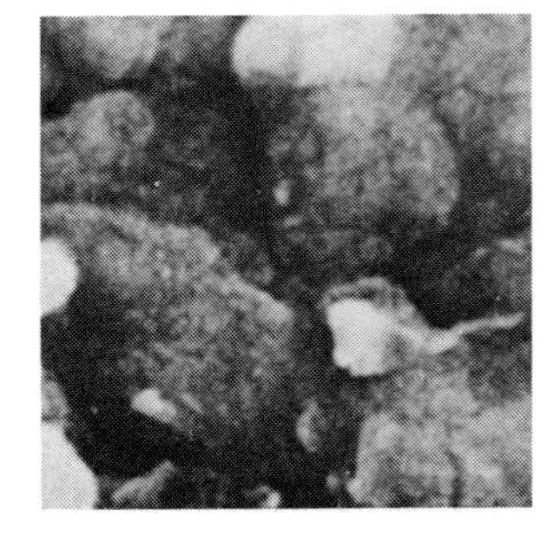

Li_8ZrO_6-NRL $Li_6Zr_2O_7$-NRL

1 μm

FIGURE 1
Microstructures of the various Li-zirconates.

Table 2. Irradiation data of EXOTIC-4.

Material	Li_2ZrO_3			Li_8ZrO_6	$Li_6Zr_2O_7$
Capsule no.	14.1	14.2	16.1	15.1	15.2
Temp. range, (°C)	350/430	480/650	380/680	450/650	470/650
Irr. time, (FPD)	97	97	97	97	97
Burn-up, (% Li)	0.13	0.13	0.13	0.14	0.15
Initial tritium release rate,					
- (μCi/min)	320	300	130	300	200
- (μCi/(min g))	5.4	5.1	3.8	11.7	6.7

3. RESULTS

In order to determine tritium release kinetics, temperature transients were performed from different temperatures levels and with various steps, both by increasing and decreasing the temperature. For each temperature step a positive or negative tritium release peak was observed. Examples are given in fig. 2. The temperature ranges and measured initial steady state tritium release rate are given in table 2. The steady state tritium release rates equal the calculated tritium production rates for all temperatures, except for Li_8ZrO_6 where steady state tritium release was not observed below 550°C.

The kinetic approach to steady state tritium release after a temperature step was analysed. The resulting characteristic time constants, τ, were determined by the following models:

a) intra-granular diffusion with constant internal tritium source and zero surface concentration.

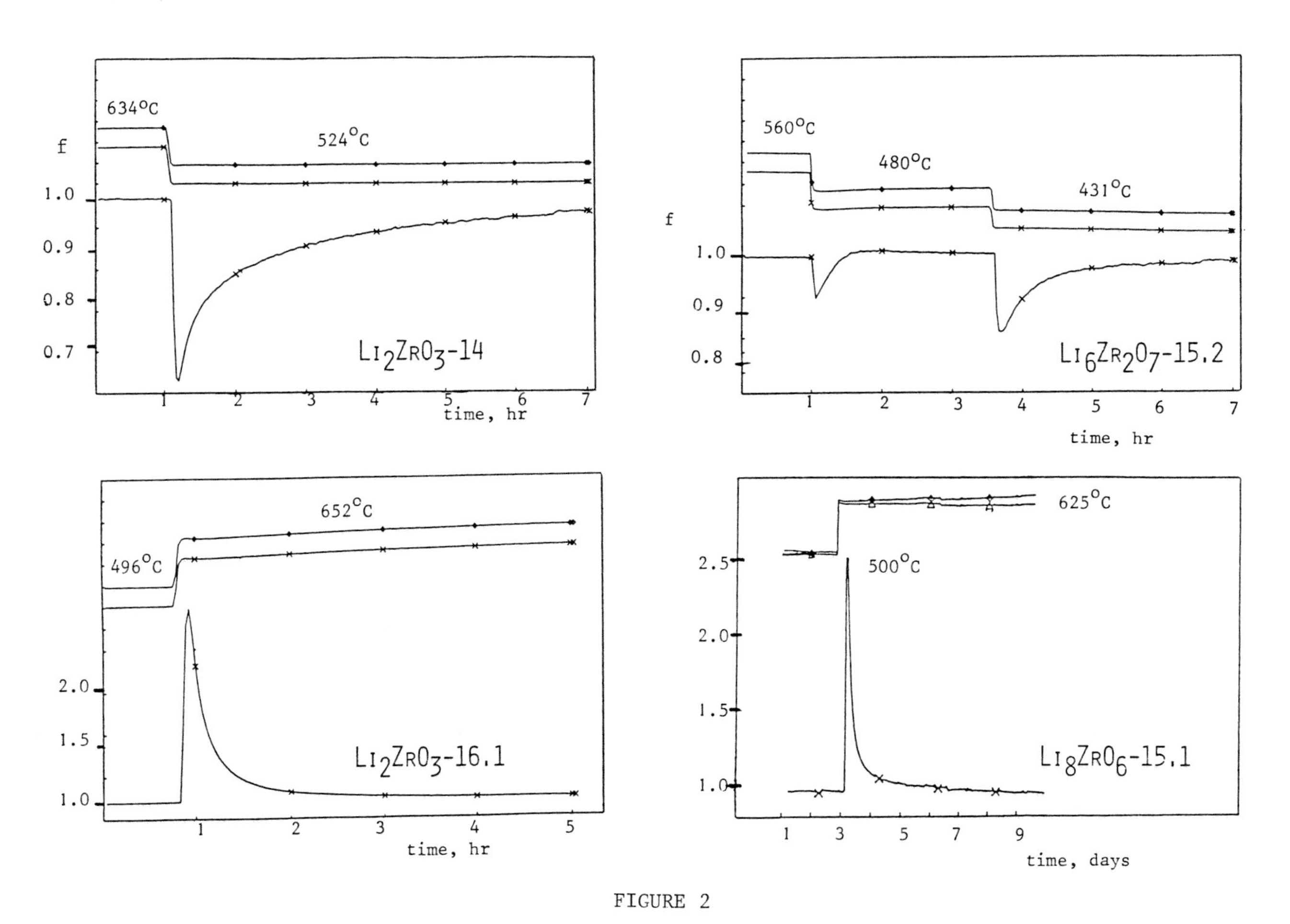

FIGURE 2

Examples of fractional tritium release, f, after temperature transients.

b) first order desorption with constant internal tritium source.

For both models the relative differential equations for a spherical geometry were analytically solved to predict the tritium release behaviour after an ideal temperature step. The solutions of both equations are given below:

$$\frac{R}{G} = 1 + \frac{6}{\pi^2}\left(\frac{D_2}{D_1} - 1\right)\sum_{n=1}^{\infty}\frac{1}{n^2}\exp\left(-n^2\frac{t}{\tau_{dif}}\right) \quad (1)$$

$$\frac{R}{G} = 1 + \left(\frac{K_2}{K_1} - 1\right)\exp\left(-\frac{t}{\tau_{des}}\right) \quad (2)$$

where,

R= tritium release rate, μCi/min

G= tritium generation rate, μCi/min

D_1 and D_2 = diffusion coefficients, cm^2/s

K_1 and K_2 = desorption coefficients, cm/s

t = time, min

τ = tritium residence time, min

and

$$\tau_{dif} = \frac{a^2}{\pi^2 D_2} \quad (3) \quad \text{and} \quad \tau_{des} = \frac{a}{3 K_1} \quad (4)$$

where,

a = grain radius, cm

τ_{dif} and D_2/D_1 were chosen as free parameters for model a) and τ_{des} and K_2/K_1 for model b). Good data fitting was obtained with the diffusion model at temperatures below 530°C. At temperatures > 530°C both models gave equal residence times. The results are given in table 3 and are shown in an Arrhenius-plot in fig. 3. The activation energies derived from these plots are given in table 4.

Table 3. Tritium residence times (τ) of Li-zirconates irradiated in EXOTIC-4, during temperature transients.

T** (°C)	Temp.*** step (°C)	τ (min)	τ (hrs)	T** (°C)	Temp.*** step, (°C)	τ (min)	τ (hrs)
Li_2ZrO_3-SCK/CEN (14.1/14.2)				Li_2ZrO_3-NRL (16.1)			
350	- 45	6262	104	369	-111	>>600	>>10
375	- 45	3624	60	376	- 17	960	16
425	+ 80	>300*	>5*	390	-180	342	5.7
470	-180	383	6.4	392	- 26	278	4.6
521	-180	117	2	418	- 15	101	1.7
524	-110	107	1.8	435	- 83	100	1.7
560	- 55	47	0.8	440	-120	24*	0.4*
588	- 82	>18*	>0.3*	518	- 55	9	0.15
620	- 50	26	0.43	538	-116	<13.5*	<0.2*
627	+106	20	0.33	570	-110	7.7	0.13
640	+116	15	0.26	652	+156	<10*	<0.17*
657	+140	9	0.16				
670	+ 30	8	0.14				
Li_8ZrO_6-NRL (15.1)				$Li_6Zr_2O_7$-NRL (15.2)			
560		1680	28	420	-170	18	0.3
560		2400	40	430	-115	25	0.4
580		756	13	430	-100	41	0.7
625		810	14	431	- 49	48	0.8
635		480	8	431	-177	26	0.4
				472	- 80	13	0.2
				480	- 81	11	0.2
				490	+ 60	<14*	<0.2*
				610	+190	13*	0.2*

* steady state not achieved or temperature not constant after the transient.

** at end of temperature transient.

*** (-) decrease and (+) increase in temperature.

4. DISCUSSION

The results given in fig. 3 indicate that, at least for the two lithium meta-zirconates under the conditions considered, tritium release is a singular activated process. However, the rate limiting process for tritium release at temperatures > 530°C is not clearly understood because of the equal tritium residence times obtained by the two models. The latter might indicate the existance of a transition region where both tritium diffusion and desorption are equally effective. However, it must be stressed that the material density, open porosity and grain size distribution as well as the existance of radial temperature gradients are not taken into account by the models. Moreover, at high temperatures the transients are much shorter and give less data points for curve fitting.

In some cases the tritium residence time could not be determined accurately because a steady state tritium release was either not achieved or the temperature was not constant after the transient. Residence times shorter than 10 minutes could not be discriminated from the time constant of the overall measuring system. The data given in fig. 3 show that the

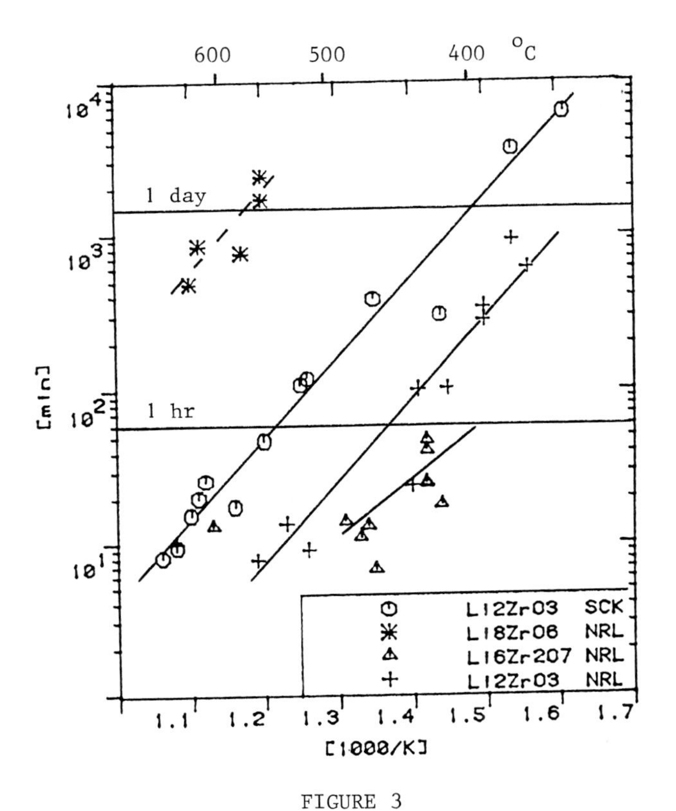

FIGURE 3

Arrhenius plot of tritium residence times, τ, in minutes.

order of tritium residence time of the various zirconates is:

τ ($Li_6Zr_2O_7$) < τ (Li_2ZrO_3-NRL) < τ (Li_2ZrO_3-SCK/CEN) < τ (Li_8ZrO_6).

4.1 Li_8ZrO_6 (caps. 15.1)

Very long times to reach steady state tritium release after a transient were observed. Most likely this can be explained by the absence of open porosity, see table 1. Below 550°C steady state tritium release was not achieved within 10 days. Most of these transients did not start from steady state release. Therefore the residence times have been estimated as 30% of the time to reach steady state for temperatures > 550°C. Just for comparison these data are also plotted in fig. 3.

4.2 Li6Zr207 (caps. 15.2)

A relatively high spread of values was observed in spite of very good data fitting. The samples of hexa-lithium zirconate show the shortest tritium residence times. Compared to Li_2ZrO_3 the density, pore size and OPV are almost equal. The average grain size,however, appears to be somewhat smaller in case of $Li_6Zr_2O_7$. This capsule was designed to operate at approximately 600°C. Therefore, a temperature lower than 420°C could not be achieved. At temperatures > 560°C the tritium residence time is < 10 min.

Table 4. Activation energies of Li_2ZrO_3 and $Li_6Zr_2O_7$ irradiated in EXOTIC-4

Material	Activation energy, kJ/mol
Li_2ZrO_3-SCK	100
Li_2ZrO_3-NRL	108
$Li_6Zr_2O_7$-NRL	~ 75

4.3 Li_2ZrO_3 (caps.14.1,14.2 and 16.1)

Fig. 3 shows that the Li_2ZrO_3-samples of SCK/CEN-Mol have longer residence times than those made by NRL-Springfields. This can be explained by the difference in characteristics of both materials, see table 1. The negative influence of increasing the material density on tritium release behaviour has been reported earlier[7]. The higher density of the Mol material consequently results in a lower fraction of open porosity. Moreover, the average grain size appears to be slightly larger compared to the Springfields material, as shown also in fig. 1. Both materials appear to have practically the same activation energy.

5. CONCLUSIONS

* The order of tritium residence times of the various Li-zirconates is τ ($Li_6Zr_2O_7$) < τ (Li_2ZrO_3-NRL) < τ (Li_2ZrO_3-SCK/CEN) < τ (Li_8ZrO_6).

* The temperature dependence of the tritium residence time can be described by an Arrhenius law. The tritium release from the lithium meta-zirconates appears to be a singular activated process.
* The derived activation energies are: 100 (Li_2ZrO_3-SCK/CEN), 108 (Li_2ZrO_3- NRL) and ~ 75 kJ/mol ($Li_6Zr_2O_7$).
* The tritium residence time is significantly affected by the material density, grain size and fraction of open porosity.
* With the diffusion model a good fit of the data was obtained at temperatures < 530°C. At temperatures > 530°C the diffusion and desorption models gave equal residence times.
* A tritium residence time of 1 day was observed at approximately 400°C for the 85% TD Li_2ZrO_3, 350°C for the 80% TD Li_2ZrO_3 and << 350°C for the $Li_6Zr_2O_7$.
* For Li_8ZrO_6 with almost no open porosity a tritium residence time of 1 day was obtained at about 560°C.

REFERENCES

1. G.W. Hollenberg, J. Nucl. Mat. 122/123 (1984) 896.
2. G.W. Hollenberg and D.L. Baldwin, J. Nucl. Mat. 133/134 (1985) 242.
3. P. Hofman and W. Dienst, J. Nucl. Mat. 141/143 (1986) 289.
4. D.L. Baldwin and G.W. Hollenberg, J. Nucl. Mat. 141/143 (1986) 305.
5. G.W. Hollenberg et al., J. Nucl. Mat. 141/143 (1986) 271.
6. H. Kudo, K. Okuno and O'Hira, J. Nucl. Mat. 154-156 (1988) in print.
7. H. Kwast, R. Conrad et al., J. Nucl. Mat. 154-156 (1988) in print.
8. K. Okuno and H. Kudo, Proc. Int. Symp. Fus. Nucl. Techn. Tokyo, Japan, April 10-15, 1988. To be published.
9. A.J. Flipot, E. Brauns and P. Diels, This volume.
10. C.N. Nilson and G.W. Hollenberg, DOE/ER -0113/2, p19, Oct. 1983.
11. P. Kennedy, Private Communication.
12. R. Scholder, D. Rade and H. Schwarz, Z. Anorg. Allg. Chem. 362 (1968) 149.
13. A. Neubert and D. Guggi, J. Chem. Thermodynamics 10(1978) 297.
14. M.S. Ortman and E.M. Larsen, J. Am. Cer. Soc. 66 (1983) C142.
15. G.W. Hollenberg and D.E. Baker, DOE/ER -0113/3, p4, May, (1984).
16. H.R. Ihle and C.H. Wu, J. Nucl. Mat. 130 (1985) 454.
17. P. Kennedy, A.J. Flipot, R. Conrad and H. Kwast, EXOTIC Annual Progress Report 1986, NRL-R-2016 (S).
18. H. Kwast, R. Conrad, J.D. Elen, J. Nucl. Mat. 133/134 (1985),246.

FUSION TECHNOLOGY 1988
A.M. Van Ingen, A. Nijsen-Vis, H.T. Klippel (editors)
Elsevier Science Publishers B.V., 1989

PREPARATION, PHASE RELATIONSHIPS AND FIRST IRRADIATION RESULTS OF LITHIUM ORTHOSILICATE DOPED WITH AL^{3+}- AND P^{5+}-IONS

Alfred SKOKAN, Dieter VOLLATH, Horst WEDEMEYER, Elmar GÜNTHER, Heinrich WERLE

Association KfK-EURATOM
Kernforschungszentrum Karlsruhe GmbH, P.O.Box 3640, D-7500 Karlsruhe, Federal Republic of Germany

A method for fabrication of lithium orthosilicate doped with Al^{3+} an P^{5+} in methanol is described. In addition the phase diagrams of the phases in question are explained. A reinvestigation of the subsystem Li_2O - Li_2SiO_3 - $LiAlO_2$ showed solubilities of Al^{3+} ions in Li_4SiO_4 which differs from literature data. Irradiation experiments showed that, solid solutions of the series Li_4SiO_2 - $LiAlSiO_4$ have a better tritium release than pure orthosilicate. The other specimens, especially the P^{5+} doped ones revealed higher tritium retention.

1. INTRODUCTION

Within the European Fusion Program the lithium aluminate ($LiAlO_2$) and the lithium orthosilicate (Li_4SiO_4) are mainly discussed as ceramic blanket materials for forthcoming fusion reactors. Recovery and release of the bred tritium are of special interest concerning the breeding rate. Tritium release is determined primarily by two processes:

- Bulk diffusion within the grains
- Adsorption-desorption on the surfaces.

For lithium orthosilicate it is assumed that surface phenomena are rate controlling /1/.

It is the aim of this paper to show whether it is possible to influence tritium release of lithium orthosilicate by doping with Al^{3+} and P^{5+} ions as the Li^+ conductivity can be substantially enhanced by addition of Al^{3+} or P^{5+} ions /2, 3/. If for tritium transport in the solid a diffusion mechanism is assumed similar to that for Li^+ the measures contributing to increase the ionic conductivity should also improve tritium transport.

Independently of a possible improvement of tritium release it is likewise of interest, with a view to application of the doped orthosilicates as ionic conductors, whether also these modified materials can be prepared by the methanol process described earlier /4, 5, 6/.

2. PREPARATION OF DOPED LITHIUM ORTHOSILICATE

Monophase lithium orthosilicate with good sintering properties will be obtained if amorphous SiO_2 is dispersed in methanol and the suspension obtained is allowed to react with LiOH. This reaction takes place in boiling methanol under reflux conditions. A milky white colored suspension is obtained. Subsequently, the methanol can be distilled off this suspension while water is added. An aqueous LiOH solution is formed which contains an organic lithium-silicon phase as a fine suspension. To obtain powders the alcohol-free suspension can be spray-dried. After spray drying a powder is available which reacts to form lithium orthosilicate at temperatures of 250 °C and above /4, 6/. This method can be used for the production in large quantities.

For the purpose of doping adequate quantities of freshly precipitated $Al(OH)_3$ prepared by hydrolysis of aluminium iso-propoxide and ammonium dihydrogen phosphate where added to the suspension after completion of the silica-LiOH-methanol reaction. In the case of phosphorous doping ammonia was distilled off together with methanol /6/.

3. PHASE RELATIONSHIPS

3.1. System Li_2O - Al_2O_3 - SiO_2

Jackowska and West /2/ described a limited solubility for aluminium in the Li_4SiO_4 phase /2/. This solubility tends by exchange of

$$Si^{4+} \leftrightarrows Al^{3+} + Li^{+} \quad (1)$$

towards Li_5AlO_4 formation thus generating solid solutions of the type

$Li_{4+x}Si_{1-x}Al_xO_4$ with $0 \leq x \leq 0.4$ /2/.

On the other hand, exchange of

$$3\ Li^{+} \leftrightarrows Al^{3+} \quad (2)$$

is possible towards $LiAlSiO_4$ formation. Then the solid solutions have the composition

$Li_{4-3y}Al_ySiO_4$ with $0 \leq y \leq 0.6$ /2/.

Because of the importance of this system, the subsystem Li_2O - Li_2SiO_3 - $LiAlO_2$ was reinvestigated with respect to ternary solutions of Li_4SiO_4. In accordance to published results two series of solid solutions were identified:

- between Li_4SiO_4 and Li_5AlO_4 following the exchange reaction (1) with a maximum amount of additional lithium of $x = 0.06$.
- between Li_4SiO_4 and $LiAlSiO_4$ following the exchange reaction (2). In this case, the maximum fraction of aluminium is $y = 0.15$.

In addition on the side of $LiAlO_2$ a range of homogeneity was found on the join $LiAlO_2$ - Li_4SiO_4 following the formula

$Li_{1+z}Al_{1-z}Si_{0,5z}O_4$ with $0 \leq z \leq 0.03$.

This solid solution is formed by the exchange of

$$2\ Al^{3+} \leftrightarrows 2\ Li^{+} + Si^{4+} \quad (3)$$

A revised isothermal section at 800 °C of the ternary subsystem Li_2O - Li_2SiO_3 - $LiAlO_2$ is given in fig. 1.

As the irradiation experiments showed a most favourable behavior of the solid solution accordingly to the exchange reaction (2) further investigations where done. From these

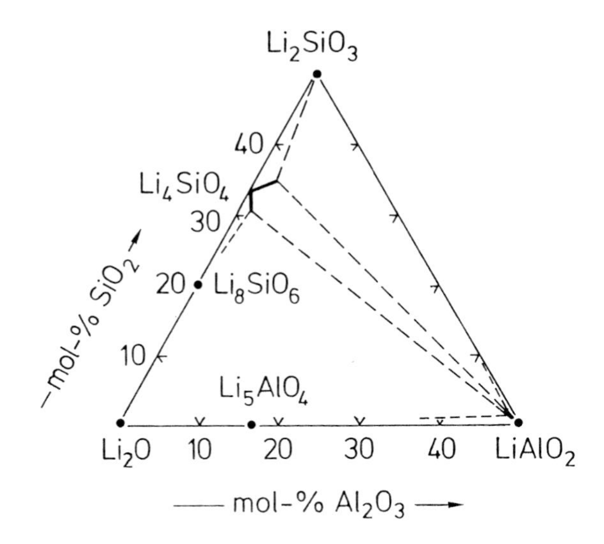

FIGURE 1
Phasediagram of the subsytem Li_2O-Li_2SiO_3-$LiAlO_2$

experiments two important results came out:

- The addition of aluminium decreases the melting point from 1255 °C for Li_4SiO_4 to 1150 °C for $Li_{3.55}Al_{0.15}SiO_4$.
- The phase transformation at 663 °C for Li_4SiO_4 is suppressed by addition of $LiAlSiO_4$.

3.2. System Li_4SiO_4 - Li_3PO_4

This system was studied in detail by Khorrasani et al /3/. In the Li_4SiO_4 - Li_3PO_4 quasi-binary system the structure of the orthosilicate is stable up to about 12 mole% Li_3PO_4. From about 57 mole% Li_3PO_4 the structure of the orthophosphate is obtained /3/. In the range of the orthosilicate and orthophosphate the following substitution takes place

$$Li^{+} + Si^{4+} \leftrightarrows P^{5+} \quad (4)$$

Accordingly, the solid solutions can be described by the formula

$Li_{4-x}Si_{1-x}P_xO_4$ with $0 \leq x \leq 0.12$

on the side of the orthosilicate and by

$Li_{3+y}P_{1-y}Si_yO_4$ with $0 \leq y \leq 0.43$

on the side of the orthophosphate. The maximum Li^{+} conductivity occurs in the range of the two-phase structure /7/. Therefore, we have prepared specimens with 10, 30 and 70 mole% Li_3PO_4.

4. TRITIUM RELEASE

4.1. Specimens and irradiation conditions

The specimens were prepared according to the processes described in chapter 2. The powder obtained, was calcined at 700 °C for 5 hours and sintered for 16 hours at 1100 °C in air. For pelletizing a pressure of 30 kN/cm² was used. The dimensional characteristics of the samples used are summarized in table 1.

TABLE 1: Summary on the properties of the pellets used

Specimen	Length (mm)	Diameter (mm)	Density (g/cm^3)	No.
$Li_{4.05}Si_{0.95}Al_{0.05}O_4$ mono phase	5.33	7.49	2.04	1
	5.64	7.49	2.01	2
	5.18	7.49	2.06	3
$Li_{3.7}Al_{0.1}SiO_4$ mono phase	5.41	7.63	1.82	4
	5.61	7.67	1.81	5
	5.50	7.66	1.81	6
$Li_{3.9}Si_{0.9}P_{0.1}O_4$ mono phase	5.35	7.54	2.05	7
	5.33	7.54	2.06	8
	5.35	7.54	2.05	9
two phase materials	5.40	7.50	1.97	10
	5.40	7.51	1.92	11
	5.41	7.50	1.92	12
$Li_{3.3}P_{0.7}Si_{0.3}O_4$	5.03	6.97	2.19	13
	5.11	6.92	2.19	14
	5.03	6.95	2.20	15
Li_4SiO_4 mono phase as reference	6.67	6.68	2.18	16
	6.65	6.69	2.17	17
	6.71	6.68	2.17	18

The specimens were loaded under helium in aluminium capsules for irradiation.

Irradiation was performed in the FR2-reactor at KFA Jülich. The temperature during irradiation was estimated to be below 100 °C. The time for irradiation was selected to give a specific tritium activity between 100 and 200 µCi/g. The large difference is caused by flux gradients within the irradiation channel.

4.2. Tritium release

Tritium release was measured in the low-activity laboratory facility made of stainless steel. Tritium water is reduced with a zinc-bed (heated to 390 °C) to avoid problems with tritium adsorption. The samples were purged with 25 STP-cm³/min He + 0.1 vol% H_2 (total impurities ≤ 1 vpm), cleaned additionally by Oxisorb (Messer-Griesheim). The water content of the purger gas at the inlet of the sample chamber was measured and found to be constant and below 1 vpm.

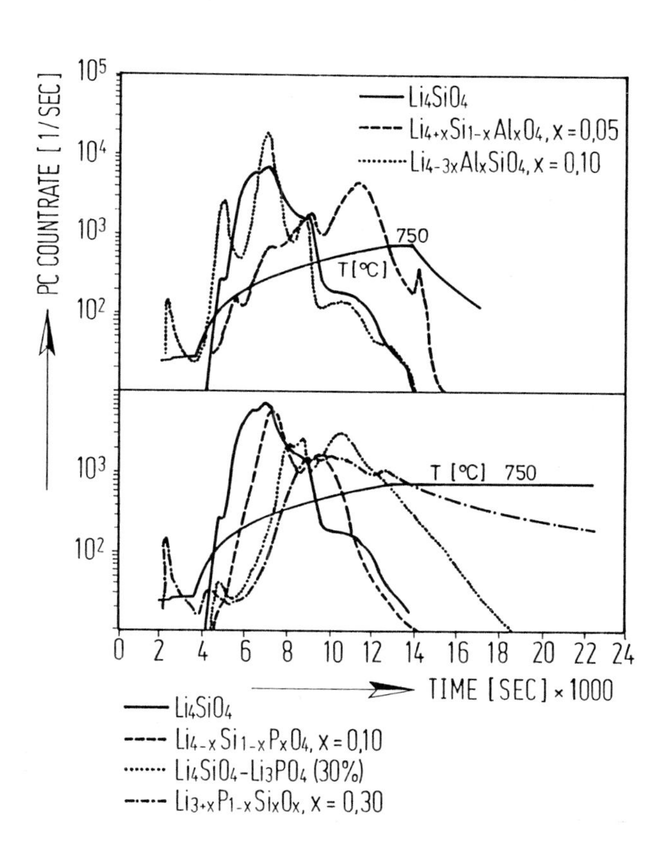

FIGURE 2
Tritium release of doped lithium orthosilicate specimens in case of continuously heated specimens (heating rate 4.8 °C/min)

For first comparison of the materials the specimens were continuously heated and the tritium release determined. The results are shown in fig.2.

If the temperature is increased linearly for diffusion-controlled release the activity should continuously increase to a maximum and then continuously decrease. For all samples, but especially for the fastest (Li_4SiO_4, $Li_{4-3x}Al_xSiO_4$) the observed release is more complex. A tentative explanation is that the release of the fast samples is mainly determined by desorption, whereas that of the slow samples is additionally controlled by bulk diffusion. It is interesting to note, that the fastest of the doped samples ($Li_{4-3x}Al_xSiO_4$, $Li_{4-x}Si_{1-x}P_xO_4$) are characterized by a deficiency of lithium. From these experiments it can be concluded, that tritium release of phosphorous doped specimens are in any case slower than those doped with aluminium. Therefore we performed for the aluminium doped specimens additional fractional release experiments. From the stepwise temperature increase tests (fig. 3) it is obvious that tritium release with $Li_{4-3x}Al_xSiO_4$ is faster, whereas that with $Li_{4+x}Si_{1-x}Al_xO_4$ is slower than that of pure lithium orthosilicate Li_4SiO_4. This is confirmed by the fractional release at the end of each temperature step, given in table 2.

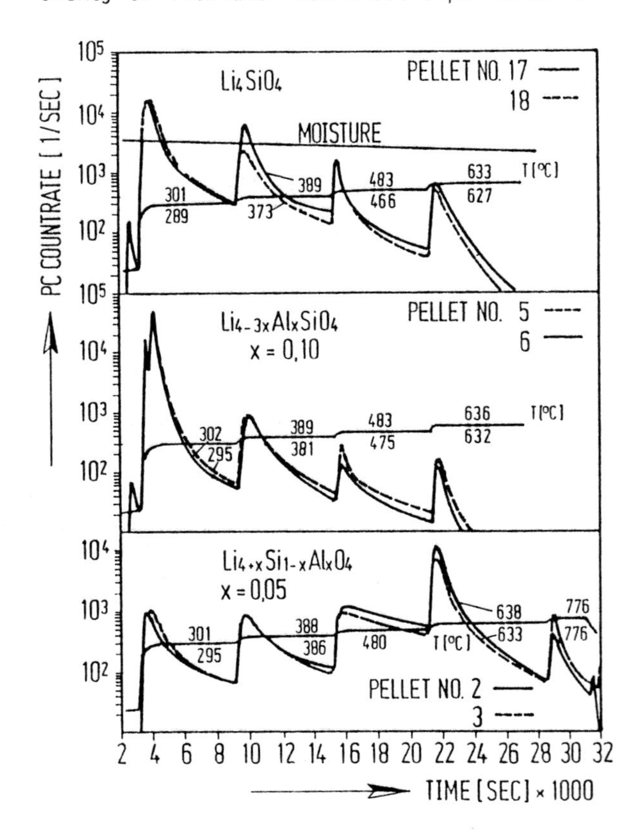

FIGURE 3
Tritium release of Al^{3+} doped lithium orthosilicate specimens during stepwise heating

TABLE 2: Fractional release at the end of each temperature step (in %)

Pellet No.	Temperature (in °C) 300	390	480	630	780
Li_4SiO_4					
17	62	90	96	100	100
18	80	93	98	100	100
$Li_{4+x}Si_{1-x}Al_xO_4$					
2	7	17	41	95	~98
3	11	23	46	95	~98
$Li_{4-3x}Al_xSiO_4$					
5	91	98	99	100	100
6	93	99	100	100	100

The reproducibility was checked by performing the same stepwise temperature increase with two different pellets and was found to be very good (fig. 3). Therefore the differences between the various types of samples can be attributed to different sample characteristics.

5. CONCLUSION

Summarizing the results presented within this paper we can conclude, that production of lithium orthosilicate doped with aluminium or

TABLE 3: Influence of doping on the tritium release as compared to lithium orthosilicate

Material	Release compared to Li_4SiO_4
$L_{4+x}Si_{1-x}Al_xO_4$ x = 0.05	much slower
$Li_{4-3x}Al_xSiO_4$ x = 0.1	faster
$Li_{4-x}Si_{1-x}P_xO_4$	slower
two phase material	much slower
$Li_{3+x}P_{1-x}Si_xO_4$	very much slower

phosphorous is not problematic if the "methanol-process" is used. Doping influences tritium release of these materials as summarized in table 3.

From table 3 we conclude that doping has an advantage only in the case of $Li_{4-3x}Al_xSiO_4$. In this case we have additional advantage of suppressing the phase transformation at 665 ^oC. As tritium release is improved especially in the case of low temperatures the use of this type of material could have advantages as compared with other lithium compounds. For final conclusion further irradiation experiments are neccessary.

ACKNOWLEDGEMENT

This work has been performed in the framework of the Nuclear Fusion Project of the Kernforschungszentrum Karlsruhe and is supported by the European Communities within the European Fusion Technology Program.

REFERENCES

1. H. Werle, J.J. Abassin, M. Briec, R.G. Clemmer, H. Elbel, H.E. Häfner, M. Masson, P. Sciers, H. Wedemeyer, J. Nucl. Mat. 141-143 (1986) 321-326.

2. K. Jackowska, A.R. West, J. Mater. Science 18 (1983) 2380-2384.

3. A. Khorrasani, G. Izquierdo, A.R. West, Mat. Res. Bull. 16 (1981) 1561-1567.

4. H. Wedemeyer D. Vollath, E. Günther, German Patent Application (1987).

5. D. Vollath, H. Wedemeyer, J. Nucl. Mat. 141-143 (1986) 334-338.

6. D. Vollath, H. Wedemeyer, Advances in Ceramics, Vol. 25, in the press.

7. Y.-W. Hu, I.D. Raistrick, R.A. Huggins, J. Electrochem. Soc. 124 (1977) 1240-1242.

FUSION TECHNOLOGY 1988
A.M. Van Ingen, A. Nijsen-Vis, H.T. Klippel (editors)
Elsevier Science Publishers B.V., 1989

DETERMINATION OF MECHANICAL CHARACTERISTICS OF γ-$LiAlO_2$ and Li_2ZrO_3 MATERIALS FOR A FUSION REACTOR BLANKET

Bernard RASNEUR

COMMISSARIAT A L'ENERGIE ATOMIQUE - CEN/SACLAY - IRDI/DESICP/DLPC/SPCM
91191 - GIF/SUR/YVETTE CEDEX (France)

Mechanical properties such as Young's modulus, ultimate compressive strength and bending strength are compared for porous materials $LiAlO_2$ and Li_2ZrO_3; the values found are similar.

FOREWORD

Data concerning the mechanical properties of the lithiated materials composing a solid breeding blanket for a fusion reactor are required to design the blanket if one wishes to retain insofar as possible the integrity of the material during reactor operation. The elastic modulus and the compressive and bending strength values are measured and compared for the material γ-$LiAlO_2$ and Li_2ZrO_3.

1. INTRODUCTION

In the present blanket designs, lithiated materials will not be subjected to large compressive stresses due to samples arrangement in their container (approximately 1 MPa maximum) but to major compressive and tensile stresses due to thermal gradients generated during reactor operation.

Contrarily to metals, ceramics generally exhibit bending or tensile strength not equal to compressive strength, which is often high, but 3 to 12 times lower. This is why the integrity of the lithiated ceramic is especially dependent on the bending or tensile strength characteristics.

The ultimate bending and tensile strengths being very similar, the bending strength is determined preferably since it is easier to measure and because bending test specimens are far easier to prepare.

The maximum acceptable tensile stress on the periphery of a cylindrical sample, for a given uniform volumetric heating, is given by the formula [1]:

$$\sigma = \frac{\alpha E}{1-\nu} \quad \frac{q''' \, \emptyset^2}{32 \, k}$$

where:

$\emptyset$ = sample diameter
α = thermal expansion coefficient
E = Young's modulus
q''' = uniform volumetric heating
ν = Poisson's ratio
k = thermal conductivity

or again:

$$\emptyset = \sqrt{\frac{\sigma \, (1-\nu) \, 32 \, k}{\alpha \, E \, q'''}}$$

The diameter of the sample can be larger for a larger σ value and a smaller E value.

σ and E values have been measured for porous γ-$LiAlO_2$ and Li_2ZrO_3 materials.

2. SAMPLE PREPARATION

The methods used here generally lead to porous materials of uniform grain and pore sizes as shown for Li_2ZrO_3 in Fig.1. This uniformity results in improved mechanical properties and a smaller scatter of mechanical values; in addition, using uniform grain size materials are the ideal case when one calculates tritium diffusion coefficients from in-pile and out-of-pile tritium release measurements.

FIGURE 1
Microstructure by S.E.M for porous Li_2ZrO_3.

2.1. γ-$LiAlO_2$ preparation

Various porous microstructures were obtained using various reactants and various reaction conditions, as shown below.

a) *0.2 µm diameter grains.*

γ-Al_2O_3 powder* + Li_2CO_3 powder
mixing, 700°C, 3 h, decomposition -->
α + γ-$LiAlO_2$ powder (stirring; 900°C, 1 h; CIP, 100 MPa) -->
sintering 1000°C, 1 h.

b) *0.4 µm diameter grains.*

γ-Al_2O_3 powder* + Li_2CO_3 powder
mixing; 700°C, 3 h, decomposition -->
α + γ-$LiAlO_2$ powder stirring; CIP, 200 MPa
-->sintering, 1100°C, 3 h.

c) *5 µm diameter grains.*

γ-Al_2O_3 powder* + Li_2CO_3 powder
mixing; 700°C, 3 h, decomposition -->
α + γ-$LiAlO_2$ powder stirring; CIP, 200 MPa
--> sintering, 1300°C, 3 h .

d) *3 µm diameter grains.*

γ-Al_2O_3 powder + Li_2CO_3 powder
mixing; 700°C, 3 h, decomposition -->
α + γ-$LiAlO_2$ powder stirring; CIP, 200 MPa
--> sintering, 1100°C, 3 h .

e) *13 µm diameter grains.*

γ-Al_2O_3 powder**+Li_2CO_3 powder
mixing; 700°C, 3 h, decomposition -->
α+$\gamma$$LiAlO_2$powder
stirring; CIP, 200 MPa -->
sintering, 1100°C, 3 h.

2.2. Li_2ZrO_3 preparation

A method similar to the method a) of preparation of γ $LiAlO_2$ was used. It allowed obtaining very uniform grains, which were relatively small (see Figure 1) given the greater sinterability of Li_2ZrO_3 compared to the $LiAlO_2$ one. The method is summarized as follows:

ZrO_2 powder + Li_2CO_3 powder
mixing, decomposition 700°C, 3 h (two times)
-->Li_2ZrO_3 powder, stirring
+ heating at 800°C for 3 h,
stirring, CIP at 200 or 400 MPa,
sintering 1000°C for 3 h.

This procedure could be simplified for industrial fabrication.

As for γ-$LiAlO_2$, the elementary grain size can be varied by slightly varying the quantity of lithium around the stoichiometry. Some of the results obtained are summarized in Table I.

* With a slight excess (2 to 5%) of Al_2O_3 with regard to the stoichiometry.

** With a slight deficit (2 to 5%) of Al_2O_3 with regard to the stoichiometry.

TABLE I

	Pressing MPa	Porosity	Ø grain µm	Ultra sound velocity ms^{-1}	Young's modulus GPa
Stoichiometric proportions	100	0.38		3600	33
	200	0.32	1.85	4000	45
	400	0.25	1.85	5000	78
With 5% deficit in lithium	200	0.20	1	4900	80
	400	0.17	1	5000	86

3. MECHANICAL PROPERTIES

3.1. Ultrasound velocity and Young's modulus

Ultrasound velocity

Ultrasound (1 MHz) velocity C was measured using the CNS PUNDIT equipment on porous samples of Li_2ZrO_3 of different porosities, obtained by varying the isostatic pressure and the sintering temperature during the preparation steps. The porosity range covered was 0.20 to 0.40.

From about 20 experimental measurements, the following equation was found for the variation of C versus porosity:

$C = C_o\ (1 - \frac{9}{7}\varepsilon)$ where $C_o = 7000\ ms^{-1}$

ε = porosity

It is recalled that for $LiAlO_2$,[2], the velocity C is not dependent on the elementary grain size; a wide pore size distribution, cracks or microcracks, as well as a lack of grain cohesion greatly reduce the value of C for a given porosity.

Static Young's modulus

The Young's modulus E was calculated from the measured ultrasound velocity by the approximate formula:

$E \simeq \rho c^2$ where ρ = apparent density

$\rho = \rho_o\ (1-\varepsilon)$ ρ_o = theoretical density

ε = porosity

For Li_2ZrO_3, the following formula was found:

$$E = \rho_o c_o^2\ (1-\varepsilon)\ (1 - \frac{9}{7}\varepsilon)^2$$

where $E_o = \rho_o c_o^2 = 203.35$ GPa

For γ-$LiAlO_2$, the following formula was found[2]

$$E = \rho_o c_o^2\ (1-\varepsilon)\ (1 - \frac{10}{7}\varepsilon)^2$$

where $E_o = \rho_o c_o^2 = 184.5$ GPa.

Table II summarizes the values calculated for c and E as a function of porosity .

TABLE II

	$LiAlO_2$		Li_2ZrO_3	
ε	c (m/s)	E (GPa)	c (m/s)	E (GPa)
0	8400*	184.5*	7000*	203.35*
0.05	7800	151	6550*	169*
0.1	7200	122	6100*	139*
0.15	6600	96.8	5650*	113*
0.20	6000	75.3	5200	90
0.25	5400	57.2	4750	70
0.30	4800	42.2	4300	54
0.35	4200	30	3850	40
0.40	3600	20.3	3400	29
0.45	3000	12.9	2950*	20*
0.50	2400	7.5	2500*	13*
0.55	1800*	3.8*		
0.60	1200*	1.5*		
0.65	600*	0.3*		
0.70	0*	0*		

* extrapolated value

3.2. Ultimate compressive strength

The ultimate compressive strength was measured on relatively small samples (dia. 5 mm, height 5 mm) which can be easily irradiated and examined after neutron irradiation. A fairly comprehensive study [2] was already made for $\gamma LiAlO_2$. Table III lists, for comparison, values of mechanical characteristics for microstructures of $\gamma LiAlO_2$ and Li_2ZrO_3 probably promising for tritium release.

TABLE III

	$LiAlO_2$		Li_2ZrO_3
Ø grain (μm)	0.5	0.7	0.8
porosity	0.24	0.19	0.20
Velocity (ms^{-1})	5520	6120	5200
Young's modulus (GPa)	61	79	90
Ultimate compressive strength (MPa)	257	357	396
Ultimate bending strength (MPa)	77	70	60

3.3. Ultimate bending strength (3 points)

The ultimate bending strength was measured on samples of cross-section 5 x 5 mm^2. The values, listed in Table III, show no great difference between the two lithiated ceramics γ-$LiAlO_2$ and Li_2ZrO_3.

4. CONCLUSION

Simple fabrication routes have led to fine and homogeneous porous $\gamma LiAlO_2$ and Li_2ZrO_3 materials.

From the mechanical characteristics measured so far i.e. Young's modulus, compressive and bending strengths, and for the microstructures investigated, $\gamma LiAlO_2$ and Li_2ZrO_3 appear to behave similarly from the standpoint of strength. Therefore, it is unlikely that mechanical strength is a determining criterion for an eventual choice between these two breeding ceramics.

REFERENCES

1. J.P. Blanchard and N.M. Ghoniem, UCLA.ENG 8606, PPG 932, February 1986

2. B. Rasneur, Fusion Technology, July 1985, Vol. 8, p. 1909

FUSION TECHNOLOGY 1988
A.M. Van Ingen, A. Nijsen-Vis, H.T. Klippel (editors)

THERMAL DIFFUSIVITY MEASUREMENTS WITH A PHOTOTHERMAL METHOD OF FUSION SOLID BREEDERS MATERIALS

M.Bertolotti,L.Fabbri,A.Ferrari,C.Sibilia,G.Suber(*),C.Alvani(+),S.Casadio(+)

Dipartimento di Energetica, Univ. di Roma, Italy
(*) Fondazione Bordoni, Roma, Italy
(+) ENEA, Casaccia, Roma, Italy

The Photothermal Deflection method is employed in thermal diffusivity measurements. A theoretical analysis is performed to reduce the influence of arbitrary parameters. Measurements on gamma-lithium alluminate samples as a function of temperature are performed.

1. INTRODUCTION

Recently the Photothermal Deflection method[2,3] has been developed for measuring thermal diffusivity of solid samples. This method is derived from the PDS technique introduced by W.B. Jackson et al[1] and consists, in its transverse configuration, of heating the sample by absorption of laser modulated source (CO_2) and detecting the deflection that the temperature field causes in a probe laser beam travelling in the air layer near the sample surface (fig.1).

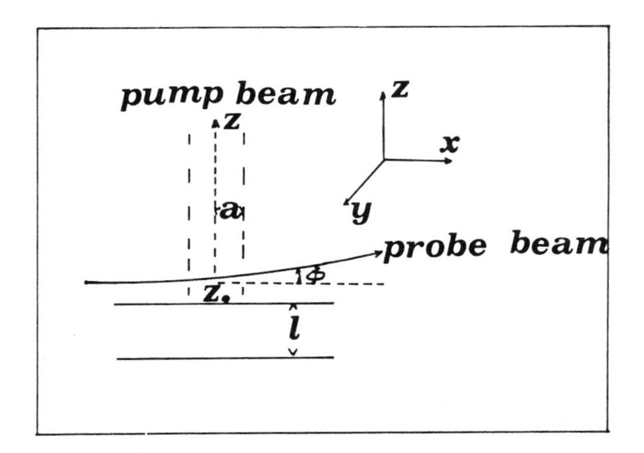

FIGURE 1
Probe beam travelling in the air is deflected by an angle Φ

Generally the following relation between the deflection angle Φ of the probe beam and the thermal field T holds [1] :

$$\Phi = 1/n\ (dn/dT) \int_{path} (\nabla T)\cdot \bar{u}\ ds \qquad (1)$$

where n and dn/dT are the refractive index and its thermal gradient respectively and $\bar{u}$ is the orthogonal unit versor to the beam path.

From equation (1) it can be seen that in order to characterize Φ the distribution of thermal field is required. Therefore, the heat transfer equations with their boundary conditions must be solved. This suggests the employment of the photothermic signal for extimating the thermal distribution in an air layer near the sample surface heated by laser source. Given a sample of known thermal optical and geometrical parameters, deflection angle Φ can be derived if one correct thermal distribution is known.

Our interest is turned to the inverse problem which consists in the determination of one particular sample parameter: the thermal diffusivity χ , but in this case all other parameters must be known.

However, for materials for which

all other parameters are unknown it is possible to avoid explicit solution of (1) by introducing an approximate expression for the deflection angle Φ . For this purpose measurements of Φ are been done as a function of the displacement y (offset) between the center of the pump beam and the probe beam.

The first step of such approximation consists in writing the deflection angle in the more general form :

$$\Phi = A(y) \exp\left[i(\omega t - \Theta(y))\right] \quad (2)$$

where A is the deflection angle amplitude and Θ the phase shift with respect to the reference modulated signal characteristic of the heating beam. This parameter, linked to hysteresis phenomenas, is influenced by several parameters; however by assuming a point-like heating source, phase shift becomes[3]:

$$\Theta(y) = -y/l_T + f(z) \quad (3)$$

where l_T is the thermal length : $l_T = (2\chi/\omega)^{1/2}$ and ω the modulation frequency of the heating beam; z is the vertical offset viz. the mean distance between probe beam and sample surface, of which f is an arbitrary function depending on several variabiles such as the thermal parameters of the sample and of the contiguous medium.

For y values for which the phase signal becomes linear, the l_T thermal length can be evaluated from the gradient of the curve, and from l_T the thermal diffusivity is derived once the angular frequency ω of the input signal is known.

2. EXPERIMENTAL SET-UP

The experimental set-up is outlined in fig. 2. It can be seen that the sample lies within an isolated system (furnace) provided with side windows which allow the two laser beams to interact with the sample.

This furnace is designed to mantain uniform sample temperature in the range 290-900 K within a 1% error.

The heating and probe beams are respectively a CO_2 (10600 nm) and a He-Ne (642.8 nm) lasers.

The detector is a Si-photodiode with its sensitive surface half shut; this detector is applied as " position detector" by alijning the probe beam so that only half of the intensity enters when no heating occurs ($\Phi = 0$), and a beam displacement is seen by the detector as an intensity change of the beam itself. The detection sistem is connected to a lock-in amplifier.

The photodiode signal ΔV depends on the deflection angle Φ through the following relation[1] :

$$\Delta V/V = (8\pi)^{1/2} (\pi w n/\lambda) \Phi$$

where V is the DC level, w and λ are respectively the spot-size and the wavelength of the probe beam.

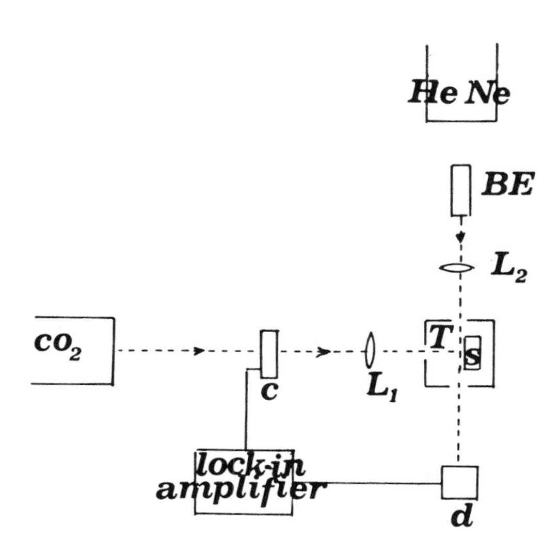

FIGURE 2
Experimental set-up

3. MEASUREMENTS OF THERMAL DIFFUSIVITY

The relationship of thermal diffusivity to temperature for a set of gamma-lithium alluminate samples has been studied.

The samples under study are divided in two classes 066 and 067 which are distinguish by different preparation methods, and within each class there are samples of different porosity.

The thermal diffusivity values are obtained from measurements of signal phase as explained in the preceding paragraph and the results are shown in fig.3,4. The porosity is expressed in term of the total density TD.

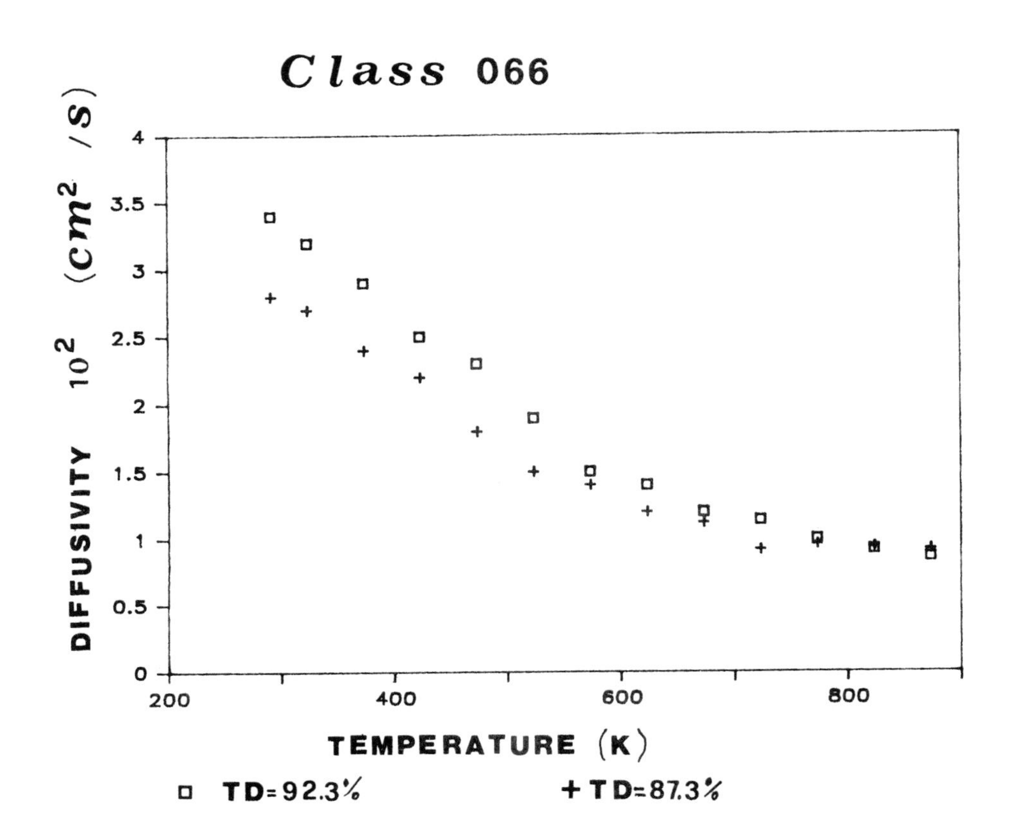

FIGURE 3

Thermal diffusivity measurements as a function of temperature for 066-samples with Total Density TD = 92.3 % and TD = 87.3 %.
The difference of the values of thermal diffusivity between the two sample is linked to the particular conduction mechanism of heat in these samples. The samples with TD = 87.3 have a room temperature value of thermal diffusivity lower than the sample with TD = 92.3. At higher temperature, the heat conduction inside the pores is not negligible and the diffusivity of the sample with low porosity decreases more slowly.

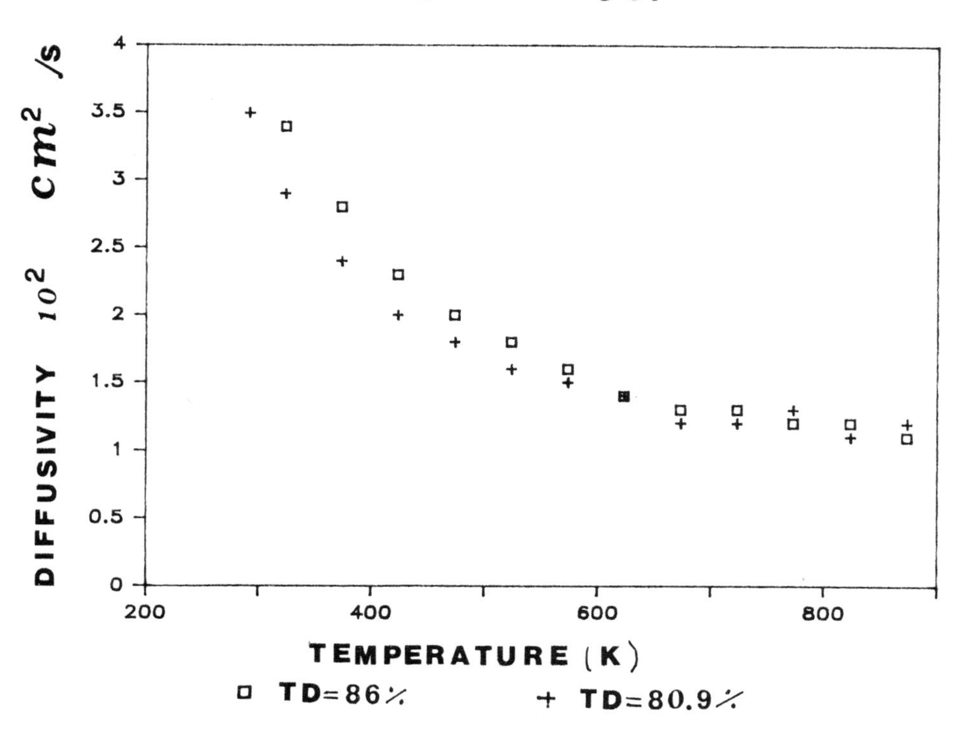

FIGURE 4

Thermal diffusivity measurements as a function of temperature for 067-samples with Total Density TD = 86 % and TD = 80.9 %.

4. CONCLUSION

The Photothermal Deflecion method was applied to thermal diffusivity measurements of solid materials. Measurements in gamma-lithium aluminate samples are in agreement with those reported in the literature[4].

It is interesting to compare our method to the most employed one for thermal diffusivity measurements: the "flash method"[5]. This last method has been employed for over 20 years during which has undergone many improvements concerning the formulation of a more "realistic" theory where actual measurement conditions were taken in account. For example, the theory of the "flash method" applies for a Dirac pulse and for samples absorbing all the optical radiation at their surface, so effects due to finite duration of the laser source must be considered. On the contrary, neither finite duration of the laser pulse nor optical trasmission through the sample obviously affect our method.

REFERENCES

1 W.B.Jackson, N.Amer et al. Applied Optics 20 (1981) 1333
2 G.Suber, M.Bertolotti et al. J. of Ther. Anal. 32 (1987) 1039
3 M.Bertolotti, L.Fabbri et al. J. of Phys.:Appl. Phys. D in press
4 B.Schultz, K.K. GmbH (1965) 2
5 A. Degiovanni, Rev. Gen. Technique 185 (1977) 470

FUSION TECHNOLOGY 1988
A.M. Van Ingen, A. Nijsen-Vis, H.T. Klippel (editors)
Elsevier Science Publishers B.V., 1989

IN SITU TRITIUM RECOVERY FROM $LiAlO_2$ PELLETS: TEQUILA-I

C. Alvani, S. Casadio, V. Violante, M. Briec*, M. Masson*

ENEA, C.R.E. Casaccia, Via Anguillarese, 301 - 00060 Roma, Italy;
* CEA, CEN Grenoble, 85 X 38041 Grenoble Cedex, France.

TEQUILA-I is the first phase of an in situ tritium release experiment performed in MELUSINE reactor, at CEN Grenoble, with almost the same facilities and irradiation conditions as used for the LISA series experiments. Three couples of $LiAlO_2$ specimens with the same density (80% of TD) but quite different microstructures (grain size) were tested in six vented capsules by thermal step cycling in the temperature range 400°-700°C. The reference He + 0.1% H_2 purging gas was used, but an oxidizing mixture containing moisture, of composition He + O_2 (37 vpm) + H_2O (115 vpm), was also successfully tested to extract the tritium from the ceramic breeders. The best tritium releasing performance was achieved by the P-type ceramic breeder, with the smallest grain size (0.3 um), for which a Tritium retention time less than one day could be measured at 450°C.

1. INTRODUCTION

Since the first miniaturized solid breeder assembly was tested by Clemmer [1], lithium aluminate has been under extensive worldwide study as one of the candidate tritiogenic materials ("breeders") for fusion reactors. However, the tritium recovery performance has not yet been well defined for all the possible variables of both the material microstructure (fabrication routes) and the in-pile testing system.

In the frame of European programme for Fusion Technology two fabrication processes were selected to obtain gamma-lithium aluminate[2] with the physico-chemical characteristics [3] which we presumed to be a reasonable compromise among the different required properties for the ceramic Breeder In poloidal Tube (BIT) of the blanket concept we referred to in ref.4.

This work deals with the first phase (I) of the in situ Tritium recovery Experiment for the QUality evaluation of Italian Lithium Aluminate (TEQUILA) ceramic breeder materials, which was scheduled for the last irradiation cycle (in June 1988) of the 30 year old MELUSINE thermal neutron experimental reactor (CEA, CEN Grenoble, France).

The main objective of TEQUILA-I was to obtain a preliminary evaluation of the material texture (microstructure) role of the tritium release properties of our specimens under conditions comparable to those of similar experiments (i.e., LISA series) performed on other ceramic breeder candidate materials. For most of these tests hydrogen was added to the helium purging gas, with the 0.1% H_2 level (in volume) actually considered as the "reference" mixture. In fact the presence of H_2 was generally found to be very effective in reducing the tritium inventory in the ceramic breeder samples under thermal neutron irradiation[1,5,6]. On the contrary, the presence of oxygen in the purging gas was found to inhibit the tritium recovery[1,5].

Some experiments in the ENEA Laboratories showed that below 600°C, hydrogen does not interact with a $LiAlO_2$ surface[7] and does not affect (such as oxygen and moisture) its ionic conductivity characteristics [3,8], while water vapor undergoes remarkable adsorption[7].

Hence, the "beneficial" role of H_2 is probably limited to "swamping" the tritium by an isotopic exchange/dilution mechanism, while the presence of oxygen, which promotes the formation of tritium oxides, holds these species on the breeder surfaces by depressing the tritium release rate.

If this hypothesis is true, the addition of water to an oxidant purging gas should also "swamp" the adsorbed Tritium oxides by isotopic exchange/dilution and by reducing the available adsorption sites for the tritiated species.

The feasibility of tritium extraction in these conditions was a further aim of TEQUILA-I, since there is strong interest in testing the tritium extraction from a ceramic breeder in an oxidizing environment to minimize (i) the tritium permeation through the walls of the gas circuitry in the BIT blankets and (ii) the isotopic dilution in the fuel reprocessing plant.

2. EXPERIMENTAL SETUP

2.1 Irradiation device and purging gas lines

The heating of the specimens under irradiation was performed in a CHOUCA furnace in which six separately vented capsules were placed at two levels. At the high level, capsule numbers 1,3 and 5 could be heated so that the specimens contained reached the required temperature (within +1°C) in the 500-700°C range. At the low level (capsules N° 2,4,6), the temperature was adjustable between 350° and 550°C.

The in-pile housing N° 80 of the MELUSINE reactor was used to irradiate the CHOUCA furnace, and the thermal neutron flux was in-line monitored by 4 collectrons placed at the aluminum box corners.

The main features of the purging gas line have been already reported[5,6]. Just down-stream of the specimen holders (after 1.45 m of heated line at 250°C, in the reactor pool), Zn-reducing reactors operated at 380-390°C to insure the full transformation of the tritium species in uncondensable reduced HT molecules. This is a very important peculiarity of the experimental assembly because it is possible to monitor the tritium produced in situ in the oxidized forms using the ionization chambers placed at several meters down-stream of the cold SS line (out of pile), thus avoiding any tritium loss due to adsorbtion phenomena.

Two purging gases were used: (i) the "reference" (R) was He + 0.1% H_2 in which the moisture level was held below 1 vpm by the cold trap; and (ii) a mixture (OW) of composition He + 37 vpm O_2 + 115 vpm H_2O by-passing the cold trap except in run 2.7 where the "dried" gas (O) was tested. In any case, the pressure on the irradiation zone was 0.15-0.16 MPa, and the flow about 2.8 l/h at 22°C and 0.1035 MPa at the ionization counter level.

2.2 Ceramic breeder targets

Three couples of three different gamma-$LiAlO_2$ specimens were inserted in the six vented capsules of the TEQUILA experimental device, following the scheme of Tab.1, where the main characteristics are also reported. The specimens were fabricated in the form of cylindrical rods (pellets of dia. 10 mm, length 30 mm) with a 2-mm-thick disk with a central hole (dia. 1 mm, depth 10 mm) to house a "K" type thermocouple was laid on top.

TABLE 1
Gamma-$LiAlO_2$ ceramic specimen identification and characteristics.

Capsule N°	1*	2	3*	4	5*	6
Specimen N°	1	2	3	4	5	6
Breeder type	P		A2		A1	
$^6Li/^7Li$ ratio	0.083		0.058		0.060	
Grain Size (um)	0.3+0.1		12+7		bimodal: 5+1 (70%) 0.2 (30%)	
Surf. Area (m^2/g)	2.6+0.2		0.2		1.06+0.2	
Density (% of TD)	80		80		80	

(*): high level position in the CHOUCA furnace

3. TEST MATRIX

Since there were marked differences in the texture characteristics of the materials under test, the tritium release transitories under thermal step variations were found to have very different time constants (see Fig.1). Because of the preliminary nature of TEQUILA-I, the experimental plan (Tab.2) was devoted mainly to characterizing the best performing tritium release breeder (the P-type) by sometimes neglecting the other specimens (A1 and mainly A2 types) placed on the same levels and for which a steady state was often not attained when the successive temperature variations (involving all the specimens of that level) were imposed.

The average thermal neutron flux was about 0.82 10^{13} n/cm^2s in the first runs, but a mean increase of 20% was detected during run 1.6, close to the value of about 0.98 10^{13} attained during the second irradiation cycle (runs 2.1 - 2.8)

4. RESULTS AND DISCUSSION

4.1 Experiments with reference purging gas

In Fig.1-a, the fractional (actual over steady-state) tritium releases (F) are plotted versus time for the 8 runs of cycle 1 for the high level specimens. The P-type ceramic breeder was found to return quite quickly to steady state (Fig.1-a,b), in much less time than that observed for the other breeder types, A1 and A2 in order, under the same conditions.

In Tab.2 reports the tritium extracted by the purging gas (recovered) with the inventory variations as measured at each run and for each specimen , the algebraic sum of this couple of values being the generated tritium.

The isothermal extractions of run 2.9 were performed out of pile to get the residual tritium content.

Tab.3 gives the tritium inventories (I, after taking into account the mass balance over all the runs) as "normalized" at the tritium production rate values (P) reported in the last row. The tritium retention times (I/P) resulting from these values are reported on the Arrhenius plot of Fig.2.

TABLE 2

TEQUILA-I test matrix and Tritium recovered +/- Tritium inventory variation (= Tritium produced) in the specimens (mCi) in each run.

RUN N°	Date	Start time	Temp. °C high-low level	Purg. gas	Specimen N° 1	2	3	4	5	6
1.1	31/5	9.14	700-500	R	7.3+0.6		6.5+1.1		5.6+0.6	
1.2		13.02	650-500	R	39.0+0.3	41.7+7.7	32.1+4.9	10.0+28.0	28.5+1.8	15.9+17.8
1.3	01/6	8.08	700-550	R	49.2+0.1	54.6-3.0	50.5-3.3	32.2+7.9	39.5-1.5	30.4+5.2
1.4	02/6	8.29	600-550	R	114.4+2.7		84.4+27.8		79.3+10.9	
	02/6	8.50	600-450	R		180.5+35.7				
1.5	04/6	17.41	550-450	R						
1.6	06/6	8.43	455 on 2	R	125.0+5.2	20.9+23.8	61.9+62.9	25.3+178	78.0+22.4	81.1+99.0
	06/6	12.44	400 on 2	R						
	07/6	9.41			n-flux variation (+20%)					
				R	57.4+1.8	32.2+28.3	29.8+23.0	7.4+36.1	35.4+5.5	22.3+19.1
1.7	08/6	8.25	600-500	R	129.1-6.3				104.5-19.9	
1.8	10/6	8.15	680 on 1,5	R	67.2-1.7	259.4-66.7	214-46.5	122+16.8	68.2-23.2	138-6.6
1.9	11/6	9.44	650-550	R	30.9+.05	163-16.9	164-35.1	202-96.6	97.3+6.2	132-32
	13/6	17.02			reactor shut-down					
2.1	16/6	13.20	650-550	R	69.6-0.65	72+1.4	60.8+0.8	52+0.2	48.3-0.8	50+1.1
2.2	17/6	15.35	650-550	OW	102+2.75	106+1.8	99.9-7.1	81.5-3.1	79.6-5.8	77.7-3.4
2.3	19/6	8.49	600-500	OW	119+2.4	115+10.4	88.5+18.9	28.2+62.5	80.8+4.6	58.4+27.5
2.4	21/6	8.48	550-450	OW						
2.5	22/6	8.23	+5°C on 5 +10°C on 2							
	22/6	8.55	600-500	OW	76.8-5.3	92-18.2	88.9-25.6	29.5+24	57.2-6.9	56.1-5.5
2.6	23/6	12.41	650-550	OW	137-3.6	154-16	145-27	204-105	101-6.9	126-32
2.7	25/6	17.28	650-550	O						
	25/6	19.20	650-550	OW	159-1.2	164-1	143-3.4	123-4.4	111-0.15	108+3.5
2.8	28/6	8.15	650-550	R	69.6-5.8	67.7-3.3	58-3.1	48.6-3	47.1-2.8	45-4.4
	29/6	9.23			end of irradiation					
2.9	29/6	10.55	550	R						
	30/6	8.11	600	R						
		18.05	end of heating		1.43	7.42	9.1	80.5	1.8	19.3
	1/7	8.35	650	R						
		12.40	670	R						
		16.00	end of eating		0.00	1.31	5.72	30.51	1.1	12.0
	4/7	8.55	670	R						
		16.03	end of eating		0.00	0.40	3.56	21.0	0.39	4.2

Preliminary considerations:

(i) The tritium extraction is characterized by retention times more or less proportional to the particle radius of the specimens. This is compatible with an extraction rate controlled more by a surface desorption mechanism than by pure diffusion in the grains[5,6].

(ii) For the P type specimens, a residence time of less than one day was measured at 450°C; this material is comparable to the best $LiAlO_2$ "breeders"[1,5,6].

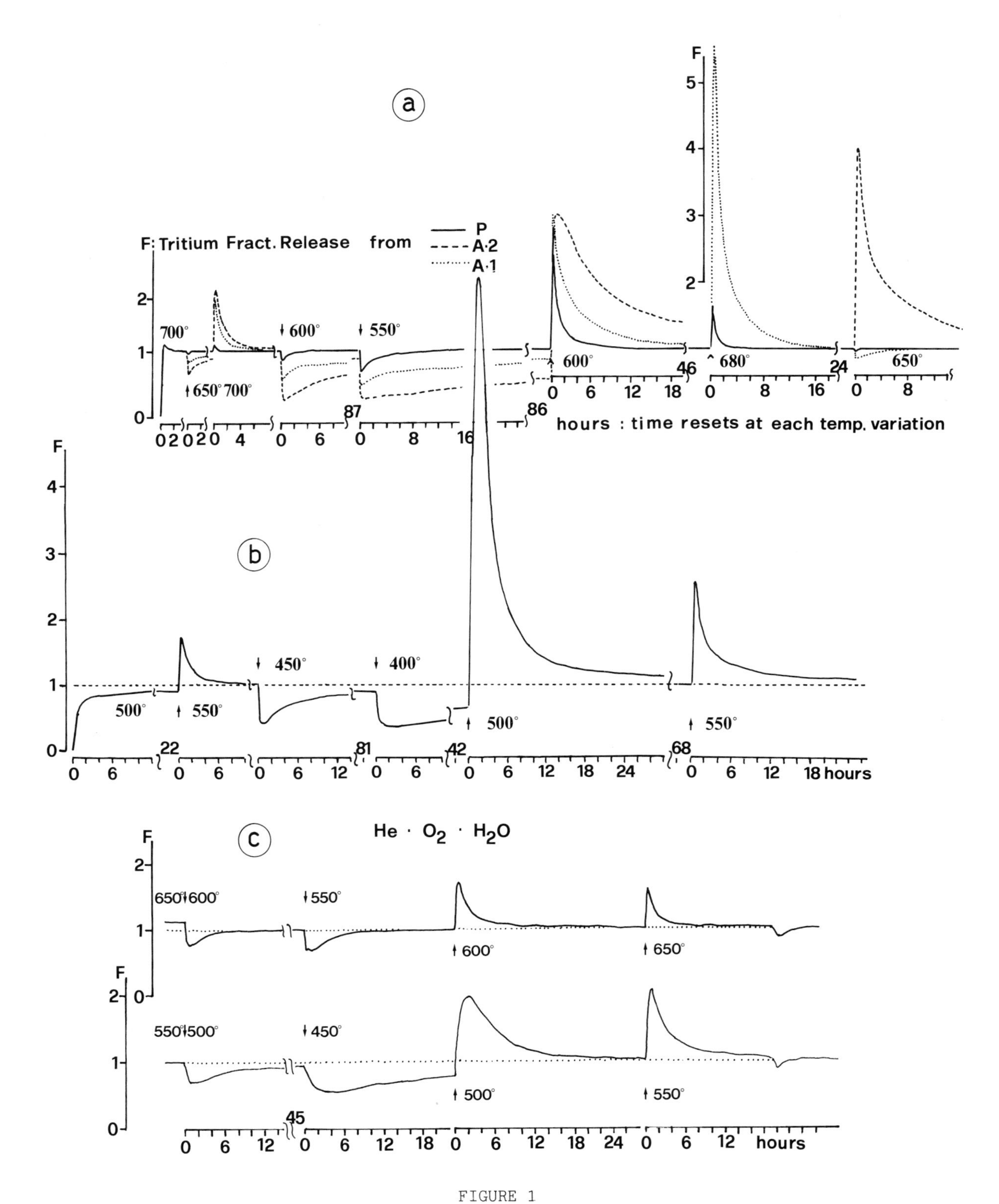

FIGURE 1

Tritium fractional release: (a) first runs at high level; (b) at low level for the P-type breeder; (c) at both levels for the P-type breeder in the moist purging gas (OW).

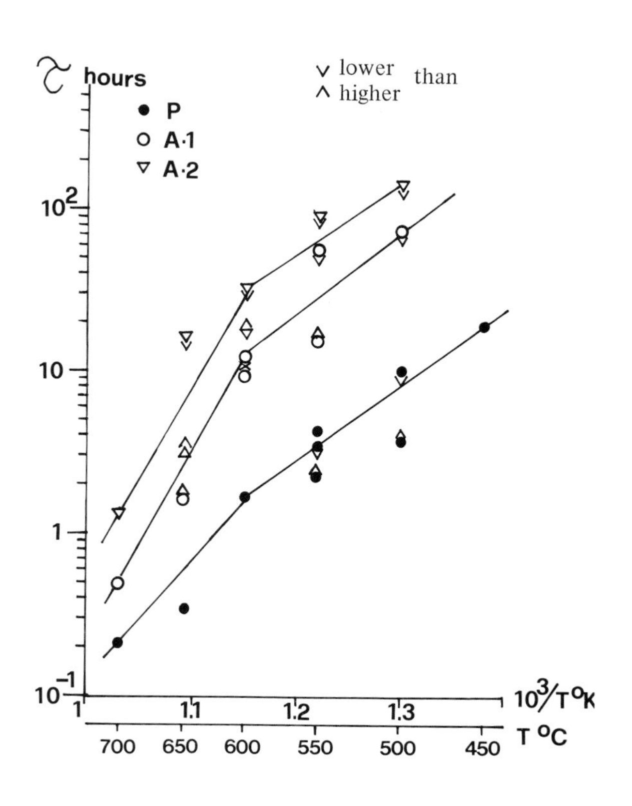

FIGURE 2

Tritium retention times vs. temperature

TABLE 3

Tritium inventories (mCi) measured on the $LiAlO_2$ ceramic specimens in TEQUILA-I

Breeder type	P		A2		A1	
Specimen N°	1	2	3	4	5	6
700°C:	0.54		2.9		0.87	
650°C:	0.88		6.8-38		2.8	
600°C:	4.2		39-73		16-21	
550°C:	10.7	4.7-9		41-170		28-102
500°C:		9.4-26				
450°C:		49.4				
Tritium prod. rate (mCi/h):	2.56	2.62	2.29	1.93	1.77	1.97

4.2 Moisture in the purging gas

Starting from run 1.9, the temperatures of 650° and 550°C were considered as references for the high and low levels.

During the purging gas variation (R vs OW, run 2.2), all the specimens except N°4 (large grains) were releasing tritium under steady state. Fig.3 shows the typical evolution of the tritium concentrations as they were recorded, including the effects of the cold trap in- and out-line (run 2.7, OW versus O) and the return

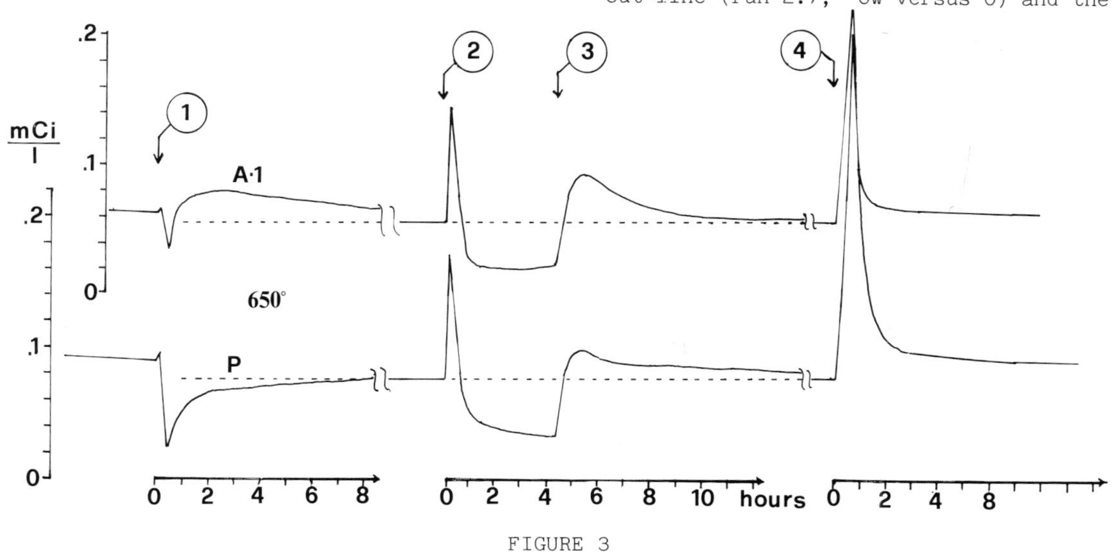

FIGURE 3

Tritium concentration in the purging gas: (1) entrance of the moist gas (R vs OW); (2) cold trap in (OW vs O); (3) cold trap out (O vs OW); (4) return to the reference gas (OW vs R).

to the "dry" reference purging gas (OW versus R, run 2.8).

The OW insertion was found to be beneficial for the A1 and A2 breeder types, while for the P type, some increase in the inventories, corresponding to 1 or 2 hours in the retention times, was observed, which is not trivial for this material considering its good performance with the R purging gas at these temperatures.

The moisture trapping in the transition OW versus O confirms the expected role of water as a "carrier" of the adsorbed tritium.

The reverse transition to the reference gas (run 2.8) was characterized by fast peaks of doubtful interpretation, so the relative inventory variations reported in Tab.2 should be considered as conservative values.

Nevertheless, the tritium release properties under thermal step variation in the moist purging gas OW (runs 2.3 to 2.6, Fig.1-c) were found to be quite similar to those measured with the reference purging gas R.

In spite of the preliminary nature of this experiment, it is demonstrated that the tritium extraction from ceramic breeder materials is feasible in oxidizing environments, using a low moisture concentration (100 vpm) in the He purging gas instead of hydrogen at a higher concentration (1000 vpm).

5. MODELLING

The boundary value problem for the tritium transport in the TEQUILA geometric assembly was numerically solved by taking into account the "local" tritium (i) generation rate, (ii) in grain diffusion (D), (iii) desorption from grain surface, (iv) in-pore diffusion, and (v) convection in the purging gas. By means of a two-parameter variance reduction between the experimental and the simulated fractional release evolution, the desorption mass transfer coefficient h and the diffusivity D could be simultaneously evaluated. The analysis was performed on the P-type breeder results.

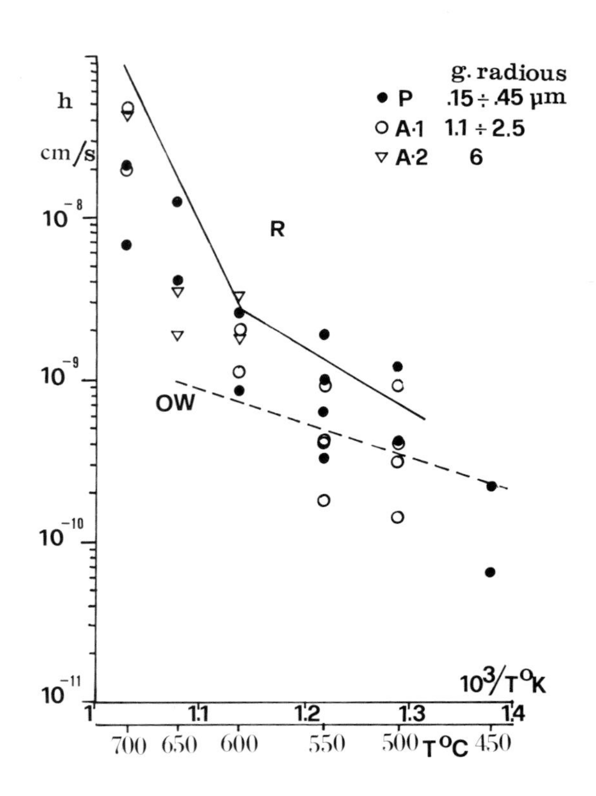

FIGURE 4
Tritium desorption mass transfer coefficients vs. temperature, as evaluated by assuming pure desorption control (points for all the specimes) and mixed diffusion-desorption control for the P-type breeder in R and OW purging gas (lines).

In Fig.4 the functions h vs temperature are drawn in an Arrhenious plot for each purging gas used. In Tab.4 the best fitting D values as "cleaned" of the surface desorption effects, hence independent of the purging gas composition, are reported. They are higher than the literature data obtained by assuming pure diffusion control[1,5,6], which, in our opinion, may induce underevaluation.

TABLE 4
Tritium diffusion coefficients for the P-type breeder material evaluated by the best fitting-computer simulation method

T(°C)	450	500	550	600	650
$Dx10^{-13}$ (cm^2/s)	0.5	1.4	2.8	3.5	5.2

To emphasize the determining role of the desorption step, the mass transfer coefficients obtained by $h = R/(3I/P)$ (pure interface resistence control) are also given in Fig.4 for all the specimens, considering as well the equivalent radius (R) coming from the Surface Area values.

5. CONCLUSIONS

Phase 1 of the TEQUILA experiment has just been completed and the resultant data are being elaborated. Phase 2 must still to be carried out in order to complete the investigation programme.

The following conclusions, however, can be reported:

(i) - The grain size determines the tritium releasing properties of the porous gamma-lithiun aluminate ceramic materials. This is in general agreement with the available results[1,5,6], but a surface tritium desorption mechanism describes our data better than the solid state diffusion one. (ii) - tritium extraction from ceramic breeders is feasible in an oxidizing environment by strongly decreasing the isotopic dilution in comparison to the actual "reference" mixture (He+0.1% H_2), and using moisture at a 100 vpm level as dopant of the helium purging gas.

(iii) - Further work is needed to optimize the tritium recovery process from the ceramic breeder materials, not only to improve the microstructural parameters, but mainly to test the effect of the purging gas chemistry.

REFERENCES

1. R.G.Clemmer et al., The TRIO Experiment, Argonne Nat. Lab. ANL-84-55 (1984).

2. C.Alvani et al., The ENEA selected fabrication processes of $LiAlO_2$ ceramic pellets for Tritium Breeding Experiments in Fusion Technology, High Tech Ceramics, ed. by P.Vincenzini (Elsevier Sci. Pub., Amsterdam, 1987) pp. 2941-2948.

3. C.Alvani et al., Physico-Chemical Properties of $LiAlO_2$ for a Ceramic type Tritium Breeding Blanket, this Volume, pp.2933-2940.

4. L.Anzidei et al., Helium Cooled Ceramic Breeder in Tube Blanket for a Tokamak Reactor: the Coaxial Poloidal Modulus Concept Design, in this Proceeding of the 15th SOFT, Utrecht, the Netherlands, Sept. 19-23, 1988.

5. H.Werle et al., J. Nucl. Mater., 141-143 (1986) 321.

6. H.Werle et al., J. Nucl. Mater., 154-156 (1988) in print.

7. M.R.Mancini et al., Hydrogen and Water Adsorption on $LiAlO_2$, ENEA Report, June 1988, in print (available on request).

8. F.Alessandrini, et al., High Temp.Electrical Characteristics of Gamma-LiAlO2 in the Proceedings of 5th Int. Conf. on High Temperature Materials, Rome, Italy, May 25-29, 1987, in print.

FUSION TECHNOLOGY 1988
A.M. Van Ingen, A. Nijsen-Vis, H.T. Klippel (editors)

IN-PILE TRITIUM RELEASE FROM LIQUID BREEDER MATERIAL Pb-17Li IN LIBRETTO 1 EXPERIMENT

Rainer CONRAD, Luigi DEBARBERIS

Joint Research Centre Petten, P.O.Box 2, 1755 ZG Petten
The Netherlands

Lithium-lead alloy samples, Pb-17Li, with lithium in natural abundance were irradiated in the HFR Petten during ~69 full power days up to a Li burnup of ~1.14%. The samples were contained in closed capsules and one plenum swept capsule, made from stainless steel (316L). The design of the irradiation facility permitted independent in-situ tritium release measurements, to study tritium permeation through the capsule walls and tritium extraction by plenum sweeping with a helium + 0.1 vol% hydrogen purge gas. The alloy was irradiated at NET relevant conditions. Temperature transients between 600 and 680 K were performed to study tritium release kinetics. The experimental results indicate that steady state tritium release by permeation was achieved at temperatures >580 K. The typical in-situ tritium residence times for temperatures between 600 and 680 K ranged from 12 to 2.2 hours. The apparent diffusion coefficient of tritium in the alloy is 3.5E-10 m2/s at 645 K. 47% of the produced tritium was extracted by plenum sweeping at temperatures >580 K, although the unfavourable ratio between gas-liquid interface to permeation areas and the absence of a permeation barrier. Tritium was released in HT form. The data seem to show that the eutectic alloy has a very low capability to confine tritium. The ongoing PIE confirms the previous statements.

1. INTRODUCTION

Lithium-lead alloys of various compositions have been proposed as blanket materials for fusion reactor tritium breeding [1]. The eutectic alloy Pb-17Li is the sole breeder material in the liquid blanket development for NET [2]. The possible advantage of Pb-17Li is that one material could serve as tritium breeding, neutron multiplier and cooling fluid. The critical issues for the choice of the eutetic alloy Pb-17Li as tritium breeding material are the selection of the tritium recovery methods, the verification of basic tritium kinetic data, the selection of the structural materials and the clarification of safety problems (polonium generation) [3]. These issues have to be investigated by in-pile tests with realistic tritium generation and dispersion in the alloy.

Before starting with the design of an in-pile loop for simulation of blanket modules, it is necessary to study under irradiation basic issues with static Pb-17Li samples contained in capsules. Up to now, only annealing experiments after irradiation were performed with Pb-17Li samples [4,5]. The Libretto experiments are designed to investigate various tritium extraction methods and tritium release kinetics with in-situ tritium release measurments of static samples. The irradiation programme Libretto is being carried out within the European Fusion Technology Programme on Blanket Technology as a joint project by JRC Ispra and JRC Petten. The irradiations are performed in the HFR Petten [6].

The first experiment, of which the irradiation results and the ongoing PIE observations are given in this paper, was performed to obtain data on:

- tritium extraction by plenum sweeping,
- temperature influence on tritium residence times and tritium release limit,
- tritium permeation through 316L containment cladding,
- safety problems (polonium generation, alloy-containment interactions, tritium retention.

2. EXPERIMENTAL

2.1 Material

The lithium-lead alloy Pb-17Li was prepared and characterized by JRC Ispra [7]. The alloy was prepared in an iron crucible from pure lead

(99.999% and Bi<1ppm) and 99.999% pure lithium. The chemical composition of the eutectic alloy was 99.26 wt% lead and 0.67 wt% lithium. The isotopic content of ^{6}Li was 7.5 wt%. All the operations were carried out in a stainless steel glove box connected to a high vacuum system and to an argon purification system. The irradiation capsules were conditioned before filling by an appropriate heat treatment, i.e. heating for 30 minutes at 1273 K under a vacuum < 1.3E-3 Pa, in order to clean the steel surfaces. Each capsule was filled with 12.5 cm^3 alloy. The capsules were closed by E-beam welding. The capsule plenum was 3 cm^3.

2.2 Irradiation facility

The irradiation was performed in a multichannel rig containing three independently controlled stainless steel capsules of 14 mm diameter and 0.5 mm wall thickness [8]. The capsules contained 120 g alloy each. The cross section of one capsule is shown in figure 1.

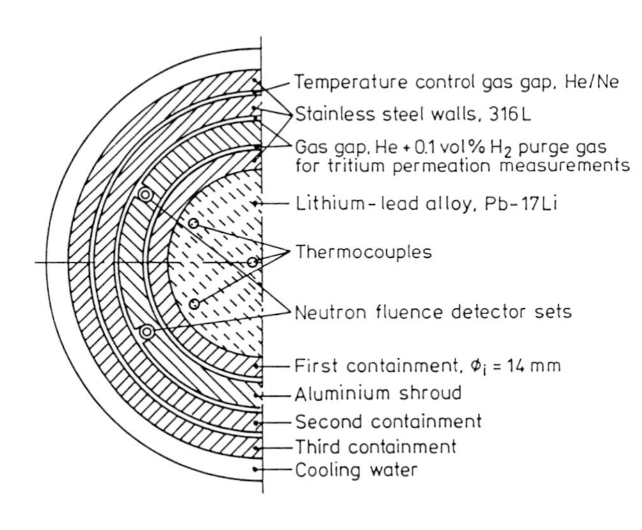

FIGURE 1
Cross section of one Libretto capsule

Two closed capsules were designed to measure in-situ tritium permeation. One plenum swept capsule was designed to evaluate in-situ tritium extraction by plenum sweeping and tritium permeation. The capsule instrumentation included thermocouples and neutron fluence detector sets. Each containment was equipped with two capillary tubes to purge the released tritium to the experimental out-of-pile system. This system was equipped with independent analytical trains for analysis of tritium content of the purge gas and for characterisation of the form of the released tritium, i.e oxidized or reduced.

2.3. Irradiation conditions

The irradiation was performed in a peripheral in-core position of the HFR Petten during three reactor cycles, i.e. ~69 full power days. The relevant process parameters of each capsule were scanned by a computerized data logger. The second containments and the plenum swept capsule were continously purged by purified helium + 0.1 vol% hydrogen, to improve tritium desorption through exchange from the alloy and/or capsule wall surface. The nuclear conditions remained constant. The samples were irradiated during the first 45 days at a temperature of 580 K. Temperature transients between 600 and 680 K were performed during the remaining 24 irradiation days to study tritium release kinetics. The radial and axial temperature gradients were <25 K.The irradiation paramters are given in table 1.

Table 1 Irradiation parameters Libretto 1

Capsule no.	1	2	3
Closed	x	x	
Sweeping			x
Irrad.time (days)	69	69	69
Temp.range (K)	560/680	560/680	560/680
Temp.gradient radial/axial (K)	15/10	21/11	16/15
T-release rate (uCi/(min g))	3.45	3.70	3.58
Burnup (% Li)	1.07	1.14	1.12

2.4. Post irradiation examination, PIE

The ongoing PIE programme of the closed capsules is focused on:

- X-ray & neutron-radiography, to visualize the alloy distribution in the capsule plenum,
- dosimetry, to confirm the calculated nuclear characteristics,

- puncturing, to determine the quantity and the composition of the plenum gas,
- scanning electron miscroscopy of alloy and cladding samples,
- Tritium retention measurements of alloy and cladding samples (in progress),
- metallography of cladding/alloy samples to examine surface interactions (in progress),
- polonium concentration measurements of alloy samples (in progress).

3. RESULTS AND DISCUSSION

3.1 Irradiation closed capsules

Initial tritium release by permeation through the capsule wall started at an alloy temperature >540 K. Steady state tritium release was achieved at temperatures >580 K.

Temperature transients were performed to investigate the influence of temperature on the tritium release. They resulted in expected tritium release response, i.e. for each temperature increase or decrease a positive or a negative tritium release peak was observed, see figure 2.

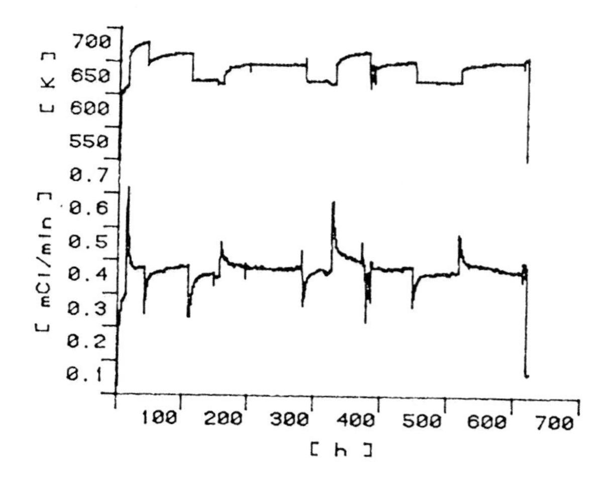

FIGURE 2
Tritium release and temperature vs. time of capsule 2 during the last cycle

Characteristic residence times, τ, were calculated in the temperature range 600 - 680 K by:

$$\tau = I / G$$

where I is the total tritium inventory in the alloy, capsule wall and plenum,
G is the tritium production rate.

The tritium inventory, I, in the capsules was calculated by the difference between production and release, assuming initial zero inventory.

The determined residence times for the different end of transient temperatures are given in table 2 and plotted in an Arrhenius plot (figure 3). These residence times are characteristic for the configuration alloy-cladding-plenum.

Table 2 Measured tritium residence times

Temperature [K]	Capsule no.	Time [h]
600	1	~12
610	1	~10
623	2	9.0
626	2	7.3
643	2	6.8
652	2	4.6
655	2	3.9
658	2	4.2
659	1	~4.0
680	2	2.2

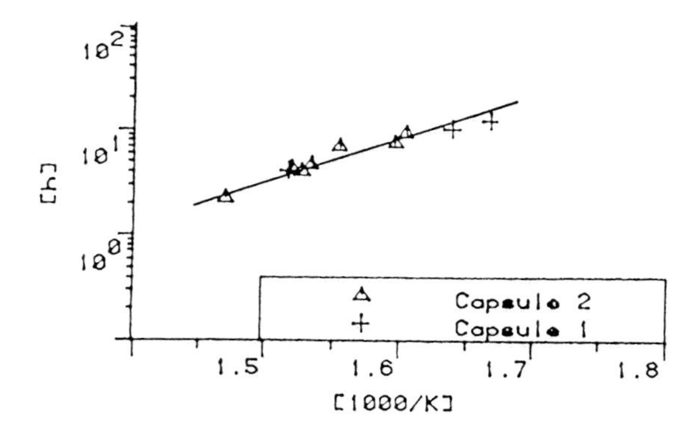

FIGURE 3
Arrhenius plot of residence times in hours

At the temperature of 680 K the in-pile measured residence time of 2.2 h is in good agreement with the characteristic radial diffusion time based on the out-of-pile diffusivity given by Reiter [9]. A good agreement at a temperature of 658 K is met with the data given by Shibuya [5]. For temperatures < 658 K the measured residence times are longer compared with data given by [9,5].

The measured tritium residence times show a single activated process with an activation energy of 75 kJ/mol, against the 20 kJ/mol given by [9] and about 5 kJ/mol given by [5] for the Pb-17Li diffusivity and 50 kJ/mol given by [11] for the AISI 316L permeability. The influence of the AISI 316L capsule wall on the residence time should not be dominant since the wall characteristic diffusion time is only a fraction of the Pb-17Li diffusion time. Using data by Reiter [9] or more recent data on AISI 316L given by Grant [10] this fraction is smaller than 0.27 and 0.18 at the lowest temperature of 600 K, respectively. However, a decrease of the diffusional properties of AISI 316L in presence of Pb-17Li has been observed by Gautsch [11], so that the lower temperature mesurements could be influenced by the capsule wall and the plenum.

An apparent tritium diffusivity of about 3.5E-10 m^2/s at the temperature of 645 K can be calculated, considering pure radial diffusion in the alloy.

3.2 Irradiation open capsule

Tritium extraction by sweeping during the initial start up was negligible at temperatures <450 K. Measurable tritium release started spontaneously after the melting temperature of the alloy was passed. Steady state tritium release rate, equal to 47% of the production rate, was achieved at temperatures >580 K. The remaining 53% of the production rate was released by permeation through the capsule wall. The observed form of tritium was HT (>99%). The careful pre-irradiation capsule conditioning resulted in a very pure downstream gas quality (dewpoint <-87°C). Rather big spikes in the sweeping tritium release signal were continously observed. Most probably, this phenomena was caused by gas bubbles (see §3.3), formed under irradiation through the thermal reaction $^6Li(n,\alpha)T$. The rising gas bubbles transported tritium into the plenum and spattered liquid alloy material onto the purge gas tubes, which became closed after 18 operating days.

Tritium extraction by sweeping was relatively efficient, considering the unfavourable ratio gas-liquid interface to permeation areas of 4% and the absence of a tritium permeation barrier. This is in disagreement with data given by Seibene et al.[12], considering pure tritium diffusion in stagnant alloy.

3.3 Post irradiation examination results

Neutron-radiography and X-ray foto's (figure 4) after irradiation confirmed the sputtering phenomena mentioned in § 3.2. Puncturing of the closed capsules at ambient temperatures revealed that the plenum contained for >99 % pure helium. The measured tritium activity was < 3 mCi in HTO form, corresponding to a production of 8 minutes. No other tritium species were observed. One capsule was cut radially into several pellets of ~2 mm thickness for further examinations. During cutting, only negligible amounts of tritium were detected by the air monitor.

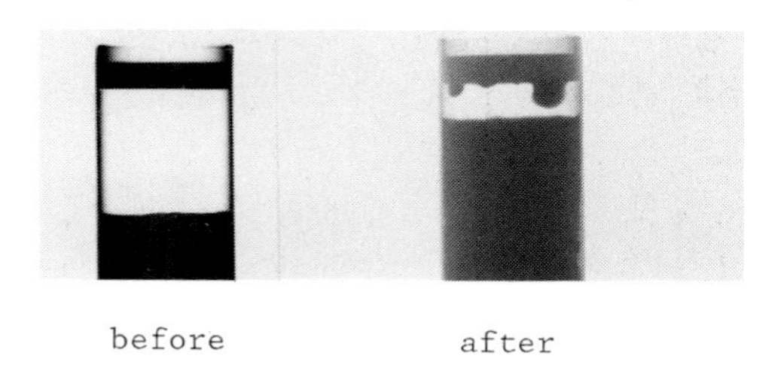

before after

FIGURE 4
X-ray before and after irradiation of capsule 2 gas plenum

The stainless steel cladding of one pellet was teared off the alloy and both pieces were investigated by scanning electron microscopy, SEM (figures 5&6). It appears from the SEM that the alloy is full of spherical bubbles with a max.diameter of 1 mm. Furthermore, the alloy is very strongly attached to the stainless steel cladding. When teared off, the alloy cracked over the equator of the bubbles. Since practically no tritium was detected during the alloy cutting and the measured helium quantity of the plenum was only half of the produced quantity, we believe that the bubbles were formed mainly by helium. Metallographic examinations of the alloy-cladding interface is currently being performed.

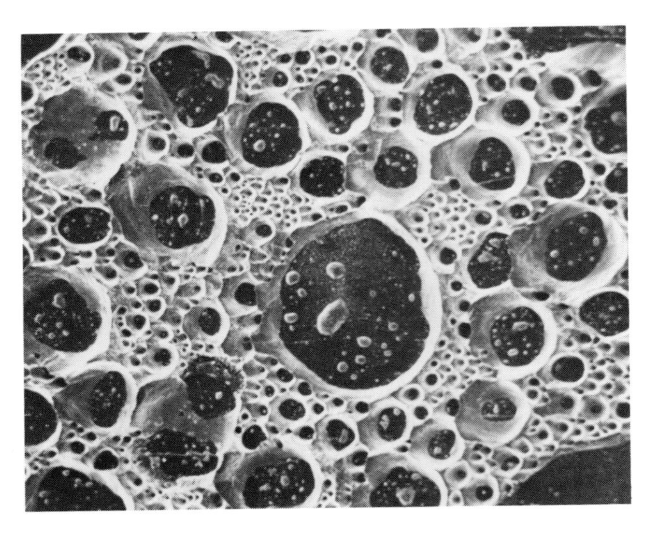

FIGURE 5
SEM of inner cladding surface (200X)

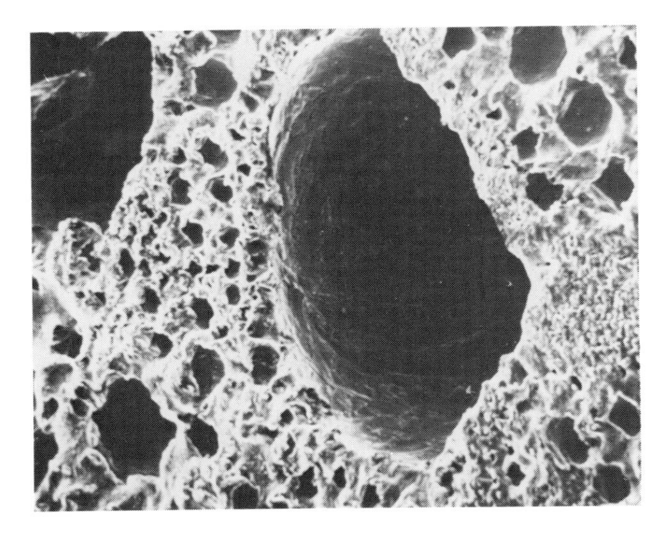

FIGURE 6
SEM of alloy surface (300X)

Small alloy samples of ~ 1 g, cut from different axial capsule positions, were investigated on retained tritium by annealing technique [13]. The measured retained quantity was ~30 μCi/g, corresponding to a tritium production of < 10 min.

The theoretical tritium inventory in the system at the end of the irradiation should be 0.9 mCi/g, or the equivalent of ~4 hours production (see table 2). Since the calculated tritium production was confirmed by the in-situ release measurements and by neutron dosimetry, we must assume that tritium permeated out of the capsule in the time period of 7 months between irradiation and PIE annealing. In fact, small amounts of tritium were always detected, when the storage container was opened.

4. CONCLUSIONS

* Steady state tritium release by sweeping and permeation was observed at temperatures >580 K.
* Tritium transport is most probably enhanced by rising gas bubbles in the alloy.
* 47% of the produced tritium was extracted in HT form by plenum sweeping at temperatures >580 K, in spite of the unfavourable ratio between gas-liquid interface to permeation areas of 4%.
* Typical tritium residence times for the closed Pb-17Li capsules were determined and a single activated process was observed in the temperature range 600 K - 680 K.
* The derived activation energy for the system alloy/cladding/plenum was 75 kJ/mol.
* Considering pure radial diffusion of tritium through the Pb-17Li a diffusivity of 3.5E-10 m^2/s at 645 K was calculated.
* The residual tritium content in the alloy was <30μCi/g, which confirms the low capability of Pb-17Li to confine tritium.
* Gas bubbles of < 1mm diameter were observed, which were most probably formed by helium.

* The gas bubbles seem to play an important rôle on tritium transport in the alloy.
* The alloy seems to be strongly attached to the stainless steel cladding.
* The calculated and measured tritium quantities are in good agreement.

REFERENCES

1. G. Casini et al., 6th Top. Meet. on the Technol. of Fusion Energy, San Francisco, March 1985

2. W. Dänner et al., Proceedings 14th SOFT (1986) p.1281

3. V. Coen, Journal of Nuclear Materials 133-134 (1985) 46-51

4. R.G. Clemmer et al., Journal of Nuclear Materials 118 (1983) 274-285

5. Y. Shibuya et al., Journal of Nuclear Materials 150 (1987) 286-291

6. H. Röttger et al., EUR 5700 EN (1986)

7. V. Coen ,private communication

8. R. Conrad et al., EUR 9515EN (1985)

9. F. Reiter et al., Proceeding 15th SOFT (1986) p.1185

10. D.M. Grant et al., Journal of Nuclear Materials 152 (1988) 139-145

11. O. Gautsch et al., Journal of Nuclear Materials 152 (1988) 35-40

12. G.Seibene et al., Proceedings 3rd Topical Tritium Technology in Fission,Fusion and Isotopic Applications (1988) in print

13. H. Kwast et al., Journal of Nuclear Materials 133-134 (1985) 246-250

FUSION TECHNOLOGY 1988
A.M. Van Ingen, A. Nijsen-Vis, H.T. Klippel (editors)

REACTION KINETICS MEASUREMENTS OF LiPb COMPOSITES WITH STEAM

H.M. KOTTOWSKI, D. DROSTE, H. DIETZ, C.H. WU*

J.E.C. Joint Research Centre, Ispra Establishment, I-21020 Ispra (VA), ITALY

* The NET Team, c/o Max-Planck-Institut fuer Plasmaphysik, Boltzmannstrasse 2, D-8046 Garching bei Muenchen, FRG.

The study on the reactions kinetics of $Li_{17}Pb_{83}$, $Li_{50}Pb_{50}$, Li_7Pb_2 and lithium with steam has been carried out as a function of temperture of samples and steam at a temperature range of 350 to 950°C.

The rate of H_2-production by reaction of H_2O with Li-Pb were determined as:

α = 0.05 g mol/ m^2.s for $Li_{17}Pb_{83}$ T > 400°C
α = 0.95 g mol/ m^2.s for $Li_{50}Pb_{50}$ T > 550°C
α = 1.2 g mol/ m^2.s for Li_7Pb_2 T > 550°C
α = 3.2 g mol/ m^2.s for Li T > 400°C

The rate of production of the alloy Li_7Pb_2 has almost the same magnitude as that of $Li_{50}Pb_{50}$.

The results have been compared and the consequences are discussed.

1. INTRODUCTION

Lithium-lead alloys of various compositions have been proposed as blanket materials of fusion reactors for tritium breeding. The compositions most likely are: "Li_7Pb_2" as a solid blanket, and "$Li_{17}Pb_{83}$" as a liquid blanket. The possible advantage of "$Li_{17}Pb_{83}$" as a liquid blanket, besides very low lithium activity, is that one substance could serve as a breeding material, neutron multiplier and cooling fluid combined. From a number of considerations of the use of Li-Pb alloys in fusion reactors, the uniqueness of this system has been predicted: good tritium breeding, low tritium inventory and high blanket energy multiplications. In particular, the lead rich region is chemically relatively stable to H_2O as compared with other blanket materials.

To optimize the utilization of the unique characteristics of Li-Pb alloys, their physical chemical properties must be known. In the previous works[1] we published results of measurements of the solubility of hydrogen in Li-Pb alloys and the activity of lithium in the Li-Pb systems. Because of the increasing interest in the use of Li-Pb alloys in fusion reaserch the interaction of H_2O with Li-Pb has now been investigated in more detail, in particular the knowledge is significant for the safety analysis.

2. EXPERIMENTAL

The apparatus and experimental procedure used for the study of reaction kinetics has been described in detail previously[2,3].

Steam is generated at atmospheric pressure in the "steam generator" and heated up before entering the test chamber to the desired inlet temperature. Behind the reaction chamber, the steam is separated from the hydrogen in a "condensor" and the hydrogen fed to the measuring roller. Temperatures, pressure in the reaction chamber and the hydrogen formation are measured simultaneously with appropriate data and acquisition devices. The test matrix shows the initial test conditions (Table 1).

TABLE 1: TEST MATRIX

Metals	In. M. temp. °C	M. Q. gr	Li-Inv. gr	In. vap. temp. °C
$Li_{17}Pb_{83}$	350	100	0.68	350
	450	100	0.68	450
	550	100	0.68	550
$Li_{50}Pb_{50}$	450	21.0	0.68	450
	550	21.0	0.68	550
Li_7Pb_2	550	6.48	0.68	550
Li	350	0.68	0.68	350
	450	0.68	0.68	450
	550	0.68	0.68	550

The investigation is performed in such a way that the Li-inventory of the test samples was the same regardless of the "alloys", which means that the total number of Li atoms are constant for all the samples. The quantity of the alloys used for the tests, therefore determined by the chosen constant Li atoms.

As initial conditions, the steam temperature was adapted to the metal temperature and kept constant during the experiment. Post-test analysis were made in order to determine the chemical composition of the residues and their mode of deposition.

3. RESULTS AND DISCUSSIONS

3.1 Reaction of $Li_{17}Pb_{83}$

The measurements at higher initial melt temperature show a large scatter, the reason for which may be attributed to the mode of reaction product deposition. The chemical analysis of the residue after reaction shows LiOH with traces of Li_2O. On the other hand, the post-test observation reveals that at 350°C and 400°C test temperature the LiOH is a compact, porous cake covering the Pb-bulk (Fig. 1). At 500°C metal temperature, which is above the melting temperature of LiOH, the hydroxide retracts and is accumulated at the edges of the test crucible. The bulk of Pb is metallic-clean.

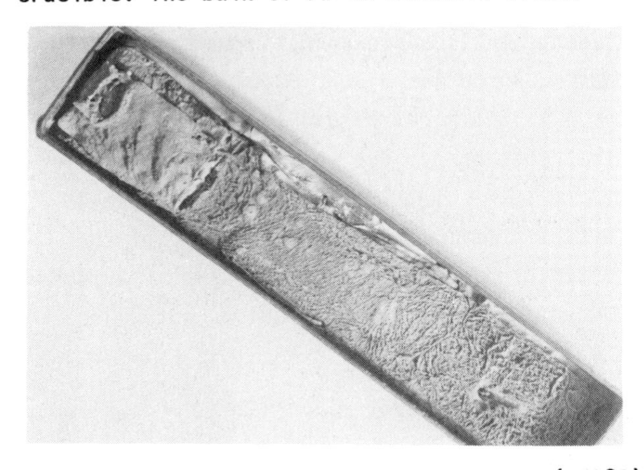

Fig. 1 $Li_{17}Pb_{83}$ sample after reaction (350°C)

The moderate reaction of $Li_{17}Pb_{83}$ with steam is also shown in Fig. 2. The metal bulk temperature, which is a function of the heat of reaction, increases only marginally after reaction inception and remains constant during the reaction.

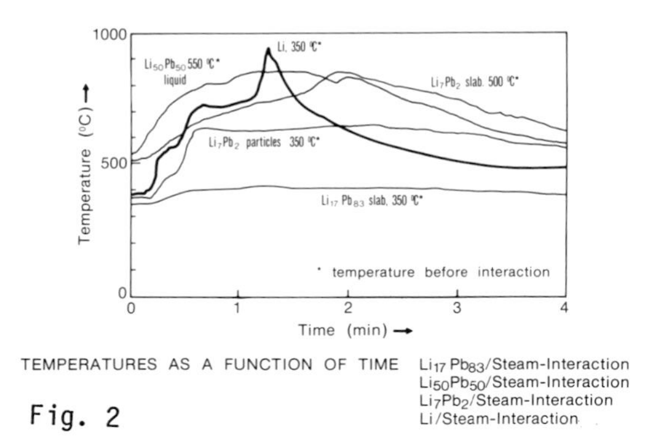

TEMPERATURES AS A FUNCTION OF TIME $Li_{17}Pb_{83}$/Steam-Interaction
$Li_{50}Pb_{50}$/Steam-Interaction
Li_7Pb_2/Steam-Interaction
Li/Steam-Interaction

Fig. 2

The influence of the mode of reaction residue deposition is illustrated in Fig. 3. At metal temperatures up to 400°C, no measurable changes in the reaction rate at the initial contact of steam with the clean metal and processing reaction has been observed. However, for metal temperatures of 450 and 500°C the initial reaction rate is significantly higher (about two to three times) than that measured at lower temperatures. Starting from the high value, the reaction rate is approaching asymmetrically the values measured for linear metal temperatures (Fig. 4). This might be due to the liquid phase of the reaction residue which retracts from the metal surface. The exponential decrease could then be a consequence of the lowering of the Li-inventory in the metal bulk during reaction. The reaction rate seems diffusion-controlled.

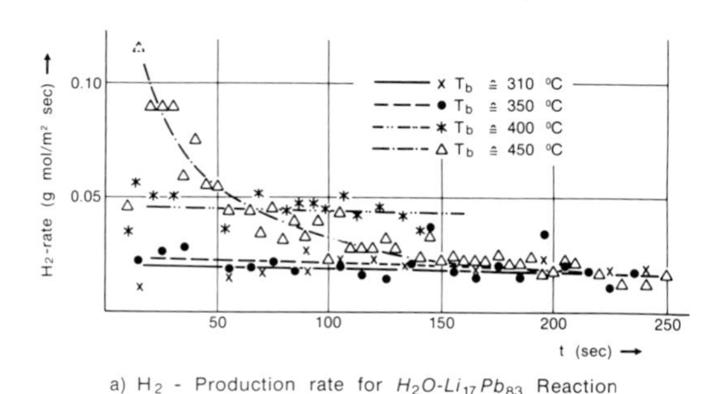

a) H_2 - Production rate for H_2O-$Li_{17}Pb_{83}$ Reaction

Fig. 3

3.2 Reaction of $Li_{50}Pb_{50}$

In contact with steam behaves similar to $Li_{17}Pb_{83}$. Since, however, the melting point of $Li_{50}Pb_{50}$ is higher than the melting point of LiOH (471°C) the molten desposit retracts from the metal surface so that the metallic surface is always exposed to the vapour stream. The reaction produces (at surface reaction) volatile aerosols which are carried with the H_2 stream and desposited downstream the sample holder (Fig. 4). This has not been observed at $Li_{17}Pb_{83}$ reactions.

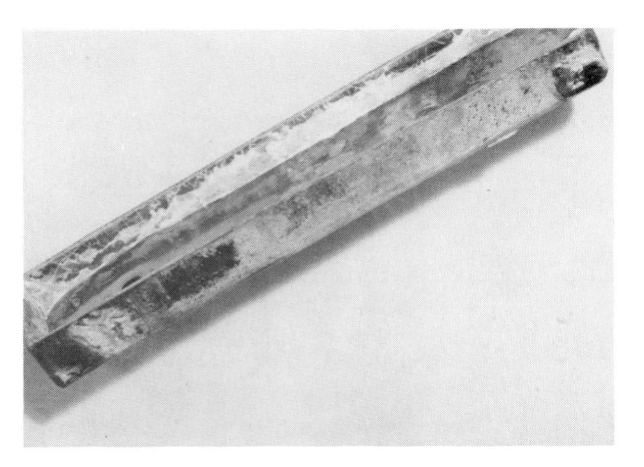

Fig. 4 $Li_{50}Pb_{50}$ sample after reaction (550°C)

The temperature scenario during the reaction is shown in figure 2. It is seen that the temperature after initial rise remains stationary for the duration of the reactions. These observations indicate that the Li-diffusion in the metal phase is the reaction dominating process. As shown also in Fig. 5, the reaction rate does not change remarkably with time during reaction.

3.3 Reaction of Li_7Pb_2

The reaction of Li_7Pb_2 with steam depends strongly on the temperature and disintegration in case of solid state. If liquid, violent reaction occurs. Compact test samples of Li_7Pb_2 (1 cm length) heatet up to 350°C, 500°C and

550 ^{0}C before floading with vapour of the same temperature showed the following result:

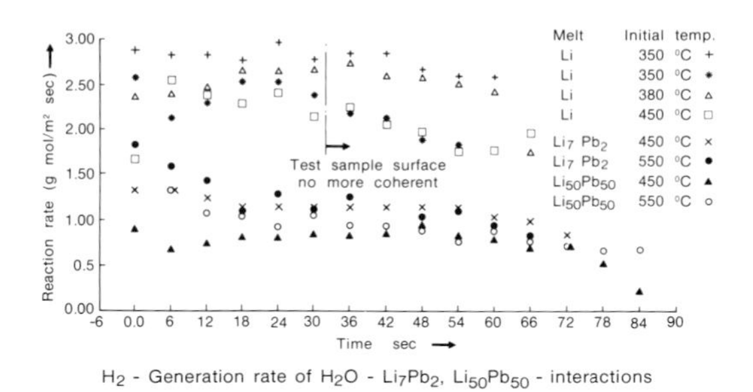

H_2 - Generation rate of H_2O - Li_7Pb_2, $Li_{50}Pb_{50}$ - interactions

Fig. 5

At 350 ^{0}C slab and steam temperature only a weak, non-sustained surface reaction was observed. At 500 and 550 ^{0}C, however, self-sustained reactions with temperature peaking of the slab at 850 ^{0}C occured (above the melting

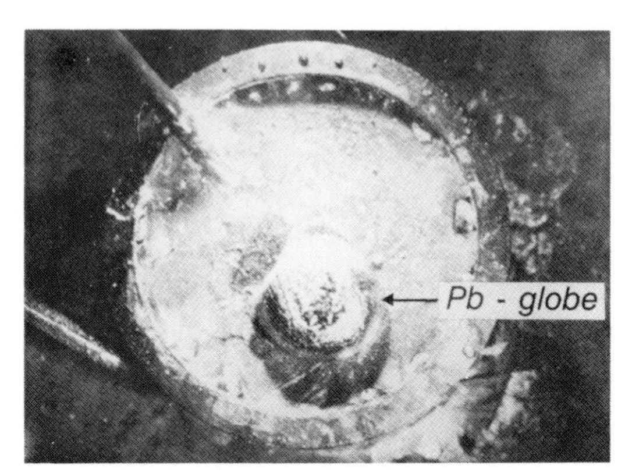

Li_7Pb_2 - After reaction (500°C initial temperature)

Fig. 6

point of Li_7Pb_2). The post-test observation shows single lead spheres embedded in LiOH (Fig. 6). This redistribution pattern of Pb-metal and solid reaction products is a reproductible feature of the Li_7Pb_2 steam reaction. Again, volatile aerosol formation has been observed. In case of Li_7Pb_2 granulates (100-500 um) self-substaines reaction can already be obtained at 350 ^{0}C initial temperature of metal and steam. These tests show also Pb-metal globules embedded in LiOH. The temperature history of the metal due to the heat of reaction is shown in Fig.2.

3.4 Reaction of Li with steam

Tests with molten Li of different initial temperatures (350, 400, 550 ^{0}C) were performed. Violent reactions were with temperature peaking up to 1000 ^{0}C (Fig. 2), depending on the initial temperature. The reaction shows also aerosol formation.

The residue of the reaction, which is mainly LiOH, is shown in Fig. 7. The black allotropy of LiOH has not been observed at Li-steam reaction.

Fig. 7 Li sample after reaction

4. SUMMARY

The measured rates of reactions for $Li_{17}Pb_{83}$, $Li_{50}Pb_{50}$, Li_7Pb_2 and Li as a function of metal "pool" temperature are shown in Fig. 8. The reaction rate of $Li_{17}Pb_{83}$ is metal temperature-dependent. It increases up to about 400 ^{0}C. Between 400 and 500 ^{0}C the mean value remains apparently constant. However, the scattering range increases up to 60 %. The tests results might be explained as follows. Above the melting point of LiOH, the molten LiOH might not wet the metal so that no protective layer is built up which could impair the reaction. The large scatter is probably caused by the partial solidification and remelting of the LiOH due to the non-uniform temperature in the reaction chamber and also the metal of nearly melt temperature of LiOH.

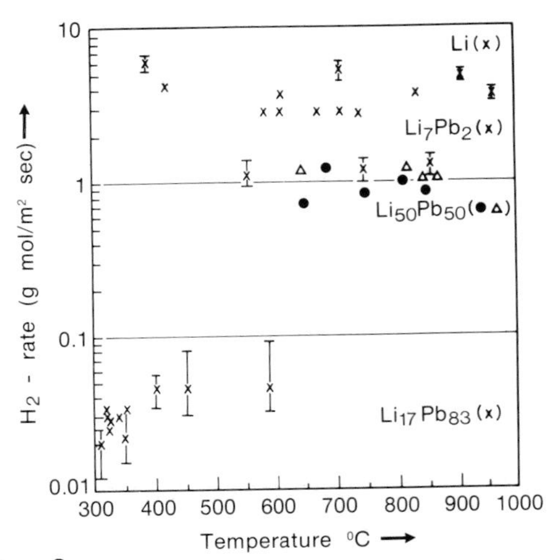

Fig. 8
Reaction rates as a function of temperature

Since the self-sustaining reaction temperature of $Li_{50}Pb_{50}$ is above the melting point of LiOH, the reaction rate seems not being influenced by the melt temperature. The reaction rate is about 3 to 4 times higher than that of $Li_{17}Pb_{83}$ at 400 ^{o}C.
The reaction rate of self-substained Li_7Pb_2-steam reactions is of an order of magnitude higher than that of $Li_{17}Pb_{83}$. It does not seem being influenced (within the temperature range measured) by the melt temperature.

The reaction rate of Li with steam is nearly an order of magnitude higher than that of Li_7Pb_2 and about two orders of magnitude higher than that of $Li_{17}Pb_{83}$.

Li $_{50}Pb_{50}$ steam interactions tend to behave like Li_7Pb_2 and Molten Li_7Pb_2 steam interactions tend to behave like Li.

Mean reaction rate constants are:

($Li_{17}Pb_{83}$) > 400 ^{o}C α = 0,05 g mol/m^2sec

($Li_{50}Pb_{50}$) > 550 ^{o}C α = 0,95 g mol/m^2sec

(Li_7Pb_2) > 550 ^{o}C α= 1,2 g mol/m^2sec

(Li) > 400 ^{o}C α= 3,2~5,0 g mol/m^2sec

REFERENCES

1. C. H. Wu
The Interaction of Hydrogen Isotopes with Lithium-Lead Alloys
J. Nucl. Mat. 122 & 123, 941 (1984)

2. H. M. Kottowski, G. Kuhlboersch, C. H. Wu, The Compatibility of H_2O with Li_7Pb_2 and its Interactions with Hydrogen Isotops, Proc. 14th Symposium on Fusion Technology, Avignon France, pp. 1173-1178, 1986

3. H. M. Kottowski, G. Grossi, G. Pautasso, Interaction of Eutectic Li-Pb and Water with Respect to Safety Aspects in the Fusion Reactor Blanket, Proc. 13th Symposium on Technology, Varese, Italy, pp. 1655-1660, 1984

FUSION TECHNOLOGY 1988
A.M. Van Ingen, A. Nijsen-Vis, H.T. Klippel (editors)

HYDROGEN RELEASE FROM LOW-Z COATED METALS

M. CAORLIN, J. CAMPOSILVAN and F. REITER

Commission of the European Communities, Joint Research Centre, Ispra Establishment
21020 Ispra (Va) - Italy

Low-Z materials are being studied, with the aim of looking into their hydrogen release characteristics. The effects of very thin TiC coatings deposited on the NET steel AISI 316L are being investigated. First measurements show a sensible decrease of the hydrogen release rate, when compared to bare steel data. In addition, surface processes are present in the experimental pressure range scanned. An interpretation of the experimental data is attempted on the basis of simple diffusion- and surface-limited hydrogen release models. Possible consequences are shortly discussed.

1. INTRODUCTION

High temperature refractory ceramics may have important fusion technology applications. For instance, tritium permeation barriers would be needed in blanket modules and coolant pipework to reduce tritium inventory in the structures and losses to the coolant and the ambient[1]. In addition, low-Z low activation resistant materials should be used as protection of plasma-facing components[2-4]. In both cases, understanding the mechanisms involved in tritum interaction with such materials is of outstanding importance in order to predict recycling, inventory and permeation.

A new experimental activity is under way at JRC Ispra on the hydrogen release characteristics of low-Z, high temperature materials for which low hydrogen release rates are expected. Titanium carbide (TiC) was chosen as representative of the ceramics class. This material has a very high melting point (3067°C), good corrosion and thermal shock resistance, good thermal stability, low vapour tension and sputtering yield[5,6]. Low diffusivity is also expected[7]. On these grounds, in order to keep measuring times as low as possible, very thin TiC layers are being studied. The NET austenitic steel AISI 316L was chosen as substrate on which TiC layers some μm thick were produced by Chemical Vapour Deposition (CVD).

In this work, first results are given on the effects of TiC on the hydrogen release rate, by comparing TiC/steel to bare steel data. The experimental method used successfully worked on testing samples of nickel, Inconel 600 and 625, of the eutectic alloy Pb-17Li and of AISI 316L[8].

2. EXPERIMENTAL

A schematic view of the apparatus is shown in Figure 1. The technique used has been discussed elsewhere[8]. It is based on pressure rise measurements carried out in a UHV vessel of known volume. Hydrogen is released after a thorough pumpdown from previously loaded samples. Loading pressures are varied between 10^2 and 10^5 Pa whereas the temperature range is 400°C - 600°C. Blank runs have to be subtracted to account for the rig inner wall outgassing.

From the final pressure attained at the end of each isothermal run the amount of gas dissolved in the sample is obtained. From the trend of pressure rise in time a value can be found for the dynamic parameter of the model used to simulate the gas release. In the case

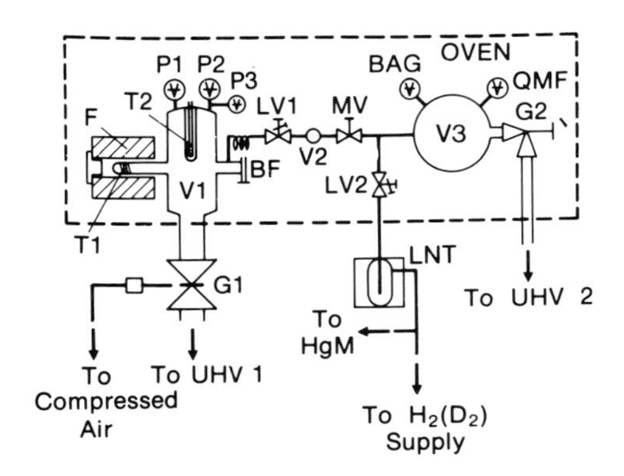

BAG	Bayard-Alpert Gauge
BF	Blanking Flange
F	Furnace
G1	Electropneumatic Gate Valve
G2	Manually Actuated Gate Valve
HgM	U-tube Hg manometer
LNT	Liquid Nitrogen Trap
LV1, 2	Manually Actuated Leak Valves
MV	Manually Actuated Magnetic Valve
P1, P2	Capacitive Manometers
P3	Spinning Rotor Gauge
QMF	Quadrupole Mass Filter
T1, T2	Pt-Resistance Thermometers
UHV	Pumping Units
V1	Experimental Chamber
V2, V3	Expansion Volumes

FIGURE 1
Schematic view of the experimental set-up.

of diffusion-limited flow regimes a value for the hydrogen diffusivity in the material is obtained.

Several CVD-TiC coated samples with coating thickness from 0.1 to 10 μm were purchased from HTM-Biel (Switzerland). The substrate was the same AISI 316L batch used for diffusion and solubility measurements[8] and was supplied in rods 60 mm long and having a diameter of 6 mm. No interlayers were deposited to increase adhesion. Nevertheless, SEM inspection showed perfect adhesion to the substrate even after mild thermal cycling from RT to 873 K. The TiC layer reproduced the substrate surface roughness and a rather sharp interface could be observed.

Hydrogen supplied by SIO-Milan had a stated purity of 99.9999%. The background pressure in the rig before each run was less than 10^{-7} Pa. According to spectrometric analysis of the gas released during each run, the peak of molecular hydrogen was dominant over the others for a factor of 1000 or more.

3. MAIN RAW RESULTS

Several runs were carried out with a 3 μm thick coating in order to have rough estimates of the loading time, i.e. of the time required to saturate a sample with hydrogen. In diffusion-limited regimes the loading time is only a function of hydrogen diffusivity in the material. For metals, hydrogen absorption in high pressure regimes is diffusion-limited. As a rule, the release time, i.e. the time required to reach equilibrium in a release run, is much the same as the loading time. At 873 K and at a loading pressure P_0 of 10^5 Pa we found a release time of about 24 hours, while at 673 K it increased to 7 days or more.

This is shown in Figures 2 and 3, where the same experimental curves for the bare steel are given. The latter come from new measurements on AISI 316L, carried out for comparison. The results obtained are in excellent agreement with a more extensive previous investigation[8].

The release is slower by nearly a factor of

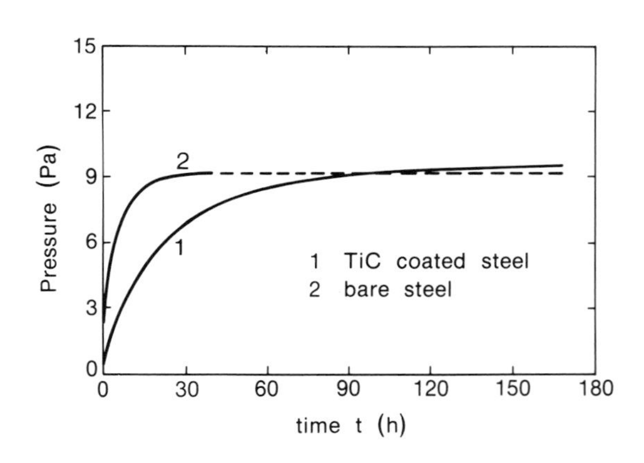

FIGURE 2
Hydrogen release from 3 μm TiC-coated and from bare steel at 673 K and for $P_0 = 10^5$ Pa.

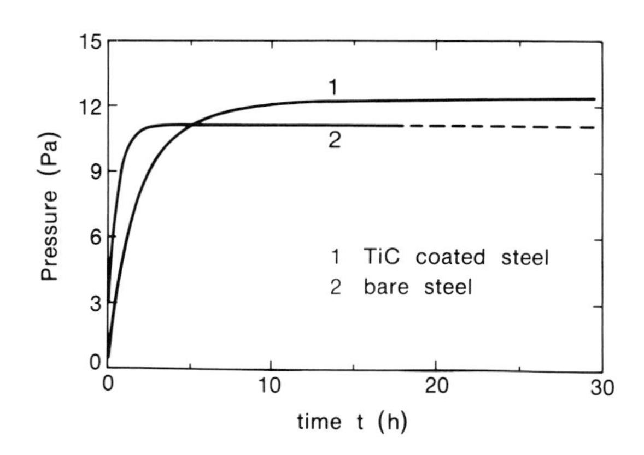

FIGURE 3
Hydrogen release from 3 μm TiC-coated steel and from bare steel at 873 K and for $P_0 = 10^5$ Pa.

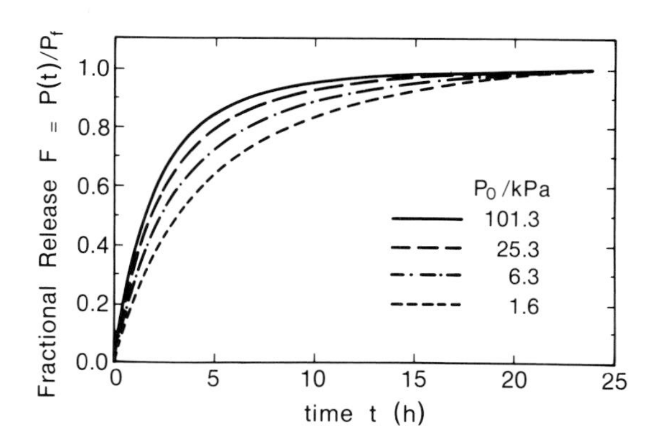

FIGURE 4
Effect of the loading pressure P_0 on the hydrogen release from TiC-coated steel at 873 K.

10 for TiC coated samples. The final pressure is higher for TiC-coated runs as compared to bare steel data, but the difference is not sufficient to give reliable estimates of the hydrogen solubility in TiC by additivity considerations. Nevertheless, the P_f behaviour outlined in Figures 2 and 3 is consistent with a very high hydrogen solubility in TiC.

Reproducibility after thermal cycling from RT to the test temperature was also assessed, in order to look into the effects of thermal expansion mismatching between substrate and coating. Satisfactory reproducibility was obtained at 673 K and 873 K.

We also looked into loading pressure effects at 873 K and found another feature in the release curves which was not observed with bare steel in the same experimental conditions. This is clearly seen in Figure 4, where the release rate sensibly decreases with the loading pressure. This might indicate a strong influence of surface processes on hydrogen release. For bare steel we also carried out a few measurements with different loading pressures. In agreement with previous studies[8] the fractional release was independent of the loading pressure, the release curves being exactly overlapping.

In addition, the final pressure P_f at 873 K varied as $\sqrt{P_0}$. This is shown in Figure 5, where the least squares straight line has a slope of 0.503 ± 0.003. Since P_f is directly proportional to the amount of hydrogen dissolved in the sample at the end of the loading phase, then hydrogen is dissolved atomically in the sample (Sieverts' law).

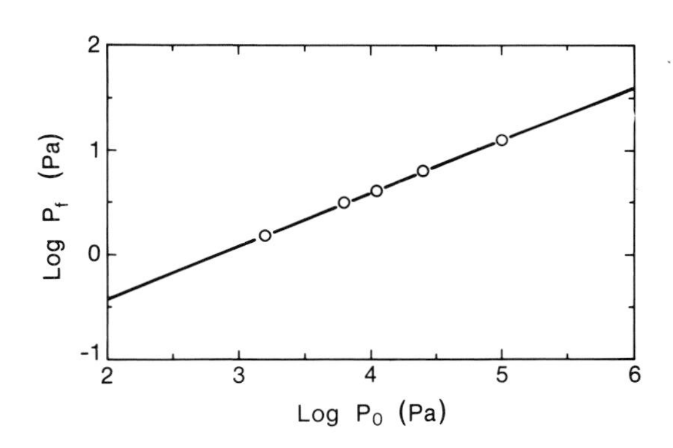

FIGURE 5
Effect of the loading pressure P_0 on the final pressure P_f at 873 K.

For hydrogen in TiC, two alternatives are possible: a) the solubility is high and goes as $\sqrt{P_0}$, or b) the solubility is not high enough to be detected with so low a volume, in which case it might also vary as P_0.

4. MODELLING

From the experimental evidence given in the previous section a convenient model for hydrogen release from our TiC-coated samples should account for surface processes and for diffusion in the bulk of the sample. Due to a quadratic term in the associated boundary condition, the differential equation cannot be solved analytically and one should resort to numerical techniques.

We have set up two separate simple models where diffusion and surface recombination, respectively, were assumed to dominate over each other. A diffusion-limited model cannot explain the patterns shown in Figure 4, but interesting information on the hydrogen flow regime in the solid can be obtained as well. These models and their analytical solutions are outlined in the following paragraphs.

4.1 Surface-limited model

In this limiting case hydrogen diffusion in the bulk of the sample is assumed to be instantaneous with respect to surface processes, so that any concentration gradient is immediately cancelled over a time scale of the order of the surface-limited release characteristic time. As a consequence the hydrogen concentration in the sample is only a function of time t, $c = c(t)$. Its variation with time is obtained by equating the hydrogen flux out of the sample with the time variation of the amount of hydrogen dissolved in it. This gives:

$$V_s dc/dt = A(k_1 P - k_2 c^2) \qquad (1)$$

to which we add a similar equation for the pressure variation in the experimental chamber of volume V:

$$(V/RT)dP/dt = A(k_2 c^2 - k_1 P) \qquad (2)$$

Here $R = 8.31\ \mathrm{J\ mol^{-1}\ K^{-1}}$, T is the gas temperature, V_s is the sample volume and A its surface area. Backflow from the gaseous phase onto the sample surface is accounted for through the "adsorption" term $k_1 P$ which is opposed to the "true" flux out of the sample $k_2 c^2$. The phenomenological surface adsorption and recombination coefficients k_1, k_2 are pressure-independent but may vary with temperature, usually according to an Arrhenius' law.

We also have the following couple of initial conditions:

$$P(0) = 0, \quad c(0) = c_0 \qquad (3)$$

The pressure law obtained by solving equations (1),(2) with the associated initial conditions (3) is expressed through the fractional release $F(t) = P(t)/P_f$:

$$F(t) = 1 + [U-(U+1)\exp(t/\tau_s)]^{-1} \qquad (4)$$

where:

$$U/P_f = \frac{[V/(K_s V_s RT)]^2}{1 + [2V\sqrt{P_f}/(K_s V_s RT)]} \qquad (5)$$

and:

$$\tau_s = V/[ARTk_1(1+2V\sqrt{P_f}/(K_s V_s RT))] \qquad (6)$$

$$K_s^2 = k_1/k_2 \qquad (7)$$

From (1) and (2) it follows that at equilibrium Sieverts' law holds. K_s is Sieverts' constant, so that the final concentration c_f in the sample in equilibrium with P_f is:

$$c_f = K_s\sqrt{P_f} \qquad (8)$$

As for the initial concentration c_0 one finds:

$$c_0 = c_f + VP_f/(RTV_s) \qquad (9)$$

Should the loading phase be long enough to saturate the sample one also finds:

$$c_0 = K_s\sqrt{P_0} \tag{10}$$

By imposing this condition in (4) instead of using P_f one can see the effect of varying the loading pressure. This is shown in Figure 6 for three different values of P_0.

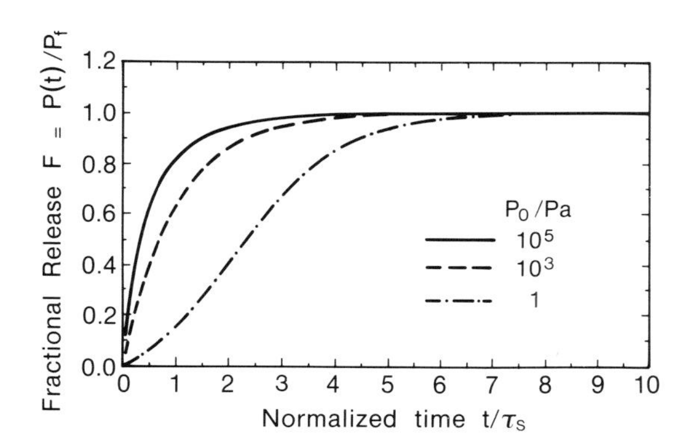

FIGURE 6
Surface-controlled simulated hydrogen release from TiC-coated steel at 873 K.

The value of the characteristic time τ_s used in this figure to normalize the time t belongs to the higher pressure curve, with $P_0 = 10^5$ Pa.

It is readily realised that this model agrees qualitatively with the experimental patterns given in Figure 4.

Extension of this model to a composite sample is accomplished by setting:

$$V_s = kV_1 + V_2 \tag{11}$$

where V_1,V_2 are the sample and coating volumes, respectively, and where a "partition factor" k has been introduced. This stems from the assumption that no hydrogen uptake occurs at the substrate-coating interface and from chemical potential continuity across it. At equilibrium k is a constant equal to the solubility ratio S_1/S_2. Since Sieverts' law holds also for the coating in this model, k is a parameter of the material:

$$k = S_1/S_2 = K_{s1}/K_{s2} \tag{12}$$

4.2 Diffusion-limited model

Basic assumptions of the model are: a) radial diffusion in the coating is the slowest step; b) the coating thickness b is small compared to the cylinder radius a, so that the coating can be approximated by a slab of the same thickness; c) both chemical potential and diffusive flux are continuous across the coating-steel interface; d) the initial radial concentration profile is flat inside the coating layer, e) the concentration at the surface is constant in time.

These assumptions yield the following set of equations:

$$\partial c_2/\partial t = D_2\partial^2 c_2/\partial x^2 \tag{13}$$

$$c_2(x,0) = c_0 \tag{14}$$

$$c_2(b,t) = c_f \tag{15}$$

$$\partial c_2/\partial x\big|_{0+} = 0 \tag{16}$$

$$c_1(t) = kc_2(0,t) \tag{17}$$

Here subscripts 1,2 refer again to steel and coating, respectively. The partition factor k is the same as in the previous section and the radial coordinate x in the coating is measured from the steel-coating interface.

The pressure rise in the experimental vessel is obtained from the diffusive flux at the gas-coating interface. The fractional release F(t) is given by:

$$1-F = [uG_1(t/\tau_d)+G_2(t/\tau_d)]/(u+1) \tag{18}$$

where the diffusion time τ_d is:

$$\tau_d = 4b^2/(D_2\pi^2) \quad (19)$$

having set:

$$u = kV_1/V_2 \quad (20)$$

for simplicity. In addition:

$$G_1(z) = (4/\pi) \sum_1^\infty \frac{(-1)^{n+1}}{2n-1} \exp[-(2n-1)^2 z] \quad (21)$$

$$G_2(z) = (8/\pi^2) \sum_1^\infty \exp[-(2n-1)^2 z]/(2n-1)^2 \quad (22)$$

The link between c_0 and P_f is again given by (9) and (11) where c_f is a priori unknown, but can be computed after Sieverts' law has been verified.

It is worth noting that the time to reach equilibrium within 1% is about $6\tau_d$.

5. RESULTS AND DISCUSSION

The experimental curves obtained were processed by means of non-linear least-squares fitting software[8], on the basis of the models developed.

The results obtained by the diffusion model at 873 K are given in Figure 7. It is evident that even at the highest pressures the release is not purely controlled by diffusion. Apparent diffusivities show a marked dependence on P_0, even stronger than for AISI 316L in a surface limited regime[9]. A trend to approach a constant value seems also present, though the actual values are still far from it. The high pressure apparent diffusivity can only represent a possibly good estimate of the "true" value, i.e. 5×10^{-16} m^2/s at 873 K. At 673 K, the diffusivity is nearly one order of magnitude lower. Then, in comparison to AISI 316L[8] the coating has a diffusivity which is lower by six orders of magnitude.

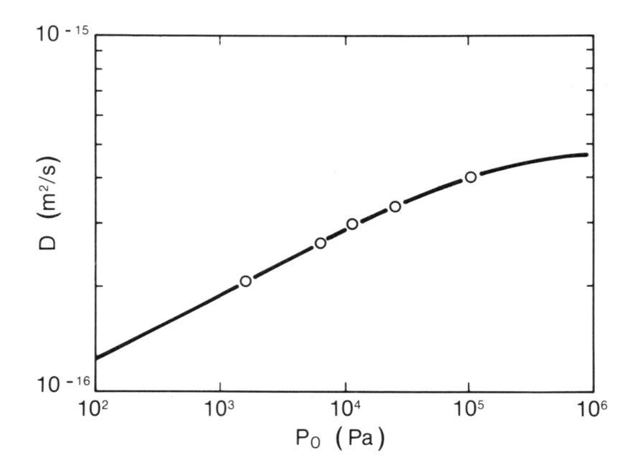

FIGURE 7
Variation of the hydrogen "apparent" diffusivity D with the loading pressure P_0 for a 3 μm TiC coating at 873 K.

Fitted-to-data curves by the surface model were rather good, reproducing the experimental ones within two percent. An example is given in Figure 8, for a coating 0.5 μm thick at 823 K.

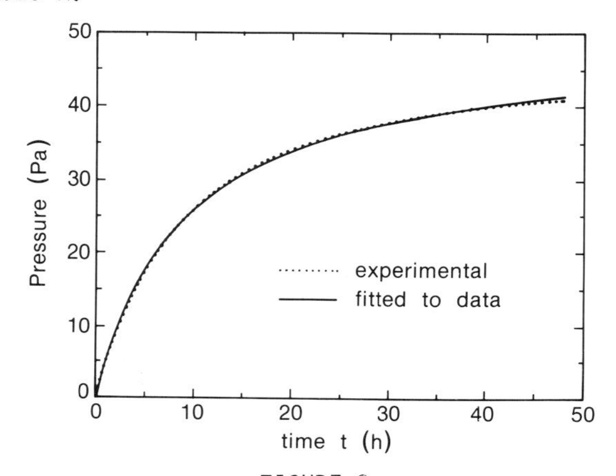

FIGURE 8
Data fitting by the surface-limited model for a 0.5 μm thick coating at 823 K and with $P_0 = 10^5$ Pa.

Figure 9 shows the results obtained at 873 K, for different loading pressures. As for D, k_1 shows a strong dependence on P_0 but the trend towards saturation, i.e. to approach a constant value as P_0 decreases, is more pronounced.

Therefore, in the pressure range covered, hydrogen release from a 3 μm thick coating is also not purely surface-limited. Instead we are

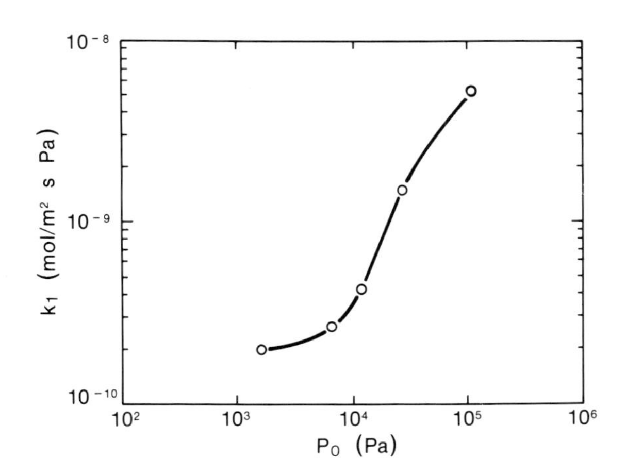

FIGURE 9
Loading pressure effect on the "apparent" surface rate constant k_1 at 873 K.

in a transition region, possibly nearer to the surface-limited edge. To check this we tried to fit the data at 873 K with P_f as only parameter, the others being kept fixed after tuning to the lower pressure curve. The fitted curves obtained were in poor agreement with experimental ones.

The lower pressure k_1 value is a good estimate for the true value, i.e. $2x10^{-10}$ mol/(m^2 s Pa). This is very low compared to bare steel data[10], as shown in Figure 10. Here the values of k_1 obtained with a completely different method[10] for a clean and a heavily oxidised AISI 316 surface are shown. The TiC data points labelled with an asterisk are to be considered only, since circles were obtained in a region nearer to the diffusion edge. The extrapolated point has been added to give a more reliable estimate of k_1 at 673 K.

As for the other parameters involved in the models, from lower pressure runs we can say that K_s is higher than for steel, at least by a factor of 200, since we obtained a value of about 12 mol/(m^3 Pa$^{\frac{1}{2}}$) against $4.98x10^{-2}$ for AISI 316L[8]. The partition factor k is about $5x10^{-3}$ in these conditions, what means that for equal volumes of steel and TiC, most hydrogen is dissolved in TiC. The same order of magnitude

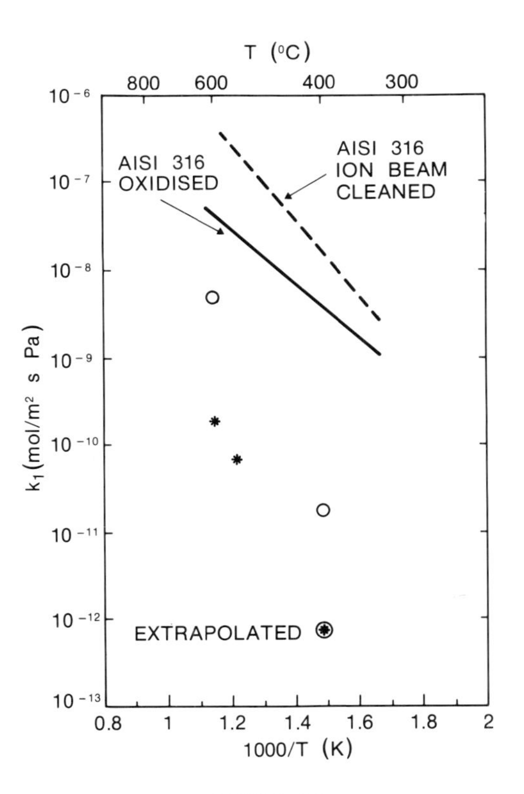

FIGURE 10
Surface rate constant k_1 for TiC compared to AISI 316 steel data[10].
$\ast$ = true k_1, surface-limited edge;
O = apparent k_1, diffusion-limited edge;
$\odot$ = extrapolation to 673 K.

for k was found with the diffusion model, too.

The value of the recombination constant k_2 may be computed from (7) to yield $k_2 = 1.4x10^{-12}$ m^4/(s mol). This also is several orders of magnitude below steel data.

Should these preliminary data be confirmed by more extensive investigations with thicker coatings, TiC could be of interest, also because of its resistance, as coating on steel structures of blanket modules and/or coolant tubes, so as to slow tritium absorption and reemission from them. This would be effective in the surface-limited regime, but much less if diffusion dominates because of high K_s which partly cancels the low diffusivity, so that rather thick coatings should have to be used. Tritium inventory would probably also be high in this case, whereas in the surface-limited regime a significant permeation reduction could be obtained with a very thin coating, eliminating inventory problems. For instance, 50 μm of TiC would absorb the same amount of tritium as 1 cm of steel. Another possibility would be to make use of a carbide where Ti has been replaced by a lower-solubility material.

6. CONCLUSIONS

Thin CVD TiC coatings on steel slow down hydrogen absorption and reemission, by lowering diffusivity and surface rate constants, whereas hydrogen solubility seems higher than for steel. This might be a source of concern only for thick coatings in the diffusion-limited region. Thin coatings in the surface-limited region might effectively reduce tritium permeation through blanket modules and coolant tubes.

ACKNOWLEDGEMENTS

The authors owe much to Dr. D.A. Blackburn (Open University, U.K.) for advice and helpful discussions and wish to acknowledge Mr. M. Mariotto (Materials Div., JRC Ispra) for sample preparation, Mrs. N. Williams (Open University) and Mr. T. Sasaki (Materials Div., JRC Ispra) for SEM work.

REFERENCES

1. M. Caorlin et al., The impact of tritium solubility and diffusivity on inventory and permeation in liquid breeder blankets, in: Proc. 3rd Top. Meet. Tritium Technology, Toronto, May 1-6, 1988, in print.

2. R.F. Mattas, J. Nucl. Mater., 141-143 (1986) 29.

3. D.R. Harries et al., J. Nucl. Mater., 133-134 (1985) 25.

4. J.M. Dupouy et al., J. Nucl. Mater., 141-143 (1986) 19.

5. F. Brossa et al., Proc. 13th SOFT, Varese, September 24-28, 1984, Vol.2, pp.1267-1273.

6. T. Yamashina, Proc. IX IVC-V ICSS, Madrid, September 26-30, 1983, pp.614-626.

7. K. Sone and Y. Murakami, J. Nucl. Mater., 121 (1984) 254.

8. F. Reiter et al., Fus. Technol., 8 (1985) 2344 and references therein.

9. F. Reiter et al., Proc. IX Congr. Naz. Vuoto, Firenze, October 7-10, 1985, Vol.1, pp.91-99.

10. D.M. Grant et al., J. Nucl. Mater., 152 (1988) 139.

FUSION TECHNOLOGY 1988
A.M. Van Ingen, A. Nijsen-Vis, H.T. Klippel (editors)
Elsevier Science Publishers B.V., 1989

THERMOMECHANICAL AND EROSION ASSESSMENT OF THE PROTECTION OF THE NET FIRST WALL

C.U. WU, E. ZOLTI, M.I. BUDD

The NET Team, c/o Max-Planck-Institut für Plasmaphysik, Boltzmannstrasse 2, D-8046 Garching bei München

Following the choices for the predesign of the NET basic machine a complete protection of the First Wall for the operational physics phase is being developed, envisaging different constructive, cooling and Low-Z material options. Presently, the most studied concepts consider mechanically attached tiles made of graphite or SiC with or without fibre reinforcement and cooled mainly by radiation or conduction through an intermediate flexible graphite layer.

This paper describes firstly recent progress in the thermomechanical assessment. The analyses are based on 2-dimensional transient, static and elastic finite element calculations and aim at a preliminary comparative evaluation of various situations (e.g. material and dimensions of the tiles, thickness of the intermediate layer, burn and dwell time) with respect to maximum temperatures, stresses, their variation in the cycle and to implications for the integrity of the assessment of the thermomechanical viability of a FW protection concept are discussed. These include effects of neutron irradiation (despite the relatively low fluence of the physics phase) e.g. thermal conductivity reduction, differential shrinkage or swelling and differential irradiation creep, statistical aspects of the graphite or SiC behaviour and effects of electromagnetic forces due to disruptions.

For the assessment the mechanisms responsible for graphite erosion: physical sputtering, chemical erosion, radiation enhanced sublimation and thermal sublimation, have been taken into account. The results show that the erosion is strongly temperature dependent, with erosion varying between 1 mm/y at room temperature to 10 mm/y at 2000 K. The tritium retention in NET protective graphite as a function of temperature has been estimated and consequences are discussed.

In addition, potential benefits and disadvantages and open questions on carbon fibre reinforced graphites as compared to fine grain graphites, particularly with regard to their thermal and mechanical properties, their response to neutron irradiation effects and industrial capabilities are reviewed.

FUSION TECHNOLOGY 1988
A.M. Van Ingen, A. Nijsen-Vis, H.T. Klippel (editors)

MATERIALS FOR THE PLASMA FACING SIDE OF THE FIRST WALL OF FUSION DEVICES - HIGH HEAT FLUX TESTS AND ELECTRON BEAM PROFILE MEASUREMENTS -

K. Koizlik, H. Bolt, H. Hoven, J. Linke, H. Nickel, E. Wallura

Nuclear Research Center Jülich, Association EURATOM-KFA, P.O.Box 1913, D-5170 Jülich, Fed. Rep. Germany

A customary facility to perform high heat flux tests on samples of candidate plasma facing materials is an electron beam device. To quantify the relation material damage/deposited energy, energy distribution under test conditions it is necessary to determine the power density profile in the electron beam. Such measurements have been carried out in two electron beam devices. Results show profiles which can deviate significantly from the ideal axially symmetrical Gaussian distribution.
In a second part, this paper summarizes results of electron beam experiments to test
- the fatigue and erosion behaviour of monolithic SiC, SiC + 5 % AlN, and AlN
- the thermal shock behaviour of graphite with and without binder
- the erosion behaviour of 2-directional carbon fiber composites (CC).
Finally, the poor thermal stability of a SiC-fiber reinforced SiC is demonstrated, and a new type of material, a bulk SiC produced by the coat-mix-process is introduced.

INTRODUCTION

One important aspect of plasma-wall interaction is the deposition of thermal energy in the surface of the plasma facing components /1/. Disregarding the fusion neutrons, the energy is transported from the plasma to the first wall by radiation, electrons, ions, and neutrals. Typical values that are discussed for this type of energy deposition are some 0.5 MWm^{-2} for the low loaded part of the first wall. About 5 MWm^{-2} are admitted to the high heat flux components during normal plasma operation, whereas during unstable plasma conditions, e.g. disruptions, the power density may reach some 50...100 MWm^{-2} during 20 ms (slow energy deposition) or even more than 500 MWm^{-2} for less than 2 ms (fast energy deposition). The surface temperatures of the plasma facing materials range - during normal plasma operation - from considerably below 1000°C to more than 1800°C, depending on the technical concept and the type of cooling.

To overcome the problems of plasma-wall-interaction under the mentioned thermal conditions the first wall shall be protected by tiles or thick (sacrificial) coatings of carbon based materials - graphites, CC, pyrocarbon - or heat conducting ceramics, either losely bonded to the first wall or brazed to metallic structural materials. These materials and material compounds have to be tested with respect to their thermal fatigue and thermal shock behaviour. Actually, customary facilities to perform such tests are electron beam machines /2/.

A necessary condition for a quantitative interpretation of test results is the information about the total power and in particular the power profile across the electron beam.

ELECTRON BEAM PROFILE MEASUREMENTS

The ideal power profile in an electron beam for material tests would be a constant value of the power density distribution across the circular beam cross section and a very sharp rise of the power density from zero to the mean value at the edge of the beam. Since this test condition cannot be realized - at least not with a stationary electron beam - the best approximation would be a Gaussian distribution of the power density. This assumption is acceptable as long as only qualitative interpretations of the materials response to the energy deposition are carried out. When more precise, quantitative evaluations are taken into account one has to know the real beam power density distribution.

Two procedures have been used for the determination of the real beam power density distribution, fig. 1:

- a current measurement of a thin beam which is masked by a tungsten aperture. For the profile measurement the beam is scanned in the x-direction while the aperture is moved in the y-direction. This method is limited to electron beams with a total current < 50 mA due to noise problems.
- the perforation of a stack of thin stainless steel or Ta foils. The diameters of the holes burned into the superposed foiles represent the resp. radial beam profile.

Fig. 2 shows a stack of perforated foils after its exposure to an electron beam. The shape of the power density distribution in an electron beam is plotted in fig. 3. All machine parameters have been kept constant except for the beam current. One measurement has been made in a beam with 60 mA beam current fig. 3 a, the other with 200 mA, fig. 3 b. Two conclusions can be drawn from these graphs:

- The power density distribution across the beam cross section is not in the least of Gaussian type, not even circular. Secondary maxima occur, and the highest power density reaches 2.5 $MWcm^{-2}$ in the case of the 200 mA beam although the average power density for a nominal beam spot size of 1 cm^2 is just 30 $kWcm^{-2}$.
- The shape of the power density distribution depends on the beam current. The inhomogeneities are more pronounced the higher the beam current is.

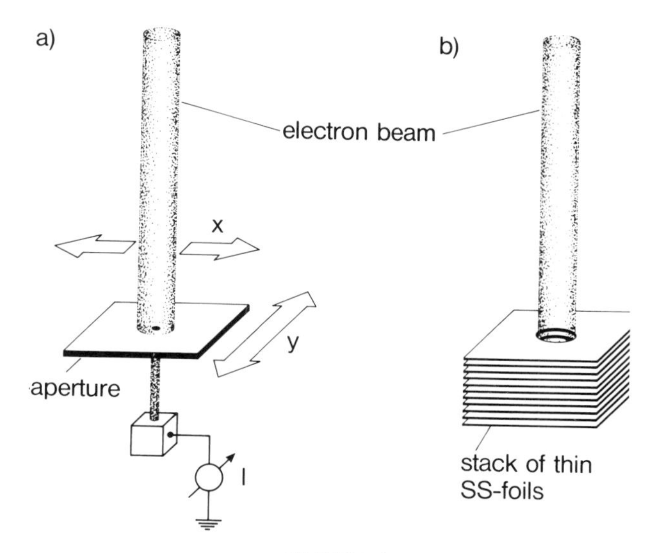

FIGURE 1
Schematic representation of electron beam profile measurements;
a) beam current measurement of a thin beam, masked by a tungsten aperture; applicable for electron beams with a total current < 50 mA
b) perforation of a stack of thin foils of stainless steel

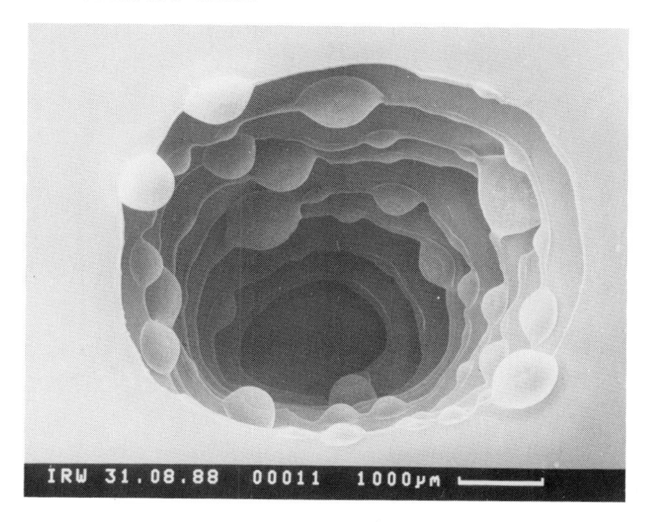

FIGURE 2
Micrograph of a stack of perforated foils after its exposure to the electron beam

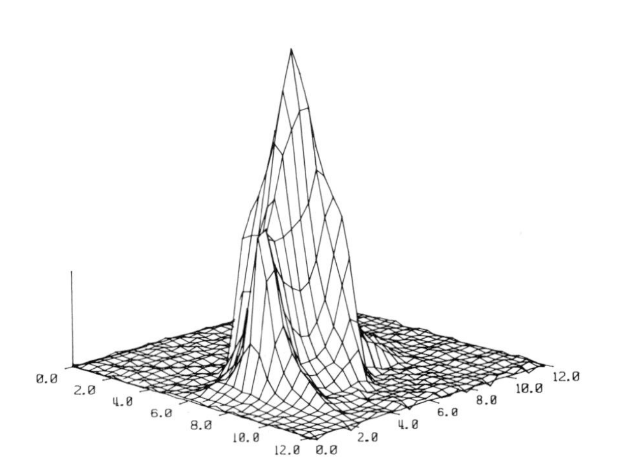

a

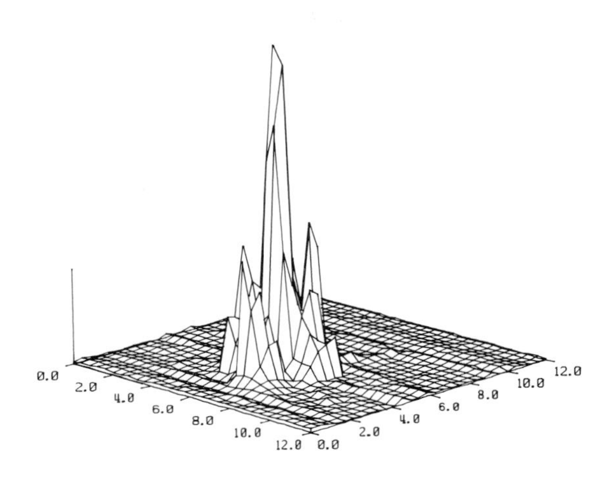

b

FIGURE 3
Shape of the power distribution in electron beams, acceleration voltage = 150 kV
a) beam with 60 mA current
b) beam with 200 mA current

RESULTS OF MATERIAL TESTS IN AN ELECTRON BEAM DEVICE

Inspite of the severe inhomogeneities of the electron beam, material tests in such a device are valid and informative, at least as far as the results are interpreted qualitatively or - with the necessary precaution and care - semi-quantitatively. The direct comparison of the behaviour of different materials is meaningful in any case /3/. This can be seen in fig. 4, which shows micrographs of the surface of two qualities of isotropic fine grain graphite after a thermal shock test with an average beam power density of 100 MWm^{-2} (equivalent to about 70 mA beam current) and 100 ms pulse duration: The binder-free quality (a) shows severe cracking, whereas the conventionally bonded graphite (b) shows only very few, small cracks which, moreover proceed only vertical to the sample surface. Additionally, fig. 4 shows the strong ellipticity of the erosion crater corresponding to the inhomogeneity of the beam profile.

Nevertheless it is surprising that the non-uniformity of the electron beam affects the material damage not much more pronounced. That might be due to a certain equalization of the energy peaks with respect to the resulting material damage, at least when the pulses are sufficiently long, may be 50 ms, a value which may be deduced from the previous electron beam tests.

As compared to graphites, CC-materials show a better crack behaviour under thermal load but at least in the case of 2-directional CC the erosion of carbon is higher than that of "good" graphites /4/. Fig. 5 shows a micrograph of an erosion crater after a single shot electron beam test, 50 MWm^{-2}, 1 s. The same is true for a SiC-fiber reinforced SiC. The special problem with this SiC-fiber material is its poor thermal stability. Already after an annealing at 1400°C for 8 h the SiC matrix recrystallizes, and the fibers begin to dissolve, fig. 6.

Finally, a new "family" of SiC materials produced by the coat-mix-process has been tested for the first time with respect to its thermal shock behaviour. The results available so far prove that the thermal shock resistance of the new material is superior to all other monolithic SiC qualities investigated so far.

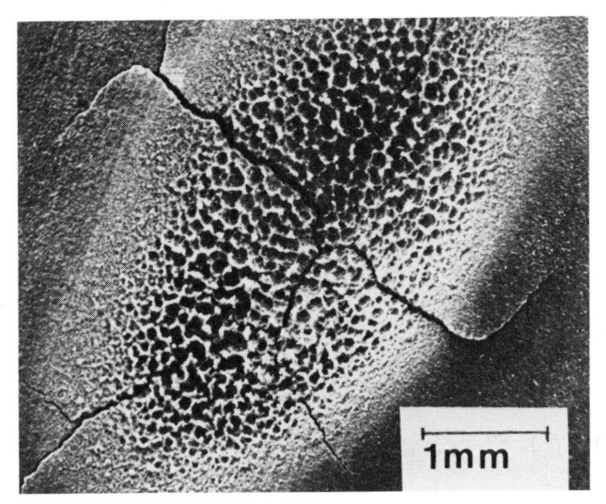

a

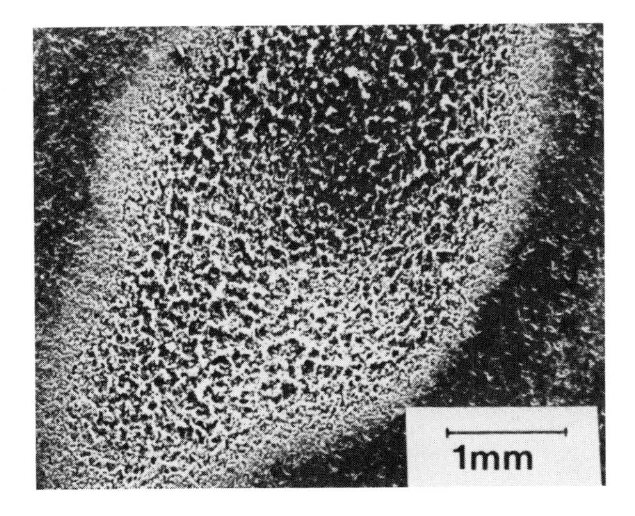

b

FIGURE 4
Micrographs of the erosion craters of two graphite qualities after a single shot electron beam test, 100 MWm^{-2}, 100 ms
a) ultra fine grain graphite, grain size 4 μm, without binder
b) fine grain graphite, grain size $\leq$ 20 μm, conventionally bonded

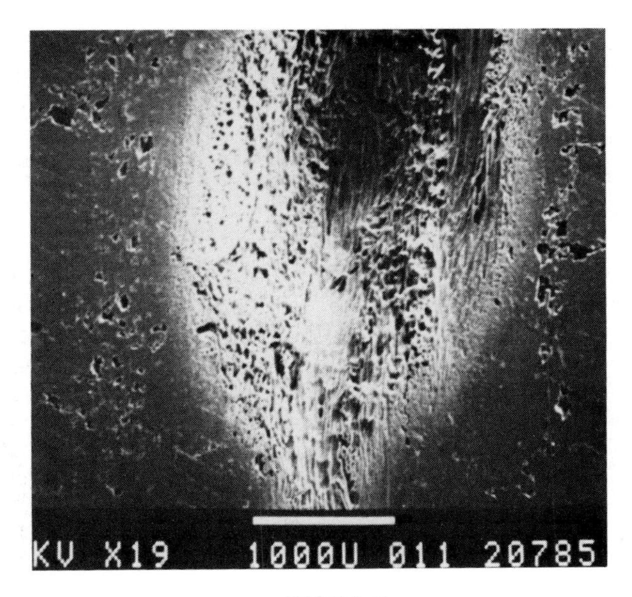

FIGURE 5
Micrograph of the erosion crater on a 2-directional CC material after a single shot electron beam test, 50 MWm^{-2}, 1 s; electron beam vertically to both fiber directions

SUMMARY

Electron beam machines are customary devices to perform tests with plasma facing materials for fusion application. The power density distribution in electron beams was determined. Two experimental procedures for this purpose are presented. The results of such measurements show

- a severe inhomogeneity of the electron beam,
- a dependence of the power density profile on the beam current.

Materials tests prove that

- conventionally bonded fine grain graphites have a better crack behaviour under thermal shock load than binder-free qualities,
- CC materials show an excellent crack behaviour but a very bad erosion behaviour, at least for 2-directional qualities,
- SiC-fiber reinforced SiC behaves similar to CC under pulsed thermal load, but has a bad thermal stability,

- Bulk SiC produced by the coat-mix-process shows a significantly better thermal shock behaviour than "conventional" SiC qualities.

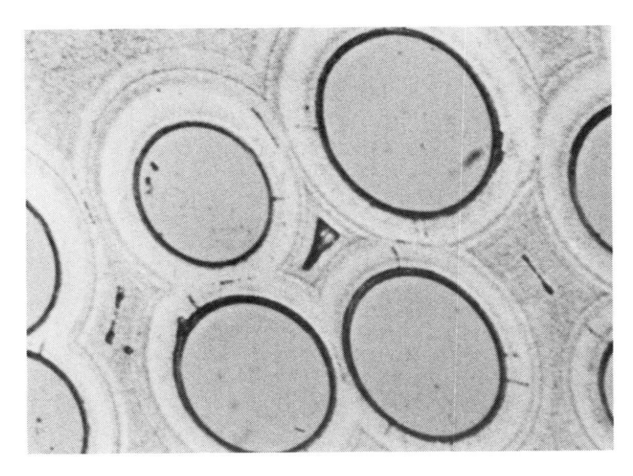

a

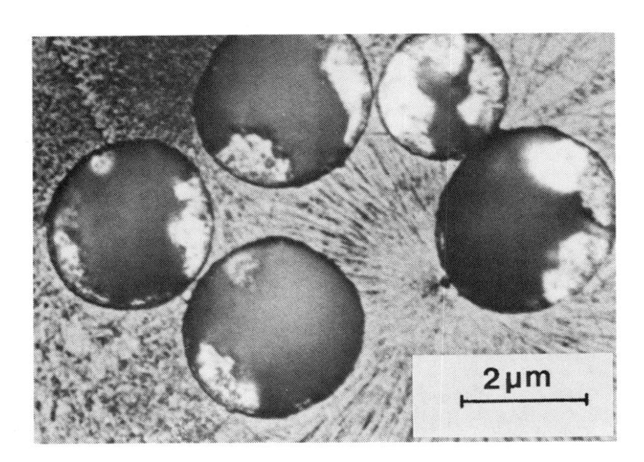

b

FIGURE 6
Ceramographic section of SiC fiber reinforced SiC
a) as produced
b) after annealing at 1400°C for 8 h

REFERENCES

/1/ 3rd Int. Conf. Fusion Reactor Materials, Abstracts, Karlsruhe, 1987 and preceeding conferences

/2/ R.D. Watson, C.D. Croessmann, J.G. Watkins, J.F. Dempsey: Sandia Report SAND 87-2452, Albuquerque, 1988

/3/ W. Delle, J. Linke, H. Nickel, E. Wallura: Berichte der Kernforschungsanlage Jülich, Jül-Spez-401, Jülich, 1987

/4/ H.H. Bolt: "Verhalten von Kandidatenwerkstoffen für Fusionsanwendungen unter thermoschockartigen Oberflächenbelastungen", Dr.-Thesis, Technical University of Aachen, Aachen, 1988

/5/ H. Luhleich, in: Jahresbericht der Kernforschungsanlage Jülich 1979/80, 39, Jülich, 1980

FUSION TECHNOLOGY 1988
A.M. Van Ingen, A. Nijsen-Vis, H.T. Klippel (editors)

EVALUATION OF HIGH TEMPERATURE BRAZES FOR GRAPHITE FIRST WALL PROTECTION ELEMENTS

I. Šmid, E. Kny*, K. Koizlik, J. Linke, H. Nickel, E. Wallura

Nuclear Research Centre Jülich, Institute for Reactormaterials, KFA-EURATOM-Association, P.O. Box 1913, D-5170 Jülich, Federal Republic of Germany

*Metallwerk Plansee GmbH, A-6600 Reutte, Austria

Four different high temperature brazes with melting points from 800 to 1865°C have been used to braze a commercial reactor grade graphite to TZM substrates. Those brazes were Zr, 90Ni 10Ti, 90Cu 10Ti and 70Ag 27Cu 3Ti (wt %).
The resulting composite tiles of 80 x 80 mm^2 with a graphite thickness of 10 mm brazed on a 3 mm TZM substrate have been tested in electron beam experiments for their thermal fatigue properties. The parameters of the electron beam testing were chosen to match NET design specifications for normal operation and "slow" peak energy deposition. The resulting damages and microstructural changes on the graphite and the brazes are discussed. Additional information is supplied on tensile test and thermal conductivity data of brazed composites. These measurements confirm that thermal contact between TZM-substrate and graphite is improved by brazing.

1. INTRODUCTION

Due to the high sublimation temperature, low-Z number and good thermal shock resistance, first wall protection structures for magnetic confinement experiments are preferably made from carbon materials. This is particularly true for plasma interactive components exposed to high heat fluxes. The bonding of a bulk carbon material to a metallic supporting structure has to meet several requirements:

- high thermal conductivity
- thermal shock fracture toughness
- good thermomechanical behaviour
- matched thermal expansion coefficients at the interface
- interface integrity even after long-term operation
- chemical and thermodynamical compatibility with the plasma
- favourable neutron irradiation behaviour (14 MeV neutrons).

In order to obtain a reliable bonding which could satisfy these needs, especially the good thermal contact across the interface, high temperature brazing was applied.

Recent investigations showed the difficulties of brazing fine grain graphites and carbon composites to stainless steel[1,2]. Pure molybdenum and its alloy TZM promise to be better candidates for substrate materials because of their attractive high temperature properties as well as thermal expansion coefficients matching the values of graphite - see table 1. For that reason considerable research work has been done at different laboratories in the past few years concerning molybdenum/graphite composites for fusion application[3-6].

2. EXPERIMENTAL

The bulk materials used were very fine grain graphite and a high temperature alloy of molybdenum (TZM). Physical properties of the bulk materials together with relevant values of the four different brazes are given in table 1. Graphite and TZM-tiles were brazed at Metallwerk Plansee, Austria, under vacuum better than 10^{-4} mbar at temperature conditions indicated in table 1. Brazes containing Ag were handled under Ar- or He-atmosphere to reduce evaporation due to the high vapor pressure of Ag at elevated temperatures.

Thermal loading experiments were carried out with an electron beam welding machine at the Nuclear Research Centre Juelich, Germany. The samples used were cut of the original tiles (80 x 80 x 13 mm^3) and had a final size of 25 mm x 25 mm with 13 mm thickness. The relevant loading parameters are listed in table 2.

To study the thermal fatigue behaviour of the brazes the power density applied was normalized to the brazing plane and corrected for losses by backscattered electrons, whose amount was estimated to be 20 %.

Loading geometry was the following: vertical beam (downwards), horizontal samples (graphite on top), placed on a massive copper plate. Usually 50 loading cycles of 1 second pulse duration and 9 s interval time as well as 5 s pulse and 5 s interval time were applied. To minimize graphite surface damages due to beam inhomoge-

Table 1 a: Properties of bulk materials used

Material	Graphite	Substrate	Molybdenum
Trade name	CL 1116 PT (fine grain graphite)	TZM (refractory molybdenum alloy)	-
Specimen geometry (mm)	80 x 80 x 10 and 25 x 25 x 10 resp.	80 x 80 x 3 and 25 x 25 x 3 resp.	-
Composition (wt %)	C	99.3% Mo, 0.5% Ti, 0.08 % Zr	Mo
Density (g/cm^3)	1.82 (2.27 theor.)	10.15	10.20
Open porosity	10 %	-	-
Expansion coefficient at RT (1/K)	$5.5 \cdot 10^{-6}$	$5.3 \cdot 10^{-6}$	$5.3 \cdot 10^{-6}$
Thermal conductivity at RT ($W/m \cdot K$)	90	126	142
Melting point (°C)	~ 3700 (subl.)	(2620)	2620

Table 1 b: Properties of brazes used

Nominal Composition (wt %)	70Ag 27Cu 3Ti	90Cu 10Ti	90Ni 10Ti	Zr
Composition after brazing wt% (EDX), metals only	30Ag 50Cu 15Ti 5Mo	83Cu 16Ti 1Mo	27Ni 3Ti 70Mo	66Zr 34Mo
Brazing temeratures (°C)	800 - 850	950 -1020	1330-1400	1520-1865
Braze thickness (μm) prior to brazing	100	100	100	100
after brazing	appr. 40	appr. 35	400	130
Penetration depth into graphite (μm)	max. 600	max. 800	max. 50	max. 100

neities, the electron beam was defocussed and rotated with 200 Hz to decrease local power density by increasing the loaded area of graphite.

The temperature at the graphite/substrate interface was recorded while loading. Temperatures above 1000°C were obtained pyrometrically, lower temperatures were measured with a Ni-NiCr-thermocouple.

To complete the information evaluated by electron beam investigation additional thermal conductivity experiments using the laser flash method as well as tensile strength tests were performed. For the later methods pure molybdenum was used instead of TZM.

Table 2: Parameters of electron beam testing

General parameters of the electron beam:	
accelerating voltage	150 kV
beam current max.	100 mA
pulse duration min.	10 ms
beam rise time	0.2 ms
Parameters for thermal fatigue tests:	
pulse duration / interval time	1 s / 9 s and 5 s / 5 s resp.
focus diameter	appr. 10 mm
frequ. of beam rotation	200 Hz
number of cycles	50
power density (in the brazing plane)	0.25 - 10 MW/m^2

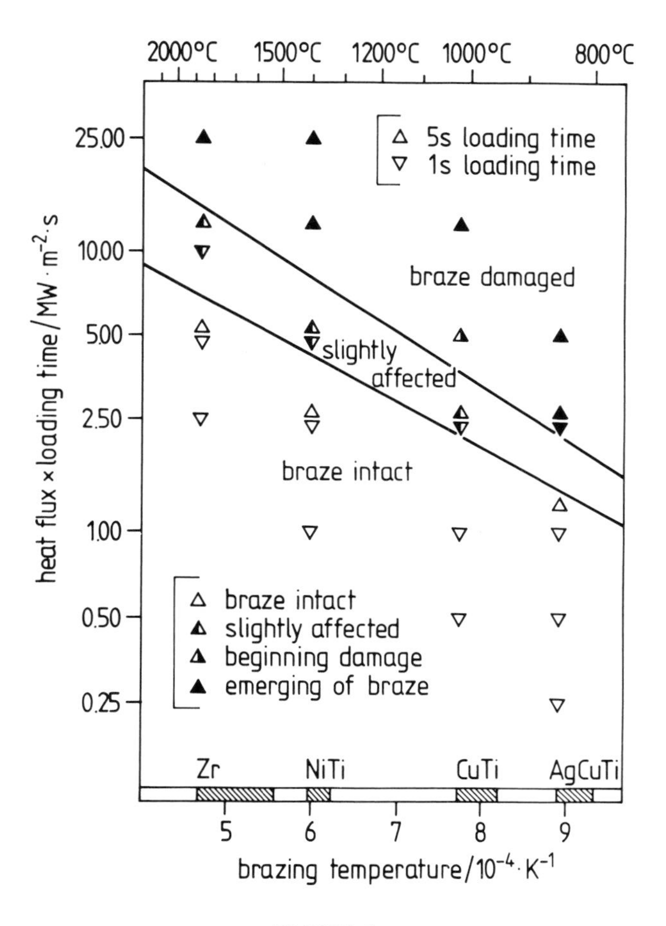

FIGURE 1

Thermal fatigue testing results for graphite/TZM composite tiles joined by different high temperature brazes. The loading experiments were carried out with a 150 kV electron beam. The indicated heat flux was normalized to the brazing plane.

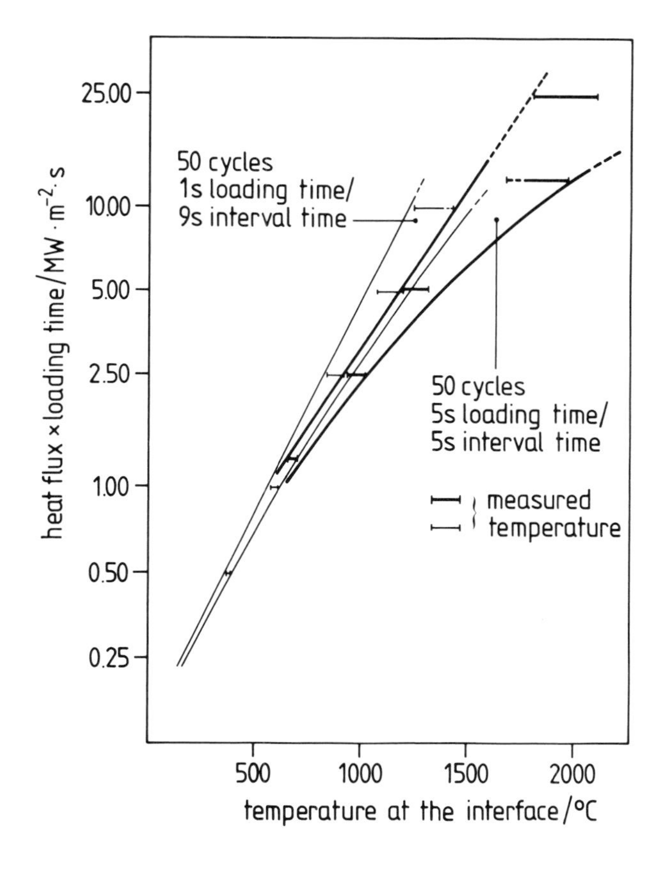

FIGURE 2

Temperature obtained at the brazing interface in relation to the thermal load at the graphite surface

3. EXPERIMENTAL RESULTS

3.1 Damage threshold

The damage threshold for all investigated composite samples has been plotted as a function of input power and brazing temperature (fig. 1). The damage threshold was defined as the first occurance of microstructural changes in the braze region or emerging of molten braze.

A change in composition does not indicate a damage in the braze. This is rather an indication of the performed brazing reaction of the active brazes and by this of the enhanced adhesion. The erosion of the graphite on the exposed surface did not usually result in a damaged interface.

The weight losses caused by the electron beam loading were recorded for each sample. Since braze evaporation occured in general together with graphite erosion the weight losses seem not suitable to qualify changes in the braze.

The temperature measured at the interface correlated only with the loading parameters and not with the braze composition. An exceeded melting temperature in any case caused at least mictrostructural changes at the interface. Since the loading experiments lasted 500 s total and the temperature quickly reached steady state, the samples were kept at elevated temperatures close to 8 min. Emerging of molten braze, evaporation in particularly of Ag and high rates of graphite erosion were observed. The results of the temperature measurements for the interface are given in fig. 2.

3.2 Microstructural changes of the brazes

Although the samples were subjected to 50 thermal cycles, crack formation was not observed. Only vertical cracks were observed crossing the NiTi interfaces already present after brazing. Cracks introduced into the exposed graphite surface did not grow into the interface region.

Fig. 3 gives a comparison of unloaded and loaded composite samples for all brazes discussed. The SEM micrographs were taken before and after the loading experiments. Intact brazes as well as the first effects in the interface are shown. The first appearing damage is different for each braze and will therefore be discussed separately.

However, in no case a complete separation of graphite and substrate appeared.

a) braze: 70Ag 27Cu 3Ti

In the brazing process a high amount of Ag is evaporated due to its high vapour pressure (see table 1). Several different phases are recognizable in the braze. 5 wt% of molybdenum originating from the substrate are found distributed in the whole braze region indicating a rapid diffusion in molten braze.

b) braze: 90Cu 10Ti

Only a negligible amount of molybdenum is found in the braze, ~1 wt% (values before loading see table 1). Neither in metallographic sections nor in SEM pictures different phases can be discriminated in the inner regions before loading. Braze decomposition has taken place resulting in a copper-rich phase of ~ 35 µm thickness (~99 wt% Cu) adhering to the substrate and a titanium-rich interlayer of ~ 2 µm thickness (~75 wt% Ti) adhering to the graphite.

c) braze: 90Ni 10Ti

Ni alloys are dissolving molybdenum along the grain boundaries - here causing an increase in braze thickness up to 400 µm after brazing and even ~600 µm after intense loading. The grains close to substrate showing a composition same as TZM are surrounded by a Ni-rich alloy. The Ni/Ti ratio appears unchanged indicating a negligible selective evaporation at higher temperatures.

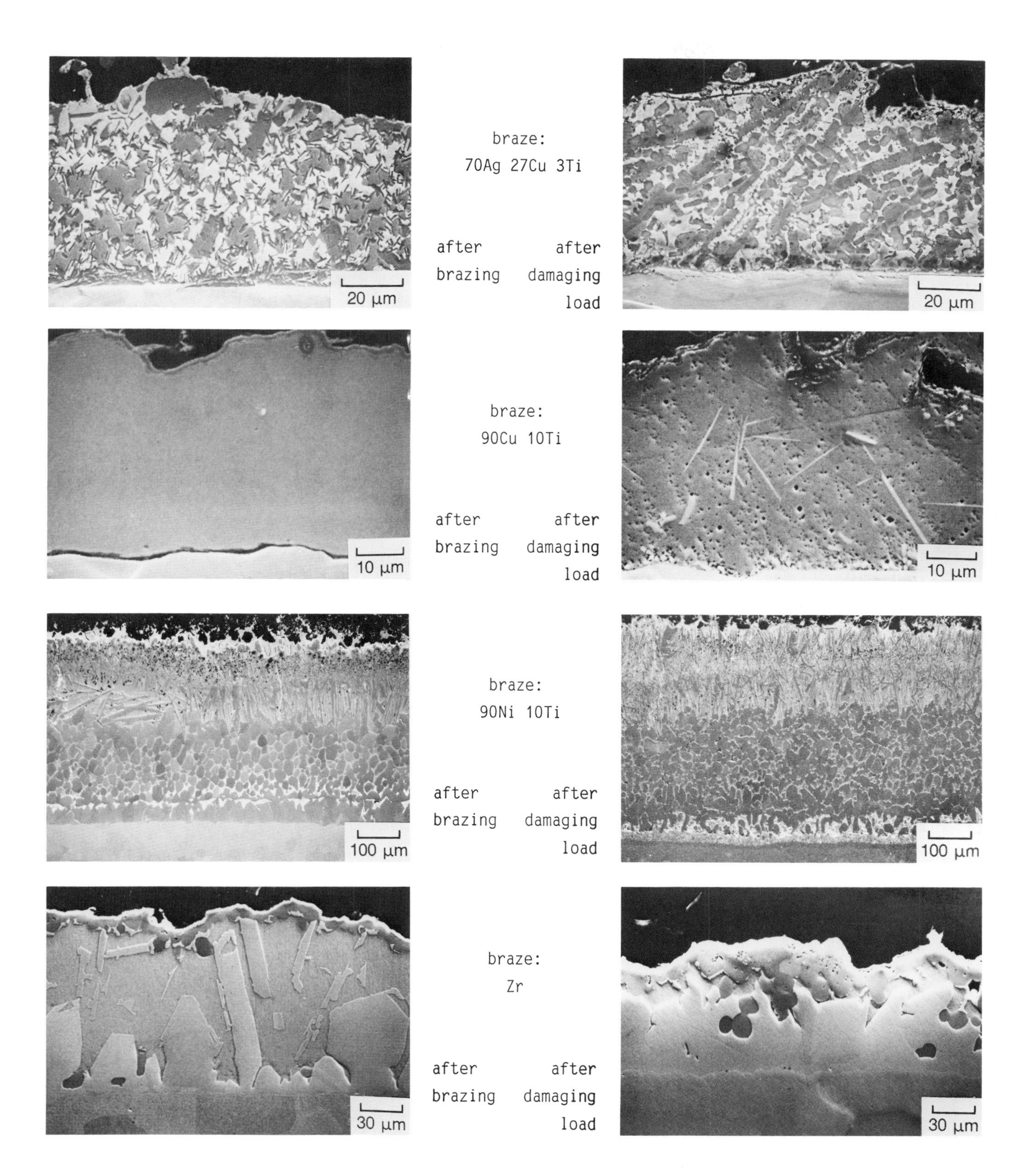

FIGURE 3 SEM micrographs of graphite/TZM composites before and after exposure to thermal loads by the electron beam

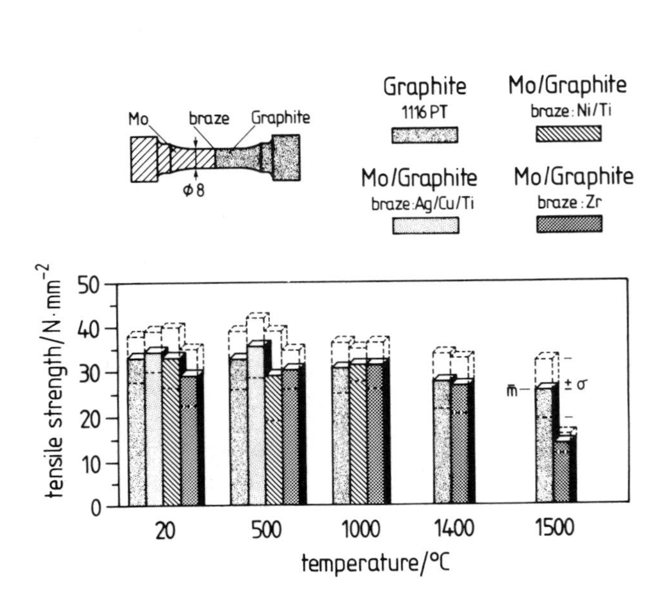

FIGURE 4

Tensile strength data of graphite/molybdenum composites in comparison to bulk graphite

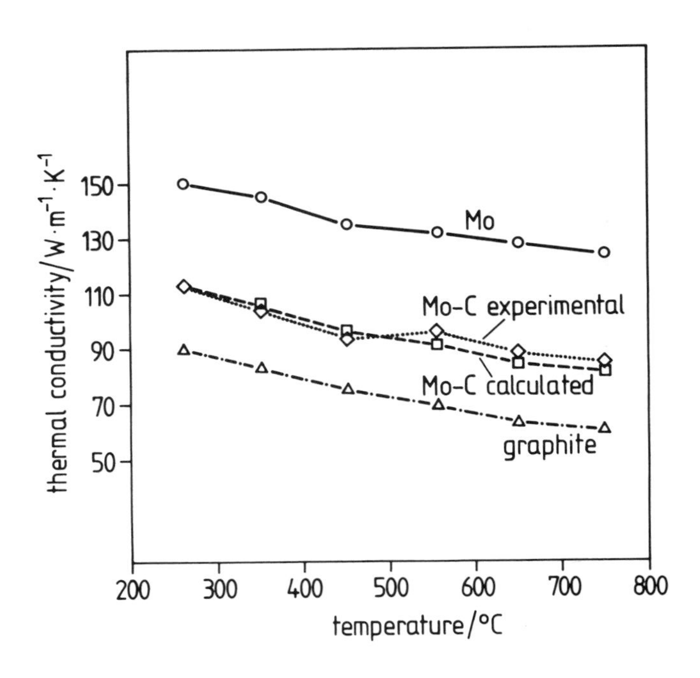

FIGURE 5

Thermal conductivity of graphite/molybdenum composites, braze: 70Ag 27Cu 3Ti

d) braze: Zr

The best high temperature behaviour of all tested brazes was obtained for the Zr-braze. Heat fluxes up to 5 MW/m^2 (1 s) in the brazing plane causing an interface temperature of 1250°C do not affect this braze in any way. Even temperatures of 1350°C caused no considerable changes in opposite to all other brazes. Higher loads resulting in an emerge of molten braze due to temperatures of ~2000°C at the interface are disabling mainly the graphite material but no complete separation of graphite and substrate was observed.

3.3 Tensile strength and thermal conductivity

Tensile strength data of brazed composites as a function of temperature and compared to bulk graphite is presented in fig. 4. The results prove the suitability of the discussed brazes up to their melting point.

Thermal conductivity data evaluated using the laser flash method at different temperatures prove the good conductivity between substrate and graphite (fig. 5). The expected enhancement in thermal contact compared to graphite is fully confirmed.

4. CONCLUSIONS

The following conclusions on brazed composites can be drawn:

- brazed graphite/TZM joints are resistent to thermal fatigue
- crack formation caused by cycling was not observed neither in bulk graphite nor in the substrate in the vicinity of the interface
- cracks were only found across the NiTi braze - however, those cracks were already present before loading
- exceeding the melting temperature of the braze resulted in microstructural changes as well as in emerging of molten braze
- at moderate temperatures (max. 800°C) all four brazes show satisfactory behaviour
- the highest temperature resistance (up to 1500°C) was observed for the Zr-braze
- 10 mm of graphite are sufficient to shield braze and substrate against temperature shocks
- brazing techniques applied are suitable to join fine grain graphite and molybdenum (TZM)
- the composites obtained withstand thermal shock loading common in fusion application.

Brazing and further development of composites was performed by N. Reheis, Metallwerk Plansee GmbH, Austria. Metallographic preparation and light microscopy have been carried out by H. Hoven, KFA-Jülich, and thermal conductivity measurements by W. Neumann, Österreichisches Forschungszentrum Seibersdorf GmbH, Austria. The research stay of I. Šmid at KFA-Jülich has been made possible by the support of the Friedrich-Schiedelstiftung für Kernfusionsforschung, Austria. All contributions are greatfully acknowledged.

REFERENCES

1. K. Ioki, M. Nayama, S. Tsujimura, M. Seki, T. Horie: Int. Symp. on Fusion Nuclear Technology, April 10-15, 1988 Tokyo, Japan

2. Y. Gotoh, H. Okamura, S. Itoh, T. Wada, Y. Karatsu: Int. Symp. on Fusion Nuclear Technology, April 10-15, 1988 Tokyo, Japan

3. G. Kneringer, E. Kny, W. Fischer, N. Reheis, R. Staffler, U. Samm, J. Winter: Fusion Technology Vol. 2 (1986) 1083-88

4. G. Kneringer, N. Reheis: Proceedings of the 11th Int. Plansee Seminar Vol. 3 (1985) 323-44

5. H.E. Kotzlowski: Fusion Technology Vol. 2 (1984) 1253-58

6. H. Bolt, M. Miyahara, M. Miyake, T. Yamamoto: J. Nucl. Mater. 151 (1987) 48-54

FUSION TECHNOLOGY 1988
A.M. Van Ingen, A. Nijsen-Vis, H.T. Klippel (editors)

DETERMINATION OF THERMAL SHOCK RESISTANCE AND FRACTURE TOUGHNESS OF GRAPHITES AS PLASMA-FACING COMPONENTS FOR FUSION REACTOR DEVICES BY ARC DISCHARGE HEATING METHOD

Sennosuke SATO, Akira KURUMADA, Kiyohiro KAWAMATA

Faculty of Technology, University of Ibaraki, Hitachi, 316, Japan

Teruhiro TAKIZAWA and Kazuhiro TERUYAMA

Hitachi Works, Hitachi Ltd, Hitachi, Ibaraki, 317, Japan

Thermal Shock resistance and thermal Shock fracture toughness of graphite to be used as Plasma-facing first wall of fusion reactor devices are quantitatively determined by applying arc discharge heating at the central area of the disk specimens. The graphites used in the measurements included IG-11 of a fine grain isotropic mold, HCB-18 of a very fine grain mesophase pitch carbon and C/C-B of a carbon fiber felt reinforced composite. The results obtained showed that C/C-B composit has outstandingly high thermal shock resistance and fracture toughness. However, contrary to expectation, HCB-18 was found to have relatively low thermal shock resistance and fracture toughness although it has higher mechanical strengths than IG-11. The measurements were also carried out for IG-11 and HCB-18 after irradiating them ($1.1\text{-}1.5\times10^{21}$ n/cm², >29fJ at 750-1000℃)in a fission neutron reactor (JMTR). The thermal shock resistances and the fracture toughnesses of these graphites were found to degrade to about 68-78% of the unirradiated values after undergoing the neutron irradiations. In this study, tile-like plates of HCB-18 and C/C-B graphites were used as the first wall materials at different positions inside a fusion reactor device, together with a fine grain isotropic graphite ETP-10, which is similar to IG-11. According to a tentative surveillance following the operations, the tiles of HCB-18 were found to be some damaged, but no damage was found for the tiles fo ETP-10 and C/C-B. Especially, the C/C-B composites, which were placed at comparatively intense positions, showed conspicuous integrity. These results agree with the values of thermal shock resistance and fracture toughness determined in this study.

1. INTRODUCTION

Plasma-facing first wall materials for fusion reactor devices are subjected to very severe and complex problems such as very high neutron and thermal flux, erosion due to sputtering by plasma particles, electro-magnetic forces in plasma disruptions,and large thermal stresses due to arcing. Selection of the material to be used in such severe environments is of primary importance.

Graphite materials have begun to be used as the first wall components for various powerful fusion research devices including JT-60 in Japan,because of the low Z (atom number), extremly high melting point, high thermal conductivity and very strong thermal shock resistance. Severities of thermal loadings for the first wall vary with positions, but are said to be subjected to thermal shocks of the magnitude, 10-100 MW/m² per 0.1-1 second, the so-called "shot". It is feared that these thermal shocks may cause cracking and erosion on the surface and/or the bulk of the wall components thereby impairing their function as first wall protection.

In this paper,test results of thermal shock resistance[1] and the fracture toughness[2] are presented for three graphites, IG-11, HCB-18 and C/C-B composite. HCB-18 and C/C-B are actually used together with ETP-10, a graphite similar to IG-11, as the first wall materials for JT-60. The tests for IG-11 and HCB-18 are carrid out for neutron irradiations in Japan Material Testing Reactor(JMTR), a fission neutron reactor. The thermal shock resistance and the fracture toughness are shown to degrade considerably due to the neutron irradiations. It is also shown that C/C-B composite reinforced by PAN carbon fiber felt has outstandingly high thermal shock resistance and fracture toughness. According to a tentative surveillance following the operations of JT-60, tiles of HCB-18 were found to be some damaged, but no damage was found for ETP-10 and C/C-B. Especially, C/C-B composites,which were placed at comparatively intense positions, showed conspicuous integrity. These surveillance results tend to agree well with the values of thermal shock resistance and fracture toughness determined in this study.

2. EXPERIMENTAL METHODS

2.1 Thermal Shock Resistance

Tile-like graphites are used as plasma-facing components inside fusion reactor devices. The distribution of stresses in the graphite tiles due to plasma heating is considered to be similar to that of transient thermal stresses in a plate of finite thickness locally heated at its surface. The theoretical analysis to obtain the exact solution to this problem is not feasible at present. Therefore, we decided to simulate it by studying the thermal stress problem of a thin circular disk which is suddenly heated at the central area.

The transient thermal stresses in a disk, which is rapidly heated at the central area, have been analyzed for adiabatic conditions of the disk surfaces by Riney[(3)]. Sato, et al[(1)] had used the results of his analysis for measuring the thermal shock resistance $\Delta (= \sigma_t k/E\alpha$, σ_t : tensile strength, k : thermal conductivity, E : Young's modulus, α : thermal expansivity)of various kinds of graphite by means of arc discharge heating.

The thermal shock fracture toughness ∇ $(= K_{Ic}k / E\alpha$, K_{Ic} : mode I fracture toughness valus)[(2)] of a disk with an edge slit was also determined. But for the actual heated disks, there are some heat transfers from the surfaces to the surrounding atmosphere during the measurments. It is thus necessary to evaluate the influence of the heat transfers on the disk by studying how the temperature distributions and thermal stresses in the disk are affected. Sato,et al[(15)] have recently analyzed the influences of the nondimenaional heat transfer coefficient on the disk surfaces on the thermal stresses as a function of Biot number, Ho.

The maximum stresses are defined as the specified maximum thermal stress S_* $(= \sigma_{\theta\theta}\max / E\alpha Q_0)$ and are used to determine the thermal shock resistance of the disk as follows (1) :

$$\Delta = \sigma_t k/E\alpha \quad (4)$$

$$= S_* \quad \beta W/\pi h(a/R)^2, \quad (5)$$

where $Q_0 = qR^2/k, q(= W/\pi a^2 h)$ is the heating quantity,W is the threshold heating power to generate cracking due to $\sigma_{\theta\theta}$ at the disk periphery and β is the heating efficiency, which is calibrated for the thermal shock testing apparatus. The value of S_* for a/R=0.3 and τ=0.25 under the test conditions in this study are 1.06 $\times 10^{-2}$ at H_0= 0.2. The H_0 for the heated disk used in this study is estimated to be about 0.2 up to 1,500℃ by convection and radiation heat transfers. To compute the heating period t_* corresponding to τ=0.25, thermal diffusivity κ of disk materisls is determined from the measurement of heat propagation in the disk by a transient converter. The values of t_* are calculated from $t_* = \tau R^2/\kappa$, where τ=0.25. Real time corresponding to shot is estimated in a range, τ=0.01-0.1, although it varies with disk size.

2.2 Thermal Shock Fracture Toughness

A disk specimen,with an edge slit of depth c previously machined,was subjected to a thermal shock as described in the preceding section. The threshold electric power W, which represents the limiting value for propagating a crack from the tip of the edge slit,was measured at about H_0=0.2. The stress intensity factor F_1 for the thermal stress distributions was calculated for different values of H_0 and c/R, with a/R =0.3 and τ=0.25. In this manner, we could determine the thermal shock fracture toughness ∇ in terms of lumped physical properties, which are related to thermal stress and fracture toughness through the equation[3],

$$\nabla = K_I ck/E\alpha = F_1 \sqrt{\pi c}\ \beta W/\pi h(a/R)^2, \qquad (6)$$

where the non-dimensional stress intensity factor $F_1=0.0147$ for $a/R=0.3$, $c/R=0.3$ and $H_0=0.2$. The heating period t_* in the arc discharge is the same 3.0s found in the preceding test of thermal shock resistance.

Sato, et al.[15] have recently analyzed the influences of eccentricity of heating area at the disk on the thermal stresses and stress intensity factors. It was revealed that stress distributions in the disk can change considerably by varying the position of heating area even with the same amount of heating.

2.3 Other Mechanical and Fracture Mechanics Properties

By using similar disk specimens, diametral compressive strength by circular anvils[3], mode I and II fracture toughnesses of the disk with a central slit were measured[7,8]. Young's modulus was determined by measurement of ultrasonic pulse speed[9]. Bending strength and relative Charpy impact energy were also measured by using circular rod specimens.

2.4 Graphite Specimenns

The graphite specimenns used in this study included an isostatically molded graphite IG-11 of fine grain petroleum coke,a press molded binderless graphite HCB-18 of very fine grain mesophase pitch carbon and PAN carbon fiber felt reinforced carbon composite C/C-B. Anisotropies of IG-11 and HCB-18 graphite were estimated to be about 1.05 and 1.1,respectively. In the structure of C/C-B, meandering carbon fibers (about 6μm in diameter) are found in coal tar pitch binder. C/C-B composite has a measure of three dimensionality due to peculiar texture of felt-like fiber.

The axial direction of the disk specimen of 20 mm in diameter and 2 mm in thickness for IG-11 and HCB-18 is chosen to coincide with the longitudinal direction of the formed bodies. Therefore the measured data for the disk specimens show the weakest properties in the transverse direction of the bodies in their slight anisotropies. The disk specimen of 30 mm in diameter and 3 mm in thickness for C/C-B was cut to the direction of thickness of the plate material of 160×180×4 mm.

2.5 Neutron Irradiations

Neutron irradiations of IG-11 and HCB-18 specimens were carried out inside two irradiation holes located in the fuel zone of Japan Material Testing Reacter (JMTR) by using two capsules, one of which was temperature-controlled in helium (80M-31U) and the other uncontrolled(82F-9U). The conditions of irradiation were$(1.1\text{-}1.5)\times10^{21}$ (>29fJ) at 750° -1000℃. The conditions of the two capsules were treated to be the same, since average data were taken for the specimens placed in the two capsules and that the differences between the two capsules are small[5].

Irradiations of C/C-B are being carried out under almost the same conditions in JMTR at present.

Round bar specimens for testing the bending strength of each graphite were also used for the measurements of irradiation shrinkage. Young's modulus,electric resistivity and relative Charpy impact absorption energy. Some disk specimens were also used for the measurements of Rockwell hardness.

3. EXPERIMENTAL RESULTS AND DISCUSSIONS

Table 1 summarizes the mean values of experimental results for specimens before and after neutron irradiation. Values in the parentheses show the ratio of "after" and "before" values. After irradiation tests for C/C-B specimens have not been completed at present.

3.1 Thermal Shock Resistance

Fig. 1 shows the thermal shock resistance before and after irradiation which were corrected by β, for two graphites and before irradiation for C/C-B. The lower and upper limits of the cross-hatched or black ranges in the figure show the limit of the fracture possibility. Below the lower limit, no cracks occurred at all, and above the upper limite cracks always occurred within the heating period of 3s. Such ranges of data may reflect the scatter of material properties and measuring conditions. Photograph 1(a) shows the typical appearances of the three graphites following the thermal shock resistance teses. No thermal shock fracture of C/C-B was observed during the heatings up to the maximum capacity (30kW) of the arc heating apparatus. Delaminations, which are apt to occur in two dimensional C/C composites of carbon cloth layer, were not at all observed during the thermal shock heatings by virtue of the peculiar felt texture.

Thermal shock resistances of the irradiated graphites degraded to about 0.68 times the values before irradiation. This value of degradation ratio was just about the same as the mean value previously obtained for four graphites, IM2-24, 7477, H327 and SMG[5]. The resistances of IG-11 graphite were stronger than HCB-18. The degradations of thermal shock resistance of graphite due to neutron irradiations can be partly attributed to the increase in both Young's modulus and thermal expansivity[13] and to the decrease in thermal conductivity[13][14]. In particular, the considerable decrease in thermal conductivity contributes greatly to the reduction of thermal shock resistance. It is noteworthy that C/C-B composite has a very strong thermal shock resistance, more than five times that of IG-11. On the other hand, the value for HCB-18 was low, contrary to expectation.

3.2 Thermal Shock Fracture Toughness

Fig. 2 shows the thermal shock fracture toughness for two graphites before and after irradiation and that for C/C-B before irradiation. Since the disk specimens have an edge slit, the electric power was considerably smaller than the data previously taken during the thermal shock resistance tests.

Typical appearances of the fracture are shown in Photo.1(b). With greater magnification, the crack propagations were found to be meandering from void to void. They were found more frequently at the boundaries of filler grains. Stationary cracks were also observed. The thermal shock fracture toughness of HCB-18 graphite was considerably smaller than that of IG-11. That is to say, HCB-18 is sensitive for thermal shock fracture,reflecting its very high rigidity. The degradation ratios were 0.78 and 0.72 times the values before irradiation for IG-11 and HCB-18 graphite, respectively. These values are fairly larger than the mean value of 0.55 previously obtained for the four graphites and are near the desirable value of 0.79 for IM2-24 graphite of gilsonite coke. Thermal shock fracture toughness of C/C-B was surprisingly high, up to about 5.7 times that of IG-11.

3.3 Other Propertie

Table 1 lists the mean values of mechanical, fracture mechanics and physical properties for IG-11 and HCB-18 graphite before and after irradiation and for C/C-B before irradiation. After irradiation, Young's modulus E of IG-11 and HCB-18 increased to about 1.5 times the values before the irradiation. Mechanical strengths of compression σ_c and diametral compression σ_{Ht} of the two graphites increased by about 20 percent due to the irradiation. The Rockwell hardnesses H_{R15X} of both graphites increased by about 10 percent over those before irradiation. However, impact absorbed energies Ec increased by only about 2 percent. Fracture toughness values K_{IC} and K_{Ic} of both graphites following the irradiation increased to about 1.3 times those before irradiation. The ratios K_{Ic} / K_{IC} =1.23-1.25 before irradiation for both graphites showed almost no change in spite of the irradiation similarly with other graphites[10].

Uniaxial tensile strengths of both graphite were deduced by our fracture criteria[11] using the results of σ_c, σ_{Ht} and K_{Ic}/K_{IC}, and they

Table 1 Mean values of experimental results of specimens before and after irradiation.

Specimen		IG-11		HCB-18		C/C-B
Irradiation		Before	After	Before	After	Before
Apparent density γ (g/cm^3)		1.75	1.75 (1.00)	2.00	2.01 (1.01)	1.77
Porosity P (%)		22.6	22.6 (1.00)	11.5	11.1 (0.97)	21.7
Young's modulus E (GPa)		7.8	11.8 (1.51)	13.7	20.5 (1.50)	26.3
Compressive strength σ_c (MPa)		70.5	89.2 (1.27)	192.1	232.4 (1.21)	66.3
Bending strength σ_b (MPa)		38.7	47.6 (1.23)	78.4	94.4 (1.20)	96.9
Diametral compressive strength σ_{Ht} (MPa)		15.3	18.5 (1.21)	37.1	44.5 (1.20)	-
Tensile strength σ_t (MPa)		23.7*	27.6* (1.16)	52.0*	61.4* (1.18)	55.4
Dimensional change(%)		-	-0.04	-	-0.14	-
Electric resistance ρ ($\mu\Omega$cm)		1150	1927 (1.68)	1208	1970 (1.63)	400
Rockwell hardness H_{R15X}		71	78 (1.10)	91	99 (1.09)	44
Impact absorbed energy E_c (J/cm^2)		0.41	0.42 (1.02)	0.47	0.48 (1.02)	1.16
Mode I fracture toughness K_{IC}(MPam$^{1/2}$)		0.78	1.01 (1.29)	0.75	0.99 (1.32)	3.44
Mode II fracture toughness K_{IIC}(MPam$^{1/2}$)		0.96	1.24 (1.29)	0.94	1.20 (1.28)	4.39
K_{IIC}/K_{IC}		1.23	1.23	1.25	1.21	1.28
Thermal diffusivity κ (mm^2/s)		48.5	26.0 (0.54)	39.3	21.7 (0.55)	56.6
Thermal shock resistance Δ (W/mm)		31.9	21.7 (0.68)	27.5	19.0 (0.69)	>194
Thermal shock fracture toughness ∇ (W/mm$^{1/2}$)		34.4	26.8 (0.78)	20.0	14.3 (0.72)	196
Equivalent crack length (mm)	$C_e=\frac{1}{\pi}\left(\frac{K_{IC}}{\sigma_t}\right)^2$	0.345	0.426 (1.23)	0.066	0.083 (1.26)	1.227
	$C'_e=\frac{1}{\pi}\left(\frac{\nabla}{\Delta}\right)^2$	0.370	0.486 (1.31)	0.168	0.180 (1.07)	-

( ) indicates the ratio (After irradi./Before irradi.).
* indicates the deduced tensile strength.

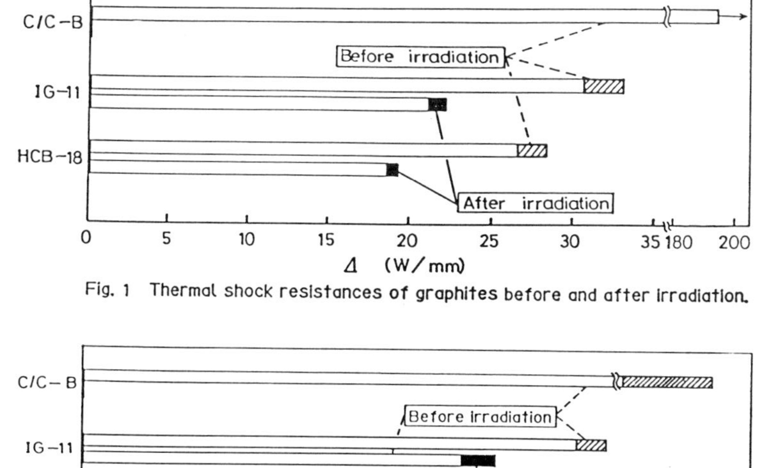

Fig. 1 Thermal shock resistances of graphites before and after irradiation.

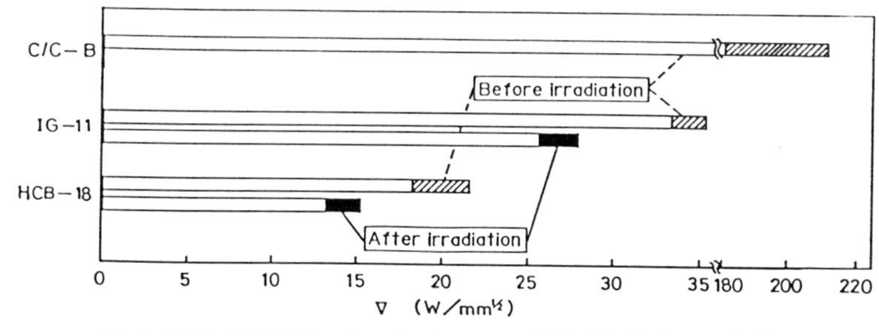

Fig. 2 Thermal shock fracture toughnesses of graphites before and after irradiation.

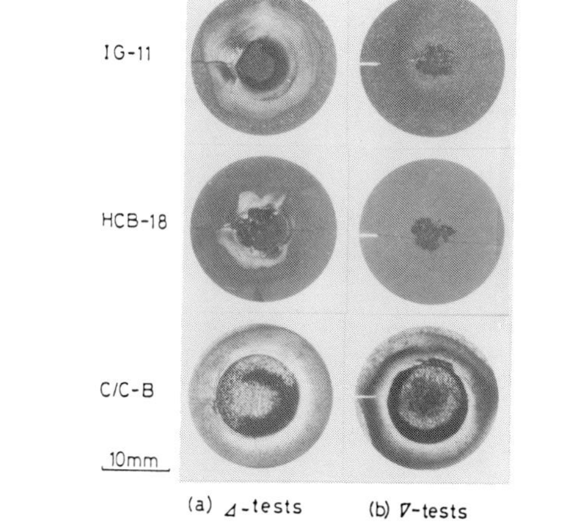

Photo. 1 Typical fractures in (a) thermal shock resistance and (b) thermal shock fracture toughness tests of IG-11, HCB-18 graphites and C/C-B composite.

are also shown in Table 1. The deduced value for unirradiated IG-11 graphite agreed well with the measured data.

Electric resistances of the two graphites increased by about 60 percent due to the irradiation. Such increases may correspond to the decreases in the thermal conductivities of the two graphites. Thermal diffusivities, which are proportional to the thermal conductivities, decreased by about 54 to 55 percent as a result of the irradiation.

Dimensional changes due to irradiation showed shrinkage of 0.04% for IG-11 and 0.14% for HCB-18. This shrinkage of IG-11 agreed with other data reported by JAERI[12]. The value for HCB-18 was remarkably large, but no tendency for worse irradiation damage was found in this study.

4. CONCLUSION

Usually, the irradiation effects on the mechanical strengths of graphites are found to be beneficial for the safety assessment. But graphite placed in a fusion reactor is subjected to severe thermal stresses due to plasma heating and disruption effects. Thermal stresses are related not only to the mechanical properties of the materials, but also to the thermal properties such as thermal conductivity and the expansivity. Fractures due to thermal stress occur when the thermal strain exceeds the strain limit that can be accommodated in the material. In practice, the thermal shock resistance and the thermal shock fracture toughness of graphite tested in this study were degraded considerably due to the newtron irradiation. Therefore, a synthetic approach is required for evaluatry the resistance of materials against thermal stress or thermal shock. Our system of disk best is not only useful for selecting quantitatively a brand of graphite for fusion reactor devices, but also very simple and convenient for carrying out capsule irradiation and post irradiation tests.

Tile-like plates of HCB-18 and C/C-B graphite were used as the first wall materials at different positions of JT-60, together with a fine grain isotropic graphite ETP-10, a graphite considered to be similar to IG-11. Accordingto a tentative surveillance, tiles of HCB-18 were found to be severely damaged after a number of shots during the operations, but no damage was found for the tiles of C/C-B and ETP-10. Especially, C/C-B composites, which were placed at comparatively intense positions, showed conspicuous integrity. Therefore these results are in good agreement with the experimental evaluations of this study.

REFERENCES

(1) Sato,S.,Sato,K.,Imamura,Y.,Kon,J.:Carbon, 13,309(1975).

(2) Sato,S.,Awaji,H.,Akuzawa,H.:Carbon,16,103 (1978).

(3) Riney,T.D.:J.Appl.Mech.,ASME-E,28,631(1961).

(4) Sato,S.,Awaji,H.,Akuzawa,H.:Carbon,16,103 (1978).

(5) Sato,S.,Imamura,Y.,Kawamata,K.,Awaji,H.,Oku, T.:Nucl.Eng.Des,61,383(1980).

(6) Awaji,H.,Sato,S.:J.Eng.Mater.Tech.,ASME-H, 101,139(1979).

(7) Awaji H.and Sato S.J Eng,Mater.Tech.,ASME-H, 100,175(1975).

(8) Sato,S.,Awaji,H.,Kawamata,K,Miyauchi,M.: High Temp.-High Press.,12,23,(1980).

(9) Sato S.and Miyazono S.Carbon,2,103(1964).

(10) Sato,S.,Awaji,H.,Kawamata,K.,Kurumada,A., Oku,T.:J.Atomic Energy Soc.Jpn,(in Japanese), 28 12 ,1172(1986).

(11) Sato S.Awaji H.Kawamata K.Kurumada A.and Oku,T.J.Nucl.Eng.and Design,103,291(1987).

(12) JAERI: Present Status of Research and Development of HTGR,(in Japanese),(1972-85).

(13) JAERI: Present Status of Research and Development of HTGR,27(1985).

(14) Binkele,L.:High Temp.-High Press.,4,401 (1972).

(15) Sato,S.et al.,unpublished.

FUSION TECHNOLOGY 1988
A.M. Van Ingen, A. Nijsen-Vis, H.T. Klippel (editors)

THERMAL DEPTH PROFILES OF ARTIFICIAL GRAPHITES AND THE ABSORPTION OF HEAT PULSES *

B.K. Bein, U.G. Bertsch, U. Krebs, H.W. Schmidt, J. Pelzl
Ruhr-Universität Bochum, Institut für Experimentalphysik VI
D-4630 Bochum, P.O. Box 102148, Federal Republic of Germany

Thermophysical properties of artificial graphites have been investigated by frequency-dependent photoacoustics. Depth dependent profiles have been found for the effusivity, $(k\rho c)^{1/2}$, the relevant parameter for time-dependent surface heating. These thermal depth profiles are related to surface roughness and porosity and are also affected by previous outgassing.

1. INTRODUCTION

In recent years, artificial graphites have widely been used as limiter materials in tokamaks. Although new materials may be required in power producing fusion reactors, where the combined nuclear and plasma heat deposition may be to high, graphite is still favourable for limiters and heat shields in near future research because of the compromise between plasma compatibility and required heat transfer properties. Thus, in tokamak devices like ASDEX UG all components facing the plasma will be armoured with graphite[1]. Apart from the steady state plasma heat deposition which e.g. on the actively cooled divertor plates of ASDEX UG will be of the order of 10 MW/m^2 during 10 s, the plate materials have to resist heat pulses after confinement instabilities or disruptions up to 10 MJ/m^2 with pulse durations of a few milliseconds.

For quantitative studies of heat removal and for the application of the IR thermography in tokamak energy balance studies[2], the thermophysical parameters have to be known. In time-dependent heating processes as observed for limiters, the combined quantities $(k\rho c)^{1/2}$ and $a=k/\rho c$, the thermal effusivity and diffusivity, respectively, are the more relevant thermophysical parameters, rather than the thermal conductivity k, mass density ρ, and specific heat capacity c, separately, as shown by the relation describing the time dependent temperature rise

$$\Delta T_s(t) = (k\rho c\, \pi)^{-1/2} \int_0^t dt'\, t'^{-1/2}\, F_s(t-t') \ .$$

Here $F_s(t)$ is the net heat flux absorbed by a thermally thick plate.

With the help of frequency dependent photoacoustics, artificial graphites without previous plasma exposure have been analysed. Here, room temperature results of thermally thick and thin samples are presented, from which the effusivity depth profile and the bulk diffusivity have been derived. To get information independent of the occasional surface roughness, the samples were prepared under equal conditions by cutting and polishing with silicon carbon paper of grain-size graduation P800 and P1200. The samples, exposed to air for a long time before, were measured with and without previous outgassing.

In Section 2 the experimental method is described briefly. In Section 3 and 4 results are presented obtained from three graphites (Carbon Lorraine 5890PT, Ringsdorff EK98, Schunk & Ebe FP159I), which are in use as wall materials in tokamaks.

* Supported by Max-Planck-Institut für Plasma-Physik, Garching, FRG.

2. EXPERIMENTAL TECHNIQUE

Frequency-dependent photoacoustics is a thermal wave method with audio-acoustic detection by which the thermal response of a solid to an oscillatory heating process is analysed. Amplitude and phase lag of thermal waves directly depend on the effusivity $(k\rho c)^{1/2}$, thermal diffusivity, and propagation distance x, as shown by the thermal-wave solution of a surface-heated semi-infinite solid[3]

$$T(x,t) = \eta\, I_o(k\rho c\; 2\pi f)^{-1/2} \exp[-(\pi f/a)^{1/2}\, x] * \cos[2\,\pi\, ft - (\pi f/a)^{1/2}\, x - \pi/4]\ .$$

Here, η is the optical absorptivity defined as the fraction of the total incident intensity I_o transformed into heat. The diffusion length of thermal waves, $\mu = (a/\pi f)^{1/2}$, can be controled by the modulation frequency f of heating, thus providing a means to measure depth-dependent thermal properties.

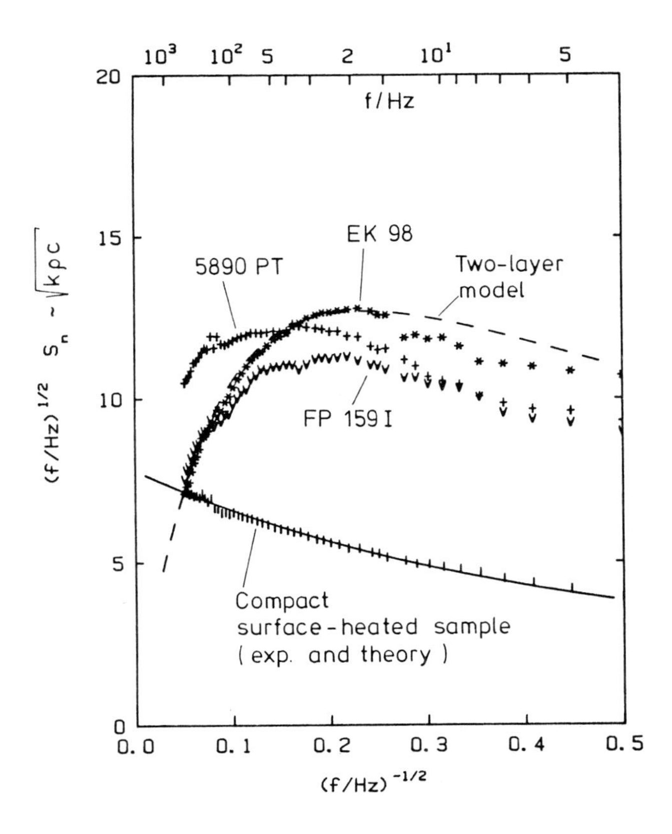

FIGURE 1

Normalized amplitudes measured for polished graphite and a compact test sample, compared to the theory for a two-layer model and a compact solid.

For a quantitative interpretation, the photoacoustic measurement device is calibrated by reference materials of known optical and thermal properties. The reliability of this normalization has already been proved in previous studies using neutral density glass and pure metals[4], where the relative errors of the amplitude measurement were only about 3%. Details of our experiment have been published earlier[4,5].

3. RESULTS OF THERMALLY THICK GRAPHITE SAMPLES

Fig.1 shows normalized amplitudes $S_n = S_r/S_s$ plotted in the form $f^{1/2}S_n$ versus $f^{-1/2}$; neutral density glass is the reference S_r. As the penetration depth from the surface is proportional to the thermal diffusion length, this form of the normalized amplitude provides information about the effusivity depth profile. The signals of the surface-heated compact sample (carbon-coated quartz glass) have been adjusted to the theoretical solution[5],

$$f^{1/2}S_n(f^{-1/2}) = (k\rho c)_s^{1/2}\ (\eta_r/\eta_s)\ (\beta_r/\rho_r c_r)$$

$$* \left[2\,\pi + 2\ (\pi\, a_r/f)^{1/2}\ \beta_r + a_r\beta_r^{\,2}/f\ \right]^{-1/2}$$

where ß is the optical absorption constant, and 's' and 'r' refer to the sample and reference. The fit yields the unknown effusivity $(k\rho c)_s^{1/2}$ as a function of the reference parameters and of the optical absorptivity, which was determined by measurements of the reflectivity. The good agreement between experiment and theory is a test for the reliability of the actual experimental conditions.

The amplitudes of the graphite samples differ from the behaviour of the compact solid by a significant increase with decreasing frequencies (Fig.1). This can be interpreted as an effusivity increasing with the distance from the surface. To lowest order, the signal amplitude of EK 98 can be approximated by a two-layer solution[6]: For small thermal diffusion lengths,

$1/\sqrt{f} \rightarrow 0$, the quantity $\sqrt{f}\, S_n \sim \sqrt{(k\rho c)}_s$ has a low value, characteristic for a thin surface layer; with larger diffusion lengths, between 50Hz > f > 20Hz, the effusivity has a maximum. For further growing diffusion lengths, 10Hz > f > 5Hz, the measured values deviate from the two-layer approximation towards a lower effusivity value.

The allover behaviour of both the signal amplitude and phase lag have satisfactorily been approximated by a three-layer model[6] where the fit parameters are the effusivities $(k\rho c)^{1/2}$ and the thermal diffusion times (l^2/a) of the three layers. These combined quantities are the data required to calculate the thermal response to time-dependent surface heating processes. Here, l and a represent layer thickness and diffusivity. The results of the fit procedure are:

- a very thin first layer with a thermal diffusion time about 20-40 µs and a low effusivity characterizing the material just at the surface,
- a second thicker layer with a diffusion time between 10-20 ms and an effusivity increased by a factor between 2 and 4 in comparison to that of the first layer,
- and the bulk region of the material, where the thermal transport is affected by the volume porosity.

Physically, the thermal depth profiles could be related to the effect of surface roughness and volume porosity on the heating process[7]:

With growing thermal diffusion lengths, the screened surface areas along the open pores, which cannot be heated directly by the incident flux, become accessible with their material to the thermal wave or pulse, thereby contributing to an increasing effective heat capacity and effusivity. When the thermal diffusion lengths finally exceed the characteristic depth of the surface roughness, the effective thermal conductivity and density are reduced by the volume porosity and thus contribute to a reduced effusivity of the bulk material.

We investigated a rough, not polished graphite sample at room temperature before and several days after outgassing. Apart from the geometric effects, we also find an important influence of the outgassing conditions on the effective thermal depth profile. The measured profile before outgassing had a very strong variation with depth, in comparison to the theoretical curves for compact solids (Fig.2). After outgassing, however, the depth variation was reduced considerably. Six days after outgassing, the profiles did not yet show the apparent effusivity maximum of the second layer between 20 Hz and 50 Hz. After further days of exposure to air the thermal depth profiles progressively returned to the original form. Since the profiles always tended to the same values at very low frequencies, the bulk properties were less or even not affected.

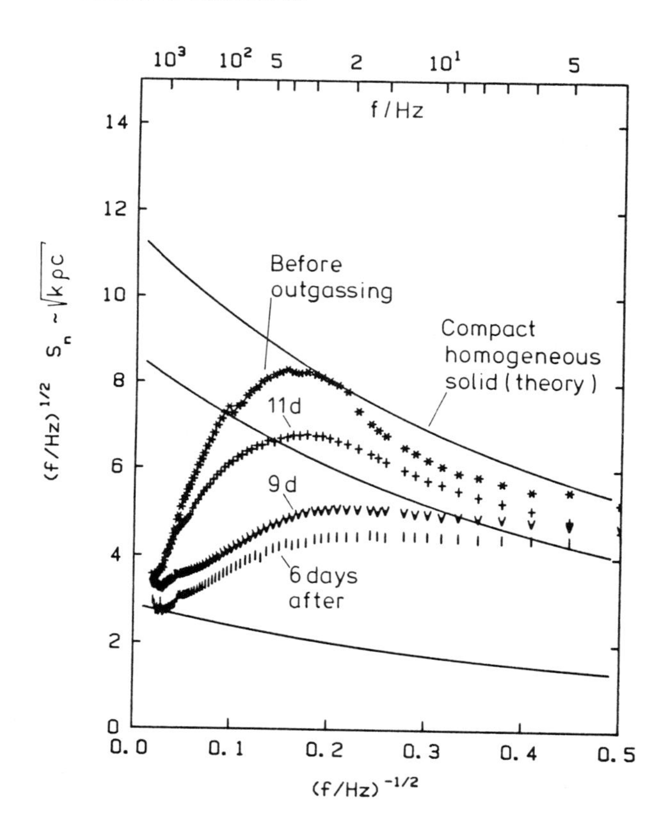

FIGURE 2
Thermal depth profiles of a rough graphite sample before and after outgassing.

The changes due to outgassing and the delayed recovery of the thermal depth profile afterwards are explained by the adsorption of gases (water vapor) at the roughness-increased large surface area, which would be in agreement with a memory effect observed after outgassing[8].

4. MEASUREMENTS OF THIN SAMPLES OF GRAPHITE

To get information about the thermal bulk properties, measurements have been performed on thin samples which were cut from the bulk material and polished (P1200). Phase lag measurements of three samples of a thickness varied between 1 mm and 1.6 mm have provided the thermal diffusivity. The results for EK 98 are adjusted in Fig.3 to the theory of compact solids of limited thickness. From the behaviour at intermediate frequencies a thermal diffusivity of approximately $a = 6.8\ 10^{-5} m^2/s$ has been derived. For high and low frequencies there are deviations between theory and measurement which are due to the roughness left over at the front and rear surface after sample preparation. Values of approximately $a = 5.8\ 10^{-5}\ m^2/s$ and $a = 5.6\ 10^{-5}\ m^2/s$ have been obtained for 5890PT and FP159I, respectively.

After outgassing at $300^{o}C$ and 10^{-2}Torr during 9 hours, the values found for the bulk diffusivity and effusivity had changed generally within a margin of 15%. The normalized amplitudes measured for FP159I before and after outgassing are compared in Fig.4 to the theory of a compact thin sample. The bulk effusivity can be derived from the signal amplitude, whereas the position of the peak as a function of $f^{-1/2}$ is determined by the thermal diffusion time l^2/a where l is the sample thickness. Again, the deviations be-

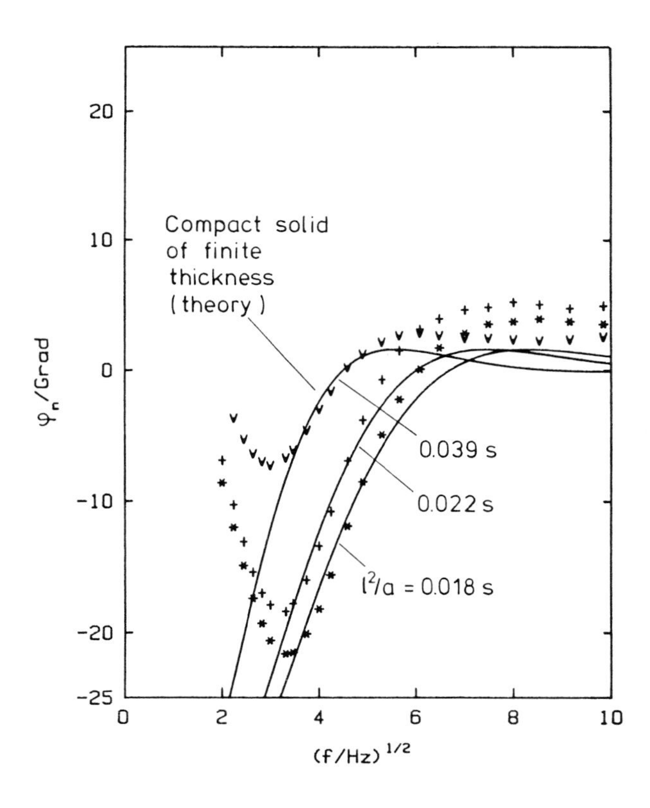

FIGURE 3
Measured phases for thin polished samples.

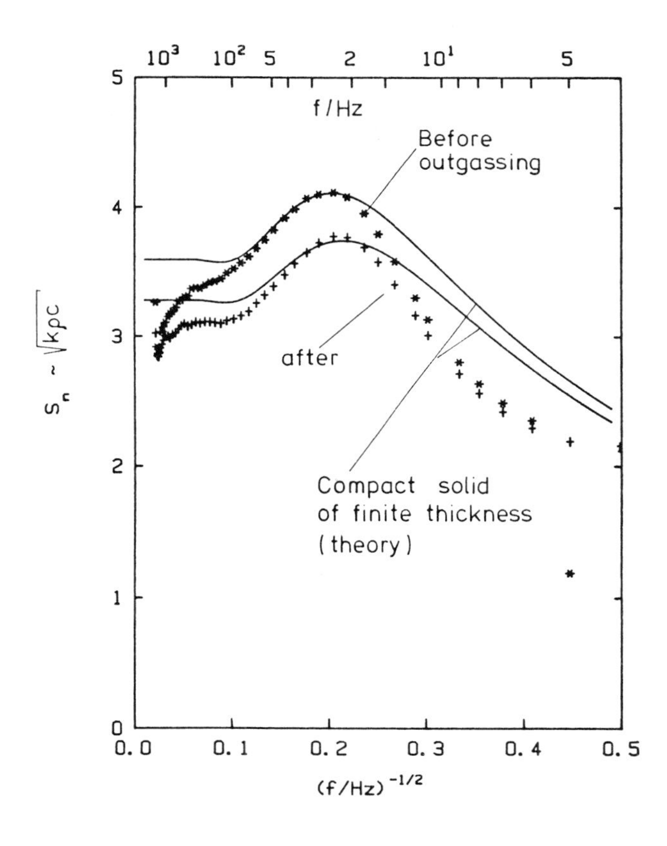

FIGURE 4
Signal amplitudes for a polished sample of finite thickness before and after outgassing.

tween experimental results and theory at high and low frequencies are related to the surface effects (roughness and gas adsorption) at the front and rear side. For the analysed graphites, the signal amplitude and thus the effusivity decreased due to outgassing. For this sample (FP 159I), the peak additionally shifted to slightly lower frequencies, which means the sample is less transparent for thermal waves, or the diffusivity slightly decreased due to outgassing.

In Fig.5 are compared two samples of equal thickness of EK 98 and 5890PT after outgassing. The higher amplitude of EK 98 shows that this sample has a higher effusivity. The position of its peak at slightly higher frequencies shows that it is thermally more transparent and thus also has a slightly higher diffusivity.

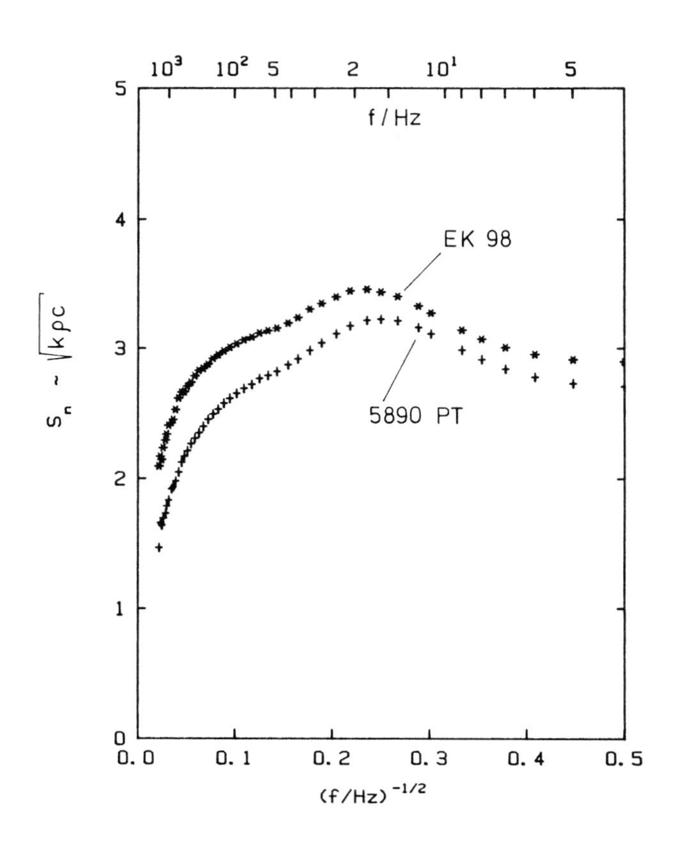

FIGURE 5
Signal amplitudes of two different graphites after outgassing.

5. CONCLUSIONS

The thermal depth profiles and bulk properties measured for different graphites at room temperature show some common features: 1) Due to the inherent surface roughness and gas content, the effusivity has a depth dependence which together with the typical thermal diffusion times can affect the absorption of heat pulses. 2) The depth variation was considerably reduced by outgassing; with time of exposure to air the original profile form was recovered again. These preliminar results stress the effect of outgassing on the thermal properties of graphite in use in tokamaks. 3) The measurement itself of the bulk properties is not unproblematic: The roughness remaining after the necessary preparation of samples from bulk material and eventually undefined outgassing in measurements at high temperatures can lead to uncertainties in the determination of the thermophysical properties.

REFERENCES

1. H. Vernickel, S. Schweizer, B. Sombach, P.D. Weng, J. Nucl. Mater., to be publ.
2. H. Würz, D. Zasche, B.K. Bein, Power Flow in the ASDEX divertor, this volume.
3. H.S. Carslaw, J.C. Jaeger, Conduction of Heat in Solids (Oxford, 1959).
4. S. Krueger, R. Kordecki, J. Pelzl, B.K. Bein, J. Appl. Phys. 62 (1987) 55.
5. B.K. Bein, S. Krueger, J. Pelzl, Can.J.Phys. 64 (1986) 1208.
6. B.K. Bein, S. Krueger, J. Pelzl, J. Nucl. Mater. 141-143 (1986) 119.
7. B.K. Bein, in: Photoacoustic and Photothermal Phenomena, eds. P. Hess, J. Pelzl (Springer Series in Optical Sciences, Vol. 58, Springe Berlin/Heidelberg, 1988) pp. 308-311.
8. J. Bohdanski, R.A. Causey, C.D. Croesmann, A.E. Pontau, J.B. Whitley, J. Nucl. Mater., to be publ.

FUSION TECHNOLOGY 1988
A.M. Van Ingen, A. Nijsen-Vis, H.T. Klippel (editors)

CARBON AND GRAPHITES AND THEIR ASSEMBLY TO METAL STRUCTURE FOR FIRST WALL PROTECTION, LIMITOR AND DIVERTOR

Michel COULON, Pierre LEHEUP, Claude ADES (Mrs)

LE CARBONE LORRAINE, B. P. 148, 92231 Gennevilliers, France

1. INTRODUCTION

LCL is involved in the fusion programmes since 1976. The earlier studies in collaboration with the CEA Saclay led to the definition of the properties of the materials required for TOKAMAK inner protection (low Z, high purity, thermal shock resistance, heat sink). The relevant grades of graphite were selected and thus the grade 5890 PT (see characteristics table I) has been chosen by a number of machines TFR, JET (fig 1), ASDEX, TORE SUPRA, RFX, VARENNES, TJ1.

	20	1000° C	2000° C
density	1.81		
total porosity (%)	20		
open porosity (%)	8		
ash content (ppm)	20		
flexural strength (MPa)	60		
tensile strength (MPa)	45	52	63
coef. of thermal expansion ($10^{-6}K^{-1}$)	4	5.3	6.1
resistance to thermal shock ($10^{4} W m^{-1}$)	8.3	3.2	2.5

TABLE I - Physical characteristics of 5890 PT

2. IRRADIATION RESISTANCE

LCL is engaged in a collaboration with the CEA to define the neutron irradiation resistance of polygranular graphites and to optimize the choice of material for future machines. It is known that neutron irradiation strongly affects the mechanical and thermal properties of graphites. A compromise has to be found between the grain size of the material and its thermal shock resistance : very small grain sizes give a higher irradiation resistance but generally decreases the thermal shock resistance mainly because of a higher CTE.

Little is known about the irradiation resistance of carbon/carbon composites and a programme has been launched to determine the influence of the texture of the fibre substrate.

3. MATERIALS WITH AN IMPROVED RESISTANCE TO THERMAL SHOCK

As the performances of the machines increase there is a need for carbon components with an improved resistance to thermal shock. It is now recognized that the carbon/carbon composites represent a considerable progress in this direction. We have developped the A05

figure 1

5890 PT tiles in JET
(by courtesy of JET)

composite which has been selected for the protection of the heavy duty area of JET (see table II).

The very high heat load reached now on limitor and divertor demands materials that combines thermal shock resistance and the highest density and heat conductivity possible in order to reduce the level of surface temperature and erosion. The pyrographite oriented with its layers parallel to heat flux seems to be the most promising material. The material has to be optimized in order to achieve the best heat conductivity with keeping a good machinability (see table II).

4. HIGH HEAT TRANSFER ASSEMBLY

Up to now most of the carbon components were used as heat sink. Steady state operation and reduced cooling time require a good heat transfer between the tiles and the actively cooled metal structure. Two types of assemblies have been developped.

4.1. Mechanical bonding with C/C bolts and nuts with a foil of Papyex as the flexible bond

An example of such a mechanical assembly studied for NET is presented on fig2. With improved grades of Papyex coefficients of heat transfer between C/C tiles and copper up to 20 - 30 kw $m^{-2}K^{-1}$ have been obtained. Such devices permit of course to replace the tiles one by one.

		Pyrographite		Composite A05	
	Unit	⊥	//	⊥	//
Density	g cm^{-3}	2.2		1.77	
Open porosity	%	0		8	
Bending strength	MPa	80		100	
CTE	$10^{-6}K^{-1}$	25	0.1	8	1
Heat conductivity	W $m^{-1}K^{-1}$	5	1500	90	270
Resistivity	$10^{-6}\Omega$ m	10^5	400		
Ash content	ppm	100		100	
Maximum sizes	mm	190 x 500 x 8		300 x 300 x 25	

TABLE II - Physical characteristics of A05 and pyrographite

figure 2

mechanical assembly studied for NET

4.2. Metallurgical bonding

Polygranular graphites such as 5890 PT can be currently brazed to metallic structural parts. Molybdenum has a good thermal expansion match (fig. 3 - Mo bolts protected by a brazed 5890 layer manufactured for RFX), but metal with strong mismatch such as stainless steel or cooper can also be brazed by mean of interdiate layers (fig. 4).

figure 3

Mo bolts protected by a brazed 5890 layer manufactured for RFX

figure 4

5890 brazed to copper

As seen above, pyrographite is a most promising material. For its application we have to solve the problem of securing a large number of thin blades (about 5 mm thick) perpendicular to their support. Brazing these blades may be the solution but a considerable mismatch of CTE has been faced which is worsened by the tremendous anisotropy of the pyrographite. We present on fig. 5 and 6 a test piece of pyrographite brazed on inconel made for the pump limiter of JET.

figure 5

test piece for JET
of pyrographite brazed on Inconel

figure 6

micrography of the brazed junction between pyrographite and inconel (studied for JET)

FUSION TECHNOLOGY 1988
A.M. Van Ingen, A. Nijsen-Vis, H.T. Klippel (editors)

HYDROGEN TRAPPING BEHAVIOR OF GRAPHITE EXPOSED TO GLOW DISCHARGE

Masahiro Kitajima*, and K. Aoki**

*National Research Institute For Metals, Tsukuba Laboratories, Tsukuba, Ibaraki 305, Japan
**National Chemical Laboratory For Industry, Tsukuba Research Center, Tsukuba Ibaraki 305, Japan

The hydrogen(deuterium) trapping of graphite exposed to the glow discharge was examined with pressure monitoring, and Raman and SIMS analyses. The exposure to the hydrogen(deuterium) glow discharge causes the refining of microcrystallite and the 'partial' amorphous formation in a surface region with the saturation of hydrogen. The saturation layer with implanted deuterium in the graphite is about 100A to the surface. The chemical trapping of deuterium due to the deuterium glow discharge is considered to be in the mode of C-D. The pressure change was discussed in terms of the ion impact desorption mechanism. The hydrocarbon formation due to the glow discharge was also described.

1. INTRODUCTION

The trapping of hydrogen and its isotopes in the first wall materials governs fuel recycle and tritium inventory. Because of its low atomic number and good high temperature properties, graphite is one of the most popular plasma facing materials. Graphite has been extensively studied with respect to hydrogenic fuel recycling[4-10].

This report describes the behavior of hydrogen(or deuterium) trapping of graphite under a glow discharge, using Laser Raman spectroscopy and Secondary Ion Mass Spectrometry(SIMS), and monitoring partial pressure changes for gases during discharge.

Laser Raman spectroscopy is a promising tool for characterization of graphite surface, since Raman scattering is very susceptible to structure change of graphite[2]. The graphite structure is discussed mainly from the Raman bands at $1360cm^{-1}$, $1584\ cm^{-1}$ and $2700cm^{-1}$. C-D mode is also discussed for deuterium trapping to a carbon.

SIMS was used to measure the depth distribution of hydrogen isotope trapped in the surface by the exposure to glow. The pressure change of hydrogen was detected to discuss the phenomenological behavior of hydrogen trapping and reemission during the discharge.

2. EXPERIMENTALS

The vacuum chamber used for the discharge experiment was a cylinder of 70cm long and 20cm in diameter made with Type 304 stainless steel. A graphite sheet of 60cm x 16cm x 5mm(IG-110, average density $1.77g/cm^3$, ash<100ppm, TOYO TANSO CO., LTD) was introduced into the chamber. The vacuum chamber was pumped to its lowest ultimate pressure, usually $2x10^{-8}$ Torr($2.66x10^{-6}$Pa) and adjusted to an initial value of hydrogen pressure P_0, typically P_0=0.1 Torr(13.3 Pa). After closing off the pumping ports and the gas leak valve, a glow discharge was initiated by applying a DC voltage between the cathode graphite sheet and the anode chamber wall, where the glow formation expanded in all the chamber and the sample was subjected to an ion flux[3,4]. In this case, the ions are accelerated through the cathode fall and are incident on the graphite with maximum energies of approximately 1keV(i.e. about 500eV/D). The pressure change in the chamber caused by the glow discharge was measured with a pressure gauge and a quadrupole mass spectrometer (QMS) in a differentially pumped by-pass.

The 530.9nm line of a Kr-ion laser was used

to induce the Raman spectrum. The 476.2nm line was used on occasion. The scattered light was analyzed with a Spex-1403 double monochrometer in the photon counting mode. The passing layer of the induced light is estimated to be an order of 500A[1]. SIMS analyses were performed with Shimadzu-ARL IMMA. Details in experimental techniques are described elsewhere[3,4].

3. RESULTS AND DISCUSSION

3.1. Pressure Monitoring

Figure 1 shows pressure change of hydrogen during and after the discharge for graphite, and molybdenum or stainless steel(SS) at ambient temperature. There are rather large pressure drops during the discharge for both cases. When the discharge is switched off, there is a little pressure increase in the case of metal of molybdenum or (SS). This shows that the hydrogens are thermally reemitted from the metal surface. On the other hand, in the case of graphite, there is no pressure change on discharge-off. This shows that the implanted hydrogen is stably trapped, leading to no thermal reemission from the graphite surface at ambient temperature.

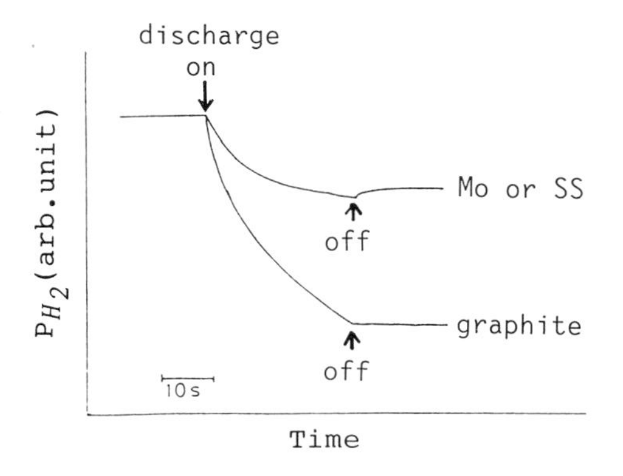

FIGURE 1. The hydrogen pressure change before and after the dischage-on or discharge-off.

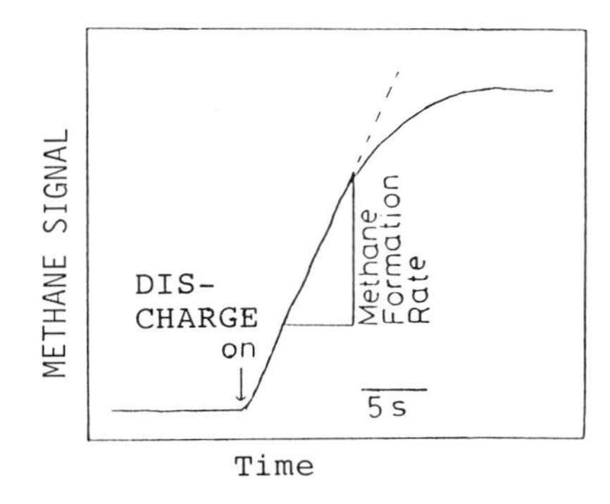

FIGURE 2. Methane formation of graphite as a function of a discharge time. Similar behaviors were observed for acethylene and ethylene formations.

The hydrocarbon formation was confirmed during the discharge, as shown in figure 2. The pressure for hydrocarbons is rapidly increase on discharge-on and reach saturation in any hydrocarbons. There were soots observed on the graphite surface after the experiment. These show redeposition of the hydrocarbons. When the discharge is initiated, the plasma contains originally little hydrocarbon. As time proceeds, the concentrations of the hydrocarbons increase and they return to the graphite surface through the ionization in the discharge. Thus, the signals of hydrocarbons reach saturation. Figure 3 shows temperature dependence of the hydrocarbon yields due to the hydrogen glow discharge, which was obtained from the slope of hydrocarbon signals at a time of the discharge-on(see figure 2). A peak value of the CH_4 yield is about $0.06CH_4/H^+$, which is similar to the case of CH_4 production from graphite under a bombardment of H^+ ions [5]. The production yield due to sub-eV H^0 is

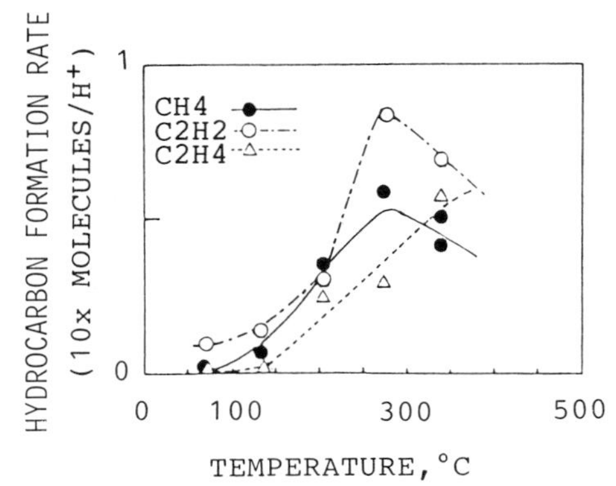

FIGURE 3. Temperature dependence of the rate of hydrocarbon formation.

about $3X10^{-4}CH_4/H^0$ [6]. The yield due to a simultaneous bombardment of the H^+ and H^0 in the presence of H_2 gas[5] is rather larger than these values, because of the 'synergistic effect': $0.4CH_4/H^+$. Thus, in the case of glow discharge, there is little synergistic effect , though the glow comprises hydrogen ions and atomic hydrogen as well as hydrogen gas.

3.2. Raman and SIMS analyses

Figure 4 shows the Raman spectra of graphites after exposure to the deuterium glow and argon glow discharges ,as well as before the exposure. The wave length of Kr-laser was 530.9nm. The Raman line at $1584cm^{-1}$ indicates the E_{2g} active mode in crystalline graphite[1,2]. The line at $1364cm^{-1}$ is an active band attributed to defects in crystal lattice of graphite, which has been pointed out first by Tuinstra and Koenig[1]. They have found that the intensity of this line is inversely proportional to the crystallite size in the direction of graphite plane. The intensity of this band allows the estimation of the crystallite size. Upon exposure to deuterium glow discharge(DG), the intensity of the line at $1364cm^{-1}$ increases(Figure 4b). A total fluence was around $4X10^{18}D/cm^2$. The relative intensity I_{1364}/I_{1584} was 0.8-1.0 after the exposure, while it was 0.38 before the exposure(Figure 4a). This intensity increase shows that the microcrystallite size becomes small resulting from lattice damage induced by deuteron under the glow bombardment. Similar behavior was observed under a hydrogen glow discharge(HG). In the case of 15keV H^+ ion bombardment[5], the two bands at $1584cm^{-1}$ and $1364cm^{-1}$ disappear and merge into a broad band with a maximum near $1525cm^{-1}$ indicating an amorphous graphite, which has been observed even at fluence of 10^{18} H^+/cm^2. On the other hand ,in the case of the glow discharge of the present work, the two lines between $1584cm^{-1}$ and $1364cm^{-1}$ still remain and a broad band is newly formed near $1540cm^{-1}$ (Figure 4b). This weak broad band may show a 'partial' amorphous formation in graphite surface layer. The ion energy in the glow is about 500eV being very small compared with the ion bombardment case(10-15keV) and thus the range is much smaller. The surface layer ,therefore, should be damaged much less. This could be a reason why in a graphite the deuterium (or hydrogen) glow discharge does not lead to the complete formation of an amorphous surface layer which

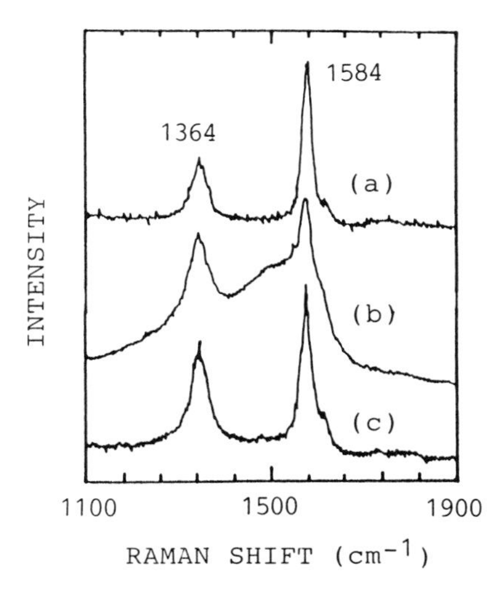

FIGURE 4. Comparison of the Raman spectra of graphite: a) before and b) after exposure to the deuterium glow discharge, and c) after further exposure to the Argon glow discharge.

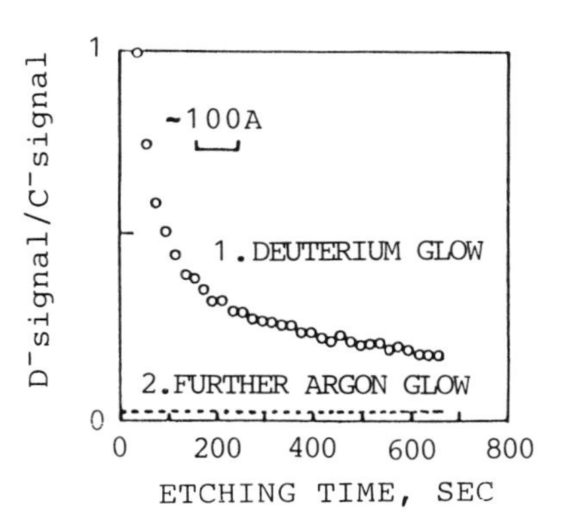

FIGURE 5. Depth profile of implanted deuterium in the graphite due to the glow discharge, using SIMS.

is observed for the 15keVH^+ ion bombardment.

Further exposure to the Ar glow discharge(AG) has almost recovered the structure before the deuterium glow(Figure 4c). The broad band between 1584cm^{-1} and 1364cm^{-1} disappeared. Figure 5 shows the SIMS depth profiling of deuterium for graphite exposed to DG and AG, corresponding to the Raman spectra in figure 4. The distribution of trapped deuterium is in an about 100A layer to the surface. AG removes this saturated layer. The intensity ratio I_{1364}/I_{1584} is still a little larger than that before DG or HG: the value is around 0.5, showing the damage production effect of AG near surface is small. From Raman and SIMS analyses, a deuterium saturation with lattice damage causes the enhancement of line at 1364cm^{-1} and the formation of a broad band near 1540cm^{-1}.

Figure 6 presents the Raman spectra with a wider frequency region 600-3400cm^{-1}. In this case a shorter wave length of Kr-laser at 476.2nm was used in order to get he information nearer surface for the graphite exposed to DG. The sharp peak at 2324cm^{-1} is the plasma line of the laser. A line at 2736cm^{-1} is also attributed to the crystalline graphite[2]. The weak line near 3250cm^{-1} is stated to correlate with the 2nd order Raman effect[2]. These two lines are weakened by the exposure to DG(Figure 6b). The intensity ratios I_{2736}/I_{1584} were around 0.37 before DG and after the AG-exposure, and 0.22 after the DG. The new feature appears near 2260cm^{-1} after the DG-exposure. This band could not detected in the spectra measured by the use of the 560.2nm laser. A very weak peak could be seen at 3100cm^{-1}(arrow 1). The ratio of peak locations expected from the reduced mass of C-H and C-D is 1.36. This value is almost equal to the ratio of these two peaks (3100/2260). The energetic ion bombardment with 15keV H^+ is reported to induce the C≡C bond[8]. If 2260cm^{-1} were due to the C≡C bond, the HG-exposure should also enhance the band at 2260cm^{-1}. Actually,however, there was no peak observed at 2260cm^{-1} in the case of HG discharge. Thus, the 2260cm^{-1} band could be due to C-D vibration showing the chemical trapping of deuterium in carbon sample. There is very weak feature seen near 2960cm^{-1}(arrow 2). This possibly show D-D mode indicating the physical trapping of deuterium, because 2960cm^{-1} corresponds in principle to a location expected from the H-H band at 4160cm^{-1} .

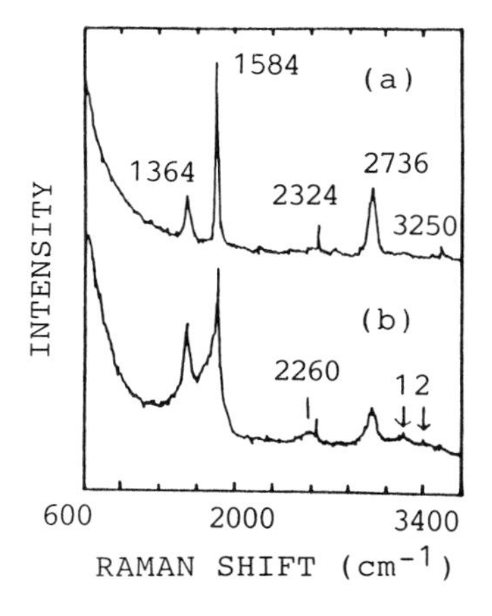

FIGURE 6. Raman spectra of a) before and b) after exposure to the deuterium glow discharge(scanning range from 600 to 3400cm^{-1}). The 476.2nm line of Kr-ion laser was used.

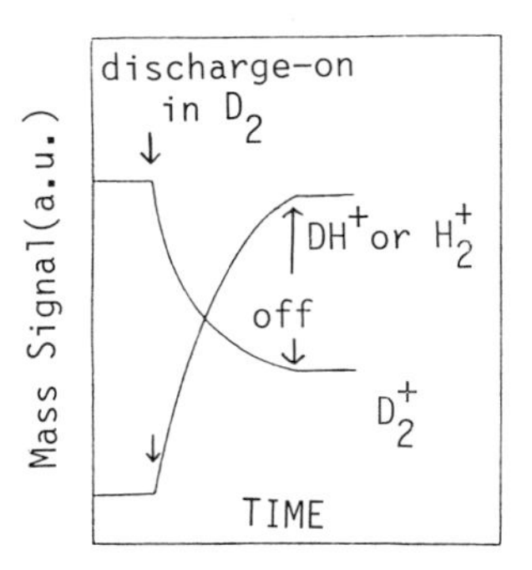

FIGURE 7. Mass signal changes of D_2^+,and DH^+ or H_2^+ due to the deuterium glow discharge for the graphite surface pre-exposed to the hydrogen glow.

3.3. Ion Impact Desorption Cross Section

The pressure monitoring ,and Raman and SIMS analyses showed the trapped hydrogens were thermally stable in the graphite. Moreover, residual gas analyses during DG-exposure for the surface pre-exposed to HG showed a signal enhancement of H_2^+ or DH^+ , only during the discharge(figure 7). It is known from these results that the dominant process of the hydrogen reemission at lower temperatures is not thermal process , but the ion impact desorption. Thus, in this section we attempt to analysis pressure decrease during the discharge, following the simplest model of the ion impact desorption, which is a similar model proposed by Braun and Emmoth[9] and Braganza et al.[10] In this model, the effective ion impact desorption cross section σ^* is defined to include the temperature dependence release mechanism also[9]. The balance equation on surface of hydrogen is given by

$$dC(t)/dt=J_0-J_0C(t)\sigma^* \quad \ldots \quad (1)$$

with a boundary condition of C(0)=0 and $C(\infty)=1/\sigma^*$, where C(t) is the areal density of the trapped hydrogen at a time t, and J_0 is the incoming atom flux. The time derivative of the solution C(t) for Eq.(1) is

$$dC(t)/dt=J_0\exp(-\sigma^*\Phi_0) \quad \ldots \quad (2),$$

where the fluence is $\Phi_0= J_0t$. The areal

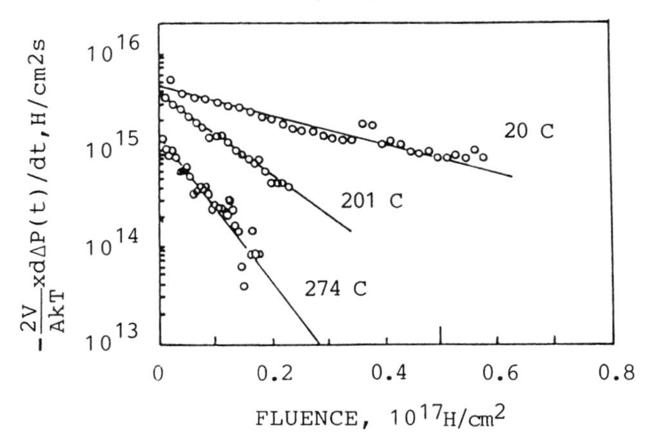

FIGURE 8. Least squares fit for the pressure change rate of hydrogen.

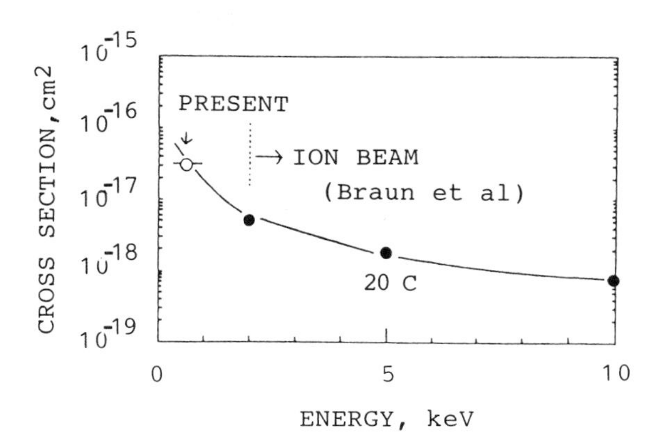

FIGURE 9. Effective ion impact desorption cross section as a function of ion energy. The present datum is on the extraporation of the curve obtained with the ion beam experiment[9].

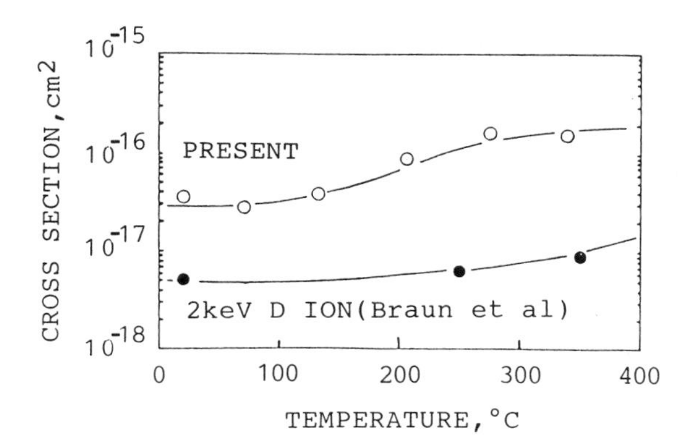

FIGURE 10. Effective ion impact desorption cross section as a function of sample temperature.

density C(t) can relate to the pressure drop of the chamber $\Delta P(t)=-AkT/2VC(t)$, where A is the surface area of the sample, V is the chamber volume and T is the sample temperature. Therefore, a slope of $\ln(d\Delta P/dt)$ vs.Φ_0 gives us the ion impact desorption cross section σ^*.

The $\ln(d\Delta P/dt)$ shows clearly a good linear dependence on Φ_0(figure 8). The desorption cross section has a trend to decrease with increasing ion energy[9,10]. The effective desorption cross section of hydrogen from the pressure change under the glow discharge is coming on an extrapolation of the curve which was obtained with ion beam experiment(2-

20keV)[9], as shown in figure 9. Figure 10 shows the effective cross section as a function of sample temperature. The cross section is enhanced above 200 C. The efficiency of the desorption process is generally not expected to be a function of the sample temperature, since ion impact desorption mechanism is kineticly, rather than thermally, activated. A possible explanation for the enhancement above 200C is thermal enhanced ion impact desorption during the discharge. The lattice sites in the surface are damaged with ion bombardment, and the implanted hydrogens are trapped on those sites (see the previous section). The binding energy of the trapped hydrogen depend on the ion energy, and there should be trap sites with different binding energies, since the ions in the glow has an energy distribution(an energy peak is about 500eV). With a temperature increase, the weakly trapped hydrogens with low activation energy could be detrapped being thermally induced by ion impact; its extreme example is adsorption of hydrogen gas. Therefore, the cross section may be enhanced with temperature increase. The thermally induced ion impact desorption mechanism has been also proposed for the desorption of graphite deuteron-bombarded to saturation[9].

4. Conclusions

The hydrogen(deuterium) trapping of graphite exposed to the glow discharge was examined with pressure monitoring, and Raman spectrum and SIMS. The following conclusions are drawn:

-The exposure to the hydrogen(deuterium) glow discharge causes the refining of microcrystallite and the 'partial' amorphous formation in a surface region with the saturation of hydrogen.

-The saturation layer with implanted deuterium in the graphite is about 100A to the surface. The chemical trapping of deuterium due to the deuterium glow discharge is considered to be in the mode of C-D.

-The dominant desorption process during the discharge could be the ion impact desorption. The ion impact cross section obtained from the pressure change curve of hydrogen gas showed its enhancement above 200C. This may be explained by the mechanism of the thermally induced ion impact desorption of the weakly trapped hydrogens.

-Some of the hydrocarbons of CH_4, C_2H_2 and C_2H_4 forming due to the glow discharge return to the cathode graphite. In the case of the glow discharge, there should be little synergistic effect on the hydrocarbon formation, though there are hydrogen ions, atomic hydrogen and hydrogen gas in the glow.

References

1. F. Tuinstra and J. L. Koenig, J. Chem. Phys., 53(1970) 1126.

2. R. J. Nemanich and S. A. Solin, Solid State Comm., 23(1977) 417.

3. M. Kitajima, M. Fukutomi, A. Hasegawa and M. Okada, J. Nucl. Mater.,141-143(1986)

4. M. Kitajima, K. Aoki and M. Okada, J. Nucl. Mater., 149(1987)269.

5. A. A. Haasz, O. Auciello, P. C. Stangerby, and I. S. Youle, J. Nucl. Mater., 128-129 (1984)593.

6. E. Vietzke, K. Flaskamp and V. Phillips, J. Nucl. Mater., 111-112 (1982) 763.

7. W. L. Hsu and R. A. Causey, J. Vac. Sci. Technol. A5(1987)2768.

8. R. B. Wright, R. Varma and D. M. Gruen, J. Nucl. Mater., 63(1976) 415.

9. M. Braun and B. Emmoth, J. Nucl. Mater., 128&129(1984) 657.

10. C. M. Braganza, S. K. Erents and G. M. McCracken, J. Nucl. Mater., 75(1978)220.

FUSION TECHNOLOGY 1988
A.M. Van Ingen, A. Nijsen-Vis, H.T. Klippel (editors)

DISRUPTION HEAT FLUX SIMULATIONS ON GRAPHITES WITH PULSED Nd:YAG LASER

J.G. van der Laan, J. Bakker and H.Th. Klippel

Netherlands Energy Research Foundation, ECN, Westerduinweg 3, 1755 LE Petten, The Netherlands

This paper describes the results of a disruption heat flux study using a pulsed high power Nd:YAG laser. Pulse durations are in the range of 0.2 to 20 ms and energy densities up to 10 MJ/m². These quantities cover the predictions for NET disruptions. The materials studied are several polygranular graphites. Results are given in terms of dimensions of the laser induced crater and erosion rate. The quantitative results are compared with FEM-calculations. The experimental values exceed the predictions with a factor of 2 for the longer pulse duration and upto an order of magnitude for the shortest pulse durations. Factors affecting this discrepancy are discussed. The erosion data can be correlated with a figure of merit of heat transfer, known as the effusivity.

1. INTRODUCTION

In the next generation Tokamaks, such as NET or ITER, protection of the First Wall and the Plasma Facing Components against plasma disruption events will be necessary. Estimates for energy densities occurring in disruptions in NET[1] are upto 2 MJ/m² with deposition times of 0.1 - 20 ms. The First Wall of NET will probably be made of austenitic stainless steel protected by carbon tiles and also the divertors will have a carbon protection. The effects of high heat fluxes on plasma facing materials are studied with high energy particle beams and lasers. Earlier work at ECN-Petten showed a marked influence of the gas environment on the resulting material damage[2]. The present paper describes experiments on six grades of graphite in a vacuum of about 10^{-2} Pa.

2. EXPERIMENTAL

2.1. Equipment

The experimental set-up for the laser heat flux experiments consists of a pulsed high power Nd:YAG laser and a vacuum chamber, see figure 1. The laser (LASAG KLS-311/321) has a multimode beam of unpolarized light with 1.064 µm wavelength. Pulse energies range from a few J to 75 J while the pulse durations range from 0.2 to 20 ms. Peak powers range from 1 to 10 kW. The beam diameter before the 100 mm focussing lense is 12 mm. The actual beam irradiated areas in the experiments range from 0.3 to 15 mm². The vacuum chamber can be evacuated downto 10^{-4} Pa. The experiments have been performed at pressures between 5.10^{-3} and 2.10^{-2} Pa. The vacuum window has to be protected by an AR-(anti-reflex)coated glass or quartz plate inside the vacuum against deposition of evaporated material which needs repeated cleaning or replacement. Absolute energy and relative power

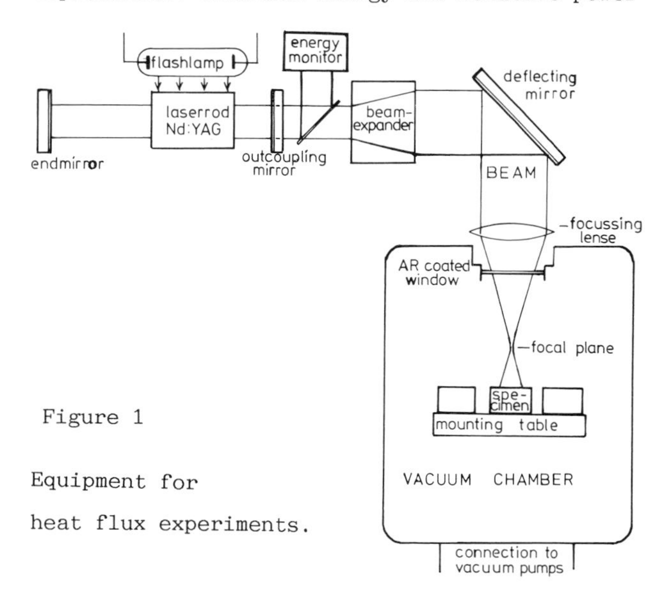

Figure 1

Equipment for heat flux experiments.

measurements have been performed with a Laser Power Monitor (OPHIR 150-A-N-DG). This device can be calibrated internally with 2% accuracy and has a 2% non-linearity. The photodiode energy monitor which measures about 2% of the laser pulse energy has been calibrated with aid of the power monitor enabling to cover a wide energy range with both devices. The total energy transmission losses by the optical components have been determined to be 15% of the initial laser values. In the vacuum chamber a specimen holder is placed with a travel distance of 125 mm. So six different specimens with a characteristic surface diameter of 20 mm can be tested in one batch.

2.2. Materials

Six different grades of polygranular graphites have been studied which cover a certain range of density and physical properties. The mean values of their main properties taken from literature values[3, 4] and manufacturers data sheets are given in table 1. All graphite surfaces have been polished before exposure to the laser beam. Specimen thicknesses are about 10 mm.

2.3. Experimental procedure

The focal point of the laser beam inside the vacuum chamber has been determined by locating the intersection point of a beam-splitted He-Ne auxiliary laser. The specimen holder containing the specimens has been placed at five different levels between the focal point and a defocus distance of 32 mm in order to control the energy density. In this way the six materials can be tested under similar conditions. All specimens have been examined by optical microscopy and scanning electron microscopy (SEM). The laser induced crater depth and diameters have been measured. The damage area has an elliptical shape, especially at increasing defocus distance.

TABLE 1. Materials and properties, see text.

Values at ~ 100°C:
ρ = density (g/cm³)
λ = conductivity (W/mK)
$\sqrt{\lambda\rho c}$ = coefficient of heat transfer or effusivity, normalized to the value for EK-47.

graphite	manufacturer	ρ	λ	$\sqrt{\lambda\rho c}$
EK-47	Ringsdorff	1.7	130	1
AXF-5Q	Poco Graphite	1.80	91	0.86
EK-98	Ringsdorff	1.86	80	0.82
5890-PT	Carb. Lorraine	1.79	82	0.82
2239	Carb. Lorraine	1.78	82	0.82
FE-219	Schunck	1.74	75	0.77

3. RESULTS & DISCUSSION

3.1. Erosion data

The resulting material damage is quantified in terms of crater depth, see figures 2a-d, and crater area, see figures 3a-d. The horizontal axis of the bar diagram shows the graphites approximately in the order of the increasing coefficient of heat transfer or effusivity $\sqrt{\lambda\rho c}$ for reasons to be given in section 3.3. The typical error in the crater depth is 2 microns for the lower values up to 10% for the deepest craters. The error in crater area determination is about 10% up to a factor 2 for the most shallow craters. No cracks were observed in all present cases due to the laser energy deposition.

3.2. Heat flux calculations

In interpreting the results reference is made to the numerical model of heating and evaporation which takes into account temperature dependent physical properties and a moving boundary[5]. Results for a one-dimensional heat flow in graphite have been published for initial temperatures of 1000°C and 1800°C[6, 7]. The effect of a 10 ms disruption has been evaluated also for an initial temperature of 30°C[7]. Results are given in table 2. To examine the influence of variations in the volumetric heat capacity this product was varied with 20%. It

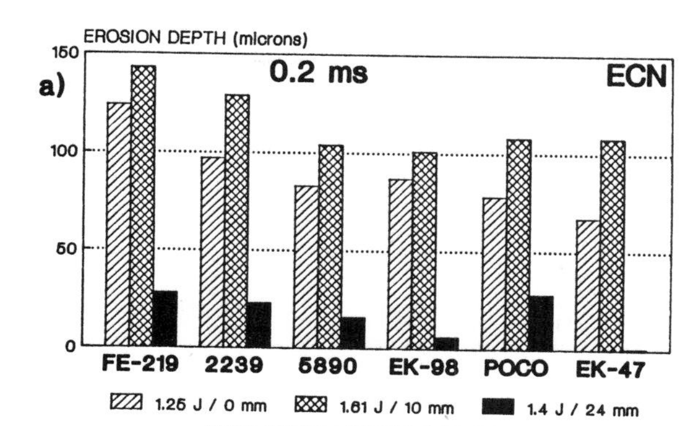

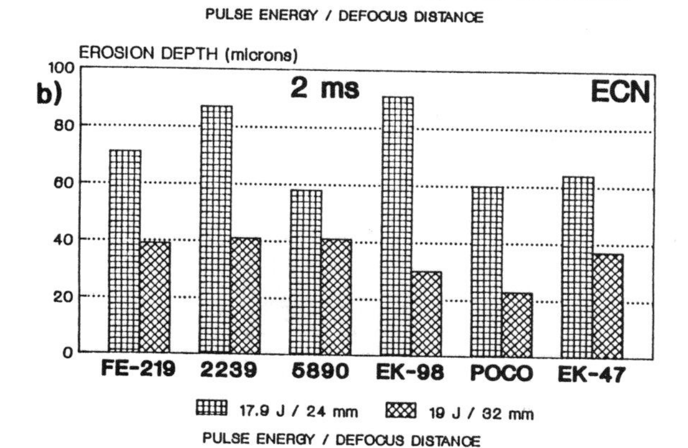

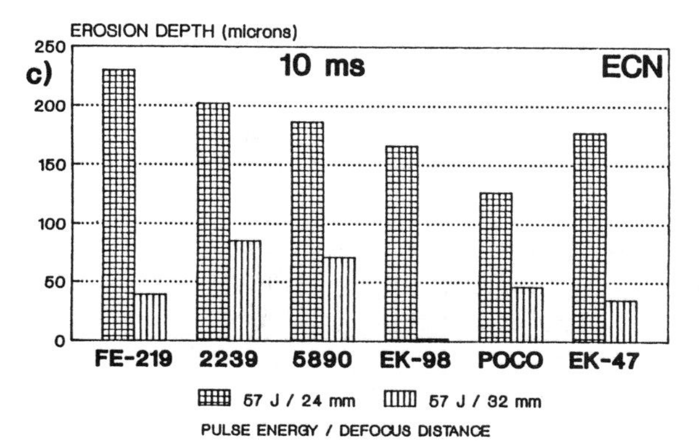

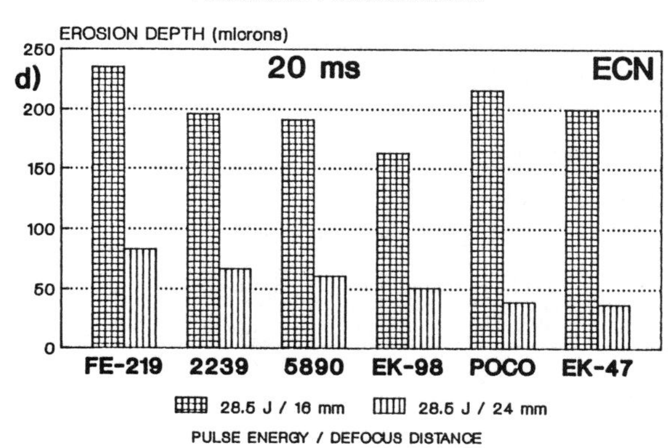

Figure 2
Depth of laser-induced erosion crater in graphites for a) 0.2 ms, b) 2 ms, c) 10 ms and d) 20 ms, at different defocus distances.

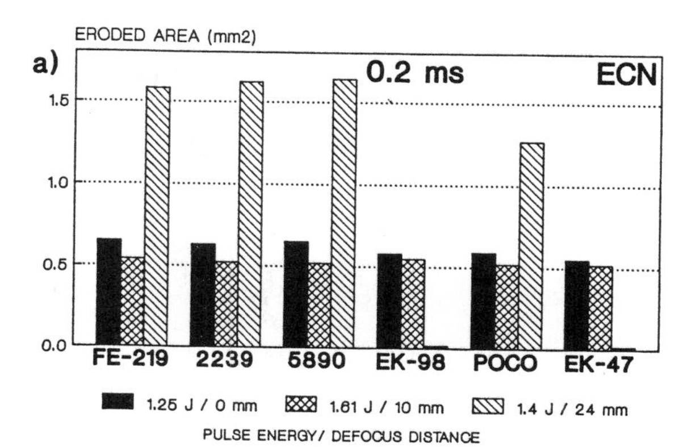

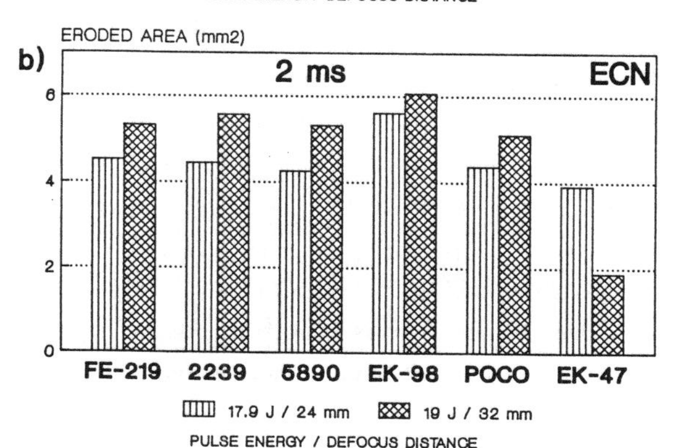

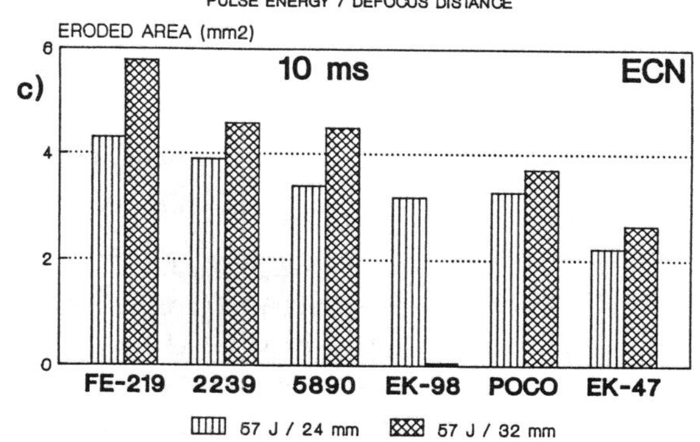

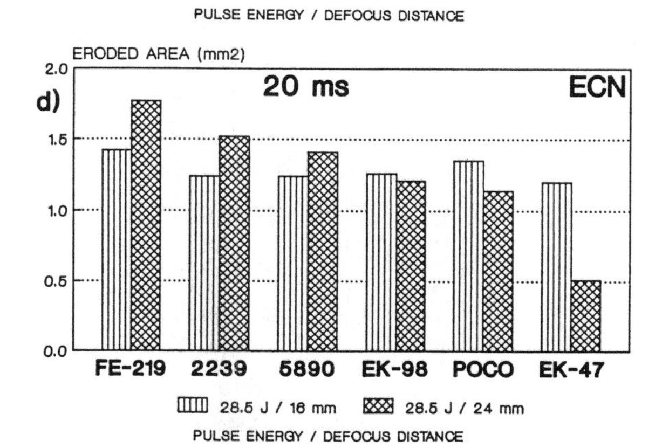

Figure 3
Eroded area for a) 0.2 ms, b) 2 ms, c) 10 ms and d) 20 ms, at different defocus distances.

appears that the final amount of evaporation varies about 20% for the 5 MJ/m^2 disruption. For the 10 MJ/m^2 disruption this is negligible (< 1 μm). A reduction of thermal conductivity with 50% increases the evaporation depth with 62% for a 5 MJ/m^2 disruption. The predictions are based on a one-dimensional heat flow. The influence of the lateral heat conduction on the laser induced crater depth is related to the ratio of beam diameter to thermal diffusion depth. For the 20 ms pulse duration at 16 mm defocus distance this ratio is about 2, for all other cases the ratio exceeds 4 which has been verified by some two-dimensional calculations to be a safe figure.

3.3. Ranking of materials

A dependence of mass loss of carbon materials on density in laser heat flux experiments has been reported by Brinkschulte et al.[8] on a similar range of graphites, Jortner et al.[9] and Evangelides et al.[10]. In a review of results from five different laboratories Bolt et al.[11] did not find such a dependence in heat flux experiments with particle beams. The present results also do not show a pronounced density dependence. Considering the previous calculations a more reasonable figure of merit may be the heat transfer coefficient or effusivity which is proportional to the square root of the conductivity and volumetric heat capacity. This figure of merit indicates the effective transfer of heat penetrating into a material. The laser induced crater magnitudes generally show an inverse dependence on this figure. It is most pronounced for long pulse duration and low energy density. This can be understood from the fact that at high energy densities and short pulse durations a larger fraction of the incident energy is consumed by evaporation, see figure 4[7]. The energy that is left for conduction is approximately the complementary fraction. The correlation of erosion with the heat transfer coefficient applied to the results of Brinkschulte et al. is equally reasonable as with the density. Also the laser experiments of Benz et al.[12] on POCO, EK-98 and CL-5890-PT show similar ranking with respect to crater depth and mass loss.

TABLE 2. Numerical calculations of the heat flux response of graphites with variation of density and thermal conductivity.

5 mm graphite thickness
10 ms disruption time
density 1.76 g/cm^3
T_0 = 30 °C
T_{max} = maximum surface temperature
δ_{ev} = net evaporation thickness
δ_u = ultimate possible evaporation thickness

				ρC+20%	ρC-20%	λ-50%
E (MJ/m^2)	1	5	10	5	5	5
T_{max} (°C)	969	3415	3580	3385	3414	3435
δ_{ev} (μm)	-	13.5	56	10.8	16.5	21.9
δ_u (μm)	18.3	91	183	91	91	91

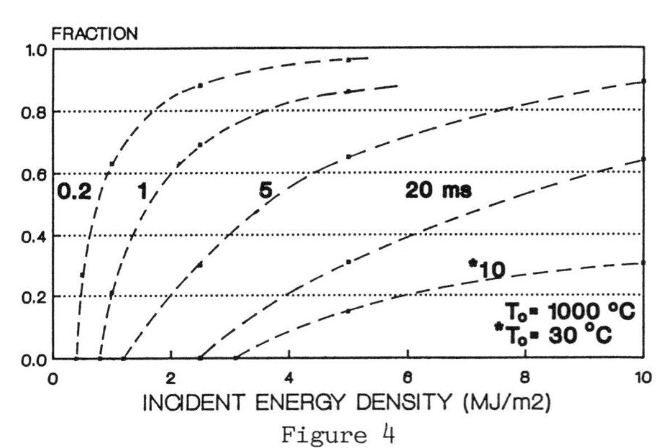

Figure 4
Fraction of incident energy consumed by evaporation versus incident energy density for several disruption durations with an initial temperature of 1000°C and for a disruption duration of 10 ms starting from 30°C.

3.4. Erosion modelling

In figure 5 the present measured erosion depths have been plotted versus the incident energy density. The transfer from energy values and defocus distances to incident energy density in this figure has been done with the assumption of a flat beam profile. In the present results total absorption has been assumed.

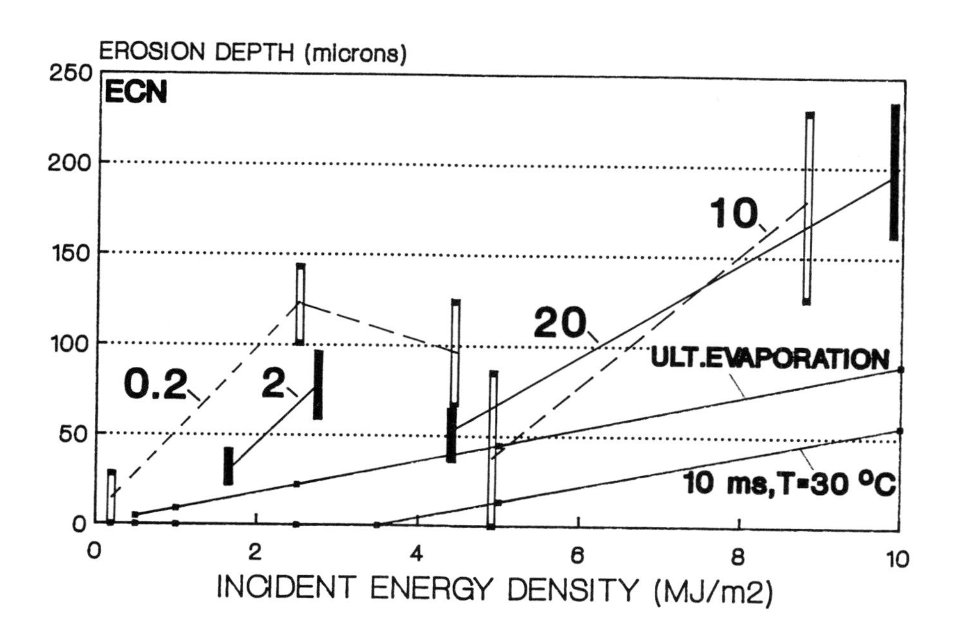

Figure 5
Present data of erosion depth versus energy density of six grades of graphite at 10^{-2} Pa.

Although low power reflection measurements indicate a 85 ± 3% absorption coefficient for the Nd:YAG wavelength the absorption may actually increase during the laser pulse[20]. The absolute values of the erosion depths exceed the predictions[6] with a factor of 10 at shorter pulse durations and a factor of 2 for the long pulses. Several contributions to the discrepancies can be identified:

a) For model calculations a sublimation energy has been taken of 63 kJ/g but values as low as 26 kJ/g have been found in literature[13, 20]. This is related to the composition of the vapor which mainly consists of C_3-molecules but C_n clusters with n up to a few hundreds have also been observed[14-16]. Equilibrium vaporization theory, as used in the model, may not apply in this rapid vaporization process[21].

b) The calculations assume a thermal conductivity λ characteristic for graphite H-451, which is comparable to EK-47, see table 1. The effect of a lower λ has been discussed in section 3.2.

c) In long pulse experiments with particle[17, 18] and laser beams[8, 14] a particle-emission process has been observed. Preferential evaporation of the binder phase, if present, and unfavourable oriented grains cause ejection of grain fragments. In the experiments presented here no plume species were collected but SEM examinations give evidence for this erosion process.

d) The model assumes:
   - a flat beam profile, which is a rough approximation of a multi-mode beam.
   - a constant power density, whereas the actual peak powers occurring in the laser spikes and hot spots range within a few ten percent from the mean value[19].

e) Absorbed gases may increase graphite erosion, as has been reported by Croesmann et al.[23] with electron-beam, and Lincoln and Covington[21], and Meyer et al.[22] with laser.

Summing up all the aforementioned effects the overall discrepancy might be of the order of 2 to 10, which might explain the results in figure 5.

6. CONCLUSIONS

a) Six grades of graphite show erosion due to a high heat flux with pulse durations in the range of 0.2-20 ms and energy densities to 10 MJ/m^2. The measured erosion is a factor of 2-10 higher than the predictions by a detailed numerical vaporization model. The erosion is increasing with decreasing pulse duration. The discrepancy in erosion is also increasing with decreasing pulse duration. The discrepancy can be partly understood from the uncertainty in model input data, the characteristics of the laser beam and material conditioning. More accurate modelling is required, because present design studies might underestimate the total erosion due to disruptions in next generation tokamaks.

b) The ranking of graphite materials with respect to their erosion behaviour to the material density is not supported by the present results. Correlation of graphite erosion to the coefficient of heat transfer or effusivity seems to be more appropriate.

ACKNOWLEDGEMENT

The technical assistance of D.S. d'Hulst and M. Zijlstra is greatfully acknowledged.

REFERENCES

1. Vieider, G., NET-team et al., paper presented at ISFNT, Tokyo, april 10-15, 1988.

2. Van der Laan, J.G., paper presented at 8th Conf. on PSI, Jülich, May 2-6, 1988.

3. Kneringer, G. et al., Proc. 14th SOFT (1986) 1083-9.

4. Delle, W. et al., Juel-Spez. 401, May 1987.

5. Klippel, H.Th., ECN-137, September 1983.

6. Klippel, H.Th., paper presented at ISFNT, Tokyo, April 10-15, 1988.

7. Klippel, H.Th., private communication, to be published.

8. Brinkschulte, H. et al., paper presented at ICFRM III, Karlsruhe, October 19-23, 1987 - also: JET-P(87)43.

9. Jortner, J. et al., Proc. 13th Int. Conf. Carbon, Irvine (1977) 411-2.

10. Evangelides, J.S., Proc. 14th Int. Conf. Carbon (1979) 251-2.

11. Bolt, H. et al., paper presented at ISFNT, Tokyo, April 10-15, 1988.

12. Benz, R. et al., Jülich-2056, April 1986 - also: J. Nucl. Mat. 150 (1987) 128-39.

13. Touloukian, Y.S. (ed.): Thermophysical properties of high temperature solid materials, MacMillan, NY, 1967.

14. Meyer, R.T. et al., High Temp. Sc. 4 (1972) 684-6.

15. Lundell, J.H. and R.R. Dickey, Prog. IAAA 56 (1977) 405-22.

16. Hastie, J.W. et al., paper presented at IUPAC 5th Int. Conf. on High Temp. & Energy-Related Materials, Rome, May 25-29, 1987.

17. Bolt, H. et al., IPPJ-AM-53, Augustus 1987.

18. Bohdansky, J. et al., Nucl. Inst. and Meth. Phys. Res. B23 (1987) 527-37.

19. Bass, M. (ed.): Mat. Proc. Theory and Practices, Vol. 3, North Holland (1983) 15-112.

20. Klein, C.A. et al., J. Appl. Phys. 61,5 (1987) 1701-12.

21. Lincoln, K.A. and M.A. Covington, Int. J. Mass. Spectrometry & Ion Phys., 16 (1975) 191-208.

22. Meyer, R.T. et al., J. Phys. Chem. 77, (1973) 1083-92.

23. Croesmann, C.D. et al., J. Nucl. Mat. 141-3 (1986) 108-12.

FUSION TECHNOLOGY 1988
A.M. Van Ingen, A. Nijsen-Vis, H.T. Klippel (editors)

THE MOZART EXPERIMENT : IN-PILE TRITIUM EXTRACTION FROM LITHIUM OXIDE, ALUMINATES, ZIRCONATES

M. BRIEC[1], J.J. ABASSIN[1], C.E. JOHNSON[2], M. MASSON[1], N. ROUX[3], H. WATANABE[4]

Commissariat à l'Energie Atomique, Institut de Recherche et de Développement Industriel, France
1 - IRDI - CEN/Grenoble, 85 X - 38041 Grenoble Cedex, France
2 - Argonne National Laboratory, USA
3 - DESICP - CEN/Saclay, 91191 Gif-sur-Yvette Cedex, France
4 - Japan Atomic Energy Research Institute, Japan

The MOZART experiment was carried out within the framework of the BEATRIX program, as part of the Commissariat à l'Energie Atomique contribution to this international collaboration program. This experiment was run during 45 days in the MELUSINE reactor at Grenoble.
Tested ceramics were Li_2O and $LiAlO_2$ from Japan, Li_2ZrO_3 from USA, $LiAlO_2$ from CEA. Influence of parameters such as temperature, sweep gas composition was investigated. Tritium residence times as a function of temperature were calculated. The comparison of the tritium release performance of the three ceramic breeders was made. This investigation has first evidenced the very good tritium release characteristics of Li_2ZrO_3 especially at low temperatures.

1. INTRODUCTION

Within the framework of the contribution of the Commissariat à l'Energie Atomique to the BEATRIX program, an in-pile tritium extraction study of Japanese and American ceramics was carried out in the MELUSINE reactor at Grenoble.

Ceramics investigated were Li_2O and $LiAlO_2$ from JAERI and Li_2ZrO_3 from ANL/HEDL, hence the name of MOZART (Melusine Oxide Zirconate Aluminate Release of Tritium) which was given to the experiment. A $LiAlO_2$ specimen from CEA was used to complete the loading ; it was therefore possible to compare the tritium release performances of ceramics of different nature.

Preliminary results are given in this paper.

2. EXPERIMENT DESCRIPTION

2.1. Samples characteristics

The samples, fabricated to required dimensions i.e. 8 mm in diameter and total column height 30 mm, are all of natural ^6Li content. Microstructural characteristics measured in the source laboratory are listed in table 1. Scanning electron micrographs made at CEA are shown in figure 1.

The microstructure (large grain, high density) of the CEA aluminate specimen was chosen in accordance with the assigned range of temperature explored i.e. the upper temperature level of the irradiation rig.

2.2. Irradiation device

The samples were mounted in the CHOUCA rig, previously described[1] and currently used for the LILA, LISA, TEQUILA experiments respectively of the CEA, KfK, ENEA programs of tritium extraction study.

It is recalled that the CHOUCA rig gives the possibility to simultaneously irradiate 6 capsules and to independently analyze the tritium released from the corresponding samples.

The temperature of groups of capsules can be varied independently. Given the diversity of the breeding ceramics to be tested, two levels of temperature (with a difference of 150°C) were selected in the MOZART experiment in order to span a sufficiently large range of temperature. The range explored was 300 - 680°C. The small

Table 1
Samples characteristics

Sample reference	S1	S2	S3	S4	S5	S6
Material	$LiAlO_2$	$LiAlO_2$	Li_2ZrO_3	Li_2O	$LiAlO_2$	Li_2ZrO_3
Source	JAERI	JAERI	ANL/HEDL	JAERI	CEA	ANL/HEDL
^6Li content	natural	natural	natural	natural	natural	natural
Dimensions, mm	8 x 30	8 x 30	8 x 30	8 x 30	8 x 30	8 x 30
Source laboratory data,						
Density % T.D.	81	81	80	80	92.7	80
Grain size, μm	1-2	1-2	bimodal 2-4/20	16	5	1
Specific surface area, m^2/g	0.7	0.7		0.05		

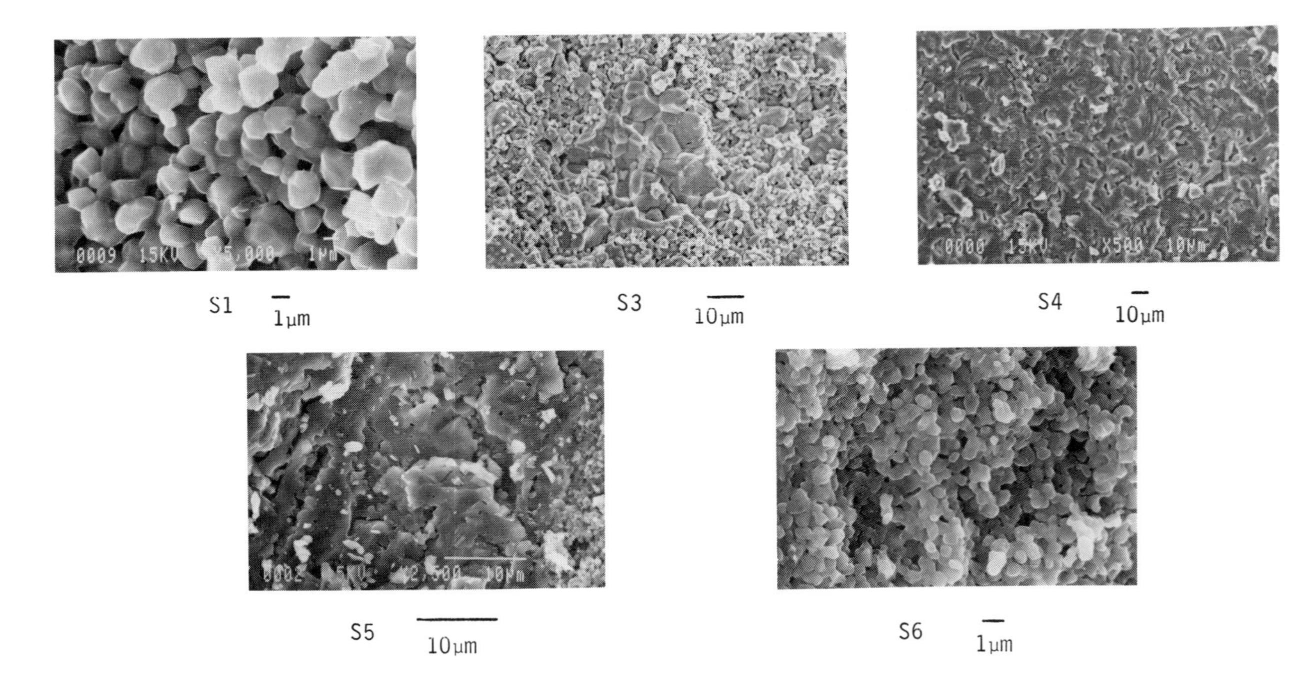

Fig. 1. Scanning electron micrographs.

grain Li_2ZrO_3, JAERI $LiAlO_2$ and Li_2O samples were tested at the lower temperature level, whereas the large grain Li_2ZrO_3, JAERI $LiAlO_2$ and CEA $LiAlO_2$ samples were tested at the upper temperature one.

In order to avoid any water adsorption on the lines, the tritiated species released were reduced using zinc beds operating at 380 - 390°C. The lines between the samples and the zinc beds are heated above 150° C, whereas the 25 meter long lines between the outlet of the irradiation device and the ionization counters are not heated.

Prior to irradiation, the samples were outgassed at 800°C.

2.3. Purge gas conditions

- sweep gas nature : helium + 0.1 % H_2, helium, helium + 0.01 % H_2
- pressure : 1.5 to 1.7 x 10^5 Pa
- mean flow rate (22°C, 1.035 x 10^5 Pa) : 2.4 l/h.

2.4. Irradiation conditions

- thermal neutron flux (unperturbed)
  1.4 to 1.45 x 10^{13} n/cm^{2}s during the first cycle
  0.85 to 1.05 x 10^{13} n/cm^{2}s in the following cycles
- irradiation duration : 45 days.

2.5. Experimental matrix

In addition to the nature of the ceramic, the parameters investigated were purge gas composition and temperature.

Runs were made in helium + 0.1 % H_2 as standard conditions, in pure helium, and in helium + 0.01 % H_2 (table 2).

In order to evaluate the effect of moisture, a rough test was made, in which moisture is added to the purge gas by removal of the cold traps used for purge gas purification. This practice only induces a transient effect and, although helpful for a scoping test, obviously cannot be contemplated for a methodical study.

3. RESULTS

3.1. Comparison of the tritium release behaviour of the ceramics

The MOZART results generally corroborate those of the LILA and VOM series for $LiAlO_2$ and Li_2O[2,3]. Thus Li_2O exhibits a good tritium release behaviour at rather low temperatures whereas in the same conditions tritium inventories in $LiAlO_2$ are high. A significant result of this experiment is the first evidence of the very good tritium release characteristics of Li_2ZrO_3 at very low (∿ 300°C) temperatures. This appears clearly in fig. 2, showing the tritium release for a temperature decrease from 500 to 350°C with He + 0.1 % H_2 as purge gas for $LiAlO_2$, Li_2O and Li_2ZrO_3.

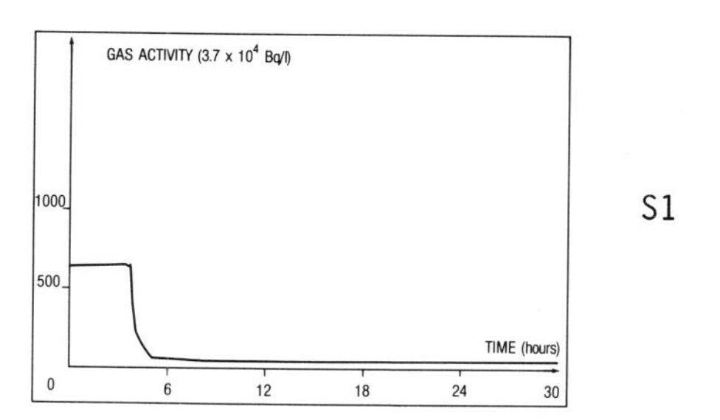

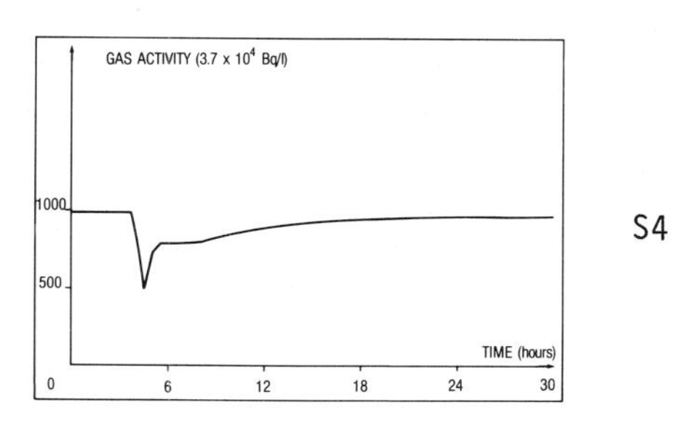

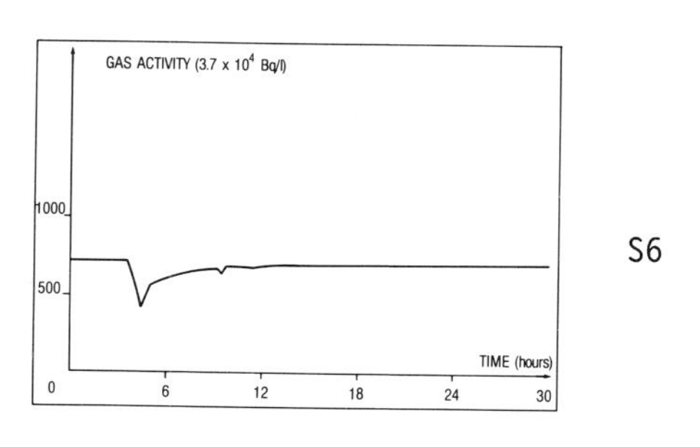

Fig. 2. Temperature influence on tritium release. Tritiume release for a temperature decrease from 500 to 350°C. Sweep gas : He + 0.1 % H_2. Samples S1, S4 and S6.

Table 2
Experimental matrix

Run	purge gas	Run purpose	Temperature, °C S1, S3, S5	S2, S4, S6
1.1	He + 0.1 % H_2	Standard conditions	650	500
1.2	He + 0.1 % H_2	Temperature decrease	600	450
1.3	He + 0.1 % H_2	Standard conditions	650	500
1.4	He + 0.1 % H_2	Temperature decrease	550	400
1.5	He + 0.1 % H_2	Temperature decrease	550	350
2.1	He + 0.1 % H_2	Standard conditions*	650	500
2.2	He + 0.1 % H_2	Temperature increase	680	530
2.3	He + 0.1 % H_2	Standard conditions	650	500
2.4	He + 0.1 % H_2	Temperature decrease	550	350
2.5	He + 0.1 % H_2	Temperature decrease	500	300
2.6	He + 0.1 % H_2	Standard conditions	650	500
2.7	He	Purge gas effect	650	500
2.8	He + H_2O	Purge gas effect	650	500
3.1	He	Purge gas effect	650	500
3.2	He	Temperature decrease	600	450
3.3	He	Temperature increase	650	500
3.4	He + 0.01 % H_2	Purge gas effect	650	500
3.5	He + 0.01 % H_2	Temperature decrease	600	450
3.6	He + 0.01 % H_2	Temperature increase	650	500
3.7	He + 0.01 % H_2	Temperature decrease	550	400
3.8	He + 0.01 % H_2	Temperature decrease	550	350
4.1	He + 0.01 % H_2	Temperature increase	550	400
4.2	He + 0.01 % H_2	Temperature decrease	550	350
4.3	He + 0.1 % H_2	Purge gas effect	550	350
4.4	He + 0.1 % H_2	Out-of-pile tritium extraction	450	310
	He + 0.1 % H_2	"	500	310
	He + 0.1 % H_2	"	550	350
	He + 0.1 % H_2	"	600	500
	He + 0.1 % H_2	"	700	600

* Thermal flux change : 1.45 x 10^{13} n/cm^{2}s to 0.95 x 10^{13} n/cm^{2}s

At temperature still lower, about 300°C, the tritium release remains significant for Li_2O and Li_2ZrO_3 as shown in fig. 3.

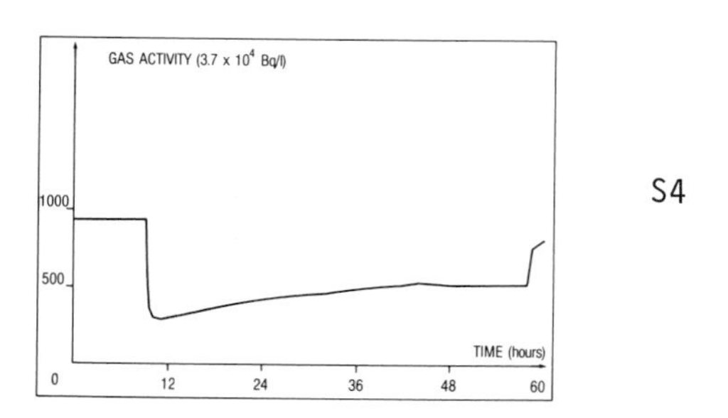

S4

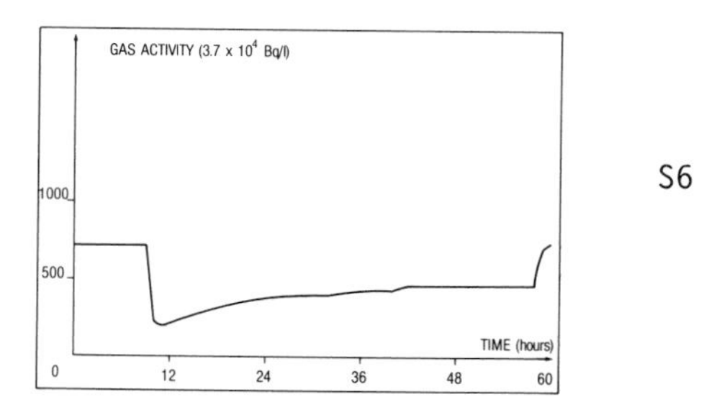

S6

Fig. 3. Temperature influence on tritium release. Tritium release for a temperature decrease from 350 to 300°C followed by a temperature increase to 310°C. Sweep gas : He + 0.1 % H_2. Samples S4 and S6.

The tritium residence times obtained with He + 0.1 % H_2 as purge gas (table 3 and figure 4) show that the minimum working temperature, defined as the value for which the tritium inventory can be limited to the production of one day, is about 500° C for $LiAlO_2$ with 1 to 2 µm grain size, 310 to 320°C for Li_2ZrO_3 with 1 µm grain size and for Li_2O.

The observed influence of $LiAlO_2$ grain size i.e. the better performance of the smaller grain size material is consistent with previous CEA studies results[2].

3.2. Effect of the purge gas composition

The experimental results show a favourable effect of hydrogen addition to the purge

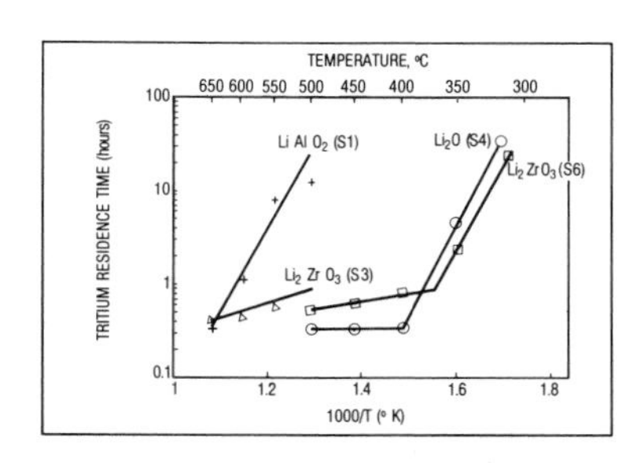

Fig. 4. Tritium residence time (in hours).

gas. Fig. 5 shows an example for Li_2ZrO_3 (sample S6) at 500° C when He + 0.1 % H_2 is replaced by pure helium. The effect is the greater the larger the specific surface area. The effect is larger with larger hydrogen contents. The same figure shows the change of tritium release at 350°C for sample S6 when He + 0.01 % H_2 is replaced by He + 0.1 % H_2. Whether the hydrogen effect depends or not on temperature is not clear.

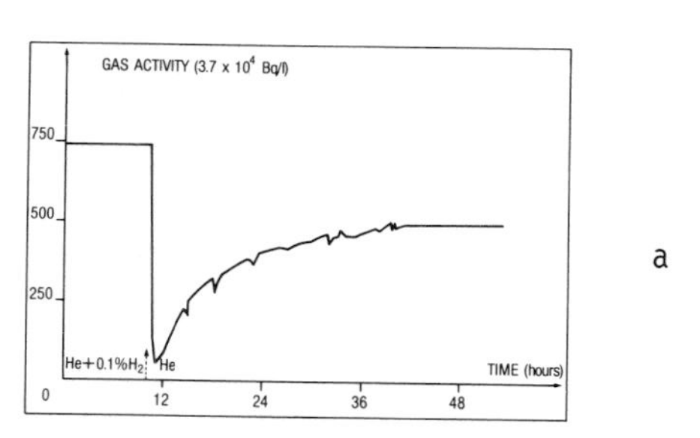

a

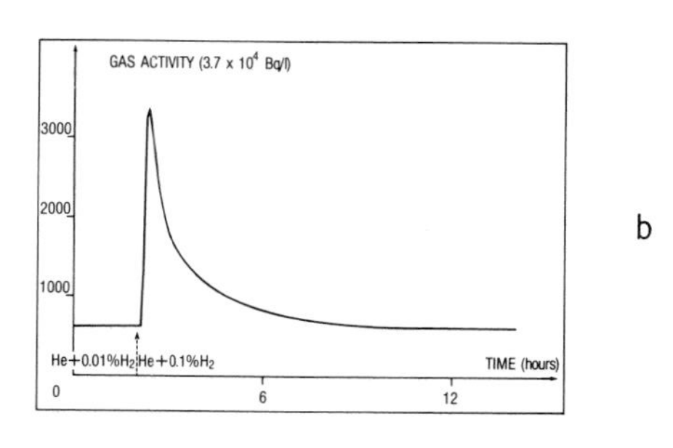

b

Fig. 5. Purge gas composition influence. Change from He + 0.1 % H_2 to He at 500°C (a). Change from He + 0.01 % H_2 to He + 0.1 % H_2 at 350°C (b). Sample S6.

Table 3
Tritium residence times (in hours)
Purge gas : He + 0.1 % H_2

T, °C	S1	S2	S3	S4	S5	S6
680	0.26		< 0.42			
650	0.35		0.42		1.28	
600	1.12		0.46		5.31	
550	7.9		0.6		> 31.3	
530				< 0.33		< 0.53
500	> 28.3	12.2	0.89	0.33	62.9	0.53
450		> 30		0.33		0.61
400				0.33		0.77
350		>> 76		> 4.53		2.34
320				< 30.3		
310				> 30.3		23.3

A short run made with water addition to pure helium indicates a favourable effect of water suggesting that water might have an effet comparable to the hydrogen one.

The main mechanisms influencing tritium release are intragranular diffusion and tritium desorption at the solid-gas interface ; their relative importance depends on specific surface area and temperature.

The beneficial effect of hydrogen addition to the purge gas, which is indicative of the contribution of surface processes, will be larger in desorption-controlled region. The experimental results show a larger effect for samples with larger specific surface areas. Although no clear temperature dependence is evidenced in MOZART, previous experiments with $LiAlO_2$[4] have shown the significant contribution of surface desorption at low temperatures.

These two experimental results infer that the controlling step of tritium release changes from diffusion to desorption when grain size and temperature decrease.

4. CONCLUSION

The MOZART experiment corroborates previous literature results concerning the tritium release behaviours of Li_2O and $LiAlO_2$.

The experiment first evidences the good tritium release performance of Li_2ZrO_3. This recent result, regarding a crucial characteristic of ceramic breeders, affirms the attractiveness of Li_2ZrO_3. Indeed, except for the activation of zirconium which may be a concern, Li_2ZrO_3 has several merits e.g. satisfactory physical properties, good compatibility with structures[5], very good irradiation behaviour[6, 7].

From the data collected so far, Li_2ZrO_3 appears as one of the most promising ceramics for water-cooled blanket concepts and for the DEMO relevant helium-cooled concept which will be tested during the technology phase of NET.

ACKNOWLEDGEMENTS

This work has been performed within the frame of the EEC program.

The authors wish to thank Dr Rasneur for supplying one $LiAlO_2$ sample and their colleagues of Division d'Etudes de Séparation Isotopique et de Chimie Physique for helpful discussions.

REFERENCES

1. E. Roth, F. Botter, M. Briec, M. Rostaing, H. Werle, R.G. Clemmer, J. Nucl. Mater. 141-143 (1986) 275.

2. M. Briec, J.J. Abassin, M. Masson, E. Roth, P. Sciers, H. Werle, J. Nucl. Mater. 154-156 (1988) in press.

3. T. Kurasawa, H. Watanabe, G.W. Hollenberg, Y. Ishii, A. Nishimura, H. Yoshida, Y. Naruse, M. Aizawa, H. Ohno and S. Konishi, J. Nucl. Mater. 141-143 (1986) 265.

4. T. Kurasawa, H. Watanabe, E. Roth, D. Vollath, J. Nucl. Mater. 154-156 (1988) in press.

5. P. Hofmann, W. Dienst, J. Nucl. Mater. 141-143 (1986) 289.

6. D.L. Baldwin, G.W. Hollenberg, J. Nucl. Mater. 141-143 (1986) 305.

7. G.W. Hollenberg, HEDL report 7643 (1987).

FUSION TECHNOLOGY 1988
A.M. Van Ingen, A. Nijsen-Vis, H.T. Klippel (editors)
Elsevier Science Publishers B.V., 1989

BLANKET FOR EXPERIMENTAL FUSION REACTOR (OTR) AND PRINCIPLES OF BASE STRUCTURAL MATERIALS SELECTION

E.O.ADAMOV, G.M.KALININ, V.V.RYBIN, A.M.SIDOROV, A.V.SIDORENKOV, Yu.S.STREBKOV, Yu.M.CHERKASHOV, V.F.VINOKUROV, V.A.IGNATOV

Research and Development Institute of Power Engineering

The paper presents the main concepts concerning the development of blanket design for the experimental fusion reactor (OTR). Its designation, design solutions and the main blanket characteristics and variables have been considered in this paper.
The X16H11M3T austenitic steel and the X20H45M4B nickel-based alloy have been chosen as the potential materials for the first wall components fabrication based on the analyses of structural material and operational conditions.

1. INTRODUCTION

The different design solutions of peaceful uses of fusion energy are being developed at present within the scope of national and international programs. One of the most promising trends is the development of reactors with magnet plasma confinement, equipped with toroidal chamber - tokamaks.

The transition from physical investigations to the development of commercial reactors of a given type is determined by a number of complicated design and technological problems to be solved. The validation of structural material selection for the reactor blanket is one of these problems. The stage of developing the commercial fusion power reactors precedes the stage of development of the experimental facilities where the physical, engineering and design solutions will be tested. The experimental fusion reactor (OTR) will permit to simulate and to study the processes changing the structural properties of materials of blanket under all operating factors.

2. OTR-FUSION REACTOR CHARACTERISTICS AND OPERATING CONDITIONS

The design developed during the last few years permitsto determine the principle design solutions for the OTR blanket, its composition, characteristics and operating conditions and to select the structural materials.

The OTR blanket is designed to solve the following problems:

- to form a plasma chamber;
- to produce the conditions for tritium breeding in amounts ($K_t > I$) required for the reactor self-sufficient and its extraction;
- to demonstrate the electricity and secondary fuel production;
- to provide the radiation shield for the electromagnet system;
- to locate the devices for additional plasma heating, the means for plasma diagnostics and fuel make-up;
- to locate the experimental channels for testing the structural components of candidate and perspective materials;
- to acquire an experience in remote and preventive maintenance of blanket;

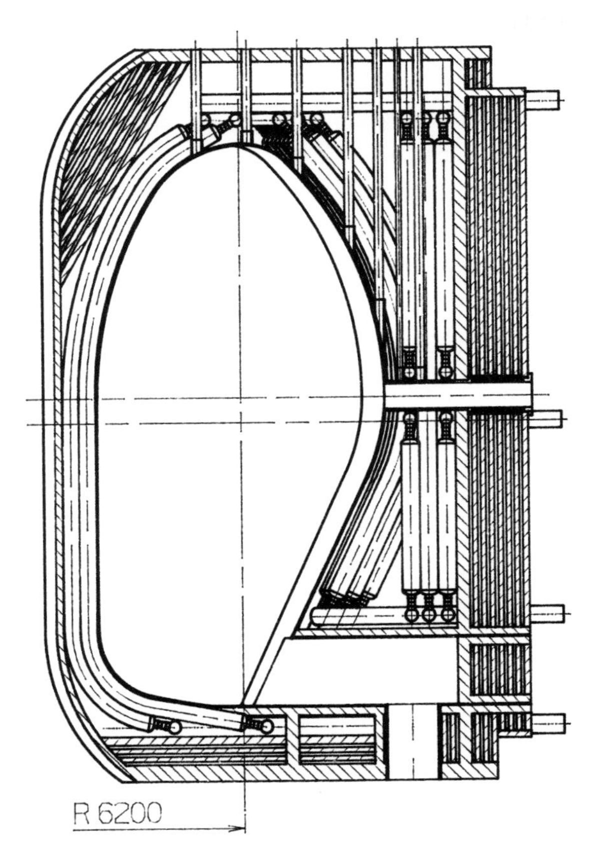

FIGURE 1
Experimental blanket sector
/cross-section/

- to ckeck the main components and facilities for reliability and safe operation.

The basic diagram of the OTR blanket is presented in Figs.1 and 2.

Blanket consists of 12 sections including:

- basic sections where the lithium zones for tritium breeding needed for reactor operation, are located (breeding ratio should be > I);
- heating sections for locating the high-frequency system for plasma heating;
- diagnostical sections where the plasma control systems and diagnoses means are located;

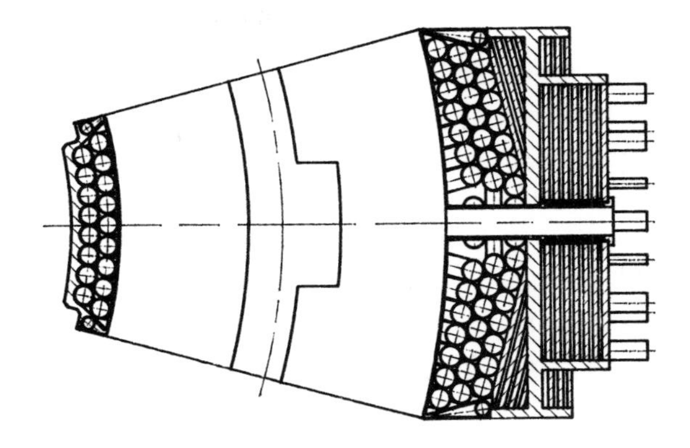

FIGURE 2
Experimental blanket sector
/longitudinal section/

- experimental sections that include the uranium zone apart from the lithium ones.

The selection of the basic structural materials is an important stage in solving the engineering problems of developing the experimental fusion reactors and depends on the operating conditions of the design elements and on the factors affecting on the materials.

The following main characteristics and parameters of the blanket are determined in accordance with the adopted concepts:

- mode of operation - cyclic	
- burn time, s	600
- pulses, s	80
- number of cycles	$3\ 10^5$
- neutron energy fluence, MJ/m^2	$1.6\ 10^8$
- maximum density of neutron flux per first wall, $cm^{-2}s^{-1}$	$5\ 10^{14}$
- maximum neutron fluence, cm^{-2}	$7\ 10^{22}$
- maximum dose of first wall radiation damage,dpa	70
- maximum heat flux, MW/m^2 per first wall area	

(limitter)	0.25
- average heat flux per first wall area, MW/m^2	0.1
- basic lithium-bearing material	$Li_{17}Pb_{83}$
- basic fuel material of experimental uranium section	UO_2
- coolant	water
- coolant temperature in basic section	50-80 °C
- temperature in experimental and lithium zones	270-285 °C

3. BASE STRUCTURAL MATERIALS SELECTION

We beleive, that the selection and feasibility of the OTR structural material operability should not be significantly differed from the adopted approach to the selection of materials used for current nuclear power reactors. The problem of material selection is, however, complicated by the fact that at present it is not possible to determine the whole complex of factors affecting the materials, and to reproduce them in experimental way. The influence of each unit factor can be analyzed at present. The analysis of the simulteneous influence of operating factors is a problem to be solved in future and is not considered here.

The possibility to use steels and alloys of different types, such as: austenitic, ferritic, of transient type, nickel-based alloys etc, has been considered.

The X16H11M3T austenitic steel and the X20H45M4B nickel-based alloys were selected as perspective materials for fabrication of the first wall elements of blanket components.

Irradiation effects significantly on the degradation of steel and nickel alloy properties (and consequently, on serviceability of the design). Despite the fact that the radiation swelling and effect of high-temperature radiation embrittlement do not occur within the operating temperature range, the neutron effect at temperatures less then 300°C results in significant decrease in plasticity. The uniform elongation can fall down almost to zero for several types of steels. Inspite of the significant increase (2-3 times) in yield stress the plasticity remains no less than 2% within the whole operating temperature range for the selected austenitic steel of X16H11M3T type and the X20H45M4B alloy, as shown from the investigations. The relationship between the uniform elongation of austenitic steels or nickel-base alloy and the neutron fluence is presented in Fig.3.

Ionic irradiation in contrast to neutronic one effects on the surface layers of the first wall materials of the blanket only. In this case thinning of wall as a result of sputtering and evaporation of materials under ionic irradiation in a pulse, at plasma disruption, the formation of unipolar arcs and scalling of surfaces due to blistering can take place. The use of diverter permits to diminish sputtering and erosion of the first wall surface and therefore to decrease the thinning of the first wall under normal operating conditions. Thinning of the wall due to evaporation will be more than 2 orders lower than sputtering.

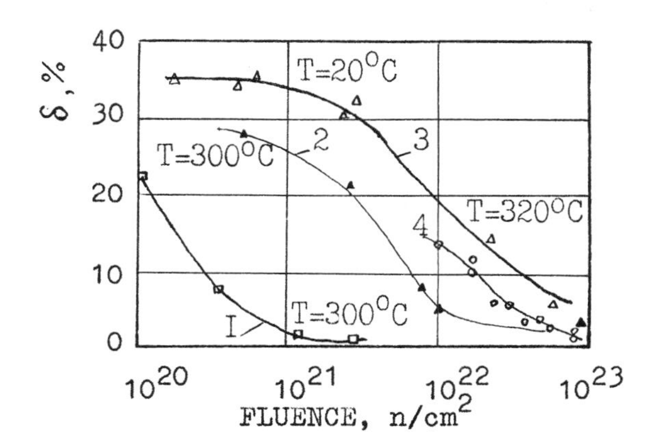

FIGURE 3
The relationship between the dose and relative uniform elongation of steels and austenitic alloys
1-X18H10T steel, 2,3-X20H45M4B alloy, 4-X16H11M3T steel

The surface erosion due to blistering and flecking is lower than in sputtering (fractions of micron per life time). Sputtering of the material due to ionic bombardment is a determining factor of wall thinning. Thinning of the first wall faced to plasma due to erosion may be decreased at the expense of various design solutions (use of protective shields, coatings, etc).

Structural materials of lithium zone elements in blanket are affected by coolant - water and lithium-bearing material-eutectic $Li_{17}Pb_{83}$.

The corrosion resistance in water does not cause a danger, due to the fact that this process has been well studied when using the austenitic steels and nickel alloys in nuclear power reactors. The uniform corrosion rate is equal to 10^{-2}g/m^{2}h in water at PH = 7÷10 and at temperatures up to 300°C for candidate materials of the X16H11M3T steel and the X20H45M4B alloy being selected.

The investigation of selected material behaviour in eutectic showed that any corrosion damages, as well as changes in short-time mechanical properties do not occur after exposing in eutectic at 340°C for 7000 hr. The corrosion rate amounts to less than 10^{-3} g/m^{2}h.

Under neutron irradiation along with the tritium forming in eutectic and diffusing into structural material of lithium zone wall of blanket, a significant amount of helium and hydrogen will be formed in it.

The calculations showed that the amount of helium and hydrogen in austenitic steel was equal to 0.07 at % and 0.3 at % respectively during the OTR life. These amounts could lead to hydrogen embrittlement along with low-temperature radiation embrittlement. However the investigations showed that the 03X20H45M4 alloy is not prone to hydrogen embrittlement with hydrogen content up to 0.5 at %. Previously, the alloy was exposed to fluence of $\sim 10^{22}$ n/cm^2, then it was subjected to electrolitic hydrogenation and mechanicaly tested. The hydrogen content was determined on samples after the mechanical tests.

Thus, the carried out analysis of effect of various operating factors showed that the X16H11M3T austenitic steel and X20H45M4B alloy could be considered as candidates for fabrication of blanket components of the OTR being developed. Moreover, the X16H11M3T austenitic steel is brought to a commercial level and it has a high manufacturability and possesses good welding properties. The X20H45M4 nickel-based alloy can be considered as an alternative candidate.

4. CONCLUSION

It should be noted that the OTR project being now developed permits to apply the available experience in developing and using the structural materials of current nuclear reactors. Low temperature permits to avoid a number of negative phenomena effecting on structural material properties-swelling and high-temperature radiation embrittlement and to decrease tritium penetration through the wall of blanket lithium zone. At the same time the low-temperature radiation embrittlement is less dangerous, especially taking into account the cyclic mode of operation (thermocyclic effect) and production of helium and hydrogen in structural materials. The study of selected material behaviour under conditions of complex effect of operating factors is the main task of further investigations on volidation of material selection.

FUSION TECHNOLOGY 1988
A.M. Van Ingen, A. Nijsen-Vis, H.T. Klippel (editors)
Elsevier Science Publishers B.V., 1989

MONTE CARLO SHIELDING ANALYSES IN THE DOUBLE NULL CONFIGURATION OF NET

Ulrich FISCHER

Association KfK-EURATOM, Kernforschungszentrum Karlsruhe, Institut für Neutronenphysik und Reaktortechnik, P.O. Box 3640, D-7500 Karlsruhe 1, Federal Republic of Germany

The shielding performance of NET reactor components has been studied in a three-dimensional analysis of a NET torus sector in the double null configuration on the basis of biased Monte Carlo calculations with the MCNP code. The main objects of this analysis are the poloidal variations of the physical quantities relevant for the radiation shielding and the neutron streaming through toroidal segment gaps. Furthermore, the relations between idealistic one-dimensional and realistic three-dimensional shielding calculations are analyzed.

1. INTRODUCTION

The design of the NET reactor provides the utilization of the blanket and the vacuum vessel as shielding components for the protection of the superconducting magnet from the neutron and photon radiation. Both components, forming the blanket/shield system, have to be optimized with respect to their shielding performance. In the past this essentially has been done on the basis of one-dimensional shielding calculations. The advanced design phase of NET, however, necessitates more detailed analyses. For treating different aspects of the shielding problems of NET corresponding activities have been started in the European Community [1,2,3,4]. In the present work a three-dimensional shielding analysis is performed using a realistic geometrical representation of a NET torus sector in the double null configuration. Emphasis is put on the appropriate representation of all relevant reactor components and the analysis of their impact on the shielding performance of the inboard blanket/shield system.

2. CALCULATIONAL PROCEDURE AND GEOMETRICAL MODEL

The shielding analyses have been performed on the basis of biased Monte Carlo calculations with the MCNP-code[5]. MCNP uses the basing nuclear data in the continuous energy representation avoiding in this way errors and uncertainties involved in the generation and application of group nuclear data. This is of great concern for shielding calculations, since the use of group nuclear data tends to underestimate the penetrating neutron and photon radiation[1].

Due to the segmentation of the torus into 48 sectors, the shielding analysis can be restricted to the treatment of a 7.5^{o} sector. Actually only one half of a sector is treated because of its toroidal symmetry. Reflective boundary conditions are applied at the lateral walls of the sector. This sector model takes into account adequately all relevant reactor components: blankets, shields, reflectors, divertors, plugs, openings and ducts. (see figs. 1, 2).

The plasma source distribution is represented by a D-shaped, exponentially decreasing probability distribution of the 14 MeV source neutrons, that is adequate for the NET double null configuration[6]; details are reported in ref.[1].

Appropriate variance reduction techniques (importance sampling with geometry splitting and Russian roulette) have been applied to follow the neutron and photon tracks through the shielding system. Typically 50000 to 150000 neutron histories have been taken into account in a shielding calculation assuring on the average a statistical accuracy of about 5 - 10 % for the relevant physical quantities scored in the spatial region of the toroidal field (TF) coil.

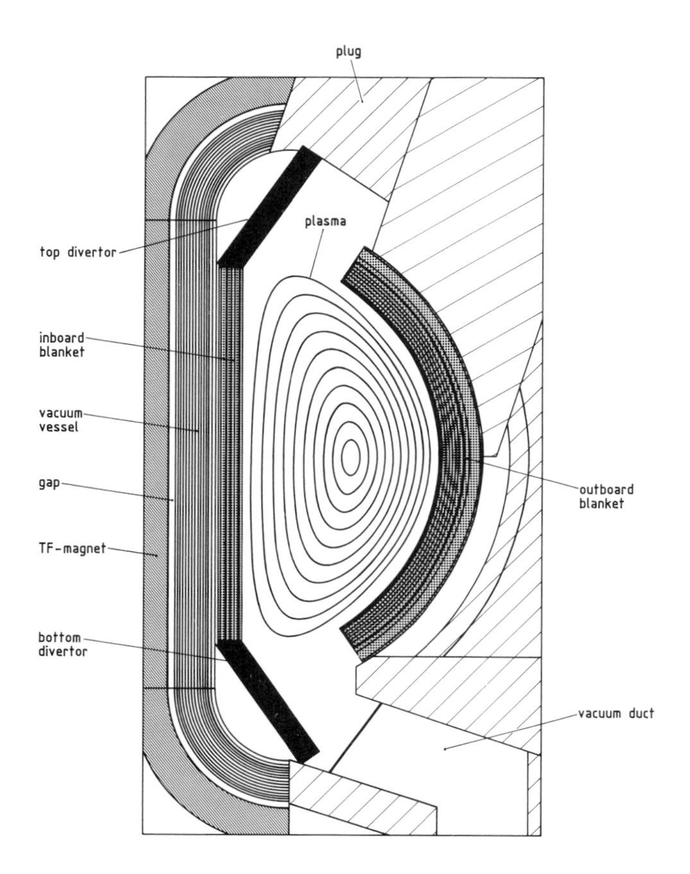

FIGURE 1
Radial-poloidal cut of the torus sector model

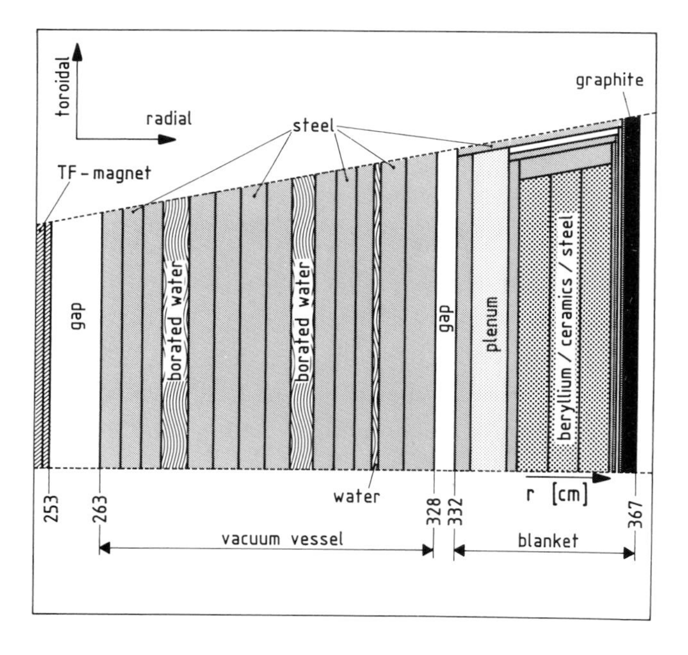

FIGURE 2
Radial-toroidal cut of the inboard segment at the torus mid-plane

3. THREE-DIMENSIONAL SHIELDING ANALYSIS OF A NET TORUS SECTOR

The KfK-design of a ceramic breeder blanket[7] and the NET-design for the vacuum vessel[8] are used in this analysis. The radial thickness of the blanket and the vacuum vessel at the inboard side is 35 and 65 cm, respectively. The results of the 3d-calculations are normalized to a fusion power of 600 MW (NET-III/DN).

3.1 Poloidal Variations

The concentration of the plasma source around the torus mid-plane leads to a strong poloidal variation of the 14 MeV neutron current impinging onto the first wall (fig. 3). The use of a plane first wall at the inboard side enhances this behaviour. Due to multiple scattering processes the poloidal profile of the total neutron flux at the first wall is somewhat flatter than that of the virgin 14 MeV neutron current (fig. 3). The poloidal peaking factor, defined as the ratio of the maximum (torus mid-plane) to the poloidal average value, is 1.82 for the 14 MeV neutron current, but 1.22 for the total neutron flux.

The poloidal variations, however, strongly depend on the radial position in the blanket/shield system (fig. 4): as seen from the central plasma region, the neutron pathways in the blanket/shield system increase as the distance to the torus mid-plane increases. Therefore the attenuation of the neutron radiation increases as the poloidal distance to the mid-plane increases. Consequently the poloidal peaking factor increases with increasing blanket/shield depth. In case of the total neutron flux the poloidal peaking factor increases to about 2.2 at the front of the TF-coil.

The radiation dose deposited in the Epoxy insulator is the most crucial issue with respect to the radiation damage of the superconducting magnet. According to the specifications by the NET-team, the Epoxy radiation dose should not exceed $5 \cdot 10^8$ rad over the lifetime of NET[9]. Taking an integral operation time of one year (for NET this is an upper limit) results in a poloidally averaged radiation dose of $3.5 \cdot 10^8$ rad. There is, however, a strong

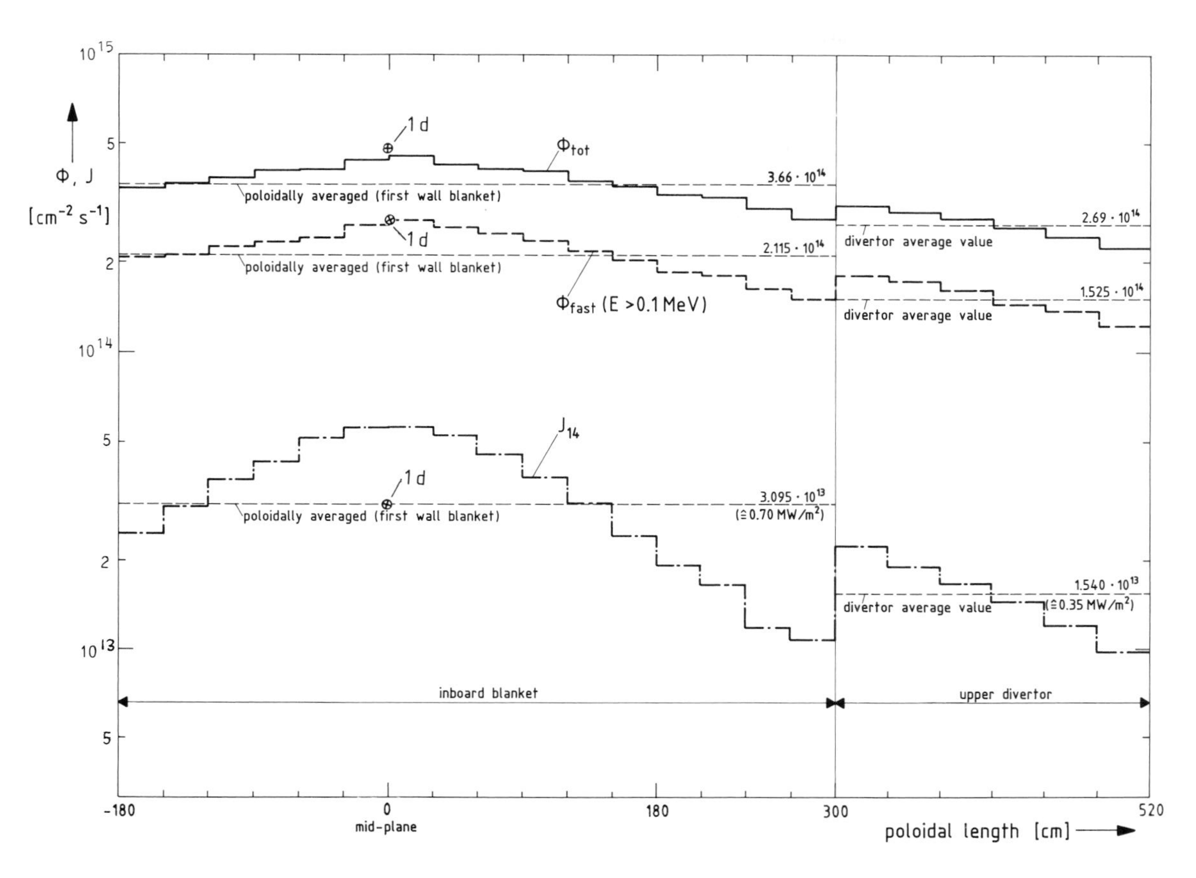

FIGURE 3
Poloidal profiles of the direct 14 MeV neutron current (J_{14}), the total (ϕ_{tot}) and fast (ϕ_{fast}) neutron flux densities at the inboard first wall

poloidal variation of the Epoxy radiation dose (fig. 5): at the torus mid-plane it amounts to about 8 $\cdot 10^8$ rad. Therefore, the dose limit is kept on the average, but it is exceeded in the central region around the torus mid-plane.

3.2 Relations Between 1d- and 3d-Calculations

One-dimensional (1d) calculations are performed in radial direction in the torus mid-plane using the same methods and data as in the case of 3d-calculations. Inboard and outboard segments are treated as cylindrical rings around the torus axis with the plasma source distributed uniformly in between. 1d-calculations are normalized to an average (i.e. inboard and outboard) neutron wall load of 1.0 MW/m^2. For a reliable comparison the neutron wall load of the 1d-calculation should agree with the averaged one of the 3d-calculation, both for the inboard and the outboard first wall. In the actual case it is 0.70 (1d) versus 0.68 MW/m^2 (3d) at the inboard first wall (fig. 3). Due to the special features of the 1d-calculation[10], the 1d-value of the total neutron flux at the first wall, however, agrees with the peaking value (torus mid-plane) of the 3d-calculation (fig. 3). Passing through the blanket/shield system, these relations no longer hold: in general the 1d-neutron fluxes agree with the poloidally averaged ones (fig.6). Consequently, the 1d-calculation underestimates the peaking values of the penetrating radiation at the torus mid-plane, although it is able, on the other hand, to reproduce them in the first wall region! Concerning the Epoxy radiation dose, this underestimation roughly amounts to a factor 2 (see fig. 5).

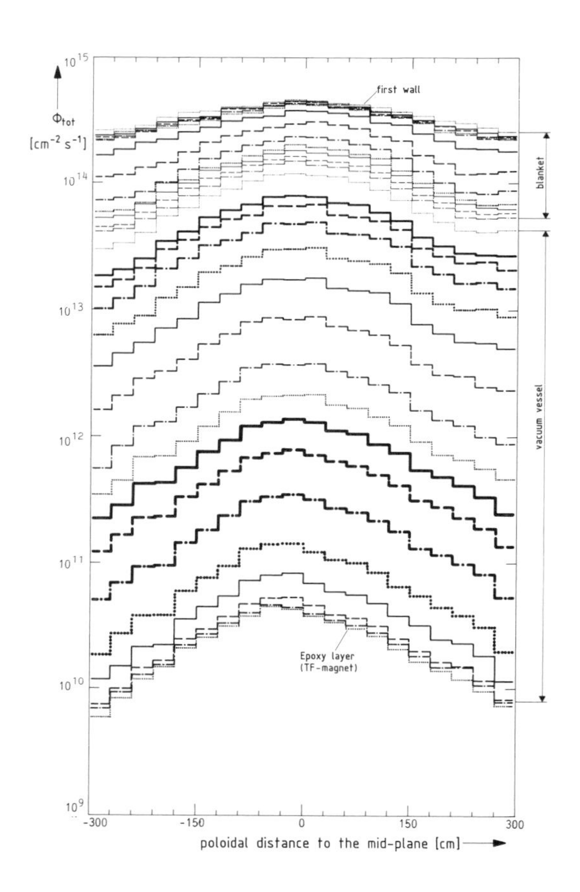

FIGURE 4
Poloidal profiles of the total neutron flux density in the inboard blanket/shield system

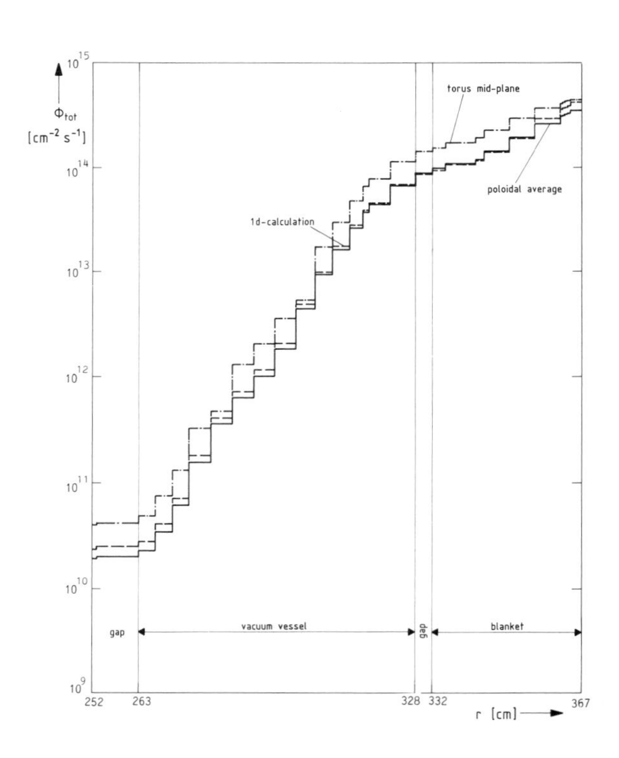

FIGURE 6
Radial profiles of the total neutron flux density from 1d- and 3d-calculations

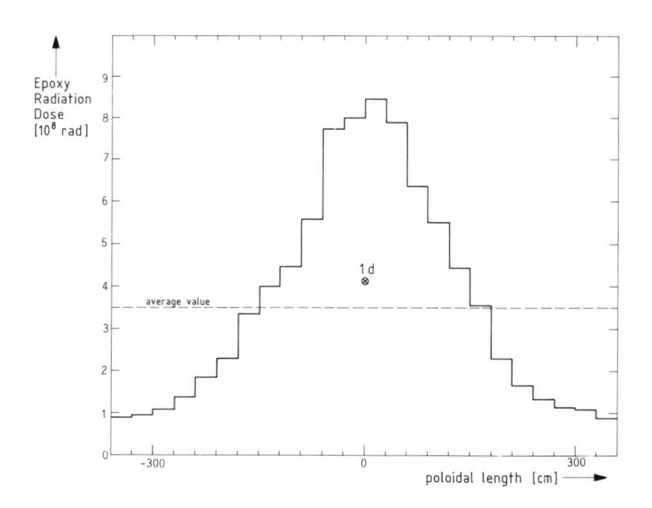

FIGURE 5
Poloidal profile of the Epoxy radiation dose

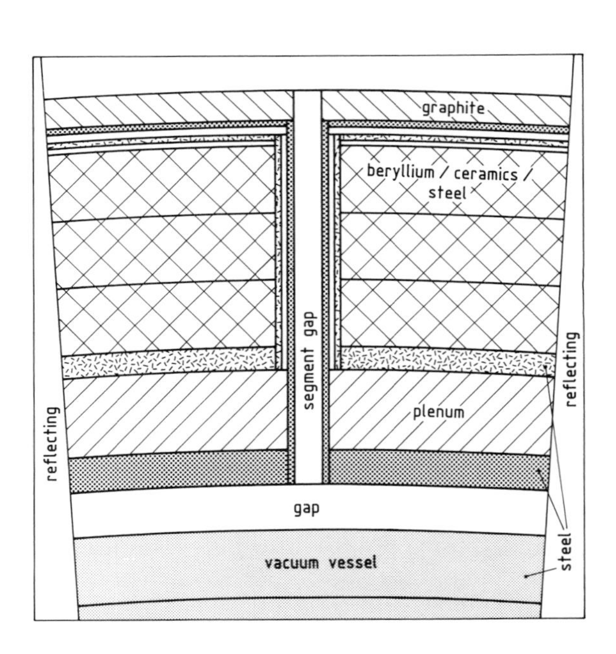

FIGURE 7
2d-sector model of the blanket segment gap

3.3 Divertor Shielding

The divertor is subjected to an average 14 MeV neutron current, that is about one half of the corresponding average value at the inboard first wall (fig. 3). It has therefore to be analyzed, if the radiation shielding in the divertor region is sufficient. In the KfK-design[7] the divertor consists of a thin (25 mm) helium cooled molybdenum plate, protected by a 5 mm thick graphite layer. The divertor plate is backed by a steel structure, enclosing the helium supply pipes. 3d-shielding calculations using the upper poloidal half of the torus sector model show, that the shielding is sufficient if the steel structure has an effective thickness of about 10 cm[1].

4. NEUTRON STREAMING THROUGH BLANKET SEGMENT GAPS

For maintenance reasons the blanket segments of NET will be separated by a gap of 20 to 40 mm width. As a consequence, there will be a neutron streaming through the segment gaps subjecting the vacuum vessel to an unshielded neutron flux at the bottom of the gaps.

The segment gap effect has been anlyzed in a two-dimensional approach. A 7.5° torus sector is modelled in cylindrical geometry taking the torus axis as symmetry axis. At the inboard side of the sector a segment gap and segment walls are inserted (fig. 7). Gap widths of 2.5 and 4 cm have been considered. Both, a breeding[7] and a shielding blanket, the latter consisting of a mixture of 80 % steel and 20 % water, have been used. The analysis is restricted to the inboard side of a torus sector, where the thickness of the blanket, and hence the depth of the segment gap amounts to 35 cm. In case of the shielding blanket, also a reduced blanket thickness of 22 cm has been considered: this refers to the extended plasma version of NET.

4.1 Radial Variations in the Blanket and the Segment Gap

The different shielding properties of the blankets (see table I) strongly affect the neutron streaming in the segment gaps.In case of the breeding blanket the total neutron flux at the bottom of the gap increases modestly due to the neutron streaming effect but strongly in case of the shielding blanket (table I). There is a weak dependence of the segment gap effect on the gap width, but a strong dependence on the blanket material composition: it is the stronger, the better is it's shielding performance.

4.2 Radial and Toroidal Profiles Across the Vacuum Vessel

The neutron streaming through the segment gap causes a toroidal peaking of the neutron flux at the front of the vacuum vessel; the toroidal profiles strongly depend on the blanket material composition (fig. 8). The corresponding peaking factors amount to 1.6 for the shielding and 1.3 for the breeding blanket; they are rather insensitive to the width of the segment gap. The toroidal profiles,

Table I: Impact of the segment gap on the total neutron flux density at the bottom of the gap

	breeding blanket			shielding blanket		
	no gap	gap width 2.5 cm	gap width 4.0 cm	no gap	gap width 2.5 cm	gap width 4.0 cm
ϕ_{tot} [10^{13} cm^{-2} s^{-1}]	8.774	11.88	13.96	1.474	3.883	5.537
attenuation factor *)	4.95	3.69	3.07	23.9	8.92	6.22
ratio gap/no gap	-	1.35	1.59	-	2.63	3.76

*) *Ratio flux at the first wall/flux at the front of the vacuum vessel*

Table II: Impact of the segment gap: neutron flux densities at the back of the vacuum vessel and Epoxy radiation dose (1 full power year operation time)

	breeding blanket			shielding blanket		
	no gap	gap width 2.5 cm	gap width 4.0 cm	no gap	gap width 2.5 cm	gap width 4.0 cm
ϕ_{tot} [10^{10} cm^{-2} s^{-1}]	2.631	3.036	3.336	0.4548	0.8062	1.179
ratio gap/no gap	-	1.15	1.27	-	1.77	2.59
total radiation dose [10^8 rad]	4.32	4.86	5.54	0.753	1.39	1.94
ratio gap/no gap	-	1.13	1.28	-	1.85	2.58

however, flatten as the depth of the vacuum vessel increases: at the back of the vacuum vessel there is no toroidal variation left.

There is, however, a smooth lifting of the neutron radiation level at the back of the vacuum vessel (table II and fig. 9). Again there is a strong dependence on the blanket material composition, but a weak dependence on the gap width. The Epoxy radiation dose is affected by the gap effect in the same way than the penetrating neutron radiation (see table II); again there is no toroidal variation.

4.3 Impact of Reduced Blanket Thickness

For the shielding blanket the reduction of the blanket thickness from 35 to 22 cm results in an enhanced neutron radiation by a factor 4.5 at the front, and a factor 5.5 at the back of the vacuum vessel. Thus the shielding performance of the shielding blanket with a thickness of 22 cm roughly corresponds to that of the breeding blanket with a thickness of 35 cm.

Consequently the effect of the blanket segment gap is considerable smaller for a blanket thickness of 22 cm than it is for a thickness of 35 cm. With a gap width of 4 cm, the neutron flux at the front of the vacuum vessel is enhanced by about 90 %. At the back of the vacuum vessel it is increased by about 60 %. The same figure is obtained for the Epoxy radiation dose.

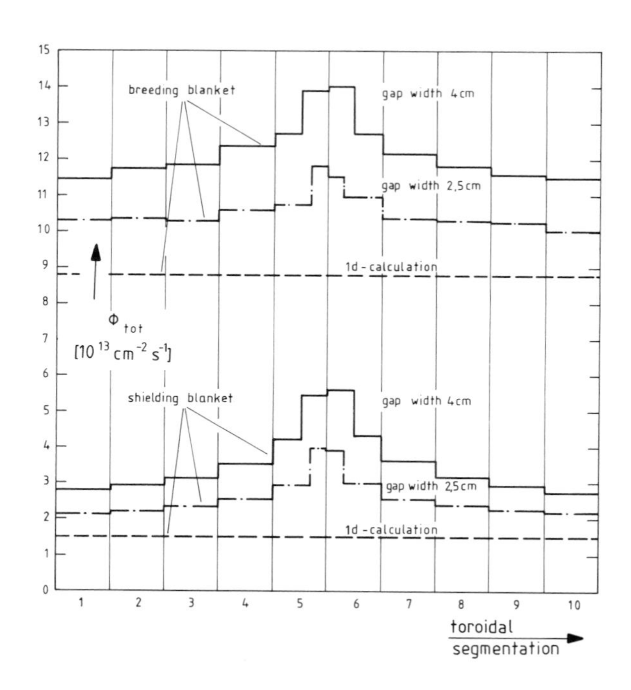

FIGURE 8
Toroidal profiles of the total neutron flux density at the front of the vacuum vessel

5. CONCLUSIONS

The poloidal variations of the neutron flux density strongly depend on the radial position in the blanket/shield system. 1d-shielding calculations are not able to reproduce this behaviour and therefore underestimate the maximally penetrating neutron radiation. For the analyzed blanket/shield system

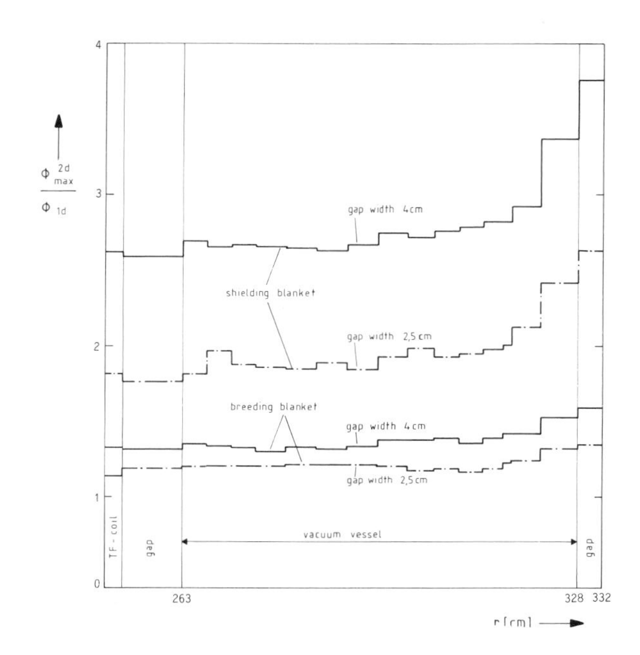

FIGURE 9
Ratio of the total neutron flux densit
without segment gap across the vacuun

this underestimation amounts to a factor 2 at the inboard side of NET.

The effect of the blanket segment gap strongly depends on the blanket material composition. It is very modest for a blanket with low shielding performance (e.g. a beryllium containing breeding blanket), but it is very strong for a blanket with a high shielding performance (i.e. a "shielding blanket"). The gap effect depends relatively weakly on the gap width, but stronger on the blanket thickness.

The gap effect may lead to a further correction factor of 2 - 3 for the 1d-calculation. Keeping in mind that the use of group nuclear data in a deterministic 1d-shielding calculation also causes an underestimation of the penetrating radiation[1] by up to a factor 2, it is realistic to correct such a shielding calculation by a factor 8 to 12, i.e. roughly by one order of magnitude.

ACKNOWLEDGEMENT

This work has been performed in the framework of the Nuclear Fusion Project of the Kernforschungs-zentrum Karlsruhe and is supported by the European Communities within the European Fusion Technology Programme.

REFERENCES

1. U. Fischer, Monte Carlo Shielding Calculations in the Double Null Configuration of NET, KfK-4411, Mai 1988

2. K.A. Verschuur et al., Assessment of the radiation shielding problems for the NET coil system, this conference

3. J.F. Jaeger et al., Shielding performance of the NET vaccum vessel, this conference

4. V. Rado et al., 3-D Monte Carlo analysis of the NET shielding system, this conference

5. J.F. Briesmeister (Ed.), MCNP - A General Monte Carlo Code for Neutron and Photon Transport, Version 3A, LA-7396-M, Rev. 2, Sept. 1986

6. K.A. Verschuur, Poloidal Variation of the NET Blanket Response Functions, ECN-87-001, Nov. 1986

7. M. Dalle Donne et al., Pebble Bed Canister: The Karlsruhe Ceramic Breeder Blanket Design for NET, Fusion Technology (in print)

8. W. Dänner, The Shielding Properties of the Vacuum Vessel, NET/87/IN-022, June 1987

9. Next European Torus, Status Report 1985, NET Report 51, CEC Brussels 1985

10. U. Fischer, Multi-dimensional Neutronics Analyses in the Double Null Configuration of NET, Int. Symp. Fusion Nuclear Technology, Tokyo, Japan, April 10 - 15, 1988

FUSION TECHNOLOGY 1988
A.M. Van Ingen, A. Nijsen-Vis, H.T. Klippel (editors)

ASSESSMENT OF THE RADIATION SHIELDING PROBLEMS FOR THE NET COIL SYSTEM

K.A. Verschuur, W. Daenner, G. Malavasi

The NET Team, c/o Max Planck Institut fuer Plasmaphysik, Boltzmannstr 2,
8046 Garching bei Muenchen, Germany.

This paper deals with the integration of the radiation shields into the complex torus geometry, with special emphasis on the penetration shields. The shielding performance of the penetration shields has been assessed using an approximate deterministic method, and solutions have been elaborated to improve the shielding at weak spots.

1. INTRODUCTION

The radiation shield required to protect the superconducting toroidal and poloidal coils from radiation damage and nuclear heat is accomodated in the space between the plasma-chamber and the coil casing. This space must be shared with other components as first wall, divertors, vacuumvessel etc. which may have competing requirements |1|. In addition this radiation shield is penetrated by ports required for plasma heating, blanket replacement, maintenance etc. These weaken the performance of the coil shield and require additional penetration shields which have to be accomodated in the restricted space between the TF coils. Further detailed evaluation of component design and changes in the design concept therefore easily may affect the radiation shielding for the coils |2|. Therefore during the development of the torus concept regularly an assessment of the performance of the shielding for the coils is required. To enable this a scheme was set up to perform approximate neutronic calculations for the penetration shields, applying 3D-kernel integration and 1D discrete ordinate. The scheme will be described shortly, and the calculations performed for the NET torus discussed.

2. NUMERICAL CONSIDERATIONS

For performing the 3D transport calculations to assess and optimize the penetration shields, the use of a 3D Monte Carlo code has certain disadvantages. Apart from the calculational effort that might be required, the statistical nature of the calculation results makes the optimization task difficult.

Therefore an approximate method is used in which the calculation is split in three steps. First the angular flux spectra at the first wall of the plasma chamber are calculated in 1D using a infinite cylinder approximation for the torus. Then a 3D kernel integration over the primary neutron source of the plasma and over the source of scattered neutrons and photons on the first wall (as produced by the first step) is performed to produce the angular flux spectra at the plasma facing surface of the penetration shield. Then finally the latter angular flux spectra are used to calculate the flux attenuation through the penetration shield with 1D discrete ordinate.

A similar approach was used to calculate the penetration shields for TIBER |4| and was reported to give sufficiently accurate results for the radiation load to the coils near the penetrations. Performing calculations for NET the same conclusion could be drawn. However, the method cannot be used always, as for example in cases where streaming phenomena are important or large contributions from leakage through slots occur. Therefore in a number of cases the calculations performed here have to be confirmed and/or extended by 3D Monte Carlo calculations.

3. DESCRIPTION OF THE SHIELD (FIG. 1)

The radiation shield for protecting the coils is essentially a layered structure of stainless steel and water that is accomodated in the vacuum vessel/shield and the shielding blanket |3,4|. See A and B in Fig. 1.

The effective thickness of shielding blanket and vacuum vessel/shield was designed to satisfy the maximum allowable radiation load to the coils as specified for the predesign phase |1|. Due to engineering considerations in a later stage the effective shield thickness of the vacuumvessel/shield was reduced, causing an increase of the radiation load to the coils which partly was compensated by mitigation of the design goal for the integral wall load from 3 Mwy/m^2 to 0.7 Mwy/m^2 |1, 2|. The main space restriction is at the inboard side, and therefore more effective options for the inboard shield were studied |5|. The most effective improvement was obtained by applying tungsten, either by replacing some of the outer steel plates by tungsten plates or by replacing the whole outer part of the shield by a heavy concrete shield in which tungsten pebbles are used as the heavy aggregate. A preliminary assessment of the engineering consequences of these concepts leads to the conclusion that a complete redesign of the vacuumvessel/shield would be required |6|. The major reason however, to doubt the application of tungsten in the inboard shield stems from safety considerations. The increase in afterheat by the use of tungsten is such that the temperature rise in the centre of the torus under total loss of coolant conditions might become a serious concern |7|. Therefore the stainless steel/water shield might remain the best option for the inboard side. The use of tungsten at the outboard poses less severe safety problems and therefore still is a useful option to improve the shielding at weak spots as will be shown later.

The shielding blanket consists of vertical removable segments with slots in between that may have a width of 2 cm. It is surrounded by the vacuum vessel/shield which is a tight bolted structure without any slots. Thus the blanket slots are fully shielded by the vacuum vessel/shield and therefore do not cause an appreciable increase of the radiation load to the coils, which is demonstrated by 2D Monte Carlo calculations |8, 9, 10|.

A more difficult situation arises at the blanket ports (F in Fig. 1). Here the vertical slots in between the blanket segments extend from the plasma chamber right to the outside forming a radiation leakage path towards the coils.

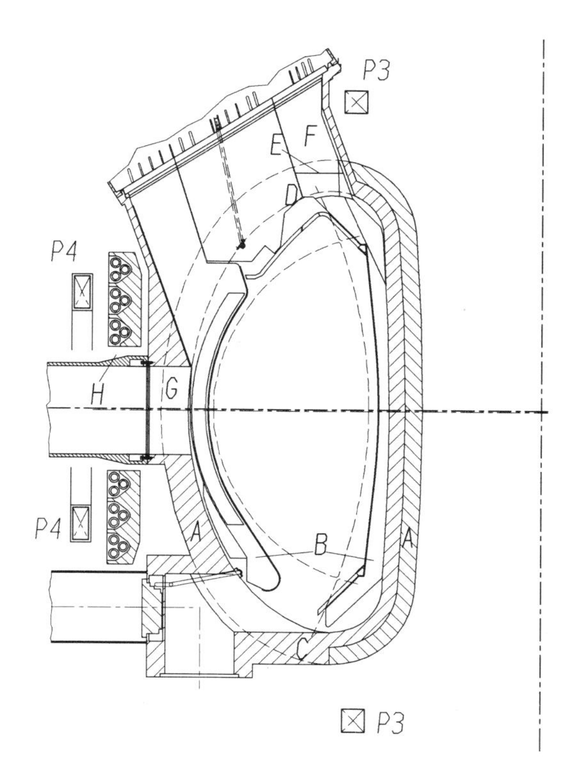

FIGURE 1
Vertical torus cross section

At the bottom of the torus the inboard and outboard blanket segments do not touch each other thus providing a gap as required for the vacuum ducts. Therefore the thickness of the vacuum vessel/shield had to be increased in that region (C in Fig. 1). A comparable situation arises at the top where part of the shielding blanket is lost to provide room for maintenance of the upper divertor (D in Fig.1).

The openings for remote maintenance are no serious concern as they will be plugged during machine operation. Diagnostic openings as discussed so far are small and should not cause serious radiation leakage. The amount of structure and coolant in the wave guides for the lower-hybrid heating occupies about 50% of the port volume where the port penetrates the shields. It can be shown easily, that therefore the wave guides do not seriously endanger the radiation shielding.

Much attention is given to the NBI-ports which are large empty ducts that cut away large pieces of the radiation shield (Fig. 4). These ports should be surrounded by penetration shields on all sides so as to protect the TF coils on its sides as well as the P4 poloidal field coils that pass above and below (H in Fig. 1).

4. CALCULATIONS

The calculations presented here are preliminar, and should be considered as a first assessment of the shielding performance of the NET torus as a whole, i.e. including ports ducts etc. To allow an acceptable safety factor it was aimed to obtain a numerical accuracy of a factor 2. Intercomparison with 3D Monte Carlo calculations is required however to obtain more certainty.

As measure for the shielding performance we use the dose in rads absorbed by the most exposed layer of epoxy/glass in the coils. This dose should not exceed a prescribed limit, which is around 10^9 rad |3|. An operation time of 4000 h during the technology phase was assumed with an average neutron wall loading of 0.7 MW/m^2 |2|.

4.1 Blanket port

A top view of the vertical blanket port is given in Figure 2. The slots that look straight into the plasma are those between the outboard lateral and central segments, those between the inboard lateral segments and the plug and those between the inboard central segment and the blanket port side wall. The latter is the more serious one as here the blanket port forms the only shield to protect the adjacent TF coil (A in Fig. 2). At this position the port wall thickness varies between 7 and 12 cm. It is constructed of two plates of 2 cm thickness with an interspace filled with coolant water and structure in a ratio which is not determined by cooling requirements there |11|. Using the approximate numerical method

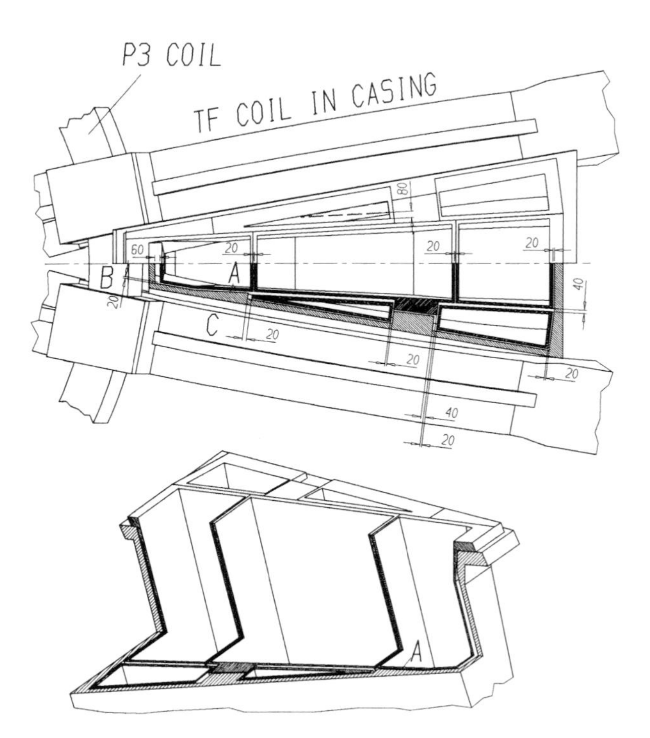

FIGURE 2
Blanket port showing slots

described the radiation dose to the epoxy/glass insulator of the adjacent TF coil was calculated assuming a blanket port thickness of 7 cm (2 cm thick side walls and 3 cm interspace). The results are as follows: Varying the stainless steel borated water ratio (SS/bH_2O) in the interspace volume from 75%/25% to 25%/75% resulted in a factor 2 reduction of the dose (as the inelastic scattering of fast neutrons towards the coils is thus reduced). With a slot widths of 2 cm a dose of $4 \cdot 10^8$ rad is calculated, which seems to be adequate. However this result will have systematic errors, not only due to the numerical approach and the geometry description used here but also due to possible errors in the nuclear cross sections used. Due to the large anisotropy of the fluxes involved it must be expected that the results will be dependent on the double differential scatter-data used. A careful further assessment of the radiation transport in this region is required using 3D Monte Carlo.

Another situtation that requires attention occurs for the poloidal field coil P3 that passes the vertical blanket port at the inboard side (B in Fig. 2). Through the same slot, i.e. the one between the inboard central segment and the blanket port side wall, the P3 coil sees the plasma directly, the blanket port inboard wall being the only protection. It was caluclated that for a 2 cm slot and a 7 cm thick inboard wall with in the interspace 25% borated water the radiation load to the coil would be $5 \cdot 10^9$ rad. Reducing the slot width to 1 cm gives a factor 2, and increasing the thickness of the inboard wall of the blanket port from 7 to 10 cm which seems feasible |11| gives another factor 1.5, which seems about adequate. However, a serious situation might occur here if (due to a change in the design goals, due to errors in the database used or otherwise) the actual dose will be higher than

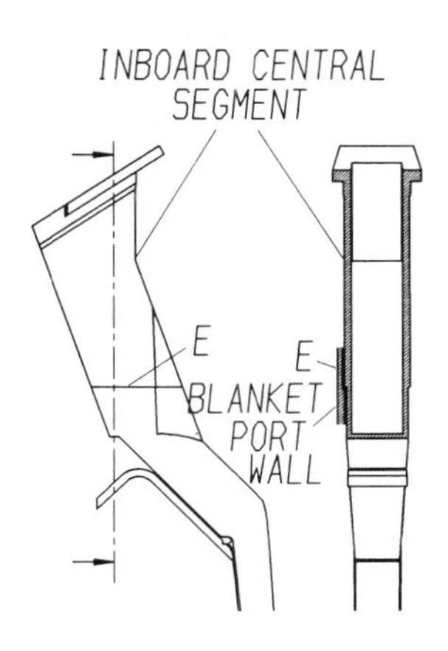

FIGURE 3
Step in slot at inboard central blanket segment

calculated here, because there is no room for additional shielding. Therefore it was advised to interrupt the slot by implementing a step. This was done with some difficulty as the poloidal dimensions of the blanket port are restricted by the presence of the TF coils, and therefore the overlap is rather small |12| as can be seen at E in Figure 3. Here again 3D Monte Carlo calculations are required to obtain sufficient confidence.

4.2 Vacuum duct

As mentioned before, due to the vacuum duct, the bottom of the torus has no shielding blanket, leaving the vacuum vessel/shield unprotected from plasma radiation (C in Fig. 1). Unless this gap can be closed otherwise, it might be considered to protect the vacuum vessel here using removable water cooled protection plates covered with carbon tiles |13|. Using our approximate method the radiation load to the bottom part of the TF coils was calculated, assuming a protection plate and carbon tiles. The 59 cm thick vacuum vessel/shield gives an epoxy/glass dose of $3.5 \cdot 10^9$ rad.

The bottom of the vacuumvessel/shield carries the whole weight of the vessel and therefore it was not assumed attractive to weaken this part of the vessel mechanically by replacing a part of the steel structure by tungsten plates. A local increase of the thickness of the vacuum vessel shield however, was feasible. It was calculated that the dose will be reduced by a factor 5 by increasing the thickness from 59 cm to 70 cm. Using the same thickness for the upper part of the vessel will compensate for the reduced shielding by the blankets behind the upper divertors.

The side walls of the duct (Fig. 1) protect the TF coils that pass on both sides. From the side walls the plasma centre is not seen directly however, so that there is no serious shielding problem here. 3D Monte Carlo calculations for the vacuum duct are being performed at Frascati |14| which will be presented at this symposium as well. A careful intercomparison of the result still has to be made however.

4.3 NBI ports

A horizontal cross section through the neutral beam injector ports where they pass in between the TF coils and penetrate through the vacuum vessel/shield and the shielding blanket is given in Figure 4. The most exposed area of the vacuumvessel/shield, point P, views plasma and plasma chamber walls over a solid angle as indicated, resulting in a neutron wall load of 0.3 MW/m^2. This is 5 times lower than the maximum wall load on the first wall at the outboard midplane. Nevertheless, a first wall and carbon tiles are assumed here to protect the vacuum vessel/shield leaving 50 cm for the actual shield to protect the TF coils. As before our approximate method was used, and it was found that for an optimal SS/bH_2O ratio of 75/25 the dose to the adjacent TF coil insulation would be $2 \cdot 10^9$ rad. As there is no space for additional shielding here, the only way to improve is to replace part of the stainless steel by tungsten. For reasons of construction at the backside a 5 cm SS plate

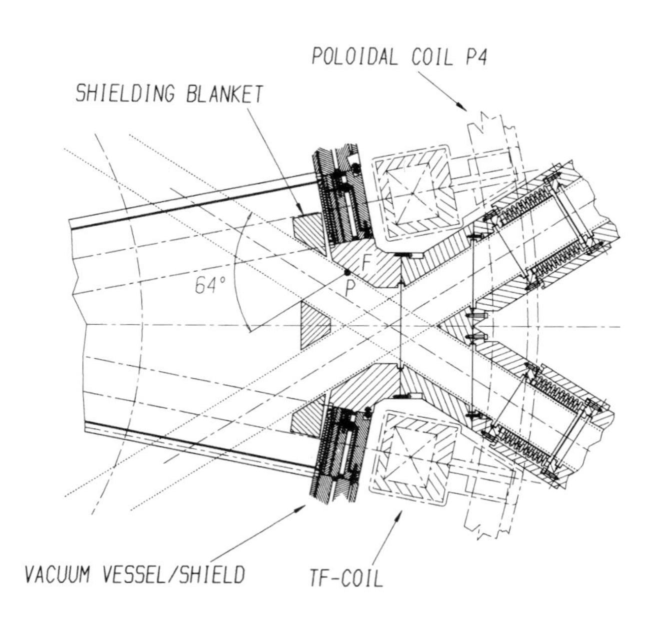

FIGURE 4
Horizontal cross section of NBI port

should be kept, and because of the heat load of several W/cm^3 at the front the first 15 cm behind the first wall was assumed to be constructed of stainless steel cooled with borated water, leaving 30 cm for tungsten plates with layers borated water. This geometry resulted in a radiation dose to the epoxy/glass of $7 \cdot 10^8$ rad for an optimal W/bH_2O ratio of 75/25. This does not look bad. However the tungsten containing vessel sector will have steel sidewalls and adjacent segments without tungsten so that the actual attenuation obtained will probably be less. To find out calculations with a better geometry description, at least 2D, are required.

An additional problem at the NBI ports are the P4 poloidal field coils that pass above and below (Fig. 5). Our calculations show that to obtain 10^9 rad for the P4 coils requires a 50 cm thick shield of an optimum SS/bH_2O mix (75/25). To properly shield the P4 coil (and the TF coils as well) the NBI port should be completely surrounded by the penetration

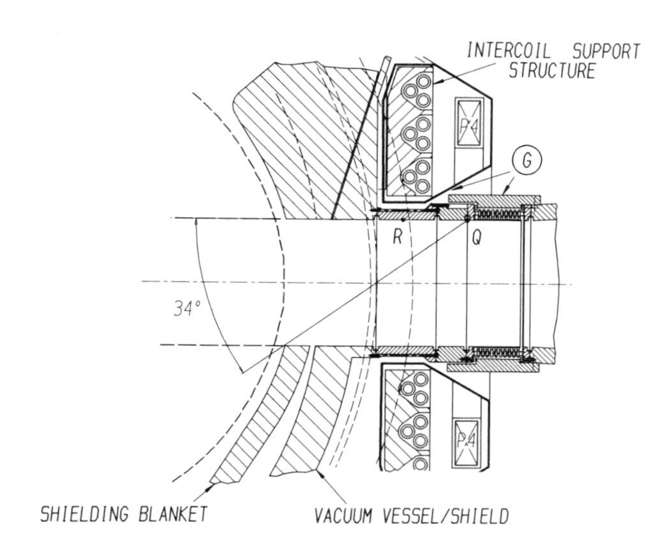

FIGURE 5
Vertical cross section of NBI port

shield. Unfortunately the intercoil support structure restricts the available space for that to 30 cm, just behind the vessel where the neutron wall load is still 0.1 MW/m^2. Therefore an improved shield with 5 cm SS/bH_2O (75/25) and 25 cm W/bH_2O (75/25) was calculated, which gives $2 \cdot 10^9$ rad to the epoxy/glass of the P4 coils. It seems acceptable to use the latter 30 cm thick shield only near the support structure, and increase the thickness behind it where there is more space.

5. CONCLUSIONS

The main conclusion of this assessment is that after the implementation of the improvements mentioned here the shielding obtained at the penetrations is of about the same quality as obtained for the bulk shielding at the inboard side |3|. However at the blanket ports the geometry is rather complex, and will therefore require more detailed Monte Carlo calculations.

In this assessment the influence of cross section inaccuracies is not included. However, as these might have a significant influence on the calculated radiation loads, sensitivity studies have been initiated.

ACKNOWLEDGEMENTS

The authors are indepted to H. Gorenflo for his invaluable assistance in the computer preparations.

REFERENCES

1. NET Status Report, December 1985, EUR/FU/XII-80/86-51 Commission of the European Communities (1986).
2. The NET Team, Fusion Technology 14,1 (1988).
3. W. Daenner et al., Status of NET Shielding Blanket Development, this Conference.
4. F. Fauser et al., Design Options for the NET Vacuum Vessel and its Resistive Elements, this Conference.
5. F. Gervaise, W. Daenner, Private Communication (1988).
6. D. Collier, G. Malavasi, Private Communication (1988).
7. J. Raeder, W. Gulden, Private Communication (1988).
8. U. Fischer, Monte Carlo Shielding Analyses in the Double Null Configuration of NET, this Conference.
9. F. Rohloff, Evaluation of Neutron Streaming Effects Through the Gaps In Between Blanket Segments, KfA Report No. KfA-ASS/NST 87-4 (1987).
10. V. Rado, Private Communication (1988).
11. F. Fauser, Private Communication (1988).
12. E. Theisen, Private Communication (1988).
13. F. Casci, Private Communication (1988).
14. V. Rado et al., 3D Monte Carlo Analysis of the NET Shielding System, this Conference.

FUSION TECHNOLOGY 1988
A.M. Van Ingen, A. Nijsen-Vis, H.T. Klippel (editors)

SHIELDING PERFORMANCE OF THE NET VACUUM VESSEL

J. F. JAEGER and J. J. ARKUSZEWSKI

Paul Scherrer Institut, (formerly EIR), 5303 Würenlingen, Switzerland

The Next European Torus (NET) is foreseen as the next step in the European development towards the controlled use of thermonuclear fusion. It has essentially two goals: the study of plasma physics with a long burning, ignited D-T plasma and the technological development of blankets, materials, tritium breeding, etc.. The neutron load is already considerable, needing detail design of the shielding blanket protecting the peripherals, more especially the super-conducting coils.

Monte-Carlo shielding calculations have been done with the MCNP code for the heat generation in the pressure vessel and for the radiation load on the outside. A very detailed 3-D geometrical model has been set up to obtain both the neutron and gamma loads on the coil insulation, as this influences both the cooling power and the life time of the coils. At present only shielding is considered, but the model is a template for later tritium breeding modules. It includes detailed representations of both the plasma as an extended neutron source and of the large vacuum pumping ducts.

Preliminary results for a plasma power of **714 MW**, show that the total heat deposited in the blanket and vessel is 810 MW, considerably more than the energy of the neutrons at birth. It corresponds to 20.0 MeV per source neutron. This high value is due to the preponderance of stainless steel and absence of lithium. The maximum currents escaping from the vessel are $3 \cdot 10^{10}$ neutrons/cm^{2}sec and nearly $9 \cdot 10^9$ photons/cm^{2}sec.

1 INTRODUCTION

In the European programme towards the controlled use of thermonuclear fusion, the Joint European Torus, JET, at Culham, is at present the largest machine investigating plasma physics. Its aim is to reach considerable α-heating from the D-T fusion reaction, but not ignition. The Next European Torus (NET) is foreseen as the next step. It has essentially two goals: the study of plasma physics with a long burning, ignited D-T plasma and the technological development of blankets, materials, tritium breeding, etc.. Initially, the machine will be run with an extended plasma, with a slightly larger plasma cross-section, and a blanket designed for shielding only, with no tritium breeding. The neutron load is already considerable, needing detail design of this shielding blanket protecting the peripherals, more especially the super-conducting coils. The power deposited in the vacuum vessel and the shielding blanket is that of a modern power station and tritium breeding will need to be considered at a later stage.

1-D deterministic shielding calculations have been done by the NET team, with the ANISN code, as these are quicker to set up and use less computer time than full 3-D ones. Done well, they give an accurate overall picture. To corroborate these, 3-D Monte Carlo calculations have been done with the MCNP code.[1] This should provide information on the poloidal and the toroidal variations, caused by large vacuum pumping ducts, division of the vessel into parallel and wedge segments, and the finite, D-shape of the plasma and vessel.[2] It is needed for the detailed thermal and thermohydraulic design.

The present calculations provide, in particular, neutron and photon currents at the outer boundary of the vacuum vessel. These are needed for the detailed design of the super-conducting magnetic coils, where the problem of heat deposition is crucial, as this has to be extracted at the temperature of liquid helium and where degradation of the electrical insulation could limit the lifetime of the coils. The model was also evolved to act as a template for later design work, e.g. when the shielding blanket is upgraded to breed tritium.

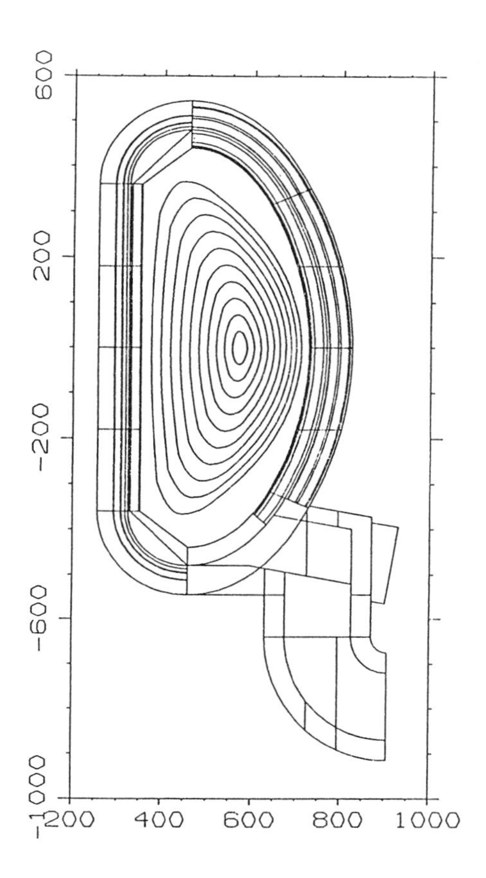

FIGURE 1: Poloidal Cross-section

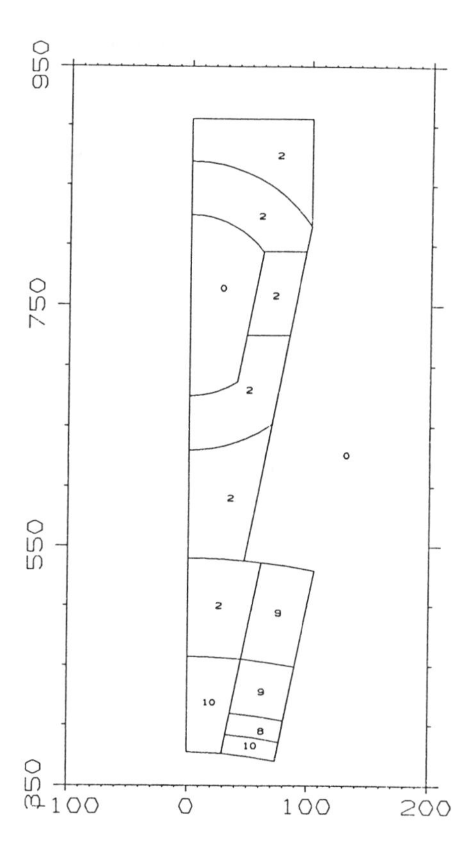

FIGURE 2: Horizontal Cross-section through the Vertical Duct

2 PLASMA SOURCE SIMULATION

The D-shaped plasma contours are described by the following set of parametric equations, which represent fairly realistically the results of the plasma physics:

$$R(\alpha, a) = R_p + a\cos(\alpha + \frac{a}{a_p}c_p \sin\alpha) + \epsilon_p a_p (1 - \frac{a^2}{a_p^2})$$

$$\begin{aligned} Z(\alpha, a) &= \frac{a}{a_p} b_p \sin\alpha \\ S(a) &= S_0 (1 - \frac{a^2}{a_p^2})^{e_{pk}} \\ 0 \leq \ &\alpha < 2\pi \end{aligned}$$

They define the contour surfaces, $(R(\alpha, a), Z(\alpha, a))$, of equal source strength, $S(a)$, for given values of the horizontal minor radius a. For the NET Extended Plasma:

R_p,	plasma major radius	= 541 cm
a_p,	horizontal plasma minor radius	= 168 cm
b_p,	vertical plasma minor radius	= 366 cm
ϵ_p,	radial plasma shift	= 0.17
c_p,	triangularity	= 0.62
e_{pk},	peaking exponent	= 4

The geometry can be defined in MCNP by surfaces up to the second order and by tori. Thus, the above contours have to be approximated by such surfaces: a rotational hyperboloid and an elliptical torus on the inboard side and a rotational ellipsoid and another elliptical torus on the outboard side. The neutron energy distribution is a Gaussian fusion spectrum for D-T fusion at 10 keV. The discrete distribution among cells is represented as a histogram of unnormalized probability density values for cell midpoints, interpolated from the $S(a_i)$ values. The angular velocity distribution is isotropic and there is no biasing.

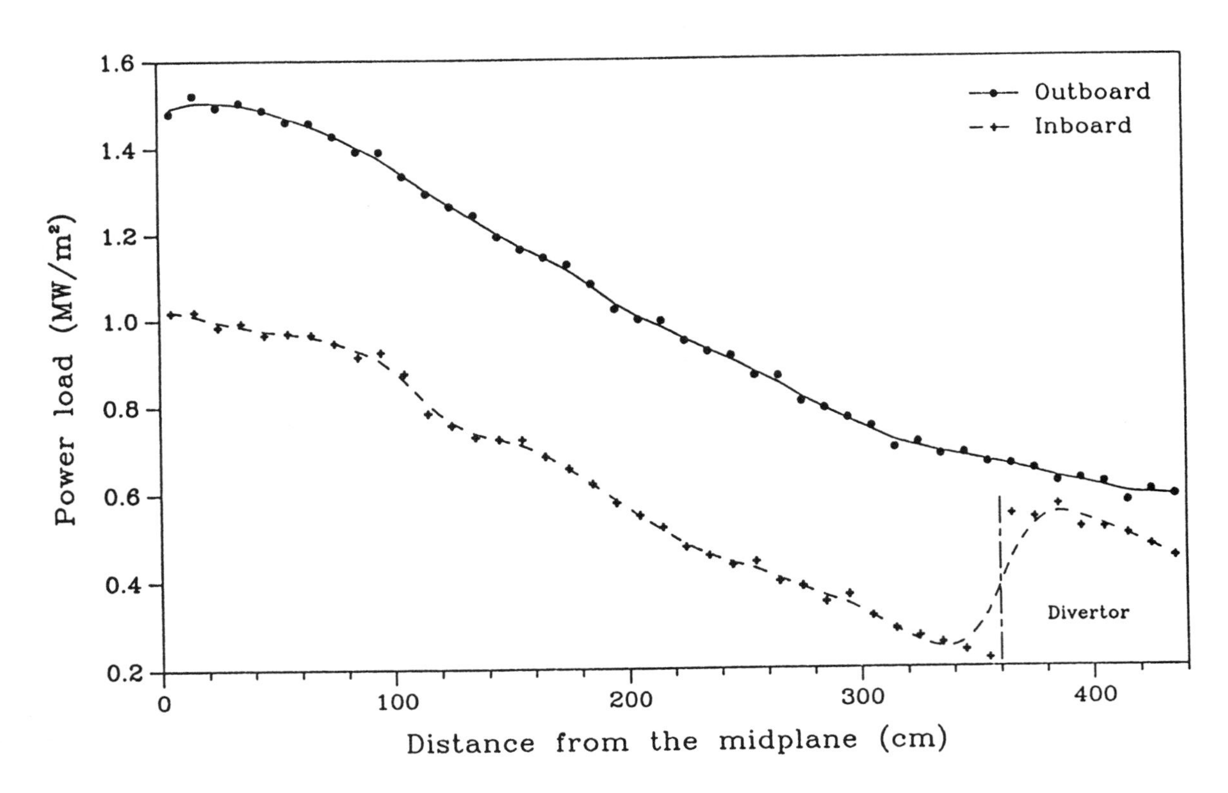

FIGURE 3: Poloidal Distribution of the Virgin First Wall Load

3 SHIELDING BLANKET AND VACUUM VESSEL MCNP MODEL

The reference profile of the internal vacuum vessel surface has been approximated by the envelope of the following surfaces: a cylinder and two circular "small" tori shifted ±360 cm along the z-axis on the inboard side; a "large" elliptical torus from the vertex down and a rotational ellipsoid on the outboard side. All the other poloidal surfaces that define both the shielding blanket and the vacuum vessel are parallel, forming a sandwich structure. The 32-fold reflective symmetry of the device allows one to compute a 11.25° sector only. The poloidal cross-section through the wedge sector is shown in Fig. 1. The lower part of the wedge sector opens onto the vacuum duct. The blanket extends into a lip in front of the duct to shield it from line-of-sight neutrons. The wedge shape of the vertical part of the duct (Fig. 2) is there to leave space for the toroidal coils. Both vacuum vessel and shielding blanket have been sectioned into 12 segments each, 6 inboard and 6 outboard. The divertors (inboard) have been somewhat simplified.

The final input to the code contains the following features:

Cross Section Data. Default MCNP libraries, compiled at the Los Alamos National Laboratory, have been used. The neutron one, BMCCS contains recommended unclassified data. Three of the nuclides ^{31}P, ^{32}S, and ^{55}Mn lack gamma production tables. However, they occur in small quantities. The gamma library, MCPL, contains cross-sections for coherent and incoherent scattering, pair production, and the photoelectric effect together with appropriate form factors, energy deposition, and fluorescence data. No (γ,n) reactions are considered.

Volume and Surface Areas. None of the volumes and surface areas required by the code could be computed analytically by the code and so had to be estimated stochastically, in a separate run.

Physical and Weight Cutoffs. An analog Monte Carlo game has been chosen for the neutrons for

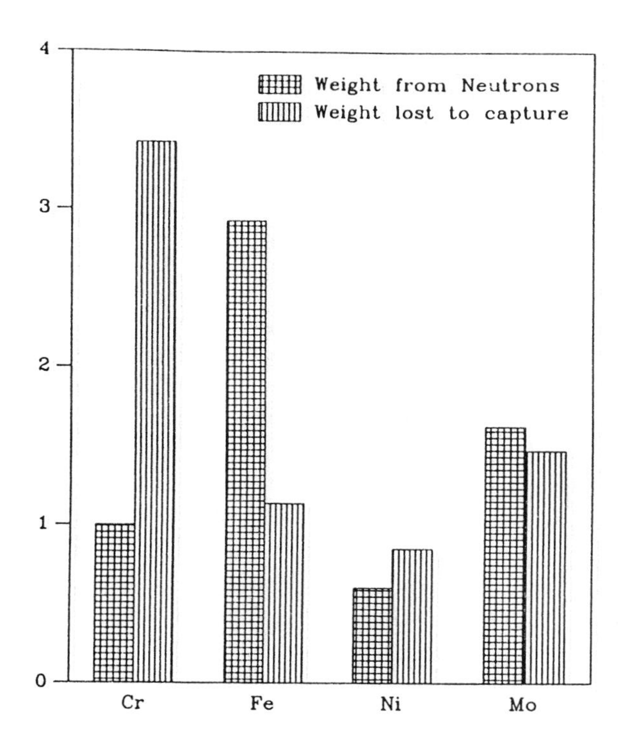

FIGURE 4: Photon Creation & Loss

the entire energy range from 0 to 20 MeV as well as for the photons from 0.001 to 100 MeV. The detailed physical treatment has been used for the photons.[1]

Weight Windows and Exponential Transform. Outboard weight windows for both neutrons and photons were generated from a series of relatively short runs with hand adjustment. As the reference tally for the outboard side, the total heat generation at the outer shell of a midboard segment was selected. One energy bin was used for both types of particles and a set of stretching parameters $p = \Sigma_a/\Sigma_t$ in directions normal to the sandwich structure was used for the exponential transform. The weight windows for the inboard and the outboard sides are different and for this reason two sets of corresponding calculations were done with different sets of weight windows.

TABLE I: Power Release in the Shielding Blanket and Vacuum Vessel

	Power Release (MW)		
	Neutron	Photon	Total
Shielding Blanket:			
– Inboard	39.6	92.9	132.4
– Top divertor	4.4	46.2	50.7
– Bottom divertor	4.0	43.4	47.4
– Outboard	163.0	354.5	517.5
– Total	211.0	537.0	748.0
Vacuum Vessel:			
– Inboard	2.3	12.5	14.7
– Outboard	7.3	41.0	48.3
– Total	9.6	53.5	63.0
Total	220.6	590.5	811.0

4 RESULTS

The total net neutron power crossing the first wall is 574.4 MW. This is close, ≈0.6%, to the total power of the virgin - uncollided - neutrons in the plasma of 571.2 MW. The power load on the first wall varies between 1.37 (MW/m^2) at the mid-outboard region and 0.67 (MW/m^2) at the divertors. The overall mean is 1.03 (MW/m^2), with a 0.2% statistical error. The virgin power load poloidal distribution is shown in Fig. 3. Done as a separate run, this fine distribution does not show the effects of back-scatter and, hence, has a much larger peaking factor ≈1.8.

The average neutron induced photon energy is 2.4 MeV, which represents indirect evidence that most of the gamma radiation is created in a capture process that leads to high energy photons.

The code produces very detailed statistics of the neutron and photon activity with respect to each nuclide, an information which is not usually available in deterministic codes. The effect of material composition on the photon production and absorption is illustrated by Fig. 4, where neutron induced weight creation and weight absorption are given for a few significant nuclides. It should be noted that neutrons are captured preferentially in iron, whereas gammas are best absorbed in chromium. This could be of importance when defining the composition of the steel(s).

TABLE II: Neutron and Photon Currents (max.)

	Currents ($cm^2 sec^{-1}$)	
	Neutrons	Photons
Outboard:		
– mid-board wedge	$1.2 \cdot 10^{10}$	$7.5 \cdot 10^{9}$
– mid-board parallel	$3.1 \cdot 10^{10}$	$8.6 \cdot 10^{9}$
– top	$6.5 \cdot 10^{9}$	$1.9 \cdot 10^{9}$
– bottom †	$7.4 \cdot 10^{10}$	$6.4 \cdot 10^{10}$
Inboard:		
– mid-board wedge	$1.8 \cdot 10^{10}$	$3.0 \cdot 10^{9}$
– mid-board parallel	$6.5 \cdot 10^{9}$	$2.1 \cdot 10^{9}$
– top †	$3.3 \cdot 10^{8}$	$1.2 \cdot 10^{8}$
– bottom †	$3.8 \cdot 10^{9}$	$7.1 \cdot 10^{8}$

† note:- statistically unreliable

The total power released is given in Table I. The overall total, 811 MW, corresponds to 20.0 MeV per source neutron, a gain of 5.9 MeV in the blanket and shield. This unusually high value is due to the fact that the principal neutron capture material is stainless steel - Fe, Cr, and Ni - all with a neutron binding energy around 7 MeV, instead of the more usual Li with only 4.8 MeV present in blankets constrained to breed tritium. Furthermore, the energy multiplication is significantly enhanced by a strong contribution from (n, xn) reactions, particularly in Fe, Mo, Cr, Ni, and Mn. These reactions create $\approx$12% extra neutron weight with an average energy loss of $\approx$0.27 MeV, as is the case with capture in ^{7}Li. Significant values of the heat deposition have statistical errors in the range 2-8%, with a few exceptions.

The maximum neutron and photon currents escaping the blanket and vessel are given in Table II. These limit the super-conducting coils design. Two regions dominate, the mid-board region which always is a maximum and the region facing the divertors. There are, however, considerable variations between upper and lower halves of the torus, some explainable, e.g. in the neighbourhood of the vacuum duct with its large loss of shielding, some less so, e.g. top and bottom, where other computational methods, for instance 2-D deterministic codes, might need to be used. Minimum values of the currents are $<$1% of the respective maximum.

5 DISCUSSION

The Monte-Carlo method using MCNP produced many good results with adequate statistical accuracy, especially heat deposition rates, in the blanket and on the outboard part of the vessel where the weight windows are well optimized. However, it is clear that on the inboard side some results are inaccurate, especially currents of escaping neutrons and photons. This is due to lack of time and the high computing costs involved in optimizing the weight windows which were synthetically constructed. This is not too disturbing, as these results are either small in comparison with other ones, e.g. power deposition, or in a region of rotational symmetry where 2-D deterministic computations can be done. One must bear in mind, however, the considerable simplifications necessary even for a 3-D model, e.g. neglect of streaming effects in the segment gaps.

6 CONCLUSIONS

The peaking factor of 1.34 on the power load on the first wall is of course reflected in the heat distribution. This is global, per segment, but should be detailed enough for design purpose. It is considerably less than for virgin neutrons but, nevertheless, shows the difficulty in modelling the neutron source for 1-D calculations adequately.

The peaking factor on the currents is much greater, but less relevant. In view of the closeness of the toroidal coils, the maximum current is more important and this is very different from that on axis. This shows the necessity of 3-D or 2-D calculations.

Another question left unanswered at this point is the sensitivity of the results to known experimental tolerances on the cross-sections.

REFERENCES

1. MCNP - A General Monte Carlo Code for Neutron and Photon Transport - Version 3A, LA-7396-M, Rev. 2 (1986)

2. J.J. Arkuszewski and J.F. Jaeger, Shielding performance of the NET vacuum vessel, PSI Report Nr.12, Paul Sherrer Institute, (formerly EIR), 5303 Würenlingen, Switzerland, (1988), also NET-Report, in print.

FUSION TECHNOLOGY 1988
A.M. Van Ingen, A. Nijsen-Vis, H.T. Klippel (editors)

STATUS OF NET SHIELDING BLANKET DEVELOPMENT

W. DAENNER, P. DINNER, B. LIBIN, E. THEISEN, K. VERSCHUUR, C.H. WU

The NET Team, c/o Max-Planck-Institut fuer Plasmaphysik, Boltzmannstr. 2,
D - 8046 Garching b. Muenchen

This paper deals with the present status of the development of the NET Shielding Blanket. It covers the description of the design concept, basic considerations for the layout, and the results from neutronics and thermohydraulics calculations for operation with water as well as an aqueous salt solution as the coolant. Finally some brief comments on the results of the associated R & D programme are added.

1. INTRODUCTION

According to the present conception, the operation of NET will start with a blanket which is optimized for shielding. It is to be designed such as to allow, together with the vacuum vessel/permanent shield, a minimum operation time of 4000 h. The average neutron wall loading is expected to be between 0.5 and 1 MW/m^2. Geometrically it has to be compatible with the requirements of the 15 MA plasma and the oblique maintenance scheme[1]. This means that in the midplane on the inboard side 220 mm and on the outboard side 385 mm of radial space are available for accommodating this component.

This paper describes the current status of the design of the shielding blanket, the results of neutronics and thermohydraulics analyses, and the progress in the associated R & D program.

2. THE BASIC DESIGN CONCEPT

The overall architecture of the Shielding Blanket is shown in Fig. 1. It is a result of a compromise between the requirements for segment removal, divertor integration, and coolant tube accommodation inside the blanket neck. The outboard segments, eigth of which include a horizontal port, are non-symmetric with respect to the midplane. The divertor target plates are attached to the inboard segments. The central and the lateral inboard segments are different in the poloidal shape in order to gain space for the tubes penetrating to the outside.

The shape of segments as indicated in Fig. 1 represents the outer surface of a segment box. Each box consists of a graphite protected first

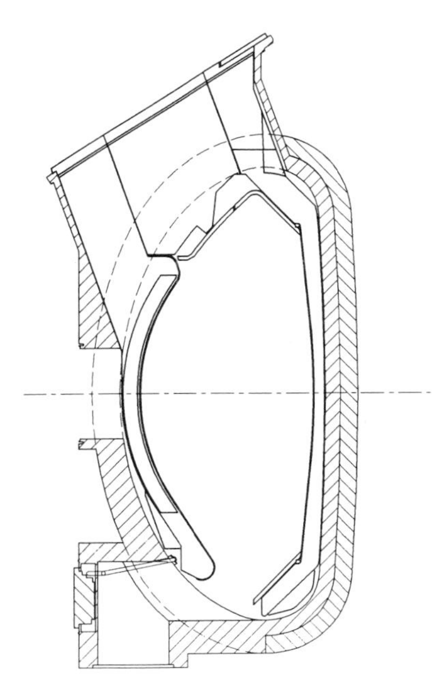

Figure 1:
NET Poloidal crossection

wall[2], two side walls, and a rigid backplate. The sidewalls of the outboard segments carry the copper plates needed for passive plasma stabilization.

From the shielding requirements point of view, the most severe problems occur on the inboard side. To adequately cope with them, a maximum amount of steel is foreseen to fill the inside of the inboard box, while maintaining a water fraction just enough for cooling purposes. For the design, a plate concept with rectangular coolant channels is envisaged as shown in Fig. 2. The plates are fabricated from two pieces and are electron-beam welded at the level of the channels. A similar design is foreseen for the outboard segments which is shown in Fig. 3. An overall plate thickness of 200 mm is sufficient to assure the same radiation level in the outboard leg of the TF coil as in the inboard one.

The dose limit prescribed for the magnet is around 10^9 rad. Experimental tests are underway to define more accurately this limit. Present estimates of the plasma performance indicate that, during the Technology Phase of operation, the wall loading will be about 0.7 MW/m^2 [1]. According to the latest estimates which include effects of the torus geometry[3,4] and those of the gaps in between adjacent blanket segments[4,5], after 4000 h of operation the dose reaches about 8.10^8 rad.

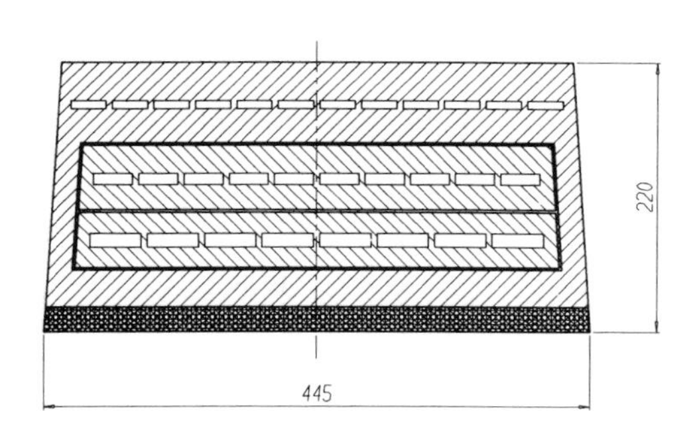

Fig. 2:
Inboard Crossection

3. DIMENSIONING OF THE BLANKET PLATES

For the dimensioning of the plates the following criteria were applied.

a) The temperature difference across the front steel part of a plate should guarantee, with a safety factor of 1.5, a satisfactory fatigue behaviour under cyclic loading conditions up to 10^5 cycles.
b) To keep the entire plate always slightly in tension the temperature difference across the rear steel part should be about 80 % of that of the front part.
c) In order to ensure gross compatability with respect to the overall deformation, the temperature differences in all plates of a box should be about the same.
d) In order to approach similar pressure drops in all plates, the coolant fraction should decrease with increasing depth into the blanket.

On the basis of a power density profile obtained from a one-dimensional neutronics calculation in which the blanket was represented as a homogeneous mixture of 75 % stainless steel and 25% H_2O, these criteria were evaluated by a computerized procedure. The resulting dimensions are summarized in Table I. The temperature differences Δ T also quoted there demonstrate that there is enough margin with respect to the indicated limit. This margin should be sufficient to account for peaking factors or even for transient overpower conditions.

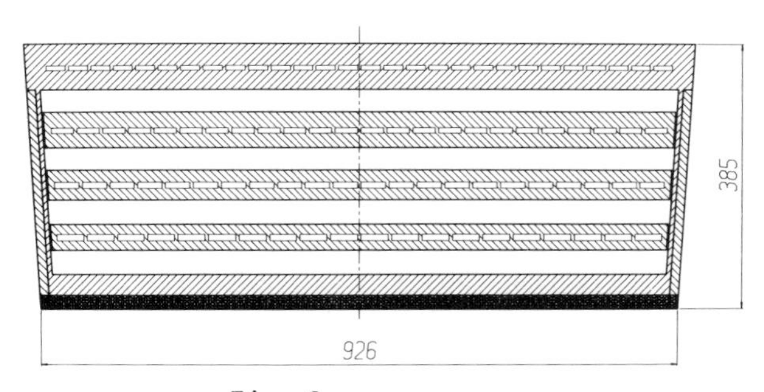

Fig. 3:
Outboard Crossection

TABLE I: Parameters of Blanket Plates

Plate No.	Thickness (mm) Front Steel	Channel	Rear Steel	Total	Temperature Difference (K)* Front	Rear	No. of Channels
Inboard							
3	27.6	6.6	32.2	66.4	74	62	12
2	20.6	9.2	23.0	52.8	74	62	10
1	16.1	11.1	17.6	44.8	74	62	8
Outboard							
1	13.7	9.2	14.4	37.3	72	59	20
2	16.7	8.6	18.0	43.3	72	59	22
3	21.1	7.8	23.3	52.2	72	59	24
4	27.9	6.8	32.5	67.2	72	59	28

* For a neutron wall loading of 1 MW/m^2

The number of coolant channels per plate, also given in Table I was determined from the channel height h, observing the relation $l/h < 4.5$ where l is the channel width. This condition assures that a maximum stress of 100 MPa is not exceeded if for baking purposes the coolant system should have to be operated at a pressure of 10 MPa.

4. THERMOHYDRAULICS IN CASE OF WATER COOLING

For the detailed plate configuration as specified in Table I, a refined one-dimensional neutronics calculation was performed. The resulting power density profile is shown in Fig. 4. To estimate the total power per plate in both the inboard and the outboard segment a relation between three- and one-dimensional calculations was used which was derived from recent results[3]. They indicate, as shown in Figs. 5 and 6, that the average power density over the entire poloidal length is about $0.81 \cdot q_{1D}$ for the inboard, and $0.88 \cdot q_{1D}$ for the

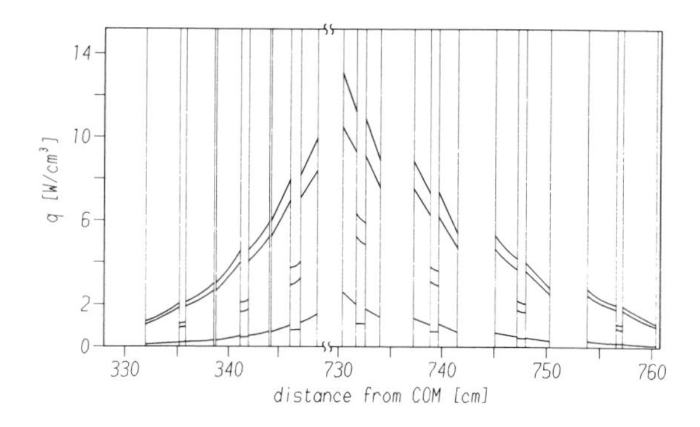

Fig. 4: Radial Power Density Profile

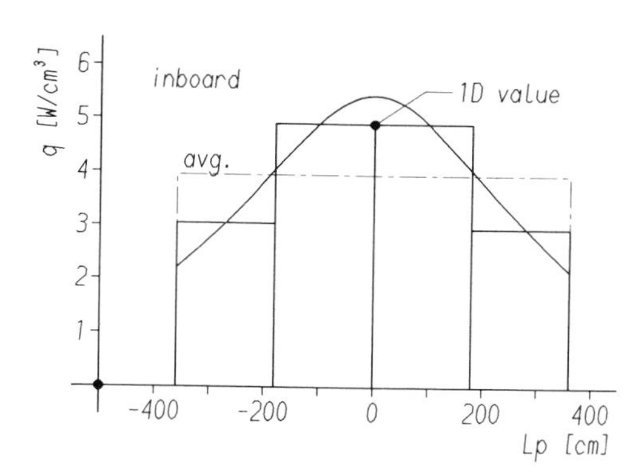

Fig. 5: Poloidal Power Density Profile Inboard Side

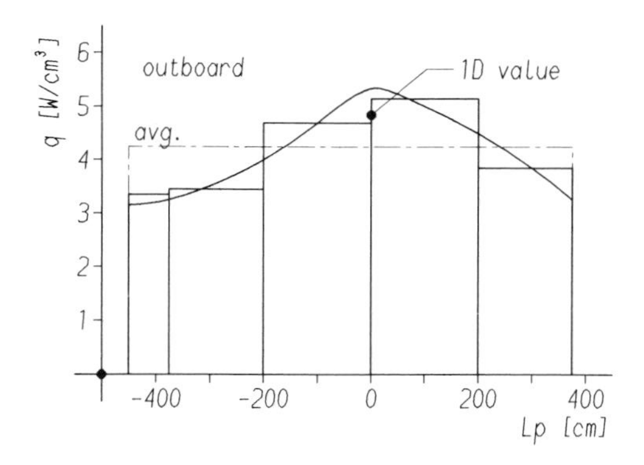

Fig. 6: Poloidal Power Density Profile, Outboard Side

outboard side, respectively, where g^{1D} are the values obtained from the one-dimensional calculation.

This estimate is, of course, a rough one and therefore preliminary because the poloidal distribution varies with blanket depth[4] and because for the inboard side the analysis is restricted to the vertical part only. Useful values for the inclined parts are not yet available. An extrapolation would suffer additionally from the unknown impact of the divertor target plates.

On the basis of the thus estimated total power the thermohydraulics parameters were calculated assuming the same coolant conditions as for the first wall[2], which are

Coolant pressure	<1.0 MPa
Inlet temperature	60 °C
Outlet temperature	<100 °C

The results are summarized in Table II. The numbers for both the coolant velocity and the heat transfer coefficient turn out to be quite reasonable. However, the goal of achieving equal pressure drop in all plates has not completely been reached yet. Appropriate tayloring of the coolant channels and adjustment of the total coolant volume fraction in the individual plates are the tools for arriving at a better balance. This procedure should, however, also include the piping outside the plates which is just now under consideration. Velocity and pressure drop on the inboard side are significantly lower than on the outboard side. This leaves some margin which is required to cope with the power contributions originating from the inclined parts of the inboard segments.

5. OPERATION WITH LITHIUM SALT SOLUTION

The Shielding Blanket as described above can be readily converted into a tritium generating driver blanket if the water coolant is replaced by and aqueous lithium salt solution. Because of the high steel and low coolant volume fraction, the achievable tritium breeding ratio is rather modest. As shown in Fig. 7, it depends on the choice of the salt, its

TABLE II: Thermohydraulics Parameters, Volumes and Weights

	Inboard			Outboard			
Plate No.	3	2	1	1	2	3	4
Power (kW)	276	472	704	2058	1585	1113	679
Mass flow rate (g/s)	1642	2809	4189	12249	9433	6621	4041
Min. Velocity (m/s)	1.68	1.94	2.29	3.47	2.86	2.21	1.56
Max. Velocity (m/s)	1.68	1.68	1.94	4.71	3.95	3.05	2.16
Pressure Drop (MPa)	0.0204	0.0179	0.0184	0.0552	0.0438	0.0319	0.0212
Min. Heat Transf. (W/cm^2K)	1.10	1.17	1.28	1.72	1.49	1.24	0.97
Max. Heat Transf. (W/cm^2K)	1.10	1.17	1.28	2.38	2.09	1.74	1.36
Water Volume (dm^3)	14.5	21.5	27.2	44.9	42.4	39.0	34.2
Steel Volume (dm^3)	163	123	97	164	205	268	372
Steel Weight (kg)	1285	968	767	1296	1620	2117	2939

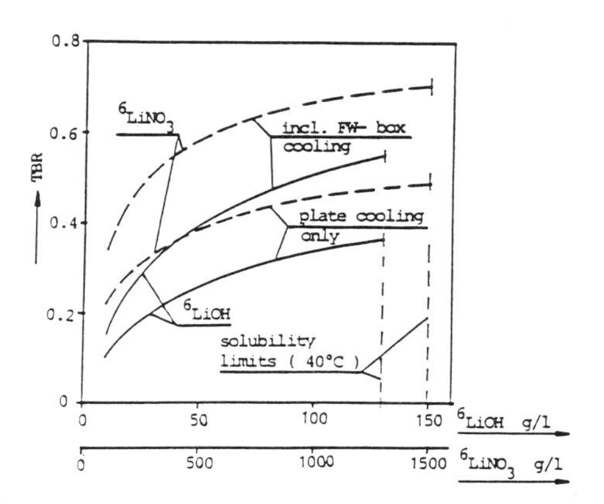

Fig. 7: Tritium Breeding Ratio

concentration in water and on the possibility to use it also for cooling the segment box. In the most optimistic case, a local breeding ratio of 0.70 is obtained from one-dimensional calculations which should reflect in a global breeding ratio in the range of 0.50.

Cooling of the Shielding Blanket with an aqueous lithium salt solution instead of water implies lower heat loads of the steel plates as can be seen from a comparison of Figs. 8 and 9. Instead, the contribution of heat dissipated in the coolant is higher. Hence, the operation conditions for the structure are relaxed. On the other hand, the thermohydraulics parameters will be affected in a negative sense. Based on estimated physical property data of the salt solution[6] is was found that both the mass flowrate and the coolant velocity will be higher by about 25 %. As a consequence, the pressure drop will increase by about 65 % and the pumping power by about a factor of 2. The heat transfer coefficient will be worse by about 30 %. Although these differences are significant they should not pose insurmountable problems. However, their impact on the design and layout of the external cooling cycle components needs to be assessed.

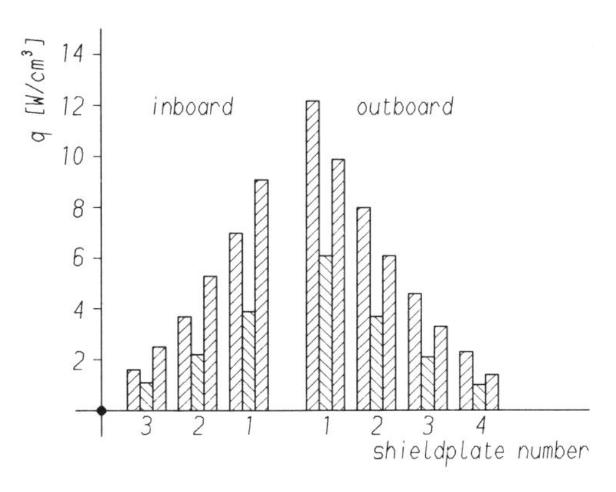

Fig. 8: Heat Deposition, Water Cooling

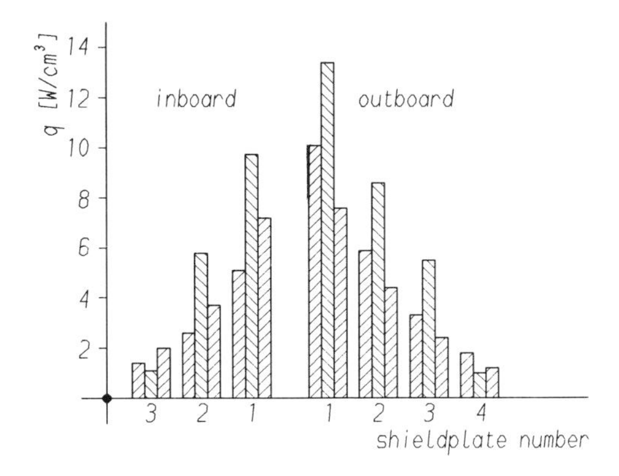

Fig. 9: Heat Deposition, Salt Solution Cooling

6. PROGRESS IN R & D PROGRAM

The R & D program associated with the Shielding Blanket development is presently focussed on the issues raised by the option of salt solution cooling. These are mainly corrosion, radiolysis, and tritium extraction.

First corrosion experiments with LiOH, $LiNO_3$, $LiNO_2$, and Li_2SO_4 solutions have been performed at the K. University of Leuven[7,8] with 316L and 1.4914 as candidate structural materials. They showed a preference for $LiNO_3$ in general. Under the NET temperature conditions, however, also LiOH seems to be applicable, while at higher temperature stress corrosion cracking might prevent its application.

The radiolysis problem has been studied so far only from a theoretical point of view[8]. To cope with it appears to be simpler with LiOH as compared to $LiNO_3$. A worst case analysis showed that it is possible to handle the free gas production by raising the coolant pressure to 1.3 MPa. An experimental program has been started to confirm the theoretical analyses.

Various possibilities for extracting the tritium from the aqueous solution have been examined which are based on the CANDU experience[9]. As a reference, a three-step process has been selected which foresees after complete separation of the salt from the water: water distillation, vapor phase catalytic exchange and cryogenic distillation. As a reasonable processing capacity 500 l/h was identified by which a tritium concentration in the range of 1 to 35 Ci/l can be maintained. This results in a total tritium inventory of some hundred grams. The equipment required is smaller than that of the CANDU Darlington facility.

7. CONCLUSIONS

The Shielding Blanket as described above has been selected as the main option for the NET basic machine. The plate concept offers optimum shielding capability and a great deal of flexibility to adapt the dimensions to any updated shielding requirements. Since it can operate at low temperatures and conservative thermohydraulics conditions, the expectation of a high reliability in case of water cooling is not unrealistic. More detailed design, feasibility, and fabricability studies will be performed in the near future to confirm these statements.

Operating this blanket with an aqueous lithium salt solution instead of water as the coolant seems to be the easiest way to provide some tritium for the machine operation. This option introduces a number of additional R & D issues which are being addressed by the current European Fusion Technology Programme. The results presently available indicate that there are also good prospects for this specific kind of operation.

ACKNOWLEDGEMENTS

The authors gratefully acknowledge the many valuable discussions held with and practical assistance in preparing this paper received from M. Chazalon, R. Toschi, M. Snykers, W. Bogaerts, A. Bruggemann and P. Gierszewski.

REFERENCES

1. The NET Team, Fusion Technology 14,1 (1988)
2. G. Vieider et al., in this conference
3. J.F. Jaeger et al., in this conference
4. U. Fischer, in this conference
5. F. Rohloff, Evaluation of Neutron Streaming Effects through the Gaps in between Blanket Segments, NET Contract 264/87-3/FU-D, KfA Document No. KFA-ASS/NST 87-4 (1987)
6. P. Gierszewski, private communication (1988)
7. W. Bogaerts et al., Application of the Aqueous Self-Cooled Blanket Concept to a Tritium Producing Shielding Blanket for NET, NET Contract 237/86-6/FU-B, K. University of Leuven, Report Jan. 1987
8. W. Bogaerts et al., Corrosion, Radiochemistry and Neutronics Studies for the NET ASCB Blanket Concept, NET Contract 261/87-2/FU-EB, K. University of Leuven, Report Aug. 1988
9. M. Galley et al., Proc. of the 3rd Top. Meeting on Tritium Technol. in Fission, Fusion and Isotopic Applic., Toronto, May 1-6, 1988 (to be published)

FUSION TECHNOLOGY 1988
A.M. Van Ingen, A. Nijsen-Vis, H.T. Klippel (editors)
Elsevier Science Publishers B.V., 1989

TWO-DIMENSIONAL CROSS-SECTION SENSITIVITY AND UNCERTAINTY ANALYSIS OF THE LBM EXPERIMENTS AT LOTUS

J. W. DAVIDSON and D. J. DUDZIAK, University of California, Los Alamos National Laboratory, Los Alamos, New Mexico 87545, USA
and
S. PELLONI and J. STEPANEK, Paul Scherrer Institute, (formerly EIR), CH-5303 Würenlingen, Switzerland.

In recent years, the LOTUS fusion blanket facility at IGA-EPF in Lausanne provided a series of irradiation experiments with the Lithium Blanket Module (LBM). The LBM has both realistic fusion blanket materials and configuration. It is approximately an 80-cm cube, and the breeding material is Li_2O. Using as the D-T neutron source the Haefely Neutron Generator (HNG) with an intensity of about $5 \cdot 10^{12}$ n/s, a series of experiments with the bare LBM as well as with the LBM preceded by Pb, Be and ThO_2 multipliers were carried out.

In a recent common Los Alamos/PSI effort, a sensitivity and nuclear data uncertainty path for the modular code system AARE (Advanced Analysis for Reactor Engineering) was developed. This path includes the cross-section code TRAMIX, the one-dimensional finite difference S_N-transport code ONEDANT, the two-dimensional finite element S_N-transport code TRISM, and the one- and two-dimensional sensitivity and nuclear data uncertainty code SENSIBL.

For the nucleonic transport calculations, three 187-neutron-group libraries are presently available: MATXS8A and MATXS8F based on ENDF/B-V evaluations and MAT187 based on JEF/EFF evaluations. COVFILS-2, a 74-group library of neutron cross-sections, scattering matrices and covariances, is the data source for SENSIBL; the 74-group structure of COVFILS-2 is a subset of the Los Alamos 187-group structure.

Within the framework of the present work a complete set of forward and adjoint two-dimensional TRISM calculations were performed both for the bare, as well as for the Pb- and Be-preceded, LBM using MATXS8 libraries. Then a two-dimensional sensitivity and uncertainty analysis for all cases was performed. The goal of this analysis was the determination of the uncertainties of a calculated tritium production per source neutron from lithium along the central Li_2O rod in the LBM. Considered were the contributions from ^{1}H, ^{6}Li, ^{7}Li, ^{9}Be, ^{nat}C, ^{14}N, ^{16}O, ^{23}Na, ^{27}Al, ^{nat}Si, ^{nat}Cr, ^{nat}Fe, ^{nat}Ni, and ^{nat}Pb. The total uncertainties obtained lie between 1.74% and 2.63%. The largest contributors were ^{7}Li (1.15%) and ^{16}O (1.35%) for the bare LBM; ^{9}Be (2.21%) for the Be-preceded LBM; and ^{nat}Pb (2.25%) for the Pb-preceded LBM.

1 INTRODUCTION

The Lithium Blanket Module (LBM) was constructed for testing on the toroidal D-T neutron source of the Princeton Tokamak Fusion Test Reactor (TFTR).[1] The LBM has both realistic fusion blanket materials and configuration and has been designed for detailed experimental analysis of tritium breeding and neutron flux spatial/spectral distributions. It is approximately an 80 cm cube, and the breeding material is Li_2O. Li_2O pellets are placed in the leading 60 cm of stainless-steel rods (the back 20 cm of each rod is solid stainless steel) which are arranged in an hexagonal array.

Due to a delay in undertaking D-T operation of TFTR, it was decided that the LOTUS facility at the IGA-EPFL in Lausanne (Switzerland) could provide a very valuable resolution of basic technological uncertainties in fusion reactor blanket physics.[2] The LOTUS experiments use the Haefely Neutron Generator (HNG) with a source intensity of $5{\cdot}10^{12}$ n/s; this source has a well defined spatial/spectral distribution with a potential for a highly accurate analysis. Using this source a series of experiments with the bare LBM as well as with the LBM preceded by Pb,

Be and ThO_2 multipliers were carried out.

A collaborative Los Alamos/PSI effort is underway to analyze the LOTUS/LBM experiments being performed by EPFL. The goals of the analysis are first, to investigate the accuracy of the most recent nuclear data on the part of the U.S. (ENDF/B, versions V and VI) and the European Community (JEF-1/EFF); and second, the adequacy of the common 1-, 2- and 3-D neutron transport and 1- and 2-D sensitivity and uncertainty methods.[3,4]

The first 2-D cross-section sensitivity and uncertainty analysis of the bare LBM presented in reference 3 indicated suprisingly large uncertainties in the calculated tritium production per source neutron from lithium in the central Li_2O rod near the LBM steell reflector. This was caused mainly by ^{7}Li, ^{16}O and ^{nat}Cr. This lead to the assumption that ^{7}Li, ^{16}O and ^{nat}Cr evaluations would need to be improved. Latter, in a further analysis, it was found that the higher uncertainties were caused mainly by incompletely included cancelling effects in the sensitivity profiles, when the COVFILS-2[5,6] covariance data library was used.

In parallel, also within the framework of a collaborative Los Alamos/PSI effort, is the development of a one- and two-dimensional cross-section sensitivity and uncertainty path of the AARE modular code system.[7,8] The main effort concentrated on the further development of the cross-section sensitivity and uncertainty code SENSIT-2.[9] Additional features were incorporated, such as the capability to calculate complex reactions (such as KERMA, dpa or He production) when the covariance data becomes available, and the reduction of the user's input using the geometry file GEOMTY, produced by TRISM, and extensions to consider all contributions to the sensitivity profiles in relation to the covariances from the COVFILS-2 library. A strategy was also developed for the one-dimensional capability. This resulted in the new cross-section sensitivity and uncertainty code SENSIBL[4,10]

The lack of suitable benchmark problems made it difficult to test sensitivity codes with a covariance library. Therefore, a benchmark problem for one and two-dimensional sensitivity and uncertainty analysis, representative of a fusion reactor blanket, was recently defined and SENSIBL was extensively tested using this benchmark (See reference 11).

As part of the recent effort, a detailed cross-section sensitivity and uncertainty analysis has been performed for experiments using the HNG with the LBM bare and also preceded with lead and beryllium multiplier plates. The work reported here is the cross-section sensitivity and uncertainty analysis performed at Los Alamos using the U.S. nuclear data base. A companion analysis will be performed at PSI using the European Fusion File (EFF)[12] when the cross-section covariance file is available.

2 CROSS-SECTION DATA LIBRARIES

The transport and reaction cross sections were obtained from libraries in the MATXS8 187 neutron group format.[13] MATXS8A, a library containing 31 isotopes and MATXS8F, containing 21 isotopes, all based on ENDF/B-V evaluations, were used for all materials. Neutron spectra (187-group) were calculated for ten regions of a 3-D model of the LBM using the Monte Carlo code MCNP.[14] The transport and reaction cross-section data were collapsed into the 74-neutron-group structure of the COVFILS-2[5,6] covariance data library using the calculated neutron spectra. The generation of the collapsed transport libraries from the MATXS libraries was accomplished with the code TRANSX-CTR.[13]

COVFILS-2 is a library of multigroup neutron cross sections, scattering matrices, and covariances (uncertainties and their correlations). The 14 materials included in the first version of COVFILS-2 are ^{1}H, ^{6}Li, ^{7}Li, ^{9}Be, ^{nat}C, ^{14}N, ^{16}O, ^{23}Na, ^{27}Al, ^{nat}Si, ^{nat}Cr, ^{nat}Fe, ^{nat}Ni, and ^{nat}Pb. COVFILS-2 was produced using various modules of the NJOY nuclear data processing system.[15,16] It is largely based on data evaluations from the ENDF/B-V library, although some minor corrections and improvements are incorporated. In cases where the covariance evaluation is missing (as in the case of Be) or judged to be inadequate, private Los Alamos evaluations are employed.[17] The 74-group structure was chosen for compatibility with the extensive, general-purpose MATXS8 187-group library which was produced including scattering reactions for which covariance evaluations are available.

3 CALCULATIONAL METHODS AND HNG/LBM MODEL

Cross-section processing, neutron transport, and the cross-section sensitivity and uncertainty analyses were performed with the cross-section sensitivity and uncertainty path of the modular code system AARE (**A**dvanced **A**nalysis for **R**eactor **E**ngineering).[7,8] This path includes the cross-section processing code TRAMIX[8] (the code based on TRANSX-CTR[13]), the one-dimensional finite difference S_N-transport code ONEDANT,[19] the two-dimensional finite element S_N-transport code TRISM,[18] and the one- and two-dimensional cross-section sensitivity and uncertainty code SENSIBL.[10] Neutron transport calculations were performed with TRISM, a computer program for solving the two-dimensional neutral particle transport equation in rectangular (x-y) and cylindrical (r-z) geometries using triangular finite elements within a general domain having curved or other nonorthogonal boundaries. The code SENSIBL was used to perform the cross-section sensitivity and uncertainty analyses. The algorithms used are based on first-order generalized perturbation theory. SENSIBL is coupled to TRISM and ONEDANT via interface files. The forward and adjoint angular fluxes as well as the geometry from TRISM (or ONEDANT) are transferred as input to SENSIBL on these files.

In the model of the HNG/LBM used for the calculations with TRISM, all of the rectangular geometry of the LBM and HNG was represented by equivalent r-z geometry. The Haefely neutron source was positioned 10 cm from the front of the LBM grill plate (or multiplier plate face) the axis. The HNG has been the model for several studies in both two- and three-dimensions[20,21]; the HNG was modeled for TRISM using 44 bands with as many as 11 material regions per band.

The LBM was subdivided along the axis into 28 bands, one in the grill plate, 22 in the Li_2O breeding zone and 5 in the reflector. In the radial direction the zones are Li_2O, cladding, the cooling channel, one zone with smeared Li_2O and cladding, and one zone with smeared Li_2O, cladding and cooling channels. The plexiglass surrounding the LBM was omitted for simplicity. In the axial direction the zones were subdivided into regions between 0.23 to 30.73 and 30.73 to 60.23 cm for the precalculation of weighting spectra with MCNP. The lead and beryllium multiplier

LBM Detector Volume		Tritium Production per Source Neutron		
		Multiplier Plate		
		Bare	Lead	Beryllium
1	T_6	$3.102{\cdot}10^{-5}$	$2.515{\cdot}10^{-5}$	$4.587{\cdot}10^{-5}$
	T_7	$1.660{\cdot}10^{-5}$	$8.381{\cdot}10^{-6}$	$1.073{\cdot}10^{-5}$
2	T_6	$1.079{\cdot}10^{-5}$	$1.010{\cdot}10^{-5}$	$1.138{\cdot}10^{-5}$
	T_7	$6.958{\cdot}10^{-6}$	$3.173{\cdot}10^{-6}$	$3.145{\cdot}10^{-6}$
3	T_6	$7.015{\cdot}10^{-6}$	$6.070{\cdot}10^{-6}$	$5.782{\cdot}10^{-6}$
	T_7	$2.600{\cdot}10^{-6}$	$1.183{\cdot}10^{-6}$	$1.119{\cdot}10^{-6}$
4	T_6	$3.620{\cdot}10^{-6}$	$2.797{\cdot}10^{-6}$	$2.481{\cdot}10^{-6}$
	T_7	$7.838{\cdot}10^{-7}$	$3.690{\cdot}10^{-7}$	$3.398{\cdot}10^{-7}$
5	T_6	$1.766{\cdot}10^{-6}$	$1.272{\cdot}10^{-6}$	$1.103{\cdot}10^{-6}$
	T_7	$2.227{\cdot}10^{-7}$	$1.061{\cdot}10^{-7}$	$9.602{\cdot}10^{-8}$

TABLE I: Calculated tritium production per source neutron in the five LBM detector volumes for each multiplier configruation.

plates were modeled as 5- and 6-cm-thick zones, respectively, adjacent to the front of the grill plate.

All TRISM forward and adjoint calculations were performed in the 74-neutron-group structure with a P_3S_8 approximation and converged to within 10^{-3}. The forward source was distributed in the copper region of the Haefely generator using a Gaussian shape on the copper surface representing the target area. A calculated source spectrum[22] was distributed into the appropriate (nine highest) energy groups. The adjoint sources for the five detector regions were specifed as the macroscopic tritium production cross section for lithium in the Li_2O detector volumes. The LBM calculational model is shown schematically in Fig. 1. The zone designations are as follows: 1 and 2, Li_2O; 3 and 4, stainless-steel cladding; 5 and 6, homogenized Li_2O and cladding; 7 and 8, homogenized Li_2O, cladding, and void; 9, grill plate stainless steel; 10, reflector steel; 11, reflector steel with void; and 12, void.

A total of 18 TRISM calculations were required, three forward calculations (for the bare LBM and for the lead and beryllium multiplier plates) and 15 adjoint calculations (five detector volumes for each of the LBM multiplier configurations). The calculated forward response, the tritium production per source neutron, is given in Table I for each of the LBM detector volumes and multiplier configurations.

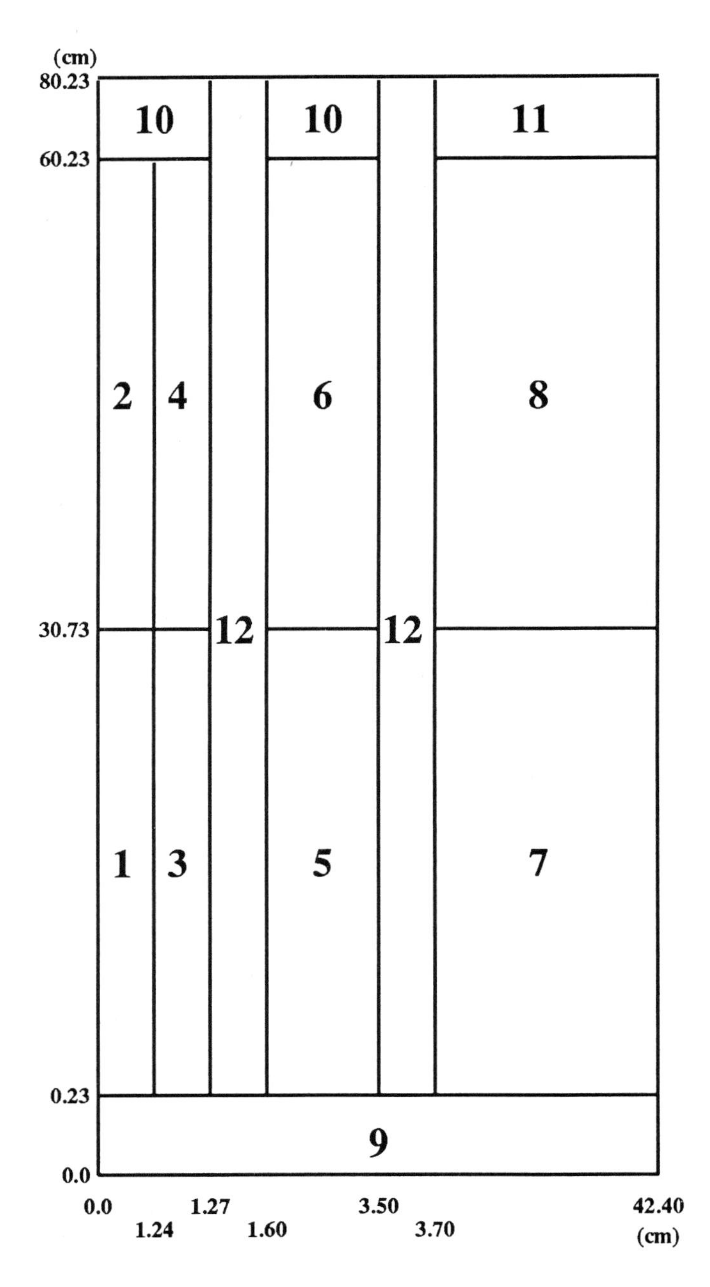

FIGURE 1: Material divisions and detector regions specified in the TRISM geometry model for the LBM.

4 SENSITIVITY AND UNCERTAINTY CALCULATIONS

The goal of the cross-section sensitivity and uncertainty analysis is to determine the uncertainties of a calculated response function, tritium production per source neutron from lithium. In the analysis reported here, the uncertainties in the calculated tritium production at five different positions along the central Li_2O rod in the LBM (see Fig. 1) were computed using SENSIBL. The detector region mid-points are at 1.0, 15.75, 29.0, 44.25, and 58.25 cm along the central rod from the rear of the grill plate. The detector volumes were 2.48-cm-diameter cylinders of length 2.0, 2.5, 3.0, 3.5, and 3.5 cm, respectively. For each of five investigated positions mentioned above, an adjoint calculation with TRISM was performed using the macroscopic tritium production cross section as the adjoint source. To check the consistency of the forward and adjoint calculations, a comparison was made between the tritium production per source neutron computed in the forward calculation for the detector region, $< \Phi, R >$, and the integral over the HNG source region of the product of the forward source and the adjoint flux for that detector, $< S, \Phi^* >$. Perturbation theory predicts that the inner product relationship $< \Phi, R > = < S, \Phi^* >$ should hold. The results of this comparison are given in Table II.

A total of 15 SENSIBL calculations were required for the sensitivity and uncertainty analysis; five for the bare LBM calculations and five each for the lead and beryllium multiplier plate calculations. These calculations required as input the forward (three sets) and adjoint (15 sets) angular fluxes calculated by TRISM. In the calculation for the beryllium multiplier plate, the MAT number 1302 was specified for ^{9}Bel (the "l" denotes the Los Alamos evaluation). There are four MAT numbers for this nuclide on COVFILS-2; 1301, 1302, 1303,and 1304. The covariance data for the (n,2n) reaction is represented as 1, 3, 9, and 27 (all) inelastic "lumps" for these MAT numbers, respectively. The response input was the macroscopic tritium production cross section for natural lithium in Li_2O collapsed from 187 to 74 groups using the weighting spectra calculated for the region in which the detector volume was located and including the appropriate ^{7}Li cross section.

LBM Detector Volume	Relative difference between inner products (%)		
	Multiplier Configuration		
	Bare	Lead	Beryllium
1	0.593	0.372	0.302
2	0.344	0.101	0.068
3	0.022	0.050	0.034
4	0.093	0.087	0.073
5	0.070	0.022	0.024

TABLE II: Comparison of the inner products of the forward $< \Phi, R >$ and adjoint $< S, \Phi^* >$ calculations for each LBM detector region and multiplier configuration.

A possible large source of uncertainty in the calculated response is the cross sections for the copper in the HNG. Because this material is not available in COVFILS-2 and since aluminum is not a HNG/LBM material, ^{27}Al was substituted for ^{nat}Cu in the SENSIBL calculations. With some knowledge of the relative covariance data for ^{nat}Cu and ^{27}Al cross sections, an estimate can be made of the contribution to the uncertainty in the calculated response from copper cross sections. The contributions to the uncertainty in the calculated tritium production in the five LBM detector volumes calculated with SENSIBL are given in Table III for all of the materials present which have data in COVFILS-2. Materials used in the calculation which did not have data on COVFILS-2 are ^{10}B, ^{11}B, ^{nat}K, ^{55}Mn, ^{nat}Cu, ^{64}Zn, ^{nat}Zr, and ^{nat}Mo; in all cases except copper, these are trace constituents.

The total uncertainties obtained lie between 1.74% and 2.63%. The largest contributors were ^{7}Li (1.15%) and ^{16}O (1.35%) for the bare LBM; ^{9}Be (2.21%) for the Be-preceded LBM; and ^{nat}Pb (2.25%) for the Pb-preceded LBM.

The larger contributions from ^{7}Li in front of the LBM, where the 14-MeV spectrum occurs, indicates bigger uncertainties in the threshold reactions, such as the inelastic scattering levels and/or in the $(n, n't)$ reaction, in comparison to other "whole-energy-range" reactions. The (n, xn) down-scattering of the source neutrons in lead or beryllium (when the LBM is preceded by lead or beryllium) makes the threshold reactions of ^{7}Li less important and leads to a decrease of the relative standard deviation.

Also, (n, xn) reactions in lead and beryllium show bigger cross-section uncertainties than other "whole-energy-range" reactions from those two materials. This leads to a larger uncertainty in the ^{7}Li tritium production in front of the LBM, where the first- and second-collision neutrons are important. Due to the neutron thermalization on the other nuclides in the LBM, the importance of the (n, xn) reactions of lead and beryllium and their contribution to the total relative standard deviation decrease towards the back of LBM.

The contributions from ^{nat}Fe are very small even at the back of LBM where ^{nat}Fe has a high concentration in the stainless-steel reflector. The same can be concluded with respect to ^{nat}Ni. The contributions from ^{nat}Cr are not large, yet still higher when conpared to those from ^{nat}Fe and considering much smaller isotopic densities of ^{nat}Cr as compared to ^{nat}Fe in the stainless-steel reflector.

5 CONCLUSIONS

The obtained uncertainties in the calculated tritium production per source neutron caused by the cross-section uncertainties are acceptable considering the uncertainties still coming from the calculational scheme and from the methods.

The obtained low uncertainties does not imply that there are no systematic errors in the evaluations. Therefore, in a next step, the calculated reaction rates will be compared with the newest measured values in the LOTUS experiments.

Also, a companion analysis will be performed at PSI using the European Fusion File (EFF) when the cross-section covariance file is completed.

REFERENCES

1. D. L. Jassby, "Overview of the TFTR Lithium Blanket Module Program," *Fusion Technol.*, 10, 925 (1986).

Material	Multiplier	LBM detector volume 1	2	3	4	5
		Relative standard deviation (%)				
^{1}H	Bare	.22	.04	.02	.01	.02
	Lead	.17	.04	.02	.02	.05
	Beryllium	.11	.03	.02	.02	.04
^{6}Li	Bare	.31	.28	.15	.06	.13
	Lead	.47	.26	.09	.08	.18
	Beryllium	.23	.19	.07	.07	.17
^{7}Li	Bare	1.15	1.09	.87	.74	.80
	Lead	.80	.86	.71	.48	.80
	Beryllium	.68	.73	.55	.47	.89
^{9}Bel	Bare					
	Lead					
	Beryllium	2.21	1.08	.94	.68	.56
^{nat}C	Bare	.07	.03	.02	.01	.01
	Lead	.06	.02	.01	.01	.01
	Beryllium	.04	.01	.01	.02	.02
^{16}O	Bare	.29	.90	1.14	1.30	1.35
	Lead	.32	.65	.75	.88	1.08
	Beryllium	.38	.54	.68	.89	1.12
^{nat}Cu (^{27}Al)	Bare	.87	.27	.18	.13	.13
	Lead	.68	.23	.17	.15	.16
	Beryllium	.59	.16	.15	.19	.22
^{nat}Cr	Bare	.08	.16	.18	.23	1.03
	Lead	.07	.17	.15	.15	.87
	Beryllium	.13	.12	.11	.12	.84
^{nat}Fe	Bare	.85	.35	.31	.30	.55
	Lead	.61	.32	.29	.29	.56
	Beryllium	.53	.25	.25	.32	.59
^{nat}Ni	Bare	.04	.11	.12	.13	.14
	Lead	.05	.08	.09	.10	.11
	Beryllium	.04	.08	.09	.11	.12
^{nat}Pb	Bare					
	Lead	2.25	1.24	.52	.37	.58
	Beryllium					
Total*	Bare	1.74	1.52	1.51	1.55	1.97
	Lead	2.63	1.73	1.27	1.14	1.82
	Beryllium	2.48	1.46	1.32	1.28	1.88

* Totals include contributions from Nitrogen, Natrium and Silicon

TABLE III: Contributions to the Uncertainty in the Calculated Tritium Production per Source Neutron in the LBM Detector Volumes for Each Multiplier Configuration.

2. S. Azam, P. -A. Haldi, A. Kumar, W. R. Leo, C. Schraoui, J. -P. Schneeberger, F. Tsang and L. Green, "Experimental Program at the LOTUS Facility," *Fusion Technol.*, 10, 931 (1986).

3. J.Stepanek, J. W. Davidson, D. J. Dudziak, C. E. Higgs, and S. Pelloni, "Analysis of the LBM Experiments at LOTUS," *Fusion Technol.*, 10, 940 (1986).

4. J. W. Davidson, D. J. Dudziak, D. W. Muir, J. Stepanek, and C. E. Higgs, "Recent Joint Developments in Cross-Section Uncertainty Analysis at Los Alamos and EIR," to be published in Proc. of IAEA Advisory Group Meet. on Nuclear Data for Fusion Reactor Technology, Gaussig, G.D.R., 1-5 December 1986 (LA-UR-87-219).

5. D. W. Muir, "The COVFILS-2 Library of Neutron Cross Sections and Covariances for Sensitivity and Uncertainity Analysis," Los Alamos National Laboratory report LA-10288-PR (1984).

6. D. W. Muir, "COVFILS-2: Neutron Data and Covariances for Sensitivity and Uncertainity Analysis," *Fusion Technol.*, 10, 1461 (1986).

7. J. W. Davidson, D. J. Dudziak, J. Stepanek and C. E. Higgs, "AARE: A One- and Two-Dimensional Sensitivity and Uncertainty Analysis Code System," 7th Int. Conf. on Radiation Shielding, 12-16 September 1988.

8. J. Stepanek and C. E. Higgs, "A General Description of AARE: A Modular System for Advanced Analysis of Reactor Engineering," Proceedings of the 1988 International Reactor Physics Conference, Jackson Hole, Wyoming, 18-21 September (1988).

9. M. J. Embrechts, "SENSIT-2D: A Two-Dimensional Cross-Section Sensitivity and Uncertainty Analysis Code," Los Alamos National Laboratory report LA-9515-MS (1982).

10. J. W. Davidson, D. J. Dudziak, J. Stepanek, C. E. Higgs, D. W. Muir, and M. J. Embrechts,

"SENSIBL: A One- and Two-Dimensional Sensitivity and Uncertainty Analysis Code of the AARE System," (to be published).

11. D. W. Muir, J. W. Davidson, D. J. Dudziak, D. M. Davierwalla, C. E. Higgs and J. Stepanek, "A Benchmark Problem Specification and Calculation using SENSIBL, a One- and Two-Dimensional Sensitivity and Uncertainty Analysis Code of the AARE System," Int. Symp. on Nuclear Fusion Technology, Tokyo, Japan, 10-19 April 1988.

12. H. Grupelaar, "Europe Sets up its Own Fusion File," Nuclear Europe, 6, 40 (February 1986).

13. R. E. MacFarlane, "TRANSX-CTR: A Code for Interfacing MATXS Cross-Section Libraries to Nuclear Transport Codes for Fusion Systems Analysis," Los Alamos National Laboratory report LA-9863-MS (1984).

14. "MCNP - A General Monte Carlo Code for Neutron and Photon Transport," Los Alamos Laboratory report LA-7396-M, Revised (April 1981).

15. P. G. Young and L. Stewart, "Evaluated Data for ^{9}Be Reactions," Los Alamos Scientific Laboratory report LA-7932-MS (ENDF-283) (1979).

16. R. E. MacFarlane, D. W. Muir and R. M. Boicourt, "The NJOY Nuclear Data Processing System, Vol. I: User's Manual," Los Alamos National Laboratory report LA-9303-M (1982).

17. D. W. Muir and R. E. MacFarlane, "The NJOY Nuclear Data Processing System, Vol. IV: The ERROR and COVR Modules," Los Alamos National Laboratory report LA-9303-M (1986).

18. J. W. Davidson, B. A. Clark, D. R. Marr, T. J. Seed, C. E. Higgs and J. Stepanek, "TRISM: A Two-Dimensional Finite Element Discrete-Ordinates Transport Code," Los Alamos National Laboratory report (Draft 1985).

19. R. D. O'Dell, F. W. Brinkley, Jr., and D. R. Marr, "User's Manual for ONEDANT: A Code Package for One-Dimensional, Diffusion-Accelerated, Neutral Particle Transport," Los Alamos National Laboratory report LA-9184-M (February 1984).

20. P. -A. Haldi and A. Kumar, "Lithium Blanket Module (EPRI/PPPL); 2-D Model for DOT-3.5 Calculations," Internal IGA-EPFL document (1985).

21. G. L. Woodruff, M. Abhold, and G. McKinney, "EPFL (Swiss) Fusion-Fission Hybrid Experiment Progress Report No. 15, January 1, 1986.

22. S. Sitaraman and G. L. Woodruff, "Monte Carlo Analysis of a High Intensity D-T Neutron Source," *Trans. Amer. Nuc. Soc.* 43 197 (November 1982).

FUSION TECHNOLOGY 1988
A.M. Van Ingen, A. Nijsen-Vis, H.T. Klippel (editors)
Elsevier Science Publishers B.V., 1989

AFTERHEAT EVALUATION FOR NET BLANKET MAINTENANCE OPERATIONS

L. Giancarli and F. Mazille(*)

CEA - CEN Saclay, DPT/SPIN, Gif-sur-Yvette Cedex, FRANCE
(*) Technicatome, Aix-en-Provence, Rue A. Ampère, BP 34, 13762 Les Milles Cedex, FRANCE

Starting from the NET-DN operating conditions and the LiPb blanket design, in the present work is analysed by means of 1D calculations the sensitivity of the thermal transient after a LOCA to some basic parameters, such as inter-structure physical emissivity and power density source. At the present stage of the design the uncertainties on the used parameters alone give a corresponding uncertainty on the maximum temperature of more than 100°C.
The afterheat production has also important consequences in the framework of maintenance operations. Using the DELFINE code, 1D blanket thermal analysis is then performed within the NET maintenance scenario, considering the detailed schedule of the successive required operations for a segment replacement. The results show that after coolant pipes disconnection (T_0 = 100°C) and LiPb draining, the segment external temperature rises just above 200°C when radiation, hall air convection and shape factors between rows are taken into account by the model. If LiPb draining is not performed freezing of LiPb is expected, expecially in the back row of tubes.

1. INTRODUCTION

The afterheat produced in the blanket of a fusion reactor, although lower than in the equivalent fission reactor, could lead to high First Wall (FW) and blanket temperatures in the case of Loss-of-Coolant Accident (LOCA) and in all downtime scenarios requiring disconnection of the coolant pipes (i.e. maintenance). The determination of the thermal transient and of the maximum reached temperature within the blanket is then necessary in order to estimate if they are compatible with the good working of the reactor.

Considering the NET operating conditions, the thermal behaviour after shutdown of a water-cooled lithium-lead blanket has been analysed. Such an analysis has proceeded in two steps. First of all, a parametric study of a 1D, three-layer model of the outer blanket in LOCA conditions has been performed. Secondly, a more recent design has been considered and a more complete model of the whole blanket has been used to estimate the maximum reachable temperature during LOCA and to analyse the effects of the afterheat production within the maintenance scenario. The objective was to determine the limitations on the maintenance operations due to the temperature transient of the blanket segment components.

2. SENSITIVITY CALCULATIONS

A simple three-layer model of the NET outboard blanket, shown in Fig. 1, has been used for sensitivity analysis. The materials fractions are given in Table 1 (MODEL 1). The afterheat sources, taken from Ref. 1, refer to a neutron wall loading of 0.95 MW/m^2 and 3 full-power years operations. Their distribution in the blanket at shutdown is shown in Fig. 2. The thermal model is completed with the assumption of having only radiation heat exchange between layers and no heat flux towards the plasma chamber (symmetry condition). The conductivity is 25 W/m°K for SS and 15.7 W/m°K for LiPb ; the specific heat 500 J/kg°K for SS and 90 J/kg°K for LiPb. Post-accidental conditions after a total LOCA in FW and blanket were assumed. The only cooling is then performed by the shielding mantained at 80°C. The 1D thermal transient analysis was performed with the DELFINE code[3], which is a finite-element code developped in Saclay permitting 1,2 and 3D geometry analysis with various kind of boundary conditions (contact, radiation, free and forced convection).

In Fig. 3 it is shown the thermal transient in the FW for various values of the physical emissivity. The maximum temperature of about 1270°C is reached after 17h from shutdown in the case of ε=0.3. Both these values decrease when ϵ increases. For ε=1.0 (black body) the maximum temperature is ∿ 820°C reached after 12h, showing the high sensitivity of the thermal transient to surface emissivity. From Fig. 4 we can see that a variation of 10 % of the afterheat sources leads to a variation of ~5% of the maximum reached temperature in the FW. It shows the necessity of having good precision in the neutronic evaluations. Finally, in Fig. 5 the temperature transient at each blanket surface (ε=0.3) is given. At some stages, more than 100°C-difference between the back and the front side of each layer is observed. The choice of the mean conductivity for each homogenized row has then an important effect on the accuracy of the results.

Table 1 - Blanket compositions assumed in the 1D neutronic calculations (1,2)

		MODEL 1		MODEL 2	
		thickness (cm)	volume fractions	thickness (cm)	volume fractions
INNER BLANKET	GRAPHITE TILES	—	—	1.2	C=1.0
	TZM SUPPORT	—	—	0.4	TZM=1.0
	FW	—	—	2.0	SS=0.66 H_2O=0.34
	FIRST ROW	—	—	12.3	SS=0.20 LiPb=0.34 H_2O=0.19
	SECOND ROW	—	—	12.8	SS=0.18 LiPb=0.31 H_2O=0.08
	BACK PLATE	—	—	6.0	SS=0.9 H_2O=0.1
OUTER BLANKET	FW	3.2	SS=0.58 LiPb=0.27 H_2O=0.15	2.0	SS=0.74 H_2O=0.26
	FIRST ROW	20.0	SS=0.11 LiPb=0.49 H_2O=0.08	17.0	SS=0.17 LiPb=0.27 H_2O=0.19
	SECOND ROW	"	SS=0.11 LiPb=0.49 H_2O=0.08	14.5	SS=0.12 LiPb=0.33 H_2O=0.06
	THIRD ROW	"	SS=0.11 LiPb=0.49 H_2O=0.08	17.6	SS=0.13 LiPb=0.45 H_2O=0.03
	BACK PLATE	—	—	10.0	SS=0.9 H_2O=0.1

3. APPLICATION TO MAINTENANCE SCENARIO

For these calculations a good temperature definition is expected. A more complete blanket model (MODEL 2) is adopted considering both inboard and outboard components (FW, LiPb tubes, back plate and coil shielding). The model is described in Fig. 6 and the compositions are given in Table 1. It refers to a recent design of NET-DN. The afterheat sources in the blanket have been determined by Ponti[2], for one full-power year and a neutron wall loading of 1 MW/m^2 (Table 2). The general thermal model is completed with assumptions similar to those given in section 2; in particular radiation exchange between layers with physical emissivity ε=0.5 for SS and ε=0.8 for graphite (inner FW) have been considered.

3.1. Temperature transient during LOCA.

If a total loss of coolant inside the FW and the blanket is assumed, the cooling being possible only through inner and outer shielding (mantained at 100°C), the maximum temperature is reached in the outer FW. Assuming no heat flux along the layers (1D), two cases have been considered, one with coupling inboard/outboard blankets through radiation exchange, the other without coupling. The results for the outer FW are given in Fig. 7 where it can be seen that the maximum tempe-

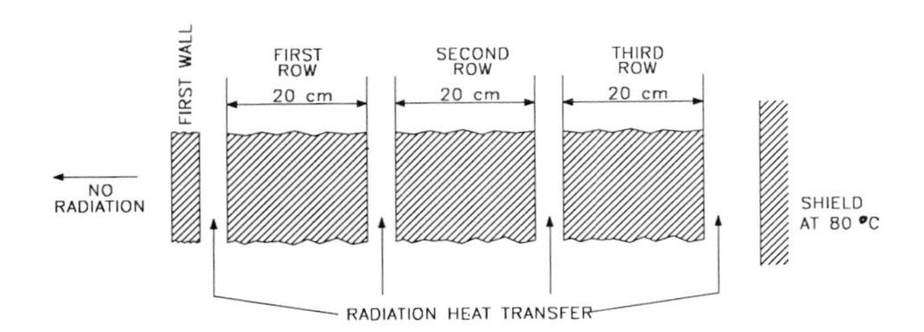

Figure 1 - NET outer blanket - model 1

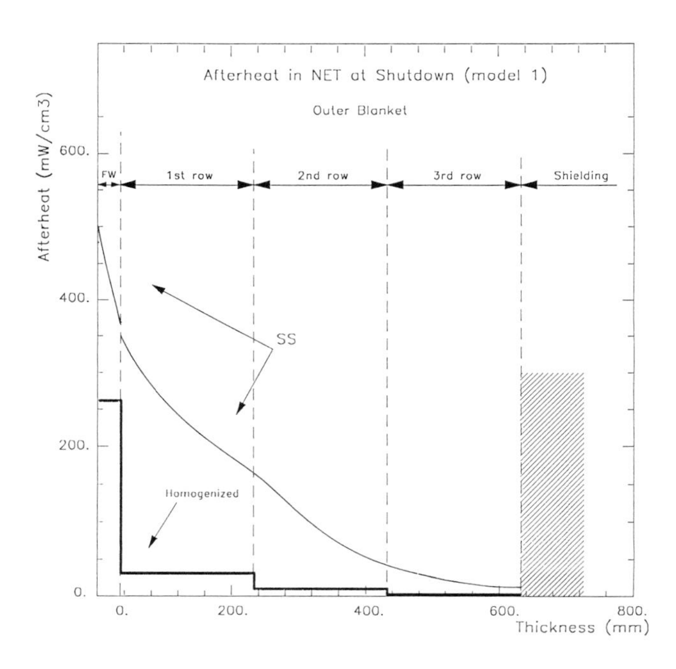

Figure 2 - Afterheat sources distribution in NEt outer blanket (model 1) at shutdown

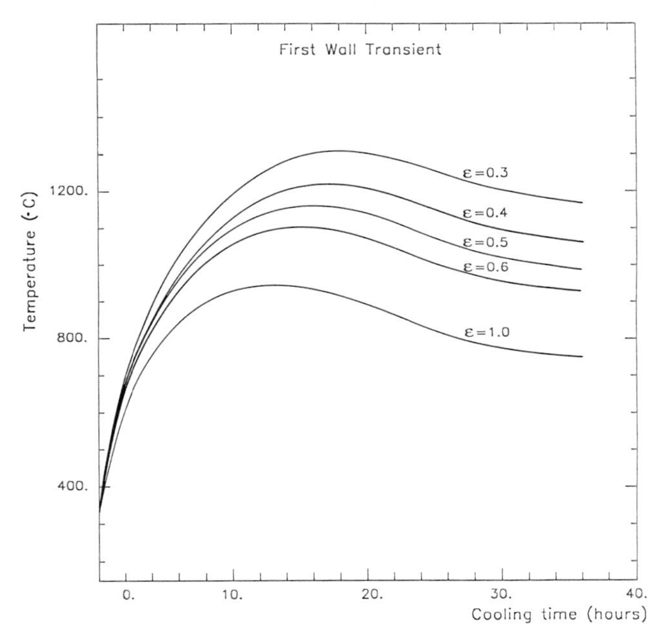

Figure 3 - Effect of surface emissivity on FW transient (model 1) after shutdown

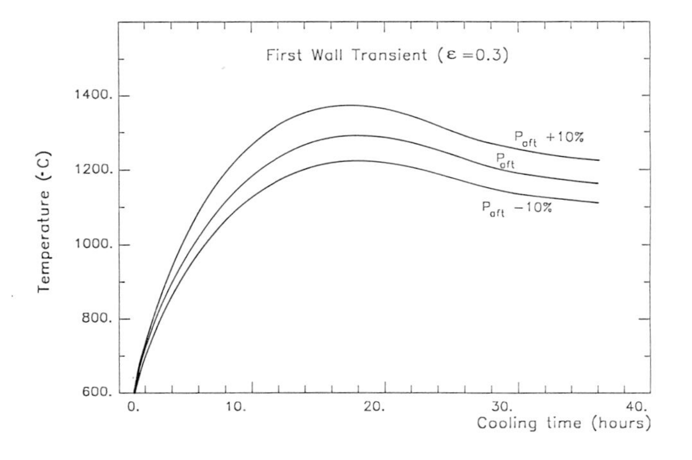

Figure 4 - Afterheat power production effect on FW transient (model 1) after shutdown

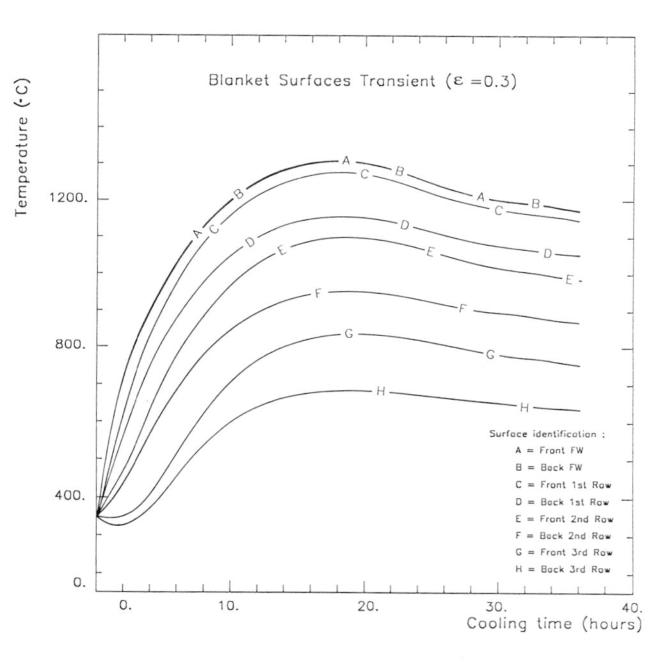

Figure 5 - Temperature transient on outer blanket surfaces in NET-model 1 (after shutdown)

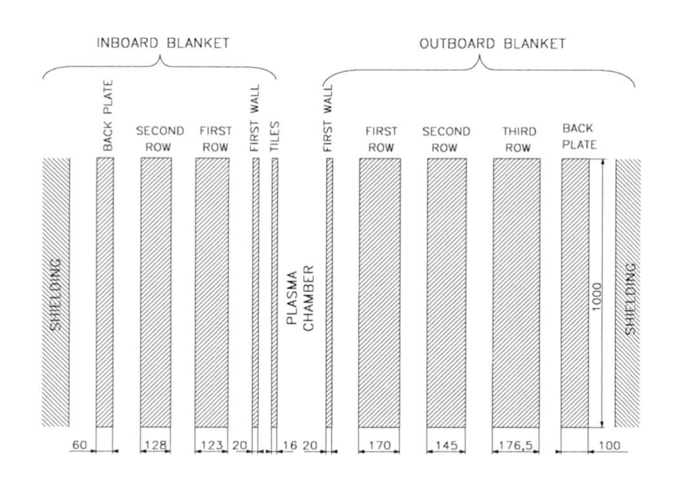

Figure 6 - NET inner and outer blanket-
model 2

rature is reached after 5h from shutdown when the blankets are coupled (∿850°C) and after 8h when they are not (∿900°C). The decrease of the time required for reaching the maximum temperature compared to the previous calculations is mostly due to the decrease of the layers thickness leading to a reduced thermal inertia.

3.2. Maintenance operation schedule

It is assumed that one sector at time will be disconnected and removed (inner or outer). Several operations have to be executed after shutdown before disconnecting the coolant lines. The required time has been estimated[4] to be ∿170h in TIC system (Tight Intermediate Containment) and ∿270h in CTU system (Closed Transfer Unit). Until then the segment could be cooled down to a given temperature and, at the same time, the afterheat sources decrease of factor 5 at least (see Table 2). The mean time required for the segment extraction is ~10h in TIC and ∿20h in CTU. In all the following calculations the afterheat sources present 170h (∿7 days) after shutdown have been assumed as heat sources.

Table 2 - Afterheat sources in NET blanket-
model 2 (2)

		Cooling times (after shutdown)						
	mW/cm³	shutdown	10 min	1 h	3 h	12 h	1 d	10 d
OUTER BLANKET	FW	317	270	225	164	84.7	76.6	67.4
	FIRST ROW	68.5	60.6	43.2	23.2	9.5	8.5	7.4
	SECOND ROW	31.9	28.5	18.7	7.9	2.1	1.7	1.6
	THIRD ROW	21.3	19.2	12.1	4.5	0.8	0.7	0.6
	BACK PLATE	14.9	14.0	11.7	8.2	3.8	3.3	2.9
INNER BLANKET	GRAPHITE TILES	—	—	—	—	—	—	—
	TZM SUPPORT	377	337	253	222	171	137	21.6
	FW	172	150	125	90.4	45.4	40.4	35.9
	FIRST ROW	59.2	53.0	37.0	19.1	7.3	6.5	5.6
	SECOND ROW	25.2	22.7	15.2	7.0	2.2	1.9	1.7
	BACK PLATE	31.7	29.1	24.3	17.4	8.4	7.4	6.5

3.3. Draining of the LiPb tubes

Some preliminary calculations have been done in order to see if the option of leaving the LiPb inside the tubes during segment removal is in any way acceptable. In particular, one must avoid LiPb-freezing in all tubes. We have examined the thermal behaviour of the third row of tubes of the outer blanket assuming that initially the temperature was set at 100°C for FWs and back plates (low T is required for maintenance operations), and 300-400°C for the LiPb tubes, the shielding being mantained at 25°C all time. The results are shown in Fig. 8. Although only radiation losses where considered, in the case C (radiation through the side walls, 2D calculation)

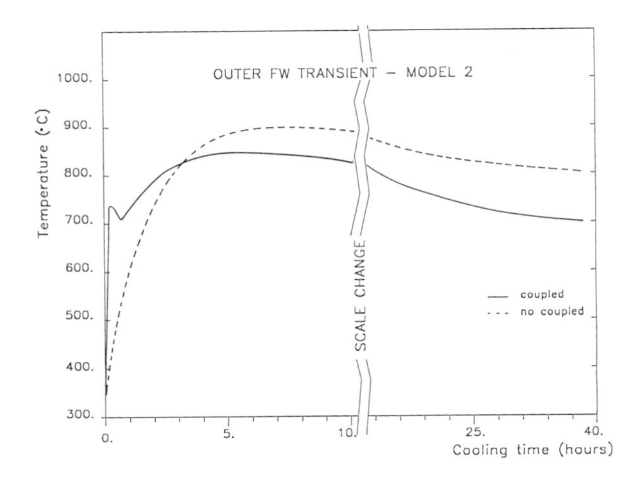

Figure 7 - Temperature transient after shutdown in NET outer FW (model 2). Effect of the coupling inner-outer blankets (radiation exchange)

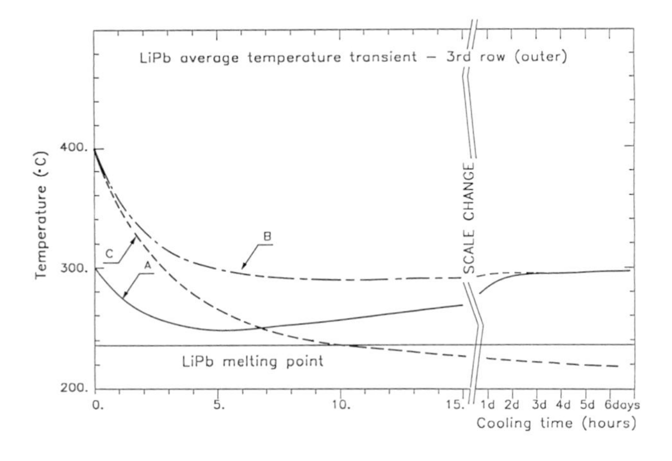

Figure 8 - Temperature transient after pipes cutting (170h after shutdown) in outer LiPb third row when (A) initial LiPb temperature = 300°C, (B) initial LiPb temperature = 400°C, (C) as in (B) plus radiation from the sides of the segment (toroidally averaged temperature - 2D calculation)

the average temperature in the row reaches 235°C, the LiPb freezing point, after only 10 hours from disconnection. The conclusion is that draining of LiPb is required anyway at least for thermal reasons.

3.4. Segment thermal transient during maintenance

The thermal transient in each segment during maintenance has been analysed, assuming the LiPb already drained. From the maintenance viewpoint the segment can be considered as a box formed by FW, side walls and back plate. Out of these surfaces, the hottest one is the FW. The temperature thermal transient in the outer FW is then analysed, even if, during maintenance, the highest blanket temperature occurs in the first row of LiPb tubes because of cooling conditions. Four alternatives have been considered. Two of them assume the segment left in the machine (A,B-neighbouring segment removed) and the two others assume removal 170h after shutdown (C,D). Referring to the results shown in Fig. 9 and Table 3 the boundary conditions for the four cases are :

A) Segment initial T=100°C, cooling through radiation towards the plasma chamber (100°C) and towards the shielding (25°C).

B) Additional air free convection in the plasma chamber (100°C, 10^5 Pa) - Correlation data taken from Ref. 5.

C) Radiation towards the hall (25°C) from FW and back plate. Air free convection in the hall (25°C, 10^5 Pa) for FW and back plate surface.

D) Same as in (C) plus account of shape factors between rows (i.e. allowance for radiation exchange through the tubes interspaces).

The outer FW maximum temperature (equilibrium temperature) is $\sim$260°C when the segment is not removed and $\sim$200°C when removed. Ten hours after pipes cutting the temperature in

Table 3 - Equilibrium average temperatures in the blanket components during maintenance for different boundary conditions (A-D)

NET inboard blanket (model 2)

⁰C	A	B	C	D
GRAPHITE (TZM) TILES	236	200	140	132
INNER FW	317	299	264	249
FIRST INNER ROW	334	322	285	263
SECOND INNER ROW	296	287	238	220
BACK PLATE	230	224	140	149

NET outboard blanket (model 2)

⁰C	A	B	C	D
OUTER FW	307	261	212	203
FIRST OUTER ROW	374	346	313	297
SECOND OUTER ROW	352	330	292	253
THIRD OUTER ROW	294	278	228	227
BACK PLATE	217	208	126	141

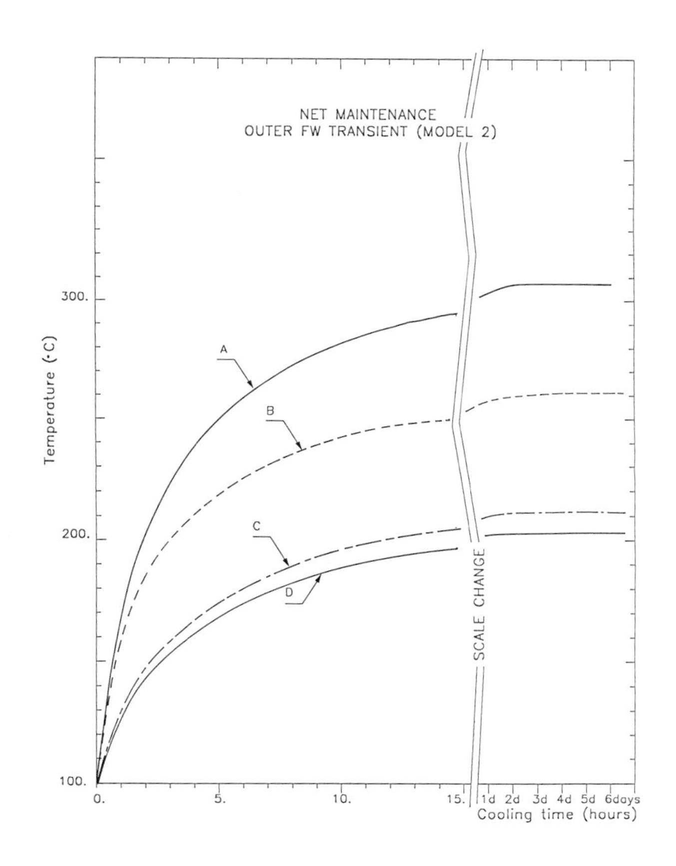

Figure 9 - Temperature transient after pipes cutting (170h after shutdown) in outer FW when (A) only radiation towards plasma chamber is considered (100°C) ; (B) radiation plus air convection (100°C, 1 atm) ; (C) radiation plus convection to the hall (25°C, 1 atm) ; (D) as in (C) plus shape factors between rows

all components is already about 90 % of the equilibrium temperature. Heat flow through the side walls was not taken into account but, although the effect would be to decrease the average temperature inside the blanket (see for instance Fig. 8), it is expected not to have important effects on the FW temperature.

4. CONCLUSIONS

In the present paper it has been demonstrated the high sensitivity of the blanket thermal transient to the physical (ε, sources) and the geometrical parameters (thickness, homogenization). In particular the maximum reached temperature in the FW in case of LOCA decreases of about 5 % in increasing the surface emissivity ε from 0.3 to 0.4, and of about 30 % if increased to 1. This effect shows the advantage of using materials and surfaces with high emissivity (corresponding to an high absorption coefficient).

The NET blanket thermal behaviour during maintenance operations has been analysed using the DELFINE code. First of all it has been shown that the draining of the LiPb is required if its solidification has to be avoided. Starting from the afterheat sources taken at the pipes disconnection time (7 days after shutdown) and considering a model which includes radiation and air free convection, it has been found that the segment will reach an external temperature (outer FW) of about 260°C if not removed from the machine and of about 200°C when removed. Because of the modelling (1D calculations) and of the assumed boundary conditions these results can be considered as conservative at least for a LiPb blanket.

ACKNOWLEDGEMENTS

The authors wish to thank Mr S. Goldstein of C.E.N.-Saclay for his advice concerning the use of the DELFINE code and Mr G. Cabaret of C.E.N.-Saclay for the technical assistance in preparing the code input data.

REFERENCES

1. P. Reynolds et al., 'Study of the Reactor Relevance of the NET Design Concept', Culham Laboratory Report CLM-R278 (1987).
2. C. Ponti, 'Activation Calculations for NET-DN', JCR-Ispra, Technical Note No. I.87.116 (1987).
3. DELFINE User Manual, CASTEM System, CEA-CEN Saclay, IRDI/DEMT (1986).
4. M. Maupou et al., 'Comparison of Vertical versus Horizontal Maintenance Approach', CEA Report (1987), to be published ; and Ph. Charruyer, personal communication.
5. F.P. Incropera and D.P. DeWitt, Introduction to Heat Transfer (John Wiley & Sons, New York, 1985).

FUSION TECHNOLOGY 1988
A.M. Van Ingen, A. Nijsen-Vis, H.T. Klippel (editors)

COMPARISON OF INTEGRAL VALUES FOR MEASURED AND CALCULATED FAST NEUTRON SPECTRA IN LITHIUM FLUORIDE PILES

Hiroshi Sekimoto

Research Laboratory for Nuclear Reactors, Tokyo Institute of Technology, O-okayama, Meguro-ku, Tokyo 152, Japan

The tritium production density, kerma heat production density, dose and certain integral values of scalar neutron spectra in bare and graphite-reflected lithium-fluoride piles irradiated with D-T neutrons were evaluated from the pulse height distribution of a miniature NE213 neutron spectrometer with UFO data processing code, and compared with the values calculated with MORSE-CV Monte Carlo code.

1. INTRODUCTION

Conventional scalar neutron spectra measured with a miniature NE213 spectrometer in fusion neutronics experiments show large oscillation and error band associated with the unfolding process;[1,2] thus it is hard to derive certain information from the spectrum structure. These large oscillations and error bands are attributed to the errors of response functions used in unfolding. The errors of absolute value and global shape of response functions themselves do not cause the large oscillation and error band, but rather these are caused by the errors of local shape of the functions. The global shape error may be small, since it comes mainly from the errors of Monte Carlo calculation and hydrogen cross-sections. The absolute value error can be reduced by using the same spectrometer both for determining source intensity and measuring integral values in the piles.[3] However the local shape error is difficult to reduce, since monoenergetic neutron sources are necessary over a wide energy range in order to determine the local structure of the shape.

The value which we actually want to determine is not the spectrum itself but its integral values, such as tritium production density, kerma heat production density and dose. The errors for these values are not sensitive to the errors of local spectrum shape, most of which are the errors of spectrometer resolution. In the present paper the tritium production density, kerma heat production density and dose are evaluated for more than 1 MeV of the neutron energy, whose weighting functions are shown in Fig. 1.[4] Since the shapes of these functions, especially for tritium production density, are similar to the response function shapes, the errors of these integral values may be expected to be small. In addition to these three integral values, straight integral values for the following four energy regions are evaluated:

- Region I (16.399-10.089MeV)
- Region II (10.089- 5.099MeV)
- Region III (5.099- 0.964MeV)
- Region IV (16.399- 0.964MeV)

2. EXPERIMENT AND CALCULATION

Experiments for two piles were analyzed. One is a 50x50x50-cm^3 bare lithium-fluoride pile, and the other is a 50x50x40-cm^3 lithium-fluoride pile with 20-cm thick graphite reflector covering all surfaces except the neutron source side. The target position of the neutron source is 4.0 cm from the front surface for the bare pile and 20.0 cm for the graphite-reflected pile. The integral values along the center line in the piles were investigated. The details of the

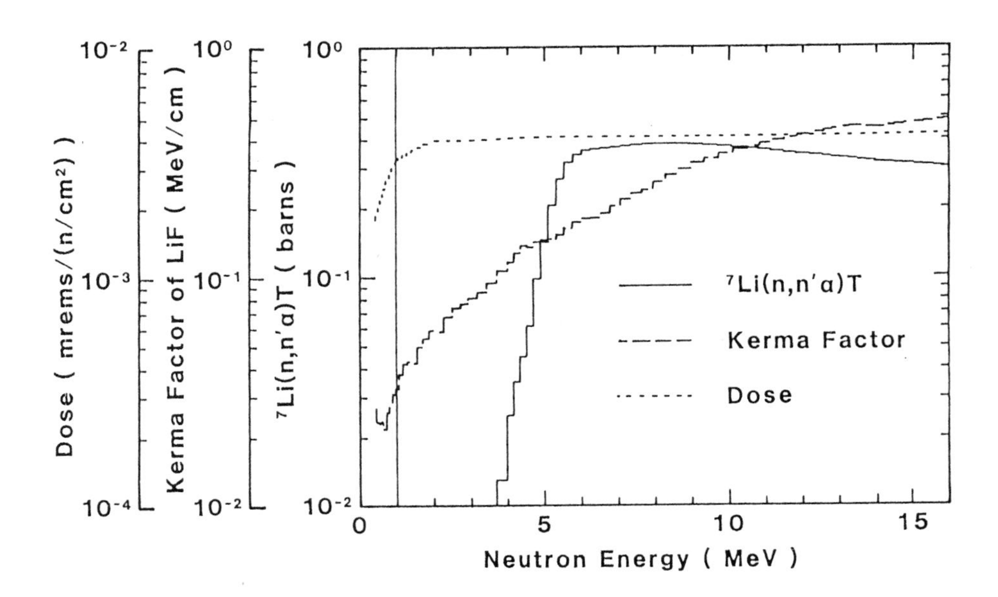

FIGURE 1
Weighting functions of integral values

piles and experiments are given in references 1 and 2.

A miniature NE213 spectrometer developed by the author[5] was employed to measure the integral values in the piles. The pulse-height data and the pulse-shape data were stored as two-dimensional data, and the neutron-gamma discrimination was executed at each pulse height. The one-dimensional pulse height distributions thus obtained were analyzed with the UFO code to obtain the measured integral values.[6] The neutron response function was measured to be almost isotropic except for the backward direction,[7] and the perturbation effect was analyzed to be small.[8] Unbroadened response functions were obtained from Monte Carlo calculations and broadened using the measured results for 14 MeV neutrons.

The integral values were also calculated with the MORSE-CV Monte Carlo code and the GICXFNS group constants[4] for comparison with the measured values.

3. RESULTS AND DISCUSSIONS

The results are shown in Tables 1 and 2 for the bare and graphite-reflected piles, respectively.

The fsd (fractional standard deviation) of the straight integral spectrum is larger for the measured value than for the calculated value except in the case of region IV for the graphite-reflected pile. The fsd of the measured integral spectrum for region II is considerably larger than for the other regions, with that for region IV smaller than the others. Compared to the measured values, the fsd's of the calculated integral spectrum are not so different from each other, though the order of increasing fsd is the same.

The fsd's of measured values for the tritium production and kerma heat production densities are considerably smaller than the straight integral spectra and dose. They are also much smaller than the calculated values. The respective fsd's of calculated values do not differ considerably among the tritium production density, kerma heat production density and dose, and the order of decreasing fsd (tritium production

TABLE 1

Integral values at distance of d from front surface of a lithium-fluoride pile
(figures in parentheses show fractional standard deviations)

(a) Integrated spectrum over region I (16.399-10.089MeV) [neutrons/cm^2/source-neutron]

d [cm]	UFO	MORSE	(MORSE-UFO)/UFO
16.4	7.244×10^{-5} (0.104)	6.761×10^{-5} (0.042)	-0.067
26.4	1.582×10^{-5} (0.139)	1.572×10^{-5} (0.054)	-0.006
31.4	8.731×10^{-6} (0.138)	7.981×10^{-6} (0.091)	-0.086

(b) Integrated spectrum over region II (10.089-5.099MeV) [neutrons/cm^2/source-neutron]

d [cm]	UFO	MORSE	(MORSE-UFO)/UFO
16.4	2.293×10^{-5} (0.366)	2.396×10^{-5} (0.068)	0.045
26.4	8.243×10^{-6} (0.310)	7.672×10^{-6} (0.071)	-0.069
31.4	4.198×10^{-6} (0.336)	3.853×10^{-6} (0.064)	-0.082

(c) Integrated spectrum over region III (5.099-0.964MeV) [neutrons/cm^2/source-neutron]

d [cm]	UFO	MORSE	(MORSE-UFO)/UFO
16.4	7.928×10^{-5} (0.147)	7.923×10^{-5} (0.042)	-0.001
26.4	2.819×10^{-5} (0.139)	2.459×10^{-5} (0.036)	-0.128
31.4	1.556×10^{-5} (0.139)	1.512×10^{-5} (0.056)	-0.028

(d) Integrated spectrum over region IV (16.399-0.964MeV) [neutrons/cm^2/source-neutron]

d [cm]	UFO	MORSE	(MORSE-UFO)/UFO
16.4	1.747×10^{-4} (0.063)	1.708×10^{-4} (0.031)	-0.022
26.4	5.226×10^{-5} (0.070)	4.799×10^{-5} (0.032)	-0.082
31.4	2.849×10^{-5} (0.071)	2.695×10^{-5} (0.046)	-0.054

(e) Tritium production density from ^{7}Li [tritium/cm^3/source-neutron]

d [cm]	UFO	MORSE	(MORSE-UFO)/UFO
16.4	1.337×10^{-6} (0.005)	1.285×10^{-6} (0.039)	-0.039
26.4	3.400×10^{-7} (0.008)	3.346×10^{-7} (0.046)	-0.016
31.4	1.826×10^{-7} (0.008)	1.721×10^{-7} (0.069)	-0.057

(f) Kerma heat production density [MeV/cm^3/source-neutron]

d [cm]	UFO	MORSE	(MORSE-UFO)/UFO
16.4	4.227×10^{-5} (0.009)	4.036×10^{-5} (0.035)	-0.045
26.4	1.069×10^{-5} (0.012)	1.019×10^{-5} (0.042)	-0.047
31.4	5.691×10^{-6} (0.013)	5.317×10^{-6} (0.059)	-0.066

TABLE 1 (Contiuned)

(g) Dose [mrem/source-neutron]

d [cm]	UFO	MORSE	(MORSE-UFO)/UFO
16.4	6.953×10^{-7} (0.053)	6.773×10^{-7} (0.031)	-0.026
26.4	2.053×10^{-7} (0.059)	1.890×10^{-7} (0.032)	-0.079
31.4	1.120×10^{-7} (0.060)	1.058×10^{-7} (0.046)	-0.055

TABLE 2

Integral values at distance of d from front surface of a graphite-reflected lithium-fluoride pile
(figures in parentheses show fractional standard deviations)

(a) Integrated spectrum over region I (16.399-10.089MeV) [neutrons/cm^2/source-neutron]

d [cm]	UFO	MORSE	(MORSE-UFO)/UFO
21.4	1.265×10^{-5} (0.086)	1.215×10^{-5} (0.036)	-0.040
31.4	3.583×10^{-6} (0.095)	3.769×10^{-6} (0.034)	0.052
41.4	1.192×10^{-6} (0.083)	1.267×10^{-6} (0.056)	0.064
51.4	5.023×10^{-7} (0.083)	4.065×10^{-6} (0.068)	-0.191

(b) Integrated spectrum over region II (10.089-5.099MeV) [neutrons/cm^2/source-neutron]

d [cm]	UFO	MORSE	(MORSE-UFO)/UFO
21.4	4.723×10^{-6} (0.252)	5.091×10^{-6} (0.047)	0.078
31.4	2.480×10^{-6} (0.150)	1.939×10^{-6} (0.038)	-0.218
41.4	8.643×10^{-7} (0.140)	8.509×10^{-7} (0.058)	-0.016
51.4	3.716×10^{-7} (0.139)	3.795×10^{-7} (0.053)	0.021

(c) Integrated spectrum over region III (5.099-0.964MeV) [neutrons/cm^2/source-neutron]

d [cm]	UFO	MORSE	(MORSE-UFO)/UFO
21.4	2.121×10^{-5} (0.041)	1.674×10^{-5} (0.028)	-0.210
31.4	7.705×10^{-5} (0.037)	6.904×10^{-6} (0.021)	-0.104
41.4	3.145×10^{-6} (0.038)	2.705×10^{-6} (0.035)	-0.140
51.4	1.445×10^{-6} (0.041)	1.213×10^{-6} (0.040)	-0.161

(d) Integrated spectrum over region IV (16.399-0.964MeV) [neutrons/cm^2/source-neutron]

d [cm]	UFO	MORSE	(MORSE-UFO)/UFO
21.4	3.858×10^{-5} (0.019)	3.398×10^{-5} (0.023)	-0.119
31.4	1.377×10^{-5} (0.017)	1.261×10^{-5} (0.019)	-0.085
41.4	5.201×10^{-6} (0.019)	4.824×10^{-6} (0.031)	-0.073
51.4	2.319×10^{-6} (0.021)	1.999×10^{-6} (0.033)	-0.138

TABLE 2 (Continued)

(e) Tritium production density from ^{7}Li [tritium/cm^{3}/source-neutron]

d [cm]	UFO	MORSE	(MORSE-UFO)/UFO
21.4	2.466×10^{-7} (0.003)	2.439×10^{-7} (0.030)	-0.011
31.4	8.237×10^{-8} (0.003)	8.184×10^{-8} (0.027)	-0.006
41.4	2.930×10^{-8} (0.002)	3.063×10^{-8} (0.044)	0.045
51.4	1.261×10^{-8} (0.003)	1.139×10^{-8} (0.046)	-0.097

(f) Kerma heat production density [MeV/cm^{3}/source-neutron]

d [cm]	UFO	MORSE	(MORSE-UFO)/UFO
21.4	7.798×10^{-6} (0.005)	7.585×10^{-6} (0.029)	-0.027
31.4	2.598×10^{-6} (0.005)	2.537×10^{-6} (0.025)	-0.023
41.4	9.087×10^{-7} (0.006)	9.084×10^{-7} (0.040)	-0.000
51.4	3.798×10^{-7} (0.007)	3.369×10^{-7} (0.044)	-0.113

(g) Dose [mrem/source-neutron]

d [cm]	UFO	MORSE	(MORSE-UFO)/UFO
21.4	1.505×10^{-7} (0.016)	1.345×10^{-7} (0.023)	-0.107
31.4	5.376×10^{-8} (0.015)	4.954×10^{-8} (0.019)	-0.079
41.4	2.028×10^{-8} (0.017)	1.887×10^{-8} (0.032)	-0.069
51.4	9.019×10^{-9} (0.018)	7.795×10^{-9} (0.034)	-0.136

density, kerma heat production density, dose) is the reverse of the measured values.

The fractional difference ((MORSE-UFO)/UFO) of measured and calculated integral values is negative in most cases. The average value is about -0.05 for the bare pile except for the region II integral spectrum and tritium production density. For the graphite-reflected pile, the fractional difference in the reflector for region I is negative large (-0.191), and the average value for region III is also negative large (-0.154).

The region IV integral spectrum and dose give similar values for the fsd's of measured and calculated values and the fractional difference between measured and calculated values, as was expected from their weighting functions.

REFERENCES

1. D. Lee et al., J. Nucl. Sci. Technol., 22 (1985) 28.

2. H. Sekimoto et al., J. Nucl. Sci. Technol., 23 (1986) 381.

3. H. Sekimoto et al., Nucl. Instrum. Methods, A234 (1985) 148.

4. H. Sekimoto, Nucl. Sci. Eng., 94 (1986) 277.

5. H. Sekimoto et al., Nucl. Instrum. Methods, 189 (1981) 469.

6. H. Sekimoto, Integral value measurements with miniature NE213 spectrometer, in: Proc. Theory and Practices in Radiation Protection and Shielding, Vol. 1 (American Nuclear Society, 1987) pp. 103-109.

7. H. Sekimoto et al., Nucl. Instrum. Methods, 227 (1984) 146.

8. H. Sekimoto et al., Nucl. Sci. Eng., 80 (1982) 407.

FUSION TECHNOLOGY 1988
A.M. Van Ingen, A. Nijsen-Vis, H.T. Klippel (editors)

STARTING EXPERIENCE ON THE USE OF Pb-17Li IN A FORCED CONVECTION LOOP

Ja. DEKEYSER, J.L. WALNIER, L. AUFFRET*

S.C.K./C.E.N., Boeretang 200, 2400 Mol, Belgium
* METAUX-SPECIAUX, Place de l'Iris 6, 92087 Paris, France

The loop MALICE started for the first time with lithium in May 1985 and had been operated during 7500 hours with pure lithium quite satisfactorily. Corrosion tests as well as magnetohydrodynamics experiments have been performed. Now the loop is filled with Pb-17Li and operated for the first time with Pb-17Li on December 11th 1987. The conversion of MALICE to the use of Pb-17Li was necessitated by the choice of the European Fusion Technology Programme in favour of this eutectic alloy, as liquid metal breeder for the Next European Torus. Pure lithium had been rejected because of its higher chemical reactivity. In this paper the fabrication technique for Pb-17Li and the first experience with Pb-17Li in the forced convection loop MALICE will be described.

1. INTRODUCTION

Mass transfer of structural materials in contact with a liquid breeder is one of the critical problems related to the use of liquid metal breeders in fusion reactors : this applies especially to the eutectic Pb-17Li. Many parameters influence the corrosion of structural materials [1]. Specific for the magnetic confinement type fusion reactors is the magnetic field, which changes the velocity profile of the liquid and laminarizes the flow. This may have a drastic influence on the extent of corrosion and mass transfer. Experiments with the Pb-17Li eutectic started now in the loop MALICE to assess the importance of this parameters.

2. PRODUCTION OF LITHIUM-LEAD ALLOYS

METAUX-SPECIAUX S.A. has recently developed a process for manufacturing Pb-17Li on an industrial scale (metric ton) and supplied various loops such as MALICE in Belgium and ALCESTE and CAMILLE at the CEA France.

The phase diagram of the binary system Li-Pb is well known and given in figure 1.

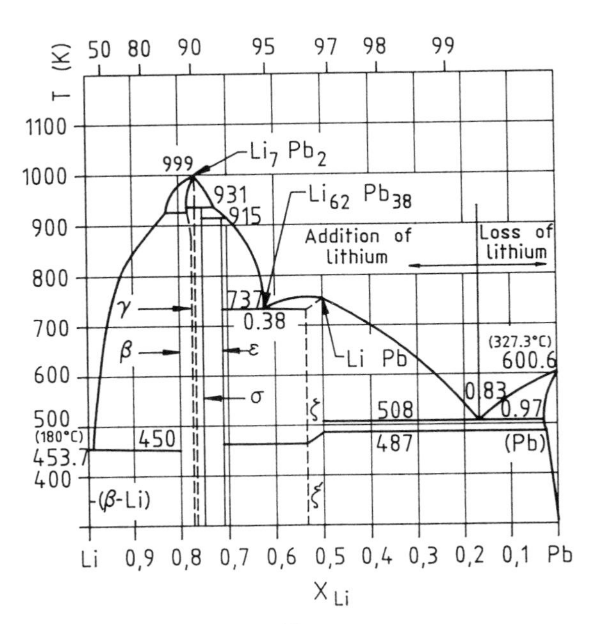

Fig. 1 : Phase diagram of the lithium-lead system.

Lithium rich alloys such as Li7PB2 show very high melting points 726°C, whereas the lead rich eutectic Pb-17Li has a melting point of only 235°C. Clearly in a nuclear loop operating at 600°C it is important to avoid high melting point components.

The fabrication technique for the Pb-17Li is as follows :

- Pure lead (free of oxides) is molten in a reactor and strongly agitated in an inert atmosphere of argon.
- In a vessel above the reactor, a stoichiometric quantity (0.69 % by weight) of alloy grade lithium is molten under argon protection.
- Using a slight argon overpressure, the lithium is transferred dropwise over an extended period of time. During the reaction, the temperature of the lead rises by over 200°C.
- The alloy is then filtered through a sintered metal mesh. The alloy can be delivered in ingot form (0.25 to 5 liters) or in bulk containers. To minimize external contamination during transfer into the nuclear loop, special containers have been developed by METAUX-SPECIAUX that can be heated and pressurized. The alloy can then be directly injected.

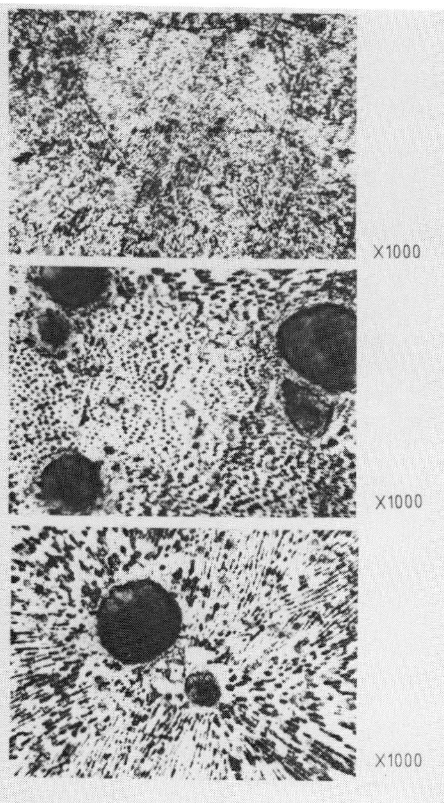

Fig.2:
Optical microscopy on three different places.

Analysis of ingot reveals that during solidification the upper part becomes slightly rich in lithium (0.73 % by weight) and the lower part slightly deficient (0.65 %). Only one melting point at 235°C has been found in accordance with the phase diagram.

Metallographic analysis of ingots in the top and bottom regions reveals two phases - one rich in lithium and the other deficient. These are shown in figure 2 at a magnification of 1000 x; from which the following comments can be made :

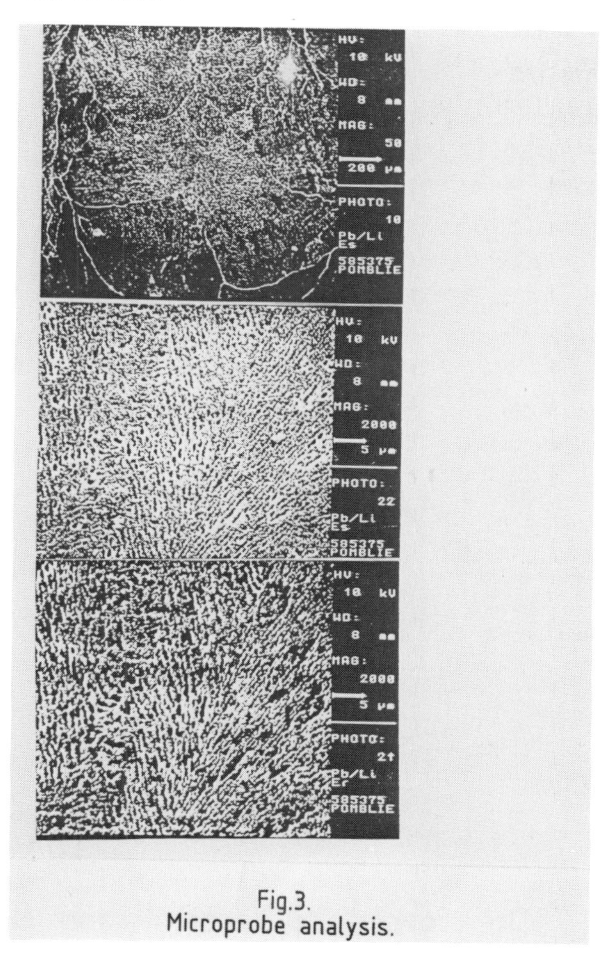

Fig.3.
Microprobe analysis.

- Grain boundaries are variable in size and indicate that solidication has occured at different intervals of time;

- Black nodules in the metallographic represent a Li-Pb phase which are dispersed in a Pb matrix.

X-ray diffraction studies confirm the presence of a Pb phase, although the melting point of lead is not found (see figure 3). Further measurements of the Pb-17Li eutectic are required in order to improve our understanding of the solid phase.

3. DESCRIPTION OF THE LOOP MALICE

The forced convection loop MALICE is used primiraly to study the influence of a strong magnetic field on corrosion and to measure the MHD-effects in the magnetic field. The loop was conceived as a "figure of eight" forced convection loop and made of stainless steel AISI 316L. It comprises two EMP-pumps in series, a Pb-17Li recuperator, an electrical heater, an expansion tank, electromagnetic flow meters, two parallel test sections with hot legs and two parallel test sections with cold legs. One of the hot legs is presently mounted inside the electromagnet. Other parts of the loop are the dumptank with the sample taker. An isometric view and the main characteristics of the loop are given in figure 4.

4. LOOP OPERATION

After 7500 hours of working with liquid lithium, the loop has been converted to the eutectic liquid metal Pb-17Li. Prior to the Pb-17Li filling some alterations of the loop has to be done, such as :

- replacing the preheating elements on the pressure measuring tubes.
- reinforcing the suspension of the loop components because of the high density of the Pb-17Li (± 9.15 kg dm^{-3}).
- improving the thermal insulation of the electromagnetic pumps because of the higher

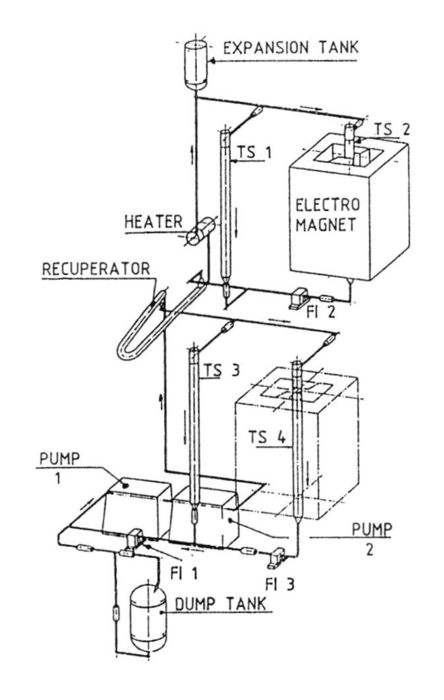

Flow (m^3 /h)	up to 1.3
Pressure head of the pump (mPa)	up to 0.3
Temperature (°C)	up to 500
Heater input (kW)	up to 50
Design pressure (mPa) at 500°C	1.2
Liquid metal volume (l)	100
Magnetic field strength (Tesla)	2
Uniformity (%)	~1.5
Dimensions (m)	1X0.1X0.1
Pipe size	
-Main circuit (mm)	⌀ 33.4/26.6
-Test sections (mm)	⌀ 88.9/77.9
Construction material	AISI 316L

Fig.4 :

The forced convection loop MALICE : a) Isometric view
b) System parameters.

melting point of Pb-17Li (235°C) versus lithium (183°C).

- cleaning of the dumptank with water in order to remove all lithium residue.

After these modifications the loop was ready to receive the Pb-17Li from the special melting tank. This tank, with the lithium lead alloy, had been supplied by METAUX-SPECIAUX France and fabricated as mentioned above. First filling of the loop was done at December 1987. Some difficulties occured during the filling operation, as a consequence of the residual lithium remained on the walls of the loop after previous dumping of the lithium. Plugs with a high melting point were formed in some valves and the temperature had to be raised till 500°C in order to remove the obstructions. This indicates that the composition of these plugs is close to the equi-atomic Li-Pb compound, which has a melting point of 482°C (755 K) (see figure 1 : LiPb-fase diagram).

When all the obstructions in the main loop were dissolved the maximum flow in the loop was about 1.3 m^3h^{-1} (instead of 8 m^3h^{-1} for lithium) with the same two electromagnetic pumps in series. During the previous tests with lithium the bellow of the valve in the first pressure tab was broken and this caused a small lithium leak. The lithium in the pressure tab was solidified and could not be dumped because the lithium valve was not movable, a content of about 1.5 l rested in the tube. During the filling operation with Pb-17Li another Li-Pb mixture was formed and even a temperature increase to 500°C could not remove the plug.

5. LITHIUM-LEAD CIRCULATION

5.1. Pumping Pb-17Li

Presently in MALICE two EMP-pumps of the conduction type are connected in series. This pumps are originally calibrated for use with sodium. In previous tests the performance curves for sodium were checked in the Li-2 loop [2] and it appeared that they can also be used for lithium at 400°C. The experimental determination of the performance of EMP-pumps with Pb-17Li has not yet been reported.

The performance of an EMP-pump [3] with lead was compared with its performance for sodium. The following formulation is given :

$$\Delta P_{Na} = f\ (Q_{Na}, E) \qquad (1)$$

where ΔP_{Na}, Q_{Na} and E are the pressure head in MPa, the flow rate in m^3h^{-1} and the applied voltage in Volt. The function f defines the family of curves in sodium at 600°C. In lead [3] at 430°C, the performances are expressed by :

$$\Delta P_{Pb} = 0.55\ f\ (5\ Q_{Pb},\ E) \qquad (2)$$

The parameters which influence the pump performance are : the specific weight, the electrical conductivity and the viscosity. The error in applying relation (2) to lithium-lead should be small : indeed the three parameters are similar for lead and for lithium-lead, and actually slightly more favourable for lithium-lead. Therefore, relation (2) has been applied to derive the performance of the conduction pumps in MALICE with lithium-lead at 400°C. The result is given by curve 3 on figure 5 whereas the resistance of the circuit with all test sections fully open is given by curve 4.

The flow resistance of the circuit for lithium-lead is calculated from the curve 2, valid for lithium, by multiplying the pressure loss by the ratio of densities.

Different settings of the conduction pumps are represented by the dashed vertical line numbers 1-3. The corresponding pressure losses in the loop and the pressure heads of the pumps cannot be given, since the pressure cannot be measured immediately before and after the electromagnetic pumps. Therefore the "true" characteristics of the pumps and the "true" resistance of the circuit cannot be plotted. The maximum possible flow rate predicted was 1.11 m^3h^{-1} while 1.3 m^3h^{-1} was practically obtained. The prediction was some 14 % lower than the actual possible flow rate.

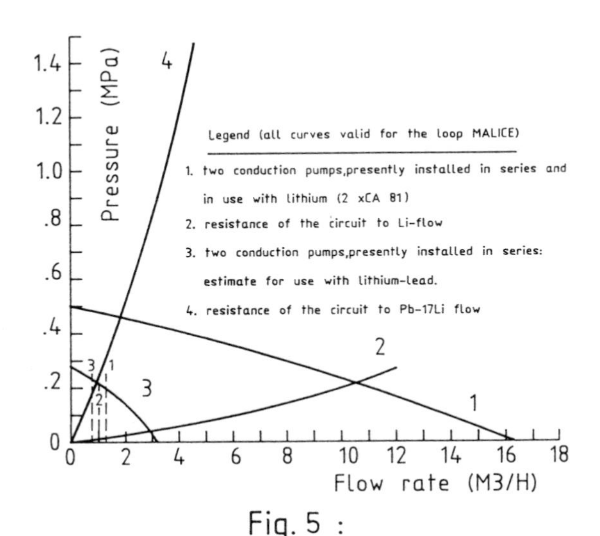

Fig. 5 :
Comparison in pump performance between Li and Pb-17Li flow.

5.2. Flow measurement

Flow measurement in the loop MALICE can be performed using :

- one of the three permanent magnet flow meters.
- the vortex flow meter.

The permanent magnet flow meter FI1 has been gauged in sodium at the temperatures 200°C and 400°C by the manufacturer. The results can be used to predict the gauging coefficient of the same flow meter at 400°C in Pb-17Li.

The flow rate has a linear relationship with the measured voltage

$$Q = K_{Na,400°C} \times E \qquad (3)$$

with Q flow rate (m^3h^{-1})

E potential (mV)

K gauging factor (= 0.473)

The formula used for the EMF of the flow meter is the following in a simple notation :

$$E = K1 \times v \times B \times d \qquad (4)$$

with v average velocity (ms^{-1})

B magnetic field (Tesla)

d inner diameter of the tube ($13.4\ 10^{-3}$ m)

The factor K1 is expressed by [4] :

$$K1 = \frac{2\left[\frac{d}{D}\right]^2}{1 + \left[\frac{d}{D}\right]^2 + \frac{\rho_f}{\rho_w}\left[1 - \left(\frac{d}{D}\right)^2\right]\left[1 + \frac{\tau}{d\rho_f}\right]}$$

with D outer diameter of the tube ($16.7\ 10^{-3}$ m)

ρ_f electrical resistance of the fluid

ρ_w electrical resistance of the tube

τ electrical contact resistance (neglectable)

Since Q ÷ v, it follows from (4) that Q ÷ E/K1. Thus the gauging factor is inversely proportional with K1.

To evaluate the influence of different liquid metals, the gauging factor for Pb-17Li must be deduced from the sodium factor :

$$K_{Pb-17Li,\ 400°C} = K_{Na,\ 400°C} \times \left[\frac{K1\ Na}{K1_{Pb-17Li}}\right]_{400°C} = 0.473 \times \frac{0.7502}{0.6074} = 0.584\ m^3h^{-1}$$

Theoretically, the vortex shedding flow meter gives a frequency only dependant of the velocity, irrespective of the temperature and of the type of fluid, provided it is used in an adequate range of Reynolds numbers (250... 250000). Thus the relationship between the frequency of detachment of the vortices and the velocity which had been obtained in lithium can be used with no correction in Pb-Li. This relationship is :

$$v = 0.02753 \times f - 0.03270\ (ms^{-1}) \qquad (5)$$

The vortex flow meter offers then an experimental means of obtaining the gauging coefficient of the permanent magnetic flow meters. From the output of the flow meter (frequency in Hertz), the velocity is computed by (5) and so the flow rate is obtained. A linear regression line represents the gauging factor of FI1. Figure 6 shows a practically linear relation between the vortex frequency and the output of FI1.

5.3. Effect of the magnetic field

Using the methods of calculation of fully developed flows outlined in ref. 5, the MHD pressure losses for lithium-lead in an annular geometry would be :

$$P = f(\phi_1, \phi_2) \times \sigma \times Vo \times Bo^2 \times L \qquad (6)$$

with P pressure losses (Pa)

$f(\phi_1, \phi_2)$ f conduction ratio of the inner and outer pipe (0.159)

σ electrical conductivity liquid metal ($7.3065\ 10^{-5}\ Sm^{-1}$)

Vo average velocity ($7.855\ 10^{-2}\ ms^{-1}$)

Bo magnetic induction (1.98 Tesla)

L total length of the magnetic field (m)

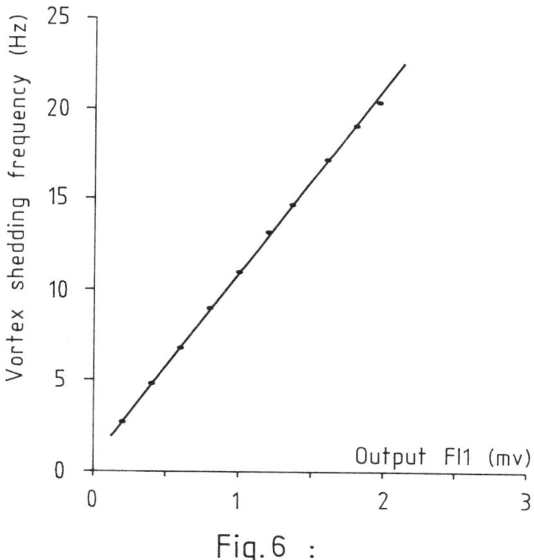

Fig.6 : Vortex shedding frequency versus voltage output of the permanent magnet flow meter FI1

The annular geometry of the test section actually installed in MALICE has the following dimensions, outer pipe of 3" sch 40 S (OD = 88.9 mm, ID = 77.9 mm) and an inner pipe of 1 " sch 40 S (OD = 33.4 mm, ID = 16 mm).

The maximum flowrate obtained with Pb-17Li in test section 2 (TS2) for a field strength Bo = 1.98 T was Q = 1.1 m^3h^{-1}. The MHD pressure loss per unit length is then obtained from (6) :

P = 36220 Pa

At the same flow rate, the pressure loss without magnetic field would be :

P = 40 Pa

Thus the Lorentz force introduces a very high pressure loss compared to the viscosity forces, even at the low velocities obtained in Pb-17Li.

However the effect of the magnetic field on the absolute change of flow rate remains rather small. This can be understood from figure 5. The characteristic of the pumps has a steep negative slope whereas the resistance of the circuit has a steep positive slope. Therefore increasing this last slope by an additional MHD pressure drop (~ 0,4 bar at Q = 1.1 m^3h^{-1}) does not change much the flow rate.

The observations are radically different from those in lithium, where the observed effect of the MHD pressure loss on the flow rate has been quite drastic.

6. CONCLUSIONS

Lithium-lead can now be fabricated on an industrial scale. It can be safely handled in industrial environments, provided that precautions are taken. Equipment needs to be of very high quality and personal fully trained.

The conversion of the liquid metal in the loop MALICE from Li to Pb-17Li has given some difficulties due to the residual lithium remained on the walls of the loop. After a temperature raise to 500°C all obstructions in the main loop were dissolved and the circulation of the liquid Pb-17Li could be started.

The gauging factors of the permanent flow meters of the loop MALICE with Pb-17Li have been obtained by two methods, nl. extrapolation from sodium and comparison with the vortex flow meter. The difference between the two results is small (5 %).

The prediction of the electromagnet pump performance and the circuit resistance in Pb-17Li was quite realistic. The predicted maximum flow rate was 14 % lower than the actual maximum obtained. The MHD pressure losses are very important relative to the viscous pressure losses. However, due to the high overall hydraulic resistance of the circuit operated with Pb-17Li, the impact of the MHD-effect on the flow rate is rather small and relative less important than with pure lithium.

ACKNOWLEDGEMENT

This work is performed under contract with the CEC Association contract Euratom-Belgium State 200-85-1 FUA B.

REFERENCES

1. Mass transfer in pure lithium and lithium-lead dynamic environments : Influence of parameters
(ICFRM-2, Chicago, USA, April 13-17.04.86)
(H. Tas, Ja. Dekeyser, F. Casteels, J.L. Walnier, F. De Schutter, SCK/CEN)

2. Performance test of an electromagnetic conduction pump
Lithium loop Li-2 (J.L. Walnier)
Internal Document

3. Experimental studies on the use of liquid lead in a molten salt nuclear reactor
(M. Broc, J. Sannier, G. Santarini, CEA; Nuclear Technology, vol. 63, nov. 1983)

4. Electromagnetic flow meter development for LMFBR (Miyagawa & al., "Specialists meeting on sodium flow measurements in large pipes of LMFBR, Feb. 1980, Interatom, Germany)

5. Theoretical and experimental studies of Annular and counter current Magnetohydrodynamic flows (LIMET 88,4th Conf. on Liquid Metal Engng. and Techn., October 17-21, 1988, Avignon, France)
(J.L. Walnier, Ja. Dekeyser, A. Falla)

FUSION TECHNOLOGY 1988
A.M. Van Ingen, A. Nijsen-Vis, H.T. Klippel (editors)

CORROSION INITIATION AND MASS TRANSFER OF STRUCTURAL MATERIALS IN LIQUID PB-17LI

P. LEMAITRE, F. DE SCHUTTER, H. TAS, Ja. DEKEYSER and W. VANDERMEULEN

SCK/CEN Materials Development Department, Boeretang 200, B-2400 Mol, Belgium

The Pb-17Li natural convection loop LELI permits to establish material damage rates and mass transfer rates in flowing Pb-17Li. Runs were performed at 450°C/413°C and 400°C/370°C. Very low corrosion rates were obtained for the austenitic AISI 316L steel and the ferritic DIN 1.4914 steel. This can probably be ascribed to the presence of a very thin and stable passivation surface layer, preventing wetting of the steel by the liquid metal. A mechanistic corrosion model, based upon the thermodynamic instability of austenitic steels at the temperatures considered and the high solubility of Ni in liquid Pb-17Li, is used to explain the surface microstructures developed in the loop tubing.

1. INTRODUCTION

Eutectic Pb-17Li has been proposed as prime candidate liquid metal breeder material in the European Fusion Technology program. Pb-17Li is attractive because of its effective tritium breeding. Furthermore, it is much less reactive with water and air than liquid lithium. However, several studies (1-3) indicate that corrosion and mass transfer are higher in liquid Pb-17Li than in liquid lithium. In the present study results are presented, obtained in the Pb-17Li natural convection loop LELI during runs 7 and 8. Corrosion results of runs 1 - 6 have been reported earlier (4).

2. EXPERIMENTAL PROCEDURE

A detailed description of the natural convection loop LELI has been given elsewhere (4). The AISI 316L loop has the shape of a parallelogram, does not include neither a cold trap nor a hot trap and contains 4 liters of liquid metal. Table 1 gives the operation conditions for runs 7 and 8.

The flow rate of the liquid metal during both runs was about 0.05 m/s. Unlike previous runs (4), the loop was not rinsed with pure lithium prior to Pb-17Li operation. Coupons of the two CEC reference materials, AISI 316L and DIN 1.4914, were exposed as polished in the hot leg and the cold leg of the loop. The non-metallic impurity content is expected to be low (5) : the concentration of nitrogen

Table 1 : Operation conditions for runs 7 and 8.

run	time (hours)	temperature (°C) cold leg	temperature (°C) hot leg
7	4000	413	450
8*	4750	370	400

* tab specimens inserted during 4000 hours only.

as measured in previous campaigns was below 10 ppm (4).

After each run the loop was dismantled for detailed analysis of the loop tubing by metallography and Scanning Electron Microscopy (SEM) in conjunction with Energy Dispersive X-ray Analysis (EDX). Weight losses were determined from the tab samples after intensive cleaning by alternate exposure to water and lithium until no further weight change was measured.

3. RESULTS

3.1. Tab samples

Table 2 summarizes the weight change data obtained. The weight losses appear to be much lower as compared to results of previous campaigns (4).

Fig. 1 shows some typical micrographs of samples exposed during run 7. Material loss occurred in the hot leg tab samples of both the austenitic 316L steel and the ferritic DIN 1.4914 steel. For the austenitic tab sample inserted in the cold leg, deposition of material as well as selective corrosion were observed, whereas for the ferritic tab specimen only deposition of material seemed to have occurred. Furthermore, the attack of the hot leg tab specimens was clearly not uniform. Some areas seem to be not affected at all.

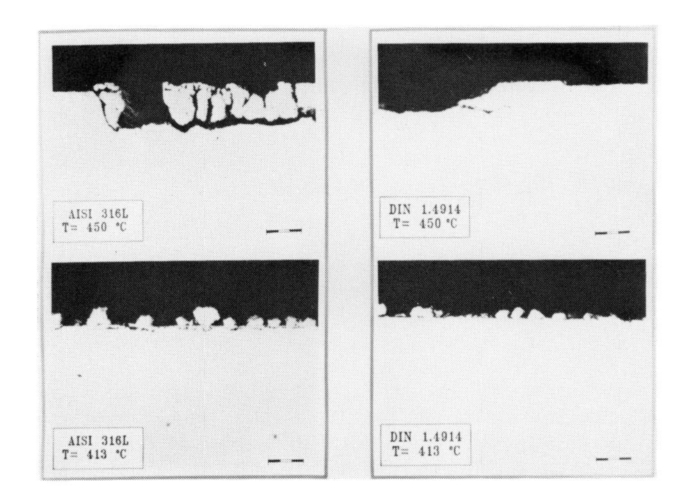

FIGURE 1
Cross-section of AISI 316L and DIN 1.4914 tab specimens after 4000 hours exposure to flowing Pb-17Li at 450°C/413°C. Notice the non-uniform attack.

The depth of the local attack in the austenitic 316L material was about 20 μm whereas previously values of 30 to 40 μm were obtained for exposure times of only 1000 hours at 450°C (4).

The hot leg samples exposed during run 8 did not show any significant attack.

3.2. Loop tubing

3.2.1. Run 7 : 4000 hours, 450°C/413°C

Table 2 : Weight change data of tab samples inserted in the cold and hot leg of the loop.

AISI 316L			DIN 1.4914		
Run/ Sample	T (°C)	Weight change (mg/m^2h)	Run/ Sample	T (°C)	Weight change (mg/m^2h)
7/8	413	+ 2.0	7/7	413	+ 2.8
7/9	450	- 10.0	7/20	450	- 3.9
7/11	450	- 6.3	7/21	450	- 3.8
8/1	400	- 1.8	8/2	400	- 0.9

Some typical micrographs of the loop tubing after run 7 are shown in fig. 2. The heating element is situated between locations 8 and 16. For each location uncleaned samples (with Pb-17Li still adhering to the surface) and cleaned specimens were investigated. Unlike the case of the tab specimens no localized attack was observed. A rather thin porous ferritic layer developed in both the cold and hot leg of the loop. The thickness of the ferritic layer as measured on cleaned specimens varied from about 6 µm in the cold leg to 20 µm in the hot leg. Comparison of the micrographs of locations 15 and 16 shows that the porosity of the surface layer is enhanced by the cleaning operation without, however, affecting the overall thickness of the layer.

3.2.2. Run 8 : 4750 hours, 400°C/370°C

Some typical micrographs of the loop after run 8 (4750 hours, 400°C/370°C) are shown in fig. 3. The heating element is situated between locations 6 and 14. The microstructure is totally different from the one observed after run 7. No ferritic layer did form (confirmed by EDX) neither in the hot leg nor in the cold leg of the loop. Etching revealed that the observed localized attack had an intergranular character.

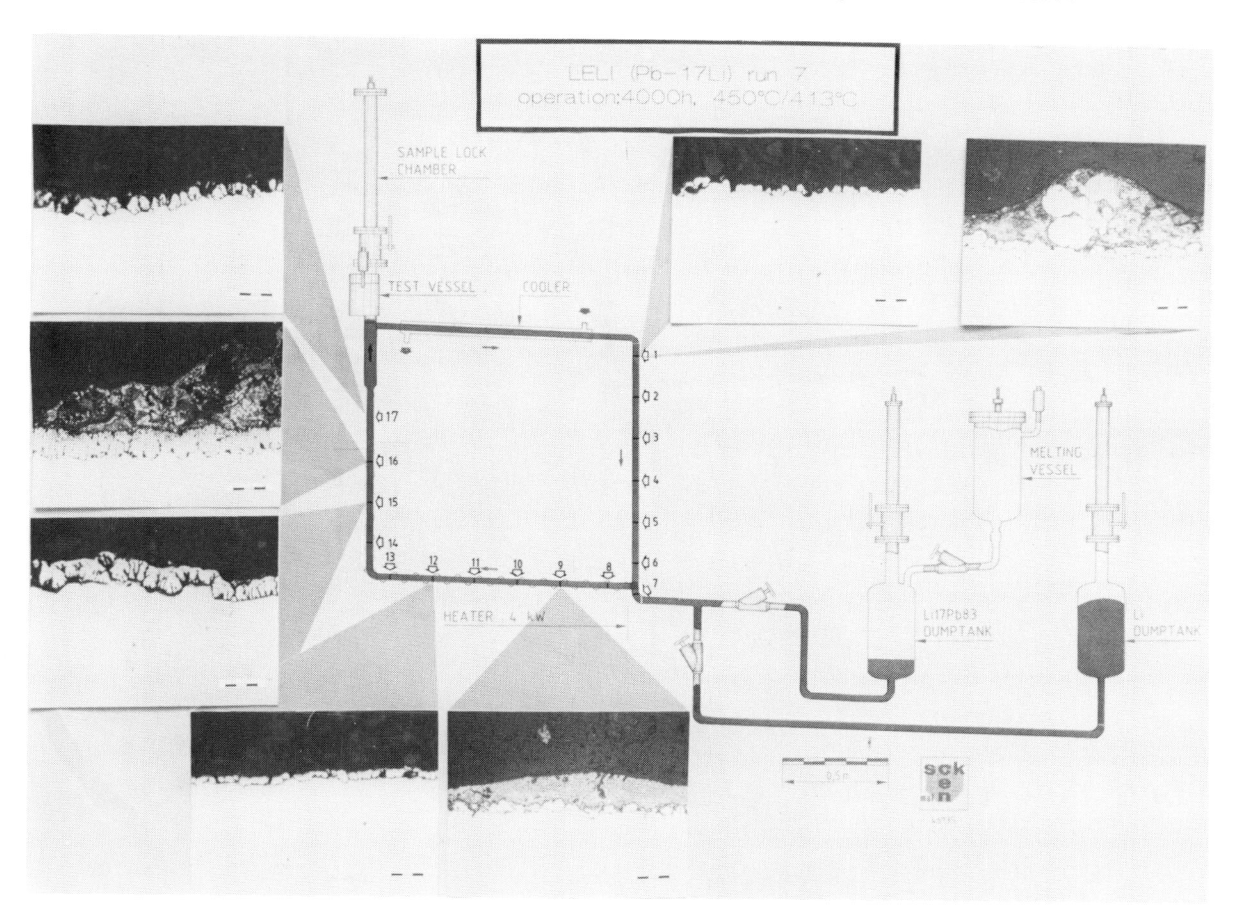

FIGURE 2

Metallographic analysis of loop tubing after run 7 (4000 hours, 450°C/413°C). Bars in the micrographs are 15 µm. Micrographs 1, 12, 15 and 17 after lithium/water cleaning; micrographs 1,9 and 16 without cleaning.

3.2.3. Runs 1 to 8 : 15864 hours

Fig. 4 shows the microstructure of the loop tubing situated immediately before and after the cooler after exposure during runs 1 to 8. This part of the loop, which is normally not dismounted for examination, has exceptionally been investigated after run 8. The total accumulated exposure time was 15864 hours, during which the loop was operated for 9338 hours at 400°C/370°C, 5176 hours at 450°C/413°C and 1350 hours at 500°C/453°C. A ferritic layer thickness of about 30 to 40 µm is observed in the part of the loop tubing situated just before the cooler (belonging to the hot leg). Within this ferritic layer brighter spots are visible pointing to the presence of austenitic islands. The presence of δ-ferrite at the grain boundaries of the austenite matrix in the weld can also be observed.

Immediately after the cooler (forming part of the cold leg of the loop) intergranular attack has occurred. No ferritic layer, however, has formed.

4. DISCUSSION

4.1. Coupons

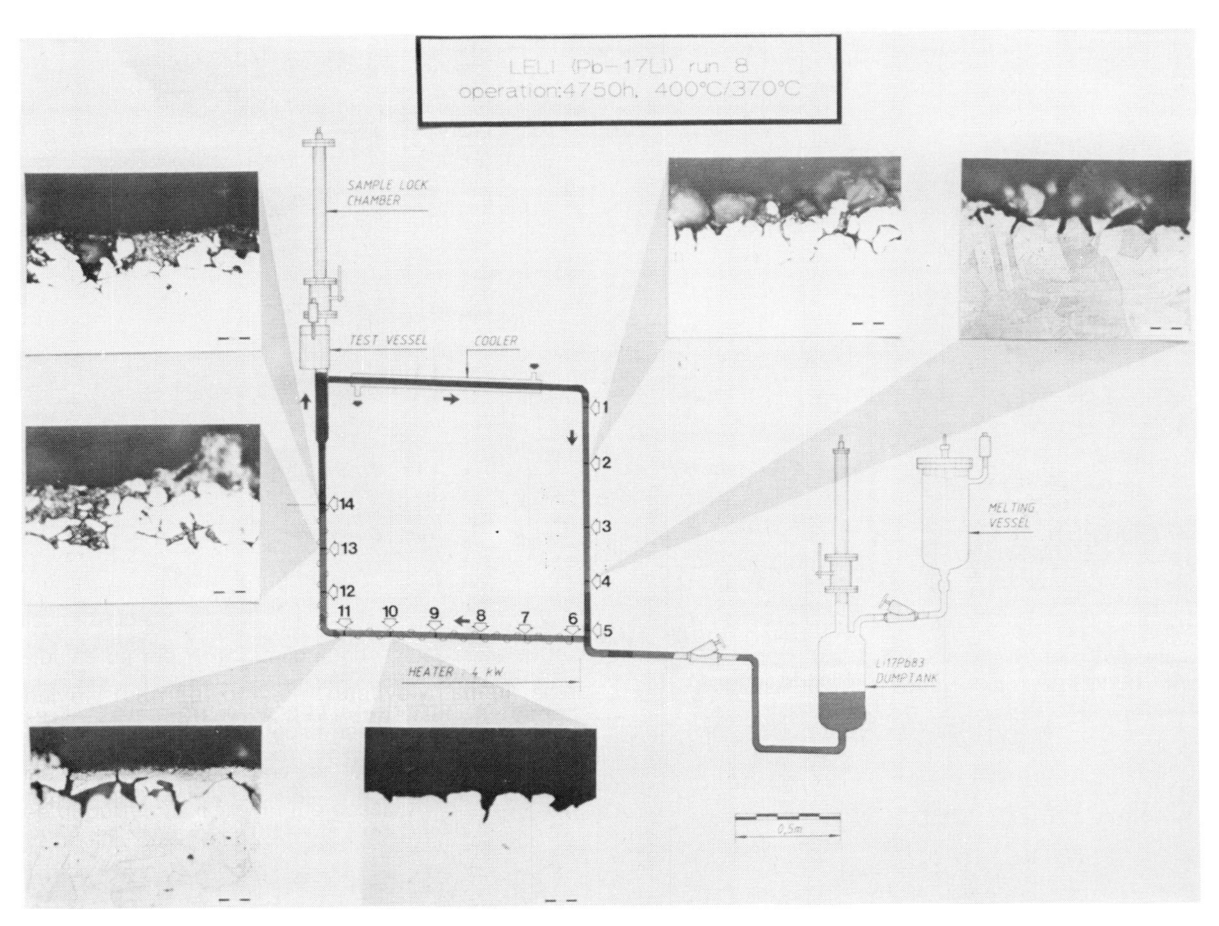

FIGURE 3

Metallographic analysis of loop tubing after run 8 (4750 hours, 400°C/370°C). Bars in the micrographs are 15 µm. Micrograph 10 after cleaning. Micrographs 2, 4 (etched), 11 (etched), 13 and 14 without cleaning.

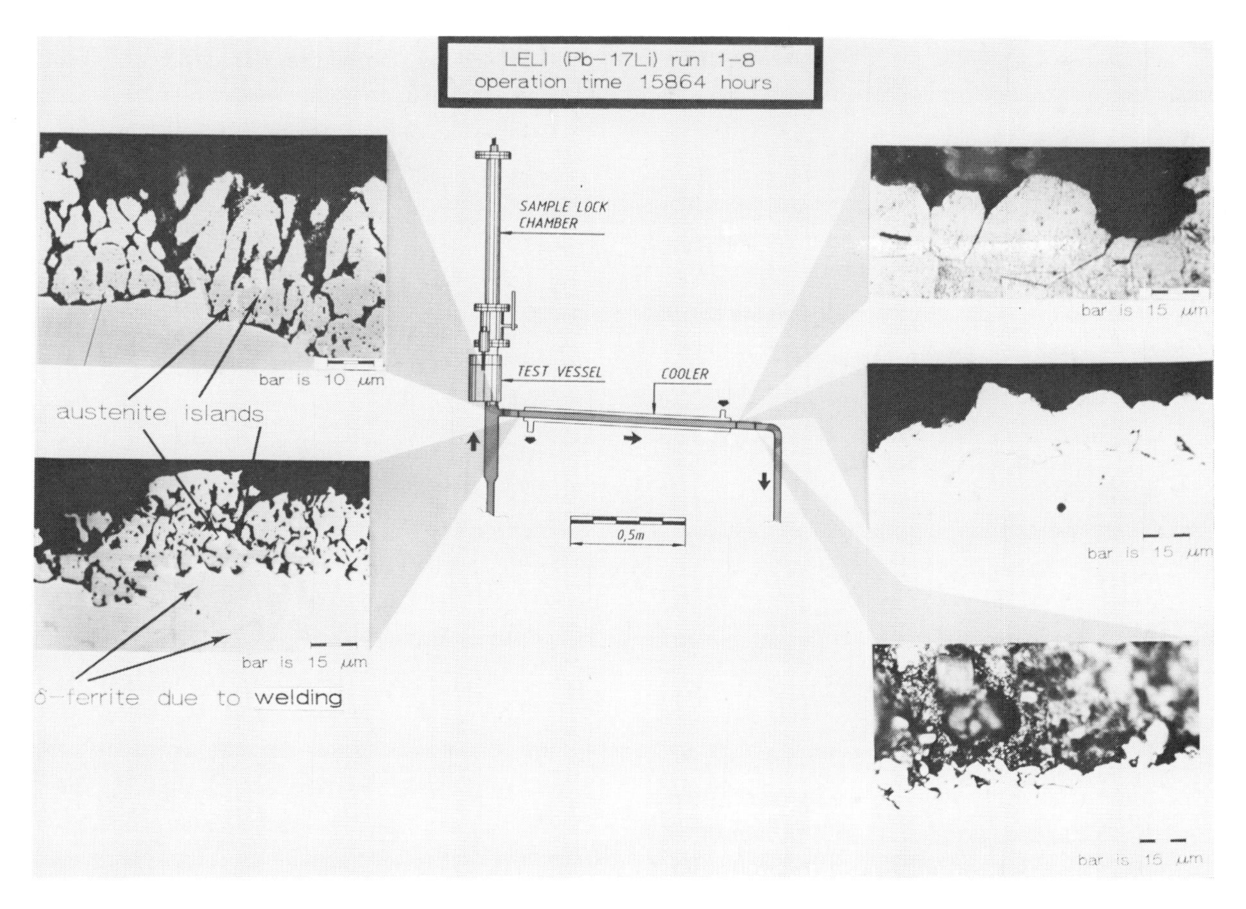

FIGURE 4

Microstructure of loop tubing situated immediately before and after the cooler after exposure during runs 1 to 8. Total corrosion time 15864 hours of which 9338 hours at 400°C/370°C, 5176 hours at 450°C/413°C and 1350 hours at 500°C/453°C. All micrographs after cleaning except for under left. Upper right micrograph after etching.

In fig. 5 the weight change data obtained are compared with data obtained during earlier campaigns i.e. runs 1-6 (4) and data reported by other investigators (6, 7-10). It appears that for both the austenitic and the ferritic steel tabs the dissolution rates obtained during runs 7 and 8 are much lower than the ones measured in previous campaigns. Microstructural analysis showed that corrosion did not occur in a uniform way during run 7 whereas during run 8 almost no corrosion was observed.

These very low dissolution rates can be attributed to a very thin and stable oxide layer covering the surface of the samples and preventing direct contact between the steel and the liquid metal. During former runs (4) the loop was rinsed with pure lithium prior to Pb-17Li filling in order to remove this layer and hence obtain good wetting. However, this caused plugging of the liquid metal filling lines. As it was argued that this procedure might also cause higher corrosion rates, it was

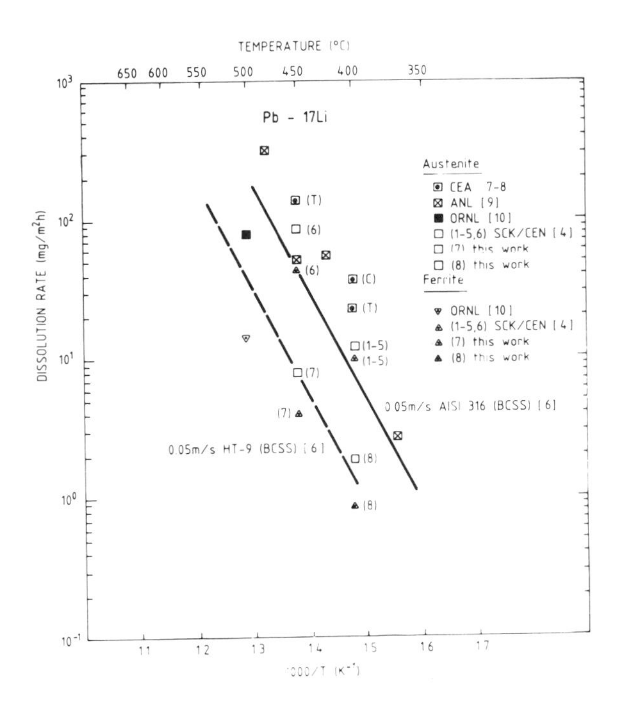

FIGURE 5
Arrhenius plot of corrosion rate data for austenitic and ferritic steels in flowing Pb-17Li. SCK/CEN results refer to 4000 h/400°C (campaign 8), 4000 h/450°C (campaign 7), 1000 h/450°C (campaign 6), 400°C (campaign 1 to 5). CEA data refer to experiments in the small convection loop "Tulip" (T) and the larger convection loop "Clipper" (C).

subsequently abandoned. The fact that the morphology of the corrosion layer of the coupons was totally different from that of the corresponding loop tubing can probably be attributed to the different nature of the oxide layer on the coupons and on the inner surface of the loop tubing. The thin passivating oxide layer on the coupons which forms immediately after polishing is apparently more resistant to wetting.

4.2. Loop tubing

A mechanistic corrosion model has been proposed by H. Tas et al. (11) for austenitic steels in contact with liquid metal, which is based on the inherent thermodynamic instability of these steels at the exposure temperatures considered and the high solubility of Ni in the liquid metal. In the model three steps are distinguished. First channels form along high-energy sites of the steel. Then ferrite forms around these channels according to the mechanism described in detail by H. Tas et al. (11) till the channels are saturated with Ni and Cr. Renewal of liquid metal re-initiates channel extension along high-energy sites so that ferrite can again be formed.

The ferritic layer observed in the loop after run 7 (4000 hours/450°C-413°C) and after runs 1 to 8 (15864 hours) just before the cooler may be formed by this mechanism. The occurrence of a ferritic layer in both the cold and the hot leg of the loop after run 7 can be attributed to selective corrosion at all points of the loop immediately after filling. The brighter spots as observed in the ferritic layer (runs 1 to 8) are most probably untransformed austenite islands. The same phenomenon has already been observed in ferritic layers exposed to liquid lithium (11, 12, 13). As far as known to the authors it is the first time it has been observed in a ferritic layer formed in contact with Pb-17Li. The presence of these untransformed austenite islands is in accordance with the mechanistic corrosion model as their presence may be linked to the inter-channel distance and the degree of solubility of Ni in the liquid metal (11).

The preferential grain boundary attack, as observed after run 8 (4750 hours, 400°C/370°C), and after runs 1 to 8 (15864 hours) immediately after the cooler may be the start of formation of channels along high-energy sites of the steel. The fact that during run 7 a ferritic layer has developed in the cold leg (T = 400°C), whereas during run 8 no ferritic layer has formed in the hot leg (also at T = 400°C) may seem conflicting. The overall temperature during run 8, however, was much lower than during run 7, probably leading to such a low solubility of Ni in the liquid metal that no formation of ferrite was possible.

5. CONCLUSIONS

The weight change data obtained on austenitic (316L) and ferritic (DIN 1.4914) tab samples inserted in the hot leg are much lower than the values obtained during previous campaigns. This can probably be explained by the fact that the loop was not rinsed with pure lithium prior to filling with Pb-17Li. It follows that a very thin but stable oxide layer prevented good wetting between the steel and the liquid metal.

The mechanistic corrosion model proposed by H. Tas et al. (11) distinguishes 3 steps :

1) formation of channels along high-energy sites.
2) formation of ferrite around these channels according to a mechanism based upon the thermodynamic instability of austenite steels at the temperatures considered and the high solubility of Ni in liquid Pb-17Li till the channels are saturated with Ni and Cr.
3) renewal of liquid metal initiating channel extension along high-energy sites so that again ferrite can be formed.

The observations which have been made concerning the microstructure of the loop tubing are in accordance with this model :

- at lower temperatures (run 8, runs 1 to 8 after the cooler) mainly intergranular attack :
  * step 1 of the model only
  * no formation of ferrite due to limited solubility of Ni in the liquid metal.
- at higher temperatures (run 8, runs 1 to 8 ahead of the cooler) occurrence of porous ferritic layer :
  * steps 1-2-3 of the model.
- untransformed austenite islands in ferritic layer, observed for the first time in Pb-17Li (runs 1 to 8, ahead of the cooler) :
  * step 2 of the model
  * presence of austenite islands is linked to inter-channel distance and degree of solubility of Ni in liquid metal.

ACKNOWLEDGEMENT

The authors are indebted to the Technology Department for loop operation and to L. Driesen for metallography. They also want to express their gratitude to M. Kaers for photography. This work is performed under contract with the CEC, association contract Euratom-Belgian State 100-82-1 FUA B.

REFERENCES

1. H. Tas, S. Malang, F. Reiter and J. Sannier, J. Nucl. Mat. 154-156 (1987).
2. O. Chopra and D. Smith, J. Nucl. Mat. 122-123 (1984), 1219.
3. H. Tas, Ja. Dekeyser, F. Casteels, J. Walnier and F. De Schutter, J. Nucl. Mat. 141-143 (1986), 571.
4. Ja. Dekeyser, F. De Schutter and H. Tas, Proc. 4th Int. ENS/ANS Conf. Geneva (1986), Vol.3, 111.
5. R. Buxbaum, J. Less Common Metals 97 (1984) 27.
6. O. Chopra, D. Smith, P. Tortorelli, J. De Van and D. Sze, Fusion Technology 8 (1985), 1956-1969.
7. M. Broc, P. Fauvet, T. Flament and J. Sannier, Proc. 4th Int. ENS/ANS Conf. Geneva (1986), Vol.3, 135-143.
8. M. Broc, T. Flament, P. Fauvet and J. Sannier, J. Nucl. Mater. 154-156 (1987),
9. O.K. Chopra and D. Smith, J. Nucl. Mater. 141-143 (1986), 566-570.
10. P.F. Tortorelli and J.H. De Van, J. Nucl. Mater. 141-143 (1986), 592-598.
11. H. Tas, F. De Schutter, P. Lemaitre, Ja. Dekeyser, Proc. LIMET Conf. Avignon 1988 (in print).
12. P. Lemaitre, SCK/CEN, Unpublished results.
13. C. Bagnall, J. Nucl. Mater. 103-104 (1981) 639.

FUSION TECHNOLOGY 1988
A.M. Van Ingen, A. Nijsen-Vis, H.T. Klippel (editors)

THE RATES AND REACTION PRODUCTS OF THE CORROSION OF 1.4914 STEEL BY LITHIUM MONOXIDE, AND THE EFFECTS OF FLOWING WATER VAPOUR.*

Richard J PULHAM, Peter HUBBERSTEY and Rick A CHAPMAN.

Department of Chemistry, University of Nottingham, Nottingham NG7 2RD, England.

Lithium monoxide was virtually inert under dynamic vacuum conditions towards 1.4914 steel (10.6% Cr) from 773 to 873K, but the rate of corrosion increased rapidly from 0.027 μmh^{-1} at 1048K to 14 μmh^{-1} at 1275K. Corrosion was via a series of successive rapid and slow steps giving nearly linear overall rates. The steel was internally oxidised giving pockets in grain boundaries in which there was Cr-enrichment, and the external surface consisted of alternate layers of Li_5FeO_4-($LiCrO_2$) and Cr-depleted steel. Experiments at 823 and 873K showed that water vapour caused a marked increase in corrosion which occurred again in successively rapid and slow steps to give a nearly linear rate of reaction. Moreover, the rate was linearly dependent on water vapour pressure rising at 823K to 0.28 μmh^{-1} on increasing the pressure to 0.97 kNm^{-2}. Above this pressure the corrosion was still cyclic but extremely rapid. At 873K the corrosion rate increased to 0.53 μmh^{-1} at 1.77 kNm^{-2} and at both temperatures the onset of severe corrosion corresponded with pressures which were large enough to convert Li_2O to a liquid LiOH phase. The corroded surface of the steel contained alternate layers of Li_5FeO_4 and ($LiCrO_2$).

1. INTRODUCTION

Solid Li_2O has potential use as tritium breeder in some fusion reactor designs so that it is necessary to assess its compatability with containment steels. During tritium breeding with oxide fuels, some $T_2O(=H_2O)$ is generated so that the potential for corrosion is enhanced, and this also merits investigation. Previously[1] we showed that there was no measurable corrosion of 1.4914 steel foils which had been immersed in Li_2O and sealed under Ar into Ni capsules at the various temperatures(K)/times(h): 1023/120, 1023/263, 1123/120, 1123/264, 1198/102 and 1198/201. It was indicated, however, that as with 316 steel, if the equilibrium:

$$M + 2Li_2O = LiMO_2 (M=Cr,Fe) + 3Li$$

is disturbed by removal of Li by vacuum, condensation or reaction with impurities, then corrosion would proceed. The present paper describes the determination of the rates of corrosion of 1.4914 steel under vacuum conditions, and also under a constant pressure of water vapour, by monitoring continuously the increase in electrical resistance of the steel[1].

*This work was sponsored by UKAEA Springfields

2. METHOD

Lithium monoxide was prepared as before[1] or supplied by UKAEA Springfields. In both cases, x-ray diffraction analysis of the powder showed the characteristic lines of Li_2O only. The 1.4914 steel (Cr 10.6 and Fe 85.9%) was supplied by KfK, Karlsruhe. Steel foils (30x8x0.3 mm) were immersed in powdered Li_2O which was contained in a ZrO_2 tube (id = 11, l = 40 mm) and heated in a tube furnace under a dynamic vacuum of 10^{-2} Nm^{-2}. The electrical resistance of the foil was determined by passing through a current of 0.3A via Ni leads (4 probe) welded to opposite ends of the foil, and measuring the potential difference (~5 mV). The thickness of the foil (t) was related to its resistance (R) by $t = \rho l/wR$ where ρ is the resistivity of the steel, and l, w are the length and width of the foil, which were both assumed to remain constant. This technique gave a continuous measurement of the undamaged steel thickness over long times.

In experiments under $H_2O(g)$, the pressure of vapour generated from liquid in an external reservoir was regulated by needle valves at each end of the assembly, and the constant pressure measured by means of an oil manometer. Although the H_2O (g) flowed continuously through the tube furnace, there was no forced flow through the Li_2O.

At the end of a corrosion experiment, the water source was removed and the assembly was cooled under dynamic vacuum to prevent LiOH formation on cooling. The corroded foil and corrodant were recovered under Ar and subjected to x-ray diffraction analysis. Subsequently the foil was mounted in bakelite, polished and examined using optical and scanning electron microscopy.

3.RESULTS AND DISCUSSION

3.1 Rate of corrosion by Li_2O at 773 to 873K

Corrosion after 160h was very slight, and close to the lower limit of detection by the resistance method. The very small increase with increasing time, which was linear, gave corrosion rates $\leq 0.003\ \mu m h^{-1}$ (Table).

Table. Rates of corrosion by Li_2O

Temperature (K)	10^2 Rate ($\mu m h^{-1}$)	Total time (h)	Comments
773	0.11	144	linear
773	0.18	144	linear
798	0.12	128	linear
823	0.31	120	linear
823	0.26	140	linear
873	0.23	92	linear
873	0.21	160	linear
1048	1.70	98	depths
1088	5.80	116	after
1118	66.5	43	40 h.
1148	83.3	40	
1273	1000-1680	6.5	min & max depth from
1275	720-1170	6.5	microscopy

An Arrhenius plot showed that there was a small increase in rate with increasing temperature (Figure 1). There was no distinction between the two sources of Li_2O, and the general inertness was similar to that towards 316 steel.

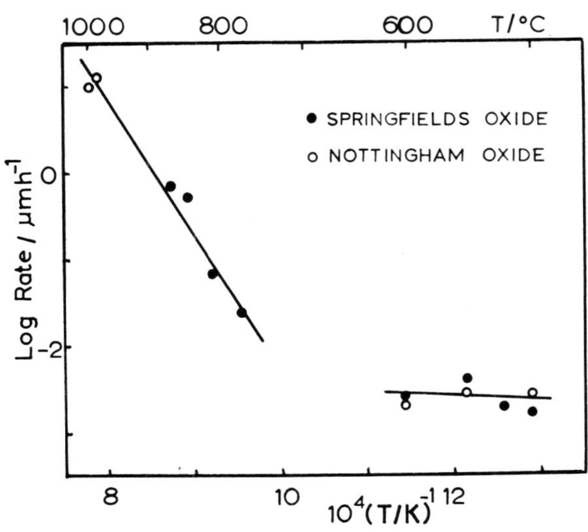

FIGURE 1
Rates of corrosion by Li_2O

3.2 Rate of corrosion by Li_2O at 1048-1275K

Corrosion was much more severe at these higher temperatures. The corrosion proceeded at each temperature in a series of alternate

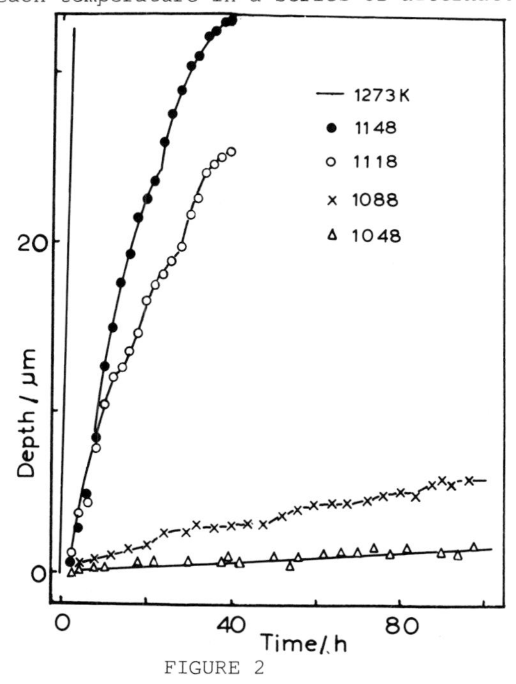

FIGURE 2
Depth vs. time for corrosion by Li_2O

rapid and slow steps (Figure 2) to give an approximately linear increase in corrosion depth with increasing time. Corrosion increased greatly with increasing temperature and the rate (defined as the depth of corrosion after 40h due the non-linearity of the process) is shown in Figure 1. These fast rates and their strong temperature dependence, which is given by the equation:

$\log(\text{Rate}/\mu m h^{-1}) = -1.6 \times 10^4/T + 13.7$; 1048-1275K

are very similar to those for 316 steel[1].

3.3 Corrosion products from Li_2O + steel

At the lower temperatures (773-873K), the foils carried a red film and x-ray diffraction showed this to be the ternary oxide Li_5FeO_4. Optical and scanning microscopy showed the presence of a compact Cr-rich inner layer of corrosion product (probably $LiCrO_2$) and a red diffuse thicker Fe-rich outer layer (Li_5FeO_4). These together were very thin (overall 2-8 μm) and the substrate steel was largely undamaged (steel loss ≦ 0.18 μm after 90 h).

At the higher temperatures (948-1275K) there was grain boundary penetration of the steel as well as external layers of corrosion products. Analysis by SEM/EDAX of the metallic parts of the corrosion zone showed (Figure 3) that there was Cr-enrichment in the damaged grain boundaries but Cr-depletion of the outer steel layer (= Fe). The non-metallic corrosion products were situated interior ($LiCrO_2$) and exterior (Li_5FeO_4 + $LiFeO_2$) to the Fe layer.

3.4 Mechanism

The results suggest that Li^+ and O^{2-} ions diffuse relatively rapidly into the steel by reaction with Cr in grain boundaries to form $LiCrO_2$. The Cr activity in the boundaries is restored by diffusion of Cr from the substrate steel on a broad front, the diffusing Cr meeting the incoming Li^+ and O^{2-} ions in expanding grain boundaries to form a layer of $LiCrO_2$. The Cr-denuded grains constitute the layer of nearly pure Fe. Surface Fe reacts on a broad

Wt. Ratios Cr:Fe

A. 0.12 Substrate steel

B. 9.3 Grain boundary

C. ($LiCrO_2$)

D. 0.49

E. 0.05 (=Fe)

F. (Li_5FeO_4)

⊢⊣ = 10 μm

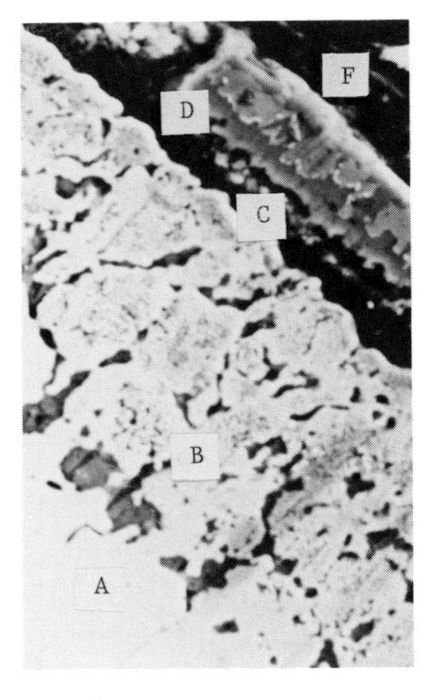

FIGURE 3
Corrosion by Li_2O at 1148K after 40 h

front with Li_2O to produce the outer red layer of Li_5FeO_4. Basically the corrosion is driven by the removal of Li(g) from the equilibrium:

$M + 2Li_2O = LiMO_2 \ (M = Cr, Fe) + 3Li$

In the case of M = Fe, further reaction with corrodant occurs to produce the red Li_5FeO_4 through which Li^+ and O^{2-} ions have to diffuse to reach the steel. The stepwise corrosion suggests the making and breaking of an inhibiting layer, probably $LiCrO_2$.

3.5 Rate of corrosion by Li_2O + H_2O (g)

The effect of flowing water vapour on the corrosion by Li_2O is shown in Figures 4 and 5 for 823 and 873K, respectively. Corrosion was stepwise again but increased nearly linearly with increasing time. Because of the stepwise nature, the rate is defined as the depth of corrosion after an arbitrarily chosen time of 30 h. This rate of corrosion increased linearly with increasing pressure of H_2O (g) as shown in Figure 6 but only up to 0.97 kNm^{-2} at 823K and up to 1.77 kNm^{-2} at 873K. The dependence of the rates on the pressures of H_2O (g) are given by:

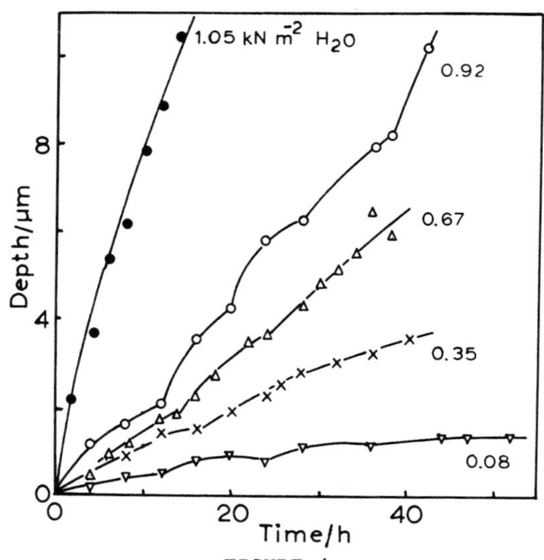

FIGURE 4
Corrosion by Li_2O + $H_2O(g)$ at 823K

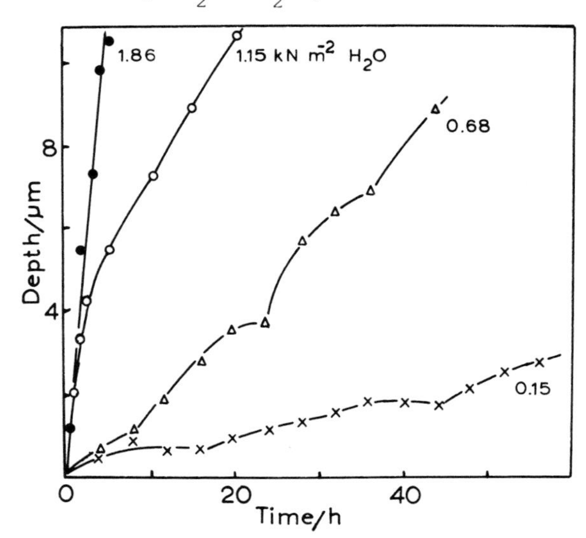

FIGURE 5
Corrosion by Li_2O + $H_2O(g)$ at 873K

Rate/μmh^{-1} at 823K = 0.29P; $0 \leqq P/kNm^{-2} \leqq 0.97$

Rate/μmh^{-1} at 873K = 0.30P; $0 \leqq P/kNm^{-2} \leqq 1.77$

Above these pressures, there was a large increase in corrosion rate to a higher and thereafter seemingly constant value. The exceptionally high rate was greater at 873 than at 823K whereas the difference was insignificant over the preceding linear region. At extremely low pressures of H_2O (g), extrapolation to $\leqq 0.04$ kNm^{-2} indicates that H_2O (g) does not destroy the inertness of Li_2O towards the steel. Moreover, H_2O (g) alone had no detrimental effect on the steel.

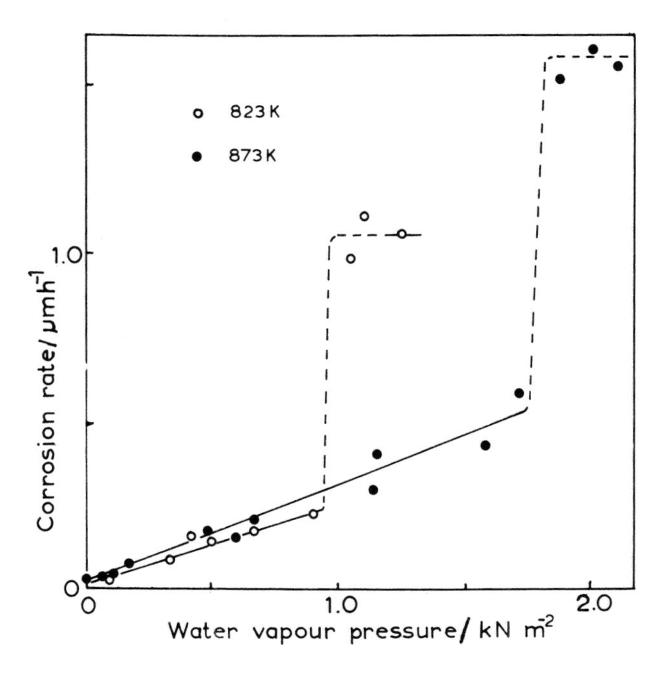

FIGURE 6
Effect of $H_2O(g)$ on corrosion by Li_2O

A foil exposed to H_2O at 873K for 40 h under 1.6 kNm^{-2} corroded quickly (20 h) to a depth of only 1.6 μm but thereafter remained intact. The surface was covered with a superficial layer of Fe_2O_3/Fe_3O_4. These data on water corrosion augment our previous work[2].

We believe that the surge in corrosion rate at the crucial pressures of 0.97 and 1.77 kNm^{-2} is due to corrosion by molten LiOH which is known to be severe[3]. Here the phase LiOH (l) forms according to:

$$Li_2O\ (s) + H_2O\ (g) = 2LiOH\ (l)$$

The equilibrium pressures[4,5] given by:

$$\log P\ H_2O\ (g)/atm = -4406/T + 3.27;\ 746 \leqq T/K \leqq 893$$

are 0.84 and 1.69 kNm^{-2} at 823 and 873K, respectively, and these are very close to our crucial pressures for enhanced corrosion. Moreover, post-corrosion examination of the corrodant which has been under these high pressures of H_2O (g) showed that part had been liquid and had flowed round the steel.

3.5 Corrosion products from Li_2O + steel + H_2O (g)

The foils carried a red corrosion product which readily spalled off. X-ray diffraction showed this to be Li_5FeO_4 + Fe_2O_3. Analysis of sections by optical and scanning electron microscopy showed the existence on the surface of narrow Cr-rich grey layers between wider Fe-rich red (Li_5FeO_4) layers (Figure 7). The substrate steel had been consumed on a broad even front rather than by grain boundary penetration. This sandwich effect seemed a more advanced form of dry Li_2O corrosion; there were more layers to the sandwich and the original Cr-depleted (=Fe) layer was now fully converted to red Li_5FeO_4.

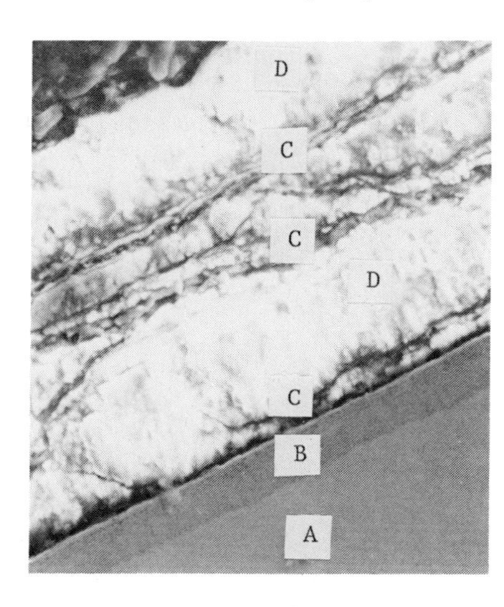

A. Steel

B. Cr-rich layer

C. $LiCrO_2$

D. Li_5FeO_4

⌴ = 10 μm

FIGURE 7

Corrosion by Li_2O + 1.2 kNm^{-2} $H_2O(g)$ at 823K after 40 h

3.6 Mechanism

Corrosion at pressures of H_2O (g) above 0.97 and 1.77 kNm^{-2} at 823 and 873K, respectively, is attributed to liquid LiOH. Below these pressures, the corrosion rate increases linearly (Figure 6) with pressure of H_2O (g) and this may be due to increased oxidation of the steel to form Cr_2O_3/Fe_2O_3 which then reacts with the corrodant Li_2O to form $LiCrO_2/Li_5FeO_4$. Alternatively, the corrosion may be due to LiOH but now as a solid solution in Li_2O, and the linear increase in rate is a reflection of the increase in OH^- ion concentration with increasing partial pressure of H_2O (g) above the solid solution. Corrosion occurs on a broad front. It appears that Fe diffuses out rapidly to form a layer of Li_5FeO_4 leaving a Cr-rich layer which is converted to $LiCrO_2$ (Figure 7).

4. CONCLUSIONS

Li_2O is virtually inert to 1.4914 steel below 873K but the rate of corrosion increases steeply from 1048 to 1275K. The layered corrosion products consist of Fe, Li_5FeO_4 and $LiCrO_2$. At 823 and 873K, no additional corrosion is caused by low ($\leqq$ 0.04 kNm^{-2}) pressures of water (reactor pressures might be *ca.* 10^{-3} Nm^{-2}). At higher pressures the rate of corrosion increases linearly with increasing pressure but there is a drastic increase in corrosion when the pressure is great enough to produce LiOH (1). The layered corrosion products are Li_5FeO_4 and $LiCrO_2$.

REFERENCES

1. R J Pulham, W R Watson and N P Young, Fusion Technology, 2, 963 (Pergamon Press, 1986).

2. R J Pulham, P Hubberstey, R A Chapman and N P Young, Proc. of the Third Internat. Conf. on Fusion Reactor Materials, Karlsruhe, Oct.4-8, 1987. J. Nucl. Materials, in print.

3. J S Collinson and R J Pulham, Proc. of Third Internat. Conf. on Liquid Metal Engineering and Technology, Oxford 9-13 April, 1984, 355.

4. M Tetenbaum and C E Johnson, J. Nucl. Materials, 126 (1984) 25.

5. N W Gregory and R H Mohr, J. Amer. Chem. Soc., 77 (1955) 2142.

FUSION TECHNOLOGY 1988
A.M. Van Ingen, A. Nijsen-Vis, H.T. Klippel (editors)

COMPATIBILITY PROBLEMS WITH BERYLLIUM IN CERAMIC BLANKETS

A. TERLAIN, D. HERPIN, P. PERODEAUD, T. FLAMENT, J. SANNIER,

CEA, IRDI, DCAEA/SCECF/SECNAU, Centre d'Etudes Nucléaires, BP.6, 92265 Fontenay aux Roses, France

Compatibility of beryllium with structural materials (316L austenitic steel and 1.4914 martensitic steel) and with tritium breeding ceramics (lithium aluminate or silicate) has been studied in contact tests between 550°C and 700°C and for durations reaching 3000 hours. Beryllium-ceramic interaction is negligeable in all the temperature range with aluminate and up to 600°C with silicates. On the other hand, noticeable interaction is observed between beryllium and 316L steel at 580°C and above. Beryllium interaction with 1.4914 steel is visible only at 650°C and above and its amplitude is lower than 316L steel one. In these two cases, the superficial layer is brittle, and adherent to the steel. Comparison between beryllium - 0.4 wt% calcium alloy and beryllium at 700°C shows that interaction with steels or ceramics is qualitatively the same but slightly weaker.

1. INTRODUCTION

Most of ceramic blankets for fusion reactors need an additional neutron multiplier. Beryllium is a prime candidate for this purpose but its compatibility with breeding and structural materials has to be prouved. Consequently, in the framework of the European Program on Fusion Technology, the interaction of beryllium with structural materials (austenitic and martensitic steels) and with tritium breeding ceramics (lithium aluminate and silicates) has been studied and results achieved in contact tests are presented in this paper.

2. EXPERIMENTAL

2.1. Materials

The two reference structural materials 316L austenitic steel and 1.4914 martensitic steel have been tested and their chemical compositions are given in Table I.

Table I - Composition of steels (wt %)

	Fe	Cr	Ni	C	Mo	Mn	Si	N
316L	65.2	17.4	12.33	0.024	2.3	1.8	0.46	0.06
1.4914	bal.	10.6	0.87	0.13	0.8	0.8	0.37	

Specimens are constituted of disks (20 mm diameter and 2 mm thickness) obtained by machining a plate with a surface state allowing to establish a good contact with the other tested materials.

Two types of beryllium have been experimented :

. unalloyed beryllium containing as major impurities oxygen (1000-5000 wt ppm), aluminium (500-650 wt ppm) and iron (300-500 wt ppm),

. beryllium - 0.4 wt% Ca alloy containing O_2 ($\leqslant$ 200 wt ppm) and N_2+Ar ($\leqslant$ 100 wt ppm).

As shown on figure 1, this alloy contains calcium-rich precipitates which can be expected to be $Be_{13}Ca$ /1/.

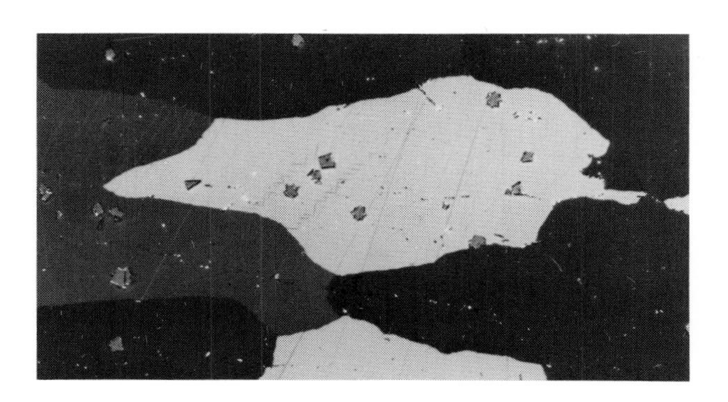

FIGURE 1
Structure of beryllium-calcium alloy

Beryllium disks (20 mm diameter and 10 mm thickness) are polished successively with grinding paper (up to 1200 grade), diamond paste (15 and 6 μm) and alumina powder in order to eliminate the cold-worked layer due to the machining of disks and to provide a good contact surface.

Lithium aluminate disks were prepared by CEA /2/ and are dehydrated either 24 hours at 600°C or 2 hours at 900°C before testing. Lithium ortho and meta silicates were delivered by KfK /3/ and are dehydrated 2 hours at 900°C.

2.2. Testing device

The tests are performed under dynamic vacuum (0.5 to 1 Pa) on stacked specimens (Fig.2). By means of a metallic ball, a weight insures a contact pressure of $1.5x10^4$ Pa.

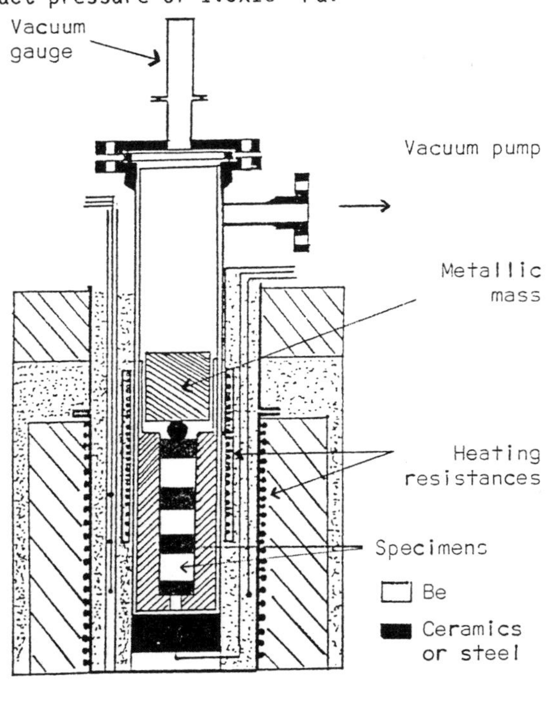

FIGURE 2
Testing device

3. RESULTS AND DISCUSSION

Operating conditions and results of the six tests carried out between 550 and 700°C and for exposure durations reaching 3000 hours are given in tables II and III.

3.1. Beryllium-steel interaction

3.1.1. Beryllium-316L steel interaction

At 550°C, no interaction is visible after 3000-hour exposure. On the other hand, the tests carried out in the 580-700°C temperature range show a beryllium-steel interaction characterized by the formation of a superficial layer adherent to the steel (Fig.3).

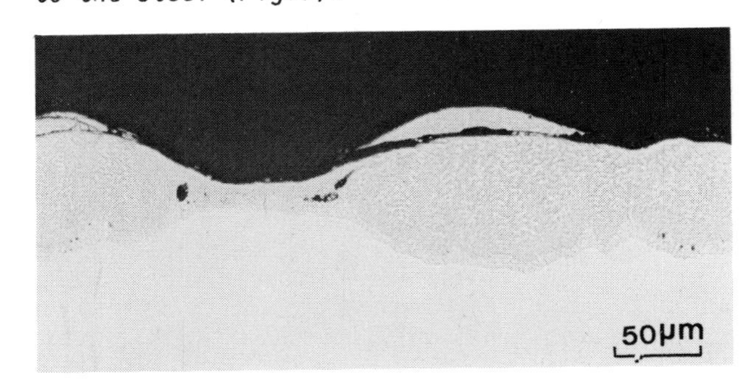

FIGURE 3
Cross-section micrograph of 316L steel after 1500 hour-test in contact with Be at 650°C

Its thickness is time and temperature dependent and can reach 100 μm after 500 hours at 700°C for example (table II). At temperatures higher than 600°C, an inner and an outer layer are distinguishable. As shown by Energy Dispersive Spectroscopy /4/, the outer layer is constituted by a brittle beryllium-rich compound whereas the inner one, about five times thinner, corresponds to a diffusion zone of beryllium in the steel.

On the beryllium side, numerous holes are visible at the surface (Fig.4).

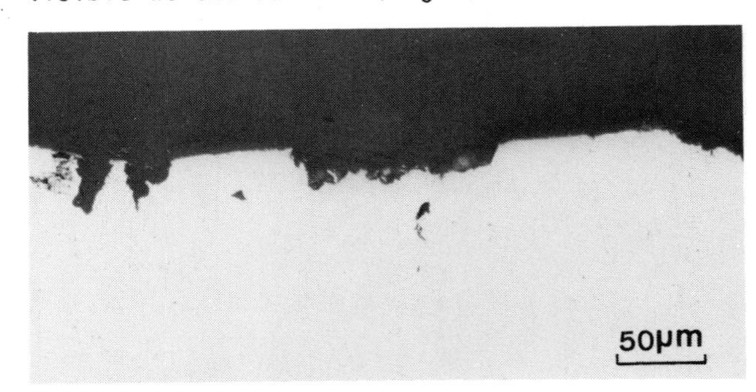

FIGURE 4
Cross-section micrograph of Be specimen after 1500 hour-test in contact with 316L steel at 650°C

Though position of theses holes is not correlated to the diffusion area into the steel, their formation is associated to the reaction of beryllium with steel. Indeed, depth and width of holes are increasing with time and temperature and they are not visible when no interaction occurs.

3.1.2. Beryllium-1.4914 interaction

At 550 and 600°C, no interaction is observed after respectively 3000 and 1500-hour exposures.

On the other hand, a brittle superficial layer, adherent to the steel, is visible after 1500 hours at 650°C or 500 hours at 700°C (Fig. 5) and corresponding thicknesses are respectively 20 µm and 30 µm (Table II).

EDS analysis has also revealed the formation of a beryllium-rich compound but no significant diffusion zone has been identified contrary to 316L steel.

Beryllium aspect is similar to that observed with 316L steel.

FIGURE 5
Cross-section micrograph of beryllium-1.4914 steel after 1500 hour-test in contact with Be at 650°C

3.1.3. Be-Ca/steels interaction

After 500 hours at 700°C, the interaction between Be-Ca alloy and 316L and 1.4914 steels is qualitatively identical to that observed with unalloyed beryllium. However, amplitude of the phenomenon appears to be lower. Layer thicknesses are respectively 65 µm and 12 µm for 316L and 1.4914 whereas they reach 100 µm and 30 µm with unalloyed beryllium (table II).

Table II - Interaction between steels and beryllium or beryllium-calcium alloy

Interaction	Temps (°C)	Duration (h)	Material	Observed phenomenon	Thickness (µm)
316L/Be	550	3300	Be 316L	No interaction No interaction	
	580	500	Be 316L	Holes at the surface (1) Discontinuous zones of Be diffusion	0-25 0-10
	600	1500	Be 316L	Holes at the surface (1) Discontinuous zones of Be diffusion	0-100 0-40
	650	1500	Be 316L	Holes at the surface (1) Discontinuous zones of Be diffusion	0-100
	700	500	Be 316L	Holes at the surface (1) Discontinuous zones of Be diffusion	0-200 0-100
	700	500	Be Ca 316L	Holes at the surface (1) Discontinuous zones of Be diffusion	0-65
1.4914 / Be	550	3000	Be 1.4914	No interaction No interaction	
	600	1500	Be 1.4914	No interaction No interaction	
	650	1500	Be 1.4914	Holes at the surface (1) Discontinuous zones of Be diffusion	0-20
	700	500	Be 1.4914	Holes at the surface (1) Discontinuous zones of Be diffusion	0-20 0-30
	700	500	Be Ca 1.4914	Indented surface Discontinuous zones of Be diffusion	0-12

(1) Not directly related to affected parts of steel surface

Surface of Be-Ca specimens appears to be more indented when compared with Be. Moreover, holes observed at the surface seem to be located preferentially in front of calcium-rich precipitates.

3.1.4. Discussion

From these experiments, the interaction between 316L and Be appears to be important and very temperature dependent. Using our data and those reported by Hofmann /5/, a tentative of prediction to one year has been made by considering two kinetics laws : parabolic and linear laws. These values are presented on the figure 6 in fonction of inverse of temperature.

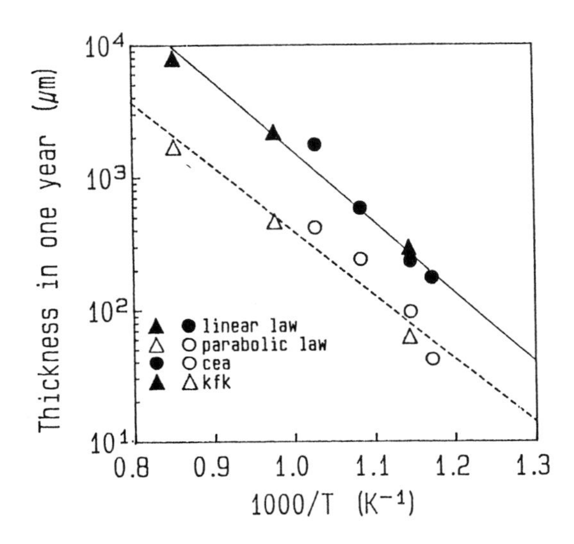

FIGURE 6
Be-316L interaction layer thickness versus inverse temperature

Whatever the law, our results and Hofmann's ones are in a good agreement. The activation energy deduced from that Arrhenius plot is about 120 $kJ.mol^{-1}$.

This prediction leads to a non negligeable interaction after 3000 hours at 550°C (36 μm and 12 μm for linear and parabolic kinetics respectively).

This is not in agreement with the experiment which shows an absence of interaction layer in these operation conditions ; such a discrepancy may originate from presence of a surperficial oxide layer on beryllium which could restrain beryllium - 316L steel interaction.

Comparison of austenitic and martensitic steel shows a better behaviour of the later one. This results corroborates Kittel's observations /6/ which evidence a larger interaction with 347 austenitic steel than with pure iron. As interaction with pure nickel is still more important, the presence of this element in austenitic steel may explain the less satisfactory behaviour of this material.

3.2. Beryllium-lithiated ceramics interaction

3.2.1. Results

Operating conditions and results are presented in table III.

For all the temperature and time range investigated (550-700°C, 500-3000h) no interaction is visible between beryllium and $LiAlO_2$.

Silicates behave similarly at 550°C and 600°C. On the other hand, at 650°C with pure beryllium and at 700°C with Be-Ca alloy, a superficial grey layer of about 10 μm thickness is locally observed on beryllium and steel specimens present in the container. This layer does not contain EDS detectable elements.

3.2.2. Discussion

Interaction of beryllium with tritium breeding ceramics appears to be very weak. The grey layer observed in certain tests is probably related to the interaction of beryllium with silicates. In this way Hofmann and Dienst /5/ have observed at 750 and 900°C an interaction of beryllium with meta silicates possibly associated to the formation of a liquid lithium-silicide phase.

From thermodynamical data, reduction of lithium aluminate and silicates by beryllium is expected. Inertia of ceramics noticed in our

Table III - Interaction between ceramics and beryllium or beryllium-calcium alloy

Interaction	Temp (°C)	Duration (h)	Observed phenomenon	Thickness(μm)
Be/$LiAlO_2$ Be/Li_2SiO_3 Be/Li_4SiO_4	550	3000	Unchanged aspect of Be surface	
Be/$LiAlO_2$ Be/Li_2SiO_3	580	500	Unchanged aspect of Be surface	
Be/$LiAlO_2$ Be/Li_2SiO_3	600	1500	Beryllium indented surface Transformed zone on Li_2SiO_3	150
Be/$LiAlO_2$ Be/Li_2SiO_3 Be/Li_4SiO_4	650	1500	Beryllium indented surface	
Be/$LiAlO_2$	700	500	Beryllium indented surface	10
Be-Ca/$LiAlO_2$ Be-Ca/Li_2SiO_3 Be-Ca/Li_4SiO_4	700	500	Indented surface	20

tests is probably related to protectiveness of the superficial BeO layer and has to be confirmed under irradiation.

4. CONCLUSION

These contact tests carried out under vacuum between 550 and 700°C point out a negligeable or weak beryllium-ceramic interaction. Practically, only silicates slightly react at 650 and 700°C.

On the other hand, beryllium-steel interaction is much more marked and becomes noticeable at 580°C with 316L austenitic steel and at 650°C with 1.4914 martensitic steel. Resulting from beryllium diffusion, a brittle layer is formed at the surface of steels. Its thickness is less important in the case of 1.4914 steel, what emphasizes the better behaviour of martensitic steels compared to austenitic materials. Unalloyed beryllium and Be-0,4% Ca alloy exhibit at 700°C qualitatively similar behaviours. However, reactivity of Be-Ca alloy seems to be slightly lower.

For the future, this study will be completed by investigating the behaviour of lithium metazirconate and studying the influence of oxygen content.

Moreover, effect of irradiation will be also investigated during an in-pile test at 550°C during 2000 hours.

ACKNOLEDGEMENTS

This work has been carried out in the frame of the Commission of the European Communities Program on Fusion Technology (Task actions B14.3 and MAT.16). The authors wish to thank the CEC and CEA authorities for giving the possibility of conducting this research and publishing the results. They are also grateful to B. HOCDE for his participation in carrying out the tests.

REFERENCES

/1/ Gmelin Handbook of Inorganic Chemistry Ed. Springer Verlag (1986)

/2/ B. Rasneur, Fusion Technology (July 1986) 1909

/3 D. Vollath, H. Wedemeyer, E. Gunther, J. Nucl. Mater., 133-134 (1985) 221

/4/ T. Flament, P. Fauvet, J. Sannier, J. Nucl. Mater., 154 (1988)

/5/ P. Hofmann, W. Dienst, J. Nucl. Mater., 154 (1988)

/6/ H. Kittel, ANL 4937 (1949) 1-29

FUSION TECHNOLOGY 1988
A.M. Van Ingen, A. Nijsen-Vis, H.T. Klippel (editors)

CORROSION OF MARTENSITIC STEELS IN FLOWING 17Li83Pb ALLOY

T. FLAMENT, P. FAUVET, B. HOCDE, J. SANNIER

CEA, IRDI, DCAEA/SCECF/SECNAU, Centre d'Etudes Nucléaires, BP 6, 92265 Fontenay aux Roses, France

Corrosion of three martensitic steels - 1.4914, HT9 and T91 - in the presence of flowing 17Li83Pb is investigated in thermal convection loops Tulip entirely made of 1.4914 steel. Two 3000-hour tests were performed at maximal temperatures of respectively 450 and 475°C with a ΔT of 60°C and an alloy velocity of about 0.08 $m.s^{-1}$. In both tests, corrosion is characterized by a homogeneous dissolution of the steel without formation of a corrosion layer. Corrosion rate is constant and very temperature dependent : the sound-metal loss of 1.4914 steel is 22 $\mu m.year^{-1}$ at 450°C and 40 $\mu m.year^{-1}$ at 475°C. Behaviours of 1.4914 and HT9 steels are very similar whereas T91 steel is about 20 % less corroded.

1. INTRODUCTION

Utilization of lithium-lead eutectic alloy as tritium breeding material requires to solve the problem of its compatibility with the structural material. Previous compatibility experiments carried out with 316L austenitic steel have revealed a noticeable corrosion by flowing lithium-lead /1-3/. The attack is characterized by a strong nickel depletion which gives rise to the formation of a ferritic layer. This feature has led to consider martensitic steels which do not contain nickel.

This paper is devoted to the presentation of the first results gained with three martensitic steels in anisothermal flowing conditions.

2. EXPERIMENTAL

2.1. Test loop

The Tulip loop is a 0.5 x 0.5 m thermal convection loop (Fig.1) built of 1.4914 steel tube (internal diameter 22.7 mm). Heating along the lower horizontal leg is supplied by shielded resistances and cooling along the upper leg by water jackets. The expansion tank is used for the introduction of the lithium-lead ingot before testing and its tightness is ensured by a flange with metallic seal and weldable lips.

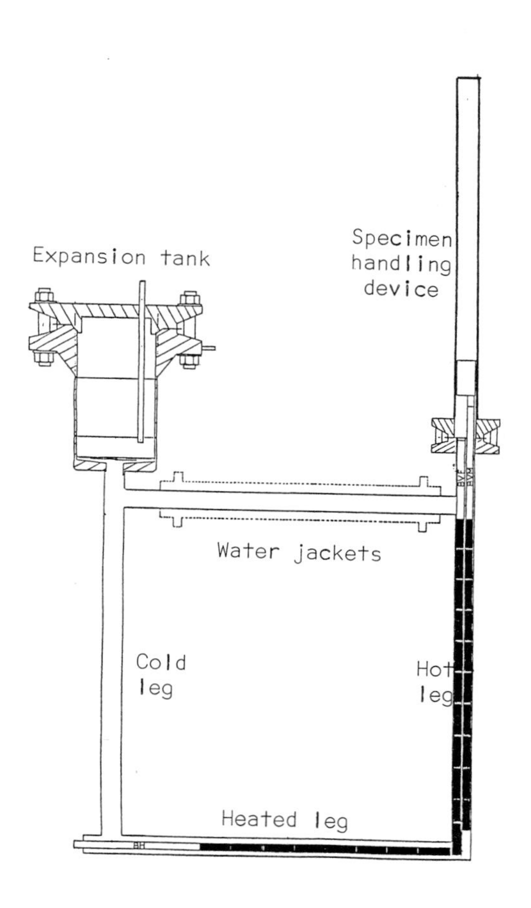

FIGURE 1
Tulip loop

Two rods of cylindrical specimens are immersed in the vertical isothermal hot leg. Above this leg, a specimen-handling device allows to partially remove one rod from the liquid alloy at given time intervals for achieving kinetics data ; a magnetic transmission avoids any tightness loss during the displacement of specimens. The second rod remains unmoved for revealing a possible downstream effect.

Another fixed rod is situated in the horizontal lower leg for determining the behaviour of martensitic steel in the heated zone.

After test, the loop is entirely dismantled.

2.2. Materials

The reference 1.4914 martensitic steel utilized as structural and also test material was supplied by KfK. After an appropriated heat treatment of normalization and tempering : (I) 2 hours at 970°C, (II) 0.5 hours at 1075°C, (III) 2 hours at 750°C, the structure is fully martensitic with a fine precipitation of niobium carbide but without ferrite

For comparison two other martensitic steels - T91 and HT9 - have been also investigated in the same metallurgical state. The composition of the different steels is given in Table I.

Table I - Composition of the steels (wt %)

Elements	1.4914	T91	HT9
C	0.13	0.1	0.21
Si	0.37	0.39	0.37
Mn	0.82	0.42	0.50
P	0.005	-	0.016
S	0.004	-	0.005
Cr	10.6	9	11.8
Ni	0.87	0.09	0.48
Mo	0.77	1	0.99
V	0.22	0.3	0.29
Nb	0.16	0.3	-
Al	0.054	-	-
Co	0.01	-	-
Cu	0.015	-	-
Zr	0.053	-	-
W	-	-	0.51

Specimens are 45 mm long and 6 mm in diameter cylinders. They are prepared by turning and simply degreased. Threads allow the assembling of the test rods.

Lithium-lead alloy used for these tests is the "standard" type alloy prepared in a casting laboratory of CEA for the European Laboratories. Its composition is given in table II.

Table II - Composition of 17Li83Pb eutectic alloy

Elements	Na	Se	Cr	Zn	As	Br	Ag	Sb	U
wt ppm	40	0.010	0.2	3	8	5	0.05	3	0.029

2.3. Operating conditions

The operating conditions of the two Tulip tests are summarized in table III.

Table III - Operating conditions of Tulip tests

Test	Hot leg temperature °C	Cold leg temperature °C	Test duration h	Velocity $m.s^{-1}$
I	450	390	1000 2000 3000	0.08
II	475	415	1000 3000	0.08

3. RESULTS AND DISCUSSION

3.1. Corrosion morphology in hot leg

At both temperatures, the three steels suffer a homogeneous dissolution. As shown on figure 2, no corrosion layer is visible. Moreover EDS analyses on cross-section do not show any chromium or iron depletion near the interface. The slight indentations observed at the surface on a depth of 5 to 10 µm are probably related to the presence of martensitic needles which may be more or less attacked according to their orientation.

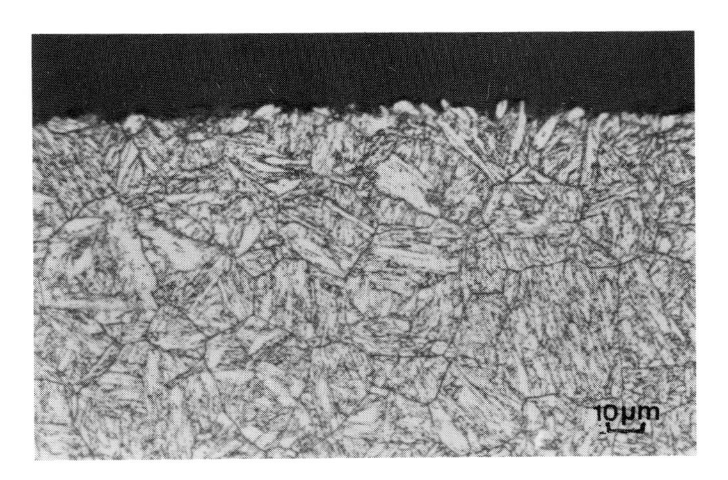

FIGURE 2
Cross-section micrograph of 1.4914 exposed 3000 hours at 475°C

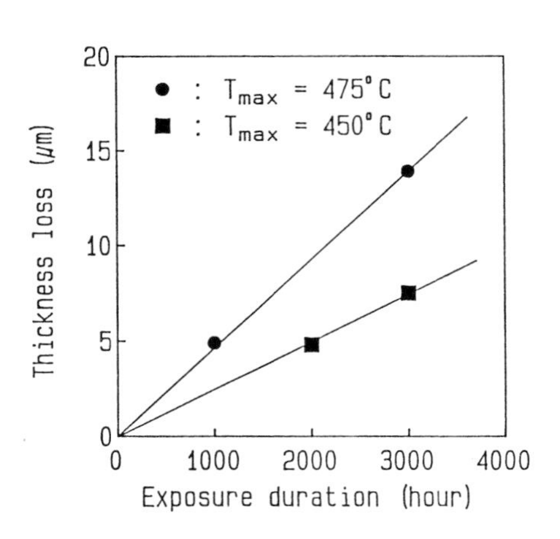

FIGURE 3
Kinetics of mass transfer

3.2. Corrosion kinetics

Weight loss and diameter variation of specimens determined after elimination of the superficial lithium-lead layer are in a good agreement and both can be used for calculating the corrosion rate. The values corresponding to the isothermal hot legs are summed up in Table IV.

Table IV
Thickness loss in the isothermal hot leg (µm)

Maximal temperature °C	Type of steel	Exposure duration 1000 h	2000 h	3000 h
475	1.4914	4.9	-	13.9
475	HT9	5.5	-	15.0
475	T91	4.7	-	11.1
450	1.4914	-	4.8	7.5
450	HT9	2.3	-	7.2
450	T91	2.4	4.7	6.4

. Influence of time

In both tests, corrosion kinetics of all the steels appear to be linear up to 3000 hours (Fig.3). Taking into account the corrosion morphology, a constant corrosion rate is not surprising and corroborates the results obtained by Borgstedt /4/ and Tortorelli /5/.

Extrapolation of the corrosion rate of 1.4914 martensitic steel leads to a thickness loss of 22 $\mu m.year^{-1}$ at 450°C and 40 $\mu m.year^{-1}$ at 475°C.

. Influence of temperature

On figure 4, the results obtained by different laboratories in monometallic loops are plotted versus inverse of temperature. The slope of the curve leads to an activation energy of 140 $kJ.mol^{-1}$ which is very closed to that achieved with 316L steel /3/.

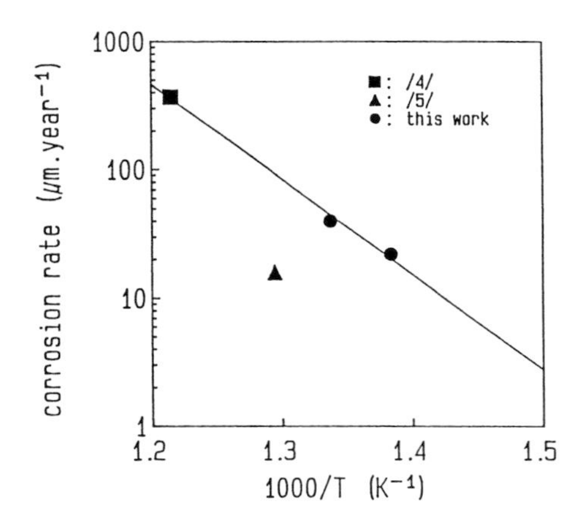

FIGURE 4
Influence of temperature on the mass transfer of 1.4914 steel in flowing 17Li83Pb

Our results are in a good agreement with Borgstedt's data /4/. On the other hand, the value achieved by Tortorelli at 500°C does not fit with this curve. This discrepancy might be due to the much lower velocity of lithium-lead in this test : 1 cm.s^{-1} compared to 8 cm.s^{-1} in Tulip loop and 30 cm.s^{-1} in Borgstedt's study.

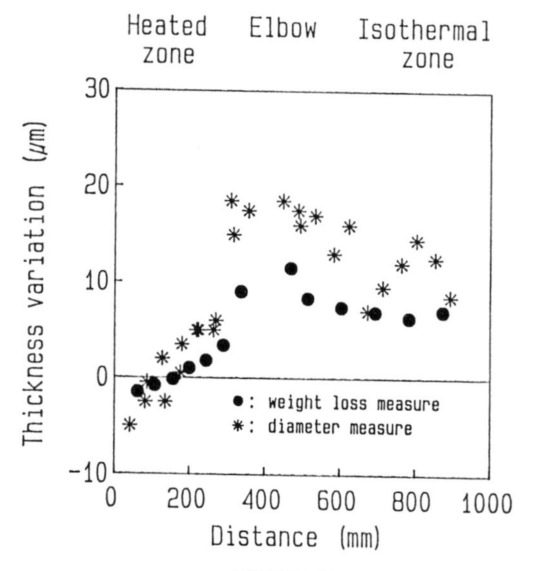

FIGURE 5
Thickness variation versus distance to the beginning of heated leg (450°C-test)

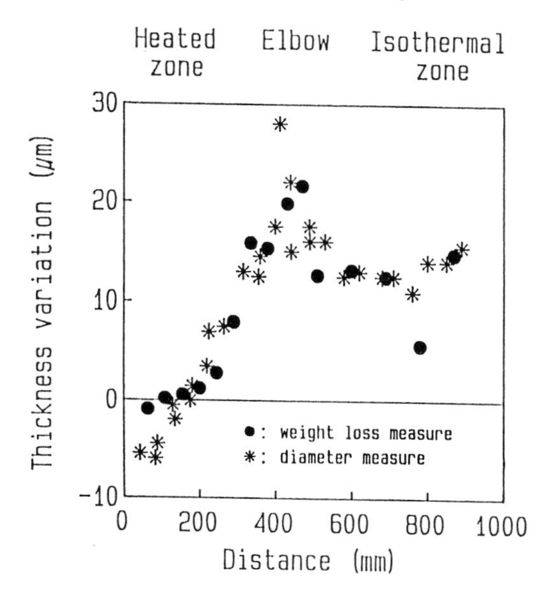

FIGURE 6
Thickness variation versus distance to the beginning of heated leg (475°C-test)

. Position effect

On figures 5 and 6 are summed up the thickness variations of specimens exposed during 3000 hours in the horizontal heated leg and in the vertical isothermal hot leg.

In both tests, curves have a similar aspect characterized by :

- a deposition zone in the first third of the heated zone ; deposits (Fig.7) are mainly constituted of iron with about 5 to 7 wt % chromium

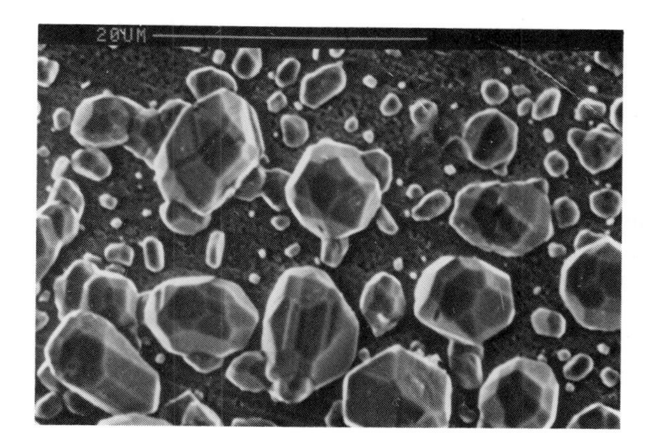

FIGURE 7
SEM image of deposits

- an increasing corrosion rate in the remaining part of the horizontal leg
- a maximal corrosion rate in the elbow
- a quasi-constant corrosion rate in the isothermal zone.

Two hypotheses may be suggested to explain the existence of a corrosion peak in the elbow :

- a downstream effect
- a hydraulic perturbation enhancing corrosion rate by reducing the superficial diffusion layer.

The sharp decrease of corrosion rate after the elbow and the quasi-constant corrosion rate in the isothermal hot leg are not consistant with a downstream effect.

So, the enhancement of dissolution by hydraulic perturbation seems to be more likely and is supported by the low corrosion rate obtained by Tortorelli at very low velocity.

. Influence of the type of steel

Though the composition of T91 steel is not very different and its corrosion morphology practically identical, this material appears to be 20 to 30 % less corroded than 1.4914 and HT9 steels which behave similarly.

4. CONCLUSION

In thermal convection loops at 450 and 475°C corrosion of martensitic steels is characterized by a homogeneous dissolution without formation of a corrosion layer.

Corrosion rate is constant and very temperature dependent. The sound-metal loss of 1.4914 steel is 22 $\mu m.year^{-1}$ at 450°C and 40 $\mu m.year^{-1}$ at 475°C. These values are four to five times lower than those corresponding to 316L stainless steel.

Interpretation of results suggests a possible effect of lithium-lead velocity on corrosion rate. This parameter is going to be investigated in Camille loop (turbulent flow) and in Celimene loop (laminar flow).

ACKNOWLEDGEMENTS

This work has been carried out in the frame of the Commission of the European Communities Program on Fusion Technology. The authors wish to thank the CEC and CEA authorities for giving the possibility of conducting this research and publishing the results. They are also grateful to H. Redureau and P. Perodeaud for their participation to the study.

BIBIOGRAPHIE

/1/ M. Broc, P. Fauvet, T. Flament, J. Sannier, J. Nucl. Mater., 141-143 (1986) 611

/2/ M. Broc, P. Fauvet, T. Flament, J. Sannier, J. Nucl. Mater., 154 (1988)

/3/ M. Broc, P. Fauvet, T. Flament, A. Terlain, J. Sannier, Proceeding LIMET 88, Avignon

/4/ H.V. Borgstedt, G. Dreschsler, G. Frees, Z. Perie, J. Nucl. Mater., 154 (1988)

/5/ P.F. Tortorelli, J.H. De Van J. Nucl. Mater., 141-143 (1986) 592

FUSION TECHNOLOGY 1988
A.M. Van Ingen, A. Nijsen-Vis, H.T. Klippel (editors)
Elsevier Science Publishers B.V., 1989

MEASUREMENTS OF THE EFFECTIVE THERMAL CONDUCTIVITY OF A BED OF Li_4SiO_4 PEBBLES

Mario DALLE DONNE* and Giancarlo SORDON

Association KfK-EURATOM. Kernforschungszentrum Karlsruhe, Institut für Neutronenphysik und Reaktortechnik, Postfach 3640, D-7500 Karlsruhe, Federal Republic of Germany

The Karlsruhe ceramic breeder design for a Demo relevant blanket is based on a concept where the breeder material is in form of a bed of 0.5 mm Li_4SiO_4 pebbles. The pebble bed is placed in 6 mm wide gaps between beryllium plates. Helium at 0.1 MPa flows through the bed and carries away the produced tritium. Data on the effective thermal conductivity of the bed and on the heat transfer coefficient at the bed walls are required for the proper design of the blanket. The measurements have been performed for a bed of 0.5 mm Li_4SiO_4 pebbles with stagnant helium. The bed in contained in an annular space between two concentric cylinders (R_1=3 and 8 mm, R_2=51 mm). The heat produced electrically in the inner cylinder flows in radial direction. The measurements have been performed for bed average temperatures up to 330°C and maximum wall temperatures of 450°C. The agreement of the obtained experimental values with general correlations is satisfactory.

1. INTROCUTION

The Karlsruhe ceramic breeder design for a Demo relevant blanket is based on a concept where the breeder material is in form of a bed of 0.5 mm lithium orthosilicate (Li_4SiO_4) pebbles[1]. A previous design was based on a mixture of Li_4SiO_4 and beryllium pebbles[2]. In both cases the pebbles are surrounded by flowing helium which carries away the tritium produced in the Li_4SiO_4 pebbles. However the helium velocity is so small (≈ 30 cm/sec)[1] that the effective thermal conductivity of the bed is not affected by the helium flow and the bed behaves like a stagnant bed[3]. The heat produced in the orthosilicate and in the beryllium is carried away by means of cooling tubes containing high pressure helium.

Reliable data on the heat transfer parameters of such pebble beds are required for the design of these blankets. If the diameter of the pebbles is considerally smaller than the dimensions of the bed, the pebble bed can be treated as an homogeneous medium and the heat transfer parameters can be reduced to two coefficients: the effective thermal conductivity of the bed and the heat transfer coefficient at the walls of the bed container. Various correlations for the two coefficients are available from the literature. Two models for the effective thermal conductivity have been considered in the frame of the present work, as they seem most relevant to the kind of bed we are proposing:

1. the model of Schlünder, Zehner and Bauer[4,5], valid for beds of pebbles of any shape and size distribution, but with the same thermal conductivity of the solid, and

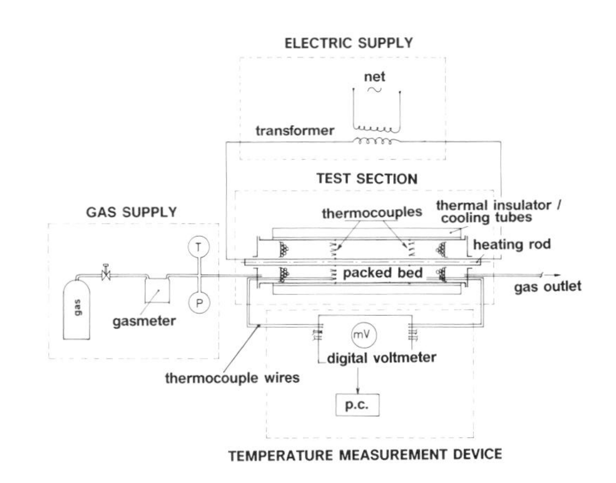

FIGURE 1
Schematic representation of the experimental apparaturs

---

* Delegated from EURATOM to Kernforschungszentrum Karlsruhe

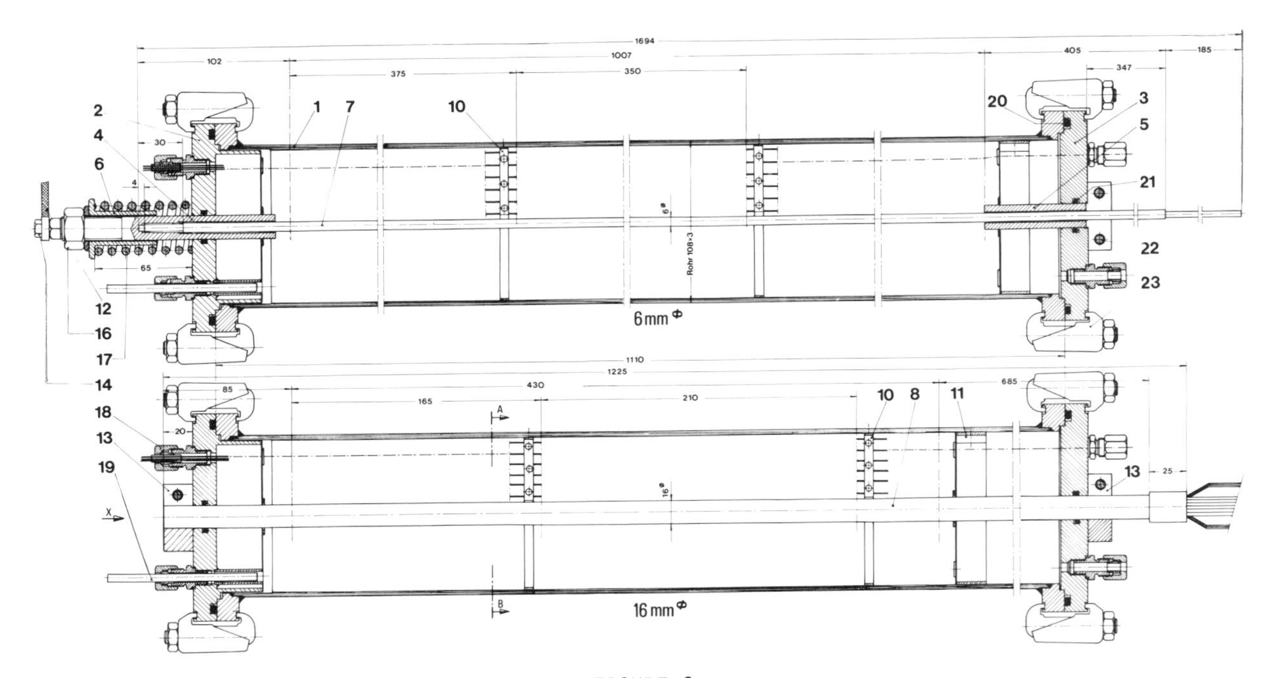

FIGURE 2
The two test sections used, with an inner cylinder of 6 mm and of 16 mm respectively (dimensions are in millimeters)

2. the model of Okazaki et al.[6], valid for beds of spherical pebbles with the same or two different thermal conductivities.

The models of Hennecke[7] and of Yagi and Kunii[8] have been used for comparison with the experimental values of the wall heat transfer coefficients obtained in the present investigation.

The use of these correlations requires the knowledge of certain empirical parameters, thus it was decided to perform some experiments to obtain more precise information on the heat transfer parameters for the beds of particular interest to us.

2. EXPERIMENTAL APPARATUS

Fig. 1 shows schematically the experimental apparatus. The pebble bed is contained between two concentrical cylinders. The inner cylinder contains an electrically heated rod. The gas (helium or argon) can flow in axial direction through the bed. Fig. 2 shows the two test sections used, which differ essentially only for the size of the inner cylinder (6 and 16 mm diameter respectively). The radial distribution of the temperature in the bed is measured in two axial positions in the central part of the test section, where the axial temperature gradients are negligible in comparison to the radial ones, by means of two banks of 32 thermocouples each, placed at various radii at four different azimuthal angles (Fig. 3). Furthermore the temperatures on the inner and outer cylinder surfaces are measured by thermocouples placed into the walls. The temperature level of the bed has been varied by cooling the outer test section surface (Fig. 4) or by surrounding it by a thermal insulation of variable thickness.

The measurements were performed with flowing or stagnant helium or argon. The investigated beds were made of aluminium oxide (1,2,4 mm in diameter), of aluminium (2 mm diam.), steel (2 and 4 mm diam.) and orthosilicate (0.5 mm diam.) pebbles. Binary mixtures of aluminium oxide pebbles of different diameters or of aluminium

FIGURE 3
View of a bank of 32 thermocouples for the measurement of the bed radial temperature distribution at an axial position of the test section.

FIGURE 4
Outside view of test section showing the water cooling tubes of the outer cylinder.

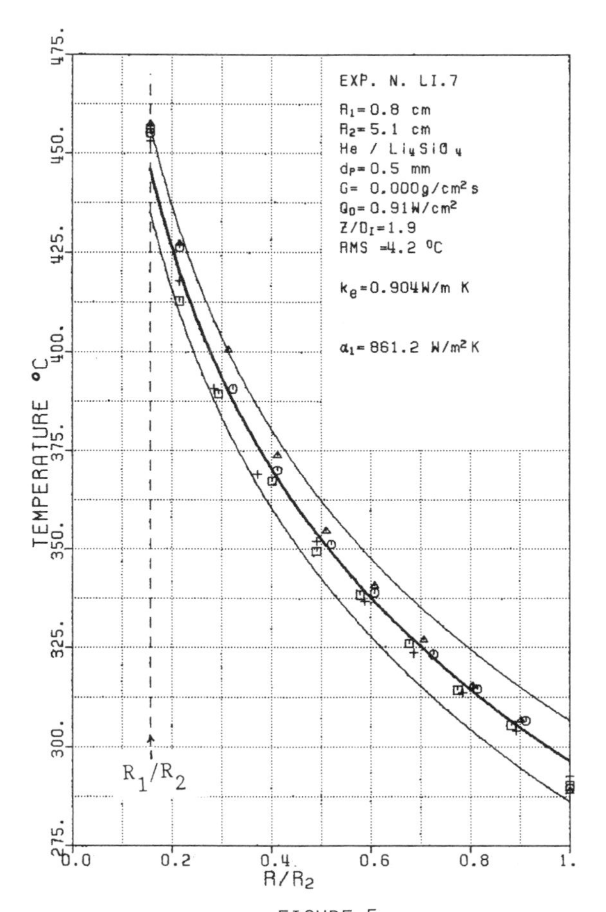

FIGURE 5
Radial temperature distribution in the bed.

oxide either with aluminium or with steel pebbles were also investigated[3].

3. EXPERIMENTAL RESULTS

The results of all experiments are given in Ref.3, here we will limit ourselves to show only those directly relevant to our blanket designs. Fig. 5 shows a typical radial temperature distribution in the bed of 0.5 mm Li_4SiO_4 pebbles in presence of stagnant helium, for a given axial position along the test section. Also shown are the measured temperatures at the inner cylinder surface ($R=R_1$) and at the outer cylinder surface ($R=R_2$). With a constant radial heat flux and bed thermal conductivity and with stagnant helium, the theory predicts a straight line in a semilogarithmic diagram temperature versus log R/R_2. Fig. 6 shows that the experimental points are correlated quite well by a straight line in such a diagram. This was always the case for all the tests with stagnant gas performed during this experiment[3]. The slope of the temperature curve allows to calculate the effective thermal conductivity of the bed. The difference between the measured temperature on the inner cylinder wall and the extrapolated temperature from the bed to the wall (Fig. 6) allows to calculate the heat transfer coefficient at the walls. This temperature difference is always smaller at the outer cylinder surface,

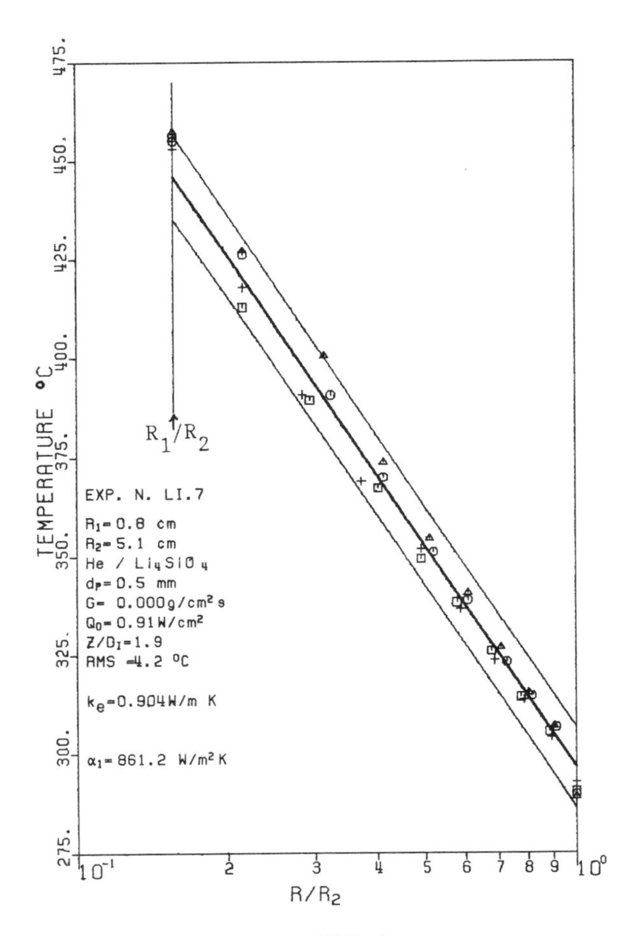

FIGURE 6
Radial temperature distribution in the bed. Semilogarithmic diagram.

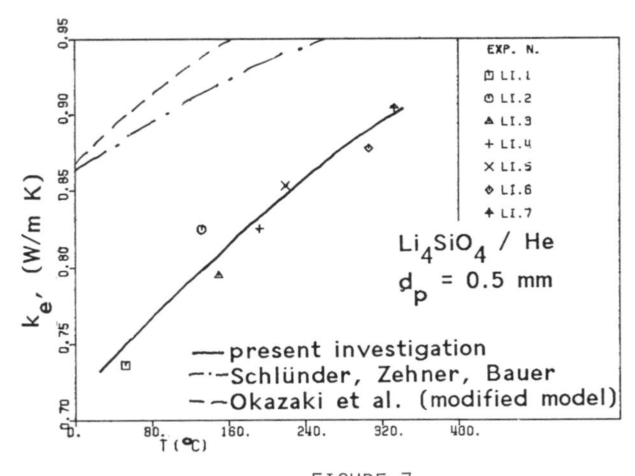

FIGURE 7
Effective thermal conductivity of the bed of 0.5 mm Li_4SiO_4 pebbles and stagnant helium.

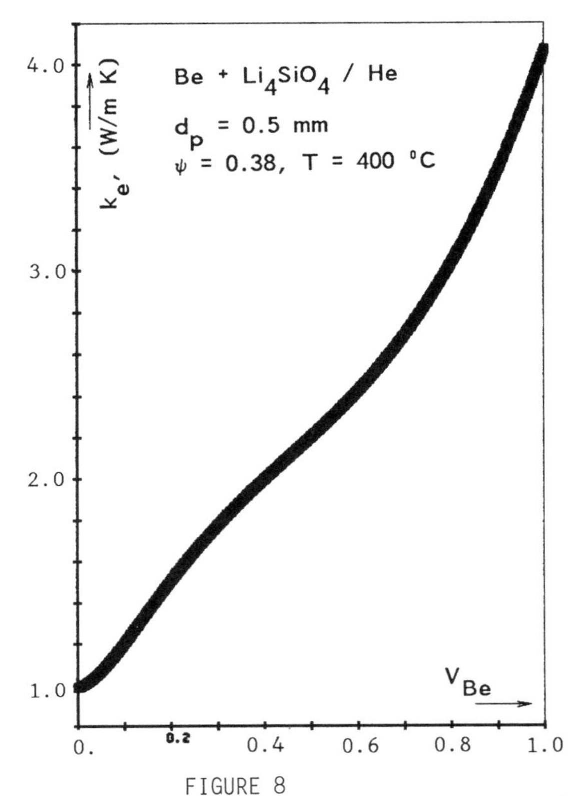

FIGURE 8
Effective thermal conductivity of a mixture of 0.5 mm beryllium and Li_4SiO_4 pebbles in stagnant helium.

so that the obtained heat transfer coefficients for the outer cylinder scattered considerally more than those for the inner cylinder and were not used for the correlation[3].

Fig. 7 shows the measured effective thermal conductivity of the bed formed of 0.5 mm Li_4SiO_4 pebbles in stagnant helium. At 400°C the thermal conductivity is about 0.92 W/mK. The agreement with the predictions of Schlünder, Zehner and Bauer[4,5] and Okazaki et al.[6] is reasonable.

Table 1 shows the measured heat transfer coefficients at the wall of the inner cylinder. In the case of direct contact between pebble bed and wall the heat transfer coefficient is about 0.45 W/cm^{2}K. Experiments were also performed with a thin (0.1 mm) stainless steel wire gauze placed between the cylinder wall and the bed, because this is the solution adopted for the present KfK design[1]. In this case the scatter of the experimental data is larger. The value 0.1 W/cm^2K is recommended.

Fig. 8 shows the thermal conductivity of mixtures of 0.5 mm beryllium and Li_4SiO_4 pebbles in stagnant helium for a bed porosity of 38% and a temperature of 400°C as a function of the beryllium volume fraction. These data have been calculated on the base of the experiments performed with beds made of mixtures of aluminium (as a simulation for the beryllium) and of aluminium oxide (for the Li_4SiO_4) pebbles and of a correlation for mixed beds suggested in Ref. 3. For a mixed bed of beryllium and Li_4SiO_4 pebbles with a beryllium pebble fraction of 80% (optimum from a neutronic point of view) the thermal conductivity increases to 3 W/mK.

ACKNOWLEDGEMENT

This work has been performed in the framework of the Nuclear Fusion Project of the Kernforschungszentrum Karlsruhe and is supported by the European Communities within the European Fusion Technology Program.

REFERENCES

1. M. Dalle Donne et al., Fusion Technology, in print.

2. M. Dalle Donne, U. Fischer and M. Küchle, Nuclear Technology 71 (1985) 15.

3. G. Sordon, Über den Wärmetransport in Kugelschüttungen, Kernforschungszentrum Karlsruhe, report KfK-4451, EUR 11391 DE (1988).

4. P. Zehner, E.U. Schlünder, Chemie-Ing.-Techn. 42, 14 (1970) 933.

5. R. Bauer, E.U. Schlünder, Int. Chem. Eng.18,2 (1978) 181.

6. M. Okazaki, T. Yamasaki, S. Gotoh and R. Toei, J. Chem. Eng. Japan, 14,3 (1981) 183.

7. F.W. Hennecke, Über den Wandwiederstand beim Wärmetransport in Schüttungsrohren, Ph-D Thesis, University of Karlsruhe, Karlsruhe, Germany.

8. S. Yagi and D. Kunii, A.I.Ch.E. Journal, 6,1 (1960) 97.

TABLE I

Wall heat transfer coefficient (W/cm^2K) for a bed of 0.5 mm Li_4SiO_4 pebbles and stagnant helium

Experiment	Li_1	Li_2	Li_3	Li_4	Li_5	Li_6	Li_7
With direct contact between pebbles and wall	0.53	0.42					
With the steel wire gauze at the wall			0.19	0.097	0.084	0.11	0.086

FUSION TECHNOLOGY 1988
A.M. Van Ingen, A. Nijsen-Vis, H.T. Klippel (editors)

HELIUM COOLED CERAMIC BREEDER IN TUBE BLANKET FOR A TOKAMAK REACTOR: THE COAXIAL POLOIDAL MODULE CONCEPT AND DESIGN

L. Anzidei, M. Gallina, L. Petrizzi, V. Rado, G. Simbolotti, V. Zampaglione

Associazione EURATOM-ENEA sulla Fusione, Centro Ricerche Energia Frascati,
C.P. 65 - 00044 Frascati, Rome, Italy

V. Violante

ENEA, Centro Ricerche Energia Casaccia,
Via Anguillara, 301 - 00060 S. Maria di Galeria, Rome, Italy

S. Bassani (ENEA Guest)

The paper presents the results of the 3-D stress analysis and neutronics of the helium-cooled ceramic breeder coaxial poloidal modules blanket for a tokamak reactor. The blanket modules form a compliant structure with acceptable values of the thermomechanical stresses. A 3-D value of the tritium breeding ratio of 1.22 has been calculated.

1. INTRODUCTION

The ENEA ceramic breeder coaxial poloidal module blanket for a tokamak reactor is shown in Figs. 1, 2. The blanket consists of tubular poloidal modules made of a central bundle of ceramic rods ($\gamma LiAlO_2$) with coaxial distribution of the inlet/outlet coolant flow (helium) surrounded by multiplier material (beryllium) in the form of bored bricks. A first study was presented at the ISFNT, Tokyo[1]. The present paper is concerned with detailed 3-D thermo-mechanical analysis and neutronics calculations

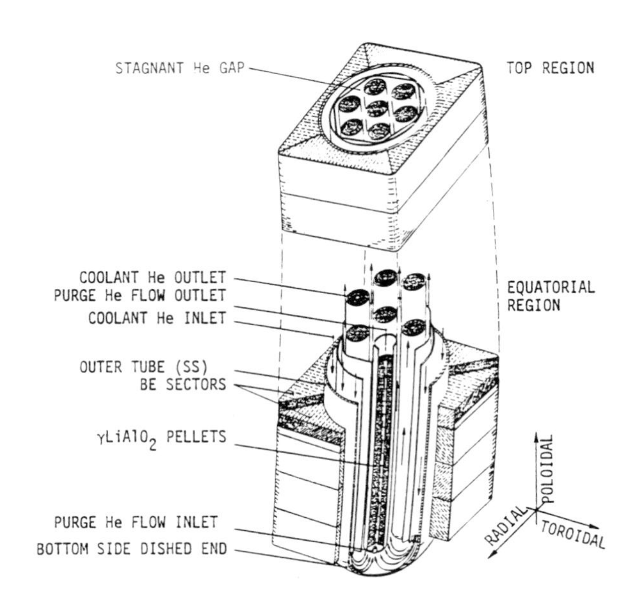

FIGURE 1
The poloidal blanket module

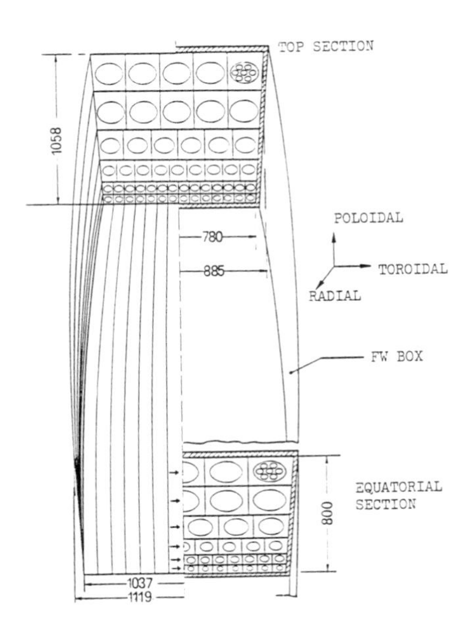

FIGURE 2
The poloidal module assembly

under the following operating conditions:

Thermomechanics

- max power density ($\gamma LiAlO_2$, Be, SS AISI316)	46,12,20 MW/cm^3
- coolant He pressure	5 MPa
- coolant He inlet/outlet temperature	250/570 °C

Neutronics

- average reactor wall load	2 MW/m^2

2. THERMOMECHANICS

The distribution of the coolant helium temperature has been calculated as a function of the average stagnant helium gap thickness s (see Fig. 1). A linear poloidal distribution of the nuclear heating has been assumed with a reduction factor of 0.56 at the top blanket section. The results are shown in Fig. 3. Accordingly, a value of 1.5 mm has been selected for the reference gap leading to a He temperature in the reverse flow zone of 445 °C, that is slightly above the minimum achievable (415 °C under adiabatic conditions). Assuming such a value of the He temperature, a thermal analysis of the cross-section of a first row outboard blanket module has been performed in the reverse flow zone by means of the DELFINE code of the CASTEM system. The resulting maximum temperatures are (see Fig. 4): 550 °C for Be, 510 °C for SS AISI 316, 630 °C for $\gamma LiAlO_2$.

The stress analysis of the pressure tube has been performed by means of the 2-D and 3-D codes INCA and BILBO, respectively, of the CASTEM system.

An important part of this study has been concerned with the definition of mechanical constraints allowing the conservation of the geometry with acceptable values of the stresses. To this aim, an axisymmetric analysis (straight tube) has been used. This has been possible because of the negligible effect of the poloidal curvature of the system. In fact, the displacements of the tube due to the conservation of its curvature lead to low flexural stresses (about 3 MPa). Secondly, the load on the tube due to the different thermal expansions of the breeder rod bundle amounts to

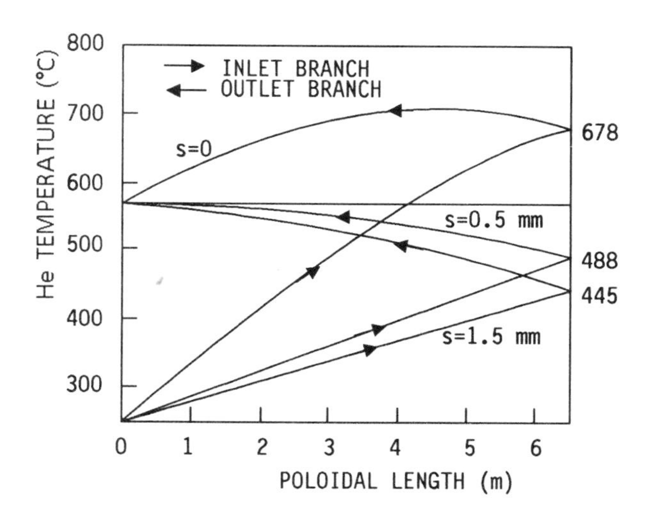

FIGURE 3
Inlet and outlet He temperature distribution along the poloidal length for different gap thicknesses

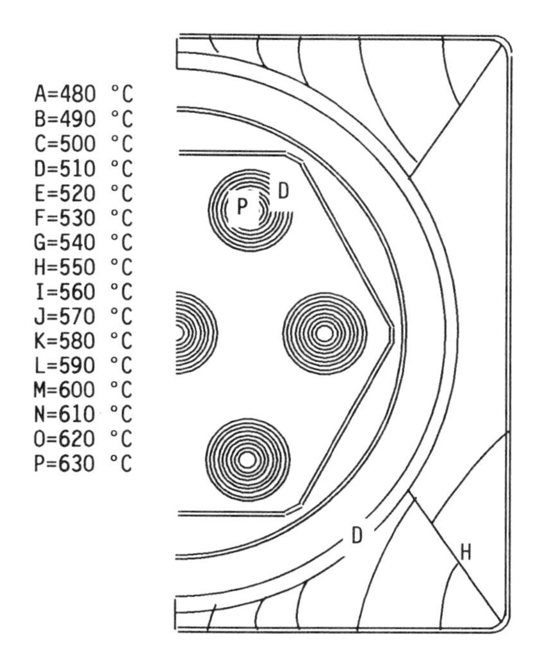

FIGURE 4
Temperature distribution in the cross-section of the first row blanket module (reverse flow zone)

1 N only. Finally, the Be bricks make up a flexible system because of their poloidal discontinuity thus not producing any load on the tube.

According to the above considerations, the poloidal blanket module is a compliant structure and, therefore, the stress analysis of the pressure tube can be performed by using as input only the pressure load and the thermal load (based on a temperature distribution obtained by means of the previous thermal analysis).

Figure 5 shows three types of constraints that have been considered. Type a constraint leads to a displacement of 3 cm, while too high stresses (300 MPa) on the bottom dished end are obtained under type b constraints. The reference case is represented by constraints of type c leading to a maximum displacement of 0.35 cm and to maximum values of the Von Mises stresses of 120 MPa in the pressure tube and 176 MPa at the bottom dished end (Figs. 6,7). The 3-D results confirm that the tube behaviour is not affected by its radius of curvature.

As a conclusion, the stresses in the pressure tube have acceptable values, while a further detailed design is required for the bottom dished end in order to reduce the stress level there (to, e.g., 120 MPa).

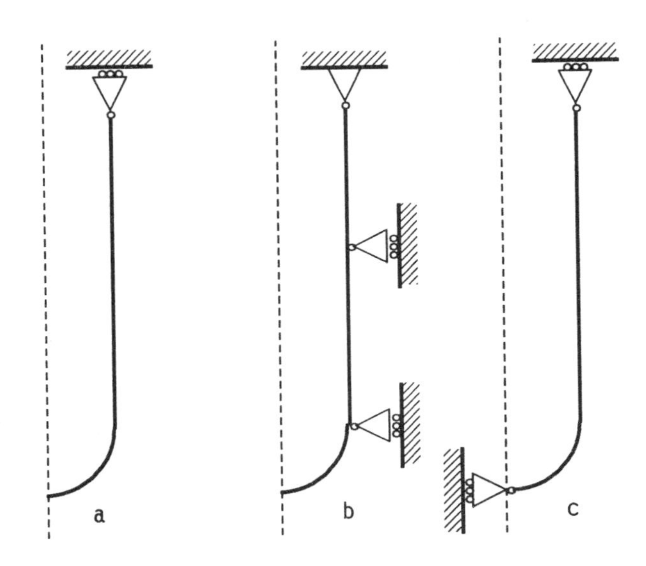

FIGURE 5
Mechanical constraints (reference case: c)

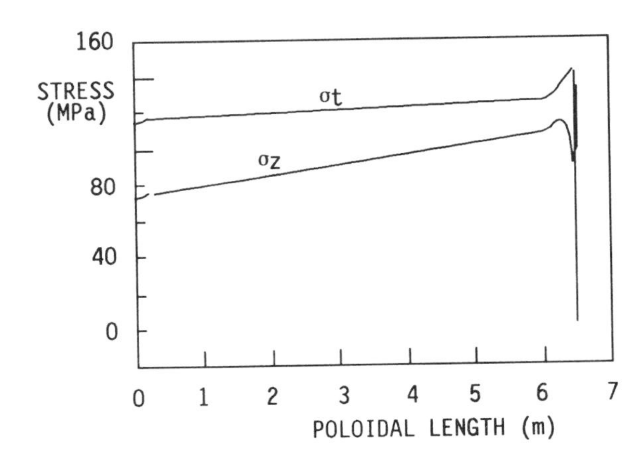

FIGURE 6
Membrane-flexure stress in the pressure tube (tangential σ_t, longitudinal σ_z)

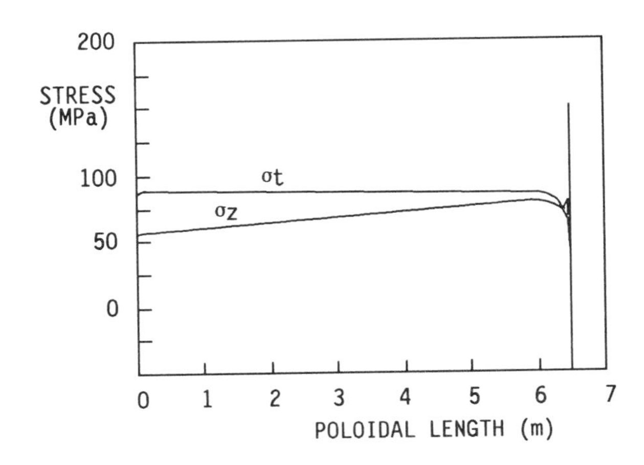

FIGURE 7
Membrane+flexure stress in the pressure tube (tangential σ_t, longitudinal σ_z)

3. NEUTRONICS

A 3-D neutronics analysis has been performed for the coaxial poloidal module blanket by means of the MCNP-3A Monte Carlo code. The fusion neutron source has been sampled from a D-shaped distribution as for the NET-DN configuration[1] but in the reactor geometry[2]. The real local features of the breeder rod bundle cannot be

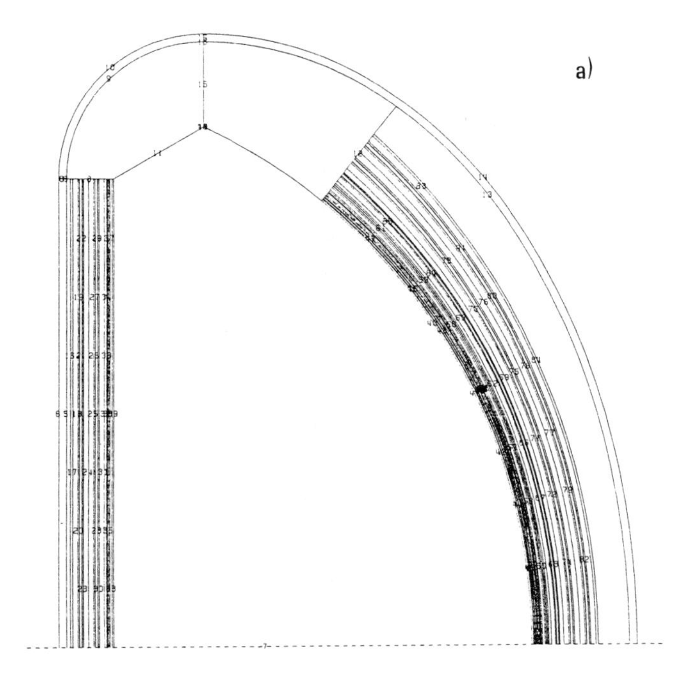

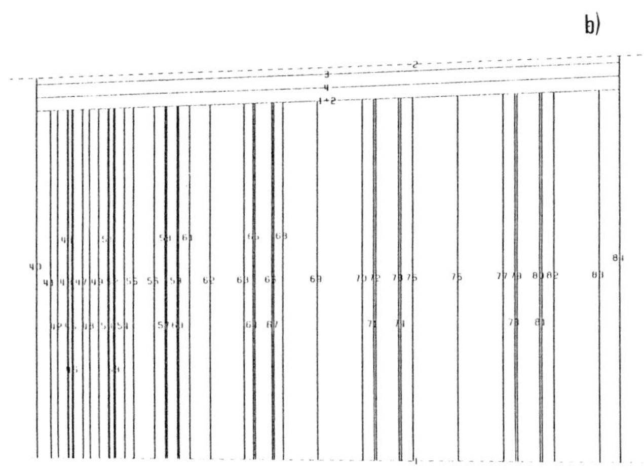

FIGURE 8
MCNP reactor geometry (poloidal and midplane outboard cross-section)

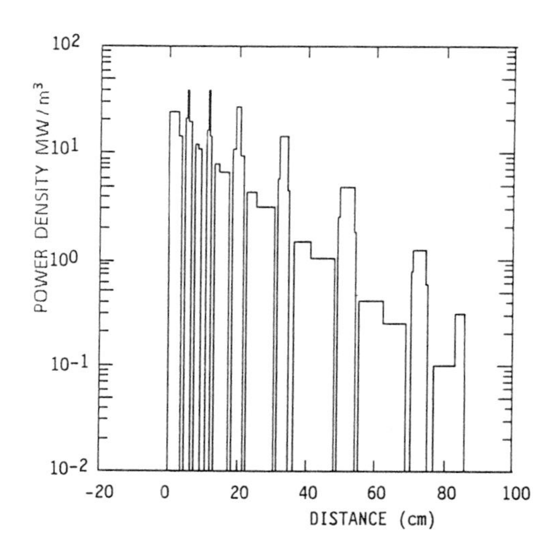

FIGURE 9
Radial distribution of the power density on the midplane (outboard)

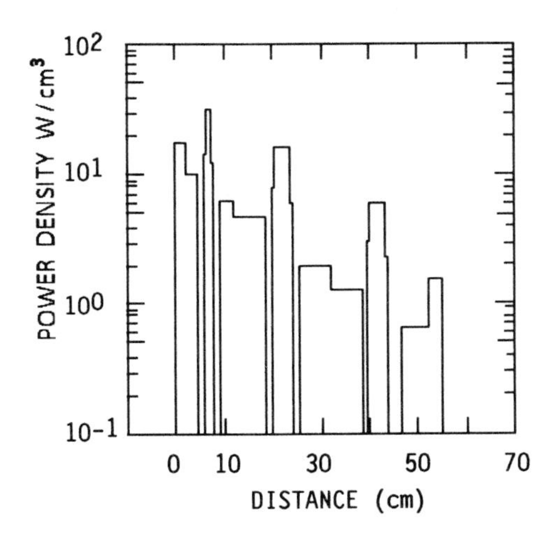

FIGURE 10
Radial distribution of the power density on the midplane (inboard)

exactly represented in the Monte 3-D toroidal geometry. A model has been adopted where the blanket features are approximately described by separate radial shells for each material and for each blanket tube row (see Fig. 8). The boundary cylinders of such shells have radii and centers so as to conserve the same area cross-section for each material on the midplane and top cross-sections according to the poloidal module assembly adopted[2] (see Fig. 2).

The reactor source intensity has been normalized to an average neutron outboard wall load of 2 MW/m^2, that corresponds to a fusion plasma power of 3350 MW. A 3-D value of the tritium breeding ratio of 1.22 has been calculated (outboard: 0.88, inboards 0.34). Power density and tritium production have been calculated for each radial shell divided into

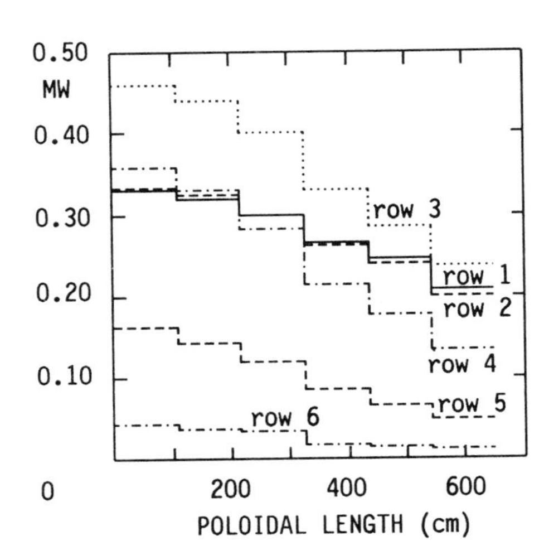

FIGURE 11
Power in the poloidal sectors for each blanket row (outboard)

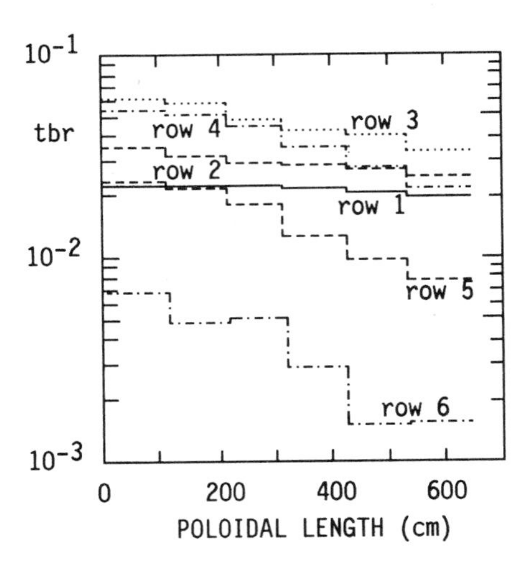

FIGURE 13
Contribution to the overall tritium breeding ratio by poloidal sectors and radial rows (outboard)

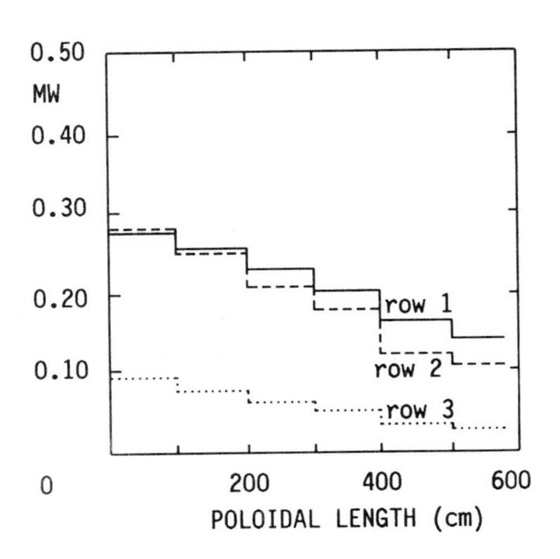

FIGURE 12
Power in the poloidal sectors for each blanket row (inboard)

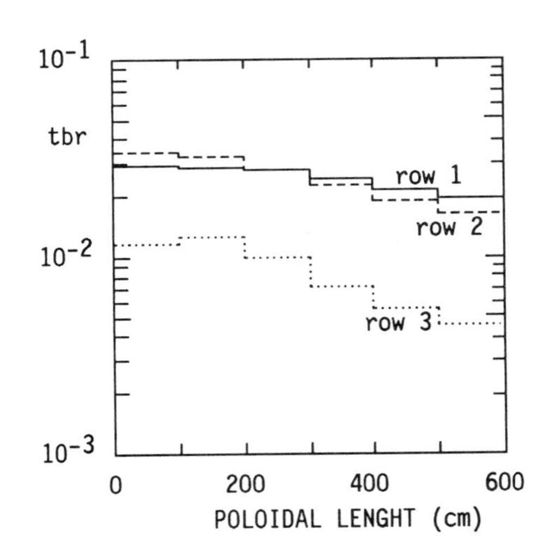

FIGURE 14
Contribution to the overall tritium breeding ratio by poloidal sectors and radial rows (inboard)

six poloidal sectors. The major results are summarized in Figs. 9-14. It can be seen that the calculated power density in the breeder material is below that assumed for the thermal-hydraulic design and based on previous 3-D calculations on *Il Mantello* blanket.

REFERENCES

1. A. Amici et al., A ceramic breeder in poloidal tube blanket for a tokamak reactor, 1st ISFNT, Tokyo, April 1988.

2. M. Gallina et al., A 3-D Monte Carlo neutronic analysis of the first wall, blanket and shield for NET, 12th Symp. on Fusion Enginnering, Monterey, October 1987.

FUSION TECHNOLOGY 1988
A.M. Van Ingen, A. Nijsen-Vis, H.T. Klippel (editors)

CERAMIC BOT TYPE BLANKET WITH POLOIDAL HELIUM COOLING

A. CARDELLA[1], W. DAENNER[1], M. ISELI[1], M. FERRARI[2], M. GALLINA[2], V. RADO[2], G. SIMBOLOTTI[2], V. VIOLANTE[2]

[1] The NET Team c/o Max-Planck-Institut fuer Plasmaphysik, Boltzmannstr. 2, D - 8046 Garching

[2] Associazione Euratom-ENEA sulla Fusione, Centro Ricerche Energia Frascati, C.P. 65, I - 00044 Frascati, Rome

This paper briefly describes the work done and results achieved over the past two years on the ceramic breeder BOT blanket with poloidal helium cooling. A conclusive remark on the brick/plate option described previously is followed by short descriptions of the low and high performance pebble bed options elaborated as alternatives for both NET and DEMO. The results show, together with those about the poloidal cooling of the First Wall, good prospects for this blanket type provided that the questions connected with an extensive use of beryllium find a satisfactory answer.

1. INTRODUCTION

For a couple of years, the NET Team , in close cooperation with ENEA Frascati, has followed a ceramic breeder blanket of the BOT (Breeder Outside Tube) type with poloidal helium cooling. Initially, the work focussed on design options suitable for application as a driver blanket for NET. These were characterized by a modest tritium breeding capability resulting from a modest utilization of beryllium multiplier. This restriction was suggested by the absence of an adequate experimental effort on this material in the European Fusion Technology Programme. Recently, the studies were reoriented towards more reactor relevant versions with a more aggressive use of beryllium. Such blankets meet more closely the requirements of test blankets for NET. This paper briefly describes the work done and the results obtained during the last two years.

2. THE BRICK/PLATE OPTION

The design option in which the ceramic was intended to be used in form of bricks or plates has been extensively described in[1,2]. A key question which could not be answered so far was what stresses will occur in the ceramic and how they compare with the material properties. Fig. 1 shows as an example[3] the calculated stress distribution in a first row brick which is loaded by a linearly decreasing power density with a maximum value of 20 W/cm^3. Stresses of > 200 MPa occur in the ceramic which are

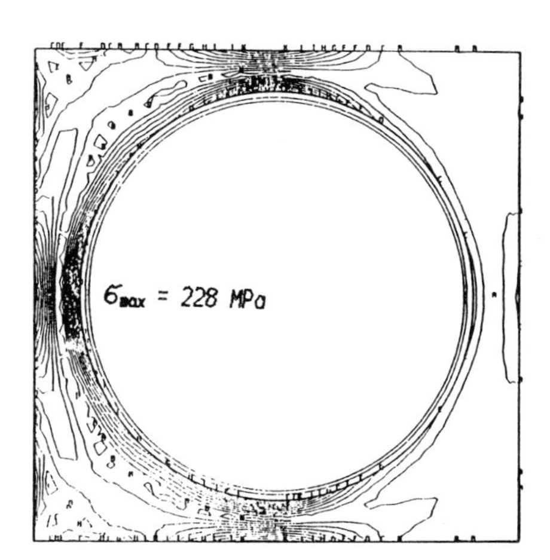

FIGURE 1: Stress Distribution in a First Row Brick/Plate

incompatible with the < 100 MPa tensile strength limit found experimentally[4]. Bricks or plates could be applicable only below 5 W/cm^3. The required improvement in strength could be obtained with whisker reinforced lithium ceramics, but an extensive experimental programme would be needed.

New comparative neutronics calculations with ANISN and MCNP allowed to resolve the big discrepancy between the two methods of about 50 % in the breeder power density in the layer closest to the beryllium multiplier: ANISN overpredicts the heat deposition if steel is mixed to the ceramic and selfshielding is neglected.

3. THE LOW PERFORMANCE PEBBLE BED OPTION

The problems arising from the mechanical behaviour of ceramic bricks/plates suggested to study a pebble bed option in parallel. The investigations resulted in the design of an outboard segment the top/bottom crossection of which is shown in Fig. 2. The pebble bed is preceded by a thin beryllium layer which is attached to the First Wall and cooled by its cooling system. The beryllium surface exposed to the pebble bed is of corrugated shape in order to better cope with the high power densities. Within the pebble bed region, coolant tubes of equal size and standard dimensions are arranged in 8 rows, the number of tubes per row decreasing according to the decreasing power density. The pitch of the tubes is kept constant along the poloidal direction. The remaining toroidal space close to the midplane region can be utilized by one or two additional coolant tubes of shorter poloidal length. The complications in the design and in particular in the containment of the pebble bed has to be weighted against the enhancement of the tritium breeding ratio. As temperature calculations have shown, an uncooled container as shown in Fig. 2 would need to be fabricated from a refractory alloy. An alternative would be an actively cooled steel container which could be part of the segment box, but would require a complicated shape of the coolant tubes.

The performance of this blanket has been analyzed in some detail. The main results are reported in the summary Table I.

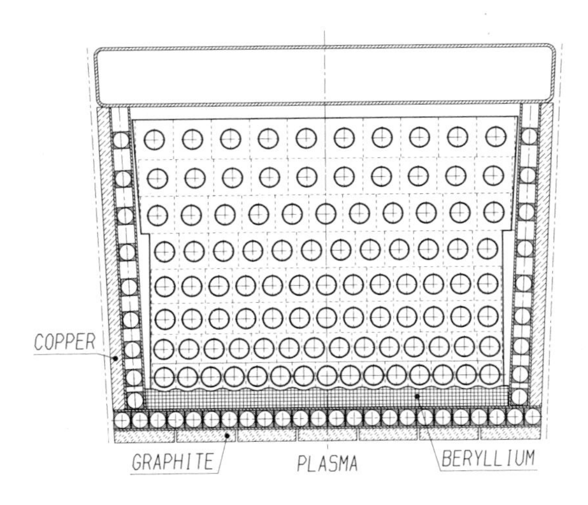

FIGURE 2: Top/Bottom Crossection

4. HIGH PERFORMANCE PEBBLE BED OPTIONS

The most significant modifications made for the design of a high performance blanket are the use of a much bigger amount of beryllium multiplier and its application for transferring the heat from the breeder ceramic to the coolant tube.

4.1 The First Approach

In the first version, the beryllium is used in the form of 50 mm thick poloidal plates (or a series of bars) fabricated in two halves and joint to the poloidal coolant tubes. The diameter and the number of tubes are the same

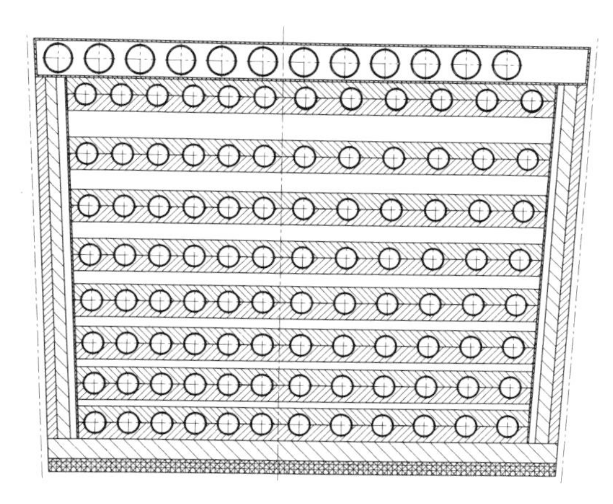

FIGURE 3: Top/Bottom and Midplane Crossection of High Performance Option - First Approach

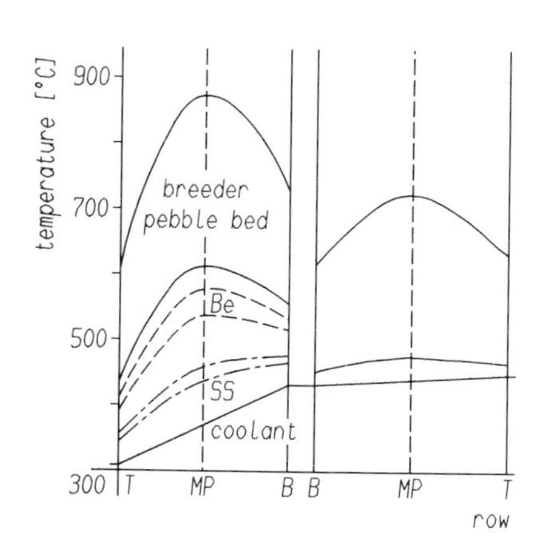

FIGURE 4: Poloidal Temperature Variation - Circuit I (Rows 1 + 8)

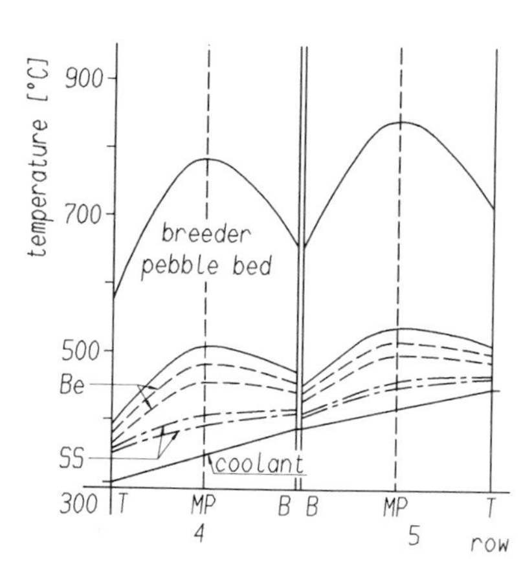

FIGURE 5: Poloidal Temperature Variation - Circuit IV (Rows 4 + 5)

in all 8 plates. Fig. 3 shows the top/bottom crossection on left and the midplane crossection on the right hand side, respectively. The pitch of the tubes varies along the polodial length. This can be afforded because the heat flow inside the beryllium is largely equilibrated as a consequence of its excellent thermal conductivity. The thickness of the pebble bed layers in between the plates increases towards the rear according to the radially decreasing breeder power density. The elimination of the coolant tubes from the pebble bed has the advantage that less breeder material is operating in the lower temperature range. Additionally, the smaller radial extension of the breeder layer allows to use an uncooled steel container at the toroidal edges. The purge gas is assumed to penetrate the bed in toroidal instead of poloidal direction, thus reducing the pressure drop.

A special feature is the use of U-tubes which allows to thermally connect the first and eighth row, the second and seventh row, and so on. This leads to a well balanced heat load distribution of all 4 coolant circuits. Figs. 4 and 5 show the temperature distribution along the poloidal length for circuit numbers I (row no. 1 and 8) and IV (row no. 4 and 5). The main parameters are summarized in Table I.

4.2 Second Approach

This approach is governed by the idea to implement the neutronically optimum ratio of 4:1 of beryllium multiplier and breeder material throughout the blanket. This requires an individual optimization of each radial layer. As a consequence, different tube diameters and a different number of tubes in each row have to be used as can be seen from

Fig. 6. As compared with the first approach this solution necessitates coolant tube manifolding and collection at both poloidal ends. The advantage is to be seen, as is shown in Fig. 7, in a well equilibrated temperature distribution in radial direction which facilitates the problems of differential dilatation in poloidal direction.

In addition to the design shown in Fig. 6 an alternative version has been studied in which the ratio of beryllium and ceramic varies from 4 : 1 in the front to 0.5 : 1 in the rear of the blanket. In this case, more tubes are required, the tritium breeding ratio is reduced, but there is a significant saving in beryllium inventory.

In a further investigation, it was demonstrated that a blanket based on this concept can also be designed for a DEMO relevant neutron wall loading of 2 MW/m^2.

Again, the characteristic parameters of the blanket versions described above, are summarized in Table I.

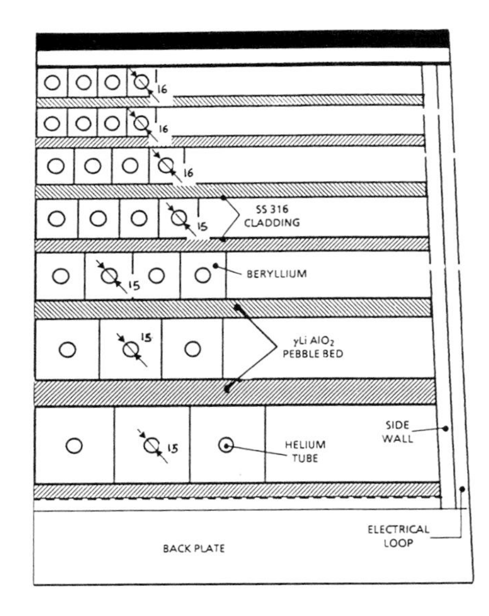

FIGURE 6: Midplane Half Crossection of High Performance Option - Second Approach

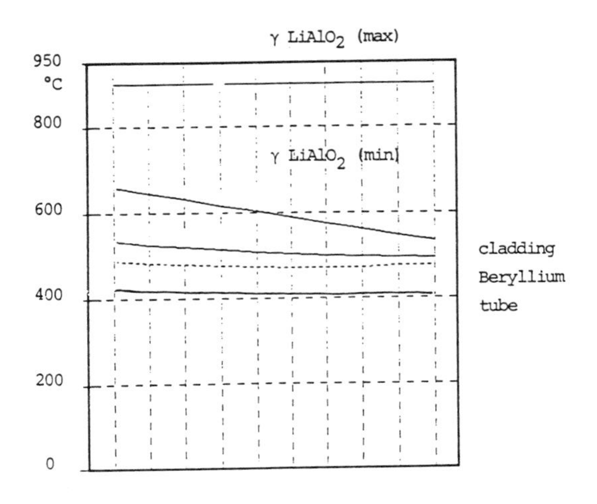

FIGURE 7: Radial Temperature Distribution

5. FIRST WALL AND TRITIUM ANALYSES

In connection with the blanket studies described above, big efforts have been devoted to first wall and tritium dynamics problems.

It has been demonstrated that helium cooling of the first wall can be accomplished with poloidal tubes under reasonable thermohydraulics conditions (7 MPa, 100/250° C) without violating the imposed conservative constraints on the maximum steel temperature (< 450° C) and the relative pumping power (< 1.5 %). This is even possible for very pessimistic assumptions for the surface heat load.

The theoretical models for analyzing the tritium dynamics problems[1] have been refined and extended and applied to the low performance pebble bed blanket in order to answer the questions about inventory and permeation. Also the helium purge gas system has been analyzed in some detail. Apart from the specific application for these blankets, the methods have been used to establish in general the test conditions for ceramic blankets in NET[5].

TABLE I: Design and Performance Parameters of Outboard Segments

	Units	Low Perf.	High Performance Blankets NET	ENEA-1	ENEA-2	DEMO
Blanket thickness	cm	65	65	65	65	85
Total poloidal length	cm	650	650	650	650	850
No. of coolant tubes	-	98+26	96	119	333	256
Be: Ceramic volume ratio	-	low	2:1	4:1	var.	4:1
Li_6 enrichment	%	30	90	75	75	75
Neutron wall loading	MW/m^2	1	1	1	1	2
Coolant pressure	MPa	5	6	6	6	7
Inlet temperature	^oC	310	310	250	250	250
Outlet temperature	^oC	400	450	450	450	450
Max. velocity	m/s	35	65	100	100	70
Relative pumping power	%	1.80	2.0	4.0	4.0	1.5
Min. breeder temperature	^oC	360	395	400	400	400
Max. breeder temperature	^oC	850	875	900	900	900
Max. steel temperature	^oC	430	480	490	490	490
Max. beryllium temperature	^oC	n.e.	580	500	500	500
Tritium breeding ratio (1D)	-	0.54	0.92	1.07	0.98	1.13

6. CONCLUSIONS

The studies on the ceramic pebble bed blanket with poloidal helium cooling have shown that it is a valid option for a reactor relevant blanket. The low number of coolant tubes and weldments on the tubes, as well as the high packing density, characteristic for the BOT concept in general, meet well the requirements of high reliability and good shielding properties. By including an inboard breeding blanket, tritium breeding ratios, as calculated by 1D methods, of 1.6 are possible. The heat can be removed with a helium coolant circuit operating under well known thermohydraulics conditions. The good prospects originating from this study need, however, to be confirmed by further work. In particular, a more detailed design is required as well as an experimental programme in which the problems of the design specific application of beryllium are adequately addressed.

ACKNOWLEDGEMENTS

The authors gratefully acknowledge many helpful discussions with M. Chazalon, C.H. Wu and V. Zampaglione and the help of H. Gorenflo in performing the calculations.

REFERENCES

1. A. Cardella et al., Proc. of 14th SOFT, Avignon, 1986, Vol 2, pp 1291 - 1297

2. W. Daenner et al., Ceramic Breeder in Poloidal Canisters with Helium and Water Cooling, Europ. Contr. to the 14th INTOR Workshop, Dec. 86, EUR FUBRU/XII-52/86/EDV 20, GROUP F, pp 25/1-18

3. M. Ferrari et al., Proc. of the 14th SOFT, Avignon, 1986, Vol. 2, pp 1231 - 1236

4. P. Kennedy et al., EXOTIC - Development of Ceramic Tritium Breeding Materials. Annual progress Report 1986 NRL Springfields Report NRL-R-2016(S) (June 1988)

5. M. Iseli et al., in this conference

FUSION TECHNOLOGY 1988
A.M. Van Ingen, A. Nijsen-Vis, H.T. Klippel (editors)
Elsevier Science Publishers B.V., 1989

LITHIUM OXIDE BLANKET DESIGN STUDY FOR A TOKAMAK FUSION REACTOR

J A DODD*, L J BAKER+, P KENNEDY‡, R S CHALLENDER#, G COAST#

* Risley Technical Services, UKAEA, Risley, Warrington, Cheshire WA3 6AT, England
+ Harwell Laboratory, UKAEA, Didcot, Oxon, OX11 ORA, England
‡ Springfields Laboratories, UKAEA, Springfields, Salwick, Preston PR4 ORR, England
# Progressive Engineering Consultants Ltd, 105 Walton Road, Warrington WA4 6NR, England

(UKAEA/EURATOM Fusion Association)

Lithium oxide as a fusion breeder material has neutronic advantages but various design limitations the implications of which are examined in this study. Blanket designs have been prepared employing helium both as the coolant and tritium carrier and operating within the temperature window of lithium oxide, with the blanket tritium inventory limited to one day's production. The designs use natural lithium with an adequate tritium breeding ratio and low radial power and burnup form factors. It is concluded that lithium oxide remains a strong contender as a reactor-relevant material.

1. INTRODUCTION

The objectives of the study [1] were to investigate the advantages and design implications of the use of lithium oxide as a reactor breeder material, to demonstrate how any known limitations can be overcome, and to provide guidance for future development of lithium oxide in terms of material properties, breeder and blanket geometry, and operating environment. These objectives were addressed within the framework of a redesign of a blanket for an existing DEMO tokamak reactor concept [2 3], using material properties gathered mainly from the EXOTIC series of fission reactor irradiation experiments [4 5].

2. LITHIUM OXIDE

Of the lithium ceramic materials which have been considered for tritium breeding the oxide is by far the best from a neutronic point of view with a high lithium density ($0.938 g.cm^{-3}$) and a low parasitic neutron absorption. It is however very reactive, absorbing water and carbon dioxide avidly, and has to be prepared and handled under clean, dry conditions.

Lithium oxide has been examined widely in the USA, Japan and in Europe and has been shown to have good thermal properties and to be relatively strong at room temperature [4]. However it loses its strength at temperatures in the region 400-600°C, becomes plastic and under irradiation swells by about 1% per 1% burn-up [6]. Consequently it must be clad.

One important requirement of any breeding material is that it should have a sufficiently wide operating temperature window. In general the lower temperature is set by the characteristics of the breeder, which define the tritium diffusion coefficient, and by the need to maintain a safe and economic tritium inventory. The upper limit on the other hand is set by the solubility of the tritoxide in the oxide, the volatility of the oxide and the compatibility of the ceramic with the clad.

In order to calculate the temperature dependence of the tritium inventory the model of Billone [7] has been used and, for a pellet of equivalent size 1cm dia, the results are shown in Fig 1. These indicate a minimum operating temperature of 600°C for a residence time of 1 day. The lower limit could be reduced to 500-550°C by using smaller pellets or by adding hydrogen to the helium. However the latter technique, which has been used in most irradiation experiments to date, leads to a mixed product ie HT + HTO instead of T_2O.

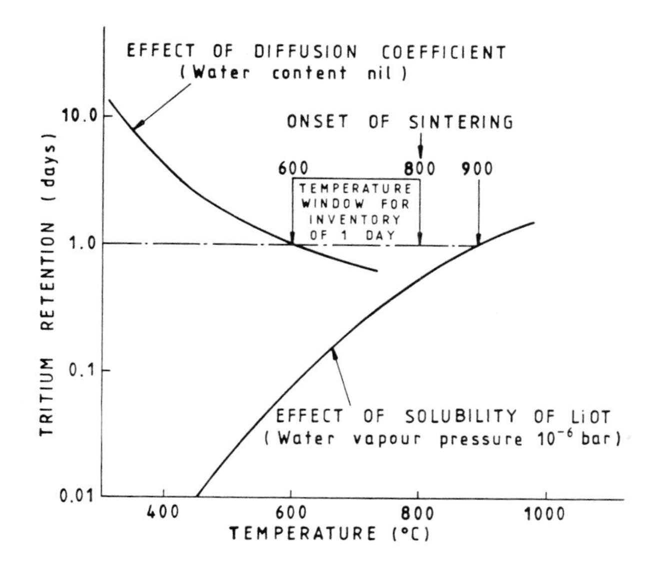

FIGURE 1 TEMPERATURE WINDOW

Lithium tritoxide solubility data[8] are also shown in Fig 1 and these indicate a maximum operating temperature of 900°C. However there are other factors to consider. Compatibility is a problem with most steels at high temperatures, or in the presence of a purge gas which removes lithium vapour from the reaction zone[9]. Volatility is however a more serious problem and at 900°C a loss of 3 weight % was observed during a single cycle's irradiation in FUBR IA[10]. Furthermore, early closed capsule experiments in America[11] showed considerable sintering and pore closure at 850°C but none at 750°C. It is considered that sintering will be more than offset by helium bubble formation. Nevertheless, to minimise the effects of volatility, incompatibility and sintering a maximum operating temperature of 800°C is proposed.

3. TRITIUM

The tritium feed to the plasma and the required blanket production in a 2500MW(T) DEMO reactor would be about 400g/day. For safety and economic reasons the blanket inventory is limited to one day's production (400g).

As in all similar gas-cooled ceramic designs tritium dispersal to the atmosphere must be limited and it remains to be shown that acceptable limits can be achieved, particularly when the main helium coolant is used as the carrier gas for the tritium. The means available, which will require further investigation, include optimisation of coolant chemistry, eg by oxygen dosing, to increase the ratio of tritiated water to tritium gas, oxidation of tritium at the exit from each breeder element, concentric ducts, oxygen dosing on the steam generator water side, and recovery of tritiated water from the helium coolant.

To maintain an acceptably low tritium loss through the steam generator tubes the partial pressure of tritium in the primary coolant must be kept below $\sim 10^{-8}$ Pa. Consequently it is necessary for the tritium to be in the oxide form as it leaves the pellets or to be oxidised before leaving the breeder element. This is incompatible with the use of hydrogen to aid tritium release, as the hydrogen would rapidly deplete the oxidiser, hence hydrogen must not be used. This necessitates a minimum breeder temperature of 600°C to obtain adequate tritium release.

4. ENGINEERING DESIGN

The constraints imposed upon the design process include, a) maintaining the lithium oxide temperature between the limits of 600 and 800°C, b) operating with helium inlet and outlet temperatures suitable for an efficient steam cycle eg 350 and 600°C, and c) handling with safety a target local tritium breeding ratio of 1.48 with the total blanket inventory limited to one day's production.

The first consideration in blanket design is to identify the most appropriate breeder

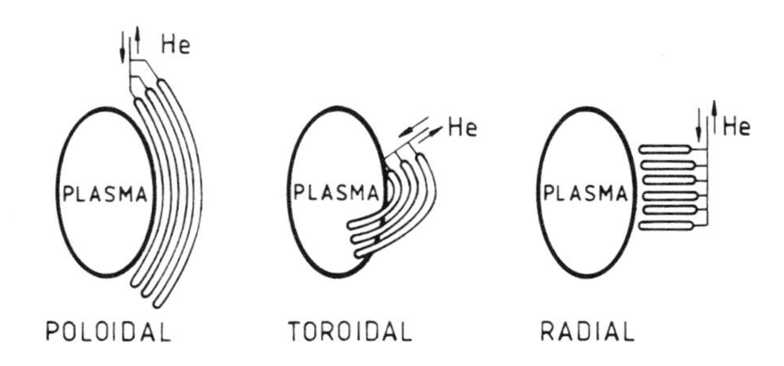

FIGURE 2 ALTERNATIVE BREEDER ELEMENT CONFIGURATIONS

element configuration. Of the three most generally accepted options shown in Figure 2 radially disposed breeder elements have been selected rather than poloidal or toroidal because the variation in heat rating from element to element is much less. The more even distribution of the coolant through the blanket is conducive to effective temperature control and, by adopting a re-entrant cooling path for the helium the manifolds are placed conveniently at the rear face of the blanket.

The concept is an array of nesting hexagonal tubes brazed together to form a strong honeycomb structure with cells facing the plasma behind the first wall, horizontal outboard of the torus and vertical above and below, with an inboard reflector. Each cell houses a cylindrical breeder element (Fig 3). Inlet helium at 350°C and 20 bar pressure flows along the outside of each breeder element through the cusp shaped interstices between the element and the hexagonal cell wall. The inlet flow is thermally insulated from the bulk of the element by a stagnant helium filled gap in order to limit the temperature rise in the inlet stream. On reaching the plasma end of the channel the temperature has increased only by about 20°C and this maintains the hexagonal structure and the end cap at acceptable temperatures. The flow is then reversed by the end cap and enters the breeder element, through which it passes back over canned annular pins of lithium oxide in a matrix of beryllium multiplier to enter the outlet manifolds at 600°C.

It is generally maintained that a separate helium purge circuit is necessary to purge the tritium product from the breeder elements. In the commercial reactor there may be over 5×10^5 pins setting a manufacturing and reliability problem for the connections. The probability of failure is high and repair impracticable, especially when the cost of an unplanned or extended shutdown of a 1500(MWe) station is in the order of £1M per day. Hence it was decided to have vented pins and use the main helium coolant as the tritium carrier gas. Elimination of hydrogen means that most of the tritium will be in the oxide form. However the

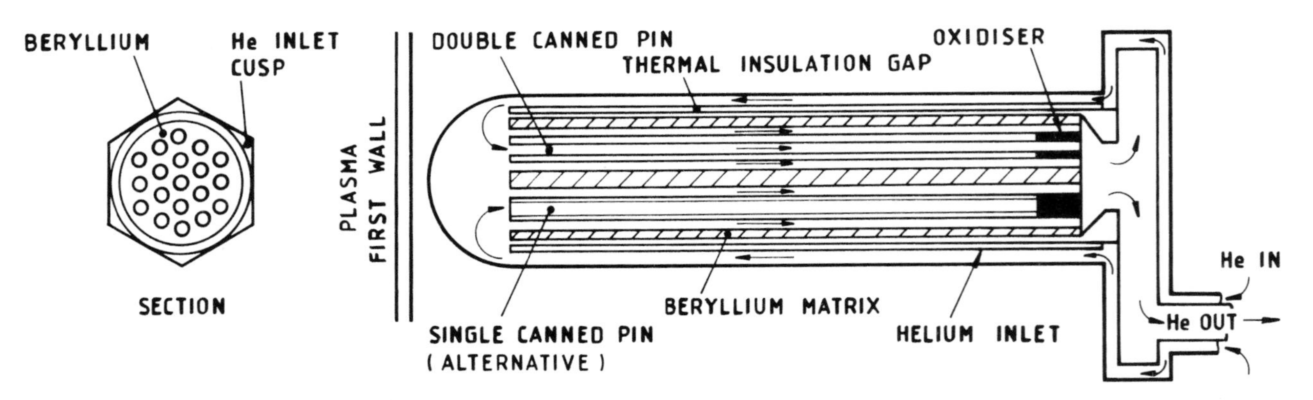

FIG. 3 RADIAL BREEDER ELEMENT

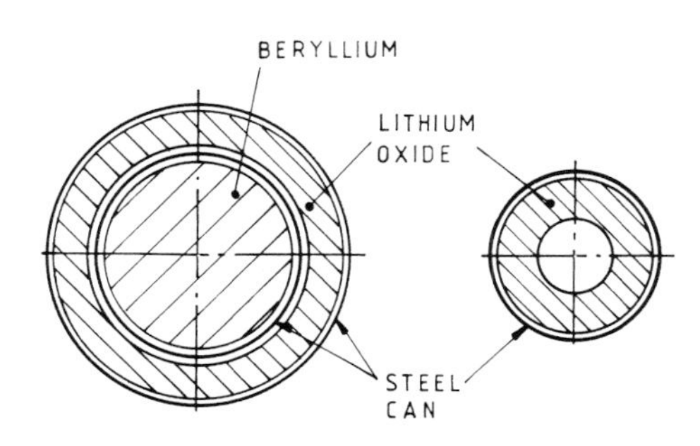

FIG. 4 TWO TYPES OF BREEDER PINS

intention is to oxidise any remaining tritium close to the point of generation. The oxidised tritium will be reclaimed from the main circuit. The problems of maintaining sufficiently low levels of tritium are well known. Much work has been done on this subject and much remains to be done.

It was decided to examine two pin configurations (Fig 4). One would be enclosed in a single can with coolant flowing over its external surface only, and the other would be encased in inner and outer cans, thus presenting two heat transfer surfaces to the coolant.

The advantages of "double" canned pins have been denied the designers of fission reactors because of the effects of differential expansion between the inner and outer cans, combined with the paramount requirement for can integrity to prevent the escape of fission products and actinides. Such products, however, are not present in fusion reactors and as this design is based on vented cans, the requirement for complete leak tightness no longer applies and the advantages of double canned pins become available.

Table 1 gives details of the main parameters of breeder elements for both single and double canned pin designs for DEMO and higher wall loading commercial reactors.

TABLE 1 MAIN PARAMETERS OF PIN DESIGNS

		SINGLE CAN		DOUBLE CAN	
Reactor		DEMO	Commercial	DEMO	Commercial
First Wall Loading	MW/m^2	2.6	5.2	2.6	5.2
Maximum Lithium Oxide Rating	W/cm^3	45	90	44.4	88.8
Channel Rating	kW	57.2	114.5	57.2	114.5
Helium Inlet Temperature	°C	370	350	350	350
Helium Outlet Temperature	°C	600	600	600	600
Circuit Pressure	bar	20	20	20.4	20.4
Pumping Power per Channel	kW	1.7	3.34	.93	8.26
Channel Length	cm	80	80	80	80
Number of Pins		12	37	7	12
Outside width of Hexagon Cell	cm	14.2	14.2	14.2	14.2
Lithium Oxide Outside Diameter	cm	1.96	1.17	3.363	2.715
Lithium Oxide Inside Diameter	cm	0.98	0.59	2.522	2.117
Lithium Oxide Surface Temp Min	°C	600	600	600	600
Lithium Oxide Centre Temp Max	°C	774	757	739	774

5. BLANKET NEUTRONICS

The neutronic study followed the design principles established for a lithium metasilicate DEMO reactor blanket[2 3] in which the volume proportions of materials were 7% martensitic steel structure, 15% helium coolant, and the balance was filled by breeder and multiplier. In the present design the breeder/multiplier space was filled by a matrix of beryllium or beryllium and graphite containing lithium oxide breeder pins, and the volume ratios and the lithium enrichment were varied during the optimisation process. The study employed a one dimensional model of an outboard first wall, blanket and shield for which, from the earlier work, the target local tritium breeding ratio (TBR) was 1.48. The calculations were normalised to the DEMO mean first wall source neutron power density of $2.6 MW.m^{-2}$, and burnup calculations assume a neutron fluence corresponding to $5.2 MW.am^{-2}$, ie a two year blanket operating life is envisaged.

Initially an all-beryllium matrix was employed with a breeder/matrix volume ratio of 1/4. Variation of the lithium enrichment from 60% ^{6}Li to the natural value produced a range of high TBRs (1.86-1.7), and the lowest peak radial values of lithium oxide power density ($44 W.cm^{-3}$) and lithium burnup (4.1%) were found using natural material. This calculation was used as the basis for the engineering design, therefore subsequent neutronic development represents an opportunity for future blanket optimisation. A disadvantage of the design point blanket configuration is that the end-of-life TBR is 1.26 (estimated by a pessimistic approximation to the spatial profile of burnt up lithium), demonstrating that a low ^{6}Li inventory leads to an undesirable decrease with time of the TBR.

Investigation of the use of depleted lithium (ie less than 7.5% ^{6}Li) with the same breeder/beryllium matrix ratio showed that the peak burnup in the oxide can be made self-limiting since it never greatly exceeds the original atom proportion of ^{6}Li. However the disadvantage of the decrease of TBR with time becomes more severe. After returning to the use of natural lithium this problem was addressed by increasing the breeder/matrix volume ratio. As this ratio is increased the TBR decreases, the target value of 1.48 being exceeded for volume ratios up to 7/3. At this limiting ratio the peak power density is $21.7 W.cm^{-3}$ and the peak lithium burnup is 1.2%, thus the radial distribution of reactions has been smoothed to considerable advantage. The much increased lithium inventory removes the problem of the temporal variation of the TBR, and the inventory of resource-limited beryllium is reduced. A further advantage is that the ^{7}Li reaction rate is maximised ($\sim 17\%$ of all breeding reactions) giving optimum use of lithium resources. At the extreme breeder/beryllium ratio of 7/3 it is unlikely that the present engineering concept is feasible, but intermediate cases (eg equal volumes of oxide and beryllium) could probably be realised using the pin/matrix concept, and the use of depleted lithium could be reintroduced yielding even lower peak/mean radial form factors and further small increases in the proportion of ^{7}Li burnt.

A final series of calculations investigated the replacement of a proportion of the beryllium matrix by graphite, a change which can be introduced with no modification to the blanket geometry. Even with the radially-smoothed reaction rate distributions obtained with natural enrichment lithium oxide used in high volume proportions, it is possible to change the rear 75% of the matrix to graphite with only a $\sim 4\%$ decrease in TBR. Thus the beryllium inventory is further reduced. To establish the viability of the use of graphite it would be necessary to investigate its influence on the coolant chemistry.

6. CONCLUSIONS

A consistent lithium oxide breeder element design concept has been obtained allowing for known properties, physical, chemical and mechanical, and its behaviour under irradiation. Particular attention has been given to temperature limitations and tritium release requirements.

The design is based on annular cylindrical oxide breeder pins, beryllium multiplier, helium coolant and steel structure. It meets a safety requirement that only 1 day's production of tritium should be retained in the blanket. The recommended conditions are to operate without hydrogen in the helium coolant, which also acts as the tritium carrier gas, and to have a lithium oxide temperature window of 600-800°C. The outstanding problems are: tritium oxidation before release to the coolant; keeping tritium partial pressure in the coolant below 10^{-8}Pa and leaktightness.

The neutronics advantages of lithium oxide as a tritium breeder material are further demonstrated in this study. Thus using natural lithium and a beryllium multiplier in a 1/4 ratio the lithium burnup is 4.1% and a local TBR of 1.7 can be achieved compared to a target of 1.48. Further optimisation has produced alternative proposals of a higher lithium oxide/beryllium ratio or replacing the rear 75% of the matrix by graphite, resulting in lower peak values of power density and lithium burnup and a lower swing in TBR through life. It is concluded that lithium oxide remains an important option for a breeder material.

7. REFERENCES

1. J. A. Dodd et al, (UKAEA report).

2. P. I. H. Cooke et al, UKAEA report CLM-R254(1985).

3. P. Reynolds et al, Transactions of the ENC '86 conference, Geneva, 3,(95-103)1986.

4. P. Kennedy et al. The preparation, characterisation and properties of lithium oxide and lithium metazirconate specimens irradiated in HFR Petten in the second and third EXOTIC experiments - Presented at the 14th SOFT in Avignon 1986.

5. H. Kwast et al. Tritium release from various solid breeder materials irradiated in EXOTIC experiments 1, 2 and 3. Presented at ICFRM 3 in Karlsruhe 1987.

6. G. W. Hollenberg et al. The effect of irradiation on four solid breeder materials J Nucl Mat 133 and 134 (1985) p242-245.

7. M. C. Billone. Mathematical models for predicting tritium transport in lithium ceramics. Advances in Ceramics (to be published).

8. M. Tetenbaum et al. An investigation of the solubility of LiOH in solid Li_2O. Fusion Technology 7 (1985) 53.

9. R. J. Pulham et al. The corrosion of steels by lithium monoxide and lithium silicate. Presented at the 14th SOFT in Avignon 1986.

10. G. W. Hollenberg. Fast neutron irradiation results on Li_2O, Li_4SiO_4, Li_2ZrO_3, Li_2AlO_2. J Nucl Mat 122 and 123 (1984) p896-900.

11. L. Yang et al. Lithium compound samples for tritium breeding application. J Nucl Mat 103 and 104 (1981) p585-590.

FUSION TECHNOLOGY 1988
A.M. Van Ingen, A. Nijsen-Vis, H.T. Klippel (editors)
Elsevier Science Publishers B.V., 1989

OPTIMIZATION OF THE COOLING OF THE LiPb BREEDER BLANKET FOR NET AND DEMO

F. ANDRITSOS
M. RIEGER

Commission of the European Communities, Joint Research Centre, Ispra Establishment,
21020 Ispra (Va) - Italy

Numerical thermal analysis of NET liquid eutectic LiPb water cooled breeder blanket, done during the past years at JRC Ispra, resulted in high temperatures and steep temperature gradients at the front part of the horizontal midplane of the outboard breeder region. Consequently, in order to attenuate this phenomenon, which can have detrimental effects on the corrosion rate of the austenitic stainless steel as well as on the allowable stress values, a new layout of the cooling tubes was proposed.
Nonlinear finite element heat transfer simulations were performed. The aim was to check and subsequently optimize the arrangement of the cooling tubes inside each breeding unit. The criterion for this was the temperature field at the horizontal midplane section of the breeder unit closer to the plasma, which is the one submitted to the highest thermal loads.
Optimized layout consisted in increasing the number of cooling tubes and repositioning them inside the breeding unit. Temperatures and gradients resulted well below the acceptable levels. The achievement of a security margin on these values for the NET heat load conditions allowed to consider the optimized solution proposed for the conditions of the breeder blanket of a DEMO reactor. For this purpose the effects of increased heat loading and different material choices were investigated. Finally, the austenitic versus Martensitic steel option is discussed.

1. INTRODUCTION

Finite element thermal analysis of NET liquid breeder blanket (Figure 1) done during the past years at JRC Ispra[1-3] resulted in high temperatures and steep temperature gradients at the front part of the horizontal midplane of the breeder. Consequently, in order to attenuate this phenomenon, a new layout of the cooling tubes was proposed by Rieger[4].

The main difference of the new from the old layout consisted in the number of tubes of the external row and the repositioning of both inner and outer rows a little further from the center, closer to the casing of the breeder.

A finite element thermal analysis of this new geometry of the NET LiPb breeder blanket was performed. The aim was to establish the temperature field at the horizontal midplane section of the breeder unit which is more close to the plasma, so as to check the new geometry and optimize it further, if necessary.

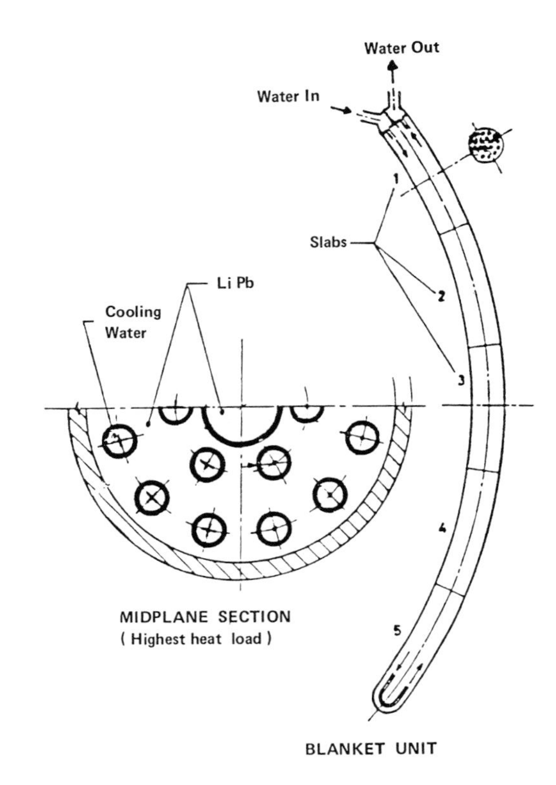

FIGURE 1
NET blanket breeder unit. Cooling tubes arrangement.

The analysis was performed through the

ABAQUS finite element package, on the AMDAHL main computer of the JRC Ispra. For discretizing the geometry of the section of the breeder, the PATRAN-G package was used on a microVAX computer. The transfer of the section geometry data to an ABAQUS file was done through the use of a purpose made routine written by Y. Crutzen of JRC Ispra.

2. ANALYSIS PROCEDURE FOR NET REACTOR

The particular thermal loading of the structure implied mirror symmetry of the analysis along the horizontal mid axis (X) of the section which, in turn, made the analysis of only half of the section possible. The upper half (that was used for the analysis) of the new proposed geometry of the breeder is shown in Figure 2. The finite element mesh used is shown in Figure 3. The elements used were quadrilateral with 8 nodes. The geometry was considered completely insulated from the exterior. Heat balance calculations for a breeder unit closest to the plasma led to midplane coolant temperatures as shown in Table 1. The heat transfer film coefficient between steel and pressurized water was considered at the value of 4.412 W/cm °C,[4]. The conductivities of both LiPb and steel were considered temperature dependent. The values from which the respective conductivity versus temperature curves were interpolated (automatically in ABAQUS CONDUCTIVITY procedure) are shown in Tables 2 and 3 for LiPb and steel, respectively.

Neutron flux induced heat generation, in-

TABLE 1
Global thermal characteristics of a breeding unit closest to the plasma, NET reactor

Volume of liquid LiPb	57 L
Volume of structure	19.7 L
Volume of coolant	19.2 L
Total heat generation	64 MW
Coolant water:	
pressure	10 MPa
velocity	6 m/s
inlet temperature	265°C
outlet temperature	294°C
Coolant temperature at midplane:	
front	272.2°C
rear	286.5°C

TABLE 2
Thermal conductivity of Li(17)Pb(83), from ref. 5, in W/cm °C

Temperature (°C)	260	280	300	310	350
Thermal conductivity	.121	.124	.138	.134	.140

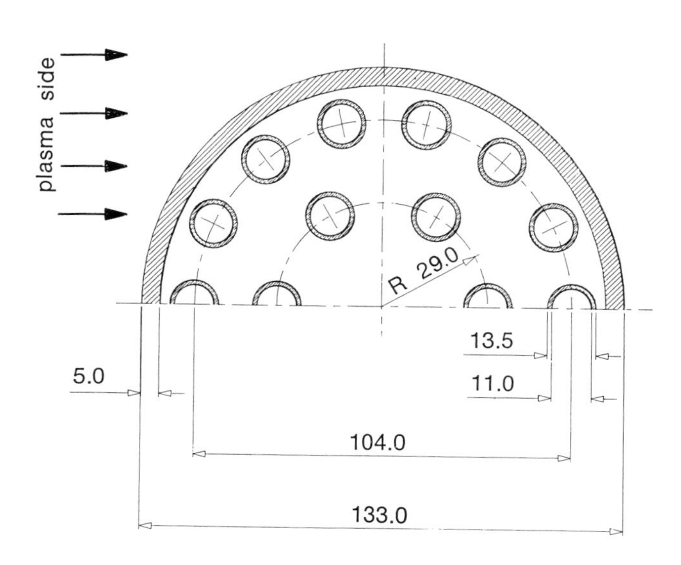

FIGURE 2
New arrangement of cooling tubes of NET breeder unit, as proposed in ref.4.

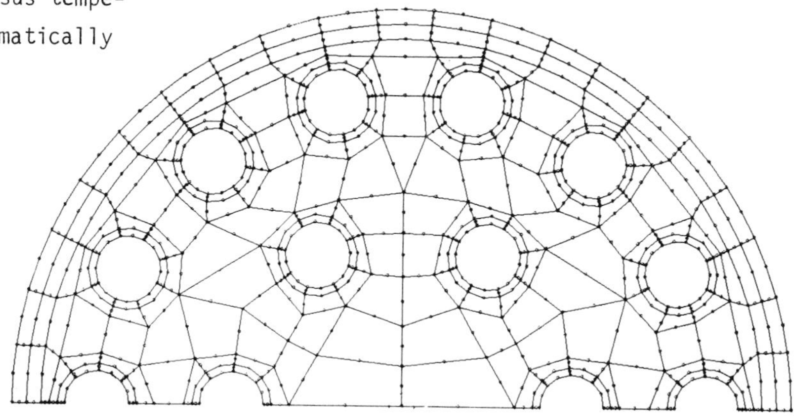

FIGURE 3
Finite element mesh, used for the analysis of the new arrangement of the cooling tubes. Generated with the help of PATRAN-G software.

TABLE 3
Thermal conductivity of AISI 316 steel, ASME code, in W/cm ^{o}C

Temperature (^{o}C)	21	149	204	315	426
Thermal conductivity	.145	.162	.170	.185	.200

side both LiPb and steel, is considered to be a function of length x. Length x is defined as the X-projection of the distance from the point O (Figure 4) closest to the plasma side. Alternatively, it is defined as a function of the X-distance from the outer casing surface facing the plasma (Figure 5). Let these two alternatives be called: heat generation, case 1 and 2, respectively.

Heat generation due to neutron flux is of different magnitude in LiPb than in steel. The above mentioned function is, therefore, dual: there exist two functions, one for LiPb and one for steel elements. These functions of x (where x is defined as above, according to the case) were interpolated from the values indicated in Table 4 (taken from ref.5) and are of the type:

A1*exp(-B1*x) + C1 + D1*x + E1*x**2 (1)

and

A2*exp(-B2*x) + C2 (2)

for LiPb and steel, respectively. The SAS numerical interpolation routines were used to produce the curves and derive the respective coefficients (A1,B1,...,C2). Their values are given in Table 5. The graph of these functions is in Figure 6.

For what concerns the ABAQUS programming, heat generation due to neutron flux was considered as a non uniform body (i.e. power/volume) heat flux. DFLUX (for distributed flux) procedure allows for the definition of the heat body flux through a user's subroutine. A FORTRAN subroutine distinguishes steel from LiPb elements calculates x according to the specific case (1 or 2, as mentioned previously) and then passes the proper magnitude of the distributed body heat flux to the main ABAQUS program.

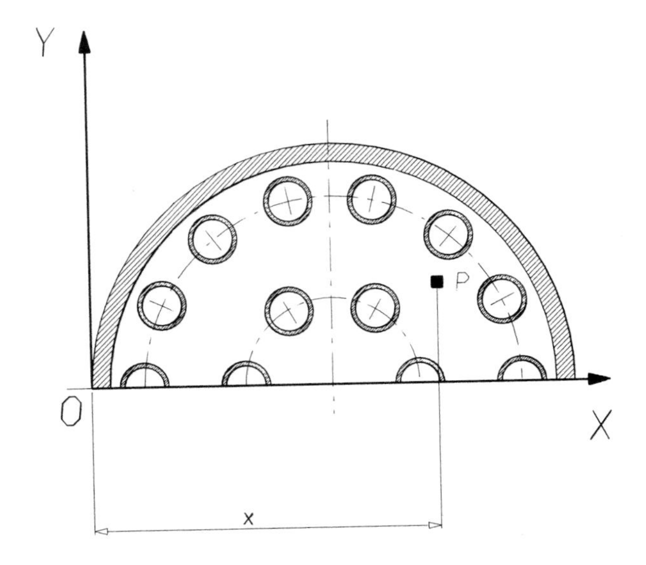

FIGURE 4
Neutron flux induced heat generation, case 1. Definition of length x.

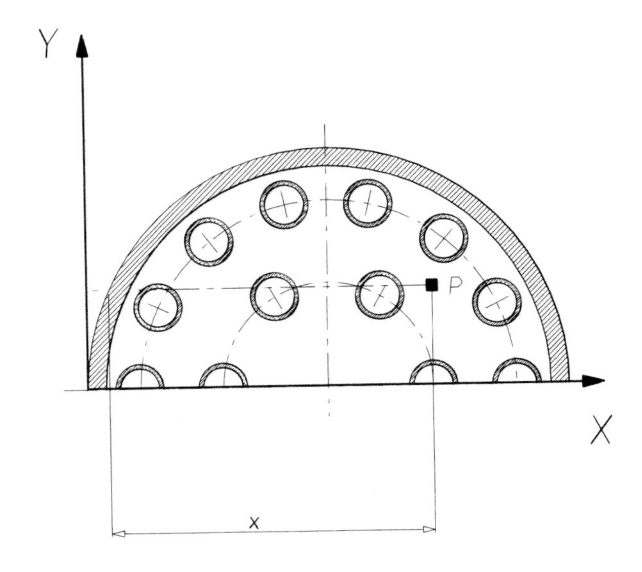

FIGURE 5
Neutron flux induced heat generation, case 2. Definition of length x.

3. EVALUATION OF THE RESULTS FOR NET REACTOR

Results are presented as isothermal contour plots. Present analysis of the new geometry gave the results shown in Figure 7. On comparing them to the results obtained in ref.1 (i.e. steady conductivity and equal temperature of 265°C on both front and rear cooling tubes), one notes the fundamental difference in the

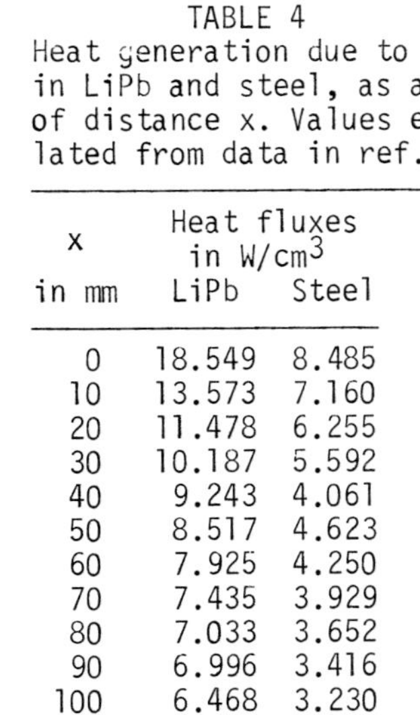

TABLE 4
Heat generation due to neutron flux, in LiPb and steel, as a function of distance x. Values extrapolated from data in ref.5.

x in mm	Heat fluxes in W/cm^3 LiPb	Steel
0	18.549	8.485
10	13.573	7.160
20	11.478	6.255
30	10.187	5.592
40	9.243	4.061
50	8.517	4.623
60	7.925	4.250
70	7.435	3.929
80	7.033	3.652
90	6.996	3.416
100	6.468	3.230
110	6.502	3.131

TEMPERATURE I.D.	VALUE
1	270
2	281
3	292
4	303
5	314
6	325
7	336
8	347
9	358
10	369
11	380

FIGURE 7
Isotherms of the new geometry, heat generation, case 1. Maximum temperature: 382.2^{o}C, center.

TABLE 5
Values of coefficients used to generate volume heat flux as a function of distance x

A1	B1	C1	D1	E1	A2	B2	C2
5.24719	.124264	13.1901	-.1203	.000537378	5.74354	.0219396	2.6437

FIGURE 6
Heat generation due to neutron flux as a function of length x, applied to both cases 1 and 2.

position of the maximum temperatures. Maximum temperatures obtained with the present analysis lie no more at the front (plasma facing) edge of the breeder but at its center. This was most probably due to the change of the arrangement of the cooling tubes of the breeder. To be sure though of the correctness of the analysis, a model of the old geometry of the breeder, with cooling water temperature of 265°C everywhere, was made and analyzed using the present ABAQUS program, as described previously. The results showed good agreement with ref.1 and confirmed the correctness of the analysis. The slight differences are due to the fact that present analysis is non linear and that neutron flux heat generation procedure is more accurate.

The fact that, with the new arrangement of the cooling tubes, the maximum temperature is at the center and not at the front of the breeder is highly desirable. The main reason is that the convection phenomena (which were not taken into account at the present analysis) will most probably lower the temperature at the center of the breeder.

As the temperature distribution at the mid-plane section of the breeder proved quite sensitive to the arrangement of the cooling tubes, a further modification was decided in order to optimize the temperature distribution of the breeder. It consisted in moving the inner tubes radially towards the center of the breeder, so as to cool it better and attenuate the temperature differences in the LiPb volume. The above optimized geometry is shown in Figure 8.

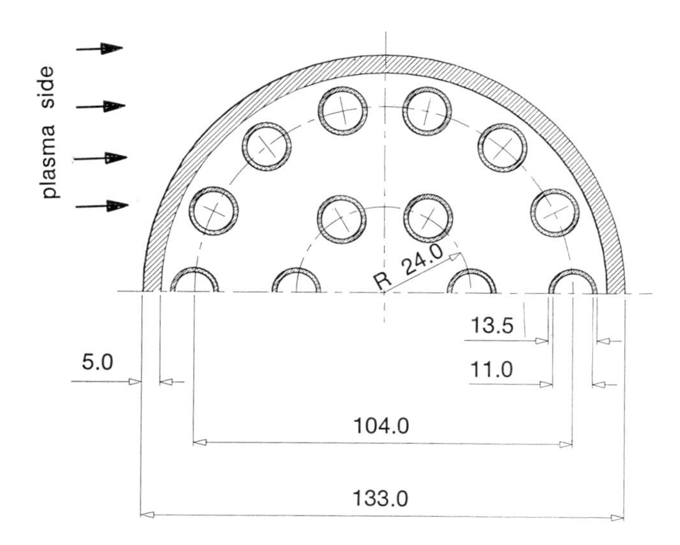

FIGURE 8
Optimized arrangement of cooling tubes of NET breeder unit. In respect to the arrangement shown in Figure 2, the internal row of tubes has been moved radially towards the center (to R = 24 mm).

In Figures 9 and 10 its temperature contours are shown for heat generation case 1 and 2, respectively. Maximum temperature has dropped by almost 40°C, in respect with the previous geometry analyzed. Maximum temperature is now around 345°C at the center and (more important) only around 330°C at the plasma facing steel casing. These results permit the consideration of this geometry at much more important thermal loads than those considered for the NET reactor.

Comparison of Figure 9 with Figure 10 confirms that case 1 of heat deposition presents only small differences from case 2. The latter leads to an insignificantly higher maximum temperature at the center as well as at the front part of the casing of the breeder.

4. ANALYSIS PROCEDURE FOR DEMO REACTOR

As thermal analysis presented above indicated, the optimized cooling arrangement of the NET liquid breeder blanket promises not only to eliminate problems of high temperature and steep temperature gradient, present in past cooling arrangements, but also to provide a large margin of security. Therefore, it was decided to test the above mentioned optimized geometry for the conditions of the DEMO reactor. DEMO operating conditions implied approximately double heat generation than in NET. This, in turn, implied higher temperatures of the cooling water at the horizontal midplane of the breeder.

A similar to the previously described finite element thermal analysis of this optimized geometry for the DEMO LiPb breeder blanket was performed. The discretization used was the same as for the NET.

Heat balance calculations for the DEMO reac-

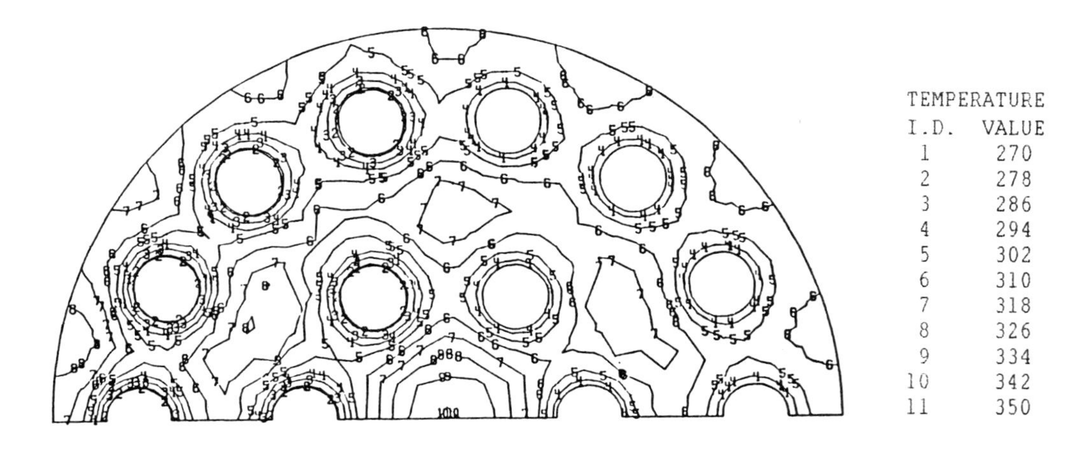

FIGURE 9
Isotherms of the optimized geometry. NET conditions. Heat generation case 1, max. temp.:344.2^{o}C center.

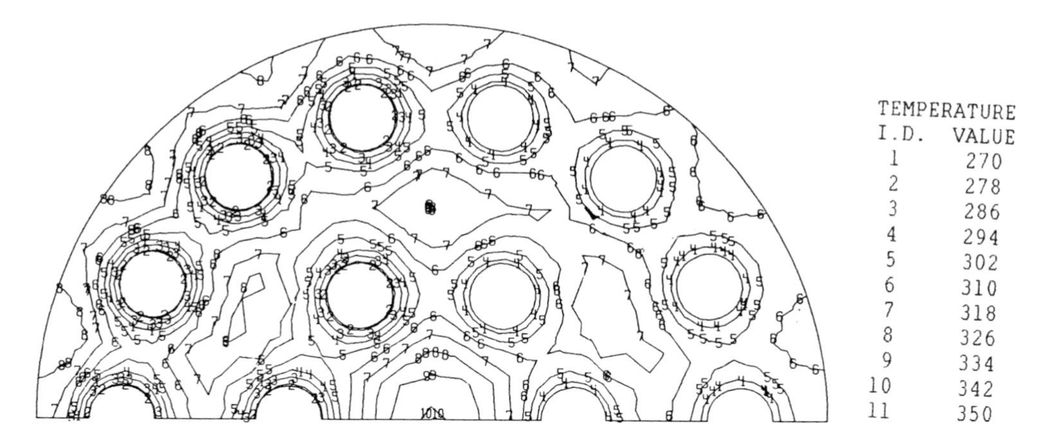

FIGURE 10
Isotherms of the optimized geometry, NET conditions. Heat generation case 2, max. temp.: 345°C center.

TABLE 6
Global thermal characteristics of a breeding unit closest to the plasma, DEMO reactor

Volume of liquid LiPb	57 L
Volume of structure	19.7 L
Volume of coolant	19.2 L
Total heat generation	128 MW
Coolant water pressure	15 MPa
velocity	6 m/s
inlet temperature	265°C
outlet temperature	322°C
Coolant temperature at midplane, front	280°C
rear	308°C

tor resulted in the values summarized in Table 6. The temperature at the midplane for the front and the rear cooling tubes was, therefore, about 10 and 20° more than at NET conditions, respectively (Table 6 versus 1).

The alternative of using Martensitic instead of austenitic steel was examined as well. The Martensitic steel ("Manet" type 1.4922) has a substantially higher conductivity than the austenitic, especially at low temperatures. The values from which the conductivity of the Martensitic steel versus temperature was interpolated (automatically in ABAQUS CONDUCTIVITY procedure) is shown in Table 7.

As neutron flux in DEMO conditions is estimated as approximately double as in NET, heat generation functions, used previously in NET conditions (formulas (1) and (2)) were multiplied by a factor of 2 and used as such in the DEMO simulation. This holds true both for LiPb and steel (as no data were available for Martensitic steel, the data available for austenitic steel was used). In Table 8 are shown the values of the coefficients of functions (1) and (2), used to simulate the heat generation due the neutron flux, in DEMO conditions.

TABLE 7
Thermal conductivity of "Manet" steel, type 1.4922, in W/cm °C

Temperature (°C)	50	100	150	200	250	300	350	400	450	500
Thermal conductivity	.2419	.2442	.2454	.2477	.25	.2512	.2535	.2559	.2582	.2605

TABLE 8
Values of coefficients used to generate volume heat flux in DEMO as a function of distance x

A1	B1	C1	D1	E1	A2	B2	C2
10.71438	.124264	26.3802	-.2406	.001074756	11.48708	.0219396	5.2874

5. EVALUATION OF THE RESULTS FOR DEMO REACTOR

Results for the optimized geometry under DEMO conditions are presented as isothermal contour plots in Figures 11 and 12, the structural material being Austenitic and Martensitic steel, respectively. Both refer to heat generation case 2.

Maximum temperature calculated was 421.5^{o}C at the center of the breeder for austenitic steel (Figure 11). It must be noted that this is a temperature of LiPb, the maximum temperature for the steel casing being less than 400°C. This temperature is acceptable.

The fact that, with the new arrangement of the cooling tubes, the maximum temperature is at the center and not at the front of the breeder is highly desirable. The main reason is that the convection phenomena (which were not taken into account at the present analysis) will most probably lower the temperature at the center of the breeder.

Comparison of Figures 11 and 12 leads to the following considerations: The use of Martensitic steel, although it does not lower but a little (-4°C) the global maximum temperature (at the center of the breeder), it lowers sig-

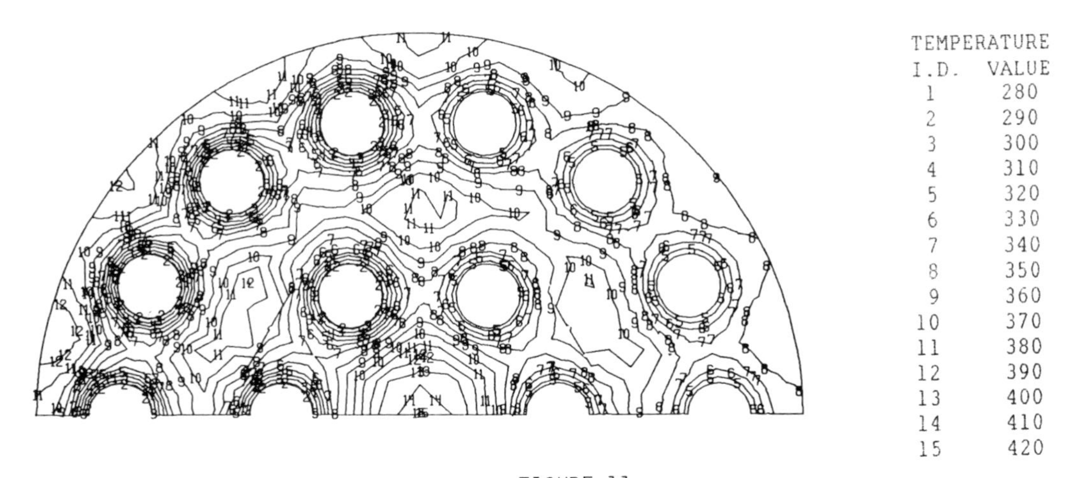

FIGURE 11
Isotherms of the optimized geometry, DEMO conditions, Austenitic steel, heat generation, case 2. Maximum temperature: 421.5^{o}C, center.

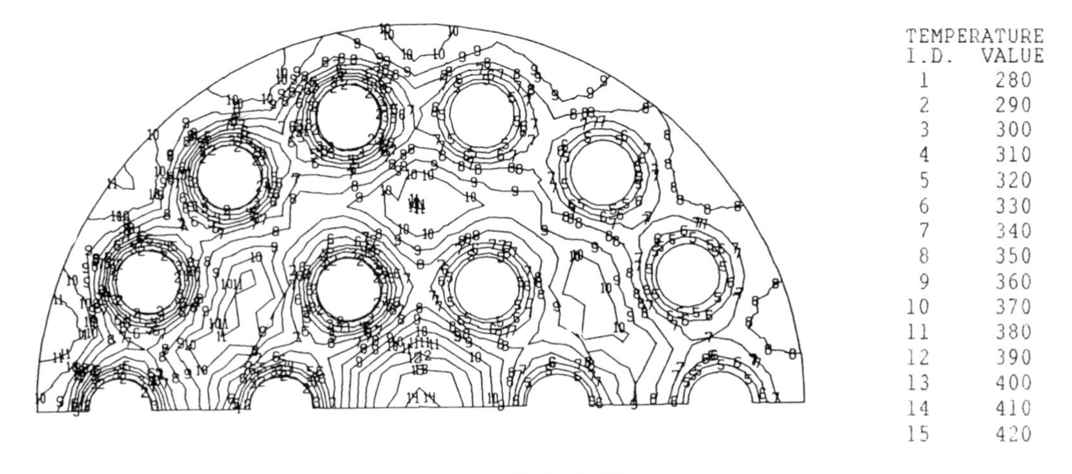

FIGURE 12
Isotherms of the optimized geometry, DEMO conditions, Martensitic steel, heat generation, case 2. Maximum temperature: 417.2^{o}C, center.

nificantly the maximum temperature at the front end of the casing (-10°C). The temperature differences between the front and the back part of the breeder casing are also attenuated by about 10°C. This represents a drop of 30% of the temperature difference and can be very significant as far as thermal stresses are concerned.

6. CONCLUSIONS

The above analysis confirmed that a breeder with an optimized arrangement of the cooling tubes such as in Figure 8, can very well withstand thermal loads of the magnitude expected at a DEMO fusion reactor.

The optimized geometry yielded a significantly higher temperature (around 420°C) at the center of the breeder than at the plasma facing edge (around 395°C). This difference is more pronounced now than in NET conditions.

The fact that the maximum temperature is at the center of the breeder section is highly desirable because of two reasons:

1) Convection (that was not taken into account in the present analysis) is expected to be more acute at the center than at the edge of the breeder. Thus, the slight temperature overestimation implied by the neglection of convection, is more pronounced at the center than near the casing of the breeder.
2) Lower temperatures near the casing imply lower thermal stresses for the structure.

The use of Martensitic steel instead of Austenitic lowers the maximum temperature at the casing significantly (about -10°C). This fact becomes even more important when the difference between the temperature at the plasma facing part and the rear part of the breeder is considered. 10°C less at the plasma facing part represents a drop of about 30% of the temperature difference between front and back of the midplane section, which is quite significant if thermal stresses are to be considered.

It is also worth noting that considering case 1 or case 2 for the neutron flux induced heat generation does not cause but minor differences of the temperature distribution. Maximum temperatures at the center as well as at the casing do not practically change.

REFERENCES

1. M. Biggio and M. Turri, Thermomechanical analysis of a liquid breeder blanket for NET using LiPb in cylindrical modules, Technical Note I.05.31.84.07, PER 799/84, JRC Ispra, January 1984.
2. J. Airola, M. Biggio and C. Piana, Thermomechanical analysis on INTOR/NET liquid LiPb breeder blanket, Technical Note I.05.07.82.133, PER 638/82, JRC Ispra, 1982.
3. W. Daenner, M. Rieger and K.A. Verschuur, Progress in design and analysis of the NET liquid breeder blanket, Proc. of 14th Symp. on Fusion Technology, Avignon, September 1986.
4. M. Rieger, personal communication, November 1986.
5. U. Jauch, G. Haase and B. Schulz, Thermophysical properties in the system Li-Pb, part II: Thermophysical properties of Li(17)Pb(82) eutectic alloy, Primärbericht, Kernforschungszentrum Karlsruhe, May 1986.
6. K.A. Verschuur, Poloidal variation of the NET blanket nuclear response function, Report ECN 87 001, Stichting Energieonderzoek Centrum Nederland.
7. G. Casini, M. Biggio, C. Ponti, M. Rieger, G. Vieider, A. Gardella and W. Daenner, First wall and blanket concepts for an experimental fusion reactor, Top. Meet. on Fusion Energy, San Francisco (USA) March 1985.
8. M. Rieger et al., Progress in blanket designs with Li17Pb83 breeder, 13th Symp. on Fusion Technology, Varese, September 1984.
9. W. Daenner, personal communication to M. Rieger, April 1987.
10. T. Munsch, Rapport de Stage, JRC Ispra, August 1986.
11. Hibbit, Carlson & Sorrensen Inc., ABAQUS User's Manual, Version 4.5(a), July 1984, USA.
12. Hibbit, Carlson & Sorrensen Inc., ABAQUS User's Manual, Version 4.6, 1987, USA.
13. F. Andritsos, Thermal analysis of NET LiPb breeder blanket, Technical Note I.87.152, PER 1420/87, JRC Ispra, December 1987.
14. F. Andritsos and M. Rieger, Thermal analysis of DEMO LiPb breeder blanket, Technical Note I.88.86, PER 1553/88, JRC Ispra, July 1988.

FUSION TECHNOLOGY 1988
A.M. Van Ingen, A. Nijsen-Vis, H.T. Klippel (editors)

NEUTRONICS BENCHMARKING STUDY OF BREEDING SHIELD FOR THE FUSION EXPERIMENTAL REACTOR (FER)

Seiji MORI, Takeshi KOBAYASHI, Hiromasa IIDA*, Yasushi SEKI*

Kawasaki Heavy Industries, Ltd. 2-4-25 Minami-suna Koto-ku Tokyo, Japan
*Japan Atomic Energy Research Institute, Mukouyama Naka-machi Naka-gun Ibaraki, Japan

Neutron flux distribution and tritium breeding ratio(TBR) were calculated with the one-dimensional discrete ordinates code using the various cross section libraries and with the continuous energy Monte Carlo code. Two types of the geometrical representation were considered; one is the homogeneous mixture of structural material, aqueous lithium salt coolant and, if any, neutron multiplier; the other is the alternating layers of the structure and the coolant. Results were compared in terms of neutron flux attenuation, integrated TBR, and energy profile and spatial distribution of tritium production rate. S_N/MC ratios of TBRs for the homogeneous model are smaller than unity by a few percents and several percents for the configurations with and without multiplier, respectively. For the heterogeneous model without multiplier, the total TBRs agree within several percents and are larger than those of the homogeneous model by 20 - 40 %. While, total neutron fluxes are underestimated with the S_N calculations by 30 - 40% compared to the MC results for both the homogeneous and heterogeneous models.

1. INTRODUCTION

Installing tritium breeding blanket at inboard region of a tokamak reactor, in general, increases space for shielding of superconducting magnets and leads to enlargement of reactor size. We have studied feasibility of blanket concepts using aqueous lithium salt[1-4] as a candidate of the inboard tritium breeding blanket for the Fusion Experimental Reactor (FER)[5]. Structure of this blanket is relatively simple compared to solid breeder blanket concepts and is easier to satisfy shielding requirements without increasing shield thickness.

This paper describes results of neutronics benchmark calculations to assess the accuracy of the multi-group 1-D S_N calculation used for the design study of FER. Results of the S_N code using different cross section libraries were compared with results of continuous energy Monte Carlo calculations. The another purpose of the study is to reveal calculational uncertainties of neutronics properties due to geometrical representations because breeding zones(i.e. aqueous salt coolant) are localized in this shield and hence it is expected that geometrical heterogeneity has large impact on tritium breeding characteristics.

2. CALCULATION METHOD

The multi-group calculation was done with the 1-D S_N code, ANISN[6] using the four cross section libraries whose major characteristics are summarized in Table 1. All the neutron cross sections are for infinite dilution. The calculational geometry is an infinite cylinder with the plasma axis. Table 2 shows the three geometrical representations and the material composisions of the shields which are followed by toroidal field coils. Two types of the geometrical model were analyzed; one is homogeneous mixture of structural material, aqueous solution and neutron multiplier if any (homogeneous model) ; the other is alternating layers of structure and aqueous solution (heterogeneous model), which is closer to the actual mechanical configuration.
The continuous energy Monte Carlo calculation was done with MCNP[13] using the ENDF/B-IV data

TABLE 1
Main Characteristics of Cross Section Sets Used in the Benchmark Calculations

	GICX40[7]	GICXFNS[9]	FSXJ3T1[10]	VITAMIN-C[12]
Nuclear Data File	ENDF/B-III[8]	ENDF/B-IV[8]	JENDL-3T*[11]	ENDF/B-IV
Number of Neutron Groups (above 1 MeV)	42(17)	135(60)	125(64)	171(59)
Processing Code	RADHEAT	NJOY	PROF.GROUCH -G/B	AMPX
Weighting function	1/E	Constant	1/E+Maxwell	Maxwell+Fission 1/E+Fusion

* JENDL-3T is a temporary file for testing the evaluated data for JENDL-3. It will be revised in JENDL-3.

library. The geometrical model is the same as that for the S_N calculation; the reflection condition was assumed at both ends of the cylinder. Neutron flux and tritium production rate were evaluated with surface crossing estimators and track length estimators.

In this study, kind of the lithium salt is not specified, and the enrichment of ^6Li is assumed to be 100%

3. RESULTS AND DISCUSSIONS

3.1. Neutron flux attenuation

Neutron fluxes are compared in Fig.1 and Fig.2 for Case 2 and Case 3, respectively, between the S_N calculations and the Monte Carlo calculation in terms of ANISN to MCNP ratios (S_N/MC ratio). The S_N/MC ratio of the outermost part of the shield is not plotted in the figure because the particle statistics is poor (fsd > 10%). The neutron fluxes calculated with ANISN are smaller than those with MCNP. Discrepancy between them increases with depth into the shield. Although the discrepancy partly originates from the difference in nuclear data fliles when GICX40 and FSXJ3T1 are used, it is attributed to the resonance self-shielding effect of the structural material elements when the same nuclear data file (i.e.ENDF/B-IV) is used. The worst S_N/MC ratios in the total neutron fluxes of Case 2(Case 3) are about 0.5(0.6) for GICX40 and around 0.7(0.8) for GICXFNS, FSXJ3T1 and VITAMIN-C. The large difference of the GICX40 case mainly comes from the inadequacy of iron cross section data of ENDF/B-III as reported in the former analysis[14]. The accuracy of the

TABLE 2
Material Compositions of the Breeding Shield for the Three Different Configurations

Components	Thickness (cm)	Compositions (Vol.fraction)
Case 1 (Homogeneous model)		
Breeding shield (I)	30	10 % SS316
		15 % AS*
		75 % Be**(Pb)
Breeding shield (II)	38	85 % SS316
		15 % AS
Case 2 (Homogeneous model)		
Breeding Shield	68	85 % SS316
		15 % AS
Case 3 (Heterogeneous model)		
First wall (I)	0.5	100 % SS316
Bank 1	0.5	100 % AS
First wall (II)	0.5	100 % SS316
Bank 2	1.94	100 % AS
Shield (I)	11.36	100 % SS316
Bank 3	1.94	100 % AS
Shield (II)	11.36	100 % SS316
Bank 4	1.94	100 % AS
Shield (III)	11.36	100 % SS316
Bank 5	1.94	100 % AS
Shield (IV)	11.36	100 % SS316
Bank 6	1.94	100 % AS
Shield (V)	11.36	100 % SS316

* Aqueous salt(^6Li : 2.464×10^{21} cm^{-3})
** 85 % theoretical density

other libraries based on ENDF/B-IV and JENDL-3T is almost identical.

Penetrated neutron fluxes behind the breeding shield are summarized in Table 3. When large fraction of beryllium(lead) is used in the shield, shielding capability degrades, i.e. the fast neutron flux of Case 1 is over five(four) times higher than that of Case 2.

Although attenuation of the 14 MeV neutron flux is not affected by the geometrical representation, the fast and total neutron fluxes of the heterogeneous model are higher than those of the homogeneous model by a factor of two (Case 2 vs. Case 3).

3.2. Tritium breeding ratio (TBR)

Calculated tritium breeding ratios are summarized in Table 4. For Case 1, the results of ANISN is somewhat smaller than that of MCNP except when VITAMIN-C is used. Discrepancy between them is around a few percents. For Case 2, the results of ANISN except the FSXJ3T1 case is smaller than that of MCNP by the maximum discrepancy of 8%. The FSXJ3T1 library gives, on the contrary, the larger TBR by 2%. Case 3 gives the larger TBRs compared to Case 2 by 20-40%. The results of ANISN agree well with that of MCNP and discrepancy is around a few percents.

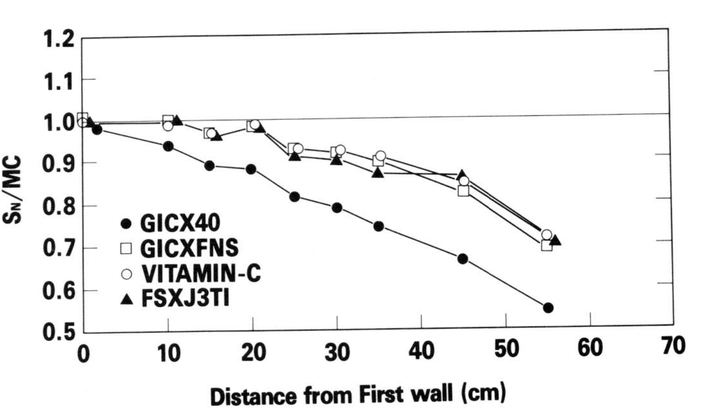

FIGURE 1
S_N/MC Ratio for Total Neutron Flux (Case 2)

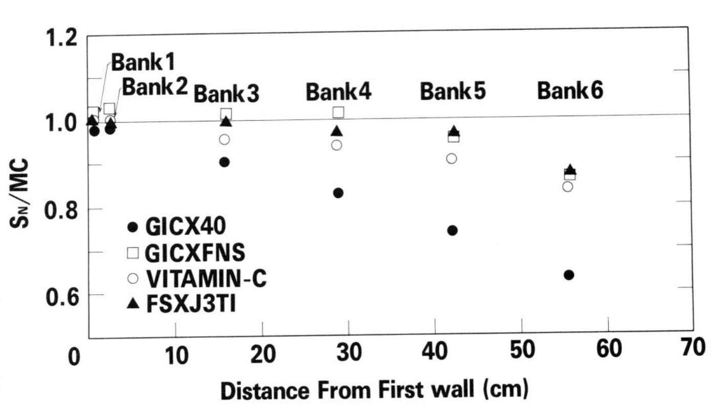

FIGURE 2
S_N/MC Ratio for Total Neutron Flux (Case 3)

TABLE 3
Comparison of Neutron fluxes behind the Shield Calculated with the Various Methods and Data (Arbitrary Unit)
(Upper row : Fast Flux, E>0.1 MeV, Lower row : Total Flux)

	ANISN				MCNP
	GICX40	GICXFNS	FSXJ3T1	VITAMIN-C	(ENDF/B-IV)
Homogeneous with Be(Pb) (Case 1)	5.31-6* (3.91-6) 1.36-5 (1.08-5)	-	6.98-6 1.82-5	8.11-6 2.08-5	9.10-6, 0.4778** 2.08-5, 0.1456
Homogeneous without Be (Case 2)	9.54-7 2.61-6	1.28-6 3.46-6	1.24-6 3.82-6	1.38-6 3.77-6	1.52-6, 0.5690 3.84-6, 0.2106
Heterogeneous without Be (Case 3)	2.24-6 5.97-6	3.50-6 8.66-6	3.34-6 8.94-6	3.46-6 8.30-6	4.85-6, 0.3254 1.01-5, 0.1478

* Read as 5.31×10^{-6} ** Fractional standard deviation

Spatial distribution of the S_N/MC ratios of tritium production rate for Case 2 is illustrated in Fig.3. The S_N/MC ratio of the outermost part of the shield is not plotted because of the poor statistics (fsd>10%). Though the ANISN results almost agree with the MCNP results in the first wall region, the difference between them increases with depth into the shield. This result is consistent with the neutron flux attenuation described above. Since the total TBR is dominated by the tritium production near the first wall, the underestimation of neutron flux due to the deep penetration does not greatly affect the total TBR . The S_N/MC ratios of the tritium production rates at each breeding bank for Case 3 are shown in Fig.4. The S_N/MC ratio near the first wall is somewhat larger than unity. The S_N/MC ratio at and behind the second bank decreases below unity in the case of G1CX40 , GICXFNS and VITAMIN-C, and is around unity in the case of FSXJ3T1.

Neutron energy profiles of TBR are shown in Fig.5. Comparison between Case 2 and Case 3 reveals that the heterogeneous configuration greatly increases the tritium production by low energy (E<1eV) neutron. The neutron balance calculated by MCNP is compared in Table 5. In the heterogeneous configuration, the neutron absorption by the elements of the structural material decreases and the absorption by ^{6}Li atoms increases, which explains the increase of TBR.

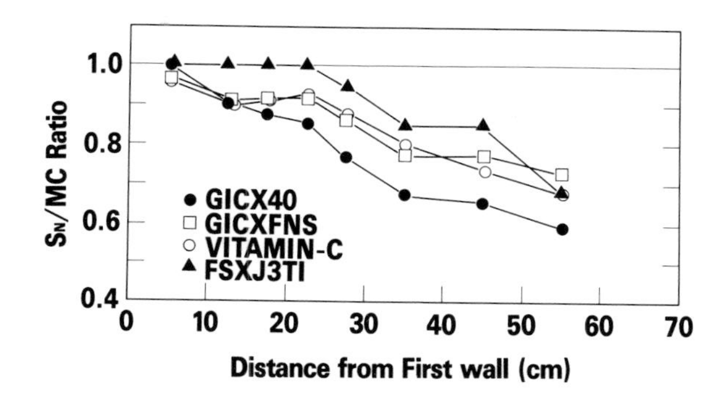

FIGURE 3
S_N/MC Ratio for Tritium Production (Case 2)

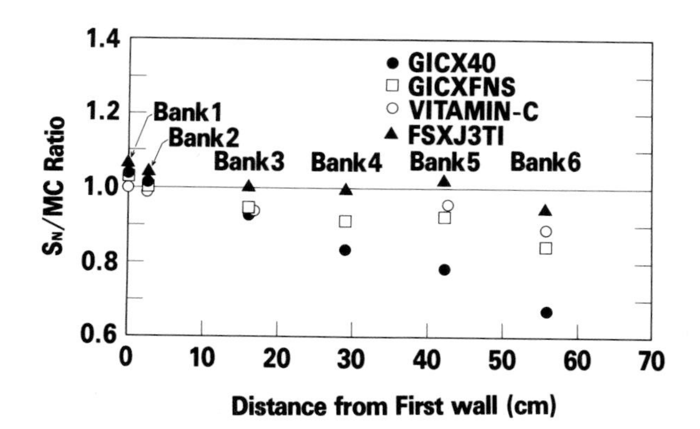

FIGURE 4
S_N/MC Ratio for Tritium Production (Case 3)

4. CONCLUSIONS

The resonance self-shielding effect does

TABLE 4
Comparison of Tritium Breeding Ratios Calculated with the Various Methods and Data

	ANISN				MCNP
	GICX40	GICXFNS	FSXJ3T1	VITAMIN-C	(ENDF/B-IV)
Homogeneous with Be(Pb) (Case 1)	1.590 (1.394)	-	1.595	1.660	1.612, 0.0087*
Homogeneous without Be (Case 2)	0.556	0.558	0.615	0.559	0.604, 0.0191
Heterogeneous without Be (Case 3)	0.756	0.765	0.794	0.762	0.784, 0.0237

* Fractional standard deviation

TABLE 5
Compaison of Neutron Balance in the Homogeneous and Heterogeneous Configurations

	Homogeneous			Heterogeneous		
	Absorption	Collision	(n, 2n)	Absorption	Collision	(n, 2n)
H	0.0057	11.95		0.0074	12.69	
O	0.0100	1.753		0.0235	0.851	
^{6}Li	0.6067	0.6565		0.7901	1.921	
Structure elements	0.6389	45.84	0.2648	0.402	32.29	0.2269
Total	1.2613	60.2	0.2648	1.2212	47.75	0.2269

not affect the TBR greatly because the tritium production near the first wall mainly determines the TBR. The discrepancy of the TBR between the different methods and nuclear data is within 10%. The neutron behavior in the homogeneous and heterogeneous configurations is rather different each other; the TBR increases, by 20-40% and shielding capability reduces by a factor of two, and hence the heterogeneous geometrical effect must be taken into account to predict the neutronics properties of the breeding shield. Further investigation (e.g. 2-D geometrical effect) is necessary to select an appropriate geometrical representation in accordance with the actual mechanical configuration.

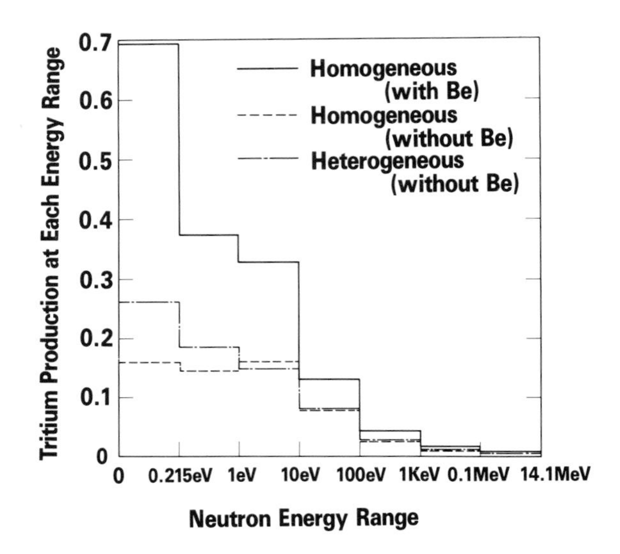

FIGURE 5
Neutron Energy Profile of Tritium Production

ACKNOWLEDGEMENT

Authors gratefully acknowledge the valuable assistance from Drs. K.Kosako, S.Matsuda, S. Tamura, M.Yoshikawa and the members of the FER design team of JAERI.

REFERENCES

1. P.Gierszewski, et al., 14th SOFT, Avignon, Sept.1986, Proceedings PP 1217-1222

2. M.J.Embrechts, et al., ANS Top. Conf. on Theory and Practices in Radiation Protection and Shielding, Knoxville, April 1987, Proceedings PP 207-216

3. G.L.Varsamis, et al., ibid, PP 226-234

4. M.J.Embrechts, et al.,Nucl.Eng.Design /Fusion 4,1987, PP 211-222

5. FER Design Team, JAERI-M 88-090, 1988

6. W.W.Engle Jr., K-1693, U.C.Corp. 1967

7. Y.Seki, and H.Iida, JAERI-M 8818, 1980

8. M.K.Drake (edited), BNL-50274, 1970

9. Y.Seki, JAERI-M 83-061, 1983

10. K.Kosako, private communication, 1987

11. JENDL Compilation Group (JAERI), Private communication, 1987

12. R.W.Roussin et al., ORNL-RSIC-37

13. J.F.Briesmeister, Editor, LA-7396-M, Rev.2, LANL, Sept.1986

14. S.Mori et al., JAERI-M 88-103, 1988

FUSION TECHNOLOGY 1988
A.M. Van Ingen, A. Nijsen-Vis, H.T. Klippel (editors)

TEST MODULE IN NET FOR A SELF-COOLED LIQUID METAL BLANKET CONCEPT

S. Malang, K. Arheidt, U. Fischer

Association KfK-EURATOM, Kernforschungszentrum Karlsruhe, Postfach 3640, D-7500 Karlsruhe, Federal Republic of Germany

The application of a self-cooled liquid metal blanket concept to the condition of a DEMO-reactor and its testing in NET is described. The neutronics analysis shows that tritium self-sufficiency can be achieved without beryllium multiplier if breeding blankets are arranged at both outboard and inboard side of the torus or, using beryllium as multiplier, with outboard breeding only. First estimates indicate that it should be possible to test all relevant features of the concept in one of the horizontal plug positions of NET.

1. INTRODUCTION

The operation of the Next European Torus (NET) will be divided into a physics phase and a technology phase. Water-cooled shielding blankets without any tritium breeding will be used during the first phase. For the technology phase "driver blankets" are needed which can provide a significant fraction of NET's tritium needs with minimal technical risk. The presently preferred solution is an aqueous self-cooled blanket concept based on lithium salts dissolved in the cooling water. The intention is to install such "low technology" blankets at nearly all blanket positions and provide a few locations only for testing reactor relevant blanket concepts. In this strategy blanket concepts using lithium ceramics or lithium containing liquid metals as breeding material are not needed for the operation of the basic machine. Therefore, they can be selected mainly based on their potential for an electricity producing power reactor.

One attractive blanket concept for a power reactor uses the same liquid metal both as breeding material and as coolant. Such a self-cooled liquid metal blanket concept is under development at the Kernforschungszentrum Karlsruhe as part of the European Fusion Technology Programme. This paper deals with the questions if this concept is relevant for a DEMO-reactor and how it can be tested in NET.

2. BLANKET CONCEPT

Liquid metals, i.e. pure lithium or the eutectic lithium-lead alloy Pb-17Li are attractive breeding materials in fusion reactor blankets. In some concepts the liquid metal is stagnant and cooled by water or helium. In other concepts the liquid breeder serves as coolant as well and is circulated for heat extraction. Such self-cooled blanket concepts are characterized by a simple design with a small number of ducts and welds, no tritium permeation problems inside the blanket and a high breeding ratio due to the large breeder volume fraction. The main problems arise from the strong influence of the high magnetic field in a tokamak on the liquid metal flow. A high pressure drop is caused in uninsulated ducts if the flow direction is perpendicular to the magnetic field and the heat transport is degraded due to the supressed turbulence. A major step in overcoming these problems was a novel flow concept employed in the ANL design of a self-cooled liquid metal blanket [1]. In this flow concept the velocity perpendicular to the magnetic field is minimized by using nearly the entire blanket cross section as poloidal duct. At the plasma facing side of the segment the flow is diverted to the toroidal direction (parallel to the magnetic field) in order to allow sufficient first wall cooling without excessive large MHD pressure drop. But even with this flow concept and a tokamak design allowing coolant acces tubes at both the upper and the lower end of the blanket segment it was necessary to taper the thickness of the duct walls and to support the blanket segment side walls by the neighbouring segments in order to stay inside given stress limits for the structural material. In fact, the interaction between duct wall thickness, MHD pressure drop and mechanical stresses in the duct walls represents the key problem in designing self-cooled blankets. A possible way to diminish this problem is to use novel flow channel inserts [2], decoupling electrically the load carrying walls from the flowing liquid metal. The present design is based on the use of flow channel inserts in all poloidal ducts. Two alternative blanket concepts have been investigated, depending on the tokamak design:

A) A blanket, shown schematically in fig. 1a, with a rather thick layer of beryllium as neutron multiplier in order to avoid the need for breeding blankets at the inboard side of the torus. This concept is suitable for a machine like NET, designed for blanket exchange in vertical direction. In this case both inlet and outlet cooling tubes have to be arranged at the top end of the torus, which makes the design of a self-cooled inboard blanket extremely difficult due to the higher magnetic field strength and the smaller space available at the inboard side of the torus.

B) A blanket, shown schematically in fig. 1b, which does not need a beryllium multiplier, but inboard breeding instead. This concept, however, requires coolant access tubes at both the upper and lower end of the blanket segments in order to allow inboard blankets without excessive large MHD pressure drops.

Both concepts are based on the following features:

- Eutectic alloy Pb-17Li as breeding material/coolant.
- Ferritic steel as structural material.

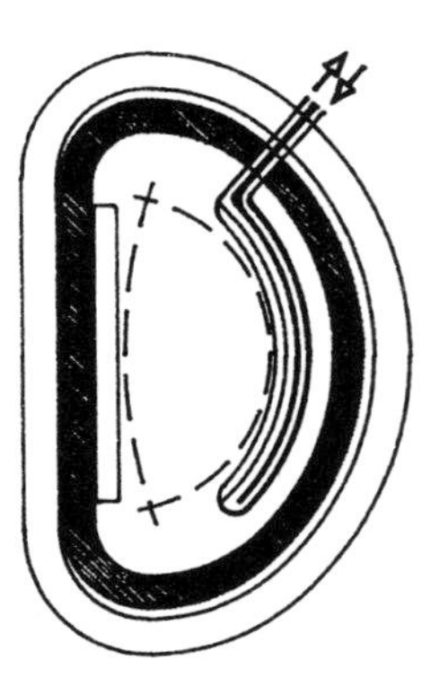

A) Outboard breeding only

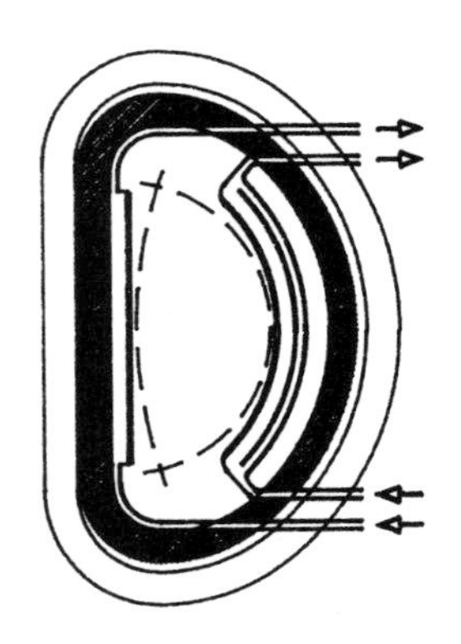

B) Outboard and inboard breeding

FIGURE 1
Arrangement of breeding blankets and coolant access tubes

- Poloidal-toroidal-poloidal flow concept with integrated first wall.
- Laminated flow channel inserts installed in all poloidal ducts.

3. APPLICATION OF THE BLANKET CONCEPT TO A DEMO-REACTOR

The goal of the blanket programme is to develop concepts for electricity producing reactors. Therefore, it has to be shown for each blanket concept that it can meet the conditions of a DEMO-reactor. These conditions are not really fixed but cover a range of required power density, thermal efficiency and life time [3].

As a first check of the reactor relevance, the NET-team has proposed the following minimum requirements for a DEMO-reactor blanket:

- Total tritium breeding ratio > 1.
- Average neutron wall loading $\geqq 2$ MW/m^2.
- Full power life time $\geqq$ 20 000 h.
- Coolant conditions for electricity production with $\eta_{net} \geqq 20$ %.

Tritium self-sufficiency clearly must be provided for a power reactor. The neutronics analysis, described in section 4, shows that this goal can be met with self-cooled Pb-17Li blanket concepts. An extrapolation of the MHD, thermal and stress analysis, performed for the boundary conditions of NET[4], to the DEMO-conditions indicates that the required power density causes no problems and that at least the thermal efficiency of a pressurized water reactor (PWR) is achievable.

The final choice between the two concepts depends on the actual design of a DEMO-reactor. An "enlarged NET-design", characterized by vertical blanket maintenance, does not allow inboard breeding due to an excessive large MHD pressure drop and, therefore, requires beryllium as a neutron multiplier. In contrast to all other blanket concepts, however, no beryllium is necessary if the reactor design is optimized for the self-cooled blanket, allowing coolant acces tubes at the upper and lower end of the blanket segments.

4. NEUTRONICS

The neutronic performance of both blanket concepts has been analyzed on the basis of Monte Carlo calculations with the MCNP code [5]. The analysis has been performed in a realistic three-dimensional geometrical representation of the DEMO double null configuration, taking the specifications of the UK-DEMO study [3] as guidelines.

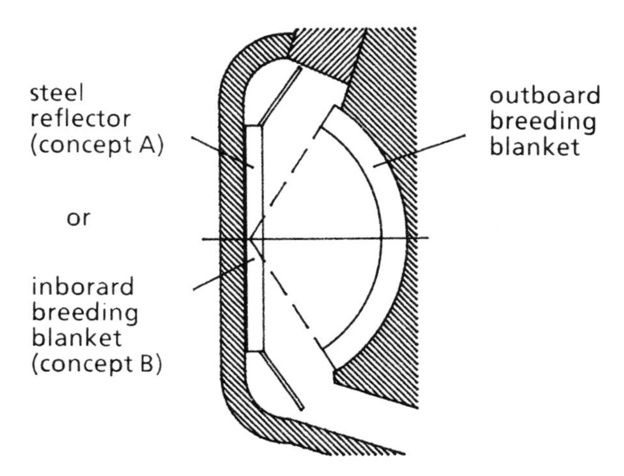

FIGURE 2
Sector model used in the MCNP calculations

Due to the segmentation of the torus into 48 segments, the neutronics analysis can be restricted to the treatment of a 7.5° torus sector. Actually only one half of a sector is treated because of its toroidal symmetry. Reflective boundary conditions are applied at the lateral walls of the half sector.

The sector model takes into account adequately all relevant reactor components: blankets, shields, reflectors, divertors, plugs, openings and ducts. Fig. 2 shows a radial-poloidal cross-section of the sector model as it is used in the MCNP calculations. Figs. 3 and 4 show the corresponding cross-sections of the blanket segments in the torus mid-plane.

The plasma source distribution is represented by a D-shaped, exponentially decreasing probability distribution of the 14 MeV source neutrons [6], adapted to the DEMO double null configuration [3].

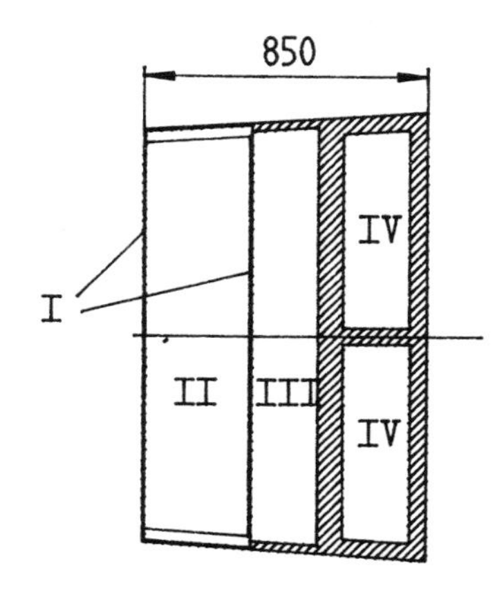

outboard blanket with beryllium multiplier

zone		I	II	III	IV
thickness [mm]		5	320	160	230
composition	steel	1	0.075	0.06	-
	Pb-17Li	-	0.100	0.94	1
	beryllium	-	0.825	-	-

FIGURE 3
Cross-Section of the blanket segments for concept A

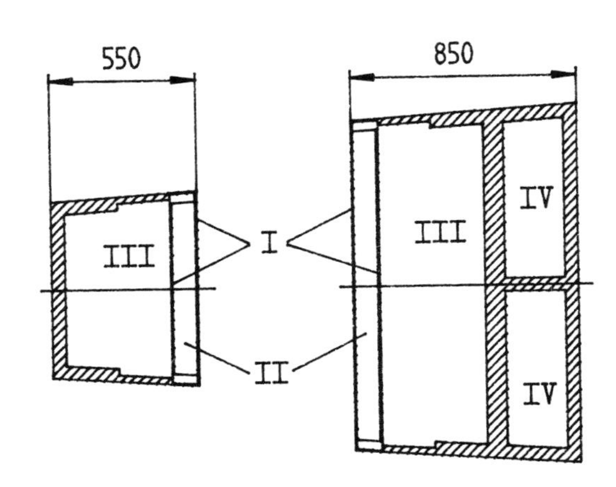

inboard blanket outboard blanket

zone		I	II	III	IV
thickness [mm]		5	90	400	230
composition	steel	1	0.1	0.06	-
	Pb-17Li	-	0.9	0.94	1

FIGURE 4
Cross-Section of the blanket segments for concept B

For assuring a sufficient statistical accuracy a large number of source neutrons has to be considered in the Monte Carlo calculations. For the calculation of global quantities, like the tritium breeding ratio, in general 10 000 to 20 000 source neutrons are taken into account providing a statistical uncertainty in the order of 1 %.

The three-dimensional analysis has been performed for both blanket concepts using alternatively Pb-17Li or Li as breeding material. Table I shows the calculated tritium breeding ratio for all cases, including also the results from one- and two-dimensional calculations.

It is obvious that tritium self-sufficiency can be obtained for both blanket concepts in the DEMO double null configuration. In concept A (no inboard breeding) the poor blanket coverage of about 60 % necessitates, however, the use of a beryllium multiplier. It can be deduced from the three-dimensional analysis, that in case of concept B (without beryllium) Pb-17Li and Li nearly posess the same breeding potential, although in a one-dimensional description Pb-17Li by far is superior to Li (table I). This behaviour is due to the higher reflectivity of Pb-17Li as compared to pure Li-metal, leading to enhanced neutron losses through openings and ducts. In concept A the reflectivity of the system is dominated by the presence of beryllium. Therefore, the (relative) neutron losses do not depend on the kind of breeding material. It is seen, however, that in this case Li is superior to Pb-17Li as breeding material, especially if it is enriched in ^{6}Li.

Table I: Tritium breeding ratio of the liquid metal self-cooled blanket in the DEMO double null configuration, results from one-, two- and three-dimensional Monte Carlo calculations

	concept **A** (with beryllium)			concept **B** (without beryllium)		
	3d	2d	1d	3d	2d	1d
Pb-17Li	1.05	1.26	1.27	1.07	1.47	1.52
Li-30	1.17	1.39	1.40	1.10	1.29	1.37
Li-nat	1.07	1.29	1.30	1.04	1.24	1.35

5. BLANKET TEST MODULE IN NET

Two different locations in NET will be provided for blanket tests:

A) Horizontal ports inside a central outboard segment

B) Complete central outboard segment

The horizontal ports are located at the torus mid-plane and allow a test module with a height of 2 m and a width of 0.65 m. Figure 5 shows the arrangement of such a test module, which can be inserted and withdrawn in horizontal direction without interfering with the cooling of the shielding blankets.

The arrangement of a complete outboard blanket segment is shown in fig. 6. This segment has to be exchanged through a large port at the

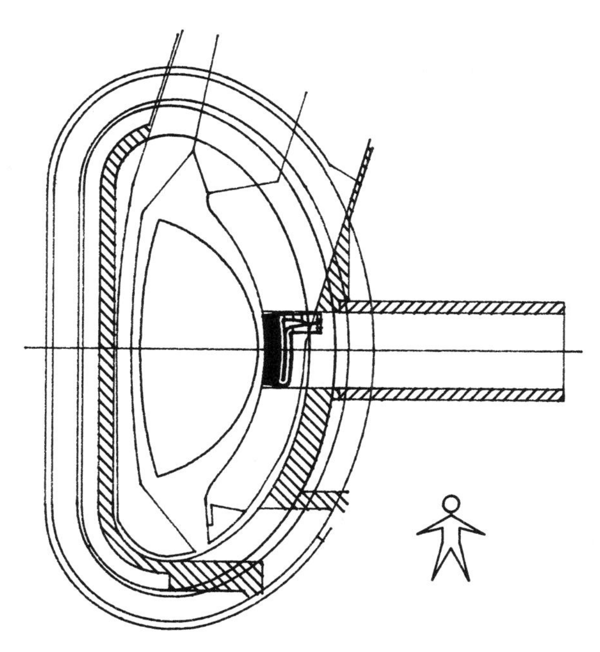

FIGURE 5:
Blanket test module in a horizontal port

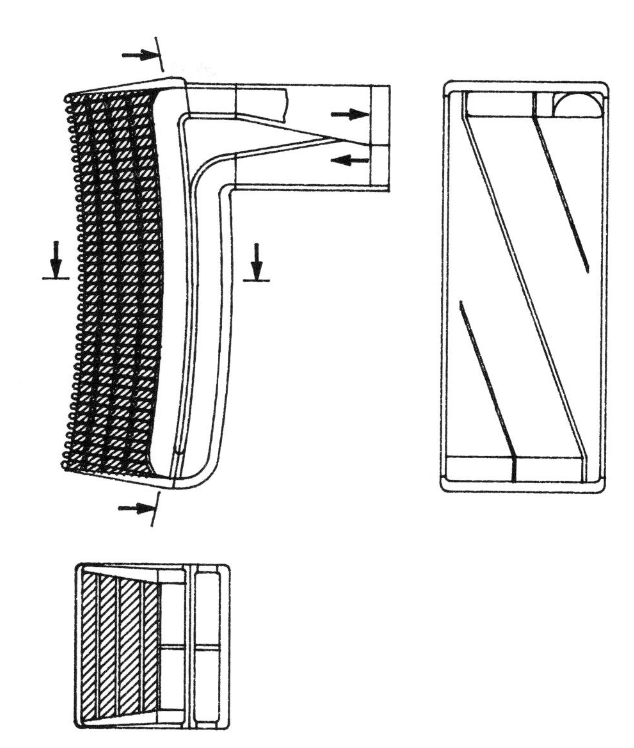

FIGURE 7
Blanket test module for Concept A

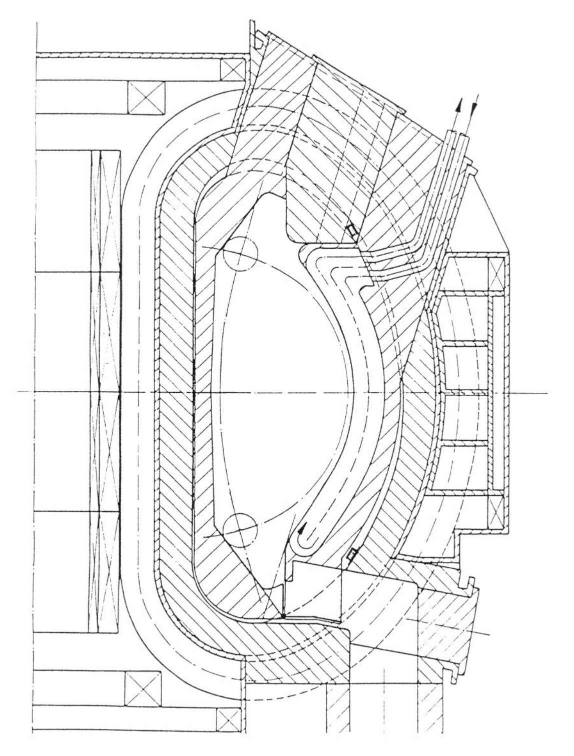

FIGURE 6:
Test blanket segment at the central outboard position

top of the torus which serves not only the test blanket but in addition 3 inboard and 2 outboard shielding segments. It is obvious that a test plug can be exchanged much easier than a segment, facilitating iterative testing of a number of test modules. One module, typical for the case of the blanket concept using a beryllium multiplier is shown in fig. 7. The height of this module is about 1/3 of the height of a blanket segment but it contains all relevant features of the blanket concept. The number of parallel toroidal front channels as well as the ratio length to width at the poloidal ducts are large enough to simulate the conditions in a blanket segment.

6. CONCLUSIONS

The requirements of a DEMO-reactor can be met by self-cooled blankets using the eutectic alloy Pb-17Li both as breeder material and as coolant.

No beryllium is needed in self-cooled blankets

if the tokamak design allows for cooling tubes entering the torus at the upper and lower end of the blanket segment.

A radial plug position is sufficient for testing all relevant features of the blanket concept. This allows first tests during the physics phase of the NET operation and facilitates iterative testing of a number of modules.

ACKNOWLEDGEMENT

This work has been performed in the framework of the Nuclear Fusion Project of the Kernforschungszentrum Karlsruhe and is supported by the European Communities within the European Fusion Technology Program.

REFERENCES

1. D.L. Smith et al.: "Blanket Comparison and Selection Study-Final Report", Vol. 1 through 3, ANL/FPP-8-1, Sept. 1984.

2. S. Malang: "Einrichtung zur Verringerung des MHD-Druckverlustes in dickwandigen Kanälen". DE-PS 3600645 (14.07.1987).

3. P. Reynolds et al.: Study of the Reactor Relevance of the NET design concept. Culham Laboratory Report CLM-R278 (1987).

4. S. Malang et al.: "Self-cooled Liquid Metal Blanket Concept, Fusion Technology (in print).

5. J.F. Briesmeister (Ed.), "MCNP - A General Monte Carlo Code for Neutron and Photon Transport", Version 3A, Los Alamos National Laboratory, Report LA-7396M, Rev. 2 (September 1986).

6. U. Fischer, "Multi-dimensional Neutronics Analysis of the "Canister Blanket" for NET", Kernforschungszentrum Karlsruhe, Report KfK-4255 (April 1987).

FUSION TECHNOLOGY 1988
A.M. Van Ingen, A. Nijsen-Vis, H.T. Klippel (editors)

THERMO-MECHANICAL ANALYSIS OF A SELF-COOLED OUTBOARD BLANKET

K. RUST, S. MALANG and G. SCHMIDT*

Association KfK-Euratom, Kernforschungszentrum Karlsruhe GmbH, Institut für Reaktorbauelemente, Postfach 3640, D-7500 Karlsruhe 1, Federal Republic of Germany
*NIS Ingenieurgesellschaft mbH, Abteilung BK, Donaustraße 23, D-6450 Hanau 1, Federal Republic of Germany

For a self-cooled liquid-metal blanket concept suggested for NET a thermo-mechanical analysis has been carried out. The purpose of the thermal-hydraulic analysis has been to optimize the flow distribution in the toroidal front channels in order to minimize the coolant-to-structure interface temperatures as well as to determine the loadings due to temperature and pressure gradients. First results of a purely linear elastic stress analysis performed for steady state conditions reflect the suitability of the present design.

1. INTRODUCTION

The key feature of a self-cooled blanket concept is the use of the same liquid metal as both tritium breeder and coolant. The viability of a concept based on the use of the eutectic alloy 83Pb-17Li as breeder/coolant material and solution annealed austenitic steel is being investigated for the Next European Torus (NET) at the Kernforschungszentrum Karlsruhe (KfK). The proposed blanket concept[1] offers the promise of a relatively simple mechanical design, a high tritium breeding capability, and a possibility to extract tritium outside the blanket as well as to control the breeder composition continuously during operation. Two main issues involved in designing a liquid-metal cooled blanket are as follows:

1.) The compatibility between the coolant and the structural material limits the allowable coolant-to-structure interface temperature.

2.) Depending on the flow direction the strong magnetic field causes a high pressure drop leading to high mechanical stresses in the duct walls and degrades the heat transport by the liquid metal due to the suppressed turbulence.

As a result of interactive effects such as flow distribution, thermal-hydraulics, magneto-hydrodynamic (MHD) pressure drop, and mechanical stress, the optimized flow concept proposed in the Blanket Comparison and Selection Study (BCSS)[2] has been improved and adapted to the conditions of NET.

2. DESCRIPTION OF THE BLANKET DESIGN

Figure 1 shows a radial-poloidal cross-section of the NET torus. The inboard side is covered by non-breeding reflector segments with integrated divertor plates. In the outboard region are 48 breeding segments arranged which can be inserted and withdrawn vertically through 16 ports at the top end of the torus. All supply lines enter through the exchange port. This avoids the need of decoupling coolant lines inside the plasma chamber. Coolant enters the blanket at the top end, flows downward in the outer part, turns 180 deg at the bottom end, and flows upward to the exit at the top end. This flow direction is perpendicular to the magnetic field and does not allow velocities high enough for sufficient first wall cooling due to the high MHD pressure drop. For this reason, the BCSS flow concept is employed in which the coolant flow is diverted into relatively small toroidal channels (parallel to the magnetic field) in the front part of the blanket in order to cool the first wall properly without excessive MHD pressure drop. Figure 2 shows the schematic of the poloidal/toroidal flow concept. The diversion is

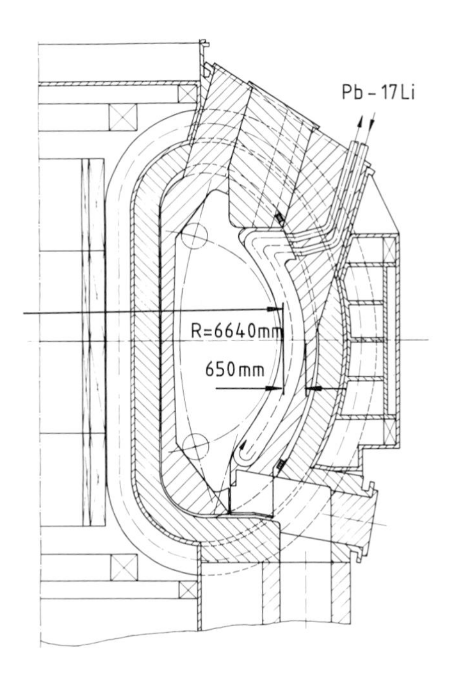

FIGURE 1
Cross-section of a NET torus

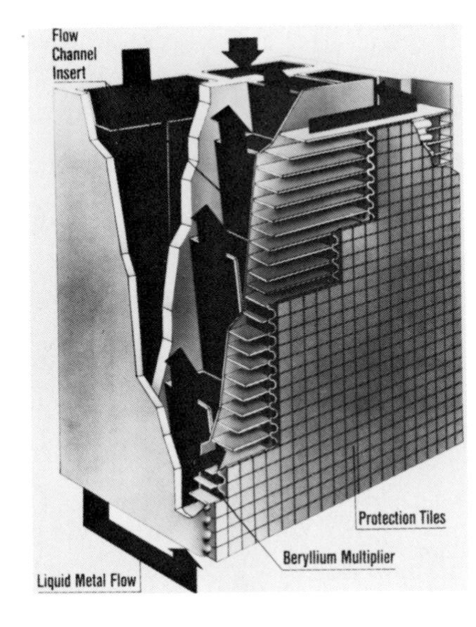

FIGURE 2
Scheme of poloidal/toroidal flow

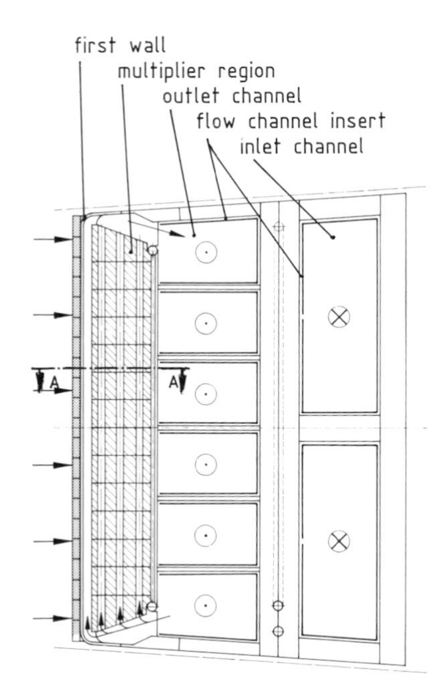

FIGURE 3
Cross-section of a blanket segment

achieved by slightly slanted walls of the return ducts. Figure 3 represents a radial-toroidal cross-section of the outboard blanket. The front part consists of a neutron multiplier region, a corrugated first wall, and a first wall protection against plasma disruptions. By the arrangement of beryllium bricks, which are fitted loosely into a web of steel structure, the need for inboard breeding is avoided[3]. The liquid metal flows through holes in the bricks. At the plasma facing surface, small graphite tiles are fitted loosely into the corrugated steel wall and transfer the heat entirely by radiation. The basic idea behind this concept[4] is to make the heat radiating surface larger than the plasma facing surface in order to lower the graphite temperature. Another design feature indicated in Fig. 3 is the use of novel flow channel inserts (FCI)[1]. The installation of FCI's in all poloidal flow channels offers the promise to diminish the key problem in designing self-cooled blankets, i.e. the interaction of MHD pressure drop, mechanical stress, and duct wall thickness. By the arrangement of FCI's, the load carrying walls are electrically decoupled from the voltage induced in the flowing liquid metal by the magnetic field. The FCI consists of a laminated sheet composed of two thin steel sheets separated by a thin ceramic layer.

3. THERMAL-HYDRAULIC ANALYSIS

The thermal-hydraulic analysis receives input from a neutronic analysis. The spatial distribution of the power density in the first wall, the multiplier region and the blanket have been determined by means of a three-dimensional Monte Carlo calculation. The volumetric heat generation in the blanket caused by the neutron flux results in a heat input to one blanket segment of about 8.2 MW. Taking into account a conservatively high value for the surface heat flux of 0.4 MW/m^2 and a surface area of 4.6 m^2, the total heat input amounts to 10 MW. The flow rate is determined by the allowable temperature rise of the liquid-metal flow between the blanket inlet and outlet. This temperature rise is limited by the allowable maximum temperature at the coolant-to-wall interfaces as dictated by corrosion considerations. An exit bulk temperature of 350 °C has been selected to limit the maximum interface temperature to about 400 °C. The inlet temperature is determined by the melting temperature, which has a value of 235 °C for the eutectic alloy 83Pb-17Li. Therefore, the inlet temperature has been set at 275 °C resulting in an overall temperature rise of 75 K. The volumetric flow rate is then 0.075 m^3/s. About 2/3 of the total blanket power are generated in the first wall and multiplier region, where the coolant flows in toroidal direction. In the poloidal channels, however, the heat-up is only 25 K. In Table I the key data of the thermal-hydraulic and MHD analyses are summarized and provide input to the stress analysis. The MHD pressure drop has been calculated for an average magnetic field strength at the location of the outboard blanket of 4 T.

	length m	average velocity m/s	bulk temp. rise K	MHD pressure drop MPa
poloidal inlet channels	7.0	0.7	5	0.25
poloidal outlet channels	7.0	0.6	20	0.6
toroidal front channels	0.8	0.9 (1.8 max.)	50	0.3 (incl. bends)
U-turn at bottom end	---	0.6-0.7	--	0.1
access tubes inlet outlet	4.0 4.0	1.2 1.2	0 0	0.4 0.4
total			75	2.05 (1.25 blanket)

TABLE I
Main results of thermal-hydraulic and pressure drop analysis

The temperature distribution in the graphite tiles, the corrugated first wall, the beryllium blocks with Pb-17Li cooling channels, and in the stainless steel structure have been calculated in order to optimize the toroidal flow channels and to minimize the coolant-to-structure interface temperatures. The optimization has been carried out by trial and error. The following assumptions have been made: a) Heat transfer from the protection tiles to the corrugated first wall by radiation only, neglecting possible hot spots; b) Slug flow with thermal conduction only because of MHD effect; c) Uniform coolant temperature at the inlet side of all toroidal cooling channels; d) Adjustment of the required coolant velocities by throttling technically possible; e) Adiabatic conditions at the interface of multiplier region and poloidal cooling ducts.

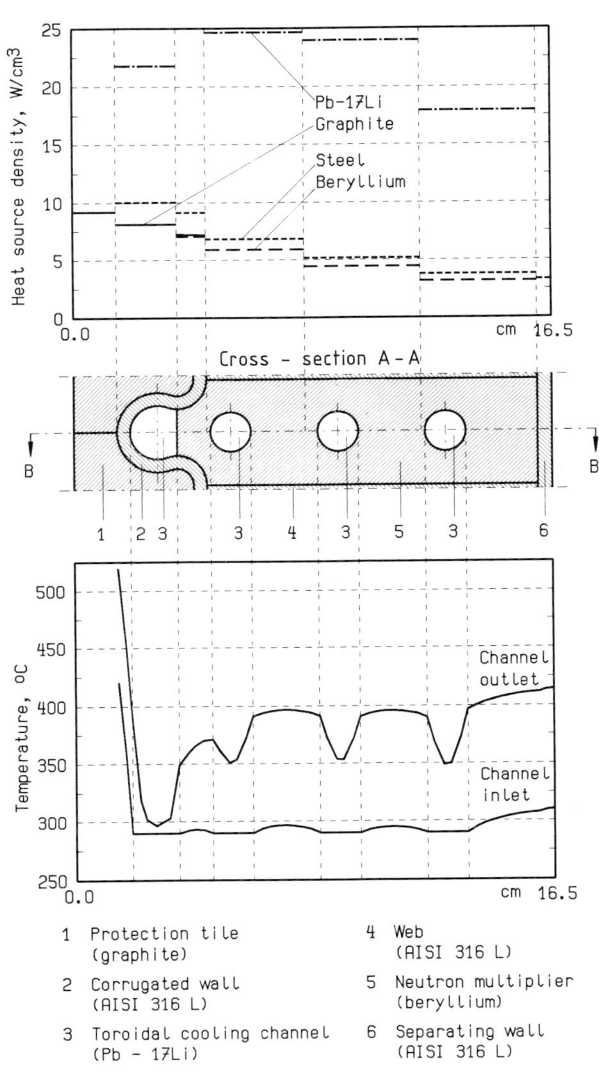

FIGURE 4
Heat source densities and radial temperature profiles at the torus midplane

Figure 4 shows in the upper part the radial volumetric heat generation in the individual materials for the torus midplane. The cross-section of the investigated geometry is shown in the middle part of the graph. The lower part of Fig. 4 shows temperature profiles along the cross-section B-B as calculated for the torus midplane. The plot shows that the coolant-to-structure interface temperatures are lower than 400 °C everywhere assuming flow velocities of 1.8 m/s in the first toroidal cooling channel and 0.36 m/s in the parallel circular channels. The maximum steel temperature is about 520 °C. The steel temperatures are higher at the torus midplane than at the top end of the blanket segment because the influence of the poloidal power profile outweighs the liquid metal temperature rise in poloidal direction.

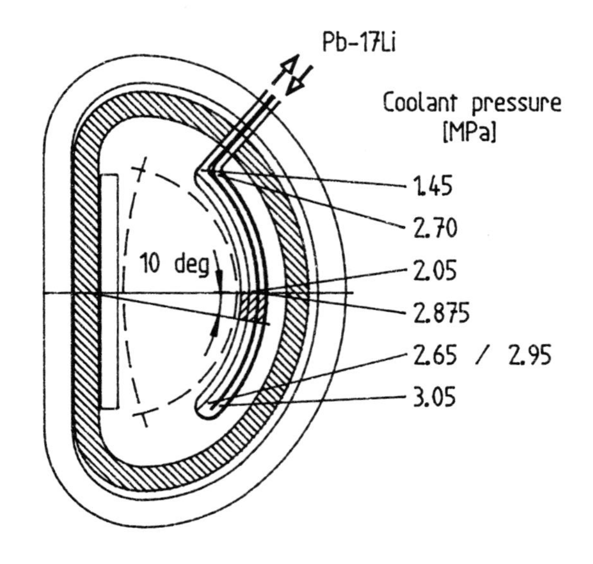

FIGURE 5
Investigated blanket sector and poloidal coolant pressure

4. STRESS ANALYSIS

Stresses in the structure result from the induced coolant pressure differences acting on the blanket walls, summarized in Table I and Fig. 5, and from the nonuniform temperature distribution evaluated by the thermal analysis and shown in Fig. 4. The main objective of the stress analysis has been to find out whether the proposed blanket structure is principally able to withstand the resulting combined stresses under cyclic operating conditions. In spite of the anticipated stresses ranging well beyond the elastic limit, an investigation on the basis of linear elastic theory has been considered to be sufficient for this purpose. The main difficulty hereby results from the rather complex design of a blanket module shown in Figs. 2 and 3. For this reason, two main simplifications have been made before creating the model for the analysis. First, instead of the complete blanket segment only a 10 deg sector as indicated in Fig. 5 has been taken as a representative part. This is justified by the arrangement of the slanted return ducts which leads to similar blanket cross sections every 10 deg in poloidal direction. The selected sector position at the torus midplane results in maximum thermal loads combined with only slightly reduced mechanical loads. The second simplification means neglecting the very weak mechanical coupling between the corrugated first wall and the remaining parts of the structure. I.e. stresses induced in this wall by thermal and pressure loading have practically no influence on the mechanical behaviour of the blanket as a whole, and vice versa. This means that the toroidal front channel structure globally behaves like an orthotropic shell whose elastic properties can be evaluated by detailed analysis of a representative cross section. Consequently, the analysis has been carried out using two different finite element (FE) models. The first one, shown in Fig. 6, by means of a fine mesh serves for the evaluation of thermal and mechanical stresses in the first wall as well as the related orthotropic shell properties. Based on these results, the complete sector has been

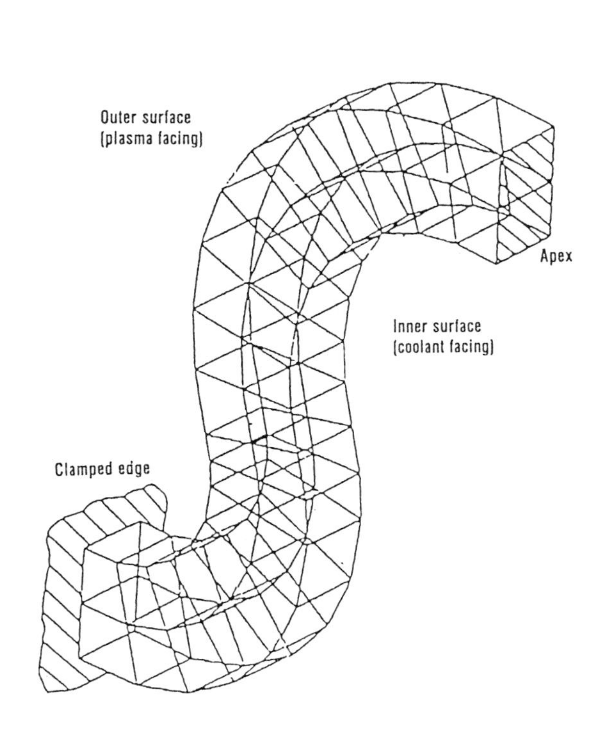

FIGURE 6
Fine FE mesh of first wall

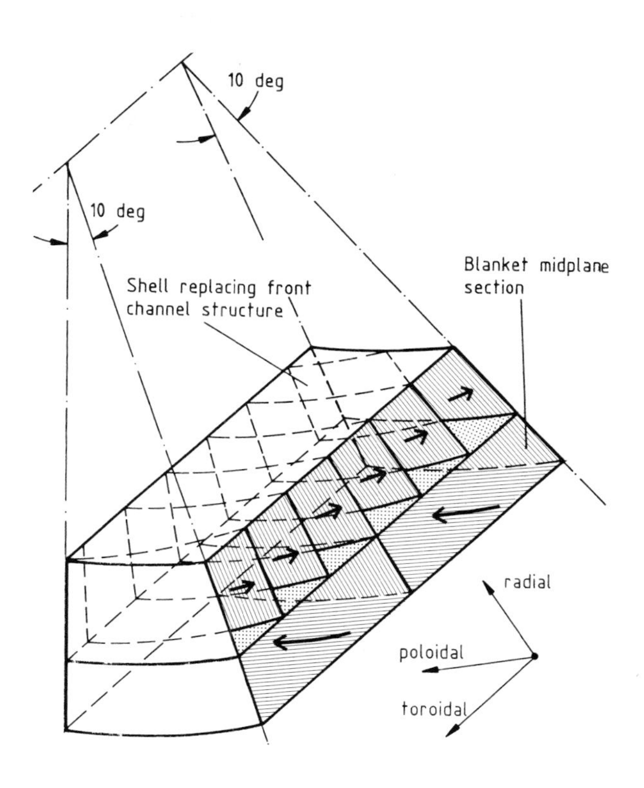

FIGURE 7
Shell model of the investigated sector

modeled corresponding to Fig. 7 as a box type shell structure which is again loaded by pressure and by thermal gradients. This structure has been analysed using a relatively coarse FE mesh. The results reveal the first wall being the most critical part of the total blanket, mainly because of high thermal stresses. Figure 8 shows stress distributions for a representative first wall section represented as rolled out. The dominating influence of thermal stresses is evident, particularly with respect to their cyclic occurence which is important under design aspects. Furthermor, it can be seen that thermal stress maxima correspond with very low mechanical stresses, and vice versa. Moreover, these maxima occur at quite different locations. Thus, thermal stresses remain the dominating ones for design. This is true too for the rest of the structure where mechanical stresses are again of minor importance as a result of sufficiently thick walls. Nevertheless, thermal stress levels in this part turn out to be considerably lower than in the first wall. So the only relevant stress contribution of some importance consists in the toroidal thermal stresses occuring at the plasma facing surface of the orthotropic shell representing the front channel structure. The data are plotted in Fig. 9. This stress component has to be superimposed to the toroidal first wall stresses shown in Fig. 8. According to the previous remarks, the resulting ASME stress intensities are mostly of secondary nature. The maximum Tresca equivalent stresses amount to about 500 MPa in the first wall, and about 175 MPa in the diagonally oriented separating wall between the hot and cold poloidal coolant outlet channels. Although the higher of these values considerably exceeds the 3 S_m limit, this is not considered to be critical for cyclic operation because there appears no possibility for the occurance of ratchetting effects.

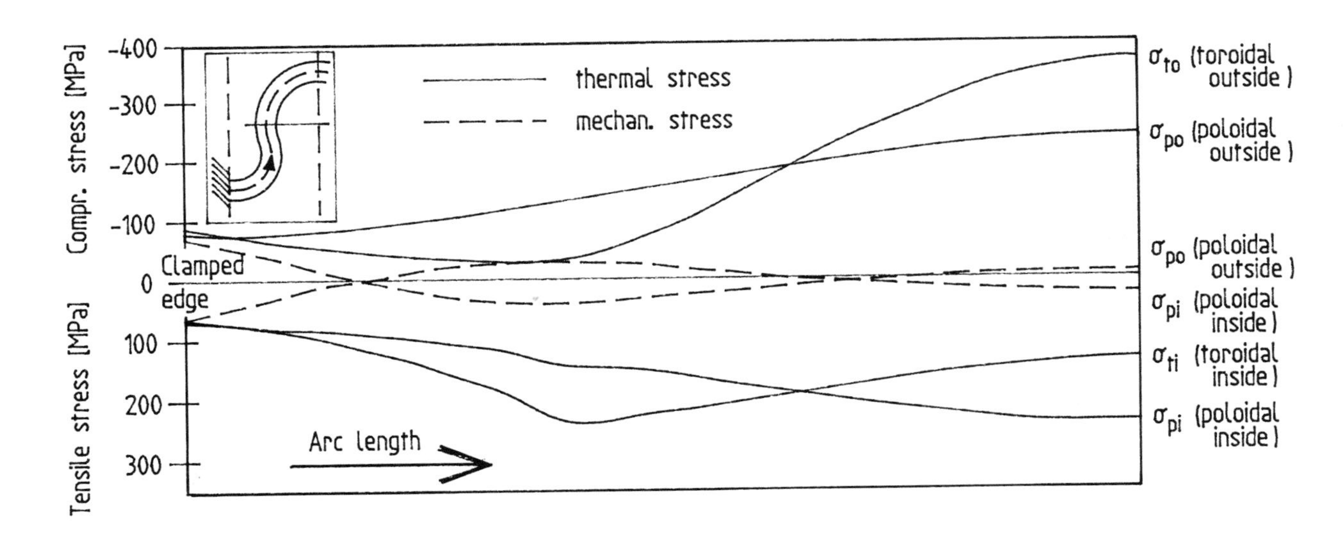

FIGURE 8
Thermal and mechanical stresses along the corrugated first wall (represented as rolled out) at the outlet side of the toroidal cooling channel in the blanket midplane

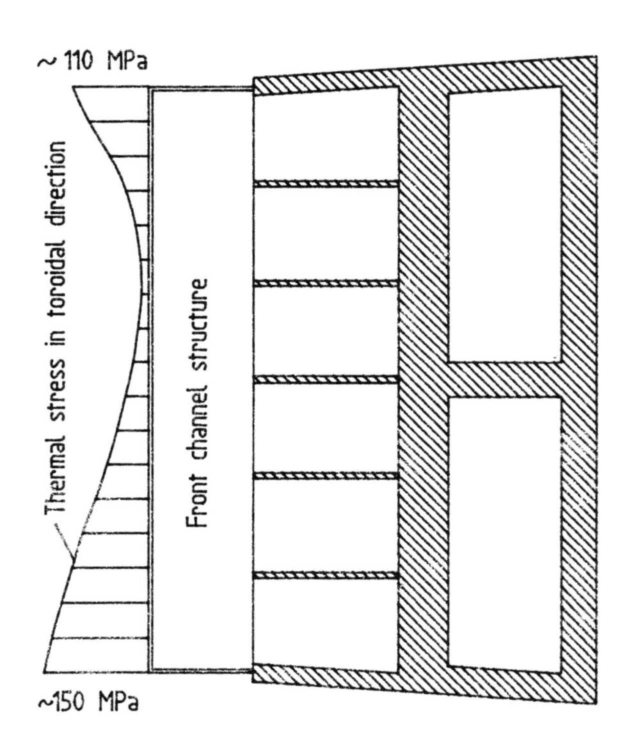

FIGURE 9
Distribution of the thermal stress in toroidal direction at the plasma facing surface of the front channel structure in the blanket midplane

The above results are valid for a blanket segment with free bottom end, where the extrapolated total displacement could reach almost 50 mm. A complete suppression of this deformation would result in a poloidal bending stress of about 82 MPa. However, this has low influence on the maximum stress intensity because the dominating thermal stresses act in toroidal direction.

5. CONCLUSIONS

According to the results of the thermal-hydraulic analysis, the chosen design dimensions and coolant velocities lead to tolerable coolant-to-structure interface temperatures.

The stresses computed for steady state conditions show that

- primary stress levels stay well within allowable limits
- secondary stress ranges do not appear to be prohibitively high for cyclic operation.

So far the overall results reflect the suitability of the present blanket design.

ACKNOWLEDGEMENT

This work has been performed in the framework of the Nuclear Fusion Project of the Kernforschungszentrum Karlsruhe and is supported by the European Communities within the European Fusion Technology Program.

REFERENCES

1. S. Malang et al, Self-Cooled Liquid-Metal Blanket Concept, accepted for Publication in Fusion Technology (May 12, 1988).

2. D.L. Smith et al., Blanket Comparison and Selection Study - Final Report -, ANL/FPP-84-1, Vols. 1, 2, and 3, Argonne National Laboratory (Sept. 1984).

3. U. Fischer, Optimal Use of Beryllium for Fusion Reactor Blankets, Fusion Technology, Vol. 13 (1988), pp. 143-152.

4. S. Malang et al., A First Wall Concept with Radiatively Cooled Protection Tiles, Fusion Technology; Proc. of the 14th Symp., Avignon, France, Sept. 8-12, 1986. in: Oxford [et al.], Pergamon Press, Vol. 1 (1986), pp. 473-478.

FUSION TECHNOLOGY 1988
A.M. Van Ingen, A. Nijsen-Vis, H.T. Klippel (editors)

SURFACE EFFECTS ON TRITIUM PERMEATION THROUGH STAINLESS STEEL WALLS

Renzo CARTA, Anna Maria POLCARO, PierFrancesco RICCI, Giuseppe TOLA, Giancarlo PIERINI(*)

Dipartimento di Ingegneria Chimica e Materiali, Università di Cagliari, Piazza d'Armi, 09123 Cagliari, Italia

(*)Commission of the European Communities, Joint Research Centre, Ispra Establishment, Ispra, Italy

An experimental study on hydrogen penetration in AISI 316 stainless steel samples and on release kinetics from the surface to an aqueous solution has been performed. The influence of the surface characteristics has been evaluated considering the effect of impurities chemisorption and of a homogeneous oxide coating.
The possibility of hydrogen segregation in the subsurface layers of the metal was verified and it was found that small amounts of chemisorbed sulphides cause a large decrease in hydrogen desorption rate. The number of sites available for hydrogen adsorption is strongly lowered by the oxide layer: so the barrier effect on tritium permeation obtained by an oxide coating of the metal surface can be justified.

1. INTRODUCTION

Tritium permeation through metallic walls is strongly affected by surface characteristics as it was pointed out by several works /1,2,3/. In fact it was verified that small amounts of oxides or impurities chemisorption, even if at low concentration, can heavily alter the kinetics of the dissociative adsorption or of the desorption with molecular recombination adding a further resistance to the global process /4,5/.

It must be noted that, if the dissociation reaction is not involved in the adsorption process (i.e. the flux entering the surface is already in atomic form), favourable conditions for high coverage degree of hydrogen or its isotopes can arouse. Consequently the dissolved hydrogen in the metal can reach a much higher concentration than the equilibrium one because the desorption reaction needs that molecular species are formed.

Moreover, when an interface metal-water is involved in the permeation process, as in the case of the cooling loop for D-T fueled machines, the process can be more complicated. Therefore, the aim of this work is to analyse the influence of the metallic surface characteristics on the permeation of atomic hydrogen generated at the surface and the kinetics of the desorption from the metal to an aqueous phase. To this end the behaviour of stainless steel AISI 316 samples with the following surface conditions has been studied.

a) clean surface.
b) inhibited surface by chemisorption of impurities.
c) surface coated by an oxide layer.

The metodology followed is based on electrochemical experimental measures that seem especially useful to obtain the proposed results.

2. SURFACE TREATMENT

The surfaces of the samples (commerial stainless steel AISI 316) were polished mechanically with 600 grade emery paper then with metallographic alumina. It was further degreased in methanol washed with distilled water and then dried with acetone in N_2 atmosphere. After this treatment, the XPS analysis showed only a negligible amount of Cr_2O_3 on the surface.

The inhibited surface was obtained dipping the sample for 1 hour in a weakly alkaline and deaerated solution of Na_2S 10^{-5}M. The XPS analysis revealed the presence of sulphide (162.6 eV) on its surface.

The oxidized surface was obtained leaving the sample, cleaned as previously indicated, in a quartz reactor with an Argon flux at room temperature. Then the temperature was risen to 800 K and the flux of Argon was replaced by an air-steam (1:1) mixture for 1 hour. Successively the sample was slowly cooled in Argon atmosphere. The XPS analysis showed the presence of an uniform layer of Cr_2O_3*Fe_2O_3.

3. EXPERIMENTAL PROCEDURE

The sample was the cathode of the electrochemical cell, adapted for electrolyte recycle, shown in fig.1.

By means of a galvanostat, a cathodic step of fixed length (t_c) and current density was applied to the sample and the open circuit potential, after the current was instantaneously (5 ns) interrupted, was recorded by means of a digital recorder and stored onto a PC IBM.

Another set of runs was performed to obtain the relationship between the imposed current density (i_{SS}) and the steady state potential (V_{SS}).

The recycle flow rate was fixed at 18 cm^3/s corresponding to a Reynolds number in the nozzle of ≈8000 because for this value no resistance to mass transfer in the liquid phase was found out in preliminary runs. More details aboutthe apparatus and the experimental procedure are reported in a previous work /6/.

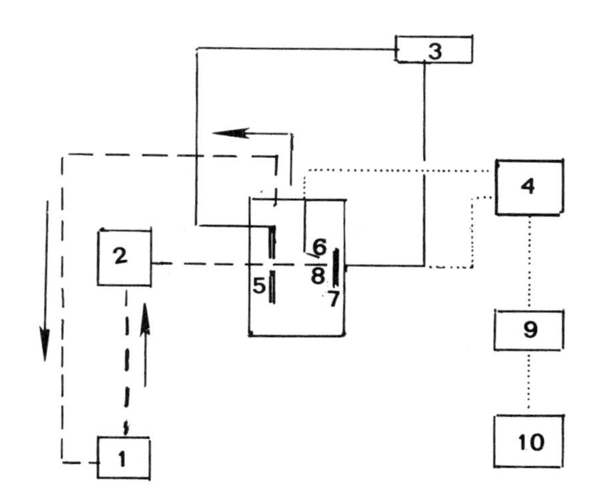

FIGURE 1
Experimental circuit
1) pump; 2) pressure equalizer; 3) galvanostat; 4) electrometer; 5) counter electrode; 6) Luggin capillary; 7) working electrode; 8) nozzle; 9) digital recorder; 10) PC IBM.
- - - electrolyte circuit; —— electric circuit

The behaviour of i_{SS} vs V_{SS} for the different examined surface conditions and a few typical potential decays are plotted in fig.2 and fig.3 respectively.

4. RESULTS AND DISCUSSION

When a step of constant cathodic current is applied to the sample, the following reactions take place:

a. Production of adsorbed hydrogen (H_A) by the proton discharge

$$H_2O + e^- \longrightarrow H_A + OH^- \qquad \text{I}$$

b. Desorption of hydrogen by molecular recombination

$$H_A + H_A \longrightarrow H_2 \qquad \text{II}$$

As a competitive step to reaction II, the ex-

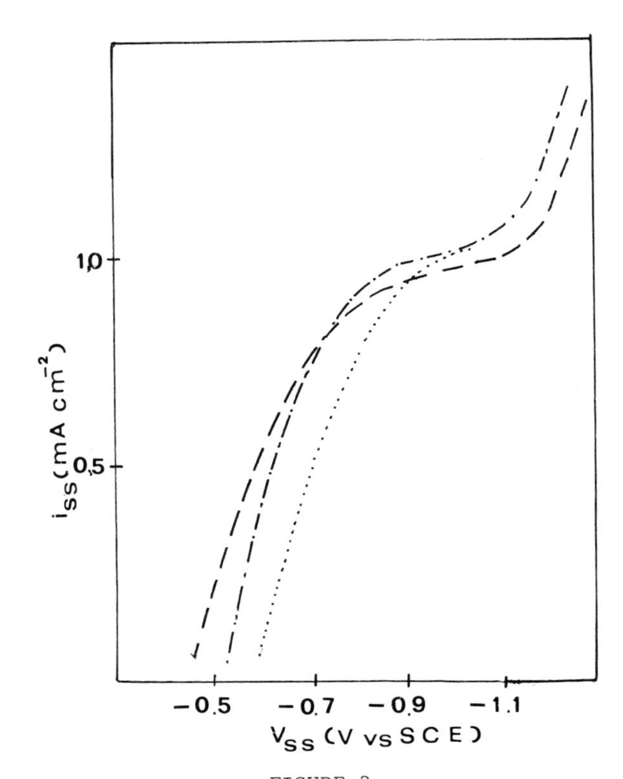

FIGURE 2
Experimental behaviour of i_{SS} vs V_{SS}
- - - oxidized surfaces; -·-·- clean surfaces
..... inhibited surfaces.

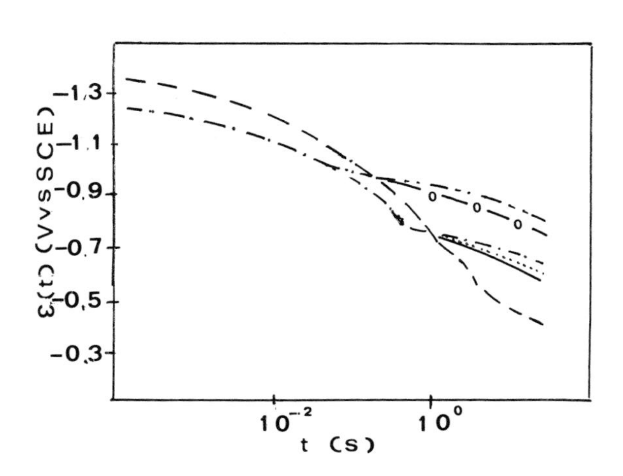

FIGURE 3
Experimental potential decays following steps of 2 mA/cm^2 for different t_c and surface conditions
oxidized - - - t_c = 60s and t_c = 15s
clean -·-·- t_c = 60s; .... t_c = 15s; —— t_c = 4s
inhibited -··- t_c = 60s; -o-o- t_c = 15s

change of hydrogen atoms between surface and interstitial sites and the diffusion into the metallic lattice can happen.

$$H_A \rightleftharpoons H\text{-}Me \Big|_{x=0} \longrightarrow \text{diffusion} \qquad III$$

Under steady state conditions the inlet and outlet hydrogen fluxes are equal. Thus, considering that they are dependent on the coverage degree θ and the kinetic con tant of reaction I is a function of the potential difference across the interface electrode-solution, for each imposed flux $J=i_{SS}/F$ a corresponding value of θ and of V_{SS} is obtained.

When the inlet flux is interrupted, the potential difference electrode-solution changes with time until the equilibrium value obtained before the cathodic step is reached.

During the potential decay, the interface electrode-solution can be represented by means of the following equivalent electric circuit/7/

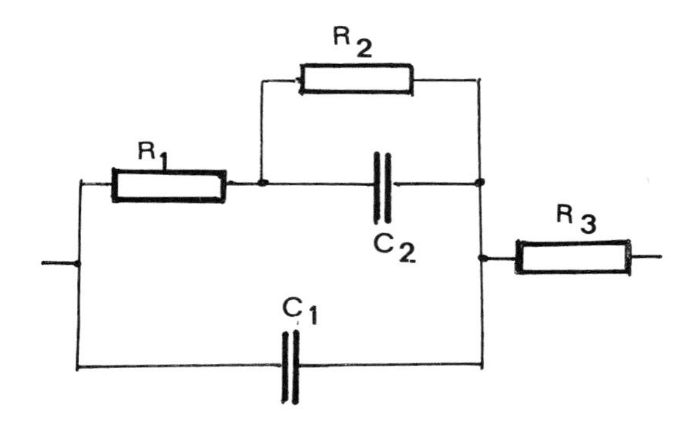

where R_3 is the electrolyte resistance R_1 and R_2 are the equivalent faradic resistances for the discharge and the recombination steps, respectively. C_1 is a capacitor having the double layer capacity C_{dL}, C_2 is a pseudocapacitor due to the charge amount Q_H that can be stored as chemical species during the cathodic charging and released when the current is interrupted.

Its capacity can be expressed as $C_\phi = dQ_H/dV$.

In a treatment based on the assumption that C_ϕ and C_{dL} are parallel elements in the equivalent circuit, C is given by

$$C = C_{dL} + C_\phi = -(i_f/(dV/dt)) \qquad (1)$$

where i_f is the current density associated with the self discharge across the double layer. C can be evaluated experimentally by dividing steady state current i_{SS} by the potential decay at the same potential /8/:

$$C = -(i_{SS}/(d\varepsilon/dt)) \qquad (2)$$

In fig.4 the values of C, calculated by means of eq.2, are plotted vs ε ; it can be seen that, for more negative potential (ε<V*) the interface capacitance is nearly constant and its small value (20 $\mu F/cm^2$) agrees with the value usually obtained for the ionic double layer.

Therefore, in this potential range, the contribution of C_ϕ is negligible because no appreciable variation of the hydrogen coverage degree occurs. As the potential moves towards its primary value, the capacitance exhibits a strong increase because θ starts to change from its value under steady state cathodic current so that the C value begins to be appreciable.

The amount of hydrogen adsorbed and absorbed in the electrode per unit of surface during the cathodic step which precedes the potential decay can be evaluated by numerical integration of the capacitance curve over the potential range of the decay curve /9/:

$$M_t = [H_A] + [H\text{-}Me] = Q_H/F = (1/F)\int C_\phi \, d\varepsilon \qquad (3)$$

In fig.5 the M_t values calculated according to eq.3, are plotted as a function of $\sqrt{t_c}$ (i.e. the duration of the previous cathodic step).

From these curves the amount of adsorbed

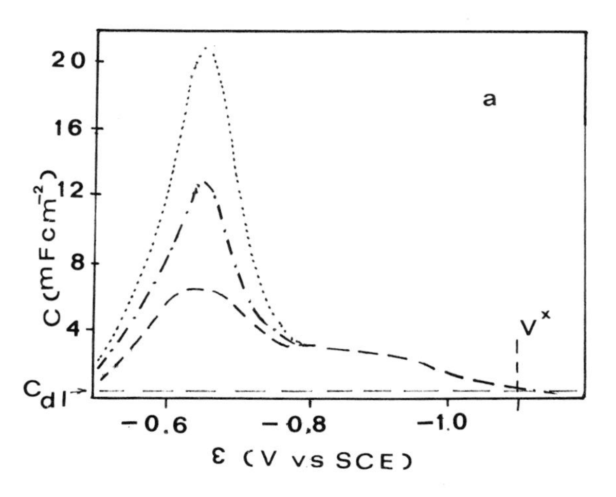

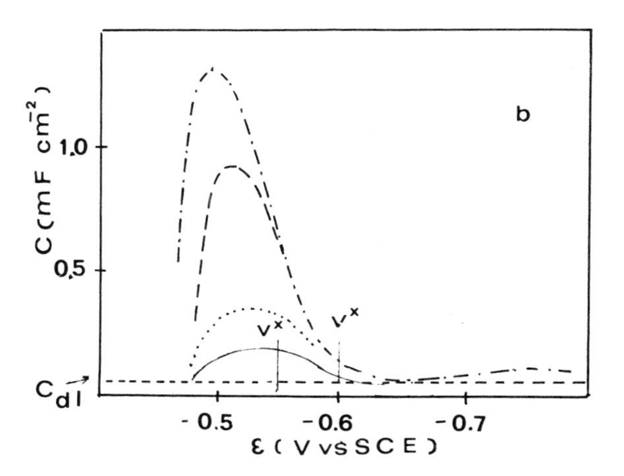

FIGURE 4

Interface capacitance for clean surfaces

a. $i_c = 2$ mA/cm^2 - - - t_c = 4s; -·-·- t_c = 15s
..... t_c = 60s

b. $i_c = 1$ mA/cm^2 ..... t_c = 4s; - - - t_c = 15s
-·-·- t_c =120s

i_c =0.5 mA/cm^2 ——— t_c =120s

hydrogen $[H_A]$ can be obtained considering the intercept at $t_c = 0$.

The values of $[H_A]$, corresponding to different J_1, for clean, inhibited and oxidized surfaces, are reported in Tab.I.

As J_1 increases, higher values for $[H_A]$ are obtained until a constant value $[H_A]_R$, depending on the surface characteristics, is achieved. Therefore, in this condition, it can be

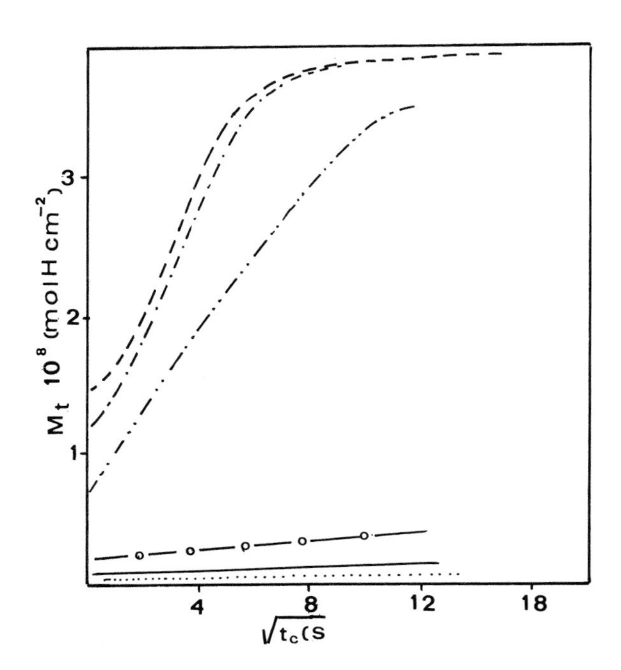

FIGURE 5

Amount of adsorbed and absorbed hydrogen as a function of $\sqrt{t_c}$ for different surface conditions and current densities

clean ..... 0.5 mA/cm^2 ; -o-o-o- 1 mA/cm^2

- - - 2 mA/cm^2

oxidized ——— 2 mA/cm^2 and 5 mA/cm^2

inhibited -··-·· 1 mA/cm^2 ; -·-·-·- 2 mA/cm^2

TAB.I

Surface	$J_1 = i_c/F*10^8$ mol/cm^2 s	H_A*10^8 mol/cm^2	$\theta = H_A / H_{AR}$
	5.18	1.4	1.0
	2.07	1.4	1.0
	1.03	0.2	0.14
	0.52	0.03	0.02
Inhibited	2.07	1.1	1.0
	1.03	0.6	0.54
Oxidized	5.18	0.1	1.0
	2.07	0.1	1.0

assumed that all the adsorption sites are saturated by hydrogen atoms (i.e. the coverage degree θ is close to unity).

Thus the θ value corresponding to each imposed hydrogen flux J_1, can be obtained as $\theta = [H_A]/[H_A]_R$. The data reported in Tab.I show that $[H_A]_R$ slightly decreases for surfaces poisoned by $S^=$ chemisorption and greatly changes when the sample surface is coated by an oxide layer.

This behaviour can be justified taking into account that the active sites of modified surfaces are partially or fully blocked by sulphide or oxigen chemisorption.

From the curves in fig.5 it is also possible to calculate the amount of absorbed hydrogen per unit surface as

$$[H\text{-}Me] = M_t - [H_A] \tag{4}$$

For samples with oxidized surfaces a nearly constant M_t value is obtained, therefore H-Me appears to be close to zero in the whole range of examined charging times.

For clean surfaces a linear trend of M_t vs $\sqrt{t_c}$ is verified in the case of small Θ values. this agrees with the hypotesis that, during the cathodic charge, the hydrogen absorption is controlled by the diffusion in the metal bulk, moreover the concentration C_0 in the first layer adjoining the surface can be considered constant and equalto the equilibrium with adsorbed hydrogen ($C_0 = [H_A]/K_V$). Considering our sample as a semi infinite medium, the equation for diffusion in one dimension can be integrated when the boundary is kept at a constant concentration C_0,the initial concentration beingzero through the medium:

$$M_t = 2C_0(D_H t/\pi)^{1/2} \tag{5}$$

Therefore, from the slope of the straight lines M_t vs $\sqrt{t_c}$, assuming $D_H = 1.1*10^{-12}$ cm^2/s at 300 K /10/, the C_0 value, corresponding to each $[H_A]$and K_V can be calculated.

For samples with inhibited surfaces or clean

surfaces charged at higher current densities (i.e. when a θ value tending to unity is achieved) the M_t curve shows a sharp rise followed by a much slower variation with time. This behaviour suggests that, owing to the high coverage degree, a remarkable concentration of dissolved hydrogen in the layer adjoining the surface is quickly established. Nevertheless, the volume expansion associated with this high concentration is not allowed in the metal bulk; so, after the narrow sub-surface region is saturated, only a much smaller hydrogen uptake can occur even if the charging times grow longer.

The hydrogen-induced phase transformation, detected in the sub-surface region (a depth of a few μm) of an austenitic stainless steel sample cathodically charged at current densities higher than a critical value, could confirm this hypotesis /11,12,13/.

A different approach, in which the kinetic equations for the elementary steps are considered, enables to evaluate the numerical value of the kinetic constant for the desorption reaction /14/. Indeed, a relationship for the potential decay can be obtained from the mass balance for the electric charge and for the hydrogen adsorbed on the electrode.

$$(C_{dL}/F)(d\varepsilon/dt) = J_1 \quad (6)$$

$$[H_A](d\theta/dt) = J_1 - 2J_2 + J_3 \quad (7)$$

In eq.6 J_1 is given by /15/:

$$J_1 = [H_A]_R K^*(1-\theta)\exp(-\alpha F\varepsilon/RT) - [H_A]_R K^*_- \exp(\alpha F\varepsilon/RT) \quad (8)$$

J_2 can be expressed as:

$$J_2 = -2K_2 [H_A]_R^2 \theta^2 \quad (9)$$

neglecting the reverse adsorption reaction, since the fluido dynamic conditions adopted in the experimental runs determine a very small concentration of molecular hydrogen in the liquid boundary layer, J_3 can be written as

$$J_3 = (K_1/\delta)[H\text{-}Me] \quad (10)$$

assuming that the decomposition of the hydrogen phase is a first order reaction /15/and putting the concentration per unit volume in the sub-surface hydrogenated region equal to $[H\text{-}Me]/\delta$, where δ is the thickness of this region.

In the early stages of the potential decay, θ does not change significantly from its initial steady state value; so dθ/dt is close to zero and the interface capacitance is equal to C_{dL}. Then θ begins to change and both eqs. 6 and 7 are required to describe the potential decay.

Nevertheless in the last stages of the transient, when J_1 is somewhat little, a simple approximation ($J_1 \simeq 0$) could be adopted:

$$(\theta(t)/(1-\theta(t)) = (K^*/K^*_-)\exp(-F\varepsilon(t)/RT) \quad (11)$$

$$(d\theta(t)/dt) = -2K_2[H_A]_R\theta^2 + (K_1/\delta)([H\text{-}Me]/[H_A]_R \quad (12)$$

Moreover, if the potential decay follows a step of brief duration and low current density, $[H\text{-}Me]$ can be considered nil and eq.12 can be analitically integrated and combined with eq.11 to give:

$$\varepsilon(t)-\varepsilon(0)=(RT/F)\mathrm{Ln}((1-\theta(0)/(1+\theta(0)(K_2+1))) \quad (13)$$

where ε(0) and θ(0) represent the open circuit potential and the coverage degree at t=0.

From the best fit of the experimental transient, the numerical value of K_2 can be obtained. Furthermore, inserting the appropriate value of K_2, eq.12 can be numerically integrated by looking for the best agreement between

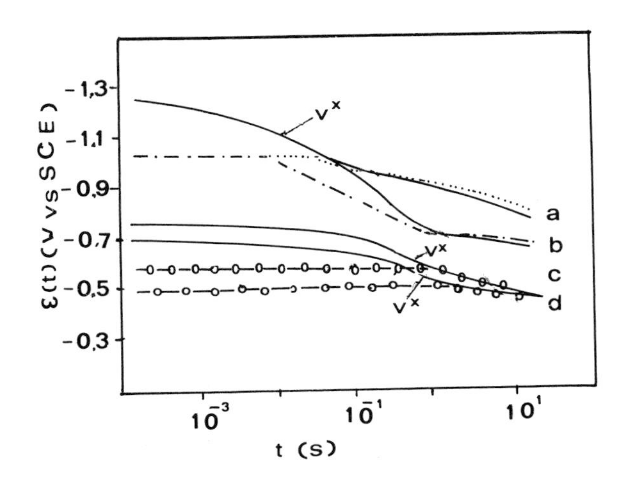

FIGURE 6

Comparison between calculated and experimental potential decays

inhibited surfaces:

curves a. i_c = 2 mA/cm^2 ——— experimental; ········ calculated

clean surfaces:

curves b. i_c 2 mA/cm^2 ——— experimental; -·-·-·- calculated

curves c. i_c = 1 mA/cm^2 ——— experimental; -o-o-o calculated

curves d. i_c = 0.5 mA/cm^2 ——— experimental; -o-o-o calculated

the calculated and the experimental transients (fig.6).

The calculated values of K_2 and K_1/δ are shown in Table II for different surface characteristics. The K_2 value obtained for clean surface can be compared with the desorption constant K_r reported in literature /17/ putting $K_r = K_2 H_v^2$. A good agreement between the K_2 value obtained in this work and that reported by Wienhold et al. /18/ at the same temperature is found.

If the surface is poisoned by $S^{=}$ chemisorption, K_2 diminishes by a factor 100 and so it can be justified that θ values higher than those obtained for clean surfaces are established for the same J_1.

TAB.II CONSTANT VALUES CALCULATED AT 300 K

K_1/δ cm/s	6	
K_2 cm^2/mol s	$10.7*10^8$	(clean surface)
	$10.7*10^6$	(inhibited surface)
K_v cm	10^{-5}	
K_r cm^4/s atoms	$9*10^{-26}$	

5. CONCLUSIONS

When an enough strong flux of atomic hydrogen hits the surface of an AISI 316 stainless steel wall, the coverage degree of adsorbed hydrogen can reach such values that a high concentration in the metallic layers adjoining the surface is established. Nevertheless this high concentration is bounded to the sub-surface region because in the metal bulk the lattice expansion caused by this concentration is not allowed. So, in this case, the high hydrogen fluxes foreseen by the diffusion equation with constant D can not be actually achieved.

As already reported in previous works /4,5/, the chemisorption of impurities in the metal surface heavily lowers the constant of hydrogen desorption. For the experimental conditions adopted in the present work a reduction factor of about 2 orders of magnitude was obtained. This reduction favours the formation of hydrides in the sub-surface layers, for the same hydrogen flux.

An homogeneous oxide layer causes a large reduction in the available adsorption sites on the surface. Consequently, an inhibition both of the adsorption reaction and of the hydrogen atoms exchange between interstitial sites of metallic lattice and surface sites can occur. Thus it can be justified that a thin protective oxide layer effectively reduces hydrogen permeation through austenitic stainless steel walls /19/.

REFERENCES

1. L.G. Earwaker, D.K. Ross and J.P.G. Farr, IEEE Trans. on Nucl. Sci. Ns-28 (1981) 1848
2. F.H. Heubaum and B.J.Berkowitz, Scripta Metall. 16 (1982) 323
3. M.R. Piggot and A.C. Siarsowsky, J. Iron and Steel Inst. 210 (1972) 907
4. D.M. Grant, D.L. Cummings and D.A. Blackburn, J. Nucl. Mat. 149 (1987) 180
5. M.A. Pich and K. Sonnenberg, J. Nucl. Mat. 131 (1985) 131
6. R. Carta, M.S. Dernini, A.M. Polcaro, P.F. Ricci and G. Tola, J. Electroanal. Chem. in print
7. E. Gileadi, E. Kirowa-Eisner and J. Penciner, in Interfacial Electrochemistry (Addison-Wesley Publishing Comp. Inc., London, 1975) p.92
8. E. Meier and B.E. Conway, J. Appl. Electrochem. 17 (1987) 1002
9. B.E. Conway and L. Bay, J. Electroanal. Chem. 198 (1986) 149
10. F. Reiter, J. Camposilvan, M. Caorlin, G. Saibene and R. Sartori, Fusion Technol. 8 (1985) 2344
11. A. Szumer and A. Janko, Corrosion 35 (10) (1979) 461
12. A.P. Butley and G.C. Shith, Metall. Trans. A 17 (1986) 1593
13. T.P. Perng and C.J. Altstetter, Metall. Trans. A 18 (1987) 123
14. D.A. Harrington and B.E. Conway, J. Electroanal. Chem. 221 (1987) 1
15. K.J. Vetter, in Electrochemical Kinetics (Academic Press Inc., New York-London, 1967) p.518
16. T. Zakroszymsky, Z. Szklarska-Smialoska and M. Smialowsky, Corrosion 39 (1983) 207
17. R.A. Langley, J. Nucl. Mat. 128-129 (1984) 622
18. P.Wienhold, R.E. Clausing and F. Waelbrock J. Nucl. Mat. 93-94 (1980) 450
19. G. Luthardt and W. Mallener, Fusion Technol. 8 (1985) 2379

FUSION TECHNOLOGY 1988
A.M. Van Ingen, A. Nijsen-Vis, H.T. Klippel (editors)

3-D MONTECARLO ANALYSIS OF THE NET SHIELDING SYSTEM

M. Gallina, L. Petrizzi, V. Rado

Associazione EURATOM-ENEA sulla Fusione, Centro Ricerche Energia Frascati,
C.P. 65 - 00044 Frascati, Rome, Italy

1. INTRODUCTION

During the first phase of the physics program, NET should be operated with an extended plasma size[1] maintaining, however, the NET-DN basic dimensions for the magnet and for the vacuum vessel shielding system[2]. This means that a reduced space is available for the blanket. Moreover, as less stringent technological conditions are desirable in this phase, a more reliable solution has been foreseen for the blanket which has been designed to fulfill only shielding and heat removal requirements and is mainly composed of water cooled stainless steel[3,4].

Within the frame of the European Task on shielding design studies (N2), a detailed 3-D neutronic analysis has been carried out for this configuration using the MCNP Monte Carlo code.

The analysis has been mainly oriented to access the shielding performances related to the radiation limits on the superconducting coils. An overall characterization of the neutron and photon field in the whole system has also been done.

Typical selected results are reported in this paper.

2. THREE DIMENSIONAL GEOMETRY MODEL

The geometry of the system, as treated by MCNP code, is shown in Fig. 1. A toroidal sector of 11.25° has been selected from the center of a magnet to the center of a pumping duct. The reflective boundary conditions at the side planes reproduce the whole system which is composed of 16 equal sectors. The slice of the geometry represented includes 3/2 of the 48 blanket modules. The void gap (2 cm) between the blanket modules, and the copper plates (1 cm) on the blanket lateral walls have also been represented. The outboard blanket has been truncated at the upper plane boundary of the vacuum pumping duct. This choice represents the worst limit condition for shielding the bottom part of the magnet. The duct across and outside the vacuum vessel has been shaped to provide a shielding thickness of 40 cm toward the magnet lateral side. The blanket and the bulk shield are composed by alternate layers of SS and water. Borated water is used, in the vacuum vessel. A more detailed description of the radial layout and material composition, according to the NET team specifications[2,4,5], is reported in Table I.

3. PLASMA SOURCE

For the extended size configuration the plasma param-eters are[1]

Major radius	RP = 541 cm
Minor radius	AP = 168 cm
Elongation	E = 2.18
Radius shift	SH = 29 cm
Triangularity	TR = 0.62

The neutron source distribution has been sampled according to the following parametric equation for the plasma contour lines[7]

$$R = RP + a \times \cos(t + TR \times AP \times \sin t) + SH \times (1 - (a/AP)^2)$$

$$I = (1 - (a/AP)^2)^4$$

where R is the toroidal radius, z the vertical ordinate and I the fusion power density.

A set of D-shaped level lines equally spaced

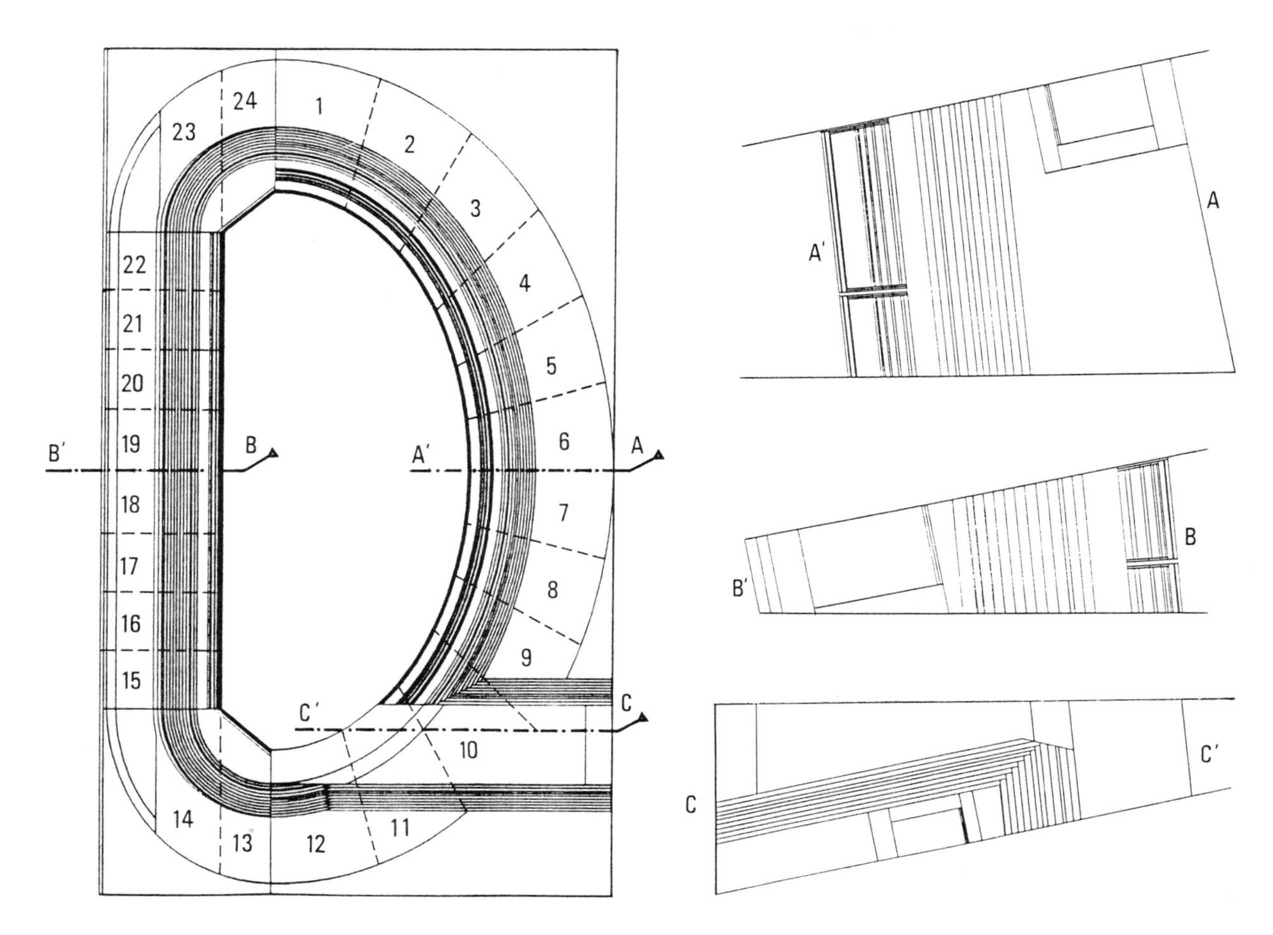

FIGURE 1
A MCNP geometry for the NET system.

TABLE I - Radial layout and material composition

Composition	INBOARD		OUTBOARD	
	R (cm)	ΔR (cm)	R (cm)	ΔR (cm)
Graphite tiles	353.5	1.5	723.5	1.5
FW: 0.74 SS+0.26 water	352	3	726.5	3
Void			741	14.5
Blanket: 0.75 SS+0.25 borated water	349	17	760.5	19.5
Void	332	16	774	13.5
Inner shield: 0.90 SS+0.10 borated water	316	21	795	21
Outer shield: 0.75 SS+0.25 borated water	295	37	825	30
Void	258	10	845	20
Inner case: SS			855	10
Insulator inmost layer	248	1.2	856.2	1.2
SC coils inmost layer	246.8	1.2	857.4	1.2
SC coils	245.6	54.1	911.5	54.1
Outer case: SS	191.5	13.5	928.3	16.8

in intensity is shown in Fig. 2a. The poloidal distribution of the neutron wall loading on the first wall profile is given in Fig. 2b. All the values reported in this paper refers to a total fusion power of 740 MW. The cosine mean value of the incident angle along the poloidal profile has also been calculated and reported in Fig. 2c.

4. COMPUTING METHODS

The 3-A version of the MCNP Monte Carlo code[6] has been used in the neutronic analysis. For the deep penetration problem related to the doses on the magnet, the self-generated *weight windows* technique has been adopted to reduce the variance. This technique is based on a space and energy dependent importance function internally estimated by the code.

The cross section libraries officially released with the code have bee used in the present case. A MCNP cross section library derived from the European Fusion File has already been produced and used in some testing cases. However it has been realized by the EFF people that the processing paths adopted to calculate the kerma factors have to be revised. Therefore, further use of the European File in routine calculations has been delayed.

5. TWO DIMENSIONAL CALCULATIONS

A 2-D model has also been adopted to analyses the same system. It consists of an infinite cylindrical sector representing the geometry of the midplane cross section. This simplified approach is normally used for parametric analyses of more detailed configurations.

In the present case, the comparison of the two models, treated by the same methods and data, can select pure geometrical bias factors to be used in extrapolating the results of a one-dimensional analysis.

The comparison is related to the normalization criterion adapted for the infinite geometry model, which is usually the wall loading on the outboard mid-plane.

Some selected result obtained from the two computing models are reported in Table II. It can be noticed that the angular distribution of the incident source neutron (represented by the mean cosine $<\mu>$) is quite different in the two cases. This can explain that, in the 2-D case, the flux-like quantities in the inner part of the blanket are over estimated and the deep penetration is under estimated.

6. THREE-DIMENSIONAL CALCULATION RESULTS

A discrete radial and poloidal distribution of the neutron and photon flux and of the power density has been calculated over the whole system. The radial layout (on the mid-plane) and the poloidal

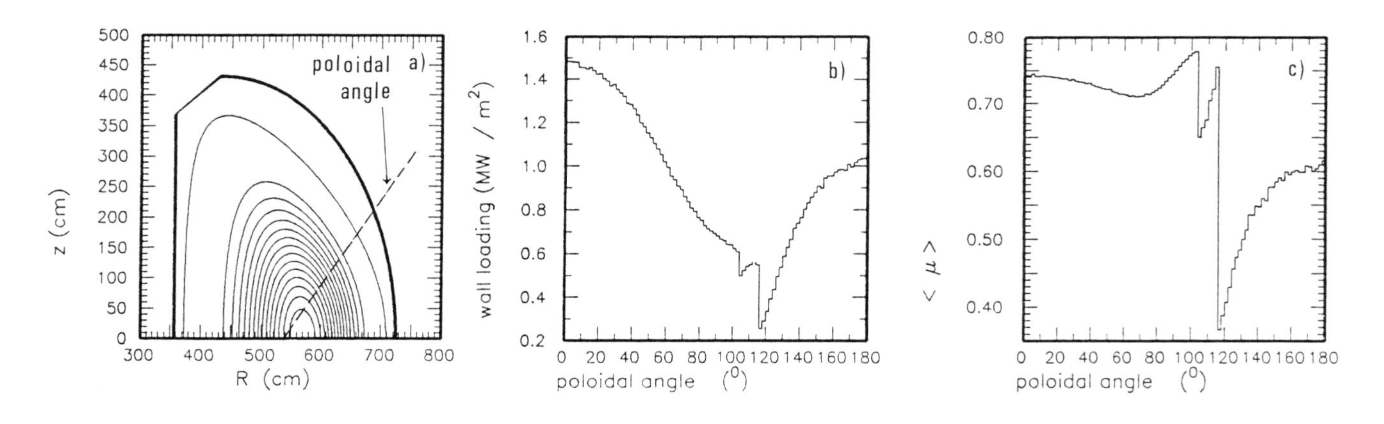

FIGURE 2
a) Plasma neutron source distribution (level lines). b) Neutron wall loading, poloidal distribution.
c) Mean cosine of the incidence angle, poloidal distribution.

TABLE II - 2-D - 3-D (mid-plane) comparative results

			OUTBOARD		INBOARD	
			2-D	3-D	2-D	3-D
Wall loading		(MW/m^2)	1.48	1.48	0.82	1.04
<μ>			0.47	0.73	0.40	0.53
First wall:	neutron flux	($n\ cm^{-2}\ s^{-1}$)	3.96 (+14)	2.84 (+14)	3.06 (+14)	2.44 (+14)
	fast flux	($n\ cm^{-2}\ s^{-1}$)	2.60 (+14)	1.94 (+14)	1.84 (+14)	1.58 (+14)
	photon flux	($n\ cm^{-2}\ s^{-1}$)	2.80 (+14)	2.04 (+14)	2.24 (+14)	1.82 (+14)
	Power density	(W/cm^3)	16.8	12.7	12.8	10.6
End of blanket:	neutron flux	($n\ cm^{-2}\ s^{-1}$)	5.51 (+13)	5.74 (+13)	6.47 (+13)	6.39 (+13)
	fast flux	($n\ cm^{-2}\ s^{-1}$)	2.55 (+13)	2.74 (+13)	3.01 (+13)	3.14 (+13)
	photon flux	($n\ cm^{-2}\ s^{-1}$)	3.96 (+13)	4.01 (+13)	4.45 (+13)	4.7 (+13)
	Power density	(W/cm^3)	2.12	2.13	2.47	2.6
SC Insulator:	neutron flux		1.13 (+11)	1.20 (+11)	8.28 (+10)	1.1 (+11)
	power density		9.33 (-4)	1.02 (-3)	1.18 (-3)	1.6 (-3)

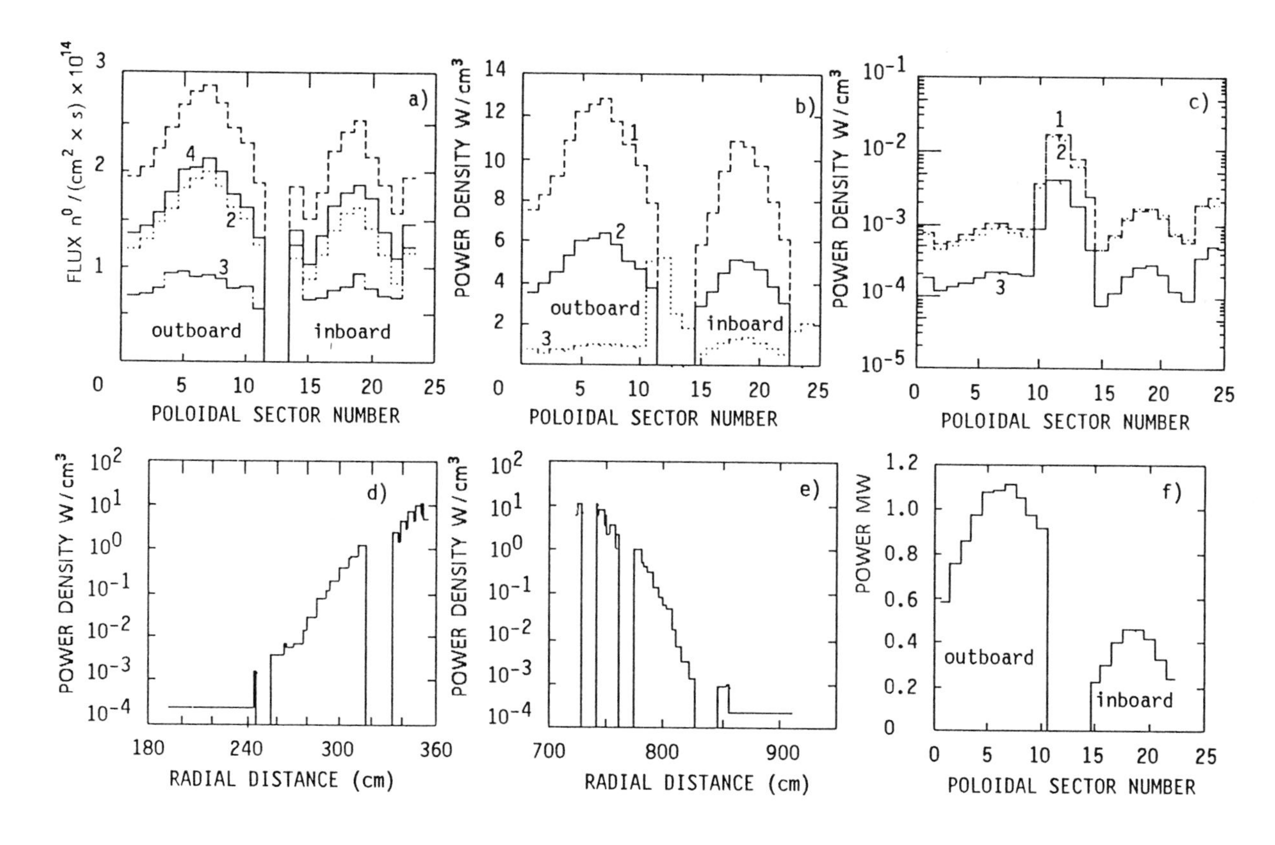

FIGURE 3

a) Poloidal distribution of radiation field on the first wall:
1. n total flux;
2. fast flux (E>0.1 MeV);
3. thermal flux (E<0.4 eV) to be divided by 10;
4. γ flux.

b) Poloidal distribution of power density:
1. first wall;
2. graphite tiles;
3. vacuum vessel (inmost layer).

c) Poloidal distribution of power density:
1. inmost insulator layer;
2. inmost SC coils layer;
3. SC coils (mean).

d) Radial (mid-plane) distribution of power density (inboard)

e) Radial (mid-plane) distribution of power density (outboard)

f) Power by poloidal sector in the blanket.

TABLE III

Power deposited in the system

Inboard blanket module	(Tot. 48)	2.84 MW
Outboard blanket module	(Tot. 48)	9.41 MW
Inboard vessel segment	(Tot. 16)	2.2 MW
Outboard vessel segment	(Tot. 16)	5.3 MW
SC magnet	(Tot. 16)	21.4 kW

Dose peak values on the magnet system

	INBOARD		OUTBOARD	
	Mid-plane	Bottom	Mid-plane	Bottom
Fast neutron fluence*(n cm^{-2})	1.07 (+18)	8.9 (+18)	1.28 (+18)	2.33 (+19)
Integral dose on the insulator*	2.8 (+9)	6.2 (+9)	1.8 (+9)	3.05 (+10)
Power density on the SC coils	1.65 (-3)	6.32 (-3)	9.15 (-4)	1.5 (-2)

* For 1-y. Operating time

segmentation which have been adopted in this case, are indicated in Table I and in Fig. 1. A selection of typical results is shown in Fig. 3a,b,c,d,e,f. Integral values of power deposition and the most significant point values for the doses are been summarized in Table III.

7. CONCLUSIONS

The irradiation limits for the magnet system during the NET operation have been specified as 5×10^{17} for the fast neutron fluence, 5×10^{8} Rad for the integral dose on the SC insulator and 10^{-3} W/cm^3 for the maximum power density on the SC coils.

In the zones shielded by the blanket and by the bulk shield, relative maximum values have been found at the mid-plane. In this points the integral dose limit on the SC insulator would be exceeded by a factor 6 in 1-year operating time.

The dose values issued by 3-D calculations overtake the estimations given by 1-D Sn methods (reported in Ref. 5) by a factor ranging from 2 to 4. This conclusion agrees with the results reported by other authors also[8].

At the bottom part of the magnet, the calculated doses dramatically exceed any reference limit by more than one order of magnitude. This effect is partially due to the very conservative representation adopted for the bottom zone which is not yet definitely designed. In the divertor region the material thickness (actually 10 cm) can be easily increased. In the outboard bottom part, the blanket truncation exposes the vessel wall to the direct view of the plasma source. In this case, a suitable shielding solution, fitting the extended plasma configuration and the vacuum pumping requirements, is not evident.

REFERENCES

1. NET Status Report, December 1987, NET Report 84.
2. NET Status Report, December 1985, NET Report 51.
3. B. Libin, NET Shielding Blanket for Extended Plasma, Rep. NET/IN/86-074.
4. W. Dänner, Shielding blanket and their nuclear performances, Rep. NET/IN/87-021.
5. W. Dänner, The Shielding properties of the vacuum vessel, Rep. NET/IN/87-022.
6. MCNP version 3A, LA-7396-M, Rev. 2, (Manual) September 1986 .
7. K.A. Verschunr, Poloidal variation of the NET blanket nuclear response functions, ECN-87-001 (1987).
8. U. Fisher, Monte Carlo Shielding Calculation in the DN Configuration of NET, KFK 4411 (May 1988).

FUSION TECHNOLOGY 1988
A.M. Van Ingen, A. Nijsen-Vis, H.T. Klippel (editors)

INFLUENCE OF IMPURITY CONTAMINATION ON SOLUBILITY OF HYDROGEN ISOTOPE IN TITANIUM

Shinsuke YAMANAKA, Yuichi SATO, Taku TANAKA and Masanobu MIYAKE

Department of Nuclear Engineering, Faculty of Engineering, Osaka University, Yamadaoka 2-1, Suita, Osaka 565, Japan

Hydrogen and deuterium solubilities in Ti-N alloys with 5 - 15 at% have been measured at temperatures of 600 - 850°C at pressures below 100 Pa. All the solubility data were found to follow Sieverts' law. For both hydrogen and deuterium, the solubility increased with the nitrogen content in titanium, while the enthalpy of solution decreased with the nitrogen content. The hydrogen solubility in Ti-N alloys was larger than the deuterium solubility. Partial thermodynamic functions of hydrogen and deuterium in Ti-N alloys were obtained by a dilute solution model and compared with those in Ti-O alloys. The tritium solubility in Ti-N alloys was evaluated from hydrogen and deuterium data. The estimated solubility of tritium was smallest, followed by deuterium and hydrogen. The quantitative expressions for the influence of impurity contamination on solubility of hydrogen isotope in titanium were derived from experimental and estimated data.

1. INTRODUCTON

Because of the high sorption capability for hydrogen isotopes, titanium has been proposed as tritium gettering and storage media[1,2]. The prime prerequisite for the successful application is an understanding of the solubility of hydrogen isotopes in titanium. During the practical operation, titanium getter will be contaminated with gaseous impurities such as O_2, N_2, H_2O, CO and CO_2, the sorption of which appears to affect the hydrogen solubility in titanium. Oxygen, nitrogen and carbon are important impurities, which are interstitially dissolved in titanium. The effect of solute oxygen on solubility of hydrogen isotope has been previously reported[3]. In the present work, the effect of nitrogen on the solubility has been studied, and the results obtained were compared with the oxygen effect to clarify the influence of impurity contamination on solubility of hydrogen isotope in titanium.

2. EXPERIMENTAL

Materials employed in the present study were Ti-N alloys with a nitrogen content of 5, 7.5, 10 and 15 at%, which were prepared by vacuum annealing of Ti/TiN mixture pellets. The Ti-N alloys were found from X-ray diffraction analysis to be alpha solid solutions possessing a hexagonal structure. Hydrogen and deuterium gases were used with purification by a hot titanium bed.

Solubility measurements for hydrogen and deuterium in Ti-N alloys were performed at temperatures from 600 to 850°C at pressures lower than 100 Pa by a constant volume method. Details of the experimental apparatus have been described elsewhere[3].

3. RESULTS AND DISCUSSION

All the data obtained for hydrogen and deuterium solubilities in Ti-N alloys were observed to obey the following relation:

$$S_U = K_U \, P_{U_2}{}^{1/2} \qquad (1)$$

where S_U is the concentration of hydrogen isotope in U/Ti (U = H or D), P_{U_2} is the equilibrium pressure in Pa and K_U is the Sieverts' constant. In Fig. 1, temperature dependence of Sieverts' constant obtained for Ti-N alloys is shown together with the results for pure titanium. It is apparent from this figure that for Ti-N alloys the relationship

TABLE 1
Parameter of Sieverts' law and enthalpy of solution

	N content (at%)	a x 10^6	b	$-\Delta H$ (kJ/g-atom)
H_2	0	9.26	5450	45.3
	5.0	7.07	5900	49.1
	7.5	5.13	6280	52.2
	10.0	4.55	6500	54.0
	15.0	3.73	6840	56.9
D_2	0	10.4	5200	43.2
	5.0	6.02	5930	49.3
	7.5	5.34	6140	51.0
	10.0	4.29	6440	53.5
	15.0	3.41	6790	56.5
T_2	0	10.0	5190	43.1
	5.0	6.72	5770	48.0
	7.5	5.24	6090	50.6
	10.0	4.55	6330	52.6
	15.0	3.62	6680	55.5

between Sieverts' constant and temperature is written by

$$K_U = a \exp(b/T) \qquad (2)$$

where a and b are constants and T is temperature in K. The enthalpy of solution ΔH (kJ/g-atom) can be obtained from the b value. The values of a, b and ΔH are listed in Table 1. The solubilities of hydrogen and deuterium at a constant pressure increase with the nitrogen content of titanium, as can be seen in Fig. 1 and Table 1. The enthalpy of solution appreciably decreased as the nitrogen content increased. For Ti-N alloys, the solubility was slightly smaller for deuterium than for hydrogen.

In our previous study, oxygen dissolved in titanium reduced solubility of hydrogen and deuterium, while the enthalpy of solution hardly changed with the oxygen content in titanium. In contrast to the behavior in Ti-O alloys, the solubility increased with addition of nitrogen to titanium, suggesting an apparent attractive interaction of nitrogen with hydrogen isotope in titanium. The solubility data for Ti-N alloys were analyzed by a dilute solution model to elucidate the effect of solute nitrogen on the solubility. The following solubility equation[3,4] was applied to our data,

$$\ln (C_U T^{7/4}/P_{U_2}{}^{1/2} A_{U_2}) = -(\overline{E}_U - E^d_{U_2})/kT + \overline{S}_U/k. \qquad (3)$$

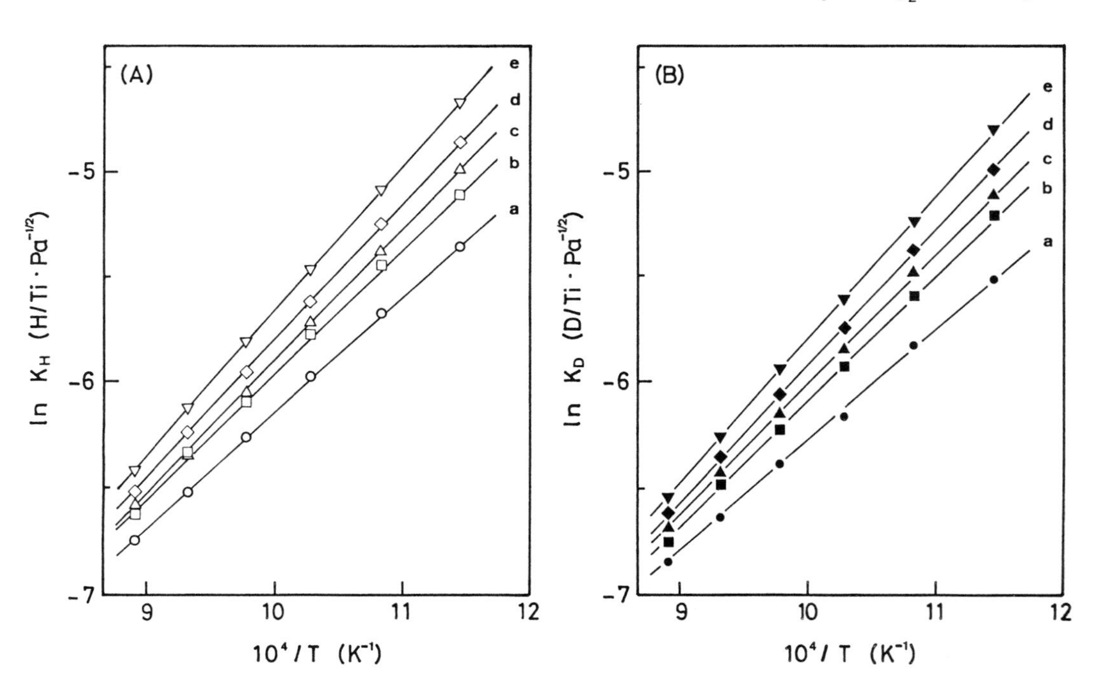

FIGURE 1
Temperature dependence of Sieverts' constant for hydrogen (A) and deuterium (B). Nitrogen content: (a) 0 at%, (b) 5 at%, (c) 7.5 at%, (d) 10 at%, (e) 15 at%

The value of C_U which shows the ratio of the number of hydrogen isotope atoms dissolved in titanium to the total number of tetrahedral interstitial sites in titanium is equal to a half of S_U value. The $\bar{E}_U$ is the energy difference between a hydrogen isotope at rest in the gas and in the solution, $-2E^d_{U_2}$ is the dissociation energy of molecule at 0 K, $\bar{S}_U$ is the partial excess entropy, and k is the Boltzmann constant. The A_{U_2} is related to partition functions of hydrogen isotope gas, the value of which was evaluated from spectroscopic data[5]. Figure 2 demonstrates the thermodynamic analysis of solubility data by eq. (3). For hydrogen and deuterium, linear relationships held between ln $(C_U T^{7/4}/P_{U_2}^{1/2} A_{U_2})$ and 1/T, independent of the nitrogen content in alloys. The values of $-(\bar{E}_U - E^d_{U_2})/k$ and $\bar{S}_U/k$ were derived from the slope and intercept of the straight line.

The changes in solubility and enthalpy of solution by incorporation of impurity atoms into titanium lattice were discussed on the basis of difference in partial energy and excess entropy between pure titanium and alloys, defined as $\delta\bar{S}_U = \bar{S}_U - \bar{S}_U°$ and $\delta\bar{E}_U = \bar{E}_U - \bar{E}_U°$ where $\bar{S}_U°$ and $\bar{E}_U°$ are values for pure titanium. In Fig. 3, changes in $\delta\bar{S}_U$ and $\delta\bar{E}_U$ values with composition are shown for Ti-N and Ti-O alloys. As evidenced by this figure, $\delta\bar{S}_U$ and $\delta\bar{E}_U$ for hydrogen and deuterium decrease with nitrogen content. The increase in the solubility in Ti-N alloys is ascribed to a net results from large decrease in $\delta\bar{S}_U$ and $\delta\bar{E}_U$, which may be due to a lattice strain effect of solute nitrogen. On the other hand, for Ti-O alloys $\delta\bar{S}_U$ decreases with oxygen content, but no marked change in $\delta\bar{E}_U$ is oberved. Site blocking effect of oxygen is likely to cause the decrease in $\delta\bar{S}_U$, resulting in the reduction of solubility in Ti-O alloys. The difference in solute effect between nitrogen and oxygen is attributable mainly to the difference in atomic size. There was no marked difference in $\delta\bar{S}_U$ and $\delta\bar{E}_U$ between hydrogen and deuterium for Ti-N and Ti-O alloys.

Influence of temperature and nitrogen content on value of ln $[(C_H T^{7/4}/P_{H_2}^{1/2} A_{H_2})/(C_D T^{7/4}/P_{D_2}^{1/2} A_{D_2})]$ must be examined to interpret the

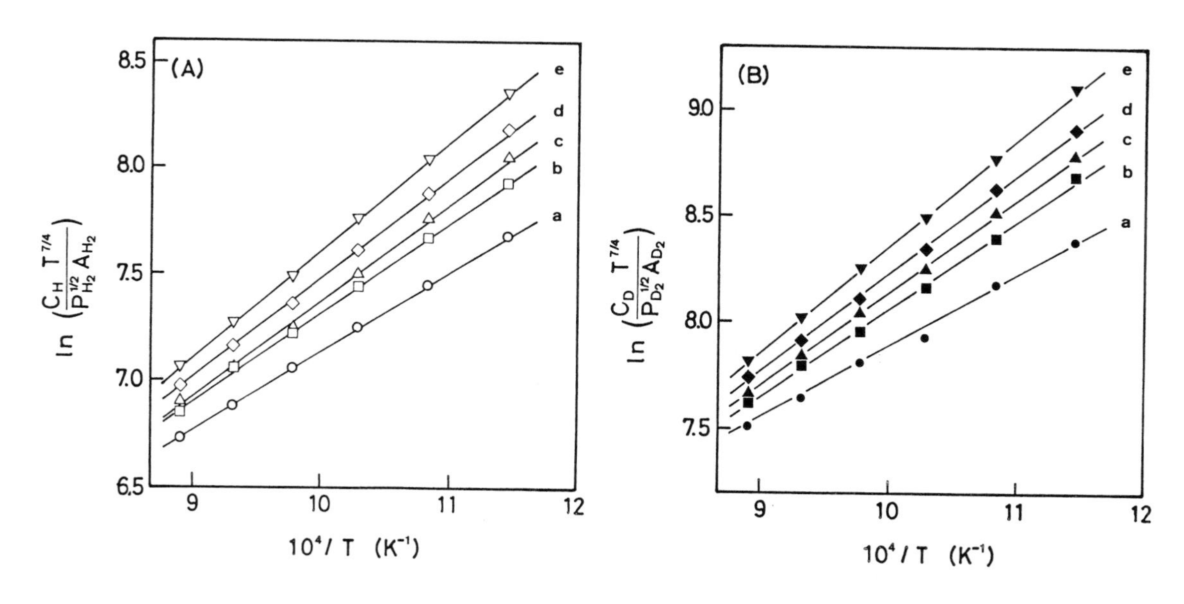

FIGURE 2
Thermodynamic analysis of solubility data for hydrogen (A) and deuterium (B). Nitrogen content: (a) 0 at%, (b) 5 at%, (c) 7.5 at%, (d) 10 at%, (e) 15 at%

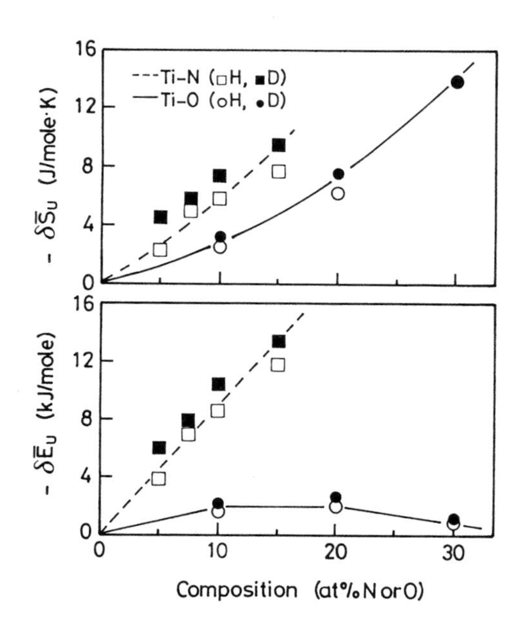

FIGURE 3
Changes in $\delta\bar{S}_U$ and $\delta\bar{E}_U$ values with composition of alloy

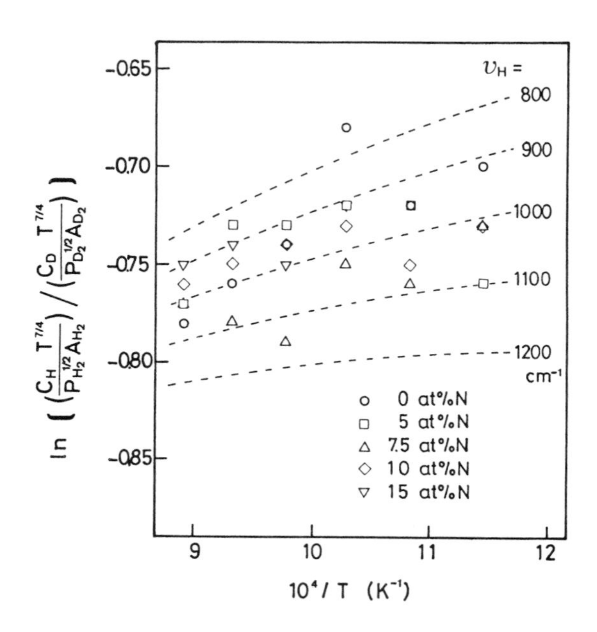

FIGURE 4
Isotope effect on solubility in Ti-N alloy

isotope effect on solubility in Ti-N alloys and to estimate tritium solubility. In Fig. 4, the values of ln $[(C_H T^{7/4}/P_{H_2}{}^{1/2} A_{H_2})/(C_D T^{7/4}/P_{D_2}{}^{1/2} A_{D_2})]$ obtained from experimental data are plotted against 1/T. There exists little correlation between ln $[(C_H T^{7/4}/P_{H_2}{}^{1/2} A_{H_2})/(C_{D_2} T^{7/4}/P_{D_2}{}^{1/2} A_{D_2})]$ and the nitrogen content in titanium. Dashed curves in this figure were theoretically drawn on the assumption that interstitial hydrogen isotope atoms in titanium behave as simple harmonic oscillators having vibrational frequencies ν_H in the range of 800 - 1200 cm^{-1}. The method for this calculation was the same as that in the previous paper[3]. The experimental values distributed around a curve for ν_H = 1000 cm^{-1}, irrespective of the nitrogen content. These results mean that the difference in solubility between the two isotopes is practically unaffected by the nitrogen content in titanium. That is to say, we are able to estimate tritium solubility in Ti-N alloys, assuming $\delta\bar{S}_H = \delta\bar{S}_D = \delta\bar{S}_T$ and $\delta\bar{E}_H = \delta\bar{E}_D = \delta\bar{E}_T$. Therefore, the value of ln $(C_T T^{7/4}/P_{T_2}{}^{1/2} A_{T_2})$ for tritium can be calculated from solubility data for hydrogen and deuterium using ν_H = 1000 cm^{-1}. The estimated values are illustrated in Fig. 5, indicating that the values from hydrogen data reasonably agree with those from deuterium data. Temperature dependence of Sieverts'

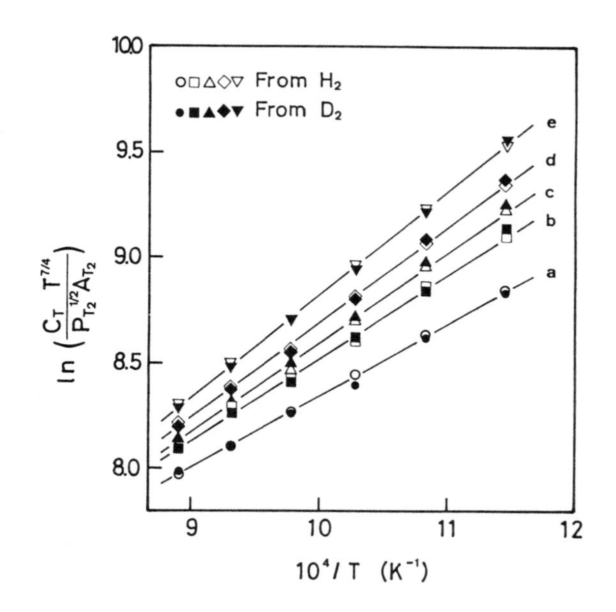

FIGURE 5
Estimation of tritium solubility in Ti-N alloy. Nitrogen content: (a) 0 at%, (b) 5 at%, (c) 7.5 at%. (d) 10 at%, (e) 15 at%

TABLE 2
Parameters of $a_{0,1,2}$ and $b_{0,1,2}$ in eq. (4)

	Nitrogen			Oxygen		
	H_2	D_2	T_2	H_2	D_2	T_2
a_0 x 10^6	9.26	10.4	10.0	9.26	10.4	10.0
a_1 x-10^2	6.80	8.56	8.04	2.84	3.47	3.04
a_2 x 10^4	18.4	27.2	25.2	0.639	2.59	1.14
b_0 x 10^{-3}	5.45	5.20	5.19	5.45	5.20	5.19
b_1 x 10^2	2.12	2.99	2.62	0.500	0.752	0.556
b_2 x-10^4	2.64	6.36	4.62	1.58	2.27	1.58

constant K_T for tritium evaluated from the ($C_T T^{7/4}/P_{T_2}^{1/2}A_{T_2}$) value was also represented as eq. (2). The values of a and b in eq. (2) for tritium are given in Table 1. The solubility and the value of $-\Delta H$ estimated for tritium increased with nitrogen content in titanium. The solubility in Ti-N alloys decreased in the order hydrogen, deuterium, tritium.

These results led to the quantitative expressions for the influence of impurity contamination on solubility of hydrogen isotope in titanium. Postulating that a and b in eq. (2) can be expressed as a function of impurity concentration x in at% as follows:

$$a = a_0 (1 + a_1 x + a_2 x^2)$$
$$b = b_0 (1 + b_2 x + b_2 x^2) \quad (4)$$

where a_0 and b_0 are values for pure titanium, we obtained the values of $a_{0,1,2}$ and $b_{0,1,2}$ as shown in Table 2. These values and eqs. (2) and (4) allow us to predict the solubility of hydrogen isotope in titanium contaminated by nitrogen or oxygen.

4. CONCLUSIONS

Solubilities of hydrogen and deuterium in Ti-N alloys of 5 - 15 at%N have been measured at 600 - 850°C at pressures less than 100 Pa. All the solubility data obtained in the present study were found to obey Sieverts' law. Nitrogen dissolved in titanium increased solubilities of hydrogen and deuterium at a fixed pressure. The enthalpies of solution for hydrogen and deuterium decreased with nitrogen content in titanium. In each Ti-N alloy, hydrogen had larger solubility than deuterium.

The partial energy and excess entropy of hydrogen and deuterium in Ti-N alloy were estimated by a dilute solution model. The increase in solubility by the presence of solute nitrogen was attributed to the marked decrease in $\delta\bar{S}_U$ and $\delta\bar{E}_U$, while the larger change in $\delta\bar{S}_U$ than $\delta\bar{E}_U$ due to solute oxygen resulted in the decrease in solubility.

The tritium solubility in Ti-N alloys was estimated from experimental data. The sulubilities were found to decrease in the order hydrogen, deuterium, tritium.

The quantitative expressions for the influence of impurity contamination on the solubility were obtained, which will offer a foundation of design criteria for tritium processing with titanium.

REFERENCES

1. N.P. Kherani and W.T. Shmayda, Fusion Technology, 8(1985)2399.

2. L.S. Krochnalneck, J.P. Krasznai and M. Crarney, Fusion Engineering and Design, 5(1922)337.

3. S. Yamanaka, T. Tanaka, S. Tuboi and M. Miyake, Fusion Engineering and Design, in print.

4. R.B. McLellan, Phase Stability in Metals and Alloys, (McGraw Hill, 1967)pp.393.

5. G. Herzberg, Molecular Sectra and Molecular Structure I. Spectra of Diatomic Molecules, (D. Van Nostrand Company, Inc., 1950)pp.530.

FUSION TECHNOLOGY 1988
A.M. Van Ingen, A. Nijsen-Vis, H.T. Klippel (editors)

TRITIUM TRAPS IN NEUTRON IRRADIATED CERAMICS FROM OUT-OF-PILE CONSTANT RATE HEATING MEASUREMENTS (EEC PROGRAM)

Fernande BOTTER, Daniel CHERQUITTE, Martine GAUTIER, Philippe MAIRE

COMMISSARIAT A L'ENERGIE ATOMIQUE - CEN/SACLAY - IRDI/DESICP/DLPC/SPCM
91191 Gif-sur-Yvette Cedex, France

Constant rate heating experiments have been performed in order to compare tritium release of pre-irradiated lithium ceramics.
Samples of sintered aluminates modified by replacement of 1.6% of the lithium by magnesium exhibit a tritium release at lower temperatures ; the profile of concentration of tritium often exhibits several peaks and shoulders, proving various kinds of tritium trapping. Many pre-irradiation examinations of the samples were pursued and seem to show electrical defects. Our first results seem promising both for the understanding of the gas release mecanism, and for the possible modification of caracteristics of retention-release of an important fraction of the generated gases, induced by doping the ceramic.

1. INTRODUCTION

Lithium aluminate γ $LiAlO_2$ is one of the leading candidates for use as a solid breeder.
However, it is necessary to have a better understanding of tritium transport and to enhance tritium release to a lower temperature range.
We investigate here the interest[1] and the influence of doping aluminate with magnesium on the tritium trapping in the irradiated solid.

2. PREPARATION OF THE SAMPLES

The preparation of sintered γ $LiAlO_2$ by dry way has been described[2]. The addition of magnesium is realised by mixing the initial γAl_2O_3 with a solution of magnesium acetate. After drying the starting product in a microwave oven, the steps of the preparation are the same as for $LiAlO_2$: mixing with lithium carbonate, decomposition during 3 hours at 700°C, stirring, cold isostatic pressing at 200 Mpa and sintering.

Aluminate containing 1.6 atom % Mg/Li is prepared. Depending on the initial proportions γ $Al_2O_3/LiCO_3$, two different micro- structures of doped material are obtained : a coarse one and a fine one respectively "5 µm and 0.8 µm" of mean grain diameter, as given by porosimetry measurements.

3. TRITIUM EXTRACTION

3.1. Experimental Procedure

Constant rate heating tritium extractions are realised on a apparatus already described (figure 1)

The usual operating conditions were :
- samples irradiation of 15 minutes in a thermal flux comprised between 2.10^{14} and 4.10^{13} $n.5^{-1}.cm^{-2}$
- observation of the tritium release, first at room temperature, then between 25 and 900°C at a heating rate of 100°C per hour under a flow rate of 0.5 cm^3/s of (Ar+0.1% H_2) sweep gas.

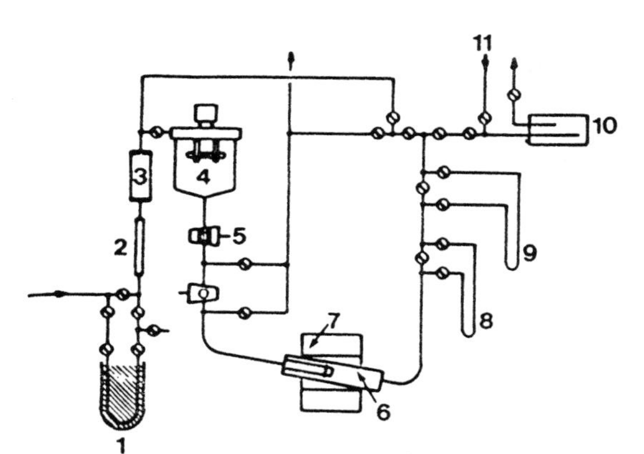

1 - Molecular sieve cold trap (-80°C)
2 - Oxygen removal device
3 - Thermal flowmeter
4 - Ampoule breaker
5 - Special grooved stopcok
6 - Quartz tube
7 - Furnace
8 - Activated charcoal cold trap (-80°C)
9 - Reducing furnace
10 - Proportional counter
11 - CO_2 (added to Ar sweep gas)

FIGURE 1 - Out-of-pile tritium extraction apparatus

3.2 Results

The curves (Figures 2 to 5) show a comparison of the behaviour of pure ceramics (aluminates and silicates) and of doped aluminates

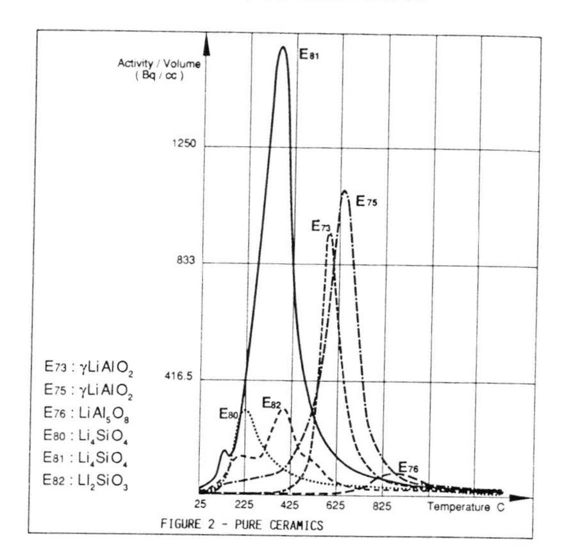

FIGURE 2 - PURE CERAMICS

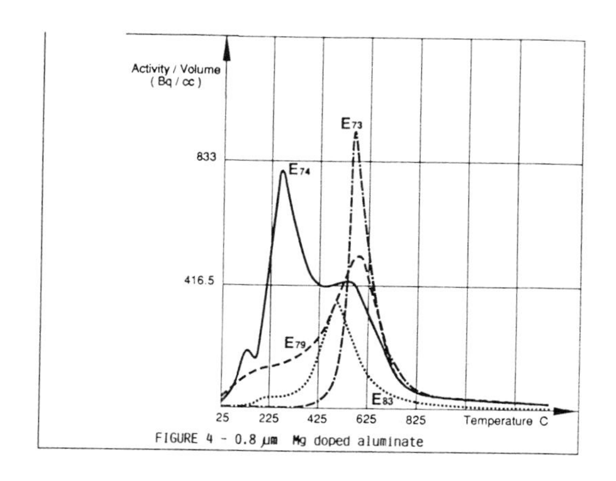

FIGURE 4 - 0.8 µm Mg doped aluminate

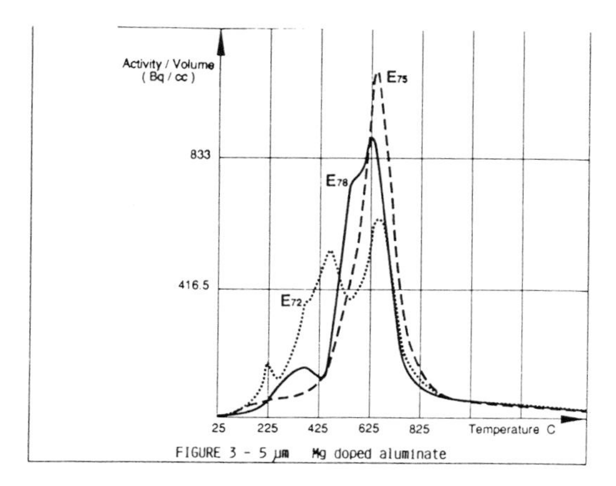

FIGURE 3 - 5 µm Mg doped aluminate

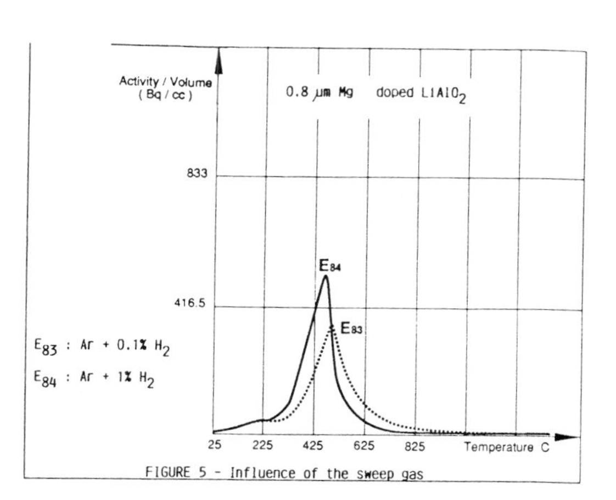

FIGURE 5 - Influence of the sweep gas

Tritium release as a function of the temperature

The last one (figure 5) indicates that the peak E_{83} is a mixture of T_2O and HT because the presence of more hydrogen in the sweep gas gives an earlier peak E_{84}, enriched in the HT species.
It is interesting to observe that the tritium release is obtained at lower temperature with the Mg doped samples than with pure $LiAlO_2$, and corresponds often to various traps of the tritium in the solid[3].

4. MATERIAL PROPERTIES INVESTIGATION

Various examinations were done on unirradiated samples. When necessary, pellets were cut or polished, always in dry conditions.

4.1. Ionic conductivity

No increase of ionic conductivity of the doped material compared to that of pure aluminate was observed.

4.2. Scanning electron microscopy (SEM)

Different Scanning Electron Fractographies are made from doped and undoped materials, covered with a carbon film. The apparatus used is generally a JSM. 880 Model.
The pictures (figures 6 to 8) show that pure lithiated ceramics present a more homogeneous microstructure than the doped material. However, some "Gypsum Flower" aspect may be observed with both materials, in various proportions. The so-called "5 and 0.8 µm" doped materials have important zones of smaller grains.

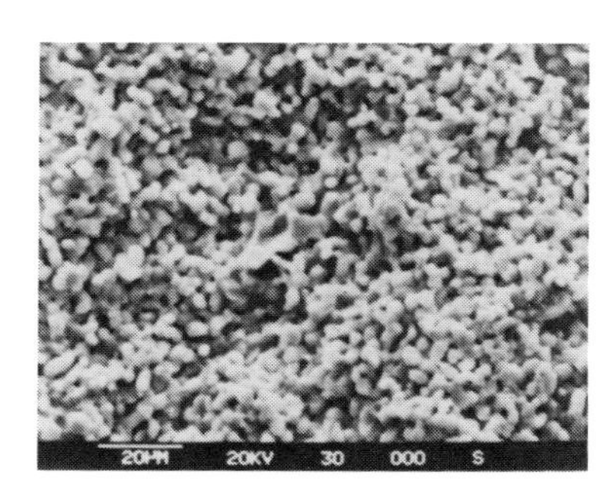

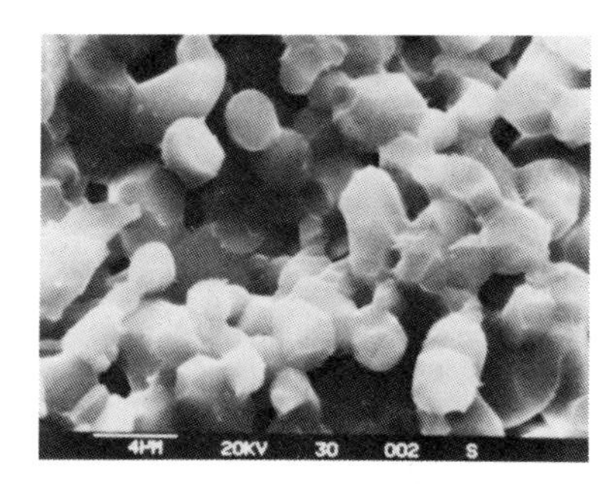

FIGURES 6
Li_2SiO_3 (CEA Origin)

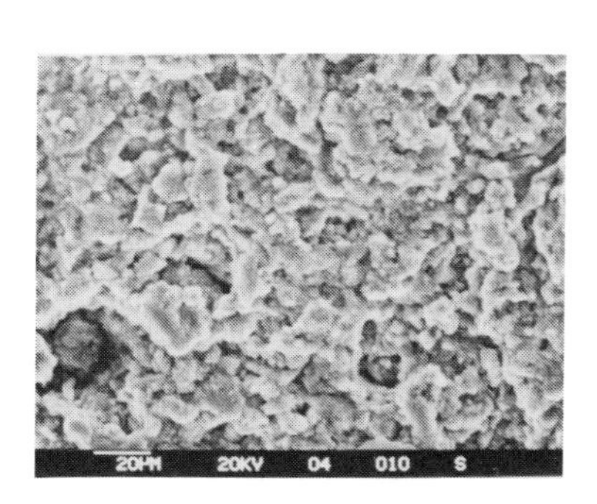

FIGURES 7
$LiAlO_2$ (KFK Origin)

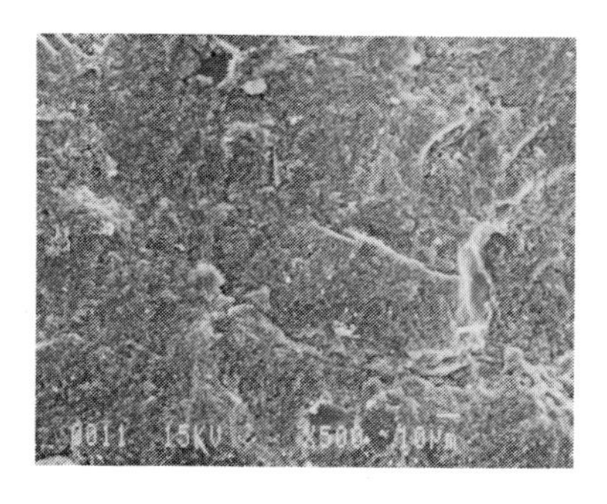

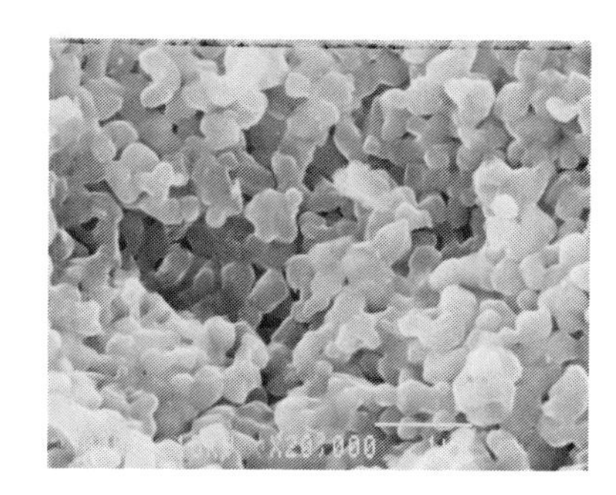

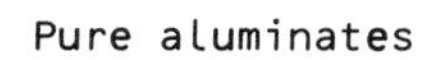

Pure aluminates

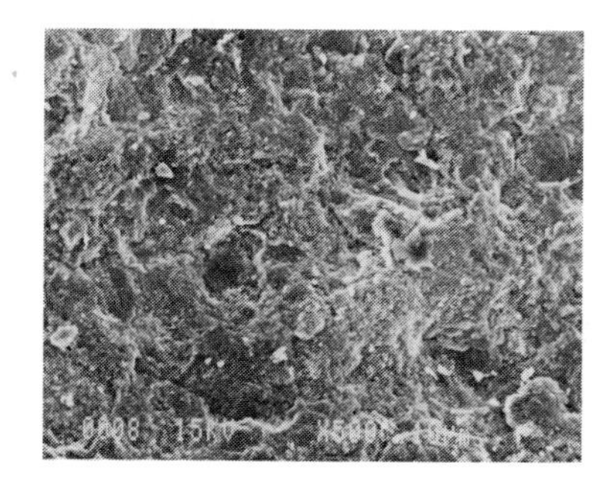

Mg doped aluminates
FINE MICROSTRUCTURE

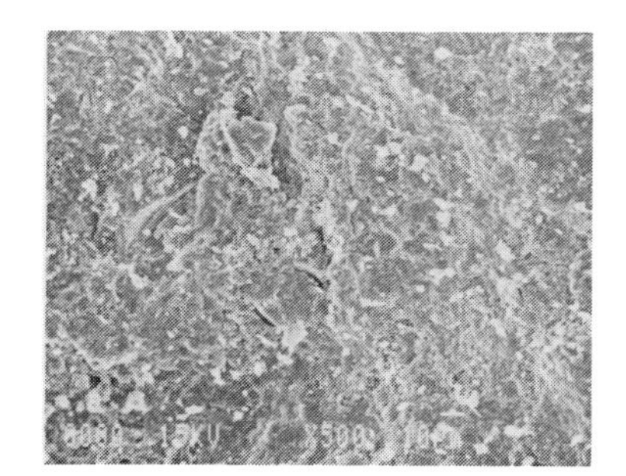

Pure aluminates

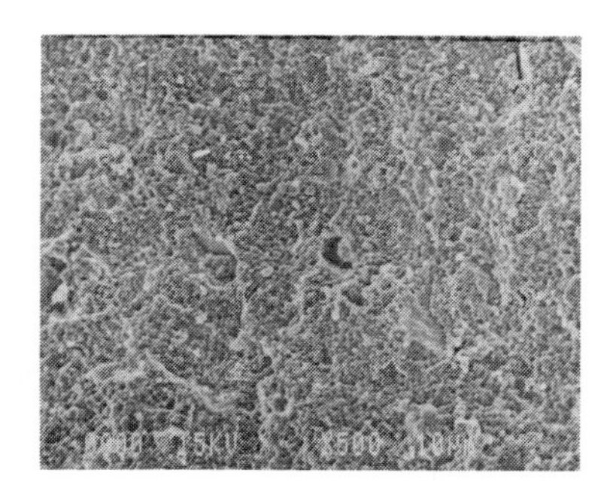

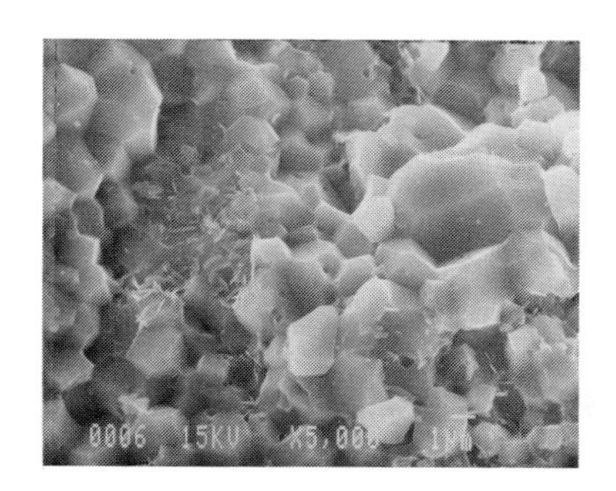

Mg doped aluminates
COARSE MICROSTRUCTURE

FIGURES 8

4.3. Electron Spectroscopy for Chemical Analysis (ESCA)

No significant change is observed when comparing pure and Mg doped aluminate : the Auger parameter is measured on oxygen and aluminium, and shows no difference in the Al-O bond.

A weak concentration of lithium is observed at the surface of the solid by ESCA, as compared to that of aluminum.

4.4. AUGER Spectroscopy

For a doped sample submitted to successive abrasions, the presence of magnesium is never shown. This proves that the magnesium is not segregated at the grain boundaries.

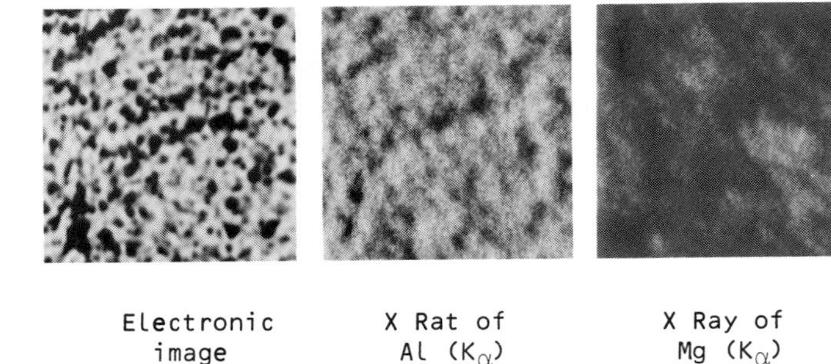

Electronic image | X Rat of Al (K_α) | X Ray of Mg (K_α)

FIGURES 9 - EPMA images on a scale of 100 x 100 µm

4.6. Charge Effects

SEM fractographies obtained with doped and undoped samples show charge effects only in the doped sample uncoated. (figures 11)

4.5. Electron Probe Microanalysis (EPMA)

Both microtextures of the Mg doped aluminate are investigated with a Castaing microprobe. The magnesium is observed in both samples, but is inhomogeneous (figure 9). The spectrum of Mg when moving the beam over 100 µm, (figure 10), again shows that the magnesium is not localized at the grain boundaries.

The examination of the sample with an ion probe seems to show less lithium and aluminum in the zones where the magnesium is located.

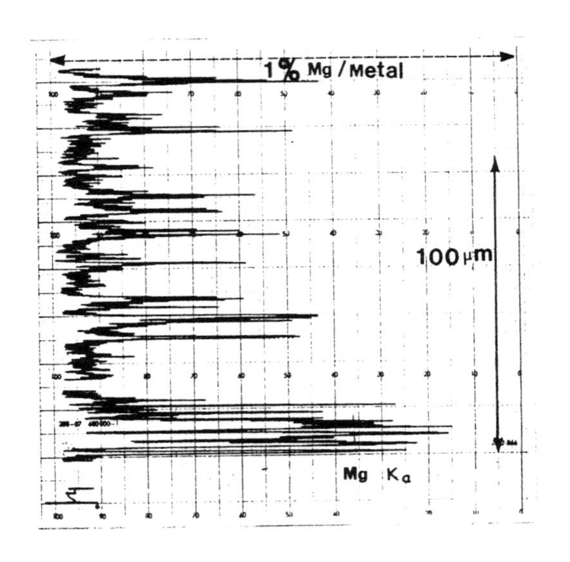

FIGURE 10 - Spectrum of the Mg (K_α) when moving the beam

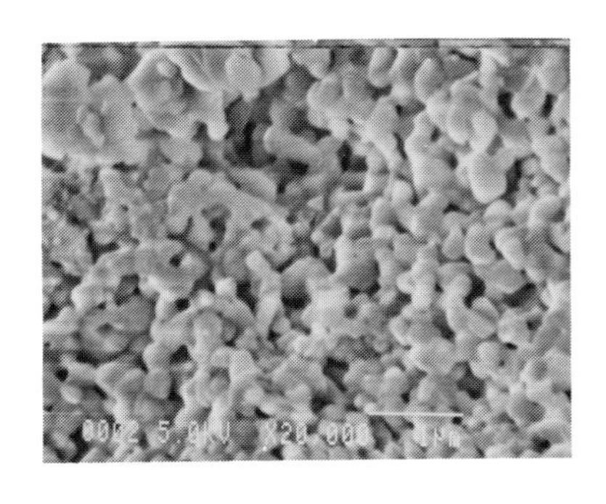

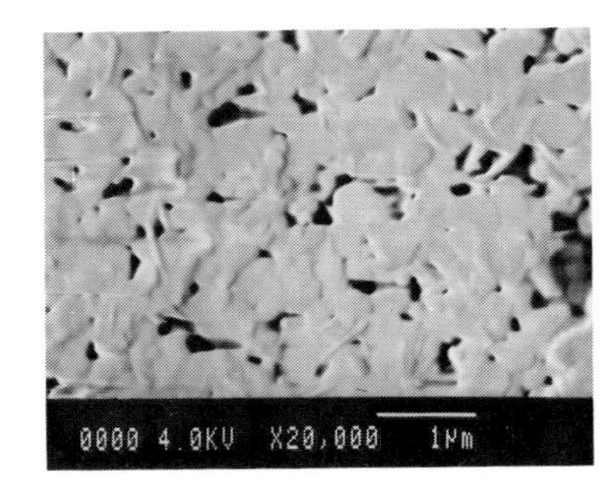

Pure $LiAlO_2$ | Mg doped $LiAlO_2$ Without Carbon Coating | With a Carbon Coating

FIGURES 11 - Charge effect on uncoated doped $LiAlO_2$

5. DISCUSSION AND CONCLUSION

The addition of magnesium seems to initiate defects in the material, which are revealed in the unirradiated state by charge effects.

In an irradiated doped material, new types of tritium traps are created, changing the tritium transport and release mechanism.

Additional preparations, examinations and tritium extraction runs should include investigations on single cristals, on both doped and undoped aluminates.

The first results seem promising both for the understanding and for the possible modification of the tritium retention or release, induced with additives to the aluminate.

ACKNOWLEDGEMENTS

This work was performed within the framework of the Europeen Community Fusion Technology Program. We would like to thank Dr F. Maurice and Mrs Zemskoff, specialists of EPMA, Mrs M. Henriot for her work in SEM, Dr B. Rasneur for providing the samples, Dr P. Charpin, J.P. Duraud and C. Le Gressus for our fruitfull discussions, and Dr E. Roth for his interest.

REFERENCES

1. J.D. Fowler, J. Am. Cer. 60, 3-4 (1977), 155.
2. B. Rasneur, J. Charpin, J. Nucl. Mat. (1988), 154.
3. F. Botter, D. Cherquitte, C. Le Gressus Letter to the Editor. in print in J. Nucl. Mat, 1988.

FUSION TECHNOLOGY 1988
A.M. Van Ingen, A. Nijsen-Vis, H.T. Klippel (editors)

TRITIUM PERMEATION INTO CONCRETE

Shigeo NUMATA and Hiroshi AMANO*

Institute of Technology, Shimizu Corporation, No.4-17, Etchujima 3-chome, Koto-ku, Tokyo,135 JAPAN
*Institute of Plasma Physics, Nagoya University, Furō-cho, Chikusa-ku, Nagoya, 464-01 JAPAN

To consider permeative behavior of tritiated water vapour into concrete, an experiment was performed using hardened cement paste and mortar specimens. It was proved that tritiated water was mixed with liquid water with which capillaries in concrete were filled, and then that some portions of tritiated water from atmosphere interacted with the water existing in solid phase.

1. INTRODUCTION

One of the most important problems for radiation protection of a fusion reactor is tritium release into the environment through the reactor building. Air containment and cleaning systems can minimize the release of tritium, and concrete walls which constitute a fusion reactor building can be considered as the final barrier for preventing tritium leakage into the environment. It, however, has been pointed out that tritiated water in the atmophere easily permeates concrete,[1] since concrete is a porous medium. Tritiated water which permeates concrete eventually leaks to the environment. In addition to this problem, re-emitting of tritium from the walls prolongs the time for the purification of the building atmosphere.[2]

An estimation of the permeability is very important in evaluating the fusion plant safety which might affect the economical feasibility of system. It is said, in general, that water in concrete exists in two states, e.g., free water and bound water. It is helpful for elucidation of the mechanism to study the interactions between permeative tritiated water and water in various states originally existing in concrete.

Present work aims to study the behavior of tritiated water permeated into concrete qualitatively. For this purpose, the state of water in hardened cement paste (h.c.p.) was first studied by the heating method. From the result, the water in h.c.p. was classified into several groups by heating temperature range. Then, the water in h.c.p. and in mortar specimen exposed to tritiated water vapour was collected in accordance with the classification and the behavior of the tritiated water was discussed.

2. CLASSIFICATION OF WATER CONTENT OF H.C.P. BY HEATING METHOD

2.1. Specimen

The h.c.p. specimen was cylindrical shape of 4mm in diameter x 3mm in height. The weight of the specimen was about 90 mg. Four kinds of cement were used to verify whether the release behavior differed from cement to cement or not. These were ordinary cement, TYPE-II cement (for use where moderate heat for hydration is necessary), fly ash cement and blast furnace slag cement. The cements were mixed at a water/cement ratio of 0.55 with pure water. After casting in moulds, they were cured in 100 % R.H. until testing. Most specimens were cured for 8 weeks, although some were cured for 16 weeks to verify the difference in behavior caused by a difference in curing period.

2.2. Experiment

To study the water release profile, the Thermogravimetric (TG) and Differential thermoanalytical (DTA) measurement were proceeded by using a thermobalance. In two experiments the general practice heat-up rate of 5°C/min was selected. The specimens were heated up to 1000°C in an argon atmosphere. The amount of released water with the rising temperature was measured with a quadrupole lens mass spectrometer.

2.3. Result & Discussion

Figure 1 shows the TG and DTA curves for ordinary cement with a heat-up rate of 5°C/min. With the TG curve, the weight loss is observed in temperature ranges below 200°C, between 450 and 550°C, and between 600 and 850°C. The DTA curve also indicates that chemical reactions occur in the same ranges. Both the TG curve and the DTA curve show similar results for other cements. The measurements were carried out more than twice for each cement, and the reproducibility of the measured results was good. When the measurements were performed with an h.c.p. specimen which were cured for 16weeks and with a heat-up rate of 2.5°C/min, no significant changes were shown in either the TG or the DTA curve.

Figure 2 shows the result with the mass spectrometer for the specimen made of ordinary cement. A large release of H_2O is shown in temperature ranges below 200°C and between 450 and 550°C. At higher temperatures, the amount of released H_2O decreases and that of CO_2 increases. These three peaks correspond to the results obtained with the thermobalance.

As the results of measurement on the temperature ranges in which large releases of water

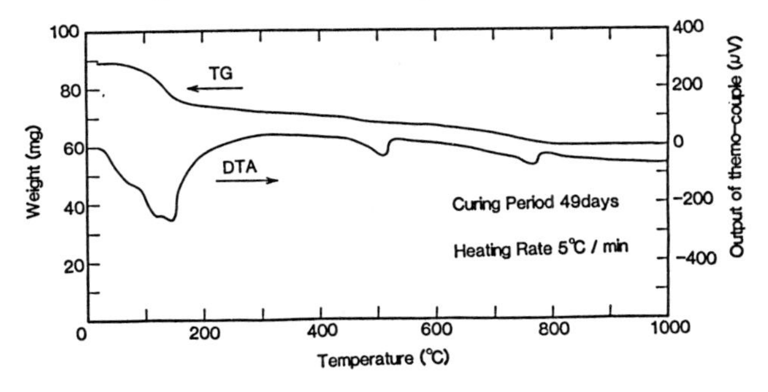

FIGURE 1
TG and DTA curves for a hardened cement paste made of ordinary cement

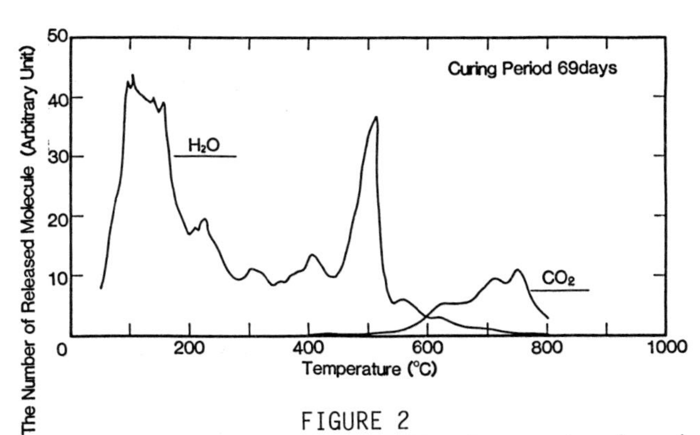

FIGURE 2
Release curve of H_2O and CO_2 from a hardened cement paste made of ordinary cement measured with a quadrupole lens mass spectrometer

occur, agree well among TG, DTA and mass-spectrometry, it seems that there may exist a fixed behavior on the release of water depending on heating temperature. So the released water is classified in accordance with heating temperature as shown in Table 1, and the state of each classified water is assumed with reference to a report on the decomposition of concrete at high temperature.[3] Group I water is assumed to be liquid water with which free spaces (pores) and capillaries in concrete are filled. Groups II, III and IV are the water conbined with constitute compounds in concrete or the constituent itself.

This classification was applied to the successive experiment.

3. PERMEATION EXPERIMENT OF TRITIATED WATER

Both of hardened cement paste and mortar specimens were prepared from a single batch of ordinary cement. The cement was mixed with distilled water and quartz sand. The specimens

TABLE 1
Classification of water distributing in a hardened cement paste by heating method.

Group	Temperature	State
I	-200°C	Free water, Capillary water
II	200-450°C	Water of crystallization
III	450-550°C	Constitute water in $Ca(CO)_2$
IV	550-850°C	Constitute water

TABLE 2
Mixture ratio of specimens (by weight)

	Water	Cement	Sand
H.C.P.	0.55	1.0	—
Mortar	0.55	1.0	2.0

were cylindrical shape of 45mm in diameter x 23mm in height, and the mixture ratio of each sample are shown in Table 2. The specimens were allowed to cure for 28 days in 100% R.H. at 25 °C (They were demoulded after one day of curing). The volume fraction of porosity is 0.33 for h.c.p. and 0.12 for mortar.

The flow sheet of the experimental procedure is shown in Figure 3. The side surface of each cylindrical specimen was coated with silicon resin to prevent the permeation from the side surface, then one end of cylinder was exposed to air at 93% R.H. loaded with tritiated water vapour (~750pCi/cm^3) maintaining at a constant humidity with the hygro-stat of the saturated solution of $Na_2SO_4 \cdot 10aq$. Another end was exposed to air at 10% R.H. The temperature was uncontrolled laboratory ambience (about 25°C). After the exposure of programmed duration, intruding tritiated water in each specimen was expelled off by heating in an electric furnace shown in Figure 4. The expelled water was fractionated into four fractions according to the previously mentioned classification. The tritium concentration in each fraction were measured with a liquid scintillation counter.

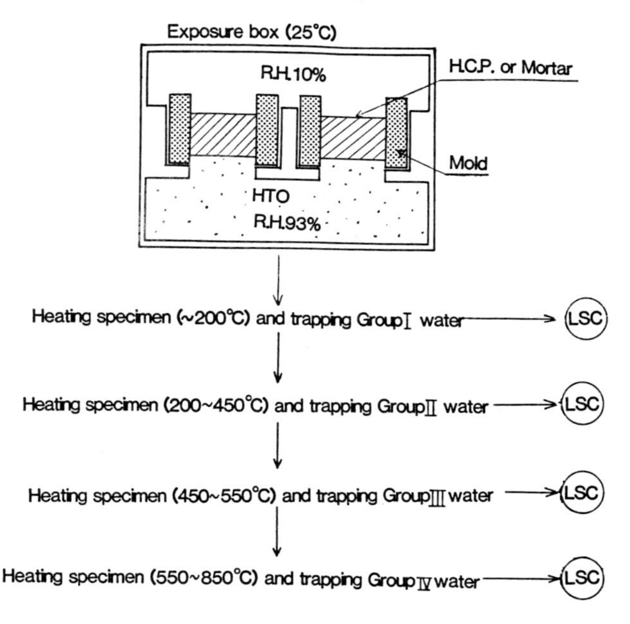

FIGURE 3
Flow sheet for experimental procedure

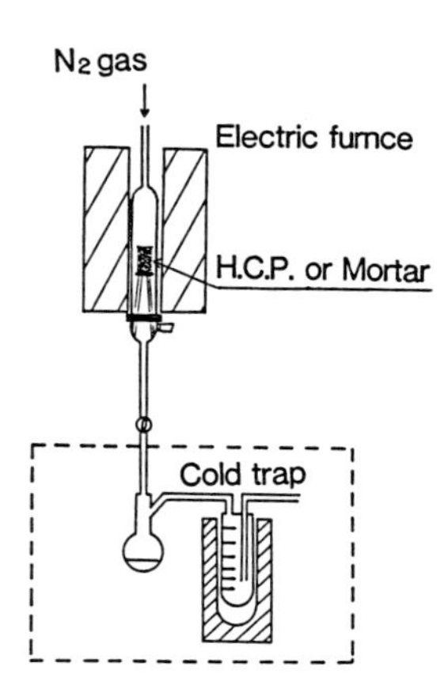

FIGURE 4
Illustration of heating apparatus

4. RESULTS AND DISCUSSION

Figure 5 shows the amount of tritiated water in an h.c.p. specimen in the form of each water

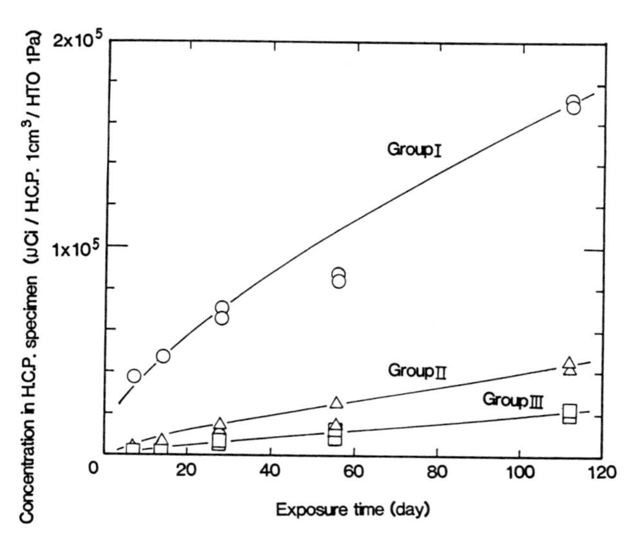

FIGURE 5
The amount of tritiated water in the hardened cement paste in the form of each water classification

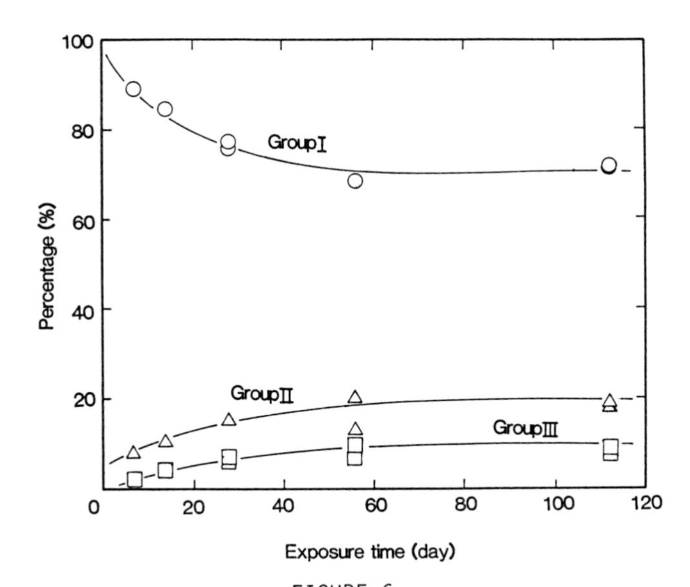

FIGURE 6
Percentage of tritiated water distributing to each classified water in the hardened cement paste.

classification. The tritium concentration in the water in Groups II, III and IV increase as that in Group I does. Some portions of tritiated water appears to adsorb on the solid and tritium in water may be exchanged with hydrogen atoms in the water of crystallization and constitute water.

Figure 6 shows the percentage of tritiated water in the form of each water. While the percentage of Group I water is more than 90% at the beginning of exposure, that of each water becomes constant at a duration of 100 days, looking like to be in a state of equilibrium. Similar results were obtained for the mortar specimens. Table 3 shows the percentage of tritiated water and that of ordinary water which

TABLE 3
Percentage of tritiated water and that of ordinary water distributing to each classified water at the equilibrium state

		I(%)	II(%)	III(%)	IV(%)
H.C.P.	HTO	71.7	18.8	9.0	0.5
	H_2O	58.1	23.2	16.4	2.3
Mortar	HTO	73.8	18.9	6.8	0.5
	H_2O	57.5	23.2	13.4	5.9

originally exists in the specimen at the equilibrium state. The percentage of tritiated water included in Group I is larger than that of ordinary water for both h.c.p. and mortar. It is assumed that tritiated water is first mixed with liquid water with which capillaries are filled, and then that tritiated water interacts with the water existing in solid phase.

The amount of water and the tritium concentration in each fraction at the equilibrium state are shown in Table 4, the difference of materials shows no difference on the tritium concentration. The absolute amount of tritium in the h.c.p. is larger than that in the mortar because of the larger water content of the h.c.p. This means that the amount of permeative tritiated water is proportional to the volume fraction of porosity which is roughly proportional to the amount of water. Since total area of capillary entrance that opens to the boundary is proportional to the volume fraction of porosity in fact, the experimental result may be interpreted that the amount of tritiated water which is mixed with liquid water increases with the volume fraction of porosity.

Figures 7 and 8 show the relations between tritium concentration in Group I water and the amount of tritium included in the water of Groups II, III and IV for h.c.p and mortar,

TABLE 4
The amount of each classified water distributing in the specimen and its tritium concentration at the equilibrium state

	Group	Water vol.	Tritium con.
H.C.P.	I	0.254	50.2
	II	0.101	33.0
	III	0.072	22.3
	IV	0.010	15.8
Mortar	I	0.103	51.0
	II	0.042	32.4
	III	0.024	20.3
	IV	0.003	13.5

Water vol.(cm^3/specimen $1cm^3$)
Tritiumcon.($\mu Ci/cm^3$)

respectively. The concentration of tritium adsorbed on the solid appears to be proportional to that in the liquid water.

5. CONCLUSIVE SUMMARY

In order to study the permeative behavior of tritiated water into concrete, a permeation experiment was performed using the h.c.p. and mortar which represented the simplest concrete systems. The concrete sample specimens were exposed to the atmosphere containing the tritiated water vapour for the designed durations, then the intruding tritiated water was fractionally collected by heating method followed by the liquid scintillation counting of tritium.

From the experimental results, followings can be said as the conclusions.

(1) Tritiated water vapour in the atmosphere intrudes into the pores in concrete and is mixed with the liquid water existing there.

(2) Then tritiated water diffuse into the liquid water with which the capillaries adjacent to the pores was filled.

(3) The exchange process between the hydrogen atoms in the liquid water and those in the crystallization and constitute water is a dominant one. It is found that this interaction is the linear-isotherm one.

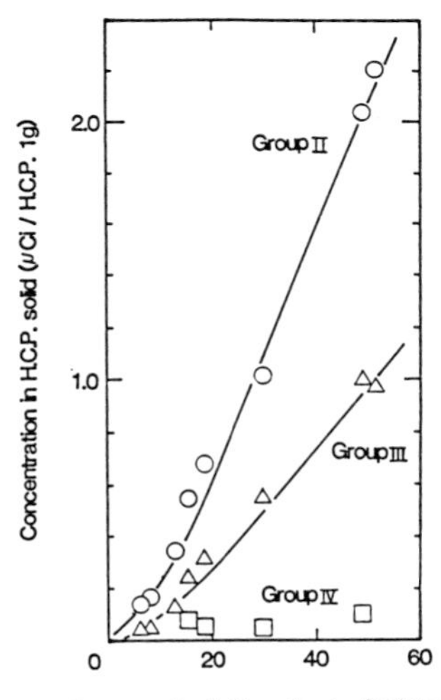

FIGURE 7
Sorption of tritium in the form of each water classification on the solid of hardened cement paste.

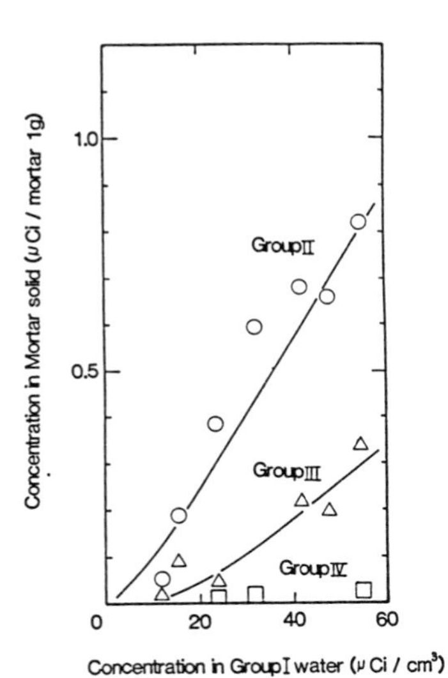

FIGURE 8
Sorption of tritium in the form of each water classification on the solid of mortar.

(4) To promote the quantitative comprehension on permeation of tritiated water into concrete, the phenomena occurring in the capillaries has to be discriminated from the phenomena in the pores which have the free paths for the outside atmosphere.

ACKNOWLEDGMENTS

The authors would like to express their sincere appreciation to Prof. M. Okamoto of Tokyo Institute of Technology for providing a thermobalance and the mass spectrometer at his laboratory. They are also grateful to Mr. A. Ashida, Mr. M. Akiyama and Mr. T. Hirata of Shimizu Corporation for their helpful supports through the experiment.

REFERENCES

1. P.A. Finn et al., Nuclear Technology/Fusion, vol.4 (1983) 389.

2. A.E. Sherwood, UCRL-80572 (1978), Proceedings of 3rd. Topical Meeting on the Technology of Controlled Nuclear Fusion, Santa Fe, May 9-11 1978.

3. U. Schneider, Behavior of Concrete at High Temperature (Deutscher Ausschuss für Stahlbeton, 1982).

FUSION TECHNOLOGY 1988
A.M. Van Ingen, A. Nijsen-Vis, H.T. Klippel (editors)

EXPERIMENTAL RESULTS ON THE HYDROGEN MASS-TRANSFER RATE IN MOLTEN Pb17Li ALLOY

A. VIOLA[1], P. LOLLI-CERONI[2] and G. PIERINI[2]

[1]Visiting Scientist of Cagliari University, Italy
[2]CEC, Ispra Establishment, 21020 Ispra (Va), Italy

Experimental work is being carried out in the Laboratory of the JRC Ispra in order to determine the kinetics of hydrogen absorption in the Pb17Li eutectic. Mathematical analysis of experimental data indicates that the overall process of hydrogen absorption in the eutectic is controlled by the heterogeneous reaction at the gas-eutectic interface with the hydrogen molecules dissociation and by the diffusion of solute hydrogen atoms in the molten Pb17Li alloy.

1. INTRODUCTION

Recent data on the H_2 or D_2 solubility in the Pb17Li eutectic[1,2] are so low that a large tritium permeation from the molten Pb17Li alloy to the cooling water system[3,4] and to the environment is expected. In order to limit the tritium losses to acceptable levels[5] it is necessary to maintain very low tritium inventory in the blanket.

The design of extraction units particularly involves the knowledge of the tritium solution rate in the Pb17Li eutectic.

Experimental work is being carried out in the laboratory of the JRC Ispra in order to determine the kinetics of hydrogen solution into Pb17Li eutectic.

2. EXPERIMENTAL PART

The apparatus, schematically represented in Figure 1, is mainly composed of the following systems: gas (H_2, He) feeding and purification, eutectic feeding to the saturator, hydrogen absorption in the eutectic contained in the saturator (Figure 2) at 673 K by bubbling a helium-hydrogen gaseous stream.

As far as the absorption tests are concerned, the H_2/He gaseous mixtures employed were fed continuously into the eutectic in the saturator, after removing O_2, H_2O impurities by means of SAES-ST 101/700 Al-Zr getters at 850 K. The hydrogen was mixed with the molten Pb17Li by

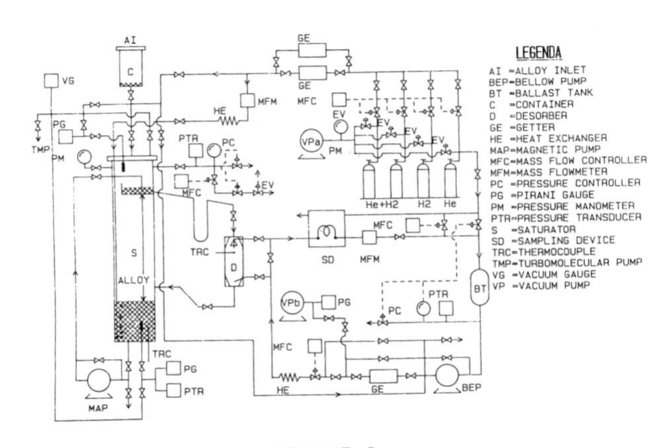

FIGURE 1
Schematic design of the hydrogen recovery apparatus from liquid Pb17Li.

bubbling the gaseous streams ($\geq$ 200 NTP cm^3 min^{-1}).

The variation of hydrogen pressure measured by the pressure sensor placed in the eutectic was recorded (Figure 3). The homogeneity of the hydrogen concentration in the eutectic was shown by the fact that by increasing the gas flow rate no variation in hydrogen pressure was observed.

3. KINETICS OF HYDROGEN MASS TRANSFER RATE

The mechanism of the overall pressure of hydrogen mass transfer rate involving the molten Pb17Li in contact with a gaseous atmosphere can be described by the following sequence of steps:

i) transport of the absorbed hydrogen by diffusion and convection in the bulk eutectic;

ii) transport of the absorbed hydrogen by

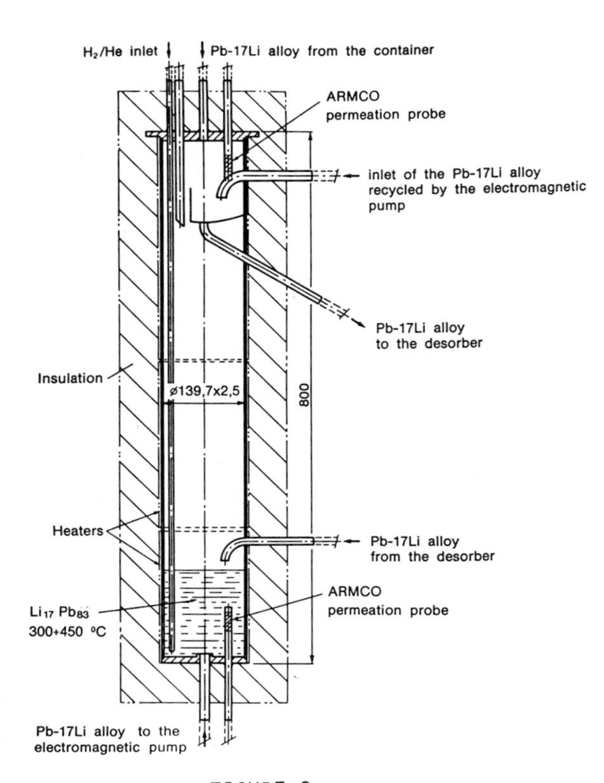

FIGURE 2
Saturator.

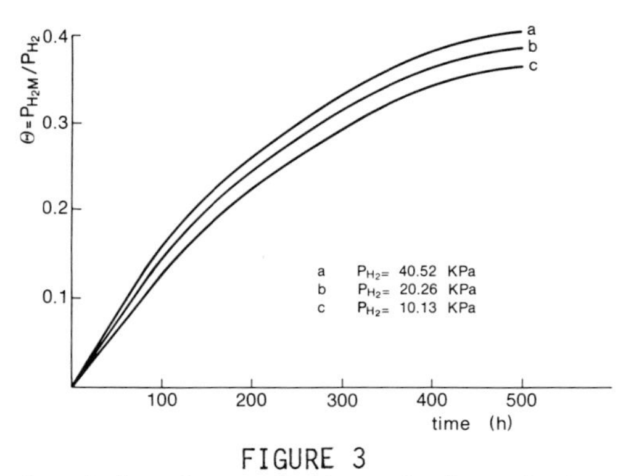

FIGURE 3
Experimental values of dimensionless hydrogen pressures in the probe device pressure vs time at different hydrogen pressures in the saturator.

diffusion through a layer of eutectic adjacent to the gas-eutectic interface;

iii) heterogeneous reaction at the interface with the hydrogen molecules dissociation or the hydrogen atoms recombination depending on the direction of hydrogen mass transfer;

iv) diffusion of the hydrogen through gas phase boundary layer to the gas-eutectic interface;

v) transport of the gaseous hydrogen by diffusion and convection from the bulk phase to the gas phase boundary layer.

Assuming the existence of a static boundary layer in the eutectic adjacent to the gas eutectic interface where the resistance of hydrogen transport in the eutectic phase is located, we obtain (step i and step ii):

$$N_{H,l} = \frac{D_l}{\gamma}\,(C_H - C_{Hi}) = K_l(C_H - C_{Hi}) \qquad (1)$$

The relationships for the mass transfer step in the gaseous phase are similar (step iv and v).

The diffusive resistance throught the gaseous boundary layer adjacent to the gas-eutectic interface can be considered negligible in respect to the liquid boundary layer, because of the low value of the thickness of this layer in the gaseous phase and the very high ratios between the coefficients of diffusion in two phases. Indeed, at 473-723 K, which is the probable range of the work temperature for NET, the order of the hydrogen diffusion coefficient is about 10^{-9} m^2s^{-1} in the eutectic phase[1,2] and it is about 10^{-4} m^2s^{-1} in the gaseous phase[6].

The third step, which concerns the breakage of the bonds between the hydrogen and the metallic elements in the liquid alloy and the reaction of recombination of the hydrogen atoms to the hydrogen molecules, can be represented by:

$$2\ H \rightleftarrows H_2 \qquad (2)$$

This heterogeneous reaction at the gas-eutectic interface proceeds by a sequence of steps which include adsorption, dissociation or recombination and desorption.

$$\begin{array}{l} H_{2g} + Pb17Li \rightleftarrows (H_2)_{ads}\ Pb17Li \\ (H_2)_{ads}\ Pb17Li \rightleftarrows (H)_{ads}\ Pb17Li + (H)_{ads}\ Pb17Li \\ (H)_{ads}\ Pb17Li \rightleftarrows H_{interface} + Pb17Li \end{array} \qquad (3)$$

Assuming then that the reaction at the interface proceeds by the sequence outlined above, the rate of this third step can be expressed by[7]:

$$N_{H,R} = K_d\, C_{Hi}^2 - K_a P_{H_2} \tag{4}$$

At equilibrium conditions and for very diluted systems, the hydrogen pressure in the gaseous phase is correlated to the hydrogen concentration in the eutectic by the Sievert's law:

$$P_{H_2}^{1/2} = S_l\, C_H^* \tag{5}$$

As the value of N_{HR} is null at equilibrium conditions and Sievert's law must be verified, Eq.(4) becomes:

$$N_{H,R} = -K_d\left(C_{Hi}^2 - \frac{P_{H_2}}{S_l^2}\right) \tag{6}$$

with $\quad K_d/K_a = S_l^2$

Now, assuming no accumulation at the interface, the kinetic equation for the overall flux can be obtained by equalizing the fluxes

$$N_{H,l} = N_{H,R} = N_H \tag{7}$$

Equalizing the relationships (Eqs.(1) and (6)), the expression of C_{Hi}, as a function of C_H and P_{H_2} can be obtained, which introduced into Eq.(6) gives:

$$N_H = K_l\left[C_H + \frac{1}{2\Gamma} - \left(\frac{1}{4\Gamma^2} + \frac{P_{H_2}}{S_l^2} + \frac{C_H}{\Gamma}\right)^{1/2}\right] \tag{8}$$

where $\Gamma = \dfrac{K_a}{K_l}$ or $\Gamma = \dfrac{S_l^2 K_a}{K_l}$

Eq.(8) thus gives the flux of hydrogen from the liquid to the gaseous phase in the hypothesis that the resistance to the mass transfer in this last phase can be considered negligible.

Generally, the kinetics of degasing of the molten metals is controlled by the diffusion step throughout the stagnant layer adjacent to the interface, even where the liquid phase is perfectly stirred[8]. In this case N_H can be calculated by the relationship:

$$N_H = -K_l\left(C_H - \frac{P_{H_2}^{1/2}}{S_l}\right) \tag{9}$$

Nevertheless, it is possible that a different situation occurs when the transfer process throughout the metal-gas interface assumes the character of a very chemical reaction[9-12] as in the case under examination. In this last case, N_H can be calculated by the relationship:

$$N_H = -K_d\left(C_H^2 - \frac{P_{H_2}}{S_l^2}\right) \tag{10}$$

4. MATHEMATICAL MODELS

The kinetic data were obtained in the form of recorded curves indicating the pressure of hydrogen in the system at each instant of time during the absorption process.
Mathematical models for the probable rate-limiting steps in the overall absorption process of hydrogen in the Pb17Li eutectic were derived in terms of pressure and time and the data correlated with both the integrated and differential forms of these relationships.

4.1 Saturator model

It is necessary to make a material balance in the saturator in order to determine C_H. As the hydrogen concentration in the liquid phase can be considered independent from the very small permeation rate into the pressure probe device, and assuming a perfect mixing of the molten liquid, the material balance referred to the hydrogen gives:

$$\frac{dC_H}{dt} = -N_H \cdot a \tag{11}$$

where N_H is expressed according to Eqs.(8),(9) and (10) by which the following expressions are obtained:

$$\frac{dC_H}{dt} = -K_l a\left[C_H + \frac{1}{2\Gamma} - \left(\frac{1}{4\Gamma^2} + \frac{P_{H_2}}{S_l^2} + \frac{C_H}{\Gamma}\right)^{1/2}\right] \tag{12}$$

or: $$\frac{dC_H}{dt} = -K_l a\left[C_H - \frac{P_{H_2}^{1/2}}{S_l}\right] \tag{13}$$

or: $$\frac{dC_H}{dt} = -K_d a\left[C_H^2 - \frac{P_{H_2}}{S_l^2}\right] \tag{14}$$

Integrating Eq.(12) at the boundary condition:

$$C_H = C_{H_0} \quad \text{for } t = 0 \tag{15}$$

Eq.(16) is obtained:

$$-\ln\frac{\theta+\frac{1}{2\Gamma C_H^*}-\psi}{\theta_0+\frac{1}{2\Gamma C_H^*}-\psi_0}+\frac{1}{2\Gamma C_H^*}\ln\frac{1-\frac{1}{2\Gamma C_H^*}+\psi}{1+\frac{1}{2\Gamma C_H^*}-\psi}\;\frac{1+\frac{1}{2\Gamma C_H^*}-\psi_0}{1-\frac{1}{2\Gamma C_H^*}+\psi_0}=-K_l at$$

where: $\theta_0 = C_{H_0}/C_H^*$; $\Gamma = \frac{S_l^2 K_a}{K_l}$; $\theta = C_H/C_H^*$ (16)

$$\psi=\left(\frac{1}{4\Gamma^2 C_H^{*2}}+1+\frac{\theta}{\Gamma C_H^*}\right)^{\frac{1}{2}} \; ; \; \psi_0=\left(\frac{1}{4\Gamma^2 C_H^{*2}}+1+\frac{\theta_0}{\Gamma C_H^*}\right)^{\frac{1}{2}}$$

For a fast rate of interphase reaction and diffusional controlling resistance in liquid phase ($\Gamma \to \infty$), Eq.(16) becomes:

$$\frac{\theta-1}{\theta_0-1} = +\exp(-K_l at) \tag{17}$$

which could be also obtained by integrating Eq.(13) at the boundary condition (15).

For a slow rate of interphase reaction the dimensionless hydrogen concentration vs time is expressed by:

$$\frac{1-\theta}{1-\theta_0}\cdot\frac{1+\theta_0}{1+\theta} = \exp(-2C_H^* K_d at) \tag{18}$$

In Figure 4 the dimensionless hydrogen concentrations in the eutectic, calculated by Eq.(16), are reported in the ordinate vs exp($-K_l$ at) for different values of Γ and P_{H_2}, respectively, equal to 10.13 kPa and 20.26 kPa.

4.2 Hydrogen concentration in the eutectic Pb17Li

It is necessary to point out that the hydrogen pressures measured by the pressure sensor do not represent directly the hydrogen pressures at equilibrium with the hydrogen concentrations in the eutectic. In order to obtain these pressures, we develop a hydrogen balance in the pressure probe device; the differential equation representing this balance is[13]:

$$\frac{dP_{H_2M}}{dt} = \frac{1}{R}\left(P_{H_2l} - P_{H_2M}\right) \tag{19}$$

where R has been assumed 5.92×10^{4} s^{-1} [13].

The hydrogen concentration in the eutectic is found from Eq.(19) utilizing the Sievert's law:

$$C_H = \frac{R\,\frac{dP_{H_2M}}{dt} + P_{H_2M}^{1/2}}{S_l} \tag{20}$$

where the (dP_{H_2M}/dt) represents the value of the derivative of the pressure-time curve (Figure 3) at each time where the hydrogen concentration in the eutectic was to be evaluated.

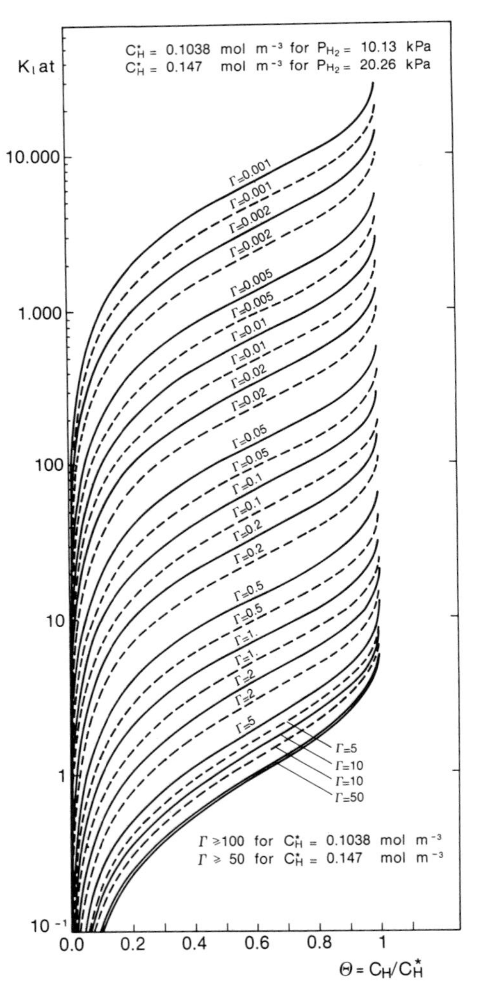

FIGURE 4
Theoretical values of dimensionless hydrogen concentration vs K_l at for different values of Γ.

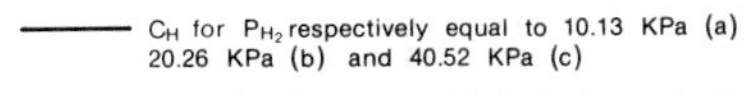

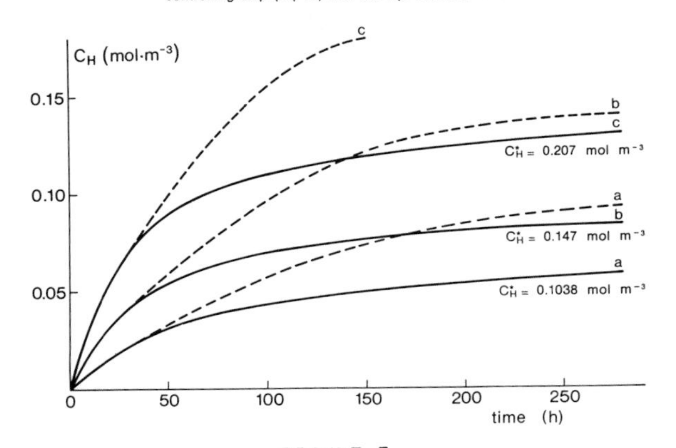

FIGURE 5
Hydrogen concentrations in the eutectic vs time calculated from experimental values of P_{H_2M} by Eq.(20).

In Figure 5 the hydrogen concentrations in the eutectic calculated from experimental values of P_{H_2M} by Eq.(20) vs time are reported.

5. DISCUSSION

As in the case of other heterogeneous systems, the mass transfer rate of hydrogen in gas-alloy dispersion is proportional to the mass transfer coefficient to individual bubbles and to the total interfacial area in the system.

As both the hydrogen mass transfer K_l and the interfacial area per unit of alloy volume, a, depend on the degree of gas dispersion in the eutectic, it is convenient to obtain experimentally the product K_la which represents the hydrogen mass transfer coefficient per unit of eutectic volume.

The differential relationships given by Eqs. (12), (13) and (14) were used to correlate data on initial rates of absorption into Pb17Li eutectic which has been vacuum degassed immediately prior to the runs. Under these conditions, Eq.(12), for a chemically and diffusion controlled process of absorptionn may be written as:

$$P_{H_2l}^{1/2} = \sqrt{R}\,\frac{dP_{H_2M}}{dt} + P_{H_2M}^{1/2} \tag{21}$$

for a diffusion controlled process in Eq.(13) may be written as:

$$C_H = \frac{R}{S_l}\,\frac{dP_{H_2M}}{dt} + \frac{P_{H_2M}^{1/2}}{S_l} \tag{22}$$

for a chemically controlled process Eq.(14) may be written as:

$$\left(\frac{dC_H}{dt}\right)_{t\to 0} = \frac{K_la}{2\Gamma} + K_la\left(\frac{t}{4\Gamma^2} + \frac{P_{H_2}}{S_l^2}\right)^{1/2} \tag{23}$$

The values of $(dC_H/dt)_{t\to0}$ are evaluated to make the derivative of the experimental hydrogen concentrations in the eutectic vs time curves at t=0 for the hydrogen pressures in the saturator equal to 10.13, 20.26 and 40.52 kPa, respectively.

If a model is applicable, then according to the respective equation, such a plot should generate a straight line having a slope whose magnitude is directly proportional to the product K_la, K_da and K_aa as being $K_aa = K_da/S_l^2$. The results are shown graphically in Figures 6, 7 and 8, respectively.

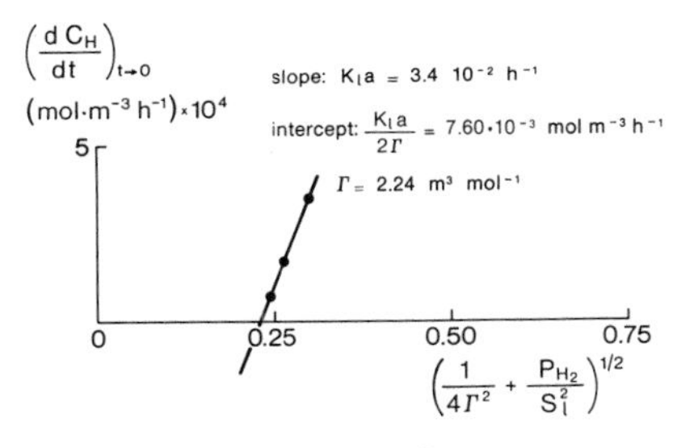

FIGURE 6
Correlation between the experimental initial rates of hydrogen absorption into Pb17Li eutectic and those foreseen by Eq.(21).

Figure 6 shows that the linear relationship predicted by Eq.(21) of the chemically and diffusion controlled process model is obeyed, but that deviation from linearity can be noted if the diffusion controlled (Figure 7) or

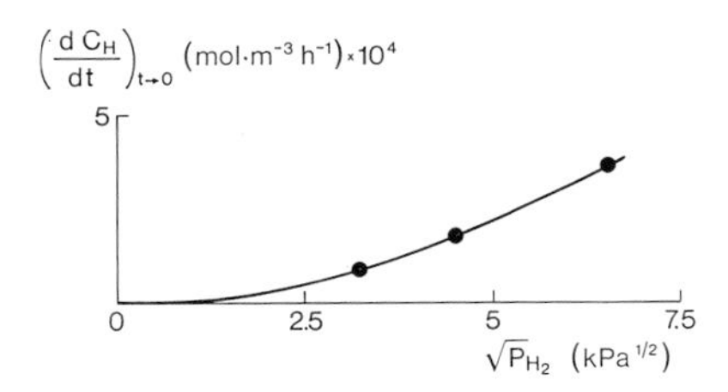

FIGURE 7
Correlation between the experimental initial rates of hydrogen absorption into Pb17Li eutectic and those foreseen by diffusion controlled model Eq.(22).

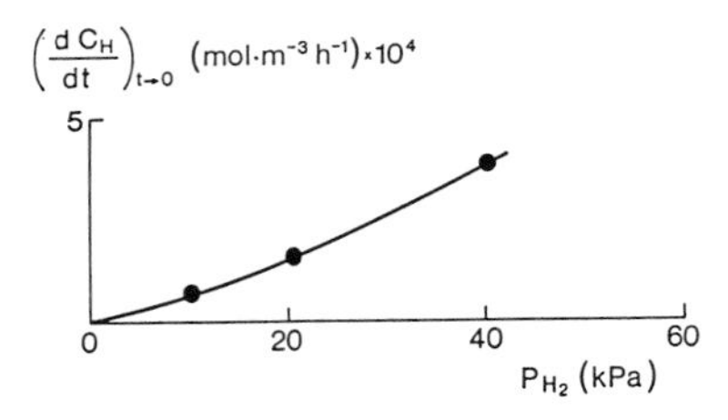

FIGURE 8
Correlation between the experimental initial rates of hydrogen absorption into Pb17Li eutectic and those foreseen by chemical reaction model Eq.(23).

the chemically controlled model (Figure 8) were applied. This indicates that the overall process of hydrogen absorption in the eutectic is represented by the heterogeneous reaction at the gas-eutectic interface with the hydrogen molecules dissociation and the diffusion of solute hydrogen atoms in the Pb17Li eutectic.

From the value of the slope of the straight line one obtains $K_l a = 9.44\times10^{-6}$ s^{-1} and from the value of the intercept of this straight line with the axis of the ordinates, $K_l a/2\Gamma = 2.11\times10^{-5}$ mol m^{-3} s^{-1} it derives $\Gamma = 2.24$ m^3 mol^{-1}. By Γ definition it derives that $K_d a = 2.11\times10^{-5}$ m^3 mol^{-1} s^{-1} and $K_a a = 2.250\times10^{-8}$ mol m^{-3} s^{-1} kPa^{-1}.

The value of $K_a a$ obtained in this work is in agreement with the value of $K_a a = 1.73\times10^{-8}$ mol m^3 s^{-1} kPa^{-1} obtained for the same system H_2/Pb17Li by means of an electrobalance by Viola et al.[14].

The results of the mathematical model (Eq. (16)) that takes into consideration a mixed control of the chemical reaction and the diffusion for values of Γ and $K_l a$ obtained by the initial adsorbing data, are reported in Figure 5. The agreement is very good during the first period of the hydrogen absorption process; for the higher values of time the deviations between the experimental hydrogen concentration in the eutectic and those foreseen by theoretical model increase, particularly when the hydrogen pressures in the saturator increase. This deviation for higher times can be caused either by hydrogen loss to the ambient from the same pressure probe device or by the fact that R (the overall resistance to the hydrogen mass-transfer from eutectic to the gaseous phase in the pressure probe device) has been considered constant[13]. Work is in progress to investigate this question thoroughly.

6. CONCLUSION

The kinetics of the global process of H_2 absorption in the Pb17Li eutectic has been analysed in the present work. The correlation of the experimental data by means of a mathematical model for the hydrogen absorption process and the examination of the results by methods which are conventionally used to determine the rate-limiting step in gas absorption processes has demonstrated that the overall solution process is controlled by the heterogeneous reaction at the gas-eutectic interface with the hydrogen molecules dissociation and by the diffusion of solute hydrogen atoms in the molten Pb17Li alloy.

NOMENCLATURE

a	interfacial area per eutectic volume unity in the gas-eutectic dispersion (m^2m^{-3})
C_H	concentration of hydrogen in the eutectic (mol m^{-3})
C_{Hi}	concentration of hydrogen in the eutectic at interface gas-liquid alloy (mol m^{-3})

C_H^* concentration of hydrogen in the eutectic in equilibrium with the gaseous phase (mol m^{-3})
C_{H0} concentration of hydrogen in the eutectic for t=0 (mol m^{-3})
D_1 hydrogen coefficient of diffusion in the eutectic (m^2 s^{-1})
k_a hydrogen absorption specific rate (mol m^{-2} kPa^{-1})
k_d hydrogen desorption specific rate (m^4 mol^{-1} s^{-1})
k_1 hydrogen mass transfer coefficient in the eutectic (m s^{-1})
N_H overall hydrogen flux from the gaseous phase to the eutectic (mol m^{-2} s^{-1})
N_{H1} hydrogen flux in the eutectic (mol m^{-2} s^{-1})
N_{HR} hydrogen flux through gas-eutectic interface (mol m^{-2} s^{-1})
P_{H_2} hydrogen partial pressure in the saturator (kPa)
P_{H_21} hydrogen pressure at equilibrium with C_H (kPa)
P_{H_2M} hydrogen pressure in the pressure probe device (kPa)
R overall resistance to the hydrogen mass transfer from eutectic to the gaseous phase in the pressure probe device (s^{-1})
S_1 Sievert's constant of hydrogen in the eutectic ($kPa^{\frac{1}{2}}$ m^3 mol^{-1})
T temperature (K)
t time (s)
γ diffusional layer of Nernst (m)

REFERENCES

1. F. Reiter, J. Camposilvan, G. Gervasini and R. Rota, Proc. of 14th Symp. on Fusion Technology, Avignon 1986.
2. P. Fauvet and F. Sannier, Hydrogen behaviour in liquid Pb17Li alloy, paper presented at the 3rd Int. Conf. on Fusion Reactors Mater., Karlsruhe (1987).
3. A. Viola, Analysis of tritium permeation through the structural material of the blanket containing liquid Pb-17Li as breeding material, 13th Symp. of Fusion Technology SOFT, Varese, September 24-28, 1984.
4. G. Pierini, A.M. Polcaro, P.F. Ricci and A. Viola, Nucl. Eng. and Design/Fusion, 1,2 (1984) 159.
5. R. Baratti, G. Pierini and A. Viola, The status of the art of tritium recovery from liquid eutectic Pb-17Li blanket material fusion reactor blanket and fuel cycle technology, Proc. of the Int. Symp. Tokai-Mura, Ibaraki, Japan, October 27-29, 1986.
6. N.D. Vargaftik, Table of Thermophysical Properties of Liquids and Gas, 2nd ed., Wiley and Sons, New York (1975) 639.
7. G. Pierini, A.M. Polcaro, P.F. Ricci and A. Viola, J. Chem. Eng. Data, 29 (1983) 250.
8. W.M. Small and R.D. Pehlke, Met. Trans., 5 (1974) 2549.
9. C. Yamauchi and K. Sano, Nippon Kinzoku Gakkaf SHI J. Jap. Inst. Metals, 33 (1969) 1249.
10. C. Yamauchi and K. Sano, Ibidem, 33 (1969) 1255.
11. C. Yamauchi and K. Sano, Ibidem, 33 (1969) 1262.
12. M.P. Gardner and M.M. Nishima, J. Phys. Chem., 85 (1981) 2388.
13. A. Viola, P.L. Lolli-Ceroni and G. Pierini, Recent experimental results of the rates of hydrogen permeation through metallic material in the presence of liquid Pb-17Li alloy; influence on the performance and design of tritium recovery equipment, ISFNT, April 10-19, 1988, Tokyo, Japan.
14. A.M. Polcaro, P.F. Ricci and A. Viola, J. Nucl. Mater., 119 (1983) 291.

FUSION TECHNOLOGY 1988
A.M. Van Ingen, A. Nijsen-Vis, H.T. Klippel (editors)

PARAMETRIC STUDY OF HYDROGEN PERMEATION RATE THROUGH Pd-Ag MEMBRANE IN THE PRESENCE OF GASEOUS IMPURITIES

J. CHABOT, J. LECOMTE, C. GRUMET, J. SANNIER

CENTRE D'ETUDES NUCLEAIRES DE SACLAY - IRDI-DCAEA-SCECF-SECNAU - 91191 GIF SUR YVETTE - FRANCE

A parametric study of H_2 permeation rate through Pd-Ag membrane in the presence of gaseous impurities, such as CO, CO_2 and CH_4 is carried out in the loop PALLAS, in a wide range of membrane temperature (100 to 450°C), H_2 pressure (0.3 to 14 kPa) and impurity concentration (0.2 to 9.5 vol.%). CO_2 and CH_4 appear to have no practical effect on the permeation rate whereas a poisoning effect may occur with CO, depending on its concentration and the membrane temperature. However the permeation rate is not affected between 250°C and 375°C.

1. INTRODUCTION

Gaseous wastes coming off the burnt gases purification plant in a fusion reactor are constituted of various impurities such as CO, CO_2, $C(D/T)_4$, $(D/T)_2O$, O_2, N_2, possibly $N(D/T)_3$, and contain also measurable amounts of deuterium and tritium. Hydrogen permeation through palladium-alloy membranes may be used to separate the residual deuterium and tritium from impurities in the waste reprocessing system. However a possible drawback of the method is the poisoning of the membranes by impurities. Though large amounts of work and data on permeation have been performed and reported, there is a lack of results concerning membranes behaviour in the presence of large amounts of impurities.

Within the framework of the European Fusion Technology Programme, a parametric study of the poisoning of Pd-Ag membranes by the most important expected gaseous impurities is under way at CEN-SACLAY, in order to determine the applicability of this method and to provide the required data for the design and development of a complete fuel clean-up system for the NET.

This paper presents permeation measurements for H_2 through Pd-Ag membranes in the presence of CO, CO_2 and CH_4 within the temperature range of 100-450°C and impurity level up to 10 vol.%.

2. RECALL OF THEORY

The permeation of hydrogen (and isotopes) through palladium-silver alloys is believed to proceed by the following steps[1] :

1 - gas phase diffusion of the hydrogen molecule to the palladium-silver surface,

2 - chemisorption and dissociation into atoms,

3 - solution of hydrogen atoms into the palladium-silver alloy,

4 - diffusion through the bulk of the metal,

5 - reemergence of hydrogen from the bulk of the metal,

6 - recombinaison and desorption of molecules from the Pd-Ag surface,

7 - gas-phase diffusion away from the surface.

In a wide range of pressures and temperatures, the diffusion through the bulk of the metal is the rate controlling step. In this case, the permeation of hydrogen is governed by Fick's and Sievert's laws and the steady-state permeation rate can be expressed by the following Richardson's equation :

$$D = a \frac{A}{t} (P_1^{0.5} - P_2^{0.5}) \exp.(-\frac{b}{RT}) \qquad (1)$$

with :

A : effective surface of the membrane, cm^2,

t : thickness of the membrane, cm,

P_1 : H_2 pressure upstream the membrane, kPa,

P_2 : H_2 pressure downstream the membrane, kPa,

T : membrane temperature, K,
R : gas constant, cal.mol^{-1}K^{-1},
a and b : empirical constants.

In the presence of gaseous impurities, the Pd-Ag accessible surface can be decreased by strongly chemisorbed species, and even a partial poisoning of the surface would significantly reduce the permeation rate. If the poisoning is important, surface reactions (chemisorption) can become the rate-controlling step and a marked deviation from Sievert's law and equation (1) may be observed.

3. EXPERIMENTAL

3.1. Apparatus

Experiments are carried out in the loop PALLAS (fig. 1) which includes a processing loop allowing to circulate the gas to be purified, an analysis circuit, a vacuum line and a feed gas line. The processing loop mainly consists of :

- a 10 litres stainless steel vessel supplied with the test-gas mixture,
- a circulation pump,
- a furnace allowing to heat the permeation cell up to 500°C,
- an aftercooler to bring back the gas to room temperature.

The membrane of the permeation cell is constituted of a palladium-23 wt.% silver tube, 10 cm long, 0.3 cm in external diameter and 0.025 cm thick. Its effective surface area is 9.4 cm^2. This tube is sealed at one end and brazed to a 316L stainless steel tube at the other one. The permeation temperature is measured by a thermocouple located inside the membrane. Operating pressure and flowrate are measured at various locations in the loop. The analysis circuit is mainly constituted of two chromatographs using thermal conductivity detectors and respectively fitted with molecular sieve and porapak column packing.

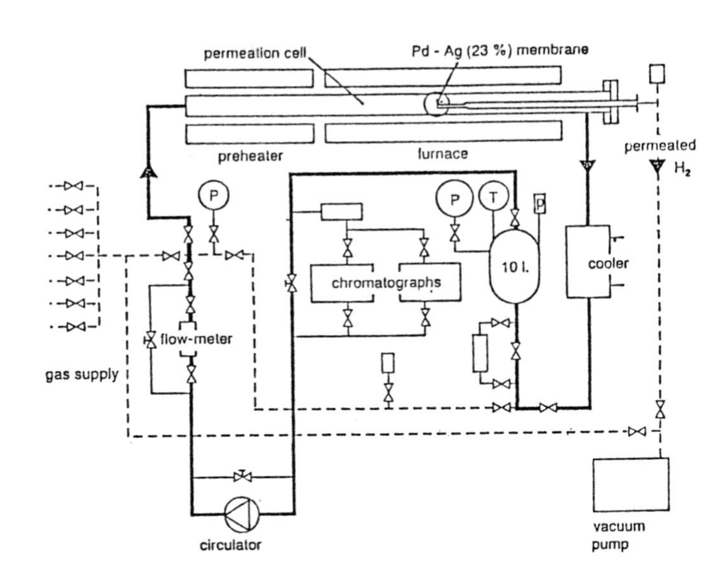

FIGURE 1
PALLAS LOOP. Experimental apparatus for permeation measurements and poisoning study of Pd-Ag membranes - Conceptual flow-sheet.

3.2. Procedure

Helium carrier gas containing a well-known amount of hydrogen and of impurities separately or in mixtures, is circulated at a 6 m^{-3}.h^{-1} flowrate through the permeation cell. The gas is preheated before it reaches the outer surface of the membrane which is kept at the required temperature. The permeated hydrogen is extracted on the downstream side of the membrane by means of the vacuum pump while unpermeated gases leaving the cell are recirculated and cooled to room temperature before analysis. The gas composition is determined at given time intervals and recorded.

4. RESULTS AND DISCUSSION

4.1. Basic experiments

The specific characteristics of the membrane were determined for He-H_2 mixtures as a function of temperature and hydrogen pressure in order to find out the validity limits of the Fick's diffusion law and Richardson's equation and also to provide the basic data necessary for checking poisoning effects of impurities.

In order to improve as-received membrane performances and to obtain a reproducible surface state for further experiments, a cleaning and activation treatment of the membrane on both sides has been defined, and afterwards applied before each run. It essentially consists in an oxidation by air at 450°C during one hour, followed by a reduction in a He-10 vol.% H_2 mixture in the same conditions of temperature and duration.

4.2. Temperature effect

The set of experimental curves obtained in the range from 100 to 450°C with an initial inloop H_2 partial pressure of 14 kPa is given in fig. 2. The specific permeation rate values calculated by considering a linear area surface dependence and the assumption of inverse membrane thickness dependence are plotted in an Arrhenius diagram in fig. 3. Log D is a linear function of 1/T with a slope change at about 150°C. It can be noted that the permeation rate does not vary greatly within the range 150-300°C.

4.3 Pressure effect

The relation between permeation flow and square-root of the membrane upstream H_2 pressure (downstream H_2 pressure being negligible) is shown in fig. 4. The permeation appears to be directely proportional to $P^{0.5}_{(H_2)}$ at pressures above 1.6 kPa and temperatures higher than 200°C.

4.4. Experimental permeation rate coefficient values

At pressures higher than 1.6 kPa and for temperatures above 200°C the semi-empirical Richardson's equation is verified and can be expressed as follows :

$$D = 9.4\ 10^{-7} \frac{A}{t} P^{0.5}_{(H_2)} \exp.(-\frac{1400}{RT})\ \text{mol.min}^{-1} \quad (2)$$

These results are in a good agreement with those reported by Ackerman[2] and Yoshida[3] at temperatures higher than 300°C.

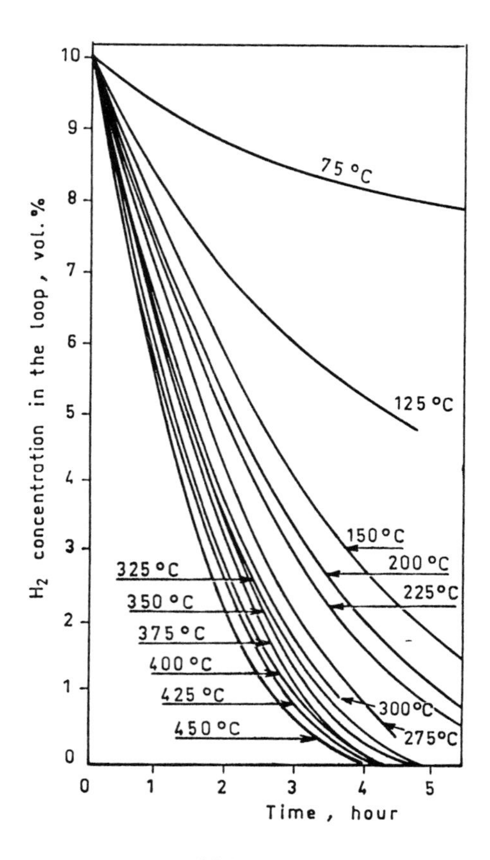

FIGURE 2
H_2 concentration in the loop versus time at several temperatures. Initial H_2 partial pressure = 14 kPa.

4.5. Effect of impurities

The poisoning effect of CO, CO_2 and CH_4 separately or in mixtures, has been investigated at different concentrations (table I).

Figure 5 shows that the presence of CO is giving rise to a poisoning of the membrane in all the studied concentration range, ie. 0.2 - 9 vol.%. For the highest concentrations (9 % and 4.3 %) the poisoning effect occurs below 300°C and is rapidly increasing when the temperature is decreasing ; it becomes practically total at 150°C. For the lowest concentrations (2 to 0.2 %) the permeation decay becomes important only at temperatures below 200°C. Practically the permeation rate is not affected in the range of 375-300°C even in the presence of high CO concentrations.

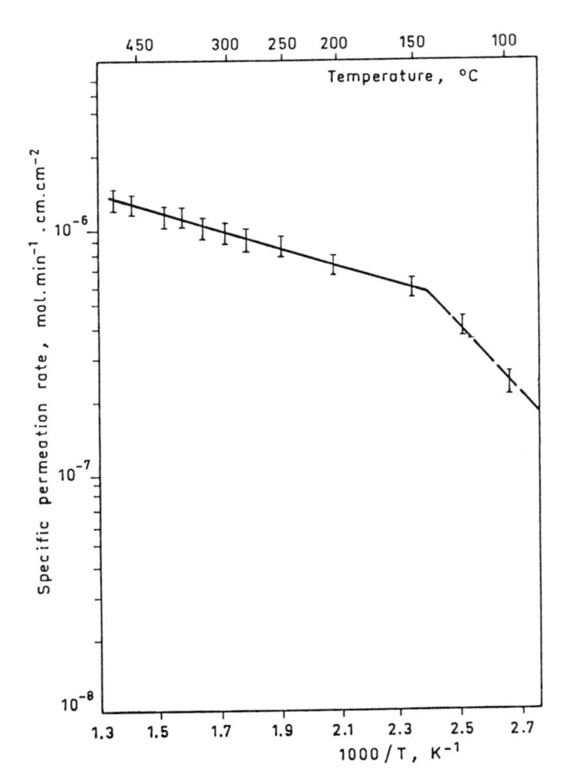

FIGURE 3
H_2 permeation through Pd-23 wt.% Ag membrane versus temperature (Arrhenius diagram). Upstream H_2 partial pressure = 14 kPa.

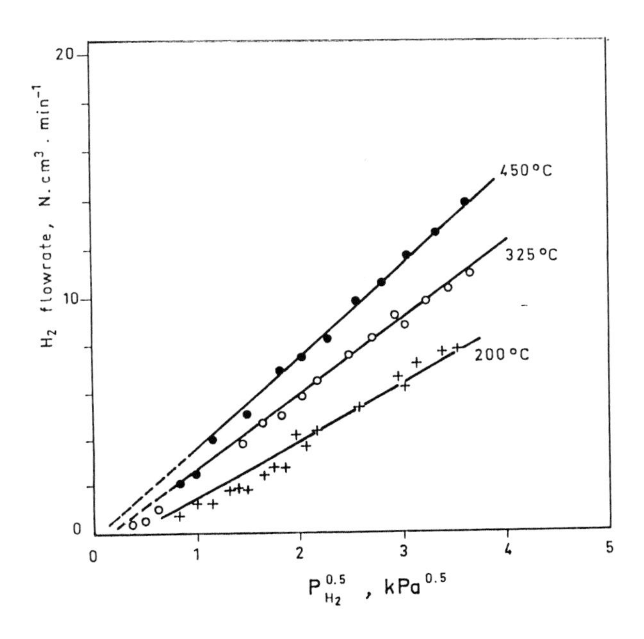

FIGURE 4
Variation of H_2 permeation versus square root of upstream hydrogen pressure for various temperatures.

TABLE I
Composition of the gaz mixtures

Ref.	impurities	gas composition (vol.%)				
		H_2	CO	CH_4	CO_2	He
a	CO	9.5	9	—	—	balance
b		9.5	4.3	—	—	"
c		9.5	2	—	—	"
d		9.5	0.4	—	—	"
e		9.5	0.2	—	—	"
f	CH_4	9.5	—	9	—	"
g	CO_2	9.5	—	—	9	"
h	CH_4+CO_2	9.5	—	4.3	4.3	"
i	CH_4+CO_2+CO	9.5	4.3	4.3	4.3	"

Above 375°C a noticeable part of CO reacts with H_2 to form CH_4, CO_2 and H_2O. The surface of the membrane probably acts as a catalyst of this conversion. Regeneration experiments show that the poisoning by CO is not permanent. Initial performances can be restored by heating the membrane under vacuum at temperatures higher than 250° or by applying the cleaning activation treatment.

As shown on fig. 6 the poisoning effect of methane and carbon dioxide, separately or in mixtures, is much less important than with CO and observed only below 150°C. When CO is added to CH_4 and CO_2 mixtures, the poisoning effects are practically identical to those observed in the presence of CO alone.

5. CONCLUSION

Based on the experimental results, the poisoning effect of carbon monoxide is a concern for the process while the presence of carbon dioxide and methane in the waste gas appears to have no practical effect on the permeation rate of hydrogen. Depending on CO concentration, temperatures of about 250-300°C appear to constitute the lower threshold for avoiding a decrease of membrane

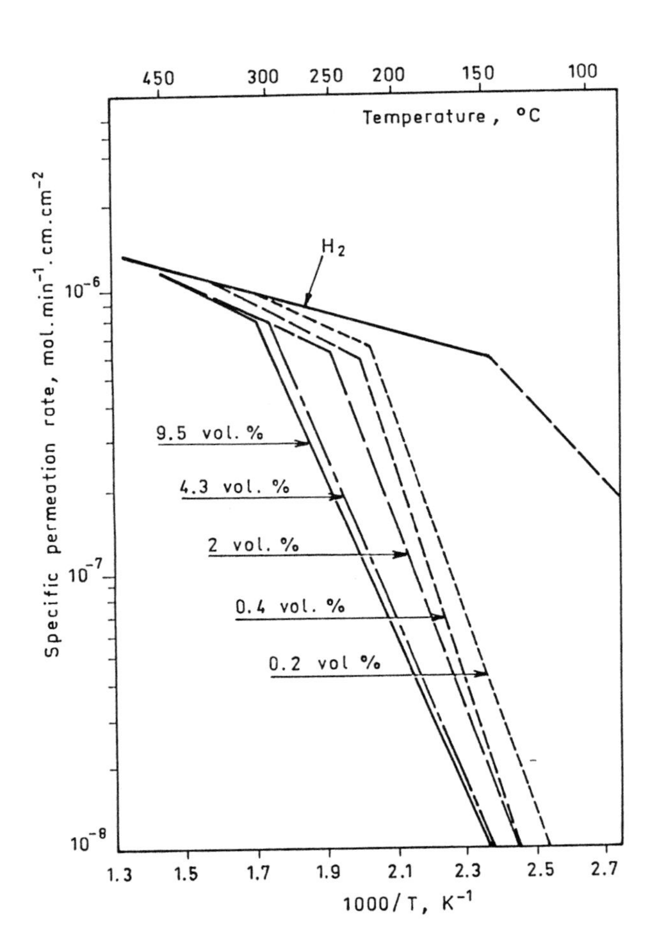

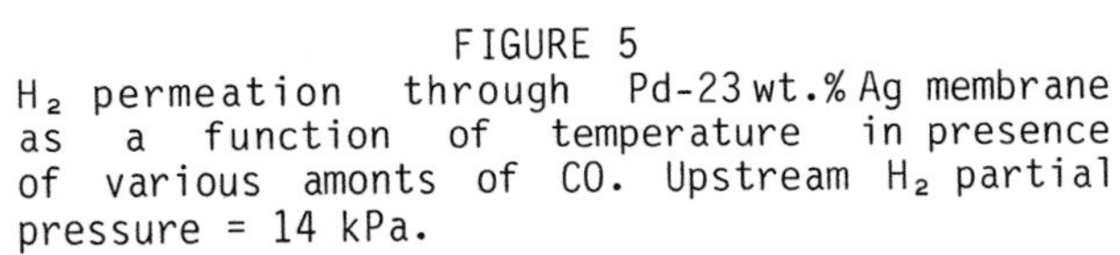
FIGURE 5
H_2 permeation through Pd-23 wt.% Ag membrane as a function of temperature in presence of various amonts of CO. Upstream H_2 partial pressure = 14 kPa.

performances. For the determination of the maximal admissible temperature, the reaction of CO and H_2 with formation of CH_4 observed above 375°C has to be taken into account. In respect to tritium recovery, this reaction is unfavourable owing to the fact that carbon-bonded hydrogen cannot be extracted by permeation. Furthermore an operating temperature as low as possible is highly desirable for limiting the tritium permeation through structural materials. This study has to be carried on by investigating the possible influence of other expected impurities like H_2O - now in progress - and NH_3.

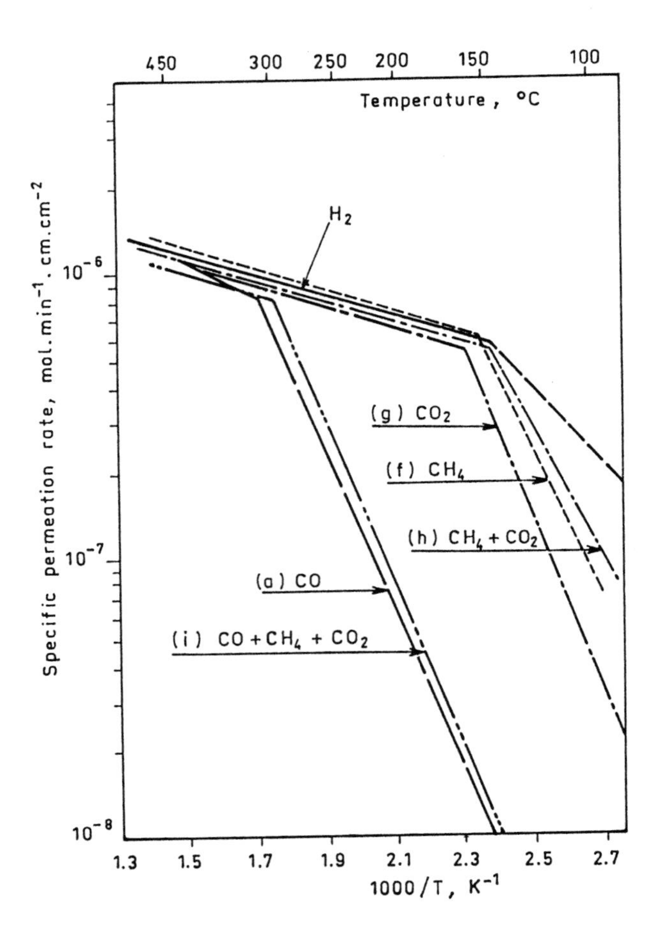

FIGURE 6
H_2 Permeation through Pd-23 wt.% Ag membrane as a function of temperature in presence of CO, CO_2 and CH_4 separately or in mixtures. Upstream H_2 partial pressure = 14 kPa.

ACKNOWLEDGEMENTS

This work has been carried out in the frame of the Commission of the European Communities Programme on Fusion Technology. The authors wish to thank the CEE and CEA authorities for giving the possibility of conducting this research and publishing the results.

REFERENCES

1. F.A. Lewis, "The palladium hydrogen system" Academic Press, New York, NY (1960).

2. F.J. Ackerman, G.J. Koskinas, J. of Chemical and Engineering Data, 17, n°1 (1972).

3. H. Yoshida, S. Konishi, Y. Naruse, Nuclear Technology, Fusion, 3, 471 (1983).

FUSION TECHNOLOGY 1988
A.M. Van Ingen, A. Nijsen-Vis, H.T. Klippel (editors)
Elsevier Science Publishers B.V., 1989

INFLUENCE OF START UP AND PULSED OPERATION ON TRITIUM RELEASE AND INVENTORY OF NET CERAMIC BLANKET

M. ISELI, B. ESSER

The NET-Team c/o Max-Planck-Institut für Plasmaphysik, Boltzmannstrasse 2 , D-8046 Garching

A first estimate for the tritium release behaviour of a ceramic breeder blanket in pulsed operation is obtained by assuming a linear steady state temperature distribution and taking into account the time constant of the thermal behaviour. The release behaviour of the breeder exposed to consecutive periods of tritium generation is described with an analytical solution of the diffusion equation. The results are compared with a simple exponential approach valid for surface desorption controlled release. The exponential model is used to simulate a blanket with aluminate as breeder material, which takes longest to reach steady state. The simulation demonstrates that a significant fraction (> 67%) of steady state can be achieved after a testing time of about one day.

1. INTRODUCTION

Full segments or modules of ceramic breeder concepts will be tested during the technological phase of NET. Discussion of testing time requirements raised the question of the influence of start up and pulsed operation on tritium release and inventory.

For diffusion controlled tritium release, an analytical solution of the diffusion equation has been applied. For desorption controlled tritium release, an exponential model analogous to that describing the transient temperature history has been used.

2. TRITIUM RELEASE BEHAVIOUR OF A CERAMIC ELEMENT

2.1 Diffusion

The tritium release is assumed to be controlled by diffusion of tritium through the ceramic grains. The temperature dependent diffusion coefficient D is expressed as follows

$$D = D_o \, e^{-Q_o/(RT)} \tag{1}$$

For diffusion controlled release, the mean residence time of tritium in ceramic material depends on temperature and geometry of the ceramic particles

$$\tau = \tau_o \, e^{Q_o/(RT)} \tag{2}$$

If the ceramic is considered to be a package of small spherical grains with radius r_g, the constant τ_o in equation (2) is given by [1]

$$\tau_o = \frac{r_g^2}{15 \, D_o} \tag{3}$$

The release behaviour of a sphere exposed to consecutive periods i of different generation rates G_i and different temperatures, is described with an analytical solution of the diffusion equation.[2] The real pulsed operating conditions are simulated by a stepwise approach, in which constant residence times $\tau_{r,i}$ and generation rates G_i are used . For the first period, the initial concentration is zero. The tritium inventory I_L $[mol/m^3]$ [a] of the Lth period may be calculated by

$$\frac{I_L}{\tau_o} = \frac{6}{\pi^2} \sum_{i=1}^{L} G_i \sum_{n=1}^{\infty} \frac{1 - e^{-X_i \frac{t'_i}{\tau_o}}}{n^2 X_i} \, e^{-\sum_{j=i+1}^{L} X_j \frac{t'_j}{\tau_o}} \tag{4}$$

$$X_i = \frac{n^2 \pi^2}{15 \, \tau_{r,i}} \tag{5}$$

The calculation is performed in terms of the time t'_i, starting from zero at the beginning of each period i. The total time at the end of a period i is termed t_i.

$$t'_i = t_i - t_{i-1} \quad \text{if} \quad i < L \quad ; \quad t'_L = 0 \text{ to } (t_L - t_{L-1}) \tag{6}$$

The release rate R_i is given by

$$R_i = G_i - \frac{dI_i}{dt} \tag{7}$$

In real operation the temperature changes during burn

and off burn time. If the mean residence time in the ceramic is much longer than the burn time, a mean diffusion coefficient is calculated by integration of the diffusion coefficient over time. These mean diffusion coefficients define the average mean residence times $\overline{\tau}_b, \overline{\tau}_{off}$ of burn and off burn . Examples of inventory and release rate behaviour during the first few pulses of start up are given in figure 1 and 2.

2.2 Desorption

With H_2 - swamping, the desorption takes place as a first order kinetic relation. The transient inventory can be calculated with the same exponential model used for temperature history (Section 3). The analogy exists between temperature and inventory. The thermal time constant is replaced by the average mean residence time $\overline{\tau}_i$ and a fictitious steady state inventory I_{FS} is introduced.(see figure 1)

$$I_{FS,i} = G_i \, \overline{\tau}_i \tag{8}$$

The corresponding release rate is calculated with equation 7.(see figure 2)

For comparison with diffusion it is assumed, that the mean residence time is the same in both cases. (Equal steady state inventory in continuous burn)

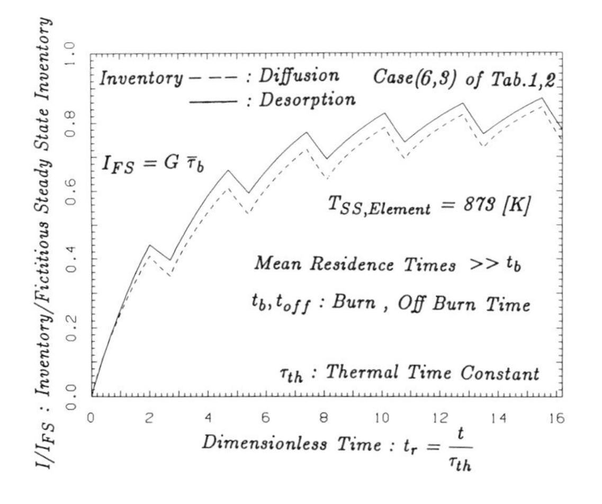

FIGURE 1 : Inventory of a Ceramic Element (Average Mean Residence Times, $\overline{\tau}_b, \overline{\tau}_{off}$)

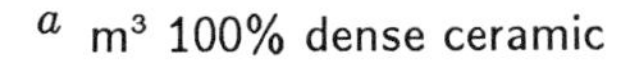

[a] m³ 100% dense ceramic

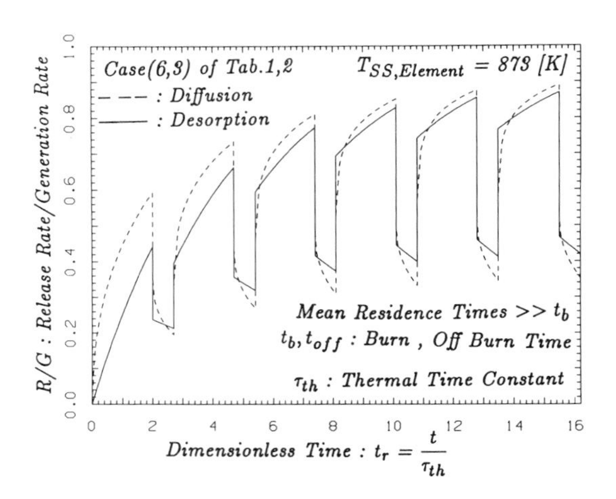

FIGURE 2 : Release Rate of a Ceramic Element (Average Mean Residence Times, $\overline{\tau}_b, \overline{\tau}_{off}$)

3. TEMPERATURE HISTORY OF A CERAMIC ELEMENT

The transient temperature is described by the well known exponential model, depending on the thermal time constant τ_{th} of the blanket concept, the steady state temperature T_{ss} of the breeder and the inlet temperature T_o of the coolant.[3]

4. TEMPERATURE DISTRIBUTION IN THE BLANKET

4.1 Steady state temperature

To obtain a first estimate, a linear distribution is assumed. The minimum breeder temperature increases axially in direction of the coolant flow and radially due to the heat generation in the breeder. Figure 3 shows a cross section of a breeder unit and the assumed linear radial temperature distribution. Limits are given by the maximum breeder and coolant temperatures. Table 2 shows the cases considered.

4.2 Pulsed operation

Each element behaves according to the exponential model mentioned in section 3.

5. RELEASE AND INVENTORY OF A BLANKET

The temperature distribution determines the fraction

Table 1 : Variation of the Material Properties

Experiment	Case Nr	Breeder	Activation Energy Q_o [kcal/mol]	Residence Const. τ_o [s]	Grain Radius r_g [cm]	Time Const. τ_{th} [s]
LISA-2	1	$LiAlO_2$	34.43	$3.95 \cdot 10^{-6}$	$4 \cdot 10^{-5}$	100
LILA-3	2	"	39.31	$9.16 \cdot 10^{-8}$	$2.45 \cdot 10^{-5}$	100
LILA-3	3	"	31.97	$1.12 \cdot 10^{-5}$	$1.15 \cdot 10^{-5}$	100
LISA-2	4	Li_4SiO_4	37.41	$3.10 \cdot 10^{-8}$	$1.30 \cdot 10^{-3}$	100
Trio[7]	5	$LiAlO_2$	35.80	$1.21 \cdot 10^{-6}$	$1 \cdot 10^{-5}$	100
Example	6	"	31.90	$0.98 \cdot 10^{-6}$	$1 \cdot 10^{-5}$	100
LILA-3	7	"	31.97	$1.12 \cdot 10^{-5}$	$1.15 \cdot 10^{-5}$	50

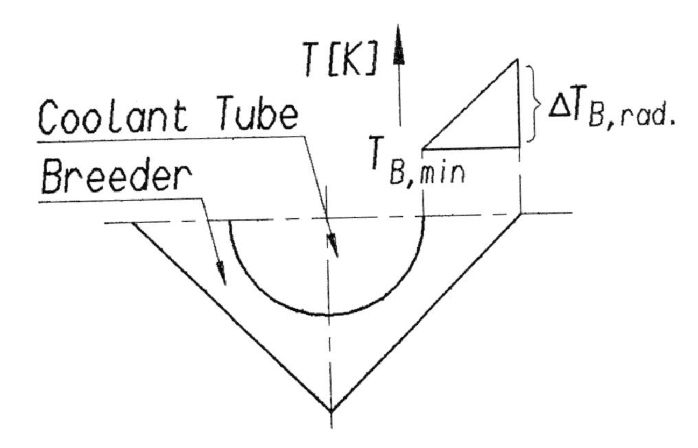

FIGURE 3 : Radial Temperature Distribution in a Cross Section of a Breeder/Coolant Unit

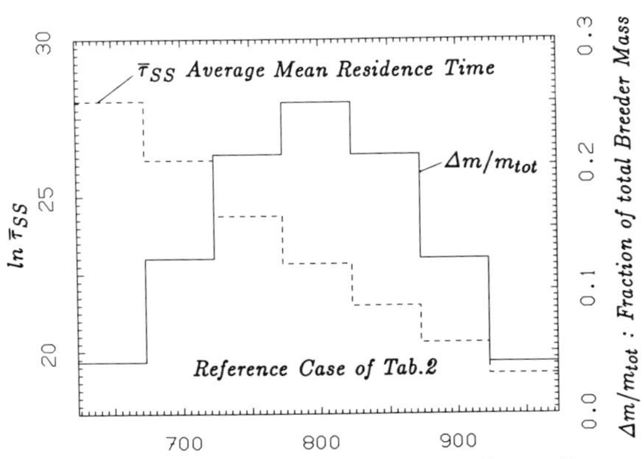

FIGURE 4 : Fraction of total Breeder Mass at the different Temperature Levels of the Blanket and the corresponding Mean Residence Time

$f = \frac{\Delta m}{m}$ of mass related to a certain temperature interval k (see figure 4). The integrated release of the whole blanket is given by :

$$R_{tot} = \sum_{k=1}^{K} f_k \, R_k(T_k) \tag{12}$$

The same holds for the inventory.

6. TYPICAL TEST BLANKET[4]

6.1 Machine operation

- Neutron wall load : 1 [MW/m²]
- Pulses a) Burn 200 [s], off burn 70 [s]
  b) Burn 300 [s], off burn 100 [s]
  c) Long Pulse of 3600 [s],off burn 300 [s]
- Continuous operating time up to 200 [h]

Table 2 : Variation of the Steady State Temperature Distribution in the blanket (Coolant outlet Temperature < 773 [K])

Case Number		Ref. 1	2	3	4	5	
Coolant inlet Temperature		573	523	623	573	623	623
Breeder Temp.	Min.	623	573	673	623	673	663
	Max.	973	923	973	873	973	903
Rad. Temp.diff. in the Breeder		200	200	200	100	200	160
Wall load [MW/m²]		1	1	1	0.5	1	0.8

6.2 Ceramic breeder material : $LiAlO_2$

The mean residence times of the in pile experiment LILA [5] and LISA [6] are used for aluminate and silicate. The values of the properties and their variations are summarised in table 1.

The transient release behaviour is estimated for

- Surface desorption (with H_2 - swamping)
- Diffusion out of a spherical grain

6.3 Blanket operating conditions

The different temperature distributions resulting from different coolant temperatures, flow rates and wall loads are summarised in table 2. Case 5 represents a change during the test.

7. RESULTS

7.1 Continuous Burn, Long Pulse

In a blanket with the reference temperature distribution, the diffusion controlled release rate of the aluminates reaches a fraction between 48 and 57% of the steady state ,whilst that of silicate reaches 82% after a continuous burn time of 3500 [s] (see figure 5). The desorption controlled release reaches a fraction of 40 to 51%. The long pulse operation with a burn time of 3600 [s] and a dwell time of 300 [s] reaches a fraction of 67% after 4 to 6 pulses and 85 to 90% after 30 pulses depending on the type of aluminate (desorption controlled release).

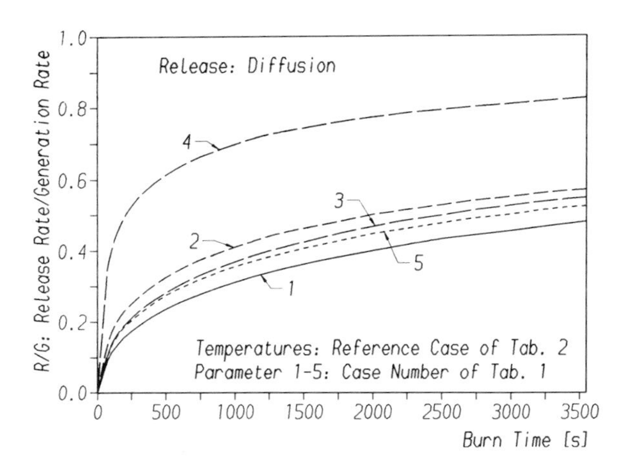

FIGURE 5 : Release Rates of a Blanket in the first long Pulse of 3600 [s]

7.2 Comparison Diffusion/Desorption

Single ceramic grains (figure 1 to 2) as well as a whole blanket (figure 6) show, that up to a significant fraction of the steady state, the release rate of surface desorption is smaller than that of diffusion. The much simpler desorption model yields conservative estimates for time requirements. The effects of changing operating conditions during tests are seen in figure 6.

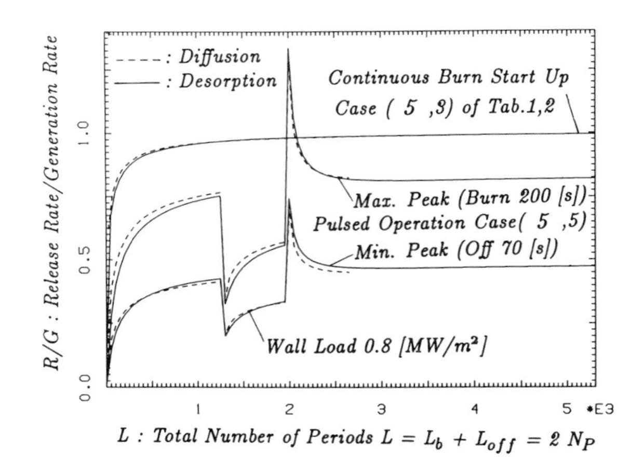

FIGURE 6 : Release Rates of a Blanket in Pulsed Operation (Extreme Peak Values) and for Continuous Burn

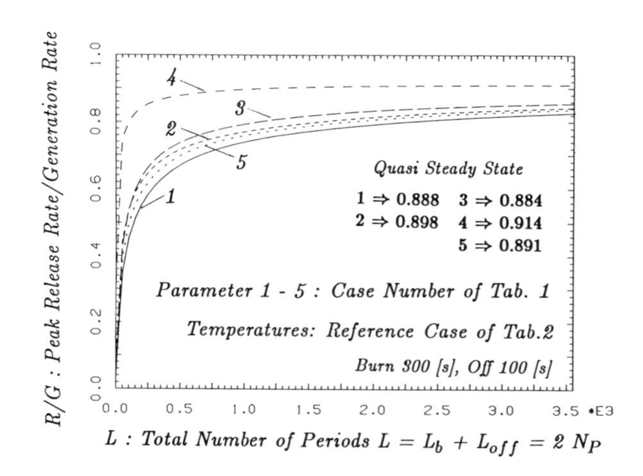

FIGURE 7 : Maximum Peak Release Rate of a Blanket

7.3 Sensitivity Analysis performed with Desorption Model

The variations of the parameters are summarized in table 1,2.

By variation of physical parameters of aluminate, the release rate reaches between 71% and 80% of the quasi steady state after 20 h and 93% to 97% after 200 h. (see figure 7)

The silicate reaches 95% after 20 h.

Figure 8 and table 3 show the effect of changes in coolant inlet temperature, wall load and thermal time constant.

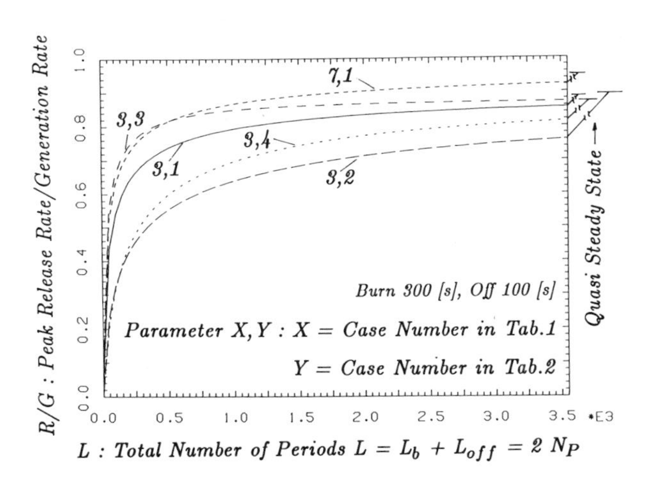

FIGURE 8 : Maximum Peak Release Rate for different Blanket Conditions.

Table 3 : Variation of the Fraction reached after Changes of Parameters

Fig.8	Parameter	Fraction of quasi steady state after 20 [h]	after 200 [h]
3.1	Reference	0.793	0.967
3.3	coolant inlet +50 [K]	0.905	0.997
3.4	wall load : 1 ⇒ 0.5	0.622	0.936
3.2	coolant inlet -50 [K]	0.571	0.853
7.1	time const. 100 ⇒ 50	0.824	0.970

8. CONCLUSION

This study gives a first estimate of the tritium release behaviour of a blanket or blanket module. Within about 1 day of pulsed operation, more than 67% of the quasi steady state is reached by a blanket with aluminate as breeder material. Uncertainties of the release data can be compensated by an adjustment of blanket parameters such as coolant inlet temperature or flow rate.

Regarding the tritium release, it seems that opration with few long pulses and relative long dwell times offer better prospects, than short pulses at high frequencies. Generally Li_4SiO_4 shows a better tritium release behaviour than the aluminates.

REFERENCES

1. K. Okula, D.K. Sze, Proc. Tritium Technology in Fission , Fusion and Isotopic Applications Dayton, Ohio, U.S.A. 1980 ,pp 286

2. J.A. Turnball, M.O. Tucker, The Release of Unstable Fission Products during variable Reactor Operating Histories , J. Nuc.Mat. 50 (1974)

3. B. Libin et. al., Testing of the Nuclear In-Vessel Components, Fusion Technology 14 (1988) 1

4. W. Dänner et al., In this conference

5. H. Werle et. al., International Fusion Technology Conference, Tokyo 1988

6. M. Briec , Tritium Extraction from solid Breeders during Irradiation, Ceramic Breeders, Progress Report 1982/86, Fusion Technology Program , Commissariat a l'Energie Atomique

7. Y. Yung et all., "Solid Tritium Breeder Materials Li_2O and $LiAlO_2$, A Data Base Review ", Fusion Technology 8 (1985)

FUSION TECHNOLOGY 1988
A.M. Van Ingen, A. Nijsen-Vis, H.T. Klippel (editors)
Elsevier Science Publishers B.V., 1989

BEATRIX-II: IN SITU TRITIUM RECOVERY FROM A FAST NEUTRON IRRADIATION OF SOLID BREEDER MATERIALS

R.J. PUIGH[a], G.W. HOLLENBERG[b], T. KURASAWA[c], H. WATANABE[c], I.J. HASTINGS[d], J.M. MILLER[d], S.E. BERK[e], R.E. BAUER[a], D.E. BAKER[a]

[a]Westinghouse Hanford Company, Richland, Washington 99352 USA
[b]Pacific Northwest Laboratory, Richland, Washington 99352 USA
[c]Japan Atomic Energy Research Institute, Tokai-Mura, Ibaraki-ken, 319-11 Japan
[d]Atomic Energy of Canada Limited-Research Company, Chalk River Nuclear Laboratories, Chalk River, Ontario, Canada K0J 1J0
[e]Office of Fusion Energy, U.S. Department of Energy, Washington DC 20545 USA

An in situ tritium recovery experiment is being fabricated for the irradiation of Li_2O in the Fast Flux Test Facility located at the Hanford Site, Richland, Washington, United States of America. Two in situ tritium recovery canisters will be irradiated with lithium atom burnups to 4%. One canister will provide fundamental data on tritium release as a function of temperature, gas composition, and flow rate. The other canister will provide integrated performance data from solid pellet specimens with large (450 °C) radial temperature gradients.

1. INTRODUCTION

In the last decade, considerable progress has been made on the development of solid breeder materials for tritium production in fusion blankets. Technological advancements in fabrication and property measurements (including fast neutron irradiation damage studies[1] and thermal reactor in situ tritium recovery experiments[2-8]) have continually yielded positive results. Past in situ tritium recovery experiments have been on relatively small samples with modest lithium atom burnup. Japan, Canada, and the United States of America have initiated an in situ tritium recovery experiment in the Fast Flux Test Facility (FFTF) reactor that incorporates high damage and tritium production rates in a fully instrumented, larger volume solid breeder irradiation. The BEATRIX-II experiment is sponsored by the International Energy Agency (IEA) and includes both vented and nonvented capsules. In 1989, two in situ tritium recovery tests will be irradiated in the FFTF reactor. This paper describes those two tests.

2. TEST DESCRIPTION

The objectives of the cycle 11 experiment are to measure the tritium release characteristics and the thermal stability of Li_2O as a function of neutron exposure, temperature, gas composition, and sweep gas flow rate. Tritium release characteristics of solid breeder materials impact the quantity of recoverable tritium fuel from fusion blankets. Thermal stability is important to ensure the maintenance of beginning-of-life blanket operating temperatures. One in situ recovery canister will contain a thin-walled (1.5 mm) ring specimen capable of incremental temperature change steps, and another will contain solid pellets with blanket relevant temperature gradients.

This in situ tritium recovery experiment has been designed for irradiation in a fast flux test reactor that will provide unique data on the irradiation performance of Li_2O. The unique experimental environment and parameters of these BEATRIX-II cycle 11 tests are:

a. *High lithium atom burnup.* Essentially homogeneous lithium atom burnups of 4% will be achieved in the Li_2O samples irradiated in 1989.

b. High total tritium production. Over 7000 Ci of tritium will be produced, measured, and recovered during the 300 equivalent full power days (EFPD) of irradiation.

c. Reactor relevant tritium partial pressures. The reference steady-state tritium partial pressure will be 50 p/m which is much higher than the other experiments (1 p/m). Higher tritium partial pressures should minimize experimental uncertainties in tritium release associated with sweep gas contaminants and surface adsorption and desorption from downstream tube walls.

d. Engineering oriented testing. The solid pellet canister will generate temperature and tritium production gradients typical of several Li_2O blanket design concepts.

The FFTF is a liquid sodium cooled, fast neutron test reactor. The FFTF is noted for its high flux (4.8 x 10^{15} n/cm^{2}s [E_{total}]; 2.9 x 10^{15} n/cm^{2}s [E > 0.1 MeV] and its ability to easily accommodate instrumented experiments. The neutron spectrum in FFTF results in a fusion prototypic cross section for neutron reactions with ^{6}Li (1-2 barns). Therefore, self-shielding is not a major limitation to sample size, and high lithium atom burnup levels can be achieved. The Materials Open Test Assembly (MOTA)[9] is a standard FFTF instrumented irradiation vehicle with over 30 instrumented canisters for materials testing. Two of these canisters will be devoted to the BEATRIX-II in situ tritium recovery tests.

The temperature change canister (Figure 1) will allow temperature changes between 460 °C and 650 °C on a thin ring of Li_2O so that in situ transient tritium release data can be obtained and compared to theoretical models. This canister's design is similar to earlier in situ tritium recovery experiments.[2-8] Temperature control is obtained by varying the helium/argon gas mixture in the gas gap between the specimen subcapsule and the reactor coolant. The exact shape and time constant associated with this transient data can then be compared with theoretical models for analysis of mechanism. Similar parametric increments are planned for flow rate (10 to 1000 cm^3/m) and sweep gas composition (helium plus 0 to 1% hydrogen). A self-powered neutron detector (SPND) will measure local neutron flux and allow normalization of the tritium production rate to the tritium release data. Thermocouples will measure the temperature gradient across the ring specimen to monitor the effective thermal conductivity during irradiation.

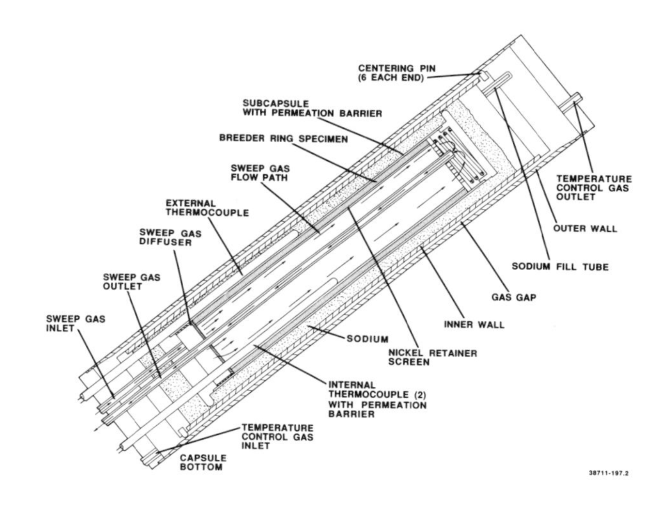

FIGURE 1
Schematic of temperature change canister

In the solid pellet canister (Figure 2), a column of solid Li_2O pellets will be irradiated. The pellets will have an initial centerline temperature of 880 °C and an outer circumference temperature of 430 °C. The solid pellet canister will provide tritium release data from Li_2O under thermal gradient conditions typical of a blanket. The large temperature gradient provides a driving force for lithium transport from the high temperature zone to the colder circumference. It also provides an opportunity to evaluate this temperature limiting phenomenon for Li_2O and the temperature stability of Li_2O during irradiation. The temperature gradient across the solid pellet will be continuously measured.

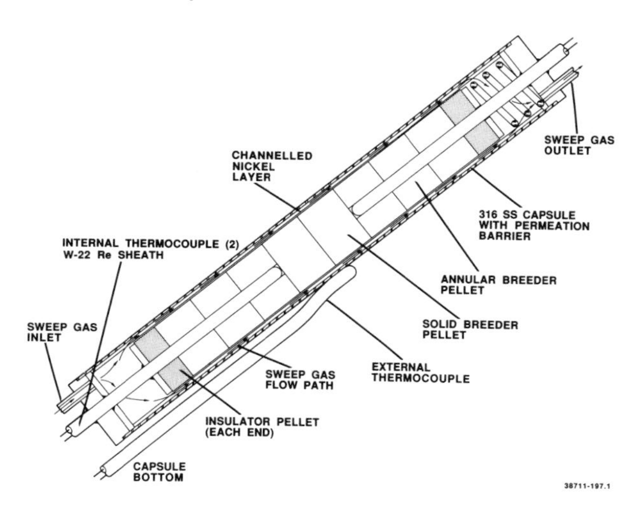

FIGURE 2
Schematic of solid pellet canister

The major characteristics of the Li_2O specimens and their anticipated environment are summarized in Table 1. The ^6Li enrichment for both specimens was chosen to obtain a lithium atom burnup of 4% by the end of the irradiation (300 EFPD). Monte Carlo calculations were performed to determine the optimum position within the FFTF core to obtain the desired lithium atom burnup and the lowest practical irradiation temperatures. The dimensions of the ring geometry specimen were chosen to minimize the radial temperature gradient and obtain the lowest achievable irradiation temperature while maintaining a temperature control capability and high tritium production levels. The dimensions of the solid pellet specimen were chosen to achieve an initial centerline temperature of 900 °C. The calculated radial temperature gradients in the two specimen types are shown in Figure 3.

The environment and test conditions of BEATRIX-II have mandated the following unique fabrication and design features:

TABLE 1
Characterization of the Li_2O specimens and anticipated environment

Characteristic/ Environmental	Ring Specimen	Solid Pellet Specimen
Dimensions		
outside diameter (mm)	18.57	17.02
wall thickness (mm)	1.50	---
length (mm)	88.90	88.90
Li^6 Enrichment (%)	61	61
Density (%TD)	85	89
Grain Size (μm)	5	5
Temperature		
inner radius (°C)	500-685	880
outer radius (°C)	460-651	430
maximum gradient (°C)	28	450
Tritium Production (Ci)	2025	5300
Sweep Gas System		
gas composition	helium + (0-1% H_2)	helium + (0-1% H_2)
flow rates (cm^3/min.)	10-1000	10-1000

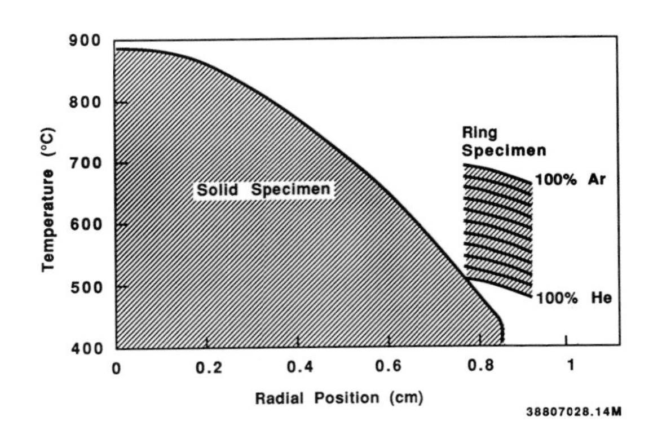

FIGURE 3
Predicted radial temperature gradients in the ring and solid Li_2O specimens

a. Tritium barriers. Tritium loss due to permeation is an issue in this experiment due to the high temperature of the FFTF coolant (350 to 450 °C). To keep errors from tritium losses below 1%, a permeation barrier was applied to the canister walls for both test specimens and to the exiting sweep gas lines between the canisters and the top of the irradiation vehicle. The centerline temperature for the solid pellet column initially will be 880 °C. To minimize tritium losses through the thermocouple penetrations, W-22%Re sheathed chromel*-alumel* thermocouples will be used. Design permeation calculations predict less than 0.3% of the generated tritium will be lost to the coolant. This prediction has been confirmed by R. Causey of Sandia Livermore.

b. Channeled nickel layer. The solid pellet canister was designed to direct the sweep gas flow away from the high temperature regions of the solid Li_2O specimen to avoid unidirectional gas phase transport of LiOT. A low density channeled nickel layer [50% theoretical density (TD)] was formed by pressing and sintering nickel powder to the inside wall of the stainless steel outer capsule. This channeled nickel layer has significant open porosity and forms the preferred sweep gas flow path around the outer circumference of the solid pellet column. It also provides good thermal contact between the pellet and the outer capsule wall.

c. Fabrication of thin-walled ring specimen with tight dimensional tolerances. Fabrication of a thin-walled ring specimen with tight dimensional tolerances was necessary to examine the temperature range of interest (460 to 650 °C). Fabrication of this fragile specimen was accomplished by isostatically pressing Li_2O to precise outside dimensions and sintering it to a net shape.

d. Highly efficient, single-pass tritium recovery system. This experiment will utilize metal getters (SAES 707)** to remove tritium from the gas stream after the measurements have been performed and before the sweep gas is dumped into the FFTF controlled atmosphere processing system (CAPS).

The sweep gas flow for the in situ tritium recovery tests is shown in Figure 4. The gas inlet system will provide the sweep gas to the in situ tritium recovery canisters. The reference sweep gas composition will be helium plus 0.1% hydrogen. The alternate gas composition will be helium with 0-1% hydrogen. The sweep gas is routed into the canisters which are mounted on MOTA. The return sweep gas lines are routed out of the reactor and to the tritium measurement system which is located within a glovebox. The sweep gas line from each canister is then split into a total tritium measurement (HT and HTO) and an elemental tritium measurement (HT) path as seen in Figure 4. The elemental tritium analysis path incorporates a molecular sieve bed prior to an ionization chamber to prevent decalibration of the ionization chamber by surface absorption of tritium. Similarly, the total tritium measurement path possesses a ceramic cell[11] for electrolytically reducing any oxidized tritium in the gas stream. The small ionization chamber volumes (100 cm^3 active volume) were specifically selected for their short time constants, which permit the measurement of short transients. High resolution is possible with these small ion chambers due to the high sweep gas concentrations of tritium (50 p/m).

To reduce tritium releases in the exhaust stream well below established limits, tritium is recovered from the flow streams with a metal getter (SAES 707). Other experiments have used ethylene glycol or molecular sieve beds, but the metal beds appear to be better suited to a nonoxidizing operation and the demonstration of tritium recovery techniques. The experimentally determined performance of the SAES 707 getter is shown in Figure 5. Detritiation factors of 2500 are required under normal operating conditions for BEATRIX-II cycle 11

*Chromel and alumel are trademarks of Hoskins Manufacturing Company.
**SAES 707 is a trademark of SAES GETTERS S.p.A.

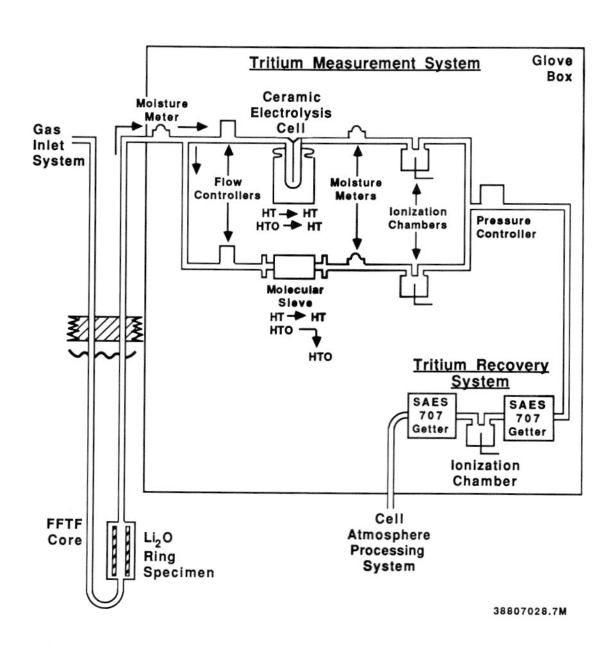

FIGURE 4
Schematic of the sweep gas flow path, analysis, and tritium recovery systems

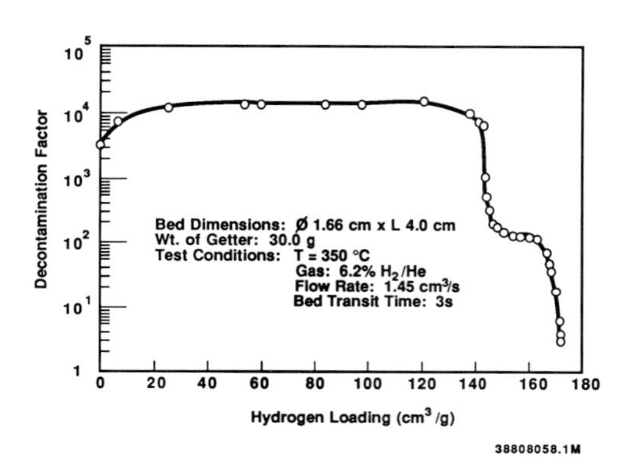

FIGURE 5
Performance of SAES 707 metal getter material

experiment. The detritiation factor degrades as the getter bed becomes saturated with the hydrogen in the sweep gas and will be replaced at periodic intervals. The hydrogen in the sweep gas improves the detritiation factor by an isotopic dilution effect.

3. CONCLUSIONS

Two in situ tritium recovery tests are scheduled to begin irradiation in FFTF cycle 11, which will start in the spring of 1989. The experiment is designed to provide tritium release and irradiation performance data on the candidate solid breeder material, Li_2O, to high lithium atom burnups. Subsequent in situ tritium recovery experiments are currently being planned for irradiation in 1990.

ACKNOWLEDGEMENTS

The inspiration and guidance of IEA leaders such as T. C. Reuther, T. Kondo, and G. Phillips is gratefully acknowledged. The tritium permeation calculations by R. Causey are appreciated.

REFERENCES

1. D.L. Baldwin and G.W. Hollenberg, J. Nucl. Mat. 141-143, (1986) 305-310.

2. R.G. Clemmer, et al., J. Nucl. Mater. 122-123 (1984) 80.

3. T. Kurasawa, et al., J. Nucl. Mater. 141-143 (1986) 265-270.

4. R.A. Verrall, et al., CRITIC-1 and On-line Tritium Release Irradiation of Lithium Oxide, presented at the Third International Conference on Fusion Reactor Materials, Karlsruhe, FRG October 1987 (to be published J. Nucl. Mater.).

5. H. Werle et al., J. Nucl. Mater. 141-143 (1986) 321-326.

6. E. Roth, et al., J. Nucl. Mater. 141-143 (1986) 275-281.

7. M. Briec, et al., J. Nucl. Mater. 141-143 (1986) 357-363.

8. H. Kwast, et al., J. Nucl. Mater. 141-143 (1986) 300-304.

9. D.L. Greenslade, et al., J. Nucl. Mat. 141-143 (1986) 1032-1038.

10. S. Konishi, et al., Fusion Technology 8 (1985) 2042-2047.

FUSION TECHNOLOGY 1988
A.M. Van Ingen, A. Nijsen-Vis, H.T. Klippel (editors)
Elsevier Science Publishers B.V., 1989

TRITIUM CLEAN-UP OF GLOVE BOX ATMOSPHERE BASED ON A ROOM TEMPERATURE NON EVAPORABLE GETTER ALLOY

G. BONIZZONI, F. GHEZZI, M. NASSI, U. PAVESI, M. SUCCI*

Istituto di Fisica del Plasma, Associazione EURATOM/ENEA/CNR sulla Fusione, via Bassini 15, 20133 Milano, ITALIA

* SAES Getters, S.p.a., via Gallarate 215, 20151 Milano, ITALIA

This work was carried out to verify the efficiency of the getter material for tritium purification in glove box. The main target of this experiment was to reach a residual partial pressure of tritium less than $4.2*10^{-7}$ mbar (corresponding to 1 mCi/m^3 as imposed by the international safety rules). Two series of measurements were carried out simulating tritium with deuterium ; the first one was performed under static conditions, in order to find the best work temperature of the getter relatively to the pumping speed. The second one was performed under dynamic conditions, in order to find the tritium residual partial pressure. Due to the sensitivity of the instrument was impossible to measure a deuterium pressure lower than 10^{-5} mbar. The Sieverts' law was confirmed to be valid also in an argon atmosphere with an error of less than one order of magnitude. Two types of getter configurations were tested: a getter cartridge pump and a getter purifier. Starting from the experimental results the proposal, for an economical tritium purification system for glove box atmosphere, is presented.

1. INTRODUCTION

Tritium catalytic oxidation techniques followed by storage on molecular sieves of the tritiated water have been recently established for the clean up systems of the glove box atmosphere. These techniques allow fast tritium removal as result of the high oxidation rates; however tritium is degraded from the gaseous phase to tritiated water, which is, from a biologic safety point of view, much more toxic. Moreover the tritium permeation increases as the temperature of the catalytic bed increases[1,2].

The purpose of this work is to study a purification system based on tritium sorption by means of a metallic getter bed other than uranium, because uranium powders formed during the hydridation process are very pyrophoric[3] in air at relatively low temperature (100 - 110 C) and, moreover, they show a low but practically constant radioactivity ($t_{1/2} = 4.5*10^{+9}$ years)[4,5].

The purification is obtained by passage of the gaseous mixture through a getter bed forming a tritium "solid solution"[6,7]. This system has the following advantages: formation of a reversible solid solution for H_2 and its isotopes avoiding the degrading of the gas; reduction of the toxicity as result of hydride formation instead of oxide formation; reduction of permeation phenomenum because the operative temperature is lower; more simple and economical plant.

The getter alloy used for this work was the Zr 70%,V 24.6%,Fe 5.4% named St 707[8] supplied by SAES Getters S.p.A. Milan. This alloy is widely used in the vacuum field; however, being inactive versus the noble gas, it is also employed when a particularly pure noble gas is required (purification in the low ppb level can be achieved).

The getter is stable in air because it is covered by a thin oxide layer. To become "active" or able to sorb the gaseous impurities, the getter must be "activated", that is heated at a temperature high enough (in case of the getter used in this work 400 C) to remove the oxide layer. Then the getter can work at any temperature, from room temperature up to the activation temperature, depending on the gas to be sorbed.

The pumping characteristics[9] under vacuum for H_2 are known and shown in Figs. 1a,1b.

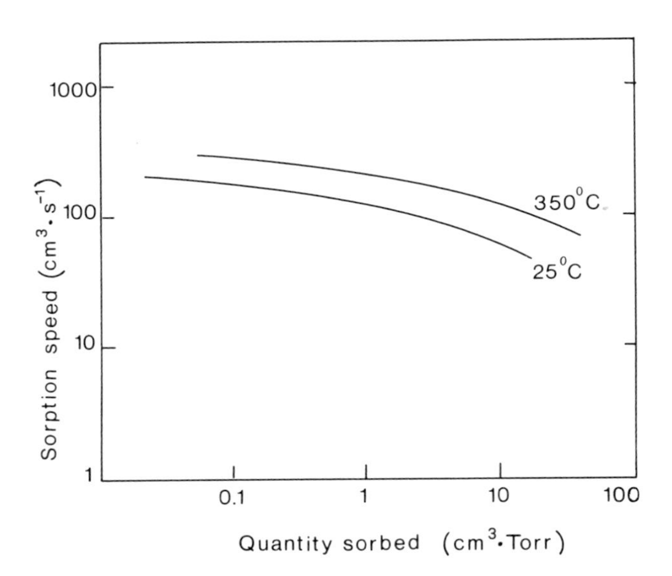

FIGURE 1a

Sorption characteristics of St 707 (100 mg, surface 50 mm^2).

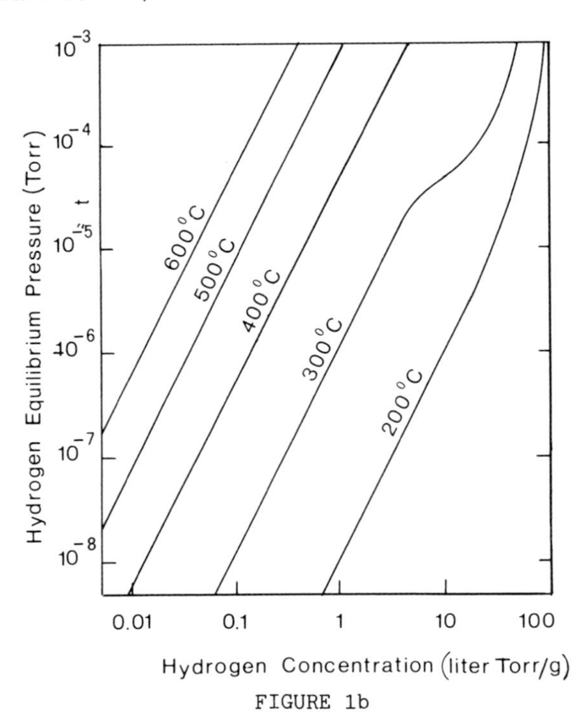

FIGURE 1b

Equilibrium pressures on St 707 getter alloy for H_2 and D_2.

The choice of the ST 707 alloy is principally related to lower working temperatures than other getter alloys which have analogous pumping characteristics for hydrogen.

To simulate the actual working conditions the getter was exposed to deuterium doped argon at 980 mbar; the Sieverts' law regulating the D_2 sorption under vacuum was compared to the data obtained under rare gas atmosphere.

2. EXPERIMENTAL SYSTEM

2.1. Apparatus

Static and dynamic tests were performed in two experimental plants (Fig.2)that consist in the following components:

- a vacuum chamber V1 of known volume for glove box simulation;
- a vacuum chamber V2 of known volume which contains the getter[10,11];
- a measurement system with a deuterium detector;
- a vacuum system;
- a circulation pump, used only in the dynamic tests.

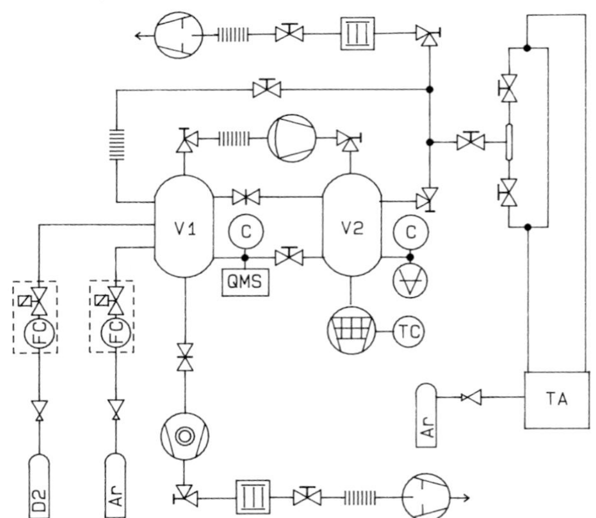

FC =mass flow controller C =capacitance gauge
TA =trace analytical TC =thermocouple
QMS =quadrupole mass spectrometer

FIGURE 2

Lay out of the dynamic system

2.2. Procedures

The systems were evacuated and baked in order to reach a vacuum of about 10^{-8} mbar; the residual gas analysis was performed by means of a quadrupole mass spectrometer. The gaseous mixture of Ar and D_2, used to simulate tritium, was introduced into chamber V1 by programmable mass flow controllers. The partial pressure values of Ar and D_2 were measured by absolute capacitance gauges. The amount of D_2 introduced in the present test system was calculated on the base of the following considerations:

- It was referred to the dimension of a real plant for tritium storage (FTU,IGNITOR)[12]. This plant is composed by a glove box volume of 4 m^3 that contains 1 g of T_2. Inside the glove box it is foreseen to mount 1000 g of St 707 alloy for the absorption of tritium.
- It is supposed that the worst accident could occur: a total tritium release in the glove box. In this condition, if the getter sorbs the total tritium amount, the corresponding concentration Q will be about 4 mbar liter/g.
- The quantity of D_2 introduced in the test system was such that the getter sorbed the same concentration Q as foreseen in the real plant. The reason of this choice is that the concentration Q, temperature apart, is the main variable in the getter D_2 sorption (as shown in Fig 1b).

The temperature of the getter was measured by a Cromel-Alumel thermocouple sited in the getter bed.

2.2.1. Static procedures

Since a small amount of gas was involved, in order to avoid considerably change in the gas pressure spills were not executed. At the beginning the sorption of D_2 was thus measured using an absolute capacitive gauge, having a sensitivity of 1 mbar. Only one spill was executed, at the end of the test, in order to measure the final deuterium partial pressure by means of a gas chromatograph.

A deuterium partial pressure of about 40 mbar was used, in order to achieve the chosen concentration Q in the 50 g of getter alloy.

2.2.2. Dynamic procedures

By opening a gate valve the mixture was expanded into chamber V2 with the getter inside. At the same time the circulation of the mixture was obtained by a diaphragm vacuum pump which allows a rate of about 10 circulations for hour. D_2 partial pressure measurement was obtained by a spill of the gas atmosphere and analysis by a gas chromatograph described later.

A deuterium partial pressure of about 9 mbar was used, in order to achieve the chosen concentration Q in the getter alloy. The initial partial pressure of Ar was about 971 mbar so that the total pressure was about 980 mbar, the same pressure of the argon atmosphere in the real plant previously mentioned. During the analysis, the total pressure slightly decreased due to the D_2 getter pumping and the spill of the mixture for each analysis; in order to maintain the total pressure of 980 mbar, fresh argon was injected from time to time.

3. DETECTION OF H_2 OR D_2 IN ARGON

The detection of trace of H_2 in argon can be performed using one of the many techniques available in the market based on gas chromatography and mass spectrometry. The main problems related to the former techniques are the sensitivity (a Thermal Conductivity detector has a sensitivity in the range of 0.1-1 ppm maximum) and the injection in the column of the sample gas considering that the argon pressure is below atmospheric pressure (about 980 mbar). The problem related to mass spectrometry is mainly the reduction of the pressure from 980 mbar to 10^{-5} mbar, the maximum pressure usually compatible with the quadrupole mass spectrometer. In fact a so great reduction of the pressure, about 8 orders of magnitude, may change the gas mixture to be analyzed. Moreover, the background partial pressure of H_2 may cover the contribution of H_2 of the sample gas. Some special techniques were developed, for example by UTI[13] (high pressure in the ionization chamber) and BALZERS[14] (the sample gas is chopped to be distinguished by the background), to overcome the problem but the lowest detectable concentration is 0.1-1 ppm.

To achieve a better sensitivity, a new type of gas chromatograph combined with a special injection system were adopted. The detector of the GC is able to measure concentration of D_2 with a practical sensitivity of 10 ppb; the detection occurs as a result of the passage of D_2 through a heated bed of mercury oxide where a reduction reaction of the oxide takes place. The resultant mercury vapour is quantitatively determined by means of an ultraviolet photometer located immediately downstream to the reaction bed. The standard rotating valve of the injection system was replaced by three valves so that the layout of the final system was arranged as shown in Fig.3.

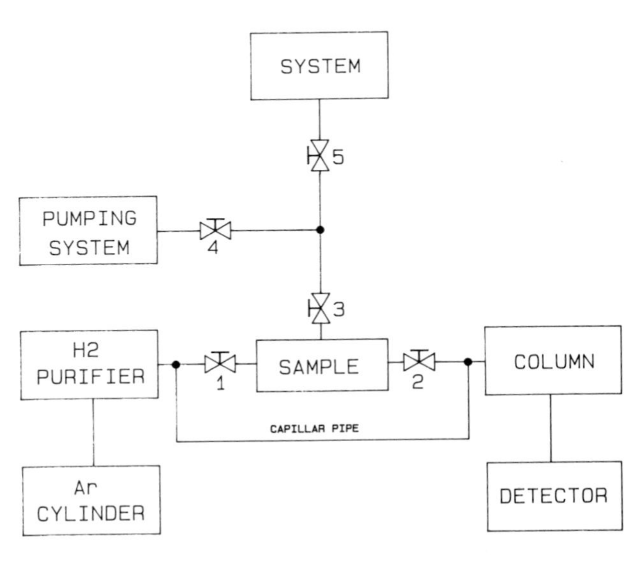

FIGURE 3

Lay out of the D_2 measurement system.

The hydrogen purifier (a catalytic combustion filter converts H_2 in H_2O) guarantees a carrier gas without H_2 (less than 10 ppb, the sensitivity of the analyzer); the bypass of the injection system allows a continuous flow through the column and the detector thus avoiding any kind of contamination.

To perform an analysis, the sample gas was forced to the sample loop, closed by the valves 1, 2 and 3; by opening valves 1 and 2, the carrier gas pushes the sample to the column and the detector because the cross section of the bypass is negligible compared to the cross section of the sample loop. The volume delimited by the valves 1, 2 and 5 is then evacuated to be ready for the next analysis. The retention time of D_2 was approximately 3 min.

A calibration of the unit was carried out introducing a known pressure of D_2 and argon up to 980 mbar. It has been possible to monitor the D_2 concentration in Ar from a partial pressure of $5*10^{-2}$ mbar down to about 10^{-5} mbar.

4. RESULTS

4.1. Static

The results obtained are shown in Fig.4.

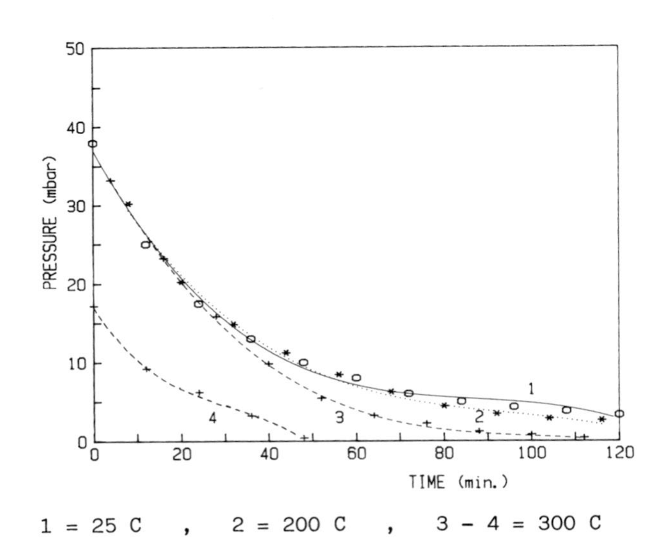

1 = 25 C , 2 = 200 C , 3 - 4 = 300 C

FIGURE 4

D_2 sorption versus time at different getter temperatures.

The curve 1 is the sorption curve with the getter material at room temperature and with a final concentration Q equal to about 4 mbar liter/g.

The measured sorption times are long, in fact a deuterium residual partial pressure of few mbar was found after two hours. Also the experiments at higher getter temperatures gave the same results. In this case the variation of the pressure due to the increase of the temperature, was experimentally measured and taken into account.

The curves 1,2 and 3 have the same behavior, within the limits of measurements uncertainty, considering the correction due to the change of the diffusion coefficient of D_2 in Ar with the argon temperature.

At these high partial pressures of deuterium, it is quite difficult to put into evidence a difference between the pumping speeds of the getter at different temperatures (this fact means that the sorption limiting factor is the diffusion speed of the deuterium through the argon, while the getter temperature has a very small influences on the phenomenum).

The values of the deuterium residual partial pressure were obtained by a measurement with the gas chromatograph 15 hours after the expansion of the gas mixture on the getter. The results showed that the values of the deuterium residual partial pressure were below the detection limit of the gas chromatograph (10^{-5} mbar) when the getter temperature was 25 C or 200 C. In the last experiment, with the getter at 300 C, the deuterium residual partial pressure was 10^{-4} mbar. This value was only about 3 times higher than the value given by the Sieverts' law for D_2 without argon[4]. The expected values from the Sieverts' law in argon pressure should therefore be affected by a maximum error of less than one order of magnitude.

The curve 4 of the Fig.4. shows the D_2 sorption with the getter at 300 C and with an initial deuterium partial pressure of 17 mbar. This test was performed in order to see if, at the same getter temperature, the pumping speed is a function of the D_2 concentration Q already sorbed by the getter.

Trying to compare the curves 3 and 4 it is possible to note that the pumping speed is not a function of the concentration Q but it depends only on the deuterium residual partial pressure, at least at relatively high D_2 pressure (=1 mbar). This behavior is easy to verify making a rigid translation along the x axis of the curve 4 until this curve overwrite the curve 3.

Other measurements were carried out with lower argon partial pressure and the getter at room temperature. In Fig.5 curve 2 shows the D_2 sorption with an argon pressure of 300 mbar while curve 3 was obtained with an argon pressure of 100 mbar.

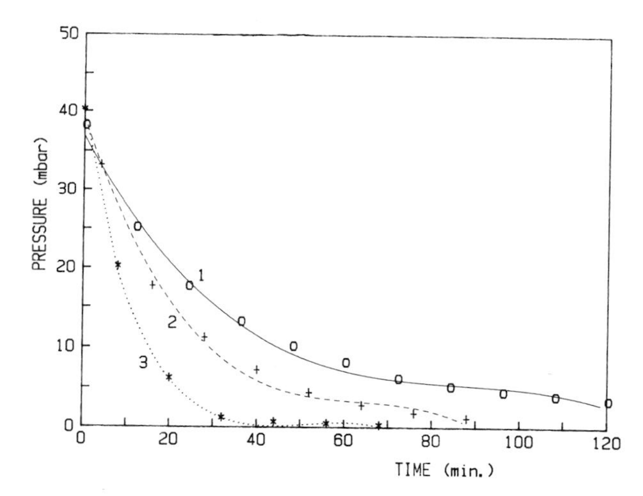

1 = 980 mbar , 2 = 300 mbar , 3 = 100 mbar

FIGURE 5

D_2 sorption versus time at different argon pressures.

Looking at the three curves, obtained at different argon partial pressures, it is obvious to see that the argon pressure is the main limiting factor concerning the sorption speed of D_2 (due to the diffusion process involved). In fact the sorption time goes down very fast, as the argon pressure is reduced; this time becomes only 0.17 sec when the argon pressure is zero.

All the tests, performed under static conditions, show that the limiting factor in D_2 pick-up, by means of the getter material, is the diffusion of D_2 in argon: the higher is the argon pressure, the higher is the time required to clean H_2 isotopes.

4.2. Dynamic

According to the results obtained under static conditions, the second series of measurements using the experimental system with a circulating pump were carried out. In these conditions the delay due to diffusion should be reduced by the forced circulation of the gas inside the system.

The final purpose was to determine the deuterium final partial pressure and to compare it with the international rule limit[15] that imposes a maximum admissible concentration of 1 mCi/m^3 of tritium in the glove box atmosphere. This concentration is equal to a tritium partial pressure of $4.24*10^{-7}$ mbar when the gas is at room temperature. This last value cannot be checked experimentally because the sensitivity limits of the gas chromatograph is only 10^{-5} mbar.

Fig.6 shows the values of deuterium partial pressure obtained with a cartridge getter pump GP 50 containing 30 g of St 707 alloy[10,11].

The dotted curve was obtained by maintaining the getter at 300 C, while the solid curve was obtained with the getter at room temperature.

After a few minutes, the deuterium partial pressure is really lower than the value measured in the first series of measurement without the circulating loop, even if the experimental system was much bigger. The last part of the dotted curve was obtained cooling the getter. It was done because the pumping speed, after about 90 min, was reducing due to the approaching of the equilibrium. It is possible to note that cooling the getter there is a further reduction in the deuterium partial pressure.

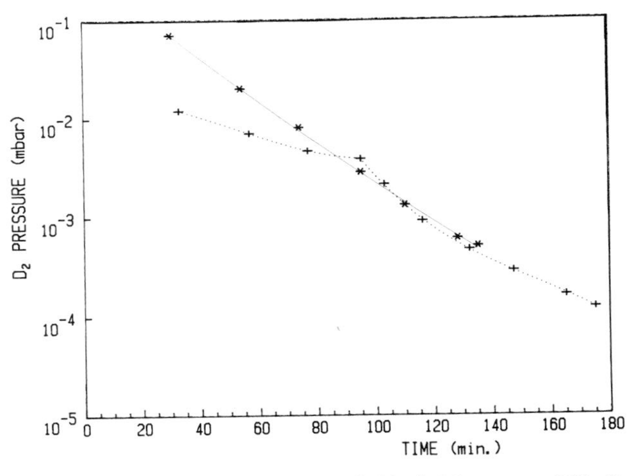

solid line = 25 C , dotted line = 300 C

FIGURE 6

D_2 sorption versus time by means of the getter pump model GP 50.

Comparing the two curves of figure 6, it is possible to see that the deuterium partial pressure is higher, during the first part of the experiment, when the getter pump works at room temperature; later on, after 90 min, this behavior is reversed. It means that the getter temperature has some influence on the sorption speed in the dynamic conditions experiment,(i.e. with gas circulation loop) since that is the sticking probability is slightly higher when the getter is hot.

However, as without argon, the lower the getter temperature the lower the deuterium final partial pressure. In any case the times required to reach the deuterium equilibrium partial pressure are more than 3 hours.

To improve the probability of collision between D_2 and the getter material a getter purifier[16], with 140 g of pills, (6 mm in diameter, 4 mm in length) was also tested. The sorption of a purifier is quite different from that of the getter cartridge: in the purifier

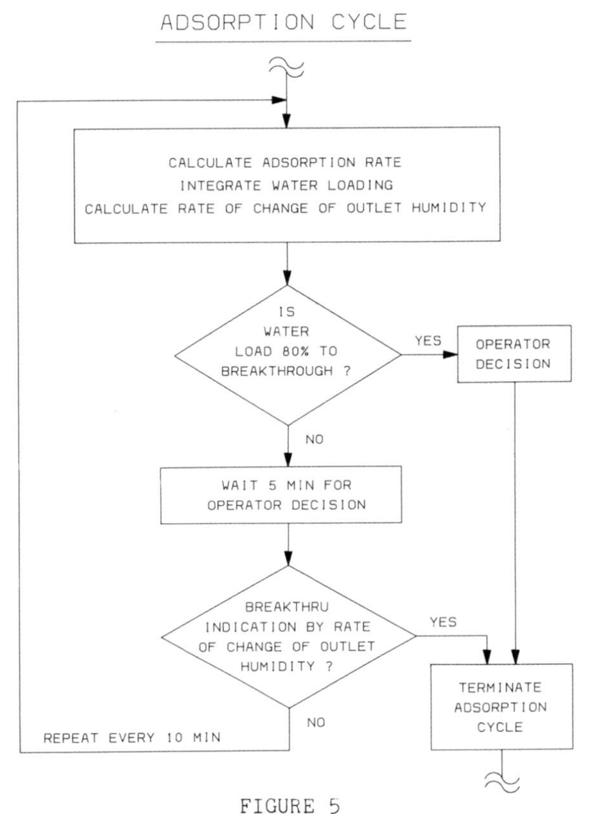

FUSION TECHNOLOGY 1988
A.M. Van Ingen, A. Nijsen-Vis, H.T. Klippel (editors)
Elsevier Science Publishers B.V., 1989

EXHAUST DETRITIATION SYSTEM FOR JET

A H DOMBRA, M E P WYKES, J L HEMMERICH, R HAANGE, A C BELL

JET Joint Undertaking, Abingdon, Oxfordshire, OX14 3EA, England

If the torus or other tritium containment is breached for maintenance or accidentally, the Exhaust Detritiation System (EDS) prevents the escape of tritium to the Torus Hall, and elsewhere, by maintaining the breached system at slightly sub-atmospheric pressure. The exhaust gas from the breached system is detritiated and discharged through the stack. The system includes catalytic recombiners for the oxidation of tritiated compounds, and molecular sieve driers for the recovery of water vapour. Provision for internal recirculation of the gas allows a fast start-up of torus detritiation operations (within two minutes) and processing of feed gas at a variable rate. An isotopic swamping technique is used, as required, to displace HTO from molecular sieve during the drier regeneration cycle. All major operations are controlled by a programmable control system.

1. INTRODUCTION

The EDS concept of preventing the escape of tritium from the primary containment presents several advantages over the conventional air detritiation method: the volume of JET Torus Hall is 28000m³, for which the detritiation system would be ten times larger, increasing the water collection rate. In addition, the Torus Hall would be contaminated with tritium.

The EDS concept, however, must meet high reliability and fast start-up requirements: the probability of sequence failure for the maximum credible release of tritium from a breached torus (Table 1) is designed to be less than 10^{-4}. The minimum EDS detritiation factor is 1000, and normally 5000.

Commissioning of the EDS is scheduled to start by mid 1990. The selected EDS supplier is Neue Technologien GmbH (NTG) in Gelnhausen, Germany.

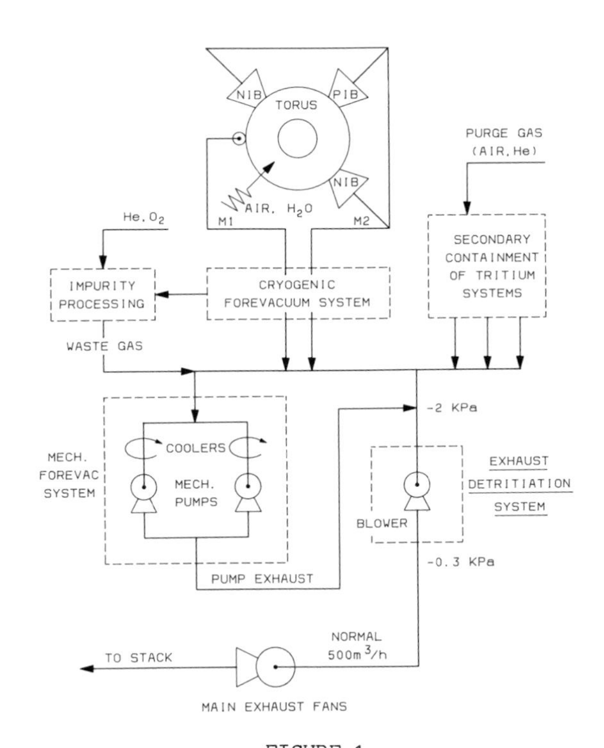

FIGURE 1
Main Exhaust Detritiation Routes

2. EXHAUST DETRITIATION ROUTES

As shown in Figure 1, the EDS provides a suction pressure of 2kPa to draw an exhaust flow, up to 500m³/h, from the torus and other tritium systems[1]. At a lower gas-source pressure, the exhaust may be delivered to the EDS by the Mechanical Forevacuum System.

2.1 Maintenance torus exhaust detritiation

During the torus maintenance periods, the EDS will maintain an air velocity of approximately 1m/s in the maintenance openings to prevent the escape of tritium and radioactive particulate. The concentration of tritium in the torus exhaust is estimated to average 1 GBq/m³, and higher if maintenance is forced without a prior glow discharge cleaning.

3. AN ANALYSIS OF TORUS VACUUM BREACH

Estimates for the worst-case air inleakage incident[2], summarised in Table 1, postulate that the following are credible events:

1) instantaneous shattering of a 100 mm - diameter vacuum window assembly, causing a choked-flow influx of air until torus and ambient pressures equalize.
2) rapid defrosting of cryopanels connected to the torus at full tritium load; the air cooling effect of the panels is neglected.

A fraction of air in the torus effluxes through the breach as it starts to expand due to heating by the First Wall.

TABLE 1
Estimates for worst-case torus air inleakage

First Wall Conditions
Temperature = constant 573K
Heat transfer area = 220m^2
Heat transfer coefficient = 5Wm^{-2}K^{-1}
Conditions at End of Air Influx
Air influx time = 125s
Air temperature = 495K
Mass of air in torus = 197kg
Fraction of hydrogen in air = 1.5%
Airborne tritium in torus = 30gT
Fraction of tritium in methane = 4%
Conditions During Air Efflux
Maximum efflux rate = 487m^3/h (STP)
Fraction of air expelled = 13.6%
Quantity of tritium expelled = 4.1gT

Analysis of the postulated events[2] is based on closed solutions of the differential equations representing the First Law of Thermodynamics applied to ingress of air into an idealized evacuated chamber. The thermodynamic processes associated with the ingress of air are illustrated in Fig. 2 for the limiting case of no heat transfer from the First Wall:
Step 1-2, represents an isentropic expansion from ambient to choked flow conditions,
Step 2-3, subsequent free expansion to vessel pressure, and
Step 3-4, compression to the final atmospheric pressure by subsequent influx of air.

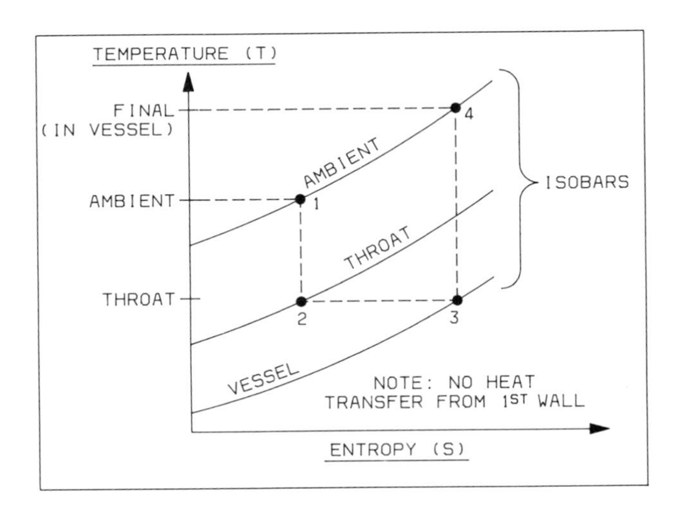

FIGURE 2
Temperature/Entropy diagram for ingress of air into an evacuated vessel

The analysis shows that the lowest air temperature at the end of the influx period (with no heat transfer from the First Wall) is given by the product of ambient air temperature and the isentropic index (γ = 1.4). The heat transfer from the First Wall increases air temperature above this value.

Table 1 sets two major EDS parameters:
Start up time of operation = 2min
EDS air flow = 500m^3/h.

4. RECOMBINER UNIT

As shown in Fig. 3, the gas entering the EDS flows through the following components of the recombiner unit before it enters the driers:

1) a chilled water cooler for control of maximum humidity: this cooler reduces the gas temperature to 6°C and condenses excess moisture.
2) a plate-fin heat exchanger which increases

the gas temperature to 450°C, in two steps.

3) an in-line heater to increase the temperature to the final 500°C. Normal power consumption is 10kW, maximum 45kW.
4) the first recombiner for the oxidation of hydrogen at 150°C to prevent ignition and permeation of tritium at subsequent gas temperature of 500°C.
5) the second recombiner to oxidize methane and the remaining traces of hydrogen at 500°C.
6) a humidifier/cooler for control of minimum humidity: this unit increases gas humidity to a dew point of -21°C to maintain the detritiation factor above 1000 if the inlet gas is very dry.

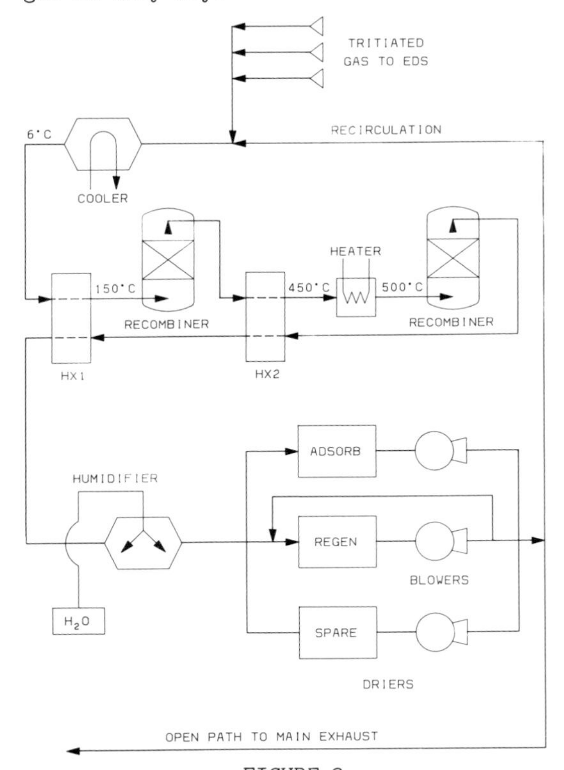

FIGURE 3
Main components of EDS recombiner unit

4.1 Gas recirculation feature

The flow through the recombiner and one of the drier units is always nearly constant 500m³/h, facilitating temperature control. As gas feeds from various sources are generally small, most of the drier flow is recirculated internally. The drier inlet dew point is -21°C, controlled by the humidifier, and the adsorption cycle time is 116 hours. This long cycle meets the two-fold requirements of maintaining the system in a ready-to-start state for torus operations, and processing waste gas from various sources. In addition, the detritiation factor for the feed gas is increased by the ratio of recirculation to feed gas rates.

TABLE 2
Main recombiner parameters

	Unit 1	Unit 2
Operating temperature	150°C	500°C
Catalyst	Pt-Pd/AlO_2	Pd/AlO_2
Catalyst bed - volume	80L	120L
- depth	0.2m	0.2m
- diameter	0.7m	0.87m
Normal pressure drop	200Pa	550Pa
Sp. heat capacity	400kJ/°C	600kJ/°C
Oxidation factor	Approximately 10^6 for hydrogen, 5000 for methane	

5. MOLECULAR SIEVE DRIERS

Figure 4 is a simplified diagram of the drier system. All three drier columns are identical. As each column operates independently, one column is on standby for reliability, instead of duplicating blowers or heaters which increases complexity[3].

5.1 Operating features

1) A change of the operating state is initiated by opening and closing interlocked valves. To rotate a drier column from adsorption to regeneration, the heater circuit valve is opened and the two outer isolation valves are closed, as shown in Figure 4. The closure of heater contacts starts the heating cycle, and lifting of the contacts starts the cooling cycle. The blower is running continuously during all operations.
2) to eliminate flow control valves, the EDS flow rates are preset with appropriate flow

orifices in the main recirculation, drier heater and blower recirculation lines. The selected blower is essentially a constant volume unit when operated at constant speed.

3) The building exhaust fans maintain a subatmospheric pressure of 0.3kPa at the EDS outlet to prevent the escape of tritium to the containment building.

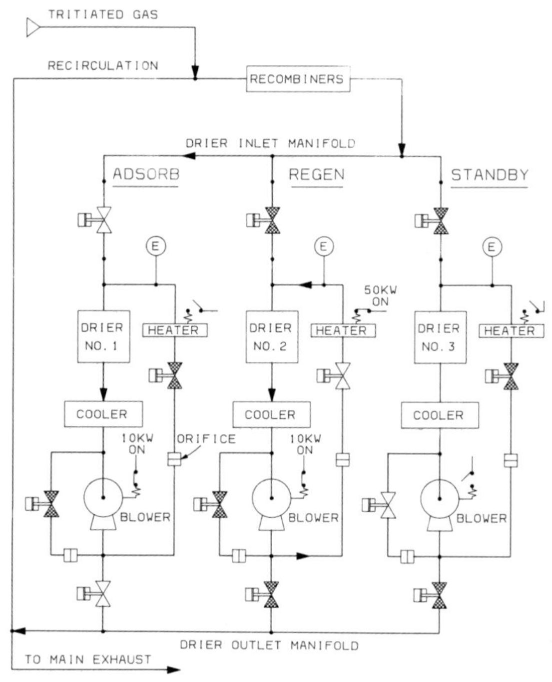

FIGURE 4
Molecular sieve drier system

TABLE 3
Main drier parameters

Weight of molecular sieve type 5A = 350kg
Bed height = 1.2m, diameter = 0.8m
Superficial adsorption velocity = 0.3m/s
Humidity reduction factor:
at inlet dew point 6°C = 8000
at inlet dew point -21°C = 1000

6. PROCESS CONTROL

All major EDS operations will be controlled by Siemens S5-135U Programmable Logic Controller (PLC). This PLC will interface with a central system for coordination of all tritium reprocessing operations, and to an existing higher level computer system (JET CODAS). Some supporting operations will be performed manually via the PLC operators console, or locally. Interlocks and trips provide mainly an independent protection against overheating or overpressure, and will be hardwired.

6.1 Auto rotation of drier columns

This is a high priority requirement for long-term operations: it assists to regenerate driers consistantly at programmed conditions, increase the adsorption capacity, and reduce operator errors.

The desired parameter for the termination of the adsorption cycle is water loading on the bed to about 80% of breakthrough capacity. This control parameter, however, depends on reliable dew point measurements which tend to present difficulties at field conditions. For this reason, the flow chart in Figure 5 calls for operator decision for the termination of the cycle before breakthrough, following review of other pertinent data.

Detection of breakthrough by the rate of change of outlet humidity[3] initiates an automatic termination of the cycle as shown in Figure 5. The controlling parameter for the termination of heating and cooling cycles is drier outlet temperature.

7. ISOTOPIC SWAMPING FEATURE

The current designs for air detritiation generally replace molecular sieve with a fresh charge following exposure to a high concentration of tritium. This method would present difficulties at JET as the EDS processes feed from various sources at different concentrations, and would require a frequent bed replacement.

The selected method for the EDS uses an isotopic swamping technique to displace HTO from molecular sieve with H_2O vapour. This technique was successfully demonstrated on industrial scale[4] and showed good agreement with the proposed model[3] for displacement of tritium by the isotopic exchange process. Estimates for the EDS based on this model show optimum conditions for the addition of water vapour to the regeneration line (point E in Figure 4) at a rate of 33kg/h, following regeneration of the bed to 290°C. The addition of 80kg of water vapour for 2.5 hours is estimated to reduce the concentration of tritium in the residual water by a factor of 170.

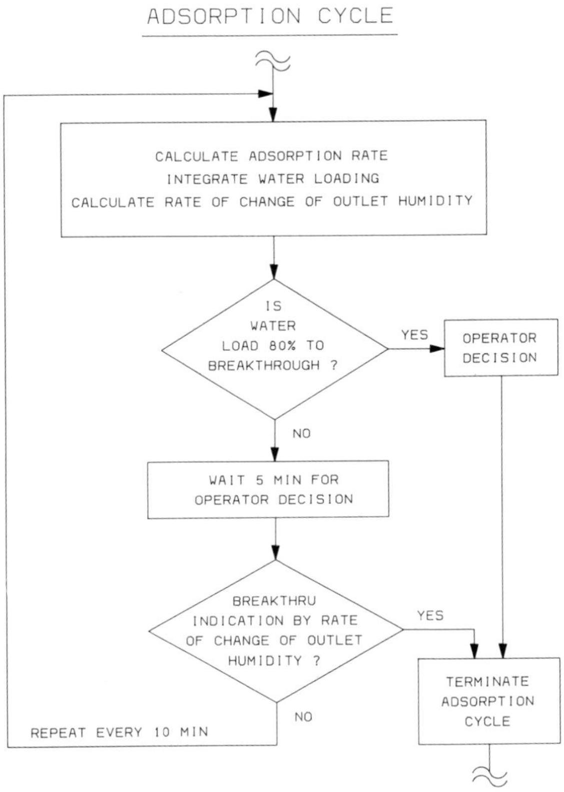

ADSORPTION CYCLE

CALCULATE ADSORPTION RATE
INTEGRATE WATER LOADING
CALCULATE RATE OF CHANGE OF OUTLET HUMIDITY
IS WATER LOAD 80% TO BREAKTHROUGH ?
YES
OPERATOR DECISION
NO
WAIT 5 MIN FOR OPERATOR DECISION
BREAKTHRU INDICATION BY RATE OF CHANGE OF OUTLET HUMIDITY ?
YES
TERMINATE ADSORPTION CYCLE
NO
REPEAT EVERY 10 MIN

FIGURE 5
Simplified sequence for termination of drier adsorption cycle.

If required, the addition of water vapour will be repeated at the end of the cooling cycle to form a tritium-free saturated section at the drier inlet, which is estimated to increase the detritiation factor during the adsorption cycle by a factor of 40.

8. FUTURE PROSPECTS

Potential advantages of the EDS design concept, if experience is satisfactory, are as follows:

Concept	Potential Advantages
1. EDS versus building air detritiation	- prevent tritium escape and contamination of building.
2. Multi-feed gas processing	- single system.
3. Two recombiners in series	- permeation and ignition control.
4. Presetting of flow rates	- eliminate flow controllers.
5. Hot recirculation	- fast start up.
6. Subatmospheric EDS pressure	- leakage control.
7. Detritiation of MS with H_2O vapour	- eliminate disposal of contaminated MS.
8. In situ isotopic swamping as required	- higher detritiation factor.

REFERENCES

1. J.L. Hemmerich, et al, The JET Active Gas Handling System - Progress Report, this Symposium.

2. M.E.P. Wykes, Safety Analysis of Torus Vacuum Breach, JET-P(88)37, 1988.

3. A.H. Dombra, W.E. Bishop and N.E. Howden, Air Detritiation Systems for Fusion Facilities Based on CANDU Developments, Proceedings 14th SOFT, 1511, 1986.

4. A.E. Everett, A.H. Dombra and R.E. Johnson, Isotopic Exchange in Air Detritiation Dryers, Proceedings Tritium Technology in Fission, Fusion and Isotopic Applications, Toronto, May 1-6, 1988.

FUSION TECHNOLOGY 1988
A.M. Van Ingen, A. Nijsen-Vis, H.T. Klippel (editors)

TRITIUM SEPARATION AND RECOVERY FROM NaK BY COLD TRAPS: RESULTS ON HYDROGEN RECOVERY FROM Na AND NaK

J. Reimann, H. John, S. Malang

Association KfK-EURATOM; Kernforschungszentrum Karlsruhe, Institut für Reaktorbauelemente, Postfach 3640, D-7500 Karlsruhe, West Germany

Tritium recovery by cold trapping in an intermediate NaK loop is considered for a self-cooled Pb-17Li blanket. The special requirements of low hydrogen loading and extremely low concentration at the outlet leads to a novel design of the cold trap. First hydrogen recovery tests have proved the feasibility of the method. Different release mechanisms were observed during recovery from NaK which suggest that tritium recovery from drained cold traps should be preferred.

1. INTRODUCTION

For a fusion reactor with a self-cooled Pb-17Li blanket an intermediate NaK (or Na) loop or a double-walled heat exchanger with NaK (or Na) flowing in the gap is required for several safety reasons. The following tritium recovery method was proposed[1,2]: the bred tritium permeates through the wall of the intermediate heat exchanger into the NaK (or Na) and is precipitated as tritide in a cold trap. The tritium is recovered by thermal decomposition of the tritide by heating up the drained cold trap and pumping off the gas at low pressure.

Figure 1 shows schematically the tritium flow sheet for the blanket system with the double-walled heat exchanger and NaK as fluid. The numbers are valid for a reactor with a thermal power of N = 1000 MW and a tritium production rate of $\dot{m}_T$ = 150 g/d. To keep the tritium losses into the steam-water loop within acceptable limits, a tritium partial pressure of p_T = $1.1 \cdot 10^{-4}$ Pa is required in the NaK-filled gap, corresponding to a tritium concentration of x_T = 0,42 appm. This requires a tritium concentration of x_{CT_O} = 0,22 appm at the cold trap outlet, which corresponds to a cold trap outlet temperature of T_{CT_O} = 40 °C for a cold trap efficiency of 1. Cold trap efficiencies in this low concentration range were not investigated up to now but are certainly significantly lower than one. A method to shift the cold trap operation into a more favourable operational regime is by adding other hydrogen isotopes, protium or deuterium (method of isotope swamping). For a protium addition of $\dot{m}_P/\dot{m}_T$ = 3 the required saturation temperature becomes 80 °C which corresponds to an outlet temperature of 64 °C for a cold trap efficiency of 0.6. Protium addition requires additionally isotope separation at the end of the tritium system. The value of $\dot{m}_P/\dot{m}_T$ = 3 is small compared to values of 10^3 - 10^5 required for tritium recovery from other blanket types[3] and, therefore, seems to be acceptable.

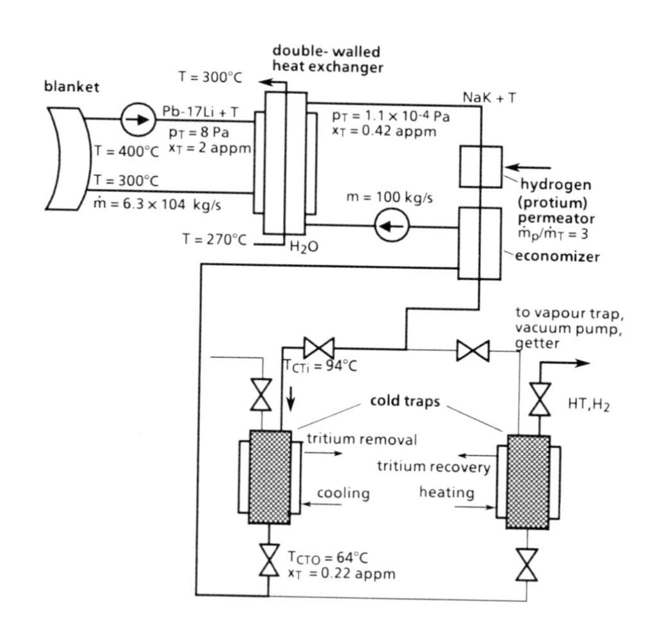

FIGURE 1
Tritium Flow Sheet (N_{th} = 1000 MW, $\dot{m}_T$ = 150 g/d)

2. REQUIREMENTS ON A FUSION BLANKET COLD TRAP

Cold traps are commonly used in liquid sodium loops for LMFBRs for oxygen and hydrogen removal. The main design goal is to achieve a high capacity (high loading of the cold trap with precipitations before replacement is required due to plugging) in combination with a high efficiency (low cold trap outlet concentration x_{CTo}). The minimum achievable outlet concentration is restricted by the relatively high melting temperature of T_{Na} = 99 °C. The cold trap outlet temperature T_{CTo} must be larger than T_{Na}, values of about T_{CTo} = 110 °C were used.

The present common experience is, that cold traps plug when only a small fraction of the theoretical capacity is reached due to local precipitation. This indicates that precipitation kinetics are still not sufficiently understood which is especially true for fusion blanket cold traps (FBCT's), where the requirements are: minimum tritium inventory and minimum tritium outlet concentration. The first requirement is met by short recovery cycles which means that the FBCT operation is characterized by low hydrogen loadings. Therefore, cold trap plugging due to hydrides does not represent a problem. However, attention has to be paid on the precipitation of impurities (mainly NaK-oxide) which are not decomposed thermally during hydrogen recovery. These impurities will accumulate and determine the operational time of the cold trap. From present LMFBR experience, the impurity deposition rates are so small that operational times are expected to be considerably longer than one year.

In order to achieve low tritium outlet concentrations, NaK appears to be better suited than Na due to its lower melting point of T_{NaK} = -12 °C. However, due to the higher saturation concentration x_{sat} for a given temperature, this advantage is somewhat compensated. Using the relationships[4] given in Table 1, a saturation temperature of T_{sat} = 63 °C is required in NaK compared to T_{sat} = 110 °C in Na, to obtain the same value x_{sat} = 0,85 appm.

TABLE 1
Hydrogen Solution in Na and NaK[4]

$\ln y = A \cdot B/T(K)$				
	Na-H		NaK-H	
y	A	B	A	B
K(wppm/√Pa	-0.921	-	-0.834	-576
pH2p(Pa)	31.28	14097	31,76	13723
xsat(wppm)	14.89	6961	15.06	6286
	x(appm) = 23x(wppm)		x(appm) = 34x(wppm)	

Therefore, NaK is only superior to Na if outlet concentrations corresponding to $T_{sat}<63$ °C can be realized. Experience with the NaK-H system is small compared to Na-H. This is especially true for the low concentration range.

Recently performed sodium experiments[5, 6] with higher concentration levels corresponding to $T_{CTo} \approx 150$ °C showed that hydrogen precipitation is dominated by crystal growth kinetics and not by diffusion. For the same temperature range, oxygen precipitation is dominated by diffusion due to the considerably lower diffusion coefficient. For the anticipated low concentrations in the FBCT, the influence of hydrogen diffusion is expected to increase and higher velocities in the cold trap compared to previous designs are proposed. In the first zone of the FBCT, concentration differences are the highest and the avoidance of plugging is of first priority. Meshless cold traps are insensitive towards plugging. In the present case, appropriate inserts should be used to increase the surface compared to an "empty" cold trap and/or other means to enhance the precipitation rate are recommended.

In the second zone, precipitation rates are

considerably lower and the surface provided for heterogeneous nucleation and crystal growth must be large. This is realized by means of wire meshs.

For hydrogen recovery, the FBCT should be drained from NaK prior to heating up the system to about T≤400 °C. This is due to the fact that otherwise for small hydrogen loadings a significant amount of hydrogen from the decomposed crystals would dissolve in the NaK which results in a much slower hydrogen release rate, compare Section 4.

Taking these considerations into account, the following design for a FBCT, as shown in Fig. 2, is proposed: The NaK with dissolved hydrogen enters the upper part and flows down in the annulus. The NaK is cooled by the return flow in

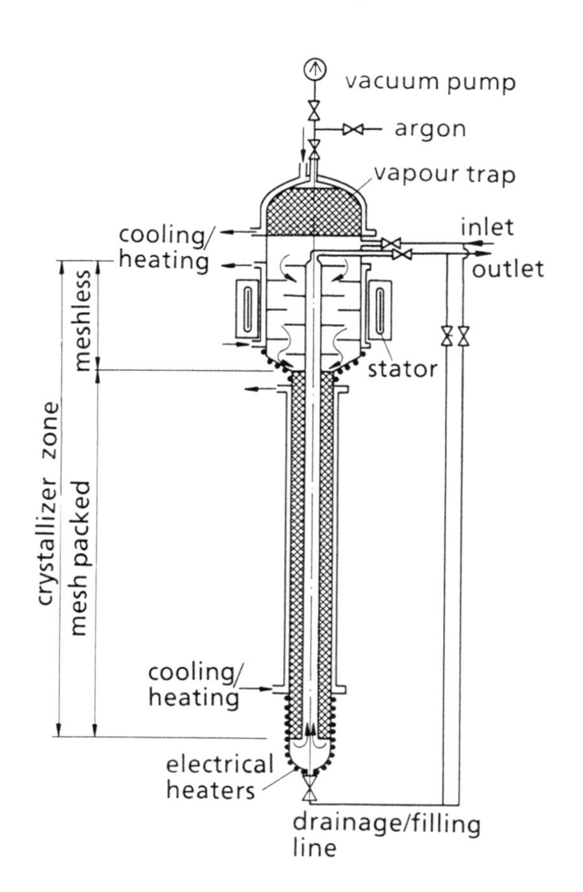

FIGURE 2
Schematical Design of Fusion Blanket Cold Trap

the concentrical tube and by external oil cooling on the vessel shell. Compared to presently used cold traps, the FBCT has a significantly larger length to diameter ratio.

Precipitation occurs in the annulus. There are two different zones: In the first zone, alternating rings of different sizes are used as inserts which provoke a meander-type axial-radial flow. Mass transfer is considerably increased by the rotating electro-magnetic field of a device similar to the stator of an electric motor[7], which superimposes a strong circumferential velocity field in the NaK. Boundary layer effects on the ring surfaces induce a secondary flow which continuously transports fluid to the surfaces (tea-cup flow). The increased mass transfer is expected to be especially effective in respect to oxygen and a considerable portion of the precipitation should occur in this zone. This flow geometry is insensitive in respect to plugging due to the large flow cross sections.

The following mesh-packed zone serves to reduce the concentration to a very low level. Through-put velocities should be as high as possible, limited by the acceptable pressure drop and the avoidance of the transport of broken crystal fragments through the exit line. Velocities of about 10 cm/s appear to be reasonable compared to some mm/s in conventional cold traps. Experiments, however, are required to justify this assumption.

To drain the cold trap, argon gas pushes the cold trap NaK through the drainage line. To initiate tritium recovery, at first the argon gas is pumped off. Then the oil is heated up and additional electrical heaters are used to heat the crystallizer region and the bottom part of the cold trap to a temperature of about 400 °C.

The hydrogen leaves the vessel at the top after having passed the integrated vapour trap which should be kept at a temperature less than 100 °C. During cold trap loading, this portion is also filled with NaK at a temperature close to the entrance temperature. Therefore, condensed potassium (and sodium) and eventually precipitated hydride is dissolved in the main stream again.

During hydride decomposition, metallic NaK, saturated with oxygen, is formed which can accumulate at the vessel bottom. To prevent plugging in this region, first hot NaK is used for filling up the cold trap from the bottom before starting the cold trapping phase.

3. EXPERIMENTAL PROGRAMME

In order to investigate the precipitation and recovery processes relevant to a fusion blanket cold trap, the WAWIK project (WAWIK = German acronym for hydrogen separation and recovery in cold traps) was started in the Nuclear Research Center of Karlsruhe (KfK).

Figure 3 shows schematically the loop for precipitation experiments which will start operation at the end of 1988. There are two experimental cold traps in parallel consisting of vertical tubes with a height of 0.6 m and a diameter of 0.1 m which can be equipped with different inserts (wire mesh, rings etc). An electric stator was built to be used in combination with one of the cold traps. The hydrogen is injected into the NaK by permeation through a coiled nickel tube 3 mm in diameter, 20 m long and with a 0.3 mm wall thickness. Hydrogen concentrations will be measured upstream and downstream of the cold traps with a hydrogen meter of the nickel membrane diffusion type.

After loading the cold trap with hydrides, the drained cold traps will be transferred to a glove box for analyzing the distributions of the precipitations. Hydrogen recovery by thermal decomposition will be performed by heating up the hydride covered cold trap inserts in an electrically heated vessel and pumping the hydrogen gas to a gasometer where the release rate is measured. This test apparatus (Fig. 4) is in operation and was used for experiments where the cold trap hydride crystals were simulated in different ways as described in the following.

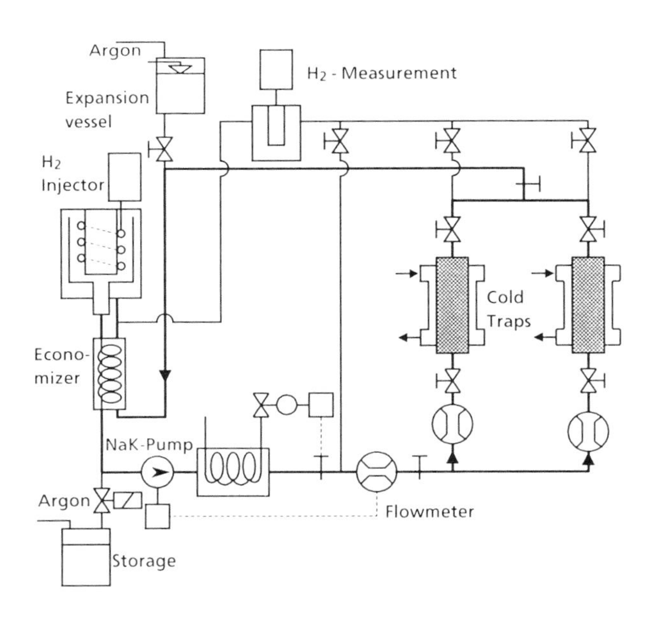

FIGURE 3
Test Loop for Hydrogen Removal

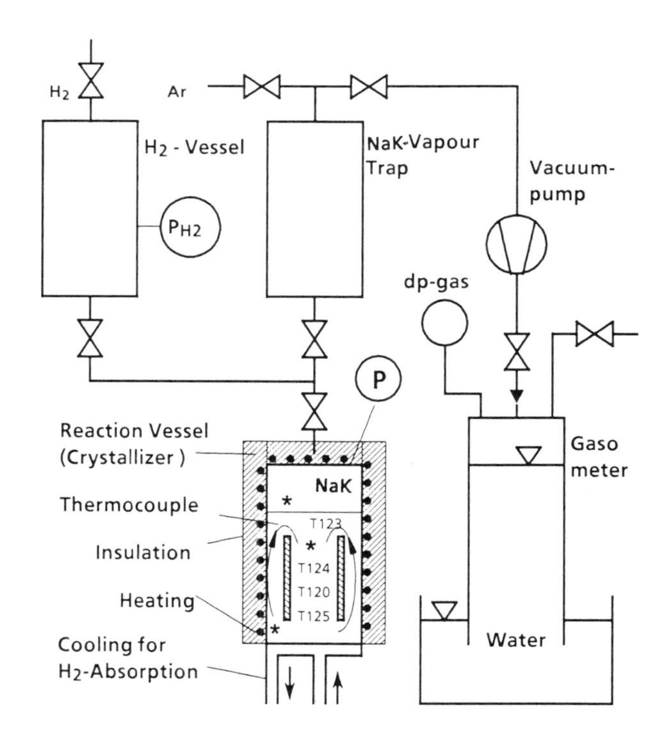

FIGURE 4
Test Apparatus for Hydrogen Recovery

4. EXPERIMENTAL RESULTS

In a first step, fine NaH-powder suspended in a basket was thermally decomposed. In a part of the experiments, Na_2O-powder was added. The reaction coefficient k was determined, given by

$$\dot{m} = k(m_o - m)^n, \qquad (1)$$

where $\dot{m}$ is the hydrogen mass flow rate, m_o the initial mass of hydrogen in the reaction vessel, m the total mass of released hydrogen at time t and the exponent n characterizes the order of the reaction. The rate constant k was evaluated assuming a value n = 1. No influence of oxygen on the release rate was found, the results agreed well with results from other workers[9]. Figure 5 contains the mean curve of the experiments in an Arrhenius plot.

The specific surface of the NaH powder might be nonrelevant for NaH-crystals in cold traps. Therefore, in a second step, NaH-crystals were generated in a vessel which can be heated at the cylindrical wall, on the top and cooled at the bottom, as shown in Fig. 4. An inner cylinder promotes a thermal convection flow which transports the sodium from close to the free surface to the bottom. Hydrogen is absorbed at the surface and crystals grow at the bottom. For hydrogen recovery, the cooling is removed and an additional heating coil at the bottom is installed.

The reaction coefficients k are significantly lower than for NaH-powder as shown also in Fig. 5. The release mechanism was clearly dominated by hydrogen bubble formation (heterogeneous nucleation).

Presently, experiments are performed with hydride crystals in NaK instead of Na. Different release mechanisms were observed[10]:

a) decomposition of hydride with hydrogen bubbling
b) Decomposition of hydride without hydrogen bubbling
c) release from unsaturated solution

Mechanism a) dominates at high release rates

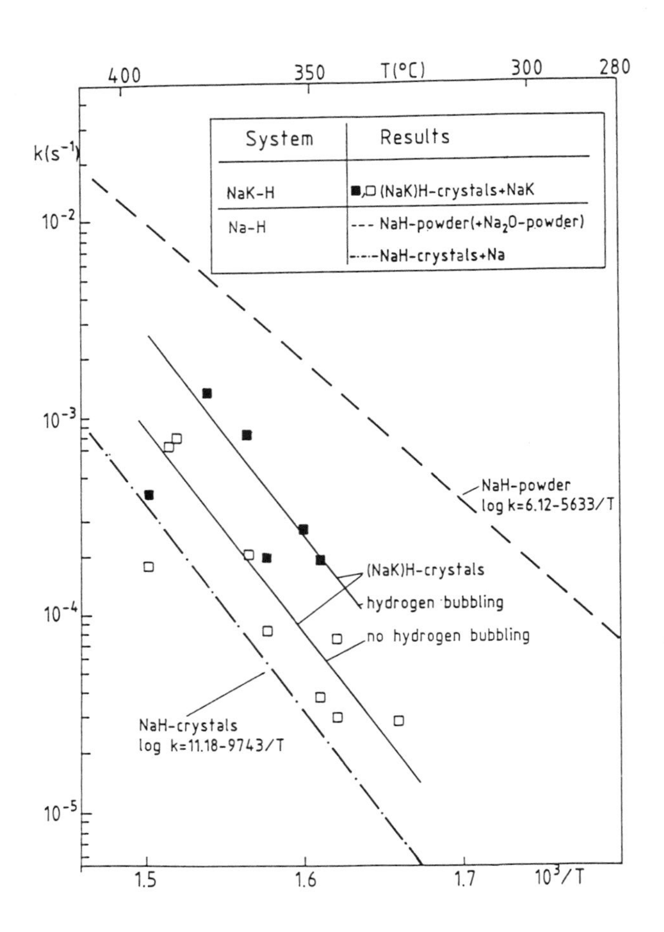

FIGURE 5
Reaction Coefficient for First Order Reaction as a Function of Reciprocal Absolute Temperature

(high temperature and/or large available hydride mass (m_o-m). The reaction coefficient k is distinctively higher than for the Na-H-system. The increase can be explained by the higher saturation pressure of (NaK)-hydride.

The reaction coefficients for mechanism b) are lower than those for a), as shown in Fig. 5. Mechanism b) occurs at low temperatures and/or small values of (m_o-m). Presently, it is not clear if the rate-determinig step is the diffusion in the liquid NaK or effects near the NaK surface.

Mechanism c) is more pronounced for NaK

compared to Na due to its higher saturation concentration. Reaction coefficients, not shown in Fig. 5, are even lower than those for mechanism b).

From present experience, it can be concluded that for cold traps with low hydride loadings, the cold trap should be drained for hydrogen recovery. Then, the reaction coefficients should be similar to those for hydride decomposition with hydrogen bubbling. Assuming conservatively the values for the Na-H system, 99 % of the hydrogen is released in less than 3 hours. Two recovery cycles per day appear to be feasible.

5. CONCLUSIONS

Requirements on fusion blanket cold traps differ from those for currently used cold traps and require a novel design. A cold trap with a high length to diameter ratio is proposed with high throughput velocities. Precipitation occurs first in a region without mesh-packing but with increased mass transfer due to an imposed velocity field by an electromagnetic field. The following region is filled with wire mesh.

Precipitation kinetics are presently not sufficiently understood for low concentrations in the NaK-H system. Corresponding experiments will be performed in the near future.

First tests on hydrogen recovery from NaK prove the feasibility of the selected method. Different release mechanisms were observed, suggesting that tritium recovery should be performed with drained cold trap. Again, more experiments are required to make quantitative statements.

ACKNOWLEDGEMENT

This work has been performed in the framework of the Nuclear Fusion Project of the Kernforschungszentrum Karlsruhe and is supported by the European Communities within the European Fusion Technology Program.

REFERENCES

1. S. Malang, K. Arheidt, L. Barleon, H.U. Borgstedt, V. Casal, U. Fischer, W. Ling, J. Reimann, K. Rust, G. Schmidt, "Self-Cooled Liquid-Metal Blanket Concept", Fusion Technology, in print.

2. J. Reimann and S. Malang, "A Study of Tritium Separation from LiPb by Permeation into Na or NaK and Cold Trapping", KfK-4105, Oct. 1986.

3. R.G. Clemmer, P.A. Finn, L.R. Greenwood, T.L. Grimm, D.K. Sze, J.R. Bartlit, J.L. Anderson, H. Yoshida, Y. Naruse, "The Requirements for Processing Tritium Recovered from Liquid Lithium Blankets: The Blanket Interface", ANL/FFP/TM-217, March 1988.

4. P. Hubbertey, "Solutions of Hydrogen in Alkali Metals", in: Handbook of Thermodynamic and Transport Properties of Alkali Metals, ed. R.W. Ohse (Blackwell Scientific Publications).

5. C. Saint-Martin, C. Latgé, P. Michaille, Journal of Nuclear Materials 151 (1988) 112-119.

6. C. Saint Martin, C. Latgé, P. Michaille, C. Laguérie, "Mechanism and Kinetics of Crystallization of Sodium Hydride in Cold Traps", Fourth Int. Conf. Liquid Metal Eng. and Techn., Avignon, France, Oct. 17-21, 1988.

7. M. Latgé, Mme Lagrange, M. Suraniti, M. Ricard, "Development of a New Cold-Trap Concept for Fast Breeder Reactors", Fourth Int. Conf. Liquid Metal Eng. and Techn., Avignon, France, Oct. 17-21, 1988.

8. M.R. Hobdell and F.J. Salzano, "A Convenient Technique for Stirring Liquid Sodium", Nuclear Appl. & Techn., Vol. 8, Jan. 1970.

9. J. Reimann, "Tritium Separation from Pb-17 Li by Permeation into NaK and Cold Trapping: First Experiments on Recovery from Cold Traps", Third Topical Meeting on Tritium Technology in Fission, Fusion and Isotopic Applications, Toronto, May 1-6, 1988.

10. R.S. Fidler and A.C. Whittingham, "A Comparison of Alternative Methods for In-Situ Sodium Cold Trap Regeneration", Liquid Metal Eng. and Techn., BNES, London, 1984.

11. J. Reimann, H. John, "Tritium Separation from Pb-17Li by Permeation into NaK and Cold Trapping: First Experiments on Hydrogen Recovery", Fourth Int. Conf. Liquid Metal Eng. and Techn., Avignon, France, Oct. 17-21, 1988.

FUSION TECHNOLOGY 1988
A.M. Van Ingen, A. Nijsen-Vis, H.T. Klippel (editors)

CATALYST STUDY FOR THE DECONTAMINATION OF ATMOSPHERES CONTAINING FEW TRACES OF TRITIUM

J. CHABOT, J. MONTEL, J. SANNIER

Commissariat à l'Energie Atomique - IRDI-DCAEA-SCECF-SECNAU
Centre d'Etudes Nucléaires de Saclay - 91191 GIF SUR YVETTE - FRANCE

The conversion of tritium at very low activity level using catalytic oxidation followed by water trapping is studied in the loop BEATRICE in order to measure kinetic parameters required for the design of the NET tritium clean-up system. Two precious-metal catalysts (Pd/alumina and Pt/alumina) are very efficient in removing tritium from contaminated gas mixtures down to a few MPC level at low temperatures, without need of isotopic swamping. However at room temperature, the trapping of tritium species on the catalyst surface gives rise to a progressive deactivation with time. Best regeneration conditions have to be determined in order to demonstrate industrial feasibility of operating at low temperatures.

1. INTRODUCTION

Tritium extraction from the atmosphere of glove-boxes containing tritium handling facilities is mandatory to minimize environmental tritium releases. Oxidation of tritium on precious-metal catalysts followed by water adsorption on molecular sieves is the most widely used method with the drawback of requiring hydrogen swamping, the necessity to regenerate molecular sieve beds and the multiplication of the amount of water formed in the system. Moreover though existing facilities are very effective in achieving the detritiation of atmospheres containing up to several curies of tritium per cubic metre using traditional catalysts, their efficiency at tritium concentrations as low as 10^{-3} ppm (89 $MBq.m^{-3}$) has not been reported and is questionable. Furthermore these catalysts are usually operated at 150-200°C.

A catalyst efficient at room temperature, without preheating and postcooling the gas stream would greatly simplify the system and improve the economic aspect of the process. In addition cold trapping of tritiated water instead of adsorption on molecular sieve beds could decrease the amount of solid wastes and make easier the tritium recovery of the formed water.

Within the European Fusion Technology Programme an experimental work on atmosphere detritiation systems is under way at CEN-SACLAY in order to determine the kinetic parameters required for the design of the NET tritium clean-up system. A two-fold goal is pursued (i) to select tritium conversion catalysts able to achieve the oxidation down to MPC level without hydrogen swamping (ii) to determine the ability to operate at low temperature, and even at room temperature if possible.

2. EXPERIMENTAL

2.1. Apparatus

Experiments are carried out in the loop BEATRICE. It includes (fig. 1) a processing loop in which the gas to be purified is circulated, a feed-gas line with a tritium injecting device, an analysis circuit and a vacuum line. The loop is located in a glove-box continuously ventilated by direct connection to a stack. The processing loop mainly consists of :

- a 250 litres stainless steel tank supplied with the tritiated test-gas mixture,
- a catalytic reactor which can be provided

with variable catalyst charges (0.003 to 1 litre) and operated at temperatures up to 300°C ; the catalyst temperature is measured by means of a thermocouple located inside the catalytic bed,

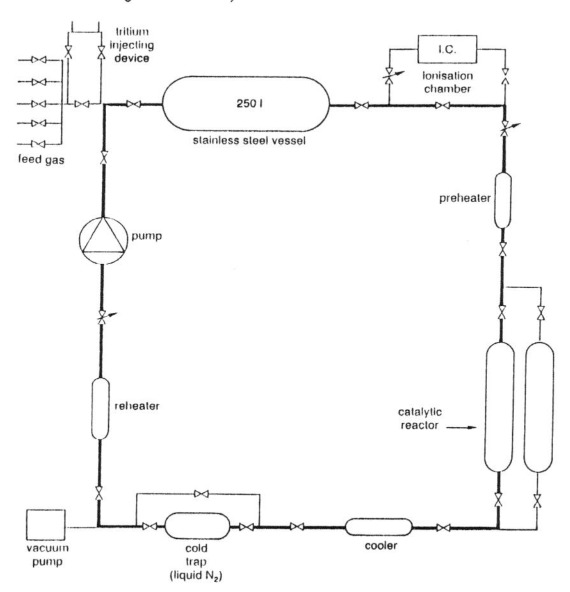

FIGURE 1
BEATRICE LOOP - Conceptual flow sheet

- a liquid-nitrogen cold trap especially designed for trapping water at very low partial pressure,
- a pump the flowrate of which is adjustable up to 10 $m^3.h^{-1}$,
- devices allowing heating or cooling of the gas before and after the catalytic reactor and the cold trap.

The analysis circuit is mainly constituted of two ionization chambers (10 litres volume) one of which is used as a reference. The analysis line is connected to the processing loop between the tank and the catalytic reactor.

2.2. Procedure

Dry nitrogen carrier gas containing up to 4000 ppm oxygen is introduced into the tank by means of the gas feed line. Invested tritium is supplied in special ampoules containing about 2.6×10^7 Bq (700 µCi). The ampoule is broken in the injecting device and tritium is swept-along into the loop by the carrier gas. Tritium is converted at the catalyst surface and tritiated water is removed in the cold trap the temperature of which is about -175/-180°C. The tritium concentration change in the loop is continuously measured and recorded. In such recirculation experiments, with the assumption of a pseudo-homogeneous first-order reaction, the tritium concentration in the loop, at time t, is expressed by the following equations[1,2] :

$$C = C_0 \exp. (- f \frac{Q}{V} t) \qquad (1)$$

$$f = 1- \exp. (- k \frac{V_c}{Q}) \qquad (2)$$

Where :

C and C_0 are tritium concentrations, $\mu Ci.cm^{-3}$, at t and t = 0 time, s,

Q flowrate, $l.s^{-1}$,

V loop volume, V_c catalyst volume, l,

k pseudo-homogeneous rate coefficient, s^{-1},

f fractional conversion.

2.3. Catalyst physical properties

The two catalysts investigated were previously selected from hydrogen experiments carried out on several available commercial catalysts at H_2 concentrations up to 100 ppm. These precious-metal catalysts are respectively made of palladium or platinum dispersed on a high-surface-area alumina substrate in pelletized form (2-4 mm diameter). Their most important physical properties are reported in table I.

TABLE I
Catalyst physical properties

catalyst	amount of active phase (%)	support surface area $m^2.g^{-1}$	support pore volume $mm^3.g^{-1}$	active surface area $m^2.g^{-1}$
Pd/alumina pellet	0.41	103	323	1.8
Pt/alumina pellet	0.46	103	305	1.3

3. RESULTS AND DISCUSSION

3.1 Room temperature experiments

Initially tests were carried out at room temperature by varying the residence time from 0.1 to 0.010 s (corresponding space velocities of 35000 to 350000 h^{-1}). Prior to each experiment the catalysts were outgassed under vacuum at 300°C. Typical data from comparative runs for two residence times with both catalysts are shown in fig. 2. After a linear part suggesting a first-order reaction the curves tritium concentration logarithm versus time become concave in agreement with already published results[1.2.] for higher tritium level experiments. The two catalysts appear to be very efficient. With a residence time of 0.1 s, the tritium level decreases to a concentration corresponding to about a few MPC value in less than one hour. Noticeably shortening the residence time leads to similar levels of residual tritium but obviously increases the duration of the tritium separation.

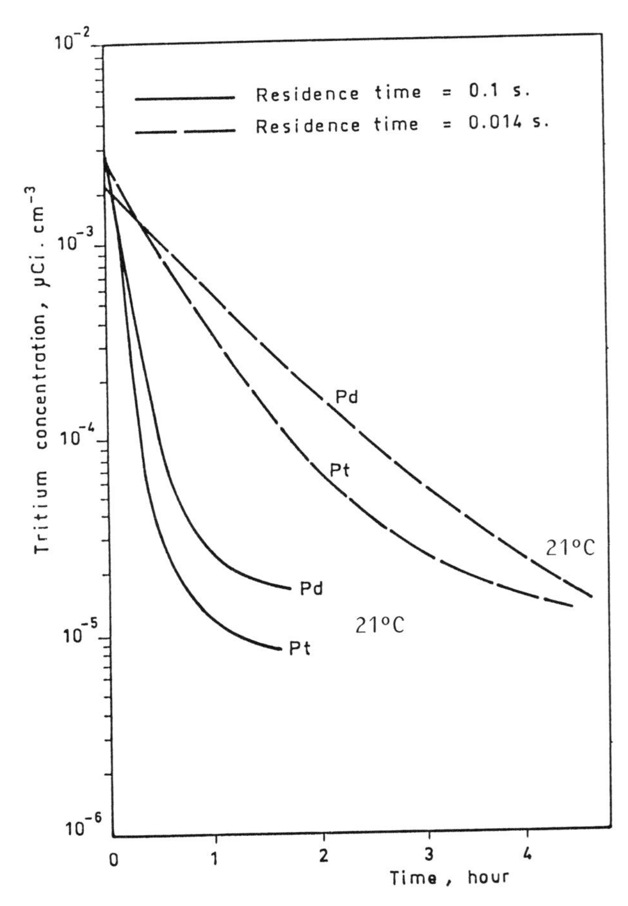

FIGURE 2
Comparison of tritium catalytic oxidation with Pt/alumina and Pd/alumina catalysts for two different residence times, at room temperature.

However when successive runs are performed with the same catalyst without outgassing between each run, a progressive decrease in the fractional conversion is observed. Simultaneously the residual concentration increases and the deviation of the curve from linearity appears earlier. Complementary experiments have pointed out the following facts :

- the phenomenon is not correlated to an eventual tritiated methane formation in the loop nor to any limitation in tritium activity measurement which could be due to the detection limit, or to a contamination of the ionization chamber (as it was observed by other workers[2.3.4.]),
- the initial performance of the catalyst can be restored by vacuum outgassing at temperatures higher than 240°C for several hours,
- only very small amounts of tritiated water are collected in the cold trap while nearly all the tritium activity (about 95 %) remains trapped on the catalyst.

Consequently, the tritium and/or tritiated water trapped at the catalyst surface when operating at room temperature would explain both the catalytic activity decrease with time and the progressive raising in the residual concentration.

3.2 Investigation at higher temperatures

Tests were performed in the temperature range of 20 to 250°C in order to find out the threshold for water desorption during the catalytic oxidation process itself.

Because of the very high catalyst efficiency

expected at higher temperatures, the catalyst volume was significantly reduced in order to lower the residence time of the gas. Consequently the conversion rate per pass of gas which corresponds to the slope of the linear part of the experimental curves was limited to make measurements possible. Prior to each experiment the catalysts were outgassed under vacuum by raising the temperature by steps up to 300°C. At the end of each run the tritium activity was measured in the catalyst and in the cold trap using a liquid-scintillation counter. Representative experimental curves and corresponding analytical results are presented on figures 3, 4 and 5 for the two catalysts. From fig. 3 it appears that in both cases the major part (80-95 %) of the tritium remains on the catalyst at temperatures up to about 70-80°C and that a significant amount of tritium is still trapped at higher temperatures. Concerning the temperature effect, the two catalysts behave similarly. The tritium decreases to a concentration corresponding to about a few MPC level for temperatures ranging from 20°C to 150°C and to the MPC level from 205°C to 240°C.

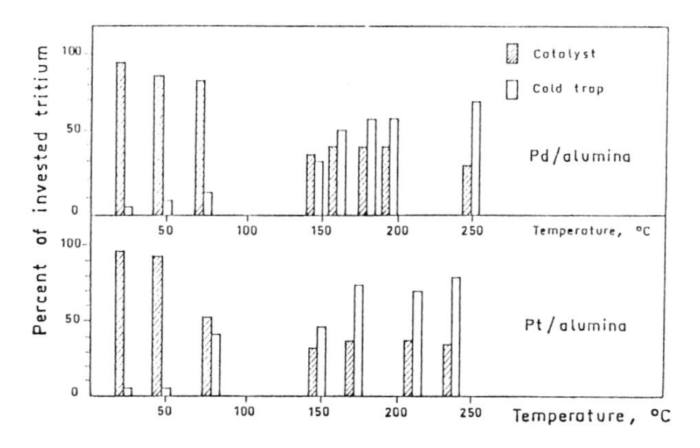

FIGURE 3
Tritium activity distribution between the catalyst and the cold trap at the end of tests carried out at different temperatures for both Pd and Pt catalysts

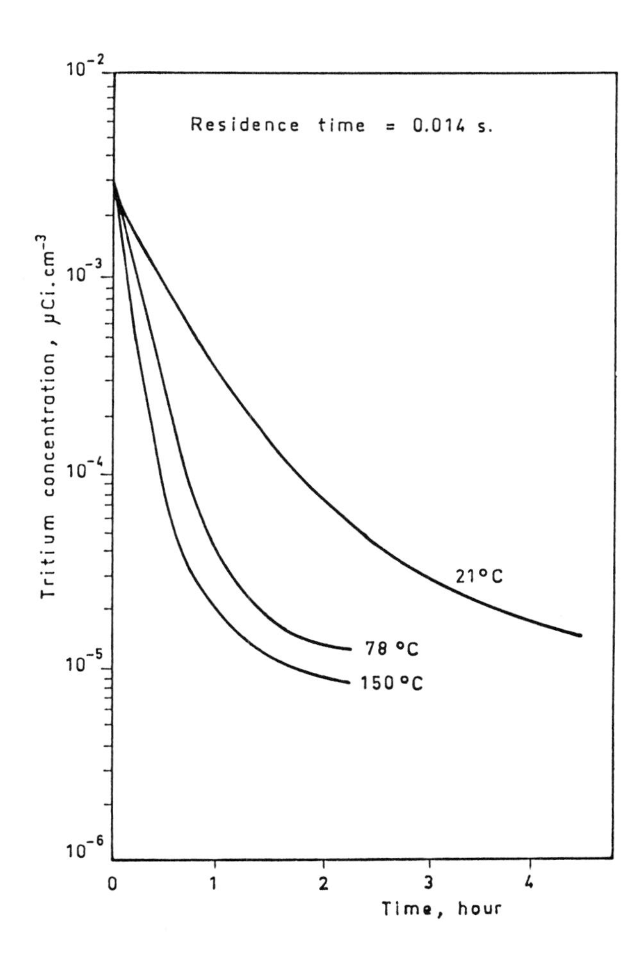

FIGURE 4
Tritium catalytic oxidation with Pt/alumina catalyst at low and intermediate temperatures.

Typical curves concerning experimental runs carried out with the platinum catalyst at low and intermediate temperatures up to 150°C and at temperatures higher than 200°C are respectively shown on fig. 4 and 5. The fractional conversions, f, determined from these curves and identical experiments with the palladium catalyst are summarized in table II.

4. CONCLUSION

Both palladium and platinium catalysts are very effective in removing tritium from contaminated gaz mixtures down to a few MPC

TABLE II
Fractional conversion, f, for different temperatures and residence times for both Pt and Pd catalysts

catalyst	temperature, °C	catalyst volume, l	flowrate $l.s^{-1}$	residence time, S.	fractional conversion
Pt/ alumina	21	$9\ 10^{-3}$	0.64	0.00140	0.18
	78	$9\ 10^{-3}$	0.63	0.0142	0.61
	150	$9\ 10^{-3}$	0.62	0.0145	0.68
	205	$3.85\ 10^{-3}$	0.403	0.0095	0.80
	240	$3.85\ 10^{-3}$	0.36	0.0106	0.84
Pd/ alumina	25	$9\ 10^{-3}$	0.64	0.0140	0.07
	75	$9\ 10^{-3}$	0.64	0.0140	0.49
	196	$3.85\ 10^{-3}$	0.34	0.0113	0.65
	250	$3.85\ 10^{-3}$	0.356	0.0108	0.89

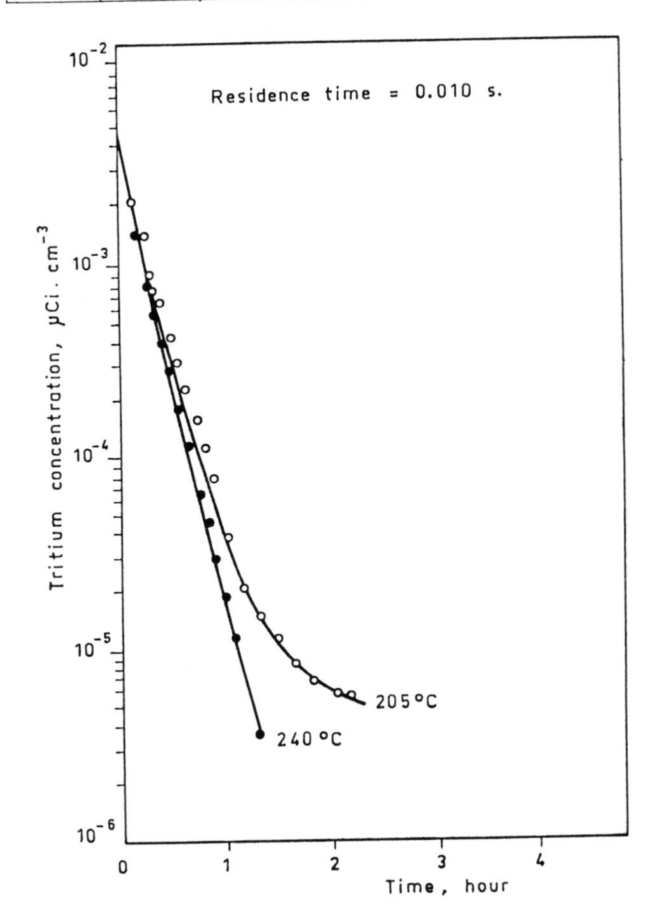

FIGURE 5
Tritium catalytic oxidation with Pt/alumina catalyst at 205 and 240°C.

level at low temperatures and a short residence time without need of hydrogen swamping. However a progressive deactivation with time may occur caused by tritium species trapped on the catalyst surface. This tritium trapping and the corresponding catalyst deactivation may be a concern both for tritium inventory and for process efficiency and cost. Room-temperature operation implies periodic regeneration of the catalyst. The best regeneration conditions depending on temperature, catalyst volume and time have to be determined in order to justify the choice of this way. However in a nuclear plant like NET it may be very useful for safety considerations to be able to start an emergency clean-up system without need of keeping the catalyst continuously hot. In this way, the catalytic bed would be heated up progressively in order to desorb the trapped species and to restore the required performances.

ACKNOWLEDGEMENTS

This work has been carried out within the Commission of the European Communities Programme on Fusion Technology (Task action T5b). The authors wish to thank the CEE and CEA authorities for the opportunity of conducting this research and publishing the results.

REFERENCES

1. O. Levenspiel, "Chemical reaction engineering. An introduction to the design of chemical reactors," John Wiley and Sons INC, New York, London, 1962.

2. A.E. Sherwood, B.G. Monahan, R.A. Mc Williams, F.S. Uribe, C.H. Griffith, "Catalytic oxidation of tritium in air at ambient temperature", Lawrence Livermore National Laboratory, Livermore, UCRL 52811, July 1979.

3. K. Yamagucchi, X. Makimoto, H. Kudo, "Detritiation of glove-box atmosphere by using compact tritium removal equipment", J. of Nuclear Science and Technology 19 [11] 948-952, November 1982.

4. G.A. Morris, "Methane formation in tritium gas exposed to stainless steel," Lawrence Livermore Laboratory, Calif., UCRL 52262, March 25th 1977.

TABLE II:

CRYOPUMP DESIGN ISSUES

ISSUE	IMPLICATION	DESIGN/DEVELOPMENT MEASURE (s)
1. Radiation field	Activation of argon	Buffer storage to allow decay
2. Tritium compatability	No organic bonds	Material test program
3. Impurity tolerance	Poisoning of cryosorbent	Avoid cryosorbent by using argon spray **or** internal separation valve and bake-out capability
4. Inventory (150 g T)	Frequent regeneration	Reliable valves
5. Magnetic field	Spontaneous regeneration	Modelling/experiment/shielding
6. Dust tolerance	Activation of materials degradation of valves	Ingress control/component design

and 500 panels have been made and subjected to mechanical and thermal cycling tests (between/50/300°C and liquid nitrogen temperature). The best combinations from the first two screenings will be subjected in 400 mm diameter panels to performance and endurance tests, to determine the optimum sorbent/bond/substrate combination for a cryopump to meet the NET duty. In addition to those fabricated in-house, sorbent samples manufactured by industry are being evaluated in this programme.

A proposed configuration of the cryopump based torus evacuation system is described in[5] (this conference). A tabulation of cryopump design issues is included as Table II.
It is considered reasonable to scale existing cryopump designs (of both variants described) to a pumping speed of 100 m^3/sec (while pumping) to give a time averaged capacity of approx. 70 m^3/s allowing for the necessary regeneration cycle. It is therefore proposed to install two such pumps on each pair of divertor ducts which will provide virtually balanced pumping of all ducts. (see figure 1)

4. TURBOMOLECULAR PUMPING OPTION

Current generation Tokamaks (JET, JT60, TFTR) are equipped with oil lubricated Leybold AG 3500 l/s turbomolecular pumps for torus evacuation. These devices are much smaller than those needed for NET. An engineering design study has been completed to assess the feasibility of extrapolation of pumping speed by an order of magnitude and modification for tritium compatability, primarily by the use of magnetic bearings[6]. Principal advantages of turbomolecular pumps are:

- low inventory
- steady state operation
- tolerance of impurities

Countering these are a range of significant issues whose specific implications and design and development measures are indicated in table III.

The table highlights the criticality of some key components, e.g. magnetic bearings, with respect to a range of the design input parameters.

Magnetic bearings need to be developed specifically for this application, where axial

loadings of up to 15 tons have been calculated for NET relevant pump designs (50.000 l/s single flow). Within the time constraints of NET it is considered reasonable to extrapolate current designs to a double flow pump with a capacity in the range of 12-15 m^3/s. This configuration features much more manageable axial bearing loads.
Installation of six or possibly eight pumps on each pair of divertor ducts would be needed (figure 4). Each pump requires an isolation valve of approx. 800 mm diameter.

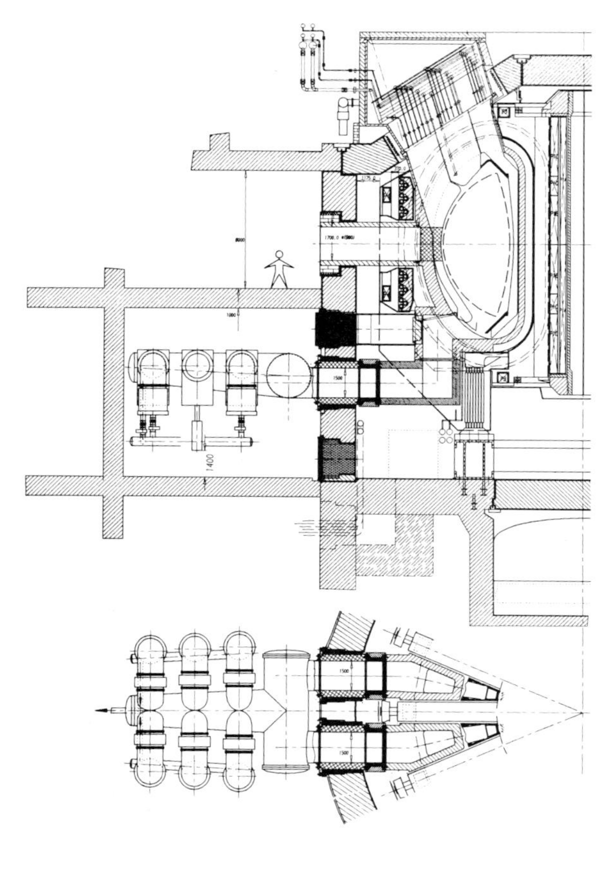

FIGURE 4
TMP System Layout

5. VALVES

For both cryopumping and turbomolecular pump options isolation valves are required upstream of the pumps. The main criteria for these valves is a tight shut off to permit maintenance or removal of pumps. The number of cycles expected of such valves will be modest, but particularly in the case of cryopumps valves larger than hitherto built will be required, probably around 1500 mm dia. Backstreaming (helium and argon are undesirable) could be reduced by maintaining the interspace between valve discs under a slight overpressure (relative to duct pressure) of deuterium. In the small clearances which might exist between disc seals and seats (due for example to degradation by dust) leakage would be of deuterium, which would be non-injurious to both upstream and downstream requirements. This feature will be investigated further as design proceeds.

For cryopumps an additional large valve (external to each pump) suitable for frequent cycling (typically every four hours) will be required for regeneration. The leak tightness requirements of these valves would be less stringent than the isolation valves.

The need for a large regeneration valve is a major drawback for cryopumps. Combination of the regeneration and isolation functions into a single valve would be a challenging design task, but elimination of one large valve per pump would be highly desirable. The feasibility of this will be reviewed as the design advances.

6. NOVEL CONCEPTS

The high required pumping speed indicated in section 1 is due partly to the limited conductance of the divertor duct system which results from the need to install conventional cryo or turbomolecular pumps some distance from the torus, outside the most intense magnetic field and where sufficient space exists for pumps, valves and remote handling devices. The required volumetric pumping speed at pump location could be reduced if the plasma exhaust pressure were boosted by an "in-line" device close to the divertor.

TABLE III: TURBOMOLECULAR PUMP DESIGN ISSUES

ISSUE	IMPLICATION	DESIGN/DEVELOPMENT MEASURE (s)
1. High gas throughput	Scale-up	Component design/replication
2. Radiation fields	Remote maintainability	Layout for accessibility Mechanical design (quick disconnect) Reliability Materials selection
3. Magnetic fields	Blade heating	Shielding Location, orientation Component design
4. Tritium compatibility	Magnetic main bearings Non-lubricated back-up mechanical bearings	Tritium compatible insulators Design for heat dissipation Minimize reliance on back-up bearings
5. Tolerance of dust	Blade imbalance/damage Bearing duty	Control of ingress Mechanical design
6. Tolerance of sudden venting	Axial bearing loads Blade bending	Double flow configuration Modelling
7. Plasma disruption	Rotor displacement	Rigid mounting/flexible duct connections
8. Seismic accelerations	Rotor may impact casing	Casing integrity

Such a mechanism is decribed by Hemmerich[7] as a so called thermodynamic transport pump. The pumping effect occurs due to the different behaviour of gas molecules striking surfaces with different characteristics. In summary fast (hot) molecules striking smooth cold surface tend to reflect specularly. Slow (cold) molecules are reemitted from a smooth hot surfaces according to a cosine law. Hot and cold molecules are reemitted from "rough" surfaces according to a cosine law. By adopting alternating hot rough and cold smooth surfaces arranged as indicated in figure 5 a multistage pump could be installed in each divertor duct. If such a device could achieve a compression ratio (and therefore effect a reduction in the pumping speed required downstream) of approximately one order of magnitude, a significant reduction in scale up of pumps and in duct and valve diameters could be realised.

The principle has been demonstrated and the next step is to build and test a multi-stage array.

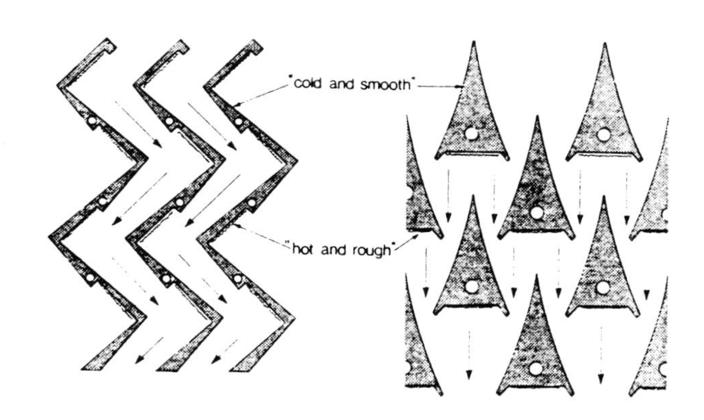

FIGURE 5
Thermomolecular Pumping Arrays

Other novel concepts which will be considered relate to enhanced regeneration of cryopumps. These include flash heating of the condensing and cryosorbent panels by microwaves, inductive or radiative means to eliminate the need to drain the respective cryostats, with an associated reduction in liquid helium consumption. However considerable design and experimental work will be required to develop a reliable system to heat uniformly the relatively large cryocondensation surfaces (20 m^2) located in large volume (10 m^3) of the cryopump body.

7. CONCLUSIONS

The mainfeatures of both turbomolecular and cryopumps have been described above. As cryopumps are more readily scaleable than turbopumps, a satisfactory level of confidence in performance predictions is achieved with larger cryopumps than turbo's. This feature is important not only for the NET duty but in meeting the long term program objective of relevance to 'DEMO' reactors. The maximum turbopump capacity which is considered achievable from currently available technology is a double flow version of 15 m^3/s. While detailed reliability studies for both options remain to be carried out it is postulated that the redundant capacity of the cryopumping system, coupled with its freedom from rotating elements will results in its being an inherently more reliable solution. Nevertheless the need for large all metal valves to cycle frequently in a tritiated and potentially dust burdened environment requires a special design and testing effort.

Also the cryogen distribution system is fairly complex and its detailed design will involve optimisation of layout and reliability.

Cryopumps are expected to deal adequately with the burn and dwell cycle duty for which purpose the system described in section 3 would be installed, however cryopumps would be less well suited to initial pump down and torus conditioning modes. Torus conditioning by glow discharge in helium would become the dominant requirement for sizing cryopumps. In addition operation in the Kundsen flow regime would probably result in cryosorption rates high enough to cause excessive cooling loads, leading to thermal instability of the cryopumps in this operational mode.

It is therefore proposed to install turbomolecular pumps sized to handle the glow discharge gas quantity in a reasonable time period (say 16 hrs. i.e. between two working days). This requires 8 (i.e. one per U-duct) TMP's each with a capacity of 10.000 l/s. Allowing for standby/repair an installed TMP capacity of 12 - 15.000 l/s is selected. These pumps would protect the CP's from the excessively demanding conditions occuring during the pump down and particularly torus conditioning while themselves not being exposed to the harshest conditions for TMP's (e.g. magnetic fields, air/water inbreaks and plasma disruptions).

The layout of the proposed hybrid CCP/TMP pumping scheme is presented in Figure 6.

The program for the immediate future will focus on resolving the issues in Tables II and III, and selection of a preferred cryopump type. In parallel the design of a 12 - 15.000 l/s double flow turbomolecular pump with magnetic bearings will be developed. This will be followed by prototype construction and testing.

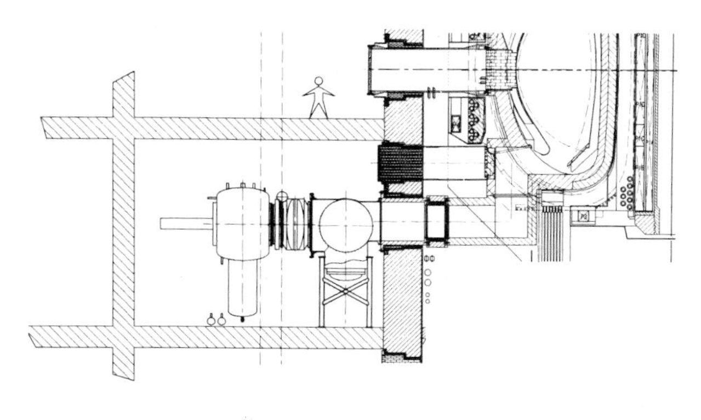

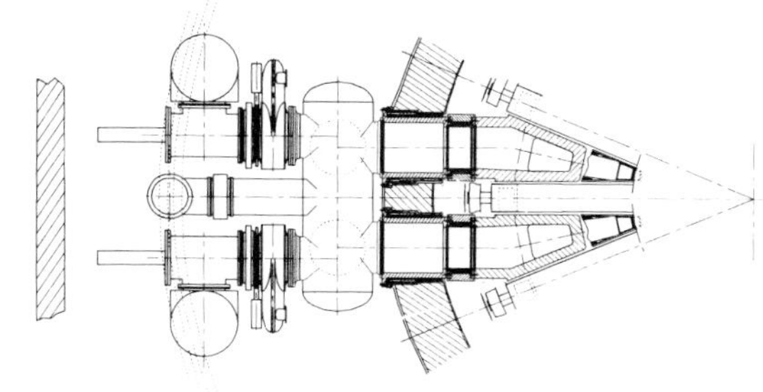

FIGURE 6
Hybrid Pumping System Layout

ACKNOWLEDGEMENTS

We would like to thank the authors of four reports prepared under NET contracts on torus vacuum systems and components, information from which is summarized and presented in this paper, i.e. Mr. H.A. Maykemper of Leybold AG, Mr. J.C. Boissin of Air Liquide, Messrs. K-A.M. Bernhardt and A. Conrad of Arthur Pfeiffer Vakuumtechnick Wetzlar GmbH and Mr. H.J. Reich of VAT AG. The valuable comments of Messrs. M. Chazalon and D. Leger of NET are appreciated.

REFERENCES

1. M. Harrison, Private Communication, Dec. 1987

2. P. Dinner et al. From JET to NET Design Steps in Tritium and Vacuum Systems, This volume.

3. D. Sedgeley et al., Conceptual Design of a Large Regenerable Cryopump, 12th Symposium on Fusion Engineering, Monterey, CA, October 1987, pp 141 - 143.

4. D. Perenic, A. Mack, Experimental Investigations into Cryosorption Pumping of Plasma Exhaust. Tritium Technology in Fission Fusion and Isotopic Application, Toronto Canada, May 1988.

5. W. Obert et al., Conceptual Design for the Plasma Exhaust Cryopumping at NET, this volume.

6. K. H. Bernhardt, A. Conrad et al., Development of Large Vacuum Pumps for Thermonuclear Reactors. Tritium Technology in Fission, Fusion and Isotopic Applications, Toronto Canada, May 1988.

7. J.L. Hemmerich, Primary Vacuum Pumps for the Fusion Reactor Fuel Cycle. T.Vac Sci Technology A6(1) Jan/Febr. 1988, p. 144-153.

FUSION TECHNOLOGY 1988
A.M. Van Ingen, A. Nijsen-Vis, H.T. Klippel (editors)

FROM JET TO NET: DESIGN STEPS IN TRITIUM AND VACUUM SYSTEMS

P.J. Dinner[1], J.L. Hemmerich[2]

[1]The NET Team c/o Max-Planck-Institut fuer Plasmaphysik
Boltzmannstr.2, D - 8046 Garching
[2]JET Joint European Undertaking, Culham, Abingdon, UK

Design requirements and design approaches taken for tritium and primary vacuum systems on JET and NET are summarised and compared. Components which can be developed from JET designs, as well as those requiring different concepts or extensive additional development for NET are highlighted.

1. INTRODUCTION

As JET moves towards tritium operation in the early 1990's, many practical engineering problems must be solved in design and construction of tritium and vacuum systems. It is important for the Next European Torus, which is currently in the pre-design phase, to maximize the lessons learned from the JET experience. In this paper, we compare the requirements of JET and NET, and consider the application of specific design elements on JET to NET.

The magnitude of the step from JET to NET can be appreciated by considering some of the basic machine parameters of the two devices[1,2] (Table 1).

Table 1
JET-NET Machine Parameters (1,2)

PARAMETER	JET	NET
Plasma Major Radius (m)	3	5.4
Plasma Minor Radius (m)	1.3	1.7
Plasma Elongation	1.7	2.2
Toroidal Field on Axis (T)	3.5	4.8
Plasma Current (MA)	5	15
Predicted Fusion Power(MW)	20-30	600-800
Flattop Pulse Length (s)	20	200+
Max. Duty Cycle (%)	3.3	80
Continuous Operation (h)	17	200

2. FUELLING AND EXHAUST SYSTEMS FOR JET AND NET

2.1 System Requirements

The principal requirements for JET and NET fuelling and exhaust systems are given in Table (2). The major difference in the design of the fuelling and exhaust loops for JET and NET are due to the need for the NET loop to function continuously. Thus the NET system must pump torus exhaust to remove He and impurities during burn, which is in fact the most demanding requirement for the vacuum pumping system. If the pumping speed provided on NET is not adequate, the He fraction will build up in the plasma. This reduces the fraction of useful plasma beta. A factor of 2 above the design value for He in the plasma may be tolerable, i.e. up to 20%. However, the modelling of plasma exhaust requirements during burn is very imprecise, since there is no experimental data-base for pumping from a divertor on a large machine. In particular, the anticipated behaviour of He in an ignited plasma is poorly quantified, and it may not diffuse to the divertor at the same rate as D or T.

Initial pumpdown requirements of the two machines are similar, however the much larger surface areas and higher temperatures of NET vacuum vessel internals poses a significant

Table 2
Fuelling and Exhaust System Requirements

PARAMETER	JET	NET
Torus:		
Volume (m^3)	200	600
Surface Area (m^2) (projected, *total*)	200, *1000*	800, *8000*
% Graphite	50	100
Ultimate Pressure @ 600K (mbar)	4.10^{-7}	<=
Pre-shot Base Pressure (mbar)	2.10^{-5}	4.10^{-5}
Leak Rate ($mbar.l.s^{-1}$)	10^{-7}	10^{-5}-10^{-6}
Effective Pumping Speed ($m^3.s^{-1}$)	6-8	250-375
Gas Flow During DT Pulse:		
Feed/Exhaust Rate ($mbar.l.s^{-1}$, *mol/hr*)	10-100, *0.5*	170, *20.5*
He (Exhaust or minority htg.)	< 5	<=
Impurity (%)	5-20	1-3
Dominant Impurities	CO, CO_2, CQ_4 Q_2O	<= Q=H,D,T
Gas Flow During GDC: (in deuterium)		
Pressure	4.10^{-3}	<=
Feed/Exhaust Rate ($mbar.l.s^{-1}$, $mol.h^{-1}$)	20, 3.3	400, 65
Impurity (%)	0.2	<=
Dominant Impurities	As in shot	<=
Neutral Beam Injectors		
Total Power (MW)	10 (160 KeV+)	50 (500 Kev-)
Torus Gas Flow ($mbar.l.s^{-1}$)	7.3	11.6
Total Pumping Speed ($l.s^{-1}$)	$1.6\ 10^7$	$1.0\ 10^8$
Gas Flow to Pumps ($mbar.l.s^{-1}$, *mol/h*)	10^3, *0.4*	10^4, 10^3
Pellet Injectors		
Propellant flow to Cryopump ($mol.h^{-1}$)	<20	<200

problem. While temperatures of 300-350 are forseen on NET, as for JET, for outgassing of the plasma facing components, this is well-below their operating temperature. Vacuum vessel and shield structures will be baked to about 180 C. Outgassing of the graphite plasma-facing surfaces will take place over several hundred hours on NET.

In addition to glow discharge in D_2 for impurity removal from the walls, recently, JET has employed frequent glow-discharge in He to degas hydrogen from the graphite tiles before startup. It is also probable that NET will use an inert gas such as He as an atmosphere inside the torus during maintenance. To cope with these conditions in the event compound cryopumps are chosen as the primary option on NET, additional pumping capacity for NET should be provided in the form of turbopumps.

JET experience with Graphite tiles is important to NET, since NET is presumed to operate with full-coverage protection from graphite tiles during the physics phase[3]. For NET, the average first-wall temperature during burn is above 1000 C, which is expected to produce fewer hydrocarbons than is the case for the lower average wall temperatures of JET. Nevertheless, graphite erosion is a serious design consideration, and the extent of redeposition and nature of redeposited material is unknown. Therefore the vacuum system must be designed to cope with substantial quantities of graphite dust, which can affect the leak-tightness of the vacuum valves as well as pose a safety hazard during open system maintenance. This dust has been studied in JET[4], and contains gamma-emitting nuclides and tritium which are also anticipated on NET. DT retention in the graphite dust may be as high as 30 at %. For this reason, as well as neutron activation and direct tritium contamination of the components, remote handling with contained-transfer is required for the removal of components to a hot-cell environment. Neutron damage is not a serious design constraint for vacuum pumps on either JET or NET. However, it is desirable that NET options be suitable for extrapolation to DEMO, which will have much higher fluences than NET.

Both continuous and time-varying magnetic fields are higher on NET than JET, which poses problems for components close to the machine. In particular the blades of turbopumps are subject to heating, and cryopumps receive an additional heat load during machine start-up and current ramp-down.

RF fields must be screened from the pumping ducts on NET to prevent re-ionization of neutral atoms in the ducts which would interfere with pumping during burn. Screening reduces conductance over open-ducts.

Fuelling on JET is performed by gas puffing both before and during shots. Currently, single-stage gas guns are installed for pellet fuelling on JET[5,6], and a 2-stage prototype gas gun[7] has obtained record velocities for pellet injection (PI), using pellets "saboted" with vespel. Plans to test the two stage gun on JET are well advanced, and JET intends to construct a 10-barrel 2-stage gun. Tritium injection via NBI is also being considered as a mechanism for central fuelling on JET. For NET, these planned experiments at JET are critical to the definition of the fuelling system. It is expected that JET will demonstrate the necessity of high-velocity PI for deep-fuelling, as well as the technical problems of achieving this with a two-stage light-gas gun. Currently, pellet fuelling is considered as an adjunct to gas puffing for fuelling on NET, to be used principally for profile control during ramp-up. If continuous deep fuelling is required, pellet velocities in the range 3-5 $km.s^{-1}$ would likely be required[8], velocities comparable to those to be explored with the JET 2-stage gas gun. However, the JET concept with saboted pellets injected by a two stage gas gun is not likely to be directly suited to continuous injection of pellets on NET, if this is required, since 1'st stage piston life is limited (10's of shots) and sabot removal is a critical operation. Concepts able to provide continuous deep-fuelling suitable for NET have been proposed[9,10], but are still speculative. The possibility exists to use NBI to provide part of the deep fuelling, but this is viewed more as a side-benefit of NBI than a deliberate fuelling approach, since high power (>100 MW) is required to provide a significant fraction (>5%) of the fuelling needs.

Pellet Injector and Neutral Beam Injector (NBI) units for NET, and their associated vacuum systems are still uncertain, and only preliminary estimates of gas loads are presented in Table 2. Pumping speeds larger than for JET are anticipated, and cryopumps will likely be used. Continuous operation will require frequent "online" regeneration to control tritium and hydrogen inventories. NET must be designed to accommodate frequent pressure surges in the torus due to off-normal operation. Thus "open-structure" cryopumps and large liquid He inventories are undesirable. To minimize the amount of water vapour and other condensible impurities pumped on the cryopumps, the NBI and PI enclosures must first be pumped down with mechanical pumps, for example by spiral-vane pumps followed by tritium compatible turbopumps (3-5 $m^3.s^{-1}$).

Total tritium inventory in JET is designed to be less than 90 g. For NET, total inventories of 2-2.5 Kg are anticipated. The design objective on NET is to keep localized inventories which could be released into secondary containment in a single process failure to less than 150 g.

2.2 System Description

The JET fast-fuelling and torus vacuum systems are described in [11,12,13]. For NET, as stated above, fuelling systems have not yet been defined, however it is anticipated that the JET fast-valve can be modified for NET by changing the valve stroke and/or feed-pressure.

High-vacuum torus pumping on JET is provided by four Leybold Heraeus turbo-molecular pumps (TMP'S), each with a nominal speed of 3500 $l.s^{-1}$. These four TMP's are located in pairs on two pumping chambers on either side of the torus, and are connected together with 2 additional TMP's on the rotary neutral injector valves to the torus fore-vacuum crown. Three TMP's- one on each NBI box and one on the pellet injector box are connected to the NBI fore-vacuum crown.

TMP isolation from the torus is accomplished via 400 mm VAT valves between the TMP's and the pumping chambers. During tritium operation, fore-vacuum is provided by cryo-accumulator panels as described in [14].

The NET primary vacuum system concepts are described in [15]. Sixteen divertor ducts are paired and connected to manifolds which can be adapted to either turbo or cryopumps. For the cryopumping option, each manifold is provided with a pair of 100 $m^3.s^{-1}$. compound cryopumps (CCP's). Pumps will be in pumping mode ~70 % and regenerating mode ~30 % of the four-hour cycle time. Thus at all times all ducts will be pumped by at least one pump. For the TMP option, each manifold will be equipped with up to six 12-15 $m^3.s^{-1}$ double-flow pumps. This would minimize the axial forces imposed on the tritium-compatible magnetic bearings during sudden-venting incidents.

For pumping speeds above 250 $m^3.s^{-1}$, which are required under some physics assumptions, CCP's would be preferred, as they provide easier scale-up to very large pumping speeds and provide higher pumping speed per unit area of duct cross-section. In the event CCP's are used to provide the main pumping duty, an additional 10-15 % of the installed capacity may be provided in the form of TMP's to provide initial system evacuation, and pumping during torus conditioning.

Novel concepts such as thermo-molecular pumping[16] are also being explored, and a "proof of principle" experiment on a multi-stage pump is planned. This concept has no "active" mechanical parts, and would permit smaller downstream ducts and pumps.

For the major options, the development issues are also discussed in [15].

3. TRITIUM PROCESSING SYSTEMS

3.1 System Requirements

The tritium processing system scheme for JET is shown in Fig (1), and that for NET in Fig (2). The JET and NET tritium systems are described respectively in [17,18,19] and [20,21,22]. The essential tritium and vacuum requirements for JET and NET have already been given in Table (2). Major points of comparison are:

- Throughput requirements for NET are more than 10 X larger than for JET. Machine size and design for continuous operation are the major factors. Furthermore, NET must operate continuously for long periods to test nuclear components, requiring a very high availability for tritium systems.
- There is no tritium breeding blanket on JET, while NET is designed to produce a significant fraction of its own tritium. This requires special circuits for the collection and purification of bred tritium.
- Total tritium inventory constraints on JET are much more restrictive than those assumed

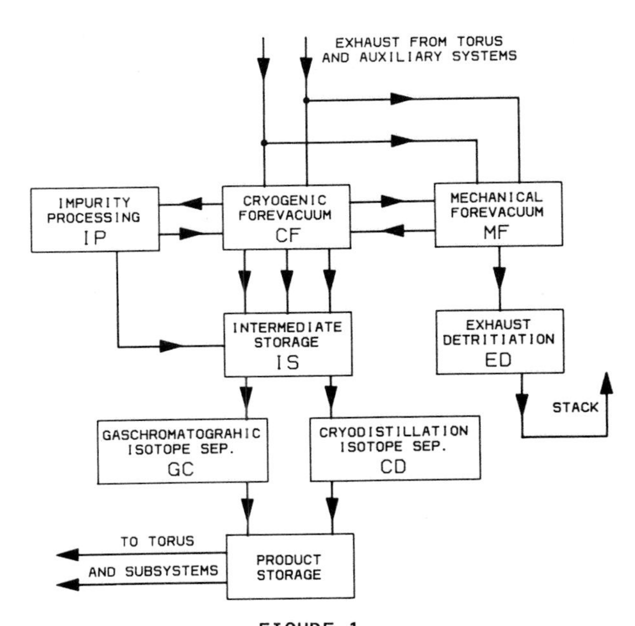

FIGURE 1
JET Tritium Processing Block Diagram

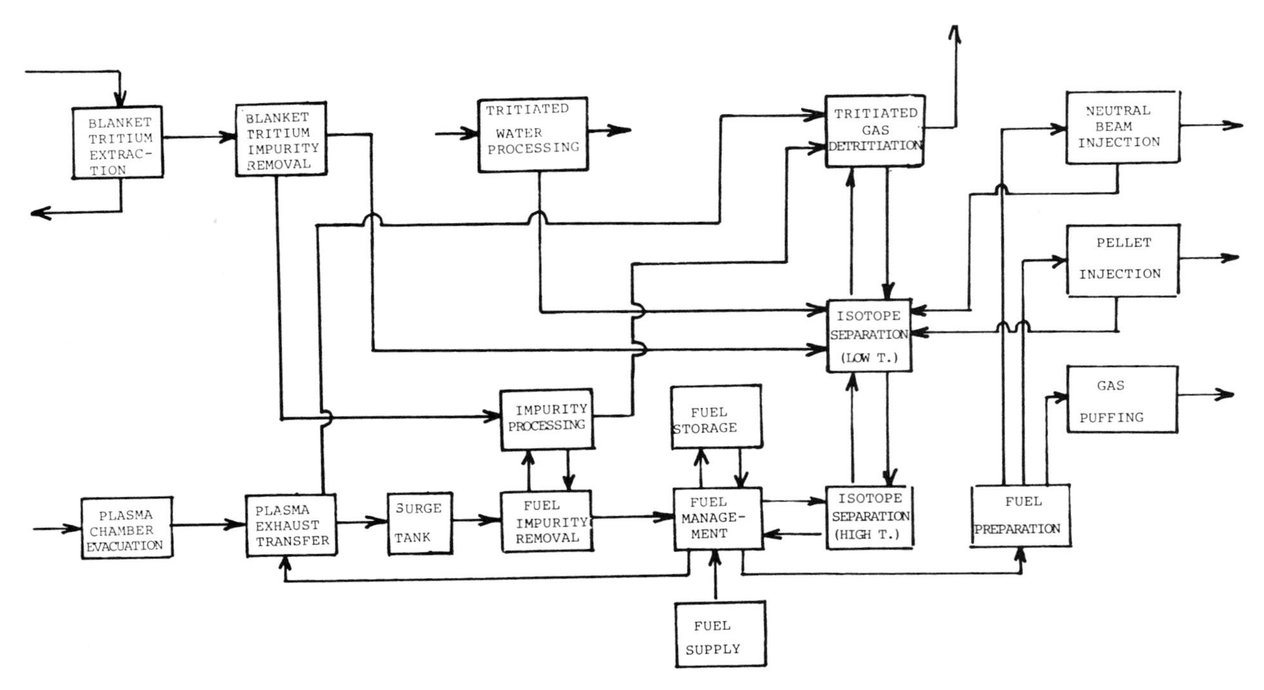

FIGURE 2
NET Tritium Processing Block Diagram

for NET. However, the large number of tritium systems and interconnections on NET requires equal or greater attention to minimizing local inventories and release probabilities.

- The larger tritium inventories required to operate NET, and potential for leakages into secondary containment require more extensive atmosphere and water clean-up systems, able to cope with both routine and non-routine releases.
- The extensive maintenance and replacement of highly activated, tritium containing, nuclear components on NET requires a number of processing circuits for the treatment of liquid and gaseous streams originating from maintenance and solid waste processing.
- The near-term in-service date for the JET tritium system requires that only proven concepts, or those relying on proven technology be employed, whereas NET can observe the results of ongoing experimental tritium activities and projects using tritium before detailed design and construction.

3.2 System Description.

3.2.1 Fuel Cleanup

Two major options are being explored for fuel-cleanup on NET. One employs permeation membranes and cracking of hydrogen impurities without producing tritiated water[23]. The second uses molecular sieve to trap impurities, which are subsequently eluted by a He carrier gas, passed over a precious metal catalyst where hydrogen containing compounds are oxidized to water, frozen out on a 160K cold trap, and reduced in an electrolysis cell[21]. The second concept employs proven approaches, or those with a broad R&D base both in European laboratories and elsewhere in the fusion community. For this reason, the second option is considered as the "reference" option for NET.

The fuel-cleanup systems for JET and the NET reference option are similar in concept. In the NET TMP-pumping option, He would be separated from DT and impurities with a modified version of JET cryoaccumulator panel (ACP) which also provides the function of a backing pump (Fig 3). For the NET compound

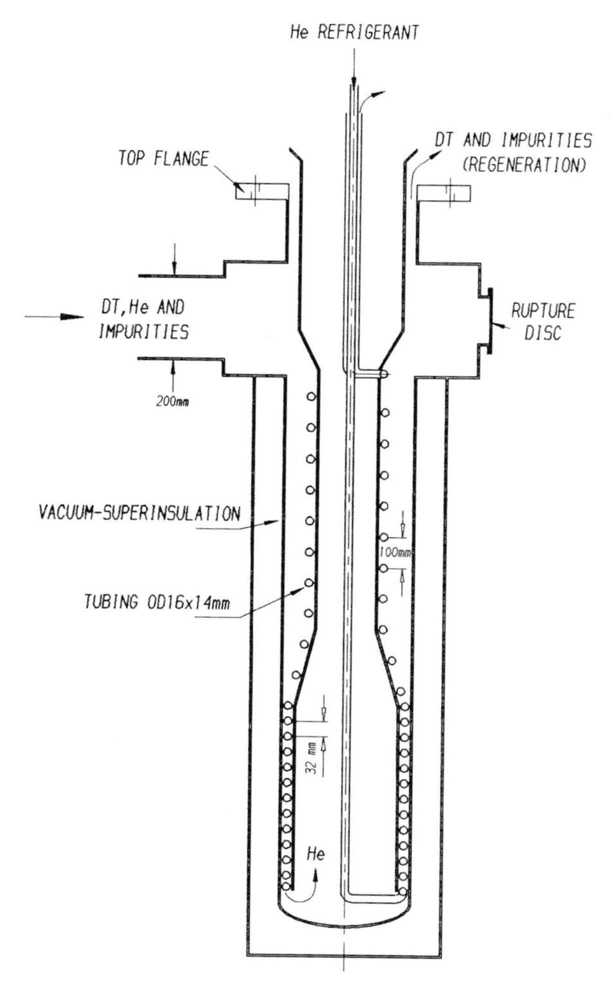

NET COMPOUND CRYOTRANSFER PUMP.

FIGURE 3
NET Cryotransfer Pump Based on JET Design

cryopump option, He is regenerated separately from DT and impurities.

The functions of other components in the JET Active Gas Handling System, and the NET Fuel Cleanup Systems can be compared. Further details of the JET components are described in 24.

Transfer Pumps

JET also uses cryotransfer to boost pressure behind the ACP's, and for transfer to Isotope Separation.

NET proposes mechanical transfer pumps: Large Normetex scroll or dry Rootes blowers (600-1200 m3/h) behind the ACP's or CCP's for He pumpoff, and SRTI circulators for transfer. Both pumps are proven in tritium service. Larger Normetex pumps are generally not suited for high pressure (1 bar+) operation, and would only be used for roughing/backing functions on NET. Metal storage beds are not used on NET for routine gas transfer, where large beds and short cycles would be required. Inventory and potential flammability of fines are undesirable features of metal beds. JET U-beds employ sintered-metal filters to control the migration of fines in the circuit. For JET-type U-beds, tests of the flammability and self-blanketing by atmospheric argon will be carried out later this year.

DT Storage

JET proposes to use cold gas reservoirs in cryogenic systems and Uranium beds elsewhere.

NET will use both approaches where appropriate. Additionally, gas-storage tanks will be used for short term storage (up to 600 g T_2). U-beds will be sized to hold the full inventory of NET ($\simeq$ 2 kg of T_2), compartmentalized to meet local inventory constraints.

Recombiner and Cold-trap

JET proposes an established process applying 800K recombiner followed by a 160 K cold trap to oxidize and trap all combustible (tritium containing) impurities.

NET will use the same approach, however the recombiner may be a two-step operation to avoid producing nitrogen oxides which are detrimental to the electrolyser.

Tritiated Water Reduction

JET proposes to use Fe and U beds at 700 K to reduce tritiated water from the cold trap, followed by adsorption of DT on a low temperature U bed.

NET currently plans to use an electrolysis cell designed for high tritium application[25,26]. Water gas shift is an alternative, particularly if high conversion efficiencies permit once-through operation of the conversion step. Although the high throughputs of highly tritiated water (several hundred cc/day) would consume several kg of U/day if this approach were used as the sole means of tritiated water reduction, NET is interested in the JET Fe and U-bed approach for "dry-out" of the FCU loop prior to maintenance on the loop, and is proposing further experiments on alternatives to U beds for water reduction on NET. JET plans to explore the alternative of the "water gas" reaction for tritiated water reduction, i.e. water reacted on carbon at temperatures in excess of 1200 K.

Cryosorbent for Impurity Removal

JET plans to use single-pass cryogenic trapping in the cryo-transfer step to separate impurities from hydrogen isotopes. The boil-off passes through fine mesh screens at T<30 K. This method reduces impurity content in the hydrogen isotope mixture to the low ppm level[24].

The NET reference approach[21] uses molecular sieve (Zeolites) for which there is a growing experimental base in Europe, Japan, and North America in fusion FCU application. The larger inventories of tritium implied (typically 50 g max for NET sieve beds) is regarded as an acceptable trade-off for the proven separation efficiency and good bed-transfer (sharp break through curves) demonstrated by molecular sieve. However, the possibility of providing a continuous process using the JET approach will be explored in the future.

Isotope Separation

JET proposes to use both cryogenic distillation (CD) and gas chromatography (GC). GC would be used to process torus exhaust (main tritium-containing feed), while CD would be used in addition for PI and NBI feeds which are larger in volume but smaller in tritium concentration.

NET flows are too large for any available GC concept. While GC inventories of tritium are much smaller than for CD, GC also requires batch operation with the added complexity of cyclic operation. The isotopic feed concentrations for NET are subject to rapid fluctuation as different feed streams are processed. Limited smoothing via buffer storage is feasible due to inventory constraints. Multiple column feeds are envisaged. In particular, the processing of large volumes of water coolant on NET for tritium removal will require a tritium "stripping" column. Preliminary design indicates a 4-column train would be preferred for NET, while a 3-column train suffices for JET. Further testing of ISS systems operation to optimise integration is required.

3.2.2 Other Tritium Systems

For JET, the only other system for processing of tritiated flows which has been designed to date is the Exhaust Detritiation System[27]. This system provides negative-pressure containment of tritium in the torus during either routine or emergency operations when the torus vacuum boundary to the containment hall is breached, for example during remote maintenance of the torus internals, or following a window-break. Designed for a flow rate of 500 m^3h^{-1} at one atmosphere, this system is based on catalytic Pd-recombiners for the oxidation of hydrogen isotopes at 150 ^oC, and methane and other tritiated compounds at 500 ^oC, followed by molecular sieve driers for the removal of tritiated water vapour from air by adsorption and isotopic swamping.

NET will employ many similar system-modules for control of oxygenated tritium-bearing atmospheres in containments and plant exhaust.

Other atmospheres on NET which must be processed include:

- Inert gas cleanup, e.g. from glove boxes
- Reducing atmospheres, e.g. from blanket tritium recovery.

Low-flow inert or reducing gas streams can be processed over metal beds (U) to extract tritium and other hydrogen isotopes. Hydrogen isotopes or isotope containing compounds in large flows must be oxidized and tritiated water recovered either with a cold trap or molecular sieve bed.

Tritiated waste-water arisings in JET are primarily due to system malfunction (e.g. water-leak into the torus) and as such are not predictable in quantity, timing and concentration. Waste water will be collected, the final disposal method depending on composition.

For NET, the major tritiated water flows would be from the Aqueous Lithium Salt Blanket tritium recovery process[28], if this is used as the "driver blanket" for NET. Typically 10,000 $l.d^{-1}$ of 10 $Ci.l^{-1}$ water would be processed to extract 10 $g.d^{-1}$ of tritium.

4.0 SUMMARY AND CONCLUSIONS

The extent to which JET tritium and vacuum components can be applied directly to NET is indicated in Table(3).

To complement the information expected from the operation of JET with tritium, a number of additional processes and components not present on JET, but which may be needed on NET, are being developed. Components include:

- Compound Cryopumps and Magnetic Bearing Turbopumps
- Advanced fuel purification systems to reduce tritium inventory, component cycling, and generation of tritium oxide. Such processes may involve using permeation, cold-trapping, and new catalysts.

Table 3
APPLICABILITY OF JET ACTIVE GAS COMPONENTS TO NET

COMPONENT	APPLY DIRECTLY	SCALES CONFIDENTLY	MAJOR DEV'T	NEW CONC.
1) Fueling fast-valve		x		
2) Low-speed pellet inj.		x		
3) High Speed Pellet Inj.				x
4) Turbopump			x	
5) NBI Cryopump			x	
6) Cryo-accumulator			x	
7) HV Isolation Valve			x	
8) Mech. roughing pump			x	
9) DT Storage on U	x			
10) Recombiner	x			
11) 160 K Cold Trap	x			
12) Trit. water reduc.				x
13) Impurity cryotrapping		x		
14) Cryo-distillation		x		
15) Ox. Atm. detrit.	x			
16) Red.Atm. detrit.		x		
17) Trit. water proc.		x		

- Reliable, low inventory, zero-leakage electrolysis cell for highly tritiated water.
- Storage approaches suitable for large inventories of tritium, such as non-U hydrides.
- Tritium extraction and recovery from breeding blankets.
- Tritium compatible diagnostics.

While only a portion of the components required for NET are utilized in JET, the practical experience with multiple, tritium containing systems required to function in a real, fluctuating operational environment will provide a unique addition to the fusion-related experience provided by tritium laboratories operating in Europe, the US, Japan and Canada.

In the course of detailed design of tritium systems for JET, JET staff have identified the need to develop and test components such as bakeable valve actuators and flow control devices which were not commercially available, or were available in a form that did not meet JET requirements. Based on this experience, it is important that NET tritium systems design proceed early to a sufficiently detailed stage to allow time to develop and test such components as may be identified for NET.

Present programs in the associated laboratories are directed towards the production of basic design information required for tritium processes. The study of basic processes over a broad range of conditions represents an efficient way to utilize such laboratory resources. However, the JET experience indicates an opportunity to enhance the value of their contribution: early identification of possible process steps, and testing of these processes and their required components, would help to focus the laboratory efforts more surely on providing results to the engineering projects which need them. A flexible process and component "screening" capability able to provide rapid feedback to designers is required.

A full appreciation of the design requirements (and capabilities) of tritium-handling systems cannot be obtained without closer contact among laboratory experimentalists, design, and operational groups. The knowledge base is dynamic, and process requirements are interwoven among many related systems. A high degree of interaction and coordination of efforts is necessary to ensure the effectiveness of assistance provided by the associated laboratories to JET and NET. For the development of tritium systems for NET, JET provides an invaluable source of experience in design, commissioning, and (in the future) operation. It is important that the steps indicated be taken so that this experience can be combined with the knowledge gained from the Tritium Technology program to produce reliable, safe, timely and cost-effective designs for NET.

ACKNOWLEDGEMENTS

The authors wish to acknowledge the assistance of other members of their respective teams in carrying out the work on which this paper is based, in particular, A.C.Bell, A.Dombra, C.Gordon, J.Gowman, R.Haange, A.Konstantellos, R.Laesser and K.Walter of JET, and Messrs. H.Brunnader, M.Chazalon, F.Fauser, H.Hopman, M.Iseli, D,Leger and D.K.Murdoch and C.Wu of NET.

REFERENCES

1. P.H. and B.E. Keen, Fusion Technology Vol. 11 (1987) 13

2. R. Toschi et al., Fusion Technology Vol. 14 (1988) 19

3. M. Chazalon et al., Fusion Technology Vol. 14 (1988) 82

4. J. Charuau et al., First Experiment on Erosion Dust Measurement in a Tokamak, in: this conference (D01)

5. P. Kupschus et al., The JET Multi-Pellet Injector Launcher - Machine Interface, in: Proc. 12th Symp. on Fusion Engineering Vol. 2 (1987) pp. 780-783

6. S.L. Milora et al., Design of a Repeating Pneumatic Pellet Injector for the Joint European Torus, in: Proc. 12th Symp. on Fusion Engineering Vol. 2 (1987) pp. 784-786

7. K. Sonnenberg et al., High Speed Pellet Development in : Proc. 12th Symp. on Fusion Engineering Vol. 2 (1987) pp. 1207-1210

8. L.L. Lengyel, Fusion Technology Vol. 10 (1986) 354

9. D.D. Schuresko et al., 'Pellet Injector Research at ORNL', Proc. 14th Europ. Conf. Controlled Fusion and Plasma Physics, Vol. IID (1987) 327

10. C.W. Hartman and J.H. Hammer, Phys. Rev. Lett. Vol. 48 (1982) 929

11. E. Usselmann et al., New Piezo Driven Gas Inlet Valve for Fusion Experiments, in: this conf. (C10)

12. E. Usselmann et al., Proc. 13th Symp. Fusion Technology (SOFT) (1984) pp. 24-28

13. M. Huguet et al., Fusion Technology Vol. 11 (1987) 43

14. E. Kussel et al., The Cryogenic Forevacuum System for the JET Active Gas Handling Plant, in: Fission, Fusion and Isotopic Applications, Toronto, 1-6 May, 1988 (to be published).

15. D.K. Murdoch et al., Vacuum Pumping Options for the NET Torus, in: this conf. (G14)

16. J.L. Hemmerich, J. Vac. Sci. Technol. Vol. 6 (1988) 144

17. J.L. Hemmerich et al., The Impurity Processing Loop for the JET Active Gas Handling Plant, in: Fission, Fusion and Isotopic Applications, Toronto, 1-6 May, 1988, (to be published)

18. R. Haange et al., General Overview of the Active Gas Handling System at JET, ibid.

19. J. L. Hemmerich et al., The JET Active Gas Handling System: Concept and Status, in: Proc. 12th Symp. on Fusion Techn. Vol. 2 (1987) pp. 1235-1238

20. P. Dinner et al., Fusion Technology Vol. 14 (1988) 178

21. P.J. Dinner and D.K. Murdoch, Integrated Plasma Exhaust Cryotransfer and Processing for NET, in: Proc. Internat. Symp. on Fusion Nuclear Technology, Tokyo, 10-19 April, 1988 (to be published)

22. D. Leger et al., The NET Integrated Tritium Systems, Proposed Concepts for Safety and Reliability Analyses, in: Proc. Tritium Technology, in: Fission, Fusion and Isotopic Applications, Toronto, 1-6 May, 1988 (to be published)

23. R.D. Penzhorn et al, 'A Catalytic Plasma Exhaust Purification System', ibid

24. J.L. Hemmerich, The JET Active Gas Handling System - Progress Report, in: this conf.

25. A. Rahier et al., Fusion Technology Vol. 8 (1985) 2035

26. D. Leger, Fusion Technology Vol. 8 (1985) 2031

27. A. H. Dombra et al., Exhaust Detritiation System for JET, in: this conf. (G10)

28. M.R. Galley et al., Recovery of Tritium from and Aqueous Lithium Salt Blanket on the NET Fusion Device, in: Fission, Fusion and Isotopic Applications, Toronto 1-6, May 1988, (to be published)

FUSION TECHNOLOGY 1988
A.M. Van Ingen, A. Nijsen-Vis, H.T. Klippel (editors)

THE SEPARATION OF HYDROGEN ISOTOPES MIXTURE ON MODIFIED ZEOLITES BY GAS-SOLID PROCESSES

A. FACCHINI[1], G. PIERINI[2], B. SPELTA[2] and A. VIOLA[3]

[1]Polytechnic of Milan, Italy
[2]CEC Ispra Establishment, Italy
[3]Visiting Scientist of Cagliari University, Italy

Experimental investigations performed at the JRC Ispra Laboratories on the possible use of zeolites and modified zeolites for the separation of hydrogen isotopes mixtures by gas-solid processes are reported. At this end the hydrogen-deuterium separation on Na- and (Ca-Na)-mordenite has been performed by gas-chromatography techniques and the following variables have been explored: flow rate, temperature, pressure and concentrations of adsorbable components, particle size of adsorbing material.
Depending on the column temperature (143 K to 173 K), resolution factors, retention times and separation factors have been evaluated. A better selection capacity of modified zeolites with respect to molecular sieves has been proved.
The theory of the moments has been applied to the adsorption of hydrogen isotopes on mordenite material. Equations are presented for the zeroth, first and second moments of the chromatographic peak in terms of the rate constants for the adsorption process. The moments obtained from experimental chromatographic data have been used to evaluate the adsorption equilibrium constants, the axial dispersion, particle-to-fluid diffusion, interparticle diffusion which may be employed to model a higher scale separation unit.

1. INTRODUCTION

Hydrogen isotopes separation is one of the key steps of the tritium fuel cycle for a thermonuclear fusion reactor. Methods for the separation of hydrogen isotopes in the gaseous state include the gas chromatography which is useful in separation for analytical processes but when employed on a large scale, has the disadvantage of not being continuous and throughput is small.

Several adsorbent materials to separate hydrogen isotopes by gas-solid chromatography have been studied in the last few years[1-4]. In a study by Panchenkov et al.[5] on the static adsorption of hydrogen isotopes, the molecular sieves including Molecular Sieves 4A show a better selection capacity as adsorbent material with respect to other materials such as Zeolites A and X, Mordenites, aluminosilicate catalysts, silica and aluminium oxides.

These results have been confirmed by Conti et al.[6]. Scope of this work performed at the JRC Laboratories of Ispra has been the experimental investigation on the possible use of modified zeolites for the separation of hydrogen isotopes mixtures by preparative gas-chromatography techniques. At this end the hydrogen deuterium mixtures separation on Na- and (Ca-Na) -mordenite has been experimentally investigated.

In recent years, processes widely adopted industrially as to the preparative chromatography techniques and pressure swing parametric pumping, have been devised which permit continuous or semicontinuous operations of chromatographic separation processes.

To model separation units at higher scale, adsorption equilibrium constants and the effect of transport processes are necessary.

The theory of the moments has been applied for the adsorption of hydrogen isotopes on mordenite material. Equations are presented for the zeroth, first and second moments of the chromatographic peak in terms of the rate constants for the adsorption process.

The moment obtained from experimental chromatographic data have been used to evaluate the adsorption equilibrium constants, the axial dispersion particle-to-fluid diffusion, interparticle diffusion.

Work is in progress considering also the separation of $H_2/D_2/HD$ mixtures.

2. EXPERIMENTAL PART

2.1 Apparatus

The main items of the apparatus, schematically shown in Figure 1, are the gas handling and purification systems, the adsorbent column, the cooling box and the gas-chromatograph.

In the gas handling systems, the inlet pressures are controlled by automatic pressure regulators. The gases are purified by passing them through traps filled with Molecular Sieves 4A at liquid nitrogen temperature. The individual flow rates of the gas components, kept constant by means of regulation needle valves, are measured by capillary flow meters. The gas phase composition was controlled by the pressure regulators and determined from the observed individual gas flow-rates.

A three-way valve allows the on-off feed of the different gas mixtures of isotopes, to a six-way sampling valve. The copper column, filled with adsorbing material and of different diameter according to the experimental runs, was situated in a cryostat which allows the temperature to be regulated to ± 0.5 K.

The inlet and outlet pressures of the column were measured by pressure indicators. The gas-chromatograph was a Varian model 1420 with thermal conductivity detector. Helium was used as the carrier gas.

2.2 Materials

The adsorbing material was constituted by Mordenite furnished by the Grande Paroisse-Montoir-France. The Na-mordenite SP and the (Ca-Na)-mordenite SP have been prepared according to the method of Pierini et al.[7]. The crystallinity was checked by X-ray powder diffraction. The Na-mordenite was converted to the (Ca-Na) form by a conventional ion exchange procedure using 0.06 mol dm^{-3} $CaCl_2$ solutions.

The Ca content in (Na-Ca)-mordenite was 70%; all the samples of modified zeolites contain 20% Al_2O_3-binder.

The zeolite was crushed, sieved and the fractions 40-60, 60-80, 80-100 mesh used to fill the copper columns. Special care was taken in the filling procedure to obtain a closed uniform packing.

To activate the zeolite surface a scrupulous dehydration has been performed, because the presence of residual water alters the selectivity in a very significant way. Therefore, after filling, the columns were dehydrated. This dehydration was obtained by heating the columns flushed by a stream of dry and purified helium in an oven with temperature increments of 2 K min^{-1} up to 693 K and maintaining fixed this temperature for 48 hours.

The complete dehydration was proved comparing the net weight loss of the column with the water content in the zeolites which was derived from a previous thermogravimetric analysis (Figure 2). The column ends were sealed with a plug of glass wool treated by dimethychlorosilane. The gases utilized in the experimental runs and furnished by SIO SpA were helium N56 (purity 99.9996%), hydrogen N56 (purity 99.995%) and deuterium (purity 99.4%). Commercial helium-hydrogen isotopes mixtures of stated composition were used for calibration purposes.

2.3 Experimental procedures

Chromatographic and breakthrough curves were measured in the apparatus schematically reported in Figure 1. When the chromatographic curves were measured, the pulses of hydrogen isotope mixtures were introduced by means of the six-way sampling valve in the helium carrier flowing through the column kept at prefixed working temperatures.

The pressure associated with a specific run was taken as the average of the measured pressures at the inlet and outlet points of the

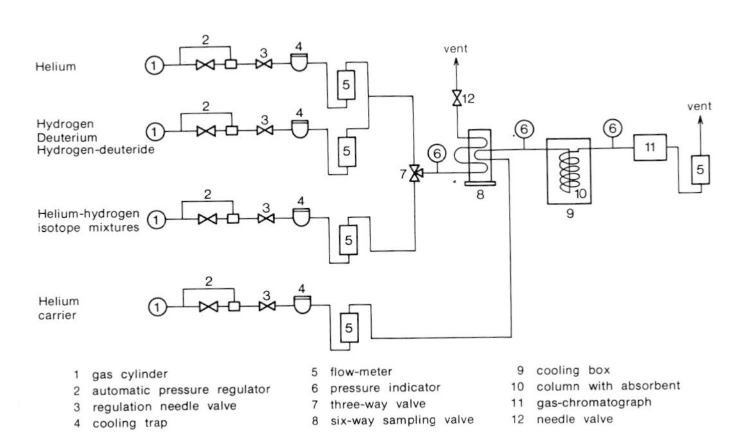

FIGURE 1
Schematic flow sheet of the absorption apparatus.

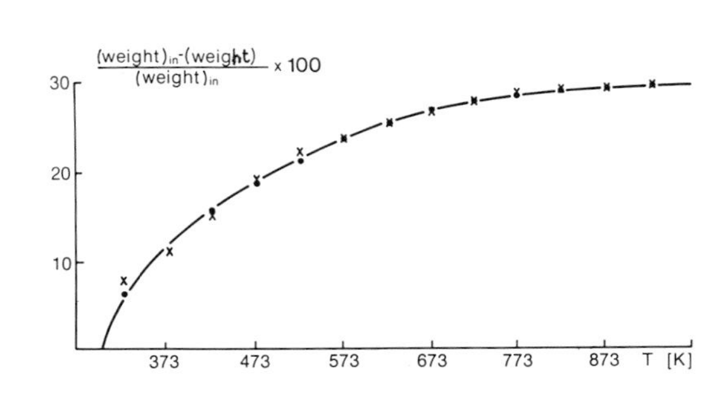

FIGURE 2
Water content removal from (*) Na- and (x) Ca-Na mordenite vs temperature measured by thermogravimetric analysis.

column. The maximum Δp measured for most of the runs was about 6% of the inlet pressure.

The measurement of the breakthrough curves was started by replacing the stream of helium, flowing at the beginning through the column, by the stream of helium-hydrogen isotope mixtures. In order to prevent changes in flow rate of the helium-hydrogen isotope mixture after switching the sampling valve the pressure in the feed line was adjusted by means of the needle valves to the value of the carrier gas pressure of the column inlet.

3. RESULTS AND DISCUSSION

In Figures 3-6 are reported the chromatograms obtained in the separation of H_2/D_2 on the Na- and (Na-Ca)-mordenite at the operative conditions indicated in the same figures. It should be mentioned that between 173 and 143 K with flow rates 25 cm^3 min^{-1} a separation is obtained on 1 m column (Figure 3), but the resolution is only complete at 143 K with helium carrier flow rates up to 80 cm^3 min^{-1} as shown in Figure 4.

Figure 5 shows as the separation of H_2/D_2 mixture can be obtained also at 163 K with helium carrier flow-rate 15 cm^3 min^{-1} and with 2.40 m column filled with Na-mordenite.

In Figure 6 the better performance of the (Na-Ca)-mordenite with respect to the Na-mordenite is shown.

Adsorption and desorption curves at 77 and 157 K, here not reported, were measured.

At 77 and 52.5 kPa the adsorption capacities of Na-mordenite for H_2 and D_2 are 3.65 and 3.95 m mol g^{-1}, respectively; at same operative conditions the adsorption capacities of (Na-Ca)-mordenite are 3.7 and 4.3 m mol g^{-1}, respectively.

At 157 K and 52.5 kPa the adsorption capacities of Na-mordenite for H_2 and D_2 are 0.12 m mol g^{-1} and 0.17, respectively; at same operative conditions on the (Na-Ca)-mordenite one obtains 0.23 and 0.24 m mol g^{-1}, respectively.

In all runs the adsorption and desorption processes are completely reversible.

Finally, to model separation units at higher scale adsorption equilibrium constants, rate constants and interparticle diffusivities for the H_2 and D_2 on the Na-mordenite are given.

3.1 Adsorption equilibrium constants

According to Eq.(17A), derived by definition of the first-moment as reported in Appendix, in Figure 7 and 8 the first reduced moment (left member of Eq.(17A)) is plotted vs L_c/v for H_2 and D_2, respectively. The values of the adsorption equilibrium constant, K_A, deduced from the slope of the straight lines from the

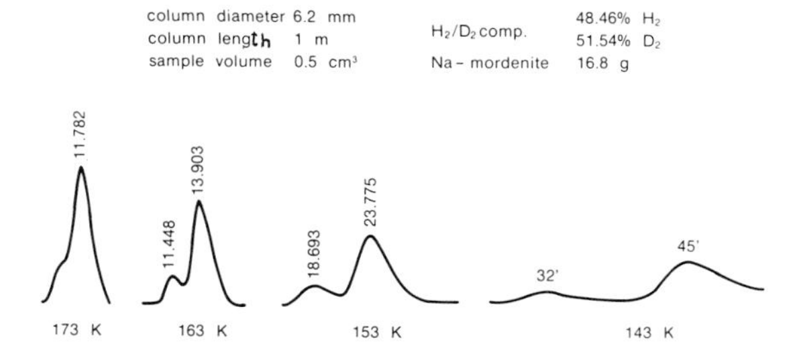

FIGURE 3
H_2/D_2 separation on Na-mordenite as a function of temperature at constant carrier flow rate (25 cm^3/min) in a column 1 m long.

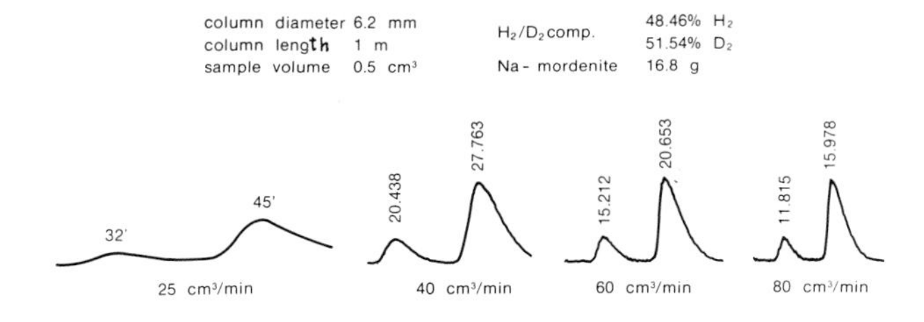

FIGURE 4
H_2/D_2 separation on Na-mordenite as a function of carrier flow-rate at constant temperature (143 K) in a column 1 m long.

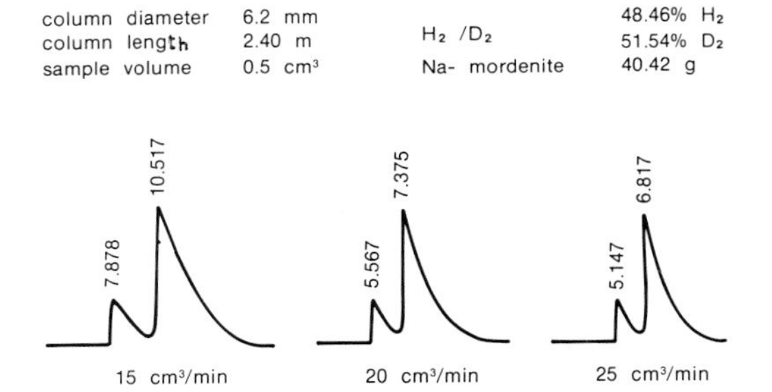

FIGURE 5
H_2/D_2 separation on Na-mordenite as a function of carrier flow-rate at constant temperature (163 K) in a column 2.4 m long.

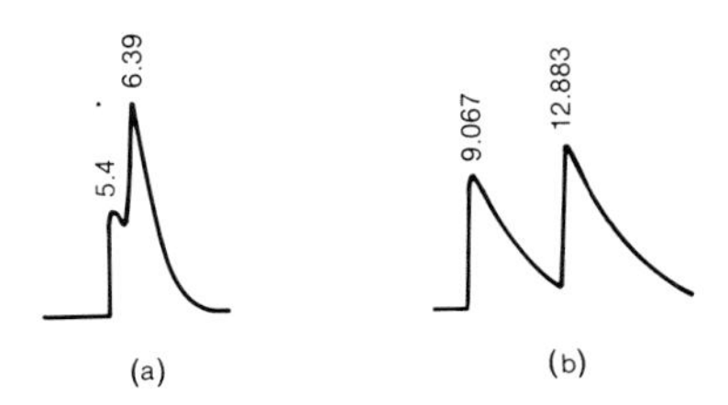

FIGURE 6
Comparison of the chromatograms obtained employing Na-mordenite (a) and (Ca-Na)-mordenite (b) in a column 2 m long, at 173 K, with 20 cm^3/min of helium carrier flow, for H_2/D_2 mixture (79.5% H_2; 20.5% D_2).

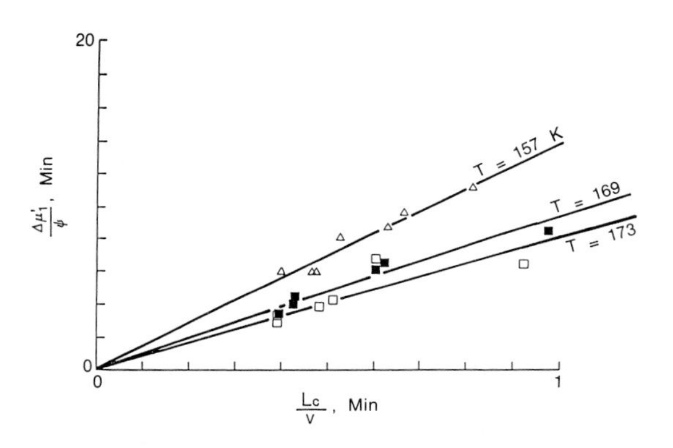

FIGURE 7
Reduced first-moments for hydrogen on Na-mordenite as a function of L_c/V at prefixed temperatures.

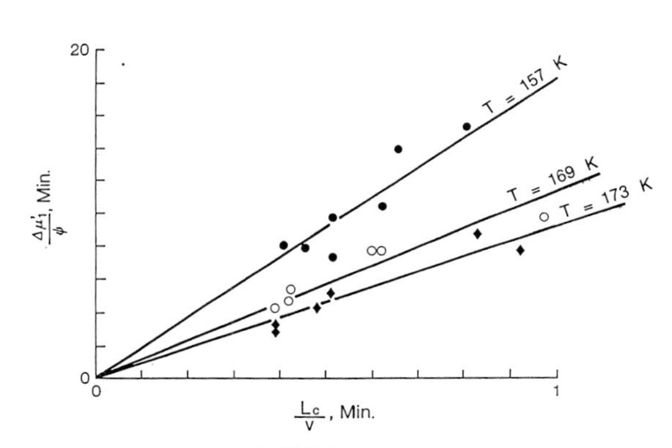

FIGURE 8
Reduced first-moment for deuterium on Na-mordenite as a function of L_c/V at prefixed temperatures.

from the property of adsorbent particles are reported in Table 1. The heats of adsorption, q, for H_2 and D_2 were calculated from the above K_A values by utilizing the Vant'off equation:

$$K_{A(T)} = A\exp\left[-\frac{q}{R_g T}\right] \quad (1)$$

Indeed, by plotting the values of log K_A vs 1/T, the q/R_g values for H_2 and D_2 can be evaluated from the slope of the straight lines as reported in Figure 9. These values of the adsorption heat are reported in Table 1: they show the physical nature of the adsorption process. In the same table the adsorption heats of H_2 and D_2 for

TABLE 1
Adsorption equilibrium constants, K_A, for H_2 and D_2 on Na-mordenite

Component	T (K)	K_A (cm^3/g)
H_2	157	15.7
	169	10.77
	173	9.09
D_2	157	20.52
	169	12.6
	173	10.45

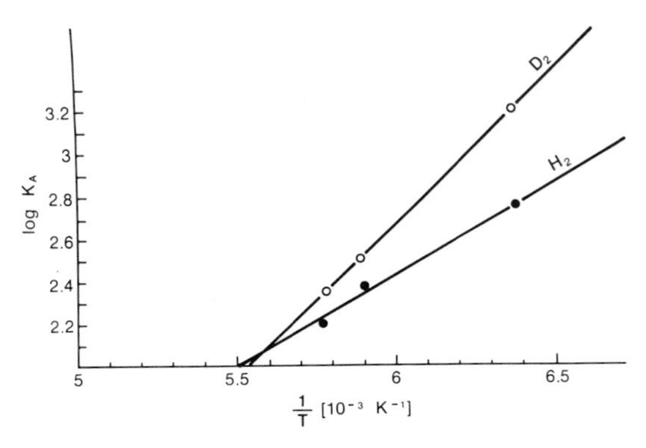

FIGURE 9
Log K_A vs 1/T to determine q/R_g for H_2 and D_2.

different adsorbent materials are reported from the literature. It can be seen that the ratio q_{D_2}/q_{H_2} is much higher for Na-mordenite than for the others.

3.2 Axial dispersion coefficient

Rearranging Eq.(13A) as follows, and as K_A (and δ_0) are known from the analysis of first absolute moment

$$\frac{\mu'_2 - t_0/12}{2(L_c/v)} = \frac{E}{\alpha}(1+\delta_0)^2 \frac{1}{v^2} + (\delta_i + \delta_a) \qquad (2)$$

the axial dispersion coefficient, E_A, can be obtained from the slope of the straight lines in Figures 10 and 11. These coefficients are reported in Table 3.

3.3 Diffusion coefficient in the macropores

Adding Eq.(15A) and (16A) and rearranging one obtains:

$$\frac{\delta_i+\delta_a}{\frac{1-\alpha}{\alpha}} = \varepsilon_m\left[\frac{\varrho_p K_A^2}{\varepsilon_m K_{ads}} + \frac{1}{15}\frac{\varepsilon_m}{1-\varepsilon_M}\frac{r_0^2}{D_i}\left(1+\frac{\varrho_p K_A}{\varepsilon_m}\right)^2\right] +$$
$$+ \frac{\varepsilon_M^2}{15}\left[1+\frac{\varepsilon_m}{\varepsilon_M}\left(1+\frac{\varrho_p K_A}{\varepsilon_m}\right)\right]^2 \frac{5R^2}{K_f R} +$$
$$+ \frac{\varepsilon_M^2}{15}\left[1+\frac{\varepsilon_m}{\varepsilon_M}\left(1+\frac{\varrho_p K_A}{\varepsilon_m}\right)\right]^2 \frac{1}{D_a} R^2 \qquad (3)$$

For the evaluation of the mass-transfer coefficient K_f this following relationship is given:

$$N_u = 2 + 0.6\, R_e^{1/2}\, S_c^{1/3} \qquad (4)$$

where:

$$N_u = \frac{2 R K_f}{D_{AB}}$$

$$R_e = \frac{2 R v}{\nu}$$

$$S_c = \frac{\nu}{D_{AB}}$$

At low Reynolds numbers Eq.(4) shows that Nu = 2 and hence

$$D_{AB} = K_f R \qquad (5)$$

Eq.(3) can also be written

$$\frac{\delta_i + \delta_a}{\frac{1-\alpha}{\alpha}} = A + B R^2 \qquad (6)$$

The term B of Eq.(6), which is the slope of the straight lines in Figure 12 and 13, has as only variable the diffusion coefficient in the macropores D_a. These coefficients are reported in Table 4.

In the term A of Eq.(6), which represents the intercept of the straight line with the axis of the ordinates (Figures 12 and 13), the adsorption rate constant and the diffusion coefficient in the micropores are included. Since K_f is given by Eq.(5) it is possible to determine K_{ads} and the effective internal diffusion coefficient.

TABLE 2
Adsorption heats of H_2 and D_2 on Na-mordenite and on different adsorbent materials as reported in the literature

Adsorbent material	Adsorption heat (kJ/mol)	
Na-mordenite	7.1	11.7
Al_2O_3	6.3	6.7
SiO_2	3.8	4.6

TABLE 3
Effective axial dispersion coefficient E_A for H_2 and D_2 on Na-mordenite

Component	T (K)	E_A (cm^2/s)
H_2	157	0.048
	169	0.055
	173	0.066
D_2	157	0.037
	169	0.038
	173	0.048

TABLE 4
Diffusion coefficient in the macropores, D_a, for the H_2 and D_2 on the Na-mordenite

Component	T (K)	D_a (cm^2/s)
H_2	157	9.4 10^{-4}
	168	4.56 10^{-4}
	173	2.56 10^{-4}
D_2	157	4.6 10^{-4}
	168	7.5 10^{-4}
	173	4.8 10^{-4}

3.4 Breakthrough curves

The rate constants K_{ads} and D_a determined from analysis of chromatographic data have been used to predict the breakthrough curves experimentally obtained. The agreement between the experimental data and predicted curves is good.

4. CONCLUSIONS

From the examination of the experimental work the following conclusions can be drawn.

1. Employing the modified zeolites Na and (Ca-Na)-mordenite SP good separations of H_2/D_2 mixtures can be achieved by gas-solid exchange at

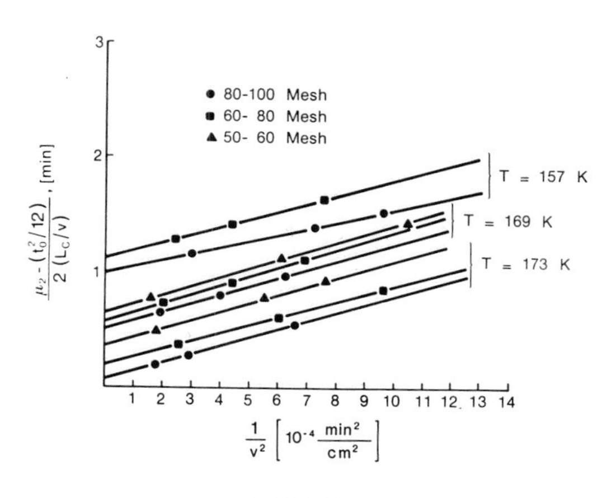

FIGURE 10
Second moments for hydrogen on Na-mordenite as a function of $1/v^2$ at prefixed temperatures and for different particle sizes.

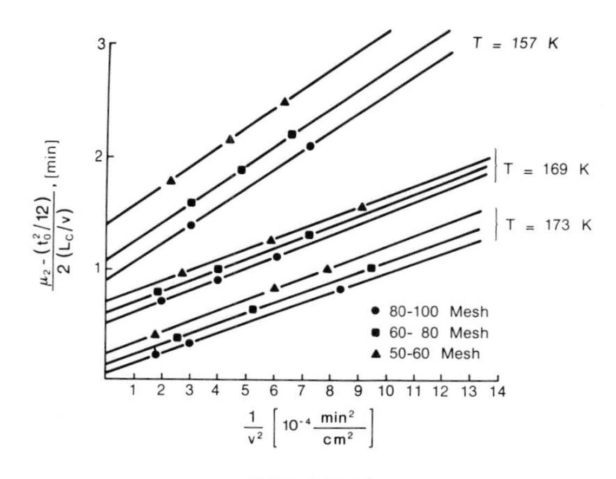

FIGURE 11
Second moments for deuterium on a Na-mordenite as a function of $1/v^2$ at prefixed temperatures and for different particle sizes.

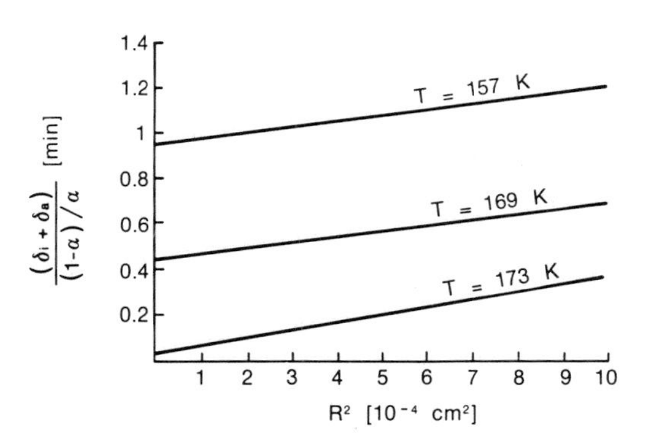

FIGURE 12
Dependence of $\frac{(\delta_i+\delta_a)}{(1-\alpha)/\alpha}$ min vs R^2 for hydrogen.

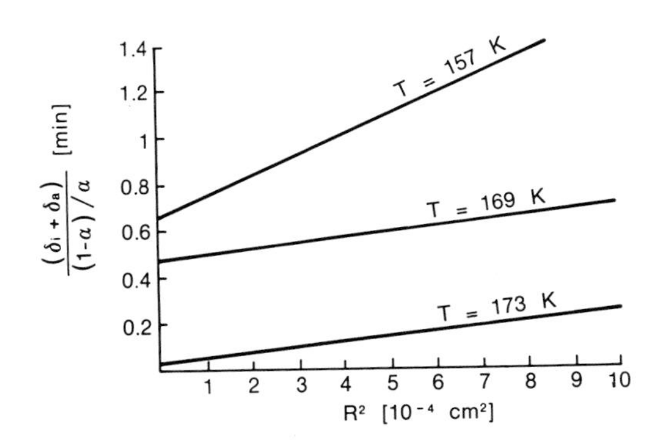

FIGURE 13

Dependence of $\frac{(\delta_i+\delta_a)}{(1-\alpha)/\alpha}$ [min] vs R^2 for deuterium.

a temperature of 160-170 K much better than those carried out at 90-120 K obtained by various authors[6,8] by using molecular sieves.

2. The values of the adsorption equilibrium constants as the rate constants and the interparticle diffusivity have been determined for H_2 and D_2 on Na and (Ca-Na)-mordenites useful for modelling gas-solid processes as the pressure swing parametric pumping, that operates in a continuous way. This has been obtained by applying the theory of moments to the experimental results.

5. APPENDIX

The set of differential equations describing the balance of an adsorbable substance in a chromatographic column (identical to that of an adsorber) is:

$$\frac{E_A}{\alpha}\frac{\partial^2 c}{\partial z^2} - v\frac{\partial c}{\partial z} - \frac{\partial c}{\partial t} - \frac{3D_a}{R}\frac{1}{\alpha}\frac{\alpha}{}\left(\frac{\partial c_i}{\partial r}\right)_{r=R} = 0 \quad (1)$$

$$\frac{D_a}{\beta}\left(\frac{\partial^2 c_i}{\partial r^2} + \frac{2}{r}\frac{\partial c_i}{\partial r}\right) - \frac{\partial c_i}{\partial t} - \frac{\rho_p}{\beta}\frac{\partial c_{ads}}{\partial t} = 0 \quad (2)$$

$$\frac{\partial c_{ads}}{\partial t} = k_{ads}\,(c_i - c_{ads}/K_A) \quad (3)$$

$$D_a\left(\frac{\partial c_i}{\partial t}\right)_{r=R} = K_f\,(c - c_i) \quad (4)$$

$$\frac{\partial c_i}{\partial r} = 0 \quad \text{at} \quad r = 0 \quad \text{for} \quad t > 0 \quad (5)$$

$$c = c_0 \quad \text{at} \quad z = 0 \quad \text{for} \quad 0 \leqslant t \leqslant t_{oA} \quad (6)$$

$$c = 0 \quad \text{at} \quad z = 0 \quad \text{for} \quad t \geqslant t_{oA} \quad (7)$$

By using the Laplace-Carson transform, Kubin[9] has solved the system of Eqs.(1) to (7) obtaining the transform s(p,z) of the concentration c(z,t), as published by Schneider et al.[10].

It is possible to obtain directly from s(z,p), explicit expression for the moments of the chromatographic curves. The nth absolute moment, μ_n, of the function c(z,t) is defined as:

$$\mu'_n = \frac{m_n}{m_0} \quad (8)$$

where:

$$m_n = \int_0^\infty t^n\, c(z,t)\, dt \qquad (n = 0,1,2,3,\ldots) \quad (9)$$

The nth central moment μ_n is defined as:

$$\mu_n = \left(\frac{1}{m_0}\right)\int_0^\infty (t - \mu'_1)^n\, c(z,t)\, dt \quad (10)$$

The evaluation of the moments from a knowledge of s(p,z) is based on

$$m_n = (-1)^n \lim_{p\to 0} \frac{d^h}{dp^n}\left[s(z,p)/p\right] \quad (11)$$

By using these equations, the first absolute and the second central moments can be expressed as[10]:

$$\mu'_1 = \left(\frac{L_c}{v}\right)(1+\delta_0) + \frac{t_{oA}}{2} \quad (12)$$

$$\mu_2 = \left(\frac{2L_c}{v}\right)\left[\delta_i + \delta_a + \frac{E}{\alpha}(1+\delta_0)^2\left(\frac{1}{v^2}\right)\right] + \left(\frac{t_{oA}^2}{12}\right) \quad (13)$$

$$\delta_0 = \frac{(1-\alpha)}{\alpha}\left[\epsilon_M + \epsilon_m\left(1 + \frac{\rho_p K_A}{\epsilon_m}\right)\right] \quad (14)$$

$$\delta_i = \frac{1-\alpha}{\alpha}\,\epsilon_m\left[\frac{\rho_p}{\epsilon_m}\,\frac{K_A^2}{K_{ads}} + \frac{1}{15}\,\frac{\epsilon_m}{1-\epsilon_m}\,\frac{\bar{r}_0^2}{D_i}\cdot\left(1+\frac{\rho_p}{\epsilon_m}K_A\right)^2\right] \qquad (15)$$

$$\delta_a = \frac{1-\alpha}{\alpha}\,\frac{\epsilon_M^2}{15}\left[1+\frac{\epsilon_m}{\epsilon_M}\left(1+\frac{\rho_p K_A}{\epsilon_M}\right)\right]^2\cdot\left[\frac{5}{K_f R}+\frac{1}{D_a}\right]R^2 \qquad (16)$$

Experimental chromatographic curves of the effluent from the adsorber can be used to evaluate the moments μ_1' and μ_2. Indeed, from Eq.(12) it follows that:

$$\frac{\Delta\mu_1' - \frac{t_{oA}}{2}}{\frac{1-\alpha}{\alpha}} = \rho_p K_A\left(\frac{L_c}{v}\right) \qquad (17)$$

where:

$$\Delta\mu_1' = \mu_1' - \mu_{1,inert}' \qquad (18)$$

The first absolute moment for an inert not adsorbable component can be evaluated by Eq.(12) with $K_A = 0$ and $t_{oa} = 0$ or it can be measured experimentally.

The first absolute and the second moment for an adsorbable component can be evaluated by means of the gas-chromatographic curves. Indeed, the time t_R, which represents the time when the maximum concentration is registered at the point of detection, is defined by:

$$\left.\frac{dc\,(L,t)}{dt}\right|_{t=t_R} = 0 \qquad (19)$$

The time t_c of the center of gravity or the mean of the peak is defined by:

$$t_c = \mu_1' = \int_0^\infty tc\,(L,t)\,dt \qquad (20)$$

The time when the maximum of the immediate concentration distribution in the column passes through the point of detection is defined by:

$$\left.\frac{dc\,(t_R, z)}{dz}\right|_{z=L} = 0 \qquad (21)$$

For large retention times, the difference between t_c and t_R depending on the longitudinal diffusion can be neglected[11]; under these conditions:

$$t_R = t_c = \mu_1' \qquad (22)$$

The second central moment (variance) depends on all factors characterising the column, its filling and the given compound transport through the column.

As the standard deviation $X = \sqrt{\mu}$ defines the width of the peak[12], the second moment μ_2 represents the square of the half-width of the peak evaluated at 0.607 of the height.

NOMENCLATURE

c	concentration of the adsorbable gas in the interparticle space (mol cm^{-3})
c_i	concentration of the adsorbable gas in the pore space (mol cm^{-3})
c_{ads}	concentration of the adsorbed gas per unit weight of adsorbent (mol g^{-1})
c_o	concentration of the adsorbable gas in the input concentration function (mol cm^{-3})
D	interparticle-diffusion parameter defined as $D_c\left(\frac{1-\alpha}{\alpha}\right)^{-1}\left(\frac{L_c}{v}\right)R^{-2}$
D_a	effective intraparticle diffusion coefficient defined as the ratio of the diffusional flux through unit area of the geometrical surface of the particle to the negative of the intraparticle concentration gradient (cm^2 s^{-1})
D_{AB}	binary molecular diffusivity (cm^2 s^{-1})
D_K	Knudsen diffusivity (cm^2 s^{-1})
E	external diffusion parameter defined as $D\,K_f^{-1}R^{-1}$
E_A	effective axial dispersion coefficient, defined as the ratio of the axial diffusional flux through unit area of the total cross section of the column to the negative of the interparticle concentration gradient (cm^2 s^{-1})
k_{ads}	adsorption rate constants (cm^3 g^{-1} s^{-1})
k_f	mass transfer coefficient (cm s^{-1})
K_A	adsorption equilibrium constant (adsorption coefficient) defined as $(c_{ads}/c)_{eq}$ (cm^3 g^{-1})

m (n=0,1,2) = integrals

M	molecular weight of the diffusing substance (g mol^{-1})
Nu	Nusselt number for mass transfer
Re,Sc	Reynolds number, Schmidt number
p	variable in the Laplace-Carson transformation
r	length coordinate in the spherical particle of adsorbent, measured center of particle (cm)
$\bar{r}_o$	average radius of the pores (cm)
R	radius of the spherical particle of adsorbent (cm)
R_g	gas constant
s	Laplace-Carson transform of c(z,t)
t	time (s)
t_{oa}	time of duration of the injection of the absorbable or inert gas in chromatography (s)
T	absolute temperature (K)
v	linear velocity of the carrier gas in the interparticle space (cm min^{-1})
L_c	length coordinate of the bed of adsorbent, measured from the inlet side (cm)

Greek letters

α	interparticle void fraction in the absorbent bed
β	intraparticle void fraction (internal porosity) of the adsorbent
μ'_n, μ_n	n^{th} absolute and central moment of the chromatographic curve, respectively (min)
μ'_{inert}	first absolute moment for the non-adsorbable gas (min)
ρ_p	apparent particle density (g cm^{-3})
ν	kinematic viscosity
ψ	defined as $(1-\alpha)\alpha^{-1}$
ε_M	macropore void fraction
ε_m	micropore void fraction

REFERENCES

1. H.A. Smith and P.P. Hunt, J. Phys. Chem., 64 (1960) 383.
2. C. Cercy and F. Botter, Bull. Soc. Chim. France (1965) 3383.
3. S. Akhtar and H.A. Smith, Chem. Rev., 64 (1964) 261.
4. D.L. West and A.L. Marston, J. Am. Chem. Soc., 86 (1964) 4731.
5. G.M. Panchenkov, A.M. Tolmachev and T.W. Zotova, Zh. Fiz. Khim., 38 (1964) 1361.
6. M.L. Conti and Lesimple, J. Chromatogr., 29 (1967) 32.
7. G. Pierini, M. Dworschak, B. Spelta and E.F. Vansant, Processing plant of exhaust plasma of fusion reactors by using zeolite compounds, Patent application No. 86902, Luxembourg, May 1987.
8. R. Vogd, KFA - Rep. ISSN 0366-0885 (February 1988).
9. M. Kubin, Collection Szechoslov. Chem. Commun., 30 (1965) 1104.
10. P. Schneider and J. Smith, AIChE Jour., 14, 5 (1968) 762.
11. E. Kucera, J. Chromatogr., 19 (1965) 237.
12. H. Purnell, Gas Chromatography, J. Wiley and Sons, New York (1962) 104.

FUSION TECHNOLOGY 1988
A.M. Van Ingen, A. Nijsen-Vis, H.T. Klippel (editors)
Elsevier Science Publishers B.V., 1989

DEVELOPMENT WORK FOR THE TRITIUM CLEANUP SYSTEM OF THE TRITIUM LABORATORY KARLSRUHE

Günter NEFFE, Eike HUTTER, Peter SCHIRA, and Ulrich TAMM

Kernforschungszentrum Karlsruhe GmbH, Hauptabteilung Ingenieurtechnik, Postfach 3640, D-7500 Karlsruhe 1

Retention of the impurities, mainly O_2, N_2, NH_3, CO_2 and CH_4, on hot uranium powder, backed by a liquid nitrogen cooled molecular sieve bed, was found to be an adequate process for the gas to be treated at the Tritium Laboratory Karlsruhe (TLK). In the KfK test facility VERDI (a German acronym for impurity retention) getter experiments were performed in the temperature range of 500 °C to 700 °C with different gas mixtures of H_2 carrier gas and the impurities (1% concentration each). As expected, O_2 is converted quantitatively to uranium oxide at these temperatures, but it substitutes C and N_2 in uranium compounds which leave the apparatus as CH_4 and N_2. The cracked CO_2 and NH_3 give rise to the formation of CH_4 and N_2, respectively. Therefore, the getter process has to be split up into operating a uranium bed at 500 °C for trapping oxygen and operating a second uranium bed at about 900 °C for hydrogen compound cracking and impurity retention. A ceramic container to be used at this temperature is in the testing phase. The molecular sieve at 77 K retains < 0.5 vpm of the impurities present at an inlet concentration of 100 vpm. Experience gathered with VERDI has been referred to in working out the concept of the TLK tritium cleanup system.

1. INTRODUCTION

The Tritium Laboratory Karlsruhe (TLK) will be equipped with a hydrogen cleanup system as part of the tritium infrastructure in order to be able to purify and recycle the tritium or mixtures of hydrogen isotopes returned from the experiments.

Gettering the impurities, mainly O_2, N_2, NH_3, CO_2 and CH_4, on hot uranium powder, backed by a liquid nitrogen cooled molecular sieve bed, was found to be an adequate process for the relatively small quantities of gas to be treated at the TLK.

The chemistry of the getter process is known; however, development work on the apparatuses is still required in order to be able to keep the residual impurities in the hydrogen below a few vpm using the said getter process.

A facility for testing the individual cleaning stages was built on an industrial scale and operated with inactive hydrogen (Fig. 1).

In the VERDI test facility getter experiments were performed in the temperature range

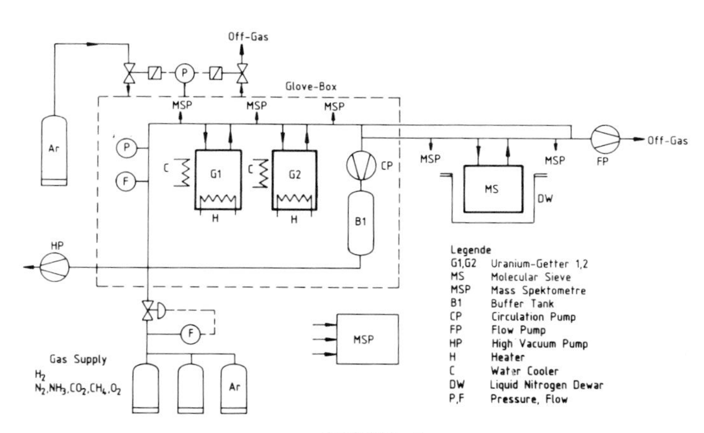

FIGURE 1
Simplified flowsheet of VERDI.

of 500 °C to 700 °C using different gas mixtures: H_2 carrier gas with the impurities (1% each) O_2, N_2, NH_3, CH_4, CO_2 as single constituents, and H_2 carrier gas with mixtures of the constituents (1% each) O_2, N_2 CH_4, CO_2 as well as O_2, N_2, CH_4, NH_3 and O_2, N_2, NH_3 and CH_4, NH_3.

As expected, all of the O_2 was retained on uranium down to less than 1 vpm at 500 °C under all experimental conditions, but O_2 substitutes

C and N_2 in the uranium compounds, and C and N_2 leave the getter stages mainly as CH_4 and N_2.

The cracked CO_2 and NH_3 give rise to the formation of CH_4 and N_2, respectively. This is a confirmation of the soundness of the concept which consists in splitting up the getter process into operating a uranium bed at 500 °C for trapping mainly oxygen and operating a second uranium bed at about 900 °C for cracking the hydrogen compounds and retaining the impurities.

A ceramic container with a low permeation rate to be used in the 900 °C U-bed is in the testing phase.

Two single-stage getter beds connected in series allowed a multi-chamber design to be studied.

A multi-chamber getter apparatus will reduce the bypassing uncleaned gas stream which results in a better decontamination factor.

2. STUDY AND DEVELOPMENT PROGRAM

The experiments to be performed in VERDI will serve to develop a process for cleaning hydrogen isotope mixtures from N_2, O_2, CO_2, CO, CH_4, NH_3 and He and to define the process parameters as well as the operating data of a facility operating on this process.

Evidence was to be provided that

- the process gas with the specified impurities (1% each of the constituents present in H_2) can be cleaned down to 1 vpm total impurities;
- a procedure is feasible which allows the process gas of any composition to be cleaned within the specified limits while avoiding that interactions of the constituents present in the gas or uranium compound phases will completely or partially cancel out the cleaning effect;
- the uranium capacity for the specified impurities can be exploited in an optimum manner by reasonable process control.

In this report, the first part of the test results will be presented which are the basis of the concept proposed for the TLK tritium cleanup system.

3. THE EXPERIMENTAL FACILITY

The VERDI facility together with its two hot metal getter stages is housed in a glovebox with argon atmosphere kept at overpressure (Fig. 2). The third stage, the cryogenic fine cleaning stage is located outside the box. The experimental facility is supplied from a gas cylinder station where premixed process gas is stored.

FIGURE 2
Test facility with the stainless steel and alumina containers.

The cleaning stages can be operated either individually or in any combination.

A getter cleaning stage consists of a water-cooled external container and the electrically heated getter container proper which is made of stainless steel (Fig. 3) or ceramic (Fig. 4). It accommodates the uranium getter metal.

The total amount of uranium is distributed among compartments of 400 g capacity each. The

FIGURE 3
Stainless steel container with one compartment insert.

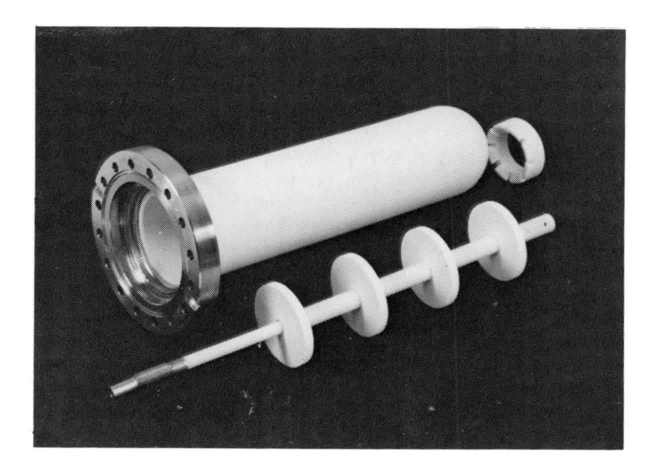

FIGURE 4
Alumina container.

number of the compartments can be adapted to the requirements of the experiments; for the time being, only single-compartment containers are used. A two-compartment container is simulated by series connection of two single-compartment containers.

The compartments are separated from each other by sintered stainless steel filters of 20 µm pore diameter. However, the end filters are always 5 µm in pore diameter.

The adsorber stage containing about 1 kg molecular sieve 5A is placed in a Dewar flask and cooled with liquid nitrogen. The process gas is passed through the cleaning stages at 1 bar pressure and 10 Nl/h flow. The cleaned gas is removed by a rotary van pump.

If necessary, the gas can be recirculated several times for cleaning.

A quadrupole mass spectrometer with a detection limit for impurities of 0.01 vpm is used to determine the composition of the gas.

4. EXPERIMENTAL RESULTS

4.1 Retention of O_2

In several series of experiments oxygen was applied to the uranium getters both alone and together with other constituents such as N_2, NH_3, CH_4 and CO_2. The getters were always kept at an operating temperature of 500 °C, and some of them had been loaded with impurities in previous experiments.

It has been confirmed in all experiments that oxygen was completely retained down to less than 1 vpm by formation of uranium oxide.

The gas constituents applied together with oxygen were either not retained at all (CH_4) or to a little extent only (N_2, NH_3).

Some of the carbides and nitrides formed in previous experiments by gettering of mainly CO_2 and NH_3 were reoxydized while CH_4 and N_2 were released.

4.2 Retention of CH_4

In a first step, methane retention by formation of uranium carbide (UC) was studied in the temperature range of 500 to 600 °C (Fig. 5) and a strong dependence on temperature was found of the concentration at the outlet of the getter stage. Above 600 °C CH_4 was completely retained (down to 1 vpm). This value is in agreement with information in the literature[1] according to which uranium powder reacts with CH_4 already at 625 ° and above.

Then, in a second step, a long-term experiment of about 43 hours duration was performed at this temperature level (Fig. 6). During this

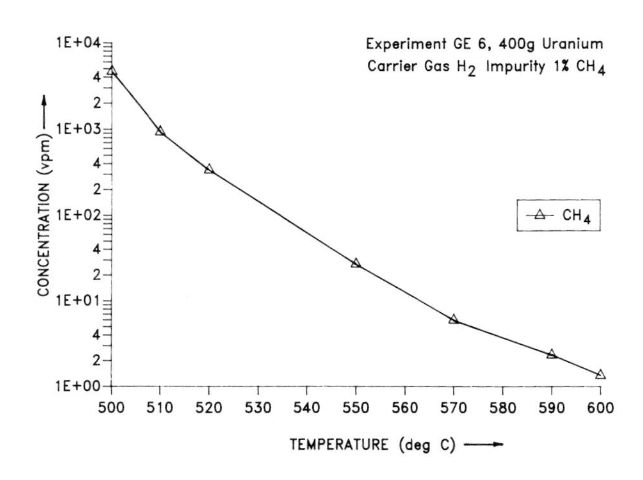

FIGURE 5
Retention of CH_4 as a function of temperature.

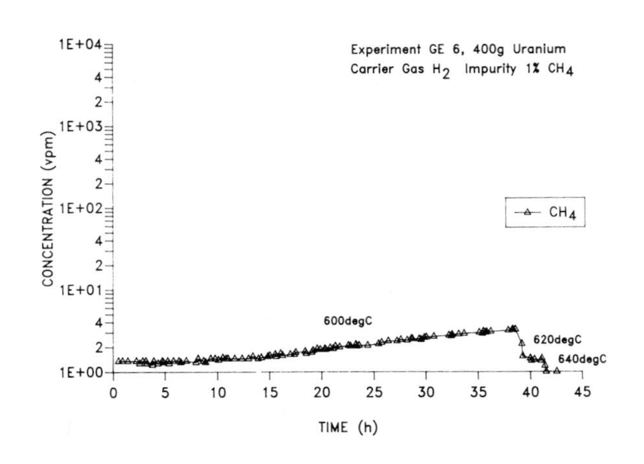

FIGURE 6
Long-term retention of CH_4.

period a slight increase was observed in the concentration of non-gettered CH_4 up to 3 vpm which is attributable to the very low retention capacity of uranium at 600 °C. By raising the temperature in two steps up to 640 °C the concentration was diminished again to 1 vpm.

It is expected that the concentration at the getter outlet will continue to rise with further loading of the uranium getter. The precise concentration plot and the related capacity of uranium as a function of the temperature will have to be determined in a later long-term experiment.

4.3 Retention of N_2 and NH_3

The retention of the nitrogen and ammonia constituents by formation of uranium carbide (U_2N_3 + UN_2) was studied in the temperature range of 450 to 700 °C (Fig. 7).

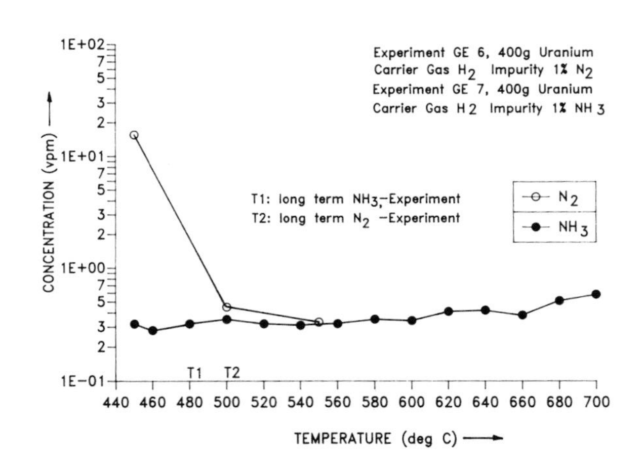

FIGURE 7
Retention of NH_3 and N_2 as a function of temperature.

For NH_3 no dependence on temperature of the initial concentration was found; complete retention was achieved as early as at 450 °C (< 1 vpm).

For N_2 a strong dependence on temperature was observed at low temperatures and the value of 1 vpm was attained at about 500 °C. The temperature values measured for complete retention of NH_3 and N_2 agree well with the values indicated in the literature[1] according to which a violent reaction takes place between uranium powder and NH_3 at 400 °C and between uranium powder and N_2 at 520 °C.

In two long-term experiments no rise was found in the initial concentration as had been found for methane. For NH_3 the concentration at 480 °C remained constant at 0.5 vpm over 40 hours; for N_2 the concentration was likewise constant at $\leq$ 0.5 vpm for 50 hours at 500 °C.

4.4 Retention of CH_4, NH_3

In another experiment made on the already NH_3 loaded uranium getter the retention was studied of a mixture consisting of NH_3 and CH_4 (Fig. 8).

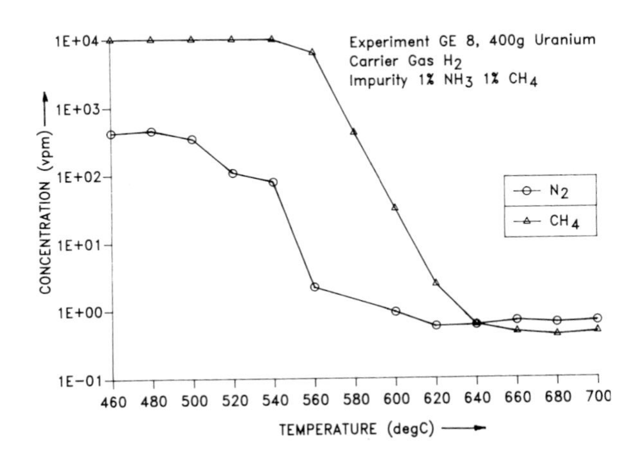

FIGURE 8
Retention of N_2 and CH_4.

No temperature dependence was observed for released NH_3; the concentration remained below 1 vpm over the entire range of temperatures.

By contrast, some N_2 was released and, again, its strong dependence on temperature was observed. Complete retention of N_2 (< 1 vpm) occurred only at about 640 °C and above.

CH_4 was not gettered at all at low temperatures (up to 540 °C) and retained completely only at 640 °C and above.

The temperature values for complete retention, due to the higher loading of uranium, were higher by several ten degrees than in the experiments involving the individual constituents.

5. CONCLUSIONS AND FURTHER DEVELOPMENT

The process engineering concept of hydrogen cleaning, as implemented in the VERDI experimental facility, is suited for use at the Karlsruhe Tritium Laboratory (TLK) within the scope of the specifications elaborated. This has been demonstrated in the experiments described here.

Splitting cleanup into two stages at different temperature levels for oxygen elimination at 500 °C, on the one hand, and nitrogen and carbon elimination at temperatures up to 900 °C, on the other hand, has proved to be the appropriate approach (Figs. 5, 7, 8).

The adsorption stage, operated with liquid N_2, is suited for subsequent fine cleaning of the process gas down to 1 vpm per constituent. However, it should serve solely to validate the result of cleanup.

Operation at temperatures up to 95 K is recommended in order to minimize the H_2 inventory.

Continuation of work at the experimental facility is reasonable and desirable on several grounds:

- Development of the apparatus design, e.g. ceramic container for a second getter stage for CH_4 and NH_3 separation at 900 °C.
- Investigation of process variants for processing different gas compositions, e.g., exhaust gas with high CH_4 and He portions as generated in experiments on the fuel cycle of a fusion reactor.
- Studies of accident conditions, e.g., component failure.

The experiments with the ceramic container will start in the near future by testing radiation heating.

REFERENCE

1. J.J. Katz, E. Rabinowitch, The Chemistry of Uranium (Mc Graw-Hill, 1951).

FUSION TECHNOLOGY 1988
A.M. Van Ingen, A. Nijsen-Vis, H.T. Klippel (editors)

THE TRITIUM SYSTEMS TEST ASSEMBLY: OVERVIEW AND RECENT RESULTS

John R. Bartlit and James L. Anderson
Los Alamos National Laboratory, Mail Stop C348, Los Alamos, New Mexico, 87545, USA

The fusion technology development program for tritium in the U. S. is centered around the Tritium Systems Test Assembly (TSTA) at Los Alamos National Laboratory. The TSTA is a full-scale system of reactor exhaust gas reprocessing for an ITER-sized machine. That is, TSTA has the capacity to process tritium in a closed loop mode at the rate of 1 kg per day, requiring a tritium inventory of about 100 g. The TSTA program also interacts with all other tritium-related fusion technology programs in the U. S. and all major programs abroad. This report summarizes the current status, results and interactions of the TSTA. Special emphasis is given to operations in May/June using large compound cryopumps that completed the fuel loop integration of all TSTA subsystems for the first time.

1. INTRODUCTION

Objectives of this project are to develop and demonstrate the fuel cycle for processing the reactor exhaust gas (unburned deuterium and tritium plus impurities), and the necessary personnel and environmental protection systems, for the next generation of fusion devices.

Related tasks include new component development and testing, operations under emergency and off-normal conditions, gathering of long-term reliability data, and operator training. To complete this mission, two major types of activities are under way at the facility. The first is the periodic operation of the integrated fuel reprocessing loop with a gradually increasing level of system integration and tritium inventory. The second major activity is the tritium testing of new components in experiments in separate gloveboxes. The latter are often done in collaboration with workers from other U. S. fusion programs and programs abroad.

2. INTEGRATED LOOP OPERATIONS

2.1 September 1986

In September 1986, a major five-day, round-the-clock operation of the TSTA integrated processing loop was carried out. The goals of the run were achieved as planned, and included the following accomplishments:

- integrated flow processing through all process systems except the compound cryopumps;
- the removal of impurities (up to 7% nitrogen) by the fuel cleanup system (FCU), though without continuous impurity addition or on-line regeneration and recovery of the captured impurities;
- the addition of 20 g of tritium to the flow loop, raising the in-process inventory to 50 g;
- the verification of improved flow control measures added to the loop;
- the production and analysis of tritium gas of 99.93% purity in the cryogenic distillation columns of the isotope separation system (ISS); and

- the training and use of personnel in different operating assignments to broaden staff experience and versatility.

2.2 December 1986

The September run was followed by a similar five-day, round-the-clock operation in December 1986. Goals and accomplishments of this run were the following:

- further development and improvement of flow control in the loop;
- elimination of unwanted interactions between the isotope separation system and effluent gas detritiation system;
- measurement of long time constants (several hours) for the isotope separation system to reach steady state after control changes are made; and
- changing from our previous 12-hour shifts to 8-hour shifts, with a two-man night shift making minimal process changes.

2.3 June/July 1987

The next major operation of the integrated loop occurred in June 1987. The goals were:

- to increase the inventory of tritium to 100 g;
- to demonstrate the removal of helium-3 from tritium decay by two techniques--gettering out the tritium on uranium beds before dumping the helium, and stripping out the tritium in the distillation columns before dumping the helium;
- to provide training for operating personnel;
- to produce and analyze high purity tritium (>99.9%) in the ISS; and
- to verify long-term, continuous addition of impurities (about 1% N_2 and 0.1% CH_4) and their removal by cold (77 K) molecular sieve beds.

The inventory was raised to 90 g and all goals except the last two were achieved before operations were concluded prematurely by a broken shaft on the commercial cryogenic refrigerator used to cool the cryogenic distillation columns. An orderly shutdown under off-normal conditions, without doses to personnel or releases to the environment, was achieved following the failure in off-the-shelf technology unrelated to tritium handling.

Following a repair and upgrade of the refrigerator by a factory representative, round-the-clock operations were resumed in July, for a period of 5 days, to add to the 4 days in June. The tritium inventory was increased to 102 g, and the last two run goals were achieved. A tritium sample of 99.98% purity was prepared in the ISS.

Impurities were added to the fuel stream continuously for 53 hours and removed by the FCU to below the limit of detection of Raman spectroscopy (low ppm levels). The impurities added were a mixture of N_2/CH_4 in 9/1 ratio and comprising 1% of the D-T fuel stream flow. The effectiveness of impurity removal was confirmed by the lack of any plugging in the cryogenic still that received flow from the FCU. On-line regeneration of the captured impurities was not part of the July operation.

2.4 February/March 1988

On-line regeneration and decomposition of the captured impurities was first done in operations of February/ March 1988.

The run was notable, not only for the eventual success, but for two abortive starts, during the weeks of February 7 and 21, that were halted by plugging lines with condensable impurities. The problem impurities were not impurities injected in the system, but were impurities that had previously leaked in or formed in the system. Nitrogen and water appeared to be the main constituents.

The problems were solved by removing all gas from the piping, heating all components, purging with helium, and evacuating over several days, before recharging the gas to the system. Although the problems caused delays, the successful recovery proved the wisdom of the TSTA design, which is one of versatility in piping and pumping capabilities.

Technical highlights of the run were:

- stable operation of the ISS over many days;
- 36 hours of addition of 0.9% N_2/0.1% CH_4 to the fuel stream and the complete removal (to the limit of detection) of these impurities using the molecular sieve beds at 77 K;[1]
- on-line regeneration of these beds, followed by catalytic oxidation of the impurities to form water, complete capture of the water by freezeout from a flowing stream, and decomposition of the captured water by reaction with hot uranium to recover tritium for reuse and
- production and offloading of 11.8 g of better than 99.98% pure tritium from the ISS into a shipping container.

2.5 May/June 1988

The most recent loop operation was a seven-day run in late May and early June. This run for the first time integrated all TSTA subsystems, including a compound cryopump.

This run involved continuous injection of fuel impurities (nitrogen and methane), cryopumping of the fuel mix of deuterium and tritium plus impurities, on-line regeneration of the cryopumps, removal of the impurities and recovery of the contained D-T, and continuous separation of the hydrogen isotopes by cryogenic distillation. All operations were successful. Cryopump results are included below.

Several weeks of tests on the cryopumps (VAC) without tritium were done to prepare for the integrated run in May. Despite minor problems, the tests achieved the calibration and readiness of VAC for system integration.

Two pumps were tested--one designed by Brookhaven National Laboratory (BNL) and one by Lawrence Livermore National Laboratory (LLNL). Torus pumping tests with these pumps were performed with D-T, D_2, H_2, He, N_2, and their mixtures.

The nominal pumping surface areas are:

	BNL	LLNL
DT condensing chevron	2400 cm^2	9000 cm^2
He sorbing panel (charcoal)	1300 cm^2	11000 cm^2

The principal results were the following:

a) Pumping speeds for the two pumps for pumping D-T at the Torus were similar despite large differences between the pumps in the areas of the cryopumping surfaces.

b) The BNL pumping speeds (in loop operation) were in the range of 2-10% of the full plasma exhaust gas flow for an ITER-size reactor. If three trains (pumping, regenerating, and recooling trains) of the VAC system are needed, 30-150 pumps will be needed.

c) Impurity effects on the pumping speed for D-T (and D_2) were investigated with 1% N_2 and 1-25% He. Little effect was seen.

d) The effectiveness of separating He/DT on both panels of the BNL pump was measured: DT gas regenerated from the condensing chevron showed no He contamination; He gas regenerated from the charcoal panel showed an acceptable level of DT contamination (10-25% DT).

3. PERFORMANCE TESTS AND MAINTENANCE OF THE ISOTOPE SEPARATION SYSTEM

Isotope separation at TSTA is done by means of an interlinked, four-column cryogenic distillation system. Individual columns of the system were operated in five-day, round-the-clock runs in October and December of 1987 and April of 1988. The purpose was to measure fundamental design parameters, such as liquid holdup in columns and the height equivalent to a theoretical plate (HETP).[2]

Repairs were made on instrumentation outside the columns, but inside the vacuum jacket. This was done by lowering the vacuum jacket. Tritium contamination inside the jacket was negligible.

4. TRITIUM TESTS OF NEW COMPONENTS

Five new components of interest to the fusion program have been tested for performance and tritium compatibility. The first four components, and organizations collaborating with TSTA, were:

a) a commercial zirconium-iron getter material for detritiation of inert glovebox atmospheres (Ontario Hydro Research Division);

b) a piezoelectric valve for fuel gas injection at the Tokamak Fusion Test Reactor (TFTR) at Princeton;

c) a ceramic electrolysis cell for recovering D-T in the reprocessing of plasma exhaust gas (Japan Atomic Energy Research Institute); and

d) a palladium-alloy membrane diffuser for purification of plasma exhaust gases (JAERI).

All four proved to be attractive and tritium-compatible items. Detailed results are given elsewhere.[3, 4, 5]

The fifth component, tested more recently, has only preliminary results available. This is a tritium-proof-of-principle pellet injection designed and built by Oak Ridge National Laboratory. Over 80 tritium pellets have been produced to date and accelerated with hydrogen propellant to speeds up to 1.4 kilometers per second. Work is continuing.

5. TESTS IN THE EXPERIMENTAL CONTAMINATION STUDIES LABORATORY (XCS)

The TSTA includes a small laboratory dedicated to studies on the contamination and decontamination of equipment, surfaces, and atmospheres exposed to tritium. In collaboration with outside programs, two new projects have been started.

1. With TFTR and the Idaho National Engineering Laboratory (INEL), studies are in progress on the effect of catalyst temperature and moisture pre-loading in the driers on the efficiency of a gas detritiation system.
2. With the Joint European Torus (JET), studies are in progress on the contamination effects and residual contamination of remote maintenance tools (welders and cutters) after use in the torus. Both studies are incomplete at this writing.

6. JAPAN ATOMIC ENERGY RESEARCH INSTITUTE JOINS TSTA

In June 1987, an international collaborative agreement was signed by the Japan Atomic Energy Research Institute (JAERI) and the U. S. Department of Energy. This agreement, Annex IV to the U. S./Japan Agreement on Fusion Energy, calls for the joint funding and joint operation of TSTA by DOE and JAERI for the next five years (till 1992), thereby doubling the size of the program. Under the agreement, JAERI will attach a four-person staff to TSTA. The first-year staff arrived at TSTA in mid June, 1987, in time to participate in the June loop operations.

7. FUTURE PLANS

7.1 JAERI-Designed Fuel Cleanup System

As part of the JAERI/TSTA collaboration, a JAERI-designed Fuel Cleanup System will be fabricated in Japan and installed in the TSTA fuel loop. The design will use palladium diffuser technology for producing a stream of hydrogen isotopes free of impurities. Installation is scheduled for 1989.

7.2 Breeding Blanket Interface

Since the inception of TSTA in 1976, a breeding blanket interface at the facility has been in the long-term plan. In 1987, the initial steps to define and develop this were taken in collaboration with blanket experts at Argonne National Laboratory.[6]

The work includes examining three leading blanket concepts (liquid lithium, solid lithium oxides, and aqueous lithium salt) to define the composition and flowrate of the tritium-bearing product stream from the extraction process in the blanket. In a fusion reactor, this product stream, after appropriate initial processing, will join the plasma exhaust stream to make up the full reactor fueling stream.

The goal is to add equipment at TSTA to demonstrate breeding blanket product processing in conjunction with plasma exhaust gas processing. The definition of stream compositions, flowrates, and processing technologies is under way. The schedule for installation at TSTA is uncertain, though at least several years away, and may depend on developments with the International Thermonuclear Experimental Reactor (ITER).

SUMMARY

The fusion program in the U. S. in tritium research, development, and demonstration has its focus at the integrated fusion fuel processing facility at TSTA. The overall TSTA program is multi-faceted, including integrated fuel reprocessing loop operations, new component testing, contamination experiments, and personnel training. The program is characterized by increasingly closer ties among elements within the U. S. and with programs abroad.

ACKNOWLEDGMENTS

The list of contributors to this work is too large and wide-ranging to include here. Organizational involvements have been indicated in the text. Thanks to all.

REFERENCES

1. R. S. Willms, "Recent Operating Results at the Tritium Systems Test Assembly Using Molecular Sieve at 77 for Purifying a Fusion Fuel Process Stream," International Symposium on Fusion Nuclear Technology, Tokyo, Japan (April 10-15, 1988).

2. T. Yamanishi, R. H. Sherman, et al., "Single Column and Two-Column Cascade H-D-T Distillation Experiments at TSTA," Third Topical Meeting, Tritium Technology in Fission, Fusion and Isotopic Applications, Toronto, Canada (May 1-6, 1988). To be published in Fusion Technology.
3. N. Kherani, W. Shmayda, and R. A. Jalbert, "Tritium Removal from Inert Gases Using Zr, Fe," Proc. 12th Symposium on Fusion Engineering, Monterey, CA, (October 12-16, 1987).
4. D. O. Coffin, S. P. Cole, and R. C. Wilhelm, "A Tritium-Compatible Piezoelectric Valve for the Tokamak Fusion Test Reactor," 3rd Topical Meeting, Tritium Technology in Fission, Fusion, and Isotopic Applications, Toronto, Canada (May 1-6, 1988). To be published in Fusion Technology.
5. R. V. Carlson, K. E. Binning, S. Konishi, H. Yoshida, and Y. Naruse, "Results of Tritium Experiments on Ceramic Electrolysis Cells and Palladium Diffusers for Application to a Fusion Reactor Fuel Cleanup System," Proc. 12th Symposium on Fusion Engineering, Monterey, CA (October 12-16, 1987).
6. D. K. Sze, P. Finn, R. Clemmer, J. L. Anderson, J. R. Bartlit, Y. Naruse, and H. Yoshida, "The Role of a Blanket Tritium System on the Fusion Fuel Cycle," International Symposium on Fusion Nuclear Technology, Tokyo, Japan, (April 10-15, 1988).

FUSION TECHNOLOGY 1988
A.M. Van Ingen, A. Nijsen-Vis, H.T. Klippel (editors)
Elsevier Science Publishers B.V., 1989

TRITIATED TARGETS MANUFACTURED BY CEA/VALDUC (FRANCE)

P. BOUCQUEY and P. GIROUX

Commissariat à l'Energie Atomique - Centre d'Etudes de VALDUC
Boîte postale n° 14 - 21120 IS SUR TILLE (FRANCE)
Tel. (33) 80.35.13.05.

ABSTRACT

The CEA manufactures neutron targets for its own needs and for outside customers. These targets are used in conjunction with linear accelerators to produce fast neutrons. This paper explains what a tritiated target is, and reviews their most useful applications. The method of manufacture of targets and quality assurance are briefly touched upon.
Finally, their uses and results obtained in the fields of neutron therapy, activation analyses and research studies are described, and main customers worldwide are listed.

1. INTRODUCTION

For more than ten years now the CEA has been manufacturing neutron targets to meet its own requirements. Its products are also marketed to French and other companies, agencies and authorities.
When bombarded with deuterons from electrostatic accelerators, theses targets generate *fast neutrons* in accordance with the following nuclear reactions :

$$ {}^{2}_{1}H + {}^{3}_{1}H \longrightarrow {}^{1}_{0}n + {}^{4}_{2}He \qquad E_n = 14.1 \text{ MeV} \quad E\alpha = 3.5 \text{ MeV} $$

$$ {}^{2}_{1}H + {}^{2}_{1}H \longrightarrow {}^{1}_{0}n + {}^{3}_{2}He \qquad E_n = 2.45 \text{ MeV} \quad E\alpha = 0.82 \text{ MeV} $$

Tritiated targets yield Reaction 1 ; deuterated targets, Reaction 2.
The high cross-section for these reactions (4.65 barns at 125 keV) enables neutron fluxes of between 10^8 and 10^{13} $n.cm^{-2}s^{-1}$ to be obtained.
This report describes the products obtained at CEA/VALDUC and the methods employed. The various applications of neutron targets are also reviewed.

2. PRODUCTS

The various standard targets produced by CEA/VALDUC are :

Small standard targets dias.	28,38 and 49 mm
Rotating targets	146 mm
Annular targets	152 mm
Cylindrical targets	128 mm O.D.
Lancelot targets	365 mm.

The quantity of tritium trapped on the target ranges from 1 Ci for the smallest to 1 000 Ci for the Lancelot targets.
CEA/VALDUC can produce targets of any other type and size to order.

3. DESCRIPTION OF NEUTRON TARGET

A neutron target (Fig. 1) consists of a metallic substrate of well defined mechanical and geometrical characteristics, part of which is first coated with a deposit of metal and finally impregnated with tritium or deuterium.

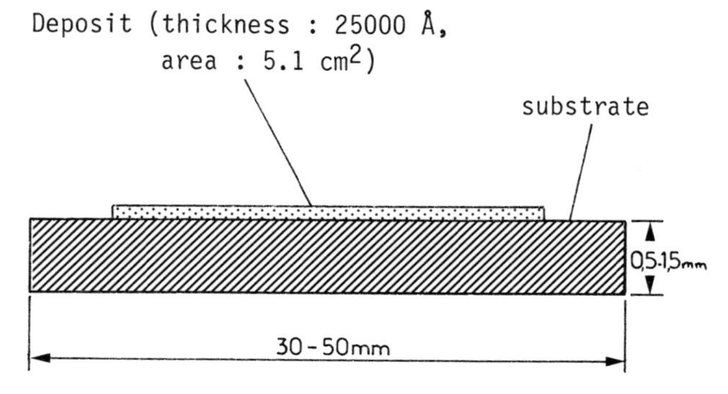

Fig. 1 : A small tritiated target

The various components of the target are not chosen at random. Years of manufacturing experience, research and development have gone into product definition and optimization.

3.1. The substrate

The substrate has to be a good conductor of heat so has to ensure effective cooling of the target. It has to combine acceptable mechanical properties with minimal sensitivity to hydrogen diffusion. Numerous materials can be employed (silver, gold, platinum, copper, aluminium, etc). As a rule CEA/VALDUC chooses to use *copper*.

3.2. Metal deposited

The metal with which the substrate is coated *by chemical vapour deposition* (CVD) has to be able to form a hydride which is both stable and able to absorb substantial quantities of tritium. It has to be resistant to sputtering under the impact of deuterons. Last but not least, it has to be tightly adherent.

The metals most frequently employed and reviewed in the literature are titanium, zirconium and the rare earths in general. In choosing between them, it is helpfull to refer to the criteria just mentioned.

3.2.1. Adhesion of deposit

Titanium and zirconium are equivalent in terms of adherent. The rare earths, on the other hand, form poorly adherent deposits and require an intermediate coating of cadmium.

3.2.2. Stability of the hydride

As evidenced by the table (Fig. 2) the hydrides formed are very stable. The dissociation pressure of the least stable (TiH_2) is as low as 8.10^{-3} torr at 400° C.

TEMPERATURE (°C)	100	200	300	400
Ti H2	10^{-10}	10^{-5}	$2\ 10^{-3}$	$8\ 10^{-3}$
Zr H2		$2\ 10^{-10}$	10^{-6}	$1.2\ 10^{-4}$
Y H2		$2\ 10^{-15}$	$8\ 10^{-10}$	$5.8\ 10^{-8}$

Fig. 2 : Hydride dissociation pressures (torr)

3.2.3. Resistance to sputtering

Investigation has shown that, in terms of ability to withstand sputtering, titanium compares favourably with the majority of the other elements studied, such as aluminium, magnesium, yttrium and zirconium.

3.2.4. Hydrogen content of hydrides

As evidenced by the table (Fig. 3), titanium is an excellent getter for hydrogen and its isotopes. Its absorption capacity (0.151 mole of hydrogen per cubic centimetre of hydrides) is far greater than that of any of the other elements listed.

HYDRIDE	DENSITY	$N_H \times 10^{-2}$
La H_2	5.14	7.27
Li H	0.775	9.68
Li D	0.883	9.8
La $H_{2.63}$	5.28	9.8
Er H_2	8.36	9.9
Zr H_2	5.61	12.05
Gd H_3	6.58	12.3
Hf H_2	11.37	12.62
Hf D_2	11.69	12.83
Y H_3	3.94	12.86
Y D_3	4.13	13.03
Er H_3	7.63	13.4
U H_3	10.91	13.58
U D_3	11.11	13.67
Ti H_2	*3.78*	*15.1*
Ti D_2	*3.94*	*15.14*

Fig. 3 : Specific gravities and hydrogen contents (mol/cm^3) of various hydrides

As we have seen, *titanium* is an excellent candidate for the production of tritiated targets, provided the working temperature is less than 300° C. For higher temperatures, rare earths should be used because of the much greater stability of the hydrides. On the debit side, the deposits are much less adherent.

4. MAKING NEUTRON TARGETS

The production of neutron targets is a five-stage process, consisting of :

- construction and surface preparation of the substrate ;
- titanium metallization ;
- impregnation with deuterium or tritium ;
- inspection/testing of the end-product ;
- packing.

4.1. Substrate construction and surface preparation

Substrates are machined from *very high purity* (OFHC) *copper*. Each substrate is degreased and buffed with felt and alumina to assist the subsequent adherence of the deposit. Finally, the last traces of contamination are removed by thorough rinsing in distilled water in ultra-sonic cleaning tanks.

4.2. Titanium metallization

Metallization with titanium is by CVD, using an electron gun. Clean high vacuum of below 10^{-6} torr is vital to the quality of the deposit and is obtained by means of cryogenic or turbomolecular pumps. To enhance the quality of the vacuum, materials with minimum tendency to outgassing are employed in the vacuum chamber.

Substrate *temperature* for the metallization operation, optimized for adherence and porosity of the deposit, is a compromise (Fig. 4) :

- at low temperature, a multiplicity of tiny islets form and grow on the deposit, yielding a coating of low adherence but high porosity ;
- high temperature favours the surface diffusion of the titanium atoms, with the formation of fewer, but much bigger, points of attachment, so that adherence is enhanced but porosity much reduced.

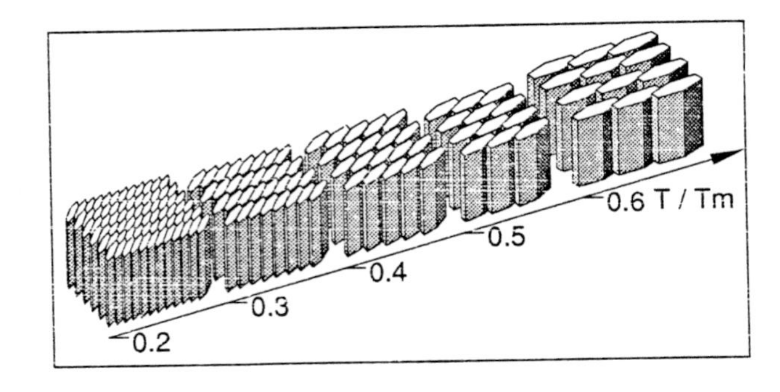

Fig. 4 : Morphology v temperature of deposit

4.3. Impregnation with deuterium or tritium

The impregnation stage determines the final quality of the target. High quality requires a homogeneous, near-stoichiometric, titanium hydride. This can be obtained by assisting diffusion of the deuterons or tritons into the crystal structure. The Ti-H phase diagram (Fig. 5) maps the conditions governing the existence of the various crystal structures, i.e. alpha-titanium (HC), beta-titanium (CC), gamma-titanium (CFC) and zeta-titanium (QFC). Since diffusion of hydrogen into titanium takes place at the tetrahedral sites, the need is to assist the formation of the *beta-phase*, with its greater number (6) of tetrahedral sites per atom of titanium.

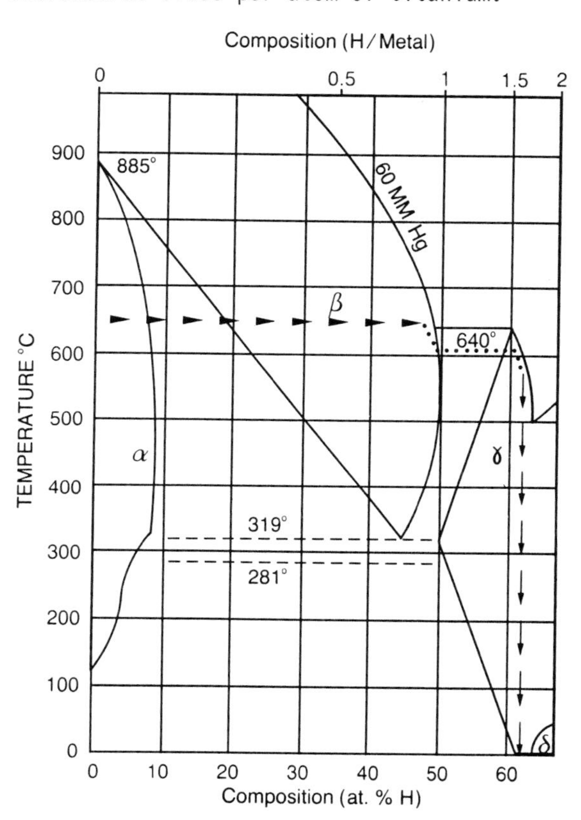

Fig. 5 : Ti-H phase diagram

After allowing a period for stabilization, the deuterium or tritium to be absorbed is introduced at a pressure and temperature calculated to promote formation of the beta-phase. Once the beta-titanium phase is saturated, the hydrogen pressure is gradually increased to assist formation of the hydride, after which the target is cooled. The QFC structure of the hydride is stabilized by heat treatment.

4.4. Inspection and testing

The process of manufacture is controlled by a *quality assurance programme*, comprising verification of the dimensional characteristics of the substrate, the quality of surface preparation, the thickness of titanium deposited by evaporation (and, after deposition, checked by weighing) and the adherence of the deposit. The quantity of tritium trapped is evaluated via an ionization chamber.

4.5. Packing

Products are packed in vacuo in a *vacuum-tight container* enclosed in a protective shell complying with international regulations governing the shipment of radioactive material. Surface contamination analyses carried out ensure against any contamination of containers.

5. MAIN APPLICATIONS

5.1. Neutron therapy

Treatment of certain types of cancer using fast neutrons (D,T, 14 MeV) has been practised since 1976 at the Hamburg Eppendorf Hospital Radiotherapy Centre in West Germany, using 600 Ci cylindrical targets yielding a flux of 10^{12} n.cm^{-2}s^{-1}. Target life is as much as twenty days or so, for an average 12 hours utilization per day. A study of several hundred patients has shown neutron therapy to have a curative action on undifferentiated tumours. No necrosis is observed in normal tissue after administration of a standard 16 Gy dose over a four-week period.

5.2. Activation analysis

In France, Pechiney's Research Centre has for twenty years been using fast neutron activation analysis (FNAA) for determinations of oxygen. In this method, the sample under test is subjected to a 10^{10} $n.cm^{-2}s^{-1}$ neutron flux produced by disc-type TiT_4 tritiated targets rated at about 5 Ci. The radiation generated by the activation reactions of the elements to be determined is detected via a photon/gamma spectrometer.

The method is suitable for the determination of oxygen in any state (dissolved, occluded or surface). It is applicable over the range 0.1 ppm to 60 %, and can be used to determine very low concentrations of oxygen in ultra-pure materials, intermetallic compounds and alloys.

5.3. Interaction of neutrons with matter

The CEA has made an experimental study of the interaction of neutrons with a thick layer of air. In this study, the air layer is simulated by a sphere 2 metres in diameter filled with liquid nitrogen. The neutron source is located at the centre of the sphere and screens of various natures are provided for energy degradation. Detectors record the time of arrival of the neutrons by the method of time of flight. The neutron source consists of the accelerator and the TiT_4 target and operates in pulse mode. 3 000 neutrons are produced per burst.

A total flux of 2.10^{11} $n.cm^{-2}s^{-1}$ is achieved. Analysis of the spectrum recorded (Fig. 6) shows the presence of 14 MeV neutrons and their energy degradation. The "blip" corresponds to 2.45 MeV neutrons (D,D reaction). This is characteristic of implantation of deuterons on the target as from a certain irradiation time.

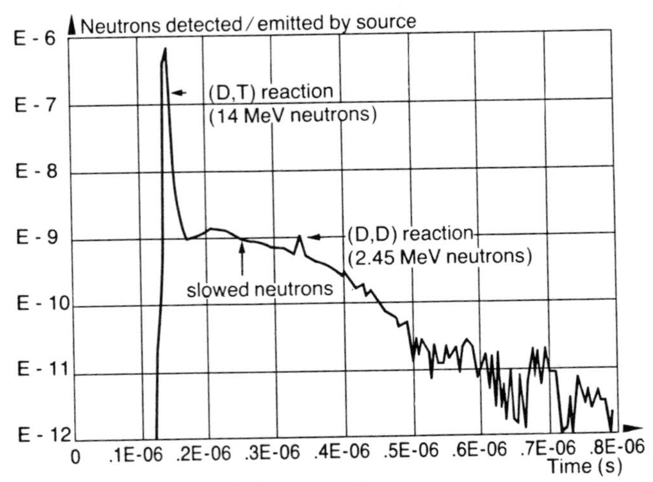

Fig. 6 : Neutron/time spectrum

6. CONCLUSION

Optimization studies and the quality assurance programme at CEA/VALDUC have significantly *enhanced target life and reliability*. The results obtained by the CEA's principal customers are highly satisfactory. *Very great safety in handling* has enabled the targets to be brought into routine use in hospitals.

7. PRINCIPAL CUSTOMERS

In France

The CEA centres.

Pechiney.

International Bureau of Weight and Measures.

National Institute of Medical Research.

Strasbourg Particle Physics Laboratory.

Nuclear Physics Institute, Lyon.

AGRILL.

IRELEC.

Elsewhere

IAEA, the UK, Germany, Belgium, Australia, New Zealand, Sudan, Iraq, Morocco, Malaysia, Turkey, Switzerland, Peru, Bangladesh, Thailand, etc.

FUSION TECHNOLOGY 1988
A.M. Van Ingen, A. Nijsen-Vis, H.T. Klippel (editors)

THEORETICAL AND EXPERIMENTAL INVESTIGATION OF CAPILLARY ELECTROLYSIS SYSTEMS

A. RAHIER, R. CORNELISSEN, A. BRUGGEMAN, P. DE REGGE

S.C.K./C.E.N., Boeretang 200, B-2400, MOL (Belgium)

In a fusion reactor environment, it is expected that highly tritiated water will be formed when tritium is extracted from the blanket as well as during the plasma exhaust purification process. As a consequence, the recovery of elemental tritium from its oxides is an essential step before recycling the fuel to the reactor. Among different basic processes that can be used for this purpose, electrolysis appears to be very promising. Therefore, SCK/CEN has developed a small dedicated cell designed to decompose 100 ml/day of pure DTO or HTO. At the present project stage, a prototype cell is available and the device has been successfully tested with light water during several thousands of hours. Active tests are planned in the near future. In the orginal concept, the liquid inventory is limited to the vertical porous gas separator. Capillarity is used as a driving force to feed the cell to avoid the use of a pump. As noted during the first experimental runs, this fact has a considerable influence on the behaviour of the electrolytic system. It has been observed that the independent variables (the electrolyte concentration, the void fraction of the gas separator, the electrolysis current and the temperature) cannot be chosen freely, otherwise instabilities may occur. In the worst cases, large amounts of mist can be formed in the anodic compartment. This particular behaviour has been theoretically investigated with the aim to allow a better basic understanding of the capillary electrolysis. A deterministic model has been developed for this purpose. The mathematical equations show clearly that the electrolyte tends to accumulate at the top of the gas separator. An equilibrium state can be reached only if sufficiently large amounts of electrolyte can flow back towards the bottom of the gas separator. This counter-flow has been taken into account by introducing a single general diffusion coefficient into the model. In a second phase, systematic experimental runs have been carried out with mock-up cells. A statistical treatment based on the maximum likelihood estimation algorithm allowed to compute the best value for the diffusion coefficient and to validate the model. Finally, acceptable ranges of the independent variables have been defined and all the subsequent experimental runs have been performed without stability problems.

1. INTRODUCTION

In the fuel cycle of future fusion reactors, technological means will be required to recover elemental tritium from its oxides. At present, hot uranium beds are the best known candidates for this purpose[1]. However, such devices present the major disadvantage of producing excessive tritiated waste.

Therefore, several alternatives are being considered among which electrolysis appears to be very promising[2]. Considerable efforts have been made in the past decade[3-7] to design and build a suitable electrolytic system for decomposing small amounts of highly tritiated water. In 1984, SCK/CEN also has started the development of a dedicated electrolysis cell, the particular merits of which are its low inventory (~ 20 ml liq. HTO), low working temperature (~ 8 °C) leak tightness and tritium compatibility of all component parts[8]. The nominal throughput (100 ml liq. HTO/day) can also be considered as an advantage when compared to the most advanced cells being presently developed elsewhere[3-4]. At present, a prototype cell is available[9] and successful long duration cold tests have demonstrated the robustness of the proposed system[10] which makes use of capillarity as a driving force to feed the cell without using any pump. Sulphuric acid is used as the electrolyte.

During the first experimental runs, it has been noted that the independent variables (i.e. the void fraction of the gas separator, the electrolyte concentration, the electrolysis current and the temperature) may not be chosen freely otherwise instabilities may occur. In the worst cases, large amounts of mist and/or SO_3 can be formed in the anodic compartment while the cell voltage can become very large (> 7 V) and/or show periodic variations with time. Searching the causes for this erratic behaviour has prompted a theoretical investigation with the aim to allow a better understanding of the capillary electrolysis.

2. THEORY

2.1. General considerations

A detailed description of the capillary electrolysis cell can be found in the literature[8-10], but for the purpose of the discussion, it is useful to recall that the cell uses a vertical planar gas separator made of porous vitreous silica. This plate is fed by capillarity through a tiny channel located at the bottom of the plate. Platinum electrodes with a suitable geometry are pressed on each side of the gas separator. There is neither catholyte nor anolyte. Liquid is present only in the gas separator and in the feed system.

Fig. 1 shows a vertical cross section of the so-called electroactive zone, namely that part of the vertical porous gas separator in contact with the electrodes. The height of this zone is referred to as h while x designates the current vertical coordinate.

As electrolysis proceeds, water is decomposed and gases are formed at both electrodes. The decomposed water is continuously and spontaneously replaced by means of capillary effects which force some liquid to raise in the plate to maintain the hydrostatic equilibrium. In particular, the water being decomposed above x is replaced by a solution of sulphuric acid flowing vertically through the horizontal cross section located at x. Since water is replaced by an electrolyte solution, and sulphuric acid losses are assumed to be negligible, the acid concentration in the upper part of the plate

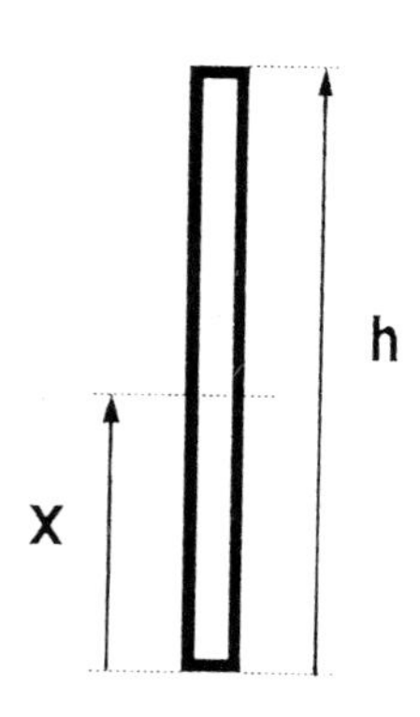

FIGURE 1
Electroactive zone of the vertical gas separator.

should increase continuously. A stationary situation can be obtained only if H_2SO_4 can find some path to flow back down. Diffusion, boundary effects along both electrodes, gradients of density as well as hygroscopic behaviour of H_2SO_4 can all combine to establish a steady-state repartition of the electrolyte along the height of the electroactive zone. However, such effects are strongly dependent on the operational conditions. Variables that can apparently be chosen freely, such as the void fraction of the gas separator, the temperature, the electrolytic current and the initial sulphuric acid concentration will all influence the flow-back rate of the electrolyte as well as the capillarity itself. Hence, it is not surprising that particular combinations of these variables will not allow to reach a steady-state but will rather lead to an unstable behaviour of the system. It can also be understood that mist and/or SO_3 can be formed when the maximum concentration of H_2SO_4 in the electroactive zone approaches 100 wt.%.

In the light of these considerations, the following points can be emphasized :

- The plate remains continuously wetted as long as the rate of electrolytic dissociation does not exceed the maximum rate of rising that can be achieved by means of capillary forces (hydrostatic equilibrium). One can find here an upper limit for the throughput of the capillary electrolysis system.

- In the normal working conditions, a concentration profile of H_2SO_4 exists along the height of the electroactive zone. The maximum concentration is expected to be found at the top of this zone. The choice of H_2SO_4 as electrolyte is, at least partially, a consequence of this fact. For instance, NaOH or KOH could easily crystallize at the top of the plate. This has also been experimentally observed.
- It is expected that the steady-state concentration profile cannot be established for all combinations of the independent variables.

2.2. Concentration profile

With the aim to characterize the concentration profile mathematically, we make the following simplifying assumptions :

- The partial molar volumes of both water and sulphuric acid are supposed to be constant at all concentrations. This means that the constituents mix with no change of volume.
- The current passing through the cell is uniformly distributed over the electroactive zone, especially along the height of this zone. Hence, the current density is constant throughout the electroactive zone.

On this basis, an obvious expression of the hydrostatic steady-state condition can be written (eq. 1). The left-hand side of this equation represents the net ascending volumetric flow rate of liquid passing through a horizontal cross section of the plate at x. The right-hand side is the volume of water dissociated above x per unit time.

$$D(x) = \left(1 - \frac{x}{h}\right) \frac{K I_e}{C_w^*} \tag{1}$$

where $D(x)$ = net ascending volumetric flow rate at x (m^3/s)

I_e = electrolysis current (A)

C_w^* = concentration of pure water ($55.56\ 10^3$ mol/m^3)

K = Proportionality constant ($5.18\ 10^{-6}$ mol/C)

Focussing then on the material balances of water and sulphuric acid respectively, equations 2 and 3 are obtained :

$$D(x)C_w^* = D(x)C_w(x) + \phi_w(x) \tag{2}$$

$$0 = D(x)C_a(x) + \phi_a(x) \tag{3}$$

where

$C_w(x)$ = local concentration of water (mol/m^3)

$C_a(x)$ = local concentration of H_2SO_4 (mol/m^3)

$\phi_w(x)$, $\phi_a(x)$ = molar flow-back rates of water and acid respectively (mol/s)

Several contributions to the dynamic flow rates (ϕ_w and ϕ_a) can be identified. An exact phenomenological description of these effects would introduce a high degree of complexity in the mathematical developments. Therefore, it has been decided to summarize all these different contributions into one single overall diffusion coefficient (eq. 4).

$$\phi_a(x) = - k_a S_f \frac{dC_a(x)}{dx} \tag{4}$$

where

k_a = overall diffusion coefficent (m^2/s)

S_f = free cross-sectional area (normal to diffusion flow) (m^2)

A trivial consequence of this assumption is that the value of k_a as derived from the experimental results should be much larger than those commonly encountered for true diffusion coefficients. This will have to be verified afterwards. Combining eq. 1, 3 and 4 yields :

$$k_a S_f \frac{dC_a(x)}{dx} = Ca(x)\left(1 - \frac{x}{h}\right)\frac{KI_e}{C^*_w} \qquad (5)$$

The electrolyte concentration profile can be calculated easily by integrating eq.5. One obtains :

$$C_a(x) = C_{as} \exp\,(\,\xi\, x(2\,h - x)) \qquad (6)$$

$$\text{with } \xi = \frac{K\,I_e}{2\,K_a\,\varepsilon\,S\,C^*_w\,h} \qquad (7)$$

where C_{as} = concentration of sulphuric acid in the feed system (mol/m³)
ε = void fraction of the gas separator
S = total cross-sectional area of the plate at x (m²).

Fig. 2 shows the shape of the theoretical normalized concentration profile as calculated from eq. 6. As expected, the curves exhibit a maximum located at $x = h$. The value of the maximum concentration is given by eq. 8.

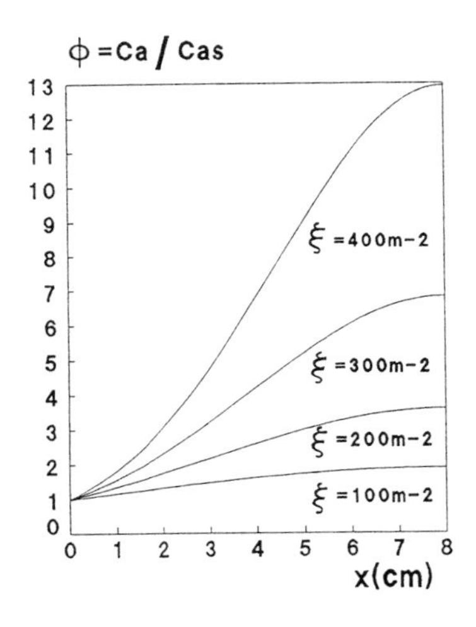

FIGURE 2

Normalized concentration profile of sulphuric acid in the electroactive zone (see eq. 6)

$$C_{a\,max} = C_a(h) = C_{as}\exp(\xi\,h^2) \qquad (8)$$

The experimental verification of the concentration profile cannot be carried out straightforward since local concentrations in the plate cannot be measured in situ. Therefore, it is useful to establish the mathematical expression of the mean concentration in the plate. This concentration can indeed be measured easily by means of common titration techniques. One has :

$$\bar{C}_a = \frac{1}{h}\int_0^h C_a(x)dx =$$

$$C_{as}\sqrt{\frac{\pi}{\xi}}\;\frac{\exp(\xi\,h^2)}{2\,h}\;\mathrm{erf}\,(\sqrt{\xi}\,h) \qquad (9)$$

2.3. Predicted behaviour of the system.

On the basis of the previous equations, it is possible to develop a theoretical criterion to predict whether or not a specified set of independent variables will lead to instability of the system. Recognizing that practical difficulties such as mist and/or SO_3 formation will surely take place when $C_{a\,max}$ approaches 18663 mol/m³ (= 100 wt.%), it can be shown that the start concentration C_{ai} may not exceed a maximum value.

Let v_p and v_s be the volumes of liquid in the electroactive zone and in the feed system respectively. If the (homogeneous) start concentration of sulphuric acid is C_{ai}, and no electrolyte is added afterwards, the following steady state material balance can be written :

$$C_{ai} = \frac{C_{as}\,v_s + \bar{C}_a\,v_p}{v_s + v_p} \qquad (10)$$

Rearranging (10) and taking (8) and (9) into account gives

$$C_{ai} = C_{a\,max}\left(\frac{1 + \eta f}{1 + \eta}\right)\exp\,(-\,\xi\,h^2) \qquad (11)$$

where $\eta = \frac{v_p}{v_s}$

$$f = \sqrt{\frac{\pi}{\xi}}\ \frac{\exp(\xi h^2)}{2h}\ \mathrm{erf}\ (\sqrt{\xi}\,h)$$

The desired criterion is obtained by introducing $C_{a\ max}$ = 18 663 mol/m^3 in (11) and by considering the right-hand side of the equation as an upper limit for C_{ai}. We obtain :

$$C_{ai} \leqslant 18\ 663\ \left(\frac{1 + \eta f}{1 + \eta}\right) \exp\ (-\ \xi\ h^2) \qquad (12)$$

Fig. 3 shows the shape of the function for η = 0.16 (mock-up cell) and for η = 1 (a value which is more realistic for the prototype cell). The curves have been calculated by anticipating the correct value of k_a. They are valid for t = 8 °C, according to the experimental conditions. Given an electrolysis current I_e and a void fraction ε, the maximum allowable start concentration is obtained at the intersection of the suitable curve with a vertical straight line drawn at the known abscissa I_e/ε.

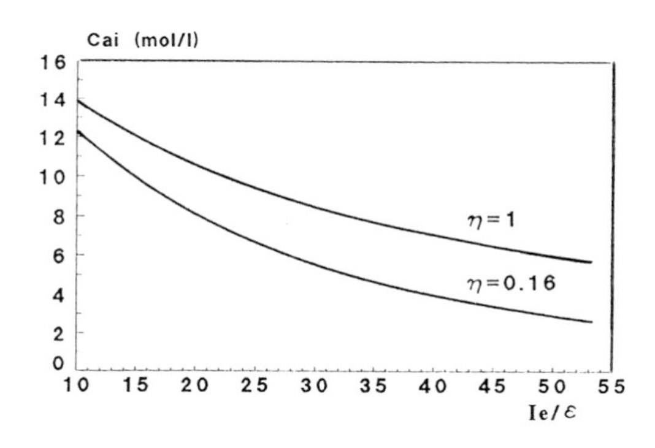

FIGURE 3

Theoretical upper limit for the start concentration of H_2SO_4 v.s. I_e/ε

2.4. Extended equation for practical use

All the abovementioned equations apply to the electroactive zone. In practice however, it is clear that this zone is only one part of the effective gas separator (Fig. 4). The latter is larger and corrections are to be made to allow an exact interpretation of the experimental mean concentration in the whole plate.

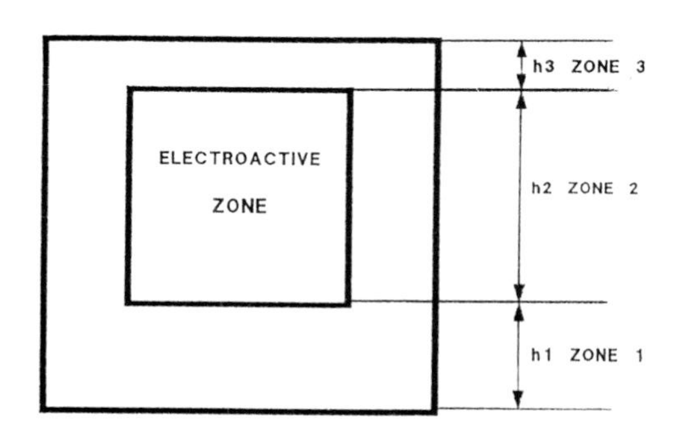

FIGURE 4

Schematic representation of the whole gas separator

For this purpose, the following assumptions are made :

- Zone 1 (Fig. 4 : In this zone, there is no accumulation of the acid. The concentration is assumed to be C_{as} everywhere.
- Zone 2 : It is assumed that the concentration profile remains valid even in the lateral non-electroactive parts of the plate. Since diffusion is the main driving force acting to flatten the eventual horizontal gradients, care must be taken to make the measurement after a sufficiently long time of steady-state operation.
- Zone 3 : The concentration in this zone is supposed to be constant and equal to $C_{a\ max}$. Note again that this condition will hold only after a sufficiently long time of steady-state working.

In these conditions, the experimental mean concentration in the plate can be described according to eq. 13 :

$$\frac{\overline{C}_{a\ exp}}{C_{as}} = \frac{1}{h_1+h_2+h_3}\left(h_1 + \frac{\overline{C}_a}{C_{as}} h_2 + \frac{C_{a\ max}}{C_{as}} h_3\right) \quad (13)$$

where $\overline{C}_{a\ exp}$ is the experimental mean concentration in the whole plate (mol/m^3)

h_1, h_2, h_3 are the heights of the different zones, according to Fig. 4 (m)

$\frac{\overline{C}_a}{C_{as}}$ and $\frac{C_{a\ max}}{C_{as}}$ correspond to the expressions given by eq. 9 and 8 respectively.

Equation 13 establishes the necessary link between the experimental data (left-hand side) and the theoretical, predicted values (right-hand side). It can be used to determine the best statistical value for k_a by using a suitable nonlinear curve fitting technique.

3. EXPERIMENTAL RESULTS AND DISCUSSION

Several experimental runs have been carried out with the aim to assess the theoretical model for capillary electrolysis. The experiments were conducted using mock-up cells (Fig. 5), a detailed description of which has already been published[10].

Each experiment comprised several sequential operations which can be briefly described as follows :

a. Measurement and adjustment of the void fraction of the gas separator

New porous plates with a void fraction ranging between 25 and 30 % were purchased. The porosity has been gradually adapted by immersing the plates in hot KOH solutions of known concentrations during controlled periods of time. Two methods were used to measure the exact void fraction[10].

b. Characterization of the start conditions

This involved the verification of the start concentration (by common titration techniques) as well as the exact determination of the volumes of solution in the plate and in the feed system respectively.

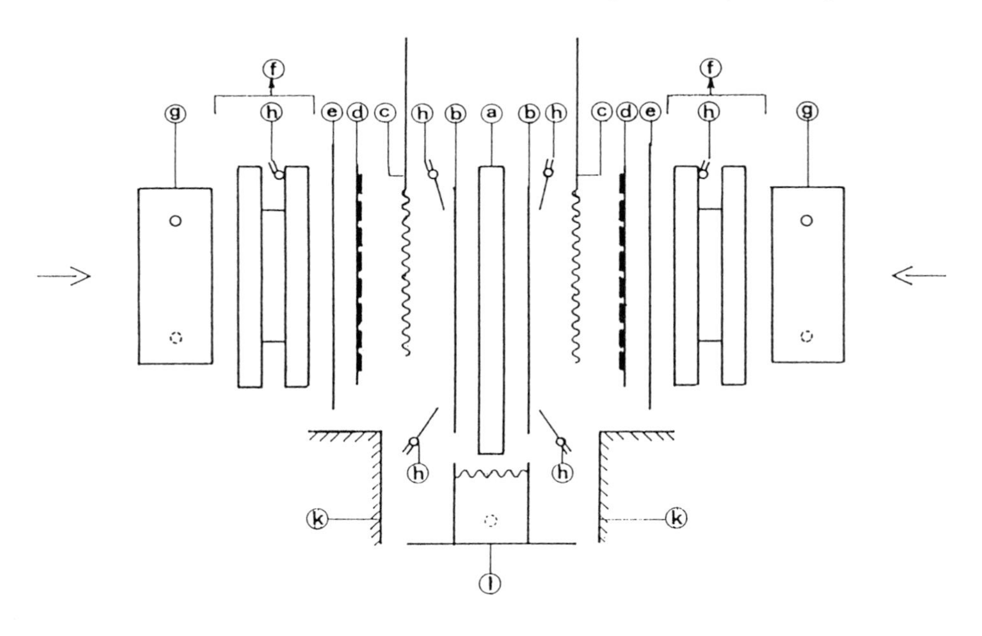

FIGURE 5

Schematic representation of a mock-up cell

a : gas separator ; b : porous spacer ; c : Pt electrode ; d : tantalum plate ; e : insulator ; f : thermoelectric heat pump ; g : heat exchanger ; h : thermistor ; k : supporting table ; l : feed vessel.

c. Electrolysis phase

The electrolysis was started by forcing the desired current to flow through the cell for at least two days while the temperature of each electrode was kept constant (8 °C) by means of thermoelectric heat pumps. Pure water was automatically added in the feed tank to maintain a constant level.

d. Analysis phase

Immediately after system shut down, the water feed was disconnected from the gas separator and the cell was dismounted to allow the determination of the acid concentration in both the gas separator ($\bar{C}_{a\ exp}$) and the feed system (C_{as}).

Although reasonable values were chosen for I_e, ε and C_{ai}, it was never sure that they would result in a stable behaviour of the system since the exact value of k_a was not yet known at this moment. In case of instability, the corresponding results were of course discarded. Table 1 summarizes the different acceptable combinations of the independent variables that have been investigated, as well as the experimental values of the ratio $\bar{C}_{a\ exp}/C_{as}$.

The results have been submitted to a statistical analysis by using a computer program based on the nonlinear maximum likelihood estimation[11-12]. We obtained the following value for k_a :

$$k_a = (36 \pm 5)\ 10^{-8}\ m^2/s \qquad (14)$$

A further analysis allows to conclude that the proposed model is sufficiently consistent with the experimental results for the purpose of predicting possible instabilities.

As expected, the value for k_a is about 2 orders of magnitude higher than common diffu-

TABLE 1 : Experimental conditions and results. All runs carried out at 8 °C

Run nr.	Start concentration C_{ai} (mol/m^3)	Electrolysis current I_e(A)	Void fraction ε	Experimental ratio $\bar{C}_{a\ exp}/C_{as}$
1	7053	6	0.34	2.33
2	2255	8.2	0.33	2.38
3	4333	8.2	0.33	4.35
4	2275	10	0.32	1.29
5	4422	10	0.34	2.15
6	4445	10	0.33	2.92
7	8166	10	0.57	1.47
8	2866	11	0.37	1.83
9	3435	11	0.37	3.00
10	2962	11.7	0.39	2.45
11	2255	12.8	0.39	1.53
12	1618	15	0.37	3.02
13	2324	15	0.37	4.46
14	2441	15	0.35	3.73
15	2866	15	0.37	4.17
16	3579	15	0.35	6.88
17	4339	15	0.37	2.97
18	6005	15	0.45	1.37
19	6640	15	0.55	1.37

sion coefficients for strong electrolytes in water. The main contribution to k_a seems to be related to boundary effects along the electrodes. As gas bubbles split at the electrodes, some liquid escapes momentarily from the plate. This liquid flows down along the electrodes and re-enters the plate at a lower ordinate, helping thereby in leveling the concentration of H_2SO_4.

As already explained, eq. 14 has been used to compute the curves of Fig. 3. For the purpose of testing the cell intensively, several other tests have been performed successfully by choosing the independent variables according to eq. 12. In each case, a steady-state behaviour could be obtained quickly. Some long-duration experiments have demonstrated the possibility to maintain the cell in continuous operation during more than 5000 hours without adding sulphuric acid.

Only small amounts of mists and a few ppm of SO_3 were detected in the oxygen stream, demonstrating thereby the usefulness and the consistency of the presented model.

4. CONCLUSIONS

A simplified deterministic model of the capillary electrolysis system has been developed successfully. A series of experimental runs allowed to compute the best estimate for the key parameter, namely an overall diffusion coefficient. On the basis of these results, a criterion has been developed which allows to predict whether or not the system will reach a steady-state working point. Subsequent experimental runs could be carried out without any problem related to instability.

ACKNOWLEDGEMENTS

We are indebted to the Commission of the European Communities for its financial support to this research.

1. E.C. Kerr, J.R. Bartlit, R.H. Sherman, Fuel Cleanup System for the Tritium Systems Test Assembly : Design and Experiments, Fusion Techn., 1 (1980) 115.

2. P.A. Finn, R.G. Clemmer, B. Misra, Tritium Management Requirements for D-T Fusion Reactors, ANL-81-32, Argonne National Laboratory, Argonne, Illinois (1981).

3. S. Konishi, H. Ohno, H. Yoshida, Y. Naruse, Decomposition of Tritiated Water with Solid Oxide Electrolysis Cell, Nuclear Techn./ Fusion, 3 (1983) 195.

4. P. Giroux, P. Boucquey, J.F. Bressieux, Electrolysis Cell for Highly Tritiated Water, Progress Report 1987, CEA, Valduc, France.

5. G. Pierini, B. Spelta, Advances in the Electrolysis of Tritiated Water for its Application to a Fusion Plasma Processing Plant, EUR-11327-EN, JRC-Ispra, Varese, Italy (1988).

6. D. Spagnolo, Fusion Fuels Technology Newsletters, First Quarter 3 (1985).

7. R.E. Ellis, Electrolysis of Tritiated Water in Solid Polymer Electrolyte Cells, MLM-2902, Monsanto Research Corporation (1982).

8. A. Rahier, R. Cornelissen, A. Bruggeman, W. Goossens, L. Baetslé, Design of an Electrolysis Cell for Highly Tritiated Water, Fusion Techn., 8 (1985) 2035.

9. A. Rahier, R. Cornelissen, A. Bruggeman, Development of an Electrolysis Cell for the Recovery of Tritium from its Oxides, Proc. 14th Symposium on Fusion Technology, Avignon, 2 (1986) 1519.

10. A. Rahier, R. Cornelissen, A. Bruggeman, P. De Regge, Preliminary Testing of an Electrolysis Cell for Highly Tritiated Water, Fusion Techn., in print.

11. A. Rahier, Obtention et Analyse Numérique de Résultats Experimentaux par des Méthodes Statistiques, 3ième Cycle en Chimie Analytique, Bruxelles (1979).

12. D.M. Himmelblau, Process Analysis by Statistical Methods (Wiley & Sons, 1970).

FUSION TECHNOLOGY 1988
A.M. Van Ingen, A. Nijsen-Vis, H.T. Klippel (editors)
Elsevier Science Publishers B.V., 1989

COMMISSIONING AND FIRST OPERATING EXPERIENCE AT DARLINGTON TRITIUM REMOVAL FACILITY

R.B. DAVIDSON / P. VON HATTEN
Ontario Hydro Darlington Nuclear Generating Station, P.O. Box 2000, Bowmanville, Ontario, LIC 3N2, 416-623-6670

M. SCHAUB Sulzer Canada Ltd. Toronto, Canada, 60, Worcester Road, Rexdale, Ontario M9W 5X2, Tel. 416/-674 20 34
R. ZMASEK Sulzer Bros. Winterthur, Switzerland, CH-8400 Winterthur Tel. 0041/52 81 37 62

INTRODUCTION

Ontario Hydro's Tritium Removal Facility is presently in it's earlier operating phase. The DTRF was built in order to reduce the average dose rate per worker and tritium emmissions in all of Ontario Hydro's CANDU reactors. As a byproduct tritium may be sold to civil users. This paper provides an overview of the system design, commissioning philosophy, program and results. Three areas of the plant are considered separately:

1. Main Extraction Process
2. Concentrated Tritium Handling
3. Support Systems

PROCESS AND DESIGN DATA

Main Extraction Process

The Darlington Tritium Removal Plant (DTRF) applies a vapour phase catalytic exchange to transfer tritium from heavy water (D2O) into a deuterium carrier stream (DTO+D2-- D2O+DT) and a subsequent cryogenic distillation (1). To make the process work, the system is supported by many more subsystems:

A feed purification system degases and purifies the entering heavy water. Dryers remove the remaining heavy water from the deuterium before entrance into the cryogenic system. Adsorbers remove any high boiling gases, mainly nitrogen and oxygen. An electrolyser produces the deuterium for initial filling and make up. A recombiner provides the capability to burn hydrogen isotopes. A draining and purging system allows evacuation and purging of process systems. A tritium immobilization system in order to safely store tritium and to fill tritium sales containers. A cryogenic refrigeration system (CRS) provides cooling of the distillation columns. A clean up system purifies the glove box argon cover gas.

An air clean up system removes hydrogen isotopes from room air in case of an accidental release. Utilities such as heating, ventilation steam, cooling water, chilled water, etc. are also required.

DTRF is designed to extract 1140 Ci/hr of pure tritium at a feed concentration of 3.4 Ci/kg D2O. A flow of 365 kg/hr of D2O is fed into the front end, where it is detritiated by a factor of 12.

The vapor phase catalytic exchange (VPCE) consists of 8 stages, each comprising an evaporator, superheater, catalyst bed and condenser separator.

Deuterium flow in the VPCE is set to 147 kg/hr. The dryer unit, consisting of two molecular sieve beds is designed to run for 24 hrs. without regeneration at a dew point of below -80 Degree C.

Regeneration is done using hot process gas. The first cryogenic distillation column has a design reflux of 300 kmoles/hr. which represents a condenser duty of approx. 100 kW. The subsequent columns havae condenser duties of 1000W, 55W and 10W respectively. Tritium purity to be reached is 99.8% T2. On the other hand, the D2 return to the VPCE is depleted in DT by factor 23 and the protium tails are designed to reach a concentration of 60% HD.

If no protium is extracted, the detritiation factor in the VPCE can be increased to 35 enhancing tritium extraction capability. This operating mode is planned to optimize extraction of T2 before steady state feed concentration is reached.

Additionally, the refrigeration system is designed to satisfy the needs of the process. It comprises two 750 kW labyrinth piston compressors and one expander turbine as moving parts, as well as a number of heat exchangers. It's refrigeration capability is designed for 6.5 kW (hydrogen liquefaction).

To burn down hydrogen isotopes, a combustion type recombiner is installed with a design flow rate of 10 MN3/hr. The elctrolyser produces deuterium at a rate of 7 MN3/hr.

Concentrated Tritium Handling Systems.

The tritium immobilization system (TIS) is designed to draw off 45kCi (4.5 gram) batches of tritium gas from the last column of the distillation system and immobilize it on a titanium getter for long term storage. During initial operation of the TRF it is anti-

DTRF SIMPLIFIED FLOW DIAGRAM

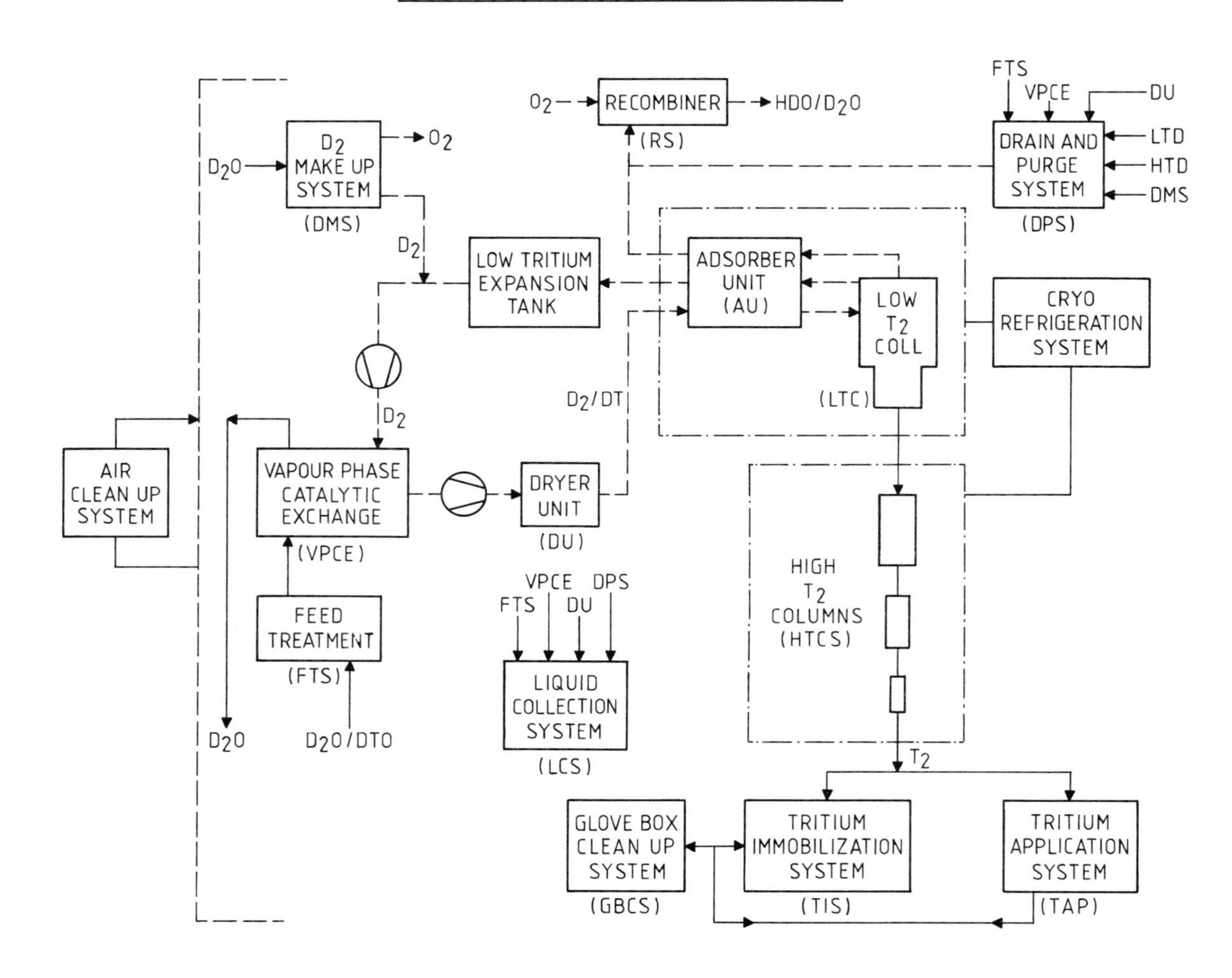

cipated that a drawoff and immobilization cycle will be carried out once every 4 hours.

The system consists of two gloveboxes each containing the necessary equipment to complete the full immobilization cycle. Thus one glovebox can be maintained in standby while the other is in service. In addition, glovebox has a tritium sales container airlock to allow filling of uranium bed and tritium gas sales containers. The gloveboxes are maintained under an argon blanket at slightly positive pressure.

Major system features include the following: A flow regulating system controls the batch drawoff. A 19.5 litre assay vessel allows accurate inventory measurements. A vacuum transfer and circulation pump set consist of a duplex metal bellow roughing and a Normetex moving spiral pump. A quadropole residual gas analyzer (RGA) unit with a turbomolecular and rotary vane pumpset verifies gas purity. Immobilized tritium containers hold 11 batches or 500 kCi of tritium on 825g titanium. Monitors for tritium, moisture and oxygen are provided on both gloveboxes and associated airlocks. A 500 kCi uranium bed provides additional flexibility in tritium sales container filling.

The titanium in the immobilized tritium containers is conditioned on site in a special oven under high vacuum. Once filled, the containers are stored in a secure vault in the lower elevation of the TRF-building.

Glovebox Cleanup System

The glovebox cleanup system is designed to purify the Argon atmosphere in the two tritium immobilization system gloveboxes and associated airlocks. It will also serve the tritium application program glovebox when it is put in service during early 1989.

The system will be operated continuously until operating experience indicates that intermittend operation is sufficient to maintain argon purity in the gloveboxes.

The major system features are as follows:

A hermetically sealed Spencer blower circulates Argon through the system at a rate of 170 standard cubic meters per hour. A mechanical filter removes particulate down to 4 micron size. An activated charcoal trap removes hydrocarbon contaminants. A catalytic recombiner recombines free hydrogen isotopes to the oxide form. The recombiner features a preheater and metered oxygen supply to achieve the recombination process. Dual molecular sieve beds knock out water vapour to 1 ppm. Dual regenerable copper oxide oxygen scavenger beds reduce the oxygen to 1 ppm. A non-regenerable nitrogen removal bed utilizing titanium pellets operates at 760C.

Support Systems

The DTRF structure is approximately 35m long by 24m wide and is an industrial type steel frame structure fitted with reinforced concrete or grating floors. The general roof of the structure is 12m high, but a tower-like structure in the cryogenic-unit area rises to a height of 38m. Approximately 260 pressure relief panels (1.4 x 2.0m) are installed in the outside walls of rooms where a hydrogen isotope release and ignition might occur.

Ten major systems are located in ten different rooms to be able to isolate the individual rooms on a tritium release. The ventilation system was designed such that airflow is maintained from tritium-free areas into areas containing tritium and then to the DTRF building stack.

Process rooms in the DTRF are provided with a hydrogen and 19 tritium monitors. The tritium monitors are required not only for alerting personnel but also for the automatic startup of the air clean up system (ACS). One monitor can perform the dual function since the monitors are provided with two setpoints.

In general the monitors are located outside the process room so that they can be maintained and calibrated in a tritium free environment. A single alarm monitors all fixed area alarm tritium monitors (FAATMs) in the Control Room.

By operating the selector switch the Operator can establish from which room monitor the alarm is initiated. Local lights and alarms are strategically located in the process rooms for audio and visual identification. The entrance doors are provided with windows so that alert lights can be observed before entering. The tritium monitor does not differentiate between elemental tritium and tritium oxide, therefore the setpoint is set for the much more hazardous tritium oxide.

The ACS will detritiate room air in the TRF building in the event of an accute leak of tritium, thus preventing its escape to the environment and its accumulation in the TRF building. Only rooms containing tritium are connected to the ACS.

The ACS is a standby system designed so that it can detritiate the larges room (VPCE room) in five hours following a spill in that room. When a high tritium alarm is received from any room, the building ventilation system dampers to that room are closed, the ACS dampers are opened and the ACS blower is started. After the tritium level in the room has been reduced to an acceptable level, the ACS is shutdown manually. One room is detritiated at a time to prevent cross contamination of rooms.

Air from any of the rooms served by the ACS passes through a filter to remove particulates before passing through the blower. The gas flow is next treated in an activated charcoal trap which removes hydrocarbon contaminants such as oil mists. An electric preheater is used to heat the steam before it enters the catalytic recombiner when the hydrogen isotope content is low. In the recombiner gaseous hydrogen isotopes are converted to their oxide forms. The existing gas/vapour is saturated with water vapour, from the swamping vaporizer, then cooled in the recombiner condenser and chilled water cooler.

Final moisture removal is accomplished in molecular sieve dryers: one operates on stream while the other is being in a regenerated mode. Finally, the dried air is returned to the room from which it originated, while decontaminated purge gases are directed to the TRF stack.

Other DTRF support systems include process steam oooo kPa (abs) and 2500 kPa (abs), electric power (4.2 kV, 600V, 220V, 115V, 48Vdc),chilled water systems, D2O detection systems and deuterium, hydrogen, helium, oxide, argon and nitrogen systems.

COMMISSIONING PROGRAM

Main Extraction Process

Commissioning of the extraction process was separated into several steps in order to prove operability.

Running the process side with helium and light water: the process was to achieve required flowrates, pressures and temperatures. System control logic was also to be proved. Using helium instead of deuterium would allow providing of system tightness with a non combustible gas and at the same time helium could also be used for the final countdown. Demineralized water was specified in stead of D2O to minimize costs when a system had to be drained.

Running the refrigeration side with helium: the goals to be achieved were running in of compressors with a non explosive gas: testing of all components including expander with non explosive gas: and establishing of performance data for the CRS, which could be used to predict performance with hydrogen.

Cooldown with helium: the goals were to prove that there are no cold leaks in the plant, and prove the piping design using a non explosive gas.

Hydrogen performance test of CRS: running the Crs only with hydrogen in order to prove its capacity.

Running the process side with deuterium and tritium: the goals were to establish rewuired refluxes in all columns, stabilize operation and detect any problems before tritium introduction.

Tracer run: running the TRF with tritium allows determination of detritiation and enrichment aspects of the performance. Sampling is easy, since the levels of tritium are low. The goals are to prove activity of VPCE catalyt, to prove sizing of VPCE and prove number of theoretical stages in columns.

Performance run: running the TRF with specified feed allows to fulfill guaranteed performance.

Before hydrogen and deuterium were loaded, the operators had been exposed to most of the aspects of the plant and could be considered fully trained at that time, all safety systems had been proven operational.

Tritium Immobilization System

Commissioning of the tritium immobilization system was divided up to into 5 phases.

Pre-operational testing involves removing the glovebox windows, cleaning the gloveboxes inspecting and cleaning as required the process piping, process flowsheet verification, inspection of all components, and instrumentation checks.

Function testing involves commissioning of vacuum pumpsets, helium leak checking of process equipment, calibration of vacuum pressure transducers, calibration of process internal volumes, commissioning of the RGA, re-installing the glovebox windows, helium leak checking the glovebox, and commissioning the glovebox argon pressure control system.

Deuterium performance testing involves initiating all operations the system is designed for, including batch draw off, assaying, analysis, immobilization, recovery, and all associated transfers using deuterium as the working gas.

Trace tritium performance testing is designed to verify the system leak tightness using deuterium with trace quantities for tritium. The entire system is to be leak tested using the installed tritium monitors and a flexible wand to reach all potential leak sites.

Concentrated tritium performance testing will verify the system operating according to design when concentrated tritium is drawn from the distillation columns. Of special interest is the level of ambient tritium concentrations that develop in the gloveboxes.

Glovebox Cleanup System

The commissioning program for the glovebox cleanup system includes the following steps:

Pre-operational checks ensure all system controls, logic and mechanical equipment is functional. Loading of the various filter and catalyst media is also carried out.

The system is purged and filled with argon. Commissioning of the hydrogen and oxygen monitors occurs at this time.

The system is started up and tested to confirm the correct functioning of the system blower, heaters and regeneration cycles.

The system is linked to the gloveboxes for integrated testing. The performance of the catalyst is confirmed with deuterium to ensure it is not poisoned. The performance of the molecular sieve dryers and the oxygen and nitrogen beds is also confirmed.

Due to the large system flowrates actual performance of the catalyst using tritium is not practical. In lieu a scaled down catalyst bed is tested under controlled laboratory conditions to determine the catalyst efficiency in the micro curie per cubic meter range, this will provide an estimate of actual system tritium removal efficiency during normal operations, and provides a benchmark to asses system performance when in service.

Support Systems

Support system turnovers from construction to commissioning staff was carried out in a detailed review. Responsible commissioning staff then issued detailed commissioning work plans to place all utilities in service and carried out commissioning acceptance tests to ensure all design criteria was met.

H2 monitor systems and FAATMs were commissioned before the H2 loading.

The ACS System commissioning program was similar to the GBCS System commissioning program.

COMMISSIONING RESULTS

Main Extraction Process

While operating with Helium and Light Water, the major design parameters (flows, temperature press.) were achieved easily, however some equipment caused problems which had to be overcome:

- diffusion pumps proved to be a problem since the 110V heaters did not work properly. Hence they were replaced with 220V heaters.

- the liquid ring compressors, a hydrauli cally standard design, were vibrating and it took 4 months to trace the source to a small orifice in the port plate and to redesign and test it.

- the control computer failed intermittently because of a system software problem which was corrected. It has to be mentioned, that the system safety is not depending on this particular computer, but on a MODICON PLC and a backup hardwired shutdown system (2) which were running without any problem all the time.

- the dewpoint of the dryer unit was not achieved until it was detected that a liquid seal line had been improperly designed. After a dewpoint of below -95 Degrees C was routinely achieved.

Running in of the CRS with helium went quite well. The automatic startup sequence had to be modified to suit all process responses. A minor problem was encountered with the sizing of valves, some of which are too small for helium operation, i.e. the achieved flowrates through the turbine were pretty low. However, it was possible to prove proper operation of the expander. The major problem during this step was leaks in a few large flanged joints. Cooldown of the process helium went much smoother than expected, and after 3 days a temperature of 24K was achieved throughout the first and largest column. No cold leaks were detected and it was possible to do some safety related interlock checking. Operator training was virtually complete at that time using on the job training after having enjoyed extensive microcomputer simulation training before commissioning start.

Filling of the process with hydrogen/resp. deuterium was a major step, since the authorities (AECB) declared this step as a licensing review point. In order to fulfill the requirements all Operators had to be trained and all the safety systems had to be operational. Progress was fast thereafter and liquid hydrogen was produced at a rate much higher than the expected 6.5. kW. The installed test heaters were too small to evaporate all the liquid hydrogen. The actual performance is estimated at close to 10 kW.

For cooldown of the largest column, deuterium circulation is required in order to avoid natural convection problems and to enhance heat transfer. Thus the dryer had been run for several days before cooldown started and after 1 day liquifation started. The plant had to run for several days to achieve the required inventory, since the electrolyser only produces D2 at a rate of 7m3/hr and a calculated inventory of 1700 NNM3 was required. We found, however, that the inventory required to achieve the reflux of 300 kmoles/hr is considerably lower. The major reason is that the holdup of the packing is smaller than assumed.

In the smaller columns reflux was achieved fairly quickly, however minor impurities had to be removed from the condenser of the smallest column. Deuterium transfer from column to column without using a compressor worked fine. The design is one of the major differences between the Grenoble and the Darlington plant. The hydrogen purification system was not available during early deuterium run. Thus we encountered problems with the hydrogen system, one filter was constantly plugging up with nitrogen snow which was very difficult to remove. Subsequent warmup revealed a residual N2 concentration of 200 ppm, which had to be purged out.

The previous steps did slightly downgrade the heavy water in use, since residual light water mixed with the heavy water. As a first step, protium extraction was performed which already allowed measurements of the performance of the upper part of the first column, and to a certain degree, the performance of the VPCE. The NTS measure exceeded the design considerably. The VPCE performance could not be measured - reliably yet. We have to wait for the results of the Tritium run. The catalytic converter (2DT--D2+T2) can only beproven when the DT concentration is high enough. At present the Tritium tracer run is in progress.

The Guarentee Run has consequently also not been carried out yet.

Tritium Immobilization Commissioning

Initial inspection of process piping revealed that the system was clean and thus methanol washing was not necessary.

Operation of the Normatex and metal bellows pumpset identified a need to better isolate these pumps from the process piping due to vibration. Isolation bellows 6 inches long were installed to effect the necessary isolation.

A significant problem developed in obtaining a leak tight system due to the large number of cajon VCR fittings in the system. This problem emphasized the need for extra care in assembling these fittings, and minimizing the number of fittings where possible.

Some valves were found to be passing across the seats due to foreign material at the vespel to seat sealing surface. Subsequently 5 micron filters were installed at the suction of the Normatex and metal bellow pumps to protect against any residual dirt in the system.

The deuterium performance test has not yet commenced.

Glovebox Cleanup Commissioning Results

Pre-operational checks have been successfully completed. No major problems were encountered.

During argon filling of the system it proved to be difficult to get the remaining oxygen concentrations down to an acceptable level. This was traced to long sections of deadheaded pipe, oxygen contamination of the argon during argon cylinder changeout, and air trapped in the molecular sieve which was being released slowly. Numerous purges were necessary to get oxygen levels below 1%

The need to upgrade the bulk argon system to a cryogenic liquid storage system from a high pressure cylinder arrangement is apparent.

System operating conditions of flow and temperatures were achieved following an extensive effort to tune control loops and regeneration sequences. This portion of the commissioning took about 4 weeks to complete.

The performance of the glovebox cleanup system has not been tested at time of this writing. However catalyst performance under laboratory conditions was determined to be in excess of 97%.

Support System

Extra catalyst will be required to reach design conversion levels in the ACS catalytic recombiner based on Lab. catalyst efficiency test which only demonstrated a 35% catalyst efficiency.

A partial powered spare DCI 4000 control system has been set up to progrom software changes, test these changes, provide limited Operator training and to provide "hot" standby spares for the process control system.

Cooling water (normal 2 Deg C during winter operations) had to be tempered to 15 Deg C t omeet Sulzer's requirements for the refrigerant H2 compressors.

Significant evaporation in the liquid nitrogen supply to the adsorber units due to heat loss will necessitate moving the nitrogen storage tank closer to the adsorber units and vacuum insulating the supply lines. The excessive vapour, estimated at 25% by weight was reducing the efficiency of the adsorber units.

Larger hydrogen and argon supplyf tanks will be installed.

IDENTIFIED PROBLEMS FOR OPERATION

In the course of commissioning, several problems arose which needed additional attention:

Some flanged connection particulary in the CRS and the cajon fittings in the tritium immobilization system are difficult to tighten.

The in-line analysers need more attention as expected, mainly the oxygen and moisture analysers. However, it is not established yet if these problems are related to the unsteady operation during commissioning.

The capacity of the drain and purge system is small resulting in long evacuation times.

Access inside the gloveboxes is difficult increasing the time required for maintenance.

Nitrogen content in the CRS is critical because of filter plugging. Using helium instead of nitrogen as purge gas may resolve the problem.

Impurities in the CRS system and the small columns have initially given some problems in the small bore tubing used in the high concentration tritium parts.

CONCLUSION

The results of the commissioning carried out till today has proven the mechanical function of all components.
Of the subsystems tested so far all have achieved or exceeded the design capacity. However some quite important steps are still to be done, such as determining of VPCE-capacity before the overall capacity can be proven. It is nevertheless expected that with further optimitation an increase in capacity by a factor of 2 to 3 can be achieved by making use of larger flowrates in the front end and in the cryogenic columns.

REFERENCES

(1) W. Riediker et al
CNS Conference, 1982

(2) S.K. Sood et al
Fusion Technology, 8 (1985) 2

(3) A. Busigin
CNS Conference, 1987

FUSION TECHNOLOGY 1988
A.M. Van Ingen, A. Nijsen-Vis, H.T. Klippel (editors)
Elsevier Science Publishers B.V., 1989

CAD-MODEL BASED REMOTE HANDLING CONTROL SYSTEM FOR NET AND JET

Klaus LEINEMANN, Uwe KÜHNAPFEL, Arnold LUDWIG

Association KfK-EURATOM
Kernforschungszentrum Karlsruhe, Institut für Reaktorentwicklung, Postfach 3640, D-7500 Karlsruhe, West Germany

For maintenance work in fusion plants a supervisory control system concept was developed, which organizes a close, problem-suited cooperation of man and machine, based on shared control and mutual help. The central module on the task control level of the control system is a real-time simulator based on a three-dimensional CAD-model. This simulator serves for planning and off-line programming of maintenance sequences, and, in the execution phase, for integrated viewing, combining TV and synthetic scene presentation. A first implementation of a geometric simulator and its integration in an overall control system was realized for JET.

1. INTRODUCTION

For fusion plants fully remote maintenance concepts are favoured for costs as well as for safety and availability reasons. The remote work in a geometrical complex and hardly observable environment with kinematically redundant manipulators calls for computer support. For these tasks, covering the whole spectrum from unpredictable to preplannable and repetitive, neither fully manually controlled tele-operation nor automatically controlled robots represent the most suitable solution. Hence, a proper organized collaboration of man and machine (that is computer) is the recommended way that best utilizes the special abilities of both partners. In our effort emphasis is laid upon the collaboration of man and machine, while conventional remote operation in nuclear environment have been mainly controlled by a person and, on the other side, industrial robot technology has concentrated on fully automatic control. Hence a *supervisory control system* concept for tele-maintenance was worked out, which organizes a close, problem-suited cooperation of man and machine, based on shared control and mutual help. The central idea of our approach in implementing such an interactive control system is to introduce geometric knowledge about the plant into the system to enhance the man-machine interface and supporting algorithms. Our first realization of these ideas, after some pre-investigations, was concentrated on a graphics attachment to the JET boom control system [1,2].

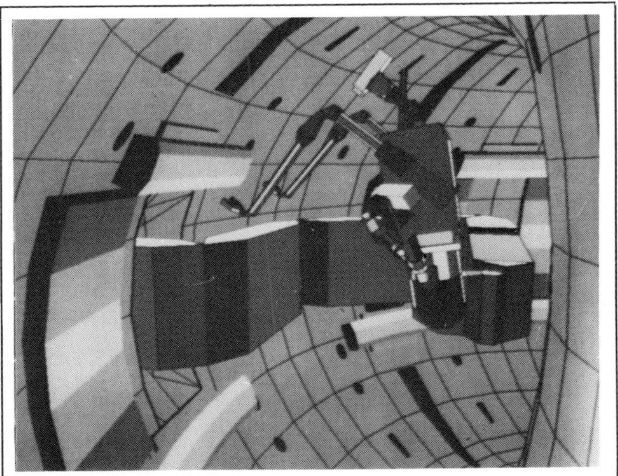

Figure 1. JET boom simulator display

2. COMPUTER GRAPHICS FOR JET

To achieve an enhanced telepresence for the operator of the JET boom, we developed the three-dimensional graphic boom simulator GBsim [3]. GBsim produces real-time synthetical images of the articulated boom, its various end-effectors, its camera arms, together with the working environment (Figure 1, Figure 2). In addition to its monitoring facilities GBsim is characterized by the following features: (1) off-line simulation and teaching, (2) TV-camera control, (3) manual control support by resolved motion and arm reconfiguration algorithms, and (4) fully three-dimensional collision detection.

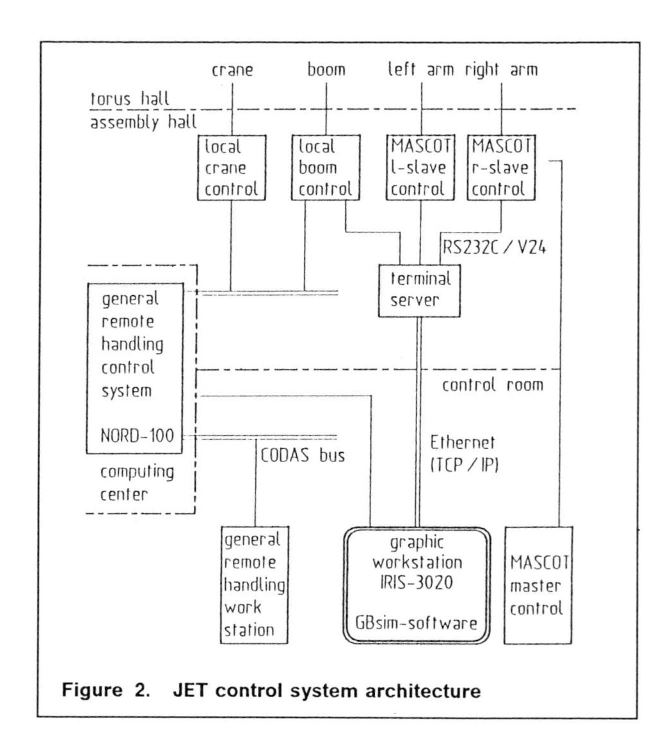

Figure 2. JET control system architecture

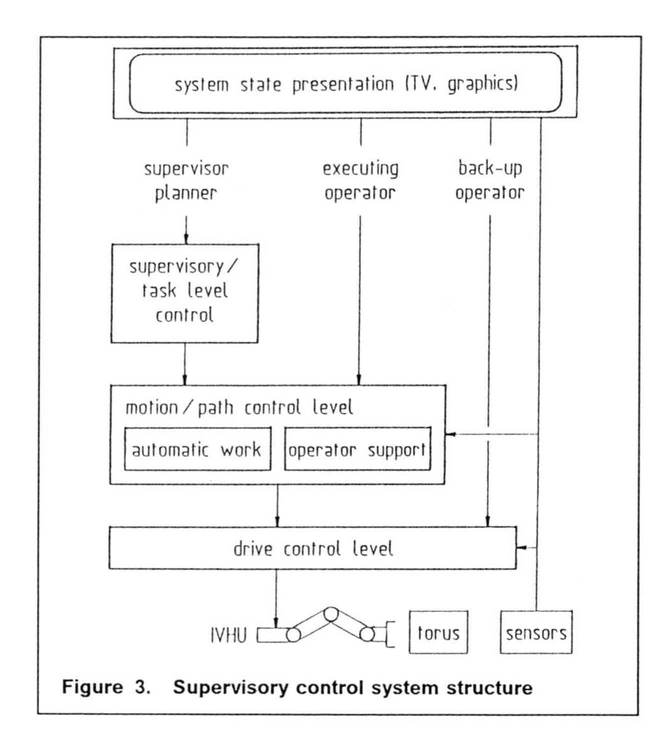

Figure 3. Supervisory control system structure

3. SUPERVISORY CONTROL SYSTEM FOR NET

A supervisory control system aims to combine conventional telemanipulation, general data processing techniques, and techniques approved in robotics. The basic principle is to have an operator in the control loop, but to provide him with the possibility of working in different roles on different control levels (Figure 3). On the highest level, the *task control* level, the operator is working as a supervisor, who has to plan, to teach or to program, to monitor, to intervene, and to learn. If necessary, the operator may work on the *motion control* level (e.g. using a master-slave manipulator) in a higher bandwidth control loop: he then acts as an executing operator. The third level, the *drive control* level should be used by the operator only in exceptional cases as a back-up solution. The general intention is to shift the actions of the operator more and more onto the supervisory level with evolving state of the art in automation. Therefore, the supervisory control system for NET has to be an *open system*, open for stepwise introduction of automation, starting e.g. with transport tasks. Man and computer are sharing control as decided by man. The functionality of the supervisory control is described in the next chapters.

3.1. Man-machine interfacing

To support the operator in execution monitoring, decision making, scene recognition, seeking for relevant informations, or diagnosis of unforeseen events the control system should provide a task oriented *state presentation*, enabling the operator to grasp the whole situation at a single glance. This is not only true for spatial informations related to the manipulators and their environment, but also for functional informations. *Integrated viewing and lighting* provide the operator with usable views of the working area, to give him the imagination of being present at the place of work. We therefore integrate standard television techniques and synthetic images (Figure 4). Integration in this sense means (1) providing both types of scene images side by side and (2) using the computer graphics system for controlling the cameras, the lights, and their carrier systems. Camera parameters may be displayed as a viewing pyramid to facilitate camera image interpretation.

The goal of an *integrated input manager* is to generalize the various input devices and to let the operator chose very special attachments between input devices and processes or end-effectors. The attachment of an input device to an end-effector for example may even vary in

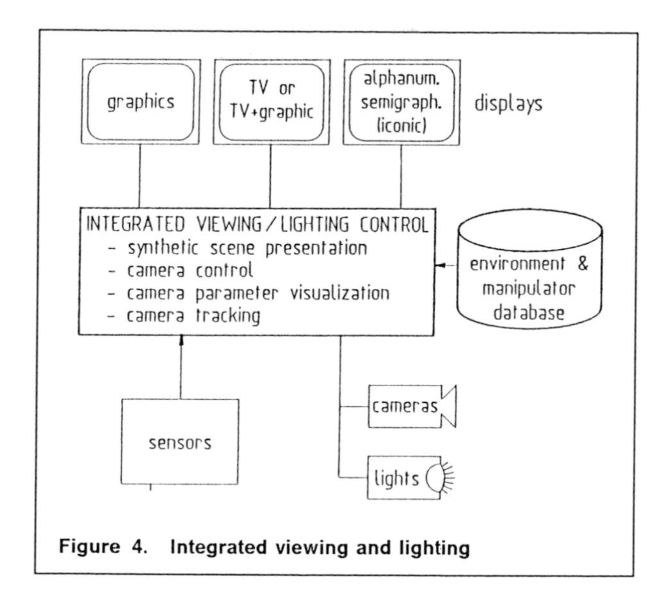

Figure 4. Integrated viewing and lighting

dependency of the task phase: contact work may better be done with a mechanical master (kinesthetic input), non-contact movements may be controlled even better by symbolic input.

3.2. Operator guidance

To guarantee a fast and safe execution of maintenance work all standard tasks like antenna exchange or tile replacement are preplanned carefully. They are decomposed into simple subtasks and actions to be done. These schedules are managed most effectively by computer like programs on the motion control level. To guide the operator they are presented textually as single instructions or perhaps graphically as action nets from an advisory subsystem.

3.3. Planning and programming

Planning and programming is done on two levels: the motion control level and the task or functional level of control. Motion or trajectory planning and programming results in programs for the motion control subsystem. This is similar to off-line programming of industrial robots, but has to include the special features of the remote handling (RH) motion controllers. But there is one additional aspect, the *ad-hoc programming*. This means, usage of off-line programming in unforeseen situations, where in conventional systems the operator would work directly on-line. In these cases fast off-line programming by off-line teaching on a simulator will make the execution safer because the solution of the handling task was simulated. This is especially helpful in situations where a solution is not obvious. The term ad-hoc programming shall subsume all the facilities required for a fast programming and testing cycle to generate immediately needed programs.

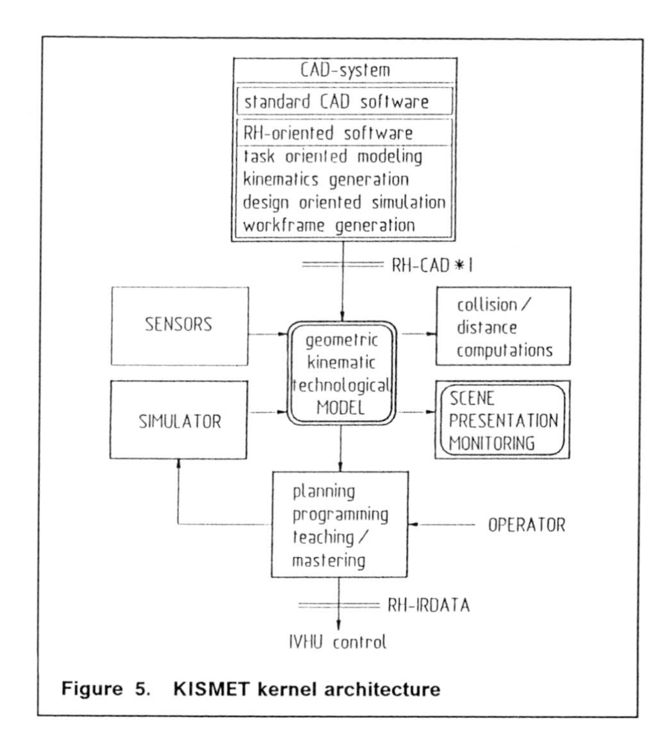

Figure 5. KISMET kernel architecture

On the task level, where no automatic execution is envisaged until now, planning means to decompose tasks into simple subtasks or action steps, including allocation of resources and coordinating communications with other RH sections. The result of this work is for example hierarchy of subtasks or an action net for concurrent activities. They are used in the execution phase to guide the operator step by step through complex work sequences which have to be done without failures.

3.4. Simulation and training

Corresponding to planning and programming two simulation and training environments should be available to simulate the proceeding process of maintenance and to simulate motions of the manipulators. An important application of a simulator is operator training to improve the task performance of man-machine cooperation by increasing the expertise of the operator. This is valuable especially for those remote handling tasks which are needed rarely, but have to be done fast and safe.

4. REMOTE HANDLING WORKSTATION

To support the supervisory tasks of the operator two basic tools were defined: a *spatial simulator* and a *functional simulator*, both implemented as special workstations, combined to the so called remote handling workstation (RHWS), which represents the supervisory level of the control system. The basic simulators will be run with different user shells, suited to the various phases of supervisory work. Our actual implementation work is concentrated on the *spatial simulator* (KISMET: Kinematic Simulation, Monitoring, and programming Environment for Telemanipulation). Figure 5 shows the architecture of the kernel. Special features of KISMET are: (1) hierarchical data structure for variable degrees of detail, (2) flexible kinematic model (trees, loops), (3) programming support by workframes, (4) enhanced collision detection. KISMET can be adapted to various applications as JET ex-vessel work and NET sections by simply exchanging the databases for the work environment and the manipulators and, eventually, a few specific modules for dedicated kinematics, operator interface, and process interface. KISMET is now implemented in a test version.

The *functional simulator* acts on data structures like task decomposition trees, action nets, instruction scripts, state graphs, or nets representing the functionality of processes (schematics). Work in this area is only in a conceptual design phase.

5. MOTION CONTROL

The integration into a supervisory control system poses special demands on the motion control, not usual in conventional controls for telemanipulation or robotics. These requirements result from the close interaction between the supervisory control level and the motion control. The central feature in this context is the possibility of *interrupting* and *influencing* running programs as a consequence of supervision. The following features are necessary: (1) abortion of a program at once or after the actual instruction, (2) interruption, performance of manual or other programmed work, and resumption after an admissibility check by a special monitor, (3) continuous manual corrections of a program (overwrite) or discrete offset processing. Investigations in subtask automation and supervision are underway in the CATROB [4] project. Other requirements are: single command execution, attaching of special options to MOVE commands (force reflection, geometric restrictions, jigs etc.), backtracking. In each working state of the motion controller user program and operating system parameters must be readable (interrogation commands) and writable for displaying the state of the controller and for adapting the behaviour of the controller to a special task phase (e.g. new payloads).

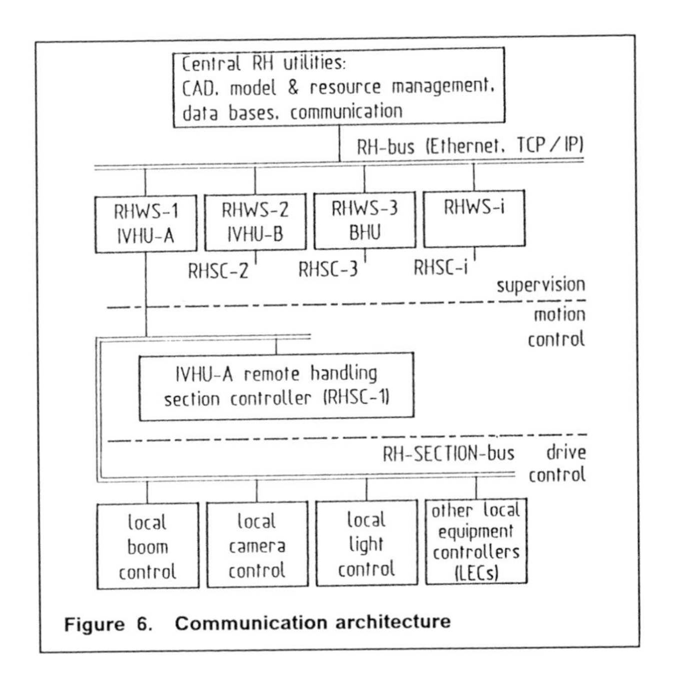

Figure 6. Communication architecture

6. COMMUNICATION ARCHITECTURE

The whole NET control system is partitioned into so-called *remote handling sections*, each dedicated to one manipulator or a group of closely related manipulators (Figure 6). The communication between sections is done via the RH bus. For this type of communication a special message transfer protocol has to be defined. The sections (e.g IVHU section) will get task-suited models by a model manager, an extension of a standard CAD-system using a standard file transfer protocol. On the task or supervisory level of the control system, each section is equiped with a remote handling workstation (RHWS, see chapter 5). The motion/path control level software will run on the remote handling section controller (RHSC), a general purpose process computer

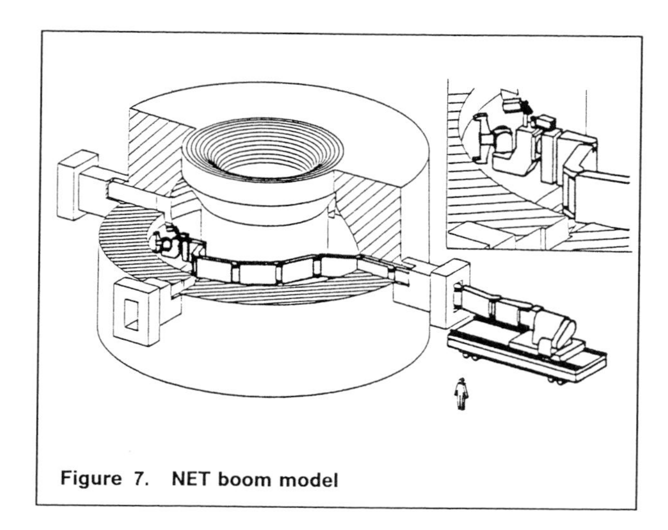

Figure 7. NET boom model

with real-time multitasking operating system. This computer coordinates the work of the local equipment controllers. A special RH section bus with real-time characteristics serves for communication in a section.

7. CAD-MODELLING FOR CONTROL

Geometric information plays a central role in the supervisory control system concept and therefore the generation and management of this information is emphasized by special tools and suited techniques. Geometric and kinematic modelling requires a qualified editor. Hence, we recommend to do it using a standard CAD system which is suited for 3D modelling. Design oriented kinematic simulations on the CAD system are supported by our ROBOT program package. The model data of the environment and the manipulators (Figure 7) is downloaded to the real-time simulator using the CAD*I data format for exchange of solid model data between CAD systems. Postprocessors generating this data format were developed for APPLICON-BRAVO and IBM-CATIA. The transferred models are used, among other things, for path programming. The resulting path geometry is submitted to the motion control system using the IRDATA format, known from industrial robotics (Figure 5). Consistency between model and reality is verified visually comparing TV pictures and synthetic images or by using a theodolite based remote surveying system [5].

8. IMPLEMENTATION FOR EDITH

The first experimental installation of a supervisory control system will be done for EDITH (Experimental Device for In-Torus Handling) by adapting the KISMET kernel to the EDITH environment for planning, programming, and monitoring the experiments. Especially the controllability of handling heavy loads will be investigated to test basic control system components and handling procedures. The task of load transfer will be broken down into subtasks to find out, where automation (by teach-in and repeat) is possible and whether special sensor support is required for working under restricted observability.

9. CONCLUSIONS

The implementation and tests of the JET graphics workstation pointed out, that a three-dimensional geometric and kinematic simulator is best suited to serve as central support unit for remote handling control systems particularly in complex work environments without direct observability. The generalisation of this tool and its extension to an integrating factor in a supervisory control system is a most promising advancement.

ACKNOWLEDGEMENT

This work has been performed in the framework of the Nuclear Fusion Project of the Kernforschungszentrum Karlsruhe and is supported by the European Communities within the European Fusion Technology Program.

REFERENCES

1. T. Raimondi, L. Galbiatti, P. Jones, Use of Teleoperators and Transporters in JET, in: Proc. IAEA meeting, Karlsruhe, in print

2. A.C. Rolfe, The Integration, Control and Operation of JET Remote Handling Equipment, in: Proc. IAEA meeting, Karlsruhe, in print

3. U. Kuehnapfel, K. Leinemann, E.G. Schlechtendahl, Graphics Support for JET Boom Control, in: Proc. Remote Systems and Robotics in Hostile Environments, Pasco, WA., (1987) pp.28-33

4. J. Benner, K. Leinemann, E. Stratmanns, W. Till, C. Walz, Automation of Remote Maintenance for Fusion Machines, this volume

5. B. Köhler, GMS - A high precision Geometry Measurement System for Large Fusion Reactor Components, in: Proc. IAEA meeting, Karlsruhe, in print

FUSION TECHNOLOGY 1988
A.M. Van Ingen, A. Nijsen-Vis, H.T. Klippel (editors)
Elsevier Science Publishers B.V., 1989

SURVEYING OF LARGE FUSION REACTOR COMPONENTS

Bernd KÖHLER

Association KfK-EURATOM, Kernforschungszentrum Karlsruhe, Institut für Reaktorentwicklung, P.O. Box 3640, 7500 Karlsruhe 1, West-Germany

For CAD-supported remote maintenance of fusion machines a surveying system has to measure the geometrical changes of the reactor components. A study of principally applicable techniques indicated that triangulation with theodolites is well suited to update the CAD-models of fusion reactor components. The remote-controlled and CAD-supported surveying system GMS (Geometry Measurement System), developed by KfK, is equipped with two digital theodolites, a laser- and a camera-theodolite, completely controlled by a computer. The data transfer CAD - GMS will be realized with the standardized IGES-interface. To show the feasibility of this draft a GMS-prototype, equipped with a single camera-theodolite, is built up presently.

1. *INTRODUCTION*

For *remote maintenance* of fusion machines exact informations about the *actual geometry* will be needed. The geometry of fusion reactor components may change during the lifetime of the machine, and for that reason a *surveying system* has to measure the *geometrical* changes in decided time intervals.

As part of the investigations of *remote handling techniques* for NET, KfK (Kernforschungszentrum Karlsruhe) presently is developing the surveying system GMS (*Geometry Measurement System*) for fusion reactor components. The GMS has to meet these main requirements :

- fully remote-controlled
- resolution 10^{-4}
  (1 mm accuracy for 10 m object distance)
- CAD-supported

The surveying system gets the nominal geometry from a *CAD-system* (Figure 1).

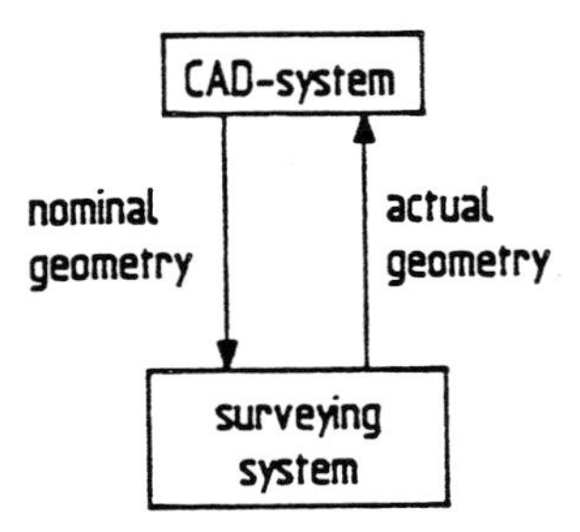

Figure 1. CAD-supported surveying system

The *actual geometry*, determined by the surveying system, is transferred back to the CAD-system and the detected geometrical changes can be considered by other *CAD-supported processes*.

A study of *different surveying techniques* have the result that the *triangulation with theodolites* is the surveying method most suitably fitted for NET[1]. This surveying technique is already approved in *industrial* use (e.g. surveying of car bodies) and with an optimum location of the theodolites a *resolution* of 10^{-5} can be obtained[2].

2. *TRIANGULATION WITH THEODOLITES*

The theodolite is a geodetic instrument which is used for measuring *horizontal* and *vertical* angles[3]. Aiming with two theodolites to the same object point, the *3-D-coordinates* of this point can be calculated with the measured horizontal and vertical angles.

The geodesy knows two different ways for the surveying by triangulation :

- the intersection
- the resection

2.1 The intersection

The theodolites are set up on a baseline appropriately with respect to the object to be measured (Figure 2). For *initialization* each theodolite is precisely aimed at the other's crosshair. This serves to zero the azimuth of each instrument. Aiming with the two theodolites to the target points A, B of the *scaling bar* the base distance can be calculated[4,5].

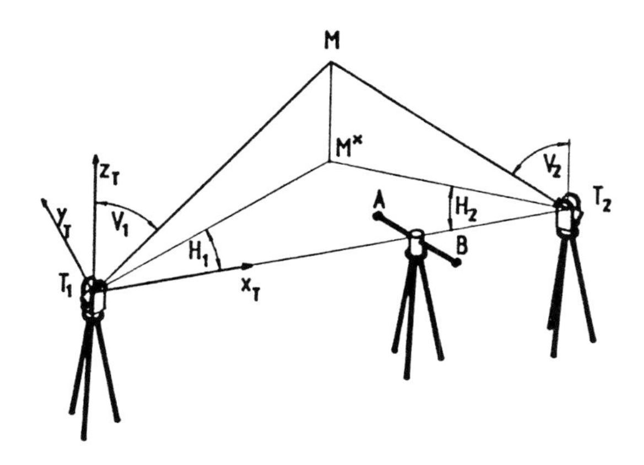

Figure 2. The intersection

After completion of the set-up procedure sightings are made to the points of interest on the object. With the measured theodolite angles the location of the object points M can be calculated [4] in the *local coordinate system* x_T, y_T, z_T.

2.2 The resection

The design of the surveying system for the *resection* looks like the system for the intersection (Figure 3). Instead of a scaling bar three target points R_1, R_2, R_3 with *wellknown* coordinates are needed (using more reference points will improve the accuracy).

Pointing with the theodolites to three *reference points* (as minimum) the theodolite positions in the *fixed*

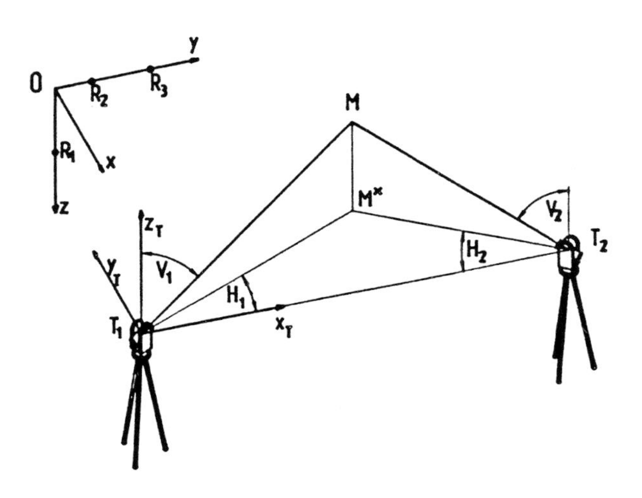

Figure 3. The resection

coordinate system x, y, z (e.g. reactor hall) can be calculated[4,5]. After this initialization the theodolites aim to the points of interest on the object. The positions of the object points in the local coordinate system are determined with the intersection method. Because of the wellknown positions of the theodolites, the *local* coordinates of the object points can be transformed into the *fixed* coordinate system.

2.3 Comparison between intersection and resection

To calculate the base distance by pointings to a scaling bar is more *accurate* then the calculation by the resection method. Therefore it will be advantageous to use the *intersection* for surveying of *relative* geometrical dimensions (e.g. diameter of bore, distances between two points). If the location of the object points are needed in a *fixed* coordinate system the *resection* method must be applied.

3. THE GEOMETRY MEASUREMENT SYSTEM (GMS)

The *surveying system* GMS consists of two digital theodolites operating in parallel (Figure 4).
To comply the required *remote control* the horizontal, the vertical and the focusing drive of the theodolites are actuated by *computer-controlled* electrical motors. For marking the aiming points one of the two instruments is equipped with a *He-Ne-laser*. The operator can observe the theodolite pointings with a *CCD-camera* mounted upon the telescope of the other theodolite.

The *nominal* geometry is made available from a *CAD-system*. Additional to the wire-frame geometry the CAD-data contain the nominal positions of the measuring points. On the one hand these CAD-data are used for building up a *graphical scene presentation* for the operator support and on the other hand for *automatically turning* the theodolites into the nominal direction of the measuring points. The actual positions of the object points are determined by the *remotely corrected* bearings of the theodolites (chap. 3.1). These results are transferred back to the CAD-system via a standardized format (chap 3.2) and the surveyed geometrical changes can be used to any *CAD-supported systems.*

3.1 The surveying process

At the beginning the operator aims with the theodolites to the *reference points* by moving a joystick. For surveying with the intersection method the operator

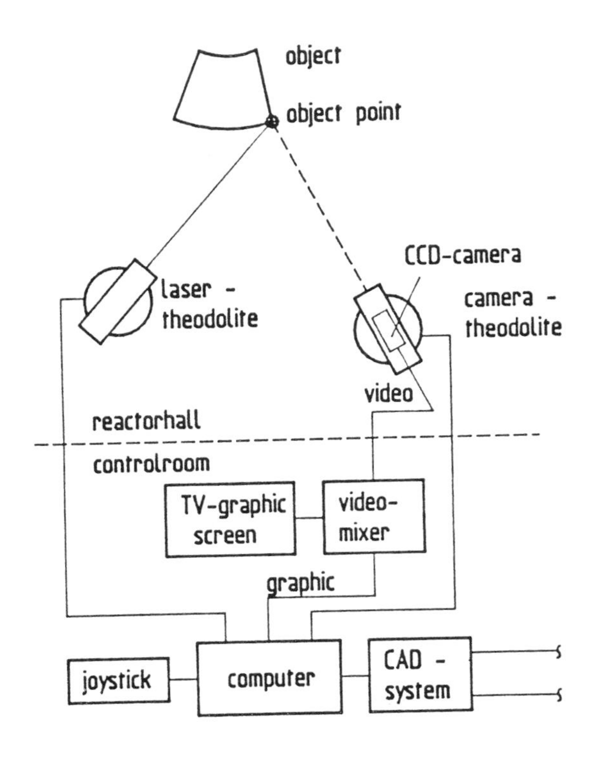

Figure 4. Configuration of the GMS

must also aim to the target points of the *scaling bar*. In this case the bearing to the reference points is only needed for turning the theodolites *automatically* into the direction of nominal points.

The theodolites automatically turn into the direction of the first object point. Using a *joystick* the operator adjusts the theodolites until they aim exactly to the *actual* position of the object point. Reading the angles of the theodolites the actual point coordinates are calculated by the computer.

The same procedure then is repeated for all other measuring points.

After aiming at the last object point the operator gets a list containing the nominal and the surveyed positions of the object points. In case the operator supposes a measurement error he can repeat the surveying of any object point.

The measured values then are transferred to the CAD-system and the surveyed geometrical changes can be used by other processes.

3.2 Data transfer CAD-GMS

For the *data transfer* between a CAD-system and the GMS the standardized *IGES-format* is employed[6]. This data format is used for the transfer of *wire-frame* geometries and the IGES-interface is already available for most of the CAD-systems. To convert the IGES-file into a data format suitable for the surveying process software for the interfaces is needed (Figure 5).

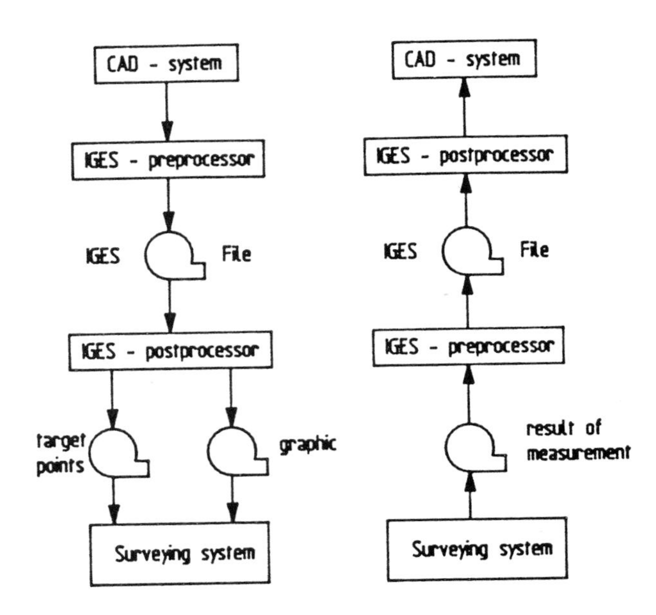

Figure 5. The data transfer CAD-GMS

For the data transfer CAD → GMS the *IGES-postprocessor* generates two data files; one containing the *nominal positions* of the measuring points and the other including the *wire-frame geometry* for the graphical support of the operator.

To transmit the surveying results to the CAD-system an *IGES-preprocessor* has to transform the result file into the IGES-format.

4. THE GMS-PROTOTYPE

The triangulation with theodolites is already employed in *industrial applications*[7], but these applications do *not* require fully remote-controlled operations. Therefore a prototype system has to be developed which is suitable for *completely* remote control.

Compared to Figure 4, a cheaper but less flexible implementation is chosen; only a single *servo-driven* theodolite (Figure 6) is used for the prototype system. The horizontal, the vertical and the focusing drive are equipped with *computer-controlled* stepping motors. A *CCD-camera* is mounted upon the theodolite telescope and a *switchable prismatic optic* allows to select the detailed image of the theodolite telescope or the complete surveying scene via a separate wide-angle objective.

Using a single theodolite, at first all the target points have to be aimed from the first theodolite position and the sighted angles are *stored* in a data file. After moving the theodolite to the second position the theodolite pointings has to be done in the *same order* and the point positions are calculated with the measured angles. For the surveying with a single theodolite it has to be guaranteed that all the target point positions are absolutely *fixed* during the whole surveying process.

The *present* development stage of the GMS-prototype is shown in Figure 7. The camera-theodolite is *completely* controlled by an INTEL-310 via RS-232 interfaces. Using joystick and keyboard the operator can *manually* control the theodolite. For controlling the surveying process and for calculating the actual point positions suitable software was developed by KfK. During the whole surveying process the operator is interactively *guided* by the GMS-software and several theodolite parameters (e.g. speed of the stepping motors) can be changed *menu-driven*. The *IGES-postprocessor* for the data-transfer CAD → GMS is already built up and successfully tested[8]. To transfer the surveying results to a CAD-system an *IGES-preprocessor* will be presently developed.

Figure 6. Computer-controlled camera-theodolite

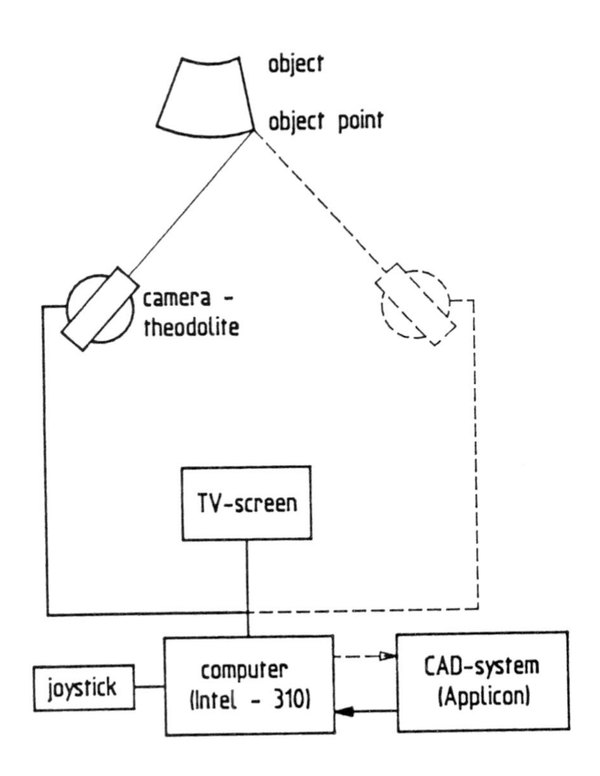

Figure 7. GMS-prototype (September 1988)

Up to now the operator is *not* supported in recognizing the target points. For *marking* the positions of the target points a computergraphic image will be selectively *superimposed* the TV-picture. This *graphical support* for the operator will be realized using an IBM-AT, equipped with a special image processing board.

After first surveying tests with an aluminium model, the GMS-prototype is already been used for two *real* surveying tasks :

• surveying the positions of thread feeders[9] which are used for dynamic surveyings of the TFTR- Maintenance-Manipulator movements
• surveying the joint positions of EMIR[10] (Extended Multi Joint Robot)
The results of these surveyings with the GMS-prototype demonstrated that the required accuracy can be complied. The surveying process with the GMS-prototype is already completely remote-controlled except the *leveling* of the theodolite. Up to now the operator has to level the theodolite *manually* using inclination adjusting screws. To realize the complete remote control either a servo driven leveling-mechanism has to be developed or the software has to be modified. For the software-solution the point coordinates and the inclination of the theodolite have to be calculated *iteratively*. It is not yet decided if the hardware- or the software-solution will be preferred. Based on approximate estimations of errors it is questionable if the required accuracy can be obtained applying the software-solution.

5. CONCLUSIONS AND PROSPECTS

The *triangulation with theodolites* is a suitable technique for surveying large fusion reactor components. The GMS is a fully *remote-controlled* and *semi-automated* surveying system which is able to *communicate* with CAD-systems via the IGES-interface. The first surveying tests show that the *required accuracy* can be obtained; for detailed statements about the resolution special accuracy tests will be needed.

Using a single camera-theodolite the final bearings to the target points has to be done *manually*. If a second laser-theodolite is used additionally the *automatic control* of the surveying process can be enhanced. The laser-theodolite *scans* the object and with the nominal CAD-data the camera-theodolite follows this laser spot. Applying *image processing techniques* the marking laser spot can be recognized and the actual point positions are calculated.

Acknowledgements

This work has been performed as a study contract for NET in the framework of the Nuclear Fusion Project of the Kernforschungszentrum Karlsruhe and is supported by the European Communities within the European Fusion Technology Program.

References

1. SCHRÖDER, J., Fernvermessung von Fusionsanlagen, Primärbericht 030503P18B (unpublished), Kernforschungszentrum Karlsruhe (1984)
2. KATOWSKI, O., RMS2000 - ein System zum berührungslosen Messen großer Objekte, Allgemeine Vermessungsnachrichten Heft 6 Juni (1985), Herbert Wichmann Verlag, Karlsruhe (1985)
3. DEUMLICH, F., Instrumentenkunde der Vermessungstechnik, VEB-Verlag, Berlin (1972)
4. KÖHLER, B., Remote-controlled surveying system for NET, Primärbericht 170101P02A (unpublished), Kernforschungszentrum Karlsruhe (1987)
5. JORDAN/EGGERT/KNEISSL, Handbuch der Vermessungskunde, J.B. Metzlersche Verlagsbuchhandlung, Stuttgart (1963)
6. SMITH, B., Initial Graphics Exchange Specifications (IGES), Version 2.0, National Bureau of Standarts, Washington (1983)
7. KARTZOW, W., Einsatzbeispiele für dimensionales Messen mit optischen Koordinatenmeßgeräten, VDI Berichte 529, VDI Verlag, Düsseldorf (1984)
8. BING, F., Entwicklung von Schnittstellensoftware zur KommuniKation eines INTEL-Rechners mit einem CAD-System, diploma work (unpublished), Institut für Reaktortechnik, University (TH) Karlsruhe (1987)
9. KÖHLER, B., Einmessen der Fadengeber für Vermessungen am TFTR-Maintenance-Manipulator, Bericht IRE/6/039/88 (unpublished), Kernforschungszentrum Karlsruhe, Institut für Reaktorentwicklung, July 1988
10. SMIDT,BLUME,WADLE, A multi-link multi-purpose advanced manipulator with a large handling hemisphere for out-door applications, 5th. int. Symposium on Robotics in Construction, Tokyo, June 1988

FUSION TECHNOLOGY 1988
A.M. Van Ingen, A. Nijsen-Vis, H.T. Klippel (editors)
Elsevier Science Publishers B.V., 1989

AUTOMATION OF REMOTE MAINTENANCE FOR FUSION MACHINES

Joachim BENNER, Klaus LEINEMANN, Erwin STRATMANNS, Walter TILL, Cornelia WALZ

Kernforschungszentrum Karlsruhe, Institut für Reaktorentwicklung,
Postfach 3640, D-7500 Karlsruhe, FRG

The CATROB telerobotic system is a prototype for remote handling equipment in hazardous environments like a fusion reactor. It consists of an industrial robot, mounted on a gantry crane with two degrees of freedom. The basic functional principle of this system is the synthesis of sensory-guided automatic and operator-controlled manual operation in order to minimize the time needed for performing remote maintenance and inspection. The goal is to achieve a high level of automation by using intelligent control strategies based on a CAD model of the environment and the information of various sensors. If manual remote handling is unavoidable the operator is supported by a graphics based simulation system. In the present state, autonomical bolting and unbolting of flanges and handling of plug connections is realized with CATROB. In the paper hardware and software architecture of the system and its sensory equipment are described, and results on the system performance are reported.

1. Introduction

Remote maintenance is one of the key issues for technologies that are applied in hazardous environments. A fusion reactor is such an example where, due to the high radiation loads, all operations for inspection, repair and replacement of components need to be performed remotely. The state of the art for this kind of work ranges from impact wrenches hanging from a crane to force-reflecting master-slave manipulators. Such devices will surely play an important role in fusion reactor maintenance. Nevertheless, it seems to be impossible to perform all the extremely complex tasks like blanket segment exchange within reasonable times completely with manually controlled handling equipment. We therefore started the *CATROB* (*C*omputer *A*ided *T*elemanipulation with *ROB*ots) project [1, 2], which aims at developing the prototype of an advanced telerobotic system as far as possible automating remote maintenance operations. It consists of an industrial robot mounted on a two-dimensional gantry crane as carrier system.

Total automation of remote maintenance in a fusion reactor environment is neither possible nor desirable for the near future, mainly due to the occurence of unforeseen events and failures. So the operator will remain as an important factor in the control loop, who supervises and eventually stops automatic task execution in order to perform manual corrections. For the CATROB system this means that a suited synthesis of manual and automatic operation is a central design goal. It is achieved by an intelligent control system following the principle of *man-machine cooperation*[3], and an elaborate man-machine interface including a graphical simulator.

For automating remote maintenance tasks two different types of operation can be distinguished. On the one hand there are the *auxiliary operations* like transport of components or control and positioning of cameras, and on the other hand the *real maintenance and repair operations* like bolting, cutting, welding or applying specialized tools. With the CATROB device it is envisaged to automate both types of operation by using different techniques.

Automation of auxiliary operations is quite simple because the needed positioning accuracy is low. So, automatic path planning for transporting the robot or camera tracking of the robot hand can be done with *nominal geometry data*. In CATROB, these data are provided by a CAD system and transfered via a standardized data format [4].

The automation of assembly operations like bolting or grasping is much more complicated. Here a very high positioning accuracy is needed. Due to inaccuracies in the nominal geometrical data (caused e.g. by thermal expansion or radiation-induced creep) and errors in positioning the carrier it is not possible to operate the robot with simple "teach and repeat" or to derive robot programs only from the nominal geometry data. To overcome this problem, the CATROB system provides for *different kinds of sensors* adapting these programs to the actual situation.

Actually the CATROB system is trained to handle various types of bolted flanges, as well as electrical and pneumatical connectors. Handling means the automatic execution of different maintenance and repair tasks like, in the case of a bolted flange, assembly and disassembly of a jumper, exchange of bolts, removal of a mother plate or insertion of sealings. In the first stage which is presently underway these components are treated under *idealized conditions* (isolated components, normal operational conditions). Furthermore, we concentrate on developing control strategies and actually do not discuss hardware related problems like radiation resistance of components and sensors.

2. Overview of the CATROB Telerobotic system

Figure 1 shows the architecture of the CATROB control system consisting of three different levels.

- The highest level called *task control level* incorporates the *man-machine interface*. On this level the remote handling operator plans the maintenance work and commands and supervises the remote handling equipment.
- On the intermediate *motion control level* of the control system operator commands like the execution of an automated subtask are translated into instructions for carrier, robot and sensors.
- On the lowest *drive control level* there are special purpose controllers for the handling equipment (carrier, robot) and two sensor systems (tactile force-torque-sensor (FTS) and vision system).

The function of these hardware and software components is described subsequently.

2.1. Task control level

Central component of the man-machine interface is the graphics based simulator GBsim[5, 6]. It is implemented on a high speed graphic workstation IRIS 3020 and has originally been developed to support the JET-boom control. For CATROB a specially adaptėd version is used supporting the following features:

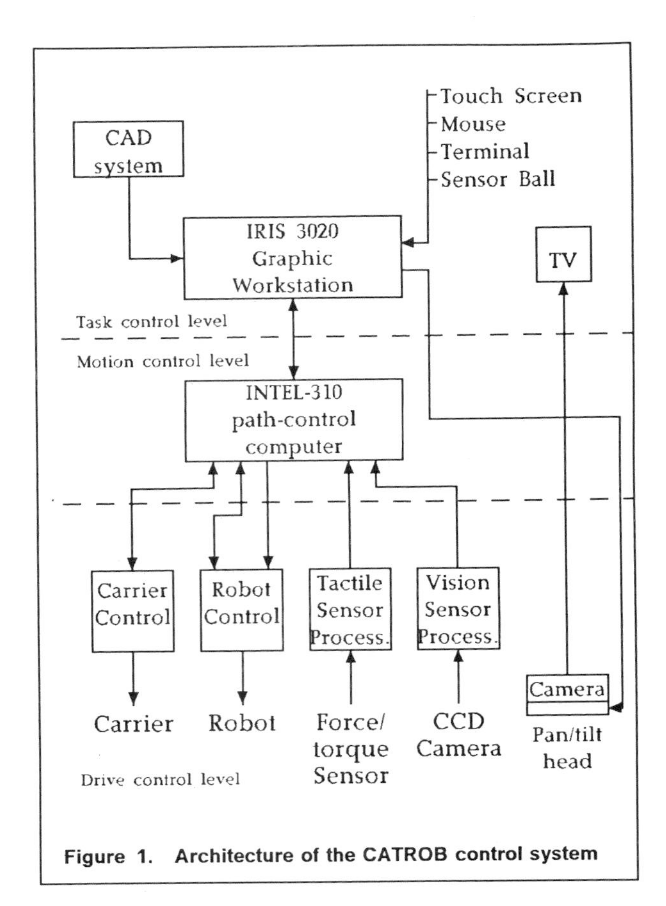

Figure 1. Architecture of the CATROB control system

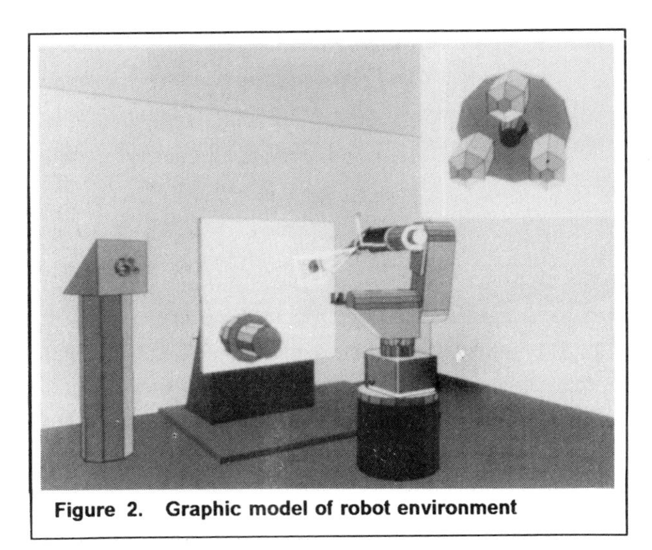

Figure 2. Graphic model of robot environment

- Graphical presentation of environment, robot and carrier (Figure 2)

- Graphics based simulation and off-line programming facility for robot and carrier
- Simulation and control of viewing cameras
- Model based collision detection

The man-machine interface supports different input devices like terminal, touch screen, and mouse. For manual robot control, a sensor ball with 6 degrees-of-freedom can be used.

2.2. Motion control level

The *path control computer* (INTEL 310 microcomputer) is responsible for coordinating and supervising the remote handling equipment, and for supervising the overall system configuration. This configuration is described by a state vector containing the state values of the different system components (input devices, graphic system, path control computer, robot and robot control, carrier and carrier control, sensors and sensor data processors), the state values of the end-effectors and of the different handling objects.

2.2.1 Execution of Autonomous Handling Modules

A remote handling subtask like unbolting a flange connection and extraction of the jumper consists of a number of different steps. For this example these are

- Transporting the robot to the flange by carrier
- Positioning of supervisory cameras
- Localizing the flange by vision system
- Applying a bolting tool with support of force-torque sensor and bolt-loosing
- Gripping and extracting the jumper flange with support of force-torque sensor
- Removing the jumper with carrier

The example shows that an automated subtask can be split into separate building blocks called *Elementary Functions (EFs).* Each EF corresponds to a single, specialized function of the handling system like robot end-effector exchange, joining with force-torque sensor or localizing with the vision system. For automatic subtask execution the EFs can be combined and form together an *Autonomous Working Module (AWM).*

Of course, the combination of EF is not perfectly free. For starting an EF, the system must be in a suited state. This is necessary in order, e.g., to prevent starting an EF "Bolting" when the robot has no bolting tool.

Therefore, every EF is described by a data record called *script* (Figure 3) where the following informations are stored:

- *IDENT:* Identification of the EF
- *OBJECTS:* Remote handling objects (like flanges, bolts or reference marks) which are *processed* by the EF
- *RESOURCE:* Resources like system components or end-effectors which are *needed* by the EF
- *STATES:* Information about system states (i.e. states of objects and resources)
  - Necessary for starting the EF (INIT)
  - Set after correct termination of the EF (EXIT)
  - Set in error case (ERROR)
- *MAX_TIME:* Maximally allowed execution time for the EF
- *PROGRAM:* Identification of a corresponding program (robot, carrier, sensor)
- *PARAM:* Information about input/output parameter of the EF.

IDENT	EF "Sensor supported bolting"				
OBJECTS	Bolt				
RESOURCE	Robot, FTS, Impact Wrench (IW)				
STATES		Bolt	Robot	FTS	IW
	Init	Loose	Active	Active	Mount
	Exit	Fixed	Inact	Inact	Mount
	Error	Not Known	Inact	Inact	Mount
MAX_TIME	10 sec				
PROGRAM	FTS-Program 'Joining"				
PARAM	Actual bolt position				

Figure 3. Example of an EF script

If the required EF's initial state is not fulfilled, the maximal time elapsed, or an error is reported by a resource, execution of the AWM is automatically interrupted. Furthermore, the operator supervising AWM execution can manually interrupt at any time. In both cases the operator afterwards has to perform manual operations in

Figure 4. Carrier system: X-Y gantry crane

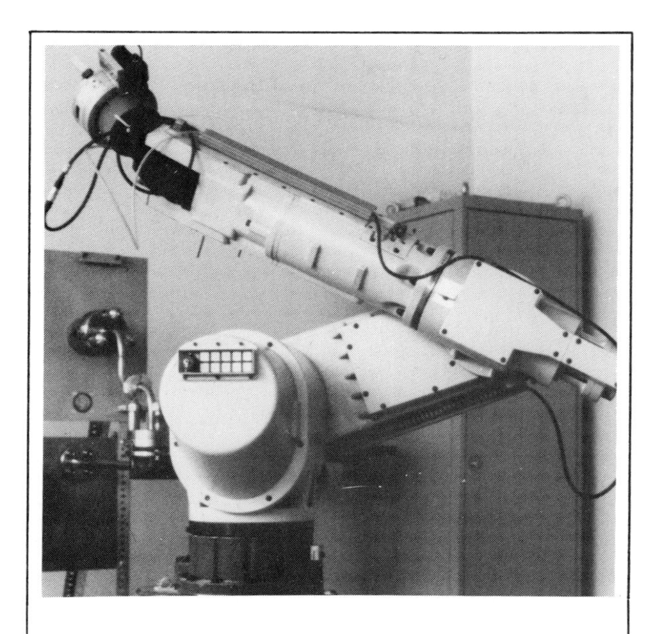

Figure 5. Industrial robot MANUTEC r3

order to correct the failure. If this has been successful, automatic execution can be started again.

2.3. Drive control level

Actually four hardware components and corresponding processors are available.

2.3.1. Carrier system

As carrier a two-dimensional gantry crane is used (Figure 4) with an operating range of 4*4 meter and a maximal velocity of 1 m/sec for every axis. It carries a payload of 500 kg, positioned with 1 mm accuracy. A positional control for the gantry crane is just under development. In the final state it will enable to control the carrier by an external computer. Among other features, continuous path control with predefined velocity and acceleration, viapoint motion with smoothing and sensor influencing the carrier motion will be realized.

2.3.2. Robot

The industrial robot MANUTEC r3 (Figure 5) with the RCM 3.2 control has been chosen. It has 6 rotatory axes, a load capacity of 15 kg, and a repeatability of 0.1 mm. The robot control is equiped with two serial interfaces to the path control computer.

2.3.3. Sensor system 1: Binary vision system

At the moment, two kinds of sensors are used in the CATROB system for recognizing and localizing remote handling objects: A binary vision system and a tactile force-torque sensor. In our concept both sensors are combined. First an object is coarsely localized by the vision system, and afterwards this result is corrected with the tactile sensor in cases where high accuracy is needed.

In the present state of the CATROB system it is not yet possible to recognize real objects like flanges with the vision system, but *artificial reference marks*, where the location of the real objects relative to a reference mark is known. This is a much easier task, because the system only has to recognize one specially selected object with high contrast. Therefore, a relatively simple binary system can be used.

The image processing algorithm follows the stereo vision principle. Two pictures of the reference mark are recorded from two different positions, for which one CCD camera is brought to two different positions by the robot ("eye in hand" principle). In the two pictures corresponding features are detected. After the correct match has been found, the mark's location is calculated by triangulation. The image processing and matching algorithm is further discussed in [1]. For the future it is planned to develop strategies to directly recognizing and localizing handling objects by using an advanced gray-level vision system with real-time image processing capabilities.

2.3.4. Sensor system 2: Force-torque sensor

The force-torque sensor is mounted between robot hand and end-effector and measures the applied forces and torques when the end-effector is in contact with the environment. After digitizing and preprocessing, the measured values are processed in a force-control algorithm superimposing the robot's normal position control loop. Here the measured forces and torques are transformed into a positional deviation, which is used for correcting the end-effector position, as long as some final state of forces and torques is reached. Presently we have implemented one force-control program for joining the nut of an impact wrench and a flange bolt [2].

3. Performance results

Though the CATROB development is not yet finished and up to now the system has not been tested under realistic conditions, some results concerning robot and sensor operation can already be reported. So it has been proven with CATROB that handling tasks of a moderate complexity like unbolting a flange or connecting an electrical plug can be performed with an automatically operating industrial robot, even if the position of the objects is not exactly known. The maintenance times for such operations are a few minutes. The performance of the used vision and tactile sensors are good, only for very delicate tasks regarding components not specially adapted for remote handling a higher accuracy is needed, which can only be achieved with specialized sensors.

4. Conclusions and outlook

In this paper the CATROB telerobotic system consisting of a mobile industrial robot guided by different sensors has been presented. It has been shown that with such a system a high degree of automation in remote maintenance can be achieved. For automating assembly tasks it is absolutely necessary to use sensors for solving the problem of vaguely known object locations. First CATROB performance tests indicate that the maintenance times with robotic equipment are much lower compared to operation with manually controlled devices like master-slave manipulators. So a system like CATROB can surely be an important completion to a master-slave manipulator in a remote maintenance system for a fusion reactor.

Next step in the CATROB development will be the integration of the carrier system. Then it will be possible to perform handling operations under realistic remote maintenance conditions. The vision system performance will be improved in order to avoid using reference marks for robot orientation. Other sensors have to be investigated for their potential of solving special handling problems.

Finally, the problem of operating a sensory guided robot in a radioactive environment has to be solved. For this purpose, special, radiation-hardened components like sensors have to be used.

References

1. J. Benner, K. Leinemann, Architecture of a telemanipulation system with combined sensory and operator control, in: Proc. RoViSeC-7, Zürich, 1988, (IFS, Kempston, Bedford, 1988) pp. 259 - 270.
2. J. Benner, C. Fischer, K. Leinemann, E. Stratmanns, W. Till, Towards more automation for remote maintenance, IAEA Technical Committee Meeting on Robotics and Remote Maintenance Concepts for Fusion Machines, Karlsruhe, 22 - 24 Feb. 1988.
3. J. Benner, C. Blume, K. Leinemann, E. G. Schlechtendahl, M. Wadle: Advanced carrier systems and telerobotics, to appear on: SYROCO 88, Karlsruhe, 5 - 7 Oct. 1988.
4. E.G. Schlechtendahl (Ed.), Specification of a CAD*I neutral file for solids, Research reports ESPRIT, project 322, CAD Interfaces (CAD*I), Vol. 1 (Springer, Berlin, 1986).
5. U. Kühnapfel, K. Leinemenn, E.G. Schlechtendahl, Graphics support for JET Boom control, in: Proc. Int. topical meeting on remote systems and robotics in hostile environments, Pasco, 1987 (ANS, La Grange Park, 1987) pp. 28 - 34.
6. K. Leinemann, U. Kühnapfel, A. Ludwig, CAD-model based Remote Handling Control System for NET and JET, this volume

FUSION TECHNOLOGY 1988
A.M. Van Ingen, A. Nijsen-Vis, H.T. Klippel (editors)

MAINTENANCE OF EX-VESSEL EQUIPMENT

J. STRINGER, C. HOLLOWAY

The NET Team c/o Max-Planck-Institut fuer Plasmaphysik, Boltzmannstr. 2,
D - 8046 Garching

A number of remote handling issues for ex-vessel equipment have been studied for the NET machine. These issues concern the remote handling of components associated with the vacuum vessel sector removal, cut/weld operations on large transfer duct connections for heating and viewing systems. The remote handling of vacuum pumping room equipment is also reported.

1. INTRODUCTION

A maintenance requirement of the NET machine is that a sector of the vessel can be removed during the lifetime of the machine. For this scenario venting of the plasma chamber and cryostat will be required.

The removal of a vacuum vessel sector will necessitate the removal of large and heavy components connected at the mid-plane of the machine. Components to be removed consist of the RH transfer duct, the TV probe equipment, the Neutral Beam duct, the Lower Hybrid Assembly and the Intercoil Structure. In order to remove these components cut/weld tooling will be required to make/break the inboard flange lip seal connections. Concepts for removing these major components, and the pumping room equipment, are presented.

2. SHIELD PLUG HANDLING

The shield plug is normally located at the inboard end of the RH transfer duct and, for normal in-vessel maintenance operations, is removed by the in-vessel handling unit (IVHU). The S/P and removal equipment is an integral part of the IVHU design and not part of this study. For maintenance operations on the machine which requires removal of the RH transfer duct, the S/P and IVHU must be removed to provide space for the installation of the transfer duct RH equipment.

3. TRANSFER DUCT HANDLING

3.1 General

Maintenance of a vacuum vessel sector will require the disassembly and partial removal of a number of components around the periphery of the machine. One of these components is the transfer duct.

Two schemes have been looked at for the remote handling of the ducts - these are 'floor mounted' and 'overhead'.

In addition, failure of the inboard lip seal welded connections will necessitate the complete removal of the duct to replace the seal.

3.2 Description

The RH equipment, as shown in Figure 1, consists of a large mandrel which is used to support the duct at the outboard flange. The duct weight is 15 tonnes and there will be an overturning moment of 75 t-metres acting on the carriage. This equipment is floor mounted.

Cut/weld tooling is located at the front end of the mandrel for making and breaking the inboard joint. Similar tooling mounted on the outboard flange is used to make/break the bellows connection.

The mandrel is supported on a carriage and roller system for installation/removal in a radial direction. The complete assembly is contained within a shielded chamber.

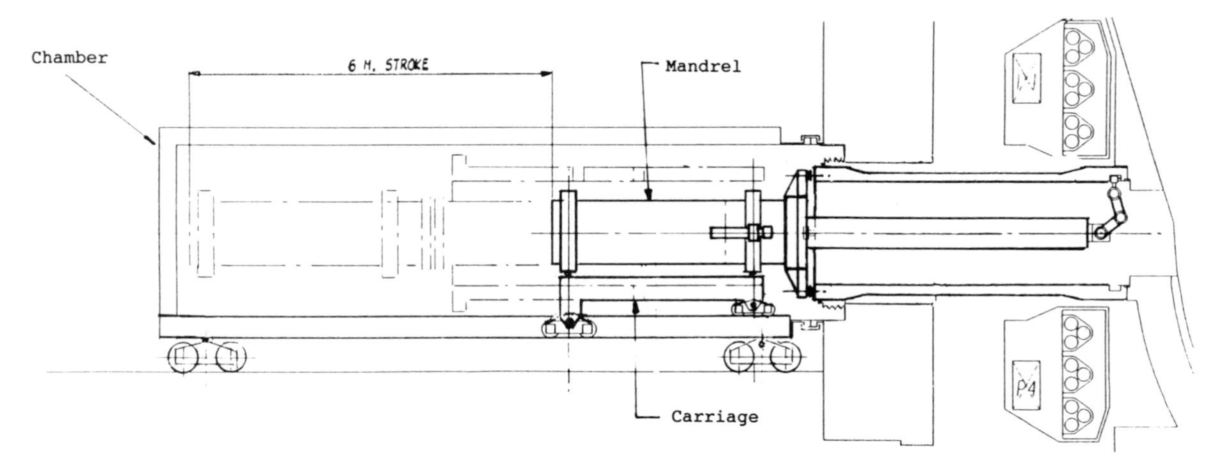

FIGURE 1
Transfer duct RH equipment

An alternative scheme for removal of the duct, using overhead mounted equipment, is shown in Figure 2.

In this arrangement a mandrel is supported by an overhead carriage. The mandrel is attached to the duct on the outboard end in order to support its weight and overturning moment. Cut/weld tooling is located at the inboard end to separate the inboard joint. Similarly, tooling is used to make/break the bellows joint.

The carriage is supported by its own roller support system which in turn is supported by overhead mounted tracks.

In order to provide sufficient space at the inboard end, the transfer duct must be retracted approximately 2 metres. This will provide adequate space for removal of various machine components.

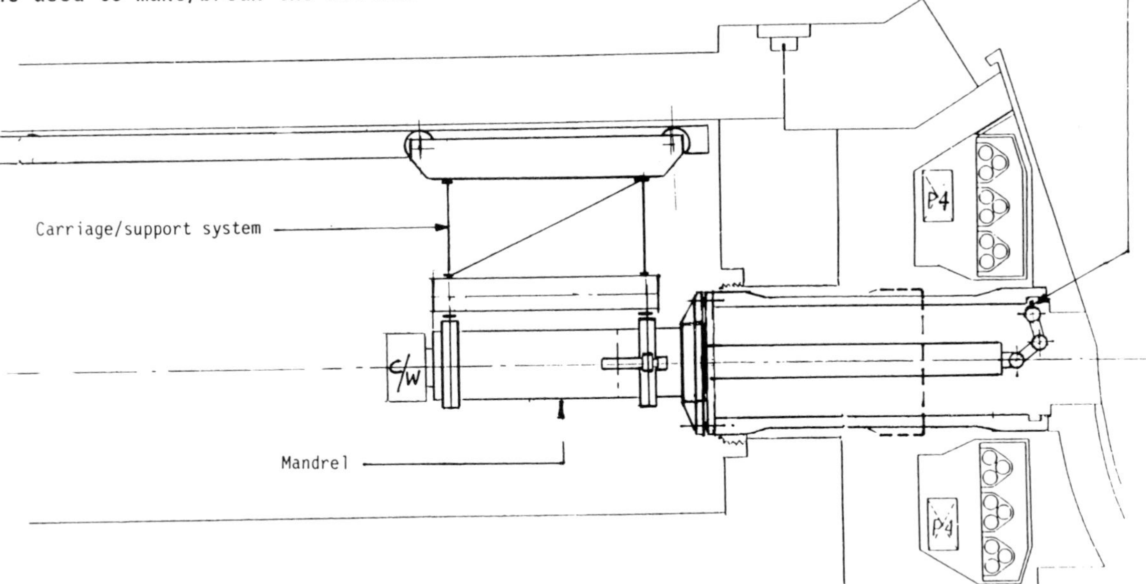

FIGURE 2
Transfer duct O/head RH equipment

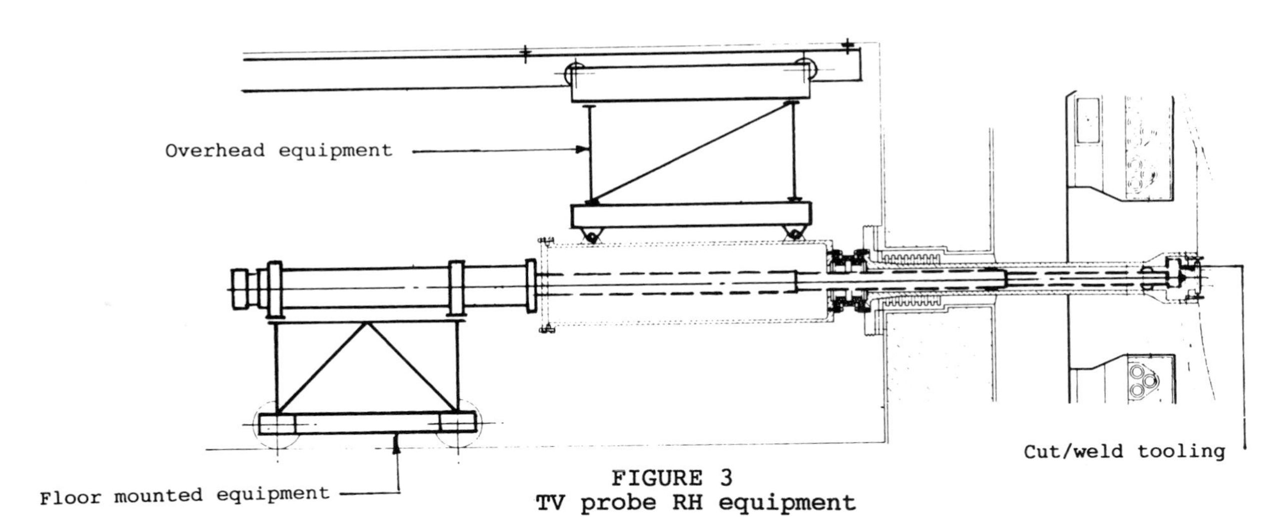

FIGURE 3
TV probe RH equipment

4. TV PROBE RH EQUIPMENT

4.1 General

The TV probe RH equipment is used to handle the probe housing and duct during various maintenance scenarios on the NET machine. The arrangement of the RH equipment is shown in Figure 3.

The equipment consists of two major components both of which are designed to be used sequentially in the handling of the probe. The major components of the system are :

a) Floor mounted equipment
b) Overhead mounted equipment

4.2 Description

Depending on the maintenance scenario the TV probe housing and duct can be removed as one piece or, the housing can be removed independently of the duct. The tooling should be adaptable to both maintenance scenarios.

4.3 Floor Mounted Equipment

This equipment consists of a carriage, telescopic tool and front-end cut/weld tooling all mounted on a carriage roller support system on the floor.

The telescopic tool supports the cut/weld tooling which is deployed down the bore of the probe duct to engage and separate the inboard joint.

In the first operational mode the telescopic tool is designed to have a reach of approximately 10 meters. In this maintenance procedure the housing and duct can be removed as a unit. In the second operational mode the telescopic tool has a reach of approximately 5 meters. In this maintenance procedure the housing is removed from the duct, therefore the stroke requirements of the tool is reduced.

4.4 Overhead mounted equipment

This equipment consists of an overhead mounted track and a carriage to support the hoist equipment. The carriage is designed to remove the TV probe instrument housing and duct (4 tonnes) as a unit, or to lift and store the housing independently of the duct.

5. INTERCOIL STRUCTURE RH TOOLING

5.1 General

Removal of a vacuum vessel sector will require the disassembly and removal of the adjacent upper and lower intercoil structures. These structures are removed using tooling deployed vertically by overhead crane. An arrangement of this tooling is shown in Fig.4.

The purpose of this equipment is to facilitate remote installation and removal of the upper and lower intercoil structures. Each intercoil structure is.7 m x 2.0 m x 2.5 m in size and weighs approximately 22 tonnes (each).

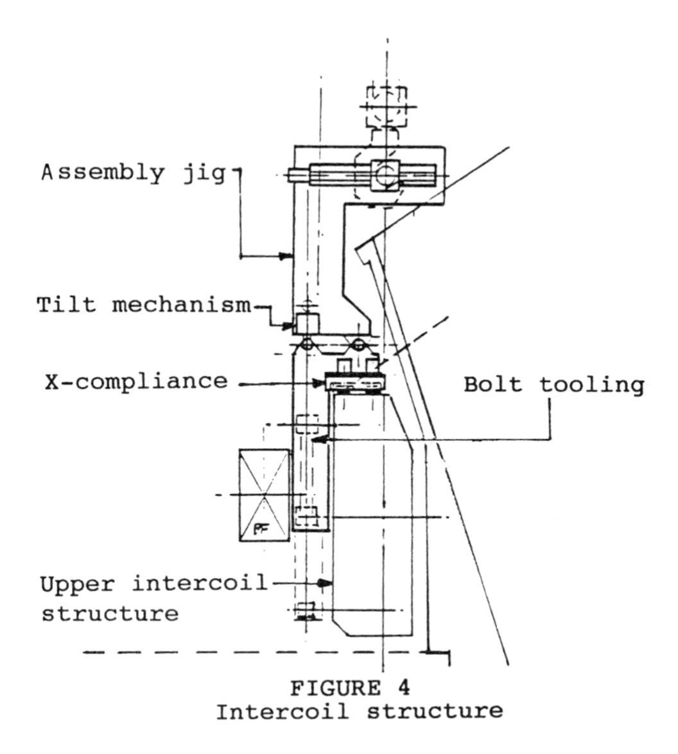

FIGURE 4
Intercoil structure

Essentially, the principle of operation of the tooling is to provide a means of initial support and positioning of the intercoil structure, utilising a fixed reference position on the upper surface of the TF coils. From this reference position, using tooling located on a fixture, fine positioning of the structure is achieved. Compliance in the fixture support mechanism is used to facilitate the assembly operation.

5.2 Installation

Initially, a combination of vertical, horizontal and tilt motions are required to position the intercoil structure in its initial position.

The assembly fixture (and intercoil structure) are deployed on guide holes located as reference points on the TF coil case.

After initial assembly, the structure is jacked radially inwards to engage the TF coil bulkhead. Lateral compliance in the Y-direction is provided by the support fixture. Once the structure is pushed home, tooling located on the fixture is used to assemble and torque the bolts.

5.3 Removal

Initially, the fixture is assembled, by overhead crane, on the guide holes located on the TF coil case. The support fixture is then attached to the upper surface of the structure and the weight of the structure transferred to the fixture.

Tooling located on the fixture is then used to unbolt and retract the studs. Following this operation tooling engages the jacking screws and proceeds to jack the structure free from the TF coils.

6. PUMP ROOM 'RH' EQUIPMENT

6.1 General

There are currently three vacuum pumping systems under study for application to NET. All of these systems will involve the remote handling of very large and heavy pumps. The assumptions made are that the pumping equipment will become contaminated, but that flasking will not be required.

Specifically, the pumping system components in these systems consist of cryopumps (4 tonnes each), TMP's (2 tonnes each), and large shut-off valves (10 tonnes each). The handling of these components, for maintenance reasons, will necessitate both overhead and floor mounted equipment (Figure 5). The RH considerations are applicable to all three arrangements currently under consideration.

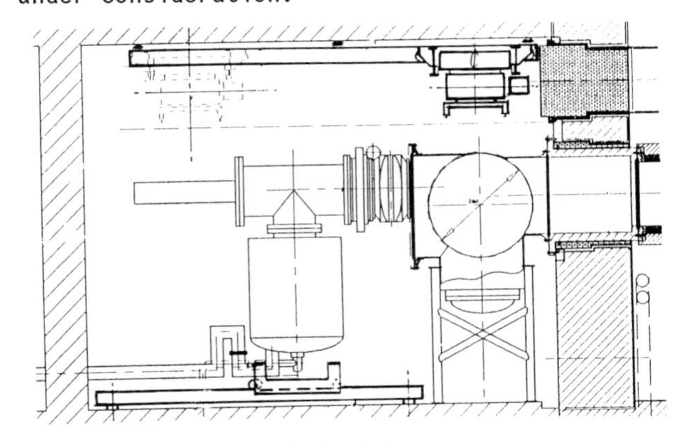

FIGURE 5
Pump room RH equipment

7. CONTROL COIL RH EQUIPMENT

7.1 General

The active control coils (ACC) are two single copper coils, segmented and placed inside the vacuum vessel to control the plasma equilibrium. The upper ACC, is removed with the blanket segments, the lower ACC, also segmented, are connected outside through lower penetrations and are removed from inside the vacuum vessel.

7.2 Equipment description

The RH equipment for maintenance of the lower ACC consists of the following major components :

a) Radially operated equipment with boom extension and front end tooling for removing the shield plug and connecting and disconnecting the control coil connectors. The equipment is overhead mounted in the pump rooms, and the connectors are reached via a horizontal access duct.

b) Vertically mounted equipment from the bridge mounted beam consisting of a strongback, rotary joint and spreader beam. This equipment is used to remove a lower ACC segment from within the vacuum vessel. The equipment is illustrated in Figure 6.

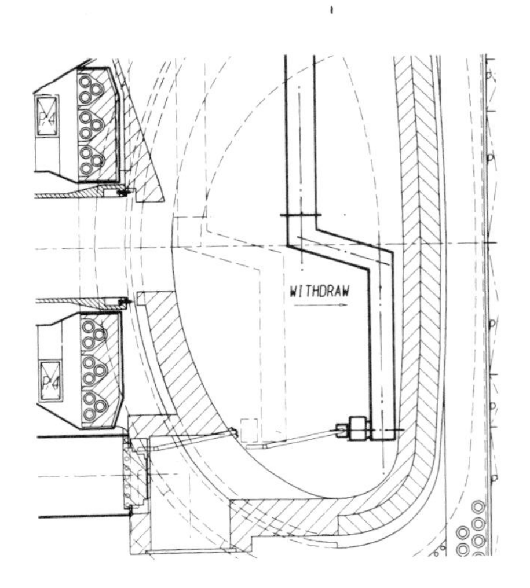

FIGURE 6
Lower ACC RH equipment

8. CUT/WELD TOOLING

There are a number of flange joint configurations currently proposed for the NET machine. These joint sizes vary from 2 x 1 m rectangular transfer ducts to the small 310 mm diameter duct used on the TV probe instrument housing. Because of the activation and contamination, remote cut/weld operations of flange lip seals will be required.

In terms of reducing the number and complexity of tools for the wide range of joint sizes, the preferred approach on NET is to standardize and simplify the tools and make these applicable to all joints.

Figure 7 shows the welding tool used on JET. This tool uses a tungsten inert gas (TIG) system and the automated trolley runs along the lip joint with no need for additional guides. The rollers bring the lips together reducing the gaps. Reliable welds are achieved on lip seals of 2 mm thick material. The driving rollers are mounted on a toe-in angle and the trolley can negotiate sharp turns.

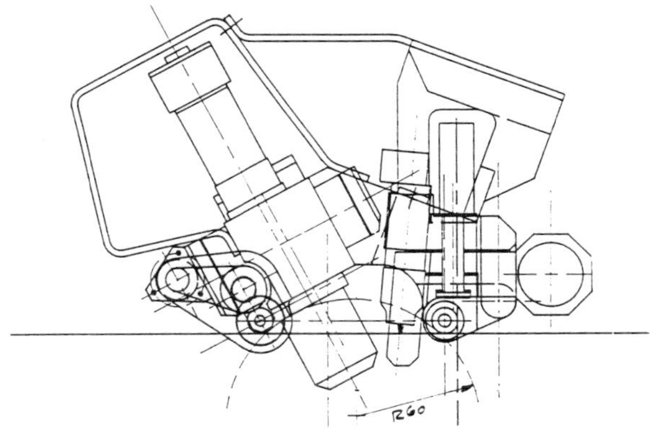

FIGURE 7
Welding tool

ACKNOWLDEGEMENTS

The authors wish to thank JET Remote Handling Groups for information on the cut/weld tooling. The authors are on assignment from the following companies in Canada:

J. Stringer Wardrop Engineering Inc., Toronto.

C. Holloway Spar Aerospace Ltd., Toronto.

FUSION TECHNOLOGY 1988
A.M. Van Ingen, A. Nijsen-Vis, H.T. Klippel (editors)

THE JET HIGH TEMPERATURE IN-VESSEL INSPECTION SYSTEM

T BUSINARO, R CUSACK, L GALBIATI, T RAIMONDI

JET Joint Undertaking, Abingdon, Oxfordshire, OX14 3EA England

The JET In-Vessel Inspection System (IVIS) has been enhanced for operation under the following nominal conditions: vacuum vessel at 350°C; vacuum vessel evacuated (~ 10^{-9} mbar); radiation dose during D-T phase 10^{6} rads. The target resolution of the pictures is 2 mm at 5 m distance and tests on radiation resistance of the IVIS system are being carried out. Since June 1988, the new system is installed in the JET machine and the first inspections of the entire vessel at 250°C have been satisfactorily done.

1. INTRODUCTION

Periodic inspections of the interior of the vacuum vessel have to be done to check for damage due to plasma disruptions. A system was developed to scan the vessel using four TV probes through small apertures in the top of the vessel without breaking the vacuum. A great effort of miniaturization was required to compress the optics and electronics into the small diameter available.

Another problem was to provide sufficient illumination, particularly since the vessel has been carbonized. The initial solution was to use high-energy flashlight and digital frame grabbers. The system was completely automatic, with microprocessor control. The vessel surface was divided into viewing areas and for each of these "named positions" the orientation, viewing and light parameters were optimized and stored on disc. The operator calls up the named positions using a keyboard or mimic diagram and the camera is pointed at the desired location with aperture, focus and flash intensity as previously chosen. In this way a series of photos are taken and can be stored on disc or tape for comparison with previous shots. Digital filters are used to enhance contrast and reduce flickering. However, even increasing the sensitivity by integrating successive flashes, the image obtained of the inner part of the vessel remained hazy because direct light impinging on the viewing glass is scattered inside it.

Recently considerable improvements have been made. Some more penetrations were made available to provide continuous lighting. Four vacuum-tight assemblies of silica light guides are illuminated by small powerful purpose designed projectors sited above the penetrations. A drawback of the original system was that the vessel had to be cooled down to below 50°C to do an inspection. After feasibility tests high temperature probes to be used at 350°C were designed and manufacture is now complete. The prototype was tested in a rig at 400°C. As with the previous model, space restrictions called for painstaking rearrangement of wiring and connectors to eliminate noise and faults. The zoom was eliminated to make room for the cooling jacket but with better lighting the system is expected to be quite satisfactory. The device is equipped with sensors to detect cooling water leaks.

If time and resources permit a study will be carried out on the feasibility of using prisms suspended in the vessel. This would obviate the optical problems connected with the cylindrical viewing glasses but would introduce others.

2. SYSTEM DESCRIPTION

1.1 Viewing probes

Each probe consists of a stainless tube terminating in a glass cylinder. The viewing probe, as shown in Fig.1, is lowered into the

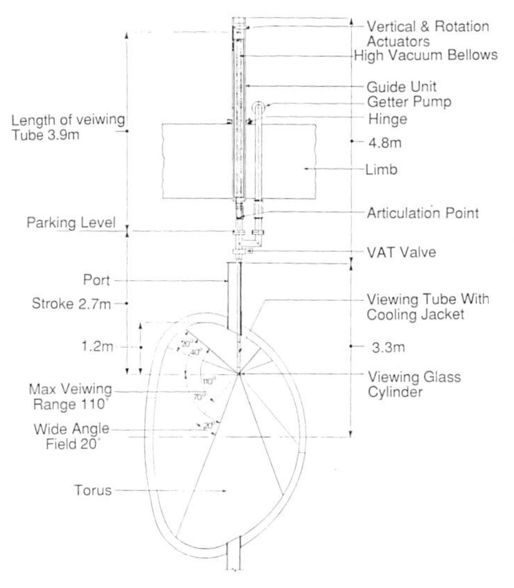

Fig. 1 Viewing Probe

vessel by a carriage driven along a guide unit which is suspended from the magnetic limb by means of a gimbal arrangement and centred with the top of the port using a hinge connection. This method of suspension allows the guide unit to comply with the thermal deformation of the vessel. The viewing probe itself is centred with the top of the port by flexible phosphor-bronze "fingers" to allow self alignment of the probe during insertion and reduce shocks due to vibrations of the port. Long vertical bellows, with 6:1 compression ratio, allow for the vertical stroke of the carriage while forming a vacuum barrier. A separation valve between guide unit and vacuum vessel is opened to insert the viewing probe after checking that the vacuum in the IVIS system matches that of the vessel.

Inside the long stainless tube a coaxial water cooled jacket has been provided to protect a high sensitivity black and white camera situated on the lower part of the probe, as well as its lenses and the motorised mechanisms for controlling iris, focus and prism tilt, from the heat radiated from the vessel wall (see Fig.2).

An external mechanism allows the cooling jacket with the camera to rotate inside the viewing cylinder. The resulting tilt and rotation of the prism allows a whole quadrant of the vessel

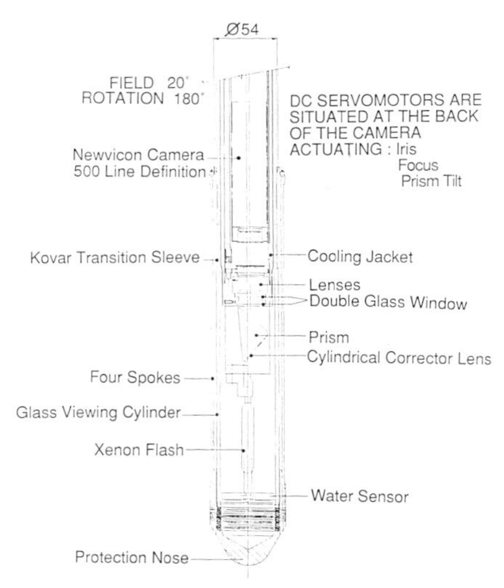

Fig. 2 Viewing Probe (Detail)

to be explored from any of the viewing positions. A flash tube, fitted at the end of each viewing probe, provides additional powerful sources of light for darker areas for single shot pictures.

A detailed description of the optics and mechanical structure of the viewing probes is reported in [1].

2.2 The cooling system

The new viewing probes are able to operate at high temperature (up to 350°C) having been isolated from the vessel thermal radiation by coaxial water cooled jackets.

The cooling jackets, shown in Fig.3, consist of two concentric cryogenic tubes 4 m long and 38.1 and 44.4 mm in diameter. In the gap between them there are 22 cryogenic tubes of 3 mm in diameter, spaced at regular intervals. The cooling water flows from the top to the bottom of the cooling jacket inside the 22 small tubes and returns to the top through the space between them. The thermal shielding has been designed for heat flow up to 2.5 KW/m².

In agreement with the design, the experimental results show that the cooling jacket keeps the working camera at less than 35°C with the entire probe fully immersed in an evacuated 4 m long oven at the average temperature of 350°C with an 18°C cooling water flow rate of 6 ℓ/min (see Table 1). While with this flow the resulting drop of pressure across the inlet and outlet hoses is only about 100 mbar, the cooling jacket is able to operate at 7 bar.

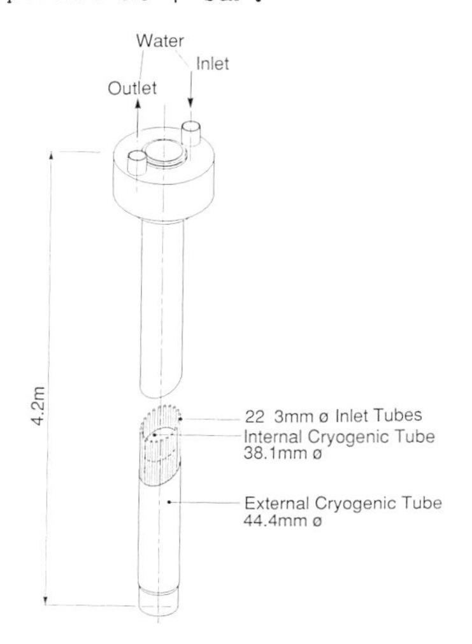

Fig. 3 Water Cooling Jacket

Furthermore, in order to reduce the heating of the camera lens due to thermal radiation coming from the front of the camera, a double glass window has been fitted between the prism and the lens and a cooling air flow (10 ℓ/min) is forced between the two glasses.

WATER FLOW RATE ℓ/min	PRESSURE DROP mbar	CAMERA TEMPERATURE °C		
		a)	b)	c)
6	100	30.5	32.5	34.5
9	250	27.5	29.0	32.0
12	500	26.0	27.0	30.0

Table 1

Camera Temperature vs Water Flow Rate at = 12°C, b = 15°C, c = 18°C

2.3 Continuous lighting system

In addition to the flash fitted in each probe a new lighting system has been developed in order to provide a continuous high intensity light source during the in-vessel inspections. The design of this system had to consider requirements of maintenance (i.e, it should not require maintenance inside the vessel) and that the vessel wall geometry not be affected.

To solve those problems a retractable 25 mm diameter 1.6 m long fused silica light guide assembly has been used. It collects the light from a powerful compact mirror arc lamp (400 W) and diffuses it in the vessel when the system is inserted.

In this design, as shown in Fig.4, only the light guide assembly is permanently installed in the vessel, while the light source is mounted in the external accessible part of the tokamak. A

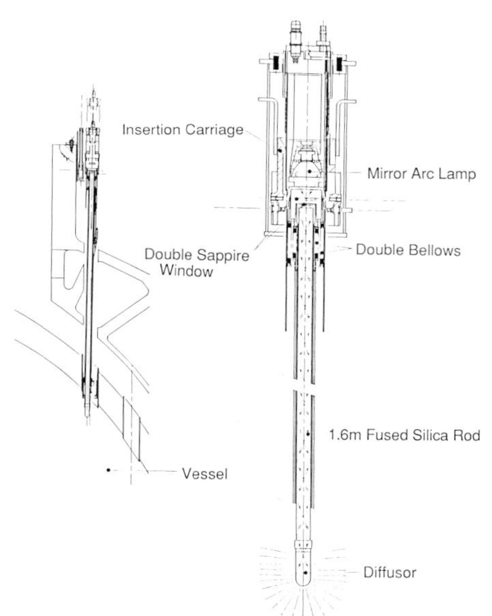

Fig. 4 In Vessel Lighting System

double bellows and double sapphire window system assures the vacuum seal of the vessel and the passage of the light into the silica rod. The penetration of the light guide in the vessel when needed, is done by a motorised trolley fitted on the top of the bellows that assures a removable mechanical connection between the lamp unit and the light guide assembly.

The efficiency of this system depends mostly on the quality of the silica rods. Imperfections, scratches or bubbles in the light guide reduce the amount of light that reaches the bottom of the silica rod where a 65 mm length of very rough surface diffuses the light all around. Experimental measurements of illuminance at 1 m distance from the diffusor of the lighting system, using a 400 W arc lamp are shown in Fig.5. This lighting system is at present the subject of a patent application.

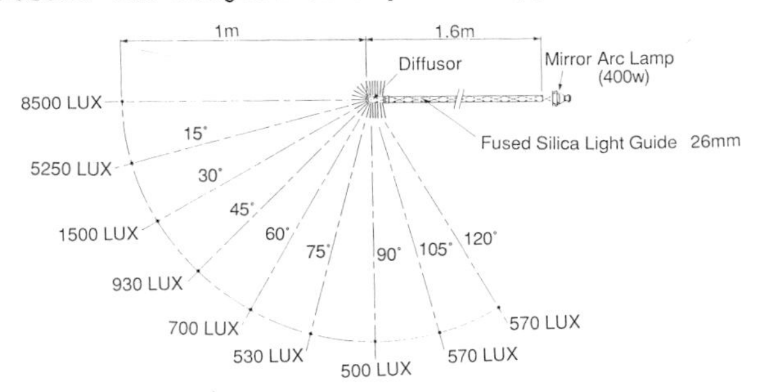

Fig. 5 Illuminance of Ivis Lighting System

3. GLASS VIEWING CYLINDER PROTECTIONS

In the present IVIS design the viewing cylinder is the most vulnerable part of the entire probe. Thermal stresses on the glass and undesirable mechanical shocks during inspections could cause the rupture of the viewing cylinder and consequently the failure of vessel vacuum seal and long delay to recover from it. Care was taken in the design in order to contain the thermal stresses in the glass particularly in the region of the welding with the metal tube. An intermediate Kovar collar whose thermal expansion characteristics are compatible with those of the borosilicate glass, had been used to join the viewing cylinder to the long stainless tube. Having assumed a linear thermal gradient from this welding of 100°C in 5 cm, a finite element calculation showed both compression and tension stresses on the glass of the order of 0.4 kg/mm^2, which compares to an allowable design tension stress for the borosilicate of 0.6 kg/mm^2, ten times less than the allowable compression stress. A metal nose pressed against the base of the glass cylinder by four spokes is used to protect the glass against rubbing on the vessel port. It also gives some benefit precompressing the glass, thus reducing the tensile stress.

Finally in order to prevent glass ruptures due to water leakages coming from the cooling jacket, a simple water sensor, based on measurements of electrical conductivity, has been installed at the bottom of the viewing cylinder.

4. RADIATION RESISTANT DESIGN

The new viewing probes (except the camera lenses) and the lighting system have been designed to be radiation resistant to withstand the above mentioned dose of 10^8 rads. To achieve that goal all components had been selected for that figure and tests on the entire system are being carried out.

Partial results of radiation test of the used TV cameras are reported on [2]. Temperature bleaching tests on the viewing cylinder show a recovery of 80% at 250°C after the glass was totally blackened by gamma exposure [3].

The fused silica rod light guides and sapphire windows of the lighting system are guaranteed by the manufacturers. The present camera lenses will be substituted with non-browning lenses during the D-T phase of the machine.

5. CONTROL ASPECTS AND OPERATION

The following movements are needed for each new probe: focus, iris, vertical stroke, prism tilt and rotation. Radiation resistant high quality miniature potentiometers have been used to provide a reference position. All movements are controlled by closed loop analogue circuits. Having continuous lighting, operations are much simpler then with the former system that had only flash light.

At present a provisional console with the aid

of PC graphics tools enables the operator to select the viewing points inside the vessel. Prism tilt and rotation are roughly shown on the PC monitor. The iris and focus are manually set in order to have the best pictures shown simultaneously on four black and white high quality monitors. If necessary a flash shot could be taken and the resulting picture grabbed for image analysis and enhancement.

6. PICTURE QUALITY AND OPTICS DRAWBACKS

In the present system good quality pictures could show a definition in the central region of up to 500 line which matches the target specification. Limitation in the quality is due to the aberrations introduced by the glass cylinder only partially corrected by an additional lens on the prism. Another drawback is glare in the glass which is particularly disturbing and strongly depends on the viewing angles.

As a result good pictures of the vessel floor and ceiling could be seen, as shown on Fig.6, taken at 250°C during the last inspection on 24.07.88, while the inner wall viewing remains partially affected by glare effects.

Fig. 6 Picture of Lower Part of JET Vessel Taken at 250°C

7. FUTURE ENHANCEMENTS

During the shutdown of the machine in July 1989 three new light units will be added to the present lighting system in order to improve the inner wall viewing. A rotating reflector will be also added on the silica rod diffusor to shadow the viewing probes from the direct light.

Feasibility tests will be carried out in order to substitute the viewing glass cylinder with a flat glass window. This solution, that requires fitting the prism in the vacuum, should solve all problem related to the glare and mechanical shocks to the glass cylinder, and will remove the optical aberrations with a strong improvement of picture quality. On the other hand problems such as the prism movements (tilt and rotation) in vacuum and at high temperature have to be solved.

A new man machine interface using advanced graphic work station will also be considered in order to facilitate the inspection.

ACKNOWLEDGEMENTS

Important contributions are acknowledged from:
Prof. W. Welford of Imperial College of London for the optics design; H. Watson and R. Govier of Culham Laboratory for the glass techniques; G. Dalle Carbonare (JET), for mechanical design of the lighting system.

REFERENCES

1. T. Raimondi, R. Cusack, L. Galbiati, "The JET In-vessel Inspection System", 14th Symposium on Fusion Technology, Avignon, France, 8-12th September 1986.

2. J. Quartly, "Television systems for inspection and maintenance in nuclear plant and the JET torus", American Nuclear Society winter meeting, Los Angeles, USA, 15-19th November 1987.

3. G. Viezzoli, "Optical behaviour of Kodial glass under irradiation and its recovery at high temperature", NET (TASK-RM3), doc. ENEA-TIB 7062(NET), Rome, 3rd June 1987.

FUSION TECHNOLOGY 1988
A.M. Van Ingen, A. Nijsen-Vis, H.T. Klippel (editors)

AUTOMATIC WELDING AND CUTTING OF JET STANDARD JOINT CONFIGURATION

P. PRESLE, L. GALBIATI AND T. RAIMONDI

JET Joint Undertaking, Abingdon, Oxon, OX14 3EA, England

The automatic welding and cutting trolleys developed for the JET standard lip weld joint are driven electrically on roller bearings which clamp pneumatically the lip (2 x 2 mm) forming the standard JET edge weld configuration. Arc voltage control is used to follow irregularities on the lip edges and geometric changes when the welding trolley moves around corners. Pulse welding makes it possible to work in any attitude. On the cutting trolley the TIG torch is replaced by a nibbler which cuts the joints ready for rewelding, all material offcuts being captured by a vacuum system. It also allows flush trimming of two joint lips, thus allowing subsequent welding even with imprecise initial alignment. To date, 250 metres of lip joint have been trolley-welded, without detectable leak, and over 180 metres cut with the cutting trolley. Typical weld penetrations in the range of 1.7 to 2.5 mm have been achieved, depending on the lip material and the shielding gas.

1. INTRODUCTION

For high vacuum applications welding is considered, in these cases, as the most reliable joining method. In general, the joining and separating operation must be repeated a number of times and motorised carriages automatically guided are envisaged. This solution is often preferred even if personnel access is allowed as it gives reliable results and avoids human errors.

The following problems are encountered:

- Remote positioning of guiding rails.
- Space required by guiding rails and interference with the design requirements of the duct.
- Precise positioning of mating surfaces.
- Space required by the motorised carriage.
- Ability for cutting and rejoining a number of times without substantially altering the relative positions of the adjacent parts.
- Complete swarf removal during cutting operations.
- Initial remote positioning of the automatic carriages.
- Position control of the welding or cutting heads along the joint.
- Ability to continue welding by closing large gaps between lips.

It is the scope of the concept hereby described to give viable solutions to the above problems[1].

2. SYSTEM DESCRIPTION

The mating surfaces of the Octant U joint, large ducts and rectangular ports, consist of two lips 2 x 2 mm, which meet on a plane perpendicular to the axis of the duct to be connected or separated. A trolley, provided with either a torch or a cutting head, is positioned on the lips. Two motorised rollers squeeze the lips onto each other, and move the trolley along by friction. They are mounted in a toe-in angle, in order to keep the trolley pressed against the joint even in an upside down attitude.

The lip joint chosen offers the following advantageous features:

- Locate and guide the trolley, thus avoiding need of rails. Initial positioning of the trolley is straightforward and can be quickly done by means of a telemanipulator.
- Provide margin for repetition of joining and

cutting operations, without varying the relative positions of the adjacent parts.

- The lips can be thin enough, even for ducts with large cross-section, to permit cutting with nibbling machines.
- The joint is totally located at one side of the separating wall of the duct, allowing total removal of the chips, avoiding the danger of them falling through the separating walls.
- The welds are kept largely free of stresses due to the flexibility of the lips.
- The disadvantage is that it is not a structural joint so that external supports have to be provided in some cases. This limits accessibility of the joint.

3. DESCRIPTION OF THE WELDING TROLLEY

The basic design is shown in Figure 1. The trolley comprises a bogey unit, torch assembly and jack system which are mounted together on a common support frame.

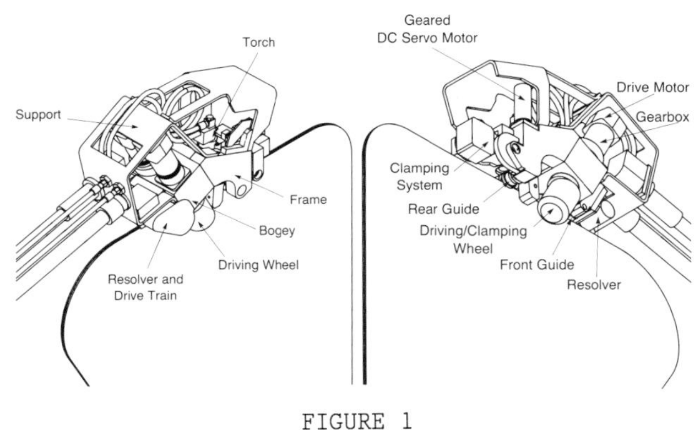

FIGURE 1
Welding trolley

The bogey is made up of two driving wheel sub-assemblies each of which consists of a wheel bearing, gearbox and electric motor all mounted eccentrically on a shaft which can be turned by a rack and pinion actuated by opposed air cylinders, pressurising of one of these cylinders will turn the shaft and bring the wheels either closer or further apart. The clamping force can exceed 1600 N for air pressure supply in excess of 6 bars. The bogey is pivoted so that the wheel axes are not perpendicular to the joint but are given a toe-in angle which makes them point inwards.

During motion of the trolley the driving wheels pull the support frame against the edge of the lips with a force equal to the inwards friction force produced between wheel and plate. Therefore the trolley will be able to ride upside down along the lips as long as the inwards friction force is superior to the trolley weight. The support frame carries two guide rollers disposed respectively at the forwards and rearwards ends of the carriage and bear on the edges of the lips. The driving wheels are mounted at a substantial toe-in angle so that they can negotiate bends of small radius while keeping the two rollers bearing hard against the lip edges.

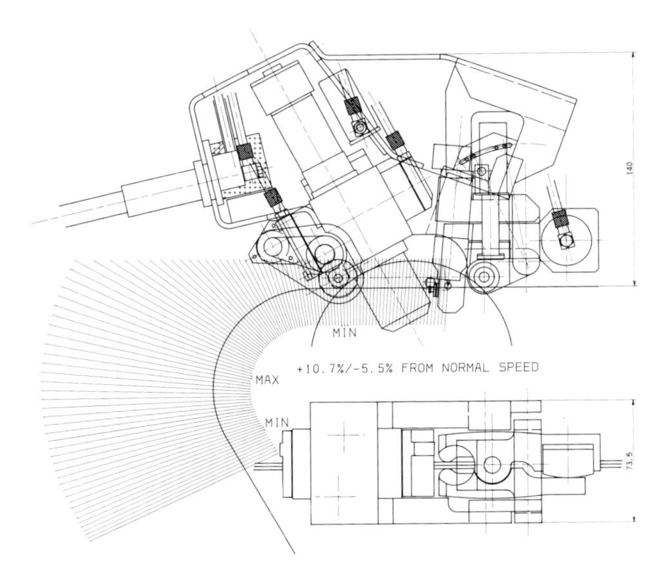

FIGURE 2
Velocity variations

Welding is done with a TIG torch mounted vertically on a pivoting arm, at the back of the trolley. Seam tracking in the plane of the lips is achieved through AVC (Arc Voltage Control described later) by a geared DC servo. The motor can raise or lower the pivot arm in order to maintain a constant voltage between the torch tungsten and edge of the lips. The position and length of the pivot arm is such as to limit velocity variation of the tungsten tip relative to the edge of the lips during straight line and curved edge welding. Optimum figures on velocity variations is 10.7% to - 5.5% ref. figure 2.

The clamping of the trolley onto the lips is normally sufficient too close the gap between lips when the components are in tolerance. Excessive gaps may occur for the following reasons: lips out of flatness tolerance, weld deformation or unflush weld caps. The trolley has the capability of closing some of these gaps with an automatic clamping system, tack welding the joint at pre-programmed locations, prior to performing the continuous TIG weld. This clamping system is mounted as part of the torch assembly pivot arm in order to squeeze the lips consistently on the edge, in both straight line and curved positions.

Constant speed is achieved through a resolver geared wheel that rides on the lip edges and this gives a true feedback of the trolley speed.

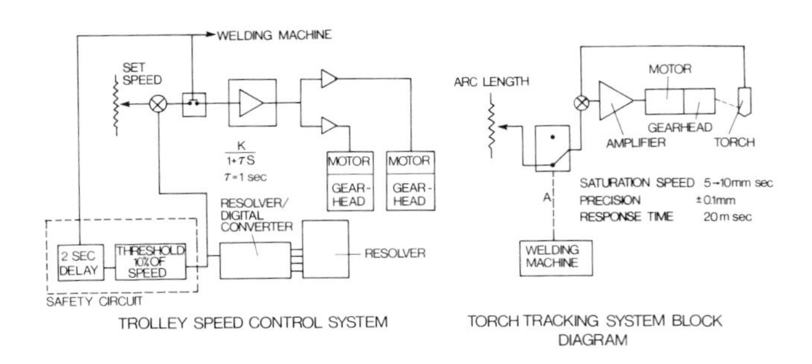

FIGURE 3
System control

4. DESCRIPTION OF THE CONTROL SYSTEM

The control of the welding trolley is incorporated in the Protig TIG welding machine[2] which is micro-processor based. The block diagram in Figure 3 shows the tracking system for the welding torch and the servo speed for the movement of the trolley.

The tracking system is an AVC synchronised with the pulsed current coming from the welding machine. The frequency of this pulse is 1.5 Hz and the duty cycle of the high current is 60% - that means 400 msec for high current and 266 msec for low current, figure 3. During the period of high current the system is switched to AVC mode using the arc voltage as a position feedback for the Servo. Instead during the low current time (266 msec) the torch is kept at the same position as at the end of the high current pulse. Using this system the weld nugget is independent of the gravitational effect.

4.1 Transporter speed control servo

Figure 3 shows also the block diagram of the velocity servo that moves the trolley at preset speed.

The resolver and the resolver to digital converter provide the position and speed signals that are processed by the welding control system which closes the velocity loop.

A safety circuit will stop the trolley when the travel speed deviates more than 10% from the set point for longer than 2 sec.

5. DESCRIPTION OF THE CUTTING TROLLEY

The basic design is shown in figure 4. The trolley comprises a bogey unit cutting head assembly and jack system on a common support.

A similar bogey design has been adopted for use on the cutting trolley. The function of the bogey assembly has been described previously in section 3.

5.1 Description of the Cutting Head

The cutting head is articulated on a support

block which is linked to the bogey body at the back of the trolley and is kept pressed against the lip edges by an air jack system.

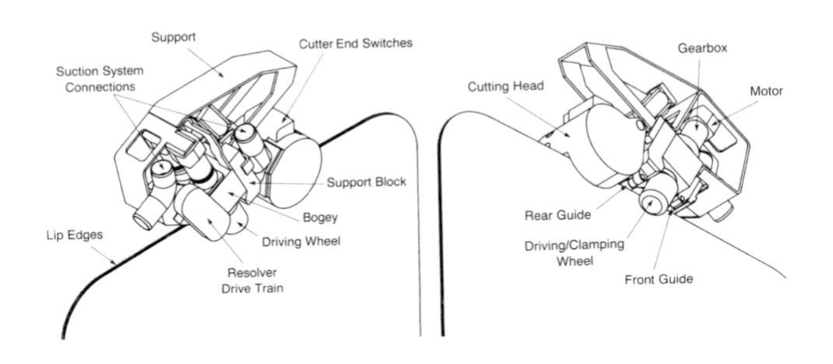

FIGURE 4
Cutting Trolley

The head is split in two, one half includes the hydraulically actuated punch, the other houses the mating die ring. Both halves are held together by a centralizing pin which also acts as a guide for the hydraulic piston, ref. fig. 5. The punch reciprocates nibbling crescent shaped chips off the welded lip as the bogey unit traverses the cutting trolley forward. After each cutting stroke, the punch is retracted by a small air cylinder incorporated opposite to the hydraulic cylinder.

The punch life is increased by partially rotating it during the return stroke. Index increments approximately of 0.5 degree are produced by the return motion of the punch in contact with the lip and linear step motion of the cutting trolley.

It is possible to index the die-ring in the housing four times to present a new cutting zone to the punch without disassembling the unit.

Cutting depth is controlled by the relationship of the punch and hub diameters and the step distance of the trolley. A maximum step equal to half the first crescent cut length will produce a minimum constant cutting depth (2 mm).

Typically lip welds can be removed by setting a 5 mm step distance which produces a constant cutting depth of 2.92 mm after 15/20 mm of trolley travel.

The punch stroke can be reduced by incorporating a spacer behind the hydraulic piston. This allows the puncher to constantly reference on top of one lip while flush trimming the other lip to the same level. Therefore subsequent welding even with imprecise initial alignment can be achieved.

During all the cutting and flush trimming sequences all material off-cuts are removed by a suction system.

5.2 Control System

The cutting tool controller is a stand alone unit based around standard Siemens micro-processor boards. The unit controls all the functions of the tool from operating pneumatic and hydraulic valves to closing the velocity loop.

The clamp and punch mechanism are operated simply by electrically driven valves with feedback derived from limit switches.

The position and velocity data are derived from a resolver as previously described. This information is used to move the trolley a known distance ready for the next cutting.

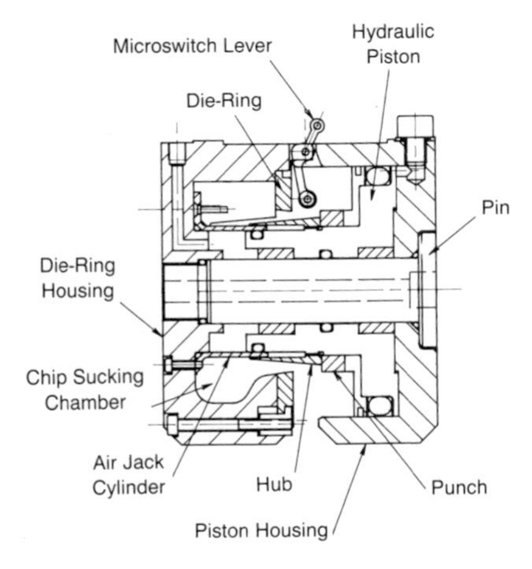

FIGURE 5
Cross-section of cutter head

6. MATERIALS

Both trolleys were handleable by the JET servo manipulator and will be made compatible with the JET remote welding system.

For reduced weight, the cutter head housing and torch welding clamping system are in titanium alloy, bogeys and frame are in aluminium alloy.

For high resistance to wear the material chosen for the cutting punch and die-ring is based on powder metallurgy sintered.

7. RESULTS

Welding - To date 250 metres of lip joints have been trolley-welded, without detectable leak. Typical weld penetrations in the range of 1.7 to 2.5 mm have been achieved, depending on the lip material and the shielding gas. Ref. fig. 6.

The welding parameters used were 60/15 amps, high/low current with 50% pulse width at 1.5 Hz, welding speed 1.7 mm/sec.

Cutting - Over 180 metres of lip welded joints have been cut during the JET machine shut-downs.

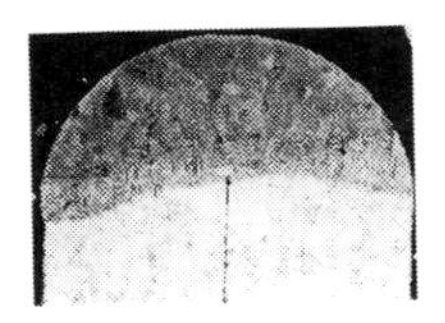

Inconel/Inconel
HYTEC 5 gas

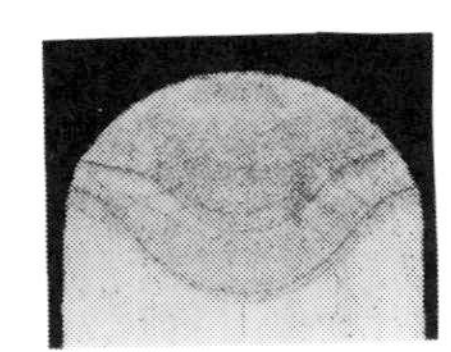

S.Steel/S.Steel
HELISHIELD 2 gas

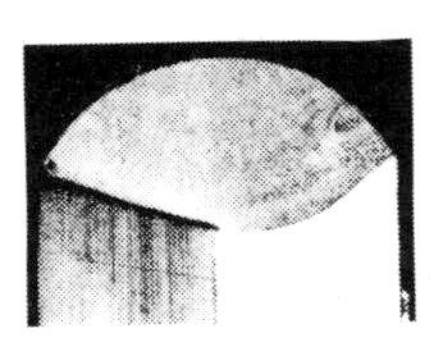

Inconel/S.Steel
HYTEC 5 gas

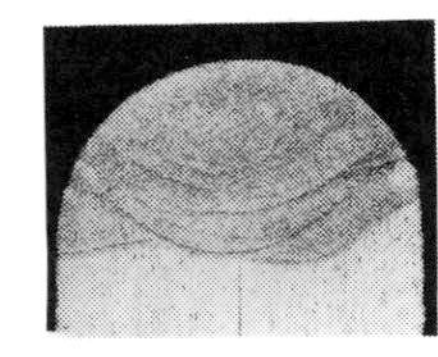

S.Steel/S.Steel
HYTEC 5 gas

FIGURE 6

Weld nugget macrograph HYTEC 5 - 95% Ar, 5% H_2
HELISHIELD2-25% Ar,75% He

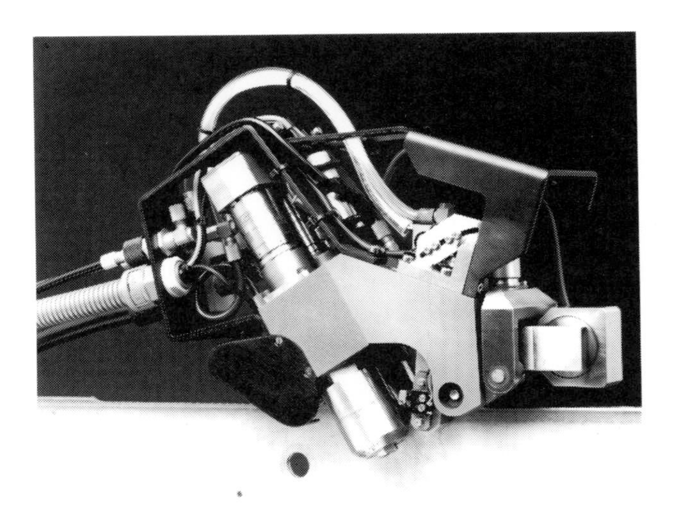

FIGURE 7
Welding trolley

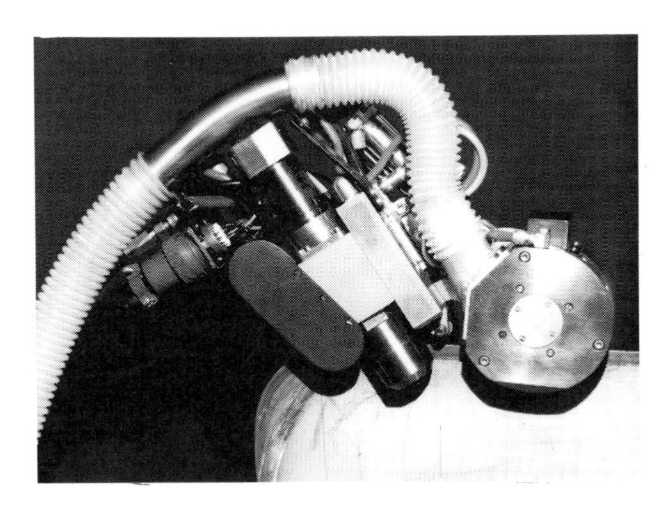

FIGURE 8
Cutting trolley

ACKNOWLEDGEMENT

The authors are grateful to G. Dalle Carbonare for his mechanical design contribution.

REFERENCES

1. S. de Burbure et al., "Remote welding and cutting for the JET project". Proc. of the Eng. Problems of Fus. Research, Chicago, 1981.
2. A. Galetsas, "The JET remotely controlled welding system", Proc. this symposium.

FUSION TECHNOLOGY 1988
A.M. Van Ingen, A. Nijsen-Vis, H.T. Klippel (editors)
Elsevier Science Publishers B.V., 1989

THE JET REMOTELY CONTROLLED WELDING SYSTEM

A GALETSAS, M WYKES

JET Joint Undertaking, Abingdon, Oxfordshire, OX14 3EA, England

A comprehensive Remote Welding System for use during the remote maintenance of JET has been specified, designed, procured and tested. The system comprises various weld tools, weld current power source, supply for tool actuators, weld gases and cooling water, a command and control interconnection system and the operator Man-Machine Interface (MMI). The welding tools which include orbital, bore, lip and tack welders some with AVC, wire feed or weave have previously been reported elsewhere. Due to the radiation environment at the workpiece all of the weld power and service supply packs are located up to 50 metres from the tool. Standard welding power packs are not usually required to generate and control welding arcs over such long distances and the device used at JET was selected only after a rigorous test programme. The command and control of the power pack from the remote control room is achieved with a standard serial link. This has been successfully implemented in the face of severe interference problems from the h.f. arc starting and weld pulses. The techniques used to overcome these problems are discussed in the paper. The control room MMI has been designed to provide the operator with real time feedback of critical parameters such as weld current and voltage, tool position and speed and wire feed speed. All control commands normally available at the power pack are available at the control room workstation. In addition the gas and cooling water flows/pressures are monitored, displayed and software interlocked for safety. The MMI also provides TV monitors for tools with weld pool viewing incorporated. This paper describes the system and its development. Its performance under remote conditions is also assessed with full scale mock-up tests.

1. INTRODUCTION

During the tritium operation phase of the JET machine, high energy neutrons will activate the machine structure to an extent which inhibits man access to it.

All maintenance, including welding tasks, will then be performed remotely. For this reason an integrated Remote Maintenance System (RMS) has been designed and it is under implementation (1).

The RMS comprises robotic arms and vehicles, servomanipulators, special end-effectors, and viewing systems. All above equipment are controlled and monitored from the Remote Handling Control Room (RHCR).

The JET Welding System (JWS) is part of the RMS and was developed in order to provide remote welding inside and outside of the JET machine.

2. OVERALL REMOTE WELDING SCENARIO

The JWS functional requirements were as follows:

- remote welding is required inside and outside the JET machine.
- the welding tasks are completely predefined.
- the JWS should have a local operating mode for the "hands-on" period of JET maintenance and welding program development.

In order to fulfil the above requirements the following design principles were adopted.

- A number of special automatic welding tools were developed according to the predefined welding tasks (2).
- A standard welding source and auxiliary equipment were modified in order to drive all welding tools remotely. One welding source is used for in-vessel welding and an

identical one for ex-vessel welding.

- The general pattern transporter, end-effector, tool was followed. Thus as it is shown in Fig. 1, the welding tools will be handled by servomanipulators mounted on the two main remote handling transporters, the articulated boom for in-vessel maintenance and the TARM for ex-vessel maintenance. All welding tools are stored in a tool box which is part of the transporter within the servo-manipulator reach. Each welding tool has a short flying lead containing all necessary weld tool services (weld current, instrument control and monitoring signals, video signals, gas, water, etc). To use a welding tool it is retrieved from the tool box by the servomanipulator and plugged into a standard RH connector on the forehead of the manipulator. All required wires and service pipes are routed from this connector through the transporter structure to the appropriate welding source or auxiliary equipment which are situated at least 50 m away in a shielded area.
- All welding sources and their auxiliary equipment are integrated into the RMS so that control and monitoring of welding is done via any RHWS.

3. THE JET WELDING SYSTEM

The JWS consists of all automatic welding tools, two welding sources (WS) and their auxiliary equipment. These include, gas distribution system, chillers, hydraulic systems and TV viewing.

The welding source is a standard programmable TIG welding source which was modified in order to fulfil the JET remote welding requirements.

The main modifications were as follows:

- a serial link (RS232) and communication software was provided for interfacing to the RMS.

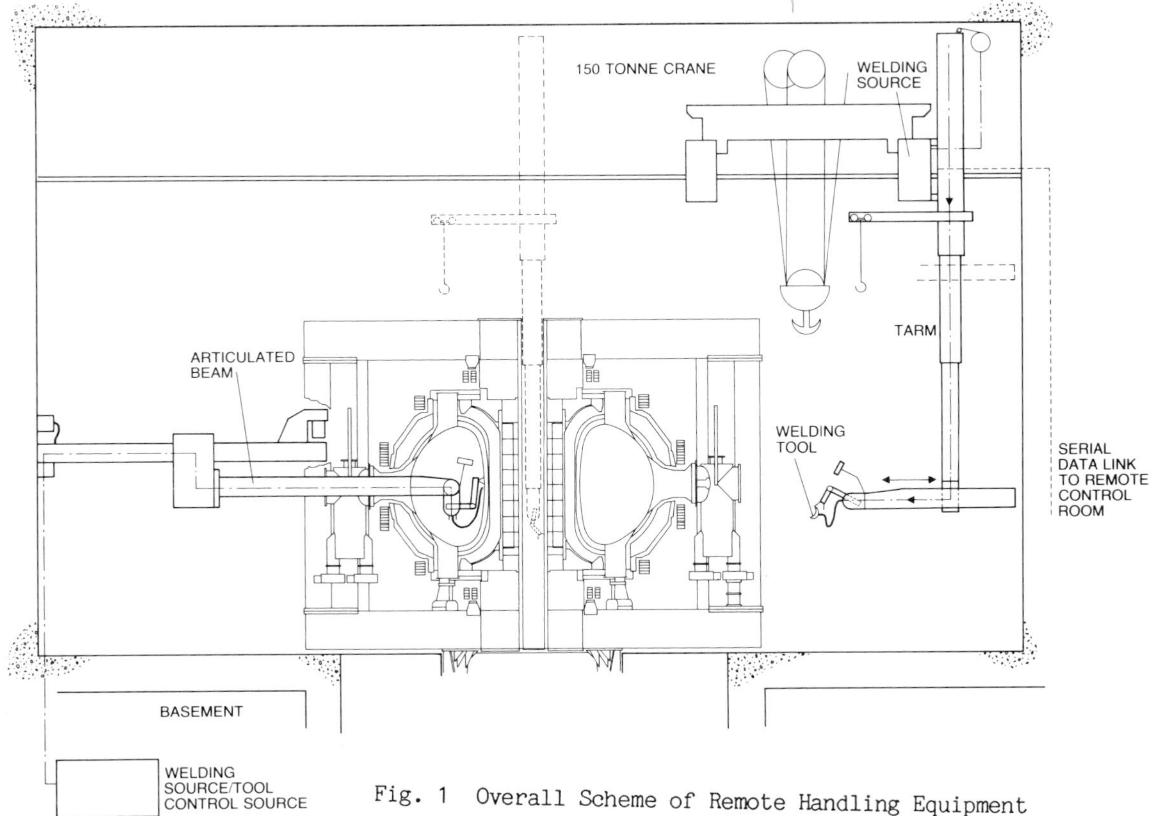

Fig. 1 Overall Scheme of Remote Handling Equipment

- remote control and monitoring of the auxiliary equipment was provided.
- software was developed in order to provide a user friendly Man-Machine Interface (MMI) on any RHWS.

The WS used was chosen because it provided reliable high frequency (50 KHz to 2 MHz) arc start over the long cable which exists between the WS and the welding tool.

3.1. Overview of welding source local operation

The welding source is a microcomputer-based (Intel 8085) welding source which is controlled locally via a standard control panel.

There are five operating modes as follows:

- STANDBY: this mode is entered on power-up.
- PROGRAMMING: this is the main operating mode which allows editing, transfer, loading and storing of welding programs.
- WELDING: during this mode execution of the required welding program occurs. In addition editing of the welding program could be achieved during the actual welding.
- WELDING SIMULATION: this mode is exactly the same as the welding mode, except that the welding current is switched off.
- MANUAL: during this mode, the electrode position and shielding gases could be controlled manually.

All auxiliary equipment (shielding gases etc) are controlled manually.

3.2. Welding source remote operation

For remote operation development work was done in two main areas:

- remote control of auxiliary equipment including TV viewing.
- integration into the RMS, including the development of a new MMI.

The welding source auxiliary equipment comprises, the gas distribution system for air and shielding gases, chiller, hydraulic system and TV viewing which is part of the Remote Handling CCTV system (1). Motorised control valves and transducers were fitted and interfaced into the RMS either directly through the welding source or indirectly through a CAMAC interface (Fig. 2). This is because some auxiliary equipment could be shared with the cutting tools. Thus the WS input/output ports were increased and new CAMAC interfaces were designed.

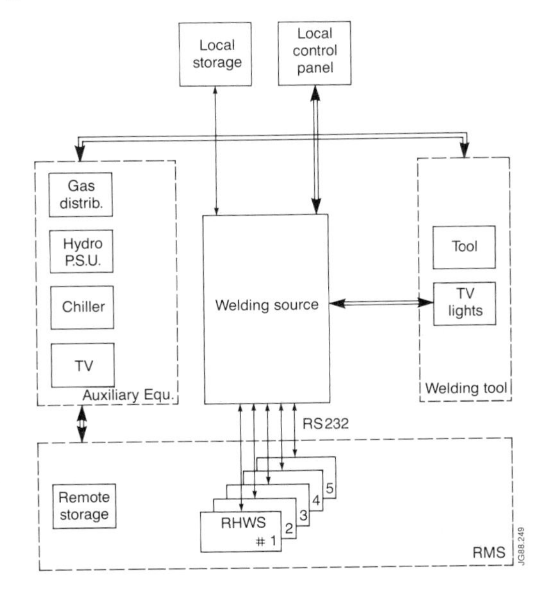

Fig. 2 The JET Welding System

Special attention was given in providing filtering for the new ports because of the strong electro-magnetic interference which is produced during H.F arc start. Also a robot deactivation output was provided for switching off all transporter actuators during the arc start period.

Because of the use of prototype welding tools and their remote requirements, some changes and additions in the welding source servoloops were necessary.

For example, the JET lip welding trolley, required a resolver feedback instead of the tacho feedback which is used by the other tools. Also it required the control of a clamp within the welding program for providing a spot welding sequence before the continuous weld.

In addition to hardware changes, software changes were made in order to accommodate the afore mentioned additional functions and also to provide safety interlocks.

The second main area of development work was the integration of the welding sources into the RMS (1). The RMS consists of the Remote Handling Control Room (RHCR) and its work-stations (RHWS), the local controllers of the various RH equipment and the interconnecting local area networks.

It was required to integrate the two welding sources in such a way as to allow control and monitoring from any RHWS, including the additional TV cameras for weld pool viewing.

Graphical displays of all main welding parameters is used for welding process monitoring and control under remote operation instead of the direct weld pool viewing which is used in the local mode operation. However, TV feedback is also provided as a complementary aid.

Because the graphical displays are the main monitoring and, their real-time operation must be guaranteed. Thus 1 sec minimum display time update rate was specified. The same delay was specified for the response to operator inputs. This is because the operator can perform welding parameter adjustment while welding to optimise the welding quality.

The above response times were achieved by using a dedicated serial serial link (RS232) between the WS and the controlling RHWS. A unit selector allows for automatic switching of the serial links between RH workstations and the two WS. For less time critical functions the CAMAC serial highway and Ethernet LAN were used.

The CAMAC was used for controlling some auxiliary units and the Ethernet for the TV camera switching.

The serial links are implemented on optical fibre because of the long distances involved and to avoid the effects of electromagnetic interference which is produced by the WS.

Communication software was developed in order to receive/send commands and data from/to RHWS.

The last area of development in the JWS integration was the design of a new MMI. This MMI was based on the standard JET RHWS hardware (Fig. 3).

The RHWS (1) was designed at JET and is a general purpose workstation allowing the control of any RH equipment.

Fig. 3 Remote Handling Workstation (RHWS)

The RHWS is a software configurable work-station, and the software guides the operator through touch panel menus.

The new MMI was designed to fulfil the following requirements.

- The RHWS operation should, as much as possible, resemble the operation and functions of the standard welding source local control panel. Thus operator training will be minimised.
- The RHWS should provide additional remote functions as follows:
  - graphic displays of main weld parameters in real time (1 sec). These are the weld current, voltage, travel speed and wire

feed. Simultaneous display of programmed and measured values with automatic alarm initiation when value out of range during welding (Fig. 3).

- display of TV pictures of the weld pool and general welding environment.
- storage of weld programs in remote computer.
- user friendly operation with the use of touch panel menus and diagnostic messages.

Both in-vessel and ex-vessel welding sources are identical and are integrated into the RMS as described above resulting in a fully remote welding system for the JET machine.

4. WELDING SYSTEM PERFORMANCE EVALUATION

An extensive program of mock-up tests was performed to prove the reliable operation of the welding source under a variety of operating conditions.

For this purpose different welding tools were connected through the RH welding umbilical through which all the required tool services are routed, Fig. 4. The welding source was connected to the RHWS through a serial optical link.

Each function was tested successfully with all available tools.

It was demonstrated that the (1 sec) response time of the new MMI (Fig. 3) was adequate for the production of high quality TIG welds. It was possible also to create a remote library of welding programs which could be downloaded to WS for use.

The error messages were found useful for identification of any problem since this is particularly difficult under remote conditions.

Adjustment of welding parameters during welding was achieved successfully. However as an operating procedure this must be avoided, since in case of a wrong adjustment it might lead to situations from which it is difficult to recover under remote conditions.

Fig. 4 Welding tool in position

6. CONCLUSIONS

In this paper the design and integration of the JET Remote Welding System (JWS) has been described.

A proven welding source was used as a basic building block. It was then modified for remote operation.

A series of prototype welding tools were designed and interfaced with the welding source. The welding source and its auxiliary equipment were integrated into the RMS thus allowing for remote welding inside and outside the JET machine.

The performance of the JWS has been tested through mock-up tests.

REFERENCES

1. A. Galetsas et al, "An Integrated Control System for Remote Handling Equipment at JET". Proc. IEEE 12th Symposium on Fusion Engineering, Oct 1987, California, USA.

2. S. Mills et al, "A Practical Experience of Using Special Remote Handling Tools on JET". Proc. IEEE 12th Symposium on Fusion Engineering, Oct 1987, California, USA.

FUSION TECHNOLOGY 1988
A.M. Van Ingen, A. Nijsen-Vis, H.T. Klippel (editors)

IN-VESSEL INSPECTION OF JT-60 FIRST WALLS

Hideyuki TAKATSU, Takashi ARAI, Atsushi KAMINAGA, Masahiro YAMAMOTO, Masatsugu SHIMIZU and Tsutomu IIJIMA

Japan Atomic Energy Research Institute, Naka Fusion Research Establishment
Mukohyama 801-1, Naka-machi, Naka-gun, Ibaraki-ken, 311-01 Japan

JT-60 in-vessel inspection system was developed to observe the interior of the JT-60 vacuum vessel during intervals between plasma discharges without break of high vacuum. This system has been successfully in use since Jan. '87. It provides much information about damages of the first walls and soundness of the other in-vessel components. Such information provides instructions for plasma operations.

1. INTRODUCTION

Periodic observation of the vacuum vessel interior is necessary to assure system integrity of in-vessel components as well as to examine the relationship between plasma operations and wall damages. The interior of the JT-60 vacuum vessel had been observed within the vacuum vessel during venting period. As venting of the vacuum vessel to the atmosphere requires lengthy time for wall conditioning and miscellaneous works (vacuum break, set-up of access manhole, pump down, leak checking, etc), venting is scheduled once a year. Therefore, very poor information had been obtained about the soundness of the in-vessel components and the damages of the first walls without in-vessel inspection system.

JT-60 in-vessel inspection system was developed in Dec. '86 and has been well in operation since Jan. '87. This system has provided much information about the damage of the first walls as well as the soundness of a number of in-vessel components. Furthermore, it can provide instructions for plasma operation.

The present paper describes the outline of the JT-60 in-vessel inspection system and the results of application to JT-60 vacuum vessel.

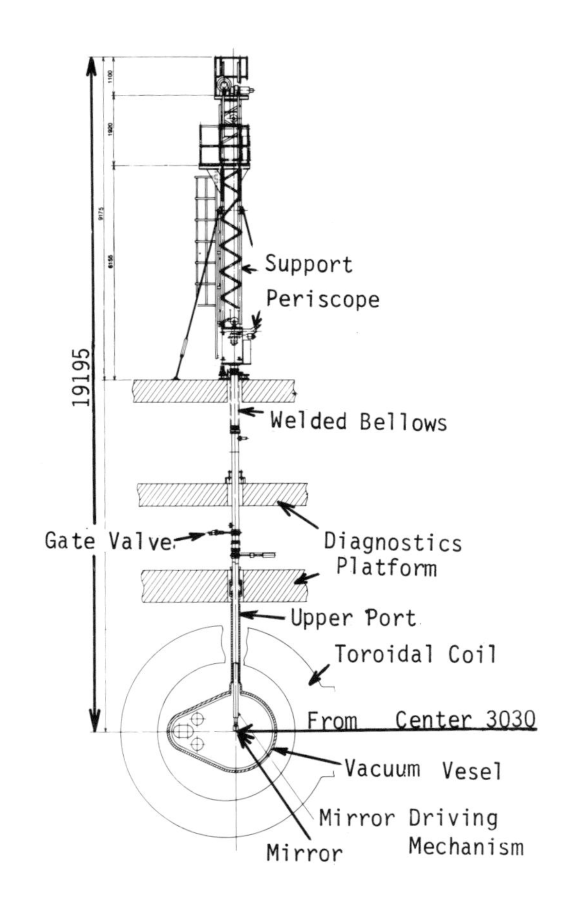

FIGURE 1
Schematic diagram of the JT-60 in-vessel inspection system.

2. IN-VESSEL INSPECTION SYSTEM

JT-60 in-vessel inspection system was developed to observe the interior of the JT-60 vacuum vessel during intervals of plasma discharges without break of the high vacuum. The objective of this system is the detection of the damages of the first walls attached on the inner surface of the vacuum vessel.[1]

A schematic diagram of the in-vessel inspection system is shown in Fig. 1. This system consists of four identical sets of perimeters including periscopes, illumination equipments, viewing systems, vacuum systems and control systems. Four perimeters can cover about 90 % of the interior of the vacuum vessel. The perimeter has a spacial resolution of 10 mm square at a distance of 5 m. Each periscope has eight halogen lamps (500 W) and three of which can be used simultaneously.

Four perimeters are placed on the diagnostics platform in the torus hall and are connected to the vacuum vessel via four vertical ports individually. High resolution pictures are provided by the periscopes, which relays images of the vacuum vessel interior with a mirror and a set of lenses. Each periscope is equipped with an inspection eyepiece, an ITV camera and a still camera. Images provided by the periscope can be observed by these viewing equipments. The pictures of the television camera are transmitted to the JT-60 control room and can be observed through a monitor television.

This system was completed in Dec., '86 and has been well in operation since Jan. '87.

3. APPLICATION

3.1 TiC-coated molybdenum first walls[2]

In-situ observation of TiC-coated first walls was carried out from Jan. to Mar. '87. In this operation period the additional heating power was increased up to 20 MW and the stored energy of the plasma reached 2.0 MJ.

After the preceding operation period (Mar. to Oct. '86), heavy melting was observed on one protection plate around sub-divertor coil casing. The appearance of this plate is shown in Fig. 2(a). This damage was thought to be caused simply by misalignment and this tile was replaced by a new one in Dec. '86. Careful observation of this tile was kept on using newly developed in-vessel inspection system during succeeding operation period from Jan. to Mar. '87.

Heavy damage was observed again on this replacement tile alone. Figures 2(b), 2(c) and 2(d) are the pictures taken on Feb. 14, Mar. 8 and Mar. 16, '87 by this system. As shown in Fig. 2(b), a half of this tile, where a slit was provided to reduce the eddy current, was distored macroscopically due to the thermal stress associated with excessive heat load concentration. Distortion of the tile disappeared and heavy melting became evident in Fig. 2(c) and melting and erosion were still in progress in Fig. 2(d). The final appearance of this tile is shown in Fig. 2(e), which was taken in the vacuum vessel during the venting period of Apr. '87. The erosion features during this operation period were similar to those during the former operation period shown in Fig. 2(a). Heavy erosion and a remarkable decrease in the thickness of the originally 5 mm thick molybdenum plate were seen.

The tile was located closest to the feeder of the lower sub-divertor coil (electron side) and the damage was concentrated on this tile alone (among 2,400 tiles) in spite of the wide variety of plasma operation. Therefore, it was presumed that this damage was caused by high energy electron bombardment, shifted due to a magnetic field error of the lower sub-divertor coil feeder. And from this result it was recommended to carry out plasma operations keeping enough gap between the separatrix line and the sub-divertor coil casing.

3.2 Graphite limiters

During the venting period from Apr. to May '87 all of the limiters and divertor plates (outer X-point) made from TiC-coated molybdenum were replaced with graphite tiles. Experiments were re-started in June '87 with new graphite limiters and divertors. During succeeding operation period until Oct. '87 operation regime was widely extended, and the additional heating power and the stored energy were raised up to 30 MW and 3.1 MJ, respectively.

In-situ observation of the graphite limiters was carried out on July 14 and Sep. 3 '87. Though a few locally damaged areas were observed, system integrity of the graphite limiters was confirmed. This was after about 1,000 shots, including conditioning ones, after re-start of experiments.[3]

The only observable damages were as follows: The edges of two outboard tiles were found to

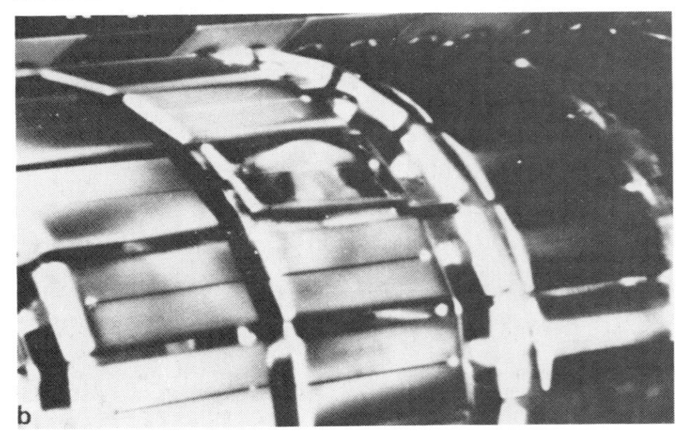

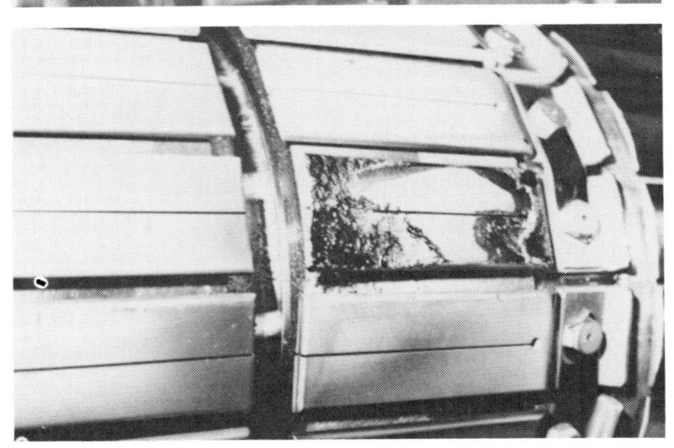

FIGURE 2
(a) Damage observed during venting period in Nov.'86 on a protection plate around the lower sub-divertor coil casing. Divertor coil feeder is located on the right-hand side. (b) Distortion of the replacement protection plate at the same location, observed during operation (Feb. 14 '87) by the in-vessel inspection system. (c) Damaged protection plate observed on Mar. 8 '87. Distortion of the plate disappeared and heavy erosion has become evident. (d) Damaged protection plate observed on Mar. 16 '87. Erosion was still continuing. (e) Final appearance of the damaged plate, which was taken during the venting period of Apr. '87.

be broken off. This damage might possibly have resulted from an excessive heat deposition associated with operation that allowed the plasma column to touch the outboard wall. This result showed an instruction for succeeding plasma operations to keep plasma column apart from the outboard wall.

Substantially eroded areas were also observed on the tiles located either at the midplane or on the edge of the rigid sectors. Eroded graphite was found to be re-deposited on the surrounding in-vessel components. Typical view is shown in Fig. 3, showing the TiC-coated Inconel liners adjacent to the rigid sector. Re-deposited graphite film and partial flake-off of the film can be seen around the fixing bolt of the liner.

A crack was observed at an innermost tile located at the midplane and on the electron side of #7 rigid sector. Figure 4(a) shows the picture of the cracked tile taken by in-vessel inspection system. Significant erosion is also evident. This might have been caused by runaway electron bombardment during a disruption. The crack seemed to propagate from the edge perpendicularly to the edge surface by 3 mm, in 45 direction by 20 mm and in parallel with the edge surface by 10 mm. Finally, the surface layer was found to have been flaked off after another one month operation. The appearance of this tile is shown in Fig. 4(b), taken in the vacuum vessel in Nov. '87.

3.3 Graphite lower divertor plates

After four month operation with graphite first walls JT-60 was shut down in Oct. '87 for

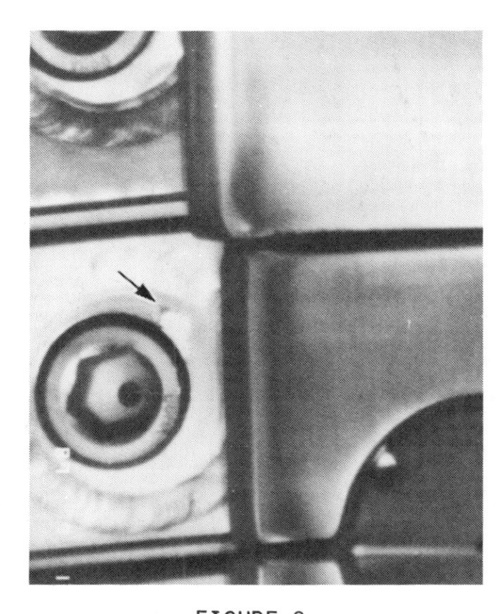

FIGURE 3
Redeposited graphite film observed on TiC-coated Inconel liner during operation. Partial flake-off of this film is also evident.

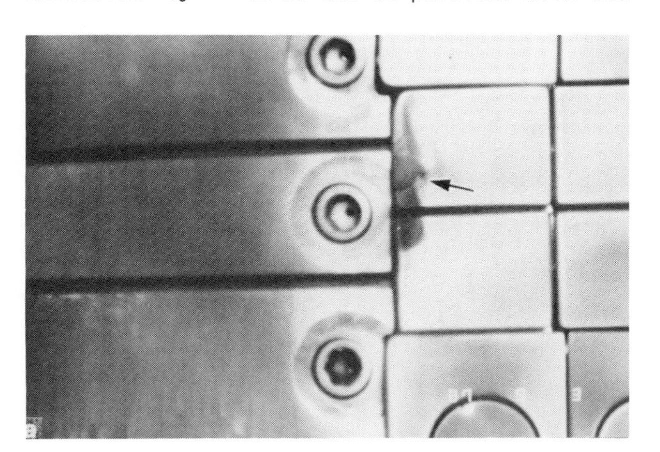

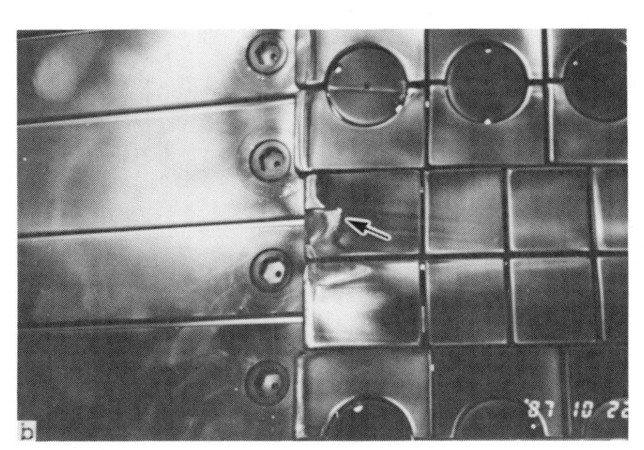

FIGURE 4
(a) A crack and heavy erosion observed on the tile located at the midplane and on the electron side of #7 rigid sector during operation (Sep. 3 '87) by the in-vessel inspection system. (b) Final appearance of the same tile observed during venting period of Nov. '87. The surface layer of this tile was found to have been flaked off. This damage might have been caused by runaway electron bombardment during a disruption.

magnetic reconfiguration (lower X-point). Associated with this change of magnetic configuration all of the TiC-coated Inconel liners on the floor were removed and new divertor plates made from graphite were installed. Plasma operations with lower divertors were started in June '88 and in-situ observation of the lower divertors was carried out on July 19 and Aug. 13 '88.

Two broad discolored zones were seen on the graphite tiles running in the toroidal direction, where separatrix lines intersected the divertor plates. Typical view of the lower divertor plates is shown in Fig. 5, which was taken on July 19. Within these zones locally eroded tiles were also observed. Zoomed-up view of the eroded divertor tiles is shown in Fig. 6.

Discoloration and erosion were more remarkable on Aug. 13 compared with July 19. As heat load conditions are going to increase during the following operation period of Sep. and Oct., careful observation by in-vessel inspection system is required.

4. SUMMARY

JT-60 in-vessel inspection system was developed to observe the interior of the JT-60 vacuum vessel. This system was completed in Dec. '86 and has been successfully in use since Jan. '87. This system has provided much information about the damages of the plasma facing components as well as the soundness of a number of in-vessel components.

TiC-coated first walls, graphite limiters and graphite lower divertor plates have been observed by this system and some damages have been found during operation. Instructions for plasma operations have been provided by this system.

ACKNOWLEDGEMENTS

The authors would like to acknowlege Drs. M. Shoju and K. Kitagawa of Toshiba Co. for cooperation of this system.

REFERENCES

1. M. Ohkubo et al., Proc. 14th Symp. on Fusion Technol. (1986) 1825.
2. H. Takatsu et al., presented at 3rd Internat. Conf. on Fusion Reactor Mater., Karlsruhe (1987) and to appear in J. Nucl. Mater.
3. H. Takatsu et al., Proc. 12th Symp. on Fusion Eng. (1987) 151.

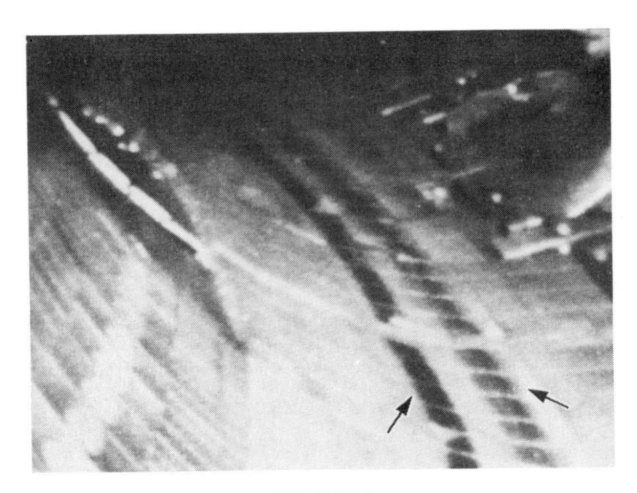

FIGURE 5
Two broad discolored zones running in the toroidal direction, which was observed during operation (July 19 '87) with lower divertor plates.

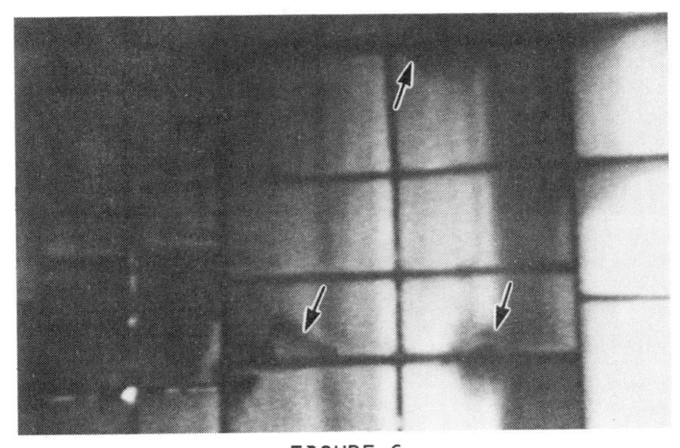

FIGURE 6
Zoomed-up view of locally eroded graphite divertor tiles.

FUSION TECHNOLOGY 1988
A.M. Van Ingen, A. Nijsen-Vis, H.T. Klippel (editors)

IN-VESSEL HANDLING UNIT CONCEPT FOR NET

Anton SUPPAN, Jörg HÜBENER

Kernforschungszentrum Karlsruhe GmbH, Association KfK-Euratom, Postfach 3640,
D-7500 Karlsruhe 1, FRG

NET in-vessel components up to a weight of 1 ton will be handled by means of handling machines which enter the vacuum vessel through four horizontally arranged ports. Out of several options KfK investigated an in-vessel handling unit based on the system of an articulated boom.
It is composed of a transport unit (carrier with articulated boom) and exchangeable work units. The boom is extendable to about 25 m and can reach any point within the torus. Carrier and articulated boom are connected by a combined yaw and roll joint. The 11 links of the boom are modular designed. A combined pitch and roll joint allows the adjustment of the end-frame. Work units investigated are an electrical master-slave manipulator unit and two devices for the replacement of divertor plates and radio-frequency antennae, respectively.
An experimental device for in-torus handling with an IVHU as a principle item will be constructed.

1. INTRODUCTION

Maintenance inside the vacuum vessel of fusion devices will only be possible by remote handling. In-vessel components as divertor plates, protection tiles, antennae and stabilization coils are supposed to be handled by an in-vessel handling unit (IVHU) based on the principle of an articulated boom. It operates on the reactor hall level.

2. DEFINITION OF REQUIREMENTS

NET requirements are based on the investigation of typical maintenance tasks, e.g. inspection of plasma facing components and their repair or replacement. Table 1 shows a summary of the requirements.

TABLE 1
Summary of IVHU-Requirements

Weight handled by special equipment	≤ 1000 kg
Weight handled by multi-purpose equipment	< 50 kg
Size of handled components	< 2 x 1 x 0.4 m
Reach of the IVHU	± 185°
Number/size of entry ports	4 /0.65 x 1,9 m
Atmosphere during maintenance	inert gas or air, 1 bar, +150°C
γ-Dose rate of components	< 3 x 10^6 rad/h

3. IN-VESSEL HANDLING UNIT

3.1. Overview

The IVHU is the basic device of the in-vessel handling system shown in Figure 1. The system consists of at least one IVHU and of one transport carrier with a telescopic arm (TC) which can enter the torus, such assisting the IVHU by transfering components and equipment. During operation of NET the entry ports are gastight closed by means of plugs. The plugs can be withdrawn and inserted by means of a special plug handling device (PHD).

In order to avoid contamination of the reactor hall the handling equipment is housed in contained transfer units (CTUs) which are permanently attached to the vacuum vessel. Transport flasks (TFs) are used for the transport of components and equipment between CTUs and storage/service areas.

3.2. In-Vessel Handling Unit

The IVHU Figure 2 is composed of a transport unit (TU) and three work units which are remotely attachable to the end-frame of the TU[1,2]. Their design depends on the maintenance tasks. The manipulator unit as multi-purpose equipment will mainly be applied for inspection tasks and protection tile replacement. Additional work units are foreseen for the exchange of radio-frequency antennae and divertor plates.

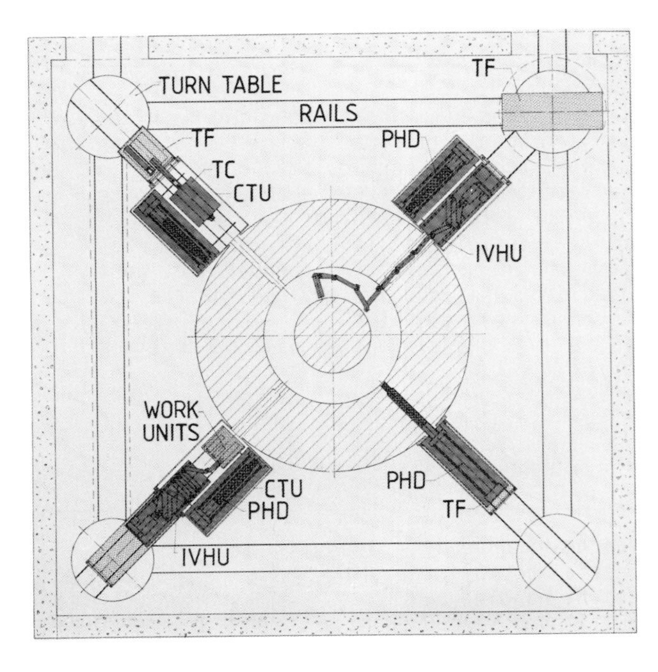

FIGURE 1
In-vessel handling system

3.2.1. Transport Unit

The TU shown in Figure 3. consists of the carrier movable on rails, the trolley guided on the carrier and movable by means of a motor driven spindle, and the articulated boom connected to the trolley. During operation, the carrier is fixed in position. The 11 boom links have box cross sections with a height of 1.35 m and a width varying between 0.35 and 0.5 m. They are connected by yaw joints. For optimization both cranked and straight links are required. The length of the deployed boom is 25 m. Any point of the torus is within its reach.

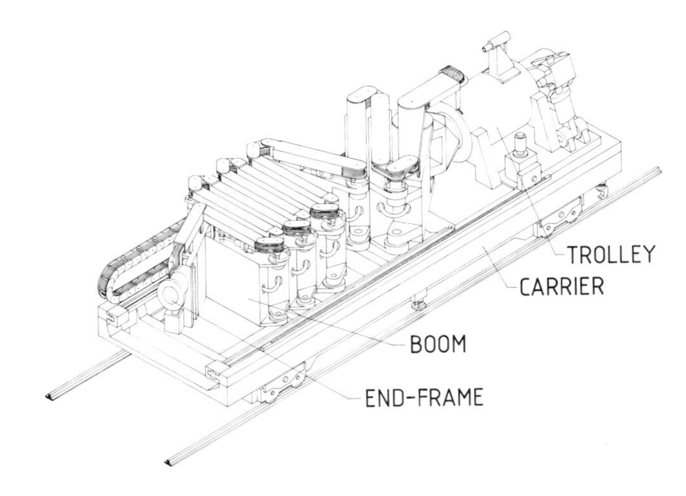

FIGURE 3
Transport unit

Figure 4 shows one of the 11 yaw joints with drive units. Each joint is split in an upper and lower sub-unit remotely replaceable as a whole from the top and the bottom, respectively.

Each link is actuated by two motor driven spindles working on a swiveling lever and allowing a joint range of 220°. In case that one unit fails the link still can be moved with reduced speed and range. The drive units are installed at the upper side of the link boxes allowing an easy remote replacement.

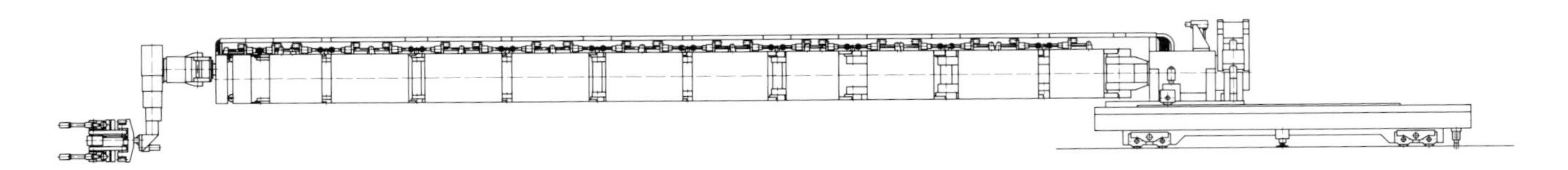

FIGURE 2
In-vessel handling unit

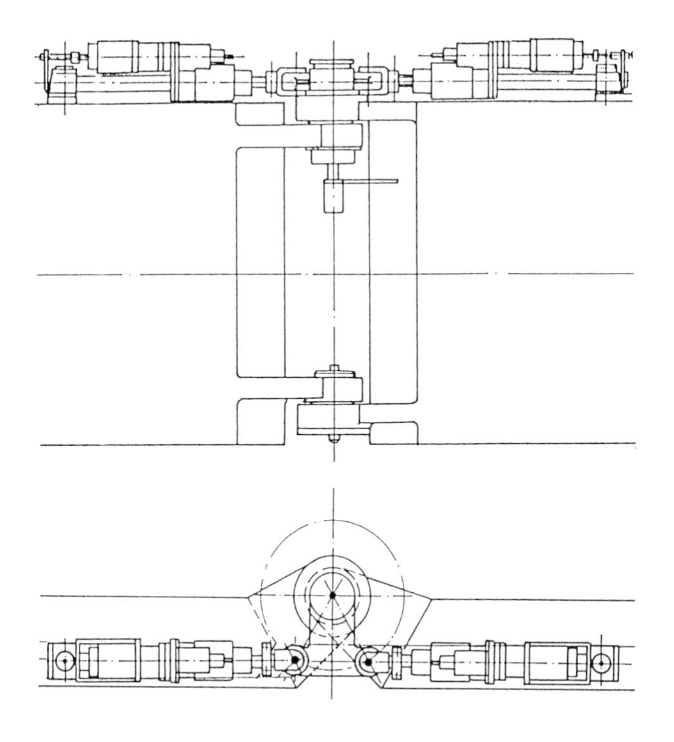

FIGURE 4
Yaw joint with drive units

Boom and trolley are linked by a combined yaw and roll joint. The latter one compesates the angular deflection of the boom with respect to the limited clearance in the port.

The tip of the boom is designed as an end-frame providing the interface for the work units. To compensate boom misalignment the frame is linked by combined spindle driven pitch and roll joints. Vertical displacement will be equalized by telescopic masts which are integrated in the work units.

3.2.2. Work Units

To perform general inspection and replacement tasks relatively small components have to be handled. For this an electrical master-slave manipulator unit is foreseen (FIGURE 5). It consists of a pair of manipulator arms and two cameras. The manipulator arms and one camera system are mounted at a common crossbar which is attached to a telescopic mast. The unit is rotatable by an angle of ± 180° making possible to reach any point at the plasma facing surfaces as well as the upper and lower active plasma stabilization coils.

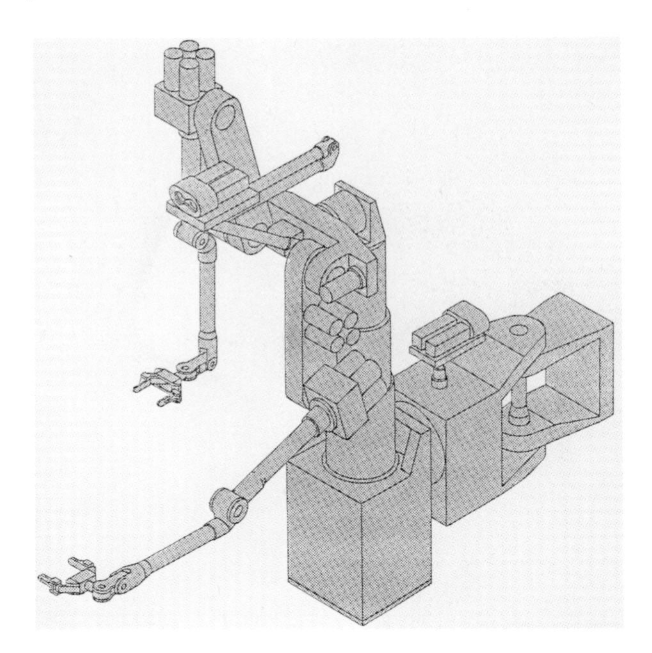

FIGURE 5
Manipulator Unit

Another handling unit is used to withdraw and insert R.F. antennae which are located in the equatorial plane of the vacuum vessel. As to be seen in FIGURE 6 this work unit consists of the gripper with a load capacity of 1 ton and two camera systems. The telescopic arm is only required to compensate vertical deflection.

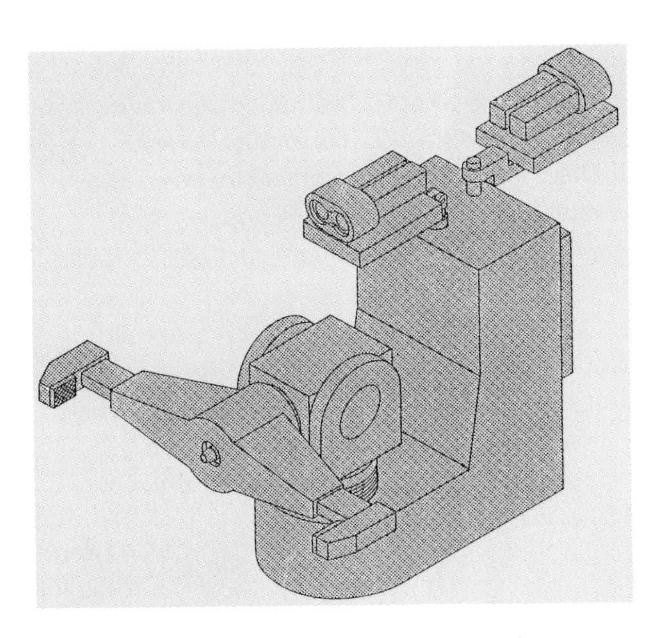

FIGURE 6
Antenna handling unit

The divertor handling unit combines four components, one gripper to handle the divertor plates, one manipulator to support replacement operations and two television systems (FIGURE 7). To locate the unit in working position a telescopic mast is required which is rotatable together with the gripper, manipulator and one camera system by an angle of ± 180°. Vertical rotation is only possible in a retracted condition to avoid excessive loads. The telescopic mast has a stroke of 2.5 m. The attachment plate for the gripper can be tilted up to an angle of 75° and rotated ±180°. A swiveling head for ± 180° is integrated in the vertical telescopic mast. In order to reduce the weight of the work unit the structure material is titanium.

The data of the transport unit and work units are given in Table 2.

3.2.3. Control System

The control system concept provides two modes of operation, an automtic mode and a manual mode.

TABLE 2
Technical data

TU-boom:		
Maximum length:		25 m
Weight:		30 tonnes
Joints:	Number	Range of operation
Jaw:	11	220°
Rotation:	1	±5°
Pitch (at end-frame)	1	±5°
Roll (at end-frame)	1	±5°
Manipulator Unit:		
Nominal/maximum load capacity of the tong:		20/50 kg
Number of motions with force reflexion:		7
Number of motions with speed control:		2
Power amplification master/slave:		1:1 to 1:10
Length of the arms:		0.8 m
Wrist length:		0.25

Divertor Plate Handling Unit:	
Gripper:	
Load capacity	1000 kg
Manipulator:	
Nominal/maximum load capacity of the tong:	50/100 kg
Number of motions with force reflexion:	7
Number of motions with speed control:	2
Power amplification master/slave:	1:2 to 1:20
Length of the arms:	0.8 m
Wrist length:	0.3 m
Divertor Plate Handling Unit:	
Gripper:	
Load capacity:	1000 kg

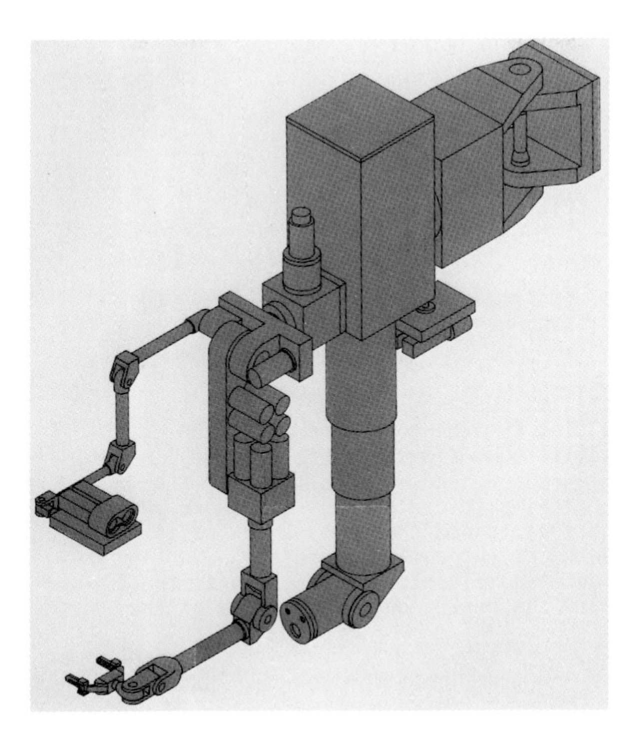

FIGURE 7
Divertor plate handling unit

4. CONCLUSIONS

The conditions for working inside the vacuum vessel are so demanding that the necessary equipment development requires assisting experiments already in an early stage. Therefore, KfK is ongoing to construct EDITH, an experimental device for in-torus handling, shown in FIGURE 8.

EDITH includes the three front links of an IVHU in full scale, linked with a support structure. The end-frame of the front link allows the attachment of work units. In a later stage the transporter is extendable to the full length of the boom. EDITH is scheduled to become the test facility for the development of in-torus remote maintenance techniques of fusion devices and serves also for component testing.

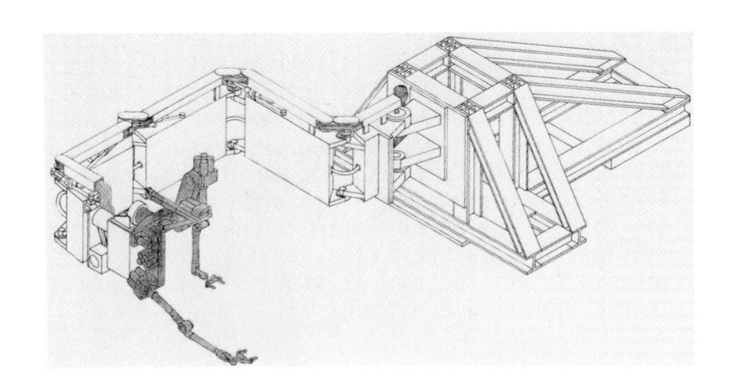

FIGURE 8
EDITH

REFERENCES

1. Suppan A. et al.: Kernforschungszentrum Karlsruhe, Handling equipment for in-vessel components - in-vessel handling unit, Primärbericht 17.01.01P09A, 1988, unpublished

2. Suppan A., Hübener J.: Kernforschungszentrum Karlsruhe, Investigations of in-vessel handling concepts for NET, Proceedings of the Technical Committee Meeting on Robotics and Remote Maintenance Concepts for Fusion Machines, IAEA, 1988, in print

FUSION TECHNOLOGY 1988
A.M. Van Ingen, A. Nijsen-Vis, H.T. Klippel (editors)

THE REMOTE HANDLING OPERATIONS ON THE NET VACUUM VESSEL DOUBLE SEALS

F. CASCI[1], F. FAUSER[1], C. HOLLOWAY[1], G. MALAVASI[1], E. SALPIETRO[1], J.E. CHAPMAN[2], R.M. HARRISON[2], J.F. FILLINGHAM[2]

[1]The NET Team c/o Max-Planck-Institut fuer Plasmaphysik, Boltzmannstr. 2, D - 8046 Garching
[2]Babcock Energy Ltd., 165 Great Dover Street, London SE1 4YB, UK

The NET vacuum vessel is made up of 16 wedged and 16 parallel segments bolted together to form a stiff toroidal structure which acts both as shielding for the coils and as vacuum tight barrier between the plasma chamber and the cryostat vacuum. Lip seals are welded between parallel and wedged segments to guarantee a continuous welded wall in front of the plasma. In order to provide an interspace for leak detection, a second seal is envisaged. No hands-on maintencance procedures will be possible on the seal, since the atmosphere inside the torus will be contaminated and the internal components activated. Therefore the welding/cutting operations on the seal will be carried out remotely. This paper reports the results of an industrial study contract placed to finalize the design of the seal and of the Remote Handling equipment.

1. INTRODUCTION

The NET vacuum vessel (V.V.) is constructed of 32 segments, 16 parallel, located in the bore of the toroidal field coils, alternating with 16 wedge shaped.

The segments are joined by mechanical connections, both bolts and shear keys, which must carry the loads induced on the structure during operation and in fault conditions.

In order to provide a high vacuum within the vessel, welded seals are used between the segments in addition to the bolted connections. Double seals provide a barrier to the tritium and a leak detection interspace.

Because the structure becomes activated and the atmosphere inside the torus contaminated, the seals must be capable of being cut and rewelded remotely, in case of removal or replacement of V.V. segments. No routine maintenance is foreseen on the seals.

2. THE DOUBLE SEAL DESIGN

A preliminary analysis based on the thermal and structural behaviour of several kinds of seal and on the Remote Handling (R.H.) requirements has permitted to identify a reference solution (Fig. 1).

The outer seal is of a "UU"-type with each half attached to either a wedged segment or to a parallel one. The inner seal, which faces the plasma, is a cover plate. It provides an interspace for leak detection and it prevents tritium contamination of the outer seal weld. The chosen material, according to the activation level, the welding requirements and the loads acting on the seal, is AISI 316LN. The high heat flux from the plasma, the pressure loads and the electromagnetic forces generated in case of a plasma disruption have been taken into account in the thermal and structural analysis of the reference seal design (see table I).

TABLE I: Main Loads on the Seal

Nuclear heat (peak)	3 MW/m^3
Radiation heat	
(cover plate only)	2.5 kW/m^2
Static pressure loads in fault conditions:	
inner volume	0-2 bar abs
outside pressure:	0-1.5 bar abs
interspace:	0-1.2 bar abs
Electromagnetic load during plasma disruption	
(peak):	2.5 bar (towards the plasma)
No. of disruptions	550

2.1 The "UU"-Seal

A seal thickness of 2 mm has been chosen. The attachement position to the segments is at the end of the seal curved section. This increases the mechanical strength and minimizes the conduction path to the heat sink.

For the most severe loadcase, 2.4 bar electromagnetic pressure, the maximum bending stress is around 180 MPa for the straight section, to be compared with the nominal 1.5 Sm (310 MPa for AISI 316LN).

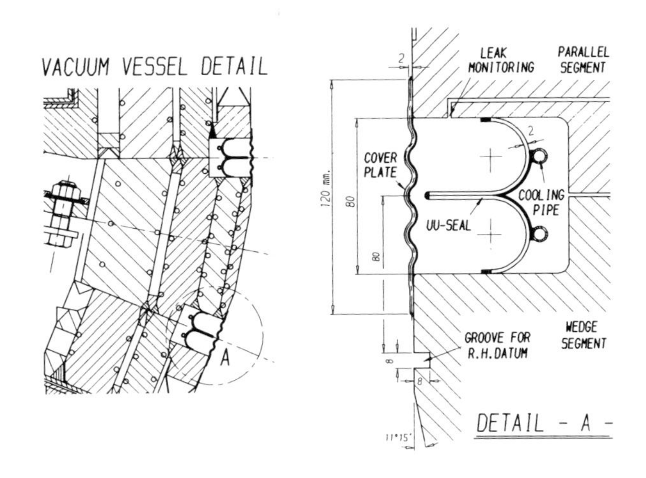

FIGURE 1
NET Double Seal

In order to maintain reasonable temperatures and thermal stresses, the heat transfer capability has been enhanced by using copper plating and a cooling pipe.

The copper plating (1 mm thick) is applied to the quarter circle section between the cooling pipe and the base of the seal lips. The cooling tube is attached to the base of the "U" and connected to the V.V. cooling system.

The maximum thermal stress, around 230 MPa at the seal top weld, is well within the 3Sm for the AISI 316LN (620 MPa) and within the endurance limits for stainless steels.

2.2 The Cover Plate

In order to minimize both pressure and thermal stresses the cover plate is 120 mm wide with 20 mm overlap on either V.V. segment. The highest load, the 2.4 bar electromagnetic pressure, has been taken as design load. The calculation shows that a 2 mm thick cover plate is satisfactory against the load if made of AISI 316LN.

In order to enhance the toroidal flexibility of the cover plate, a corrugated cross section has been preferred. A design with five undulations permits, according to the ASME fatigue curves, a toroidal displacement of about 0.3 mm for 10^5 cycles. Temperature control on the cover plate permits to limit the tritium permeation and the thermal stresses.

Because of the R.H., no cooling pipes can be directly attached to the cover plate, therefore the heat conductivity must be increased by using a 0.5 mm thick copper plating in the central portion of the outer face. Thermal stresses are well within the 3Sm limit, also in the case of a corrugated cover plate.

3. THE O-RING SEAL OPTION

The use of an O-ring placed behind the "UU"-join has also been investigated as an alternative solution to the cover plate. Possible materials for the Helicoflex seal liner are silver, aluminium and copper.

Major problems are envisaged due to the huge radial force, in the order of 1200 t for a 20 mm diameter seal, which is required to slide the wedged segment between the parallel ones and compress the seal. This solution would require special devices to push simultaneously the outboard and the inboard of the wedged segment, while reacting the force on the surrounding structures. A more detailed study, including tests, is envisaged in the next months to assess the feasibility of this approach.

4. THE REMOTE HANDLING OPERATIONS

4.1 The "UU"-Seal

Each half of the seal is welded to the V.V. segments in the workshop and no remote operations are required for this weld. The JET trolley[1] (fig. 2) has been foreseen to cut/weld remotely the "UU"-seal. These operations must be repeated up to three times before replacement of the whole seal is required. This means that due also to assembly tolerances and initial trimming, the seal centre leg can be shortened of 18 mm. The trolley will be mounted on the central leg of the seal without needs of further registration.

4.2 The Cover Plate

As for the "UU"-seal, up to three cuts/welds should be accomodated within this design.

Seal removal will be performed by milling cutters to be used in tandem. Neither cutting fluid nor lubrication is allowed inside the torus. Therefore suitable cutting parameters (i.e. feed rates, cutting speed, ...) have been chosen on the basis of tests (see table II).

The swarf removal during cutting will be based upon suction methods, using heads

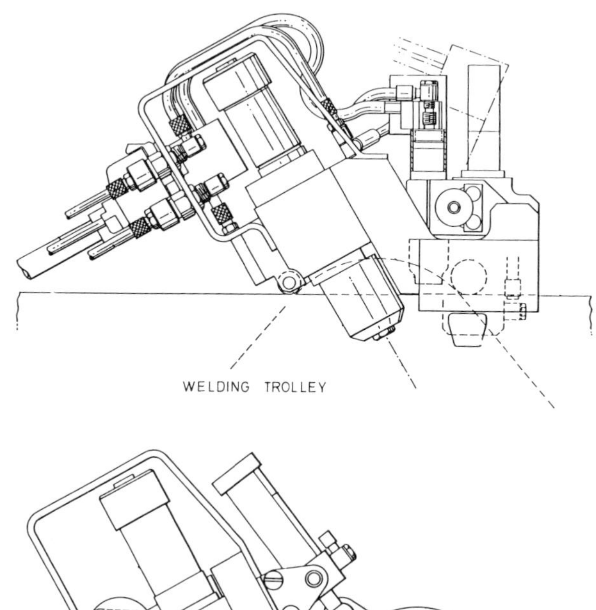

FIGURE 2
JET Welding and Cutting Trolleys

enclosing the milling cutters. Welding will be performed using the TIG method with a wire fill technique.

Tandem heads will be used to carry out simultaneously the welding on both sides. An Arc Voltage Control (AVC) is foreseen to keep the arc length constant. A groove in the V.V. will be used as tracking datum for the seal placing and to control deviations during operations. Different tools will perform the final closing transversal weld of the strip.

The handling and the positioning of the cover plate inside the vessel is by far the most difficult remote operation to be performed.

TABLE II
Results of Cutting Trials

Test No.	Feed per min.	Speed RPM	Remarks
1	0.762m	800	After machining 7.2m, tool edge showing signs of wear
2	0.508m	400	ditto
3	0.762m	1000	1 pass of 0.4m
	1.0m	1000	1 pass of 0.4m
	1.2m	1000	28 passes of 0.4m After machining 12m, the tool edge showing signs of wear
4	1.2m	1050	After machining 14.4m, tool edge worn out

Three options have been evaluated according to key parameters as reliability, control, mass, maintainability, etc.

The first approach foresees the use of a rigid mast carrying the pre-coiled strip stored on a drum. Although easy to handle, this solution would require a very sophisticate control and no multiple tools would be permitted.

The second option consists of an overhead gantry carrying a mast and a manipulator. The operating cycles, the control and the high number of transversal welds are the big disadvantages of this approach.

The third solution consists of a large frame to be inserted into the V.V. carrying both the equipment and the strip. This option has been chosen as the reference one because of the large number of advantages in respect to the other solutions (fig. 3).

5. THE REMOTE HANDLING MACHINES

The following description will focuse on the equipment for the cover plate, since the JET trolley will be used for the operations on the "UU"-seal. The cutting head is an assembly of three basic components: the trolley, the tilting frame and the cutting head (fig. 4).

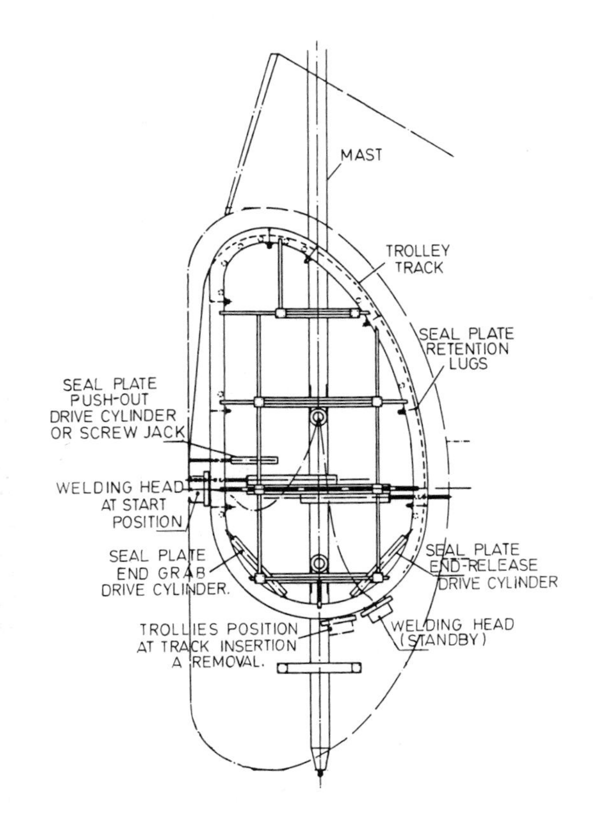

FIGURE 3
Large Frame

The trolley provides the motion with and electric motor connected to two pinions which engage with racks built into the frame assembly. The tilting frame caters for angular variations between cutting head and drive trolley as the machine moves along the VV profile. The cutting head consists of two

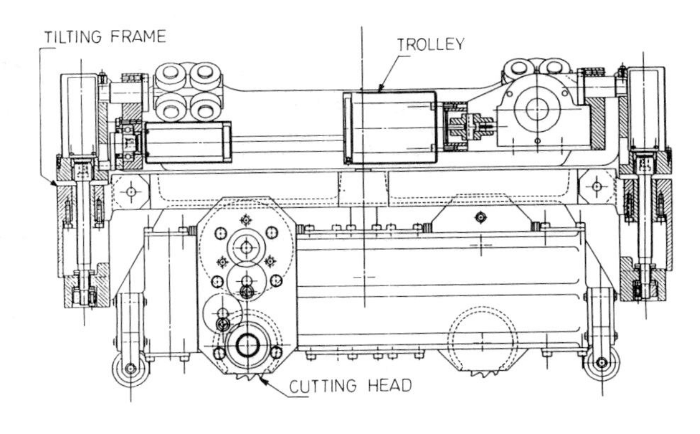

FIGURE 4
Cutting Head Assembly

milling cutters powered by air motors through individual gear boxes. The cutters can be advanced and retracted and a pantograph caters for segment to segment mismatch.

The estimated total weight of the three subassemblies is 65 kilos. The same basic concepts have been used for the seal welding machine: trolley, tilting frame and head (fig. 5). The welding torch is mounted oblique in order to guarantee a full root penetration and its alignment is given by a seal edge tracking roller located to precede the arc. Two welding heads are carried by the machine head assembly.

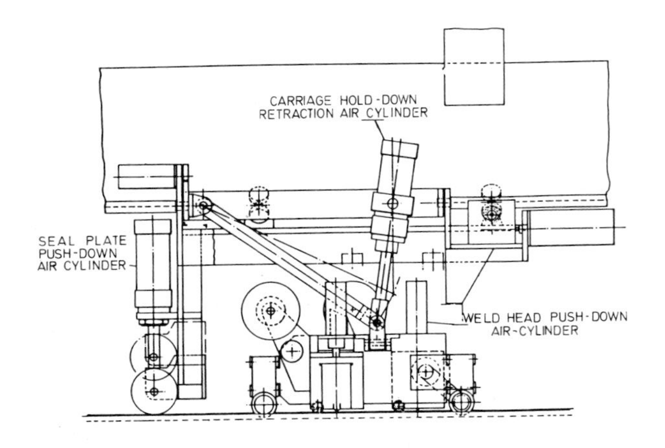

FIGURE 5
Welding Machine

6. HYDROGEN-DOPED WELD TRIALS

There is some concern that hydrogen (tritium) implanted into the seals and vessel may cause problems during the rewelding of the seals after a maintenance operation. Such a risk does not exist for the cover plate, since the weld is every time performed on a material which is not directly exposed to the plasma. However, for the O-ring option, the "UU"-seal weld would be exposed to the plasma and potentially vulnerable to hydrogen damage. Welding trials have been performed on samples of 316L which had been previously doped with hydrogen.

Electrolysis has been used to charge the 316L 2 mm thick strips with residual volumetric hydrogen levels of around 0.5 cc H_2/cc metal. The specimens were welded together autogenously along the charge edges using manual TIG.

Both dye penetrant and microscope examinations have revealed no effect on welding.

REFERENCES

1. S. de Burbure, L. Galbiati, T. Raimondi, Remote Welding and Cutting for the JET Project, Proceedings of the 9th Symposium on Engineering Problems in Fusion, Chicago, Nov. 1981, Vol. 2, p. 1138.

FUSION TECHNOLOGY 1988
A.M. Van Ingen, A. Nijsen-Vis, H.T. Klippel (editors)
Elsevier Science Publishers B.V., 1989

A Dynamic Simulation To Study NET In-Vessel Handling Operations

Dr. Patrick T.K. FUNG

Spar Aerospace Limited, Remote Manipulator Systems Division, 1700 Ormont Drive, Weston Ontario, M9L 2W7, Canada

Abstract

The inspection, maintenance and repair of the Next European Torus (NET) fusion machine will require the extensive use of remote handling equipment to minimise the human exposure to the high radiation environment. The use of efficient manipulators will reduce the NET downtime by reducing the preparation time for entry into the controlled area and by performing the task with reasonable dexterity and speed, consistent with safety. A high fidelity simulation is a valuable tool to assist in the manipulator design, operations, trajectory planning, parameter optimisation and system verification.

A manipulator simulation package called ASAD was originally developed by Spar for space manipulator applications. It is now being adapted to simulate the In-Vessel Handling Unit for the NET program. This terrestrial version of ASAD has been named ASAD_T. Spar, through the services of the Canadian Fusion Fuels Technology Project, is under contract to the NET program for the performance of this activity. This paper describes the capabilities and underlying assumptions of ASAD_T, along with a description of the simulation development of the NET in-vessel manipulator.

1. Introduction

In the late seventies, a non-real time manipulator simulation program named ASAD was developed by Spar [1,2,3], through funding from the National Research Council of Canada, for the remote manipulator on the US Space Transportation System program. Since then, ASAD has undergone an extensive validation process against other simulations, test data and actual flight data. The performance of ASAD has been accepted by NASA and used for many preflight and post-flight analyses. ASAD, particularly the kinematic model and the arm control algorithms, was originally configured for the Space Shuttle Remote Manipulator. For the purposes of simulating the NET in-vessel handling manipulator, ASAD has now been modified to accommodate terrestrial manipulator models. The modified version of ASAD is named ASAD_T.

In addition to its dynamic simulation capability, ASAD_T also has facilities for basic manipulator requirement analyses and pure kinematic control simulation. The dynamic simulation facility of ASAD_T is capable of simulating terrestrial manipulators consisting of up to seven rotary joints and includes models of the flexible dynamics of the drive system and structure. For basic manipulator requirements analyses and kinematic control simulation, ASAD_T can handle up to three manipulators connected to a common base, each with up to ten rotary or translational joints. The package includes a number of control algorithms, both at the joint level and at the manipulator level. If a new control algorithm is needed, the modular structure of the package allows the substitution of alternative control algorithms. An extensive use of parametric data to describe a manipulator and its control system permits the modelling of a wide variety of manipulator configurations.

2. Program Description

ASAD_T is a menu-driven program which consists of eight principal menus. Menu 0 allows the user to terminate the program execution. Menu 1 is for the direct selection of any submenu. Menus 2, 3 and 4 are for defining manipulator system parameters, initial conditions and reference frames respectively. Menu 5 consists of various functions for basic manipulator requirement/capability analyses. Menu 6 is for post-simulation data processing. Menu 7 is for defining and executing a kinematic simulation or a dynamic simulation. The type of simulation is defined through a parameter in a Task Definition file.

The parameters of the dynamics models and the joint servos are stored as block data. An input array is set up to store the parameters. If any of these parameters needs to be changed, the input array index and its new value are added to an input file which is used to update the default values when the simulation is executed.

The inputs of the kinematics models and arm control are divided into six groups with the data of each group stored in a user-named file with a designated extension. The groups of the inputs are:

(a) Manipulator Topology And Geometry,

(b) Initial Conditions,

(c) Important Reference Frames,

(d) Task Definitions,

(e) Control System Parameters, and

(f) Simulation Parameters

There are a large number of simulation output variables. The user can select the output variables to be stored in several types of output files for various analysis purposes such as active display, printout, plotting and post-mortem animation.

3. Kinematic Simulation

The execution of this software skips the manipulator dynamics and joint servo modules, and yields ideal joint rate responses. Kinematic control simulation, unlike dynamic simulation, does not normally reflect the actual characteristics of a manipulator system, but it requires much less computing time than dynamic simulation. Therefore it can be used for testing and tuning of a new control algorithm prior to detailed dynamic simulation. The kinematic simulation is also a valid theoretical verification method for arm control algorithms which are developed from kinematics models.

For more complex manipulator applications, task and trajectory planning is vital. Planning usually requires repeated simulations in order to yield optimal solutions. In this case, an efficient method would be to use pure kinematic control simulation in the initial trial-and-error analysis, followed by a dynamic simulation to verify the results.

4. Dynamic Simulation

The dynamic simulation facility of ASAD_T is capable of simulating a manipulator with up to seven rotary joints. Any two consecutive joints are connected by a rigid or a flexible structure, which is modelled as a beam with uniform mass and stiffness distribution. The flexibility of the joint about the two cross axes are represented by two massless torsional springs. The flexibility about the drive axis is included in the drive train model.

One end of the manipulator is assumed to be connected to a rigid base and the other end may be connected to a payload which can be modelled as a rigid or a flexible structure. The structural flexibility between the manipulator and the base/payload is represented by a torsional spring. A schematic model of the flexible manipulator is shown in Figure 1.

The dynamic equations are mainly based on [4]. The Hooker-Margulies type of approach is used to derive the equations of motion for an open chain of elastic bodies. The translational and rotational equations of motion, which include the body elastic deformations, are first derived for each free body in the chain. Next the equations for deformation of each elastic body are derived in terms of generalized coordinates. Each beam element shown in Figure 2 is considered to be inextensible along its longitudinal axis but to be flexible along its two lateral axes. The rotary inertia and the shear deformation of the beam are considered to be negligible and are therefore ignored. The hinge forces/torques are then eliminated analytically from the assembled equations. The dynamics of the servo system and the drive train are modelled separately. The system equation is of the form:

$$A(x)\ddot{x} + Kx = f(\dot{x}, x, u, u_o) \qquad (1)$$

where x is the system state vector, x and x are the first and second time derivatives of x, A is the system inertia matrix, K is the system stiffness matrix, u is the driving torque vector generated by the drive train, u_o is a vector which includes the disturbance forces/torques such as gravity, f is a force/torque function with x, x, u, and u_o as variables. The arm natural frequencies and mode shapes are computed from the non-zero submatrices of A and K. They are then used to transform Equation (1) into an equation in arm mode generalised coordinates. The transformed equation is partitioned into active (low frequency) and passive (high frequency) modal coordinate equations. The active modal coordinate equation is integrated using an extended Euler's method, while the passive modal coordinate equation is treated quasi-statically by ignoring the acceleration and velocity terms. The integrated active and computed passive coordinates are finally transformed into physical coordinates for display purposes. The above modal truncation technique can significantly reduce computing time, yet maintain simulation accuracy.

5. Joint Servo And Drivetrain Simulation

The simulation model is based on the assumption that control of the manipulator is derived from an actuator, a drivetrain and perhaps a rate and position servo loop at each joint of the manipulator. The model of the joint servo to be used in a particular application is usually unique, at least in some aspects. However, components such as motors, standard controller types, tachometers, and position sensors often have similar models from one application to another. For this reason, the design of a simulation model of a particular joint servo, such as the NET IVHU servo, is usually left to the user who can choose from available component modules in the simulation library. The user provides the code to call these modules in an appropriate sequence to form the overall servo. Where a model of a particular component does not exist in the library, the user may create one and add it to the library. All models of components are made as general as possible by using input data to set parameters.

An existing model for a typical servo and drivetrain is shown in Figure 2. The drivetrain model is basically a single stage model of a geartrain. This model has undergone extensive validation against hardware tests of multi-pass spur gear and planetary gear drivetrains with extremely good results. The ability to shape the drive stiffness nonlinearity and to set the forward drive and backdrive efficiency by means of simulation inputs allows considerable flexibilty in tuning the model for a particular application. The model can also be used for non-backdrivable drivetrains by setting the backdrive efficiency to zero or to a low value.

The integration step of the servo and drivetrain simulation is set separately from that of the arm dynamics simulation. The former is usually smaller than the latter. The output of the servo and drivetrain model is a control torque input to the dynamics model.

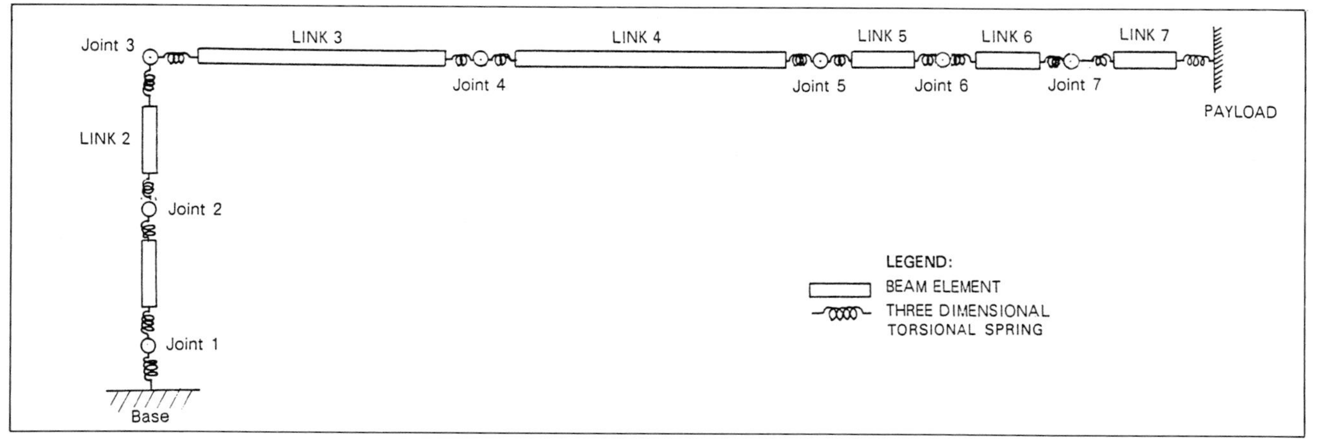

Figure 1 ASAD–T Joint Flexibility Schematic Model

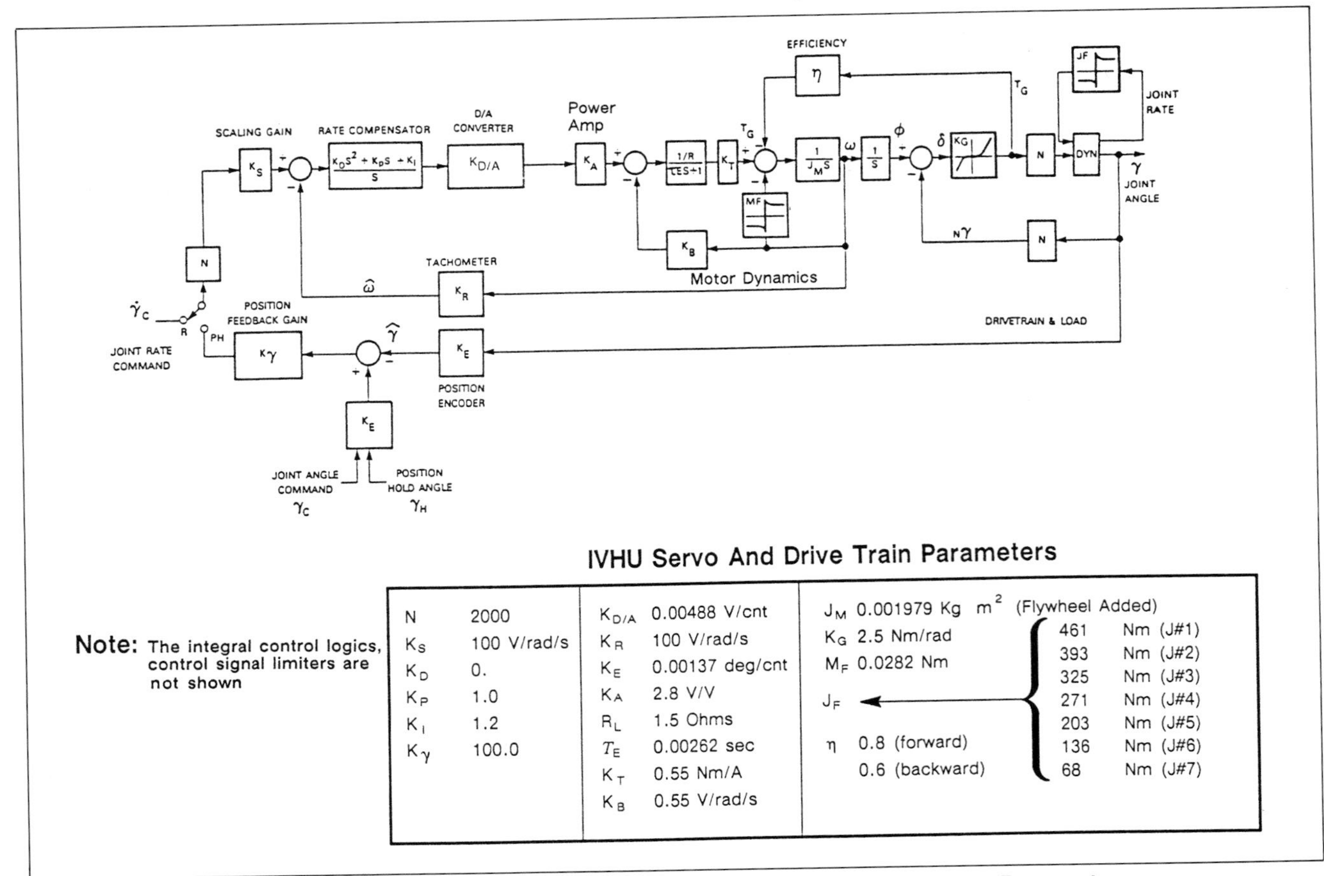

IVHU Servo And Drive Train Parameters

N	2000	$K_{D/A}$	0.00488 V/cnt	J_M	0.001979 Kg m^2 (Flywheel Added)
K_S	100 V/rad/s	K_R	100 V/rad/s	K_G	2.5 Nm/rad
K_D	0.	K_E	0.00137 deg/cnt	M_F	0.0282 Nm
K_P	1.0	K_A	2.8 V/V	J_F	461 Nm (J#1); 393 Nm (J#2); 325 Nm (J#3); 271 Nm (J#4); 203 Nm (J#5); 136 Nm (J#6); 68 Nm (J#7)
K_I	1.2	R_L	1.5 Ohms	η	0.8 (forward)
K_γ	100.0	T_E	0.00262 sec		0.6 (backward)
		K_T	0.55 Nm/A		
		K_B	0.55 V/rad/s		

Figure 2 Typical Servo and Drive-Train Model and IVHU Servo Parameters

6. Operation And Control Simulation

Operational control modes of a manipulator must be modelled and programmed into the simulation to duplicate those of the particular control system design being simulated. As a result of simulation models developed in the past, there are a number of modes of operation and control features available in ASAD_T which can be of use in developing manipulator control methods.

The present available manipulator control algorithms permit operation of the manipulator in five modes of operation:

a) Single Joint Drive Mode - The user specifies the joint to be commanded by entering the joint number and the command rate and direction. The joints not selected remain in Position Hold mode.

b) Manual Augmented Mode - This mode simulates human-in-the-loop operation by applying simulated operator velocity commands in the task space (translational and rotational).The Jacobian inverse (pseudoinverse with cost weightings in the case of redundant manipulators) is used in the transformation of these commands from the task space to the joint space in the resolved rate algorithm.

c) Automatic Lock-on Mode - In this mode the user specifies a final position and orientation of the manipulator. The program calculates the errors between the current position and orientation and the target position and orientation, and then commands the manipulator to move in a 'straight line' towards the target point. The angular velocities and the linear velocities are scaled such that both the manipulator orientation and position reach their targets at the same time. Around the target point, position and orientation 'washout spheres' are provided to reduce the magnitudes of the commands to produce a smooth deceleration to the target. Small position hold spheres in the neighbourhood of the target point also exist so that the manipulator can switch to joint Position Hold Mode.

d) Automatic Fly-by Mode - This mode is similar to the Automatic Lock-on mode, except that the target point is not the termination point for the simulated manipulator motion. Instead, when the manipulator approaches the target within the radii of 'fly-by' spheres in position and orientation, the control algorithm executes the next trajectory in the Task Definition file. The next trajectory could, for example, be commands of the similar manoeuvre to another fly-by target or an Automatic Lock-on manoeuvre to a termination point. A common use for the Automatic Fly-by Mode is to construct a complex and complete non-stop manipulator trajectory from a starting point to a terminal point.

e) Position Hold Mode - The Position Hold mode of control uses a position loop around each joint servo to hold the joint at a setposition. It can be implemented as a submode of the Single Joint mode, Manual Augmented mode, and the Automatic Lock-on mode, any of which can be automatically switched to Position Hold mode when specified quiescent conditions are satisfied.

7. Validation

Prior to being adapted to terrestrial applications, ASAD was subjected to an extensive validation process. The joint servo and drivetrain dynamics were validated against a single joint model of a manipulator. This model, previously, was validated against engineering model hardware of a manipulator joint driving a variety of dummy loads. The complete multi-joint dynamics simulation was validated against other similar non-real time simulation models, independently developed by NASA and other NASA contractors, and against Spar's real time simulation, SIMFAC. It has also undergone extensive comparison with onorbit flight results, and is now considered by NASA to be the standard against which other simulations of the shuttle orbiter manipulator are validated.

ASAD has now been extended to create the terrestrial manipulator simulation program ASAD_T. Those aspects of the program which are affected by the extension require re-validation. This task has begun using results from tests of the Tokamak Fusion Test Reactor (TFTR) Maintenanance Manipulator built by KfK (Kernforschungszentrum Karlsruhe).

8. NET IVHU Simulation

8.1 IVHU Test Model Description

Following completion of validation of ASAD_T, it is planned to simulate the NET In-Vessel Handling Unit (IVHU) to optimize its design and control performance. The IVHU is to be used for remotely performing maintenance tasks inside the NET fusion reactor. The current concept of the IVHU consists of a 9.5 metre long transfer boom, to which a ten degree-of-freedom manipulator is connected. The first six rotary joints are of the yaw type, followed by two pitch joints, another yaw joint, and finally a roll joint. This particular arrangement of joints allows the IVHU to enter the reactor torus and position itself in various service locations (Figure 3).

Since ASAD_T is limited to simulate the dynamics of manipulators with up to seven joints, the two pitch joints and the roll joint of the IVHU Test Model will not be simulated. They will be treated as passive segments of a link. Also the many hardware subsystems attached to the IVHU booms will not be simulated. ASAD_T will be used to model and simulate the kinematics, dynamics and controls of the IVHU manipulator only.

8.2 Simulation Runs

As a means to demonstrate the capabilities of ASAD_T, several simulation runs have been performed. The runs include the simulation of kinematics, flexible dynamics, joint servos and modes of operation of an early concept of the IVHU.

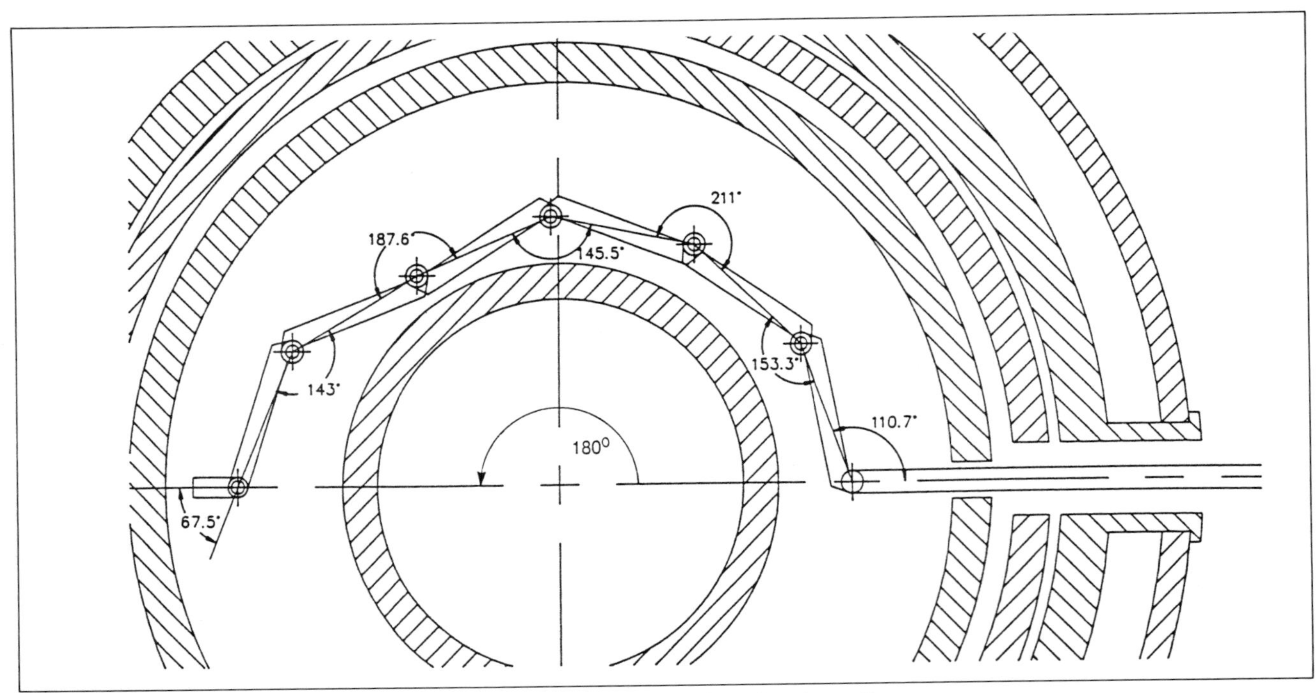

Figure 3 IVHU 180° Service Location

Three runs are performed separately to simulate Single Joint Drive, Manual Augmented (Resolved Rate) and Automatic Fly-by and Lock-on Modes control. In each run, the first 10 seconds of simulation time are used for the IVHU dynamics to settle under the application of gravitational load.

Run 1 simulates Single Joint Rate Mode of control of Joint No. 1(nearest to the base), while the other joints are in Position Hold Mode. The fully extended IVHU is used as the initial configuration. In order to reduce flexible dynamics oscillations, the joint rate command signal is ramped up gradually to 0.4 deg/sec in 1 second and then remains at this rate for 29 seconds.

Run 2 simulates Manual Augmented Mode of control. The IVHU is initially folded and then commanded to move longitudinally (along the Z-axis) for 80 seconds at a speed of 25.4 mm/sec. This is followed by an 80-second lateral (along the X-axis) command at the same speed.

Run 3 simulates Automatic Fly-by and Lock-on Modes of control. The IVHU is initially folded and then commanded to reach a target point via three intermediate spheres in space. The Automatic Fly-by Mode is used to guide the IVHU to pass through the intermediate Fly-by spheres, and the Automatic Lock-on Mode is lastly used to manoeuvre the IVHU to the final target point.

8.3 Simulation Results

The simulation results of Run 1 are given in Figures 4 and 5. Figure 4 depicts the response of Joint 1 to the rate command of 0.4 deg/sec at this joint. The response shows a 10 second period of position hold, followed by a transient rise which reaches a steady state in about 10 seconds. The effects of flexibility and friction are apparent in the high frequency components of the response. Figure 5 shows the joint angle response of Joint 2 to the position hold command. The effects coupling from joint 1 motion causes this joint to overshoot about 0.065 degrees.

Figure 6 shows the simulation results of Run 2 in the form of the X-Z position of the IVHU tip. The total response comprises two trajectories generated by two consecutive velocity commands to the manipulator tip.

Figure 7 shows the simulation results of Run 3. In this run the initial and final positions are the same as those of Run 2, but in this case, automatic control is used to drive the IVHU via three intermediate flyby points in the X–Z plane.

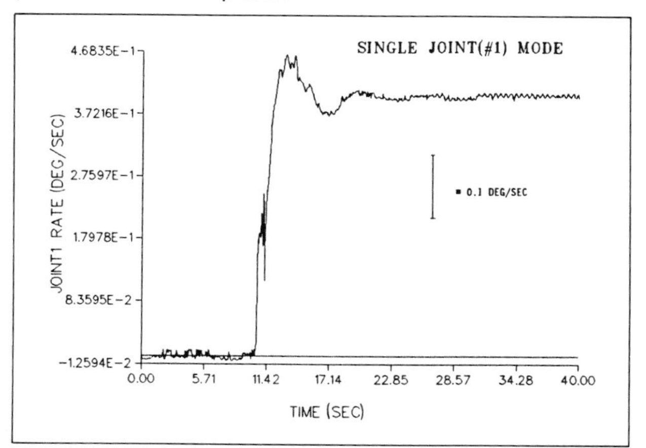

Figure 4 Run 1, Joint 1 Response to Rate Command

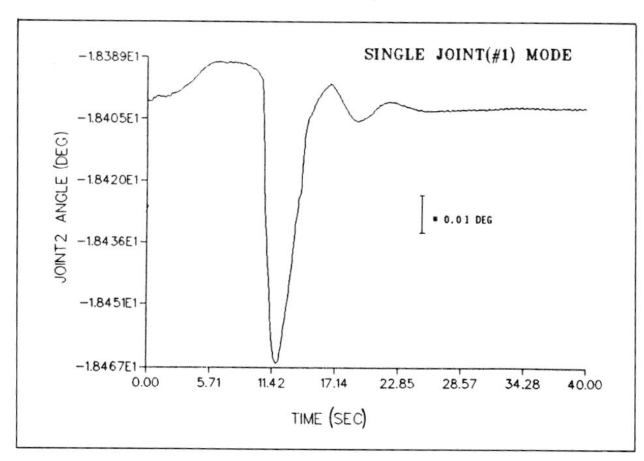

Figure 5 Run 1, Joint 2 Position Response to Joint 1 Rate Command

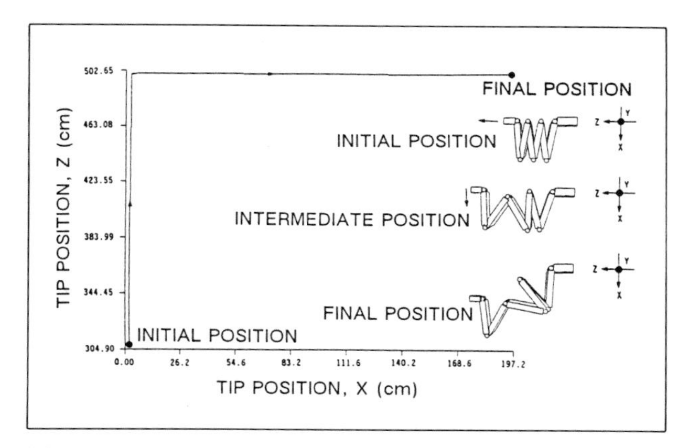

Figure 6 Run No.2 Responses to Two Orthogonal Tip Velocity Commands in Manual Augmented Control Mode

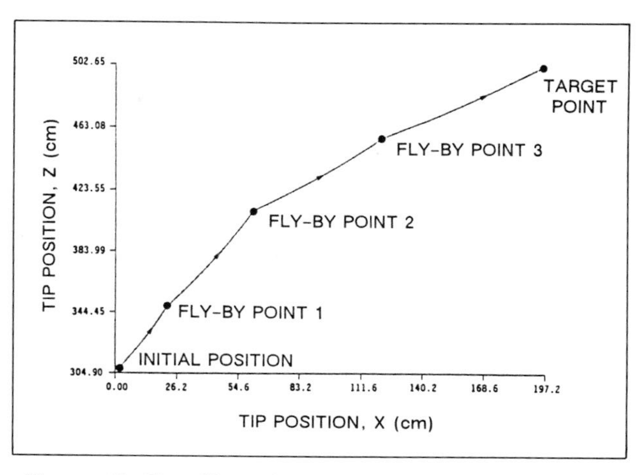

Figure 7 Run No.3 Combined Automatic Fly-by and Automatic Lock-on Response in Automatic Control Mode

9. Conclusions

A dynamic simulation package for simulation of terrestrial manipulators, ASAD_T, has been developed for application to the design and analysis of the NET in-vessel handling unit. In this application, the following aspects of dynamics and control will be studied:

(a) *Determination of the dynamic responsesof the joints and boom.*

(b) *Simulation of in-vessel operational activities.*

(c) *Determination of efficient trajectories and movements of the handling unit necessary to reach a given point within the torus.*

(d) *Optimization of the NET IVHU joint servo control system.*

Acknowledgements

The author wishes to thank the NET program and the Canadian Fusion Fuels Technology Project for their funding and assistance in extending ASAD to the terrestrial version, ASAD_T. He also wishes to thank KfK for their assistance in planning and developing a validation test program and performing tests on the TFTR Maintenance Manipulator, and the National Research Council of Canada for their permission to extend the use of ASAD to the simulation of terrestrial manipulators. Finally the author wishes to thank Mr C.P.R. Trudel and Mr F. Rakhsha of Spar Aerospace Ltd for their contributions to the development of ASAD_T.

References

[1] Nguyen, P.K., Ravindran, R., Carr R., Gossain, D.M., Doetsch, K.H., "Structural Flexibility of the Shuttle Remote Manipulator System Mechanical Arm", Proc. AIAA Guidance and Control Conference, San Diego, California, U.S.A.(1982).

[2] Carr, R., Nguyen, P.K., Ravindran, R., Trudel, C.P.R., "ASAD - A Non-Real Time Simulation Program for the Shuttle Remote Manipulator System", Proc. Summer Computer Simulation Conference, Washington D.C., U.S.A.(1981).

[3] Ravindran, R., Sachdev, S.S., Aikenhead, B., "The Shuttle Remote Manipulator System and its Flight Data", Proc. 4th International Symposium on Space Technology and Science, Tokyo, Japan(1984).

[4] Hughes, P.C., Motion Equations for a Flexible Articulated Controlled Manipulator Arm, Dynacon Report 75-05-10(1975).

FUSION TECHNOLOGY 1988
A.M. Van Ingen, A. Nijsen-Vis, H.T. Klippel (editors)

AN IMPROVED POWER SUPPLY SYSTEM CONSISTING OF DC MFG SETS AND CAPACITOR BANKS FOR HT-6M(II) TOKAMAK DEVICE

Pan Yuan, Xu Jiazhi, Liu Baohua, Huang He, Peng Jiafu, Yao Xiuluan, Hu Suhua, Chen Erlian, Liu Yanqin, Xu Yuzhen, Zou Jianhua, Li Xiaobin, Zhuang Ming, Liu Fuguo

Institute of Plasma Physics, Academia Sinica, Hefei, P.R.China

One of the main objectives for HT-6M(II) experiment is to study the feature of the plasma with high-power heating density and the plasma current driven by LHW and the combination of ICRH and LHCD. Because all of the magnets are non-watercooling and the current density already reaches $63A/mm^2$, in order to extend the pulse duration. the wavefront and wavetail of the pulse must be shortened as much as possible so as to increase the ratio of effective pulse length to whole equivalent pulse length. For this reason, we have developed a power supply system which consists of DC MFG sets and capacitor banks. This combined power supply can be operated in long pusle discharge mode (> 150ms) or in short pulse discharge mode (> 50ms) and the operation modes can be changed easily. This paper describes the design and the characteristices of this power supply system.

1. INTRODUCTION

One of the main objectives for HT-6M(II) experiment[1] is to study the feature of the plasma with high-power heating density and the plasma current driven by LHW and the combination of ICRH and LHCD. For this task, we have prepared additional heating facilities, i.e. 1MW ICRH source, 100-200KW ECRH source, 100KW LHW source, and 0.3MW NBI, the pulse length of them are all 30ms. Consequently, we must prolong the flat-top pulse length of HT-6M(II) from the previous value 30ms to 150ms.

The main parameters of HT-6M(II) device[1] are: R=65cm, a=20cm, B_T=150KG, I_p=150KA, T(flat-top) =150ms

The main parameters of HT-6M(II) magnets are:

Magnet	TF	OH	VF
Coil resistance at 75°C (mΩ)	63.5	25.5	28
Coil inductance (mH)	16.25	4.37	8.42
Mutual inductance(mH)		1.43	
Max. operation current(KA)	15	± 17	6/12
Max. operation current density (A/mm^2)	63	61	20/40
Allowed max. current (KA)	19	20	16
Allowed max. voltage to eartn (kv)	10	10	10
Temperature limitation (°c)	100	100	100
Pure copper weight (kg)	1500	800	1220

According to the requirement of physical experiment, the HT-6M(II) operation modes are:

1). Long pulsed OH experiment T(flat-top) > 150ms.

The DC MFG sets should be put into operation.

2). Short pulsed OH experiment T(flat-top) > 50ms.

Only the capacitor banks are put into operation.

3). The additional heating or/and current drive experiment.

The vertical field must be increased in the additional heating process.

4). Cleaning discharge including high power pulsed dischargecleaning, Taylor discharge cleaning, 400Hz discharge cleaning, glow discharge cleaning and ECRH discharge cleaning.

Up to now, our main power souces which can be utilized are Four DC MFG sets and 3.5MJ capacitor banks.

The main data of DC MFG set are:

Type	ZMF-500/150
Peak power (MW)	20
Max. voltage (v)	500(no load)/400(rated load)
Max. Current (KA)	50
Armature resistance (mΩ)	1.7
Armature inductance (μH)	15
Driving motor power (KW)	630
Total GD^2 ($T.M^2$)	770
Speed (r.p.m)	368 (no load)

The main data of capacitor are:

Type	My-50/150
Limited value of charge voltage (kv)	4.5
Capacitance per unite (μF)	150

2. Design of the power supply system

2.1. Construction of the power supply

As stated above, the current density of HT-6 M(II) magnets is very high and the thermostability of the magnets is very limited. In order to prolong the flap-top duration, the raise time and fall time of the pulse must be shortened as much as possible. Howerer, if these magnets are only powered by DC MFG sets, to realize this is impossible because of their low voltage. In addition, taking account of the required operation modes, finally, we decided that a combined power supply with two DC MFG sets and all of 3.5MJ capacitor banks is adopted. The main principle circuit of the power supply system is shown in Fig.1.

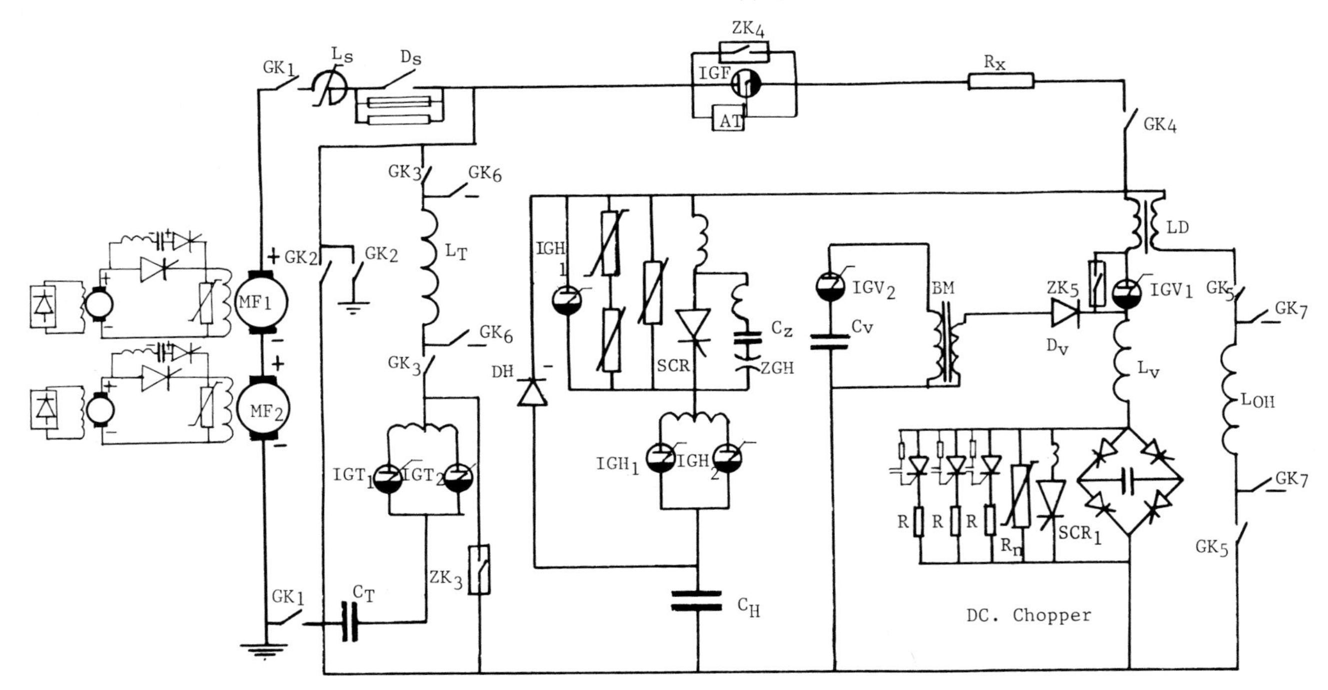

FIGURE 1: The Main Principle Circuit of the Power Supply System for HT-6M(II)

2.2. Long pulse operation

Both of the DC MFG sets and capacitor banks are put into operation. Refer to Fig.1, the disconnecting switches GK_1, GK_3, GK_4, GK_5, are switched on, GK_2, GK_6, GK_7 are turned off. A typical waveform of long pulse discharge is shown in Fig.2.

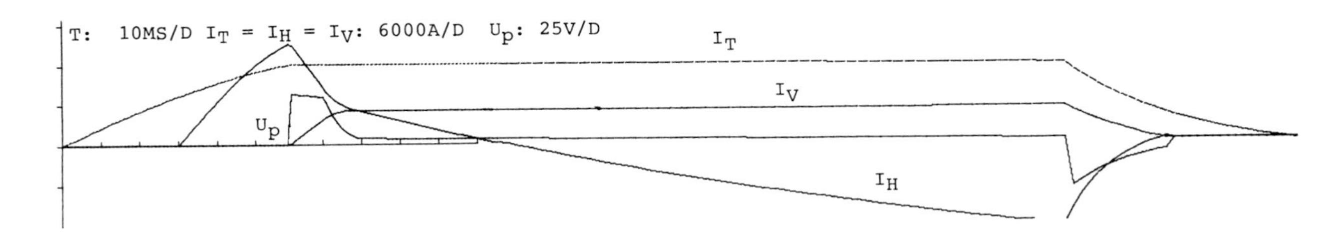

FIGURE 2: A Typical Waveform of Long Pulse Discharge

This waveform is obtained by means of computer calculating. The voltage of two DC MFG sets in series is 2× 500V, both capacitor banks C_T and C_H are charged to 4.5KV. So far, the long pulse has not yet been put into operation.

After a pulse discharge, the relation between flat-top length, equivalent square wave length, adiabatic temperature raising and loop voltage is as follows:

Magnet	TF			OH		
Flat-top pulse length (ms)	137	150	200	137	150	200
Equivalent square wave length (ms)	183	195	244	48	53	83
Adiabatic temperature raising (°C)	4.20	4.46	5.57	1.07	1.18	1.84
Mini loop voltage (v)				2.0	1.9	1.5

From the above table, we can see that the most heated component is TF, so it must be monitored in whole experiment period. When the temperature goes beyond a limited value, for example 100°C, the control microputer will automatically switch to the non-maskable interrupt and the operation is stoped.

2.3. Short pulse operation

The DC MFG set are not put into operation, all of the magnets are powered only by capacitor banks. The disconnecting switches GK_2, GK_3, GK_5 are switched on, and GK_1, GK_4, GK_6, GK_7 are turned off. The vacuum switches ZK_3, ZK_4, ZK_5 and fast DC breaker DS as well as ignitron IGF are all put out of action, but all the rest will be put into service. The operation process is basically the same as the long pulse operation process. The typical waveform of a short pulse discharge only with ohmic heating is shown in Fig.3.

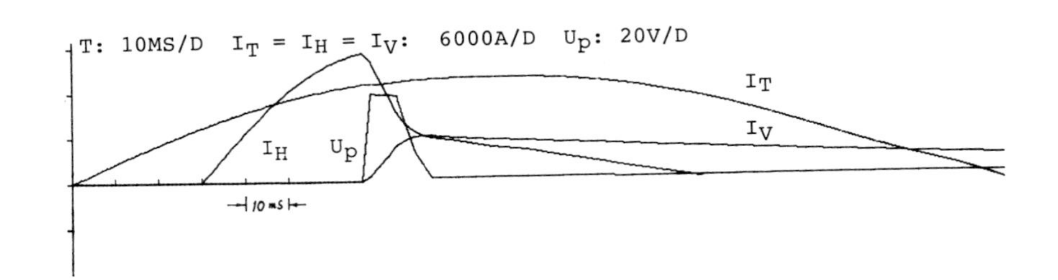

FIGURE 3(a): The Calculated Waveform of a Short Pulse Discharge

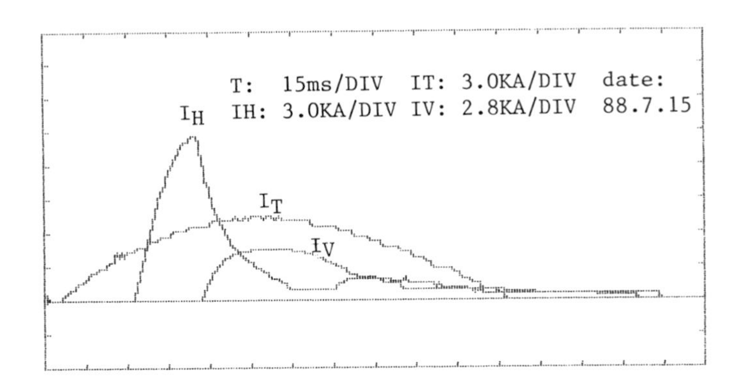

FIGURE 3(b): Experimental Waveform of Three Fields

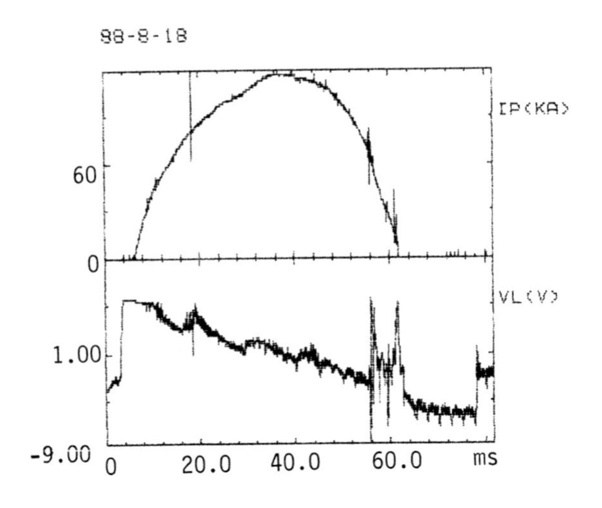

FIGURE 3(c): Waveform of Plasma Current and Loop Voltage

3. THE CHARACTERISTICES OF THIS COMBINED POWER SUPPLY SYSTEM

The characteristices of this combined power supply system can be summarized as follows:

1) In long pulse operation, the pulse front of TF, which is the most heated component, is set up by the combination in series of capacitor bank C_T and two DC MFG sets (in series), but the pulse flat-top is sustained by only two DC MFG sets, while the bank C_T is crowbared by the vacuum switch ZK_3. After flat-top, the pulse will be fallen rapidly by means of de-exciting the field of the DC generators, through a ZNR (Z_nO nonlinear resistor) network. In this case, the Max. de-exciting time T_d<500Ms.

2) The OH magnet is pre-magnetized by the capacitor bank C_H. A 8KV/20KA composite thyristor DC breaker[2], which was specially developed for the commutation branch of OH circuit and has been operating for over 10000 shots, interrupts the pre-magnetizing current and forces it into a ZNR-LR (Z_nO nonlinear resistor-linear resistor) network. The ZNR-LR network combined with backswing voltage on the capacitor bank C_H are adopted for establishing a counter e.m.f. across the OH magnet to induce an initial loop voltage which initiates the breakdown and builds up the front of pulse current. This circuit can realize the effectively utilizing of the V-S value of OH transformer. The stack of diods is for making full use of the energy storage in capacitor bank C_H (in short pulse operation). In addition, the high peak voltage which arises from LR (if only LR is adopted)

can be avoid. And thus the insulation level of OH magnet and other apparatus can be relatively low, otherwise, some multiple commutation circuit have to be adopted especially in some major Tokamak device[3]. About this, Dr.P.Bellomo etc[4]., have the same point of view. Moreover, it is possible that a relatively idealized waveform of loop voltage can be obtained by means of this circuit. And the E/P value, where E is the electric field in V/cm and P the pressure in torr, will be controlled in a reasonable range in which Tokamak can be successfully operated. And this is very important to decrease the runaway electrons and to establish rapidly the toroidal current as well as flux sufaces[5].

3) The primary establishment of VF is set up by OH magnet which acts as a pulse inductive energy storage source. In this case, the capacitor bank for VF current front establishment can be saved. When the OH current has been commutated from the thyristor DC breaker into the ZNR-LR network, the voltage on the capacitor bank C_H will swing back, and the counter e.m.f. across the primary side of OH transformer will not only induce the plasma current, but also forces the current passing through the OH magnet to pass the VF magnet. It will synchronously set up the pulse front of plasma current and VF current. Once VF magnet current close to the OH magnet current, the ZNR-LR network will be in closed state and then the two magnets will be powered simutaneously by DC MFG sets (in long pulse operation) or by another capacitor bank C_V (in short pulse operation). Consequently, the pulse will get into the flat-top phase.

4) The feedback system is composed of feedback coil, decouple transformer (LD) and an adjustable network (DC·chopper) in series with the VF branch. (see Fig.1). The feedback coil is powered by a middle-frequency generator of 400Hz and 400V which is followed by a SCR convertor[6]. The adjustable network is essentially a passive chopper. The novel property is dominated by the ZNR (R_n) which is used for stabilizing the counter e.m.f. when the ZNR is switched into VF branch. We can adjust the rasing rate of VF current by means of switching the ZNR into the VF branch or bypassing it with the resistor R or thyristor SCR_1. In the same time, the resistor R also act as an current limitation resistor in long pulse operation. The adjusting action of the decouple transformer LD for the FV current rasing is realized by changing the decouple degree.

Finally,We must give some explanation. UP to now, the ZNR in the commutation branch of OH circuit has not yet been put into operation, And also, the VF upgrade for additional heating has not yet been finished.

REFERENCES

1. HT-6M Team, "The HT-6M Project and Design", Research Report, ASIPP, 1982.
2. Huang He, Pan Yuan, Xu Jiazhi and Liu Baohua, "A 20KA-8KV Composite Thyristor DC Breaker", Research Report, ASIPP, 1987.
3. N.Miya, et al., "JT-60 Ohmic Heating Power Supply", Fusion Technology, Vol.2, 1980.
4. P.Bellomo, J.Calpin, R.Cassel and H.Zuvers, "Plasma Striking Voltage Production", 9th Symp. on Enger. Problems of Fusion Research, Proceedings Vol.2. 1981.
5. S.A.Eckstrand, R.D.Bengyson and J.E.Benesch, "A Simplified Ohmic Heating Circuit for Tokamaks", IEEE Trans. on Plasma Science, Vol. Ps-10, No.3. P207 (1982).
6. Zhou Yongcheng, et al., "The Radial Position Feedback Control System for HT-6M Device", Research Report, ASIPP, 1985.
7. Zou Jianhua, Pan Yuan and Guo Wenkang, "An adjustable VF Power Supply for HT-6M(II) Device", Research Report, ASIPP, 1987.

FUSION TECHNOLOGY 1988
A.M. Van Ingen, A. Nijsen-Vis, H.T. Klippel (editors)

MODIFICATIONS AND OPERATIONAL EXPERIENCE OF JT-60 POWER SUPPLIES

T. Matsukawa, K. Omori, S. Takahashi, S. Omori, T. Terakado, J. Yagyu, Y. Omori, K. Miyachi, A. Takeshita, Y. Ikeda, H. Ichige, S. Nagaya and T. Akiyama

Department of JT-60 Facility, Naka Fusion Research Establishment, JAERI, Naka-machi, Naka-gun, Ibaraki-ken, 311-01 Japan

R. Shimada

Tokyo Institute of Technology, Ookayama, Meguro-ku, Tokyo, 152 Japan

JT-60 power supplies have been under operation from 1985 with some components of them being modified in shutdown period of JT-60 to satisfy the operational requirements in the plasma experiments. Especially, in 1987, to achieve the extended plasma current operations of 3.2MA in limiter and 2.7MA in magnetic separatrix configurations, the output current capabilities of JT-60 power supplies were enhanced by installing a booster thyristor convertor and so on. Here, such operational experiences, including relevant modifications, of JT-60 power supplies are described. And also, the present status of JT-60 power supplies for the operation in 1988 is mentioned.

1. INTRODUCTION

JT-60 power supplies include the poloidal and toroidal field power supplies (PFPS and TFPS) and the motor-generator system for the additional heating equipments (NBI, ICRF and LHRF). PFPS consists of five thyristor convertor systems; namely, ohmic heating (PSF), vertical field (PSV), horizontal field (PSH), quadrupole field (PSQ) and magnetic limiter (PSM) circuits, which energize the corresponding five poloidal field coils of JT-60 independently of one another. PSF provides the needed volt·sec for driving the plasma current with interrupting the coil current by the vacuum circuit breaker (VCB) in breakdown period. PSV and PSH control the each coil current to keep the plasma radial and vertical positions respectively. PSQ has the role to control the shape of plasma cross section, which means the clearance between the plasma surface and the first wall. PSM energizes the magnetic limiter coil, installed in the vacuum vessel, to generate the magnetic separatrix configuration with the single null point in the outer side. All the thyristor convertors of PFPS are equivalent 12 or 24 pulse one, and some of them are connected in anti-parallel to enable themselves to control the coil current in both positive and negative directions. The electric power for PFPS is supplied

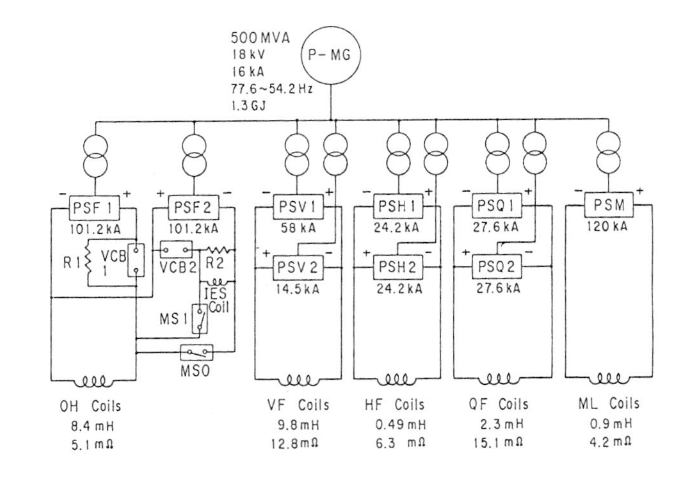

FIGURE 1
Schematic diagram of the poloidal field power supplies on April 1985

by a motor-generator, which can yield 500MVA as peak power and 1.3GJ for one cycle operation. Figure 1 is the brief scheme of PFPS, where the rating output voltage and current of each thyristor convertor, the rating coil currents and the poloidal coil inductance values are also shown.

TFPS has two 24-pulse diode convertor systems; one receives the electric power from the 275kV utility power network directly, the maximum power allowed being 160MW, and the other is energized by a motor-generator (215MVA-18kV-4GJ). Both of them are connected in series at their DC output terminals, Figure 2 shows the skeleton of TFPS with main parameters of the major components of TFPS.

The motor-generator for the additional heating equipments (NBI, ICRF and LHRF) has the capability of 400MVA-18kV-2.6GJ for the simultaneous and 10sec operation of NBI (20MW) and RF (10MW; IC+LH) systems.

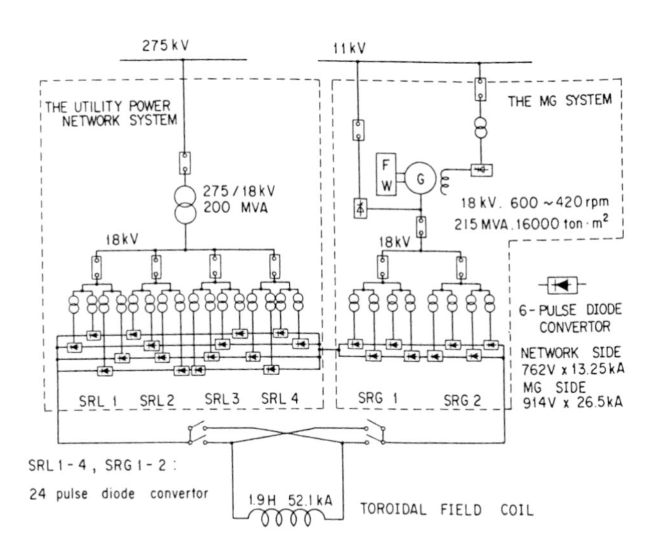

FIGURE 2
Schematic diagram of the toloidal field power supplies

2. OPERATION IN F.Y. 1986

From the first plasma operation in 1985, JT-60 has successfully progressed its own operational region. In the initial operations in 1985, the tokamak discharges of JT-60 were established upto 1.6MA plasma current in limiter and magnetic separatrix configurations, and the power supplies' operational scenarios in such stages were not so in crucial level. Some modifications, however, were done to get much better plasma performance. One of them is changing the pulse length from 5sec to 10sec.

In F.Y. 1986, the additional heating experiments with NBI, ICRF and LHRF were started, and the plasma currents reached the rated values of 2.7MA in limiter and 2.1MA in magnetic separatrix configurations with 10sec pulse length. As the main purpose of the first 3.5 cycles' experiments was to establish joule-heated target plasmas for additional heating, various operational parameters of PFPS were tried. The output currents of PFPS were reached to the following value in operation.

	Current in operation (kA)	Nominal output current (kA)
PSF	+85.4 ~ -92.0	± 92.0
PSV	+26.1	+58.0 ~ -8.7
PSH	+17.2	± 22.0
PSQ	+25.0	± 25.0
PSM	84.9	120.0

The typical waveforms of PFPS output currents and voltages in this phase were shown in Figure 3, which was the well-operated case of 2MA flattop current-10sec pulse length plasma in magnetic separatrix configuration.

In last 11 cycles' experiments with NBI and RF, PFPS was operated in the

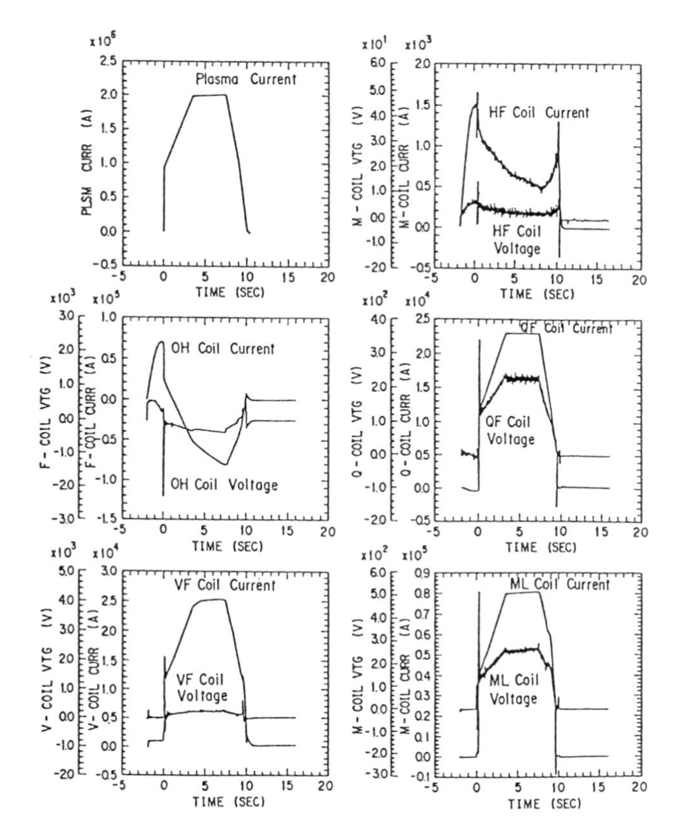

FIGURE 3
Waveforms of plasma current and currents and voltages of 5 poloidal fielid coils

same region as previous operations. But, from the point of OH coil current control, the plasma current drive experiment by LHRF resulted in different OH coil current swing from joule-heating experiment. Figure 4 shows that OH coil current changed its own direction of swing (OH coil was recharged), while the plasma current was sustained constant by LHRF.

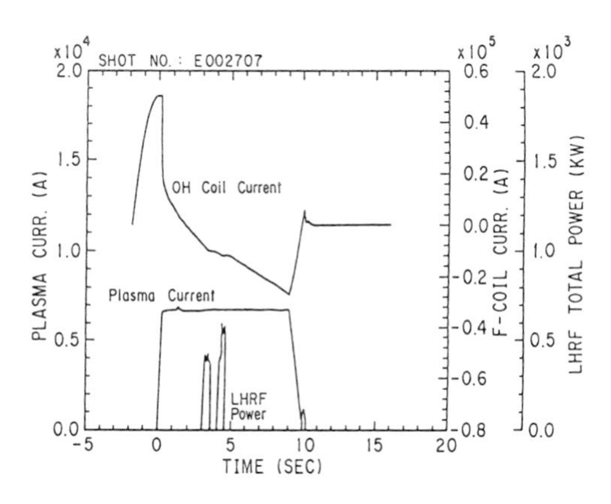

FIGURE 4
Waveforms of plasma current and OH coil current when LHRF drives the current

3. EXTENDED OPERATION IN F.Y. 1987

Based on the plasma experimental results in the previous two years, the extended plasma operations, which mean 3.2MA discharge in limiter and 2.7MA in magnetic separatrix configurations, were planned in F.Y. 1987 to achieve much higher plasma characteristics. Considering the margins of hardwares, the following modifications were done.

a) The maximum toroidal field should be strengthened upto 4.8T (rating value : 4.5T), corresponding to increasing the toroidal field coil current to 55.2kA (rating value:52.1kA). The software programe of TFPS control system and the setting of TFPS protection relays were readjusted to allow 55.2kA operation. And also the tap-changer's positions of the transformers for diode convertors were changed to keep the power distribution from the network and MG appropriately.

b) To drive the plasma current upto 3.2MA, PSF should swing the OH coil current from +92kA to ~ -110kA (rating value : ± 92kA). The output current control method of PSF was modified to suspend the circulating current control of thyristor convertors when the coil current reached -50kA. So that PSF can provide the OH coil current equal to the maximum output of the thyristor convertors.

c) To keep the clearance between the plasma surface and the first wall in the magnetic separatrix operation, a booster thyristor convertor of 10kA DC output was installed and connected in parallel to PSQ. It was designed that the booster convertor should be operated as 10kA current source, so the inductor (92kA-8.00mH) was connected in series to the convertor to avoid the DC current fluctuation caused by the plasma disruption.

d) The dummy load coil (92kA-8.00mH) used for the power tests of PFPS, was converted to the decoupling inductor and connected in series to PSV. It has the role to suppress the vertical field coil current fluctuation induced by the plasma disruption, which makes the mechanical stress of PF coils larger than the allowable value.

Figure 5 is the scheme of PFPS for the extended plasma operation, and the successful 3.2MA plasma current waveform is shown in Figure 6(a) with the corresponding PF coils' current waveforms in Figure 6(b).

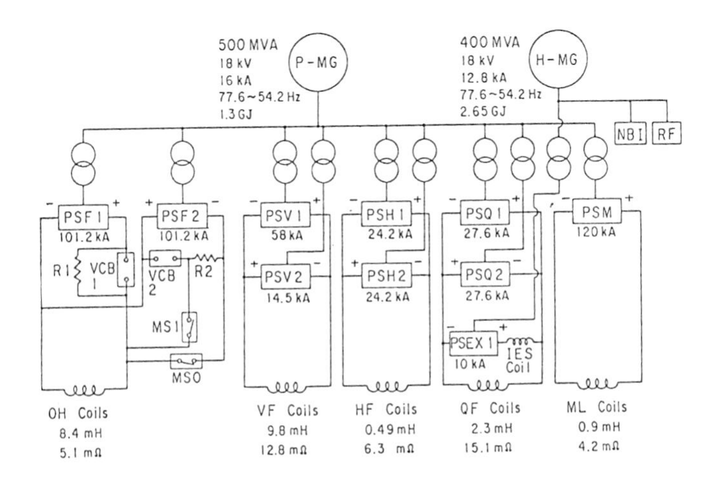

FIGURE 5
Schematic diagram of the poloidal field power supplies in FY 1987

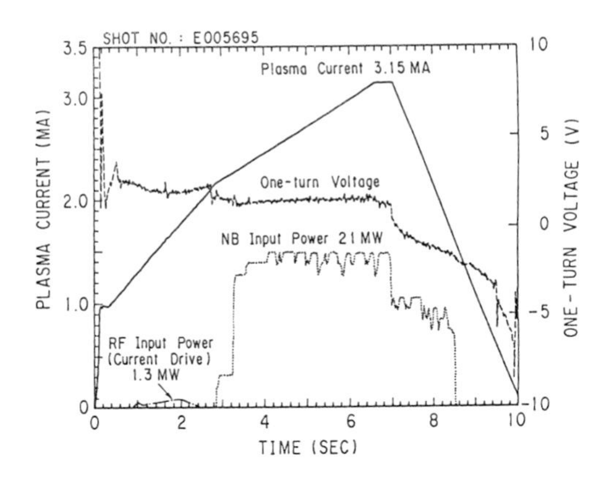

FIGURE 6(a)
Waveforms of plasma current, one-turn voltage and the power of NBI when plasma current is 3.2 MA

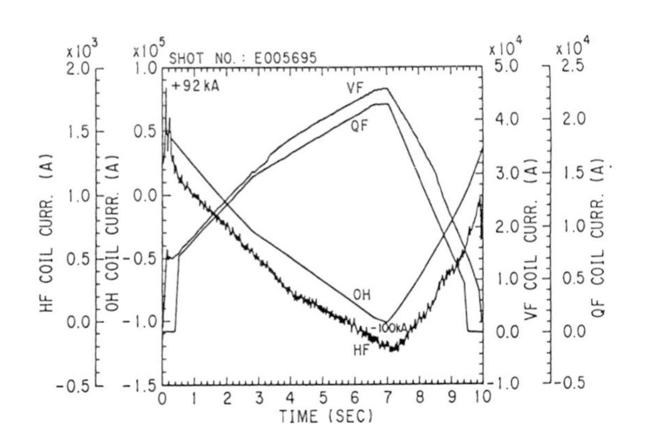

FIGURE 6(b)
Waveforms of coils' currents of OH, VF HF and QF when plasma current is 3.2 MA

In F.Y. 1987, PFPS and TFPS were operated in the following region;

	Current in operation (kA)	Nominal output current (kA)
PFPS		
PSF	+92.0 ~ -106.8	± 92.0
PSV	+46.4	+58.0 ~ -8.7
PSH	+12.9	± 22.0
PSQ	+33.3	+35.0 ~ -25.0
PSM	106.2	120.0
TFPS	55.2	52.1

4. RELIABILITY ANALYSIS ON JT-60 POWER SUPPLIES

As stated above, 3 sets of large motor-generators, that's total power is 1,115MVA, are provided in JT-60 power supply systems. And particular accident for MG has not occurred until now. The running hours of each MG is following.

	Running hours (hours)	Numbers of operation (times)
PFPS		
F.Y. 1985	432.3	76
F.Y. 1986	1211.9	118
F.Y. 1987	854.1	77
Total	2498.3	271
TFPS		
F.Y. 1985	443.1	63
F.Y. 1986	1362.8	121
F.Y. 1987	884.5	74
Total	2690.4	258
MG for AH		
F.Y. 1985	674.4	88
F.Y. 1986	2843.5	219
F.Y. 1987	1145.3	105
Total	4663.2	412

Some faults occurred in operation, and they can be assorted roughly CAMAC system errors and machine errors. The former has module or crate failure as hardware trouble and communication error as software trouble. And, as an example of typical troubles for the machine, unsatisfactory operation for VCB, destruction of thyristor element and the carbon brush abrasion of the MG slipring, etc. occurred. Every trouble was managed suitably, and machine was restored to former condition without delay. Numbers of fault and their repair hours for the past three years are following.

	Numbers of fault (times)	Repair hours (hours)
F.Y. 1985		
PFPS	27	13.9
TFPS	17	11.8
MG for AH	-	-
Total	44	25.7
F.Y. 1986		
PFPS	83	53.7
TFPS	23	8.0
MG for AH	31	7.8
Total	137	69.5
F.Y. 1987		
PFPS	64	44.3
TFPS	23	14.5
MG for AH	21	6.0
Total	108	64.8

5. CONCLUSIONS

As summarized above, the operational enhancement of JT-60 power supplies in 1986 and 1987 was successfully progressed. Specially, the extended operation in 1987 was achieved with some modifications on the main circuits and the reliable margins on output capabilities of JT-60 power supplies.

The analysis of operational experience of JT-60 power supplies, including some examples of faults, showed the availabilities and reliabilities of the power supplies system.

The operation for the new experimental regime of JT-60 in 1988 was already started, and it was also planned to use JT-60 power supplies in shutdown period for energizing the proto-type superconducting magnet in the test bed of the laboratory.

FUSION TECHNOLOGY 1988
A.M. Van Ingen, A. Nijsen-Vis, H.T. Klippel (editors)

A 220 MVA TURBO-GENERATOR FOR THE TCV TOKAMAK POWER SUPPLIES

A. Perez, I. M. Canay*, J.-J. Morf**, J.-D. Pahud***, R. Seysen****, J.-J. Simond*

Centre de Recherche en Physique des Plasmas, Association Euratom - Confédération Suisse,
Ecole Polytechnique Fédérale de Lausanne, 21 Av. des Bains, CH - 1007 Lausanne / Switzerland

A new 220 MVA, 120 Hz, 4 pole turbo-generator will be used as a pulsed power source to supply the toroidal and poloidal power supplies of the TCV tokamak, which is being built at the Ecole Polytechnique Fédérale de Lausanne in Switzerland. The paper describes the particular requirements of the TCV poloidal power supplies and the main electrical and mecanical features of the turbo-generator and its principal auxiliaries.

1. INTRODUCTION

The utility line supplying the EPFL (50 kV, Sn = 40 MVA, Ssc = 300 MVA) is too weak to directly feed the power supplies of the TCV tokamak. A motor generator is therefore necessary to deliver the peak power and energy needed by the toroidal and poloidal coil systems of TCV.

Because of the special requirements for the supplies of the poloidal coils, such as high frequency response, and being no longer limited by the public network characteristics, the TCV power supply system is designed with some innovative and optimal solutions. As a result, the specifications of the motor generator present some very special aspects.

2. POWER REQUIREMENTS FOR THE TCV TOKAMAK

The aim of TCV (Tokamak à Configuration Variable) is to explore the domain of highly elongated plasma shapes. In order to initiate, shape and stabilize such plasmas, 16 poloidal (shaping) coils are placed around the vessel. Two ohmic heating coils will provide a flux swing of 3.4 V sec. and the toroidal coil a field of 1.5 tesla.

Each coil will be fed by its own power supply, bi-directional in current and voltage for the shaping and ohmic coils, and bi-directional only in voltage for the toroidal coil.

The growth rate of the plasma vertical position may be very high, which implies a high cut-off frequency of the shaping power supplies. However, the vacuum vessel prevents the penetration of the flux at high frequencies. Its second pole, for a radial flux generated by two coils placed at the vessel external corners, is situated between 200 and 300 Hz. Therefore, a voltage amplifier with a cut-off frequency higher than 300 Hz would not help very much in plasma stabilisation.

Network commutated thyristor rectifiers supplied with a 50 Hz voltage have their cut-off frequency between 50 and 100 Hz, which is lower than allowed by the vessel. On the other hand, power supplies like chopper, GTO or transistors, which have cut-off frequencies much higher than allowed by the vessel are very expensive for the required power. Therefore, the best technical and most economic solution is the use of network commutated thyristors supplied with a 100 Hz to 200 Hz voltage.

Auxiliary coils with lower power requirements will be used inside the vessel to control high growth rate instabilities.

Regarding the generator, 200 Hz and hundreds of MVAs seem to present important technical problems. However, a frequency of 120 Hz can be delivered, for example, by a turbo generator running at 3600 rpm with a 4 pole rotor.

All 19 rectifiers will be supplied by the generator. Figure 1 show the total apparent, active and reactive power to be delivered by the generator. The commutation

* ASEA BROWN BOVERI AG, Baden, Switzerland / ** DE-EPFL, Lausanne, Switzerland /
*** CERN, Division SPS, Geneva, Switzerland / **** LHPP-EPFL, Lausanne, Switzerland

of the toroidal and ohmic rectifiers, because of their high pulse power (50 MVA), can generate large voltage perturbations. In order to reduce interference between the rectifiers, the generator unsaturated subtransient short circuit power must be higher than 1000 MVA at maximum frequency. This leads to very large short circuit currents, particularly at the minimum allowed frequency.

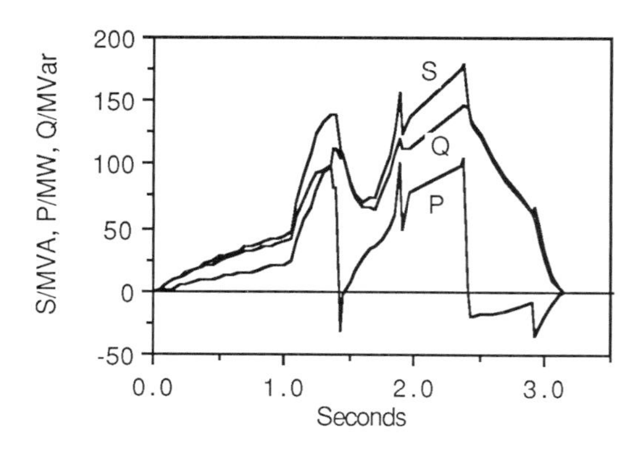

FIGURE 1
Total power requirements for the TCV power supplies

The TCV plasma will be initiated with 10 to 12 V/turn, i. e. by the voltage inversion which can be generated with the ohmic rectifiers. This inversion will produce a significant load change (as shown in figure 1) for the generator, which must not lead to excessive voltage fluctuations. During the entire pulse, the generator voltage must be kept within +/- 2% of its nominal value.

The standard pulse shown in figure 1 corresponds to a spent energy of 100 MJ. By limiting the minimum working frequency to 80 % of the maximum, the stored energy of the generator must not be less than 300 MJ. This value is usually available in the standard turbo generators of 50 to 100 MVA continuous power range. Therefore, the use of a flywheel is not necessary.

All these characteristics will be fulfilled by a turbo-generator which is presently under construction by Asea Brown Boveri in Baden. This machine is a variation of a standard type and its characteristics are summarized in Table 1. Figure 2 shows in perspective the machine, its foundation and the lube oil plant.

TABLE 1 Generator main characteristics

Type	Turbo, 4 pole, air cooled
Duty cycle	3 seconds/5 minutes
Nominal power	220 MVA
Nominal voltage	10 kV
Voltage fluctuation	< 2 %, for the figure 1 pulse
Short circuit power	> 1100 MVA (120 Hz)
Initial speed	3600 rpm
Frequency range	120 Hz - 90 Hz
Max. stored energy	418 MJ
Extractable energy	183 MJ (120 Hz - 90 Hz)
Total weight	146 tons

3. GENERATOR DESIGN

3.1. General

The machine is a 4 pole synchronous turbo-generator, which is derived from ABB standard air cooled turbo-generators. Its design is mainly influenced by the high short circuit power requested and the very low voltage fluctuations which are allowed. Complying with these two requirements, which are interdependent, leads to an active volume which is largely sufficient to comply with the requested stored energy without an additional flywheel.

Based on the power curves shown in figure 1, simulations have been carried out to verify the behaviour of the machine when submitted to the pulsed load, and to

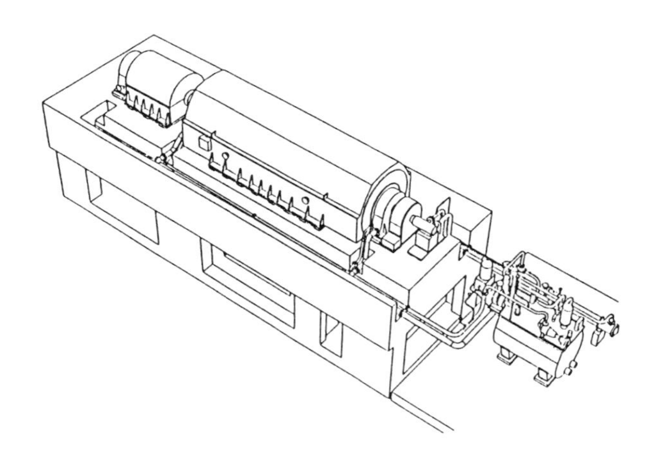

FIGURE 2 Generator perspective view

ensure good reliability. The electrical, mechanical and thermal stresses have been predetermined with accuracy. Consequently, the stator end-winding support and the rotor damper winding are designed for the particular kind of operation. Some results of these simulations are shown in figure 3.

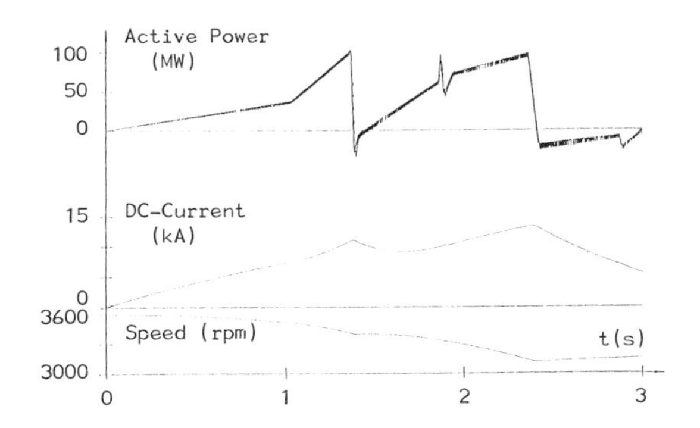

FIGURE 3 Simulation of the generator behaviour during the standard pulse. The second curve shows the DC-current delivered by a single rectifier carrying the load of all the TCV power supplies.

3.2. Stator

The complete stator consists of the housing and the assembled stator core and windings. The housing, of fabricated construction, is split horizontally in two parts. The stator core is built up with a large number of varnished, low loss lamination packets, with intermediate radial cooling slots. At the ends, the complete core is held under permanent compression by ring-shaped press-plates with press-fingers.

The winding is a two-layer bar winding using Roebel bars, insulated in ABB's Micadur^R-Compact system. All insulation materials comply with Class F specifications. The supporting beam of the stator end-windings are strengthened and attached to the housing.

3.3. Rotor

The rotor body is machined from a single high-quality alloy-steel forging. Longitudinal, rectangular slots are milled in the surface of the winding zones to carry the rotor windings as shown in figure 4.

The rotor winding (field winding) consists of hollow rectangular conductors of hard-drawn silver alloyed

FIGURE 4 Rotor view

copper. All the insulation materials used in the rotor comply with Class F specifications.

The damper winding is formed by wedges regularly distributed around the entire circumference. In the winding zone, these are winding slot wedges, and in the pole zone, wedges with the same profile are placed in slots especially milled in for them. The wedges are made from a copper-nickel alloy with good conductivity. Each is a single piece without interruptions along the length of the rotor. A complete damper cage is formed by the wedges together with the two end-bells due to the centrifugal forces. The function of the damper winding is to provide a low resistance path for currents caused by fields rotating relative to the rotor, thereby keeping that energy from having destructive effects elsewhere.

The retaining rings (end-bells), made of a high-quality austenitic stainless steel, hold the end windings firmly in place and protect them from deformation due to centrifugal forces. They are shrunk on the body of the rotor in an overhung configuration and are axially restrained, fixed by means of a snap-ring type lock.

The rotor driving end is provided with a half coupling to allow for the possible addition of a flywheel which would double the stored energy of the group.

3.4. Bearings

The rotor is supported in plain journal bearings carried in the bearings' pedestrals. In order to minimize the distance between bearings, the pedestrals are incorporated in the housing ends. The operating range for the pulses is situated between the second and the third critical bending speeds.

3.5. Excitation

The simulations mentioned under 3.1. have confirmed the need for a rapid static excitation system with fast response and high voltage ceiling value (1500 V).

Sliprings and brush gear are used to transfer the excitation current from the excitation rectifier to the rotating field winding. The sliprings are self-ventilated.

3.6. Cooling

The generator is air self-ventilated. The closed system has two symmetrical circuits, corresponding to the two axial fans. The air-to-water coolers are located below the machine in the foundation.

Due to the low duty cycle, the generator cooling presents no particular problems. Due to the special design of the rotor circuits, the machine losses have been kept at a low value. Therefore, the air flow provided by the axial fans is reduced in comparison to standard machines for network operation, in order to minimize the ventilation losses.

4. GENERATOR PLANT

The single line diagram of the generator plant is shown in fig. 5. The main auxiliaries are fed by the 20 kV

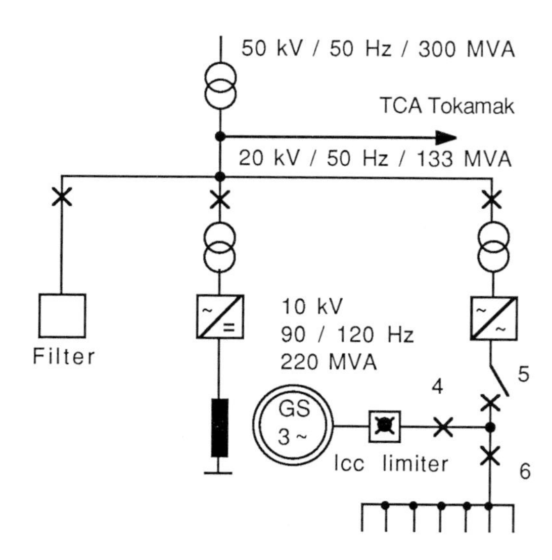

FIGURE 5 Single line diagram of the generator plant

mains of the EPFL dedicated to the TCV and the TCA, the tokamak presently in operation at the CRPP. All equipment is installed in a single building whose width is only 7 m, situated beside the TCV tokamak building. Figure 6 shows a cutaway view of the generator building.

4.1. Excitation rectifier

The rotor winding will be supplied by a rectifier delivering 4500 A and 1500 V. To prevent excessive current harmonics in the 50 Hz mains, a 12 pulse rectifier (series connection) is chosen. Two cast-resin transformers of 3.7 MVA each (3.5 sec / 5min) will supply the rectifier. The transformers have voltage taps in order to reduce the reactive power consumption and the mains voltage fluctuation if a lower ceiling voltage is sufficient to maintain the generator voltage fluctuations within acceptable limits and according to tokamak pulse requirements.

4.2. Starting convertor

To start the generator and to speed it up between pulses, a static frequency convertor is chosen to drive the generator as a motor. A design with a starting motor coupled to the rotor shaft would have needed a gear because of the maximum rotor speed (3600 rpm). Consequently, the machine overall length would have been significantly increased. The mains side of the convertor consists of a 12 pulse rectifier in order to reduce the current harmonics on the 50 kV mains.

The convertor will start the machine up to its maximum speed within 12 min. and speed it up from 80 % to 100 % in 5 min. During the speed-up phases, the convertor works at maximum and constant active power in order to reduce the reactive power consumption on the 50 kV mains. The convertor whose outputs are 2500 V and 512 A at 120 Hz is supplied by two cast-resin transformers of 1.2 MVA each.

4.3. Harmonics filter

As already mentioned, special care must be taken to prevent excessive voltage perturbations on the local 50 Hz network and to be in agreement with the new rules applied in Switzerland. Despite the use of 12 pulses rectifiers for the excitation system and the starting convertor, the harmonics they will generate exceed the maximum values allowed by the new rules. Consequently,

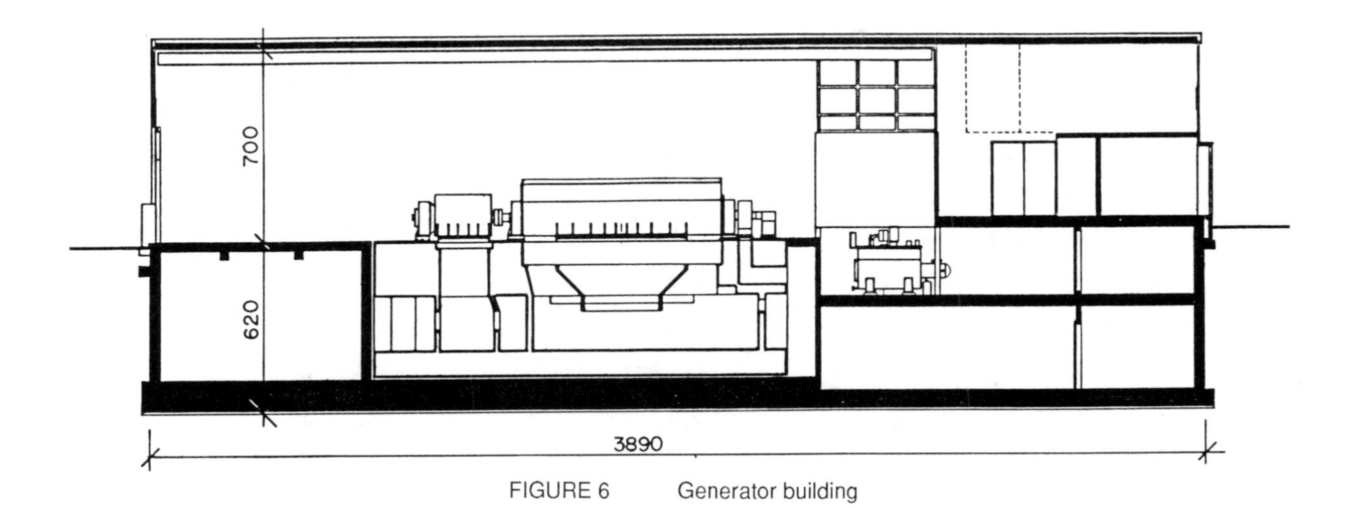

FIGURE 6 Generator building

a harmonic filter will be connected to the 20 kV input to the generator plant. It consists of three branches, 5th, 7th and 11th-13th harmonics. Special care has been taken to avoid disturbing the network remote controls. Due to the poor short circuit power of the local network, the voltage fluctuations generated by the excitation rectifier (6 to 7 MVA) during the tokamak pulse, should be just within the limits imposed by the rules.

4.4. Generator switchgear and breakers

Because of the high short circuit power of the generator at the minimal and unusual frequency of 90 Hz (higher than 1500 MVA), a short-circuit current limiter explosive fuse is chosen to protect the generator and the 10 kV network up to the primary of the rectifier transformers from the consequences of a short circuit. The breaker no. 4 (fig. 5) allows the interruption of the electrical faults occuring below the secondaries of the rectifier transformers. It can interrupt fault currents up to twice the generator nominal current. Breakers no. 5 and 6 are only used as contactors and they are never closed simultaneously. Breaker no. 6 connects the generator to the tokamak rectifier transformers. It is open between pulses and is operated at approximately zero voltage. All the three circuit breakers are air blast types.

4.5. Generator control

The whole generator plant will be controlled and supervised by a full electronic and programmable system of the same type as currently used by ABB in power generation plants. The machine can be controlled from a local console and directly from the TCV control system through a digital interface. The plant status will also be available on the TCV control console.

4.6. Oil plant

The main lube oil pump is driven, without an intermediate gear, by the rotor shaft. An auxiliary lube oil pump, AC driven, is only used for generator start-up and shut-down. An emergency lube oil pump, battery driven, is used in case of mains failure. The oil temperature is controlled by a thermostatic valve bypassing the oil/water heat exchanger.

5. CONCLUSIONS

The direct supply of the TCV tokamak power supply system from the utility network is not possible and an intermediate storage of the required energy is needed. Because of the space constraints and economical considerations, a turbo-generator solution has been chosen. This choice allows not only the fulfilment of the requirements of the tokamak power supply system, but also guarantees good reliability, flexibility and independence. Additionally, it will be possible in the future to increase the extractable energy to 360 MJ with an additional flywheel. The turbo-generator, although derived from a standard model, has several distinct features required by the particular kind of operation. The machine, which is under construction in the ABB factories in Baden (Switzerland), should be ready for operation on schedule in summer 1989.

FUSION TECHNOLOGY 1988
A.M. Van Ingen, A. Nijsen-Vis, H.T. Klippel (editors)
Elsevier Science Publishers B.V., 1989

COMPASS ELECTRICAL SYSTEMS DEVELOPMENT AND COMMISSIONING

D.C. EDWARDS, P.M. BARNES, J.H. HAY, D. RICHARDSON, S.E.V. WARDER AND L.A.E. VAN LIESHOUT*

UKAEA CULHAM LABORATORY, ABINGDON, OXON, OX14 3DB, U.K.
(UKAEA/Euratom Fusion Association)

ABSTRACT

The COMPASS machine is in the latter stages of construction and all the electrical and control systems have been installed. This paper covers the comprehensive development and commissioning programme including the testing of power supplies into dummy loads which is now complete. As a result Culham have confidence that reliable operation of the systems will be achieved.

1. INTRODUCTION

In common with other tokamak machines COMPASS has complex electrical and control systems requiring extensive testing and commissioning. The paper outlines the special tests carried out on major items of equipment, in particular tests on high power thyristor power supplies which proved necessary to ascertain compliance with specifications especially in respect of dynamic control system optimisation.

Development of the poloidal field energy transfer system required a programme of testing to produce two high power opening switches of which Culham had no previous experience.

The current status is that all power supplies have been operated into dummy loads in local control and are now being commissioned in turn from the Machine Control System.

Extensive power supply testing and development has been justified by the need for commissioning modifications and has resulted in confidence that the final working systems will be both reliable and in line with physics requirements. The transfer system development has demonstrated that the opening switches will work reliably and that the system confirms circuit design calculations.

2. POLOIDAL FIELD POWER SUPPLY SYSTEM

This has been previously described in some depth[1] and can be divided into three main types of power supply equipment: Thyristor convertors, high power transistor amplifiers and capacitor bank and inductive energy storage transfer system.

The outputs of these items are connected to the COMPASS machine windings via a series of linkboards in order to achieve the various modes of operation required.

The convertor and amplifier equipments required special testing to comply with comprehensive performance specifications, and the transfer system required a six month development programme.

2.1. Thyristor Convertors

The Magnetising, Equilibrium and Shaping Field convertors were designed and manufactured by Holec Projects BV., the following table gives a short list of their respective ratings:

Power Supply	kA (0.5s)	Voltage Range	Imposed kV	Crowbar kV
MFPS	12	+650/-435	+8.0/-2.2	+12.0
EFPS	16	+265/-175	+1.0/-2.2	+ 2.4
SFPS	10	+540/-360	+2.2/-2.2	+ 4.8

* L.A.E. VAN LIESHOUT - HOLEC PROJECTS BV, 7550AG HENGELO, THE NETHERLANDS

Each 12-pulse unit consists of two six-pulse thyristor bridges in parallel, naturally cooled and without series thyristor fuses. The bridges are paralleled via air cored reactors and blocking diodes.

All units underwent comprehensive convertor-frame works tests (without transformers); MFPS also underwent the following combined equipment tests at the Kema test laboratory in Holland:

a) Faulty thyristor test: One bridge thyristor was replaced by a copper block. The firing pulses were released at $\alpha = o°$ causing an asymmetric peak of 45kA through the thyristor receiving the first firing pulses. This test proved that only that thyristor was damaged; the ACCB cleared the rms symmetrical short circuit current of 21kA after 110ms.

b) Other tests included feedback loop response, dynamic regulation at fast current ramp-up/down and mains voltage variation.

In the works a convertor-frame underwent an imposed current test to simulate an MFPS thyristor bridge firethrough from the fully charged Tokamak start capacitor bank. Each bridge arm was tested up to a peak current of 32kA by discharging a test capacitor bank.

The three month site test programme covered full operation of the convertors and included:-

a) Heatrun for 22 hours at continuous current followed by 2 hours of pulsed current operation. The electronic thryristor junction temperature analogue output signals were checked against the calculations.

b) Small and large signal dynamic response tests in voltage control and the checking of current limiting operation.

c) Verification of protection co-ordination and static regulation curves.

d) Imposed voltage tests to simulate normal and fault conditions during Tokamak operation. These applied voltages were up to +15kV (MFPS) to check crowbar protection operation and -8kV to check firethrough of bridge arms.

2.2. High Power Transistor Amplifiers

The three amplifiers each rated 250kW (50V, 5kA) for 5s pulses, were designed and manufactured by Spitzenberger and Spies.

Each amplifier uses conventional audio amplifier techniques, considerably modified to take account of the higher ratings. Comprehensive protection circuitry is fitted in order to prevent a catastrophic cascade type failure in the 4800 output transistors.

To demonstrate reliable operation from such complex electronics, considerable works and on-site testing was undertaken. In order to reduce possible problems from 'infant mortality', the works tests included a 100 hour burn-in at the full continuous rating; the output of the amplifier was returned to the ac mains supply via a matching transformer, thus saving about 9MWh of electricity. During burn-in tests fewer than 10 failures occurred in over 16000 output and driver stage transistors.

Routine tests of available output voltage and current, harmonic distortion and bandwidth showed the amplifiers to be within the required specification. It was noted that there was a significant reserve of output voltage and the harmonic distortion was nearly 100 times lower than specified.

The modular construction of the amplifiers allowed installation, commissioning and site testing to be completed in seven days. The site tests included the discharge of a 0.2F 2kV capacitor bank across the amplifier output terminals to simulate the impulse expected when a plasma disruption occurs. The crowbar thyristors connected across the amplifier output terminals automatically fired and prevented damage to the amplifier.

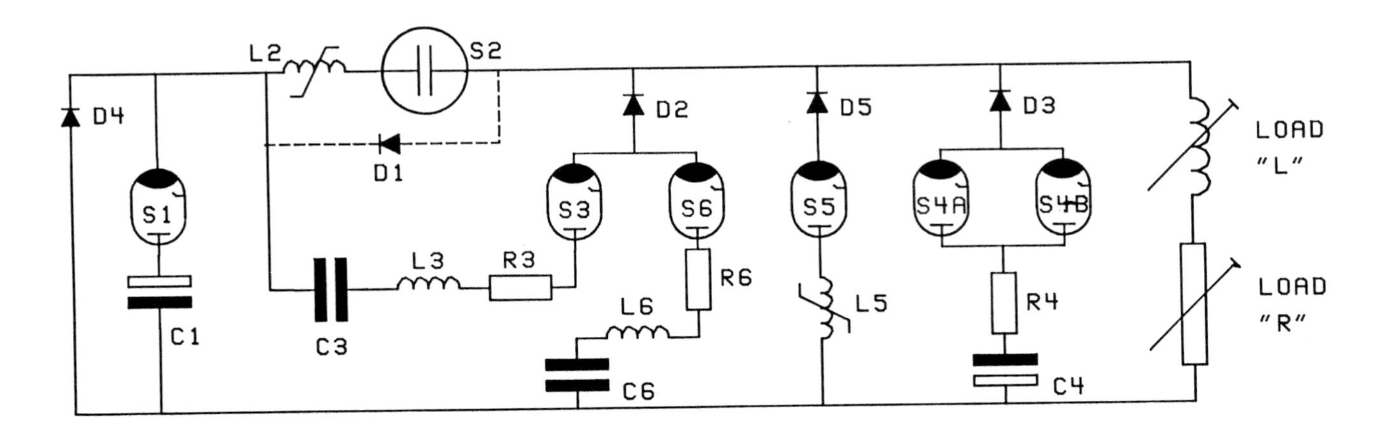

FIGURE 1

2.3 Transfer System

Figure 1 shows a simplified diagram of the development circuit. The sequence of operation has been described previously[1].

The application of inductive energy storage required the development of two opening switch systems. The first switch (S2) has to open when carrying 20kA and is subject to a re-applied voltage of about 6kV and is based on a vacuum interrupter. The second opening switch (S5) breaks 15kA at 5kV and ignitrons are used for this duty. The opening of the vacuum switch and commutation of the current into S4A establishes the plasma and drives the current up to about 200kA. For shaped Tokamak, S5 then removes the loop voltage whilst the required plasma cross-section is produced; S5 is then opened to re-establish the loop voltage and drive the plasma current up to its maximum value.

The development circuit employed a reduced start bank C1 of 0.2F in place of the final 1.1F bank. The machine windings were represented by a dummy load circuit with variable resistive (8.6-35mΩ) and inductive (0.23-2.73mH) components.

The complete system was enclosed in a temporary HV test area. An 8-channel signal memory recorder was used in conjunction with 14 Rogowski coils with 6 integrators, one coaxial shunt and 3 voltage dividers to capture waveforms within the area and display results via an optical fibre link in the control room above. The temporary control system comprised cam controllers for slow timing of capacitor bank chargers and a 12-channel fast timing system for firing ignitrons and the vacuum interruptor via an optical fibre interface.

Figure 2 shows a typical load current pulse and vacuum interrupter contact gap with the firing sequence of each switch superimposed.

The majority of development testing concentrated on obtaining interruptions in the vacuum switch S2. The circuit variables were limited to counterpulse amplitude and time constant, contact gap at point of interruption and load parameters.

In order to obtain the correct peak load current a dummy load of L/R = 0.55mH/16mΩ (34ms)was used and to obtain the correct waveshape a dummy load of L/R = 2.73mH/45mΩ (60ms) was used.

Initial testing of the counterpulse bank C3 showed that an impedance of L3/R3 = 15μH/ 120mΩ gave good results up to the maximum voltage of 8kV at a peak current of 32kA. It was also found advantageous to add a diode in

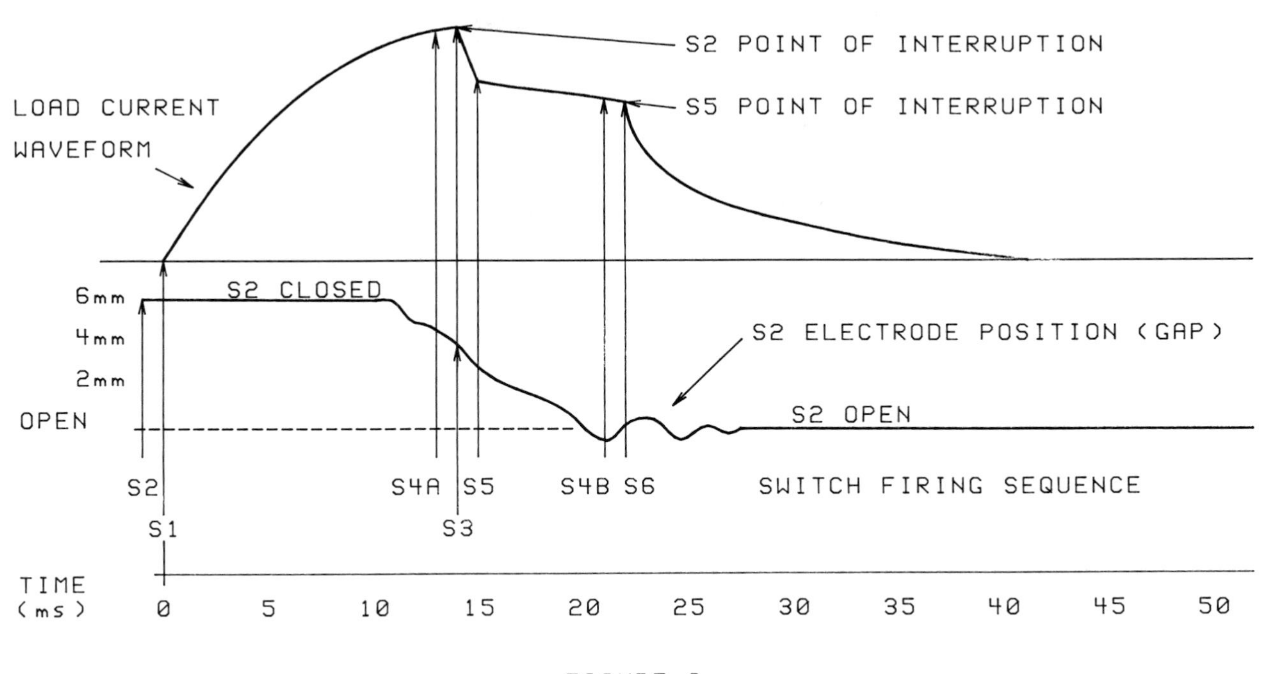

FIGURE 2

series with the ignitron to prevent current reversal.

The vacuum interruptor actuator mechanism is adjustable for opening speeds between 0.3 and 0.9m/s (excluding the dead time prior to contact separation) and gaps of 1 to 10mm. Since the counterpulse time constant was very short it was possible to adjust the timing of the counterpulse to coincide with a particular contact gap. Too small a gap (<3mm) resulted in unreliable operation and a large gap meant that more energy would be dissipated in the arc during interruption. The optimum gap was found to be 4mm at interruption with an opening speed of 0.7m/s.

It was also found important not to close the switch too soon after interruption since the contact surfaces can remain plastic for many seconds and reclosure under this condition causes spreading of the contact material and subsequent reduction in useful life.

The majority of testing was carried out to ascertain the area of reliable operation, and Fig 3 shows this relationship. The minimum counterpulse current is only slightly higher than the load current and excess counterpulse does not prevent interruption over most of the area. However, above a certain level of load current the counterpulse becomes more and more critical until eventually the switch will not interrupt at all, or fail to interrupt on a reliable basis.

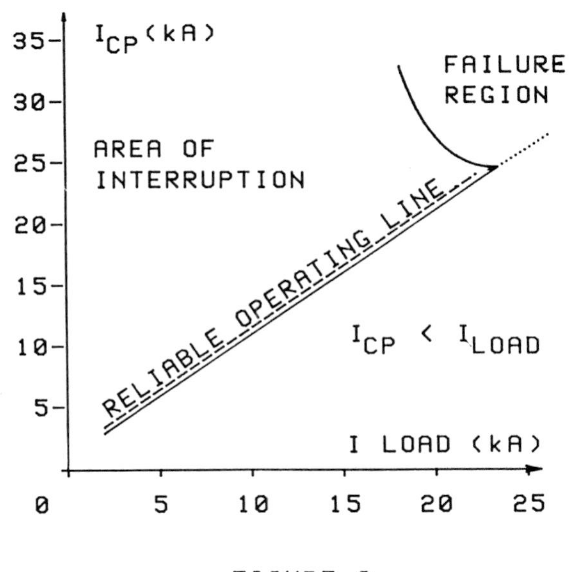

FIGURE 3

As the counterpulse current forces the main current through zero, a change in slope about the zero occurs due to a saturable reactor, which is only effective when the current is below about 200A. For reliable interruption the minimum time for a current zero produced by the counterpulse was 200μs.

The current passes through zero twice during operation (once in each direction) normally with an interruption at the first zero, however, at high load currents with high counterpulse, interruptions occurred at the second zero. This effect was random and accompanied by failures to interrupt.

The limiting value of current could not be explained by the manufacturers as it is around 23kA and the vacuum switch is normally used in ac systems and has been type tested to 40kA rms symmetrical.

The reapplied voltage was adjusted in the circuit by changing the value of transfer resistor R4; a maximum of 6.2kV was measured. The value of the re-applied voltage had no effect on the limiting value of interruption current.

The diode D1 across the interruptor was intended to divert excess counterpulse current, however, the relative impedance of the diode and the arc in S2 meant that current would not commutate out of the switch into the diode and the idea was abandoned.

The second interrupting switch S5 comprises two EEV-BK496 ignitrons in parallel. A special version with an internal baffle was recommended, however this proved to be less reliable than the standard type often causing re-ignition up to 7ms after interruption.

Sharing of current between the ignitrons was within 10%. Many tests were carried out to establish the area of operation as for the vacuum switch. No limiting value of interruption current was established within the range required. For reliable interruption the minimum time for a current zero produced by the counterpulse was 700μs.

Another area of development was the investigation of the final turn off of ignitron S4B. Failure to turn off would mean a short circuit across MFPS and the destruction of the ignitron. In order to obtain a natural zero at the end of the pulse the value of C4 was adjusted and tests repeated for various values of R4. The monitoring of ignitron S4B to establish a definite off state required careful measurement of the arc voltage using an automatic dual range divider system.

3. TOROIDAL FIELD POWER SUPPLY SYSTEM

3.1 The toroidal field (TF) power supply for Tokamak mode operation consists of two, parallel connected, 24 pulse transformer/controlled rectifier sets designed and manufactured by Thorn-EMI Electronics Ltd.

Ratings		
	Open Circuit voltage	450V
	Current (continuous)	25kA
	Current (5 second pulse)	52kA
	Current (½ second pulse)	92kA
	Load L/R	2.5mH/2.85mΩ
	Composite error	< ± 0.5%

Works tests included operating the complete equipment, from a 33kV ac supply, into a short circuit at 92kA. Temporary control system modifications were necessary for this test.

At site each set was first commissioned separately to its rated 46kA into a representative dummy load. The two sets were then run in parallel up to the 46kA rating of the load with satisfactory current sharing.

The complete system was used to test a single prototype TF coil up to full current (92kA). The rectifiers were run in full conduction to avoid control loop instability (severely mismatched load), current was adjusted with the transformer tap changers.

3.2 Reverse pinch operation requires a low current system (up to 2250A = 0.05T), with the facility to reduce the current to zero within a few msec. A 360V 10kJ capacitor bank is discharged into the load via thyristors and a 2kV thyristor counterpulse system is used to commutate the current into a SiC resistor to produce a controlled current decay.

Development tests showed that the system provides 2400A rising in 18msec with a natural decay time of 260msec. Satisfactory commutation was achieved at any point on the current waveform with a reduction from 2250A to zero in 7msec, lower currents gave shorter decay times.

The maximum load voltage developed during switch-off was 2100V and therefore the stress on the centre-point earthed TF coil will only be 1050V.

4. CONTROL SYSTEM

The Machine Control System (MCS), a hybrid network of computer and programmable controllers, has been described previously[2]. Following extensive works acceptance testing, the system has been installed and is being used in the commissioning of major power supplies.

The rigorous acceptance tests required realistic models of the equipment to be controlled. The control requirements were specified using a UKAEA developed software package, the Requirements Definition System (RDS), which is described in an accompanying paper[3]. One element of this package is the ability to develop software models of the equipment. These models were separated out and additional software added to enable them to drive analogue and digital inputs and outputs, thus the programmable controller logic could be effectively tested by driving the models in a sequential mode. The ability to use these models for the works acceptance tests allowed for immediate software corrections and reduced post-installation testing.

Electromagnetic compatibility testing of a control cubicle was carried out at Culham using a coil to generate magnetic fields of 1-2mT with rise times of 80msecs. Attenuation of field levels by the cubicle enclosure was measured and programmable controller operation monitored for errors. The tests demonstrated the ability of the control cubicles to operate satisfactorily in the COMPASS electromagnetic environment.

The control requirements were frozen at contract placement, but now that the system is installed and operational, the requirements are being re-assessed for a system revision. Software changes will be required to the central and subsystem controllers and the operator interface (touch-screen displays); significant spare memory space was allowed for the revision when selecting the control processors. Hardware modifications were allowed for by in-built spare I/O module capacity, using replaceable mosaic tile control panels and by running spare plant cables. The system revision will be completed and tested ahead of machine commissioning.

REFERENCES

1. HAY J.H. et al, Proc. 14th SOFT, Avignon 1986

2. RICHARDSON D. and WATKINS J.E., Proc 14th SOFT, Avignon 1986

3. ENDSOR R. et al, Proc 15th SOFT, Utrecht, 1988

FUSION TECHNOLOGY 1988
A.M. Van Ingen, A. Nijsen-Vis, H.T. Klippel (editors)
Elsevier Science Publishers B.V., 1989

THE RFX MODULAR AC/DC CONVERTER SYSTEM

Elena GAIO, Alvise MASCHIO°, Roberto PIOVAN
Paolo BORDIGNON*, Ambrogio BOSELLI*, Sandro TENCONI*, Giordano TORRI*

Istituto Gas Ionizzati - Associazione EURATOM-ENEA-CNR - Corso Stati Uniti 4, 35020 Padova, Italy
° Università di Trento - Dipartimento di Ingegneria - 38050 Mesiano TN, Italy
* Ansaldo Sistemi Industriali e Ansaldo Ricerche, Viale Sarca 336, 20126 Milano, Italy

The modular system of converter units, on which the RFX power supply plant is based, is described in the paper. The main features of the system are highlighted, both in the power and control system; the main design choices and constuction aspects are also illustrated.

1. INTRODUCTION

The AC/DC thyristor converter system was designed in the RFX experiment to feed the toroidal and poloidal windings during the pulse directly from the Italian 400 kV grid, with a total installed power of over 400 MVA. In particular, it will be used for charging the magnetizing winding, for controlling the vertical field in the plasma and for sustaining the plasma current and the toroidal magnetic field at the wall during the flat top phase.

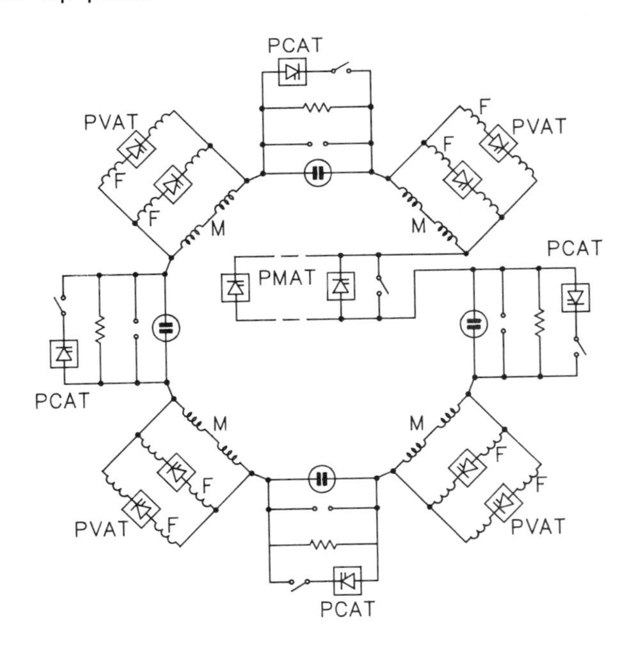

Fig. 1 - Poloidal scheme

A modular system of converter units was a basic choice since the beginning of the project, mainly to make easier reconfigurations of the power supply system that could be required in the future to investigate RFP different plasma configurations[1,2].

2. THE CONVERTER SYSTEM SPECIFICATIONS

The converter system, which has been designed for a number of 10^5 operations, is split in four groups, as shown in figg. 1 and 2, where the poloidal and toroidal circuits are represented.

The different groups have the following duty:

- PMAT feeds the magnetizing windings (50 kA, 1.35 kV load voltage);
- PCAT sustains the plasma current (1 kV maximum voltage required);
- PVAT controls the magnetic field in the plasma region (5 kA maximum current);

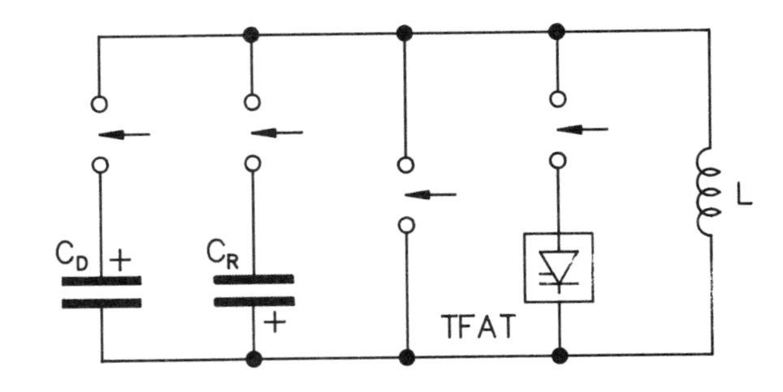

Fig 2 - Toroidal scheme

- TFAT sustains the toroidal magnetic field at the plasma vessel (50 kA maximum current).

The maximum values of the power extracted from the grid during the pulse are 100 MW and 200 MVAR for active and reactive power respectively.

The duty cycle, related with the duration of the pulse, is 5 s every 600 s.

2.1. Unit structure

The design of the system was based on two different types of units, whose rating and number are shown in Tab. I.

	Type A	Type B
Load voltage	1350 V	1350 V
DC current	12.5 kA	6.25 kA
Number of units	12	8

Tab. I - Rating of the converter units

The current rating is increased of a factor 1.3 in case of operation with a duty cycle of 0.5 s every 600 s.

In figg. 1 and 2, which refer to the nominal configuration, PMAT and TFAT groups are made up of four A units each, the four PCAT consist of one A unit each and the eight PVAT of one B unit each.

The converter units are split in two subunits, which work with 6-pulse operation (fig. 3). They can operate in series (type A only) or in parallel in a 12-pulse operation; in addition they can operate also independently.

The power module operates without any connection to ground; the nominal insulation level, chosen taking into account the maximum voltage in case of fault (35 kV), is 12 kV AC.

2.2. Transformer

To avoid any interference between the two subunits, each one has its own transformer; both are put in the same tank. Due to the particular operation and fault conditions, the transformers have been designed to withstand 10^4 short circuits: 90% of them at the DC terminals of the units, 5% inside the thyristor bridges and 5% at the low voltage terminals of the transformers.

As a consequence, the thermal design power of the windings is very close to the peak power rating, even if the low duty cycle could allow a low ratio between them.

The secondary windings of the transformers are insulated to ground and to the primary ones for a nominal level of 24 kV AC, as a consequence of possible fault conditions[3].

2.3. Control system

The control system has to follow a modular approach too: each subunit has its own control system, equal to all the others. In series and in parallel configuration only one regulator controls both the subunits. Voltage and current sharing has to be assured by controlling the thyristor firing angle in series and parallel operation respectively. Open and closed loop operation can be chosen, and in the latter case voltage or current feedback can be selected.

2.4. Protection

The analysis of the fault conditions induced by a malfunction of the other components of RFX

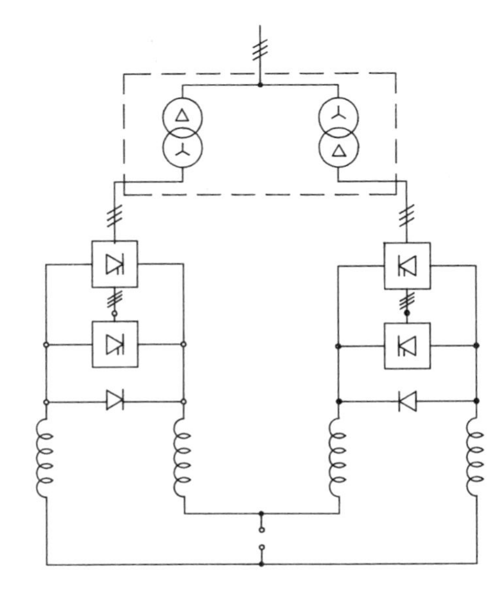

Fig. 3 - Structure of the type A unit

circuits, as well as the consequences of sudden plasma terminations (disruptions) had a strong impact on the design of the converter system. In particular, thyristor crowbars with series resistors had to be connected in parallel to each converter subunit. A special impulse test was requested to verify the dynamic protection capability of the thyristor crowbar; a impulse test voltage (35 kV crest, 10 μs rise time) is applied between the DC output terminals, with the converter working at its rated voltage.

Due to the limited space available, the results of this analysis and the design criteria of the crowbars can not be included in this paper and will be presented elsewhere[3].

3. BASIC ELEMENTS OF THE CONVERTERS

In the AC/DC converter units the same "basic bridge" has been used, whose parameters are shown in Tab. II.

Load voltage	1350 V
No-load voltage	2000 V
DC current	3125 A
Insulation to ground	12 kV rms
Duty cycle	5 s/600 s
Number of operations	10^5
Thyristor number per bridge	6
Cooling	Natural air

Tab. II - Parameters of the basic bridge.

Owing to this choice, the B unit is formed by two basic bridges in a 12-pulse arrangement; the A unit is formed by four basic bridges, again in a 12-pulse arrangement.

3.1. Choice of thyristor and diode

Tab. III shows the parameters of the thyristors and diodes used in the basic bridge.

	Diameter mm	U_{RRM} kV	I_{AVE} kA
Thyristor	75	4.4	1.9
Diode	75	2.4	4.0

Tab. III - Thyristors and diodes in the bridge

The current sharing between the two bridges in parallel in the A subunits was obtained by utilizing cables of equal resistance between each bridge and the transformer.

Because of the very high current capability demanded in case of fault conditions, the blocking voltage of the free-wheeling diode was chosen equal to 2.4 kV. Therefore two diodes in series are used for the free-wheeling. Moreover, in order to avoid special selections of diodes, the current sharing between the free-wheeling paths of the bridges in parallel in the A subunits is assured by a small resistor connected in series to the diodes.

3.2. Firing systems

The high insulation level requested and the need to avoid parasitic couplings between the converter and the control equipment require the use of optical fibre connections both for transmitting the feedback signals (i.e. the DC voltage and current) and for giving the firing command to the gate drive amplifiers of each thyristor.

During the 120° of conduction of the thyristor, the gate is maintained fired by means of a long tail current. A device included in the amplifier is driven by the voltage across the semiconductor and prevents any current from flowing into the gate when a reverse voltage is applied.

3.3. Mechanical and thermal design

The bridge described above is conceived as a module which can be transported or handled without disassembling any of its parts. The structure, very simple, consists of a metal frame acting as a supporting base and as a potential reference. All the components are installed over insulators; every floating point is far from the frame 150 mm at least. The semiconductors are arranged in 3 stacks, grouping 2 diodes and 3 thyristors respectively. The solution adopted allows a full accessibility to every device of the bridge (fig. 4).

The cubicle is open frame, protected only on

Fig. 4 - Bridge module assembly

the front by a metal panel; in this way, there is no need of recirculating the air inside by means of fans.

3.4. Transformers

The primary winding of the trasformers is of extended delta type, with a phase shifting of ±7.5°, to allow, by means of a proper connection to the grid, the 24-pulse operation. The secondary windings of the two transformer of the same unit are delta and star connected respectively, for the usual 12-pulse operation.

The windings of the transformers are built with controlled work-hardening copper which is stressed with one third of the maximum σ. This gives a substantial safety margin, which is requested by the fatigue problems related to the high number of operations and short circuits expected. In addition, the blocking system of the windings has been carefully designed to withstand the repetitive axial stresses.

4. CONVERTER CONTROL

The converter control will be performed both in remote mode, during normal operation, and in local mode, during tests and commissioning. The remote control will be operated through the central control system (SIGMA) from the central control room. The local control will be performed by the operator from the local control cubicles LCC.

Due to the electromagnetic interference, the cubicles, one for each converter, are grouped in the local control room in the Power Supply Hall, which is a completely screened area. All the signals from (to) the converter module to (from) the local control cubicles are transmitted via optical fibre links (fig. 5); in particular V/F and F/V converters are used, which assure the required 10 kHz bandwidth.

For test and commissioning operation a high monitoring level is required; in addition to all the state signals, all the measurements and detailed alarm signals, required to well understand the converter operation, are displayed in the LCC.

In table IV the total amount of signals bet-

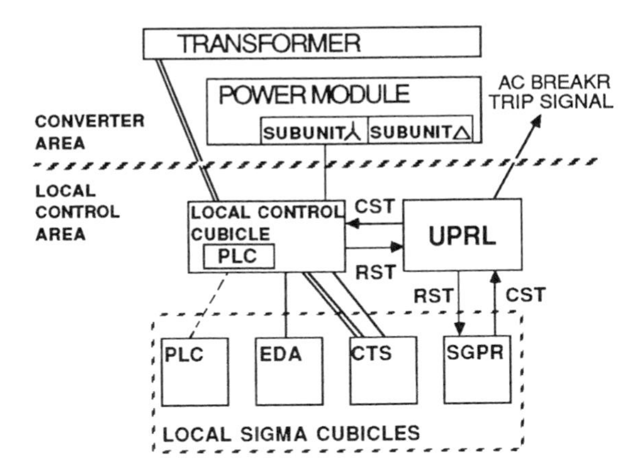

PLC = PROGRAMMABLE LOGIC CONTROLLER
EDA = ENGINEERING DATA ACQUISITION
CTS = CENTRAL TIMING SYSTEM
SGPR = GLOBAL FAST PROTECTION SYSTEM
UPRL = LOCAL FAST PROTECTION SYSTEM

RST = SOFT TERMINATION REQUEST
CST = SOFT TERMINATION COMMAND

── = FIBER OPTIC
══ = PAIRS
--- = PLC to PLC BUS CONNECTION

Fig. 5 - Signal links in the control system

	SIGMA cubicle ⇕ LCC	Power module ⇕ LCC
Unit A (12)		
Commands	13x12	24x12
State signals	24x12	4x12
Measurements	7x12	15x12
Fault signals	9x12	14x12
Unit B (8)		
Commands	13x8	12x8
State signals	24x8	4x8
Measurements	7x8	13x8
Fault signals	9x8	14x8
	1060	1028

Tab. IV - Converter control signals

ween the converter module and the local cubicle and between this one and the SIGMA cubicle is given.

Signal transmission between the SIGMA cubicles and the LCC is not particularly critical because the cables are laid inside a screened area; therefore standard solution are chosen, which utilize industrial components.

SIGMA controls separately the converter units utilized in the poloidal and toroidal subsystems; therefore, in case of reconfiguration of the converter units, provision has been taken in order to enable a quite easy reconnections of all signals which have to be sent to different SIGMA control cubicles.

In case of fault each unit is able to put itself in free-wheeling, to trip its AC circuit breaker and to send a protection signal for soft termination request (RST) to a local protection unit (UPRL), one for each of the four converter groups (fig. 5). The UPRL provides to handle the RST request from the faulty converter unit, to send a free-wheeling command to the other units of its group and to send an RST signal to the central fast protection system (SGPR), which in turn sends a soft termination command (CST) to the other UPRL, with different sequences, and stops the RFX operation.

RFX can operate with a reduced number of converter units, due to the modular design chosen. Therefore provision has been taken for a proper procedure of exclusion of units from the communication with the Central Timing System (CTS) and, if necessary, with the SGPR.

5. CONCLUSIONS

The converter system is presently under installation and the final commissioning is expected to start at the beginning of 1989. The type tests have been completed and the results are according to the specifications.

REFERENCES

1. G.Rostagni et al. : "The RFX Project: a Design Review", Proc. of the 13th Symp. on Fusion Technology, Varese, 1984, pp. 189-201.
2. I.Benfatto, R.Piovan, P.Tenti: "Transient Analysis of the AC Network During the RFX Pulse", Proc. of the 5th IEEE Pulsed Power Conference, Arlington, 1985, pp. 182-185.
3. E.Gaio, A.Maschio, R.Piovan, S.Tenconi: " Thyristor Crowbar for RFX Converter Unit Protection", to be published..

FUSION TECHNOLOGY 1988
A.M. Van Ingen, A. Nijsen-Vis, H.T. Klippel (editors)
Elsevier Science Publishers B.V., 1989

OTR MAGNETS POWER SUPPLY SYSTEM

E.V.KORNAKOV, F.M.SPEVAKOVA

D.V.Efremov Scientific Research Institute of Electrophysical Apparatus, Leningrad, USSR

OTR magnets comprise both the coils forming the poloidal field and the toroidal field coils. All these coils are superconding and this mainly determines the requirements to the electric power supply system. The most complicated is PF coils supply system, realizing the complex scenario of plasma discharge formation. TF electric power supply scheme is, to a great extent, determined by the necessity of energy emergency discharge in case of normal phase appearing in a superconductor.

1. TF COIL ELECTRIC POWER SUPPLY SYSTEM

TF coil is designed for the rated current 25 kA, energy store 45 GJ, insulation to the ground - 10 kV, maximum velocity of current ramp-up is 3.5 A/s. Specified velocity of current ramp-up can be provided by 0.5 kV source. Taking into account the voltage drop at power supply system services, the source voltage is chosen to be equal to 0.8 kV. To improve the diagnostics of the normal phase appearing in coil superconductor, there is a filter at power supply output limiting the ripples to the amplitude of 0.5 V for 300 Hz frequency and 1 V for 50 Hz.

Power supply system scheme is mainly determined by the method of energy discharge when the normal phase appears in the coil.

For a conductor, chosen nowadays the admissable time constant of energy discharge is chosen to be 17 s, what corresponds to the total magnitude - 8.5 of output resistors and the voltage in the loop - 215 kV. To reduce the voltage to the ground, the coil is divided into 12 sections, each of them being shunted by two resistors, connected in series through the diode (Fig.1). The midpoint of seriesly connected diodes is tapped to the ground. The coils sections are connected through the breaking devices. In such a scheme the coil voltage to the ground doesn't exceed 2 kV. The diodes introduced into the output loop, prevent the currents disbalance appearing in coils sections at current ramp-up.

Water cooled mechanical commutators for 25 kA - current and 20 kV - voltage, developed at D.V.Efremov Institute, can be used as breakers for energy discharge.

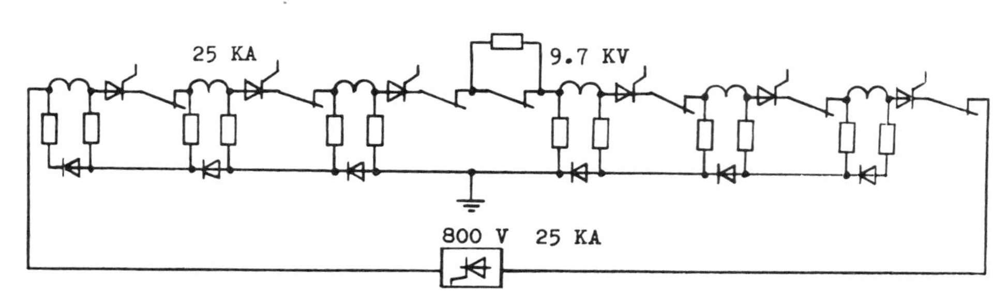

FIGURE 1

2. PF COIL ELECTRIC POWER SUPPLY SYSTEM

Significant pulse length (600 s) at comparatively small time interval between the pulses (80 s) is the specific feature of PFC power supply system. Poloidal field is generated by 17 coils, 12 of which form the central solenoid.

Developing the electric power supply system, the scheme in which each coil is feeded from a separate source, except 12 coils of the central solenoid, has been chosen. Such an approach provides the experiment flexibility and makes the consequencies of the possible emergency conditions less serious.

In plasma current drive scenario at initial moment of discharge formation it's necessary to apply high voltage to some PF-coils, this voltage being significantly reduced with plasma current rise. Energy storage seems reasonable sometimes because of such supply sources pulsed operation.

As it is known, to reduce poloidal magnetic field magnitude (which variation induces eddy emf), the negative initial flux is created in a groupe of coils. These coils are used as inductive energy storages, generating power, necessary to ensure the specified velocity of field rising at plasma discharge start-up. Energy is discharged when the resistor is introduced into the coil loop. After this, the currents are varied by thyristor converters in accordance with the required laws. These converters also provide the current ramp-up in the coils for the period of negative storage in the interval between the operating pulses.

In the most PF-coils bidirectional current should be induced, for this aim reverse converters are necessary. In the central solenoid supply system the currents of different polarity are approximately equal and the thyristor converter power is rather high. Therefore use of unidirectional converter with the mechanical reverse for current polarity change seems expedient from economical viewpoint. In the rest coils the currents of opposite polarities differ by a factor of ten, approximately, and to induce the reverse current, converters of comparatively low power are used.

In case of reverse power supply use, the time when the current polarity is changed and during which the coils are not under the voltage, should be minimum. For this purpose, the resistor is connected in parallel with the converter for a time of current reversal. Thyristor converter, operating as an invertor at current drop, doesn't conduct the current when the resistor voltage is less than that of the converter. After converter current vanishing and reverser operating, the converter again operates as a rectifier. The coil current drops to zero under the converter voltage and begins to rise in opposite direction. The resistor current in this case rises abruptly and therefore it should be switched off when the reversal is terminated. In time interval, when the converter doesn't conduct the current, the resistor voltage is applied to the coil.

PF-coils are designed for 40 kA - current and voltage to the ground - 20 kV. Powerful thyristor converters, usually, are formed of several modules. As a module, 6-pulse bridge converter is mostly expedient to be taken. To reduce the harmonics of rectified current and converter voltage and current, consumed from a.c. source, 12-pulse conversion scheme is more preferable. Such a scheme can be created by two 6-pulse modules due to 30 el.degree shifting of 3-phase system of a.c. voltage systems feeding these modules and connecting 6-pulse bridges in series or in parallel through the separating reactor.

One of PF supply system specific feature is the voltage necessary reduction with current rising. Such operation conditions attained due to converters control angle increase result in higher reactive power and current consumed from the network and rise of rectified voltage harmonics. To reduce this effect converters segmentation and change of sections connection from series to parallel is used in PF-coil power supply system. Use of reconnection scheme permits to lower the specified power of a.c. sources and thyristor converters sections. For sections commutation the additional commutators are needed. The electric power supply system energy characteristics can be improved using thyristor converters asymmetric control. In this case in each pair of the bridge converters, connected in series and in parallel through the separating reactor and feeded by a.c. voltage, coincident in phase, the control angles of the first bridge cathode groupe and of anode groupe of the second one are changed. All other valves control angles are kept constant: under rectifying conditions - close to 0 el. degree, under invertor - to 180. With such control method in the interval of the control angle variation from 60 up to 180 el.degrees, current is conducted simultaneously by two opposite bridge arms. The current in this anode transformer winding is reduced and the sections of zero voltage appear on each bridge rectified voltage curve. Variation of rectified voltage average value from maximum to zero occurs at control angle change from 0 to 180 el. degrees (Fig.2).

Fig.2 shows the simplified scheme of the power supply system of the central solenoid, 12 coils of which (coil 1 - coil 12) are connected in parallel branches, six coils in each. The reverse thyristor converter 5...8, providing equal currents in the central solenoid coils parallel branches is connected to one of the branches (between coil 9 and 10). The reverse thyristor converter 5...8, providing equal currents in the central solenoid coils parallel branches is connected.

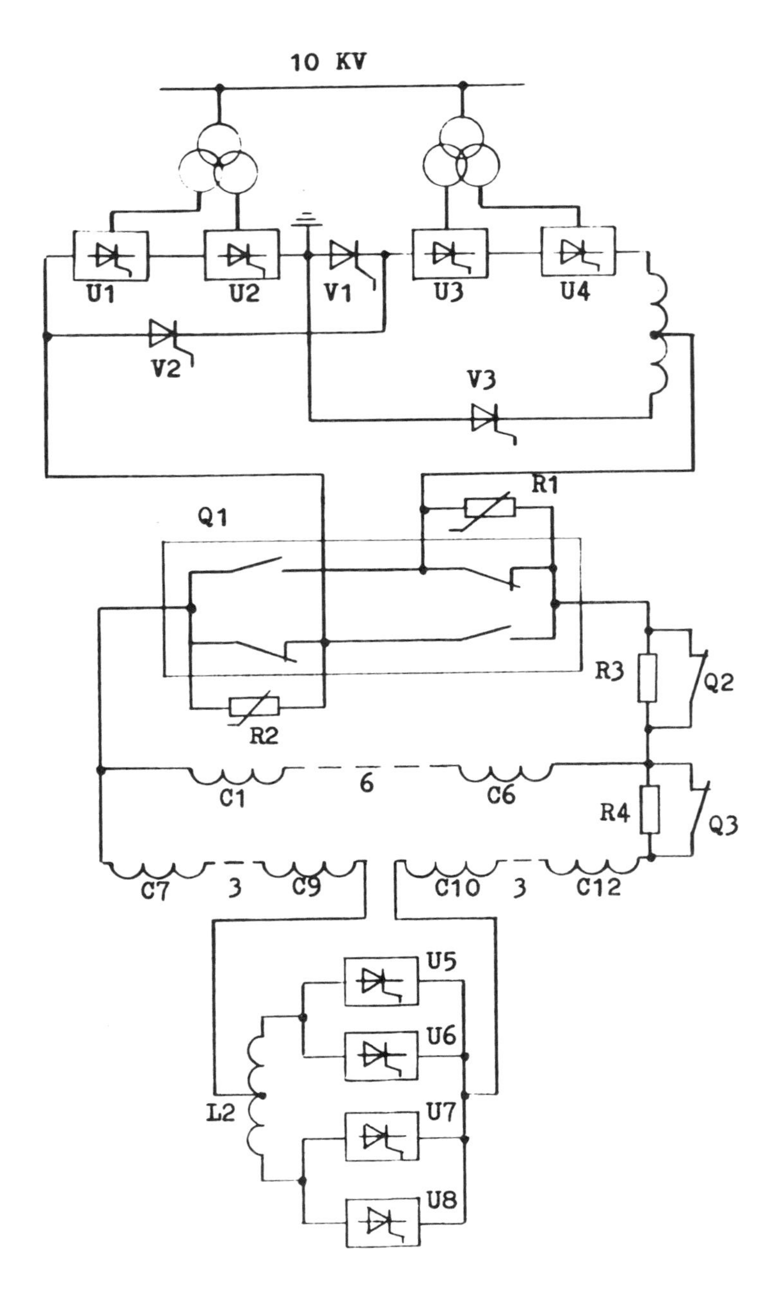

FIGURE 2

Coils 1-12 are feeded from the segmented thyristor converter 1...4 through the mechanical reverser. When energy is stored in coils, the converter sections are firstly connected in series using the thyristor switch 1. According to the accepted scheme of transformer windings connection, the converter rectified voltage is 12-pulsed one. After current ramp-up to the value equal to half maximum, converter sections 3,4 operate as an inverter, thyristor switch doesn conduct the current any more, thyristor switches 2 and 3 and converter sections 3,4 are opened in rectifying regime. Current transformation under this regime (in case of sections 1,2 and 3,4 parallel connection) is provided by the separating reactor-1. After current ramp-up to maximum, the reverser breakers operate and the resistors 1,2 are introduced into coils loop, providing high voltage - 3,6 kV appearing in the coils. The coils current begins to drop. When current is reduced down to half value, the switch 1 is opened, the converters 1...4 operate as rectifiers for a short period of time and there is no more current in switches 2,3. Then these converters again operate as invertors with all sections being connected in series. After the central solenoid current reversal and plasma current ramp-up to the preset maximum value, the converter required voltage is significantly reduced and section 1...4 are again connected in parallel. Coils 1-12 energy is discharged after burning cycle termination using converters 1...4 and breaker 2, introducing the resistor 3 into the coils loop. Sometimes emergency energy discharge is carried out when breaker-3 operates using resistor 4. PF coils 13 and 16 are feeded by unidirectional current, the scheme is given in Fig.3 (coil 13).

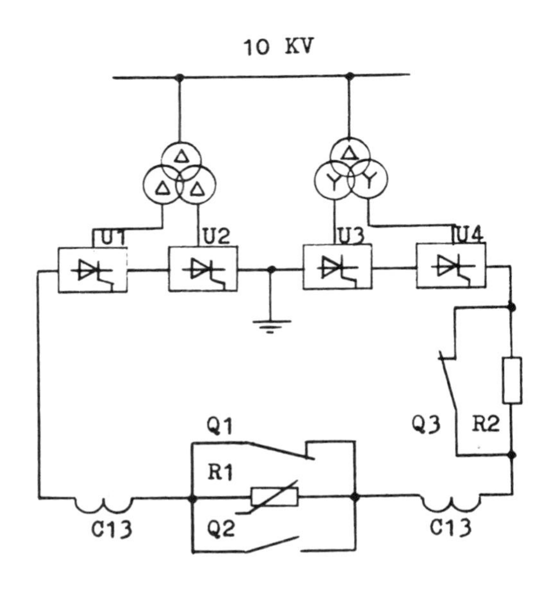

FIGURE 3

Energy storage in a coil and current control are performed by converters 1...4. In a process of control the angles of cathode groups control U1 and U3 and that of anode groupes 2 and 4 are varied. The other valves control angles are close to zero when the converter operates as a rectifier and discretely rises up to the limit when it operates as an invertor. High voltage at plasma current generation is provided by breaker-1, which is shunted then by closing device-2. Emergency energy discharge is provided by breaker 3, introducing the resistor 2 into the loop. In power supply system of coils 14,15 and 17, the reverse thyristor converters, each comprising

4 sections, are used what permits asymmetric control and change of sections series connection for the parallel one.

The total power consumption (a.c. source) isn't more than 500 MWt.

OTR magnets feeding directly from the network seems possible provided the reactive power compensation.

FUSION TECHNOLOGY 1988
A.M. Van Ingen, A. Nijsen-Vis, H.T. Klippel (editors)
Elsevier Science Publishers B.V., 1989

TEST FUSION REACTOR (OTR) ELECTROMAGNETIC SYSTEMS

V.A.BELYAKOV, L.B.DINABURG, N.I.DOINIKOV, V.V.KALININ, E.V.KORNAKOV, A.I.KOSTENKO, S.V.KRASNOV, YU.M.KRIVCHENKOV, A.A.MALKOV, I.F.MALYSHEV, N.A.MONOSZON, V.P.MURATOV, I.I.SABANSKY, S.N.SADAKOV, YU.V.SPIRCHENKO, G.V.TROKHACHEV, G.F.CHURAKOV, O.G.FILATOV, V.G.YAKUBOVSKY

D.V.Efremov Scientific Research Institute of Electrophysical Apparatus, Leningrad, USSR

1. INTRODUCTION

Construction of a test fusion reactor is an essential step of thermonuclear power engineering, making provisions for fusion reactor high-temperature plasma investigations and the development of the main engineering solutions, necessary for industrial power installations creation. OTR is a tokamak, aimed at demonstration of possible reliable and safe electric energy and nuclear fuel production on the base of thermonuclear fusion reaction. Here D-T reaction ignition, tritium self-supplying and specific energy load and fluence, close to those, necessary for industrial devices, should be attained. Also all the main reactor systems, including the most expensive and complex - superconducting electromagnets - should be optimized and tested on OTR.

OTR main parameters have been chosen taking into account the following suppositions:

- the discharge reference scenario is based on inductive plasma current ramp-up and maintaining; the burn stage duration is 600 s, the cycle duty factor - not less than 0.9;
- impurities control using single-null poloidal divertor;
- moderate plasma elongation - 1.5, the average neutron load to the wall - 1 MW/m^2, safety factor at the boundary - q = 2.1;
- physical limitations on plasma density: Murakami parameter m = 18; on ultimate pressure: Trojon parameter = 3.1, scaling for energy confinement time - T-11 (T-10);
- engineering limitations: magnetic field at superconductor - 12 T.

2. OTR MAIN PARAMETERS

Major radius:	
plasma, m	6.3
chamber, m	6.2
Minor radius:	
plasma, m	1.5
chamber, m	1.7
Plasma elongation (along the magnetic surface F = 0.95)	1.5
Magnetic field at chamber axis, T	5.8
Plasma current, MA	8.0
Burn time, s	600
Dwell time (between the nominal power levels), s	80
Pulses number	$3 \cdot 10^5$

The inductive scenario of plasma current ramp-up and maintaining has been taken as reference one. Burn phase duration is taken to be 600 s; for plasma current ramp-up, discharge and Ohmic heating coil recharging 80 s are necessary. Thus, the cycle duty factor in OTR is approximately 0.9.

3. CONFIGURATIONS

All OTR-structures are divided into 3 groupes, each having the specific temperature level and independent support system: superconducting magnets, cryostat and blanket (Fig.1).

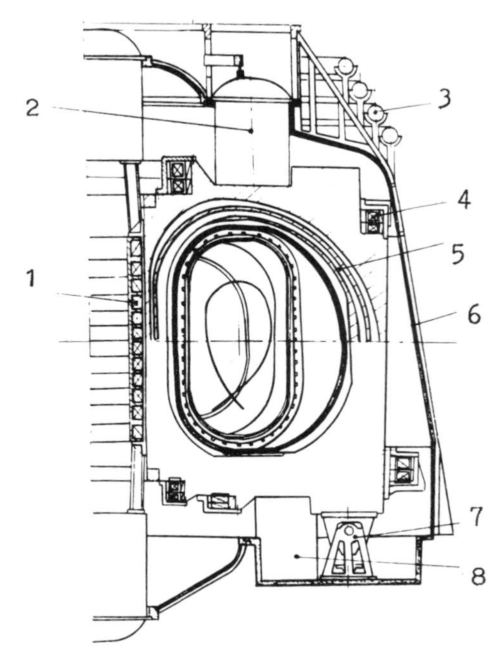

FIGURE 1
OTR electromagnetic system:
1 - central solenoid; 2 - current leads and helium lines unit; 3 - nitrogen and helium cooling pipes; 4 - outer PF-coils; 5 - toroidal field coil (TFC); 6 - cryostat wall; 7 - TF-coil mechanical support; 8 - blanket module support

EMS configuration and construction are mainly determined by the whole reactor layout.

The requirement on blanket sectors and RF antenna replacement by a single straight horizontal motion and the divertor plates - by azimuthal and radial motions, and the active zone elements - by a straight vertical motion was made the basis of the given project. Assembly and replacement of all the abovesaid elements should be performed without cryostat vacuum break i.e. through the access ports to the blanket.

All EMS superconducting coils are located in a common cryostat (vacuum volume) with the outer casing and inner toroidal cavity, connected with 12 horizontal and 24 vertical ports.

All PFC superconducting coils are located outside TFC and are stationary fastened to its support structure.

Assembly joints of all blanket replaceable elements are located around the torus or in its upper part and are easily accessible for diagnostic and maintenance equipment.

TF coils dimensions and number are chosen on condition of removable blanket sectors disassembly ($^{1}/12$ of torus) by a straight radial motion. The requirements to the magnetic field ripples are automatically satisfied in this case.

PF ring coils position choice is limited by the condition of indivisible in height torus blanket disassembly. As a result, location of large ring PF-coils isn't optimal.

Normal coils for plasma vertical position control system are located directly at the outer surface of blanket replaceable elements.

4. TF-COILS DESIGN

TF-coil comprises 12 winding blocks and 24 intercoil support structures. Each block (Fig.2) consists of two semi-blocks in which two superconducting D-shaped coils are located. The latter are made of different conductors: outer - for 7 T - field, inner - for 11.6 T. At torus inner radius all TFSC blocks form closed vault-structure, taking the centering forces, acting for it. The overturning moments, acting on TFSC, are taken by outer upper and lower supporting belts (formed by TF-coils cases and intercoil support structures) and also due to friction in a vault region. In each TF-coil upper part vacuum-tight block of cryogenic lines and current leads is located.

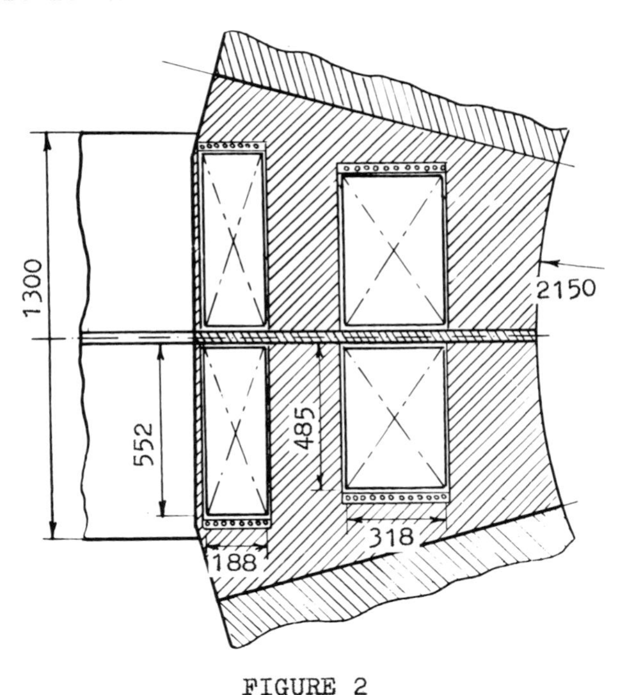

FIGURE 2
Block of TF coil

5. THE CENTRAL SOLENOID

The central solenoid comprises 12 coils seriesly connected into 3 groupes. Each of the coil consists of two concentric sections. For coils winding the supporting strip of variable thikness is supposed to be used, which should take the rupture forces. The axial compressing forces, acting to the extreme coils are transfered to TF-coil cases.

6. THE RING COILS

All PF ring coils are of the pancake winding type and have the discrete support structure, providing their uniform fixing to that of TFC. Minor ring coils are made of NbTi - based conductor and the major coils - on Nb_3Sn - base. Each PF ring coil has a separate block of cryogenic lines and current leads.

7. EMS COILS MAIN PARAMETERS

1. TF coil:

Coils number	12
Maximum field, T	11.6
Operating current, kA . . .	20
Current density in a conductor, kA/cm^2.	3.44/4.0
Winding bore height, m . . .	10.8
Winding bore width, m . . .	7.9
Superconductor	Nb_3Sn
Stabilizing material	Cu
Cooling	forced-flow
Winding to ground voltage, kV	6
Stored energy, GJ	45
Conductor mass, t	1000
Cases and intercoil support structures mass, t	5000

2. PF coil:

Total volt-seconds, V.s . .	210
Superconductor	NbTi/ Nb_3Sn
Maximum magnetic field, T. .	8
Maximum field variation rate, T/s	3.5
Plasma breakdown voltage, V	40(0.3s)
Operating current, kA. . . .	40
Current density in a conductor, kA/cm^2	2.3
Winding to ground voltage, kV	20
Stored energy, GJ.	9

PFC conductor mass, t 700
PFC support structures mass,t 300

8. THE SUPERCONDUCTING COILS

The conductors for inner (Fig.3,a) and outer (Fig.3,b) TF-coil sections are soldered compositions of the copper tubes (cooling channels) and flat superconducting cables enclosed in a stainless steel conduit. Superconducting cable comprises composite Nb_3Sn conductors and copper wires. Cooling tubes copper is also the stabilizing metal. Use of these tubes for cooling gives an advantage over the conductors option accepted in some other projects, as in this case vacuum-tightness of the conduit isn't necessary. For a.c. losses level reduction in the conductors, the insulating gasket is inserted into the centre of the superconducting cable and "coupling" losses are significantly decreased there. Besides, the adjacent copper tubes surfaces are oxidized. The total energy losses in TF-coil conductors are $1{,}5 \cdot 10^6$ J for the normal operating cycle and $1 \cdot 10^6$ J at plasma current disruption.

The "cable-in-conduit" type conductor is the most suitable for PF-coils. This type of conductor is widely used in other similar projects (for example, NET, FER, TIBER). Alongside with the low energy losses in variable magnetic fields, sufficiently high stability margin for thermal disturbances at high average current densities in a conductor is provided. The disadvantage of such conductors is high hydraulic resistance and, respectively low helium flow velocity (0.1-0.2 m/s), what results in possible appearing of several plasma current pulses for a time of helium full change in a conductor channel.

For the central solenoid (maximum field isn't more than 8 T) and the adjacent to it ring coils with minor winding radius NbTi conductor has been chosen (Fig.4,a). To increase helium circulation velocity, two central helium channels with low hydraulic resistance are provided. The space between the central helium channels walls and outer vacuum tight conduit is filled with 12 multistaged cables made of superconducting composite strands (3^4 strands). There is stagnant helium in these strands region and this helium may come to the central channels through the openings in their walls. At cooling branch length 700-800 m, helium is totally changed in a cooling channel for one plasma current pulse duration. For the rest coils Nb_3Sn conductor is used (Fig.4,b). It is a cable obtained by multistaged cabling of composite ($6 \cdot 3^5$) and copper (3^5) strands of 0.7 mm diameter, enclosed in a vacuum-tight stainless steel conduit. Composite conductors have the resistive barriers to reduce coupling energy losses. Copper quantity in both conductors has been chosen in such a way that winding maximum temperature will not exceed 80 K at quench energy discharge with winding to ground voltage 20 kV.

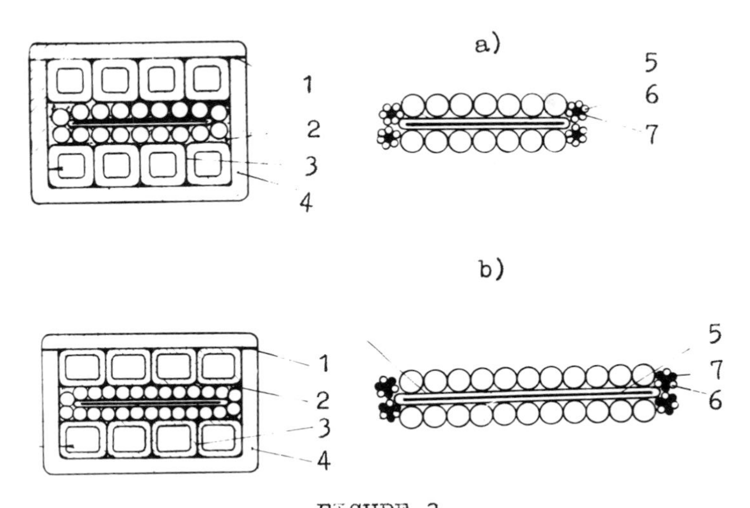

FIGURE 3
TFC superconductor:
a) for inner coil; b) for outer coil.
a: 1 - soldering or welding; 2 - solder;
3 - oxide insulation; 4 - casing;
5 - insulating gasket; 6 - superconducting wire (SW); 7 - copper;
b: 1 - soldering or welding; 2 - solder;
3 - oxide insulation; 4 - casing;
5 - insulating gasket; 6 - superconducting wire (SW); 7 - copper

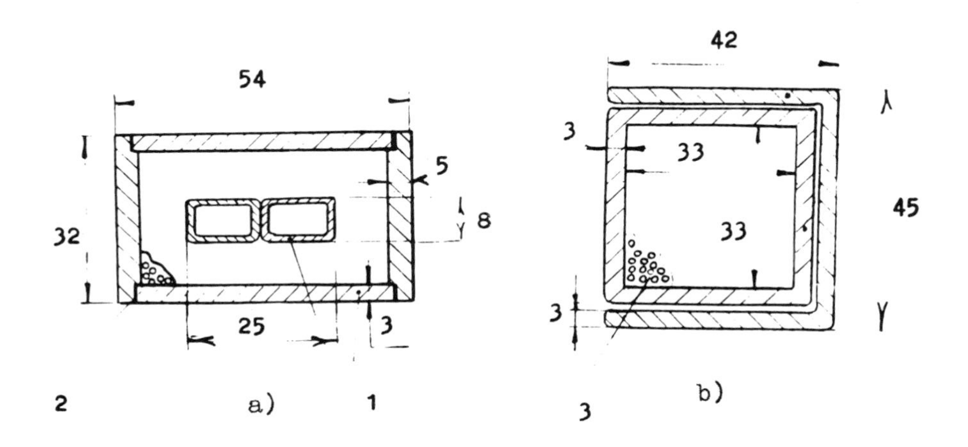

FIGURE 4
PFC superconductor:
a) NbTi based conductor; b) Nb_3Sn based conductor.
1 - stainless steel casing; 2 - composite strands located in a stagnant helium; 3 - composite strands

FUSION TECHNOLOGY 1988
A.M. Van Ingen, A. Nijsen-Vis, H.T. Klippel (editors)

THE SULTAN-III PROJECT

A. della Corte, G. Pasotti, M.Ricci, N.Sacchetti, M.Spadoni
Associazione EURATOM-ENEA sulla Fusione, Centro Ricerche Energia Frascati C.P. 65, 00044 Frascati, Italy
G. Dal Mut, G. Spigo, G.Veardo
Ansaldo Componenti, via N. Lorenzi 8, 16152 Genova, Italy

J.A.Roeterdink, J.D.Elen, A.C.Gijze, W.M.P.Franken
ECN, Westerduinweg 3, 1755 ZG Petten, The Netherlands

E.Aebli, I.Horvath, B.Jakob, C.Marinucci, P.Ming, G.Pasztor, G.Vécsey, P.Weymuth
Paul-Scherrer-Institute (PSI), 5234 Villigen, Switzerland

1. INTRODUCTION

The main task in the area of magnet technology is to develop a toroidal magnet system for NET with superconducting coils for about 8m times 11m bore for generating maximum field of 10-12T, highly reliable in the Tokamak environment at a stored energy level of tens of Gigajoules. A significant present step in this task is the International Large Coil Task Project (LCT), initiated in 1977 as a joint experiment by the USA and its partners EURATOM, Japan and Switzerland. The experience gained in the LCT programme has to be extended in coil size by factor 3× and from 8 to 12 Tesla in max. field strength. Ultimately, integration of coils at that size into a Tokamak configuration will require the most extensive development effort.

A single full size TF coil for NET represents an investment of the order of 40 MUC. Therefore, it is very probable that the full size coil fabrication and tests will only be possible as a first phase of the final commissioning of NET, with the sequence of coil fabrication being influenced by successive performance tests of the first set of coils. For such an undertaking prior to final design, risks need to be reduced significantly by implementing of a comprehensive development programme closing the gap LCT-NET in several areas as:

- design studies,
- high field conductor development including also conductors for the more advanced objective of pulsed superconducting OH coils,
- coil technology verification tests,
- full size conductor tests,
- full size and/or subsize pancake tests.

ECN, ENEA and PSI, presently engaged in a collaboration on the development and testing of high field forced flow superconductors for fusion devices, are now preparing conductors and insert coils in order to demonstrate the feasibility of s.c. technology for high current, high field applications based on a react and wind technique. Their programme includes the modification of the existing SULTAN facility[1].

2. SPLIT PAIR CONFIGURATION

The main component is a superconducting magnet system with a free bore for testing of conductor loops or coils in a background field of 12 Tesla at 0.58 m diameter after realization of the SULTAN Phase II stage. Upgrading of SULTAN-II by split coils replacing the 8 Tesla section will enhance the flexibility of the test facility considerably. This upgrading represents a relatively cheap solution to fulfill the special needs required by the NET programme. In order to test fullsize NET conductors a 12 Tesla magnet system with at least 4 m inner bore or a smaller magnet system with split coil configuration is needed. It is proposed to construct a

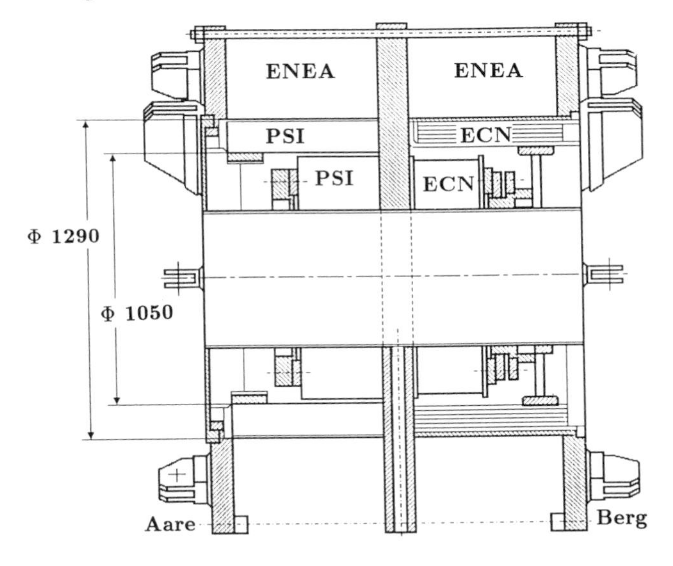

FIGURE 1
Arrangement of split coils for SULTAN-III
(dim. in mm)

split coil facility using as many of the existing magnets of SULTAN-II as possible. In fact only one of the present NbTi coils has to be replaced by two coils, due to the fact of having in the new facility, called SULTAN-III a splitted magnet system. The "new" intermediate coils must have a higher current density and as a consequence to withstand a higher magnetic field than the "old" NbTi coil. The new magnet will have therefore to be built by A-15 conductors. The splitted-III facility will provide a possibility to test short samples from industrial pilot productions of full size NET-conductors which could be extended to testing of larger coiled samples (pancakes of 2.0m diameter).

3. ANALYSIS OF MAGNET CONFIGURATION

Field: The magnetic field distribution has been calculated using the code GFUN. The iron parts of the facility have been divided into 350 elements, the maximum allowed by the code. Figure 2 shows the radial distribution of the field modulus, at the center of the gap, produced by the magnet system alone and in presence of a test pancake energized. In Figure 2 the field distribution at the inner radii of the coils in presence of a test pancake is reported. The field has been calculated for each coil along the generatrix on which the field modulus reaches its maximum value. Since the self-field of the test pancake decays rapidly along the Z direction, it has only a limited influence on the peak fields showed in Figure 2.

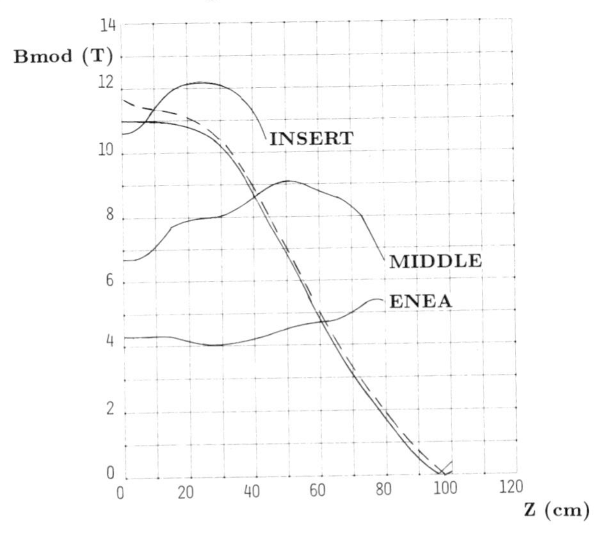

FIGURE 2

Field distribution in the gap and max. field at the inner radii of the SULTAN-III coils

(a) in presence of a energized test pancake

(b) magnet system alone

In fact, without the test pancake the peak field on the A15 insert lowers from 12.2T to 12T while there is practically no influence on the rest of the coils.

Forces: The forces acting on each coil of the magnet, in presence of a test pancake, have been calculated using the local value of field and current. Each coil has been divided into 256 elements, the field being calculated in the middle point of each element. The total forces F_X and F_Z acting on each coil are schematically reported in Figure 3. The X axis is the vertical one. There are no net forces in the Y direction because of the symmetry. It is worth noting the huge centering force acting between the test pancake and the magnet system due to the asymmetric position of the test pancake as shown in Figure 3. This force necessitates the application of a rugged central flange and the application of a 10mm stainless steel liner inside the ENEA 6T coil. The liner has to prevent possible radial movements in the pancake wound 6T solenoid.

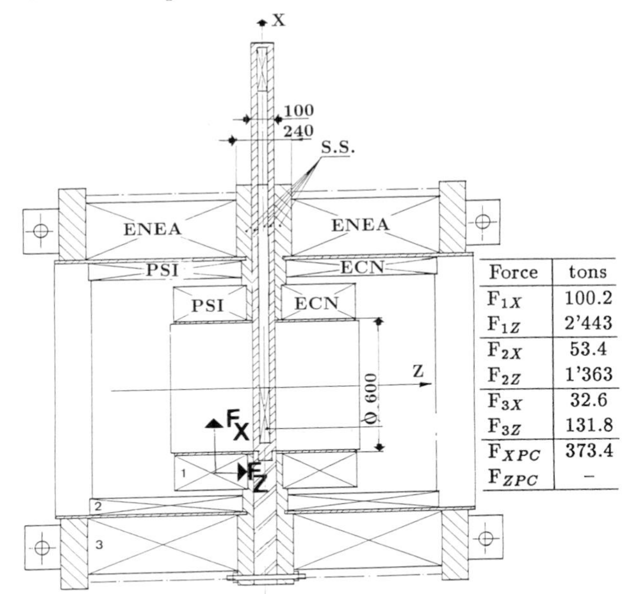

Force	tons
F_{1X}	100.2
F_{1Z}	2'443
F_{2X}	53.4
F_{2Z}	1'363
F_{3X}	32.6
F_{3Z}	131.8
F_{XPC}	373.4
F_{ZPC}	–

FIGURE 3

Forces acting on the different coils

In absence of a test pancake, the axial forces reported in Figure 3 have practically the same values but, in the vertical direction, the six coils of the magnet are subjected only to a total downward force of about 3 tons. This force is generated by an asymmetric distribution of the iron masses with respect to the middle plane of the magnet.

Stresses: The mesh of the SULTAN-III facility with the imposed boundary conditions consists of 448 isoparametric elements with 20 nodes and reduced integration (C3D20R) and of 136 interface elements (INTER 8) placed around the two coils in order to separate these two components from the flange and the stainless steel cylinder. The mesh represents only one half of the SULTAN facility because the magnetic loads and the geometry are symmetric with respect to the xy plane. The displacements of the flat side of the flange are fully contained. In the corners of the two coils near the corner between flange and cylinder, separate but coincident nodes have been introduced in order to have distinct normals of contact. The coincident nodes have also been constrained to have the same displacements and rotations by means of the multi-point constraints relations (MPC5). Three different materials have been considered in the model: stainless steel and two different equivalent materials for the two coils. The stainless steel and the ENEA coil have been supposed isotropic while the 9T coils have been assumed orthotropic. The loads due to the Lorentz force have been applied to the model as body force.

As a result the peak value of the von Mises stress are:

- on the ENEA coil 80 MPa
- on the liner 200 MPa
- on the 9T coil 180 MPa
- on the flange 300 MPa

These values are well below the allowable stresses for the liner and the flange. As far as the 9T and ENEA coils are concerned, the calculated stresses are rather high but still within the capabilities of these windings.

Energy Discharge: The magnet system is divided into 3 subsystems for powering and energy discharge viz. the 12T section (6 kA current), the 9T section (12 kA current) and the 6T section (4.5 kA current).

In case of detection of a normal zone in one of the coils the power supplies are switched off from the magnet system and dump resistors are switched into the circuit.

The new 9T coils will operate at a current density being about twice as high as in the other coils. As the dissipated energy inside a conductor during quench is about proportional to $\int j^2\,dt$, the energy of the new 9T coils has to be discharged faster from the new coils than from the existing coils at the detection of a normal zone inside one of the magnets. This enables to keep the hot spot quench temperatures for all the magnets at about the same level. A faster discharge means on the other hand that a large fraction of magnetic energy will be transferred to the other coils.

Figure 4 gives an impression of the current decay in the 3 subsystems. It is shown that the current in the 6T section rises initially from 4.5 kA up to 5.4 kA. The hot spot temperatures in the coils are respectively 34K (12T), 42K (9T) and 43K (6T).

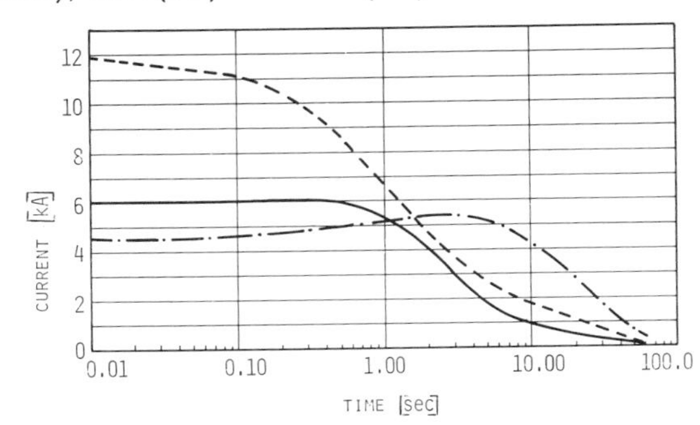

FIGURE 4
Fast discharge of SULTAN-III coil system; current decay ofthe three subsystems

4. 6T MAGNET

The present ENEA 6T coil is a pancake wound solenoid, where 40 double pancakes are stacked together[4]. In order to reach the split pair configuration, the 6T coil will be divided into two halves. The new mechanical structure has been designed in a way to use as much as possible of the existing components. The main additional new element is a system of two stainless steel central flanges and a spacers. The flanges are necessary to obtain two separate semi-coils. The spacer, 10cm thick, will be inserted between the flanges to withstand axial compressive forces. A central hole in the spacer will allow radial introduction of the sample holder for conductor tests. This assembly of rather thick stainless steel elements necessitate its own cooling circuit.

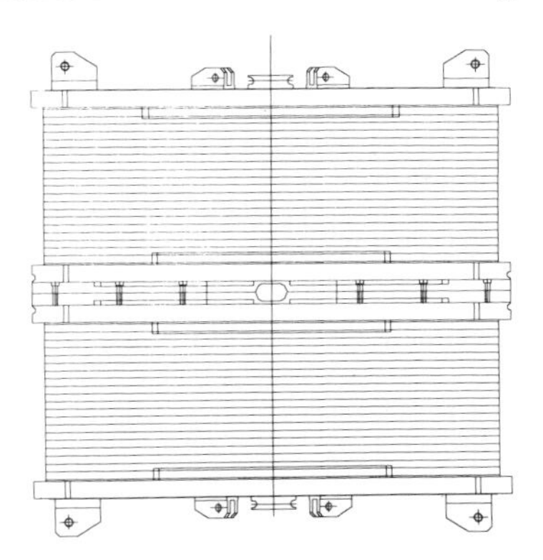

FIGURE 5
The splitted magnet with the central flange. The hole to insert test cables is shown

For this purpose a pattern of canal series will be obtained by machining the flanges and the spacer to get some rectangular quarries closed by stainless steel helium tight welded strips. Once the two halves of the coil will be reassembled, the internal surface will have to be machined in order to fit for coupling with the liner. On the other hand surface layers with a reduced roughness are required to lower the friction heat generated between the stainless steel liner and the ENEA coil by movements produced by the compression Lorentz forces. The use of special antifriction materials is also being considered.

The new general view of the magnet showing the hole where the superconducting cable will be inserted for tests is given in Figure 5.

5. DESIGN OF THE 9T COILS

Starting point for the 9T coils is the use of the 12T niobium-tin composite conductors as developed for SULTAN stage II[2,3]. Use of these pre-reacted conductors in the new 9T coils saves costs for development testing and production of the conductors.

These conductors have been designed for a transport current of 6 kA at 12.2 T and 4.5K. At 9T, being the maximum field on the winding package of the new middle magnets, the conductors can carry a transport current of 12 kA taking into account the same margins to the critical current. The critical current will be about 18 kA (9T, 4.5K).

Lorentz-forces acting on the conductors are roughly a factor of 2 larger than for the 12T coils. It will be clear that additional structural material is necessary to support the conductors, therefore 2 stainless steel strips are soldered to the conductors. Both coils will be layered: 10 layers of 24 turns each constitute the winding pack of 240 turns. The outer stainless cylinder will be used to support the coils.

Both coils will be cooled by supercritical helium. In the ECN 9T coil each layer will have its individual coolant inlet, located at the outer radial side-plane of the coil, coolant outlets are on the transition from one layer to another so that 2 layers have one common outlet. In the PSI coil only one helium inlet and outlet are foreseen at the outer radial plane.

In the ECN coil 5 individual conductor lengths (about 180 m long, being sufficient for 2 layers) are used. In the PSI coil only one conductor length of about 900m is used. For both coils the electrical connections are located at the outer radial side-plane. Coil characteristics are given in Table I; a global sketch of the ECN coil is given in Figure 6.

TABLE I
Coil Characteristics

		ECN	PSI
Operational current (9T,4.5K)	kA	12	12
Critical current (9T,4.5K)	kA	18	18
Number of windings	-	240	240
Number of layers	-	10	10
Number of windings per layer	-	24	24
Ampere-turns	MA-turns	2.88	2.88
Inner diameter winding package	m	1.088	1.064
Outer diameter winding package	m	1.270	1.275
Length of winding package	m	0.675	0.676
Average length of 1 winding	m	3.71	3.68
Total conductor length	m	891	886
Overall current density (wind.package)	MA/m^2	46.8	40.3

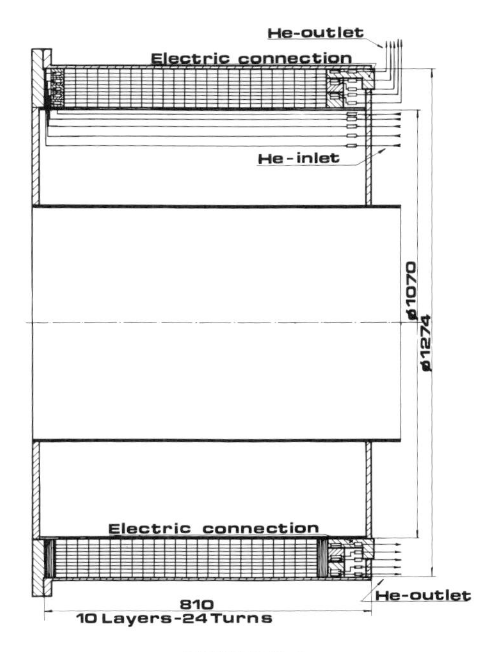

FIGURE 6
Global sketch of the ECN coil

PSI Conductor Concept: The concept for the PSI conductor is based on a flat cable of Nb_3Sn strands cooled by forced flow of supercritical helium. The conductor, designed to carry 12 kA at 9T and 4.5K is shown in Figure 7. It consists of a pre-reacted flat cable sandwiched between two copper strips and positioned between two cooling tubes. To support the hoop stress of about 190 MPa acting on the conductor in the coil, two reinforcing stainless steel strips of 1.8 mm each are added to the assembly. All seven conductor components are joined together in one operation by soldering. The superconducting cable design is based on a Nb_3Sn strand of 0.6 mm diameter made by the internal tin or bronze technique. To reach the specified critical current, a cable design was made using 112 strands in two cabling stages.

ECN Conductor Concept: The conductor is based on Nb_3Sn material produced by the ECN powder process. Strand material to be used is 1mm in diameter. Thirty-six strands are cabled to a flat Rutherford cable, having a thickness of 1.84 mm and a width of 18.1 mm. The cable is enclosed in copper necessary for additional stabilization and current by-pass in case of normal zones. Two copper cooling channels are present. The equivalent diameter of the coolant channels is 2.75 mm. Helium massflow in the conductor is 2.5 g/s. Average pressure drop over 1 conductor layer is 60 kPa. Two stainless steel strips (thickness 1.5 mm) are placed on top and bottom of the conductor as reinforcement. The cable is heat treated for 48h at700°C in argon atmosphere. After heat treatment the cable is soldered to the other components of the composite conductor. The 7 components of the conductor are soldered together in only one soldering run. To have a good bonding between copper and stainless steel the stainless steel strips are pretinned.

TABLE II
Characteristic dimensions composite Nb_3Sn conductor

	ECN		PSI	
Cross section	mm^2	%	mm^2	%
Nb_3Sn + Powder+ Nb	11.31	4.7	8.0	2.87
Copper	109.84	45.5	97.0	34.88
Bronze + Ta	–	–	7.8	2.8
Stainless steel	78.00	32.3	94.6	34.0
Insulation	21.12	8.8	25.6	9.2
Helium	15.12	6.3	25.1	9.02
Solder	5.81	2.4	15.3	5.50
Voids	–	–	4.7	1.69

Table II gives the dimensions of the two composite conductors and the distribution of the different materials.

6. MEASUREMENT OPTIONS

Short Sample Tests: In order to reach high repetition rates for sample measurements a special sample insert unit allowing fast sample insertion without warming up the whole magnet system is foreseen. This unit - similar to a telescope - will be mounted together with the sample holder on top of the facility. The insert unit has an overall length of 6m and a diameter of 0.5m.

A superconducting transformer will be used as current supply for the short samples. Such a solution eliminates large and power consuming current leads. The transformer is specified for currents up to 50 kA. It will be mounted on top of the sample holder in the insert unit to avoid large and complicated current feed throughs (see Fig. 7).

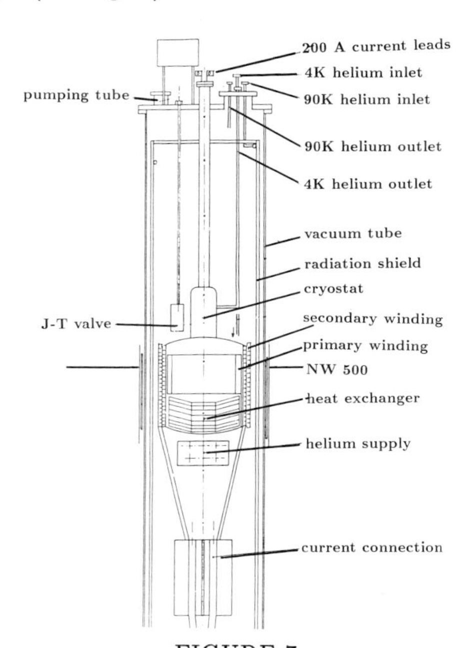

FIGURE 7
Top of sample holder with cryostat and superconducting transformer

Pancake Test: The facility will be modified in such a way that also small pancakes of NET-conductors can be tested in the midplane gap of the magnet system. A free gap region of 100 mm is reserved for the windings of the pancake. For the design of the split coil system a "standard" double pancake is taken into consideration. This standard double pancake has $2 \times 9 = 18$ turns; the inner diameter of the pancake is 1.6 m and the outer diameter is 2.1 m. The width of the double pancake is 76 mm. The conductor current taken into account for the design is 16 kA. The position of the standard pancake is given in Figure 3. The maximum field in the gap region is 11.7 T; the self field of the pancake (0.7T) is included in this figure.

REFERENCES

1. J.D.Elen, N.Sacchetti, G.Vécsey, 12 Tesla Split Coils for the SULTAN Test Facility, Project Proposal, January (1988)
2. J.A.Roeterdink, M.W.Brieko, A.C.Gijze, H.V. Mertens, IEEE Trans.Mag. Vol.24, (1988),1429
3. G.Pasztor, B.Jakob, I.Horvath, P.Ming, G.Vécsey P.Weymuth, IEEE Trans.Mag. Vol.24 (1988),1086
4. G.Pasotti et. al., IEEE Trans.Mag. MAG-17, (1981), 2007

FUSION TECHNOLOGY 1988
A.M. Van Ingen, A. Nijsen-Vis, H.T. Klippel (editors)

CONSTRUCTION AND TEST RESULTS OF THE FTU TOROIDAL MAGNET

A. Pizzuto, A. Cecchini, S. Migliori, M. Gasparotto

Associazione EURATOM-ENEA sulla Fusione, Centro Ricerche Energia Frascati,
C.P. 65 - 00044 Frascati, Rome, Italy

G. Dal Mut, A. Maragliano

ANSALDO COMPONENTI, Via Lorenzi 8, Genoa, Italy

The FTU toroidal magnet was constructed by ANSALDO in Genova. This paper illustrates the procedures adopted during the main manufacturing phases when particular care had to be taken to realize a component capable of the performance aspected.

1. INTRODUCTION

In order to obtain a component able to withstand the very severe operating conditions, the FTU toroidal magnet[1] was manufactured using very precise tools, respecting close tolerances and performing a large number of tests to check the materials and component quality at the various construction stages.

The main interesting procedures of the toroidal magnet manufacturing phases were the following:

- Fabrication of the TF coils;
- Assembly of the coils in the steel casings;
- Cooling tests;
- Torus assembly.

2. FABRICATION OF THE TF COILS

Each TF coil (in total 24) consists of 42 wedge-shaped copper turns insulated from each other by 41, 0.22-mm-thick, glass-fabric epoxy sheets; one thick filler wedge; one prefabricated ground insulation piece in which the copper structures for the passive plasma control are fastened; and an external ground insulation which matches the steel casings very closely.

The copper turns, made of 100% hard ETP copper with a 290-MPa yield, are bonded with interturn insulator using the vacuum impregnation technique, as illustrated in reference 1.

The filler wedge is made from a thick tube manufactured using the filament winding method with an angle of about 30° in order to obtain an axial thermal contraction lower than that of stainless steel.

The prefabricated ground insulation component, shown in Fig. 1, was produced by vacuum impregnation of dry E-glass cloth. Four hollows of rectangular

FIGURE 1
Ground insulation prefabricated piece

cross section are formed in its toroidal region, where the passive control coils are to be inserted.

The external ground insulation was made by applying several layers of pre-impregnated B-stage cloths on the coil outer region (see Figs. 2 and 3), providing the necessary overlapping with the prefabricated piece installed in the internal region. The final shape was achieved pressing these layers against the TF coil, using a special toroidal mould, in two phases: the first at room temperature and the second at 70 °C.

FIGURE 2
TF coil partially ground insulated

FIGURE 3
TF coil: hydraulics and electric connections

FIGURE 4
Calibration of skins for the coil matching

Finally, in order to limit the thermal stresses in the components, the coils were polimerized at 90 °C for 24 hours.

3. ASSEMBLY OF THE COILS IN THE STEEL CASINGS

This procedure, which must allow good cooperation between the coils and the steel casings at 77 K, was performed in two steps:

1) During the first step, each TF coil was matched perfectly with its steel casing, filling the gaps with charged epoxy (see Figs. 4 and 5).
2) The two coils of each toroidal magnet module (in total 12) were then cooled down to -50 °C in about 1.5 hours by streaming LN2 in the coil channel. At this point, the module precompression system (see Figs. 6, 7 and 8) was actioned for a predetermined quantity originating a stress of ~ 50 MPa in the steel.

4. COOLING TESTS

It is very important for the TF coil to have a mean cooling rate of ~ 3 K per minute between 80 K and 110 K to allow the required duty cycle of the FTU machine, so cooling tests were carried out to check this. Each module was inserted in a cryostat and cooled down (10

FIGURE 5
TF coil finished

FIGURE 7
Particular of the precompression system

FIGURE 6
Precompression system

FIGURE 8
Pre-compression phase

K/minute) using LN2 rain on the steel casing and LN2 forced flow on the coil. When 77 K was reached, the coils were heated with a DC current up to the magnet typical service temperature and then cooled down as fast as possible, checking the time taken.

All the tests gave repetitive values which were in accordance with the technical specifications.

5. TORUS ASSEMBLY

In order to be able to begin delivery of the FTU toroidal magnet before its completion, the assembly procedure was accomplished using a step-by-step method. Thus, it was necessary to use a suitable tool (reference table) on which

FIGURE 9
Torus assembly

the angular and radial positions of each module had to be defined. First, modules 1, 2 and 3 were set on the table, the intermediate spacers (which assure the magnet continuity) were positioned and machined properly to obtain a good contact, and then the holes for the final pins were bored. Once the fourth module was positioned and fixed, the second was ready for delivery. These operations were repeated until the twelfth was fixed to the first one (see Fig. 9).

6. CONCLUSIONS

The manufacture of the FTU toroidal magnet was completed at the Ansaldo works in Genoa. The first module was delivered to ENEA in October 1987 and the rest in February 1988.

REFERENCES

1. G. Dalmut, A. Pizzuto et al., Manufacturing of LN2 Cooled Toroidal Coils of FTU, 14th Fusion Technology Proceedings, Avignone, September 1986, pp.1663-1668.

FUSION TECHNOLOGY 1988
A.M. Van Ingen, A. Nijsen-Vis, H.T. Klippel (editors)

THE FTU POLOIDAL SYSTEM: MANUFACTURING AND TESTING

L. Lovisetto M. Gasparotto, A. Pizzuto

Associazione EURATOM-ENEA sulla Fusione, Centro Ricerche Energia Frascati,
C.P. 65 - 00044 Frascati, Rome, Italy

G. Dal Mut, M. Marin, G. Tacchi

ANSALDO COMPONENTI, Via Lorenzi 8, Genoa, Italy

In order to reach the expected performance, the FTU tokamk will have to operate at very high levels of magnetic fields for at least 10000 experimental shots. The toroidal magnet and the poloidal field windings will undergo very high stresses due to thermal and electromagnetic loads. After a survey of the poloidal field winding requirements, the paper reports the main manufacturing process and equipment used for the construction of the coils; the mechanical and electrical tests are also illustrated.

1. INTRODUCTION

The entire load assembly of the FTU machine[1] is kept during operation at liquid nitrogen temperature (-196 °C) to maintain the required power and the energy consumption in the toroidal and poloidal windings within reasonable limits and to take advantage of the higher mechanical properties of the structural materials (stainless steel and copper) at cryogenic temperature.

The manufacture of the poloidal system (central solenoid T1 excepted) has been completed at the Ansaldo works in Genoa. The FTU poloidal system includes four sets of poloidal field windings: ohmic heating coils (transformer -T), vertical field windings (V), plasma horizontal position feedback windings (F) and plasma vertical position feedback windings (H); a decoupling transformer composed of 17 identical pancakes and the thermal insulated bushings connecting the windings inside the cryostat at 77 K with the busbars installed in the FTU Hall.

2. POLOIDAL WINDINGS

2.1. Requirement

The T-windings used to start the discharge (plasma breakdown voltage up to 40 V), provide most of the flux swing (5.1 Vs) required to raise and keep the plasma current flowing according to a preprogrammed waveform. The T-windings have been arranged to give a very low value of the stray field (≤ 50 G) in the plasma region with a null on the plasma centerline.

The V-windings provide the preprogrammed equilibrium field (up to 0.65 T) and contribute to the flux variation during the rise of plasma current with 1.3 Vs.

The F-windings, together with a set of simple passive circuits, are capable of correcting horizontal plasma column displacements due to error in preprogramming or to plasma soft disruptions, producing a vertical field up to ± 750 G. The F coils are positioned to have a zero mutual inductance with the plasma. Furthermore, in order not to perturb the plasma current by magnetic coupling between the feedback coils and the V- and T-coils, an external decoupling transformer has been designed to compensate for the mutual inductance.

The H-windings, together with the above mentioned passive circuits, are capable of correcting plasma column vertical displacements. The position of the H-coils has been chosen to produce a horizontal field up to ± 200 G uniform within 2% in the equatorial plane of the plasma

TABLE I - Electrical parameters of FTU poloidal windings

		WINDINGS			
		T	V	F	H
Self and mutual inductance (mH)	T	64.5	12.6	3.81	0
	V		33.4	3.48	0
	F			4.52	0
	H				47
Resistance at -196 °C (mΩ)		11.7	7.0	6.3	64
Magnetic field vs current (G/kA)		--	272	63	222
Maximum current (kA)		25	25	12	1.2
Magnetic field index on plasma axis		--	0.41	-1.05	--
Copper weight (t)		4.54	2.6	0.55	0.77

region. The electrical parameters of the four sets of poloidal windings are resumed in Table I.

2.2. Design

The FTU magnet is designed as a monolithic structure and the poloidal field windings are supported by the magnet through supports which are adequate for transferring the vertical electro-magnetic loads while allowing the thermal expansion of the windings with respect to the magnet during the different operation phases of FTU. Each copper coil has been designed to individually stand the electromagnetic radial forces[2].

The central transformer solenoid (T1) manufactured by INDELVE, is similar to the FT one[3]. It consists of 612 turns made up of twelve, specially built enamelled conductors (3×8.8 mm^2) in parallel.

The other 14 poloidal field coils are made of ETP hard-drawn copper tapes of rectangular cross section with a surface roughness better than 0.4 μm and indirectly cooled with liquid nitrogen by a layer of hollow conductors located on one or both sides of the windings (see Fig. 1).

This solution permits compact windings without damage being caused to the insulating material by impurities in the hydraulic cooling circuit.

The copper tapes used for the windings fulfil the following requirements at room temperature and liquid nitrogen temperature:

	20 °C	-196 °C
- electrical resistivity (Ωm)	$\leq1.7\times10^{-8}$	$\leq0.24\times10^{-8}$
- yield limits (MPA)	≥ 260	≥ 300
- ultimate tensile strength (MPA)	≥ 290	≥330
- elongation at fracture (%)	≥ 20	≥ 30
- hardness (Vickers)	≥ 90	

The electrical insulation between turns, between the cooling hollow copper conductors and vs ground are made of layers of half-lapped glass fibre tape interleaved with kapton tape (50 μm thick), wound on the same helix. A glass fibre tape 80 μm thick is interposed between adjacent turns and between the winding and the layer of cooling conductors. The nominal thickness of interturn insulation and between the winding and the cooling conductors is 600 μm and the minimum thickness of the ground insulation is 2.5 mm.

The liquid nitrogen to cool the coils enters and exits from the hollow copper conductors through two manifolds placed at 180° and made of insulating material to avoid induced currents on the cooling circuit. In order to minimize the voltage between the hydraulic circuit and the coil, the external and internal hollow copper conductors have been electrically connected to the external and internal turns respectively of the coil itself.

2.3. Fabrication

In order to minimize the thickness of the electric insulation between the winding and the hydraulic cooling circuit for reducing the

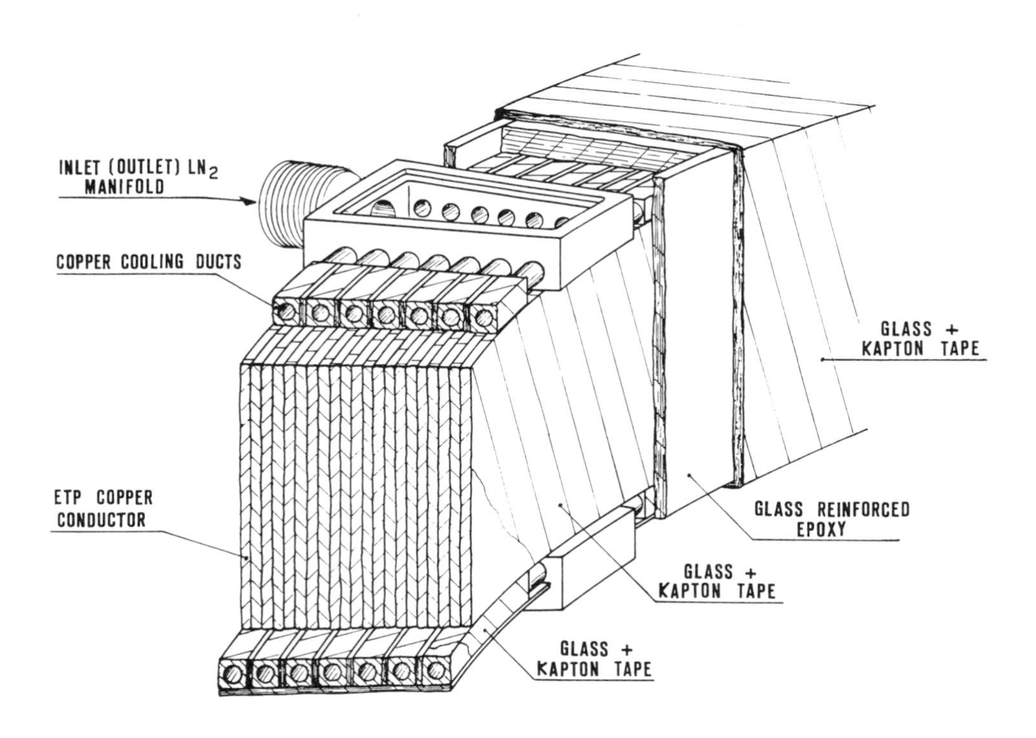

FIGURE 1
Simplified sketch of the coil construction

cooling time of the poloidal coils between two successive shots, each copper tape was machined so as to obtain a flatness within $\pm$ 0.1 mm. The copper tape was then inserted into an automatic assembly line (see Fig.2) installed in a clean area where the following operations took place:

- removal of sharp edges;
- cleaning, sand blasting, coating with DZ80 primer and baking of the copper conductor;
- application of one layer of half-lapped fiber glass (80 μm thick, 30 mm wide) interleaved with kapton tapes (50 μm thick, 28 mm wide) (see Fig.3);
- winding of the conductor within the required tolerances.

The hollow copper conductors for the hydraulic cooling circuit were also prepared and electrically insulated on the same assembly line.

The coupling of the hollow copper conductors to the manifold was done so as to guarantee

FIGURE 2
Automatic assembly line for the poloidal field coils

FIGURE 3
Forming the insulation on the poloidal field coil conductor

FIGURE 4
Layer of hollow conductors for the LN_2 cooling

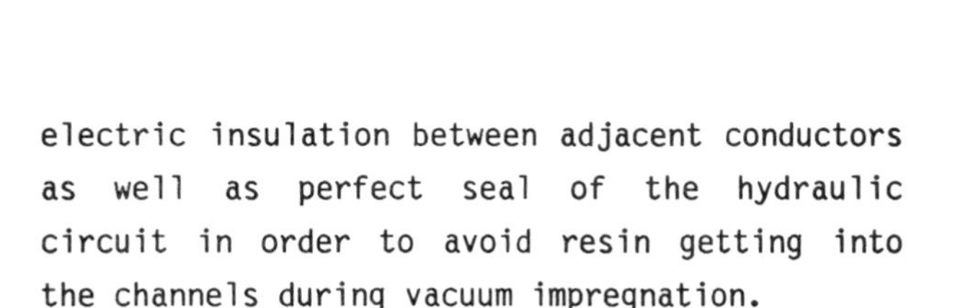

electric insulation between adjacent conductors as well as perfect seal of the hydraulic circuit in order to avoid resin getting into the channels during vacuum impregnation.

The winding, including the hydraulic circuits (see Fig. 4), was ground insulated by at least six layers of glass-kapton and then inserted into a mould and vacuum impregnated with an epoxy resin particularly suitable for operation at liquid nitrogen temperature (see Fig. 5).

2.4 Tests

The manufacturing methods and the automatic assembly line previously described were checked and set during the construction of the prototype coil cooled by a double hydraulic circuit. The coil was electrically tested between turns up to 500 V/turn and versus ground up to 35 kV DC and 25 kV AC. Cooling tests were carried out on the same prototype, using first one hydraulic circuit, then the other, and, finally, both. The temperatures of the hollow copper conductors were measured using thermocouples, and the average temperature of the winding was

FIGURE 5
Impregnated poloidal field coil

obtained from the measurement of the electric resistance. Uniform cooling was found in the hydraulic circuits, and a cooling rate from 105 K to 95 K in about 5 min was measured in the copper winding using both circuits.

The coil was then installed in a fatigue test machine and submitted to tensile and compressive cyclic load tests at liquid nitrogen temperature (77 K) for 20.000 cycles with a maximum load of ± 100 kN, corresponding to a maximum stress on the copper of 160 MPa. At the end of the tests, the prototype coil was once again submitted to the electric and cooling tests already described, without any appreciable variations being found.

The dimensional, electrical and cooling tests on the prototype coil were also carried out on the other 14 poloidal coils; the cooling time from 105 K to 95 K for each coil was less than 10 min, which is the scheduled time between 2 successive shots on the FTU machine.

3. CONCLUSIONS

The poloidal winding system was delivered to the Frascati Research Center in July 1988 after it had been preassembled by Ansaldo to check the electrical connections between the coils, test the electric insulation and measure the self- and mutual inductances between the T,V,F and H systems.

REFERENCES

1. R. Andreani and the FTU Project Team, The FTU Frascati Tokamak Upgrade, Proc. 14th Symp. of Fusion Technology, Avignon, 8-12 September 1986, pp. 149-159.

2. R. Andreani et al., The FTU Magnetic Structure, Proc. 8th Int. Conf. on Magnet Technology, Grenoble, 5-9 September 1983, pp. C1-143, C1-156.

3. Frascati Tokamak Upgrade, ENEA Report 82.49 (1982) Frascati, Rome, Italy.

FUSION TECHNOLOGY 1988
A.M. Van Ingen, A. Nijsen-Vis, H.T. Klippel (editors)
Elsevier Science Publishers B.V., 1989

THE RFX TOROIDAL FIELD WINDING: DEVELOPMENT TESTS AND MANUFACTURE

F. BELLINA, P.P. CAMPOSTRINI, A. STELLA*, J.L. BORNE**, J.P. GARY**

Istituto Gas Ionizzati (Associazione EURATOM-ENEA-CNR), Corso Stati Uniti 4, 35020 Padova, Italy
*Universita' di Padova - Dipartimento di Ingegneria Elettrica, via Gradenigo 6/a, 35131 Padova, Italy
**Jeumont- Schneider, Rue de l' Industrie 27, 59460 Jeumont, France

The RFX Toroidal Field Winding consists of 48 coils placed around a thick toroidal aluminium shell. For assembly and easy maintenance the coils are demountable into two halves along a diameter. Electrical continuity and mechanical strength are provided respectevely by demountable joints connecting the corresponding half turns of the coil, and by a clamping belt designed to withstand the electromagnetic expansion loads during the pulse.This basic choice has caused a number of unusual mechanical and thermal problems, and their solution has required accurate calculations and the manufacture of prototypes, to finalize the technology, and the setting up of an extensive testing programme. The results of the tests performed and the most significant adjustments to the preliminary design, as well as the most significant technological solutions adopted are reported in the paper.

1. INTRODUCTION

In a Reversed Field Pinch (RFP) the toroidal and poloidal field components present similar maximum amplitude; the toroidal magnetic field required is about one order of magnitude lower than in a tokamak with the same plasma current and similar aspect ratio. This also means that the forces acting on the Toroidal Field (TF) Coils and the overturning moment due to the interaction between their current and the equilibrium field are much lower than in tokamak of similar size.

On the other hand in a RFP even a local deviation from the ideal theoretical configuration may lead to instability conditions: this requires a large number of coils, in order to minimize the toroidal field ripple, due to the discretization.

Studies on the RFP configuration have been so far extensively carried out on relatively small size experiments only; it is then quite hard to establish today precise operating conditions for RFX, so that a wide degree of built-in flexibility in the possibility of setting-up the working parameters has been the leading criterion of the machine design. In addition it is quite reasonable to expect that, along the RFX experimental life, some of the plasma system components will have to be modified according to scientific needs and experimental evidence.

As far as TF winding is concerned, these requirements are met by providing a wide range of possibilities in the number of turn setting and by demountable coils, which can be opened at the equatorial plane of the machine in order to get an easy access to the vacuum vessel.

In spite of the relatively low value of the toroidal magnetic field, the design and manufacture of the winding has raised a number of unusual problems, mainly due to the basic design choice to get each coil demountable into two halves. The feasibility of the preliminary design had to be proved not only by means of theoretical models and calculations, but also with a number of tests on samples and prototypes, in order to finalize both the design and the costruction technology.

Tab. 1. Main Parameters of the Toroidal Field Winding

Parameter	Value
Number of coils	48
Turns per coil	8
Coil mean diameter	1.24 m
Copper section	22x24 mm^2
Peak toroidal field	0.7 T
Peak current	18.3 kA
Total ampereturns	7 MAt
Thermal parameter i^2t	300 MA^2 s
Peak voltage to earth	3.5 kV
Peak voltage per turn	219 V
Turn ratios	32-64-96-128-192-384

The coils are presently being manufactured by Jeumont-Schneider Company in Jeumont (France).

2. DESIGN PARAMETERS

The RFX Toroidal Field winding consists of 48 coils of 1.24 m diameter, evenly distributed around the torus and partially inserted into grooves machined in a thick aluminium shell [1-2]. Each coil consists of eight turns.

The winding has to produce a maximum toroidal bias field $B_0 = 0.7$ T and a reversed field $B_w = -0.44$ T; the corresponding currents in the coils are $I_0 = 18.3$ kA and $I_w = -11.5$ kA (Fig. 1).

To obtain the required current waveform, the winding is first fed by a 5 MJ, 7 kV capacitor bank and then by AC/DC converters, to sustain the current during the flat-top. The main winding parameters are summarized in Tab.1.

The coils are permanently connected in series, in groups of four, by means of busbars close to the coil terminals, to form twelve sectors. The sectors are connected to a collector system by means of cables.

The thermal parameter i^2t considered for the winding design is 300 MA^2s every 10 minutes: for its evaluation, not only the nominal current waveform shown in Fig. 1 has been taken into account, but also a number of different experimental conditions, such as the "pulsed discharge cleaning" operation (where the winding is subjected to lower current applied for a longer time), and possible future upgrades of the machine. No forced cooling of the conductors is necessary, since the heat generated is removed mainly through the aluminium shell, which is water-cooled. The mean thermal power transmitted to the shell is about 30 kW and the maximum temperature rise of the winding is 32 °C at the machine full performance .

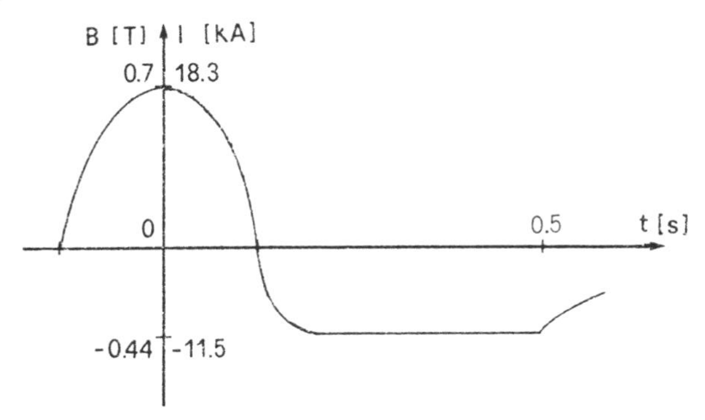

FIGURE 1
Wafeform of the winding current and of the magnetic field on the axis

3. MANUFACTURING DESIGN

3.1. Coil manufacture

Each coil is made of two identical half-coils and presents the turns arranged on two layers. The transition from each turn to the next are located at the outermost end of the half-coils (Fig. 2).

Very narrow dimensional tolerances have been specified to get the coils to tightly fit into the grooves on the shell; moreover the coils are designed to be completely interchangeable.

The conductors are made with oxigen-free high conductivity copper with silver, ASTM C10700, in a half-hard temper condition.

During manufacture each copper bar, having a 22x24 mm^2 cross-section, is first rolled and cut

FIGURE 2
Picture of a prototype half-coil end.

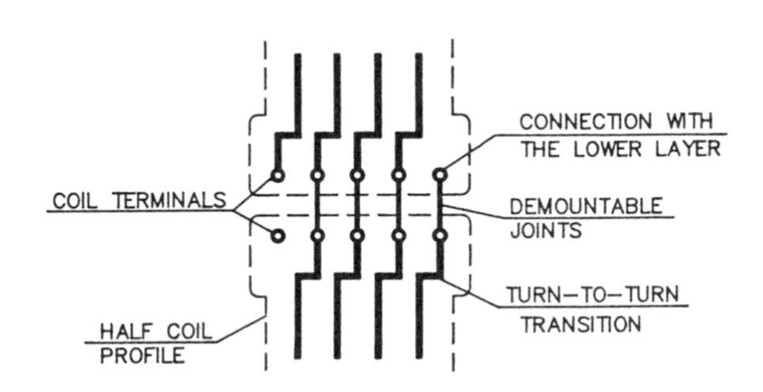

FIGURE 3
Sketch of the turn-to-turn transitions of the outer layer.

lengthwise to form an half turn; a suitably shaped copper block is then brazed at one end, to make the turn-to-turn transition, as sketched in Fig. 3.

The brazing is made by means of induction heating; a water cooling system is applied around the bar to minimize copper annealing. Due to the particular shape and to the small size of the block brazed, it is not possible to subject the joint to the usual tensile test, so that an alternative acceptance procedure had to be developed. Each brazing is examined by ultrasonic; the acceptable level of defects has been established by comparison with a preliminary qualification test made on several specimens, which have been subjected to tensile stress, X-ray, and ultrasonic.

The insulation is made with glass ribbon impregnated with epoxy resin under vacuum. The copper bar is sandblasted, cleaned and then primed with DZ 80®. The fibre-glass ribbon, 0.25 mm thick and 40 mm wide, is wrapped half-overlapped to form the insulation between turns, layers and to earth. The final thickness of the insulation is 1 mm between turns and 3 mm between layers and to earth.

The half-coil end represents a critical aspect of the coil manufacture, not only as far as the brazing process is concerned, but also because of the little space available for the demountable joints. Mechanical and electrical problems were both considered in detail during design: the need to have a large contact surface conflicts with the need of providing an insulation distance long enough to prevent flashovers. The solution represented in Fig. 2 has been finally adopted: all the contact surfaces are gold plated and the end of the conductors are insulated with suitably shaped glass fabric, which provides insulating tiles between turns, layers and to earth.

Two prototypes of half-coils have been manufactured and subjected to mechanical and electrical tests, in order to prove the soundness of the design and of the manufacturing process.

The first prototype was cut to get several specimens suitable to subject the insulation to rupture and fatigue shear tests. The shear strength obtained, which is the most critical, is above the minimum required by the RFX specifications (5 MPa for 10^5 cycles).

Dielectric tests were successfully performed on the second prototype: the values of the working voltage, the test voltage and the breakdown voltage are summarized in Tab. 2.

3.2. Demountable joints

Each demountable joint consists of two solid copper blocks with a short stack of 85 flexible copper sheets (0.2 mm thick) welded in between. The two blocks are designed to be bolted to the half turn ends, as represented in Fig. 4.

During a machine pulse, the demountable joints not only are subjected to electromagnetic forces but also have to cope with a relative movement (about 1 mm) of the two half-coils concerned, due to their thermal expansion.

Because of the little space available, the filling coefficient chosen for the sheet stack represents a compromise between flexibility and heat generation. The demountable joints are mainly cooled through the coils

Tab.2. Normal operating and test voltages for the prototype of half-coil

Voltage	Nominal V (peak)	Test 1' AC V (r.m.s.)	Breakdown AC V (r.m.s.)
between turns	220	2000	8000
between layers	1750	5000	23000
to earth	3500	10000	24000

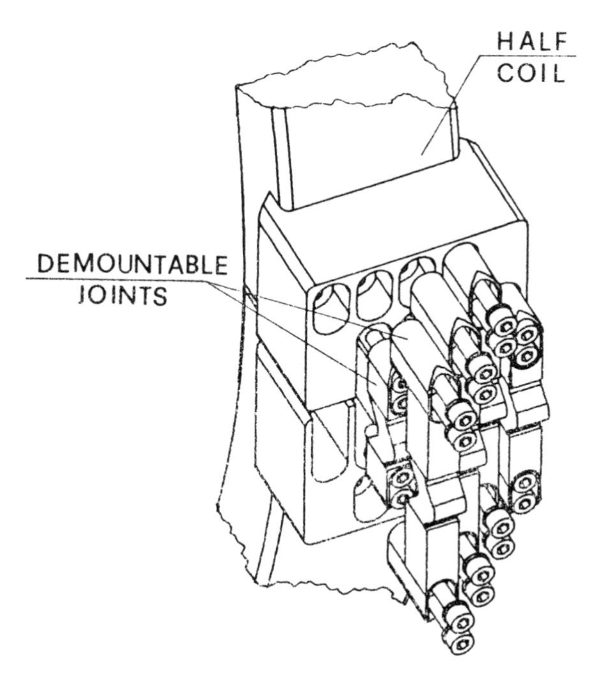

FIGURE 4
Expanded view of the connection between two half-coils.

and their peak value of temperature rise is always kept below 50°C.

With the required value of the filling coefficient no brazing process can be used to join the stack to the solid blocks: tests performed showed that the flexibility of the joint is fully jeopardized by capillarity effects of the melted alloy between the sheets. Electron-beam welding has been adopted instead: using this method, not only a good standard level of welding is achieved, but a reduced copper annealing is obtained as well. The electron-beam welding is performed in a vacuum vessel (pressure about 10^{-5} mbar) with an electron gun, maximum power 30 kW (60 kV - 500 mA). The pieces are fixed on a numerically controlled table having two axes of displacement. The displacement speed is adjusted to be 1 cm/s.

The demountable joints are obtained by cutting a long welded piece as shown in Figure 5. Electroerosion is used to avoid metal swarfs to be produced which not only would fill the space between the sheets, with consequent stiffening of the sheet stack, but also might be dangerous for the integrity of the insulation of the

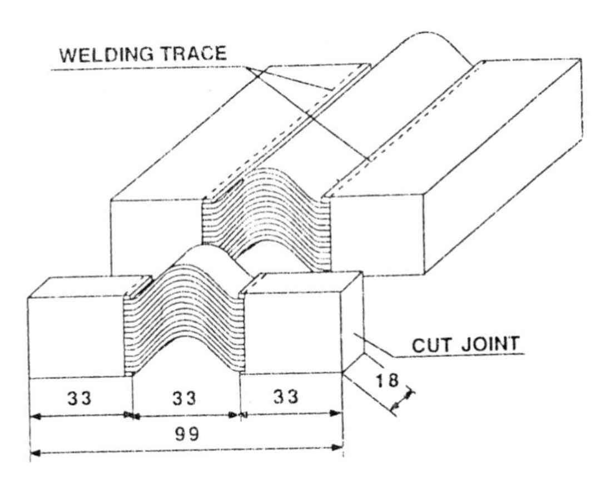

FIGURE 5
Cutting of the demountable joint from the long welded piece.

assembled equipment. Moreover, using electroerosion also prevents side cracks in the sheets which strongly would reduce their fatigue strength.

Since the breakdown of even a single demountable joint would have harmful consequences on the machine[3], the reliability of the components has to be checked carefully. Therefore, to do this, severe preliminary and routine tests have been planned. Several demountable joints selected among the production will be subjected to a pulsed electromagnetic force test: the joint will have to withstand current pulses in presence of a magnetic field. Prototypes of the joint will be subjected in addition to current pulses and mechanical fatigue tests based on impressed deformation. The values of the test levels, compared with

Tab. 3. Normal operating and test conditions for the demountable joints.

	Normal operation	Test
Electromagnetic Force (kN/m)	20	35
Thermal parameter i^2t (MA^2s)	300	600
Impressed deformation (mm)	1.1	1.7

the nominal working conditions, are shown in Tab. 3.

3.3. Clamping system

A clamping system keeps each coil fastened into its groove, to withstand the electromagnetic forces and to avoid any detatchment from the bottom of the groove itself [4].

The clamping system is composed by two belts, made of rolled plates of AISI 304 stainless steel, 8 mm thick (Fig. 6). These belts are connected at their outer end by two high-strength tie-rods (outer mechanical connections) and, at the inner end, by a demountable frame made of Inconel 625® (inner mechanical connections). Using this solution access to the demountable joints is made possible even with the coil assembled and fastened to the shell.

Two stacks of spring washers are inserted between the tie rods and the belts, to recover the decrease of pretension due to the thermal expansion of the belts.

Both the inner and the outer mechanical connections are provided with glass-epoxy insulating breaks, designed to withstand the maximum voltage per turn.

Since pretension is applied from the outer mechanical connections only, a too high friction coefficient between the belts and the coil would cause an untolerable decrease of pretension along the belts; then low friction layers are inserted in between the belts and the coils, which also are given the function of compensating for little geometrical mismatches.

Preliminary tests have been carried out with several types of materials: the most promising arrangement consists of overlapped layers of teflon® and polyurethane. However, due to the complexity of this problem, the final choice will be made according to the results of the tests made on the prototype coil.

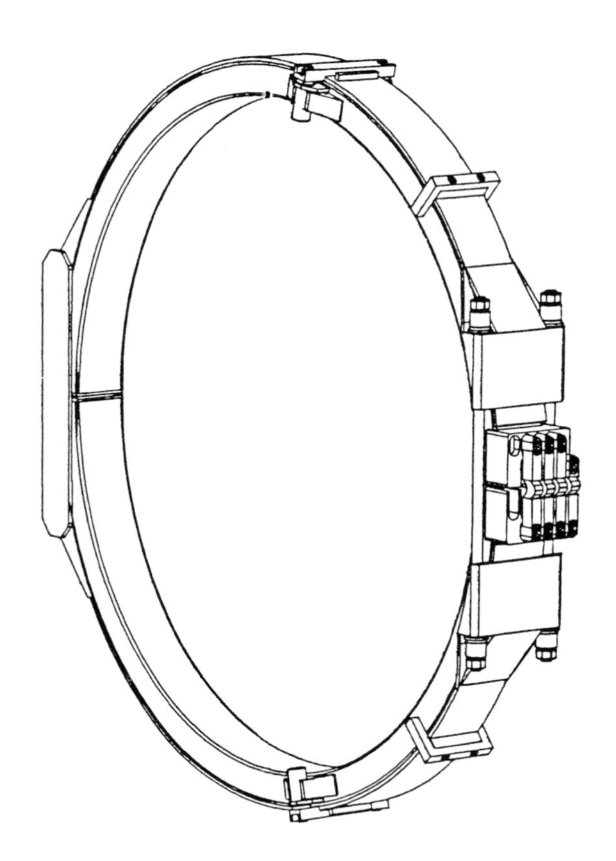

FIGURE 6
Drawing of a coil with its clamping system.

4. CONCLUSIONS

The operational flexibility pursued into the RFX Toroidal Field winding design is an important and qualifying feature of the machine. However, the penalty to be paid for such a flexibility is represented by some relatively new problems to be solved and a large number of additional components, such as the clamping belts and the demountable joints, which had to be developed and severely tested.

REFERENCES

1. G. Malesani, G. Rostagni, "The RFX Experiment", Proceedings of the 14th Symposium on Fusion Technology, Avignon, France, 1986, pp.173-184.

2. F. Bellina, M. Guarnieri, A. Stella, "The RFX Toroidal Field Winding Design", IEEE Transactions on Magnetics, vol.24, no.2, pp. 1252-1255, March 1988.

3. F. Bellina, P.P. Campostrini, G. Chitarin, M. Guarnieri, A. Stella, "Fault analysis and Protection of the RFX Toroidal Winding", Proceedings of the 12th Symposium on Fusion Engineering, Monterey, California (USA), 1987, pp. 315-318.

4. F. Bellina, P. Zaccaria, "Full 3D Stress Analyses of Complex Shaped Electrical Busbars Under High Electromagnetic Load", to be presented at the 15th MSC/Nastran Users' Conference, Rome, October 20-21, 1988.

FUSION TECHNOLOGY 1988
A.M. Van Ingen, A. Nijsen-Vis, H.T. Klippel (editors)

ASSESSMENT OF THE STRUCTURAL DEFORMATIONS CAUSED BY A POSTULATED LOCAL BREAK OF A TOROIDAL NET COIL

E. Wehner, R. Krieg, R. Meyder

Kernforschungszentrum Karlsruhe GmbH, Institut für Reaktorentwicklung, Postfach 3640, D-7500 Karlsruhe, Federal Republic of Germany

As a postulated accident of the magnetic coil system of NET the complete melting of a toroidal coil cross section is assumed while the current remains unchanged. Then under the existing electromagnetic load conditions the free coil ends will bend outward. In this paper the resulting bending deformations are calculated using the computer program EFFI to describe the electromagnetic forces and the program ABAQUS for the structural mechanics problem.

1. INTRODUCTION

1.1 Background

It is very unlikely that a local loss of the superconductivity and a resulting structural failure may propagate over the whole cross-section of a toroidal coil of NET, leading to a coil break. However, a reliable proof is rather difficult and not available, so far. Therefore in this paper a cut of a toroidal coil perpendicular to the winding pack is assumed and the resulting deformations are calculated.

On the basis of the results it should be possible to evaluate, whether important structures e.g. coolant pipes or other equipment, can be damaged. In addition it should also be possible to evaluate, whether plastic joints occur, which may be able to cause a rupture at another location of the coil circumference, leading eventually to a missile. If it turns out that the mechanical damage is quite limited possible difficulties of the proof that a failure propagation can be ruled out will not be a crucial problem for safety assessment.

1.2 Coil design

The overall layouts of the toroidal coils and of the poloidal coils within the NET machine are shown in the Figs. 1 and 2. Both sets of coils are enclosed within a common cryostat. Under operation conditions the temperature is given by the liquid helium (4 K).

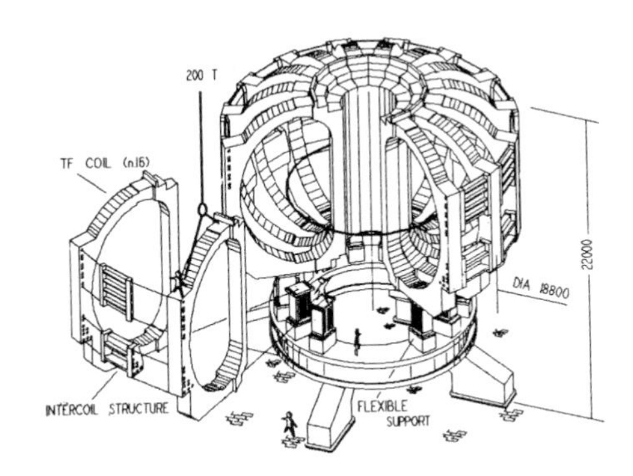

Fig. 1 : Toroidal coil system

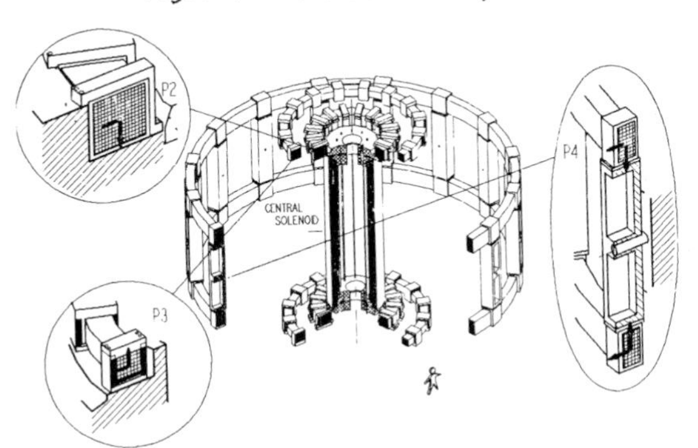

Fig. 2 : Poloidal coil system

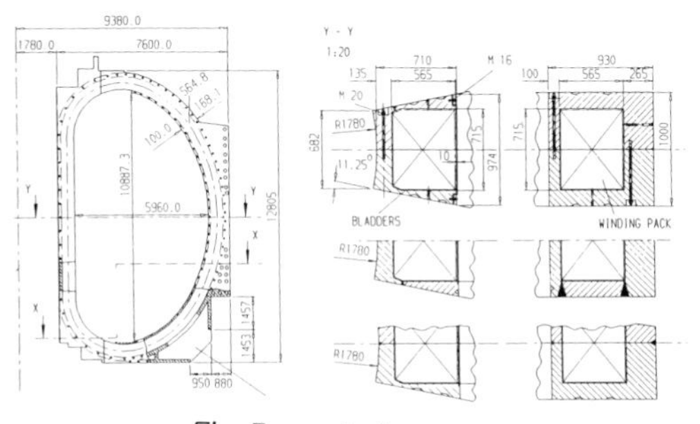

Fig. 3 : Coil case

The toroidal coils are individually supported against gravity by flexible multiplate supports which allow easy radial movement but are stiff in other directions. The magnetic forces are supported at the inside by a vault formed by the noses of the coils and at the outside by an intercoil structure in the region of the outer equator.

Each toroidal coil consists of a winding pack contained inside a stainless steel (316LN) case, Fig. 3. The case thickness varies around the coil perimeter.

2. MAGNETIC FORCES

It is assumed that one toroidal coil is cut over the total cross-section at one location of the perimeter. Nevertheless, the nominal current shall be maintained.

The loading on the toroidal coils is considered to be only due to magnetic forces, since gravitational forces are negligible compared with these.

The magnetic forces have two sources. Firstly, there are the loadings due to the self field of the toroidal coils. Summed over the winding pack cross-section, this leads to a bursting force confined by the toroidal coil and a centring force towards the machine axis. The toroidal coils will remain charged for long periods, and so these forces can be considered to be steady. Secondly, there are the fields of the poloidal coils. They produce force distributions normal to the planes of the toroidal coils. These forces undergo cyclic changes, since the poloidal coils are pulsed for each burn cycle of the machine. In this upper bound assessment it will be assumed that the maximum current occurs at the same time.

The magnetic forces were determined using the computer program EFFI [ref. 1] - a code for calculating the electromagnetic field, forces and inductances in coil systems of arbitrary geometry. The assumed currents of the poloidal and toroidal coils are given in Table I.

Table I : Assumed currents for the EFFI–calculation

name of the coil	current (A)
P1B (solenoid)	30.E+06
P1C (solenoid)	10.E+06
P1D (solenoid)	25.E+06
P3	12.E+06
P4	8.E+06
toroidal coils	8.E+06

The magnetic forces, calculated with EFFI and prepared for the structural analysis are shown in Fig. 4. The magnetic forces acting at the poloidal coils were not considered.

radial forces distributed over the coil perimeter (kN/mm)

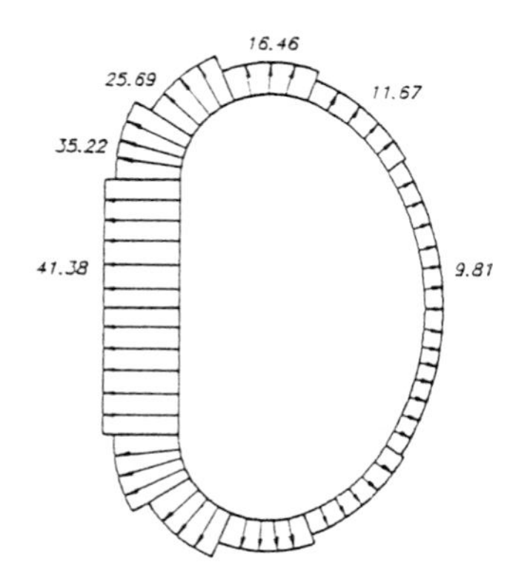

out of plane forces distributed over the coil perimeter (kN/mm)

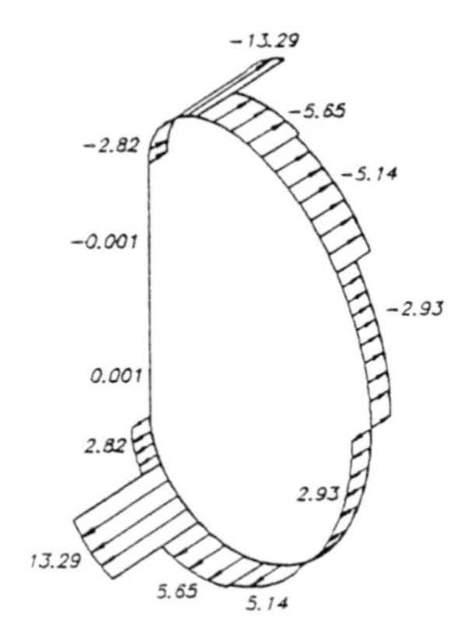

Fig. 4 : Coil loadings

3. STRUCTURAL ANALYSIS

The drawings of the toroidal coil are shown in Fig. 3. The finite element model used for the structural analysis includes seven toroidal coils, two poloidal coils and the intercoil structure in order to describe the interaction with the neighbouring coils. It is shown in Fig. 5. The coil 4 is assumed to be cut over the total cross-section at one location of the perimeter.

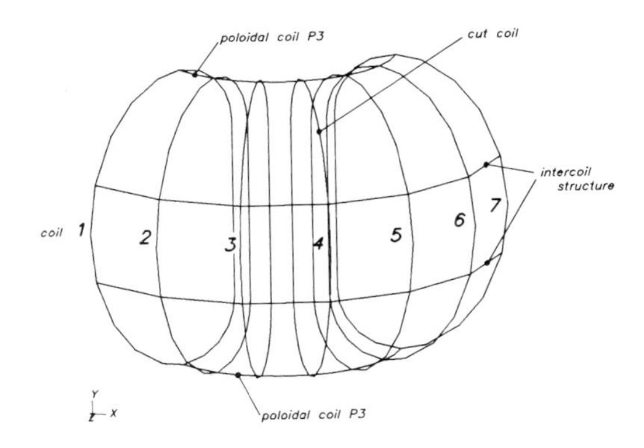

Fig.5 : Finite element model

The boundary conditions which are assumed for the coils 2-6 are shown in Fig. 6. For coil 1 and 7 rigid clamping was assumed at all nodes. The poloidal coils P3 are rigidly fixed to the toroidal coils.

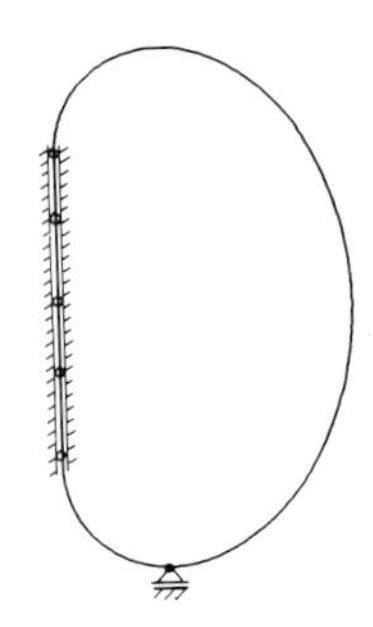

Fig.6 : Boundary conditions

The model was set up using the pre-processor FEMGEN [ref. 2], an interactive mesh generator. The calculations have been carried out with the finite element program ABAQUS [ref. 3]. Nonlinearities due to the elastic-plastic material behavior and large geometric changes (but small strains) were included. However, dynamic effects have not been considered. To display the results the interactive graphics post-processsor FEMVIEW [ref. 4] was used. It should be mentioned, however, that FEMVIEW is not able to plot curved elements, instead the nodes are interconnected by straight lines.

Based on the given geometry the cuts were assumed at five different locations as shown in Fig. 7.

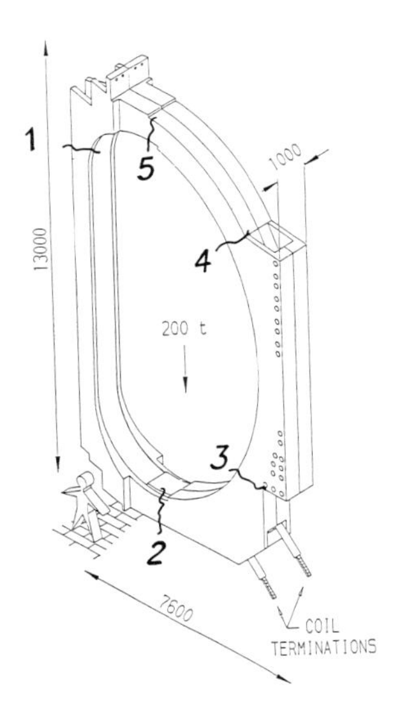

Fig.7 : Assumed cut locations

Because the stiffness of the coil case is much higher than the stiffness of the winding pack, in first calculations the winding pack of the toroidal coil was neglected. The stiffness of the winding pack of the upper poloidal coil, however, has an important influence. Since this coil has a complicated structure consisting of different materials which is difficult to model, a homogeneous ersatz model was used. In order to describe an upper and lower bounding case, first material properties of copper and then material properties of steel were assumed. For the coil cases the assumed stress-strain curve is based on the data of stainless steel AISI 316LN.

Table II : Summary of the results

location of the cut	1	2	3	4	4	4	4	5	5
toroidal coil winding pack	neglected	neglected	neglected	neglected	considered	neglected	neglected	neglected	considered
materail of the poloidal coil	copper	copper	copper	copper	copper	steel	copper with steel case	steel	copper
displace-ment of the cut coil Scaling factor=1	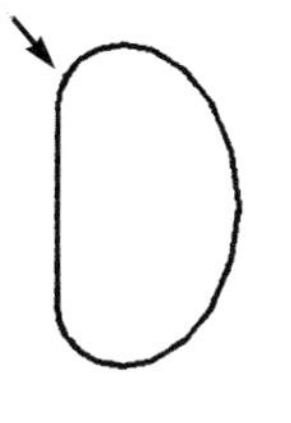	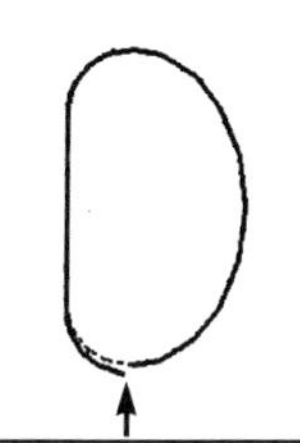	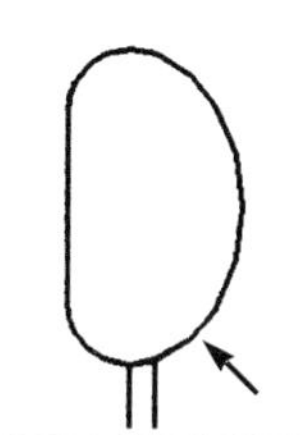	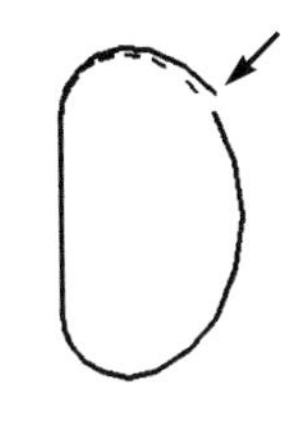	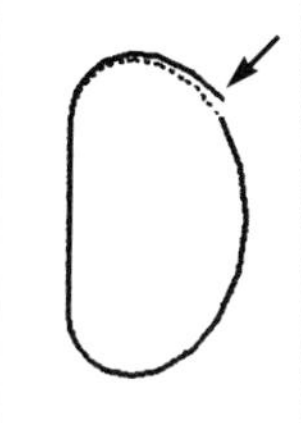	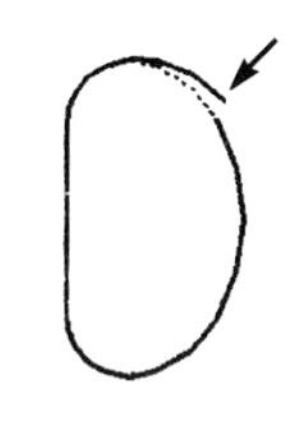	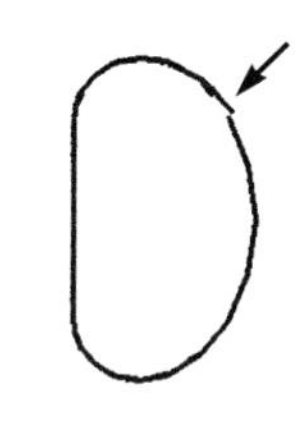	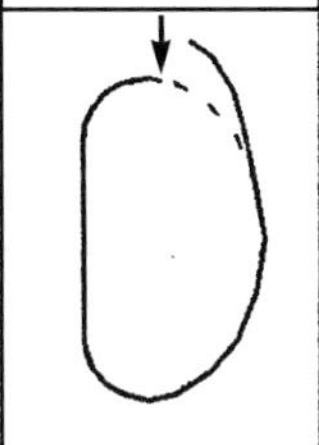	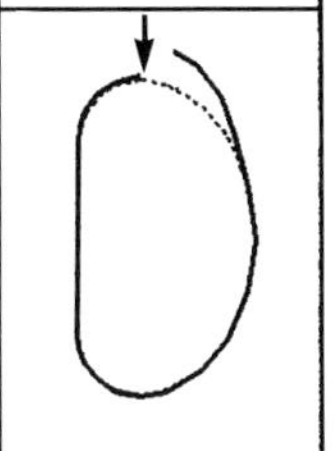
max. load (% of the nom.load)	100	89	100	32.5	38.1	100	100	100	100
max. displ. (mm)	109	454	70	695	708	749	265	2210	1660

4. RESULTS AND CONCLUSIONS

The results are summarized in Table II. It shows the deformations of the cut coil. As an example, for the fourth case (see fourth column in Table II) the calculated deformations of the complete model are given in Fig. 8. In some cases the load could be increased only to a certain percentage of the nominal value of Fig. 4 (see fifth line in Tab. II), because the calculations stopped due to numerical problems.

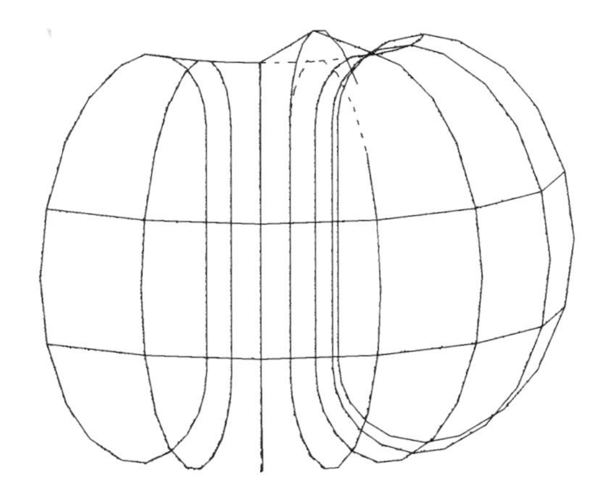

Fig.8 : Displaced structure (Scaling factor = 3)

Five different cut locations were analysed. The cut location 4 and 5 were found to be the most critical. For the cut location 4 the destructive consequences are considerably reduced, if the upper poloidal coil P3 is surrounded by a steel case. For a cut at location 5 the deformation can be reduced by a local stiffening of the toroidal coil (see Fig. 9).

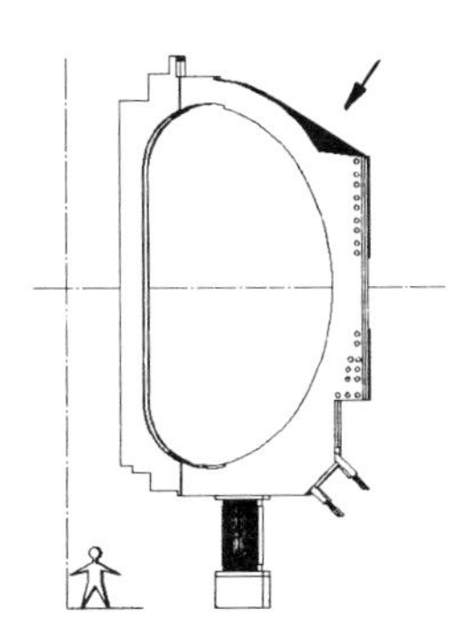

Fig.9 : Proposed design modification

The analysis with consideration of the toroidal coil winding pack shows that the stiffening influence of the winding increases the load carrying capacity by 10 % only.

Finally, some essential shortcomings should be mentioned. Long before a coil is cut, the control system intervenes and switches off the circuit. The current through this coil will then increase by a factor of 1.0 to 1.5. This provides an additional increasing of the load and of the deformations of the coil. Secondly, the load is not static but dynamic. This increases the deformations, too. As a consequence of the assumed accident, the coil will have a higher temperature, which reduces the stiffness and therefore again increases the resulting deformations. Therefore the presented results may allow only for a rough assessment.

Nevertheless, it may be concluded that even for a local break of the toroidal coil the resulting mechanical damage is limited. For more precise statements additonal analyses without the above shortcomings are required. As indicated before, these may be necessary if the failure propagation analysis turns out to be not sufficient.

REFERENCES

1. EFFI - User's Manual, Steven J. Sackett, University of California, Livermore

2. FEMGEN - A General Finite Element Mesh Generator 1984, IKO-Software Service GmbH, Albstadtweg 10, 7000 Stuttgart 80

3. ABAQUS - User Information Manual, Update to version 4.5, Hibbitt, Karlsson & Sorensen, Inc.

4. FEMVIEW - User Manual, IKO-Software Service GmbH, Albstadtweg 10, 7000 Stuttgart 80

5. N. Mitchell, Structural Assessment of the NET Toroidal Field Coils, EUR-FU/XII-80/87/69, March 1987

6. G. Malavasi, N. Mitchell, E. Salpietro, Progress in the NET Toroidal Field Coil Mechanical Design, 14th SOFT, Avignon, 1987

FUSION TECHNOLOGY 1988
A.M. Van Ingen, A. Nijsen-Vis, H.T. Klippel (editors)

STRUCTURAL ANALYSIS OF THE NET TOROIDAL FIELD COILS AND CONDUCTOR

N. MITCHELL[1], D. COLLIER[1], R. GORI [2]

[1]The NET Team, c/o Max-Planck-Institut fuer Plasmaphysik, Boltzmannstr. 2, D - 8046 Garching b. Muenchen

[2]Istituto die Scienza e Tecnica delle Costruzioni, Padova University, Italy

The NET toroidal field coils will utilise A15-type superconductor at 4.2 K to generate fields up to 11.5 T. The superconductor strands themselves are sensitive to strain, which causes degradation of their current carrying capacity, and thus the detailed behaviour of the coil conductor must be analysed so that the strain can be minimised. This analysis must include the manufacturing processes of the conductor as well as the normal and abnormal operational loads. The conductor will be insulated and bonded by glass fibre reinforced epoxy resin, with limited bonding shear strength, and the overall support of the complete coil system must be designed to reduce these shear stresses, The coils will be subjected to pulse loads form the poloidal field coils, and analysis of the slip between the various coil components, such as conductors and the coil case, giving rise to frictional heating and possible loss of superconducting properties is another important factor, which has been investigated by a number of stress analyses. The manufacturing, thermal and normal magnetic loads on the coils and the analysis leading to the proposed structural design are described. In addition to the normal operating conditions, there is a range of abnormal load conditions which could result from electrical or mechanical faults on the coils. The effect of these potential faults has been analysed and the coil design modified to prevent catastrophic structural failure.

1. INTRODUCTION

The layout and design of the toroidal field coils of the NET machine are reported in a separate paper (1). Broadly they consist of sixteen D shaped coils arranged toroidally. The coils are superconducting, and consist of a winding pack and steel case. The winding pack consists of turns of the superconductor, stabiliser and liquid helium coolant contained within a steel jacket and insulated/bonded with vacuum impregnated glass fibre reinforced epoxy.

The derivation of this coil design from the structural viewpoint has already been reported (2,3). The current set of analyses can be considered as the third iteration in the development of the design. The first analysis used a finite element analysis with a simple beam and shell model of the coils to look at a wide range of layouts and support methods and select a rough optimum. The second stage, performed after initial layout studies, modelled the coils and conductor down to the fine details of conductor and superconducting strands (4) and confirmed that the range of design criteria relating to the coil case, insulation and superconductor (5) were satisfied. Some detailed layout optimisation and verification of the finite element model was also performed.

This third iteration is being performed after the completion of design and feasibility studies of the coils (6,7). It can thus look at some of the critical issues coming out of these studies, at the range of operating loads that the coils may experience, possible fault conditions and manufacturing/assembly problems. These studies generally use finite element stress analysis (8,9) of the coils or their subcomponents with a wide range of different models, together with magnetostatic calculations to determine the loads (9,10). The TF coil layout is shown in figure 1. The coils are formed together into a central vault which carries the centering forces. The friction between coils in the vault is reinforced by shear keys at the top and bottom of the inner leg. A shear box consisting of intercoil structure and a reinforced coil leg supports the coils in the region of the outer equator and also acts as a belt constraint against bursting forces in the outer coil leg.

2. TF COIL UNDER NORMAL OPERATING CONDITIONS

2.1 Model

The finite element model of the coil is shown in figure 1 (11). The model falls into three parts. The coil case, including the central vault and a part of the outer intercoil structure, is modelled using 20 and 16 noded isoparametric isotropic elastic elements. The outer intercoil structure, for the part that connects individual coils, is represented by 3 noded triangular plate/shell bending elements. Each node has 6 freedoms and both bending and translation are linked to the main coil. The winding pack is represented with 20 noded orthotropic brick elements where the material axes are aligned with the local direction of the winding pack.

The coil is part of a sixteen-coil toroidal assembly and thus 16 fold symmetry conditions are applied at the boundary of the outer intercoil structure and within the central vault.

The internal magnetic loading on the coil is generated from the Hedo program (10). The resultant loading conditions for the out of plane forces are shown in figure 3. Additional loads on the coil arise from the outer poloidal field coils which are supported on the toroidal field coil and fully constrained by it. These coils cause additional vertical and radial loads on the toroidal field coil above and below the central vault.

The coil is supported by a leg, flexible only in the radial direction, under the lowest part of the coil. This support is a complex plate structure and exerts some restraint on the lower part of the coil. It is represented here by beam elements whose properties, equivalent to the actual support, are derived by detailed finite element analyses in (11).

To look at the effect of separation between coil case and winding pack, the original model with case and winding rigidly joined was changed to allow slip and gaps at the interface to be shown (12).

It is desirable to model the effect of frictional slip at the surface by limiting the shear stress to the value (friction coefficient) x (normal pressure) but this is not possible with the ADINA program (8) used for the analysis. It is possible to define either zero friction or infinite friction.

ADINA can be used to model the winding pack/case interaction in two ways. In the first the two interacting surfaces are covered with target and contactor contact surfaces. In the second, the winding pack outer surface is shrunk by an arbitrary 5 mm and all corresponding case/winding pack nodes are linked by 2 noded truss "gap" elements. These have a very high stiffness in compression but zero strength in tension. The elements are generally aligned normal to the winding pack surface, and as they have no bending strength, the winding pack is free to slip relative to the case. The contact surfaces allow a large displacment analysis to be performed, whereas the truss gap elements are limited to a small displacement calculation (since the truss

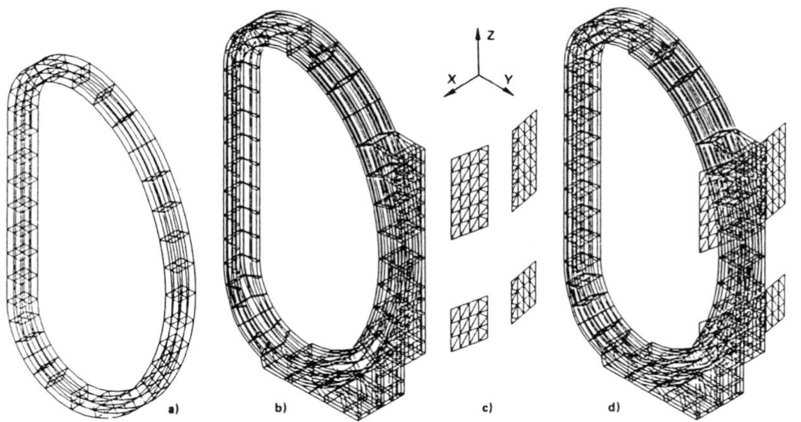

Fig.1. The Finite Element Model. a) Winding package; b) Steel coil case; c) outer intercoil structure; d) Complete model.

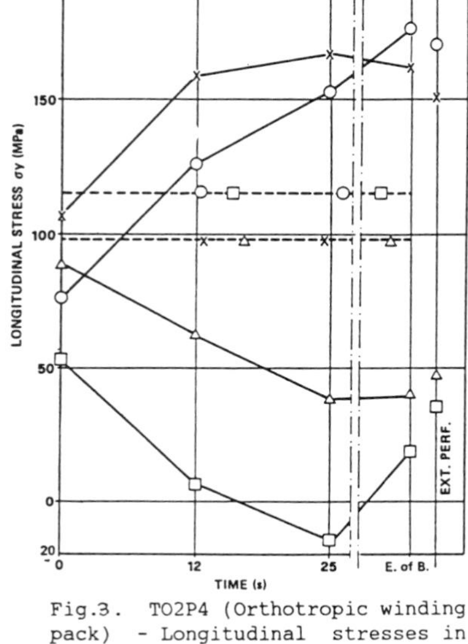

Fig.3. TO2P4 (Orthotropic winding pack) - Longitudinal stresses in case at group 1 during burn and with extended performance.

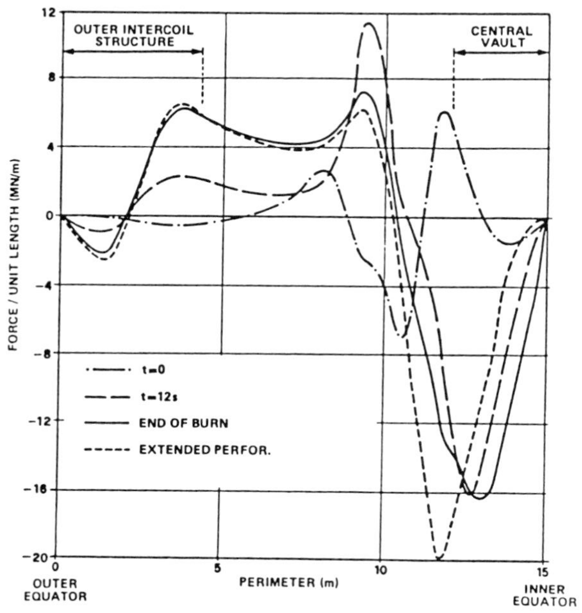

Fig.2. Out of plane forces on a toroidal field coil.

elements must keep their original alignment).

For the large displacement contact surface problem, the loads were imposed in 3 steps, since in general the coil stresses are load path dependent. However comparison with analyses where the loads were applied in a single step showed that for this case (with zero or infinite friction) the effect is negligible.

For gap problems, the preloading due to either precompression from the case or differential thermal contraction can be very significant. Here, we assume no preloading and no differential contraction.

2. RESULTS

The full presentation of the results from this analysis requires considerably more space than is available here. We therefore attempt to pick out some of the main features and summarize the peak stress levels. For the model with rigid bonding between case and winding, figure 3 shows the longitudinal stresses at the top part of the coil (where it is unsupported by intercoil structure). The build up of bending stresses due to out of plane forces is clear. Figure 4 shows the displaced shape of the coil at the end of burn condition. The peak stresses for this model are as follows:

Peak tensile stresses in:

1) Coil case outside central vault	150-250 MPa
2) Inside the central vault	250 MPa
Von Mises stress	460 MPa

in vault	
Longitudande tensile strain on winding	0.15 %
Peak shear stress in winding (across insulation)	25 MPa
Minimum friction coefficient between coils in vault	0.22

The friction coefficient between coils in the vault is rather high. To reduce this and to prevent any loss of vault rigidity, the vault has been extended at the top and bottom with the coils joined by bolts/shear keys through flanges, accessible from above and below (1). These keys provide overall strength in the event of non-symmetric coil currents arising from fault conditions. They can also be used to precompress the vault structure.

Examination of the winding pack surface stresses shows substantial areas with either normal tension or high shear. The level of these is such that separation or slip of the winding is rather difficult to prevent. It is better to allow such slip and provide a low friction surface to accomodate it.

When slip/gaps between case and winding are included, there are substantial modifications to the stress patterns in the outer part of the coil. The overall maximum displacement of the coil, for the rigid, contact surface and truss models are shown in table 1 for end of burn. Comparing the contact surface models with the original rigid winding/case model, it is clear that the out of the plane deflection has been substantially decreased (by 4 mm), as has the maximum upward motion (by 3 mm). The sum of the inward and outward (y direction) motions is however substantially larger in the contact surface model (by up to 7 mm). This is an artificial effect due to the limited number of contact surfaces that are used.

These changes are in line with what may be expected. The contact surface model has a case-

TABLE 1

PEAK COIL DISPLACEMENTS AT END OF BURN

	Displacement (mm)			
	Rigid coil/WP	Contact Surfaces Zero friction	Contact Surfaces Infinite friction	Truss gap elements
Out of plane (x direction) at	-26.6 x = 450 y = 7117.8 z = 5244.6	-16.5 x = 357 y = 7754.3 z = 4384.2	-16.3	-16.2 x = 352 y = 7750 z = 4381
(note: in + x direction value is 0 due to support position)				
Upward (z direction) at	9.1 x = 450 y = 6639 z = 5461.4	6.0 x = 0 y = 6639 z = 5461.4	5.2	7.9 x = 352 y = 1943 z = 3876
Outward	4.88 x = 500 y = 9179 z = -3493	6.4 x = 357 y = 5921 z = 5236	6.1 x = 357 y = 7107 z = 4500	9.4 x = 352 y = 5924 z = 5240
Inward (- y direction) at	- 3.3 x = 493 y = 2480 z = -440.7	- 9.2 x = 357 y = 2198 z = -2203	- 5.6 x = 357 y = 2198 z = 1322	- 3.6 x = 352 x = 2475 z = 2203

winding pack assembly that has a reduced stiffness to bending. The coil outside the central vault has an initial out of plane rotation imposed by the torsion of the central vault which is reduced by the out of plane forces. With a more flexible coil, these forces correct the initial rotation over a shorter length of coil and the maximum out of plane displacement is reduced.

The reduction in the upward displacement is a combination of the out of plane effect (since this out of plane movement has also a torsional component along th conductor direction) with a change in coil in-plane behaviour (i.e. due to the TF coil self field forces). The coil has become less stiff, which allows the outer intercoil structure to move further outward. The coil radial distortion increases and the vertical distortion decreases.

Fig. 5 shows where gaps occur between the contact surfaces on the case and on the winding pack. At the bottom of the coil there is a gap in the outer part of the case for the current leads and the winding pack shows a tendency to

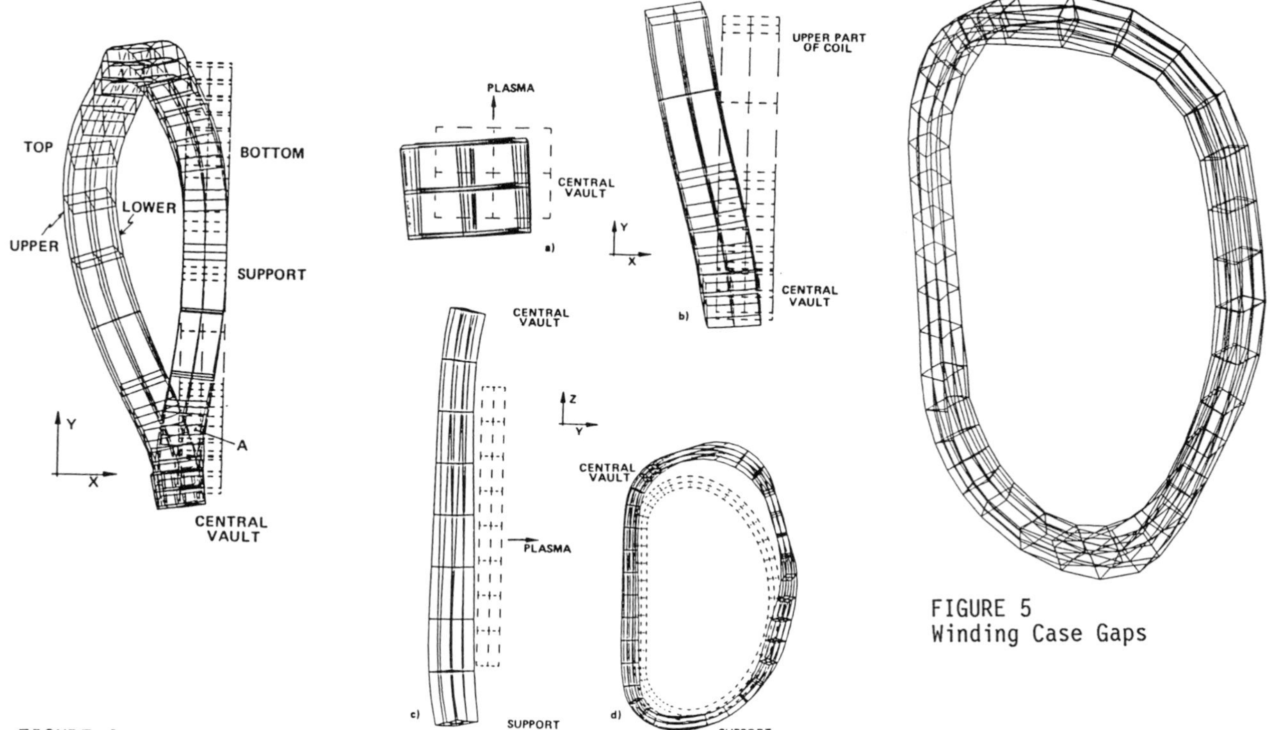

FIGURE 4
Displacement Plots

FIGURE 5
Winding Case Gaps

"bulge out" through this. Elsewhere, the winding pack is pressed out against the outside of the case as would be expected from the magnetic bursting forces. There is a gap almost everywhere in the inner bore. Inside the central vault the transverse loading keeps the sides of the case in contract with the winding pack, but outside the vault gaps appear at the sides.

The shear stresses within the winding are substantially relieved by the slip/gaps. The peak value drops from 25 MPa to 17 MPa. This is little affected by the friction coefficient.

The reason for the reduction seems to be a tendancy for the winding pack magnetic forces to be transmitted by direct pressure rather than shear due to the gaps at the sides. The peak case stresses are not significantly changed by the slip/gaps, although the distribution is altered. It is also possible to assess frictional heating between case and winding, significant for superconducting coils. The maximum heat rate appears to be about 100 - 200 w/coil averaged over a burn cycle, and the actual value is likely to be much less than this.

In conclusion it can be noted that any detailed assessment of the TF coil stresses must take account of the possiblity of slip/gaps between case and winding.

3. THERMAL COOLDOWN STRESSES IN THE TF COIL

Due to the composite nature of the winding pack, it has a different thermal contraction to the case over the temperature range 300- 4 K. The steel case contracts by 0.306 %, and the winding by 0.284 % in the longitudinal direction and 0.345 % in the two transverse directions. This thermal contraction has been modelled with the winding-case model of the previous section, with rigid bonding between the two using the ANSYS code (9). The deflections on cooldown (taking the coil support as fixed) are:

-Inward movement of vault 5 mm.

-Vertical hight reduction of top of coil 46 mm.

-Inward movement of outer internal structure 28 mm.

The winding pack tends to push out on the case and pull away at the sides. The winding pack tensile stresses of up to 20 MPa normal to the case suggest that a gap will open.

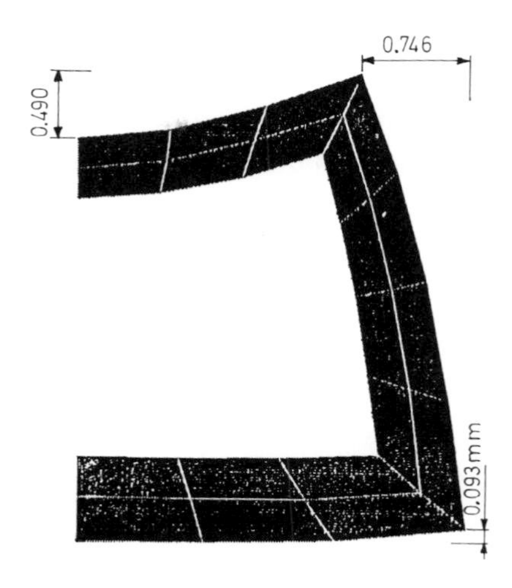

FIGURE 7
Distorted Shape 0.5 m Radius

4. BENDING OF THE CONDUCTOR JACKET DURING WINDING

During the winding of the coil, the conductor is plastically deformed from a straight section to the D shape of the coil. This causes a deformation of the conductor crossection (keystoning) which needs to be controlled as it can cause non uniform areas in the insulation. A nominal limit of 0.1 mm deformation is taken as acceptable. For the 40 KA wind and react conductor (6), which has the largest crossection, an elastic-plastic finite element analysis has been performed (13).

The model shown in figure 6 using the ANSYS code (9) consists of a 0.2 m long half section of the conductor jacket. Symmetry conditions are applied at the half section bondary. The end conditions needed to bend the jacket length into a constant radius needed considerable investigation in order to avoid spurious end effects. The final condition selected forced one end section to remain plane while imposing a rigid rotation on the other. The model of the conductor bending thus neglects:

a) Any surface pressure deformation if the conductor is bent by pulling around a curved surface

b) Any work hardening or deformation due to intermediate storage on a drum

Both of these effects may be expected to decrease the keystoning deformation.

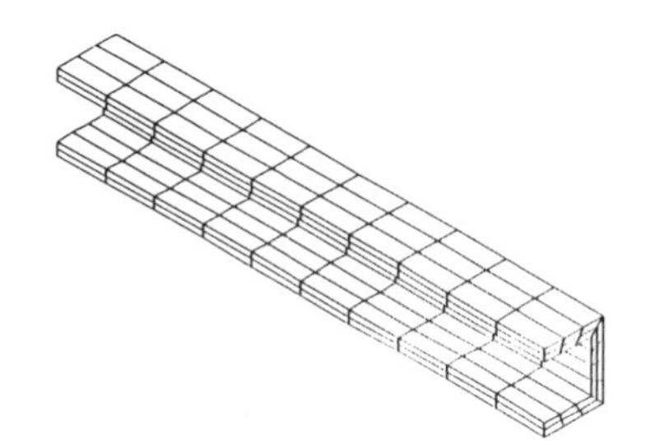

FIGURE 6
Conductor Jacket Model

The conductor has been deformed from a straight section to a 0.5 m radius, with intermediate resutls at 1.0 m radius. Figure 1 shows the deformed shape at 0.5 m radius. The lower, convex outward surface is that which is in longitudinal compression. The maximum section deformations are 0.75 mm for a 0.5 m radius and 0.33 mm for a 1.0 m radius. For the TF coil, with a minimum bending radius of 2.5 m, the deformation will be slightly larger than the nominal acceptable limit. However when the effect of tension during winding is added, it is expected that the deformation will be acceptable.

5. FAULT ANALYSIS

Under electrical short conditions, non uniform currents can arise in the coils system. As an example, the case with zero current in one TF coil has been analysed using the finite element model shown in fig. 8. This configuration causes high sideways forces on the coils adjacent to the failed one (226 MN compared to 260 MN centring force). Winding pack shear stresses, case bending stresses and intercoil structure stresses in the outer part of the coil increase, but not to critical values (60 MPa, 400 MPa, 150 MPa respectively).

Most of the sideways forces are carried inside the central vault and by the support leg. Further examination of the results is required but it appears that the stress levels are within the ultimate limits for the material for the critical adjacent coils.

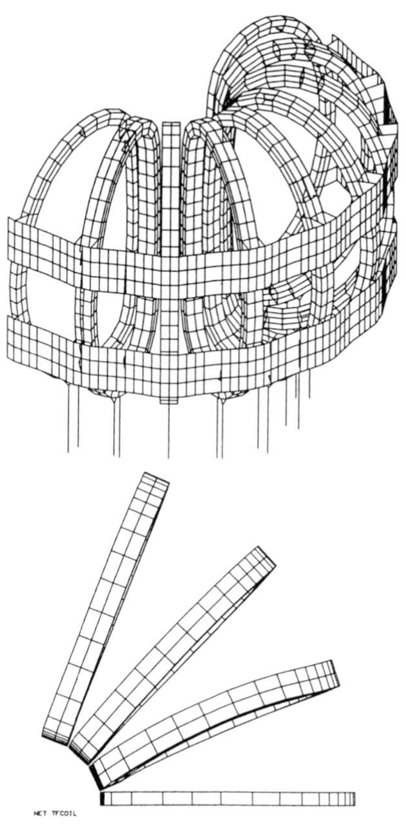

FIGURE 8
Coil Model and Displaced Shape

5. CONCLUSIONS

The results from the various structural analyses of the toroidal field coils confirm the overall structural feasibility of the proposed design under both normal and faulted conditions. More detailed analyses are underway to look more thoroughly at the effects of case-winding slip and with realistic friction coefficients and to consider in more detail the impact of fault conditions, giving rise to non-symmetric coil currents.

REFERENCES

1. R. Poehlchen et. al., Design Status of the NET Toroidal Coils, Paper presented at 15th SOFT, Utrecht, 1988
2. NET Status Report, Dec. 1987, Eur-FU/XII-80/88-84
3. NET Status Report, Dec. 1985, Eur-FU/XII/80/86-51
4. N. Mitchell (editor), Structural Assessment of the NET Toroidal Fiedl Coils, March 1987, EUR-FU/XII-80/87-69
5. N. Mitchell, Mechanical and Electrical Design Criteria for the ITER Magnets, 9/6/88, ITER Internal Report ITER/IL/MG/1/8/3
6. Final Report on a Fabricability Study of the NET Toroidal Coil System, May 1988, ABB Zuerich Report No. HISM 20376
7. Final Report on a Fabricability Study of the NET Toroidal Coil System, Aug. 1988 ANSALDO, Genova
8. ADINA Users Manual, Adina Engineering Inc., 1984
9. ANSYS User Manual 4.3, Swanson Analysis Systems Inc.
10. P. Martin, H. Preis, Programmbeschreibung und Benutzeranleitung zum Magnetfeld-Computer Programm, HEDO 2, Max-Planck-Institut fuer Plasmaphysik, Rep. No. 34, 1977
11. R. Gori, N. Mitchell et al., Stress Analysis of a NET Toroidal Field Coil Under Normal Operating Conditions, Part I, NET Internal Report NET/IN/87-07
12. N. Mitchell, H. Gorenflo, R. Gori, Stress Analysis of a NET Toroidal Field Coil Under Normal Operating Conditions Part II, NET Internal Report, NET/IN/87-055
13. B. Schrefler, P. Zaccaria, R.Gori, Bending of the NET Cable in Conduit-Conductor During Coil Winding, University of Padua, NET Contract Report Aug. 1988

FUSION TECHNOLOGY 1988
A.M. Van Ingen, A. Nijsen-Vis, H.T. Klippel (editors)

RESULTS OF THE STRAIN MEASUREMENTS FOR THE SWISS LCT-COIL

J.A. Zichy, B. Jakob, C. Sborchia, G. Vécsey
Paul-Scherrer-Institute (PSI), CH-5234 Villigen, Switzerland

The Swiss LCT-coil was one of the six toroidal field magnets tested over a period of 22 months in the International Fusion Superconducting Magnet Test Facility (IFSMTF) at Oak Ridge National Laboratory (ORNL) USA. Each of the D-shaped 2.5 × 3.5 m·m bore superconducting magnets achieved 9 T in the full-array configuration. The coil tests included single-coil and full-array experiments with and without pulsed ac-fields, tests in both torus configurations above rated current and runs with extreme out-of-plane loads. This paper presents the systematized results of the strain measurements obtained during the whole test period. Comparison with calculations, made in the design phase of the coil, are also included.

1. INTRODUCTION

The Large Coil Task (LCT) was a collaboration between the US, EURATOM (EU), Japan (JA) and Switzerland (CH) under the auspices of the International Energy Agency (IEA), to fabricate and test six D-shaped superconducting (SC) toroidal field (TF) coils of 8T field [1]. The test facility IFSMTF at ORNL was built by the US as were three TF magnets, General Electric (GE), General Dynamics (GD) and Westinghouse (WH). Each of the other participants contributed one coil. Three coils were bath cooled with boiling liquid helium and the other three were forced-flow cooled with supercritical helium. Five magnets used NbTi and one US coil, made by WH, Nb_3Sn as SC material. The design and fabrication of the CH coil was carried out by Brown Boveri and Cie., Ltd. (BBC Baden, Switzerland) in close cooperation with and under the management of the Swiss Institute for Nuclear Research (SIN). The testing of the Swiss coil was the responsibility of SIN, performed jointly with the LCT staff and the other foreign participants. In the meantime both BBC and SIN merged with others. The former is now called "ASEA Brown Boveri" (ABB Zurich-Oerlikon, Switzerland) and the latter is known as "Paul Scherrer Institute" (PSI).

2. THE SWISS LCT-COIL

The conductor was a fully transposed copper stabilized three stage square cable, made from 17 μm thick NbTi filaments and filled with high strength CdZnAg solder. The supercritical helium circulated at pressures higher than 1 MPa in a seamless copper tube. The winding consisted of 11 double pancakes each subdivided in 2 pies. The 22 pies were in series electrically and parallel circuited hydraulically. The conductor was wrapped with fiber glass tapes. The winding was vacuum impregnated with epoxy and cured at elevated temperature. The box-type bolted case was in five parts subdivided; inner and outer ring, two side plates and a connection box on the top. Hot rolled plates and forged blocks were used for the structural components and sheet metal for the connection box, all made of 316N/316LN stainless steel [2]. The design was based on extensive Finite Element Method (FEM) calculations [3].

The separate cooling circuits of the winding and case each had a set of sensors rendering a continuous calculation of the He-mass flow rate and heat loss for both circuits [4,5]. Five temperature compensated strain gauges were installed on the case at strategic locations, see Fig. 1. The ac signal conditioning drifted slowly, therefore only the relative strain was reproducible. The magnetic induction on the horizontal midplane in the straight leg was used as reference field.

3. MOUNTING OF THE COILS

The coils were installed in a cylindrical stainless steel vacuum vessel, of about 11 m diameter and 10 m height, and attached to a hexagonal bucking post on its axis. The coils were set 60° apart, clockwise in the following sequence; GE, GD, JA, CH, EU and WH. A key on the outer ring of the straight section and machined areas on top and bottom of both side plates served as mating surfaces between coil and supporting structure. The coil key fitted into a vertical groove on the bucking post and the upper and lower collars held the six coils. The coils were supported by torque rings against out-of-plane forces clamping together their outer corners at top and bottom [6].

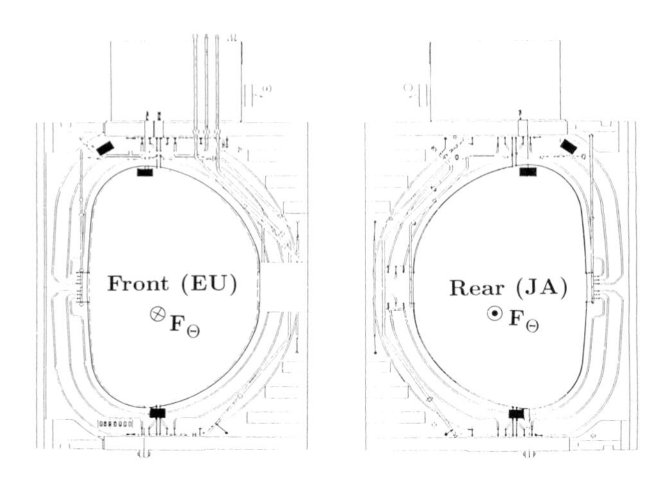

FIGURE 1
Schematic view of the Swiss coil case with the location and direction of the strain gauges and of the out-of-plane load marked.

To simulate the ac losses caused by vertical transient fields, a liquid nitrogen cooled copper Pulse Coil, movable on a circular track in the bore of the coils, was used [7]. The trapezoidal pulses with a repetition rate of 150 s had a linear rise (1 s), variable length (15 s used) and an exponential fall off (time constant 1.2 s). The maximum field created at the reference point of the coil was 0.14 T.

4. SINGLE-COIL TESTS

In the first series of single-coil tests each magnet was energized up to its rated current to demonstrate stable operation, and to verify the performance of the facility. In a second test series the coils were charged to higher currents to verify the safety margins applied in the design. The CH coil reached its rated current of 13 kA in a self field of 5.77 T without training in the first test series. In the second series three runs were needed to energize and operate the coil stably at 120% current in a self field of 6.92 T. The stored energy was 86.7 and 121.5 MJ in the first and second test series, respectively.

The strain gauge in the upper apex of the inner ring reached 450 $\mu\epsilon$ at maximum current in both test series. The two symmetrically arranged pairs of gauges on the D-shaped side plates read, in the first test series, strains up to 100 and 300 $\mu\epsilon$, respectively. The slope of the strain versus current squared was for these sensors nearly constant in the first runs. The smallest strain was measured where the curvature of the winding is smallest, verifying the sound design of the structure. In the second test series, the two sensor pairs still measured a modest strain of 150 and 425 $\mu\epsilon$, respectively, as shown in Fig. 2. The strains near the lower apex displayed the expected linear increase, however at the smallest curvature of the winding three distinct slope regions were seen.

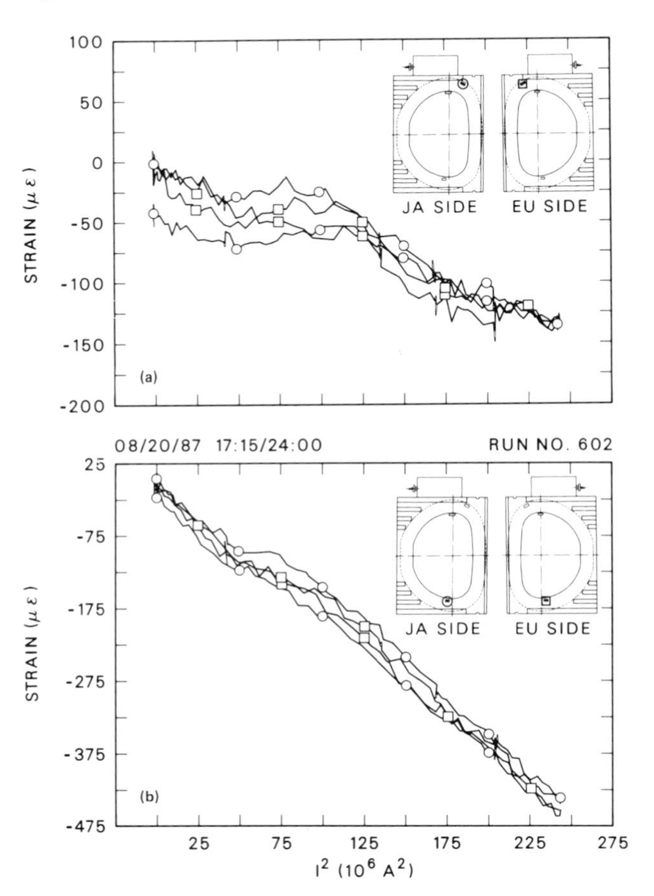

FIGURE 2
Strain versus current squared on the symmetrically arranged gauges on the side plates of the coil in a single-coil test above rated current.

5. FULL-ARRAY TESTS

In the full-array test each coil demonstrated its performance at rated current in an 8 T field on the conductor. The actual current setting was chosen such that the out-of-plane force on the CH coil was only 0.2 MN at rated current. The stored energy rose to 114.4 MJ in the CH coil and reached 510.7 MJ in the torus, the smallest amount in any of the full-array tests. To ensure that the load on the coils rises smoothly with current, the current ratios in the six power supplies were kept constant, while ramping the torus. Except for the WH coil, the current ratio was only asymptotically approached. This resulted in a discontinuity of the out-of-plane force at each hold during a charge up.

In Fig. 3 is plotted the strain versus current squared for the gauge in the upper apex of the inner ring, measured during a single-coil and a full-array run. The strain in the inner ring was in the full-array runs about 30% smaller than in the single-coil runs. The slope changes in both tests at about 60% of the rated current. The second change, at about 79% current, is not observed in the full-array run. This is explained by the centering force, which changes the load distribution on the inner ring in a full-array run in a different way as in a single-coil test. Since the out-of-plane load was very small, the corresponding pairs of strain gauges on the side plates displayed identical readings for the whole ramping cycle. The strain was the same in all runs, and for rated current reached at the smallest curvature of the winding 75 $\mu\epsilon$ and at the lower apex 200 $\mu\epsilon$.

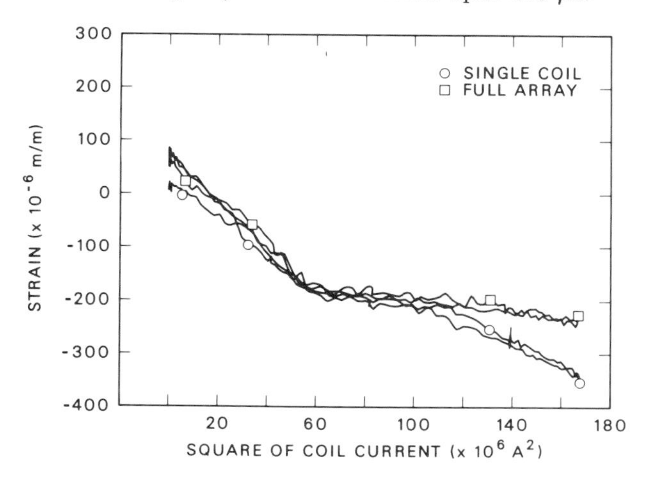

FIGURE 3
Comparison of strains measured on the inner ring in the upper apex, in single coil and full-array torus configuration, plotted versus current squared

6. EXTENDED CONDITION TESTS

In the High Field test five coils were energized to maximal 90% and the test coil to about 110% current. In the Symmetric and Maximum Torus tests the coils were energized to create at their reference point 8.5 T and at least 9 T field, respectively.

For its High Field test the setting of the CH coil was chosen to reach 8.8 T at 110% current and to limit the out-of-plane force on the coil to 0.8 MN. The torus was successfully energized creating a peak field of 8.96 T on the conductor. However, about 4 minutes later, while the current in some other coils was still settling, a quench occurred. It was preceded by a sudden flux change, triggered most probably by a sudden movement of the torus detected in all coils. The flux change in the coils started earlier than the normal zone in the CH coil. The strain on the five gauges was approximately the same as in the full-array tests, however, the signals on the corresponding symmetric gauges did not coincide. On two gauges attached to the torque rings the strain versus current squared displayed changes in slope at the same current as the two gauges near the upper apex on the side plates of the coil, see Fig. 4. The break in the slope of the strain data reflects the change of the out-of-plane force transmitted by the torque rings.

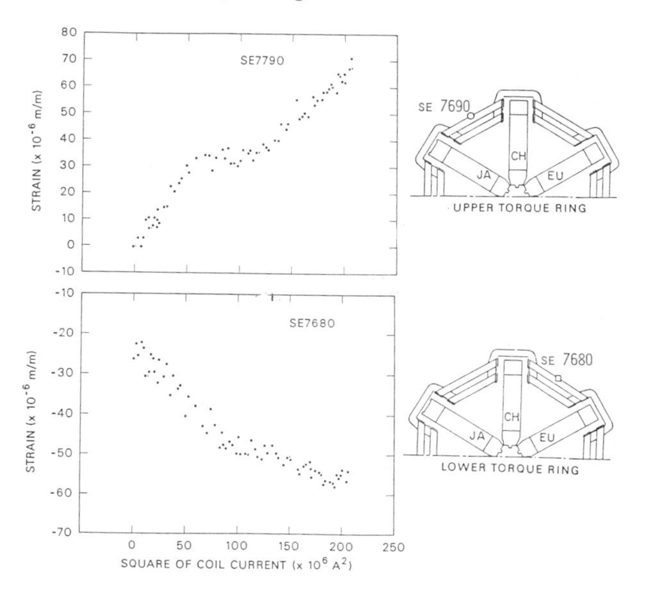

FIGURE 4
Strain versus current squared on the torque rings of the test facility during the High Field test.

In the Symmetric Torus test the strain was on all gauges approximately the same as in the High Field test. In the Maximum Torus test the signals of the two gauges on the JA side of the coil were nearly linear versus current squared and the gauge at the lower apex reached 300 $\mu\epsilon$. The other two gauges on the EU side displayed the usual pattern with three slope regions. The asymmetric reading of the symmetrically arranged sensors on the side plates originated probably from the weak clamping of the coil by the supporting structure, since the out-of-plane force was only - 1.1 MN, therefore not high enough to deformate the side plates. During the maximum torus test some large voltage spikes were detected again simultaneously on several coils, created by movements of the whole torus.

7. OUT-OF-PLANE LOAD TESTS

In these tests, with only five coils energized, the mechanical strength of the case in a failure mode of the torus was checked, since the out-of-pane load was high on the two coils next to the idle coil. The CH coil was tested in three different runs, two times up to 50% of the full load with either the JA or EU coil idle and up to 80% load with the EU coil turned off. In the first two tests 70.7% of the rated current, corresponding to 50% of the design force, was obtained, creating at first a negative then a positive out-of-plane load of 11.6 MN. Since in the two consecutive runs the direction of the out-of-plane force changed without affecting the coil performance, these tests proved that the winding was well fixed in the case. In the third run up to 89.5% current, 80.6% of the maximum out-of-plane force was reached, loading the coil with 18.6 MN.

The strain on the upper apex of the inner ring had a similar form versus current squared as in the other full-array test, with a break in the slope at 54% current. The maximum strain reached only 200 $\mu\epsilon$ in the third test. The strain of the symmetrically arranged gauges on the side plates displayed the expected antisymmetric signals, see Fig. 5. As shown on the left side of this figure the strain in the second and third run agreed well. The measured and calculated strain did agree reasonably well at low current on the inner ring and on both side plates near the bucking post, however not at the lower apex of the coil. These results are explained by the assumption that the rear side of the case was not well clamped in the torque ring at higher current in the coil. All five gauges displayed hysteresis not seen earlier, however, no irreversible changes were seen in later runs.

The strain measured on the torque rings in the third test is shown in Fig. 6. The signals of both gauges display discontinuities at approximately 65% and 84% of the rated current in the CH coil. The sudden change of the strain on the beams of the torque rings at 65% current was also observed in the second run, however, not in the first test. This implies an asymetric behaviour of the supporting structure. Simultaneously with the discontinuities in the strain on the torque ring large voltage spikes were registered for several coils. These spikes were crearly created by movements of the coils in the supporting structure. The inspection of the facility after warmup confirmed this hypothesis, since gaps of several millimeter width were found between coils and structure.

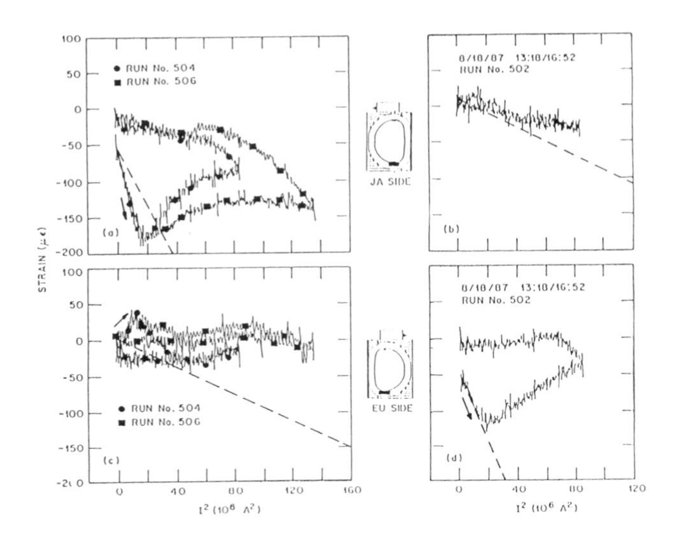

FIGURE 5
Strain versus current squared during the Out-of-Plane Load tests on the symmetrically arranged gauges at the lower apex of the coil. The graphs on the right show data for the first and on the left for the second and third runs. The dashed lines mark the calculated strain.

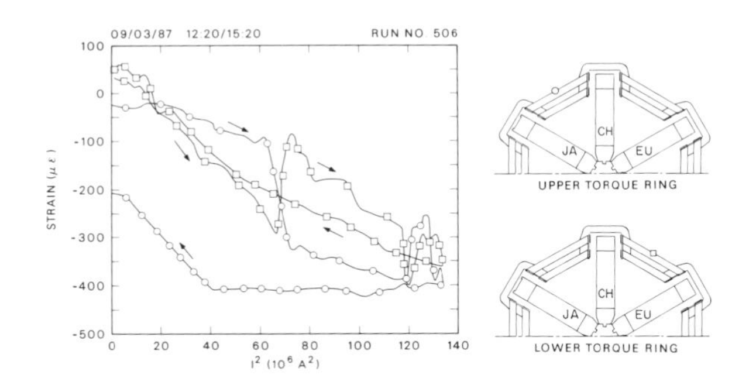

FIGURE 6
Strain versus current squared during the third Out-of-Plane Load test on the torque rings of the test facility.

8. PULSE COIL TESTS

The tests were conducted at constant current under stationary conditions in the coil by triggering a train of transient field pulses. The measurements, performed at rated current, in single-coil and full-array configuration, yielded at the highest pulse field, an ac power loss of 4.4±0.6 and 9.0±1.0 W in winding and case, respectively.

The two symmetrically arranged strain gauges near the lower apex showed a decrease in strain of approximately 15% during the pulse sequence, see Fig. 7. The other gauges did not show any appreciable change. However, peaks were superimposed on the background signal of each pair of gauges installed on the side plates. The peaks were antisymmetric, i.e., when the gauge on one side plate had a positive peak its symmetrically arranged mate displayed a negative peak. The resolution of the data was not good enough to make quantitative statements.

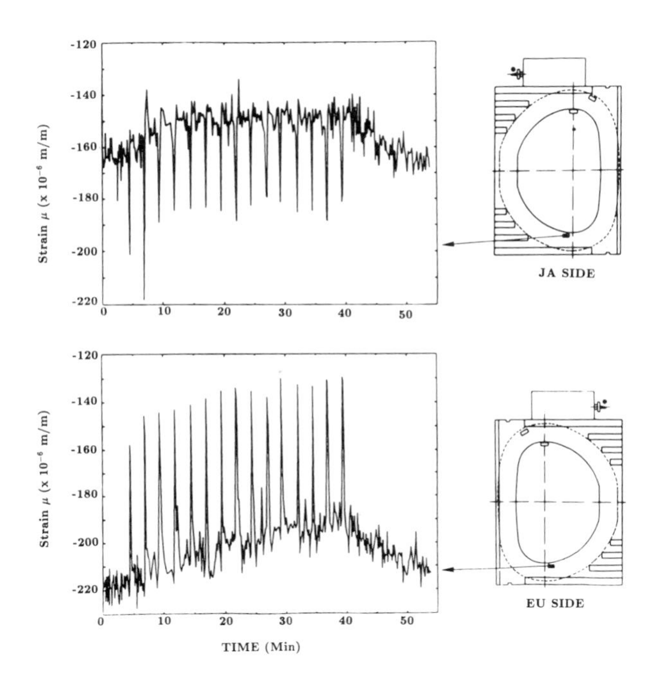

FIGURE 7
Strain versus time on the symmetrically arranged gauges at the lower apex on the side plates, while a train of 15 ac pulses were triggered on the Pulse Coil.

9. CONCLUSIONS

The measured strain gauges data on the Swiss LCT-coil confirmed that the results of the FEM calculations were properly implemented in the design. It was further shown that

- The comparison of calculated and measured strain indicated deficiencies in the fastening of the coil as confirmed on dismantling the facility.
- The coil was neither designed nor instrumented for the large disturbances induced by movements of the whole supporting structure. On the other hand stable coil operation was always achieved when the position of the structure was settled. The strain data taken during runs in which irregularities appeared strongly indicate that aborted runs were caused by insufficient stiffness of the torus.

The evidence presented strongly suggests that in a future superconducting fusion reactor structural movements must be suppressed, and a firmer clamping of the coils achieved, by improved design techniques.

ACKNOWLEDGEMENTS

The assistance of all participants of the LCT program during the tests is gratefully acknowledged. The efforts of the operating crew should be mentioned specially, since their engagement was essential to perform the tests.

REFERENCES

1. P.N. Haubenreich et al., Productive International Collaboration in the Large Coil Task, 12th Symp. on Fusion Engineering, 1987, Monterey, CA.
2. J.A. Zichy et al., IEEE Trans. Magn., MAG-23, p. 819 (1987)
3. A. Segessemann et al., Proc. of 12th Symp. on Fusion Eng., Vol. 2 (1983), p. 1139.
4. J.A. Zichy, Adv. in Cryog. Eng., Vol. 33, p. 1053 (1988)
5. J.A. Zichy et al., IEEE Trans. Magn., MAG-21, p. 245 (1985)
6. C.C. Queen et al., Proc. of 8th Symp. on Eng. Problems of Fusion Research, p. 1183 (1979)
7. R.O. Hussung et al., Proc. of 9th Symp. on Eng. Problems of Fusion Research, Vol. I, (1981), p. 166.

FUSION TECHNOLOGY 1988
A.M. Van Ingen, A. Nijsen-Vis, H.T. Klippel (editors)

ANALYSIS OF THE MECHANICAL BEHAVIOUR OF THE EURATOM LCT COIL UNDER THE EFFECTS OF OUT-OF-PLANE LOADS

S. Gauss, A. Maurer*, W. Maurer, A. Ulbricht, F. Wüchner

* Siemens AG, ZFE TPH 23, Research Laboratories, Postfach 3240, D-8520 Erlangen, FRG
Kernforschungszentrum Karlsruhe, Institut für Technische Physik, Postfach 3640, D-7500 Karlsruhe, FRG

The Euratom LCT coil was investigated under different load cases with different boundary conditions in TOSKA in Karlsruhe and in the IFSMTF in Oak Ridge. The experimental results were compared with finite element model calculations. The coil carried successfully out-of-plane forces up to 27 MN at expected equivalent stress levels < 200 MPa. Extreme load cases like single coil tests and tests under out-of-plane forces (one coil without current in a torus) showed deviations from the finite element model. The results indicated some backlash in the torus structure and changes in the mechanism of force transmission. Both effects led to observed nonlinearities in displacement and stresses which were not reflected by the finite element model. Problem areas for future and larger toroidal magnet systems were identified.

1. INTRODUCTION

The Large Coil Task was an international magnet technology experiment under the auspices of the International Energy Agency (IEA) for the development of superconducting toroidal field magnets for fusion[1]. The main experimental activity was concentrated in the years 1985 - 1987. In this experiment the Euratom LCT coil was one of the six test coils as constribution of the European Community in the frame of the European Fusion Technology Program. Besides the electrical and thermohydraulical properties of the coil the mechanical behaviour was of great interest from the following point of views:

- The coil winding itself was a sophisticated composite of different materials with anisotropic material properties exposed to a central overall forces up to 57 MN.
- The winding was enclosed by a stainless steel structure where special care was taken to transmit the force from the winding to the case.
- The coil itself (winding and case structure) was imbedded in the supporting structure of the toroidal configuration of magnets.

This was a basic mechanical arrangement of the magnets in the toroidal configuration. All test load cases of the test program were analyzed by a finite element model[2]. The coil was equipped with a suitable instrumentation which allowed the comparison of the measured and expected mechanical behaviour. Especially the analysis of extreme load cases like a free standing single coil test and the exposure to out-of-plane forces demonstrated clearly the limits of the used finite element model and the resolution of the instrumentation. The mechanical properties of the coil during exposure to out-of-plane forces between 3 to 27 MN in two directions are the subject of this contribution.

2. PREVIOUS MEASUREMENTS, ANALYSIS AND CONCLUSIONS

The coil was tested as free standing single coil, as single coil clamped against a central bucking post, in a toroidal configuration and under the torque moment of a poloidal field. While the test results in the toroidal structure not only in single coil but also in toroidal configuration were in a fairly good agreement[3] the free standing single coil showed the most discrepancies[4]. Some corrections used of material data based on the experimental results with cross checks to winding sample tests led to an agreement between measured and calculated results[5]. Especially the measurements of overall data like the deformation of the D shape and the gap between the flanges of the coil case showed nonlinearities and considerable deviations from the calculated results. The attempt to improve the finite element model led only partially to a better agreement[6]. Especially the nonlinearities were not represented in a successful manner within the model. Due to the limited resolution of the instrumentation no further input of model modifications or changes of boundary conditions was possible. The expectations were therefore directed to special extreme load cases which could lead to a better understanding.

3. THE OUT-OF-PLANE LOAD BEHAVIOUR

The exposure of a coil to lateral forces has to be assumed as a failure mode in a toroidal magnet configuration if one coil is without current. Therefore this load case was implemented in the test program of the LCT (Fig. 1).The Euratom LCT coil was exposed to out-of-plane forces in positive and negative direction in the torus coordinate system. In the first out-of-plane load test Euratom-General Electric coil (EU-GE) where the Westinghouse coil (WH) were operated without current the force direction was positive (towards the Swiss (CH) coil) and reached 22.9 MN before the five-coil array was dumped after a quench in the CH coil (Fig. 4). The decay of the out-of-plane forces calculated from the current decay of the coils demonstrated that no higher out-of-plane load appeared in any coil other than the test coils (Fig. 2). But it is remarkable that the Japanese-coil (JA) was exposed to an increasing out-of-plane force of about 12 MN within 10 s during the dump due to the lower discharge time constant of the General Dynamics/Convair (GD/C) coil. Operating the CH-coil without current the Euratom coil experienced a negative out-of-plane force of 26.6 MN (towards the WH coil) in the Euratom-Japanese (EU-JA) out-of-plane load test (Fig. 4).

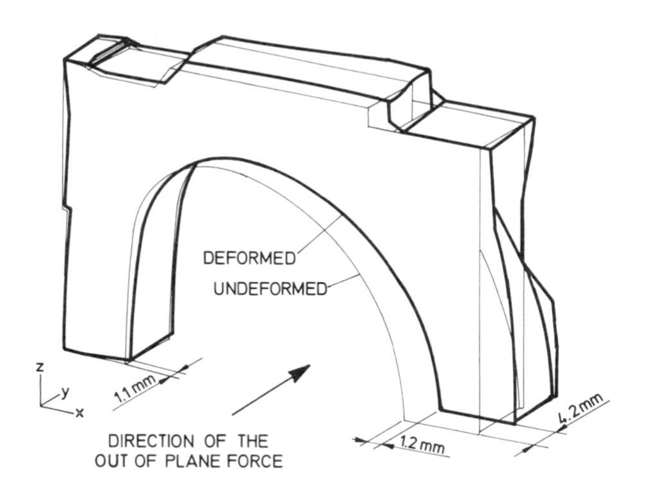

FIGURE 1
The EU LCT coil under the effect of out-of-plane forces derived from the finite element calculations.

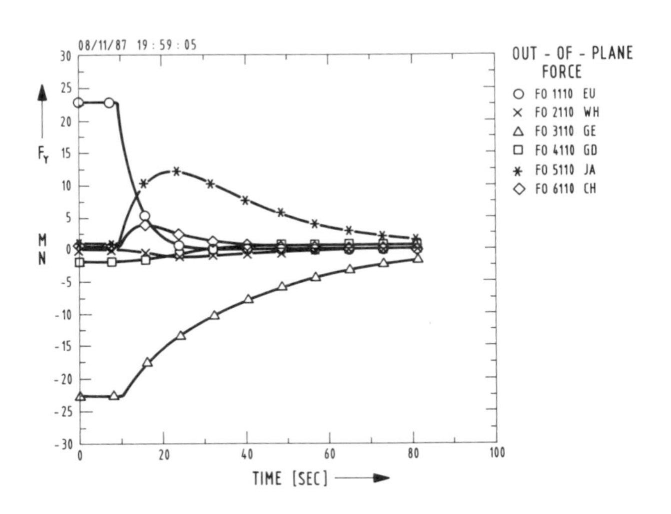

FIGURE 2
Time dependence of the out-of-plane forces during the full torus dump in EU/GE out-of-plane load test. Different discharge time constants cause considerable transient out-of-plane force for the JA coil.

3.1 Von Mises equivalent stresses

The basic stress measurements were done by strain gauge rosettes mounted on the coil case

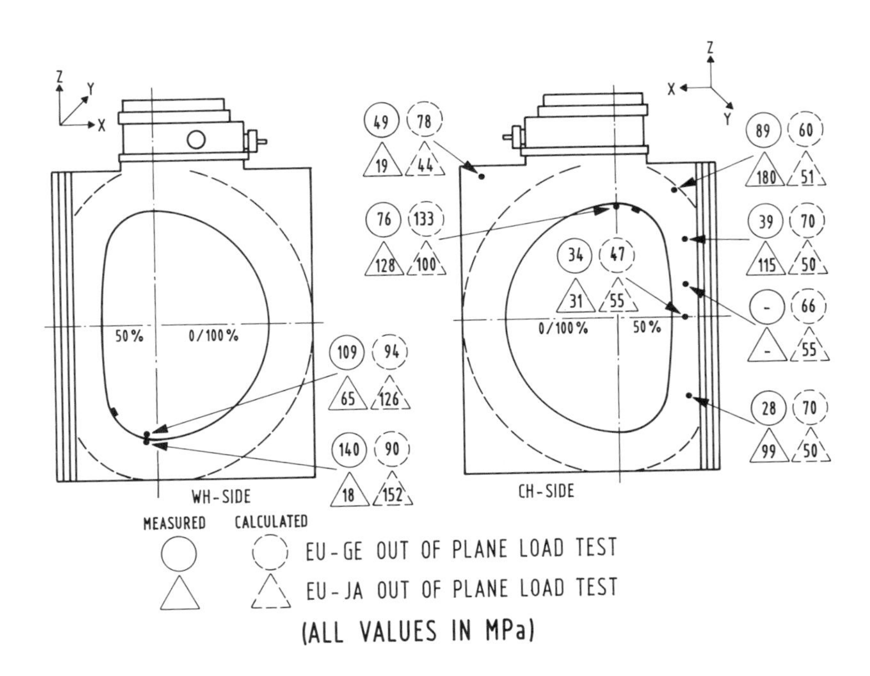

FIGURE 3
Comparison of measured and calculated equivalent (von Mises) stresses for both out-of-plane load tests.

surface. From these strain measurements the von Mises stresses were calculated for comparison with the results obtained from the finite element analysis (FEM) during the design phase[2]. Because of the different the maximum currents reached in both out-of-plane load tests which also different from the current used in the structural analysis, measured and calculated von Mises stresses were made comparable by using the rated current of J_o = 11.400 A as reference (Fig. 3). The peak stresses measured during the EU-GE out-of-plane load test were lower than the stresses during the EU-JA out-of-plane load test. The original unscaled maximum values are 135 MPa for the EU-GE test and about 175 MPa for the EU-JA test. The values predicted by FEM calculations are in the same order of magnitude. But in general they agree better with the measurements of the EU-GE test. In the EU-GE test the equivalent stresses measured on the coil case during current ramping up were linear in

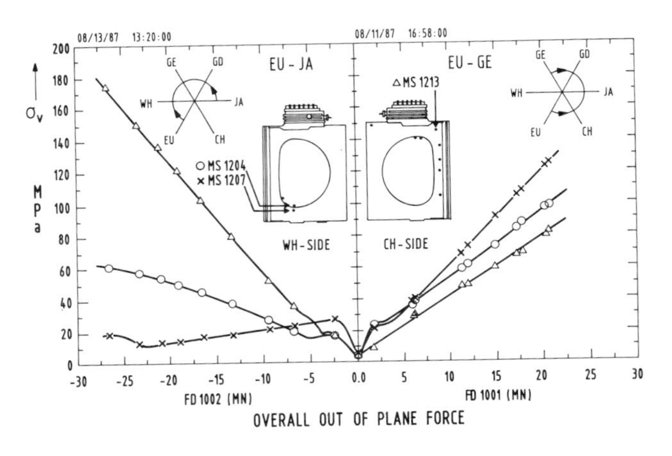

FIGURE 4
Linearity check of the equivalent stresses for both out-of-plane load tests.

current square within the measuring accuracy . A completely different behaviour was observed in the EU-JA test. The equivalent stresses were partly nonlinear in current square which led to the conclusion that the boundary condition between coil case and torque ring support structure as well as between winding and case had changed. This fact can be demonstrated clearly by comparing the measurements from both tests choosing the strain gauge rosettes MS 1207 und MS 1213 where the von Mises stresses are derived from (Fig. 4).

Two high field torus runs were performed during the experimental programm. In one run the out-of-plane force acting on the EU coil in negative direction was rather low about 3.4 MN. In another run the out-of-plane force was 10 MN in the negative direction generated by a slight change in the current distribution of the torus. The winding experienced in the last load case higher pressure to the side plate than before. Some of the measured equivalent stresses show two slopes as a function of the current square (Fig. 5). This demonstrated impressively that the mechanism of force transmission was different for both load cases. This cannot be described by a simple linear finite element model.

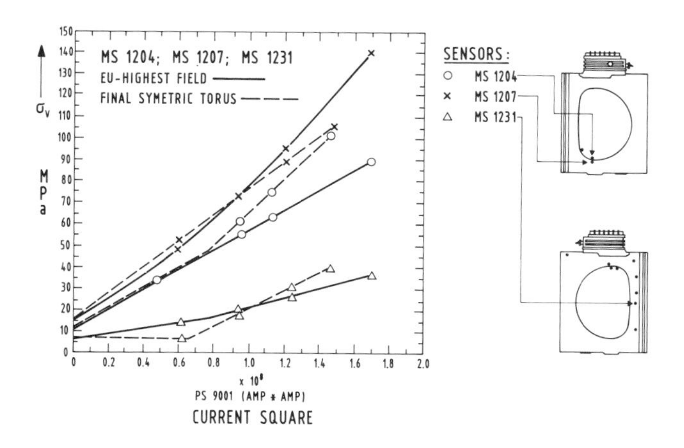

FIGURE 5
The linearity of the equivalent stresses in a full torus with a low (EU highest field test) and a high (Final Symmetric torus test) out-of-plane force. The out-of-plane force led to nonlinearities.

3.2 Displacements between winding and coil case

The Euratom LCT coil was equipped with four displacement transducers on each side plate which measured the gap between winding and side plate. During out-of-plane load tests it was observed for all displacement transducers that the winding was pressed against the side plate in the direction of the out-of-plane load . This fact was demonstrated in Fig. 6 for two displacement transducers on the both side plates in opposite positions. The measured

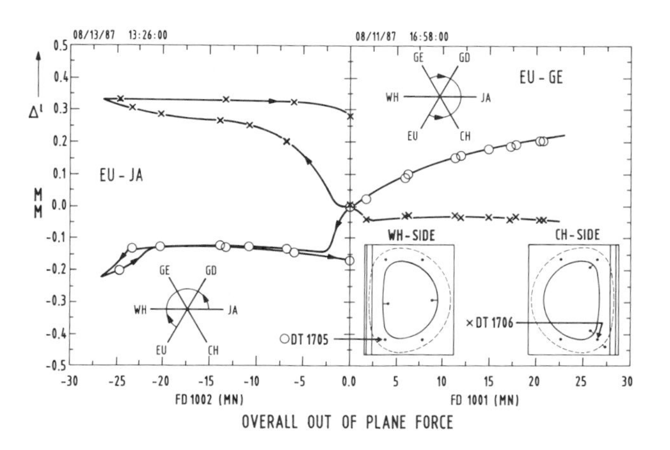

FIGURE 6
Demonstration of hysteresis effects and the local character of gap deformation between winding (wdg) and coil case (cc) for both out-of-plane load tests. The winding was pulled against the corresponding side plate.
(+: gap wdg-cc enlarges, –: gap wdg-cc reduces)

values were in all cases a superposition of coil case and winding deformations. Therefore interpretation was difficult. In the EU-GE out-of-plane load test the gaps could not be a global movement of the winding. There was no correlation between them. One gap increased nearly linearly while the other one had a constant value. Therefore both gaps were more determined by local deformations. Similar statements were also valid for the EU-JA out-of-plane load test. The strong hysteresis observed here demonstrated some kind of plastic deformation in the epoxy filled stainless steel bladders caused by the non machined case side plates and the winding. The reduction of the gap above - 22 MN was probably caused by coil case deformation. It was completely reversible. All gap changes were about or less than 0.3 mm.

4. DISCUSSION

All facts described above indicated that the stresses in the coil case were partly no longer linear in the forces applied. The used finite element model did no longer describe the behaviour of the coil correctly. Especially extreme load cases like single coil testing and out-of-plane load tests led to deviations from the calculated values. An explanation can be changes in the mechanism of force transmission between coil case and winding and also changes in boundary conditions at the supporting area of the coil to the facility. Not only equivalent stresses but also displacements showed hystereses effects. This demonstrated that irreversible processes in the mechanism of force transmission took place. The two slopes of the equivalent stresses in Fig. 5 can be interpreted as a change of the mechanism of force transmission at a certain current or field. Changes of the boundary conditions between coil and the facility should have no impact on the linearity of the finite element model assumed that the coil behaved like a solid body. The change of the boundary conditions between coil and facility probably causes changes in the mechanism of the force transmission between winding and coil case.

Measurement of stresses in the facility indicated some backlash and also hystereses effects which was a sign that the boundary condition between coil and facility changed.

5. CONCLUSION

The coil carried successfully out-of-plane forces up to 27 MN with acceptable equivalent stresses of about 200 MPa as expected. The gaps between winding and case were less than 0.3 mm and were mainly determined by local deformations. The winding was well fixed in the case. No global movement of the winding in the case was observed. Besides these results the analysis of the measurements allowed conclusions which could be useful for future magnet systems. The agreement of calculations and measurements was rather good for specified toroidal load cases. For single coil tests and out-of-plane load testing the measured and calculated results showed differences. The measurements demonstrated that the finite element calculations were no longer linear in forces. The resolution of the installed instrumentation was not sufficient to get further information for improvements of the finite element model. It seems doubtful that this is possible at all. In a first attempt the boundary condition between the facility structure and the coil should be made backlash free by pretension and suitable design principles. The constant boundary conditions between coil and facility will lead to a better analysis of the mechanism of force transmission between winding and coil case. This boundary conditions are much more complicated and need further experimental data for getting better finite element models. The experiment showed clear that a backlash free support structure is mandatory for a safe long time operation of a toroidal system with cycling loads.

ACKNOWLEDGEMENTS

This work has been performed within the framework of the IEA-Implementing Agreement for a Programme of Research and Development on Superconducting Magnets for Fusion Power, Annex I, "Large Coil Task," between Japan, the U.S. Department of Energy, EURATOM, and Switzerland. Thus, the reported results could be achieved only by the common effort of all participants involved. Especially recognized is the effort by the ORNL group responsible for operation of the facility.

This work has been performed too within the

framework of the Nuclear Fusion Project of the Kernforschungszentrum Karlsruhe and is supported by the European Communities within the European Fusion Technology Program.

REFERENCES

1. P.N. Haubenreich, P. Komarek, S. Shimamoto, G. Vecsey, Productive International Collaboration in the Large Coil Task, 12th Symp. Fus. Eng., October 12-16, 1987 Monterey, California

2. A. Maurer, Structural Analysis (Siemens) of the Euratom Coil for the Large Coil Task, Proc. MT-7, IEEE Trans. on Magn., MAG-17 (5) Sept. 1981, 2043-2046

3. A. Maurer et al., Investigation of the mechanical properties of the Euratom LCT coil by tests under different boundary conditions, Trans. 9th Int. Conf. Struct. Mechanics in Reactor Technology/Lausanne 17-21 August 1987, A.A. Balkema, Rotterdam 1987, p. 55-60

4. W. Herz et al., Testing of the Euratom LCT Coil in the TOSKA facility, Proc. 13th Symp. Fusion Techn., Varese, Italy, Vol. 2, 1984, p. 1487

5. A. Maurer et al., Effect of Azimuthal Dependence of Radial Young's Modulus on the Mechanical Behaviour of the European LCT Coil, Proc. 9th Int. Conf. on Magnet Techn. (MT-9), Zürich 1985, p. 428

6. A. Maurer et al., Improvements in the FEM-Model of the Euratom-LCT Coil and Comparison with Results Measured in TOSKA, Proc. 14th Symp. on Fusion Technol., Avignon, France, 1986, p. 1637

FUSION TECHNOLOGY 1988
A.M. Van Ingen, A. Nijsen-Vis, H.T. Klippel (editors)

ARCING EXPERIMENTS FOR MAGNET SAFETY INVESTIGATIONS

K.P. JÜNGST, H. KRONHARDT, M. OEHMANN* and J.S. HERRING**

Association KfK-EURATOM, Kernforschungszentrum Karlsruhe, Institut für Technische Physik, P.O.Box 3640, D-7500 Karlsruhe, Federal Republic of Germany

Investigations of the behaviour of superconducting magnets under fault conditions are of considerable significance for safety and availability of fusion systems. This report concentrates on one fault of major importance, the electrical arcing. Experiments were performed with the TESPE six-coil set. Main aspects were the behaviour of high-current arcs under various boundary conditions and the interaction of arcs with the magnet system. Arcing was initiated across the terminals of one coil or the entire coil set in special helium gas filled chambers outside of the TESPE cryostat, thereby avoiding early magnet destruction. After initiation the arc burned sustained by the stored energy of the magnet.
Different chamber geometries and gas pressures were applied to simulate both free burning arcs and arcs within windings.
Measured arcing voltage characteristics were used to calculate the current distribution within the magnet system.

1. INTRODUCTION

Disturbances occurring during the operation of large superconducting magnets, e.g. in fusion power plants, not only impair the availability of the plant, but also may raise safety problems. The main hazard potential of superconducting magnets exists in the large amount of stored energy, should this energy be released in an uncontrolled way in an accident[1]. One such incident, which will be mentioned in more detail below, is the electric arc[2]. The arc initiation may be of electrical, mechanical or thermal kind such as voltage breakdown, conductor breakage or conductor melting, respectively[3].

2. EXPERIMENTS

Arc experiments were performed to study the behaviour of one magnet or of the entire inductive system shorted by an electrical arc as well as the behaviour of the arct itself and the corresponding interactions.

The experiments were carried out using the six-coil superconducting torus TESPE[4]: operating current 7 kA, stored energy 8.3 MJ, all coils in series with one dump resistor. The test facility allowed to ignite an arc across one coil or across the entire system. Figure 1 shows the TESPE network with these two arcing possibilities. In order not to endanger the TESPE facility at an early stage of investigation, the arcs were initiated between two copper electrodes in external chambers filled with helium gas. The energy dumped in the arc comes from the magnet system which supplies the driving potential during fast discharge. High voltages were avoided by using ignition aids to initiate the arc. The three methods of arc initiation used were meant to simulate moreover the possible ways in which an arc can occur in the magnet system: (1) By using a high voltage pulse an electrical breakdown was simulated. (2)

* KfK, HVT/EA, **Visiting Scientist from INEL, EG & G, Idaho Falls, Idaho 83415-3523

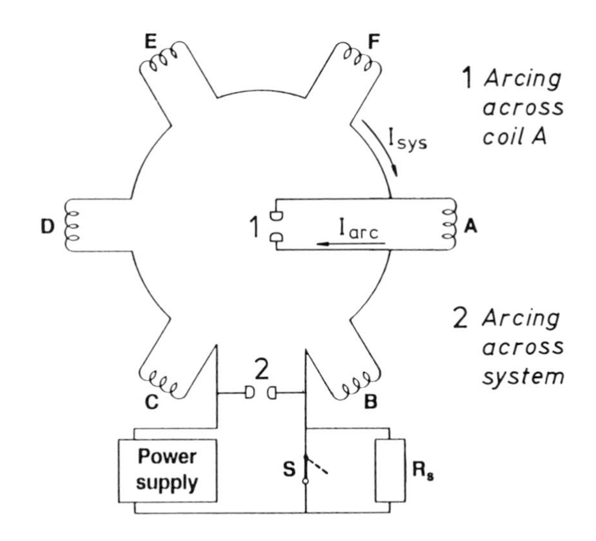

FIGURE 1
Network of TESPE for arc investigations

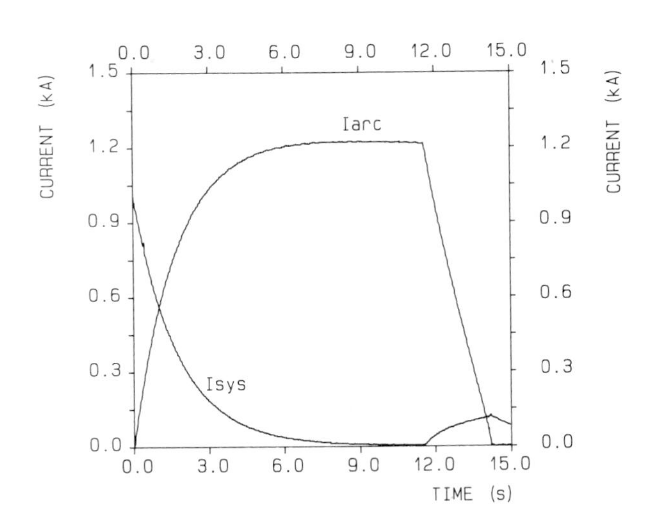

FIGURE 2
Current distribution for a system with a short-circuited coil followed by an arc.

The production of metal vapour e.g. by melting of leads or conductors, which reduce the resistivity of the gas, was simulated by copper wire vaporization. (3) Mechanical conductor breakage initiating an arc was simulated by electrode separation after a deliberate short-circuiting of the coil A.

2.1 Behaviour of the system with a short and an arc across one coil

If a short circuit occurs across one of the coils during a regular fast system discharge, the current increases in the short-circuit by induction. This increase in current depends mainly on the inductive coupling coefficient of the system and the short-circuit resistance. The maximum attainable current enhancement in one coil of TESPE, i.e. for zero short-circuit resistance is calculated as 126,5% with the program HEDO[5].

The current distribution for a system discharge from about 1000 A with a short circuit across coil A followed by the arc is shown in Fig. 2. The system current I_{sys} decreases according to the magnitude of the dump resistance and of the magnet coupling coefficient. The short-circuit current I_{arc} rises reaching a peak level of 121,5% of the initial current in the system. Separation of electrodes initiates the arc supplied by the stored energy of the coil. Now the arc current decreases linearly as a function of the roughly constant arc voltage and causes partial re-coupling of energy into the main system. The system current I_{sys} continues to rise until I_{arc} has become zero. Thereafter a usual discharge begins.

2.2 System's coupling influence on arc's behaviour

The aim of the following experiments was to study the system's coupling influence on arc's behaviour and its characteristic during system discharge. After start of a regular fast system discharge an arc was initiated by a high voltage pulse across coil A. A typical example of the curves of the system current I_{sys}, the arc current I_{arc} and the arc voltage V_{arc} is shown in Fig.3. The system was discharged from 3000 A. The shape of the arc current I_{arc} results from the superposition of the two processes occurring consecutively in Fig. 2.: The system couples energy into the short circuit, produced by the arc across coil A, and simultaneously energy is dumped in the arc itself.

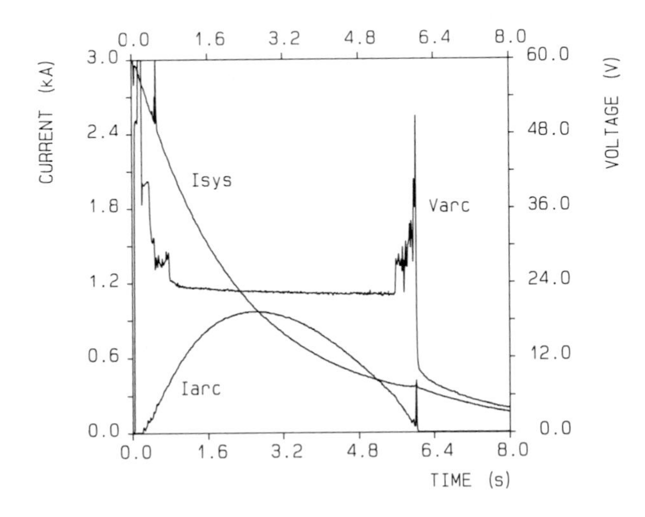

FIGURE 3
Curves of system current I_{sys}, arc current I_{arc} and arc voltage V_{arc} for an arc across one coil initiated by high voltage pulse.

The arc voltage is approximately constant (except for very low currents) amounting to roughly 22 V. Similar values were obtained in other experiments carried out at different system currents and hence different arc currents, whose peaks were between 350 A and 1800 A, as long as the gas pressure in the arc chamber was kept at low values (here 40 mbar).

A continuous lengthening of a free burning arc has a substantial influence on its characteristics. Arc lengthening e.g. by melting or vaporization of electrodes leads to increase of the burning voltage, if the metal vapour concentration can be kept low e.g. by large surrounding volume. In experiments an increase of initial pole distance by a factor 2.7 leads to a rise of the arc voltage from 28 V to 52 V. The general characteristic of arc current persists but the higher voltage accelerates the current decrease and shortens the burning duration.

2.3 Arc across the total magnet system

The behaviour of inductive magnet systems with an arc across is different from that with arc across one of the coils. The inductive coupling of energy does not occur in this case, thus I_{arc} has an other shape. In Fig. 4 current

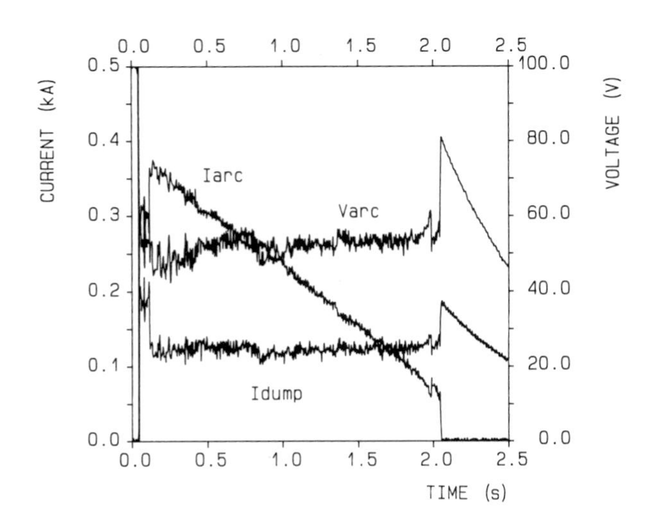

FIGURE 4
Currents and voltage for an arc across the total magnet system in parallel to the dump resistor.

and voltage characteristics for an arc across the entire system and parallel to the external dump resistor are shown.

The arc voltage limits the current I_{dump} through the dump resistor by its magnitude. The drop of the arc current I_{arc} is dominated by the arc voltage only. Therefore the arc voltage determines the global discharge behaviour of the magnet system.

The magnitude of V_{arc} is mainly influenced by environmental conditions. Here the arc was burning within a confining chamber, which simulated a part of coil windings. So the gas pressure, cooling by surrounding walls and available space influenced the magnitude of the arc voltage .

2.4 Efficiency of a discharge stop

Earlier it was expected that the stop of discharge, i.e. reducing the inductive voltage to zero, would eliminate the driving potential for the arc leading to extinguishment.

Realization of stop of discharge is possible by short-circuiting the dump resistor. In case of an arc e.g. across the total system this measure short-circuits the arc, too, and leads to the desired result.

However, in cases like an arc across one coil with no own dump resistor or for arcs in the current path, stop of discharge does not stop the arc. As can be seen from Fig. 5 a discharge stop by shorting of dump resistor (at t = 1,62 s) interrupts the current decrease in the main system und thus the coupling of energy into the arc circuit. But the arc does continue to burn at the same voltage. The voltage required is continued to be supplied by inductance present in the arc circuit, i.e. in this case the arc characteristic determines the voltage in the coil. The arc current decreases linearly as a function of the constant arc voltage. The arc continues to burn after discharge stop until the current in the coil I_{la} becomes equal to the system current I_{sys}. Only then the arc will be extinguished.

3. NUMERICAL CALCULATIONS

In case of quiet behaviour of the arc, thus burning with constant voltage, the current distribution can be calculated analytically. Such a simplified calculation yields already good agreement with measured shapes.

For more realistic voltage shapes better results can be obtained by using the network program MSCAP with the arc voltage as input parameter. Comparisons of corresponding calculations and experimental data will be

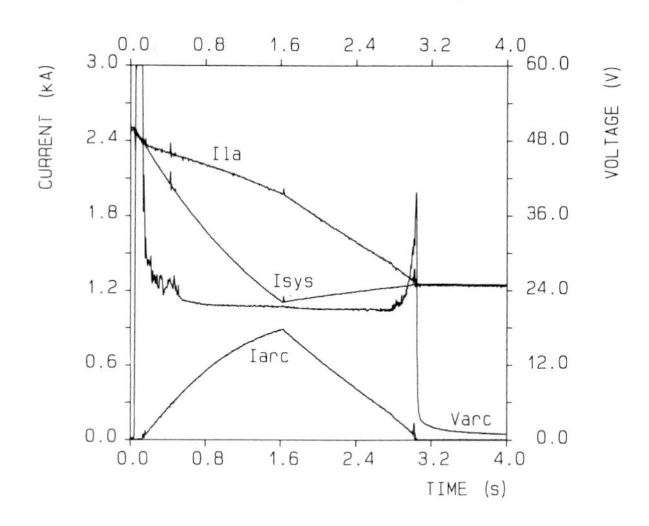

FIGURE 5
Current distributions in case of discharge stop at t = 1.62 s.

presented elsewhere[7]. MSCAP was specially developed to analyze non-linear and fast response transients of large scale magnetic systems on normal and off-normal operating conditions, including coil quenchs, coil distortion, internal coil shorts and arcs. The code includes time-, current- or voltage-dependent voltage supplies, current supplies, resistors, capacitors and inductors.

4. CONCLUSIONS

The main parts of the voltage characteristics were found to be basically the same for all experiments with stationary arcs. In detail, numerical values depend on environmental conditions such as gas pressure and flow, arc length, available space, heat capacities, material melting and vaporization. For free burning arcs at low pressure the arc voltage is found typically between 20 V and 30 V and is approximately independent of the current in the kA-range. However, voltages up to 200 V and pressures up to 30 bar were observed for arcs confined to winding-simulating small volumes.

Measures for extinguishment are effective, if the voltage supplied to the arc can be reduced below burning voltage, e.g. by short-circuiting the arc. A deliberate stop of system discharge does not extinguish the arc; this measure can be advantageous, nevertheless, due to stop of further inductive energy transfer to the arc.
For the calculation of current distributions in magnet systems with arcs, typical arcing voltage characteristics have to be determined first experimentally. In addition to the presented work a few further experiments are on the way.

ACKNOWLEDGEMENTS

This work has been performed in the framework of the Nuclear Fusion Project of the Kernforschungszentrum Karlsruhe and is supported by the European Communities within the European Fusion Technology Program. The authors acknowledge support from the U.S. Department of Energy under DOE Contract DE-AC07-76ID01570. Preparation of components and assistance in the measurements by H. Kiesel, G. Obermaier, J. Seibert, and E. Süss is gratefully acknowledged.

REFERENCES

1. F. Arendt, P. Komarek, Potential failure and hazards in superconducting magnet systems for fusion reactors, Nucl. Technology/ Fusion, Vol. 1, Oct. 1981

2. W. Finkelnburg and H. Mäcker, Elektrische Bögen und Thermisches Plasma, in Handbuch der Physik, Bd. XXII, S. Flügge, Hrsg., Springer-Verlag, Berlin (1956)

3. K.P.Jüngst, Zur Sicherheit von supraleitenden Magnetsystemen für die Fusion, KfK-Nachrichten, Jahrg. 19, 1/87,31-36

4. K.P.Jüngst, G.W.Leppelmeier, W. Geiger, P. Komarek, M.M.Steeves, First results of the TESPE-S magnet system safety experiments. Proc.14th SOFT, Avignon, Sep 8-12, 1986

5. H. Preis,Berechnung des magnetischen Feldes, der Kräfte und des Betriebsverhaltens großer Spulensysteme für Fusionsexperimente, IPP III/24, April 1976, IPP München

6. H.G. Kraus, J.L. Jones, Hybrid finite difference/finite element solution method for nonlinear superconducting magnet and elctrical circuit breakdown transient analysis. Int. J. of num. Methods in Engineering (1986), Vol. 23 pp. 1003-1022

7. J.S.Herring, K.P.Jüngst, J.L.Jones, H.G.Kraus, MSCAP Simulations of TESPE magnet safety transients, to be presented at 8th Topical Meeting on Technology of Fusion Energy, Oct 9-13,1988, Salt Lake City, Utah

FUSION TECHNOLOGY 1988
A.M. Van Ingen, A. Nijsen-Vis, H.T. Klippel (editors)
Elsevier Science Publishers B.V., 1989

OPERATIONAL CHECK AND CONTROL OF THE ASDEX UPGRADE POLOIDAL FIELD COILS

M. Pillsticker, G. Klement, F. Werner
Max-Planck-Institut für Plasmaphysik, EURATOM Association
D-8046 Garching, Fed. Rep. Germany

Abstract
The poloidal field coils of ASDEX Upgrade consisting of ohmic heating, vertical field and control coils, have to be protected. To prevent overstress by forces or thermal effects, we measure and control the timely variable currents of all coils and voltages, temperatures of the coils (Cu) and the cooling medium, water flow within all cooling loops, geometrical positions and movements of the coils and holdings, and mechanical maximum stresses of all coils. The physical quantities mentioned will be measured, whereas all mechanical stresses and coil temperatures will be calculated on line being based on current measurements and compared with critical values.

1. INTRODUCTION

The poloidal field (PF) coil system of the ASDEX Upgrade tokamak consists of ohmic heating (OH) coils for producing and heating the plasma and compensating plasma losses, vertical field (V) coils for producing the plasma equilibrium, plasma shape, and divertor configuration, and vertical field control (Co) coils for correcting the plasma position and movements to prevent instabilities. All five OH coils (fig. 1) are electrically connected in series and fed by a power supply generating extremely steep currents (max. 50 kA) and voltages (max. 30 kV). The eight V coils are each fed by a separate power supply; the coils are exposed to maximum currents of 60 kA of moderate steepness. The six Co coils are also each fed by a separate power supply with different maximum currents (CoA and Co OH 16 kA, CoI 30 kA). The slope depends on the control conditions and can be as high as $5 \times 10^6 A/s$.

When specifying the terms "PF coil control". We have to distinguish between normal coil current control within given limits under experimental conditions, called "plasma feedback and position control", and control at the superior command level, called "PF coil protection". It is the latter that this report is concerned with.

2. FOUNDATIONS

All of the coils are normal-conducting and are usually operated with pulsed currents. DN (double-null) operation requires equal pulse currents in all coils of the same type and yields a symmetric plasma shape with a divertor configuration. The SN (single-null) configuration is asymmetric and requires different pulse currents in all of the coils. Like DN, the L (limiter) condition can be realized by having equal pulse currents in coils of the same type. For all operational conditions mentioned, the pulse time will be less than 10 s. The heating energy of a normal pulse is too high for it to be transferred to a cooling medium during the current pulse time. Nearly 95% of the heating energy of a current pulse causes a maximum coil temperature rise of 65 degrees. The temperature will be reduced to normal during a cooling time of nearly 10 min by water flow within the bores of the coil conductors.

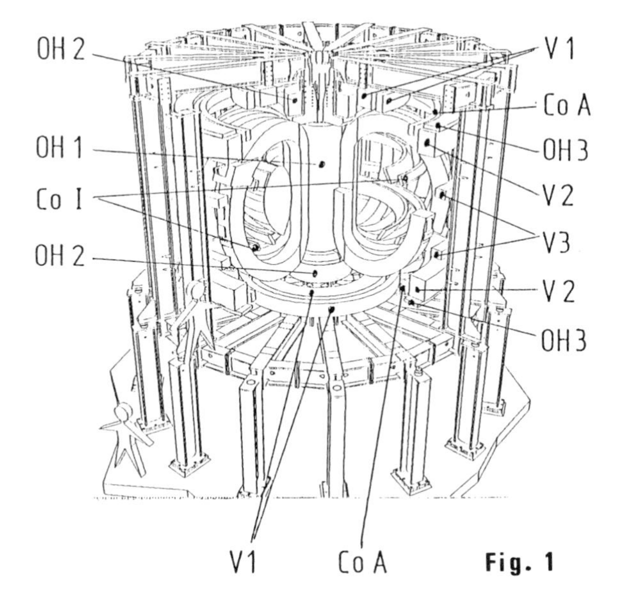

Fig. 1

Furthermore, we defined other operational conditions, such as current drive, pulse cleaning, etc., which ensure continuous service of the PF coils. Here the heating energy of the coils can be permanently transferred to cooling water. The coils consist of several cooling loops; a cooling loop in any case is a coil element such as a cylinder or pancake. This is a necessary condition to enforce uniform temperature gradients all over a coil cross-section when directly applying cold water of minimum cooling-down temperature to a hot coil. Moreover, it is important that the temperature and water flow in all of the coil cooling loops has to be controlled. It is stresses produced by these factors, not the physical properties of the insulation that limit the maximum temperature value in a coil.

Maximum current values are defined because of stresses in the current lead-ins and contact problems with the connections. Electromagnetic stresses within the coils should not be controlled by defining upper limit currents for the coils. In this case some well-known experimental conditions would give extreme PF currents and define the utilization factor of the tokamak to a very low value. Furthermore, new, as yet unknown, experimental conditions would produce riscs affecting the safety of the PF coils. To avoid riscs and a low efficiency of the tokamak, the momentary currents should be controlled and should be the basis of permanent stress control during pulse time.

All of the PF coils are manufactured with a high voltage insulation and are tested at 25 kV (V coils) and 50 kV (OH and Co coils). Even extreme conditions cannot induce voltages high enough to destroy the V and Co coils. Because the OH coils are part of a switching network generating steep current rise, the OH coils must be protected against too high voltages.

3. CONTROL FUNCTIONS

We must consider functions which are controlled at very short time intervals (less than 50 ms) and at long time intervals (150 to 200 ms). All controls concerning the cooling procedure will be executed at long time intervals with a programmable logic controller (PLC). All controls influenced directly by current measurements will be carried out at short time intervals with a VME computer (fig. 2).

3.1 PLC control in detail

3.1.1 Water flow: $Q_{min} < Q_{act} < Q_{max}$,
$Q_{max} = 1.1(k_p \cdot \Delta\ p_{act})^{0.571}$
$Q_{min} = 0.975(k_p \cdot \Delta\ p_{act})^{0.571}$

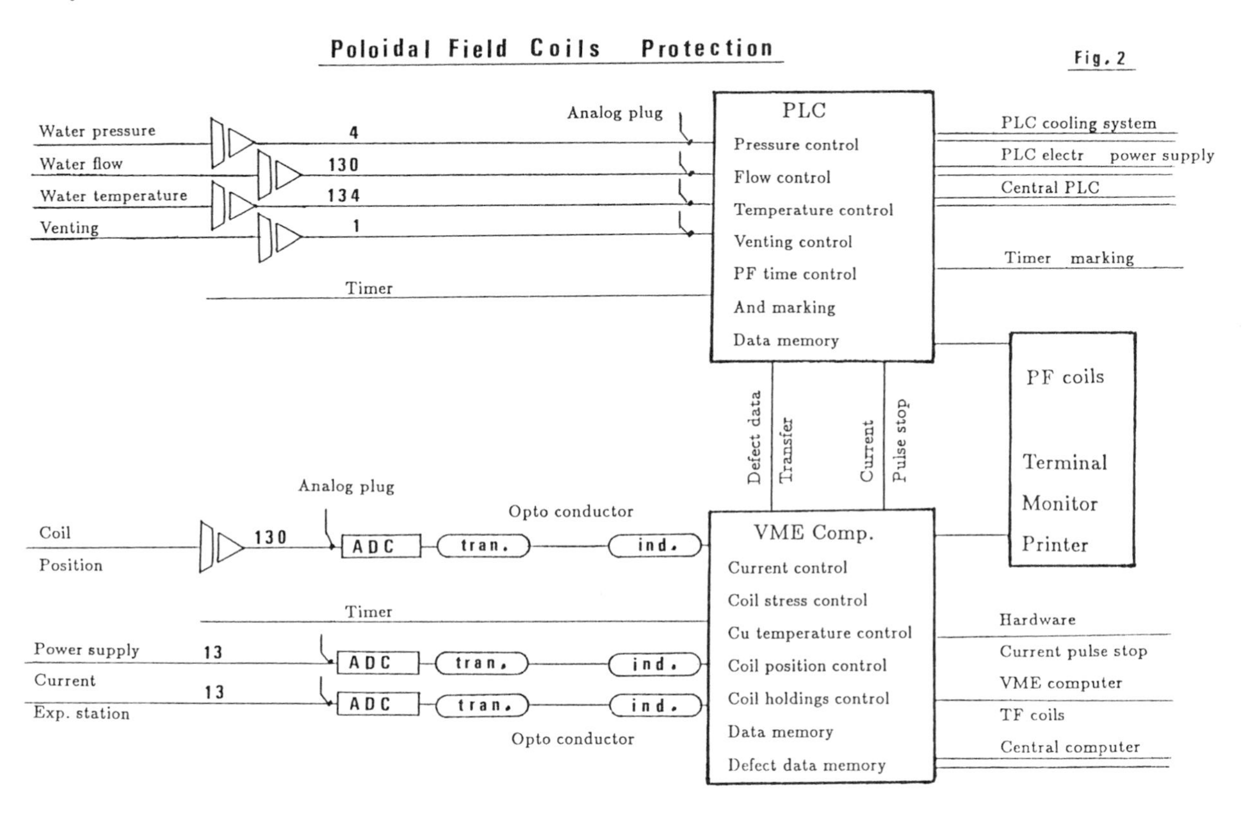

The factor k_p has to be defined on the basis of exact measurements, $k_p = Q^{1.75}/\Delta\ p$, for each of the 130 cooling loops.

3.1.2 Water pressure: $\Delta\ p_{loop} = p_{flow} - p_{return}$, $\Delta\ p_{act} > \Delta p_{min}$; $\Delta\ p_{min} = 2.5 MPa$.
The pressure measurements will be carried out at the flow and return pipes before branching to the PF coils in the upper and also lower coil regions.

3.1.3 Water temperature: $\partial_{flow} < 293K, \partial_{return} < \partial_{max}$, where ∂_{max} are coil-specific values. The cooling-down criterion is given when $(\partial_{flow} + 2°) > \partial_{return}$. The temperatures have to be measured in each of the 130 cooling loops.

The control functions mentioned above have to be fulfilled in a positive manner for all the cooling loops. One negative response is sufficient to stop the operational procedure and necessitates testing of the system. In the different phases of the operation time the control functions must be submitted to special criteria (fig. 3): The phase of current pulse preparation requires control of the water flow and pressure and a low water temperature. During the current pulse the water flow, water pressure and maximum water temperature will be controlled. During the cooling-down phase the water flow and pressure have to be controlled; the water temperature sets the mark for the end of cooling-down of the PF coils. When the PF coils are controlled in accordance with the criteria mentioned, mechanical stresses due to thermal effects will not destroy the coil materials.

3.2 VME-computed control in detail
3.2.1 Currents: $I_{act} < I_{max}$ means protection for the current lead-ins of the coils. The measurements will be carried out in the experimental power station and be compared with values from the power supply. The current values are available in both analog and digital form for calculation procedures in the VME computer.

3.2.2 Stresses: $\sigma_{v\ act} < \sigma_{v\ max}$ will be controlled at intervals of less than 50 ms, which is the time needed to compute $\sigma_{v\ act}$ on the basis of instantaneous currents measured at the beginning of the computing procedure. During the computing time the PF coil currents normally change their values, so that the computed result of $\sigma_{v\ act}$ would no longer be up to date if the current values measured at the beginning are not corrected to the values expected at the end of the computing time. We have to consider the maximum possible current change in all coils at the time of computing and to add these values to the measured ones. For all OH and V coils we get $I_{act} = I_{meas} + \Delta I_{comp.time\ max}$; here the maximum values of ΔI are only a few per cent of the maximum coil currents. For the Co coils ΔI can reach the values of $I_{coil\ max}$, so that in any case it is $I_{coil,max}$ that must be used for computing $\sigma_{v\ act}$, not the measured instantaneous currents.

The comparative stress $\sigma_{v\ act}$ is a function of the tangential, vertical and radial tensile stresses and at the shear stresses:

$$\sigma_v = f(\sigma_{tan};\ \sigma_{rad};\ \sigma_{ver}\ \tau_{tan};\ \tau_{rad};\ \tau_{ver}).$$

For critical well-known regions in all PF coils the tensile stresses can be calculated when the current-dependent loads (electromagnetic forces) of three dominating spots of each PF coil cross-section are known. We have to consider 24 coils and/or - elements. The matrix equation

$$F_{1,2,3} = J \cdot B_{1,2,3} \cdot J \cdot E$$

gives the radial forces for spots at the internal (index 1) and the external (index 2) radii, and the vertical force for the centre (index 3) of all coils. B is a constant magnetic field unit matrix that already takes the mutual coupling of the coils into account and consists of 24 elements in rows and columns. E is a 24-row 1-column unit matrix. J has to be generated by measured current values and is a 24-row 24-column diagonal marked matrix:

$$\sigma_{p,b,s} = K_v \cdot F_3$$

(with indices: $p \rightarrow$ pressure, $b \rightarrow$ bending, $s \rightarrow$ support), and

$$\sigma_{T,R} = K_R \cdot F_{(1,2)}$$

with indices $T \rightarrow$ tangential, $R \rightarrow$ radial) will give the important tensile stresses of the coils.

The defined matrices consist of:

K_v $\rightarrow$ 72 rows and 24 columns;
F_3 $\rightarrow$ 24 rows and 1 column;

$K_R \rightarrow$ 48 rows and 48 columns;
$F_{1,2} \rightarrow$ 48 rows and 1 column;
combined F_1 and F_2 matrix

The matrices K_v and K_R were generated on the basis of exact stress calculations, so that the results, σ, obtained with these matrix equations are in very good approximation to the exact results. Only the solution of the matrix equations makes it possible to calculate stresses fast enough to compute on line within a control procedure. Extensive investigations have afforded convincing evidence that a $\sigma_{v\ act}$ calculated by just tensile stresses gives a good stress control criterion without having to consider shear stresses. The reason can be seen in the fact that the coil construction and holding principle does not allow critical shear stresses.

3.2.3 Condcutor temperatures: $\partial_{cu} < \partial_{max}$ affords fast control of the critical coil temperature. ∂_{cu} has to be calculated in steps of less than 50 ms (equal to σ-computing):

$$\partial_{cu} = \partial_{flow} + \sum_{i=1}^{n} \frac{1}{\alpha}\Big(exp(\rho_o \cdot j^2 \alpha \cdot \Delta t_i)/(\gamma \cdot c)\Big)$$

It is not possible to measure ∂_{cu} because of the high voltage insulation, which cannot be destroyed by installing sensors for measuring.

3.2.4 Coil positions: Positions should be controlled in the vertical and radial directions. The controls provide information on a) thermal and electromagnetic stresses (indirectly) and failures within the coils; b) the absolute positioning of the coils within the experiment, and c) the conditions of the holding facilities. Different criteria have to be met, such as deformation controls for different diameters of each coil

$$\Delta D = \Delta R_1 + \Delta R2 < 1.1 D_o(\alpha \Delta \partial + \sigma_T / E),$$

position controls for different radii of each coil

$$R < R_{max},$$

and several vertical position controls of each coil

$$Z_{min} < z < z_{max}.$$

These controls will be carried out at intervals of less than 50 ms, so that the basis for ΔD control is the computed values of $\Delta\partial$ and σ_T.

4. REACTIONS TO CONTROLLED VALUES

Positive control results do not interrupt experimental cycles. Negative control results cause different reactions within the phases of the cycles. Failures within the cooling system (pressure, flow, or temperature) necessitate a water cooling stop in the coils concerned for all phases of experimental cycles. If a failure occurs during the current pulse time, it is additionally necessary to set all coil currents to zero under normal operation conditions. When currents, stresses, conductor temperatures, or coil movements are found to be too high, it is only the currents of all coils that must be immediately reduced to zero, whereas the water flow continues.

5. SENSORS

For measuring the physical data we use sensors giving analog current or voltage values transferred by isolation amplifiers to the analog/digital convertors of the PLC or VME computer.

All currents will be measured with shunts in the coil circuits and in the power station.

The water flow is measured by inserting standardized orifice plates in pipelines and recording their differential pressures being analogons to the flows. The differential pressures of each orifice plate, chosen between 0.03 % and 0.05 MPa, will move a disc forming part of a capacitor within a convertor system and produce a change in capacitance and also in the output signal of the convertor. The maximum distance between the orifice plate and the coverter can be 20 m, so that the electronic convertor can be placed within low magetic field regions beside the tokamak. Several other systems of flow measurement were tested but were found to be unserviceable within high magetic fields.

The pressure will also be measured within low magnetic field regions. The pressure bends a membrane which is surveyed by a wire strain gauge. The change of the wire resistance being equivalent to the pressure is a measure influencing the output signal of the convertor.

The sensors for measuring the temperatures are "PT 100" resistors. Their resistance values are dependent on the temperature so that the voltage change is a measure of the temperature if the current flow in the

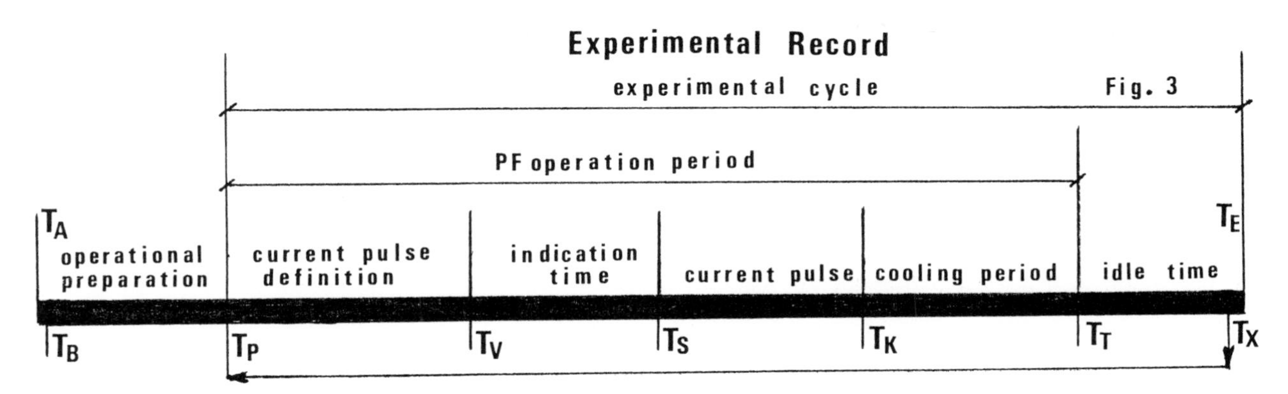

resistor is constant and the heating energy of the current can be neglected. This way of measuring is independent of the magnetic field of a tokamak.

Coil movements will be expressed in voltages if using potentiometers with tracers which will be set correspondingly to the distortions of the coils. We observed that no influence is exerted by tokamak magnetic fields on the measurement. The resolution of the measurement is a question of the mechanical precision of the potentiometers and is in the region of 0.01 mm.

6. CONCLUSIONS

The PF coil protection necessitates the control of coolant data like flows, pressures, temperatures, and currents including current-dependent values like mechanical coil stresses, the orientation of the coils, and the conductor temperatures. The coolant data control with a PLC will be performed within long time intervals ($\sim$ 200 ms), and in case of a defect the cycle of operation has to be interrupted. All the currents and current-dependent values must be controlled within short time intervals ($<$ 50 ms) to make possible a high efficiency of the tokamak. A VME computer or a transputer will calculate the current-dependent values and execute the controls. Illicit results immediately stop current increase in all PF coils and reduce the currents to zero within normal time intervals.

REFERENCES

W. Köppendörfer et al.,
The components of the ASDEX Upgrade tokamak system and the assembly;
15th SOFT, paper A02.

A. Wieczorek et al.,
Electric power circuit for plasma stabilization in ASDEX Upgrade;
15th SOFT, paper B56.

H. Bruhns et al.,
The control system of ASDEX Upgrade;
15th SOFT, paper K02.

FUSION TECHNOLOGY 1988
A.M. Van Ingen, A. Nijsen-Vis, H.T. Klippel (editors)

COIL PROTECTION CALCULATOR FOR TFTR

R. J. Marsala, J. E. Lawson, R. G Persing, T. R Senko and R. D. Woolley

Princeton University, Plasma Physics Laboratory, Princeton, N. J. 08543, USA

1. Overview

A new coil protection system (CPS) is being developed to replace the existing TFTR magnetic coil fault detector. The existing fault detector sacrifices TFTR operating capability for simplicity. The new CPS, when installed in October of 1988, will permit operation up to the actual coil stress limits by accurately and continuously computing coil parameters in real-time. The computation will be done in a microprocessor based Coil Protection Calculator (CPC) currently under construction at PPL. The new CPC will allow TFTR to operate with higher plasma currents and will permit the optimization of pulse repetition rates.

The CPC will provides real-time estimates of critical coil and bus temperatures and stresses based on real-time redundant measurements of coil currents, coil cooling water inlet temperature, and plasma current.

The critical parameter calculations are compared to prespecified limits. If these limits are reached or exceeded, protection action will be initiated to a hard wired control system (HCS), which will shut down the power supplies.

The CPC consists of a redundant VME based microprocessor system which will sample all input data and compute all stress quantities every ten milliseconds. Thermal calculations will be approximated every 10ms with an exact solution occurring every second. The CPC features continuous cross-checking of redundant input signal, automatic detection of internal failure modes, monitoring and recording of calculated results, and a quick, functional verification of performance via an internal test system.

2. Present Protection System.

The present TFTR field coil control systems are implemented in a hybrid fashion including dedicated analog feedback electronics (the Plasma Position and Current Control System, or PP&CC) and a Power Conversion Computer (PCC). The present protection, which is implemented in analog electronics, calculates I^2T and |I| for each coil system and commands the coil power supplies to shut down when necessary to limit these quantities. Cool-down of the coils is simulated by a single exponential decay of the integrated I^2T. This simulation is fairly good for air-cooled busses and cables, but is inaccurate for TFTR's water-cooled coils. A compromise has been made in selecting the I^2T threshold and decay time constant so that a single channel protects both the air-cooled and water-cooled components. This greatly reduces the system capability.

Feedback control algorithms in the PCC do not explicitly take into account these constraints on I^2T and |I|. Although operations people specify the desired plasma position and current as functions of time, the resultant PP&CC commands to drive current through the EF and OH coil systems are also dependent on factors which are difficult to predict (performance of auxiliary heating, gas injection, plasma behavior). For that reason, the PCC contains its own additional I^2T and |I| protective algorithms in software, which can also command coil power supplies to shutdown. The software I^2T and |I| limits are set slightly lower than the analog hardware protection, so that the protection circuits are not routinely challenged on many "shots".

3. The New CPS

The CPC accepts signals representing the coil currents, plasma current, coil cooling inlet water temperatures, and mode signals. All input data is sampled and all stress calculations are completed every 10ms. The 10ms rate was chosen so that the worst case stress would not exceed 3000 psi. Stress overshoot is due to the rise in current between the time the actual current passes the threshold value and the termination of the current rise by the shutdown mechanism and is a function of the sample rate, the time for computation, and the delays in the input filter and the shutdown hardware.

The calculated stress and temperature values are compared against limits stored in ROM, and if the limits are reached or exceeded, a command to the HCS is issued which will cause the power supplies to shut down and bypass rectifiers to fire, effectively shorting the coils. This protective action is known as a Level 1 fault.

3.1 Hardware

The CPC consists of two redundant RFI equipment racks, each capable of stand alone operation. Each rack contains a Input/Output (I/O) chassis, a VME based processor chassis, and two CAMAC crates and an IBM clone computer used for testing the CPC.

3.1.1 Redundancy Requirements

The CPS will no longer directly limit I^2T and |I|. Instead, the CPC limits will be on coil stresses and temperatures. Because these quantities are more difficult to calculate, and because TFTR is not now changing its PCC, it is not practical to include algorithms for coil stresses and temperatures in the PCC. Since operations people cannot accurately predict EF and OH current waveforms in advance, CPC initiated terminations of a shot will not be uncommon. To prevent single-point CPS failures from allowing shots to terminate too late, TFTR will be provided with two redundant CPS's. The test system will be used daily to check CPS operation.

3.3.2 Hardware Description

The I/O chassis accepts all analog input signals, filters the inputs and upon command from the microcomputer, holds the value for digitizing. The chassis also accepts digital signals, buffers the signals and provides any decoding necessary. A control panel on the I/O chassis allows the operator access to vary a limited subset of the operating parameters as well as monitor the status of the CPC's input, output, and internal parameters. The interfaces between the CPC and the Level 1 Fault lines of the HCS and between the CPC is also accomplished in the I/O chassis. The chassis also provides internal checking of the microcomputer in order to detect computer failures.

The Processor chassis consists of a M68020 microcomputer module, a memory module, a serial input module, two 16 channel analog to digital, two 16 channel digital to analog, and three 48 bit digital input/output modules all operating on a VME bus.

The testing system consists of a MS DOS computer interfacéd to the CAMAC crates by a IEEE-488 bus. All of the test input and output signals are supplied from or sourced to the CAMAC equipment. Much of this equipment is also used by CICADA (the TFTR data acquisition system) for data acquisition. The tester runs a canned test program which applies inputs to the CPC, analyzes the results, and provides a pass/fail indication. With additional software the tester can be used for detailed trouble-shooting of the CPC.

3.2 Input Description

Coil currents - The coil current signals used for all calculations are provided by Hall effect current transducers. In order to assure the integrity of the protection current signals, the outputs from each of the redundant current transducers used for current control are also transmitted to the CPC. The CPC compares the control signals with the protection signals and creates a Level 1 fault if the two signals deviate by more than a predetermined amount.

Plasma Current - The magnitude of the plasma current, I_p, is obtained by integrating the outputs of two redundant Rogowski coils. This provides redundant outputs proportional to I_p. The Ip signals are digitized upon command from the CPC and transmitted together with parity and validity bits from the personnel unsafe Data Acquisition Room (DARM) to the CPC located in the personnel safe Field Coil Power Conversion (FCPC) area via a fiber optics link. The CPC accepts the serial bit stream, checks for validity and compares the two redundant values. If all checks are passed, the Ip data is used in the magnetic stress calculation. If any of the checks fail, a Level 1 fault will be issued.

Inlet Water Temperature - The computation of coil temperatures requires an accurate measurement of the coil system inlet water temperature. The water temperature is monitored by redundant sensors. The temperatures are transmitted as analog signals to the CPC where the CPC checks the values against each other and against expected limits. If the checks are passed, the CPC uses the data in coil temperature computations.

Water Flow - The coil temperature and thermal stress calculations depend upon a minimum water flow in the coils. If the flow is less than this minimum, the computation will under estimate the coil temperatures. Redundant signals from the water flow switches located on each coil will be transmitted to the CPC indicating that the flow for all coils is above the minimum value. A level 1 fault is issued if either flow signal indicates insufficient flow.

Modes - The mode signals include both operational modes and coil connection modes. The two operational modes are high power pulsing (HPP) and pulse discharge cleaning (PDC). The allowable stress limits are reduced in the PDC mode.

There are two common coil connection modes used on TFTR which affect the OH and EF coil systems. In Mode A the OH6 coil is left open, while in Mode B the EF1 coil is removed from the EF system and inserted along with the OH6 coil into the OH system. Since the algorithms and coefficients differ for the two modes, the CPC must be provided with the mode information. The mode is decoded from limit switches and relays located on the switches and bus links in the coil power system and transmitted to the CPC as two mutually exclusive signals. The CPC uses the decoded mode information to select the proper algorithm and coefficients.

The Variable Curvature, VC, coil can be inserted into the EF system in either polarity or left unconnected. The CPC is provided with three mutually exclusive inputs indicating the status of the VC coil. The CPC uses the decoded mode information to select the proper algorithm and coefficients.

Internal Support Structure Temperature - The maximum allowable stress allowed is a function of the temperature of the internal support structure (ISS). When the structure is cool, the CPC allows high power pulsing at the maximum stress levels. If the ISS is warm, the CPC limits the stress values to a lower level. If the structure is hot, no pulsing is allowed. The three temperature levels (cold, warm, and hot) are transmitted to the CPC from a separate system as three mutually exclusive signals.

3.3 Software Description

Every ten milliseconds the software will do the following steps:

1. Strobe the track and hold amplifiers in the I/O chassis
2. Read the 32 analog inputs from the analog to digital converters
3. Read the digital inputs
4. Convert the data into engineering units

5. Read the plasma current
6. Check the inputs for validity
7. Check for machine cycle time constraints
8. Do 1/100th of the slow thermal model
9. Do the fast thermal model
10. Do the stress calculation
11. Compare the temperatures and stresses against limits
12. Write the analog outputs to the digital to analog converters
13. Write the digital and Fault outputs
14. Wait for the 10ms interrupt.

The 68020 microprocessor with a 68881 floating point processor operates with a 16.7MHz clock. Currently the execution time is 8.6ms out of the total 10ms available. All calculations are done is 64 bit floating point arithmetic. The majority of code is written in "C". The RAM usage is about 22K bytes, while the ROM usage is divided into 29K for the program and 34K for look up tables and other constants

3.4 Functional Description

a) Thermal Simulation of Air-Cooled Busses- The temperatures of three air-cooled busses are calculated by solving the differential equation of the form:

$$\tau \frac{dT}{dt} T + T = KI^2$$

where T is the characteristic temperature of the cable or bus, t is time, τ is the cool-down time constant of the cable or bus, and I is the electrical current flowing in the cable or bus.

The differential equation solution is performed in real-time, using a difference-equation updated every 10ms.

b) Thermal Simulation of Water-Cooled Coils -The temperatures of the water-cooled coilsare computed by solving partial differential equations of the following form:

$$\frac{\partial T_c}{\partial t} = a_1(1-\alpha T_c)\, I^2 - a_2(T_c - T_w) - a_3(T_c - T_e)$$

$$\frac{\partial T_w}{\partial t} = a_4(T_c - T_w)\, I^2 - V_{H20}\frac{\partial T_w}{\partial x}, \quad \frac{\partial T_e}{\partial t} = a_5(T_c - T_e)$$

for $0 \le x \le L$

with the boundary condition

$T_w|_{x=0} = T_{inlet}$ where:

T_c is the copper temperature (a function of position, x, and time, t),

T_w is the water temperature (a function of position, x, and time, t),

T_e is the epoxy insulation temperature (a function of position, x, and time, t),

T_{inlet} is the cooling inlet temperature, (a function of time, t),

V_{H_2O} is the water flow velocity from x=0 to x=L,

α is the temperature coefficient of copper resistivity,

L is the length of the coil,

I is the current through the coil,

and a_1, a_2, a_3, a_4, and a_5 are assorted geometry dependent constants.

The numerical solution is obtained by solving difference equations derived from the partial differential equations (using a finite number of nodes for each coil). The fast thermal analysis, which includes the I^2R heating of coil copper, the thermal variation of copper resistivity, and the local thermal heat transfer between copper and cooling water, provides an accurate model of the heat-up of the coil during a TFTR pulse, but is not accurate during the long cool down cycle between pulses. The fast thermal analysis predicts the coil hot spot temperature only, and does not predict the detailed spatial variation of temperature in the coil. The slow thermal analysis, which is completed every second provides a detailed finite element analysis of the epoxy, copper, and water which varies from 16 nodes for the EF1 coil to 31 nodes for the EF4 coil. This slow model uses accumulated data from the fast simulation to model the I2R heating, thermal variation of resistivity, and thermal equilibrium between water and copper. The slow model also includes the effect of motion of the cooling water and heat flow between copper and the epoxy insulation. A temperature predictor adds an additional temperature increment (representing the energy stored in the inductance of the coil) to the computed copper temperature (at the outlet node). This stored energy would cause further heating of the coil if the coil were shorted as a result of a fault.

c) Stress Calculations - The total stress on the TFTR coils is the sum of the predicted thermal stress and the calculated magnetic stress.

The thermal stress is calculated by multiplying the simulated copper temperatures by coefficients stored in the microprocessor ROM. The effect of a coil's temperature on its own epoxy shear stress is obtained by multiplying the temperature rise of its outlet node by a coefficient. For multiple coils in a stack, the effect of the temperature of one coil on the epoxy shear stress of another coil is obtained by multiplying the average copper temperature by a suitable coefficient.

The vertical and radial magnetic stress forces are calculated for each of the 18 PF coils as the product of coil current times a weighted sum of the currents in the PF coil systems and plasma current. Geometric weighting coefficients for the coil systems are stored in the CPC ROM. Geometric coefficients from the plasma current are directly dependent on the plasma position. The position information is not an input to the CPC because of financial and other practical considerations. Instead the plasma current, I_p, is divided by the TF current, I_{TF}. The ratio is used as an index to look up the plasma field and inductive coupling coefficients in a ROM table, for each coil. The table was created using the worst-case geometry for each ratio.

The epoxy insulation shear stress is normally limited to 2750 psi. During pulse discharge cleaning operation, the stress is limited to a lower level.

d) Level 1 Faults - Level 1 faults are issued whenever any of the coil currents, the plasma current or any of the computed stresses or temperatures exceed upper limits which is stored in ROM.

4. Expected Improvements

The new CPS will allow TFTR coils and busses to be driven to their maximum allowable limits. The total stress is composed of the sum of thermal and magnetic components. Thus when the coils are cool, as during OH precharge, more magnetic stress can be tolerated allowing a higher precharge coil current. The OH current is presently limited to +21.5KA during precharge and -21.5KA during reconduct with a thermal limit of 1.45E9 A^2sec. The new CPS will permit a +28.0KA precharge and a -21.5KA reconduct assuming an I_{EF} of 24KA, an I_p of 2.5MA, and a maximum stress of 2750psi. This produces a 22% improvement in the total OH current swing and should produce an approximate 22% increase in the maximum plasma current.

While there will be similar improvements in the allowable limits for the other (EF, VC, HF, and TF) coil systems, the operation of TFTR at this time is limited by the OH system capabilities. If other mechanisms of current drive were added to TFTR, the CPS would allow the other coil systems to be driven harder, producing a much longer shot duration.

Acknowledgement

This work was supported by the Department of Energy under contract No. DE-AC02-76CH03073.

FUSION TECHNOLOGY 1988
A.M. Van Ingen, A. Nijsen-Vis, H.T. Klippel (editors)

DESIGN AND FEASIBILITY ASSESSMENT OF THE NET CENTRAL SOLENOID

N. MITCHELL[1], G. MALAVASI[1], J. MINERVINI[1], R. POEHLCHEN[1], E. SALPIETRO[1], E. WARSCHEWSKI[1], J. RAUCH[2], H. BENZ[2], TH. ROMAN[2], D. SALATHE[2], A. SEGESSMANN[2]

[1]The NET Team c/o Max-Planck-Institut fuer Plasmaphysik, Boltzmannstr. 2, D - 8046 Garching b. Muenchen
[2]ABB Zuerich, Werk Oerlikon, CH - 8035 Zuerich

The central solenoid of the NET machine occupies a critical region in the centre of the machine and exerts a significant influence on the overall size of the tokamak. It is therefore desirable to exploit this coils up to its technical limits, considering the materials and manufacturing possiblities. Following an initial layout produced by NET a detailed feasiblity study was performed by industry in collaboration with NET. Subsequent to this study, the solenoid design and the machine operating scenarios were iterated to produce full consistancy between the two. This process has resulted in a 13 m high cylindrical solenoid stack comprising 3 independant pairs of coil sets. The coils will utilise A15-type cable in conduit superconductor with a conductor current of 40 kA. A wind and react manufacturing method is proposed for the coils due to the tight conductor bending radius of about 1.5 m. The coils are insulated with vacuum impregnated glass fibre reinforced epoxy, to give a terminal voltage and ground voltage limits of 20 kV. The stability and cooling properties of this conductor have enabled a coaxial copper insert coil, originally proposed to reduce Bmax and dB/dt on the superconductor, to be eliminated. The conductor is now the subject of a development programme.

1. INTRODUCTION

The development of the NET central solenoid has been an iterative process involving three main aspects, firstly the currents required to form the plasma scenarios, secondly the design concept for the coils and conductor and thirdly the analysis of the mechanical, electrical and thermal problems of the coils. Rather than presenting the development as an historical process we will show the final design choice and its analysis, indicating where appropriate the reasons that led to this choice.

The starting point for the solenoid design follows from the broad design philosophy for NET as it relates to the poloidal field (PF) coils

- These coils must be individually removable without disturbance to the rest of the machine, particularly the toroidal field (TF) coils.
- The load paths within the individual coils must be clearly defined and the coils themselves should be designed to allow well defined analysis procedures to be followed.
- The coils should be superconducting so that the application to a fusion reactor of superconducting technology can be demonstrated.

The first two requirements led to the choice of a straight solenoid stack, structurally self supporting and able to be withdrawn vertically from the machine. The third requirement does not directly influence the design since for a machine with the operating requirements of NET, superonducting coils offer clear advantages over normal conducting in terms of cost and performance[2]. Earlier NET concepts had a copper insert coil inside the main solenoid stack to boost performance, but detailed design and analysis led to its elimination and replacement with extra turns of the superconducting coils.

A final factor which has a major influence on the design comes from the plasma configuration. NET is required to produce a range of plasma shapes with both inductive and non-inductive plasma current scenarios. It is readily shown that this requires separate control of individual current "blocks" within the solenoid, and that there can be substantial differences in current between these blocks. Further, to make most effective use of the solenoid, the coils must all have a dual function in generating both the fields to confine the plasma and the flux to drive it.

2. CENTRAL SOLENOID DESIGN

The overall layout of the solenoid[1] is shown in fig. 1. The parameters are summarized in table 1. The coil stack has an overall height of 13 m and is subdived into 13 1 m high units. These are grouped into coils blocks as follows:

Table 1

Main Coil Data

Conductor Current	40 KA
Maximum field	12 T
Maximum Field Gradient	2.5 T/s
Cooling Length	134 m
Total Solenoid Weight	760 t
Inner Coil Radius	1.18 m
Outer Coil Radius	1.68 m
Total height	15.8 m

3-1-5-1-3, so that the centre coil is 5 m high, 2 x 2.5 m each side of the machine equator. The vertical support of the coils is provided by a segmented insulated cylinder (to reduce eddy current losses) running down the inside of the stack and connected by flanges to the coils at top and bottom. This structure allows precompression of the stack.

The superconductor used for the coils is described in a separate paper (6), but basically consists of a ternary Nb_3Sn type superconductor formed into a cable in conduit type conductor. The conductor has a thick

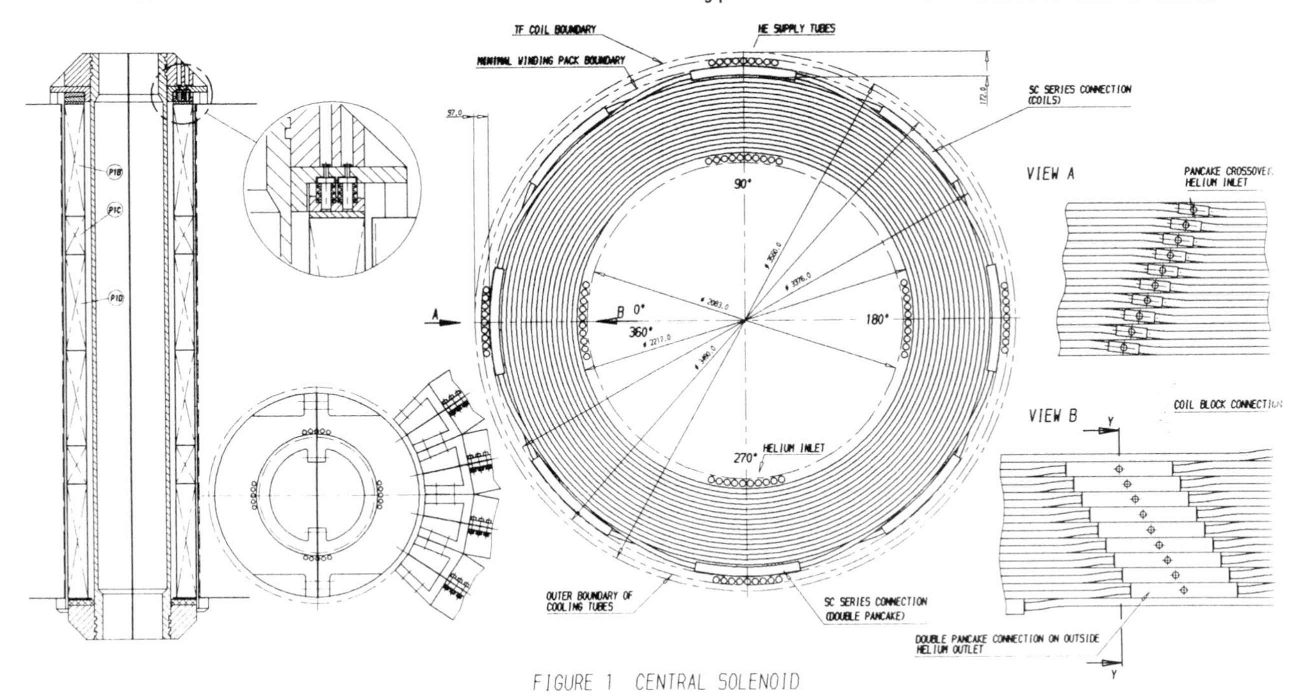

FIGURE 1 CENTRAL SOLENOID

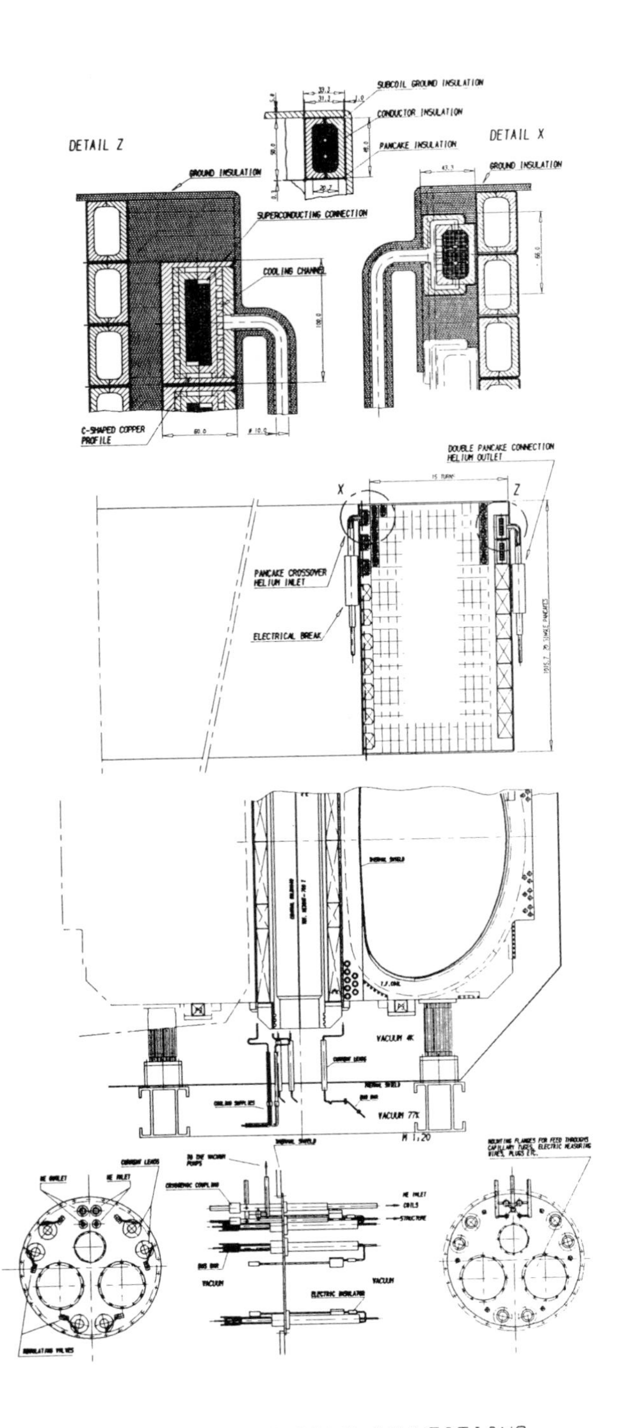

FIGURE 2a SUPPLY CONNECTIONS

(5 mm) steel jacket which transmits both vertical and hoop forces in the coils. The conductor current is 40 kA maximum. A high current is necessary to reduce voltages on the coils during the plasma scenario but the current is limited by the need to reduce deformation of the conductor crossection during winding and by stress intensification areas in a jacket with a larger crossection. A cable-in-conduit type conductor gives a good superconductor stability in the changing poloidal field due to good superconductor-helium contact, and the relatively short cooling channel lengths (150 m) allow effective heat extraction with the helium with reasonable pressure drops (0.06 MPa) and mass flow rates.

The conductor is insulated with a special "R" glass insulation before winding and reaction to form the Nb_3Sn superconductor (700°C for at least 8 hours). Tests have shown that this glass can resist the reaction temperature without degradation of its mechanical properties. After reaction, the coil is vacuum impregnated with epoxy resin. The coils are wound as double pancakes to enable the electrical connections to be placed in the outer low field region. Additional interpancake and ground insulation are added after winding. The helium inlet is on the inside at the crossover in the double pancake, and the two helium outlets correspond with the double pancake electrical terminals on the outside.The principal design of the electrical and helium connections is shown in fig. 2. The final geometrical form of the conductor ends are prepared before reaction of the conductor. After reaction the strands are soft soldered to the copper profiles and welded under compression into a steel jacket.

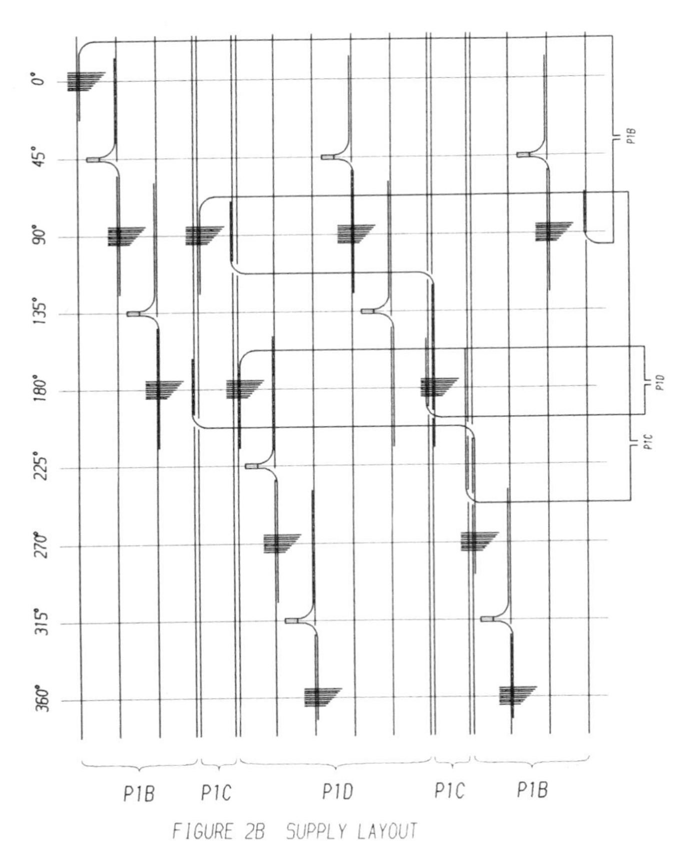

FIGURE 2B SUPPLY LAYOUT

3. SOLENOID ASSEMBLY

The solenoid is assembled outside the machine and then installed as a single unit. The assembly procedure is as follows

a) Assembly of bottom flange and central support tube
b) Mounting of the individual coil blocks onto the bottom flange and consecutive shiming of the mating surfaces
c) Assembly of the top flange
d) Prestressing of the stack
e) Series connection of the coils, installation of coolant supply and instrumentation.

4. COIL CONFIGURATION

The original PF coil positions and currents for NET were derived before the design and analysis of the coils (8); and have now been rederived. With the previous arrangement of the P1 coils substantial outward vertical forces occurred on the P1B, C coils (over 500 MN). These were generally caused by the P1B coil carrying a current of opposite sign to the P1D, especially at the start of the burn and during heating. The effect was compounded by the high current in the P2 coil (between P1B and P3) attracting P1B and by the current in P1C, adjacent to P1D, being of opposite sign.

The central stack was therefore reconfigured, reducing the height of the P1C coil by 50 cms and increasing that of P1B by 150 cms. During heating and burn the current in P1C has been kept to near zero. This separation between coils with opposing currents reduced helps to reduce the forces to acceptable values. The increase in size of P1B enables the current from P2 to be transferred to P1B (with some corrections using the coil P3). The P2 coil has been eliminated, further reducing the vertical outward forces (to under 70 MN).

The new current scenarios for the start up and burn portions of the technology phase (plasma current 10.8 MA) and physics phase (15 MA) plasma configurations are shown in figure 3, as the plasma develops from initiation to an elongated limiter plasma to a double null low

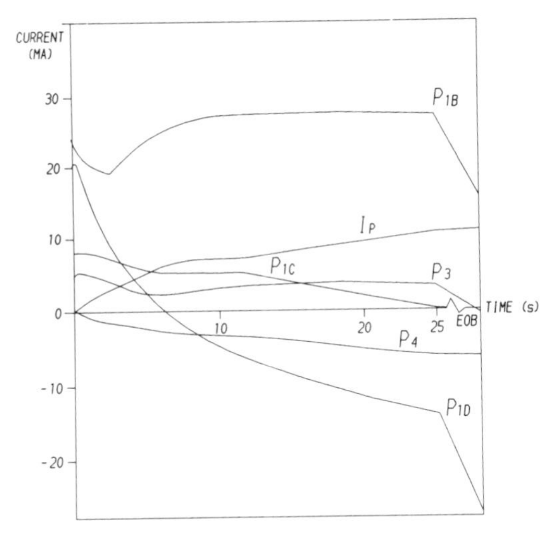

PF COIL CURRENT VARIATIONS DURING INDUCTIVE START UP OF AN 11 MA D.N.PLASMA

FIGURE 3

beta plasma on the inner equatorial limiter, to a "free" low beta double null plasma and to start and end of burn[5]. The configurations at start of burn also include an RF assisted scenario, where RF current drive during start up is provided and the burn length is limited only by the central solenoid mechanical limits.

The timescale for the start up process is driven by the need to keep the power transfer rates reasonable. The scenario shown in figure 3 requires a peak total input power to the coils of under 150 MW. The peak voltage on the P1 solenoid coils is 1-20 KV on each coil of an up-down pair, reached during the first 3 seconds of start up. The dB/dt values on the coils peak at about 2.5 T/s over the first 5 seconds of start up. The plasma initiation is performed with RF assistance to reduce the resistive part of the breakdown loop voltage. Thus a loop voltage of about 10 - 15 V (0.3 V/m) is required for plasma initiation, corresponding to an initial plasma current ramp rate of 1 - 1.5 MA/s. Previously, it was considered that NET would require a loop voltage of 30 - 35 V for initiation, which gives higher dB/dt and makes the superconductor design more severe. This was avoided by the use of a normal conducting start up coil inside the superconducting stack. The reduced dB/dt at start up allowed the replacement of this coil by extra superconductor.

5. COIL ANALYSIS

5.1 Operating Stress Analysis

The limiting hoop and pressure stress has been established using a detailed 2-D axisymmetric stress analysis of the P1 solenoid (1). The elements of this stress analysis are given in figures 4 and 5. The P1D coil was initially represented by global elastic properties which produced the global stress levels shown in figures 5 (a) and (b). The lower part of the coil was then analysed in detail using the mesh shown in figure 4, with boundary displacements imposed from the global analysis. The mesh is refined down to a single conductor. In this analysis it is assumed that all conductors are vertically aligned. A 3-D analysis of the crossover/terminal region is also required. The conductor jacket total Von Mises stress is shown in figure 5c. The superconductor cable within the conduit is assumed to make no contribution to the coil strength. These analyses enable the stress concentration factors of the global (average) stress levels to be derived and applied to the other P1 coils. The generalised stress limits are shown in table 2 for the coil bursting and pressure forces. Total stresses are limited to a Von Mises value of 600 MPa and a fatigue limit is given in reference (9). It should be

FIGURE 4 FE MESH

FIGURE 5

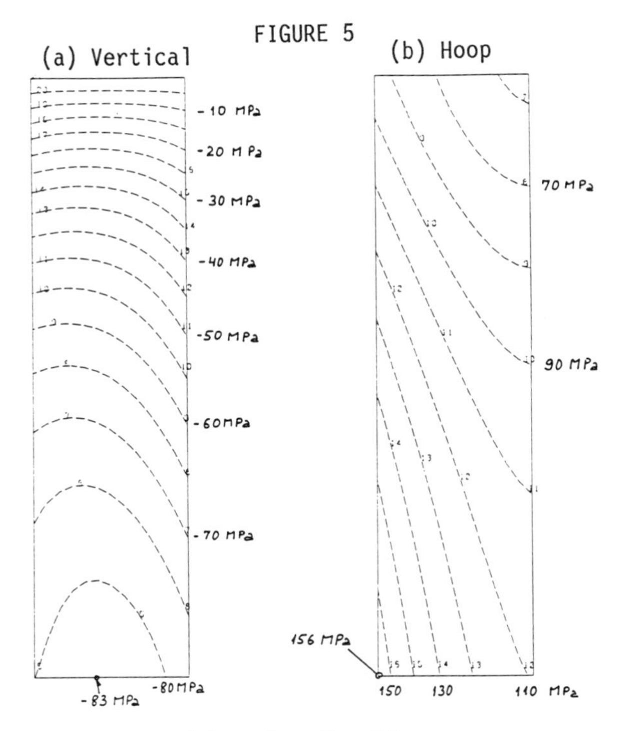

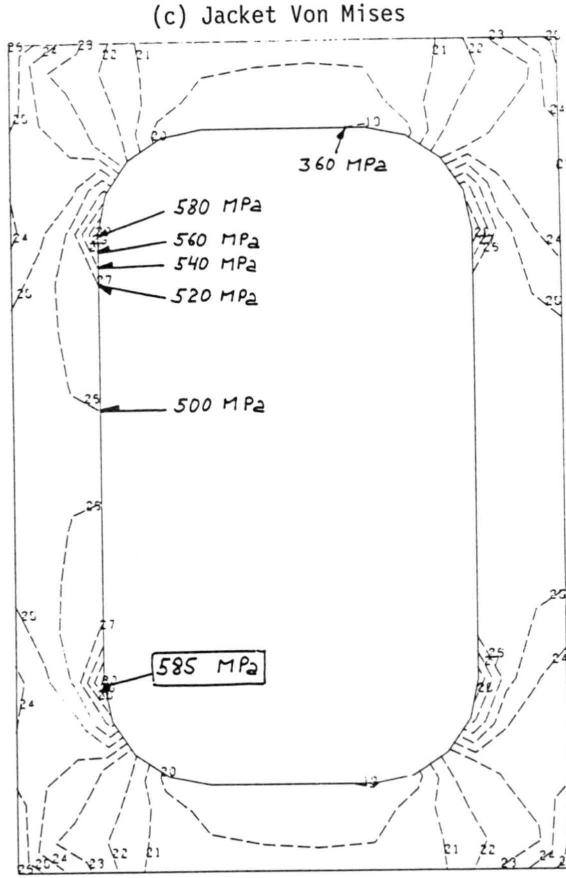

TABLE 2

Generalised Stress Limits in Central Solenoid

Applied Loads: Average hoop 130 MPa
Average Pressure -80 MPa
(peaking factor of hoop is 1.2 globally)

Stress limits on jacket peak values (reference 9)
Hoop (tensile) < 470 MPa (fracture criterion)
Pressure and hoop (Von Mises criterion) < 600 MPa

Calculated peak stresses on jacket
Hoop (tensile) 400 MPa
Von Mises 585 MPa
(components: -425 pressure and 280 tensile)

Stress concentration factors to peak jacket stresses
Average hoop (H) 3.1
Average pressure (P) 5.3

Limits: $3.1\ H < 470 \ = H < 152$ MPa

$$28\ P^2 + 9.6\ H^2 + (5.3\ P + 3.1\ H)^2 < 600$$

The Von Mises Limit assumes that pressure and hoop stress peaks are coincident, whereas they are not. It can therefore be considered pessimistic.

noted that this limit applies to the peak stress in the jacket as determined by the finite element analysis. This peak stress (fig. 5c) occurs over a very limited region of the jacket and can thus be taken as fairly conservative.

The scenarios shown in figure 3 give stresses that are at these limiting values.

5.2 Precompression of the Solenoid Stack

The central solenoid stack carries a range of currents in its 3 sets of superconducting coils. These create magnetic forces which can lead to compression or to outward separation of the coils in the vertical direction, in addition to the hoop (bursting) forces. While these forces occur it is necessary that the coils remain in contact with each other, since gaps can lead to rotation of the coil sets, sudden impact forces and wear of the insulation. To achieve this contact, the coil stack must be precompressed using the central support inside the coils.

The coil has been analysed using a one dimensional model of the coil and support (3), with the magnetic forces calculated numerically. The load cases considered are

a) At maximum transformer prebias, with peak compression of the solenoid coils

b) At start of burn with RF assisted start up, when peak outward forces on the PIB coils occur.

The necessary room temperature precomression of the stack is calculated. The model includes differential thermal contraction between support and coils and poissons contraction of the coils due to the hoop strain. A number of potential structural materials were considered, and the results are summarised in table 3. The

Table 3

Precompression of the Solenoid

Support Material	Typical Cryo. Alloy Name	E GPa (20K)	TYS 297 K	(MPa) 20 K	% Contraction 273 - 4 K	Support Pretension 4K	273 K
Al	2219/T89	80	383	512	- 0.35 %	200	175
Fe	316 LN	210	278	614	- 0.30 %	230	210
Ni	Inconel x-750	210	1000	1100	- 0.222 %	230	290
Ti	Ti-5Al-2.5 Sn	110	760	1270	- 0.152 %	230	330
Fe	304L	210	406	522	- 0.275 %	230	260
Ni	Incalloy 9 x A	210	961	1070	- 0.2 %	230	300

support should have a thermal contraction greater than that of the coils, since otherwise the precompression stress at room temperature is rather high. With a steel jacket on the conductor, the most appropriate materials for the support are steel or aluminium. However, the aluminium gives substantial extra loading with load case a) in order to satisfy load case b). This is also true for Titanium. The pretension on the steel required to satisfy load cases a) and b) is roughly similar. Thus steel seems the best support material.

If an Incolloy jacket material can be used, the coil thermal contraction reduces to -0.22 %. In this case the steel support preloading at room temperature is reduced to about 110 MPa, the remainder being provided by the differential contraction. Thus an incalloy jacket offers advantages as regards the overall coil design when used with a steel support structure.

6. CONCLUSIONS

The central solenoid for NET has been designed and analysed in sufficient detail to confirm its overall feasibility. Manufacturing and assembly problems have also been considered. Since the solenoid is one of the most critical machne components, iterations of the overall plasma/coil configuration as well as the coil design have ensured that the solenoid operates to the limits of, but not beyond, its mechanical capability.

Aspects which should be considered further, although not an overall feasibility issue, are relative radial movements between coils in the stack under differential hoop forces, and detailed stress analysis of the pancake crossover regions in the coils.

REFERENCES

1. Final Report of the Feasibility Study of the NET DN Inner Poloidal Field Coils, BBC Report HIMS-20-313, Zuerich, May 1987.

2. R. Poehlchen, J. Rauch et al., The Central Solenoid of NET, 14th SOFT, Avignon, 1986.

3. N. Mitchell, Precompression of the NET Central Solenoid Coils, NET Internal Note 81/IN-042, 1987.

4. N. Mitchell, Poloidal Field Coil Configuration for the NET Reference and Extended Plasma Options, NET Internal Note 87/IN-016, 1987.

5. NET Status Report, NET Report No. 84, Dec. 1987.

6. J. Minervini et al., Conductor Designs for Superconducting Poloidal Field Coils of NET, 15th SOFT, Utrecht, 1988.

7. Final Report of the Feasiblity Study of the NET Toroidal Field Coil System, ABB Report HISM 20 376, Zuerich, May 1988.

8. N. Mitchell, Plasma Equilibria Capabilities of the NET Poloidal Field Coil System and the Impact of RF Current Drive, NET Internal Note STA/02-1-87.

9. N. Mitchell, Structural Assessment of the NET Toroidal Field Coils, NET Report No. 69, May 1987.

FUSION TECHNOLOGY 1988
A.M. Van Ingen, A. Nijsen-Vis, H.T. Klippel (editors)

CONDUCTOR DESIGN FOR SUPERCONDUCTING POLOIDAL FIELD COILS OF NET

J. MINERVINI[1], R. POEHLCHEN[1], M.V. RICCI[1], E. SALPIETRO[1], N. MITCHELL[1], A. TOROSSIAN[2], B. TURCK[2], P. BRUZZONE[3], H. BENZ[3], J. RAUCH[3], P. BLASIO[4], S. CERESARA[4]

[1] The NET Team c/o Max-Planck-Institut fuer Plasmaphysik, Boltzmannstr. 2, D - 8046 Garching
[2] CEA/CEN de Cadarache, B.P. 1, F - 13108 St. Paul-les-Durance
[3] Asea Brown Boveri, Werk Oerlikon, CH - 8050 Zuerich
[4] Europa Metalli/LMI, Via Borgo Pinti 99, I - 50121 Florence

A programme has been launched to develop conductors suitable for the poloidal field coils of NET. This programme is being carried out in the associated European Laboratories and through direct development in industry. Three conductor designs are described which differ in the basic strand, cabling pattern, cooling channel design, method of jacket fabrication and method of inserting the cable in the jacket. An R & D programme is described which is intended to solve all the fabrication and operational problems, e.g. strand sintering, jacket material selection and fabrication, insulation development, conductor manufacturing line development, etc. An intensive programme to measure the mechanical, electrical and thermohydraulic properties of the conductor is also underway.

1. INTRODUCTION

The design of NET requires that the superconducting poloidal field coils operate efficiently and reliably under very demanding conditions of pulsed magnetic fields and stress[1]. The design of the central solenoid, for example, must satisfy the requirements of peak field of 12.5 T, peak stresses of 600 MPa, and pulsed field rate of about 40 T/s during plasma disruption[2]. The peak fields and stresses in the other PF coils are not as high as in the central solenoid, but the physical size of the coils (approx. 20 m for P4) results in difficulty with cooling long lengths of conductor and problems in manufacturing and handling[1].

The preferred conductor for this application is the forced-flow cable-in-conduit type because it can be made into a rigid winding pack with inherent electrical and mechanical advantages. The superconductor for the central solenoid will be made from Nb_3Sn due to the high magnetic fields. The small bending radius (1.2 m) and high operating current (40 kA) dictate that the wind and react process be used. In the wind and react technique the unreacted cable is first welded into the jacket, then the conductor is insulated, wound into coil form, and then reacted. The coils are epoxy impregnated after the heat treatment.

The other PF coils can use NbTi superconductor due to the lower magnetic fields. Again the preferred configuration is 40 kA cable-in-conduit because this gives a conductor with low AC loss, excellent stability, and good transient and steady-state heat transfer properties.

For both types of conductor the size, operating current, and mechanical and magnetic loads require that a programme be undertaken to

- develop the manufacturing techniques required to fabricate a conductor suitable for the PF coil
- perform testing and measurements in order to establish a data base and to qualify the conductor

- establish a basis for estimating the cost of producing the full amount required for the PF coil system.

This programme was launched in the middle of 1988 and includes participation by the associated European Laboratories and by direct contracts to industry. CEA/Cad is contributing to the development of a NbTi conductor for the equilibrium field coils and a Nb_3Sn conductor for the central solenoid. Direct industrial manufacturing experience will be gained from a conductor being developed by Asea Brown Boveri (ABB), Switzerland, and separately a conductor by Europa Metalli - LMI (EM-LMI), Italy. Further contribution to the programme comes from PSI/Villigen, ENEA/Frascati, ECN/Petten and KfK/Karlsruhe through participation in the testing programme.

2. CONDUCTOR DEVELOPMENT PROGRAMME

The R & D programme to be carried out is quite extensive and includes all sub-components up to final assembly. Much data already exists for individual components, e.g. superconducting strands, but there is still much to know about the interaction of the components or their characteristics in the operating conditions of NET. The programme is focused on developing a short length (ca 15 m) of full-size (40 kA) wind-and-react conductor. The purpose is to establish the manufacturing techniques, including quality control, and to create a data base for the conductor and components through extensive tests and measurements. Short lengths of sub-size conductor will be used for development of fabrication methods and sub-component testing.

2.1 Superconducting Strand

The strand for the central solenoid should be a multifilamentary composite wire of Nb_3Sn and stabilizing copper with a non-copper critical current density specification of 600 A/mm^2 at 4.2 K and 12.5 T. It should have a low AC loss, both hysteresis and coupling. Commercially available wire will be used for this programme. Future work will focus on strand optimization.

2.2 Cable Configuration

The cable configuration must satisfy the following (often competing) requirements:

- low AC loss by use of small strands and full cable transposition
- large wetted surface for good cooling
- large hydraulic diameter
- well distributed strand to strand contact points to avoid stress concentrations and limit the maximum transverse stress[3]
- mechanical stability for cable manufacturing and insertion in jacket
- precompression to limit strand motion and mechanical (frictional) hysteresis.

2.3 Strand Isolation

If the strands are pressed together during the reaction heat treatment it is likely that they will sinter together. This would result in a significantly increased AC coupling loss and possibly critical current degradation because the cable will act as a monolith. A surface treatment will be used that prevents sintering, reduces AC losses, is cheaply applied to long lengths and has excellent bonding strength so that it will not flake off after repeated loading cycles. The use of nickel or hard chrome plating will be investigated.

2.4 Jacket Material

The jacket material selection is very critical because it must have the excellent mechanical properties of a cryogenic grade steel in base and weld metal after being subjected to the superconductor reaction heat treatment (central solenoid conductor only). Standard commercial grade, stainless steels, such as 316LN do not fulfill these requirements after the heat treatment. The design goals for the ideal jacket material are listed in Table I. It is also desirable for the jacket material to have a coefficient of thermal expansion (COE) closely matched to that of the Nb_3Sn in order to avoid degradation of the critical curent by differential longitudinal thermal strain. Some materials being considered for the jacket are the Japanese cryogenic steels JK1, JKA1, and JN1, the iron-nickel alloy INCOLOY 908, and 316 LN modified with 0.05% Nb.

2.5 Cable/Jacket Assembly

The process of cable insertion in the jacket must not damage the cable. It is desirable to maintain a small amount of precompression in the cable after reaction in order to have load transmitted to the jacket with a minimum of strand motion. On the other hand, this compression should not be so great that the critical current is degraded due to transverse stress.

2.6 Current Termination

A means of transferring both current and helium to/from the conductor must be devised that gives a low electric and hydraulic resistance and uniform current distribution in the cable.

2.7 Electrical Insulation

A fibre reinforced epoxy resin insulation system will be developed so that the reinforcement fibres can withstand the reaction heat treatment. It must be demonstrated that the jacket-epoxy bond and insulation shear strength is equivalent to that achieved in standard insulation sytems for magnets where the reinforcing fibres have not been subjected to a heat treatment (at least 50 MPa shear strength at 4K).

TABLE I: Design for the Ideal Jacket Material

Property	Value
Yield stress (0.2%)	>1000 MPa
Tensile elongation	>10 %
Threshold for crack growth	>10 MPa$\sqrt{m}$
Fracture toughness, K_{IC}	>130 MPa$\sqrt{m}$
Fatigue life	>$2\cdot10^5$ cycles
Weldability by laser	good - excellent
Superconductor reaction Heat treatment (650 C/200 hr·750 C/50 hr)	compatible
Cobalt content	low or zero
Magnetic hysteresis	low

Note: all properties evaluated at 4 K for both base and weld metal

2.8 Production Line

Since only a short length of conductor is to be developed it is not required to establish a true production line. However, it must be demonstrated that the full-size prototype conductor can be produced in a manner feasible for producing lengths up to about 400 m.

3. CONDUCTOR DESIGNS

The elements of the development programme are being addressed by CEA, ABB and EM-LMI in different degree and by different means. CEA, working through the European Technology Programme, bases their design on experience gained during development of the POLO conductor with KfK[4]. This design uses a superconducting cable made of NbTi inserted in a conduit and cooled by supercritical helium. This conductor is dedicated to the large external coils (P4). These coils (20 m in diameter) will have to be manufactured on site and the use of NbTi as a superconducting material which does not require heat treatment could reduce significantly the overall cost. The possibility of using Nb_3Sn or NbTi/HeII with the same design is also being investigated for the central solenoid.

EM-LMI will address all the programme elements listed above with the exception of the insulation development which will be performed by ABB. ABB will also address all the programme elements and in addition will do a complete characterisation of the jacket material.

The conductor designs presented here are

preliminary and subject to change due to the developmental nature of the programme.

3.1 ABB Design

The conductor design to be produced by ABB is shown in Fig. 1. It consists of a 588 strand multistage cable surrounded by a steel jacket. The basic Nb_3Sn strand is made by the internal tin diffusion process. It will be supplied by Teledyne Wah Chang, USA. The guaranteed non-copper critical current density for this wire is 600 A/ mm^2 at 12.5, 4.2 K, E_c = 0.5 v/ cm. The strands will be coated with a hard chrome layer for strand isolation and the first cable stage will be braided. This is expected to give a good distribution of strand crossing points yielding a sub-cable with good mechanical stability, large helium transparence and minimization of transverse stress concentrations. The remaining cable stages are made by twisting and compacting.

The jacket will be made in two halves of U-profiles which will be butt welded to form the full length of jacket halves. The cable will be inserted continously between the jacket halves with a moderate and controllable amount of pre-compression. The jacket will be continuously laser welded. Corner welds are avoided in this design. The U-profiles can be provided in nominal lengths of 12 m but in the future the draw lengths can be extended to 20-30 m.

The nominal critical current for this conductor is 106 kA (588x181A). However one must allow for degradation from the single strand critical current due to several factors; a) strand or filament damage or breakage during cabling and conductor fabrication, b) longitudinal strain due to differential thermal contraction between jacket and cable, and c) longitudinal and transverse stresses due to operational electromagnetic forces. The total degradation of critical current must not decrease more than 37% of the initial value to

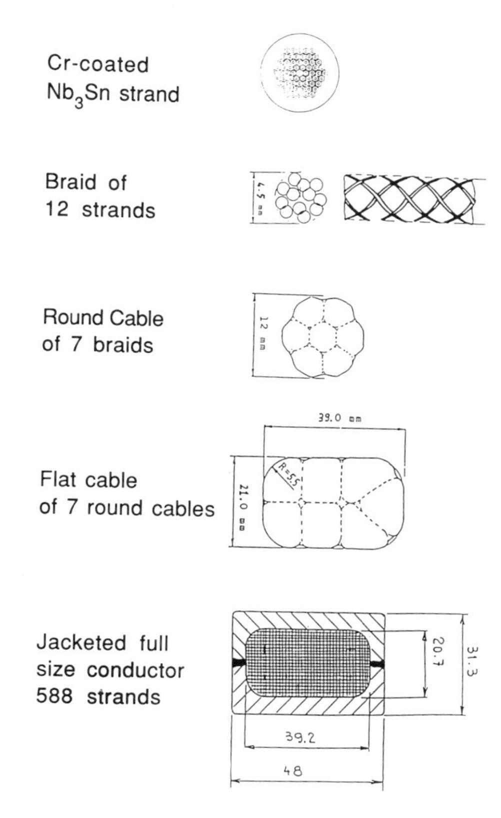

Fig. 1: ABB 40 KA Nb_3Sn Conductor

give an operating to critical current ratio of 0.6. This then leaves a temperature margin for the operational heating loads plus some stability margin.

3.2. EM-LMI Design

The 40 KA conductor design of EM-LMI is shown in Fig. 2. It is also a multi-stage cable but here the subcable stages are all made by twisting until the 4th level where the five 3rd level stages are cabled around cooling tubes made of either Cu-Ni or steel. The 5th level is cabled and shaped before insertion in the jacket.

The final strand has not been chosen yet but it will probably be similar to the ABB strand, although different in size. Both chrome and nickel plating are being considered for the strand isolation. The configuration of cabling around tubes gives two beneficial effects. The first is that the 4th level cable has a better mechanical stability. The second is that the tubes provide a large hydraulic diameter so that a large amount of heat can be removed with small pressure drop. Enough interstitional helium remains in the annular cable space to provide transient stability. The possibility of perforating the tubes to provide communication between the two helium spaces is being investigated.

The jacket profile is a hollow cross-section without any longitudinal welds in lengths 10-30 m long. These lengths are butt or scarf welded to form the full length. Then the compacted and aspected cable is pulled through the jacket. The conductor is given a small final compaction.

A particular advantage with this arrangement is that all longitudinal welds are eliminated

	NR STRANDS	% VOID FRACTION
STEP 1ST 3Nb (2%Ti)Sn STRANDS 0.74 TWISTED AND COMPACTED (ø0.74, ø1.32)	3	5.7
STEP 2ND NR 3 STEP 1ST ELEMENTS TWISTED AND COMPACTED (ø1.32, ø2.36)	9	11.5
STEP 3RD NR 3 STEP 2ND ELEMENTS TWISTED AND COMPACTED (ø2.36, ø4.20)	27	16
STEP 4TH NR 5 STEP 3RD ELEMENTS CABLED AROUND TUBE AND CALIBRATED (ø4.20, ø11.20)	135	37.7
STEP 5TH NR 8 STEP 4TH ELEMENTS CABLED AND COMPACTED (R6, JACKET, 18.00, R2, 45.50, 4.70)	1080	40

Fig. 2 : EM-LMI 40 KA Nb_3Sn Conductor

resulting in reduction of many of the quality control and leak checking problems.

3.3. CEA Design

The conductor design by CEA is shown in Fig. 3. The jacket is made out of hot drawn hollow tubes with a circular central hole and an external square shape. The lengths of about 8 m are machined at both ends then automatically butt welded by TIG process. Each weld will be carefully tested at the different fabrication stages. Welding the jacket in this fashion allows the use of a well known, reliable method.

The full length of the jacket, about 400 m, remains straight and the cable is pulled into the jacket with a force of about 4t. A pulling test performed recently on a length of 400 m with a clearance of less than 2 mm has not shown any significant problems.

The cable is made out of NbTi wire, the diameter of which should be as large as possible in order to reduce the production cost and also to get small strand displacement. These considerations have to be balanced by the efficiency of the cooling capability of the strand assembly and by the level of AC losses. Two designs having the same cross section for

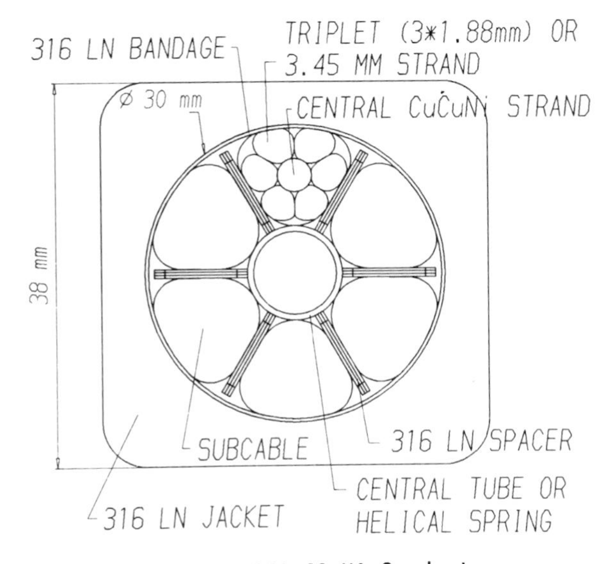

Fig. 3 : CEA 40 KA Conductor

the cable and subcable are investigated simultaneously with a wire of 3.45 mm in diameter and with a triplet made out of 3 wires of 1.88 mm. Both look acceptable but the final choice will be made on the basis of cost evaluation, overall current density and also in taking into account possible utilisation of the same wire diameter for a cable made from Nb_3Sn superconducting material. Stainless steel spacers are located between subcables in order to limit coupling currents. Both sides of the spacers are provided with radial notches to allow helium circulation. The central tube does not play a mechanical role. Its purpose is to ease the assembly of subcables and spacers and to increase the hydraulic diameter. A bandage is put around the cable to keep all the subcables in compression and to protect it during pulling into the jacket. This bandage can also balance the residual torque of the cable. The last operation consists of a mechanical compaction by drawing of the jacket around the cable in order to suppress all clearance.

4. CONDUCTOR QUALIFICATION AND TESTING

A rather comprehensive series of tests and measurements will be carried out on conductor components and the full-size conductor. Mechanical measurements will include properties of the jacket material (base metal and weld, 4 K) and mechanical properties of the cable and conductor assembly (e.g. elastic constants, mechanical hysteresis, bending tests). Critical current will be measured on the strand and sub-cables as a function of the B,T, $\varepsilon_{/\!/}$ and $\sigma_{\perp}$. The AC losses will be measured on these components also. Full-size conductor testing possibilities are limited due to the large conductor size, high current and field, and large mechanical loads that need to be simulated. It is planned to measure the full-size conductor critical current in the SULTAN III facility[5].

5. SCHEDULE

The industrial development is just beginning. The programme runs until the end of 1989 when the short lengths of full-size conductor will be delivered. Test results on components and sub-size cables will begin to be available this autumn and continue throughout the programme as development proceeds. CEA's effort is a continuing part of the European Magnet Technology Programme and their results will come in parallel with the industrial results. At the end of this programme it is hoped to narrow or merge the conductor designs for further evaluation and production of a long length. The next step would be production of a double-pancake followed eventually by fabrication and testing of a model solenoid by 1993.

REFERENCES

1. E. Salpietro, et al., Fusion Technology, 14 No. 1 (1988) 58.

2. N. Mitchell, et al., Design and Feasibility Assessment of the NET Central Solenoid, 15th SOFT, Utrecht, this volume.

3. W. Specking, F. Weiss, and R. Fluekiger, Effect of Transverse Compressive Stress on Ic up to 20 T for Binary and Ta Alloyed Nb_3Sn Wires, in Proceedings of the 12th Symposium on Fusion Engineering, Monterey, CA (1987).

4. C. Schmidt, U. Jeske, and E. Specht, A Low Loss Superconducting Cable for Poloidal Field Coils, in: Proceedings of the 14th SOFT, Avignon, France, Vol. 2, 1986, pp 1787-1793.

5. N. Sacchetti, et al., The SULTAN III Project, 15th SOFT, Utrecht, this volume.

FUSION TECHNOLOGY 1988
A.M. Van Ingen, A. Nijsen-Vis, H.T. Klippel (editors)

3-D EDDY CURRENTS EFFECTS IN THE NET VACUUM VESSEL STRUCTURE

L. BOTTURA[1], E. COCCORESE[1], R. ALBANESE[2], R. MARTONE[2], G. RUBINACCI[2]

[1] The NET Team c/o Max-Planck-Institut fuer Plasmaphysik, Boltzmannstr. 2
D - 8046 Garching b. Muenchen

[2] University of Salerno, Istituto di Ingegneria Elettronica, I - 84100 Salerno

Most of the mechanical structure of NET will operate in a strongly changing magnetic field. The magnitude of the field depends on space and time, as well as on the particular condition of operation (normal, abnormal, faulted). These variations induce eddy currents in the passive structural components whose effects are many, such as electromagnetic forces, heat generation by Joule effect, gap voltages, deformation of the magnetic field lines, penetration delays of the electric and magnetic field, etc. Aim of the paper is to predict some of the eddy currents effects which could have an impact on the design of the vacuum vessel structure proposed for NET. In this respect, the most worrying operating condition is the occurrence of a major plasma disruption, where the time-varying magnetic field is produced by the rapid disappearance of the plasma current, even accompanied by macroscopic movements of the plasma column. The calculations have been carried out by means of CARIDDI, a 3-D FE eddy current code. The results here reported are mainly large scale effects and global quantities, such as total currents in relevant direction, total magnetic energy and ohmic power, forces resultants. The analysis is parametrized on some of the relevant characteristics of the plasma disruption scenario as well as on some geometric parameters, in order to be able to derive design guidelines for the vacuum vessel.

1. INTRODUCTION

The design of the structures surrounding the plasma in a fusion reactor is considerably affected by the effects of the eddy currents circulating in case of a major plasma disruption. In this event, the magnetic energy lost by the plasma during the current quench impacts on the surrounding structures producing heavy electromagnetic loads.

From the electromagnetic viewpoint, the phenomenon can be regarded as an eddy current problem driven by the time-varying magnetic field of the plasma current; the interaction of the currents induced in the metallic structure with the existing magnetic field produces the searched distribution of body forces.

As a difference to the present day tokamak devices, fusion reactors present much more complex geometries. In particular the thin shell approximation is not satisfactory in the case of a reactor, where thick structures are required for heat removal and neutronic shielding.

As a consequence, for purposes other than scoping studies or simplified conceptual designs, fully 3-D eddy current codes have to be employed and integrated among the design tools of the reactor structures. In order to keep reasonable the CPU time the number of elements is limited to a few hundreds; this implies the necessity of introducing simplifications when modelling the structure. However this simplification has to be done with great care, due to the long range nature of the electromagnetic phenomena.

Aim of the present paper is to predict some of the eddy current effects which are expected in the vacuum vessel proposed for NET in case of a rapid disappearance of the plasma current.

After a brief description of the mathematical approach used in the calculations, the schematization of the vacuum vessel structure is discussed; some of the results achieved are then presented and commented.

2. EDDY CURRENT AND FORCE CALCULATIONS

The calculations were performed with the aid of CARIDDI[1], a 3-D computer code capable of solving a transient eddy current problem in a non magnetic conducting region and fully 3-D geometry.

The underlying mathematical model is based on an integral formulation[1] of the quasi-stationary Maxwell equations, where the unknown current density is deduced by its vector potential. The geometry of the structure, following classical Finite Element techniques, is specified by means of 8 nodes brick elements (generated by a suitable preprocessor like PATRAN-G), filling only the electrically conducting regions of the structure.

The main output of the code is the time evolution of the three components of the current density in each element of the structure.

A separate section of the code is devoted to the calculation of the magnetic field (both static and dynamic) in the elements; the main output of this section is the time evolution of field and body forces in the elements.

For purposes other than the stress analysis of the structure, the code provides the time evolution of a number of integral quantities, such as:

- magnetic energy of the eddy current
- ohmic power dissipated in the structure
- magnetic field in selected points
- total forces and torques acting on specified pieces of the structure.

3. STRUCTURE SCHEMATIZATION

The presence of the first wall/blanket and divertor plates has been ignored when modelling the structure, for the following reasons:

- the stage of the design for these structures is not as advanced as for the vacuum vessel
- to allow the increase of the discretization level of the vacuum vessel
- these structures are dynamically well decoupled from the vacuum vessel, because of their shorter time constants.

The CAD produced drawings used as source for the schematization of the vacuum vessel are shown in Fig. 1. It can be seen that each of the 16 modules has a poloidal plane of symmetry; this allows a great saving in the number of elements necessary for the schematization. Two different meshes have been created:

- an up-down symmetric mesh which models only 1/64 of the whole structure, and does not consider the penetrations (mesh I in Fig. 2)
- a non-symmetric mesh, modelling 1/32 of the vessel, with vertical access port, horizontal port and pumping duct (mesh II in Fig. 2).

The concept followed in both schematizations is to consider the vacuum vessel as a series of closed boxes (i.e. the rigid sectors) connected by resistive strips (i.e. the elastic elements)

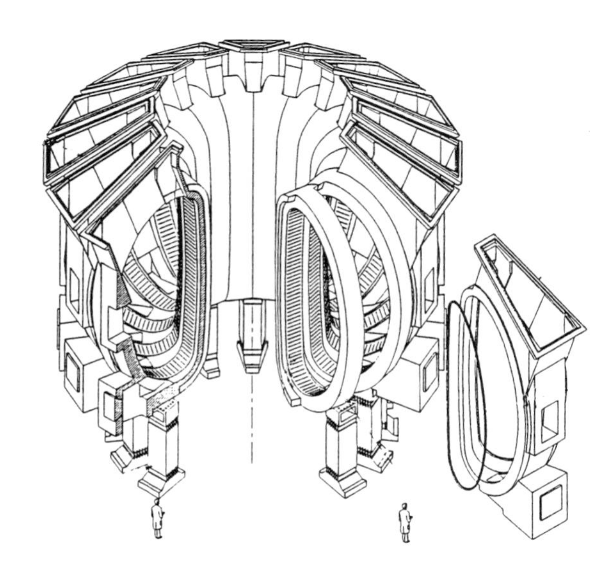

FIGURE 1
3-D view of the NET vacuum vessel

providing the toroidal continuity. The thickness is subdivided into 4 layers in mesh I and in 5 layers in mesh II.

Fig. 3, showing the outer equatorial cut of mesh I, illustrates this concept. In this figure two net currents can be discerned: the NET toroidal current I_t, flowing through the elastic elements, and the current I_b

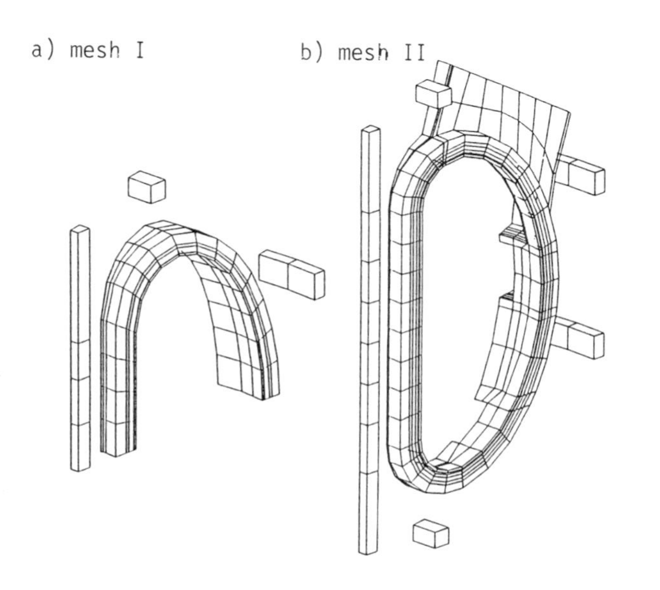

FIGURE 2
Finite element meshes used for the calculation of the eddy currents. 1/64 and 1/32 of the vessel are modelled in mesh I and mesh II respectively.

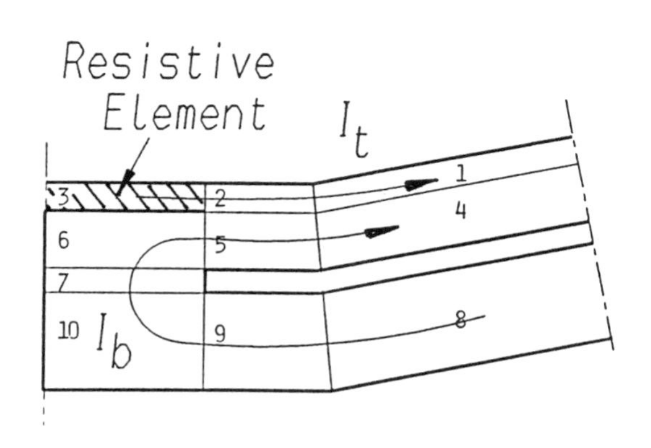

FIGURE 3
Horizontal cross section of mesh I on the equatorial plane. Thick lines stand for electric break. Note the definition of toroidal and box currents I_t and I_b.

circulating through the box-like structure of the rigid sector. The one turn toroidal resistance of the vacuum vessel was taken to be 0.2 m Ω. Both meshes include the presence of the superconducting PF coils, which have been assumed to be short-circuited and up-down series connected. It is assumed that the central solenoid is made of three independent pairs of coils, resulting in three indepedent currents: the first coil, denoted as P1D[2] is modelled with 2 elements. The remaining two, denoted as P1C and P1B[2], are modelled by 1 element each. This schematization implies that the current density is uniform in the coils; this is consistent with the assumption of having each coil made up of several series-connected turns evenly distributed in the cross-section. In a similar way coils P3 and P4 were modelled.

4. RESULTS

The results reported here refer to a disruption where the plasma current decays linearly to zero with a constant profile and no movement. The three following cases have been analysed:

- case A: NET reference equilibrium configuration at flat top (I_p = 10.8 MA) and decay time τ_D of 20 ms;
- case B: as case A but τ_D = 10 ms
- case C: NET extended performance configuration (I_p = 15 MA) and τ_D = 20 ms.

Figs. 4a, 4b and 4c report the time evolution of the net toroidal and box current flowing in the structure as shown in Fig. 3 for the three cases considered in up-down symmetric mesh I. The toroidal current I_t is responsible for the radial force acting on the modules of the vacuum vessel. The global effect of this force, due to the interaction with the vertical field, is to push the vacuum vessel inwards.

The main 3-D effect is the appearance of a significant box current I_b; this current would be the only one circulating in the vessel if the modules were totally insulated each other. The magnitude of this current is determined by two contrasting mechanisms. The first one is the spreading of the toroidal current I_t into two parallel paths across elements 1-4 and element 8 of Fig. 3 respectively; this mechanism tends to produce a current I_b of opposite sign with respect to I_t. The second mechanism, which results to be prevalent, is due to the time varying flux through the cross sectional area of the box structure (encircled by elements 4-5-6-7-10-9-8). This flux depends mainly on the decaying plasma current and the simultaneously growing toroidal current I_t. Obviously, during the early phase of the current quench, the box current I_b is expected to be negligible, because of the shielding effect of the toroidal current I_t which flows through the plasma facing wall of the vacuum vessel.

At the end of the disruption the toroidal current I_t attains its peak value and then starts decaying. On the other hand, the box current I_b keeps growing after the disappearance of the plasma. This behaviour is explained by the fact that the time constants of I_t and I_b are rather different (typical values are 20 ms and 200 ms, respectively). The consequence is that the electromagnetic forces in the vacuum vessel act for a time longer than the plasma disruption time constant. Figs. 5a, 5b and 5c report the time evolution of the ohmic power, magnetic energy and radial force on the whole vacuum vessel respectively.

For case C (15 MA plasma, 20 ms disruption), the complete analysis was repeated using the non-symmetric model of the vessel (mesh II). When the presence of the access ports is

a) case A (11 MA, 20 ms)

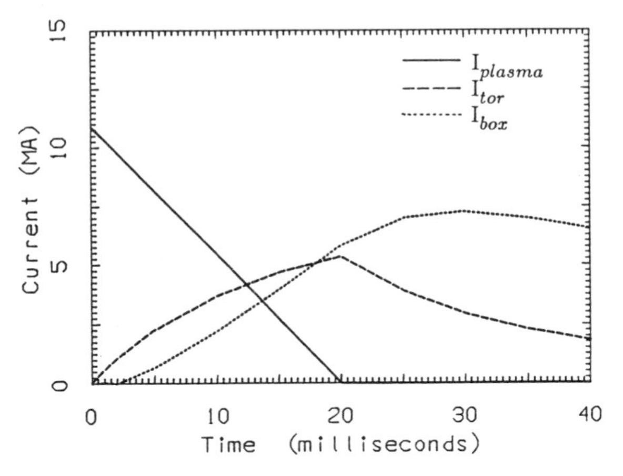

b) case B (11 MA, 10 ms)

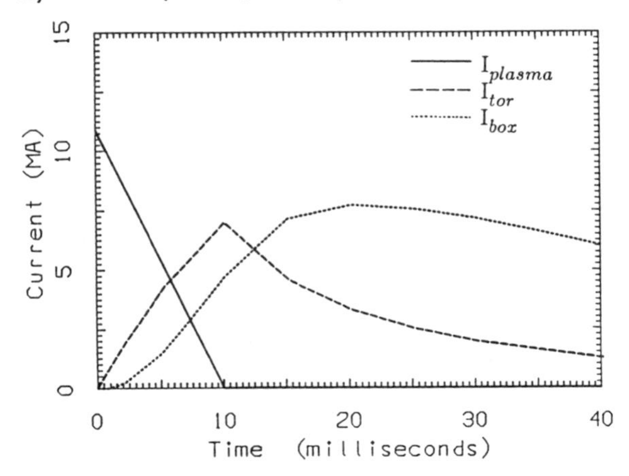

c) case C (15 MA, 20 ms)

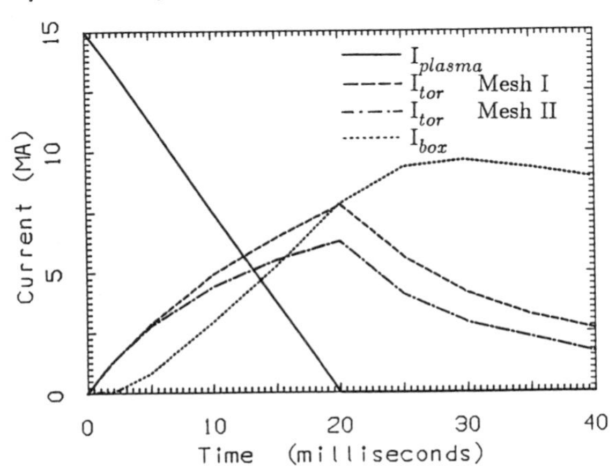

FIGURE 4
Time evolution of plasma, toroidal and box currents during the plasma disruption cases A, B and C.

a) Magnetic Energy

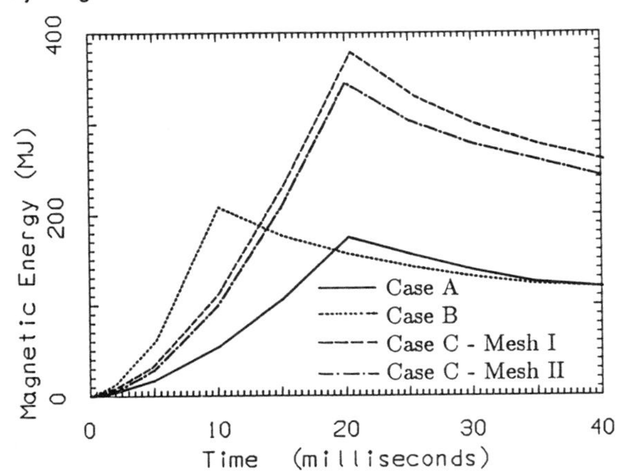

b) Ohmic Power

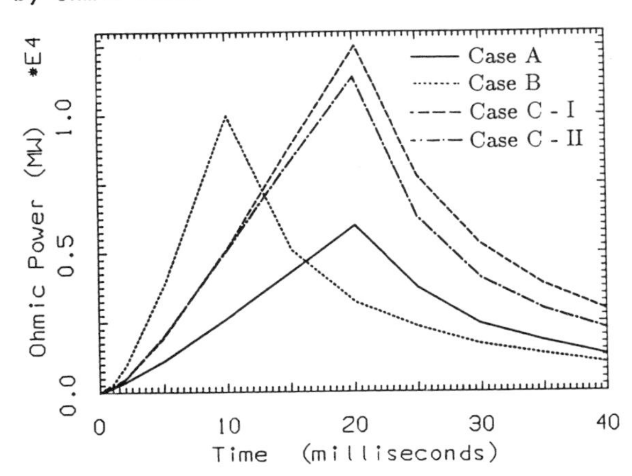

c) Radial force

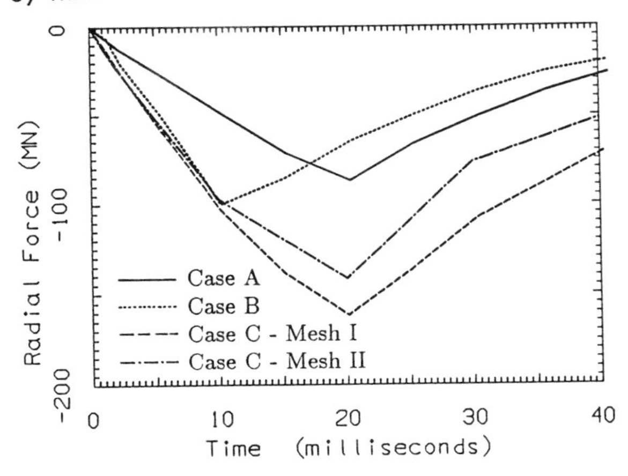

FIGURE 5
Time evolution of magnetic energy, ohmic power and radial force in the whole vessel during the disruption cases A, B and C.

considered, the electromagnetic response of the structure has a much higher degree of complexity due to the loss of up-down symmetry. The toroidal current is forced to flow in the connections around the ports in the outer part of the vessel, while remaining almost symmetric in the inner ports (see Fig. 6). This causes the toroidal resistance to increase and the net toroidal I_t current to decrease, as shown in Fig. 4c in comparison with the analogous solution obtained for mesh I.

The current distribution in the elastic element (in Fig. 7 for the two meshes at 2 and 20 ms during the disruption case C) shows a clear difference in the magnitude, but only slight changes in distribution. This is due to the low resistive paths around the ports that

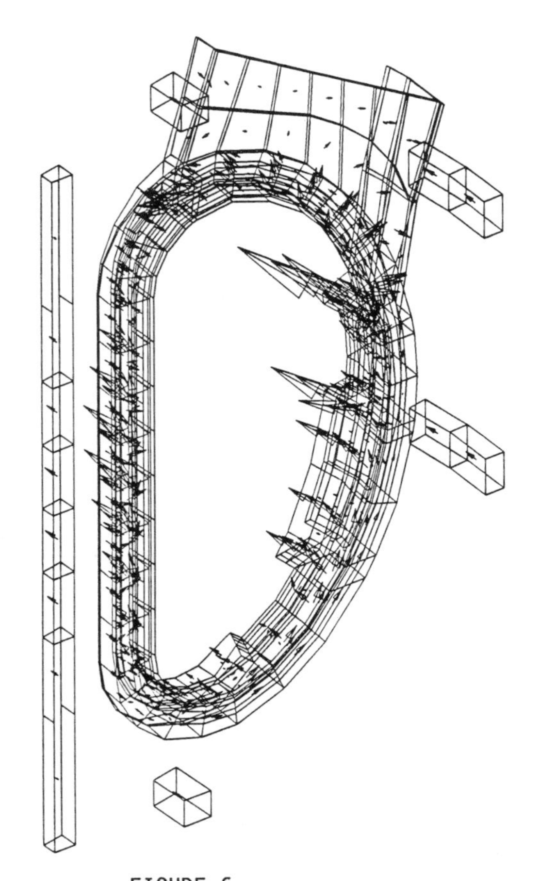

FIGURE 6
Map of the element current densities in mesh II at t=20 ms during a plasma disruption, case C.

redistribute the current in the resistive strip.

The smaller toroidal current gives a smaller ohmic power, magnetic energy and radial force (see the curves relative to case C in Figs. 5a, 5b and 5c), but the more interesting effect of the non symmetry is the presence of a net vertical force, downward directed, of non negligible magnitude (ca. 10 MN) that must be equilibrated by the supports of the vessel (in Fig. 8). The appearance of this force is due to the non symmetric location of the "centre" of the toroidal current and the consequent interaction with the poloidal curved field.

The effect of the toroidal field interaction with the non symmetric current patterns is also remarkable in the creation of a high shear in the connection of the segments[3].

5. CONCLUSIONS

The effects of the eddy currents generated by a major plasma disruption in the NET vacuum vessel have been investigated. The use of a fully 3-D capable computer code has allowed the recognizement of a 3-D current pattern (the box current I_b) coupled with the major toroidal mode. This current must be taken into account in the detailed structural analysis of the vessel, since it is heavily excited and its time constant is much longer than the time constant of the toroidal current. In addition, the reduction of the degrees of freedom of the problem obtained using a vector potential based formulation allowed the study of a larger dimension model that includes the effects of the vessel penetrations. It was shown how the main effect of the penetrations are the decrease of the net toroidal current and the creation of non-symmetric type of forces (vertical force, shear sollecitation) of relevant magnitude, while on other global parameters, such as ohmic power, magnetic energy and radial force, only minor changes are observed.

REFERENCES

1. R.Albanese, G.Rubinacci, Integral Formulation for 3D eddy current computation using edge elements, IEE Proc. &, vol. 135, part A (1988), pp. 457-462.

2. E. Salpietro et al., Fusion Technology, 14 No.1 (1988), pp. 145-155.

3. D. Collier, Thermal and Mechanical Analysis of the NET Vacuum Vessel, Presented at this conference, Paper D 19.

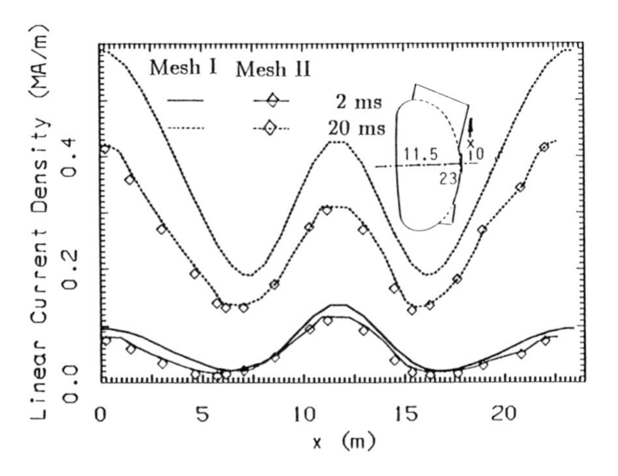

FIGURE 7
Linear current density in the resistive strip of meshes I and II as a function of the curvilinear coordinate x (defined in the inset figure) at two instants (2 and 20 ms) during a plasma disruption (case C).

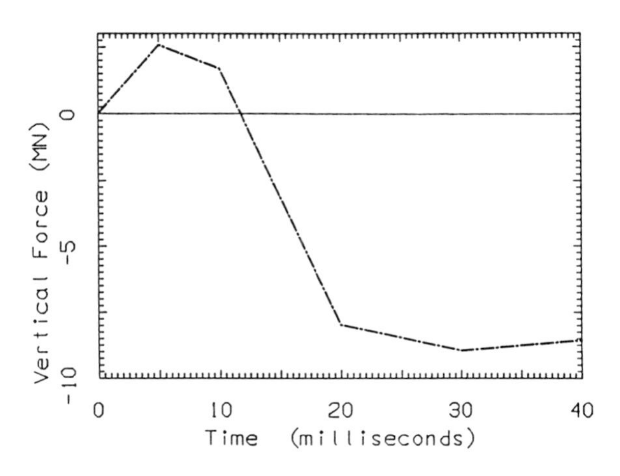

FIGURE 8
Net total vertical force on the vessel during a plasma disruption (case C, mesh II).

FUSION TECHNOLOGY 1988
A.M. Van Ingen, A. Nijsen-Vis, H.T. Klippel (editors)

HIGH VOLTAGE INSULATION AND TESTS OF CRYOGENIC COMPONENTS FOR THE SUPERCONDUCTING MODEL COIL POLO

G. Schenk, S. Förster, U. Jeske, G. Nöther, K. Schweikert, A. Ulbricht, V. Zwecker
Association KfK-EURATOM
Kernforschungszentrum Karlsruhe, Institut für Technische Physik,
P.O. Box 3640, D-7500 Karlsruhe, Federal Republic of Germany

Future superconducting magnets for fusion Tokamaks with stored high magnetic energy and fast flux change need high voltages in the order of 10 to 50 kV for safety discharges, ohmic heating ignition and plasma positioning.

The insulation material has to support the electric strength, the thermal and mechanical stresses and the pressure or vacuum condition in the presence of helium between 300 and 4,2 K.

In collaboration with industry suitables insulation materials have to be selected and to tested for this specific application. Special components, like high voltage and high current breaks, instrumentation feedthroughs, LHe breaks and other bushing elements have to be developed.

In the frame of the poloidal field coil development POLO at the Institut für Technische Physik (KfK-ITP) a cryogenic high-voltage (HV) laboratory was recently installed. There we are equpided with AC, DC and pulsed HV-supplies up to 140 kV, pool- and flow-cryostats at Helium-temperature for component testing. For insulation test we use the non destructive method of partial discharge analysis, the most specific test procedure. In order to measure the electric charges beginning from 0,1 pC, we installed a HF-tight Faraday cage.

In the POLO-programm already the conductor insulation and the coil insulation have been tested between 300 and 4.2 K. The partial discharge of the components insulation material were tested during mechanical and thermal cycling as well as under pressure and vacuum conditions.

1. INTRODUCTION

Ten years ago many investigations of the insulation material of superconducting cables for power transmission up to 220 kV were made [1]. Generally, the dielectric strength of liquid or supercritical helium, as well as vacuum was used for insulation in combination with spacers or wrapped insulation materials. For low dielectric losses any polar insulation material was unsuitable. Mechanical forces and thermal stresses were insignificant.

The development of fast dischargable superconducting magnets for fusion machines require HV-insulation for the coil and the components [2,3]. This insulation has to withstand thermal cycling, high electric fields due to magnet charging [4] and current variations, under vacuum and high pressure Helium gas conditions.

Up to now there exist no systematic investigation of such insulation materials which simultaneously meet all these requirements. The most commonly used insulating material today is a composit material of a fiber glass, anorganic granulates and epoxy resin [5]. The specific composite material can be choosen only if the dielectric, mechanical and thermal requirements are known. It is necessary that the development, fabrication and testing of the components are closely linked. For this reason, a HV-laboratory for cryogenic application was recently installed at the Kernforschungszentrum Karlsruhe in the Institut für Technische Physik. Of special importance is a non-destructive testing method inorder to survey the insulating material under working conditions.

2. CRYOGENIC HV-LABORATORY

Besides the standard cryogenic and HV-equipment (Fig. 1) a special measuring equipment is required. The insulation characteristics must be measured and analysed long time before any electric breakdown takes place.

FIGURE 1
HV-Laboratory-equipment

Such non-destructive testing methodes could be:

- Insulation resistance under HV-stress
- Determination of the dissipation factor tg δ
- Measurement of the partial discharge (PD).

The last testing technique has recently grown up in importance [6,7] and is based on a standard [8].

Partial discharges - practically measured in units of pC - result in a measuring signal of 10^{-5} V in the presence up to 10^5 V on the test object. It is evident that noise signals must be smaller and that the HV-supply has to be free of harmonic components. A Faraday cage (of about 150 m^3 volume) attenuates the electromagnetic noise penetration by 120 dB. Important filtering, line conditioning and grounding methods have to be choosen carefully in order to avoid adding noise when the HV is applied.
A partial discharge test circuit, where the measuring impedance is in serie with the test object, has been choosen, because only this circuit is not influenced by the parasitic capacitances of the required test cryostat. The minimum partial discharge, beyond the noise level, which can be measured is 0.2 pC.

A calibration of the object and the related installation including the partial-discharge detector, were made before each measurement.
The HV-equipment - the modules of Messwandler-Bau, Bamberg - supplies AC up to 100 kV, DC up to 140 kV and pulses 250/2500 to 100 kV peaks.

The standard PD-cycle starts at 0-voltage increasing at 500 V/s to the maximum voltage and decreasing again to 0 voltage. The max. voltage then was either a nominal voltage or was limited by the full scale of the partial discharge measure, i.e. 200, 2000 or 20000 pC. Such cycle datas are aquired stored and plotted as shown in figure 2 by a HP 9000 computer.

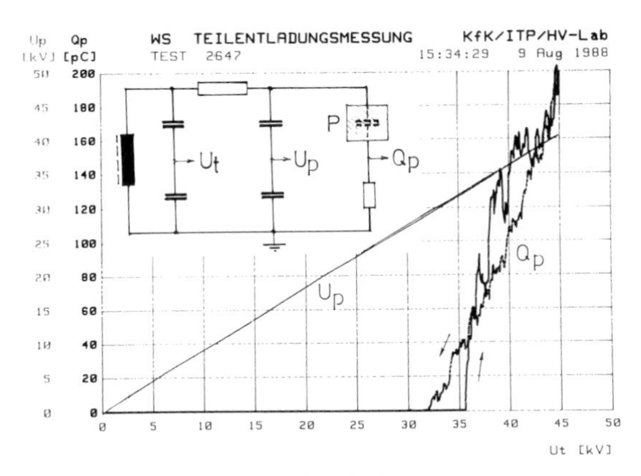

FIGURE 2
Standard test circuit and measuring plot

The voltage change has to be absolutely smooth in order to avoid any fault PD-signal by voltage transitions. The cryogenic equipment consists of LHe bath (200 ltrs) and flow cryostats. The time of cooling down depends on the test object and the heat exchange of the cryogenic system. Generally, cooling down and warming up rates are 30 K/h. One standard test cycle consists of

- test object installation, calibration,

initial test
- cooling down
- low temperature test
- warming up
- final test - object dismounting or continuation.

A reasonable time for a full cycle is 2 days. The minimum number of cycles for statistic evaluations is 5, i.e. 10 working days or 2 weeks. Additional tests, i.e. sharing stress or ageing behaviour, require even more time.

3. BASIC INVESTIGATIONS

The HV insulation of the superconductor cable at 4 K represents only one of the problems. Major attention has to be paid to all elements which connect the superconductor from the 4 K to the 300 K environment, where helium gas at low temperature and low pressure might appear. Moreover, fault conditions such as a superconductor quench, high helium over-pressure or vacuum breakdown have to be considered.

These are the conditions determining the choice of the insulating material. Basic investigations should analyse the dielectric behaviour (from partial discharge ignition up to break down) taking into account the parameters: vacuum pressures, temperatures and gases (Paschen characteristics). In order to verify the physical parameters, a relative homogeneous electric field strength, obtained by using the 25/75 electrodes IEC 243, has to be applied.

In a first test serie the partial discharge ignition voltage must be determined in respect to the mentioned parameters, and can than be listed as material characteristics.

A second test serie concerns the ageing of the insulation material as a function of the applied voltage and the total pulse number.

The basic material investigation helps to design the different components quite accurately. However the manufacturing process requires complete tests of prototype components and as final elements.

4. COMPONENT TEST

The standard test program of a prototype component consists of a series of 5 cycles in temperature and pressure, whereas the partial discharge measurement is taken only at the beginning and at the end of the cycles. The partial discharge standard cycle was described in part 2.

If fissures result from thermal dilatation, generally, this can be observed at the beginning of cooling down. It was found advantageous - reducing time and effort - to make 4 cycles with LN_2 and only the last cycle with LHe.

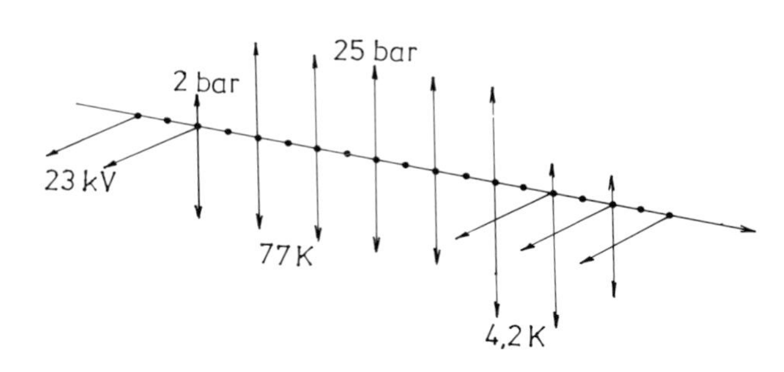

FIGURE 3
Schematic graph of the standard test series

If the PD-curves at the beginning and at the end are identical, there is a strong indication that no alteration has influenced the insulating material of the component, probably the leak test would also confirmed this.

In the case of not identical PD-curves, the leak test might confirm a crack in the insulating material. If not, the possibility of other PD mechanisms such as, surface PD, corona or treeing have to be considered.

By analysis of the 50 Hz PD-distribution the

mechanism of the discharge can be indicated. Prenature ageing of the composite material can be seen by a change in the PD hysteresis, (difference between inception/extinction voltage). For individual localisation of the PD, acoustical and optical detection methods should be applied. Computer added pulse and delay analysis are under development [9].

Strong HV which might lead to irreversible influence of the insulating material should be avoided during the standard tests cycles. It should be reserved to a special ageing test.

5. AGEING TEST

The lifetime of any dielectric component under HV-stress is a function of the applied voltage and the number of pulses. Poloidal field coils of the actual tokamaks have to support typically 10^5 HV-pulses. An ageing test series should have 10^5 to 10^6 pulses (17 min to 3 h of 50 Hz) under typical operating conditions.

Starting from the nominal high voltage U_n one proceeds by 1.2.. 1.4.. 1.6.. 1.8 up to 2 U_n and more, up to the breakdown voltage, if appropriate. The breakdown might indicate the weakest part of the component, but statistical significant breakdown tests can, hardly, be made.

6. RESULTS

HV components having been developed in the POLO program and tested are:

- superconductor insulation
- superconducting coil insulation
- radial breaks (cable end-insulator)
- axial breaks (LHe-conduits)
- current feed-throughs (cryostats)
- instrumentation feed-throughs
- cables for instrumentation
- insulating amplifiers

As a representative component the radial break is shown in figure 4. First, the electric

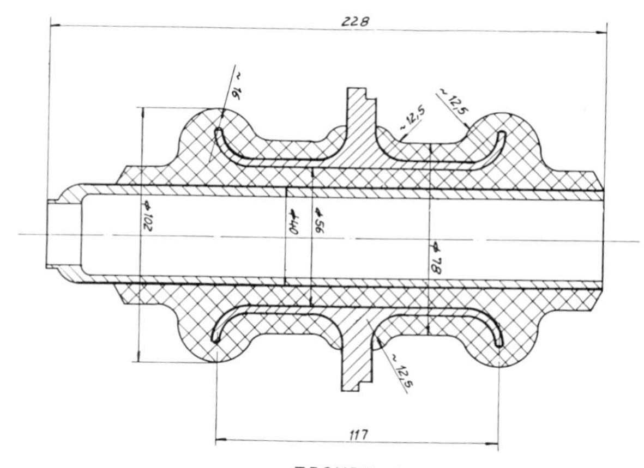

FIGURE 4
Radial HV-Break

field strength were computed for different shapes, in order to find a suitable final geometry [10]. After that, the prototype fabrication were discussed with three different manufacturers. Different breaks, with and without potential guiding sheets, varying the fibre structure by filaments, tape or sheet winding, different epoxy resins and filling granulates and vacuum or wet impregnation have been produced and tested.

All conceptions have positive and negative qualities. Best PD behavoir results from vacuum impregnation (void free), which, however, could be problematic under thermal stresses (fissures).

Wet impregnation shows better mechanical characteristics under thermal stresses, but also shows relatively early PD ignition.

The best overall results were obtained by the following process:

- Prestressed wet winding impregnation
- alternating filament and tape fibers
- metal surfaces were grooved and sandblasted
- use of filling granulates with matching thermal expansivities
- use of resins with low temperature hardener for avoiding non-uniform polymerization.

It is difficult to detail the manufacturing process, since a large part depends on artistic-ability. The finally choosen component fullfills the previous demands not perfectly, but quite adequately.

The partial discharge measurements at 293 K of a radial break before thermal cycling (figure 5a) is identical with the measurements after cycling, which indicates that no fissuration occurs. The PD measurements at 4.2 K (LHe) (figure 5b) show that the PD inception voltage increases by a factor of ~ 3. This is an indication that HV-discharge problems are less significant at LHe temperatures.

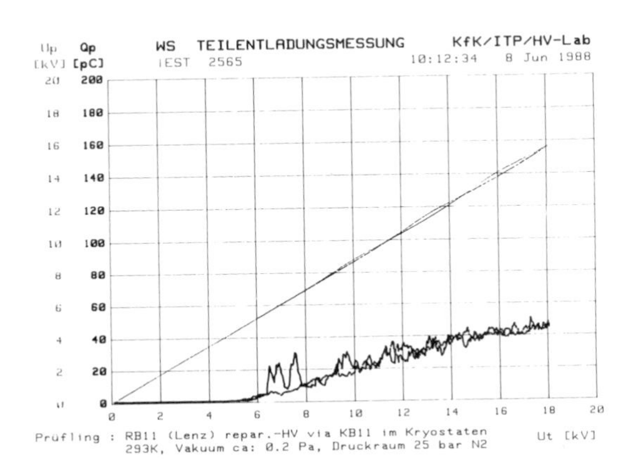

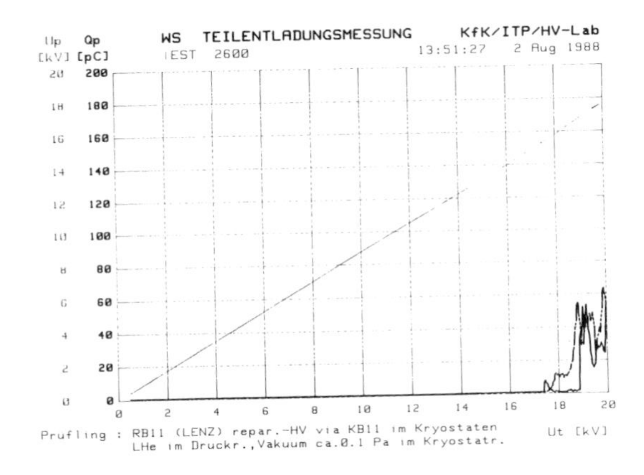

FIGURE 5
PD-measurement of the radial break
a) at 293 K (before and after thermal cycling)
b) at 4.2 K (with LHe filled)

ACKNOWLEDGMENT

The authors wish to express their gratitude to Mr. H. Katheder and the whole group of cryogenics at the ITP for their permanent support. Our sincerest thanks to Mrs. Y. Han, IEE-Academia Sinica, Beijing, for their kind help during the starting phase of the HV-laboratory.

This work has been performed in the framework of the Nuclear Fusion Project of the Kernforschungszentrum Karlsruhe and is supported by the European Communities within the European Fusion Technology Program.

REFERENCES

1. B. Fallow, A review of the main properties of electrical insulating materials used at cryogenic temperatures.
Int. Conf. on Magnet Technol., Frascati 1975

2. P. Komarek, A. Ulbricht; Aspects of Superconducting Energy Storage Systems for Supply of Fusion Magnets, Proc. Int. Symp. on Superconductive Energy Storage, Osaka 8-10 October 1979.

3. A. Ulbricht, A resistive superconducting power switch with a switching power of 40 MW at 47 kV, Cryogenics, Vol. 19, Oct. 79, pp. 591-601.

4. Technical Specifications, Large Coil Program Rev. E.Nov. 8, 1976 ORNL, Oak Ridge, USA.

5. M.N. Wilson, Superconducting magnets Oxford Science Publication 1983, p.66.

6. P. Osvath, W. Zaengl, H.J. Weber, Measurements of Partial Discharge, Bulletin SEV/VSE 76 (1985) 19.

7. Fifth International Symposium on High Voltage Engineering (ISH), Group 41 and 42 24-28 Aug. 1987, Braunschweig Germany.

8. Partial discharge measurements IEC Publication 270 (1981).

9. I.D. Gassaway, P.B. Jacob, C.A. Vassiliadis, P.H. Reynolds, Computer-aided Partial Discharge Measurement and Recognition,
5th ISH, Braunschweig, Paper 41.03, 1987.

10. P. Deister, B. Schaub, Feldberechnungen an einem Potentialaltrenner im Rahmen des Projekts Poloidalfeldspule, Hochspannungsinstitut, Universität Karlsruhe 1987.

FUSION TECHNOLOGY 1988
A.M. Van Ingen, A. Nijsen-Vis, H.T. Klippel (editors)

FABRICATION OF A 15 kA NbTi-CABLE FOR THE 150 T/s HIGH RAMP RATE POLO MODEL COIL

S. FÖRSTER, U. JESKE, A. NYILAS
Kernforschungszentrum Karlsruhe, Institut für Technische Physik, Postfach 3640, D-7500 Karlsruhe, FRG

A low loss cable for 15 kA current is under fabrication using Cu/CuNi/NbTi strands with outer CuNi shells. The conductor length will be 4 x 150 m. This conductor will be wound to a 3 m diameter POLO model coil. Coil and all components like conductor connections and cable terminations will be operated under high ramp rate pulses from 80 T/s (simulation of a plasma disruption in Tore Supra) up to 150 T/s in the TOSKA facility at KfK in 1989. The project aim is the development of conductors and components for the outer poloidal field coils for NET. As intermediate step, the operation of an 8 m diameter coil in Tore Supra is forseen at 1991. The paper reports on the current fabrication experience.

1. INTRODUCTION

The European Fusion Technology Programme is orientated towards NET. It was therefore our understanding that a development programme for poloidal field coils in fusion reactor test facilities has to consider NET as goal.

But on the other hand, we were convinced, that it is not possible to go immediately by one development step into such a big, complex machine like NET.

We have started our work in 1983 at an early stage of the NET design with rather rough knowledge on parameters like size, induction B, field change rates $\dot{B}$ (start up, plasma position control and disruption) and associated mechanical forces, high voltage conditions and transport currents[1]. Our first practical application should be the operation of an 8 m ø PF coil in TORE SUPRA 1989. However, a small, already existing machine has its own constraints different from the planned NET. So we had to design PF coil, cable, cryostat, current leads, high voltage breaks and other components specifically for TORE SUPRA[2]. We tried to stay as "NET relevant" as ever possible. But a gap between the needs of NET and TORE SUPRA could not be avoided.

2. THE CONDUCTOR DESIGN AND DEVELOPMENT

The POLO conductor design can be seen in Fig. 1 and Fig. 2. It has the main components: superconducting round cable, protection tube, and outer steel conduit, which are made in separate industrial fabrication steps.

Round cable

- basic strand: ø 1.25 mm, NbTi in CuNi matrix (Table 1)
- subcable (6 NbTi x 1 CuNi), twist pitch 40 mm
- compaction of the subcable from ø 3,75 mm to 3,65 mm ø
- subcable insulation, preimpregnated with resin (Glass/Kapton/Glass, δ = 220 μm, b = 8 mm or CuNi, δ = 265 μm, b = 8 mm)
- inner cooling tube (13 x 1 mm) with Kapton insulation 2 x 80 μm, coated with B-stage resin adjacent to the tube surface)
- quench detection wires (2 x AWG 22(1), 1 mm ø with Kapton insulation)
- round cable
  (ø 21,4 mm with GKG subcables, lp = 229 mm)
  (ø 21,8 mm with CuNi subcables, lp = 256 mm)
  (2 x Kapton insulation δ = 80μm)

Protection tube

- tube forming (26 x 0,55 mm) in a roller set with 8 stations; sheet perforated with

2 mm ø holes every 0.5 m to avoid uncontrolled He reservoirs; stainless steel 1.4311 (~ 304 LN)
- TIG welding
- compaction to 22.4 mm outer diameter by

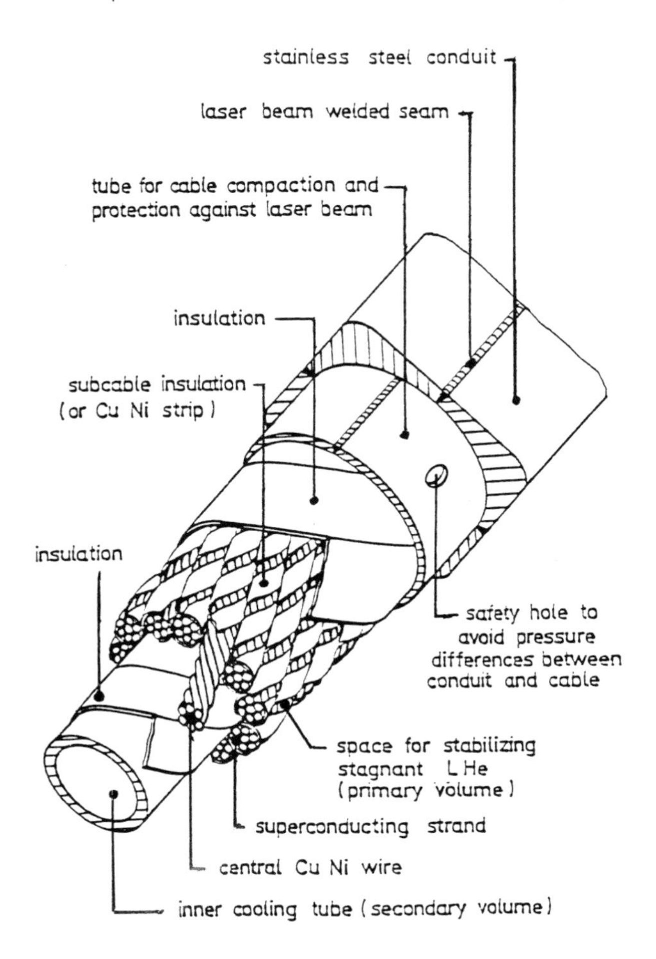

FIGURE 1
View of the POLO conductor

drawing (1/3) and swaging (2/3)

Conduit
- quarter section (stainless steel 1.4311)
- U-shaped section laser welded out of two quarters

Final assembly
- round cable and two U-shaped section laser welded by two beams.

2.1. The round cable

The development of strand and conductor was started with development contracts at VAC, Hanau and Cablerie Dour, Belgium to produce different samples of 6 m length.

2.1.1 The NbTi/CuNi/Cu-strand

The mixed matrix strand developed by VAC (Fig. 3) is a one stage strand and for that reason supposed to be cheeper than the conventional two stage designs with preextruded honeycomb elements in circular arrangements (Alsthom, VAC). The advantages of the used strand should be: - the large areas of Cu, which are not affected by heat treatments (RRR between 170 and 200) - the good thermal conductivity to the surrounding surface. Whereas the rather large distance from the center of the filamentary element via a CuNi barrier to the stabilizing Cu might be considered as disadvantage, if the transport current is near the critical current. For the POLO conductor, 57 km of this strand have been produced, with a mean value of critical current

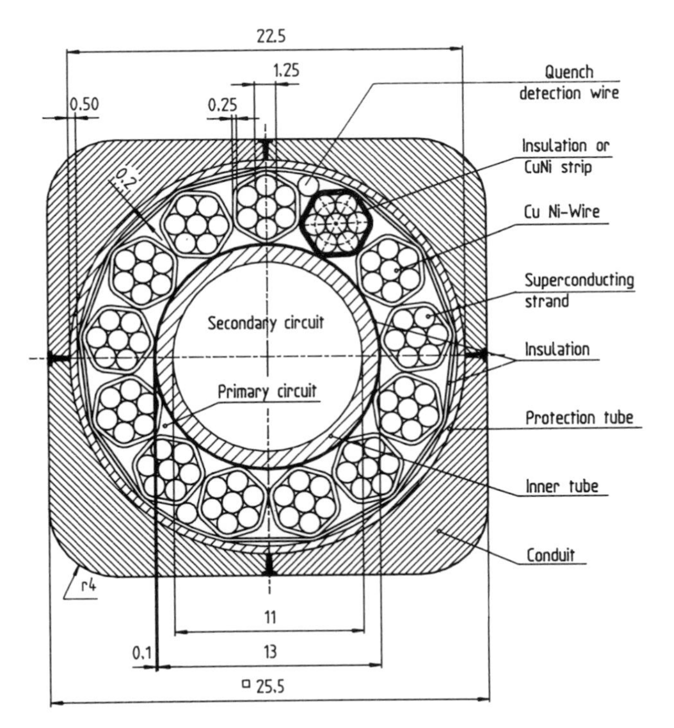

Figure 2
Cross section of the POLO conductor

in the final conductor of I_C = 262 A at 5 T, 4,2 K (j_C = 1.9 x 10^5 A/cm^2 and a mean RRR = 186).

Figure 3
POLO NbTi/Cu/CuNi strand

2.1.2 The Subcable

The subcable configuration 6 NbTi strands around 1 CuNi wire gives good mechanical contacts of each strand to the other (line contact) and on the other hand low ac-losses due to the high contact resistance between adjacent strands.

In the subcable manufacturing a preforming device for each strand is used in the cabling machine to avoid elastic spring back effects. After cabling the subcable is drawn through a die and slightly compacted from 3.75 mm to 3.65 mm. The gaps between the strands could be kept in the range of 10 μm by this method.

TABLE 1
Summary data of the POLO strand

ratio NbTi/Cu/CuNi	1 : 5 : 3
diameter	d = 1.25 mm
critical current	
(specified minimum)	I_c ≧ 220 A (4.2 K, 5 T)
filament diameter	d_{fil} = 10 μm
number of filaments	N_{fil} = 1638
coupling loss time constant	τ_c ≦ 0.2 ms
twist pitch	l_p = 10 x d = 12.5 mm
thickness of CuNi 10 shell	δ_w = 70 μm
tensile strength	σ_{02} = 520 MPa
ultimate strength	σ_u = 680 MPa

2.1.3 Subcable insulation

TORE SUPRA might have very fast plasma disruptions with τ = 5 ms and associated field change rates of 80 T/s. The conductor should not show transition to the normal conducting state in this case. We have therefore chosen a Glass/Kapton/Glass (GKG) insulation for the subcables. This insulation covers 70 % of the subcable surface. 30 % are left free for He penetration and heat transfer surface. The thickness of the strip is 220 μm after cabling

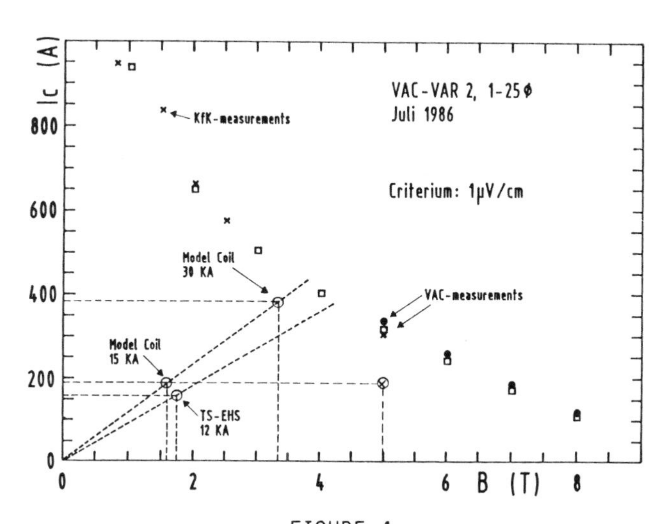

FIGURE 4
Critical current of the VAC NbTi/Cu/CuNi strand versus applied field. The operational points of the Model Coil and the TORE SUPRA EHS coil are indicated by arrows (5 T design point encircled.

and has to prevent electrical contacts between adjacent subcables. The insulation is preimpregnated with epoxy resin which will be cured together with the later coil impregnation. This process aims to fix the strands sufficiently against mechanical movements.

Two length of the conductor have plastic reinforced insulation, whereas the other two length of the conductor have CuNi strips instead of GKG. With the CuNi, the final conductor will have higher ac-losses, but recent measurements on cable samples show that even TORE SUPRA might operate with this type of conductor[3]. Anyhow, a CuNi strip is supposed to be the appropiate NET solution.

2.1.4 Round cable

2.1.4.1 The two cooling systems

It seems to be preferable to keep the temperature in the PF coils everywhere constant. (There exist divertor coils which have the highest field at the outside). Thats one reason why we have considered a two phase He cooling circuit. On the other hand we did not want to allow gas bubbles at the strand surface. So we decided to use two separate channels. The primary channel with LHe surrounding the strands at a pressure of 4 bar and a secondary channel with two phase He at 1.2 bar in the central tube. In this system, the primary He is more or less stagnent and the secondary has, even at very long length, a low pressure drop[4].

2.1.4.2 The inner cooling tube and different cooling concepts

The inner cooling tubes (⌀ 13 x 1 mm) have been tested to be vacuum tight to a limit of 10^{-6} mbar l/s. In the case of only one cooling system, e.g. for NET the tube could be perforated to allow He convection. In this latter configuration, the cable could be operated either with superfluid He at 1.8 K in high field applications with NbTi or with supercritical He between 4 and 10 bar.

2.1.4.3 The use of the inner tube for the quench detection system

The central cooling tube is wrapped and adhesive bonded with two layers of Kapton to improve the heat conduction. If the conductor is wound to a coil, the insulated central tube is a cowound coil too, which is tightly coupled to the superconductor. By balancing the resistive parts and compensating the inductive parts of both "coils", a very low level of quench detection voltage should be achievable despite the very high field ramp rates (23 kV discharges in 1-2 s in TORE SUPRA).

A second task for the Kapton is to avoid electrical contacts between tube and the GKG insulated strands. Otherwise high local coupling currents might be induced.

Table 2
Summary data of the POLO-conductor

	dimensions [mm]	cross section [mm^2]	material density [Kg/m^3]	E-modulus [GPa]
stainless steel				
inner tube	⌀ 13 x 1	37.7		
protection tube	⌀ 22.45 x 0.5	34.5	7900	205
outer shell	⌀25.5 x 25.5	240.6		
strands	⌀ 1.25		8900	135
subcables	⌀ 3.66	118		
insulating				
inner tube	2 x 80 μm	6.6		35
subcable	240 - 250 μm	30	2000	
round cable	2 x 80 μm	11		
conductor	1.2	141.6		35
Helium				
inner tube		95		
primary channel		63.5		
overall dimensions including conductor insulation	27.9 x 27.9	778.41	4920	108
weight			3.83 kg/m	
hydraulic diameter of primary channel	0.55			

2.1.4.4 Quench detection by co-wound insulated wires

Two Kapton insulated wires AWG 22(1) of 1 mm ⌀ are co-cabled with the CuNi-type subcables (see Fig. 2). If later, for conductors with higher allowable coupling loss time constants (e.g. NET), the Kapton insulation on the central tube would be left out, one wire could be used for quench detection, the other being redundant. In the model coil the performance of the different systems shall be compared.

2.1.4.5 Round cable and its insulation

4 x 150 m round cable have been manufactured in a conventional cabling machine with a capstan to pull the cable through. As two different types of subcable wrappings have been used, the outcoming geometries slightly differed from each other, but without affecting the further manufacturing process. The round cable is wrapped with two layers of Kapton to prevent shorts in the GKG type and to lower the coupling losses in the CuNi type of cable. Here too, if higher coupling loss time constants are allowed, the Kapton can be left out. For the model coil this insulation should give an additional quench detection system between outer conductor conduit and the round cable. The cable design thus has a high degree of flexibility.

Always two round cables have been wound on one transport drum with 3.6 m inner and 4.2 m outer diameter to be used in the further fabricational steps.

2.1.5 Protection tube

The choice of the GKG insulation system inside the cable, which is very sensitive to mechanical treatments, forced us to enclose the round cable in a thin walled protection tube. Drawn through a die, this tube could give a certain compaction to the cable, sufficiently high to avoid critical strand movements, and nevertheless, leaving enough He-volume to keep the quench pressure sufficiently low. The tube serves to adjust the outer diameter of the round cable to the jacket inner diameter. Later it gives a protection against the heat of the laser beam during the jacket. The tube is sheated with a steel sheeth of 0.5 mm thickness continuously around the round cable in a conventional tube forming station. The tube is TIG welded with 0.8 m/min, while the insulation system is sufficiently shielded. Thus, the diameter of the tube had to be about 26 mm. Using a linear capstan, a die, and a swaging machine in the same production line, the tube is compacted to 22.4 mm outer diameter.The reduction is shared among drawing and swaging by 1/3 and 2/3 respectively.

In the prototype manufacturing we had successfully produced conductor pieces of 25 m length. Unfortunately, the first go over 150 m with the original cable failed after 65 m due to tube buckling. It is now tried to overcome the difficulties by optimizing the welding procedure and by enhancing the tube wall thickness to 0.55 mm (0.7 mm clearly would give safer conditions, but this is not possible in the present design).

2.1.6 Replacing of the protection tube by two layers of stainless steel strips

Faced to the manufacturing problems, which were not expected due to the successfull tests on short lengths, it was necessary to search an alternative solution.

We now can propose to replace the tube by wrapping on the cable two layers of stainless steel strips of 15 - 20 mm width and 0.25 mm - 0.3 mm thickness. These two layers can be immediately put onto the cable in the cabling machine during the manufacturing of the round cable. The compaction can be done with an additional step using a swaging machine, through which the cable is pulled by the capstan. This new method has been tested up to 5 m length with the original cable. Bending tests have shown, that the outer Kapton insulation is not damaged by this operation. In cables without the need for an outer Kapton insulation, the new method is completely without any risk.

2.2 Conduit manufacturing and laser welding

Looking to NET conductor sizes with currents

in the range of 40 kA and unit lengths of more than 500 m, we had to emphasize, that the steel jacket manufacturing would be one of the biggest efforts. With expected wall thicknesses up to 5 mm for TF and PF conductors the TIG welding technology has been considered not to be adequate due to the high energy input into the jacket material during welding with the associated problem of distorting the geometry of the conductor. Minimum energy input at high welding velocities (2 m/min and more) give a big potential advantage to the laser welding[5]. We decided to develop for NET the continuous laser welding technology with two beams operating simultaneously either using two CO_2 lasers (1,5-3,5 kW each) or beam sharing of one high power laser (finally 6 kW) with operating time in the range of hours. It has been decided to demonstrate the potential of this new technology with the POLO conductor jacket welding. Two development contracts had been placed to VAC. One for the laser welding with beam sharing for full length manufacturing, the other for the jacket development.

2.2.1 The steel section manufacturing

When we started our efforts we only found one manufacturer (ISOPROFIL, Mannheim) who was willing to cooperate. But as we had only small quantities to order (less than 2 to), we had to decid to produce at first quarter sections of 6 m length out of a 15 mm ø wire and then to weld two quarter sections to one U-shaped section. The main unknown step for the steel section manufacturer has been the production of long lengths, which need take up spools of 2 m ø, and 1 m width after the final drawing bench instead of the usual cutting shear. The production was possible up to 700 m length without difficulties.

The prefabrication of the sections had been done with stainless steel 1.4301. The surface of the section was ferritic coated, what is commonly used to facilitate the section drawing process. This coating did not influence the laser welding, but as it showed ferromagnetism, it had to be removed before the final drawing

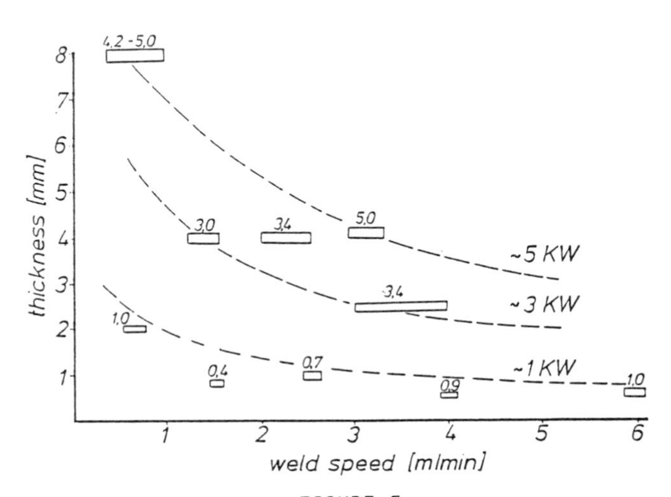

FIGURE 5
Thickness of welds as function of the welding velocity with the laser power as parameter.

step in a pickling bath. After this, there were serious problems during laser welding, which at first had been originated to the beam sharing device. Then we found in our institute, that the surface of the section was not clean enough. After removal of the tiny quantities of (unknown) material with abrasive paper, the laser welding could be performed as expected with good quality of the weld.

The required geometry of the edges of the steel section can be met by conventional drawing steps in the section manufacturing. But the cleanliness of the surface is the most important parameter to be controlled.

2.2.2 Laser-welding

In a separate development contract, VAC had overtaken the task to develop the beam-sharing of one laser primary beam (vertical) into two secondary beams (horizontal) which were focussed into the welding slits of two U-shaped sections. As there were no lasers available at the site, the decision had been taken to develop the beam sharing in a laser workshop where lateron the final conductor production could take place. The beam sharing device was thermally unstable and we had to limit the welded lengths. The maximum welded length has been 50 m of prototype jacket welding (without cable) which already has been used for coil winding tests. The maximum welded length (prototype length) of a complete conductor has been 25 m.

Unfortunately the beam sharing difficulties coincided with bad weld quality. So the beam sharing development was stopped and two lasers have been taken for the further tests to identify whether the problems came from the laser or from the steel section in its specific arrangement. The operating costs per hour doubled by this fact, and the beam sharing still looks attractive under this point of view.

During the course of work, new measuring systems for the beam quality had become available on the market and with this, a quick and precise adjustment of the beams (diameter, focus location, energy distribution) was possible. It could be demonstraded that the lasers themselves have not caused the bad weld quality.

The jacket will finally be welded with two CO_2 lasers of 1.5 kW and 6 kW beam power, operated at 1 kW each and 1.8 m/min.

After cleaning the rest of available material for prototype manufacturing, it was possible to get 9 m complete cable which is now vacuumtight better than 10^{-8} mbarl/s. Before the final manufacturing of the 4x150 m conductor can be started, the thermal stability of the optical systems of the two lasers will have to be checked in a long time running test.

2.3 Transportation and vacuum testing

Each conductor will be pancake-wound onto a drum with 3.2 m inner and 4.2 m outer diameter. The drums are transported to the coil manufacturer and serve as supply drums during coil winding. For the production of the four conductor lengths, a total number of 9 drums with 4.2 m outer ø is necessary. Special end pieces are welded to the conductor ends which allow the testing of the vacuum tightness of the conductor wound on the drum. The procedure of the Euratom LCT conductor testing is adopted here[6]. The specified vacum tightness is 10^{-6} mbarl/s.

2.4 Quality control

Due to a lot of technical problems, which had to be solved, time and money have not been available to develop specific quality control methods. This has to be done in a next step.

2.5 The manufacturing lines

Schematics of the manufacturing lines can be seen in Fig. 6 to Fig. 9. Fig. 7 containes the

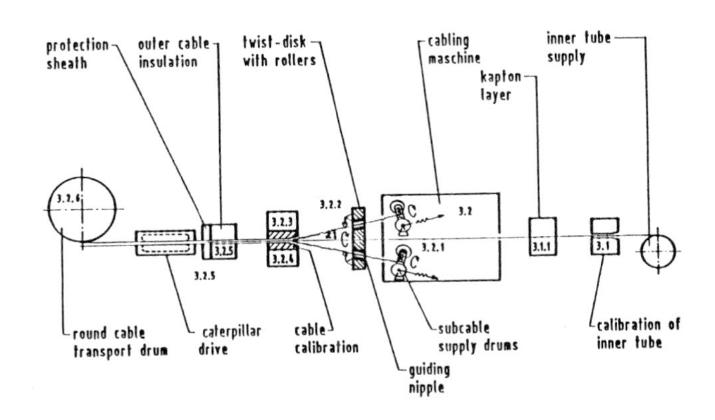

FIGURE 6
Cabling line to fabricate the round cable (with or without steel protection strips).

new protection strip compaction in comparison to the tube sheeting, welding and compaction.

3. CONCLUSION

The manufacturing of the superconductor for POLO will be continued.

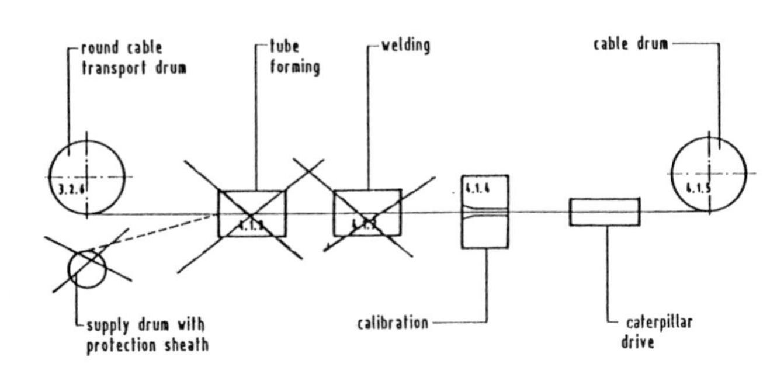

FIGURE 7
Welding line to enclose the round cable with the protection tube and simplification of the line for cables with the protection strips.

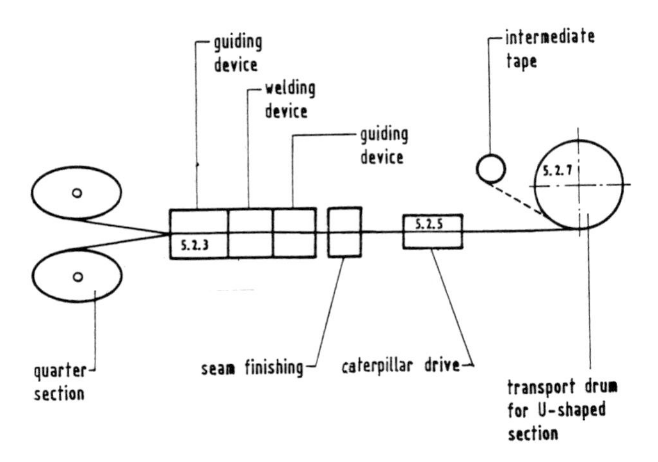

FIGURE 8
Welding line of the U-shaped section of the outer conduit (with one laser).

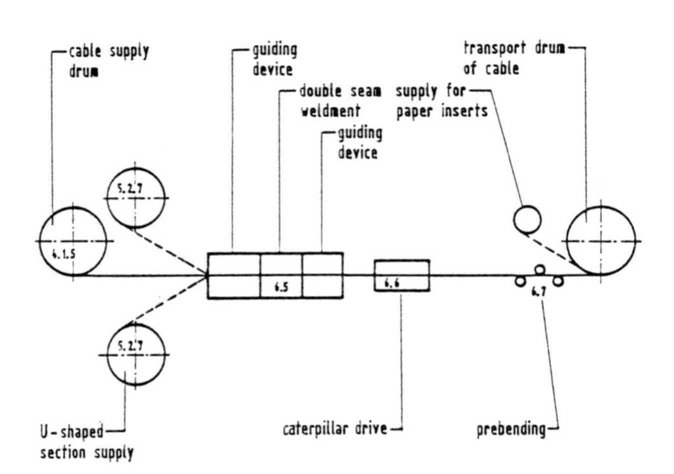

FIGURE 9
Assembly line of round cable and outer U-shaped sections (with two lasers).

ACKNOWLEDGEMENT

This work has been performed in the framework of the Nuclear Fusion Project of the Kernforschungszentrum Karlsruhe and is supported by the European Communities within the European Fusion Technology Program.

We thank the industrial companies and their staff, Cablerie Dour, VAC, Isoprofil, Heraeus Industrielaser for their engagement in the conductor development and manufacturing.

REFERENCES

1. H.G. Dittrich, S. Förster, A. Hofmann, U. Jeske, H. Katheder, A. Khalil, H. Krauth, W. Lehmann, W. Nick, A. Nyilas, A. Ulbricht, Development of Superconducting Poloidal Field Coils for Medium and Large Size Tokamaks, Proceedings of the 10th Symposium on Fusion Engineering, IEEE Cat. No. 83 C11 1916-6, Philadelphia, Dec. 6-9, 1983

2. S. Förster, G. Friesinger, U. Jeske, H. Katheder, G. Nöther, A. Nyilas, G. Schenk, C. Schmidt, L. Siewerdt, M. Süßer, A. Ulbricht, F. Wüchner, A 2 MJ, 150 T/s Pulsed Ring Coil. Status of the Design and the Test Arrangements, Proceedings of the 9th Magnet Technology Conference, UT-9, September 9-13, 1985, Zürich

3. C. Schmidt, Stability of Poloidal Field Coil Conductors: Test Facility and Subcable results, Proceedings of the 12th International Cryogenic Engineering Conference, ICEC 12, July 12-15, 1988, Southampton, England

4. H. Katheder, U. Süßer, Results of Flow Experiments with 2-Phase Helium for Cooling of Superconductors, Proc. of the International Cryogenic Engineering Conference, ICEC 12, July 12-15, 1988, Southampton, England

5. A. Nyilas, Laser Beam Welding of Advanced Superconducting Cable Conduits, Magnet Technology 1986, Proc. of the 14th Symp. on Fusion Technology, Sept. 8-12, 1986 Avignon, Pergamon Press

6. H. Katheder, K. Lennermann, Zur Dichtheitsprüfung sehr langer enger Kanäle (Innengekühlte supraleitende Spule), Vakuum-Technik, 33. Jahrgang (1984), Heft 2

FUSION TECHNOLOGY 1988
A.M. Van Ingen, A. Nijsen-Vis, H.T. Klippel (editors)

FERROMAGNETIC INSERTS FOR THE RIPPLE CORRECTION IN NET

M.V. Ricci[1], N. Mitchell[1]

[1]The NET Team, c/o Max Planck Institut fuer Plasmaphysik, Boltzmannstr. 2, D-8046 Garching bei Muenchen (FRG)

The possibility of generating large plasmas in NET might be limited by the high values of the toroidal field ripple. A possible means of ripple reduction is to include pieces of ferromagnetic material in the shielding structure. This paper presents the results of 3-D calculations that show how inserts of reasonable size can reduce the ripple to acceptable levels. The influence of the different dimensions of the ferromagnetic sectors is also shown.

1. INTRODUCTION

In a torus made of a finite number of coils the magnetic field values in a plasma cross-section are not the same if different toroidal planes are considered.

The largest differences are found in the outer plasma region when comparing field values on a cross-section contained in the vertical plane of symmetry of a coil to the values at the corresponding points on a cross-section in the middle of two coils.

If B_M and B_m are the field values at two corresponding points (same R and Z, different PHI, see Fig. 1), we define the ripple (in percent) by:

$$\text{RIPPLE} = 100 \, \frac{B_M - B_m}{B_M + B_m}$$

This ripple must be limited to avoid excessive alpha particle losses and first wall loads.

This limit is presently assumed to be 1.5%.

With the present configuration of the NET toroidal magnetic system (16 D-coils, see Fig 1), the ripple is well below this limit for an 11 MA plasma, but goes up to more than 3% at the border for a 15 MA plasma.

Means to reduce the ripple are: 1) increasing the number of the toroidal field coils (TFC), 2) increasing their size.

However, undesired drawbacks are: a reduced accessibility to the interior of the reactor in the first case and a rise of the overall size and cost in the second.

A third solution is the use of ferromagnetic inserts inside the TFC. The results of a study on the electromagnetic effects of the use of ferromagnetic materials within the NET machine were presented earlier [1]. The possibility of decreasing the ripple by using martensitic inserts in the vacuum wall/shielding structure was there suggested.

This paper presents the results of computer calculations which illustrate the applicability of this solution.

Since a high precision is required in these magnetic field computations, both in the absence and in the presence of iron, different models have been generated and several 2-D and 3-D codes have been used (POISSON, MAG3DWF, GFUN, TOSCA) and the results compared. The data here presented are those obtained with TOSCA[2].

Fig. 1 shows the coordinate system R, Z, PHI, the 11 MA and 15 MA plasma separatrices, the shield region (dashed) and the coils cross-sections.

The magnetic field values have been computed in the plasma region along the radius R at different values of Z (0, 1, 2, 3 m, see the dotted lines in the plasma cross-section in Fig. 1), for several values of PHI (from 0 to 11.25 deg, or 15 deg in the case of 12 coils).

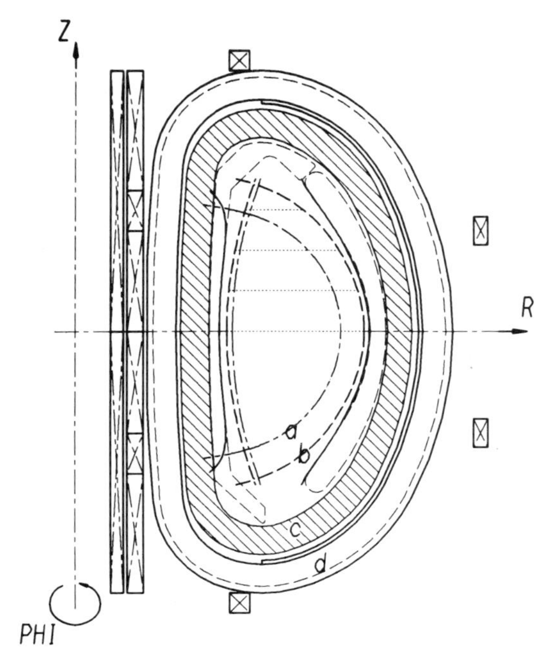

FIGURE 1
Cross-section in the plane of a TF coil.

Note that the four lines, ending at both extremities at the plasma border, have different lengths. This must be remembered when observing the following figures.

The 16 TF coils of Fig. 1 generate the field ripple which is shown in Fig. 2.

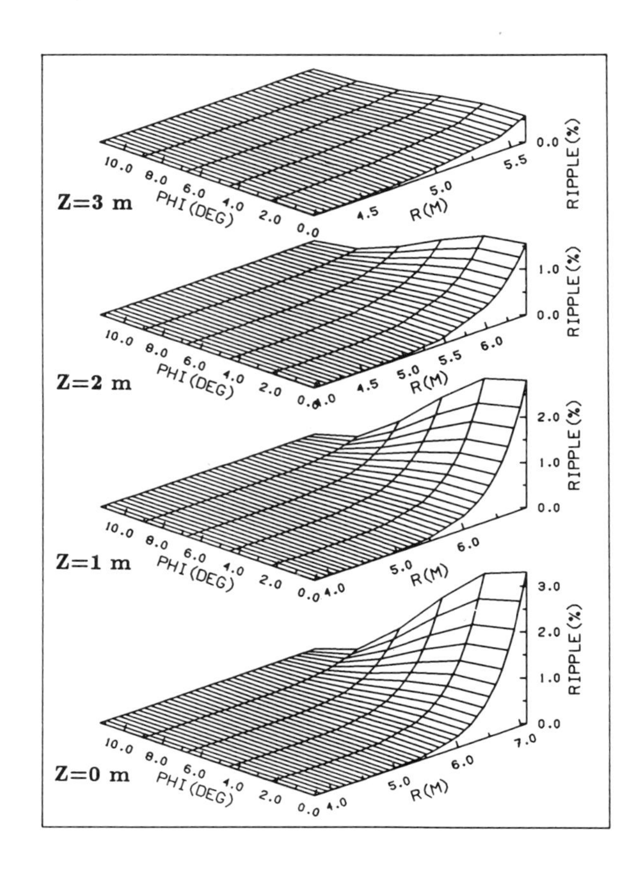

FIGURE 2
Ripple distribution in the 15 MA plasma region with no iron inserts.

The ripple is always zero at PHI=11.25 deg because field differences are always refered to this plane. It can be noticed that, as previously mentioned, the largest differences are found at the outer plasma border for values of Z lower than 2 m.

If the 11 MA plasma is considered, ripple values are always below 1%.

The field in the plane of a coil is obviously higher than in the plane between two coils. The idea for correcting the ripple is this: if a small portion of the shield structure underneath each toroidal coil is made of ferromagnetic iron (see Fig. 3), part of the magnetic flux will "prefer" to go through the iron, thus lowering locally the field.

It is clear that to adopt this effect to reduce the ripple to the desired value in the correct region, the iron inserts must be accurately dimensioned.

For these calculations, we have introduced in the codes the magnetic properties of a good commercial iron. The possibility of using more sophisticated materials will be considered in the future.

2. RESULTS

2.1. 16 coils

Fig. 4 shows the effect of rather small inserts. By comparison with Fig. 2 it can be seen that the reduction of the ripple is appreciable at Z=0 and Z=1 m, small at Z=2 m.

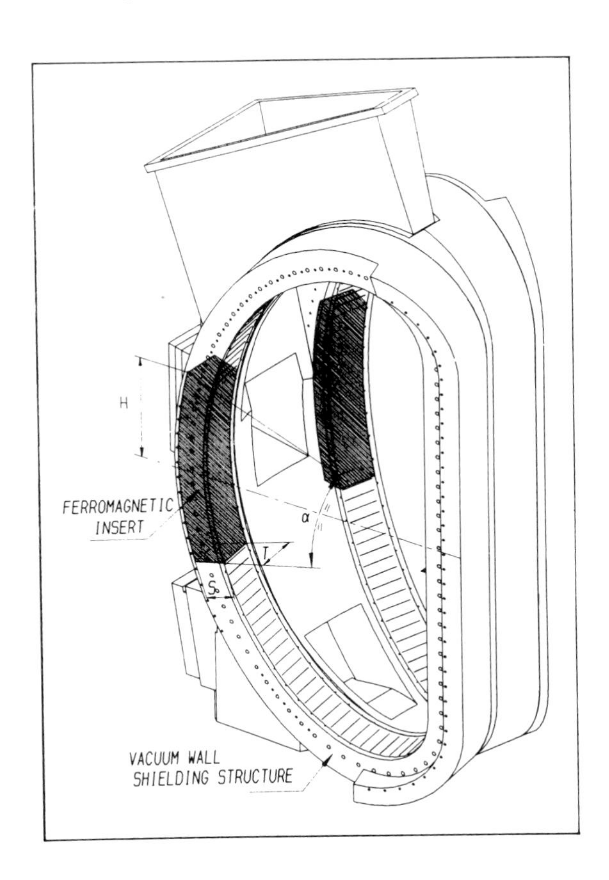

FIGURE 3
Schematic drawing of the inserts located in the shielding structure. Dimensions: S= radial thickness, T= toroidal width, H= height.

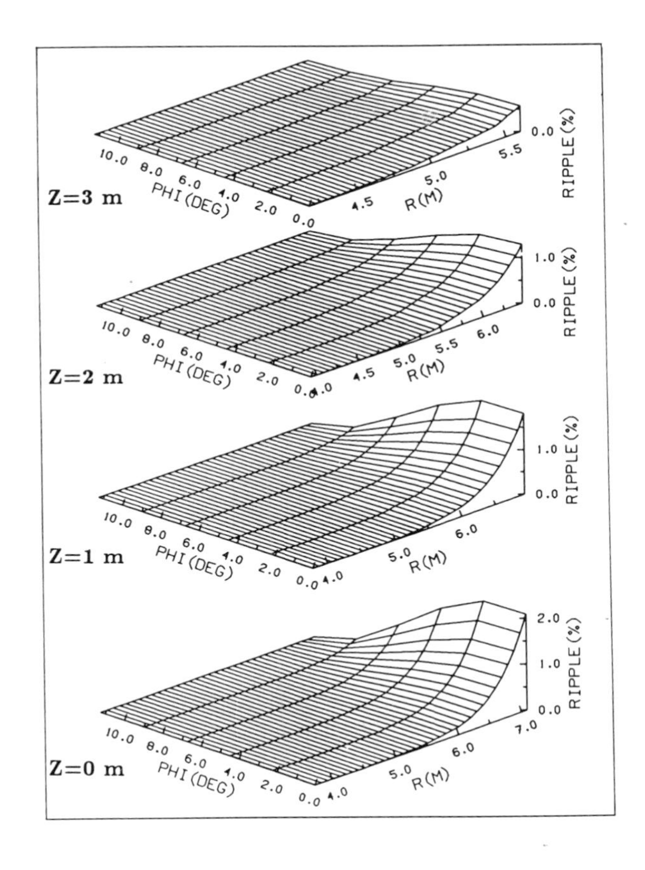

FIGURE 4
Ripple distribution in the presence of inserts of small cross-section (S = 25 cm, T= 64 cm) and limited poloidal extension (H= 224 cm).

The effect of increasing the poloidal extension of the inserts can be clearly seen in Fig. 5. The ripple reduction is higher than in the previous case at Z=2 m. At the other locations the correction is, as expected, quite the same.

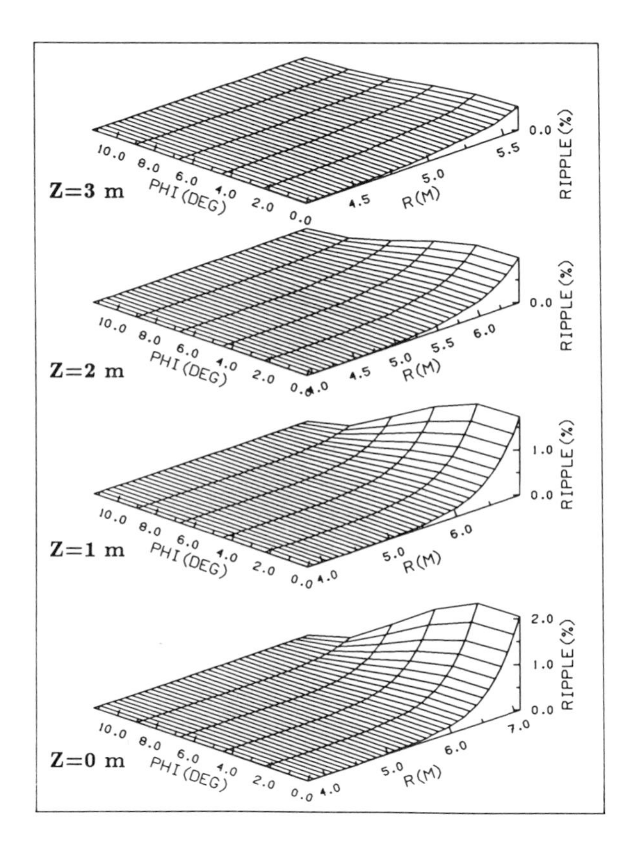

FIGURE 5
Ripple distribution in presence of inserts with the same cross-section as in Fig. 4 but more extended poloidally (H= 328 cm).

Fig. 6 shows how the ripple can be reduced to a value lower than the required 1.5% by using heavier inserts.

2.2. Lower fields

As long as the iron is saturated, its effect is independent from the field, hence at lower fields the inserts of Fig. 6 produce an

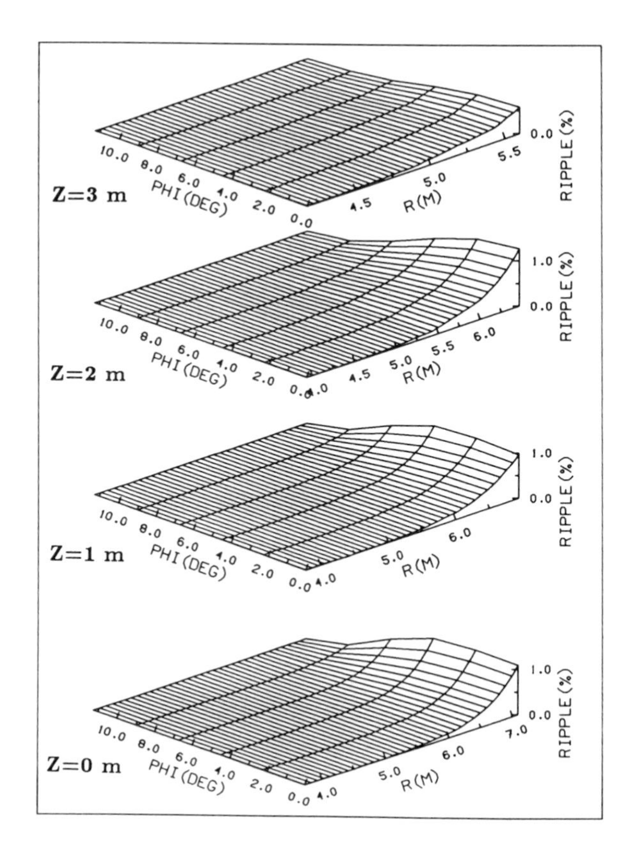

FIGURE 6
Ripple distribution in the presence of inserts of large cross-section (S= 42 cm, T= 90 cm) and limited poloidal extension (H= 174 cm)

overcorrection. Calculations show that down to about one half of the maximum operating field, the ripple is still acceptable (see Fig. 7).

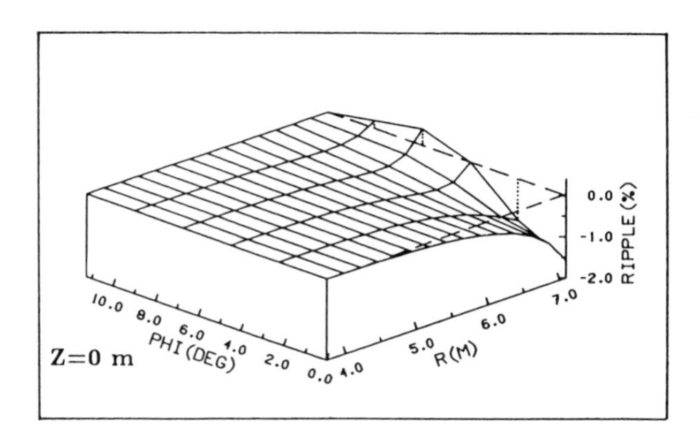

FIGURE 7

2.3. 12 coils

If the number of coils is reduced to 12 but their dimensions are left unchanged, the ripple goes up to a maximum of 8%. Inserts of reasonable size are unable to produce enough correction. The calculations demonstrate that the coils must be enlarged by about 60 cm in the radial direction.

2.4. Forces

Since the inserts are in a field gradient, they experience a net centring force.

The various codes give rather different values of this force so more work is necessary to clarify this point. The upper limit is given by POISSON and is of the order of 2 MN on each insert of Fig. 4. This force must be taken into account when integrating the inserts into the shielding structure, but no major problems are foreseen.

3. CONCLUSIONS

These calculations shows that iron inserts of reasonable size can effectively reduce the ripple in the outer plasma region. However, the necessity of this correction, and in this case the optimization of the dimensions of the inserts can only be stated by precise calculations of the alpha particle losses in a given field distribution.

AKNOWLEDGEMENTS

We would like to aknowledge G. Pasotti (ENEA Frascati) for his valuable help in the use of TOSCA.

REFERENCES

1. N. Mitchell, G. Pasotti and M.V. Ricci, Electromagnetic Effects of the use of Ferromagnetic Materials within the NET Machine, 14th SOFT, Avignon, Sept. 1986

2. J. Simkin, C.W. Trowbridge, Three-dimensional Non-linear Electromagnetic Field Computation using Scalar Potentials, IEE Proc. 127 B, no. 6, 1980

FUSION TECHNOLOGY 1988
A.M. Van Ingen, A. Nijsen-Vis, H.T. Klippel (editors)
Elsevier Science Publishers B.V., 1989

MAGNETIC FIELD AND LOOP VOLTAGE RIPPLE IN THE RFX PULSE: EVALUATION AND REDUCTION CRITERIA

Elena GAIO, Roberto PIOVAN, Vanni TOIGO

Istituto Gas Ionizzati, Associazione EURATOM-ENEA-CNR, Corso Stati Uniti 4, 35020 Padova, Italy

The ripple analysis of the magnetic field and loop voltage in the RFP experiments have been carried out by means of simplified equivalent schemes of the toroidal and poloidal machine circuits. An elementary plasma model has been utilized to estimate the diffusion of the magnetic and electric field inside the plasma region. An estimation of the ripple in the RFX experiment was then performed to evaluate the influence of some machine design choices.

1. INTRODUCTION

The large fusion research experiments in the magnetic confinement line require a considerable amount of energy and power to produce the plasma current; static power converters are so utilized to feed the toroidal and poloidal windings. The converter output voltage shows a high harmonic content whose value depends on the operation condition and converter configuration.

The converter voltage ripple produces fluctuations in the voltage which gives rise and sustains the plasma current. Moreover it produces fluctuations in the toroidal component of the magnetic field near the plasma vessel.

In the Tokamak experiments it was seen that, for the typical converter frequency range, the field penetration and the influence on the plasma behaviour can be neglected. In the RFP machines the difficulty to evaluate the ripple effects, in particular in the external plasma region where complex phenomena determine the plasma behaviour, stimulates to a first analysis to evaluate which is the ripple on the plasma due to the converters and what technical choices can influence its value. Using a simplified approach and neglecting the interaction with the plasma edge the diffusion inside the plasma was analysed in order to get the needed information for the technical choices.

2. CIRCUIT ANALYSIS FOR RIPPLE EVALUATION

The circuit parameters, which can strongly influence the fluctuations of the magnetic field on the plasma edge and of the loop voltage, are: the number of converter pulses, the liner resistance and any inductances in series to the converter. In order to simply evaluate the influence of these parameters, equivalent electric schemes for the poloidal and toroidal machine circuits have been analysed.

General schemes for RFP machine circuits are shown in fig. 1 and 2; the schemes refer to large machines, with magnetic energy stored, which windings are fed by thyristor converter units at

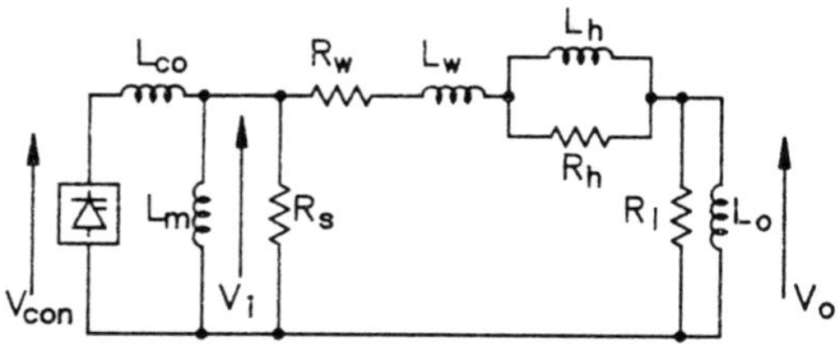

Fig 1 - Poloidal equivalent scheme

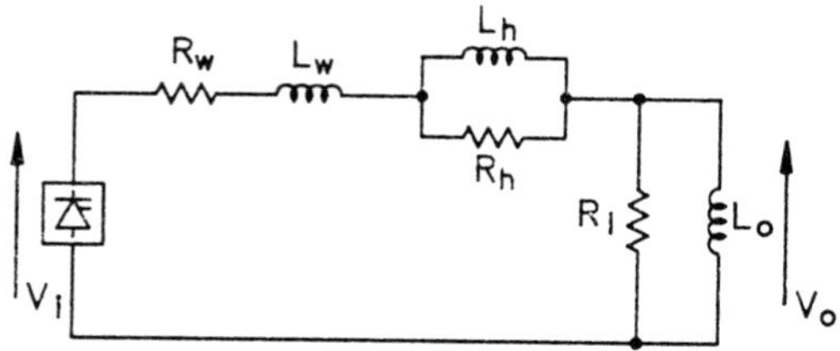

Fig. 2 - Toroidal equivalent scheme

least during the flat-top phase.

In the schemes R_w is the winding resistance and L_w is the inductance due to the space between the windings and the liner; in the toroidal circuit it can also include any inductance in series to the converter. The loop R_h and L_h is a first order approximation to take into account the phenomena due to the penetration of the magnetic field into the shell. In both schemes R_l is the resistance of the liner in the toroidal and poloidal direction respectively and L_o is the inductance related to the plasma region. In the poloidal scheme L_{co} is the converter inductance, L_m is the magnetizing one and R_s is the transfer resistance.

The frequency range considered for this analysis is that typical of the converter operation, which can vary from 300 to 1200 Hz. In this frequency range the two schemes can be simplified considering that the impedance of the parallel elements L_h and R_h can be neglected. Moreover the resistance R_w is always negligible.

In the poloidal scheme, disregarding the high magnetizing impedance L_m, the attenuation effect of L_{co} - R_s path has to be evaluated. This can be made by considering an equivalent generator with series impedance; the generator value V_i is given by:

$$|V_i|=1/\sqrt{(1+\omega^2\tau_s^2)}\ |V_{con}|$$

where $\tau_s=L_{co}/R_s$ and ω is the ripple frequency. The generator series impedance can be neglected, being much smaller than the other impedance.

The two schemes can so be reduced to that of fig. 3. The voltage V_o on the plasma side is given by:

$$|V_o| = \frac{1}{\sqrt{(1+\lambda)^2 + \omega^2\tau^2}}|V_i|$$

where $\lambda = \frac{L_{eq}}{L_o}$ and $\tau = \frac{L_{eq}}{R_l}$

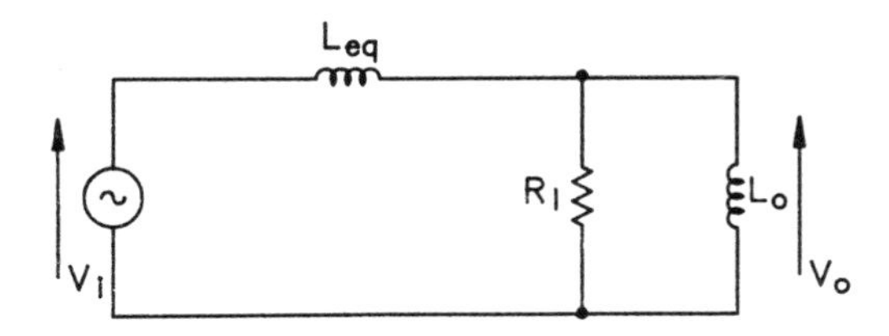

Fig. 3 - Simplified equivalent circuit

As the inductance L_o depends on the ripple frequency, an analysis has been made fixing the frequency to the most significant values: 300, 600, 1200 Hz, which correspond to six, twelve and twentyfour converter pulse operation.

The pattern of $|V_i|/|V_o|$ versus τ is drawn in fig. 4 for a frequency value of 600 Hz. The range of τ and λ were chosen taking into account the typical parameter values of RFP machines.

The figure shows that the ripple attenuation increases by reducing the liner resistance. Moreover it increases with the frequency and this effect is particularly appreciable because of the corresponding decrease of the equivalent plasma inductance due to the lower penetration thickness. These solutions for ripple reduction give however appreciable results only for λ values smaller than one.

The evaluation of V_o in the poloidal circuit gives the value of the toroidal electric field ripple

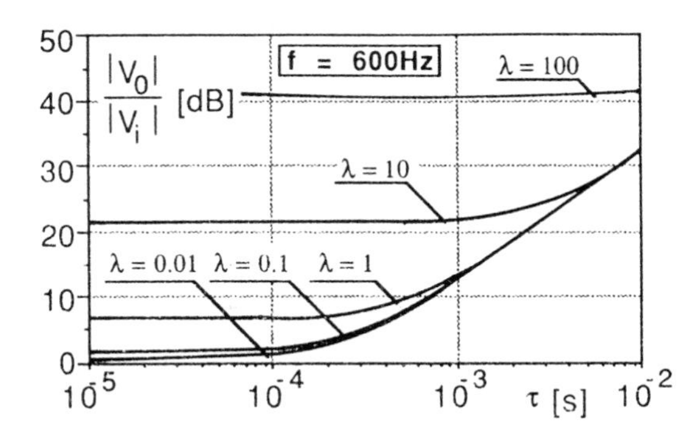

Fig. 4 - Converter voltage ripple attenuation due to the circuit

at the plasma edge; the toroidal magnetic field ripple in the same region can be evaluated from the analysis of the corresponding circuit considering that the current flowing in L_o, easily obtained from V_o, is related to this field component.

3. RIPPLE DIFFUSION INSIDE THE PLASMA

To estimate the magnetic and electric field ripple diffusion inside the plasma a simplified model was assumed, where the plasma is considered like an axisimmetric conductor with non-uniform resistivity along the radius a.

The plasma resistivity $\rho(r)$ was derived from the Spitzer[1] formula. A parabolic profile for the density n was assumed, with a decrease to one tenth at the plasma edge with respect to the axis value. The profile shape used for the temperature, elsewhere introduced[2], is:

$$T(s) = T(o)\left[1 - c_1\left(1 - e^{-c_2 s}\right)\right]$$

The same hypotesis of $T(a) = T(0)/10$ is considered; the value of c_2 that gives profiles in agreement with experimental observations is equal to 3.5 and the resulting temperature distribution is shown in fig. 5.

This model does not consider the complex plasma behaviour near the liner and the anomalous plasma resistivity, which can increase the losses of a factor two or more. Therefore this model can be used to get indications on the field diffusion in the plasma region but more complex models must be considered if an analysis of the ripple effect on the plasma behaviour was performed.

The diffusion equation, if the normalized coordinate $s=r/a$ and the normalized time $t^*=t/a^2$

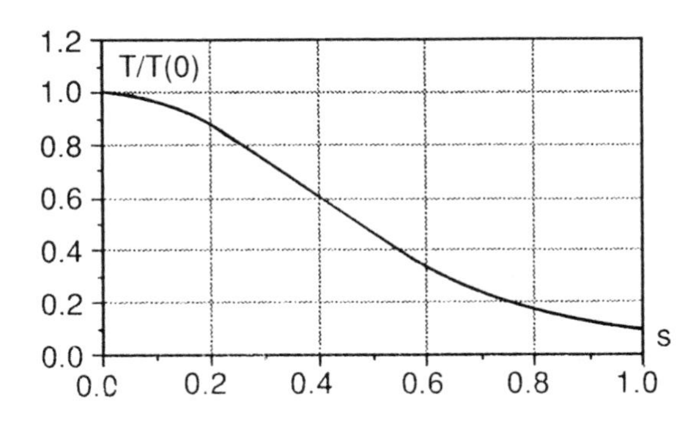

Fig. 5 - Normalized temperature profile

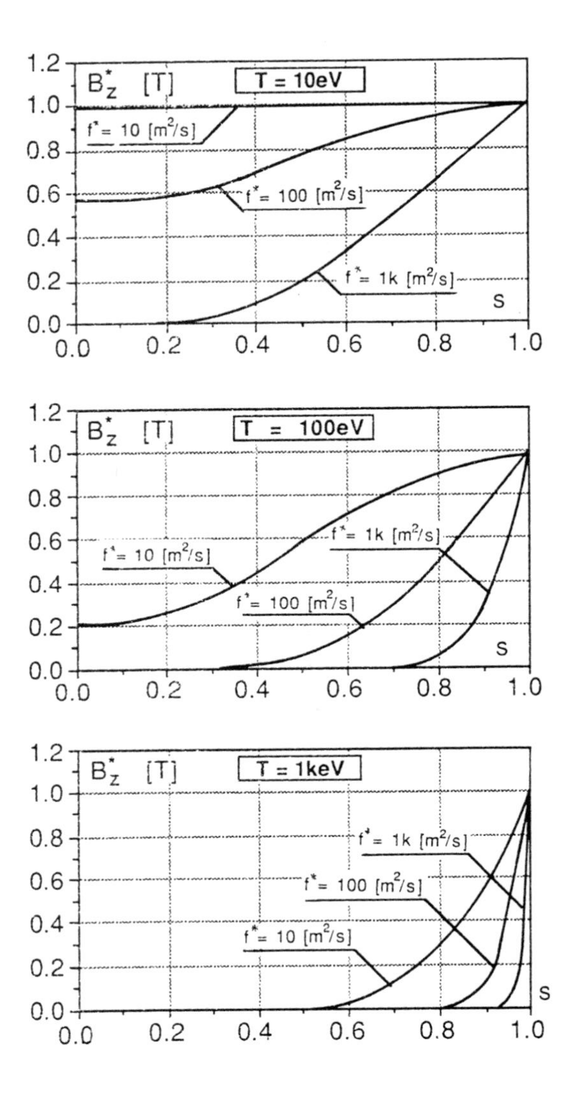

Fig. 6 - Bz* radial distribution with various temperatures and frequencies

are introduced, becomes:

$$\frac{\partial^2 B_z}{\partial s^2} + \left(\frac{1}{s} + \frac{1}{\rho}\frac{\partial \rho}{\partial s}\right)\frac{\partial B_z}{\partial s} = \frac{\mu}{\rho}\frac{\partial B_z}{\partial t^*}$$

where the effect of resistivity radial variation is included.

The solution of the differential equation was calculated by a finite difference method, considering a sinusoidal magnetic field with frequency f and unitary amplitude at the plasma edge. The field penetration was seen little sensitive to the plasma density, being the relationship logarithmic, so that a typical condition with a density on the axis of 10^{20} m^{-3} was chosen. The equation solution can be written in a stationary condition as:

$$B_z(s,t^*) = B_z^*(s) \quad \sin(\, 2\pi f^* t^* - \phi(s)\,)$$

where $f^* = a^2 f$ is the normalized frequency.

In fig. 6 the waveforms of $B_z^*(s)$ at three given frequencies and temperatures are shown.

A penetration thickness δ, which is adimensional, was defined by approximating $B_z^*(s)$ with an exponential function $e^{(s-1)/\delta}$; therefore δ is the value which minimizes the function:

$$F = \int_0^1 \left(\ln B_z + \frac{s}{\delta}\right)^2 ds$$

In fig. 7 δ is plotted vs. frequency for few temperatures. If δ is lower than one the ripple can have effect only in the outer plasma region.

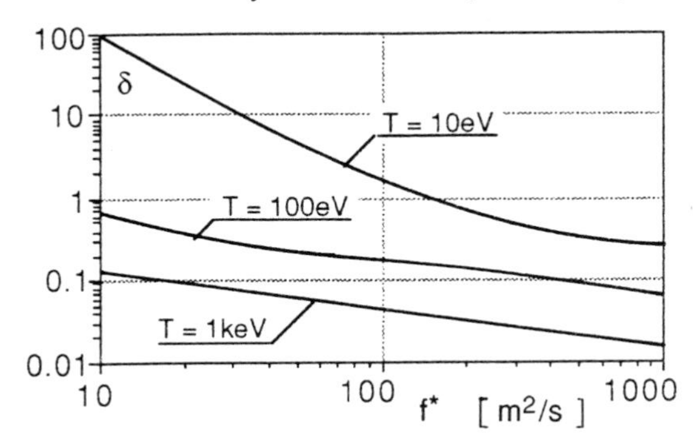

Fig. 7 - B_z^* penetration thickness inside the plasma

The same analysis described above was carried out for the plasma voltage ripple; in this case the toroidal current density j_z is considered and the diffusion equation becomes:

$$\frac{\partial^2 j_z}{\partial s^2} + \frac{1}{s}\frac{\partial j_z}{\partial s} + \frac{1}{\rho}\left(\frac{\partial^2 \rho}{\partial s^2} + \frac{1}{s}\frac{\partial \rho}{\partial s}\right) J_z = \frac{\mu}{\rho}\frac{\partial j_z}{\partial t^*}$$

This diffusion equation would be of the same type of the torodal field one if the resistivity was constant but it assumes a different form with non-uniform resistivity.

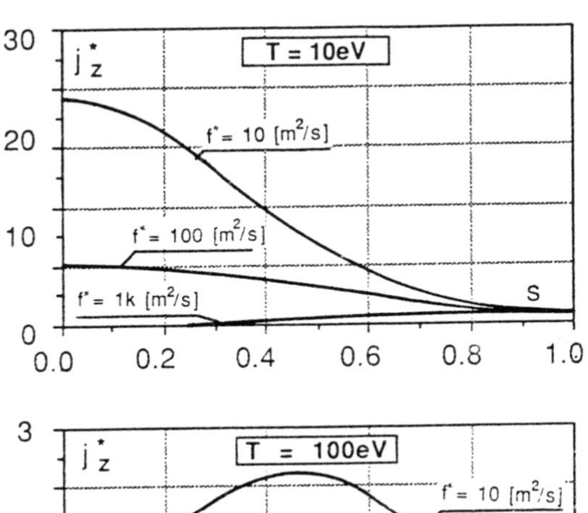

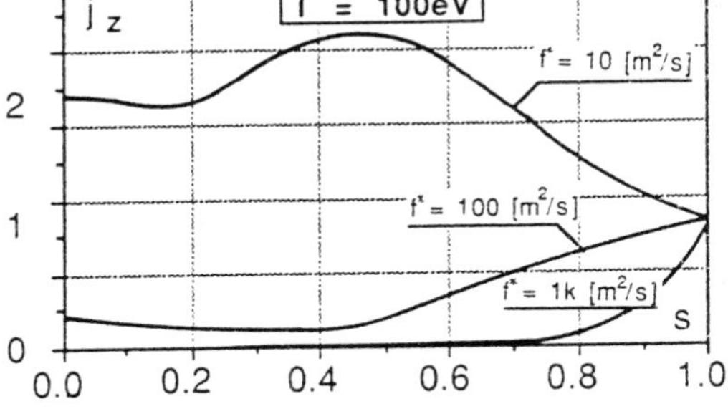

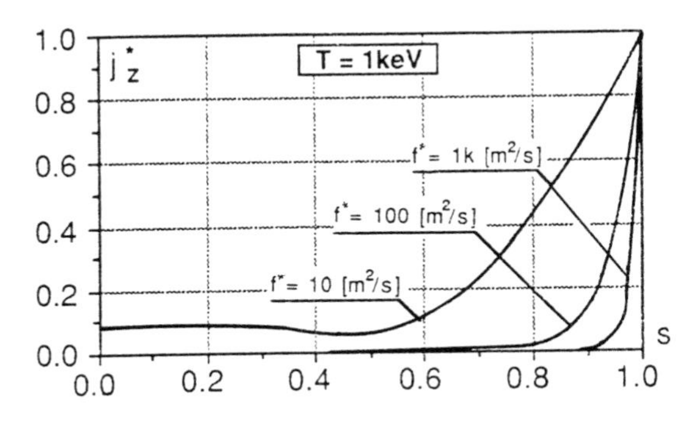

Fig. 8 - j_z^* radial distribution with various temperatures and frequencies

In fig. 8 the waveforms of $j_z^*(s)$, which correspond to $B_z^*(s)$, are shown for a unitary current density at the plasma edge; an increase of the current density in the inner plasma region occurs for the low frequency range; this can be explained because the current density ripple diffuses quickly into the central region of the plasma where the resistivity is lower.

With this model it has been analysed the results obtained in the ZT 40M RFP experiment; a voltage and current ripple in the windings was induced during the application of a current drive technique named Oscillating Field Current Drive (OFCD)[3]. It was demonstrate that the plasma current is affected through an enhanced plasma resistance by this ripple; moreover a strong dependence of the plasma behaviour from the phase-displacement between the poloidal and toroidal circuit ripple was noticed.

Considering the plasma radius of 0.2 m and a modulating frequency of 2 kHz the penetration thickness of the toroidal field is of the order of half radius. The experimental evidence is that the central plasma region is affected from the ripple and, in spite of the high semplification adopted, the model is sufficient to justify these results.

4. RIPPLE ANALYSIS FOR RFX EXPERIMENT

In the RFX machine, thyristor converter units are utilized to supply the windings of both poloidal and toroidal circuits[4]. The units operate at 12-pulses but the transformer connections also allows 24-pulse operation when more than one unit are used in series or parallel configuration.

During the flat-top phase, a group of converter units in the poloidal circuit provides to sustain the plasma current and another group in the toroidal one provides to maintain the toroidal component of the magnetic field. Both these groups operate at 24 pulses.

The evaluation of the voltage ripple on the plasma in both toroidal and poloidal circuits has been carried out by utilizing the simplified schemes; in the analysis it was taken into account also inductances, in series to the converters, needed to limit fault currents. In the toroidal circuit a configuration with four converter units in parallel and a turn ratio of 32, which maximizes the ripple on the liner, is considered.

The results found are an attenuation of 18 dB for the voltage V_θ in the toroidal circuit and of 29 dB for the voltage V_Z in the poloidal one.

In order to check the results obtained with the

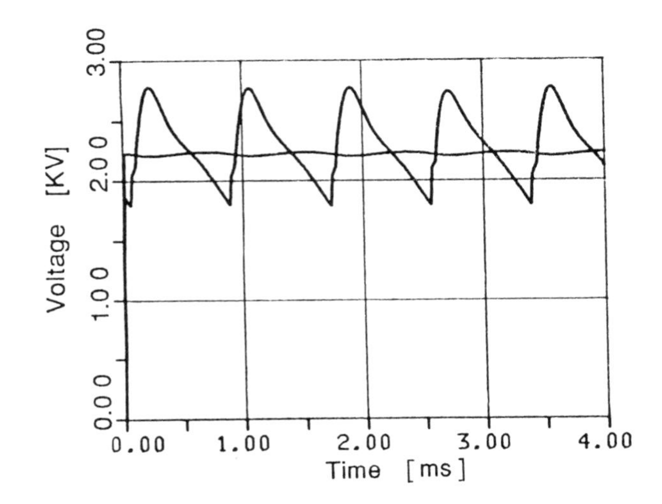

POLOIDAL CIRCUIT

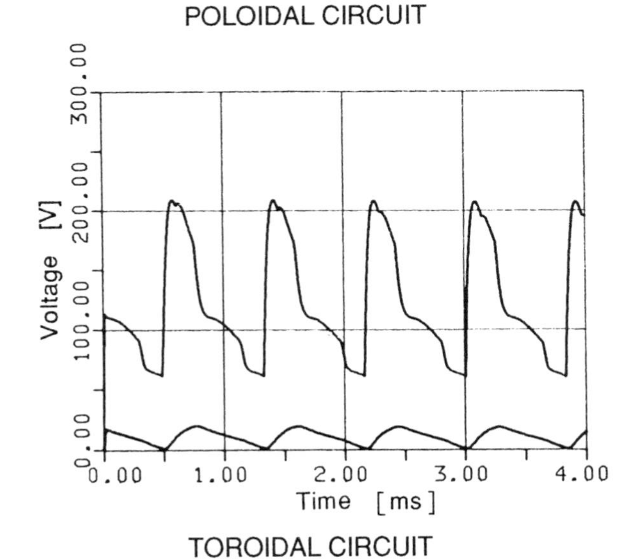

TOROIDAL CIRCUIT

Fig. 9 - a) Converter output voltage
b) Liner voltage multiplied by the windings turn ratio

simplified equivalent circuits, simulations of the complete RFX toroidal and poloidal circuits have been performed. These simulations include all the impedances of the AC grid, the converter transformers, the bridges and the complete circuits with all mutual inductances. In fig. 9 the waveforms of the output converter and liner voltage are shown for poloidal and toroidal circuit. The attenuation values found are in optimum agreement with the previous ones.

A capacitor in parallel to the converter units is foreseen for operation purpose. These capacitors give a significant additional attenuation to the ripple value on the plasma side; simulations results give 33 dB in the poloidal circuit and 24 dB in the toroidal one.

The resulting peak to peak ripple value of V_Z is of 0.13 V, which corresponds to an electric field of 10 mV/m. Considering an axis plasma temperature of 100 eV, the edge current density ripple is lower than one per cent of the current density on the axis.

The toroidal magnetic field at the liner shows a ripple of 0.14 mT which also is about one per cent of the value at the liner. The penetration thickness is 4 cm.

The ripple values and the penetration thickness of B_Z and j_Z in the RFX machine show that some choices, even if justified by other reasons, give a considerable contribution in the ripple reduction. In particular 24-pulses operation for the converters, the reduction of the liner resistance from 6.7 to 1.1 mΩ and the introduction of the capacitor banks produce the most significant benefits.

5. CONCLUSIONS

The ripple analysis for the RFX experiment, made with a simplified circuit and an elementary model of the plasma, has shown that the fluctuations induced by the converter system are limited to the external plasma region and have an amplitude of the same order of the natural plasma level.

ACKNOWLEDGEMENTS

We wish to thank Prof. G. Rostagni, Prof. G. Malesani and Dr. V. Antoni for their useful discussions and suggestions.

REFERENCES

1. L. Spitzer "Physics of fully ionized gases", Interscience Publ., New York, 1967.

2. K.A. Werley, R.A. Nebel and G.A. Wurden, Phys. Fluids 28,1450 (1985).

3. G.A. Wurden et al., "RFP Experiment: Results from ZT-40M and ZT-P",Int. School of Plasma Phys., Varenna (1987), pp.159-182.

4. G. Rostagni et al., "The RFX Project: a Design Review", in Proc. of 13th Symposium on Fusion Technology, Varese 1984, pp. 189-201.

FUSION TECHNOLOGY 1988
A.M. Van Ingen, A. Nijsen-Vis, H.T. Klippel (editors)

DESIGN STATUS OF THE NET TOROIDAL COILS

R.Poehlchen, L.Bottura, H.Katheder, G.Malavasi, J.Minervini, N.Mitchell, M.Y.Ricci, E.Salpietro[1)]
J.Rauch, H.Benz[2)]
G.Dal Mut, M.Perrella[3)]
A.Di Meglio, I.Nebuloni[4)]

[1)]The NET Team, c/o Max Planck Institut fuer Plasmaphysik, Boltzmannstrs. 2, D-8046 Garching bei Muenchen, Germany.
[2)]ABB, ASEA Brown Boveri AG., CH-8050 Zuerich, Switzerland.
[3)]ANSALDO Componenti SpA., Via Nicola Lorenzi 8, I-16152 Genova, Italy
[4)]Franco Tosi Industriale SpA., Piazza Monumento 12, I-20025 Legnano, Milano

The Toroidal Field Coil System consists of 16 superconducting coil windings, their coil casings and the intercoil structure. All of these components are located inside a common cryostat vessel and will therefore be at a temperature of about 4.50 °K during operation of the machine. The 16 coils are arranged in a toroidal configuration in order to provide a magnetic field for the confinement of the ring shaped plasma. The inner legs of the D-shaped coils form a vault which is subjected to the centering forces that are caused by the toroidal field itself. The interaction between the poloidal field and the toroidal currents creates Lorentz Forces which are perpendicular to the TF coil plane. Intercoil structure and vault have to resist these forces.
The huge size of the coils in combination with the fact that an A15 conductor material has to be used require techniques that are somewhat beyond the present state of the art. Therefore, a conductor and magnet development program has been launched. The development studies carried out by Associated Laboratories in cooperation with the NET Team have resulted in several forced flow cooled composite conductors. Furtheron, full size conductor samples were manufactured and two subsize conductors were manufactured and wound into two 12 T model coils. Proposals for the manufacture of the coil winding, the power supply and quench protection system, the cooling system and the instrumentation have been worked out in the course of these studies. To ensure the feasibility of the coil two study contracts have been placed with industry. This report will stress the most difficult aspects of the coil manufacture, the assembly of the winding in its steel casing and the assembly of the 16 coils with the intercoil structure to a toroidal configuration.
The results of the thermomechanical and electromagnetic analysis (e.g. eddy currents in coils case, stress, a.c. losses) - will be reported and their impact on the design of the TF system will be explained.

1. INTRODUCTION

The principal layout of the Toroidal Coil System and the underlying design consideration have been described rather recently |1|. Therefore this paper will be restricted to the latest developments: Industrial feasibility studies of the Toroidal Coil conductor, the Toroidal Coil itself and the Toroidal Coil Assembly have been completed these days. Together with a more detailed structural analysis |9| these studies result in:

a) a better knowledge on the manufacturing aspects and the assembly aspects.
b) a feedback on the basic design of conductors coil winding, coil case and intercoil structure which was done by the NET Team in cooperation with the associated laboratories |5,6,7,8,10|.
c) a first design of details of the coil winding and the interfaces between winding pack/coil case, between coil case/intercoil structure and between the individual coil cases in the vault area.

2. THE CONDUCTORS FOR THE TOROIDAL FIELD COILS.

All study contracts placed with research laboratories and industry on the toroidal coil conductor and the toroidal coils have resulted in proposals which use

- Nb_3Sn as superconducting material (alternatives could be ternary alloys, $(NbTa)_3SN$, $(NbTi)_3Sn$)
- a forced flow of supercritical Helium through the individual conductor as means of cooling.

While all earlier proposals are based on a React and Wind (R&W) process for the manufacture of the conductor and the coil winding a more recent proposal is based on a Wind and React process. In the first case the cable is reacted (heat treated at about 700°C for several days in order to produce the Nb_3Sn) on a spool and has to be straightened afterwards for application of stabilizer, steel jacket and the wrapping with glass tape. If a W&R process is used the conductor (cable and steel jacket) will be assembled first and then wound into double pancakes which have already the final coil shape. Then the double pancakes will be subjected to the reaction treatment. The strain sensitivity and therefore the choice of a R&W or a W&R process have considerable implications on the conductor and coil design and manufacture. No comprehensive list of advantages and disadvantages of the two methods can be given here but a list of all problems connected with the W&R process can be found in |13|. Moreover the NET Team has initiated conductor development and testing programs |5|, |6|, |7|, |8|, |14| which will lead to more manufacturing experience and test results for both types of conductors and will therefore influence to a large extent the final choice. While the question of feasibility will be answered this way the implications on the manufacturing process and the layout of the TF conductor and coil are already quite clear and are therefore a major point of discussion in this presentation. For the conductor a R&W process means that the strands of the cable - in other words the Nb_3Sn material - have to be located as close as possible to the neutral axis in order to minimize the strain degradation due to bending operations after the reaction treatment. This leads to a limit of the operating current of the conductor. The higher the operating current the wider the conductor has to be. Finally the steel jackets will get to weak mechanically. If a W&R process is chosen the SC strands can be distributed over the whole cross sectional area inside the steel jacket with the Helium flowing through the interstices in between the strands. The operating current can be much higher and therefore the quench discharge faster, the cooling path shorter or the winding process simpler (one in hand winding possible). But as outlined in |13| the conductor itself poses several problems which have to be solved in the conductor development program |14|. The main problems are: mechanical degradation of the steel conduit during the reaction treatment, electrical degradation of the SC due to a difference in the thermal expansion coeff. of the conduit and the Nb_3Sn, energy release due to strand movements. Fig. 1 shows the various conductor proposals with a list of parameters. Of these the number of electrical series connections per coil, the cooling path length, the total length of the conductor show already some implications on the Toroidal Coil layout and manufacture.

3. THE TOROIDAL COIL AND THE COIL CASE.

3.1 Coil Winding

In case of a W&R process each double pancake will be wound from a single length of conductor, each double pancake will be reacted individually then all will be stacked to a coil, electrically connected, wrapped with dry insulating tape for the ground insulation and finally vacuumimpregnated. A vacuumimpregnation process of the whole coil in one step (VPI process) with a suitable Epoxy resin system will result in a mechanically strong bonding

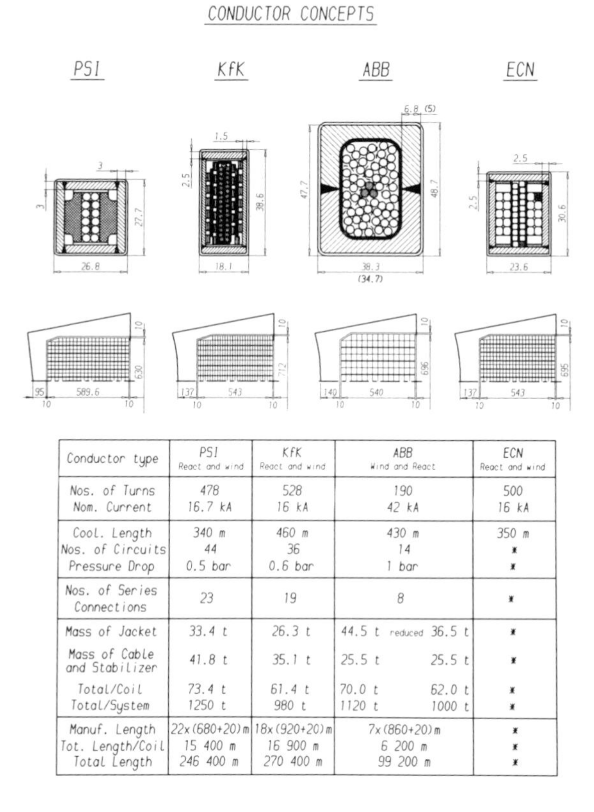

Conductor type	PSI React and wind	KfK React and wind	ABB Wind and React		ECN React and wind
Nos. of Turns	478	528	190		500
Nom. Current	16.7 kA	16 kA	42 kA		16 kA
Cool. Length	340 m	460 m	430 m		350 m
Nos. of Circuits	44	36	14		*
Pressure Drop	0.5 bar	0.6 bar	1 bar		*
Nos. of Series Connections	23	19	8		*
Mass of Jacket	33.4 t	26.3 t	44.5 t	reduced 36.5 t	*
Mass of Cable and Stabilizer	41.8 t	35.1 t	25.5 t	25.5 t	*
Total/Coil	73.4 t	61.4 t	70.0 t	62.0 t	*
Total/System	1250 t	980 t	1120 t	1000 t	*
Manuf. Length	22x(680+20)m	18x(920+20)m	7x(860+20)m		*
Tot. Length/Coil	15 400 m	16 900 m	6 200 m		*
Total Length	246 400 m	270 400 m	99 200 m		*

Fig. 1: Conductor Concepts and their Impact on the Winding Layout

between the turns and pancakes and in an insulation with a high electric strength (20 KV to ground e.g.) - as it has been achieved for large toroidal copper coils but also for large S.C. coils. In case of NET a certain degradation of the insulation by irradiation has to be taken into account - for that reason a test pogram has been started by NET.

In case of a R&W process the reacted conductor has to be unwound from storage spools |7| and then wound into double pancakes. For the rather small operating current a two in hand winding is more suitable in order to restrict the manufacturing length of the conductor to 300-400 meters. The remaining manufacturing steps are identical to those of the W&R process.

Figure 2 shows details of the winding cross section. Figures 3,4,5,6,7 illustrate the difference in layout and manufacture between a coil to be manufactured by a R&W process or by a W&R process. These drawings are self explanatory. What should be stressed here is that for a W&R process - see Figs. 3 and 5 - the electric joints can be made without prior bending of the reacted conductor, the access for making the joints is easy and the number of joints required is very much reduced (Figs. 1 and 4). Fig. 6 illustrates that in case of a R&W process welding has to be carried out in order to close the steel jacket after joining the SC cables. This welding takes place in the vicinity of cable and stabilizer which contain a considerable amount of soft soldering material with a low melting point.

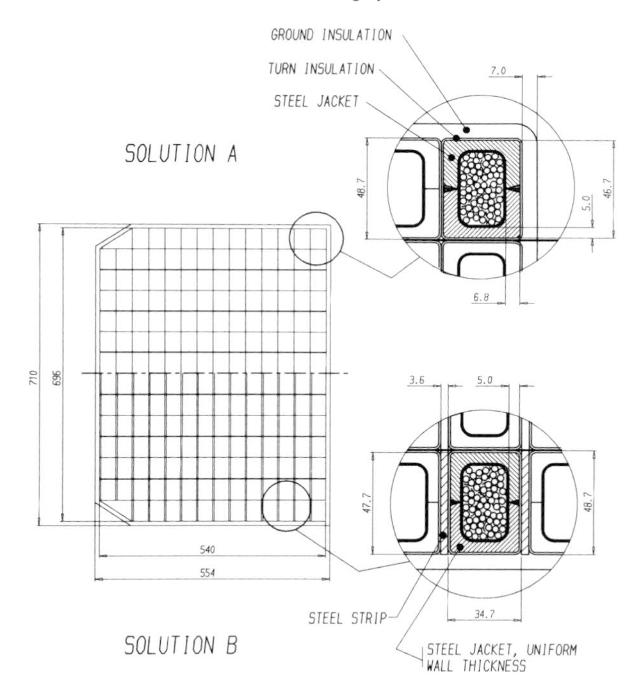

Fig. 2: Winding Pack Details

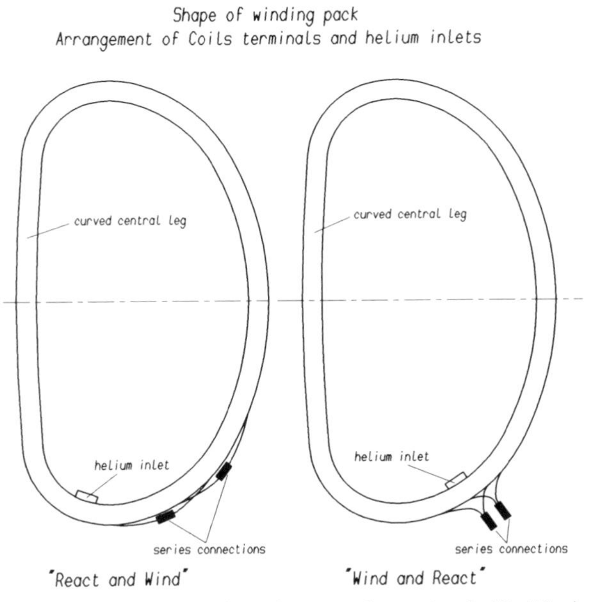

Fig.3: Electric Series Connections in Coil Winding, Location and Schematic Layout

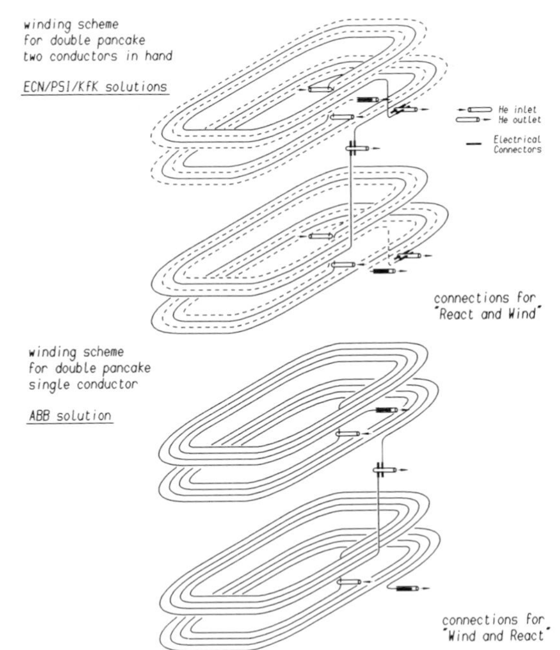

Fig.4: Winding Scheme for a Double Pancake

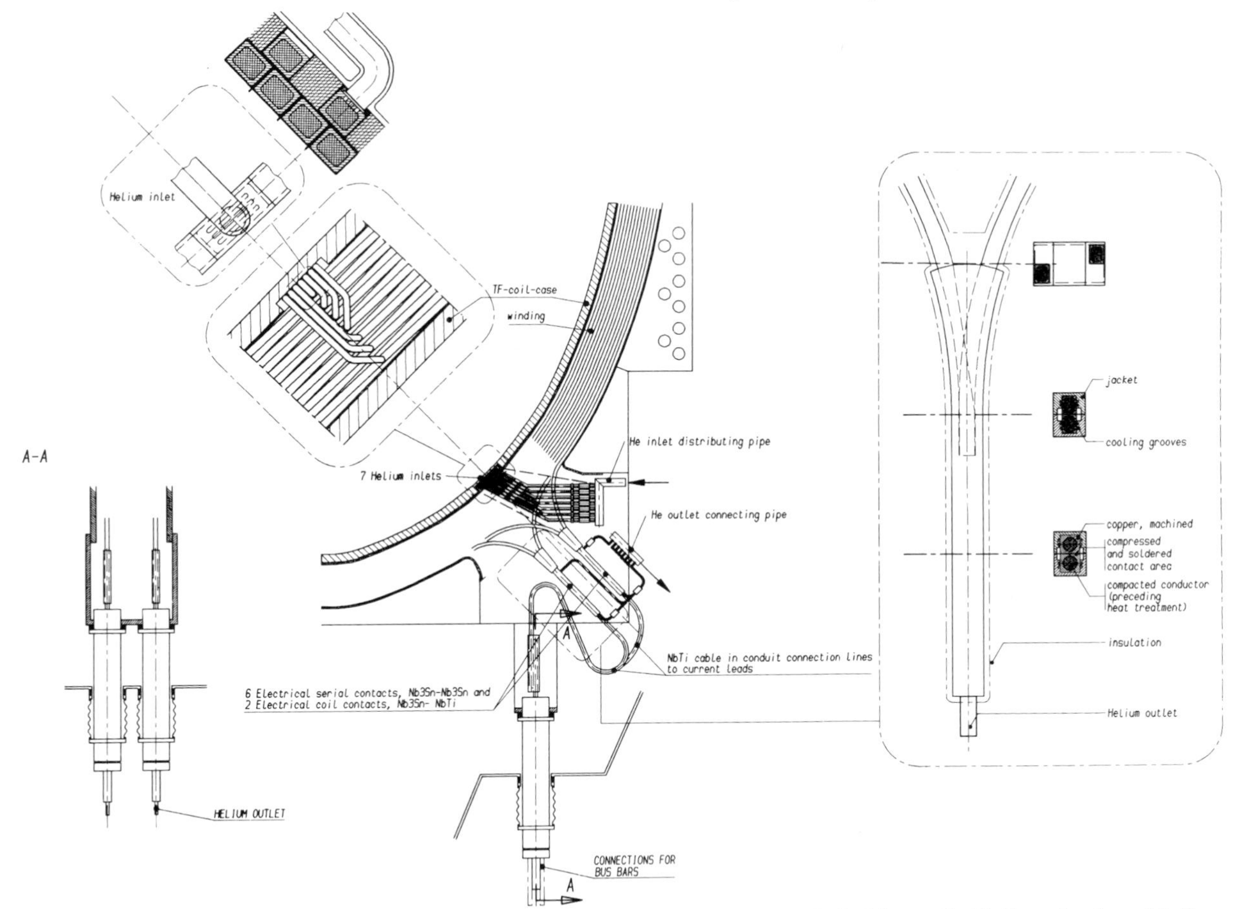

Fig.5: Details of the Toroidal Coil Winding for a Wind an React Solution: Coil Terminals, DP Connections, Helium In- and Outlets, Current Leads.

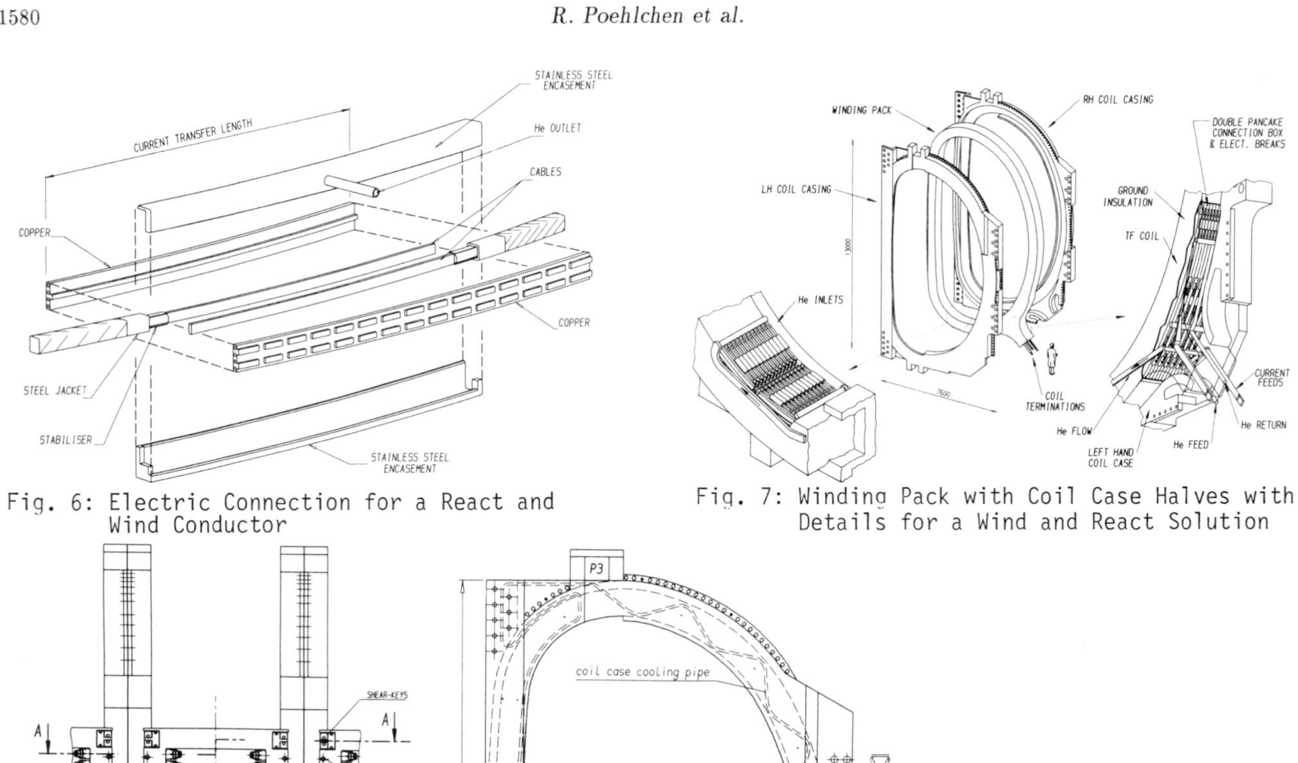

Fig. 6: Electric Connection for a React and Wind Conductor

Fig. 7: Winding Pack with Coil Case Halves with Details for a Wind and React Solution

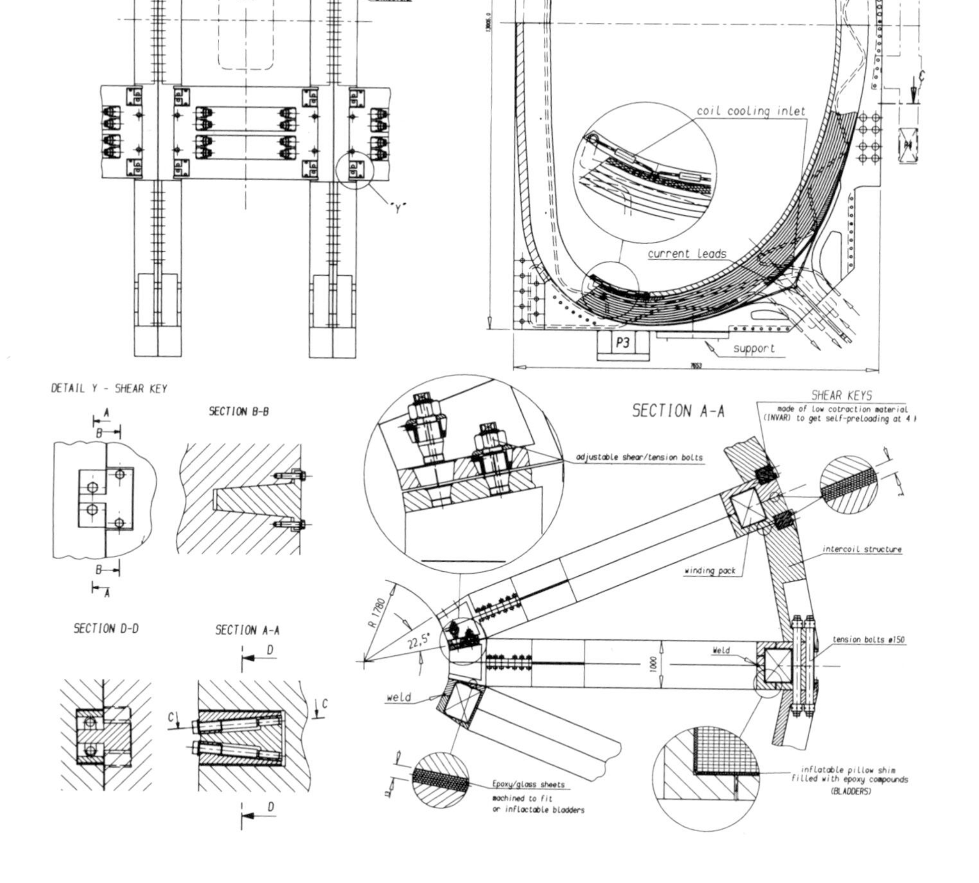

Fig. 8: Details of the Assembly of the Tor.Coils -Joining of Coils in the Vault Area and Joining of Coils and Intercoil Structure.

3.2 Coil Case and Assembly of the Coil Winding in its Case.

The coil case is shown in Figs. 7 and 8. It consists of two halves which will be partly bolted together and partly welded. Both halves will be electrically insulated from each other, at least in the critical areas, which occur in the neighbourhood of the Poloidal Coils P3. An eddy current analysis was carried out by NET for the coil case. With the halves insulated against each other except at the straight portion, 4.3 KW are generated in the 16 coil cases during a 300 sec. cycle and for a 11 MA plasma current. With no insulation at all between the coil halves there would be 24.4 KW generated in the 16 coil cases. In the vault area the tilting moment on each Toroidal Coil due to the interaction with the Poloidal Coils will be taken by friction between the coil cases and shear dowels at the lower and upper ends of the case, see Fig. 8.

In the circumferential direction, the case halves have to be welded together from six separate pieces for manufacturing reasons. For the highly stressed straight inner part a 316 LN material, forged and heat treated will have to be used. The other parts can be stainless steel castings or structures welded from hot rolled stainless plates, (316LN, 316 L or 304L depending on the stress level, see |2|, |9|). During cool down of the magnet and during operation the coil case has to be cooled. The cooling channel is indicated in Fig. 8.

Fig. 9 illustrates how a tight fit of the winding in the coil case will be achieved. Between winding pack and steel case inflatable bladders (pillow shims) will be inserted on all sides. These bladders consist of stainless steel sheets (about 0.4 mm thick) welded together at the edges. During assembly of the winding in the case the bladders are filled under pressure with a mixture of resin and a

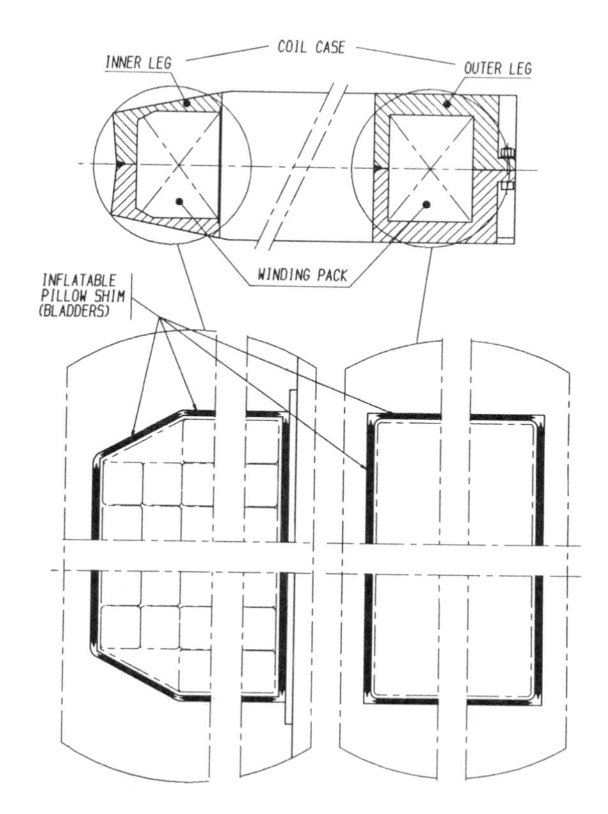

Fig. 9 : Assembly of Coil Winding in its Coil Case.

mineral filler. The same method has been used sucessfully for the assembly of the winding of the EURATOM LCT coil in its steel case (compressive strength measured for this application at 4 ^{o}K : ~ 500 N/mm^2).

4. INTERCOIL STRUCTURE

The Intercoil Structure has - together with friction and shear dowels in the vault area - to resist the overturning moment acting on the Toroidal Coils. Stainless steel shear boxes are placed between the outer coil legs and joined through the - locally - thick walled coil case by tension bolts - see Fig. 8. The manufacture of the Intercoil Structure does not pose any manufacturing problem.

5. FLEXIBLE SUPPORTS FOR TOROIDAL COILS

Each of the Toroidal Coils is resting on a support which allows a movement in the radial direction via a bending of the thin steel plates - see Fig. 10. A detailed stress analysis has been carried out and results of the max. von Mises Stresses and max. displacements are given in Fig. 10. For further details see |11|.

6. ASSEMBLY OF THE TOROIDAL COIL SYSTEM.

The assembly of the Toroidal Coils to a toroidal configuration poses mainly two manufacturing and assembly problems:

- to achieve a good matching of the large contact areas (between intercoil shear boxes and coil cases) and to achieve at the same time a precise positioning of the 16 coils,
- to achieve at these interfaces a sufficiently strong joint by means of shear dowels and bolts.

Industrial studies have resulted in detailed assembly procedures |2, 3|. There is no room to outline these procedures here. But as solutions to the two problems indicated above the following has been proposed: Either we use resin filled bladders as it is foreseen for the assembly of the winding in the coil case to achieve a perfect matching. Or Epoxy glass sheets have to be machined according to measurements - see Fig. 8. In case of the application of Epoxy/glass plates a procedure for the sequence of assembly, gap measurements, machining and final assembly has been worked out - see |2| and |3|. For point two drilling operations in situ should be avoided. Therefore adjustable shear dowels (see Fig. 8) have to be used. Another solution would be to cast a filled Epoxy around the dowels.

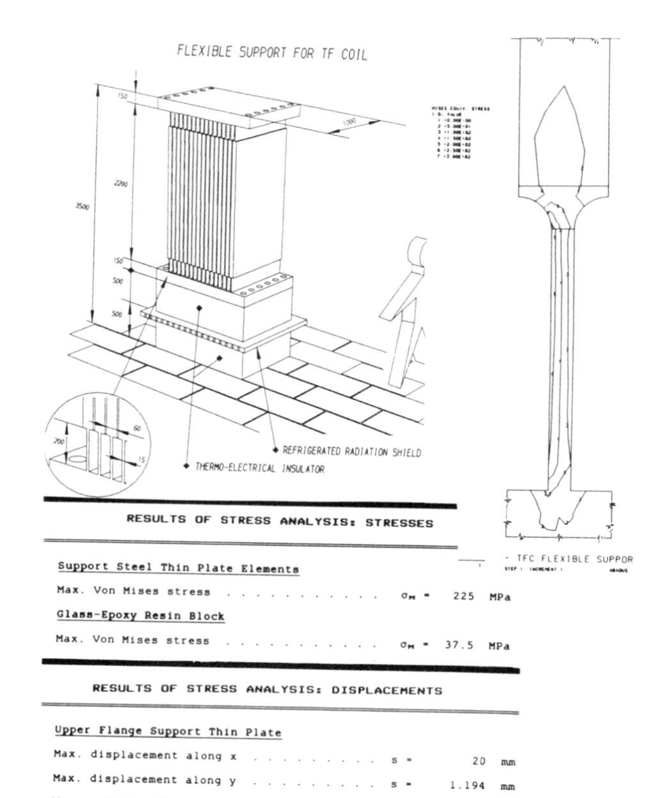

Fig. 10: Flexible Coil Support

REFERENCES

1. "Next European Torus Basic Machine", E. Salpietro et al., Fusion Technology, July 1988, Vol. 14, No. 1, pages 59-70.
2. "Final Report on the Fabricability Study of the NET Toroidal Coil System", J. Rauch et al., Rep. No. HISM 20376, Zuerich, May 1988.
3. 2nd Progress Rep. "NET Toroidal Coil Feasibility Study", Dal Mut et al., December 1987, ANSALDO, Genova.
4. 2nd Progress Rep. on "Concept Verification and Predesign of the NET Machine Support", A. Di Meglio et al., F. Tosi Legnano, Sept 1988.
5. "NET Toroidal Field Coil Design Studies, Final Report", C. Marinucci et al., PSI Villigen, Nov. 1987.
6. "Final Report on NET Tor. Field Coil Design", J.A. Eikelboom et al., ECN Petten 1987
7. "An A 15 Conductor Design and its Implications for the NET II TF Coils", Final Report, R. Fluekiger et al., KfK Karlsruhe 1984.

8. "Final Report on the TF Coil Design Study", Fus/Tecn/Superc 7/87 & 8/87 & 11/87, R. Cerreda et al., ENEA Frascati 1987.
9. "Structural Analysis of the NET Toroidal Field Coils and Conductor", N. Mitchell, 15th SOFT, 1988, Utrecht.
10. "Progress in the NET Toroidal Field Coil Mechanical Design", G.Malavasi et al, 14th SOFT, Avignon 1986.
11. "Final Report on the Flexible Toroidal Coil Support Analysis and Design", J. Nebuloni et al, F. TOSI INDUSTRIALE, Legnano, September 1988.
12. "Eddy Current Calculation in the NET TF Coil Casing", L. Bottura, NET Internal Note 87-57/1987.
13. "Problems of the Nb_3Sn Cable in Conduit Wind and React Conductor", R. Poehlchen, J. Minervini, NET/INT/87-034, 1987.
14. "Superconductor Designs for the Central Solenoid of NET", J.Minervini, 15th SOFT 1988, Utrecht.

FUSION TECHNOLOGY 1988
A.M. Van Ingen, A. Nijsen-Vis, H.T. Klippel (editors)

TRANSIENT THERMAL ANALYSIS OF THE NET TF COIL DURING NORMAL OPERATION

L. BOTTURA[1], J. MINERVINI[1]

[1] The NET Team c/o Max-Planck-Institut fuer Plasmaphysik, Boltzmannstr. 2, D - 8046 Garching bei Muenchen.

The thermal behaviour of the NET TF coil has been investigated in normal operating conditions. The detailed temperature map in the most critical cooling paths for two TF conductor designs were computed, considering the heat influxes due to nuclear heating, conduction and radiation, AC loss and the heat removal through the flowing helium. Using these temperature distributions and a detailed magnetic field map in the pancakes, it was possible to compute the minimum temperature margin in the whole coil. The location of the minimum temperature margin is computed to be the same for both designs, in the first turn of the winding. Its magnitude is mostly determined by the nuclear heat deposited in the coil. For the considered heat loads the computed margin appears sufficient, but the simplifications in the model and the uncertainties in the heat influx can lead to optimistic results.

1. INTRODUCTION

The TF coils of the NET machine will be subjected to several thermal disturbances coming from normal and abnormal operating conditions, such as nuclear heat, AC loss, conduction and radiation from the environment, sudden conductor motion and cracks in the epoxy, frictional heat production and Joule-Thomson effect in the helium.

These thermal disturbances will change the conductor temperature during a burn cycle, depending on their distribution in time and space and on the conductor properties.

Since one of the aims of the operation of the TF coils is to achieve the maximum availability as possible, it is necessary to maintain the temperature in the superconductor as far as possible from the current sharing temperature T_{cs} where the conductor begins to Joule heat. The conductor must operate everywhere with the largest temperature margin as possible.

At given operating conditions, winding pack geometry, conductor properties and dimensions, the location and time of the minimum temperature margin can be computed. To safely operate the coil this minimum temperature margin must be greater than the maximum temperature increase expected due to any disturbance during the operation. To know the temperature margin in the coil a detailed, transient analysis of the temperature must be performed for each helium flow path, considering the spatial distribution of the heat disturbances, the field distribution and the real properties of the conductor. This analysis must take into account the heat accumulation phenomena that is due to the repetition of the burn cycles with period smaller than the helium transit time in a flow path.

2. COIL AND CONDUCTORS GEOMETRY

The geometry of the NET TF coil is extensively described[1]. The dimensions, operating conditions and flow characteristics for the PSI and KfK conductors[2,3] are briefly summarized in Tab. I.

3. HEAT LOADS IN THE TF COIL

The thermal loads have been divided into steady state and transient heat fluxes, depending on the time scale of their variation.

Table I: Relevant dimensions and properties for the PSI and KfK conductors

	Pancakes	
	PSI	KfK
Double pancakes	12	10
Turns/pancake	20	28
Technique	2-in-hand	2-in-hand
Length (m)	302+302	385+385

	Conductors	
	PSI	KfK
Ext. dim. (cm)	2.68*2.77	1.81*3.86
A_{Metal} (cm^2)	4.257	3.593
A_{Helium} (cm^2)	0.498	1.076
Hyd. diam. (cm)	0.32	0.253
p_{in} (atm)	5.0	6.0
T_{in} (K)	4.5	4.7
$\dot{m}$ (g/s)	3.7	5.0
T_c (K)	$10.8*(1-B/28)^{1/2}$	17-B/2
J_{op}/J_c (T_{in}, B_{max})	0.426	0.492

The steady state heat loads considered were:

a) Nuclear heating. The NET reference double null 10.88 MA plasma was considered for the neutronic flux calculation, leading for both conductors to a maximum of 3.6 mW/cm^3 in the inboard side of the inner straight leg and average value of 500 W in one TF coil. An exponential decay described the nuclear heating distribution along the flow path in one pancake. The conservative assumption of constant nuclear heating throughout the burn cycle was done.

b) Conduction and radiation. No detailed analysis was previously performed on the the amount of heat that reaches the coils due to conduction heat flux or radiation. We assumed here that a total of 200 W (uniformly distributed in space) is deposited in the TF coil winding pack due to these two effects.

c) Joule-Thomson effect. The Joule-Thomson coefficient is tabulated in the helium property tables, and was taken into account during the temperature calculation in the helium.

d) Frictional heating in the helium. This kind of heat input is really significant only for large pressure drops in the flow path.

The only transient heat influx considered was:

e) AC loss. The amount of AC loss deposited in the TF coil conductors depends strongly on time and space. The computed detailed maps[4] have been used, leading to the cycle averaged values of 123 and 206 W/coil for the KfK and PSI designs respectively.

An example of the distribution along the flow length of some of the heat fluxes considered

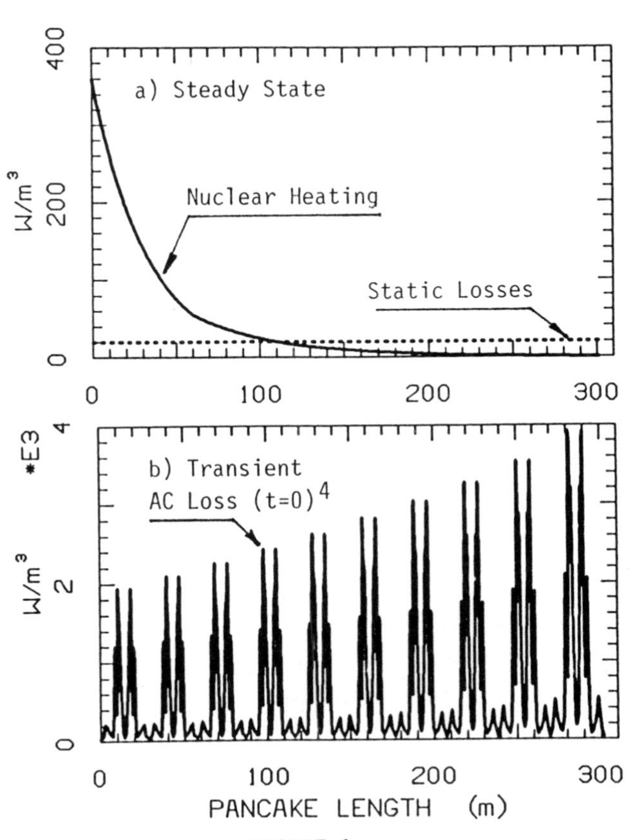

FIGURE 1
Steady state and transient heat loads in a PSI flow path

here is given in Fig. 1, where, for the PSI flow path, the nuclear heating, conduction and radiation fluxes and AC loss are plotted.

4. MODEL FOR THE ANALYSIS

4.1. Magnetic field

The magnetic field along each pancake has been computed by interpolation of a field map consisting of about 4000 points in one TF coil at 12 times during a plasma burn cycle (0 to 290 s). The magnetic field was interpolated at about 1000 points in each pancake as a function of the developed pancake length and time[4].

4.2. Current sharing model

From the field interpolated maps the spatial and temporal dependence of the current sharing temperature was known through:

$$T_{cs} = T_{op} + (1 - I_{op}/I_c) (T_c - T_{op}) \qquad (1)$$

where T_c, I_c are the critical temperature and current and T_{op}, I_{op} are the operating temperature and current, respectively. In reality, given the small relative change of the magnetic field absolute magnitude during a burn cycle, the current sharing temperature distribution was assumed to be the same for the whole cycle, fixed at the initial value.The distribution of T_{cs} shows a periodic behaviour coming from the similar periodicity in the magnetic field[4].

4.3. Thermal model

Each flow path in the PSI and KfK pancakes was modelled as a 1-D pipe, with the flowing helium thermally coupled to the conductor. No temperature gradient was considered in the cross-section of the conductor, and a constant mass flow rate has been assumed, since the time scale of the heat deposition is very large[5]. The pipe wall represented the composite structure of the conductor, with an equivalent density, thermal conductivity and heat capacity. The thermal conductivity in the longitudinal direction, in particular, was taken simply as that of copper, since this is usually several orders of magnitude greater than the other conductivities.

For the helium, the relevant properties (density, heat capacity, thermal conductivity, Joule-Thomson coefficient) were computed using interpolation tables created from the NBS computer properties package, since the use of this last would have slowed the computation too much.

5. RESULTS OF THE CALCULATION

The global analysis of the temperature margin in the coil requires subsequent results of the magnetic model, of the AC loss calculation, of the current sharing model and of the thermal model. Since the field and AC loss calculation was already described elsewhere[4], we deal here only with the thermal and current sharing models.

The detailed thermal analysis can be restricted to the worst pancake, that is, where the heat influxes are higher, the magnetic field is higher (lower T_{cs}) and the enthalpy availability of the coolant is lower. The AC loss has a maximum in the centre of the winding pack, while the nuclear heating is uniformly

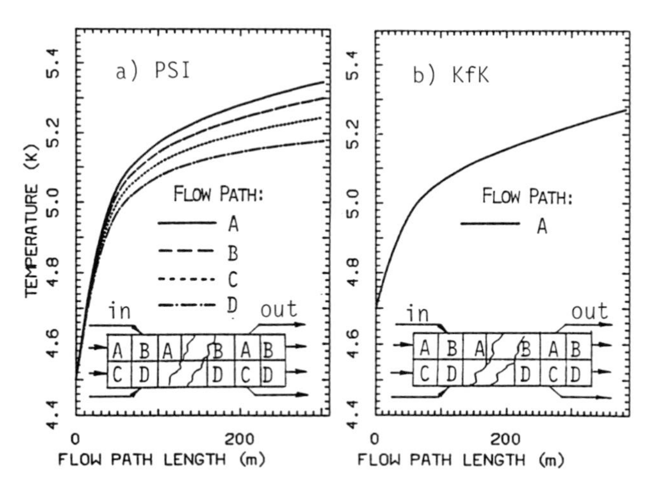

FIGURE 2
Steady state temperature in the PSI and KfK flow paths

distributed. In addition, the magnetic field has also a maximum in the centre of the winding pack. Therefore, only the central pancake of the winding packs has to be considered.

A calculation of the steady state temperature distribution in the 4 flow paths of the central pancake for the PSI geometry was done (flow paths A,B,C and D of ref. 2), showing that the flow path with the inlet closest to the plasma and the highest inlet pressure (i.e. flow path A) is in the worst condition (see Fig. 2a). The whole thermal analysis can be reduced to the transient temperature calculation in this flow path for both PSI and KfK geometries.

Starting from the steady state temperature distribution in the flow path A (Figs 2a and 2b), the transient analysis was performed for a time interval of 900 seconds, that is, about 3 plasma burn cycles. An example of the temperature distribution obtained at the third burn cycle is reported in Fig.3 for the first two turns of the PSI flow path A.

The effect of the transient heat influx due to the AC loss is the creation of small temperature peaks of about 0.2 K height for the PSI conductor that drift with the speed of the

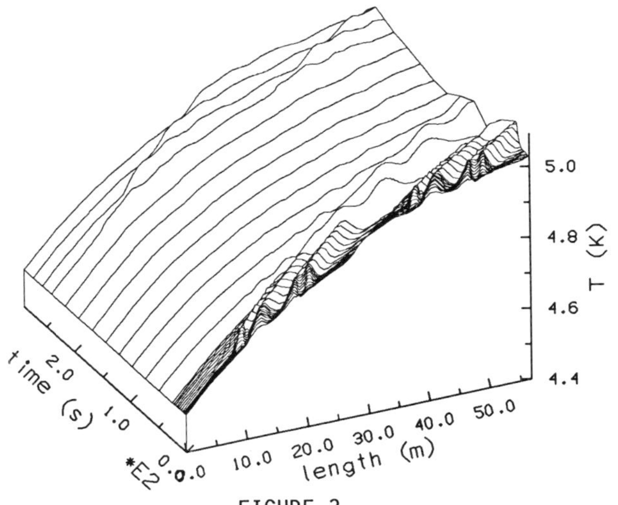

FIGURE 3
Transient temperature in the first two turns of the PSI flow path A during a plasma burn cycle

steady state flowing helium. Due to the lower instantaneous value of AC loss in the KfK conductor, the magnitude of these peaks is lower (less than 0.05 K), and can be neglected. Therefore for the calculation of the temperature margin in the KfK pancake, only an average steady state value of AC loss was used, reducing the computational cost by a factor 100.

The field map in the two geometries was used to compute the local value of T_{CS}, as given by (1). Comparing T_{CS} to the instantaneous value of T, the minimum temperature margin was found in the first turn for both geometries. The comparison between T and T_{CS} is reported in Figs. 4a and 4b for both geometries.

In Tab. II the minimum temperature margin is reported for the PSI and KfK TF winding packs. The margin seems adequate for the normal operation of the coil, but some assumptions in the model might have given a too optimistic result:

a) the model is only one-dimensional, and neglects completely the temperature gradients in the cross section of the conductors. In reality the two conductors considered here have highly resistive barriers that decrease the heat removal capability to the coolant. The effect of these barriers is surely negligible on the time scale of a plasma burn cycle, but can be appreciable on the time scale of the AC loss during the early start-up phase (0 to 0.3 s)[5].
b) the amount of heat entering the winding pack has some basic uncertainties, such as the neutron streaming effect that increases the nuclear heating, or the amount of radiation and conduction to the coil casing and the environment.
c) other transient heat inputs have been neglected such as increased AC loss due to extended plasma scenario, plasma disruption or other abnormal operation.

Additional margin is required for disturbances that are not easily quantifiable such as epoxy cracking or solder breaking.

6. CONCLUSIONS

The analysis of the temperature of the NET TF coils during several operation cycles has shown that the temperature distribution is mostly determined by the steady state type heat inputs (nuclear heating, conduction and radiation to the environment, frictional losses in the helium), and only slightly affected by the AC loss transient heat flux. This is due to the low average value of the AC loss, and to the high time constant of the heat deposition that produces low temperature gradients between wire and helium. For the same reason, the magnitude and location of the minimum temperature margin are mostly governed by nuclear heat, which has the highest average value in the coil of all the heat inputs considered and an exponential distribution along the pancakes. The normal operating heat fluxes investigated here do not produce instabilities in the TF coils designs examinated, but the analysis must be extended to include more severe or abnormal operating conditions. Additionally these results should be supported by experimental measurements.

Table II: Summary of the most critical condition for PSI and KfK conductors

		PSI	KfK
x (from inlet)	(m)	18.1	18.1
Temperature	(K)	4.82	4.87
Field	(T)	10.96	10.80
T_{cs}	(K)	6.90	8.20
T_{margin}	(K)	2.08	3.33

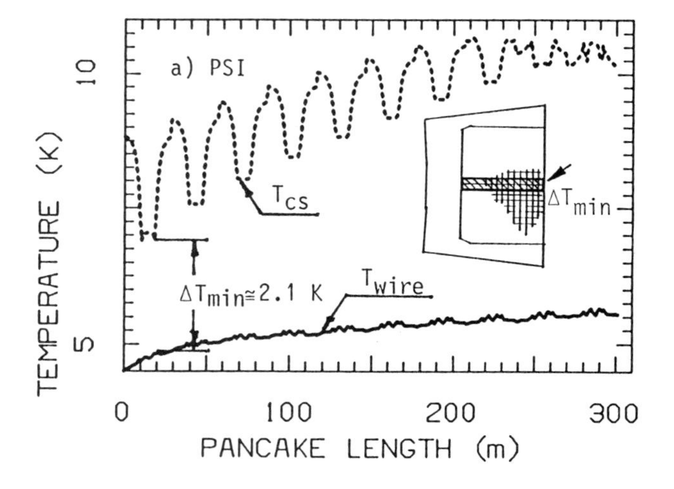

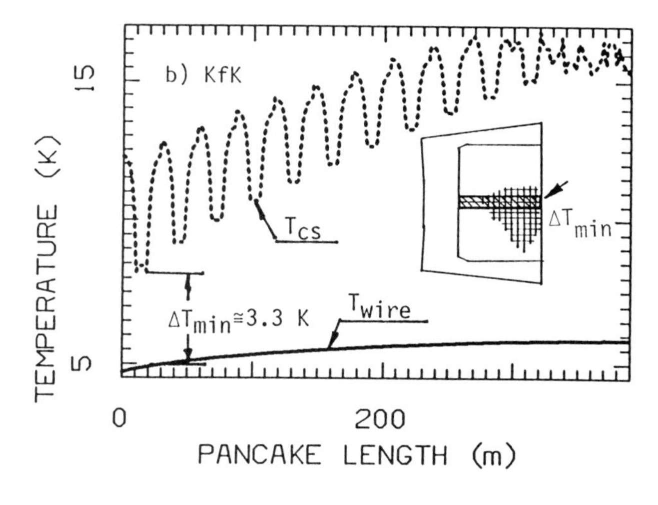

FIGURE 4

Figure 4. Comparison of T and T_{cs} in the most critical flow paths for PSI and KfK geometries.

REFERENCES

1. Salpietro E. et al., Fusion Technology, 14 No.1, pp.58-81, (1988).

2. Marinucci C. et al., Thermo Hydraulic Analysis of the SIN Conductor for the NET Toroidal Field Coil, IEEE Trans. Mag. 24, pp.1315-1318, (1987).

3. Fluekiger R. at al., An A15 Conductor Design and its Implication for the NET TF Coils, European Community Commission Rep. EURFU XII 361-85/37, (1985).

4. Bottura L., Minervini J., Detailed Distribution of AC loss in the NET TF Coils, presented at ASC, San Francisco, (1988).

5. Bottura L., Transient Thermal Analysis of Force Cooled Conductor, presented at ICEC 12, Southhampton, (1988).

FUSION TECHNOLOGY 1988
A.M. Van Ingen, A. Nijsen-Vis, H.T. Klippel (editors)

STATUS OF THE DEVELOPMENT OF THE KFK NET TOROIDAL FIELD CONDUCTOR

R. FLÜKIGER, U. JESKE, P. KOMAREK, A. NYILAS, P. TUROWSKI, A. ULBRICHT

Kernforschungszentrum Karlsruhe, Institut für Technische Physik,
P.O. Box 3640, D-7500 Karlsruhe, Federal Republic of Germany

The 16kA/12T Nb_3Sn conductor development carried out at the Kernforschungszentrum Karlsruhe (KfK) for the superconducting NET toroidal field coils is described. Most design principles are derived from those previously used in the Euratom-LCT coil, but have been adapted to the particularities of Nb_3Sn, taking into account ac losses, nuclear heating, thermohydraulic and stability considerations. A particular care has been given to the effects of transverse mechanical stresses on the critial current density of Nb_3Sn strands during the operation under full load.

The conductor, designed for a react and wind process, consists of a flat Nb_3Sn cable soldered within a rectangular core confined by very thin CuNi sheets. For electrical and thermal stabilization, Roebel processed rectangular copper sections are soldered on the flat core faces, the space between two neighbouring copper sections constituting the He channels. The whole system is surrounded by a steel jacket. In the present advanced stage of the work major emphasis is put on the manufacturing of the different components and their assembling under the aspect of industrial production of long lengths.

1. INTRODUCTION

In accordance with the present technological experiences large superconducting toroidal field coils in tokamak reactors will be cooled by forced flow supercritical helium and will be based on high field superconductors. The KfK design for the NET-TF Nb_3Sn conductor is based upon the reference magnet design of the NET machine[1]. Considering this reference design the TF-coils will be operated at 16 kA transport current generating a magnetic flux density of 12 T at the conductor. In order to achieve this high magnetic field the present technology refers to the A15 Nb_3Sn material for the conductor. The "react and wind" technique is chosen. A first evaluation resulted in the conductor design study[2], performed for the NET-team. Due to several modifications and further technological progress in past, the design was changed considering the superconducting core region, the electrical stabilizing unit and the stainless steel jacket. However, the main superconductor principles such as forced flow cooling, predictable thermohydraulics, effective force transmission, mechanically fixed superconducting strands to avoid disturbance energy created by mechanical movement and the "react and wind" philosophy were maintained. The main design features of the KfK-TF conductor can be summarized as follows:

- A transport current of 16 kA at 12 Tesla magnetic flux density and 4.2 K.
- Low ac losses induced by pulsed operation of PF coils by using Nb_3Sn strands with external bronze.
- Extremely flat Nb_3Sn cable to minimize the bending strain.
- Forced flow supercritical helium flowing through defined internal cooling channels.
- Stainless steel jacket as a conduit withstanding high transversal and compressive stresses.

According to the above given guidelines the present status of the superconductor designed at KfK is given in Fig. 1.The design can be roughly divided in three major components.

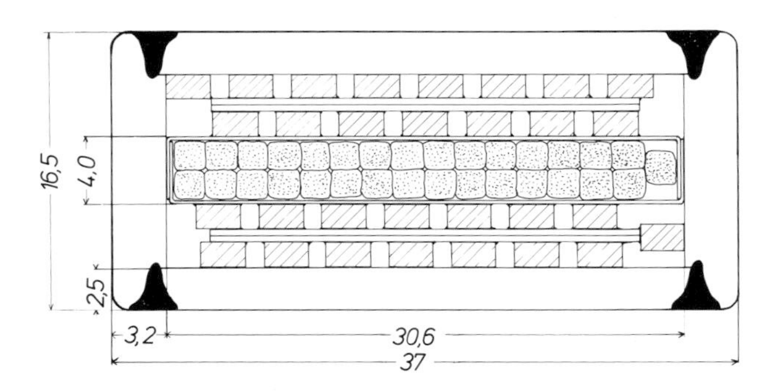

FIGURE 1
Cross section section of NET/KfK TF superconductor concept with its main dimensions.

These components are: The core, the electrical stabilizer and the stainless steel jacket. The development philosophy is based on:

- A step by step progress to a full size superconductor development on the basis of an industrial manufacturing program of a relevant subsize conductor.
- Introducing of the manufacturing experience and conductor test results collected during the subsize development to the full size NET-TF superconductor in earliest stage.

2. DETAILS OF CONDUCTOR DESIGN

2.1. Superconducting core

The rectangular superconducting core consists of a Nb_3Sn flat cable, soldered within a thin sheated CuNi conduit with external dimensions 4.0x30.6 mm^2. The flat cable consists of 31 internally stabilized Nb_3Sn strands. The strand diameter is 1.92 mm and can be routinely produced by the available bronze route process. The large diameter of 1.92 mm leads to a number of ca. 50 000 filaments with an approximate diameter of 4µm. The strands are composed of 433 Nb/CuSn elements at a first stage which are bundled again to 114 elements (total of 433x114 = 49362 filaments). The internal stabilizing copper is placed at the center of the strands, the required diffusion barrier during the reaction heat treatment consisting of Ta.

The 31 strands are cabled with a standard planetary cabling machine with 100% backtwist. The transposition pitch is in the same sense as the twist pitch, the transposition length being about 300 mm. In contrast to former the design where a ceramic insulated core strip was incorporated at the center of the cable to minimize the losses, the current status is to produce the cable without an electrical insulated core strip. The reasons for the elimination of this core strip are:

- Cabling of the Nb_3Sn strands on a core still bears manufacturing risks due to the high elastic spring back effect of the Cu-13wt.%Sn bronze. After the cabling process the core strip is subjected to high compressive stresses in axial direction, thus enhancing the risk of cable collapsing after removal from the cabling machine, e.g. during the reeling of the cable on the reaction heat treatment drum.
- Removal of the core strip was beneficial in this design, because of the reduction of the core thickness. The latter became necessary because of the increased jacket thickness of the cover sheets. As will be seen later this sheet thickness was no more force withstanding due to the increase of transversal stresses from 70 MPa (former design value) to 140 MPa in the current magnet reference design. A second beneficial effect of the core thickness reduction is the further reduction of the bending strain.
- The current situation of the ac loss calculations for this cable along with some experiments shows that the ac losses of the cable may still be acceptable without an insulated core strip. A final decision will be made after the first ac loss measurements to be carried out soon.

After the calibration to a rectangular shape the Nb_3Sn cable itself will have a dimension of 29x3,45 mm. This cable will be subjected to a reaction heat treatment process. During the subsize cable development program the heat treatment process turned out to be one of the most important key problems. The first reaction heat treatment with a ~ 100 meter length subsize cable (12,45x1,95) was carried out with a 1200 mm ø drum of the Swiss PSI laboratory under pure Argon atmosphere. The drum material (Inconel 600) and the core strip (Duratherm) of the cable have integral thermal expansions between 300 K and 1000 K of approximately 1.1% and 0.6%, respectively. This non-matched condition of thermal expansion coefficients and the volume increase of the strands during formation of the Nb_3Sn layer led to unacceptable cable distortions after the heat treatment process (In the present case, the cable was reeled on the drum as a layer winding). The final analysis showed that the cable dimensions after the process were increased ~ 0.4% in all dimensions (volumetric effect). This physical phenomenon must therefore be be considered in overall conductor design. In order to tackle these problems thouroughly a program was set up early this year which comprises the fabrication of a special vessel of ~ 1000 ø x 500 mm in which several drums (in pancake or in layer winding type) can be placed. The unit is already completed and still in this year three types of cables (subsize) will be reaction heat treated under various winding conditions. These cables have identical Nb_3Sn strands but different core strips, i.e. a bronze core strip of ~ 0.4 mm thickness with and without Al_2O_3 insulation, a bronze strip in hard condition and no core strip at all, which would be the preferred solution. The ac loss measurements on flat cables with and without core strips will give the answer concerning the minimum coupling losses. The results of these tests will later be considered in the full size cable design.

Taking the current cable dimension of 29x3.45 mm and adding the 0.4% volumetric increase one can fulfill the design goal to insert the cable after the reaction heat treatment process into the two CuNi10 U-type sections. These U-sections of 0.2 mm thickness give with their legs a defined closure during the soldering process, therefore the dimensional stability can be guaranteed. The soldering will take place in a vertical ceramic mould, electrically heated. This device is already completed and the first dummy tests gave positive results. Figure 2 shows a schematic view of this soldering unit, which is designed in such a way that soldering of the subsize cable as well as of the full size cable is possible.

As solder material the well characterized Sn58Pb39 In3 with a melting point of ~ 235°C has been chosen. In order to check the solder performance, double lap shear tests were carried out with a material combination of copper/copper (OFHC) and a lap area of 200 mm^2. The lap shear strength results are 23/26 MPa at 300 K and 30/48 MPa at 4 K, respectively. The ultimate strain capacity of the ~ 50 μm thick solder layer was found to be ~ 0.1%, thus giving enough operational safety margin. The complete core will have a dimension of 30.6x4 mm^2 according to Fig. 1. If necessary an increase of the dimension in width of some tenths of a mm is still possible without harming the overall design. This aspect will be important at a later manufacturing period.

2.2 Electrical stabilizer

The two electrical stabilizers at both sides of the core (see Fig. 1), (which should carry the transport current during a fault condition), comprise rectangular 1/2 hard (RRR ~ 100) copper sections, which have been Roebel processed on a sandwich core strip with an electrical insulation at the center (state-of-the-art). The sandwich core consists of two copper plated strips of Ni80Cr20 glued together by adhesive insulation tape. The sandwich will be covered on both surfaces with the same solder (Sn58 Pb39 In3) as used in the superconducting core. The overall dimensions of this electrical stabilizer will be 30.6x3.75 mm^2. Each stabilizer consists of 15 copper sections of 2.6x1.5 mm^2. The Roebeling process, in particular the absence of electrical contacts between neighbouring Cu sections at the edges (important for

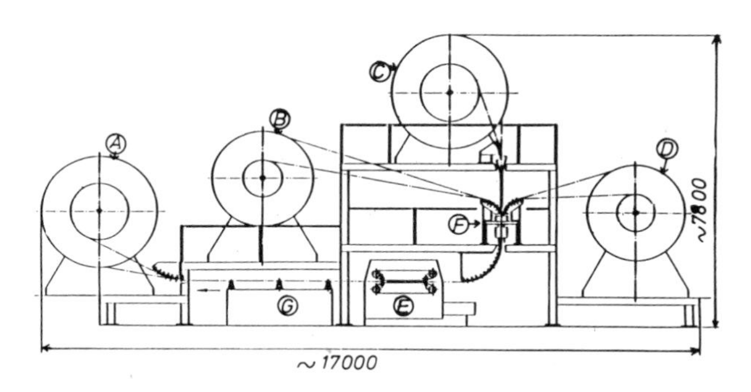

FIGURE 2
Present core soldering line (courtesy of Vacuumschmelze GmbH., Hanau).
A) Take-up reel of the completed core 1500 ø - 3000 ø mm (working range)
B) Pay-off reel for CuNi U-section No. 1 1000 ø - 2600 ø mm
C) Pay-off reel for prereacted superconducting cable 1500 ø - 3000 ø mm
D) Pay-off reel for CuNi U-section No. 2 1000 ø - 2600 ø mm
E) Linear capstan
F) Vertical soldering station
G) Dimension control plus quality control (All dimensions in mm)

ac losses) has been already successfully demonstrated in short lengths of ~ 3 meters. Calculations with the boundary condition of 16 kA and 15 s discharge time constant have shown that the Cu cross section in the stabilizer units together with the Cu cross section in the Nb_3Sn strands can carry the transport current in the quench case without reaching non permissible "hot spot" temperatures (~ 100 K). The manufacturing of such cables is well demonstrated and the tolerances are very low, in the range of ± 10 μm for the completed cable as indicated by the manufacturers report[3] of the LCT cable.

These two stabilizers will then be joined with the prereacted core on an assembly line by soldering. The solder will be Sn50/Pb32/Cd18 with a low melting point of ~ 185°C. By correct temperature adjustment the soldering can be performed without a melt down risk of the already soldered core region. The unit lengths can be collected on a take up reel ready for the final jacketing process. This solder was also tested to have an appropriate data base for the present design. Again the 4 K values are higher compared to the ambient values. The results show lap shear strength values of 17/33 MPa at 300 K and 24/28 MPa at 4 K, respectively. The straining capacity of this solder is also in the same range as the previous one (~ 0.1 %).

2.3 Steel jacket

The steel jacket as given in Fig. 1 is an essential part for the mechanical integrity of the conductor in the coil winding. It consists of two rectangular cover sections and two special T-shaped side sections. The jacket must withstand all mechanical loads during the magnet operation. According to the NET reference design the radial coil stresses and the toroidal coil stresses are -40 MPa and -140 MPa, respectively.[1] The hoop stresses are +140 MPa. The jacket sheet thickness along with the conductor aspect ratio give thus the mechanical performance of the conduit. The jacket material will be a high strength 316 LN type stainless steel, which was already successfully used in the past. Figure 3 shows schematically the jacket and the stresses acting on it. Before going on to detailed mechanical finite element calculations, a first estimation of the critical loads shows that the chosen jacket geometry may be capable to withstand the operational loads. The toroidal stress of -140 MPa results in a compressive stress of ~ -460 MPa in the 2.5 mm thick cover section. The materials 4 K yield strength (~1200 MPa) is far above this value.

The toroidal load of each side section is calculated to be 1150 N per unit (mm) of conductor length. Taking the cover section of 37 mm length as a column, the calculated critical load is ~ 1800 N per unit (mm) of conductor length. Considering the weldments plus the winding pack stiffening effect, a buckling phenomenon may be therefore outruled. The side section with its 3.2 mm thickness is subjected to a deflection under the toroidal stress of -140 MPa. The maximum deflection assuming a non supported beam of 11.5 mm length is ~ 60 μm. This value will drop down considerably by the mechanical supporting of this beam by the weldments. The deflection of the cover section is accordingly the more severe one due to the length of 30.6 mm. Considering the radial stress of - 40 MPa the deflection yields a value of ~ 1.7 mm. This means that the interior of the jacket will be subjected to considerably compressive stresses. The smaller stress of -40 MPa will increase to ~ -70MPa due to the stress transmission (see Fig. 3). This - 70 MPa will therefore act as a radial stress on the superconducting core. Due to the presence of solder and the correlated distribution of stresses in the core, the compressive stress of 70 MPa is the effective stress acting on the Nb_3Sn wires. This is strongly different from other configurations, e.g. the cable in conduit, where the effective compressive stress at the crossover points is much higher. According to Ref. 4 transversal compressive stresses of the order of 70 MPa at 12 T lead, however, to a decrease of J_c by

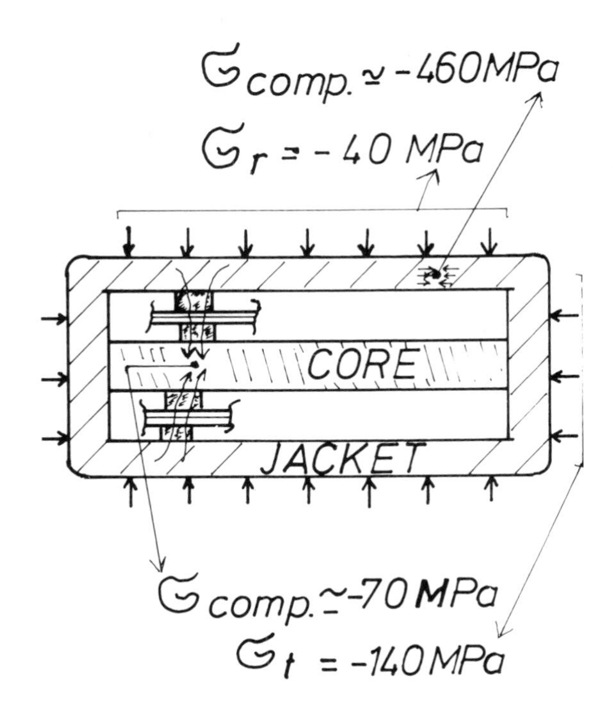

FIGURE 3
Conductor jacket and the stress acting on it during the operation.

~ 25%. In order to maintain a sufficient safety margin for the overall critical current, it is thus necessary to use Nb_3Sn wires with considerably higher critical current densities than originally thought. A detailed study is actually undertaken in our laboratory in order to decide whether the envisaged bronze route process for the production of Nb_3Sn wires should be replaced by more advanced processes yielding higher critical current densities. A detailed stress distribution calculation accompanied an integral electrical test is necessary to determine the effect of local stress enhancement and the corresponding current degradation of the superconducting core.

The present jacket design has been also improved considering the former jacket configuration[2], which used tack welded spacer sheets. A major manufacturer could be persuaded to give a bid on the fabrication of the side

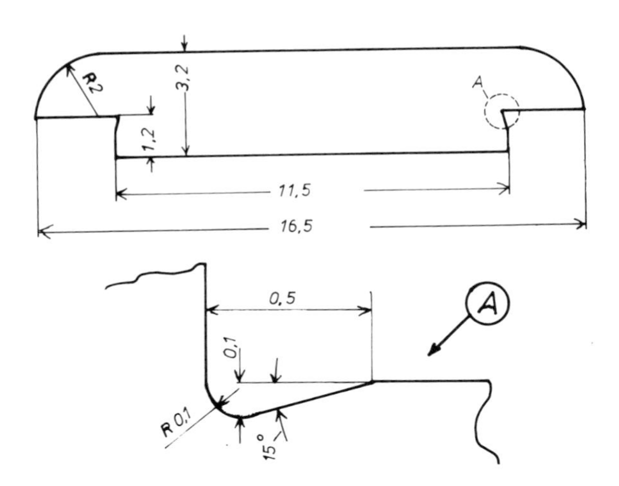

FIGURE 4
Side section of the conductor jacket manufactured by continuous forming process. A) according to DIN 509.

section by a forming process for the necessary long lengths. Figure 4 shows the section in detail. The geometrical tolerances and the evaluation of manufacturing risks for long lengths of such a section represents one major milestone before going on to a detailed final jacket design.

A key technology of the jacket design is the continuous laser beam welding of the four seams. The deep weld profile of ~ 2.8 mm forces to switch on to laser beam weld process, because of the low heat input of this technique. Details of this weld procedure are given in ref. 5. According to the recent short dummy welding tests on short lengths the temperature rise of the interior (important because of the soldered stabilizer) has been proven to be negligible[6]. At present we follow the following manufacturing route: Prefabrication of the U-shape with two simultaneously working laser stations of ~ 4 kW power range. Reel on of the produced U-shaped configuration on a > 3.5 meter ø take up reel. Closure of the U-shaped section after the cable assembling in one step, again with two laser beam working stations. The success of this process must be proven as early as possible with representative lengths.

3. OUTLOOK

In spite of the substantial progress, further development must be done before the manufacturing of full length, full size conductor. Concluding the subsize conductor development early in 1989, tests and the measurements will supply further informations. These informations can be taken as a basis for the next iterative procedure for a detailed full size length. At the moment the bronze route for manufactured Nb_3Sn wires is still preferred. Alternative production processes, in particular the internal Sn diffusion process still suffer from being not available at the required long lengths. In the case of the necessity of higher current densities, additional development work must be performed to achieve long strand lengths.

The jacket welding process with two laser stations does not suffer from the laser technology itself but from the actual non-availability of laser welding equipments at the traditional SC manufacturers site. Additional investments have therefore to be foreseen early in the manufacturing preparation phase.

Design, fabrication and test of the conductor, as well as the electrical connections plus the terminations can be evaluated with subsize type conductors. Also the electrodynamical behaviour (ac losses, stability) can be assessed by the measurements of the subsize conductor.The gained experience can be transferred to the full size conductor at the early stage. Each manufacturing step has to be developed separately until the production step is under control and suitable methods of quality insurance are the state-of-the-art.

4. ACKNOWLEDGEMENT

The authors would like to express their gratitude to Dr. Pasztor and Mr. Jakob of PSI - Swiss Laboratories for supplying of the reaction heat treatment drum for our subsize heat treatment process. For the hot spot calculations the authors thank Dr. C. Schmidt (KfK). This work has been performed in the framework of the Nuclear Fusion Project of the Kernforschungszentrum Karlsruhe and is supported by the European Communities within the European Fusion Technology Program.

5. REFERENCES

1. NET Status Report No 84 Dec. 1987 EURFU XII-80/88-84

2. R. Flükiger et al., An A15 Conductor Design and its Implications for the NET-II TF Coils, EURFU XII/361-85/37

3. R. Bezouska et al., LCT Entwicklungsarbeiten zur Herstellung eines Supraleiters mit forcierter Kühlung, 1982 VAC Abschlußbericht VAC BA 99371

4. W. Specking, W. Goldacker and R. Flükiger, Adv. Cryo. Eng., Vol. 34 (1987) 569

5. A. Nyilas, Laser Beam Welding of Advanced Superconducting Cable Conduits, 14th Symp. on Fus. Techn. Proc. SOFT Conf., Avignon, F, Sept. 8-12, 1986

6. J.M. Plaum et al., Development Toroidal Field Conductor for NET, ECN-87-158 presented at the 10th Int. Conf. on Magnet Techn., Boston, Sept. 23-26, 1987

FUSION TECHNOLOGY 1988
A.M. Van Ingen, A. Nijsen-Vis, H.T. Klippel (editors)

FURTHER USE OF THE EURATOM LCT COIL

A. Hofmann, P. Komarek, A. Maurer*, W. Maurer, G. Ries*, B. Rzezonka**, H. Salzburger*, Ch. Schnapper*, A. Ulbricht, G. Zahn

Association Euratom - Kernforschungszentrum Karlsruhe, Institut für Technische Physik, Postfach 3640, D-7500 Karlsruhe, FRG
* Siemens AG, Research Laboratories, Postfach 3240, D-8520 Erlangen, FRG
** Interatom, Postfach, D-5060 Bergisch-Gladbach, FRG

The Euratom LCT coil ran sucessfully through the LCT test programme in the International Fusion Superconducting Magnet Test Facility (IFSMTF) at Oak Ridge, USA. The coil was operated successfully at 9 T with 15 950 A as single coil and at 9.2 T in the toroidal configuration.
In a first series of discussions on testing of NET model coils wound by NET toroidal field coil conductors it was proposed to use the background field of two LCT coils in a cluster arrangement. The available space in the TOSKA facility at Karlsruhe is limited by the size of the existing vacuum vessel. Therefore the use of only one LCT coil as background coil is investigated. The missing ampere turns of the second LCT coil shall be compensated by the operation of the Euratom LCT coil at temperature and current levels promising the obtainment of a field strength between 10 T to 11 T. It was investigated by finite element calculations that a suitable mechanical structure keeps the stress levels at acceptable limits. The electrical and thermohydraulic properties of the different components of the coil were assessed considering a 20 kA operation. Nothing was found which prevented such an operation current.
In a suitable two coil arrangement LCT coil-NET-TF-model coil ("TWIN") obtainable stress and field levels were investigated and optimized.

1. INTRODUCTION

The Euratom LCT coil was the contribution of the European community to the manget technology experiment "Large Coil Task". Five of the six coils used NbTi, only one coil Nb_3Sn as basic superconducting material. The test results indicate that before the use of Nb_3Sn in reactor size coils further steps of development are mandatory[1]. Since 1985 the design for the next European fusion experiment NET is proceeding. For the magnet system 11 - 13 T field strength is needed. Therefore conductor designs and development steps were performed by the European superconducting magnet laboratories (KfK, PSI, ECN, ENEA). Three toroidal field (TF) coil conductors were designed in react and wind technique[2,3]. The big quantity of this conductor and the quality needed for long time operation of such a magnet system require careful steps in the industrial development and verification tests. Therefore the necessary testing steps of such a conductor were defined by the involved laboratories. Besides the basic investigation of subsize conductors two testing steps seemed to be necessary: test of samples in meter range and tests of long pieces of industrial fabricated conductors[3]. The last test should also include the impact of the winding technique and all techniques needed for the fabrication of a coil. In a study possible test configurations were evaluated[3]. The test conditions were compared in relation to the electrical, mechanical and thermohydraulic coil loading. In order to save length of NET prototype conductor a background field will be an advantage. Therefore the favoured arrangement included two LCT coils as background coils. An upgrading of the existing TOSKA facility in KfK-Karlsruhe was proposed.

Combined with obtained LCT test results an increasing of the operation parameters of one background coil and saving the other one seemed to be one possibility for simplification of the test configuration. This will be the subject of this paper.

2. PROPOSED TEST CONFIGURATION FOR NET-TF-MODEL COIL TESTING

The industrial fabricated NET prototype TF conductors have to be tested in a model coil in order to check the fabrication line from the conductor to the coil. In a collaboration of the European superconducting magnet laboratories two possible testing configuration were investigated for that purpose. One configurations called "Cluster" used two LCT coils as background coils. The TF model coil is mounted between the two LCT-coils (Fig. 1). The other one known as "Solenoid" was assembled from three circular coils. Two coils should be manufactured from TF conductor. The central coil between them should be manufactured completely from OH (ohmic heating coil) conductor. The both test configurations were optimized to reach as near as possible the electrical and mechanical values of NET-TF and OH coils (Tab. 1).

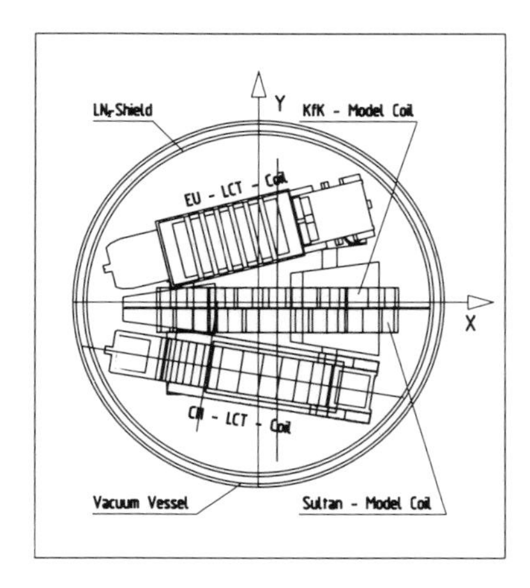

FIGURE 1
The Cluster configuration. Two LCT coils with the NET-TF model coil in the middle.

To save conductor length the operating current of the NET-TF conductors had to be increased from 16 kA to 22 kA. Both facilities reach the required NET fields, but not all the pressure and stress levels. The main characteristics are compared and the solenoid solution seems to be the technically simpler one from an engineering point of view determined by modifications of the existing facility, installation, maintenance, and operation of the facility. However, the risk to reach the goal of the test is much higher because all pancakes made from different conductors must work together to get the design values. The OH conductor cannot be tested up to the required $\dot{B}$ = 3 T/s in the neighbour hood of the TF solenoids ($\dot{B}$ = 1 T/s. The Cluster Test Facility has a higher operational flexibility compared with the Solenoid Test Facility. It copes easier with a failure of a magnet. The performance of the LCT magnet is already proven. Therefore the risk to reach the test goals for NET is lower in the Cluster Test Facility.

Costs were estimated and a time schedule was worked out. The cost of the Cluster was about 30 % less than for the solenoid but still high enough to look intensively for simplifications. The test programme and procedure were also outlined.

3. SUMMARY OF TEST RESULTS OF THE EURATOM LCT COIL

The coil was tested in single coil and toroidal configuration. High out-of-plane forces and loading with poloidal pulsed fields transients were included in these tests. The test results were extensively presented[1] and shall only be summarized briefly: The coil was stably operated at 9 T max. field as single coil and at 9.2 T in toroidal configuration. The temperature difference to the expected current sharing temperature was between zero and 0.05 K. From this point of view the coil was operated at its short sample critical currents. Insulation properties of the coil were excellent. The coil was dumped 24 times in total with and without induced quenches.

The coil was operated beyond the region of the cryogenic stability. A total flow stoppage over 10 minutes at rated current had no impact on the stability. Pressure increase during dump tests with and without intentionally induced quenches remained moderate. A sufficient buffer volume did avoid any losses of helium. High current dumps, generated max. temperatures of about 10 K in the

Table 1: Comparison of Values required by NET and attainable Values in the Cluster and Solenoid Test Facilities. All Values given in this Table are from FEM Calculations.

	NET		Solenoid-Configuration S1			Cluster-Configuration C6		TWIN (1)
	TF-Coil	OH-Coil	KfK-Coil	OH-Coil	Sultan-Coil	KfK-Coil	Sultan-Coil	NET-TF Model Coil
Maximum Field [T]	11.4	11.5	12.27	11.62	11.54	12.1	12.0	11.85
Current [kA]	16	40	22	50	22	22	22	22
Axial Pressure [MPa]	140	100	66.4	57.9	66.5	27.4	26.3	48/52**
Radial Pressure [MPa]	40	10	43.3	39.1	42.1	65.1	64.3	62/82**
Hoop Stresses [MPa]	140	200	209	222	185	310.7	309.9	210/170**
Shear Stresses [MPa]	30	30	19.3*	18*	20.7*	26.8	28.8	14/11**

* For this calculation it was assumed that the pancake is supported by a rod at two areas and without a case.

** First number for 25 mm thickness of the coil case/second number for 50 mm thickness.

(1) This configuration will be treated in section 5.

winding and 15 K in the coil case structure. Recooling needed not more than 1 - 1.5 h. Only about 2 % of the stored energy were dissipated in winding and case.

During the test program the coil exposed to about 57 MN overall central forces, overturning moments by poloidal field transient of 3.2 MNm and out-of-plane forces up to 27 MN. Measured stress levels of the equivalent stresses reached their highest values of 285 MPa in high current single coil tests, which demonstrated an healthy design of the coil. The winding was well fixed in the coil case structure and a global movement was not observed[4].

The test results obtained were very encouraging to proof the extensions of the operation range above the reached limits.

4. EXTENSION OF THE OPERATION RANGE OF EU-LCT-COIL TO (10-11) T AT 1.8 K

The well developed technology of NbTi conductors recommends this material to be used at the 1.8 K temperature level for reaching higher fields . This was practised successfully with smaller coils[5] and mid size magnet systems[6]. The possibility of operating also large internally cooled coils was investigated and looked promising by implementing He II forced flow[7]. Estimations show, that the typical heat load in reactor coils can be removed with acceptable mass flow rates and pressure drops[11]. The qualification of the Euratom LCT coil for operation at 1.8 K and at field levels of 10 - 11 T was checked concerning its mechanical, electrical and thermohydraulic properties.

Mechanical properties: In a first attempt suitable attachment of supporting structure was investigated by finite element calculations.

- Doubling of the thickness of the stainless steel coil case.
- Deformation of the coil was prevented by a D shaped disk in the inner bore.

In the first case a reinforcing of the two halfes of the coil case by ripping and in the second case an additional external mechanical structure was represented in the finite element model. The results showed clearly that the first method for reinforcement would be near the boundary (case 512 MPa) or would exceed considerably stress limits (shear stresses $\sigma_{r\phi}$) and is therefore not applicable (Table 2).

The addition of external structure in the coil bore brought a release in the stress levels. A further improvement was obtained by usage of an experimental measured radial Youngs modulus. The investigation of shear stress level measurements

Table 2: Calculated stress levels for reinforcement of the Euratom LCT coil by coil case ripping and an external structure for 21 kA operation current.

		Ripping	External structure		External structure		
			$E_r = 2.7$ [GPa]	$E_r = 4.6$ [GPa]	Typical 18 [kA] Loading	Upper Stress Boundaries	Safety Factor
Coil case	Δl_{max}	9.13	1.68	1.68	1.23		
	σ_v [MPa]	512	283	283	208	500	2.40
Winding	Δl_{max}	5.8	5.86	4.2	3.1		
	$\tau_{r\phi}$ [MPa]	58	38	32	23.5	~ 50	2.12
	σ_r [MPa]	- 70	- 70.4	- 60.0	- 44.1		
	σ_z [MPa]	- 66.7	- 67.2	- 66.0	- 48.5		
	σ_ϕ [MPa]	391	338.0	296	217	350	1.61

Δl: max. displacement of the inner contur of the outer leg in [mm]
σ_v: equivalent stresses (v. Mises)
$\tau_{r\phi}$: shear stresses
σ_r, σ_z, σ_ϕ: radial, axial and azimuthal stresses

leads to better understanding and to more realistic shear stress levels[8]. The finite element calculations show that the Euratom LCT coil can be operated at 18 kA current level (typical current in the TWIN configuration) with sufficient safety margins if the coil is reinforced by a suitable structure.

Electrical properties: According to the test results the coil can be operated at higher currents at suitable choosen operation parameters if the mechanical integrity of the coil is guaranteed. Beside the magnetic characteristic and stability also current carrying components of the coil have to be reviewed in design and test results. These are the current leads and the conductor joints between the pancakes. Each copper bar of the current leads was shunted by superconductors which total critical current carrying capacity at 2 T was about 25 kA. The effectivness of this superconductors was demonstrated in the tests, were the temperature increase at the copper bars was less than 0.3 K near 16 kA current. The mechanical support of the current leads was above all determined by the proof pressure level of the coil case and can withstand therefore the 50 % higher Lorentz forces.

The estimated joint losses for a 20 kA operation led to losses of 116 mW per joint in a field of 1 T. These results were derived from prototype investigations[9]. During tests the losses of the 13 joints were about 1 W within the measuring accuracy and could not be analyzed. The dump voltage can be increased to about 5 kV if suitable switches and signal conditioning are available.

Thermohydraulical properties: For a quench at 20 kA the generated heat per length reaches 1.7 kW/m and is nearly doubled compared to the maximum current of about 16 kA (0.97 W/m) used in the tests at ORNL. The heat flux at the copper surface reaches about 1 W/cm^2 at 20 kA. Therefore the thermohydraulical properties were carefully investigated with the computer code used for the design[11]. The calculations were done with slightly changed operation parameters to avoid the singularities at the He- λ point (Table 3).

TABLE 3
Operation parameters assumed for thermohydraulic calculations

Inlet pressure	0.3 MPa	Discharge voltage	2,5 kV
Inlet temperature	2.5 K	Dump delay	1 s
Mass flow rate per channel	1.2 g/s	Current Average field	20 kA 6 T
Discharge time current	15 s	Max pressure inlet, outlet	1,0 MPa

The assumption that a full conductor length goes normal results in an unacceptable maximum pressure of 17 MPa. The temperature of the conductor and the helium remains below 60 K. The pressure maximum is reached after 2.5 sec. An increase of the discharge voltage to 4.5 kV ($\tau = 10$ s) has therefore no major effect on the pressure maximum.

More realistic is the assumption of a normal region of one third of a turn (~ 3 m) as experienced by the tests. Calculations were performed for a 3 m normal region in the middle and at the beginning of a cooling path. The pressure increases very rapidly (1 s for the quench in the middle, 0.2 s for the quench at the inlet) to its maximum till the relief valves open at 1.0 MPa (Fig. 2 a-b). Maximum temperatures were 60 K or 22 K, respectively. A comparison of calculated and measured pressure increase with conditions used in the tests is demonstrated in Fig. 3. The measured pressure increase was much lower than predicted. This can be explained by the volume of the ventlines acting as buffer.

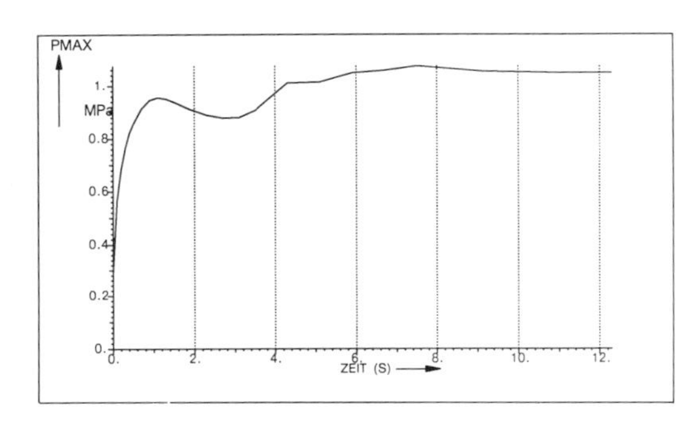

FIGURE 2b
The maximum pressure as function of time.

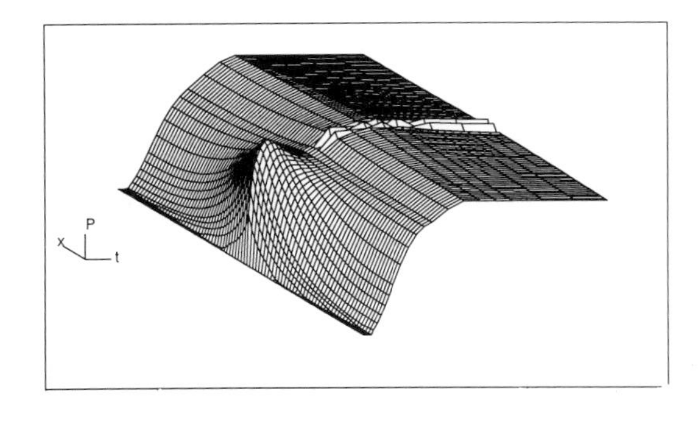

FIGURE 2a
The pressure profile for a 3 m quench in the middle of the cooling path.

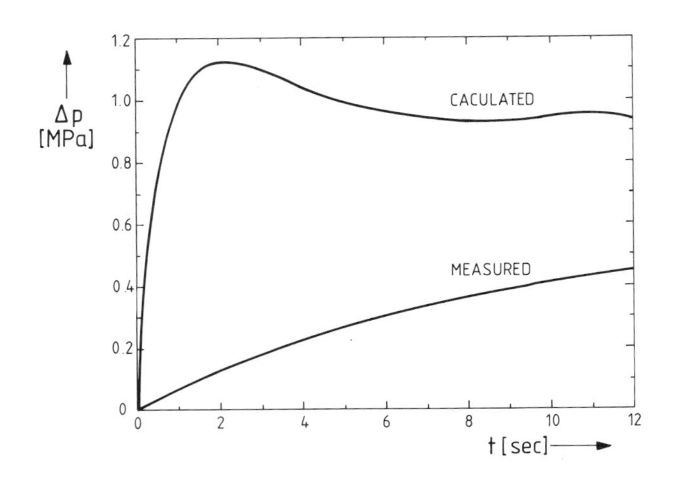

FIGURE 3
Calculated and measured pressure increase. The differences can be explained by the buffer volume of the warmer ventlines.

The results indicated that a realistic quench at 20 kA can be controlled without excessive pressures and temperatures in the conductor. The estimation of coil case temperatures after a dump were in the range of 27 K. Different operational temperatures of case and winding demand some care to avoid thermal shorts in the inlet region of the coil. Instrumentation can be used with the extension of the calibration of some kinds of sensors. If a satisfying mechanical support structure reinforcing the D shaped coil can be designed and built, there is no major obstacle to extend the operation parameters up to 20 kA at 1.8 K.

5. THE TWIN TEST CONFIGURATION FOR NET MODEL COIL TESTS

Investigations showed that an operation of the EU-LCT coil at 1.8 K can save one LCT coil in the Cluster in order to reach the field level of about 11 T

Table 4: Comparison of the operation parameter of the Cluster and TWIN

		CLUSTER C6		TWIN	
	Units	NET-TF Model Coils (KfK/SULTAN)	EU-Coil	NET-TF Model	EU-Coil
Minimum bending radius for TF Model coil	[m]	1.0	not important	1.0	not important
Temperature (Inlet)	[K]	3.5/3.5	3.5	3.5	1.8
Conductor current Total coil current	[kA] [MA]	22 4.488/4.84	13 7.644	22 9.856	18 10.584
Magnetic field at reference point of winding pack	[T]	12.1/12.0	8.3	11.69	10.33
Force in x-direction (centering force) Force in y-direction (out-of-plane force)	[MN] [MN]	- 68.7 1.5	37.5 - 86.6	- 18.5 115.2	18.5 - 115.2
Stored self energy Total stored energy	[MJ] [MJ]	34/38 51	134 2	138 50	253 7
Rotation angle to x-direction	[degree]	0/0	14	0	9

in the NET Model coil (Table 4). The test configuration fits better into the existing vacuum vessel. A disadvantage is the increasing out-of-plane force as well for the NET-model coil (100 times) as for the LCT coil (1.3 times) which needs further investigation. The electrical data are comparable with the Cluster test. The mechanical stresses and pressures are in most cases higher (Table 1). The first step toward the TWIN test facility will be the upgrading of the Euratom LCT coil for 1.8 K in the TOSKA facility.

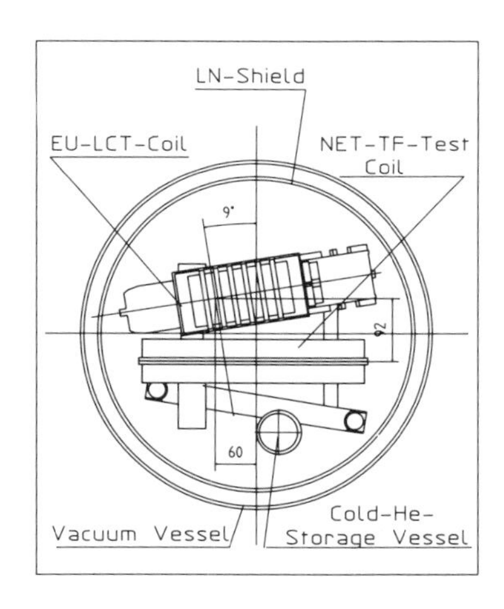

FIGURE 4
The TWIN configuration in the TOSKA vacuum vessel.

6. CONCLUSIONS

The performed investigations for the Euratom LCT coil look very much promising to operate the coil at 1.8 K up to 10 - 11 T. The proposed cluster for NET model coil testing can be reduced then to a TWIN configuration which fits better into the existing TOSKA facility.

ACKNOWLEDGEMENTS

The authors acknowledge the collaboration with the NET Team and the other European Laboratories.

This work has been performed in the framework of the Nuclear Fusion Project of the Kernforschungszentrum Karlsruhe and is supported by the European Communities within the European Fusion Technology Program. Parts of the work were performed as a study contract for NET.

REFERENCES

1. P.N. Haubenreich et al. The IEA Large Coil Task. Development of Superconducting Toroidal Field Magnets for Fusion Power; Fusion Engineering and Design 7 (1988) North-Holland, Amsterdam.

2. R. Flükiger et al., An A15 Conductor Design and its Implications for the NET-II TF Coils, KfK 3927, Juni 1985, KfK-Karlsruhe.

3. J. Erb et al. NET Model Coil Test Possibilities, KfK 4355, Nov. 1987, KfK-Karlsruhe.

4. S. Gauss et al., Analysis of the Mechanical Behavior of the Euratom LCT Coil under the Effects of Out-of-plane Loads, Contribution at this conference.

5. P. Turowski, N. Brünner, S. Förster, A 12 T NbTi Solenoid at 1.8 K with a Clear Bore of 290 mm Diameter, Proc. MT-9, Zürich, Sept. 9-13, 1985.

6. B. Turck, Tore Supra: a tokamak with superconducting toroidal field coils, Proc. ASC, San Francisco CA, Aug. 21-25, 1988.

7. A. Hofmann, A. Khalil, Considerations on Magnet Design Based on Forced Flow of He II in Internally Cooled Cables, Proc. 14th SOFT, Avignon, France 8-12 Sept. 1986, p. 1811.

8. A. Nyilas, private Communication.

9. C. Albrecht, H. Marsing, H.W. Neumüller, Electrical Joints for the European LCT coil, Journal de Physique, Colloque C1, supplement an no 1 Time 45 jauvier 1984, p. C1 607-610.

10. G. Ries, Quench induced Helium flow and recovery in the Euratom LCT coil, Proc. MT-7, IEEE Trans. Mag. Vol. 17, No. 5 (1981) 2097.

11. A. Hofmann, A Study on Nuclear Heat Load Tolerable for NET/TF Coils Cooled by Internal Flow of Helium II, KfK 4365, February 1988, Kernforschungszentrum Karlsruhe.

FUSION TECHNOLOGY 1988
A.M. Van Ingen, A. Nijsen-Vis, H.T. Klippel (editors)

DESIGN CONCEPT OF CRYOGENIC SYSTEM FOR THE PROTO TOROIDAL COIL PROGRAM

T. Kato, E. Tada, T. Hiyama, K. Kawano, H. Yamamura*, M. Sato**, M. Hoshino, N. Ito, J. Yoshida***, and S. Shimamoto

Japan Atomic Energy Research Institute (JAERI), Naka Fusion Establishment
Naka-Machi, Naka-gun, Ibaraki-ken, 311-01, JAPAN

Design study of cryogenic system for the Proto Toroidal Coil Program has started at JAERI, aiming the proto type of the cryogenic system for the Fusion Experimental Reactor. The system is designed to have the refrigeration/liquefaction capacity of 10 kW at 4.5 K or 3000 l/h and to supply the supercritical helium (0.1 MPa, 3.7 K) of 1000 g/s for the test coils in the Proto Toroidal Coil Program. The design is also investigated to achieve the energy efficiency of better than 1/400. Cryogenic pump system is selected for circulation of supercritical helium instead of elevated temperature pump system.

1. INTORODUCTION

A large scaled, advanced cryogenic system is indispensable for the Fusion Experimental Reactor (FER) of a tokamak type which is the most advanced in plasma heating and confinement among several types. Recent study shows required capacity of the cryogenic system as follows:

Refrigeration capacity at 4 K - 100 kW
Cool down capacity - 1200 kW at 100 K
(Cool the 10,000-ton-coils within 300 hours)
Supercritical helium supply - 6000 g/s,
(0.5-1.0 MPa, 3.5-4.5 K)

For realization of such a large capacity, three or four units of cryogenic system have advantage. It is, accordingly, possible to change the number of operating units depending on

heat load. Four units are, for example, required in the coil cool-down operation but three units are enough to maintain steady-state heat load. The 30-kW cryogenic system is, consequent-

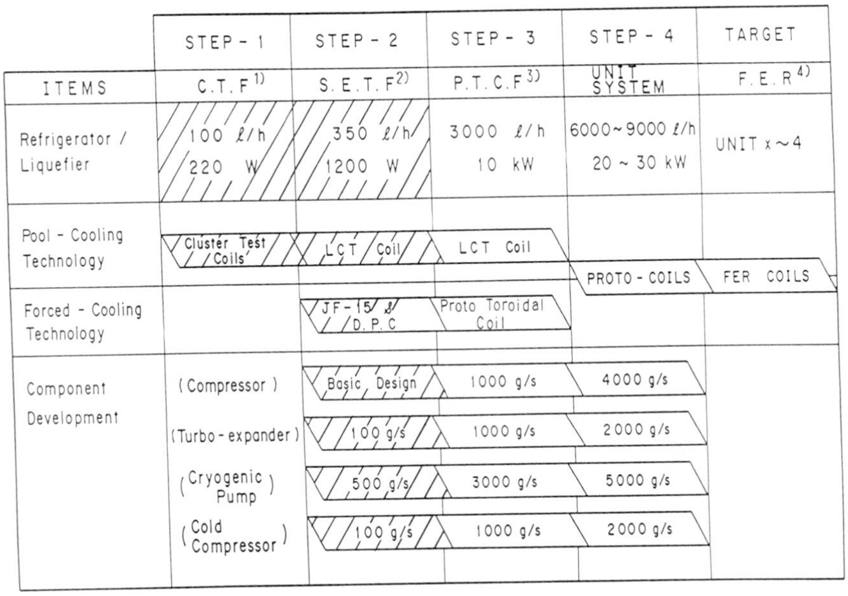

	STEP - 1	STEP - 2	STEP - 3	STEP - 4	TARGET
ITEMS	C.T.F [1]	S.E.T.F [2]	P.T.C.F [3]	UNIT SYSTEM	F.E.R [4]
Refrigerator / Liquefier	100 ℓ/h 220 W	350 ℓ/h 1200 W	3000 ℓ/h 10 kW	6000~9000 ℓ/h 20 ~ 30 kW	UNIT x ~4
Pool - Cooling Technology	Cluster Test Coils	LCT Coil	LCT Coil	PROTO - COILS	FER COILS
Forced - Cooling Technology		JF-15/ℓ/ D.P.C	Proto Toroidal Coil		
Component Development (Compressor)		Basic Design	1000 g/s	4000 g/s	
(Turbo - expander)		100 g/s	1000 g/s	2000 g/s	
(Cryogenic Pump)		500 g/s	3000 g/s	5000 g/s	
(Cold Compressor)		100 g/s	1000 g/s	2000 g/s	

1) Cluster Test Facility
2) Superconducting Engineering Test Facility
3) Proto Toroidal Coil Facility
4) Fusion Experimental Reactor

Hatching parts show the programs which were already carried out.

Fig. 1. Cryogenic system developing plan at JAERI

* On leave from Kobe Steel Co.,Ltd.
** On leave from Ishikawajima-Harima Heavy Industries Co.,Ltd.
***On leave from Hitach,Ltd.

ly, required as a unit system.

On the other hand, the Proto Toroidal Coil Program[1], which is a program of developing the superconducting toroidal coil for FER, is under consideration and the program requests the cryogenic system with the refrigeration power of around 10 kW at 4.5 K. According to our development plan of the cryogenic system as shown in Fig. 1, the cryogenic system of the Proto Toroidal Coil Program is located as step 3, just before the proto type system and it has the good scaling factor for the proto type system. In this paper, cryogenic capacity requirement, refrigeration cycle and cryogenic pump system will be introduced as the design concept of the cryogenic system of the Proto Toroidal Coil Program.

2. COOLING REQUIREMENTS FOR THE SYSTEM

The Proto Toroidal Coil System is composed of five coils: the Proto Toroidal Coil (test coil) which will be tested and verified as a proto type of toroidal field coil for the FER, two LCT coils[2], and two LCT insert coils. The LCT and LCT insert coils are used as backup coils to produce more than 12 T at a part of the test coil. In the design study of the Proto Toroidal Coil System, a forced-flow cooled type is selected as a cooling method for the test coil and the LCT insert coils. The LCT coils are a pool-boiling cooled type. Heat load of these coils is estimated to be around 500 W: the test coil of 200 W, two LCT coils of 200 W, and two LCT insert coils of 100 W. Since evaluation of the effect of nuclear heating on the superconducting toroidal coil is one of the purpose of this program, the nuclear heating up to 2 kW could be applied to the test coil. Cooling conditions for all coils are listed in Table 1. The forced-flow cooled coils request inlet temperature of 3.7 K and total mass flow rate of around 1000 g/s with pressure drop of 0.2 MPa. The total heat loss to achieve these conditions is estimated to be around 5 kW, which will be discussed in section 4. Finally, the total required refrigeration power is 7.5 kW and including design margin, the required refrigeration power of the Proto Toroidal Coil System is determined to be 10 kW and the design guide lines are as follows:

Refrigeration capacity	10 kW at 4.5 K
Cool down capacity	100 kW at 100 K
Energy efficiency	better than 1/400
Supercritical helium flow	1000 g/s

Table 1. Tentative cooling conditions for each coil

	TEST coil	LCT coil	LCT insert coil
cooling method	Foreced cooling	Bath cooling	Forced cooling
Inlet Temp.	3.7 K	4.2 K	3.7 K
Inlet pres.	1.0 MPa	0.1 MPa	1.0 MPa
Flow path	22	-	15
Flow rate per path	6 - 25 g/s	-	6 - 25 g/s
Pres. drop	0.2 MPa	-	0.2 MPa

3. REFRIGERATION CYCLE DESIGN

Separated Brayton cycle is selected as a basic cycle of the Proto Toroidal Coil Cryogenic System because of the following reasons;

i) to protect the turbine from the back pressure in case of a coil quench.

ii) to protect turbine circuit from impurities.

iii) to optimize the pressure of the JT valve regardless of the turbine operating pressure.

iv) to achieve high controllability.

The various separated Brayton refrigeration cycles, from case 1 to case 4 as shown in Fig. 2, are considered. Carnot efficiencies were calculated for each case to compare them each other. The turbine efficiency of series turbine configuration is assumed to be slightly higher than that of the parallel configuration because of lower expansion ratio in series configuration. The calculation results are plotted in the same figure and the following comparisons could be pointed out;

i) Double JT valve cycle has higher efficiency than single JT valve cycle by around 20% resulting in an advantage of non-linearity of helium property.

ii) Carnot efficiency of parallel turbine configuration is slightly higher than that of series turbine configuration.

iii) The case 4 has the highest carnot efficiency (24.9%), in which the turbine capacity is determined to optimize the refrigeration efficiency and the same turbine is added in case of liquefaction to improve the refrigeration efficiency.

The cycle of the case 4 is consequently selected. The detailed refrigeration and liquefaction cycle of the system are calculated to take

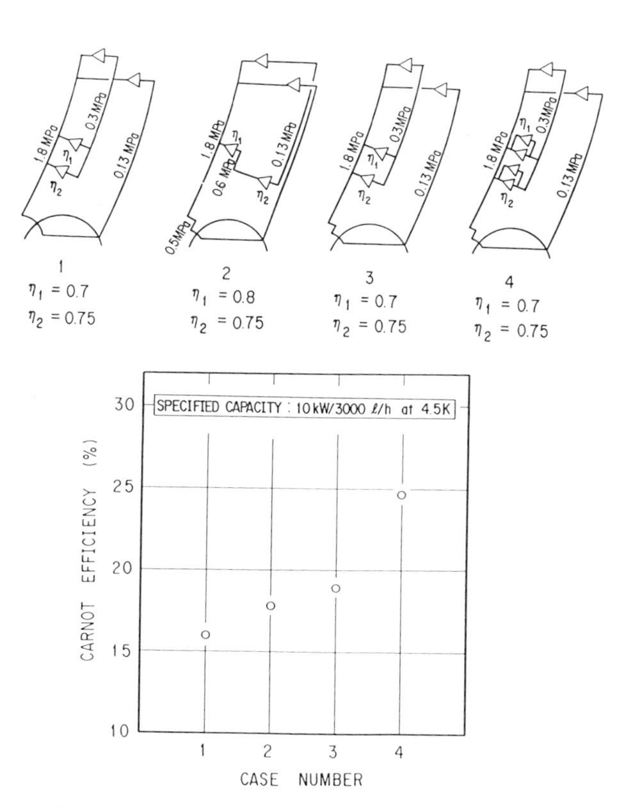

Fig. 2. Investigated refrigeration cycles and their carnot efficiency

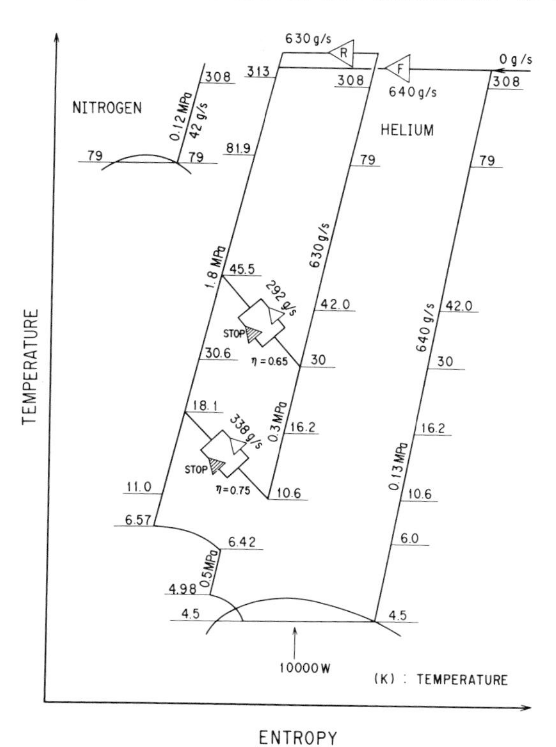

Fig. 3. Refrigeration cycle diagram

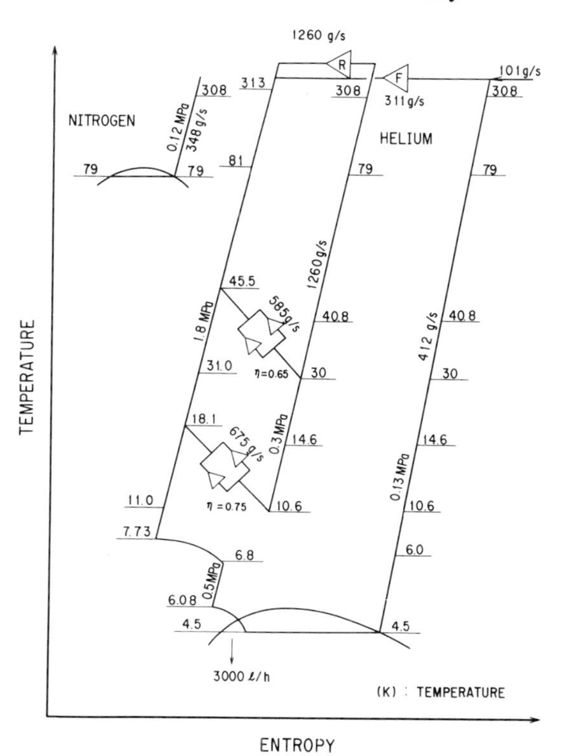

Fig. 4. Liquefaction cycle diagram

account of the heat inleak into the heat exchangers and the turbine efficiency as shown in Fig. 3 and Fig. 4, respectively. The carnot efficiency is 23.5 % and the energy efficiency is 1/289.

4. CRYOGENIC PUMP SYSTEM

A large scaled supercritical helium circulation system (cryogenic pump system) is one of the key components for the cryogenic system which drives the forced-flow cooled magnet system. The cryogenic pump system should have the following functions;

i) to circulate efficiently a large amount of supercritical helium with the pressure head of a few hundred kPa.

ii) to generate the supercritical helium below 4.2 K with the operating pressure of around 1.0 MPa as coil inlet conditions.

A helium pump system is required to perform functions described above. Compared with a room temperature pump system, a cryogenic pump system, which is composed of a cryogenic circulation pump and a cold compressor, has following advantages;

i) The pump system should be compact, then the total system should be compact even to drive a large amount of helium flow.

ii) The total efficiency of the cryogenic system should be higher if the pump efficiency is higher than 40%[3].

iii) The piping line under subatmosphere region should be short length, then the system should be more reliable so as to avoid sucking air.

JAERI has already developed such a cryogenic pump system in the Demonstration Poloidal Coil Test Facility[4] which is the world largest helium circulation pump system and composed of the circulation pump, which supplies flow rate up to 500 g/s, and the cold compressor, which produces subatmosphere helium down to 3.7 K. These pump system achieves an excellent adiabatic efficiency of more than 60 % even in the maximum operation design. In particular, such systems could be applied to the Proto Toroidal Cryogenic System to scale them up.

Figure 5 shows the block diagram of the cryogenic pump system with the calculated mass and heat balance at the several process points. The helium flow is 1000 g/s, the coil inlet temperature is 3.7 K, and the pressure head of the pump is 0.2 MPa. In this condition, the pump loss for circulation and evacuation is about 2.4 kW and about 2.6 kW with the same pump efficiency of 0.65, respectively and the details are shown in Table 2. Liquid helium load of 300 l/h and heat load of 250 W is used for cooling of the current leads and the pool-boiling cooled coils, respectively. The flow rate required to satisfy the total heat load of the pump system, around 540 g/s shown in Fig 5, is less than the flow rate of about 640 g/s in case of the 10 kW refrigera-

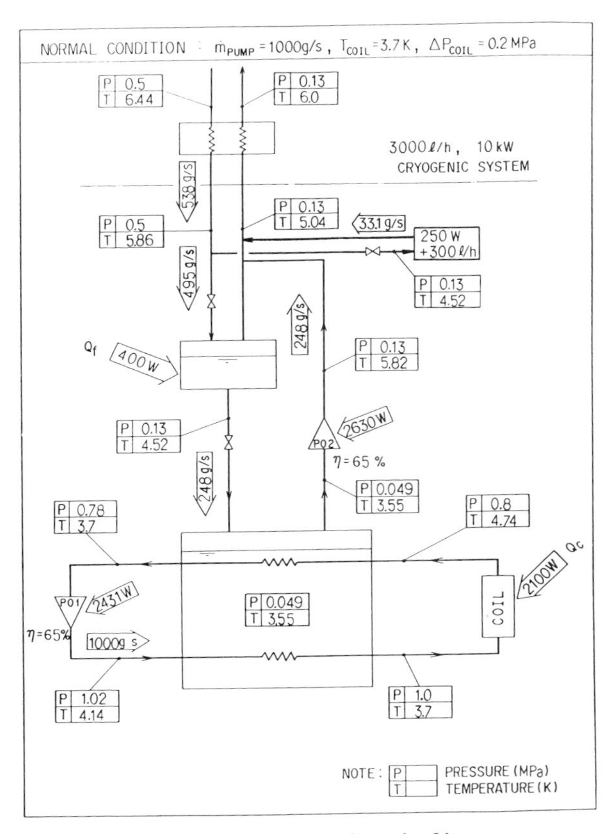

Fig. 5. Pump system block diagram

tion as shown in Fig. 3. The system can consequently operate with this pump condition. In case of the standby mode, the pump load is about half of the design load as shown in Table 2.

5. CONCLUSION

The cryogenic system for the Proto Toroidal Coil Program is the 10-kW refrigerator and has good scaling factor for the 30-kW cryogenic system which is a unit system for the Fusion Experimental Reactor. For realization of not only this system but also the 30-kW refrigerator, JAERI has investigated advanced cryogenic technology and key components of cryogenic system such as forced-flow cooling technique, cryogenic circulation pump, cold compressor, turbine expander and so on. In particular, the turbine expander and the circulation pump of 10-kW size is under developing. Fabrication of the Proto Toroidal Coil Cryogenic System, consequently makes sure in near future as shown in the bird's eye view of Fig. 6.

Table 2. Calculated heat loss for the pump system

	Design condition	Standby condition
	Flow :1000 g/s Temp.:3.7 K Pump head:0.2 MPa	Flow :500 g/s Temp.:4.2 K Pump head:0.1 MPa
Heat loss Circulation pump Cold compressor Forced-type coil Pool-type coil Miscellaneous	 2431 W 2630 W 2100 W 250 W 400 W	 562 W 610 W 2100 W 250 W 400 W
Total heat loss	7811 W	3922 W
Requested mass flow rate (g/s)	540 g/s	300 g/s

ACKNOWLEDGMENTS

The authors would like to thank Drs. S. Mori, M. Yoshikawa, and M. Tanaka for their continuous encouragement on this project. Contribution work from Isikawajima-Harima Heavy Industries Co.,Ltd., Kawasaki Heavy Industries Co.,Ltd. and Kobe Steel Co.,Ltd. are gratefully acknowledged.

REFERENCES

1. K. Koizumi et al., Design of a test coil and the test facility for Proto Toroidal Program, this volume

2. S. Shimamoto, Development and experiment of LCT superconducting toroidal coil, J. Nucl. Sci. Technol., 25[7] (1988) 557.

3. A. Bejan, Refrigerator-recirculator system for large forced-cooled superconducting magnet, Cryogenics 17 (1977) 97.

4. E. Tada et al.,Performance test results of cryogenic pump system for demonstration poloidal coil, IEEE Trans. Magn. Vol. 24 (1988) 1003.

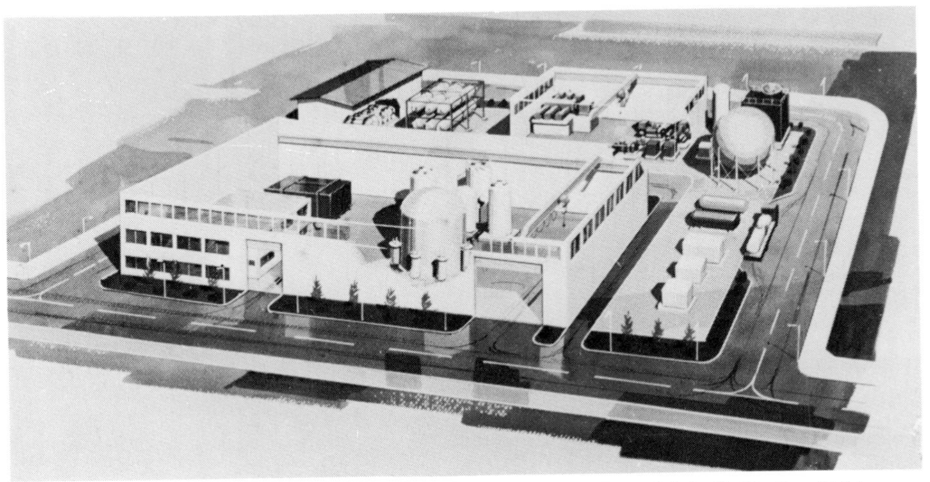

Fig. 6. Bird's eye view of the Proto Toroidal Coil Facility

FUSION TECHNOLOGY 1988
A.M. Van Ingen, A. Nijsen-Vis, H.T. Klippel (editors)

DESIGN OF A TEST COIL AND THE TEST FACILITY FOR PROTO TOROIDAL COIL PROGRAM

K. Koizumi, K. Yoshida, H. Nakajima, Y. Takahashi, M. Nishi,
H. Tsuji, K. Okuno, T. Kato, I. Itoh*, Y. Wachi**, H. Mukai**,
T. Ando and S. Shimamoto

Japan Atomic Energy Research Institute (JAERI)
Naka Fusion Research Establishment

Naka-machi, Naka-gun, Ibaraki-ken, Japan 311 - 01

JAERI started the design study for the Fusion Experimental Reactor (FER), named "Proto Toroidal Coil Program". Key objective of this program is the development and demonstration of 12-T, 30 kA, forced-flow cooled, conductors and coils, which satisfies technical specifications of toroidal coil for the FER. Based on the 12-T high field technologies established by the Test Module Coil (TMC) in the Cluster Test Program at JAERI, and on the 8-T, large-scale magnet technologies demonstrated by the Large Coil Task (LCT), a test coil for the Proto Toroidal Coil Program was designed. Since it is difficult to test a test coil at 12-T, only with the backup field of LCT coils, whose maximum field is limited to 8 - 9 T, the additional insert coils which has the maximum field of 12-T, are employed for the backup coil system. This paper describes the latest design status of a test coil and the test facility for the Proto Toroidal Coil Program.

1. INTRODUCTION

Japan Atomic Energy Research Institute (JAERI) is developing large superconducting coils for tokamak fusion machines. Our present goal of the development is to establish the construction techniques required for the Fusion Experimental Reactor (FER)[1] which is under designing as the next large tokamak machine in Japan. Applications of superconductivity both for DC toroidal field (TF) coils and for pulsed poloidal field (PF) coils is indispensable to realize long pulse operation. For this purpose, JAERI has been making systematic development both for TF coils and for PF coils. Regarding the TF coil development, JAERI has been making effort through the Cluster Test Program[2] at JAERI and the Large Coil Task (LCT)[3] under the auspices of IEA. In the former program, aiming at the demonstration of high field technologies more than 10 T, the Test

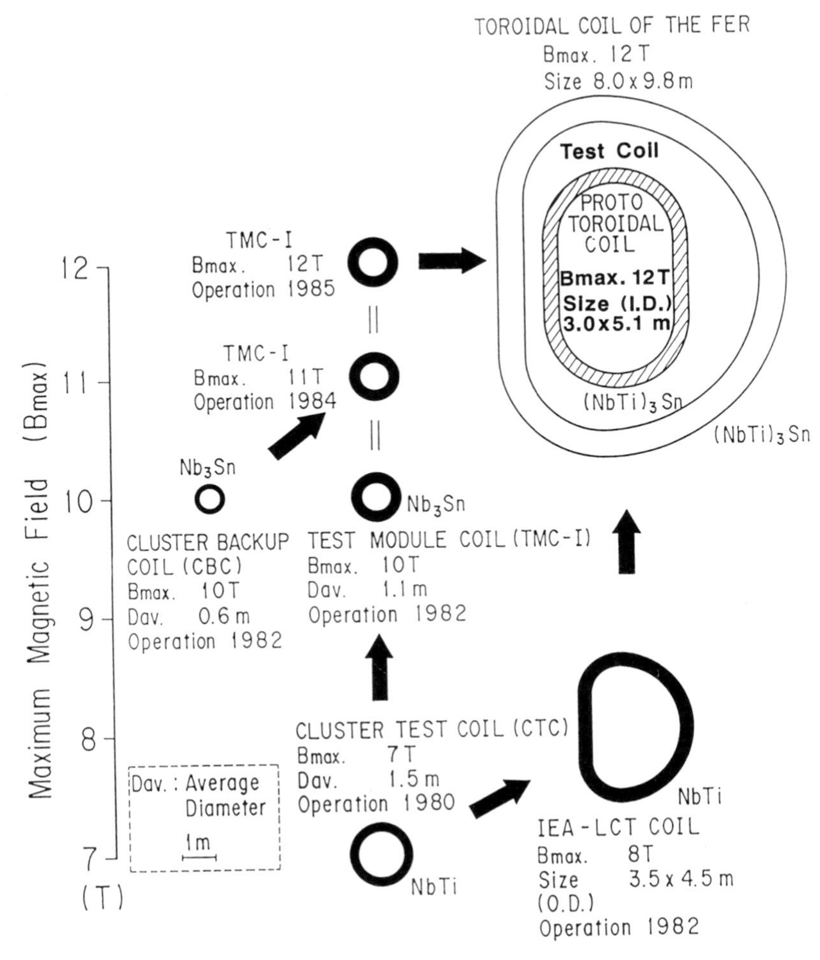

Fig. 1 TF coil development program in JAERI

* On leave from Fuji Electric Co., Ltd.
** On leave from Toshiba Corporation.

Module Coil (TMC), a 60 cm bore Nb_3Sn magnet, achieved 10 T, the specified value, in 1982 and 12 T in 1985. In the final operation, the TMC coil achieved 12.2 T with the winding current density of 33.5 A/mm^2. In the latter program, the Japanese LCT coil demonstrated a stable coil operation up to 9.1 T. Also for the development of PF coil, the Demo Poloidal Coil (DPC) program[4] is now in progress at JAERI.

Based on these progress, JAERI started the design study of new development step for the FER, named "Proto Toroidal Coil Program". Major objectives of this program are summarized in Table 1.

Table 1 Major objectives of the Proto Toroidal Coil Program

(1) Development of high performance TF coil conductor (12 T, 30 kA, and forced-flow),
(2) Development of manufacturing techniques of high rigidity, high voltage and stable coil,
(3) Fabrication of real conductor and coil as a proto-type conductor and TF coil for the FER
(4) Evaluation of conductor and coil performance required to the FER specifications in the test facility for the Proto Toroidal Coil

Table 2 General requirements from the FER

Maximum Magnetic Field	12 T
Rated Current	30 kA
Winding Current Density	30-40 A/mm^2
Maximum Dump Voltage	20 kV
Margin Factor of Critical Current	2
ENVIRONMENT	
Nuclear Heating Rate	1 mW/cc
Pulsed Field	20 T/s - 100 ms
CONDUCTOR	
Pulsed Loss Time Constant	< 100 ms
Stability Margin	50-100 mJ/ccst
Hot Spot Temperature	100-150 K
SUPERCONDUCTOR	$(Nb\text{-}Ti)_3Sn$
Jc at 12 T - 4.2 K	700 A/mm^2
Critical Axial Strain	0.7 %
Critical Transverse Pressure	50 MPa
STRUCTURAL MATERIAL	JCS*
Yield Strength	1.2 GPa
Tensile Strength	1.6 GPa
Fracture Toughness	200 MPa/m
COOLING CONCEPT	
Helium Inlet Temperature	4.0-4.5 K
Helium Inlet Pressure	6-10 bar
Maximum Inner Pressure	50 bar

* JCS : Japanese Cryogenic Steels[5]

2. DESIGN OF TEST COIL

2.1. REQUIREMENTS

Specifications of the conductor and test coil for the Proto Toroidal Coil Program have to satisfy those of the conductor and the TF coil of the FER. However, from view point of plasma engineering, the requirements from the FER design are not fixed yet. Therefore, plasma engineering staffs and magnet staffs discussed and specified the general requirements as shown in Table 2. Most of the items in Table 2 can be tested by a small test sample. However, mechanical and thermal performances of a large superconducting coil can be verified only by practical manufacturing and testing.

2.2. CONDUCTOR DESIGN[6]

Based on the general requirements listed in Table 2, three different candidates of conductor (TMC-FF type, Preformed Armor CICC type, and Advanced Disk type) are selected for the proto type. Cross-sectional views of three candidate types are shown in Fig. 2, Fig. 3, and Fig. 4.

Basic concepts of each type are as follows ;

(1) TMC-FF type : Hollow cooling type, concept of "Stranded cable in Cu stabilizer" employed for the TMC coil
(2) Preformed Armor CICC type : Reacted cable-in-conduit with thin first conduit in thick, preformed second conduit (armor)
(3) Advanced Disk type : Semi-reaction-after-winding technique and disk reinforced

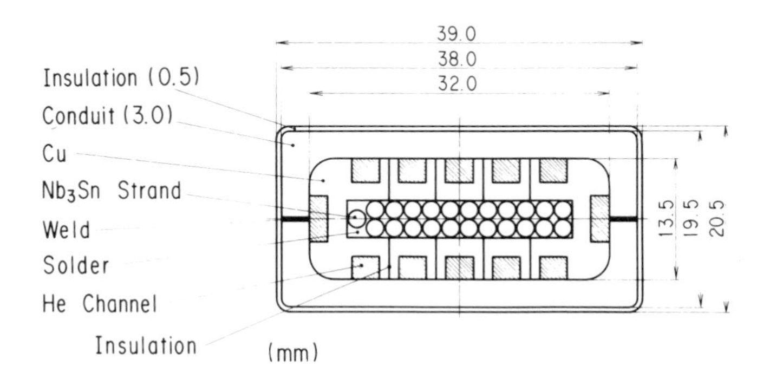

Fig. 2 Cross-sectional view of TMC-FF type conductor

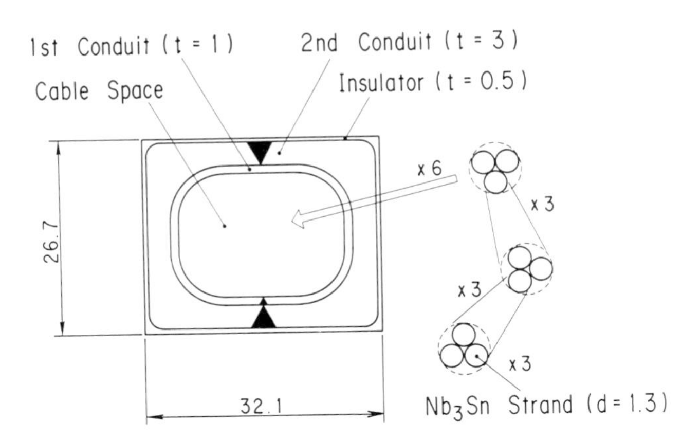

Fig. 3 Cross-sectional view of Preformed Armor CICC type conductor

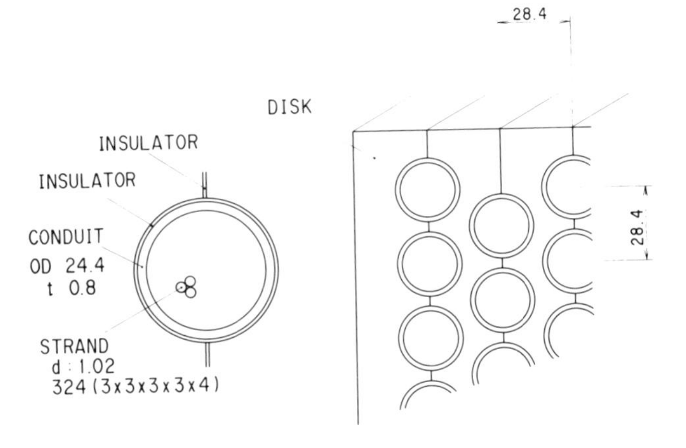

Fig. 4 Cross-sectional view of Advanced Disk type conductor

Table 3 Major advantages of three candidate conductor types

(1) TMC-FF type
- a) Low degradation of Nb_3Sn during cabling
- b) Grading technique of Nb_3Sn and NbTi can be applied in lower field (less than 5 T)
- c) Easy pancake joint by soldering

(2) Preformed Armor CICC type
- a) Difficulties during winding procedure are removed because two halves of armor are preliminary formed into coil shape by milling
- b) Nb_3Sn does not suffer from bending stress during winding
- c) Material for the armor is free from heat treatment for Nb_3Sn formation

(3) Advanced Disk type
- a) High rigidity of winding
- b) Epoxy impregnation for the insulation between conductors is not necessary
- c) Circular shape of conduit is effective to reduce the degradation of Nb_3Sn

The major advantages and design parameters of each conductor are summarized in Table 3 and Table 4, respectively. However, each conductor still has many items to be developed and verified. Therefore, JAERI and several Japanese industries started the practice manufacturing and verification test in April 1988. The final conductor type employed for the Proto Toroidal Coil will be decided based on the verification test results.

Table 4 Major design parameters of three candidate conductors

1. TMC-FF type conductor	
Winding Current Density	35.8 A/mm^2
Cross sectional Area	
Total	800 mm^2
SC	87 mm^2
Cu Stabilizer	251 mm^2
Helium Cross Section	85 mm^2
Stainless Steel	309 mm^2
Insulator	59 mm^2
Joule Generation Density	25.3 W/cc-He
Pulse Loss Time Constant(⊥)	60 ms
(winding average) (//)	0.8 ms
Helium Inlet Pressure	6.0 bar(4.2K)
Pressure Drop	0.22 mbar/m
2. Preformed Armor CICC type	
Winding Current Density	35.0 A/mm^2
Cross sectional Area	
Total	853.5 mm^2
SC	79.8 mm^2
Cu Stabilizer	149.1 mm^2
Helium Cross Section	155.0 mm^2
1st Conduit	85.6 mm^2
2nd Conduit (Armor)	322.8 mm^2
Insulator	61.2 mm^2
Joule Generation Density	26.5 W/cc-He
Pulse Loss Time Constant	< 3 ms
Helium Inlet Pressure	10 bar(4.2K)
Pressure Drop	3.9 mbar/m
3. Advanced Disk type	
Winding Current Density	37.0 A/mm^2
Cross sectional Area	
Total	807 mm^2
SC	107 mm^2
Cu Stabilizer	158 mm^2
Helium Cross Section	143 mm^2
Structure	360 mm^2
Insulator	39 mm^2
Joule Generation Density	24 W/cc-He
Pulse Loss Time Constant	< 1 ms
Helium Inlet Pressure	10 bar(4.2K)
Pressure Drop	10 mbar/m

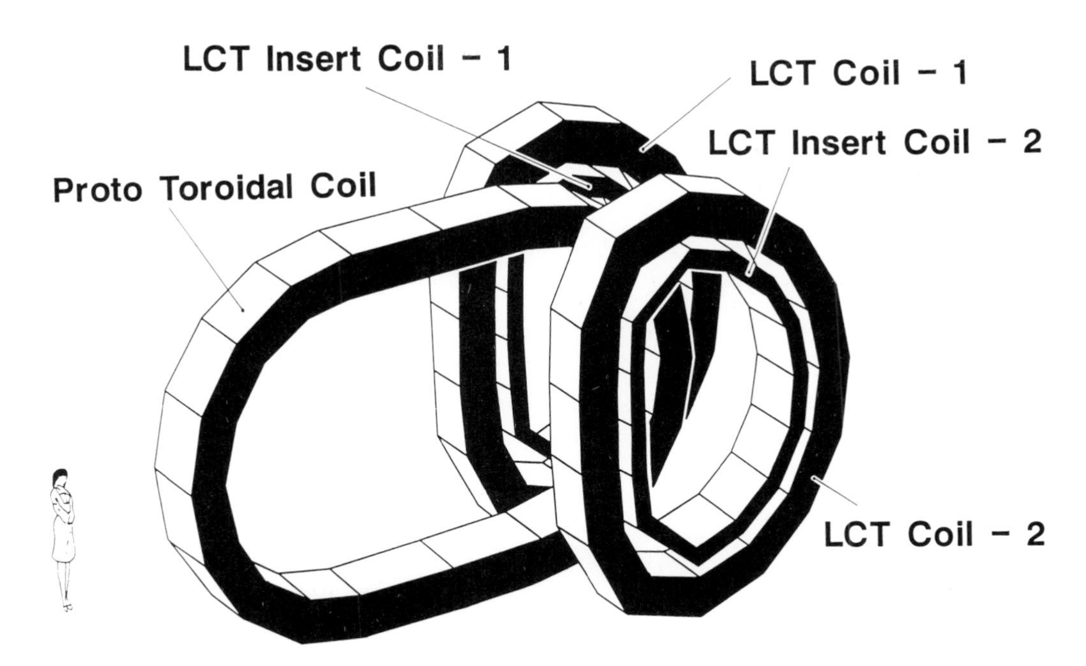

Fig. 5 Test configuration of the Proto Toroidal Coil

2.3. DESIGN OF PROTO TOROIDAL COIL

The Proto Toroidal Coil is requested to have adequate coil size to establish the mechanical design technique and construction techniques as a proto-type TF coil for the FER. Major design parameters of the Proto Toroidal Coil are shown in Table 5. So as to increase the stress level of the Proto Toroidal Coil in 12 T operation, coil shape of a race truck with the inner bore of 3.0 m x 5.1 m is employed for the test coil. Almost all parameters except the coil size, coil shape, and ampere turns, satisfy the general requirements for the FER listed in Table 2.

Table 5 Major design parameters of the Proto Toroidal Coil

Design Magnetic Field	12 T
Superconductor	$(NbTi)_3Sn$
Operation Current	30 kA
Critical Current	60 kA
Winding Current Density	> 35 A/mm^2
Coil Shape	Race Truck
Coil Size (I.D.)	3.0 m x 5.1 m
Minimum Bending Radius	1.5 m
Ampere Turns	10 MAT
Conductor Length	300 m/Pancake (600 m in double pancake)
Insulation Test Voltage	20 kV

3. TEST FACILITY

3.1. TEST CONFIGURATION

Test facility of the Proto Toroidal Coil is required to have adequate capacity to simulate the serious operating conditions of the FER TF coil. Also the initial cost required for the construction of test facility is the key issue to carry out the program. Therefore, JAERI employed the LCT coils in the design of test system. Test arrangement of the Proto Toroidal Coil is shown in Fig. 5. Test coil is placed in the parallel configuration. Since it is impossible to test the Proto Toroidal Coil at 12 T with the backup field of the LCT coils only, whose maximum field is limited to 8 - 9 T, the additional insert coils, which have a maximum field of 12 T, are also employed in the coil system. Additional coils, installed into the inner bore of the LCT coils, are called the LCT insert Coils[7]. In this test configuration, the LCT coils and the LCT insert coils are charged up to 60 %, and 100% of their rated current, respectively. The operating current of the LCT coils is limited to reduce the magnetic force in the out-of-plane direction. The peak field of the

Proto Toroidal Coil is 12.0 T in the midplane of inner bore. The peak field of the LCT coils and the LCT insert coils is 5.7 T, and 9.0 T, respectively. The out-of-plane force acts to the LCT coil is 33 MN. This value is 1.3 times larger than the value experienced in the six-coil test. In an extended test condition, in which all the coils are operated at the 100% of their rated current, the peak field of 13.2 T and 12.0 T is expected to the test coil and the LCT insert coil.

3.2. STRESS ANALYSIS OF TESTING SYSTEM

The mechanical stability of the backup coil system is one of the key issues to decide test configuration, because the LCT coils are not originally designed for this test configuration. On the contrary, the Proto Toroidal Coil is required to have the higher stress level to simulate the serious loading condition of the FER. Therefore, mechanical behavior of the LCT coils and the stress level of the Proto Toroidal Coil have been analyzed by using the three dimensional FEM model of SAP-V2. Tentative support structures between the test coil and the LCT coil are attached to the LCT coil. So as to estimate the more realistic stress level in the LCT coils, the equivalent winding stiffness obtained by the advanced calculation[8] at JAERI was applied for stress analysis.

The maximum displacement and the Tresca stress obtained by the calculation are summarized in Table 6. The maximum displacement

Table 6 Maximum displacement and Tresca stress calculated by the FEM analysis

	Test Coil	LCT Coil	LCT Insert Coil
Current Rating	100 %	60 %	100 %
Maximum Disp.(mm)			
X : Radial	5.0	2.0	2.2
Y : Out-of-plane	-0.6	-1.3	-1.3
Z : Vertical	16.7	-0.5	0.6
Max. Tresca Stress (MPa)			
Conductor	473	65	121
Coil Case	779	167	288

and Tresca stress in the LCT coil are lower than the values appeared in the six-coil test. On the contrary, the maximum Tresca stress of the Proto Toroidal Coil conductor is 473 MPa. This value is close to the stress level of 500 MPa, which is estimated as the maximum conductor stress of the FER TF coil in the real operating condition. Therefore, the Proto Toroidal Coil can simulate the serious loading condition of the FER under the magnetic field of 12 T.

4. SUMMARY OF THE PRESENT STATUS

As a further development program of TF coil for the FER, design study of the Proto Toroidal Coil Program has set in JAERI. As the first step of the development, three candidate conductors were selected for the test coil of this program. JAERI and several Japanese industries started the practice manufacturing of proto-type conductor in April 1988. A design study of the 10 kW cryogenic system[9] for the Proto Toroidal Coil Program is also in progress at JAERI.

ACKNOWLEDGEMENTS

The authors would like to thank Drs. S. Mori, M. Yoshikawa, and M. Tanaka for their continuing encouragement on this work. Also the authors would like to thank Fuji Electric Co. Ltd., Hitachi Ltd. and Toshiba Corporation for their manufacturing of the proto-type conductor.

REFERENCES

1. M. Kasai et al., Japanese contribution to IAEA INTOR workshop, JAERI-M 88-010 (1988)
2. T. Ando et al., Proc. of 11th Symp. on Fusion Eng. (1986) 991
3. K. Okuno, et al., IEEE Trans. MAG-24. No. 2 (1988) 767
4. H. Tsuji et al., IEEE Trans. MAG-24 No. 2 (1988) 1303
5. H. Nakajima et al., Advances in Cryogenic Engineering Materials Vol.34 (1988) 173
6. K. Yoshida et al., Design of the proto-type conductors for the fusion experimental reactor, presented to 1988 Appl. Super. Conf.
7. K. Koizumi et al., Proc. of 12th Symp. on Fusion Eng. Vol.2 (1087) 1035
8. S. Shimamoto et al., Cryogenics Vol.25 (1985) 539
9. T. Kato et al., Design concept of cryogenic system for the PROTO Toroidal Coil Program, paper in the conference

FUSION TECHNOLOGY 1988
A.M. Van Ingen, A. Nijsen-Vis, H.T. Klippel (editors)

RELIABILITY AND AVAILABILITY ANALYSIS OF TWO ALTERNATIVE EVACUATION SYSTEMS DESIGNED FOR THE NEXT EUROPEAN TORUS (NET)

Naeem A. Tahir

A B A S GmbH, Schirmerstr. 2a, 7500 Karlsruhe, FRG

This paper discusses the reliability and availability issues in case of two different evacuation system designs which have been proposed for the Next European Torus (NET). One of these designs uses turbo molecular pumps while the other employs cryogenic pumps to evacuate waste products from the torus after every fusion cycle. The aim of this paper is to assess and compare the feasibility of the above two designs from the reliability and availability point of view. A detailed failure mode analysis has been carried out for these two systems and appropriate mathematical models have been developed to calculate their respective reliabilities. Using these mathematical models an extensive parameter study of the system reliability has been carried out over a given range of the component reliabilities. This parameter study shows that the maximum value of the turbo molecular pump system reliability is 96 % while the corresponding value for the cryogenic pump system is only 81.6 %. The target value for the system availability is 99.9 %. This requires that the system mean repair time should be < 48 h for the turbo pump system and < 9.8 h for the cryogenic pump system. These calculations show that from the reliability and hence the availability point of view the turbo pump system is much more feasible compared to the cryogenic pump system. However if the system mean repair time is > 48 h, appropriate modifications must be made to the turbo pump system design to increase its reliability accordingly.

1. INTRODUCTION

This paper discusses the reliability and availability of two different evacuation systems designed by Leybold-Heraeus GmbH, Hanau, for the Next European Torus (NET). One of these system designs uses turbo molecular pumps whereas the other employs cryogenic pumps to evacuate waste products from the NET machine. The above two systems are schematically shown in figures 1 and 2 respectively. Figure 1 shows one of the eight subsystems of the turbo pump design and each of these subsystems is connected to the torus by two pipes. Each subsystem consists of three identical "upper branches" connected in parallel and these upper branches are connected in series with a "lower branch". Each of the upper branches consists of a "big value" V_B with a reliability, R_{VB} and has a diameter 1.5 m. This big valve is connected in series with a turbo pump P_T which is also connected in series to a small valve V_S. The reliabilities of P_T and V_S are denoted by R_{TP} and R_{VS} respectively and diameter of V_S is 0.6 m. The lower branch consists of three roughing pumps P_R connected in series and they all have same reliability R_{RP}. The entire system thus contains 24 "upper branches" and 8 "lower branches" and the reliability of these branches is denoted by R_{up} and R_{lo} respectively, where

$$R_{up} = R_{VB} \times R_{TP} \times R_{VS} \quad (1)$$

and

$$R_{lo} = (R_{RP})^3 \quad (2)$$

There also is another valve V_{by} which is called a "bypass valve" and is identical to V_S. This valve is opened at the start (initial operational phase) to create a low pressure of 10^{-3} mb by operating the roughing pumps so that the turbo pumps could be operated which only work if the pressure is of the order of 10^{-3} mb or low. This initial operational phase last for about 100 h after which this bypass valve is closed and it remains closed during the rest of the operational time (main operational phase). The reliability of the system in the initial operational phase has been discussed in reference 1. In this paper we will only consider the main operational phase in which V_{by} is kept closed. Similary figure 2 shows the logic diagram of one out of the eight identical subsystems of the cryogenic pump design. It is seen that the torus is connected to a middle chamber by two cryogenic pumps P_{C1} and P_{C2} by two big valves V_{B1} and V_{B2} respectively. Each of these valves has a diameter 1.5 m and a reliability denoted by R_{VB} while the cryogenic pump reliability is denoted by R_{CP}. The middle chamber is also connected to a set of three roughing pumps (connected in series) by a bypass valve V_{S3} which is only opened during the initial operational phase. During this phase the pressure in the system is reduced to 10^{-4} mb after which the cryogenic pumps can be operated. As before, the initial operational phase will not be considered in this paper. Interested reader may consult reference 2. The two cryo pumps P_{C1} and P_{C2} work alternately. First one of these pumps say the left hand side (LHS) pumps in all the eight subsystems are connected to the torus by opening the corresponding big valves. The waste products from the torus are sucked into the pump and are deposited in the pump in a frozen form. One operational cycle for such a pump is 2 hours after which it is disconnected from the torus by closing the valve while the RHS pumps are connected to the torus by openig the RHS big valves. The LHS pumps are then regenerated and this regeneration cycle lasts for about one hour. In this manner the RHS and the LHS units exchange their function alternately. We may divide such a subsystem into three units, namely, two upper units and one lower unit with reliabilities

$$R_{up} = R_{VB} \times R_{CP} \times (R_{VS})^3 \quad (3)$$

and $$R_{lo} = R_{R1} \times R_{R2} \times (R_{R3})^3 \quad (4)$$

where R_{R1}, R_{R2} and R_{R3} are the reliabilities of various roughing pumps shown in figure 2. For simplicity we may assume that $R_{R1} = R_{R2} = R_{R3}$.

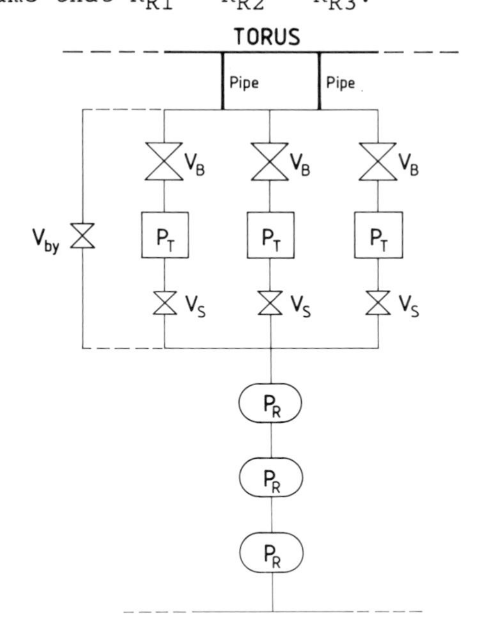

FIGURE 1
Logic diagram of a turbo pump subsystem

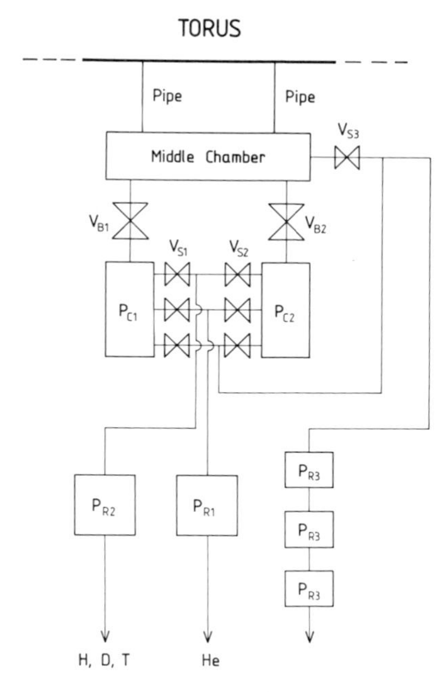

FIGURE 2
Logic diagram of a cryogenic pump subsystem

2. RELIABILITY CALCULATIONS

A detailed failure mode analysis has been carried out for the above two systems assuming that a loss of up to 25 % of the system will not affect the system operation. Applying the above constraint, amplitudes of all those possible system modes which do not stop the system operation and hence contribute to the system reliability have been evaluated and the corresponding system reliability has been calculated by adding the amplitudes of these modes. Details about these calculations are given in references 1 and 2.

2.1. Turbo pump system reliability

The logic diagram of one of the eight subsystems of this design is shown in figure 3. The purpose of these calculations is to check sensitivity of the system reliability to variations in the component reliabilities. A parameter study of the system reliability has therefore been carried out over a wide range of the component reliabilities given below.

R_{VB} is assumed to lie from 0.90 to 0.94 while R_{TP} is considered to vary from 0.95 to 0.98. Also R_{VS} is varied from 0.97 to 0.99 and R_{RP} is given values from 0.95 to 0.99. Substituting the component reliability values over the above range in equations 1 and 2, the corresponding range of values of R_{up} and R_{lo} is 0.83 to 0.91 and 0.86 to 0.97 respectively.

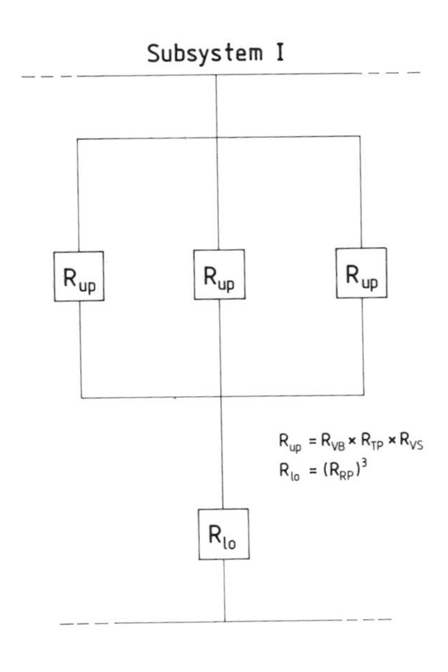

FIGURE 3
Block diagramm of reliabilty of turbo pump subsystem

In Figures 4-6, the system reliability and amplitudes of different system modes are plotted as a function of the upper branch reliability for three different values of the lower branch reliability, R_{lo}. These three values of R_{lo} correspond to R_{RP} = 0.95, 0.97 and 0.99 respectively. It is seen from these figures that the system reliability, R_S varies from 0.46 to 0.96 over the above range of the component reliabilities. These minimum and maximum values of R_S correspond to the lowest and the highest values of the component reliabilities. It is also interesting to note that the amplitudes of lower order modes R_{M1}, R_{M2} and R_{M3} which represent no component failure, a single component failure and two component failure respectively, increase with increasing values of the component reliabilities. The amplitudes of the higher order

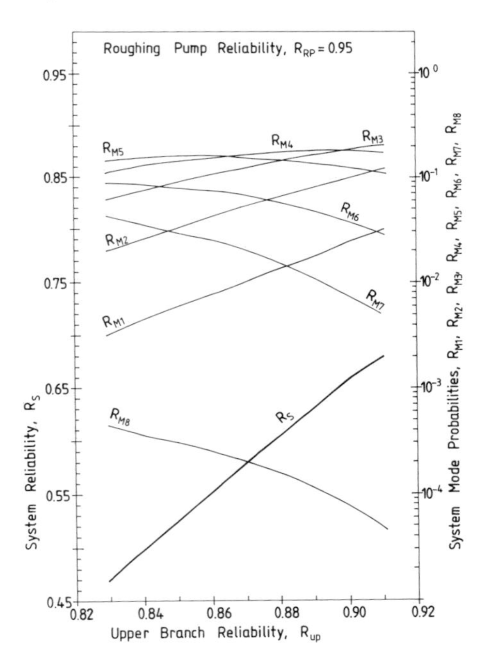

FIGURE 4
Turbo pump system reliability, R_S and amplitudes of various system modes vs upper branch reliability, R_{up}, keeping R_{RP} = 0.95

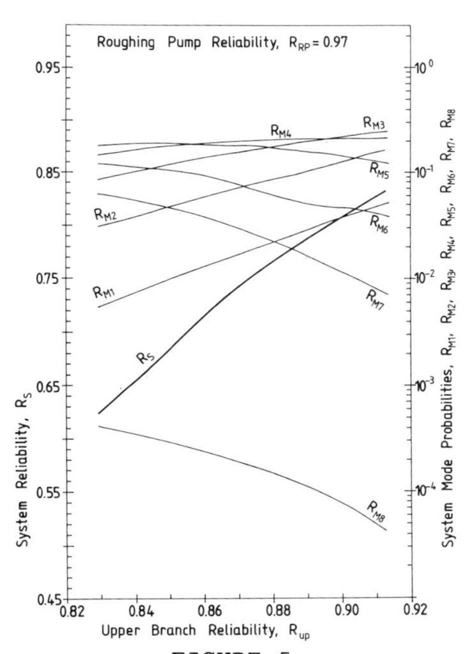

FIGURE 5
Same as in figure 4 but for R_{RP} = 0.97

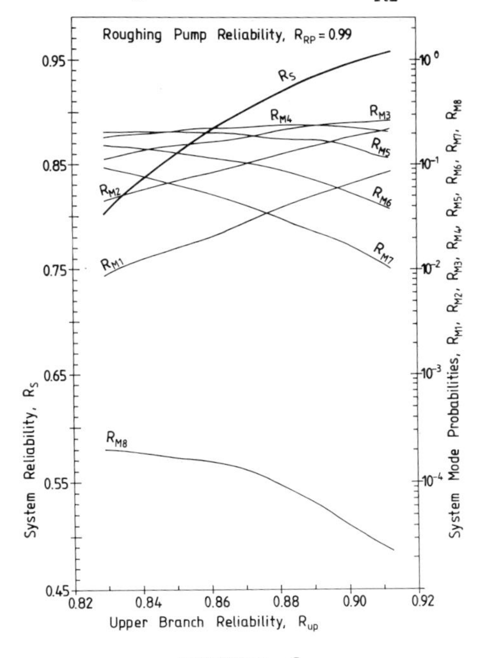

FIGURE 6
Same as in figure 4 but for R_{RP} = 0.99

modes, on the other hand, decreases as R_{up} is increased. This is because if the components have high reliability the probability of failure decreases.

2.2. Cryo pump system reliability

The logic diagram of reliability of one out of eight subsystems of this design is shown in figure 7. In this case we assume that the cryogenic pumps have the same reliability as the turbo pumps in the previous design. Also the same parameter ranges are used for the reliabilities of other components as in the turbo pump design. The corresponding ranges for the reliabilities of the upper and lower units calculated by equations 3 and 4 respectively, are, R_{up} = 0.78 - 0.89 and R_{lo} = 0.77 - 0.95. The reliability R_S, of this system design is shown in Fig. 8 as a function of R_{up} for three values of R_{lo} which correspond to

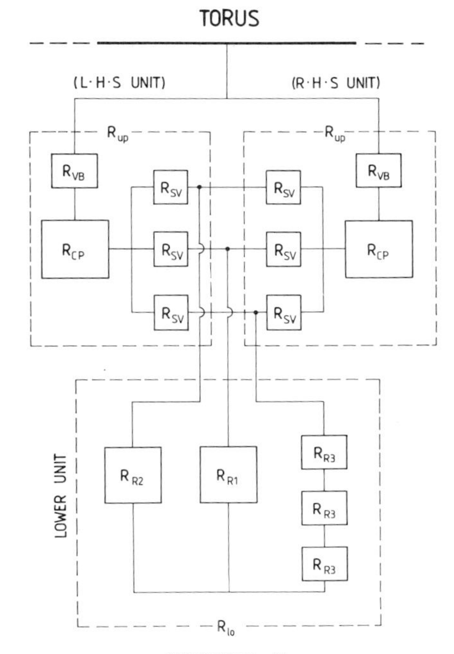

FIGURE 7
Block diagram of reliability of a cryogenic pump system

R_{RP} = 0.95, 0.97 and 0.99 respectively. It is seen that in this case R_S varies from 0.17 to 0.81 which shows that the turbo pump system is much superior to the cryogenic pump system.

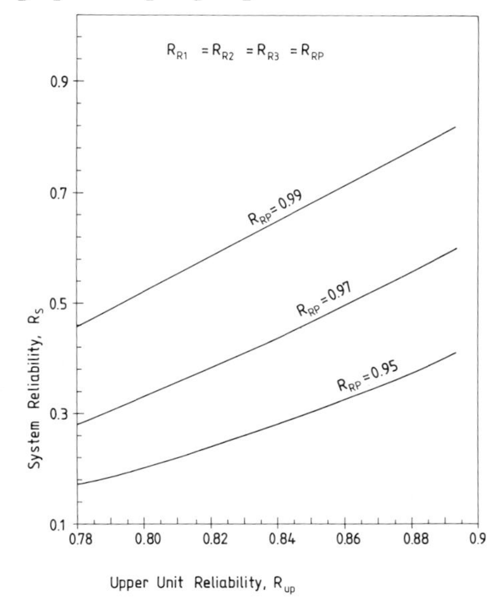

FIGURE 8
System reliability, R_S vs upper unit reliability, R_{up} for R_{RP} = 0.95, 0.97 and 0.99.

3. SYSTEM AVAILABILITY CONSIDERATIONS

In figure 9 is plotted the mean time between failure (MTBF) and the system failure rate as a function of system reliabilty. These curves have been plotted assuming a constant failure rate for the system, as discussed in reference 3. It is seen that the best values of the failure rate and MTBF for the turbo pump system are 2.04×10^{-5}/h and 4.89×10^{4} h respectively, which are obtained for maximum system reliability of 0.96. The corresponding values for a cyro pump system are 1.017×10^{-4}/h and 9.836×10^{3}h respectively. In figure 10 is plotted the system availability, A_S as a function of R_S for different values of mean repair time, T_m. The NET design demands a system availability of 99.9%. In case of a turbo pump system with R_S = 96%, this value of A_S is possible provided T_m < 48 h. In case of a cryogenic pump system, with R_S = 81.6 %, the above availability value is only possible if T_m < 10 h. This shows that from the reliability and availability point of view, the turbo molecular pump system design is much superior to the cyrogenic pump design. Details about the mathematical models used for the above calculations will be published in reference 4. One operational cycle should last for 2000 h and system life time is 25000 h.

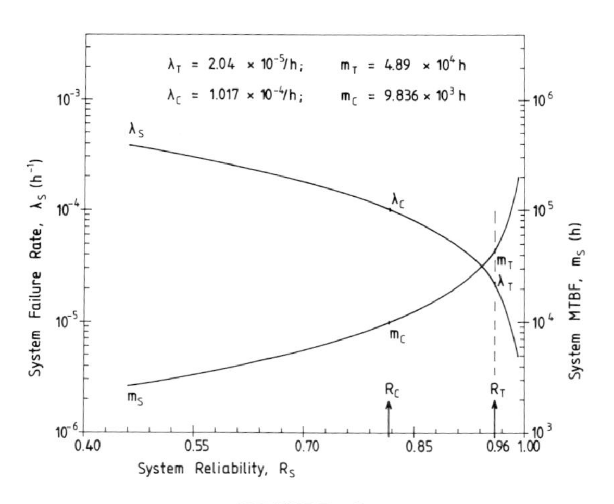

FIGURE 9
System MTBF and failure rate vs system reliability, R_S

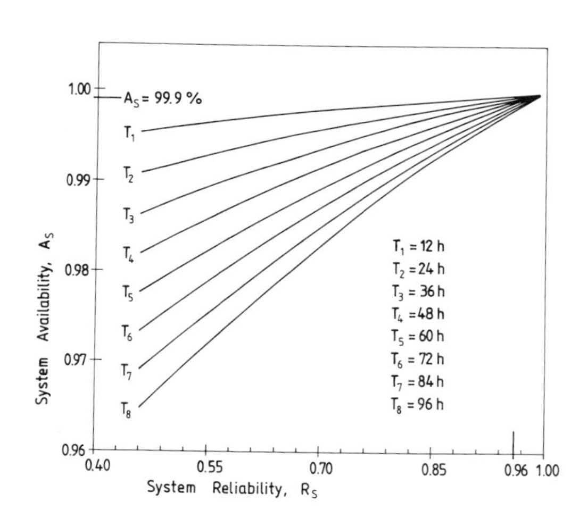

FIGURE 10
System availability, A_S vs system reliability R_S, for different mean repair times

ACKNOWLEDGEMENTS

This work was carried out by ABAS GmbH as a subcontractor to Leybold-Heraeus GmbH who were given a contract for this work by the NET project. I wish to thank Mr. P. Forscht, Dr. R. Bünde and Mr. H. Maykemper for their support. I also wish to thank Mrs. Reichert and Mrs. Berninger for typing this manuscript.

REFERENCES

1. N. A. Tahir and P. Forscht, A parameter study of reliability of an evacuation system designed for the Next European Torus as a function of component reliabilities using a static reliabity model, ABAS GmbH Rep. (1987).

2. N. A. Tahir, A reliability study of cryogenic pump evacuation system designed for the Next European Torus, ABAS GmbH Rep. (1987)

3. E. J. Henley and H. Kumamoto, Reliability engineering and risk assesment (Prentice Hall Inc., Engelwood Cliffs, 1980)

4. N. A. Tahir, Reliability and availability considerations in thermonuclear fusion devices, To be submitted to IEEE Trans. on Reliability (1988)

FUSION TECHNOLOGY 1988
A.M. Van Ingen, A. Nijsen-Vis, H.T. Klippel (editors)
Elsevier Science Publishers B.V., 1989

RELIABILITY AND AVAILABILITY STUDIES FOR THE NEXT EUROPEAN TORUS (NET)

K. Chaharbaghi, A.J.H. Goddard, R.S. Sayles and K. Khodabandehloo*

Centre for Fusion Studies, Imperial College of Science and Technology, London, SW7 2BX, United Kingdom.

R. Bünde

The NET Team, c/o Max-Planck Institut fur Plasma Physik, Boltzmannstrasse 2, Garching bei Munchen, Federal Republic of Germany.

The design requirements for the Next European Torus (NET) include a range of availability targets over physics and technology phases of operation. In order to simulate the dynamic, cyclic operation of NET, discrete change simulation software has been employed, which allows the prediction of availability and the study of the influence of subsystem failure. An 'avalanche' strategy has been developed for the systematic improvement of unavailability drivers.

1. INTRODUCTION

The aim of reliability and availability studies for NET is principally to help demonstrate that physics and technological objectives can be met within overall target calendar times. In terms of reliability and availability, there is a major step from the performance of current large fusion experiments to the availability implications of the NET machine. NET must meet, during its technology oriented operating phase, a range of targets: targets for neutron fluence to the first wall, selected availability targets over timespans between several months and one year and specified availability during individual blanket tests.

Bünde[1,2] has described the breakdown of the NET plant, for the purpose of functional analysis, illustrated in Figure 1. Main functional sections are labelled A-Y; the nuclear heat system is broken down further into subsections (AA-AQ), the TORUS system (AA) into assemblies (AAB-AAF). The plant component identification scheme (PCIS), which makes use of this functional breakdown, is described in Ref.1.

Figure 2 gives a simplified plan of both physics and technology orientated phases which are envisaged to meet the objectives of NET and which extend overall for thirteen calendar years. This paper is restricted to the technology oriented phase. It is currently envisaged that during this phase 4 test campaigns will be required; Figure 2 indicates that each campaign is preceded by a major refurbishment period. During each campaign, five successful test runs are required to be performed. Such a test run would be considered successful when a given integrated operating time (200 hours in this case) for neutron production, with a minimum availability (80% in this study) and with single down times not exceeding a selected maximum. Thus several attempts might be made to complete a test run before all desired criteria were met. Additional overall targets need to be borne in mind; the total exposure of first wall materials to energetic neutrons of about 1 MWa/m^2 over the two phases, and also availability of 25% over times between several months and one year should be demonstrated.

A measure of success in meeting these targets has been defined and determined. This is referred to as the target success ratio (TSR). It is defined as:

$$TSR = \frac{\text{demanded calendar time to complete the planned 4 test campaigns}}{\text{predicted calendar time to complete 4 test campaigns}}$$

A value of this ratio equal to or greater than unity implies that the specified targets will be attained.

2. SIMULATION TECHNIQUE

Previously used "static" availability assess-

*Present address: University of Bristol, U.K.

ment techniques were limited in various ways: for example availability could only be estimated in terms of failure rate and mean down time. The new technique described in this paper based on an advanced discrete change simulation technique of which the software is known as DSSL (Dynamic Systems Simulation Language)[3], which has previously been applied in the study of manufacturing systems[4,5,6,7]. This technique allows the construction of an analogue of the system under study whose time dependent characteristics are as similar to those of real life as possible.

This may be set in motion to follow the cyclic operation of NET and yield measures of effectiveness which cannot be offered by other techniques.

Using DSSL, the systems are modelled with the aid of transformation logic diagrams. These highlight the way in which entities (for example NET operating state, NETOS) flow through the simulated system when it is set in motion. Figure 3 shows the symbols used in the construction of such diagrams.

Figure 3(a) contains the title sub block which is the "entity's reference name (e.g. NET operating state) while the property sub block gives the status of the entity at the particular point in the diagram.

The simplest transformation activity, which assumes perfect reliability, is shown in Figure 3(b). The activity is described by "transformation definition" sub blocks. "Duration" sub block represents the time taken for the transformation while "transform if" sub block contains the conditions, which, when satisfied, start the transformation.

Figures 3(c) and (d) show activities which may be disabled. In addition to the three sub blocks outlined above they contain "enable if" and "disable if" sub blocks. The "enable if" sub block holds the conditions, which when satisfied, enable a disabled activity. The "disable if" sub block contains the condition which if satisfied disables the activity. The letters 'K' nad 'T' show whether or not an activity is re-engaged at the same point after having been disabled. The content of arrow heads represent the number of entities transferred between sets and activities or vice versa.

3. APPLICATION

Only an outline description of the application of DSSL to the NET system is possible here; reference should be made to more detailed applications. The simulation has included only major sub systems; Table 1 shows reliability data while Table 2 shows the time sequence, including refurbishment before a test campaign and the sub systems which must be fully operational during the course of each operation.

Figures 4(a) and (b) show the transformation logic diagram of NET in a simplified form. Activities representing the refurbishment or preparation phase are named PERFRM 1-8, while the five test runs, each involving 200 hours neutron production, are represented by activities PERFRM 9-13. If a particular test run is completed but its percentage availability falls below that required then the operating state is transformed to the pretransformation state of the test run via activity RESTART.

In order to treat the reliability of individual sub systems, failure and repair transformation activities (referred to as FAIL and REPAIR) change entities SSOC - sub system operating condition - from fully operational to non-operational and vice versa. This is represented for one sub system in Figure 4(c). Duration of operations is indicated by DURTN. Failure rate and mean down time data for all sub systems are used, with the aid of exponential random sampling, to determine times between two failures and down times.

4. RESULTS AND DISCUSSIONS

From the failure logic of a series system, the steady state availability of NET can be represented via the relationship:

$A = 1/[1 + \sum_{j=1}^{n} G(j)]$ with $G(j) = FR(j) \times MDT(j)$ where $FR(j)$ and $MDT(j)$ $\{j=1,2,...,n\}$ are the sub system average failure rates and mean down times respectively. The sub systems with the highest value of G have the greatest influence upon availability. G for any sub system can be improved by reducing the value of failure rate and/or down time. Achievement of the required four test campaigns over seven years, specified for the technology oriented phase (i.e. TSR $\leq$ 1) hinges not only on the attainment of an acceptable level of plant availability but also on the correct definition of the rejection criteria for test runs halted because of a sub system failure. The reason for the latter is that, following the repair of a failed sub system, continuation and completion of the test run may or may not result in the 80% availability specified.

Taking the tolerable down time for NET to be 24 hours (i.e. semi-completed test runs will be rejected if the down time of the failed sub systems exceeds 24 hours) and using the reliability data given in Table 1. NET has been simulated. The simulation predicted a value of 0.28 for TSR, obviously requiring improvement in assumed sub system reliability.

Study of Table 1 shows that sub system AC has the greatest G value. Figure 5 shows TSR predicted using a range of G values for sub system AC. Reducing the value of G for sub system AC initially improves TSR but, the effect on TSR diminishes,and there is no effect on TSR below a value of 0.15. Below 0.15, AC has become a very minor unavailability driver. There are other sub systems, such as ABA and AG with G values of 0.40 and 0.21 respectively, that are then heavily responsible for plant unavailability. Similarly, as indicated by Figure 6, reducing the G value for ABA alone does not give any noticeable improvement in performance, as sub system AC continues to have the greatest influence. Figure 7 shows that reducing G values of sub systems AC and ABA simultaneously results in an improvement in TSR which is much larger than those resulting from the previous two cases combined (compare Figures 5,6 and 7). However, once again at lower G values of sub systems AC and ABA the improvement becomes less marked when their G values pass below G value of other sub systems.

This leads to the conclusion that in planning to realise significant improvement in performance, sub systems with similar magnitude and values of group members reduced simultaneously. This is akin to the operation of an avalanche, and has been termed an 'avalanche strategy'. The strategy for NET performance assessment operates in a similar manner in that it initially lowers the G value of the sub system which is the greatest. When the G value of this sub system reaches the G value of another sub system, they join together to form a group. The G value of all the members of the group are then lowered simultaneously. This procedure is repeated until the desired performance level is attained. Figure 8 shows the improvement strategy in graphical form. Its application to NET performance has been simulated and results are summarised in Figure 9; it justifies the 'avalanche' approach. It also demonstrates that having a maximum value of G equal to 0.10 will enable NET to achieve the specified targets (i.e. TSR = 1.27). A maximum value of G equal to 0.10 is achieved when the G value of sub systems AC, ABA, AG and AF are lowered to that of sub system AQ.

Another major consideration in the analysis of NET for performance reliability is the specification of an appropriate value for the tolerable down time (currently set at 24 hours). If the specified value is too small then it will result in rejected test runs which would have achieved the 80% availability target if they were allowed to continue. On the other hand if it is too great then too many test runs having availability of less than 80% will result. These test runs rould then have to be repeated. Figure 10 demonstrates the sensitivity of NET performance to the tolerable down time for the avalanche strategy cases considered in Figure 9. For all cases Figure 10 clearly shows that the initial value of 24 hours dselected for the tolerable down time results in a better TSR than for an alternative choice of 5 hours.

In the simulation of differing strategies which have been considered in this paper the same independent random number stream[8] has been used to characterize the reliability of each sub system. This ensures a consistent sensitivity analysis with respect to different variables of interest. However, the strategy that has resulted in a satisfactory value for TSR (i.e. TSR = 1.27 in Figure 9) should be studied further, to determine whether a distribution can be fitted to the predicted values of TSR. Using 50 differing independent random number streams, the strategy has been simulated 50 times. Normal probability plotting[9] has been used to study the predicted values of TSR. Figure 11 gives the cumulative frequencies of the TSR values where the percentage cumulative frequency, (Px_i) has been computed via the relationship

$$P(x_i) = \frac{\text{number of observations } x_i}{\text{total number of observations} + 1} \times 100$$

where x_i represents different values of TSR predicted by simulation. Data points in Figure 11 lie roughly on a straight line, which leads to the conclusion that they are approximately normally distributed. The value of TSR whose cumulative frequency is 50% equals 1.14 and the difference between the two values of TSR whose cumulative frequencies are 50% and 84% is 0.18. These values represent respectively the mean and standard deviation of the distribution and indicate that the strategy described earlier should enable NET to achieve the required targets for neutron production.

5. CONCLUSION

This paper has put forward an integral approach for analysing the performance and reliability of the Next European Torus. Hitherto, the work has mainly concentrated on the development of a tool for realistic examination of the prospects for NET to achieve the neutron production targets over the 7 years specified for the technology oriented phase. In future, the technique devised and presented should be exploited extensively to help:

(a) determine expectation values for differing parameters that affect NET performance

and

(b) establish the most effective way of improving reliability and operating parameters, which are technically and economically feasible, in order for NET to achieve the required targets.

REFERENCES

1. R. Bünde, Plant component identification scheme (PCIS) for NET (Version 3, April 1988), NET/IN/88/005.

2. R. Bünde, Reliability and availability assessment during the design of the Next European Torus, 12th Symposium on Fusion Technology, Monterey, Cal., U.S.A. 1987.

3. K. Chaharbaghi, Introducing DSSL II, CROBAS Technical Report RCR/86/2, Imperial College of Science and Technology, 1986.

4. K. Chaharbaghi and B.L. Davies, Manufacturing systems simulation using DSSL, CAD, 18(5), 242-256, 1986.

5. K. Chaharbaghi and B.L. Davies, The analysis of flexible manufacturing systems for reliability. Int.J.Advanced Manufacturing Technology, 1(4), 79-100, (1986).

6. K. Chaharbaghi, B.L. Davies and H. Rahnejat, Application of simulation in planning the design of flexible assembly systems. Proc. of the Fourth Conf. on U.K.Research in Advanced Manufacture, I.Mech.E., (1986).

7. K. Chaharbaghi, B.L. Davies, H. Rahnejat and P.J. Dobbs, An expert system approach to discrete-change simulation, Int.J. of Operations and Production Management, 8(2), 14-34 (1988).

8. K. Chaharbaghi and B.L. Davies, Numerical evaluation of a multi-stream pseudo-random number generator, Applied Mathematical Modelling, 11(3), 219-228, (1987).

9. C. Chatfield, Statistics for technology, (Chapman and Hall, 1983).

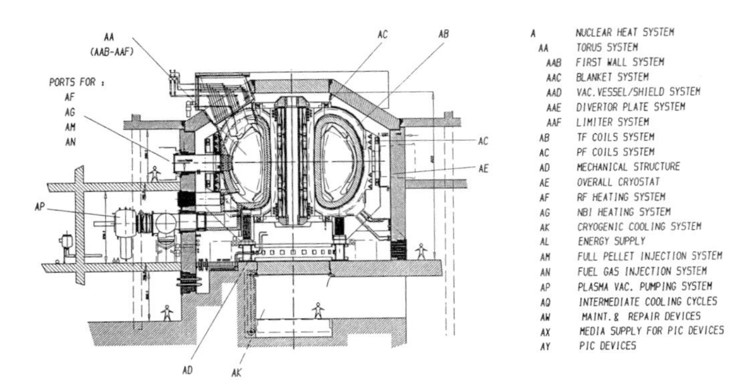

Fig. 1 (a) Breakdown of nuclear heat system

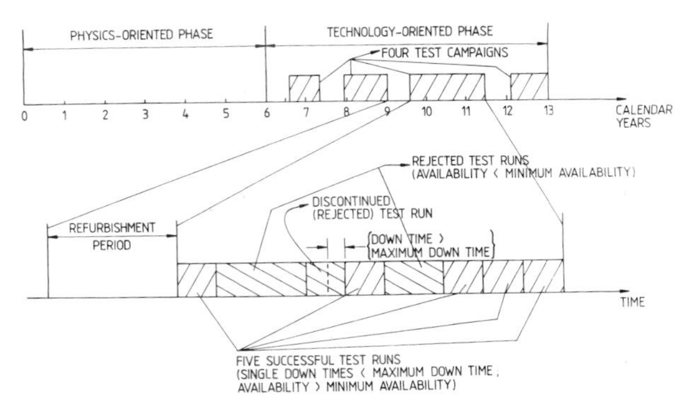

Fig. 2 Time chart for physics and technology oriented phases of NET

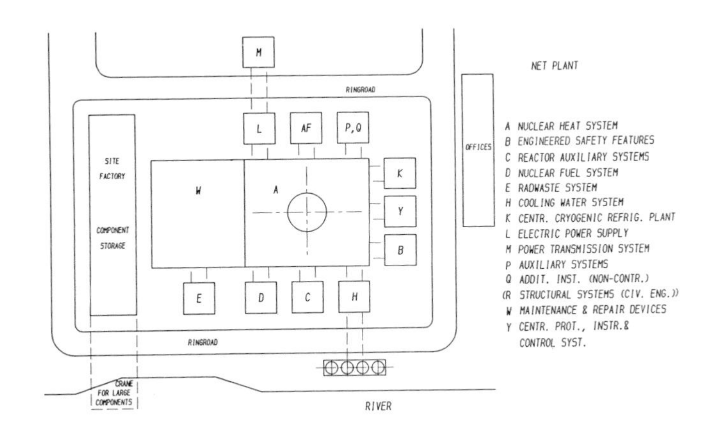

Fig. 1(b) Breakdown of NET plant into functional sections (A-Y)

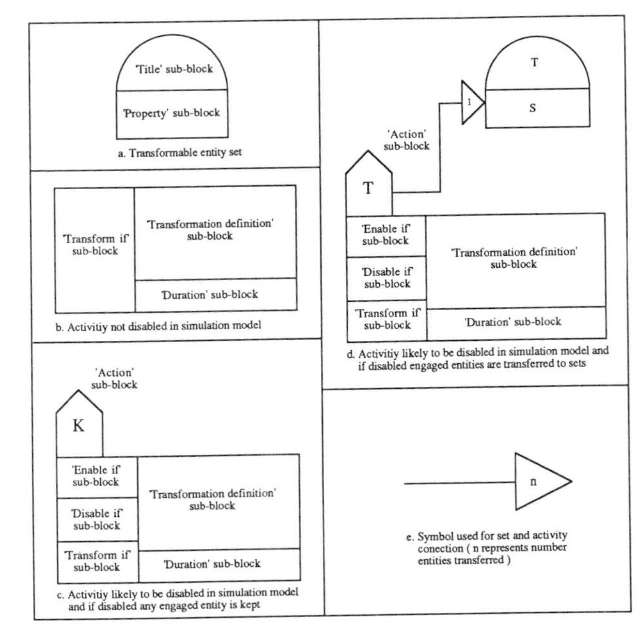

Fig. 3 Symbols used in construction of transformation logic diagram for NET

Table 1 Subsystem reliability data

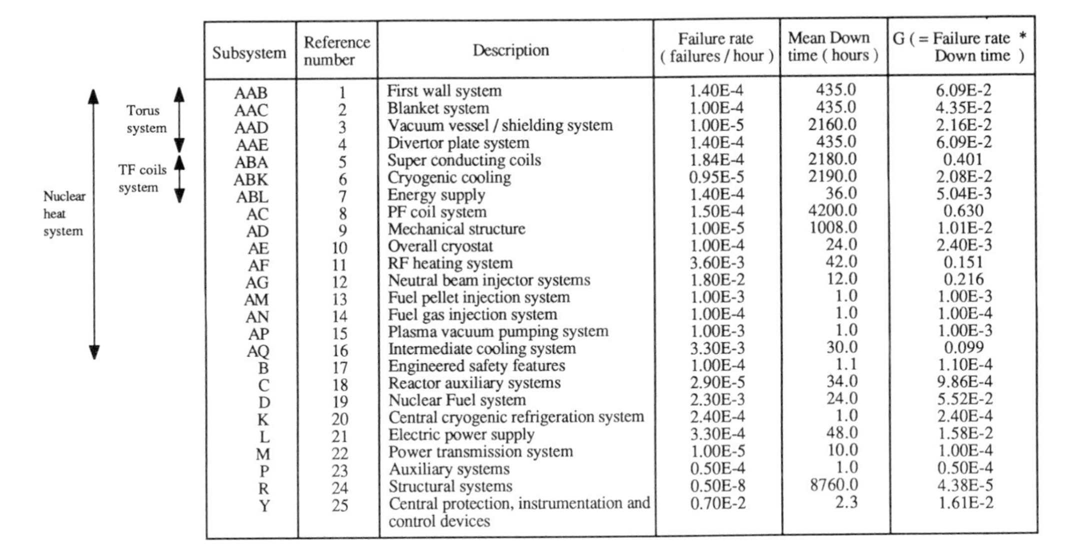

Subsystem	Reference number	Description	Failure rate (failures / hour)	Mean Down time (hours)	G (= Failure rate * Down time)
AAB	1	First wall system	1.40E-4	435.0	6.09E-2
AAC	2	Blanket system	1.00E-4	435.0	4.35E-2
AAD	3	Vacuum vessel / shielding system	1.00E-5	2160.0	2.16E-2
AAE	4	Divertor plate system	1.40E-4	435.0	6.09E-2
ABA	5	Super conducting coils	1.84E-4	2180.0	0.401
ABK	6	Cryogenic cooling	0.95E-5	2190.0	2.08E-2
ABL	7	Energy supply	1.40E-4	36.0	5.04E-3
AC	8	PF coil system	1.50E-4	4200.0	0.630
AD	9	Mechanical structure	1.00E-5	1008.0	1.01E-2
AE	10	Overall cryostat	1.00E-4	24.0	2.40E-3
AF	11	RF heating system	3.60E-3	42.0	0.151
AG	12	Neutral beam injector systems	1.80E-2	12.0	0.216
AM	13	Fuel pellet injection system	1.00E-3	1.0	1.00E-3
AN	14	Fuel gas injection system	1.00E-4	1.0	1.00E-4
AP	15	Plasma vacuum pumping system	1.00E-3	1.0	1.00E-3
AQ	16	Intermediate cooling system	3.30E-3	30.0	0.099
B	17	Engineered safety features	1.00E-4	1.1	1.10E-4
C	18	Reactor auxiliary systems	2.90E-5	34.0	9.86E-4
D	19	Nuclear Fuel system	2.30E-3	24.0	5.52E-2
K	20	Central cryogenic refrigeration system	2.40E-4	1.0	2.40E-4
L	21	Electric power supply	3.30E-4	48.0	1.58E-2
M	22	Power transmission system	1.00E-5	10.0	1.00E-4
P	23	Auxiliary systems	0.50E-4	1.0	0.50E-4
R	24	Structural systems	0.50E-8	8760.0	4.38E-5
Y	25	Central protection, instrumentation and control devices	0.70E-2	2.3	1.61E-2

Table 2 Operating requirements for each test campaign

Operation number	Description	Duration (hours)	Subsystems that must be fully operational
1	Shutting down for refurbishment	168	—
2	Scheduled maintenance, repair and checking	5208	—
3	Filling up appropriate systems with working media	168	AAB, AAC, AAD, AAE, ABK, AE, AQ, B, C, D, K, L, M, P, R and Y (Subsystem Group 1)
4	Cooling down central cryogenic plant, central valve box, TF coils, PF coil, structure, cryostat and pellet injection system	504	AAB, AAC, AAD, AAE, ABK, AE, AQ, B, C, D, K, L, M, P, R, Y, AF, AG, AM and AN (Subsystem Group 2)
5	Vacuum pumping of plasma and RF heating system and operating torus system and, main and intermediate cooling systems	336	AAB, AAC, AAD, AAE, ABK, AE, AQ, B, C, D, K, L, M, P, R, Y, AF, AG, AM, AN and AP (Subsystem Group 3)
6	Operating TF super conducting coils and energy supply	12	AAB, AAC, AAD, AAE, ABK, AE, AQ, B, C, D, K, L, M, P, R, Y, AF, AG, AM, AN, AP, ABA, ABL and AD (Subsystem Group 4)
7	Operating PF coil and energy supply, equilibrium control coils, RF heating, cooling and gas supply, and gas puffing inlet	156	AAB, AAC, AAD, AAE, ABK, AE, AQ, B, C, D, K, L, M, P, R, Y, AF, AG, AM, AN, AP, ABA, ABL, AD and AC (Subsystem Group 5)
8	Operating RF heating generator, transmission, launching and energy supply systems, and fuel pellet injector	0.00417	AAB, AAC, AAD, AAE, ABK, AE, AQ, B, C, D, K, L, M, P, R, Y, AF, AG, AM, AN, AP, ABA, ABL, AD and AC (Subsystem Group 5)
9	Test run 1	200	AAB, AAC, AAD, AAE, ABK, AE, AQ, B, C, D, K, L, M, P, R, Y, AF, AG, AM, AN, AP, ABA, ABL, AD and AC (Subsystem Group 5)
13	Test run 5	200	AAB, AAC, AAD, AAE, ABK, AE, AQ, B, C, D, K, L, M, P, R, Y, AF, AG, AM, AN, AP, ABA, ABL, AD and AC (Subsystem Group 5)

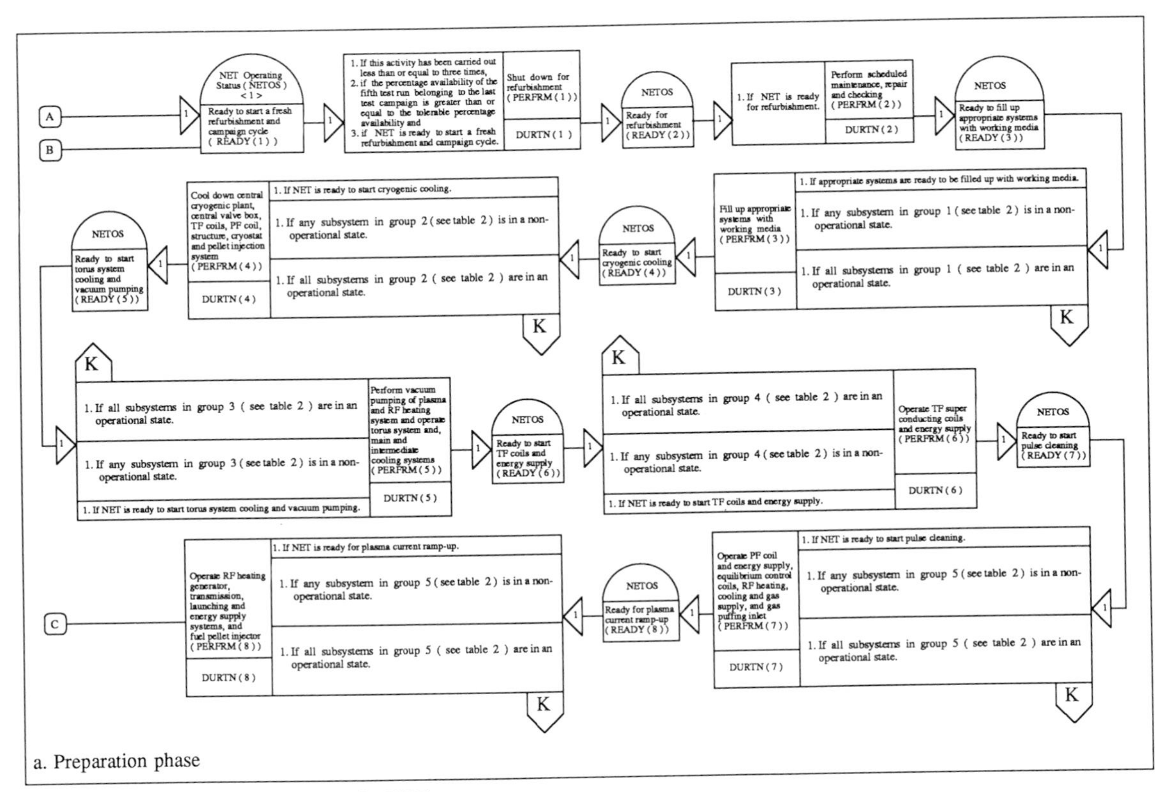

Figure 4. Transformation logic diagram for NET

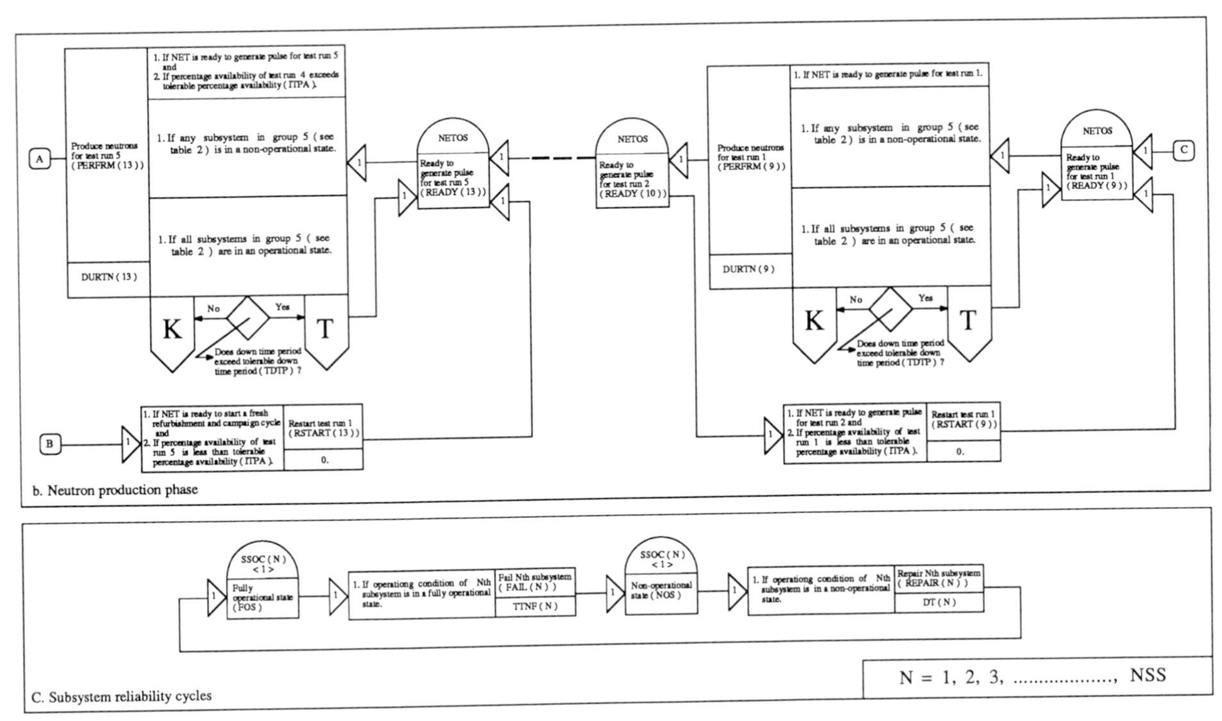

Figure 4. (Continued)

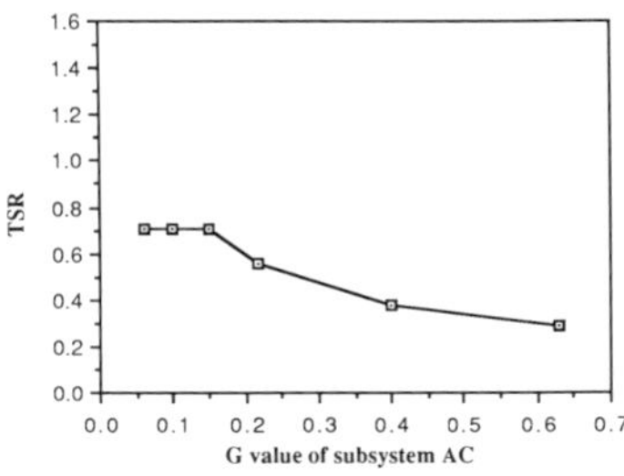

Fig. 5 Effect of reducing failure parameter G for subsystem AC, upon TSR

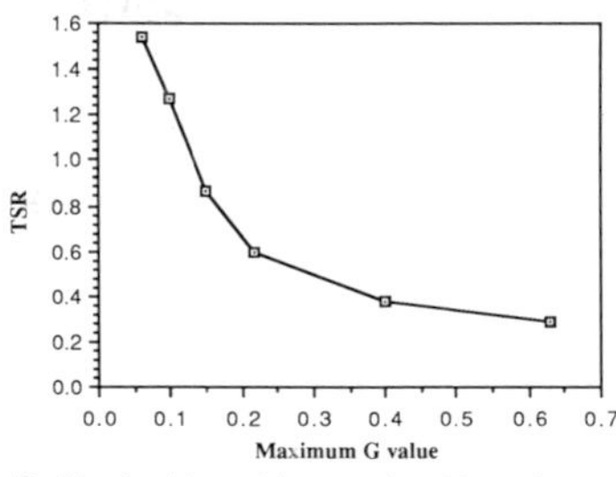

Fig. 9 Illustration of the application of systematic subsystem performance assessment to NET

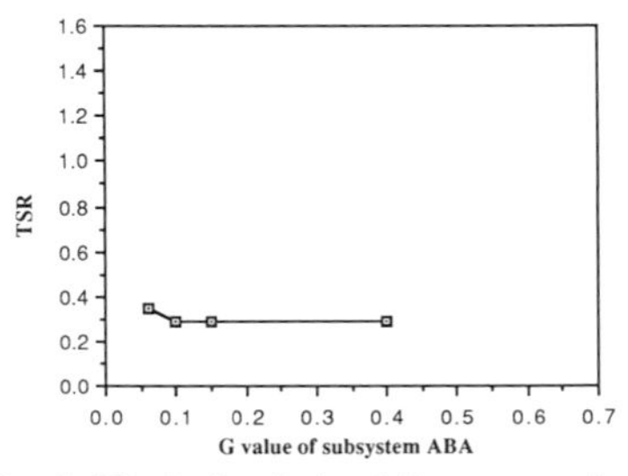

Fig. 6 Effect of reducing failure parameter G for subsystem ABA upon TSR

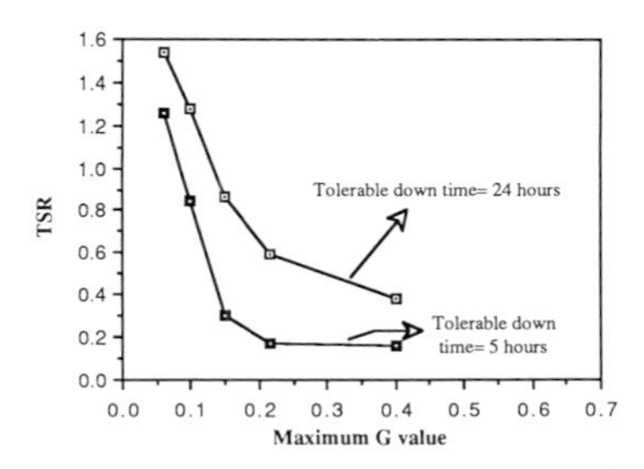

Fig. 10 Sensitivity of NET performance to tolerable down time

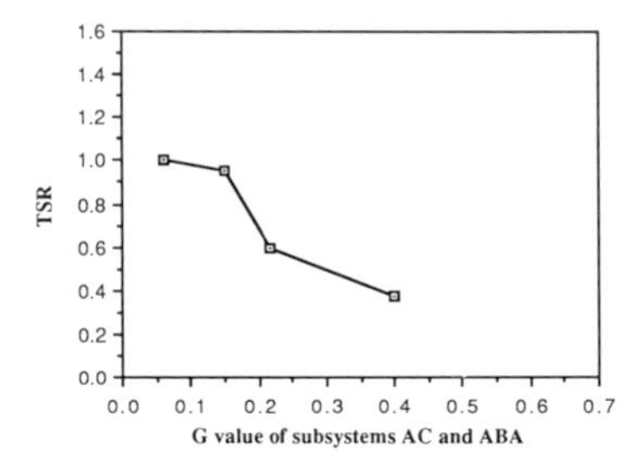

Fig.7 Effect of simultaneous reduction in G for subsystems AC and ABA, upon TSR

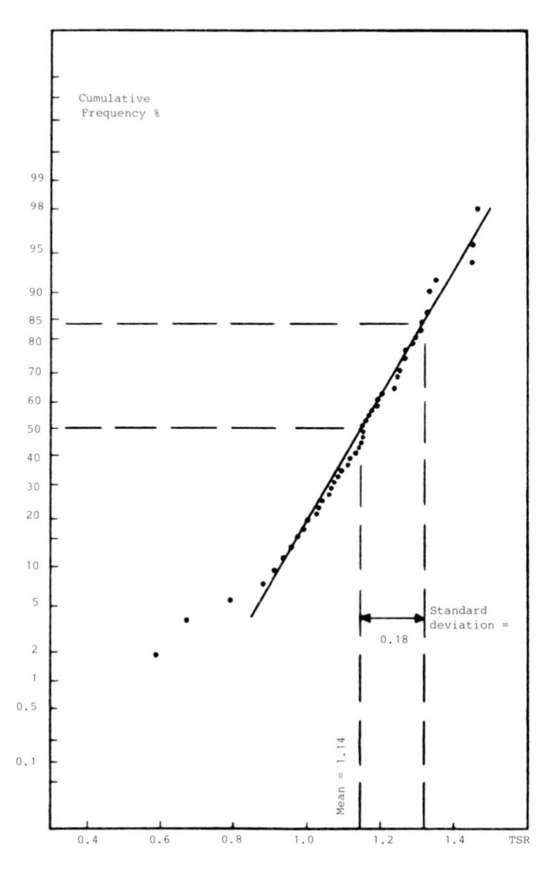

Fig. 11 Cumulative frequency distribution of TSR values, using independent random number streams

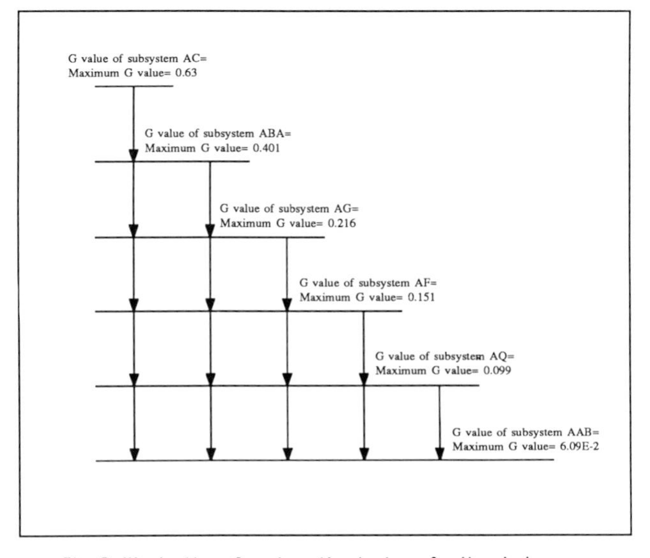

Fig. 8 Illustration of systematic strategy for the study of the influence of subsystem performance upon TSR

FUSION TECHNOLOGY 1988
A.M. Van Ingen, A. Nijsen-Vis, H.T. Klippel (editors)

DESIGN,TRANSIENT ANALYSIS AND AVAILABILITY ANALYSIS FOR THE NET COOLING SYSTEM

J. STICH[1], B. LIBIN[2], R. BUENDE[2]

[1] Technische Universitaet Muenchen, now KSB AG, D-8570 Pegnitz, Federal Republic of Germany
[2] The NET Team c/o Max-Planck-Institut fuer Plasmaphysik, D-8046 Garching, Federal Republic of Germany

A cooling system for NET is designed for separate cooling trains for first wall together with divertor, outboard blanket, inboard blanket, and vacuum vessel / shield. A thermal transient analysis for pulsed operation shows that the thermal inertia of the coolant and the materials masses of the cooling system reduce the temperature changes in the primary cycles such that no specific thermal storage is required. A Failure Modes, Effects and Criticality Analysis (FMECA) results in a mean expectation value for the cooling system availability of about 92%.

1. INTRODUCTION

Since a preliminary design of the NET cooling system [1] and corresponding reliability and availability (R&A) analyses [2],[3] were performed the NET design has been reviewed and the reqirements for the cooling system have become more specific. The earlier analyses had shown that the cooling system, although being conventional in technology, in context with a fusion machine might contribute a non negligible percentage to the unavailability of the plant. Therefore it was necessary to update the design, to consider the question of necessity of a thermal storage to cope with the pulsed operation, and to redo the R&A analysis now applying the procedure of the Failure Modes, Effects and Criticality Analysis (FMECA) which meanwhile had been introduced as the basic element of the NET R&A programme. This work is described in detail in [4], its highlights are presented in this contribution.

2. NET COOLING SYSTEM

The system is designed for the option of pressurized water cooling of the in-vessel components at low temperatures as described in [5]. As far as the blanket is concerned it comprises only the shielding blanket. The cooling of breeding blanket test segments has to be considered separately in context with the R&A analyses of these segments. Hence, their unavailability contribution has still to be added.

The design is based on the power input distribution to the various in-vessel components as given in tab. 1 [6].

Table 1. Input power distribution to in-vessel components (values in MW_{th})

	outboard	inboard	total
First wall (panel & box)	140	60	200
Divertor plates		80	80
Shielding blkt	350	120	470
Vac. vessel/shld	50	50	100
Total NET	540	310	850

The cooling circuits are shown in fig. 1 where temperatures, pressures and throughputs are given for each cycle. The cooling system is mainly characterized by the following features:

* The in-vessel components are grouped into first wall together with the divertor plates (cooled by two parallel cycles), inboard blanket, outboard blanket (cooled by two parallel cycles), and vacuum vessel/shield. Each of these is cooled by a separate cooling train which comprises primary cycle, secondary cycle and water cooling system. Each water cooling system has a separate water intake, the water discharges to river and/or cooling tower are combined. The separation has the advantage that a failure in one of the trains does not cause failing of other trains.

* Each primary cycle includes an emergency heat exchanger which transfers in a separate secondary cycle up to 2% of the

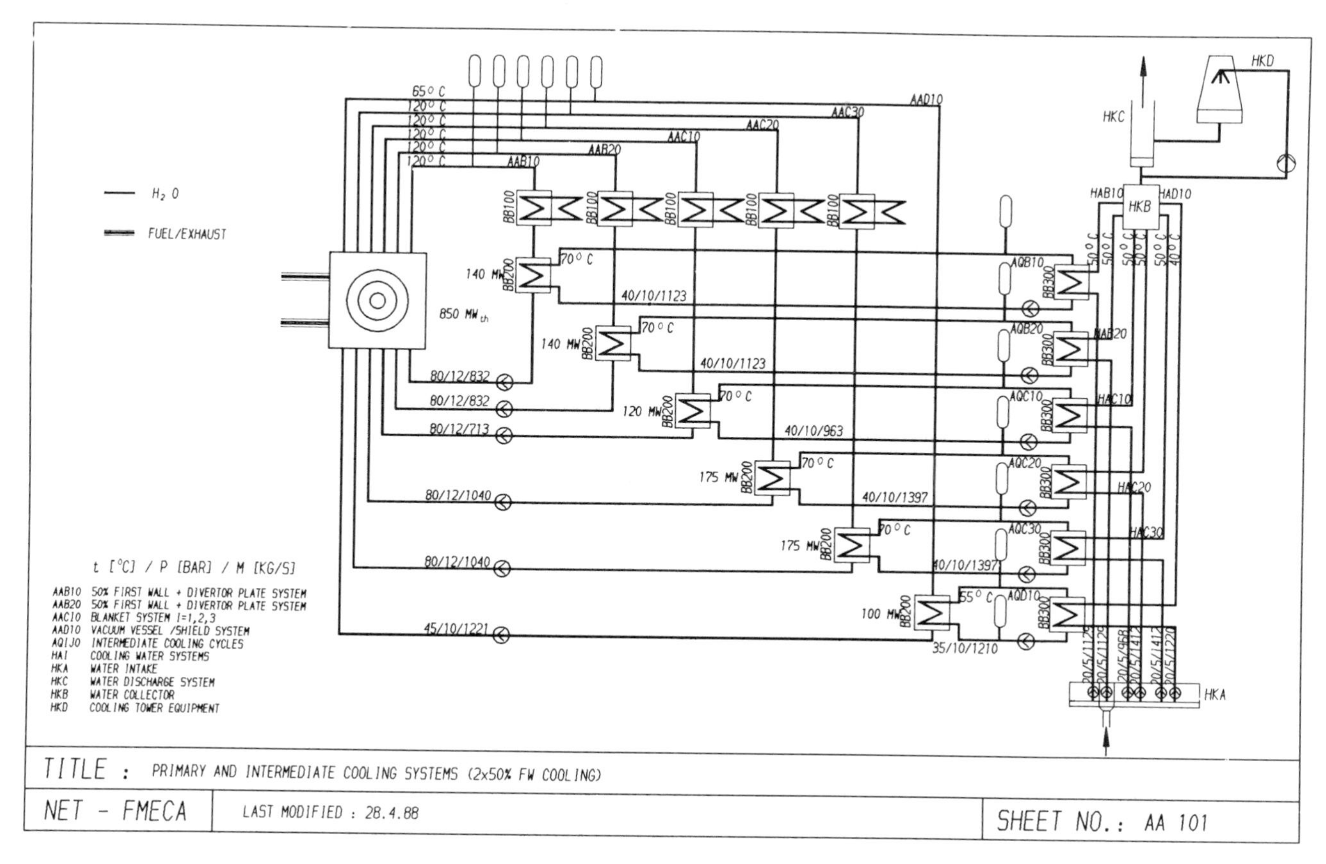

FIGURE 1
NET cooling system

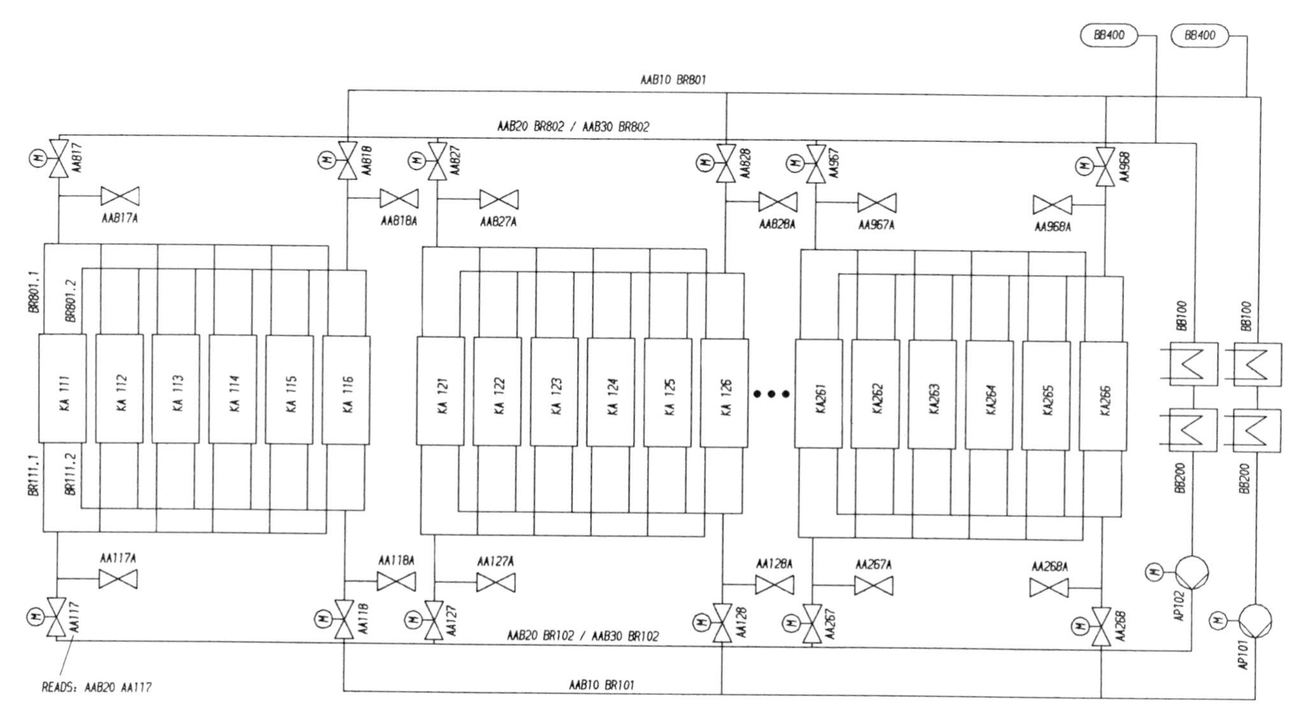

FIGURE 2
Cooling scheme of first wall

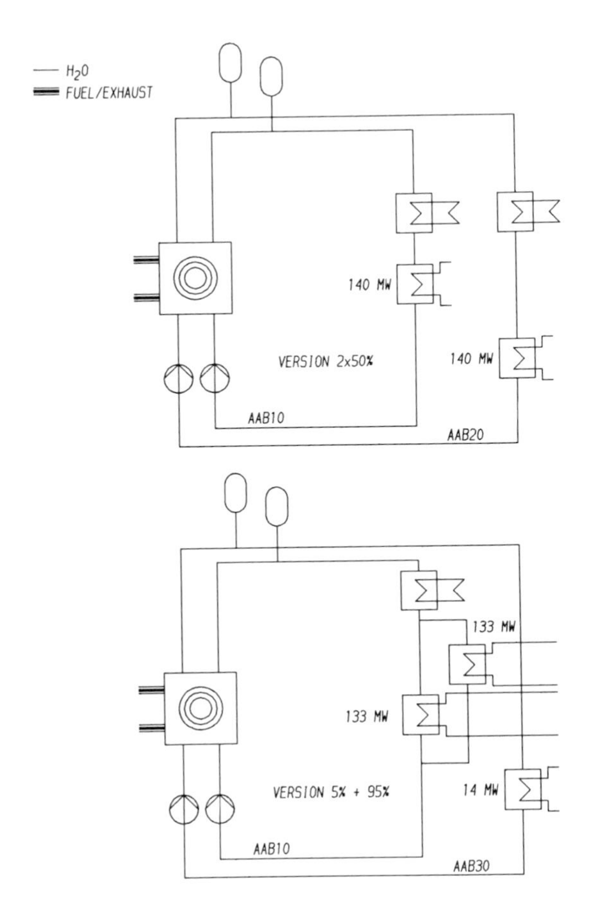

FIGURE 3
Alternative first wall cooling schemes

thermal power to an aircooled heat exchanger. The components of these cycles are arranged such that this power will be transferred to the outside by natural convection also in case of a loss of primary cycle pumping power supply.

The primary cooling scheme of the first wall is represented in more detail in fig. 2. Each of the 16 sectors (only the first two and the last one are shown) of the NET machine comprises 6 segments which are connected to two independent parallel cooling cycles for the purpose of redundancy in the cases of LOFA and LOCA. For the power distribution among these two cycles two options are under consideration (fig. 3): 2 * 50% or 95% + 5%. As was shown in [4] this distribution has only a minor influence on the availability of the plant due to the very similar number of vital components. Hence the decision for one of the two options will be dominated by safety considerations.

3. THERMAL TRANSIENT ANALYSIS

The primary cooling cycles of the first wall are mostly affected by the power input variations due to the pulsed operation of the machine and, hence, would be those which require in first order thermal storage. Therefore the transient analysis was performed for this cycle. The target of the analysis was to determine the temperature of the coolant at the outlet of the first wall and to decide on the necessity of thermal storage on the basis of the size of the occurring temperature variations.

A set of differential equations was set up from the energy balances and the heat flux conditions for the graphite tiles, the steel first wall, the cooling pipes, and the cooling medium itself. The first wall coolant outlet temperature thus was determined as a function of the coolant inlet temperature, the surface heat flux, the volumetric heating, the geometry, the coolant mass flow, and time. The calculations were based on the nominal burn time of 200s and an off-burn time of 70s, the geometry of the first wall was taken as described in [5], the thermodynamic data of the cooling cycle are those given in fig. 1.

The result is shown in fig. 4. Already after three pulses the outlet temperature has reached a quasi-steady state the temperatures varying between 93 deg C and 67 deg C. This corresponds to about 25% of that temperature which would have been reached in the case of real steady-state i.e. in non-pulsed operation. These fluctuations are sufficiently low so that no thermal storage is needed.

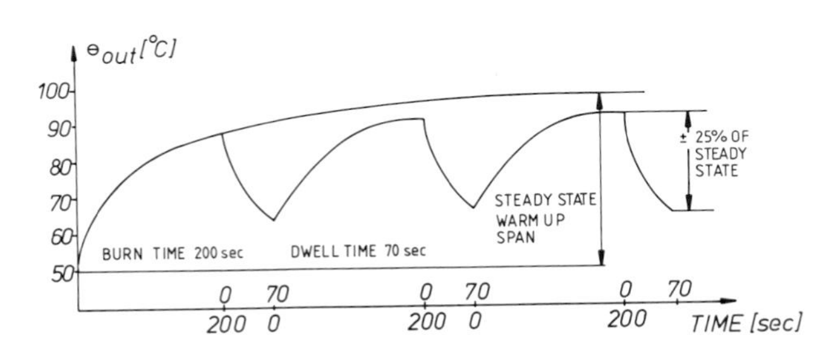

FIGURE 4
First wall coolant outlet temperature as a function of time

4. FAILURE MODES, EFFECTS AND CRITICALITY ANALYSIS (FMECA) OF THE COOLING SYSTEM

The overall NET reliability and availability (R&A) programme, the procedure and status of which are being described in [7], requires to perform so-called FMECA for all plant components partly for identifying the most effective measures for improving

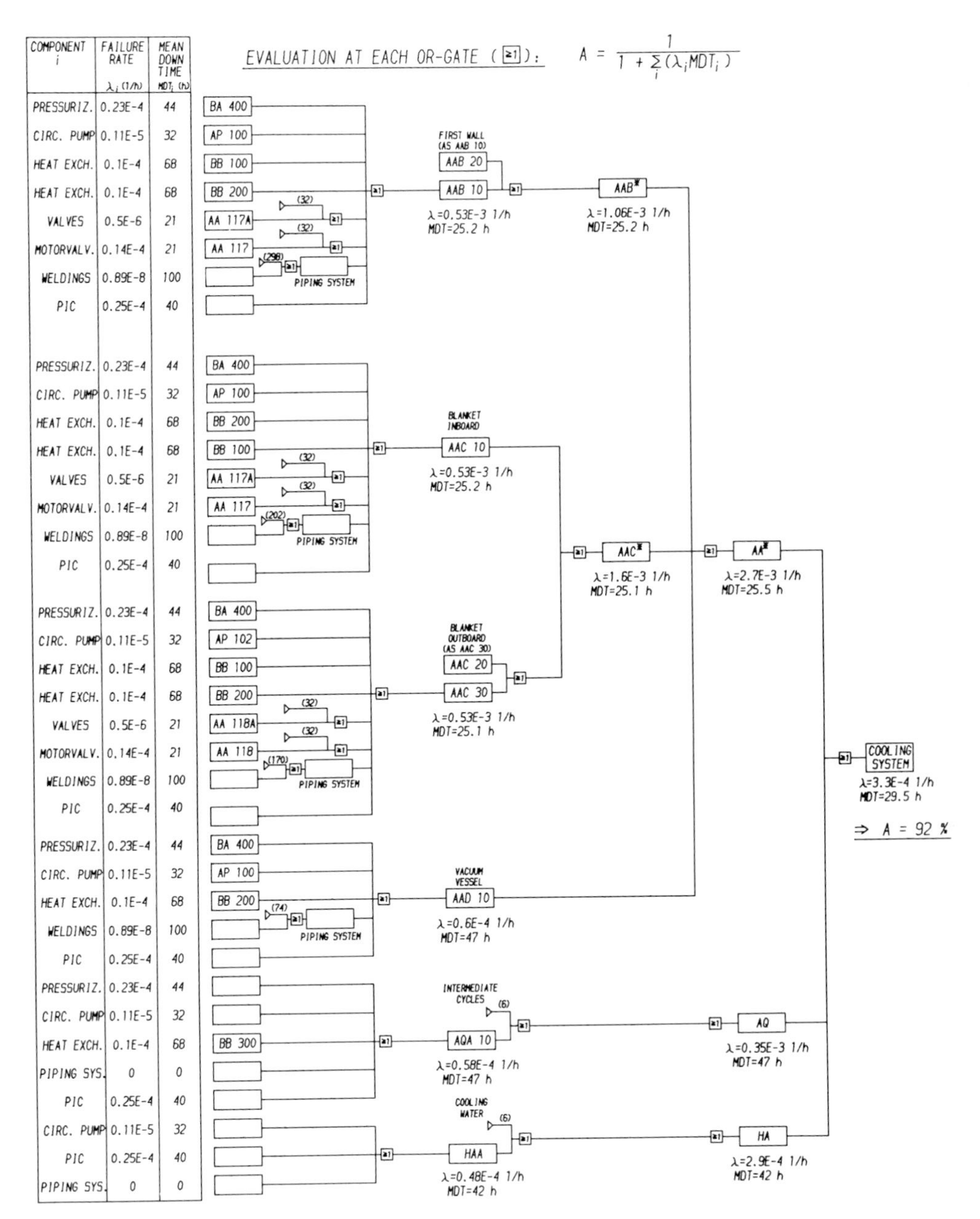

FIGURE 5
Cooling system fault tree

R&A as early as possible in the design process and partly for verifying components R&A assessments which have been made, indicating that the unavailability contribution of a component can be expected to be low. Although the cooling system applies conventional technology earlier studies ([2], [3]) indicated that the many parallel cooling cycles envisaged, the considerable number of headers required by the breakdown of the in-vessel components in sectors and segments, and the possibly large number of valves represent at least the potential for a non-negligible contribution of the cooling system to plant unavailability.

The steps of performing an FMECA are: Functional analysis, set-up of an outage logic, data assessments of subcomponents, fault tree evaluation, and the identification of reliability actions.

The *functional analysis* of the overall cooling system is represented at least in rudimentary form by the circuits shown in fig. 1 and 2. Figure 1 contains also basic information on the operating conditions and fig. 2 which is valid for the first wall primary cooling cycle thus being representative for the degree of detail required to sufficiently characterize all single cycles, i.e. identifying all basic components like pumps, valves, heat exchangers, tanks and pipes.

The *outage logic* is represented by a fault tree (fig. 5). It links the occurrence of a basic component failure, a so-called primary event, to the outage of the cooling system, the so-called top event. The latter then in turn causes an outage of NET. To simplify the later evaluation the fault tree was kept as a series system, thus containing only OR-gates as logic connections.

The *data assessment of subcomponents* was fully based on the extented operating experience with cooling cycles in Pressurized Water Reactor (PWR) power plants [2], [3]. The data of [2] are based on world wide experience, whereas [3] specificly refers to French PWR plants.

The series system *fault tree evaluation* is also represented by fig. 5. It results in a failure rate of $3.3\ 10^{-3}$ per hour and a mean down time of the entire NET plant due to such a failure of about 30 h representing a cooling system availability of 92 %. These values are to be considered as expectation values based on PWR experience transferred to the NET cooling cycles which comprise very similar components, however, in a different arrangement.

The identification of *reliability actions* focusses on the largest contributors to unavailability, the primary cycles. They contribute about 70 % of the total cooling system unavailability what can easily be seen from the comparison of the products of failure rate and mean down time for primary (AA), secondary (AQ) and cooling water (HA) cycles in fig. 5. In further going backwards in the primary cycle part of that fault tree it becomes evident that the largest contributing basic components are the motor-operated valves. Accordingly reliability improving actions should refer to the reduction of failure rate of these components and to a reduction of their number.

ACKNOWLEDGEMENT

The authors gratefully acknowledge the very helpful informations, discussions and the support by M. Chazalon, J. Raeder and G. Vieider, all members of the NET Team.

REFERENCES

[1] R. Buende, Preliminary design of a reference cooling system for NET, NET/IN/85-035, 1985.

[2] J. Huber and P. Kafka, Evaluation of the data base for the "Deutsche Risikostudie Kernkraftwerke", Report GRS-A-1305, 1986.

[3] H. Djerassi and J. Rouillard, Availability of nuclear plant components of NET, Report CEA MC/100-284,1987.

[4] J. Stich, Design and availability assessment of the cooling system for NET, Diploma thesis, Technische Universitaet Muenchen, 1987.

[5] M. Chazalon et al., Next European Torus in-vessel components, Fusion Technology (1988) 82.

[6] G. Vieider, personal communication.

[7] R. Buende, Reliability and availability issues in NET, 15th Symposium on Fusion Technology, Utrecht, Sept. 1988, to be printed in Nuclear Engineering and Design / Fusion.

FUSION TECHNOLOGY 1988
A.M. Van Ingen, A. Nijsen-Vis, H.T. Klippel (editors)
Elsevier Science Publishers B.V., 1989

MACHINE UTILISATION AND OPERATION EXPERIENCE WITH JET FROM 1983

B.J. GREEN, P. CHUILON, B. NOBLE, R. SAUNDERS, D. WEBBERLEY

JET Joint Undertaking, Abingdon, Oxfordshire, OX14 3EA. UNITED KINGDOM.

The operation of JET commenced in June 1983 and is scheduled until the end of 1992. This seemingly long period is actually quite limited when compared with the time needed to implement and commission the planned machine enhancements, and pursue research and developments which result from the experiments. There is an ever-present urgency to make the best use of the machine. 1983-84 was a learning period and only in 1985 was it felt worthwhile to adopt double-shift day working. Data has been compiled and analysed for utilisation of the machine, delays in terms of time lost and systems involved, and number and frequency of machine pulses. This paper presents an overall picture of machine availability and utilisation. It describes the JET operational arrangements and the experience of system faults. Finally, it draws conclusions and identifies lessons learned which may be relevant to the next stage of fusion power development.

1. WHY IS THIS TOPIC IMPORTANT?

The cost of plant involved in the JET experiment is high and hence JET must be used as often and as successfully as possible to maximise the operation information (for both plasma and machine systems) in the limited life of the project.

JET is generating unique plasma information for the international fusion data base because of JET's ability to operate over a wide range of plasma conditions.

The machine systems information (capability, reliability) is required to assist in the design of future controlled fusion machines and in the assessment of reactor reliability, environmental impact and economics. Again JET's size and subsystem capabilities can make unique contributions. In particular, operation experience with JET will be important when considering realistic availability targets for the next step (e.g. NET, ITER etc).

2. WHAT IS A USEFUL MEASURE OF OPERATION PERFORMANCE?

In an experiment like JET, it is important to distinguish between operation performance and scientific programme performance. In this paper we will concentrate on operation performance. The chosen measure of this performance is the number of successful tokamak pulses (TS) achieved i.e. pulses for which all systems performed successfully. For a meaningful measure of performance these TS pulses must have been achieved in the implementation of an agreed scientific programme.

Each JET pulse is classified as follows:

TS =Successful tokamak pulse - all systems for the production of plasma were working, (account not taken of the quality of the discharge).

TF =Failed tokamak pulse - the specified plasma pulse was not achieved because of some system or operator fault.

CS =Successful commissioning pulse - typically dry run, single system power test. No attempt to produce a plasma.

CF =Failed commissioning pulse - the specified pulse was not achieved because of some system or operator fault.

The measure of scientific programme performance must be measured in some other way e.g. the achievement of specific goals in plasma parameters. For JET, the cumulative number of fusion neutrons and the $n_i \tau_E T_i$ value have been chosen.

3. IMPROVEMENT OF JET OPERATION PERFORMANCE

Since JET operation began on June 25 1983 there has been a continual review of performance and as the programme has progressed, certain measures have been taken for improvement. Basically there are two main ways to improve machine performance:

i) maximise the available operation time,

ii) reduce the delays in operation.

Relevant JET experience will now be described.

3.1 Time available for operation

JET began in a basic configuration and the strategy was to continually add systems to bring it up to its extended performance state ("phased exploitation"). Further, JET is an experiment and subsystems will be altered in the light of operation experience. Installation and commissioning of new and modified equipment takes some time. Thus, since 1983, the operation of JET has been broken by significant periods of shutdown (Fig.1). The relative time devoted to machine operation in 1983/84/85 was not great but improved markedly in 1986. Once again in 1987 the relative time devoted to machine operation was reduced (Fig 2).

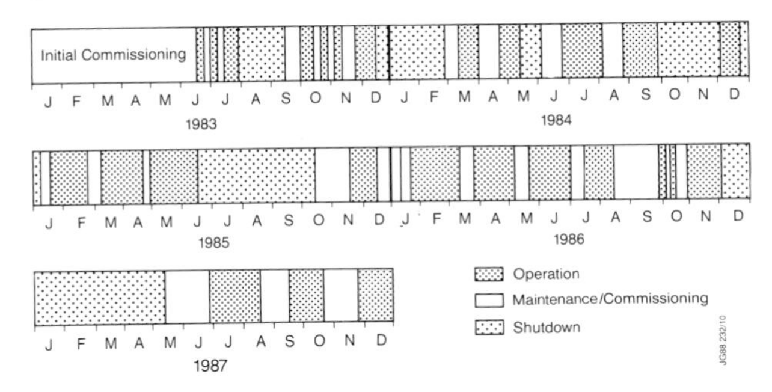

Fig.1 JET Operation Calendar 1983-1987

JET has sought to improve operation by having regular maintenance/recommissioning intervals between operation periods. The extent of plasma operation which is possible before a maintenance/ recommissioning period is required has not been systematically studied, however the pattern of 6 to 7 weeks operation broken by a 1 to 2 week period has been successful.

Often the installation and commissioning of new equipment takes place in this period and usually determines its duration. Sometimes the commissioning of new equipment can spill over into plasma operation periods and this reduces the time available for the experimental progr-

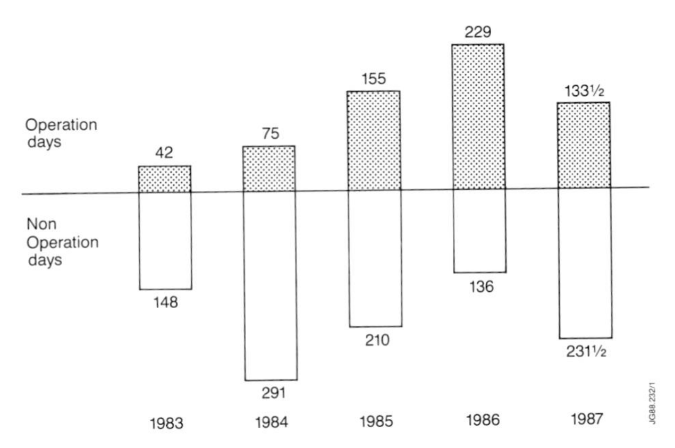

Fig.2 Numbers of Operation Days 1983-1987

amme. While all difficulties cannot be foreseen, the installation and commissioning of new equipment must be properly planned; and this certainly includes the timely preparation of commissioning procedures, implementation of an acceptable man/machine interface and proper operation documentation.

JET has also increased the time for machine operation by introducing double-shift day and weekend working. In both cases the JET Project has been somewhat constrained by the limited manpower available for operations. The improvements made are shown below:-

1983/84	Some weekend operation. Single-shift days (10 to 12 hrs). Up to 60 hr/week.
1985	Monday commissioning: single-shift day. Tuesday to Friday: double-shift (16 hr) days. 74 hr/week. Single-shift days for maintenance and commissioning.
1986	Monday to Friday: double-shift days. (80 hr/week). Some weekend operation.
1987	Double-shift days attempted for maintenance and commissioning. Some weekend operation. Monday commissioning reduced.

A significant part of each day can be lost to operation because the power supply re-energisation and isolation procedures are complex and time-consuming. Although they have been improved, they still take up 1½ to 2 hours of tokamak operation time. Nevertheless the average number of hours available for pulsing in a 16 hour double-shift day including 1 hour for meal breaks has been increased to nearly 13 hours (Fig. 3). This time is affected strongly by commissioning and recommissioning requirements following shutdowns.

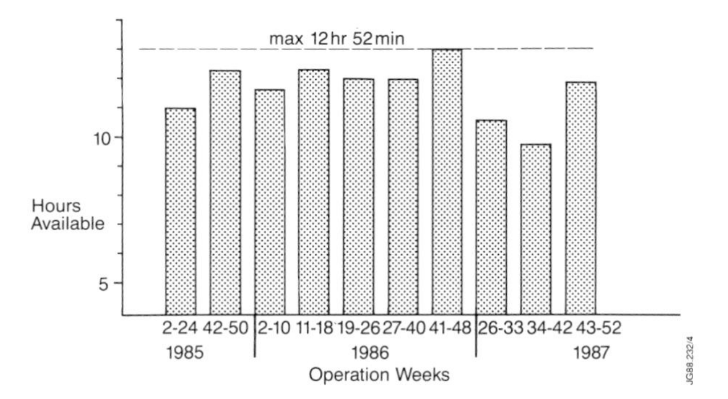

Fig.3 Average Daily Hours Available for Pulsing

3.2 Delays In Operation

Faults which occur during operation days are recorded by the Engineer in Charge of operations (EiC) in a shift report. They are identified by subsystem, any associated delay to operation is noted and the action taken or required is outlined. Not all system faults cause actual operation delay and some faults are "hidden" by others. This fault recording is continually being reviewed for improvement. Other fault reporting systems (e.g. Power Supplies and CODAS) are used primarily to specify remedial and maintenance work. These fault recording systems are narrower in scope than the EiC system but provide more detail for the subsystem involved as they include faults not "seen" by operations. The EiC fault reporting provides data for the analysis of frequency and delays due to subsystem faults. In this way the major sources of lost operational time can be identified and appropriate action defined.

An analysis of faults indicates that the delay is typically 20% of the total time available for operation (Fig.4). The analysis of the subsystems responsible for delays (Fig.5) clearly shows that two subsystems dominate i.e. Power Supplies and CODAS. Other systems which have significant fault delays are: Vacuum Systems, Neutral Beam, Operator and Water/Air.

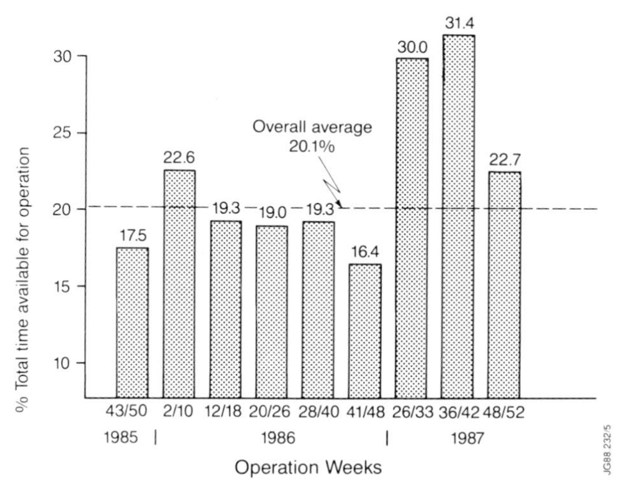

Fig. 4 Relative Time Lost due to All Faults

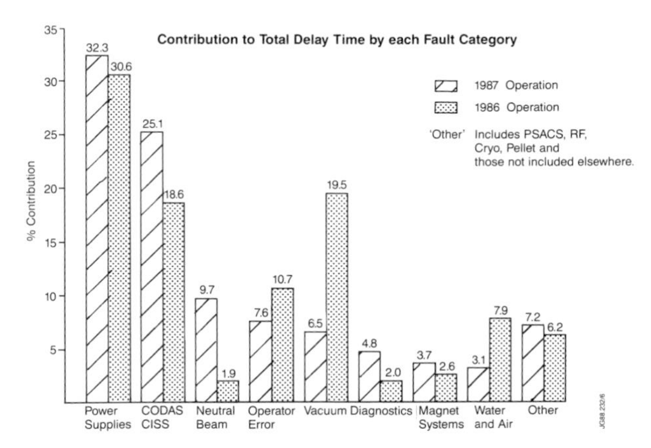

Fig.5 Relative Delays for Each Fault Category

An analysis of the number of faults for each subsystem (Fig.6) reveals a different picture in that CODAS and Operator categories dominate. This means that the average time per fault occurrence (Fig. 7) is quite small for the CODAS and Operator categories but is large for Power Supplies and for Vacuum Systems faults. This is because, in both of these categories, troubleshooting requires electrical isolation before the fault-finding can commence. In many cases the remedial work involved is lengthy (e.g. welding to repair vacuum leaks, replacement of electrical switchgear) and almost always needs testing before operation can be restarted.

Some measures have been taken to reduce the time lost in identifying and rectifying systems faults:

i) trouble-shooting procedures have been improved,

ii) up-to-date documentation has been provided,

iii) written procedures have been prepared.

The availability of 'expert' staff is also important - the JET situation where design staff has been involved with the installation, commissioning and operation of equipment has shown clear advantages over the industrial practice of separating the design, installation, commissioning and operation phases with handovers between different staff teams before each new phase.

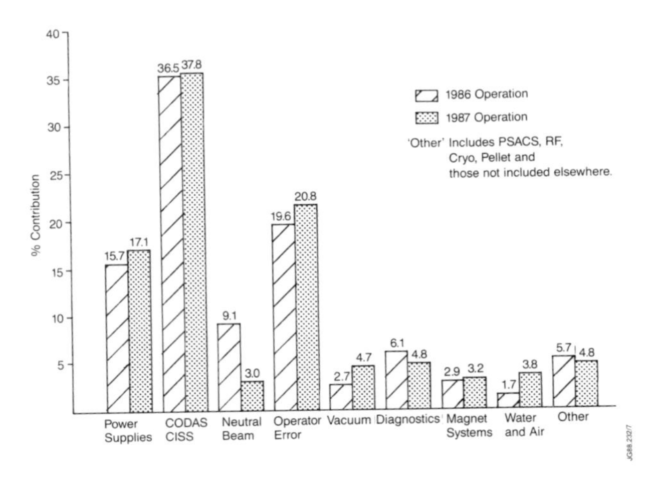

Fig.6 Relative Number of Faults per Category

It is difficult to quantify system operation reliability because of the continual changes and new modes of operation. However, an analysis of JET power supply subsystem faults (M. Huart et al SOFT 14) showed how regular maintenance has increased reliability. The reduction in frequency and severity of faults in some subsystems clearly demonstrates the benefits of:

a) simple system control and protection,

b) more automation of operator activities,

c) improved man/machine interface,

d) improved operator procedures,

e) stricter control of system interventions.

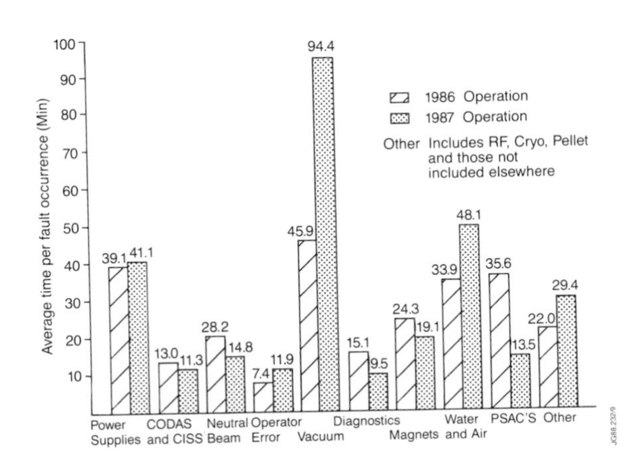

Fig.7 Average Time per Fault for Each Category

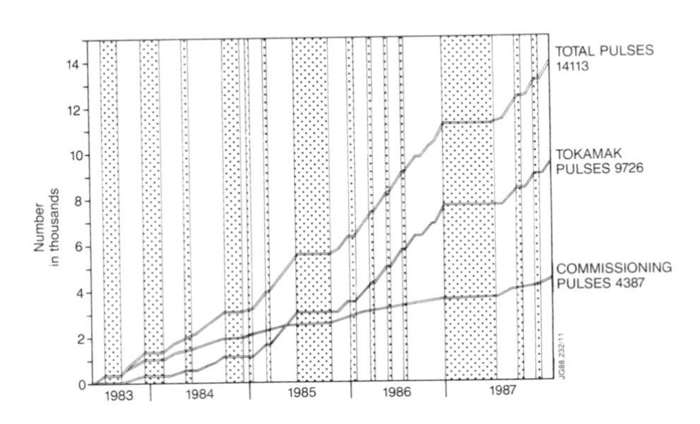

Fig.8 Cumulative Totals of JET Pulses

4. PULSE PERFORMANCE

The growth of the cumulative numbers of JET tokamak and commissioning pulses is shown in Fig.8. Overall, the relative amount of commissioning has decreased but the average number of pulses per year has not yet reached the design target of 5000. The improvement of machine operation discussed above can be seen in Fig.9. The number of tokamak pulses per week has increased and the weekly number of successful tokamak pulses is regularly between 100 and 150.

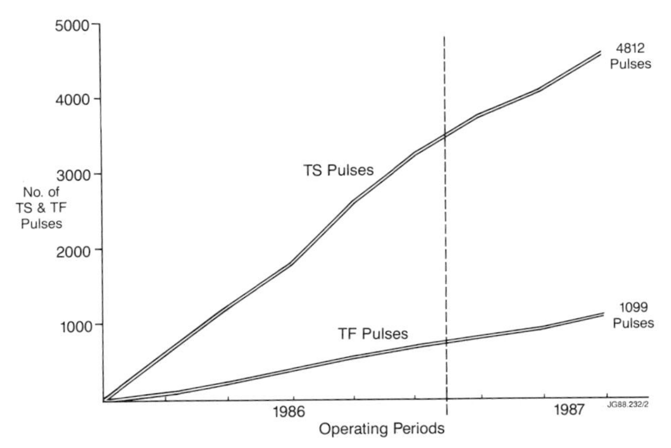

Fig.9 Tokamak Pulses for 1986/87

5. CONCLUSIONS

After a learning period (1983/84), routine double-shift day JET operation began in 1985. Operation performance has been continually improved by:

i) increasing the time devoted to operation e.g. double-shift days, weekend work, optimum preparatory and operation activities

ii) reducing the time lost due to system faults e.g by design simplification, control of interventions, regular maintenance, optimisation of trouble-shooting procedures, automation of many operation activities

iii) optimising the organisation and planning of operation (e.g. use of back-up programmes) and closer co-ordination of installation, commissioning and operation activities.

Further improvement could be made in all three of the above areas.

6. LESSONS FOR THE NEXT STEP

The Next Step (NET, ITER etc) will be an experiment like JET only more complex, with even greater demands on the "quality of operation performance". Certain experience gained in the operation of JET appears relevant:-

i) in the setting of availability targets, adequate time should be allowed for:

a) upgrading systems (if, like JET the phased exploitation strategy is used)

b) maintenance and repair (note the impact of radioactivity)

c) regular recommissioning.

ii) adequate manpower is required; where possible design staff should be used for installation, commissioning and operation. Proper training of staff (and system documenation) are required.

iii) at the system design stage, proper thought should be given to commissioning and operation (in particular the man/machine interface e.g. machine protection interlocks, alarm handling, control software). Autonomous control is important as it allows commissioning in parallel.

iv) strict configuration control should be applied (to ensure access for remote handling and to facilitate trouble-shooting).

v) design attention should be given not only to new technology and novel items but also to "standard" things e.g. piping, cabling, cranes, shielding elements.

vi) particular design attention should be given to systems which appear likely to cause the greatest delay e.g. the control, monitoring and data acquisition system and the power supplies; also the location and repair of vacuum leaks.

vii) system operation should be automated as much as possible but controls kept simple

viii) operation should be organised so that responsibilities are clearly defined.

ix) planning must be thorough so that machine operation time is optimised.

x) interventions on machine systems should be strictly controlled.

FUSION TECHNOLOGY 1988
A.M. Van Ingen, A. Nijsen-Vis, H.T. Klippel (editors)

APPLICATIONS OF RELIABILITY TECHNIQUES IN THE EARLY DESIGN OF AN INSTRUMENTATION AND CONTROL SYSTEM

Henrik AID, Jan COLLÉN, Kurt PÖRN, Stephen DINSMORE*

Studsvik AB, 611 82 Nyköping, Sweden

Probabilistic Availability Assessment methodology is being used to ensure that the availability targets defined for the Next European Torus (NET) are met with high probability. The present project involves an assessment of the unavailability contribution of the Protection, Instrumentation and Control (PIC) system which was defined as the plant instrumentation and the electronic devices used to collect, interpret and react to the signals from the instrumentation.

The first study, which included a preliminary design of a PIC system based on information from the computer-based PIC system at the Joint European Torus (JET) project, was performed in three phases. In the first phase, instrumentation for coherent units such a superconduction coils, cryopumps and radio frequency heaters was identified and the unit unavailabilities were calculated. The second phase quantification was carried out at the system level and included the expected number of first phase units and all local control PIC devices. The third phase included all the systems and the global PIC devices such as data transmission lines and the central computers.

The study was completed during 1986-87, and generated a structural overview of the instruments and the signals and control paths from all the plant equipment, and gave a first indication of potential unavailability drivers in the NET PIC system.

The objective of the second study, which is being undertaken during 1988-89, is to supplement and update the results of the original study, as well as to adjust the entire analysis to the specific architecture of the PIC system currently being developed at NET. A major task is to describe in more detail the characteristics and failure modes of each PIC device and to assess the corresponding failure rates and failure consequences.

The general experience of these studies is that a structured and well-documented tool such as provided by any systematic availability assessment is particularly important for new and very complex technical endeavours. This applies especially to projects where many different organisations contribute at various levels.

1. INTRODUCTION

An early oversight study[1] indicated that the Protection, Instrumentation, and Control (PIC) System could be a major contributor to plant unavailability in the Next European Torus (NET). In the light of the existence of a well-defined and operational computer-based PIC system at the Joint European Torus (JET), a more detailed study of the NET PIC system has been undertaken[2].

The PIC System was defined as the plant instrumentation and the electronic devices used to collect, interpret, and respond to the signals from the instrumentation (Figure 1). An instrument was defined as a device which transforms a physical quantity under observation to a usable electronic signal. An electronic device was defined as a device which inputs more than one instrument signal, interprets, transmits, and manipulates them; and

*Battelle Institute, Am Römerhof 35, D-6000 Frankfurt Am Main 90, Federal Republic of Germany

responds by generating control signals which are sent back to the plant's equipment and/or to other electronic devices.

Besides the availability assessment, the objective of the study was also to develop a preliminary design of the PIC system, to identify potential problems, and to suggest solutions. This required the identification and characterizations of the systems whose functions are so vital that signals from their instruments would interrupt the pulse cycle.

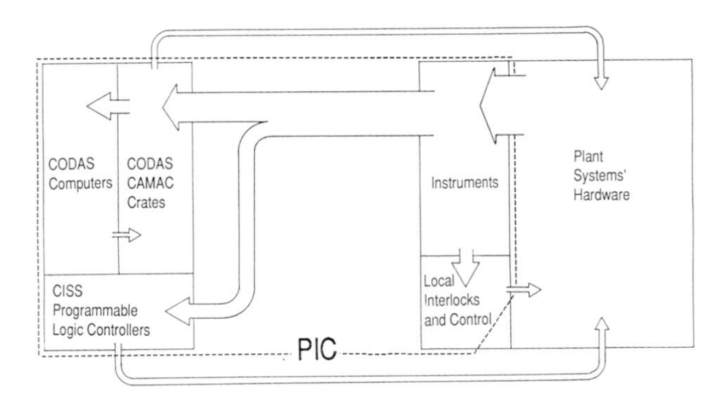

FIGURE 1
Schematic overview of plant parts and data links included in this study.

2. APPROACH OF THE STUDY

Initial identification of the sub-systems was performed using an event tree (Figure 2). Each of the pulse states requires that specific systems should be operational. These systems were identified by meeting with design engineers at NET and JET, by reviewing the main supervisor control program of JET's CODAS system, and by reviewing the input to JET's CISS safety subsystem. The systems identified were further reviewed to determine the required operational mode (the success criteria) and any required support system.

The information was used to model the PIC requirements of major components. The component models were used to build up system level models according to the specifications for the NET design. A clear distinction was made between system component failure and system instrumentation failure as illustrated in Figure 3. Terminology and labelling proved to be a problem during the course of the study. Terms such as device, equipment, component, and assembly are interchangable and consistency of use was difficult. Another reason was the fact that the PIC system, unlike other NET subsystems, is a plantwide system included in most other subsystems.

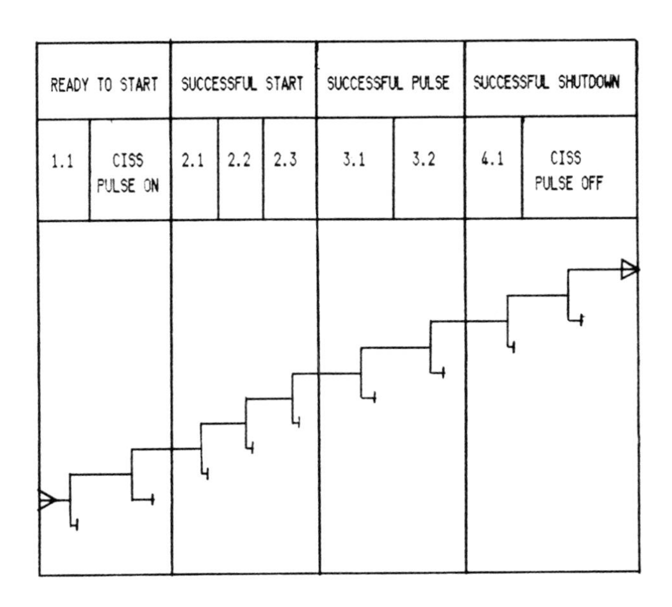

FIGURE 2
Event tree for a single pulse. Any deviation from the top path places the plant in an unavailable state.

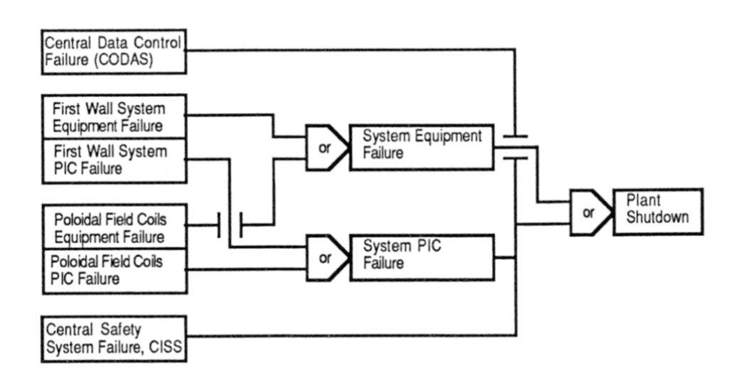

FIGURE 3
Example of system level subdivision into process and PIC equipment.

The work was organized in system tables where each unit was identified along with the physical parameters which must be monitored. These tables were subsequently turned into fault trees such as Figure 4. During quantification the number of independent points where the parameter must be monitored was estimated. Voting logic was not included in this study so the failure rate for unit monitoring was approximated as being the sum of the failure rates for single instruments.

3. DATA AND QUANTIFICATION

Two types of failure modes exist.

- Incorrect indication of an out-of-bounds condition when none exists (spurious operation).
- Failure to indicate an out-of-bounds condition when one exists.

To isolate the PIC system's contribution to plant unavailability the assumption was made that no other equipment would fail so only the first failure mode was of interest, i.e. spurious operation. It was further assumed that the instruments were set-up in the "fail safe" mode so that the vast majority of physical failures would produce spurious alarms. The often reported "all modes" failure rate was subsequently used for almost all PIC devices. Software and/or data transmission errors were not explicitly taken into account.

For the purpose of this study, a mean down time (MDT) of 1-3 hours per PIC device failure was used for the most common devices and 5-10 hours for complex.

For non-accessible instruments inside the Torus the MDT could be of the magnitude of a month. A general assumption on the provision of "adequate" redundancy or remote repair capability (resulting in a negligible contribution to the overall unavailability) was therefore made for such instruments.

Due to the lack of proven fusion technology as a reference source much work remains to be done with respect to data input. References 4 and 5 are examples of two data sources used in this study.

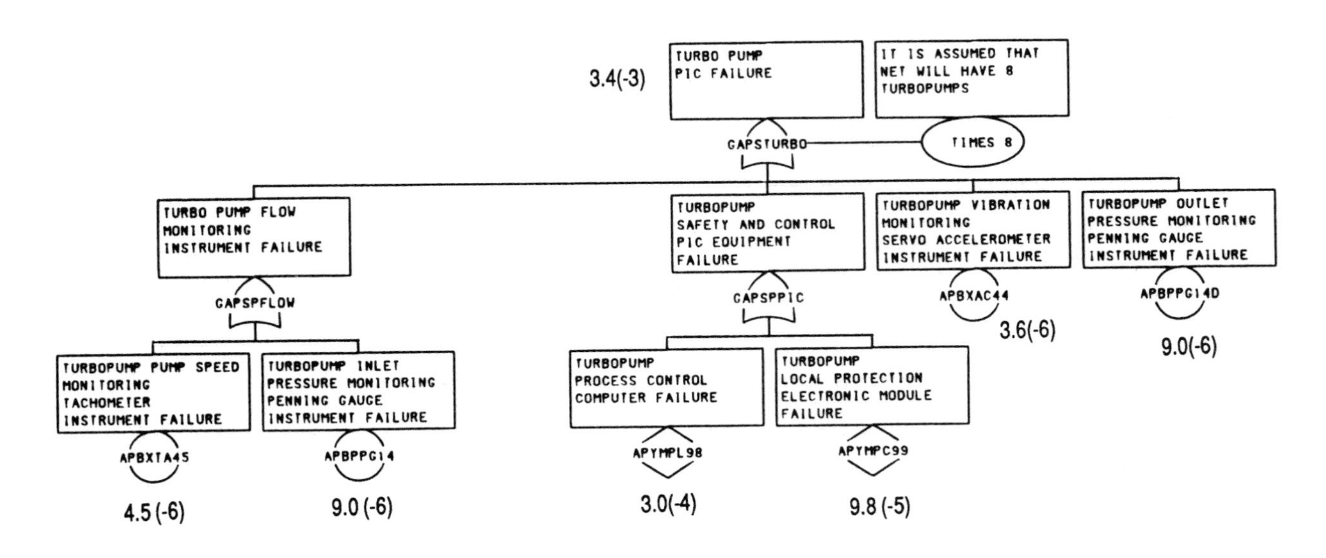

FIGURE 4
PIC fault tree for a turbo pump. Unavailabilities are given for all events.

The calculations used in this study are simple, straighforward, and in line with those discussed in Reference 1. The equations below are quite accurate up to 1 % unavailability and reasonably accurate up to 10 %.

λ_i = failure rate of the i'th component (/hr),

τ_i = down time associated with the i'th component's failure (hr), and

U_i = unavailability of component i.

If the down time of component i is a small fraction of the operation time under consideration, then U_i can be expressed as:

$$U_i = \frac{\lambda_i \tau_i}{1+\lambda_i \tau_i} \cong \lambda_i \tau_i \qquad (\lambda_i \tau_i < 0.01)$$

Defining a "unit x" as a series grouping of components, the unavailability of unit x (U_x) can be calculated as:

$$U_x = \sum_i \lambda_i \tau_i$$

where the subscript i ranges over the components in unit x. For the purpose of reporting and to facilitate subsequent calculations we define the following,

$$\lambda_{eq,x} = \sum_i \lambda_i$$

$$\tau_{eq,x} = \frac{U_x}{\lambda_{eq,x}}$$

where $\lambda_{eq,x}$ is the equivalent failure rate for unit x and $\tau_{eq,x}$ is the equivalent down time.

These equations were used to calculate the unavailabilities, equivalent failure rates, and equivalent down times given in Table 1.

TABLE 1
NET PIC Study Major System Results, Instrument and Electronic Device Failure

SYSTEM	Unavail-ability	Failure Rate (/hr)	Repair and Down Time (hr)
Central Control	3.9(-2)	3.6(-2)	1.1
Plasma Position and Control	3.1(-2)	3.5(-3)	8.9
Poloidal Field Power Supply	2.9(-2)	9.4(-3)	3.0
Toroidal Field Coil	1.5(-2)	5.9(-3)	2.6
Poloidal Field Coil	1.1(-2)	4.1(-3)	2.6
Radio Frequency Heating	7.5(-3)	2.5(-3)	3.0
Blanket Cooling	7.2(-3)	2.2(-3)	3.3
Vacuum Pumping	6.1(-3)	1.9(-3)	3.2
Cryostat and He and N2 Distribution	4.0(-3)	1.6(-3)	2.5
First Wall	3.8(-3)	9.9(-4)	3.8
Fuel Pellet Injection	2.7(-3)	9.9(-4)	2.7
Divertor Plate	2.1(-3)	7.0(-4)	3.0
Protection Interlocking	2.0(-3)	1.2(-3)	1.7
Pulse Termination Network	5.9(-4)	2.7(-4)	2.2
Gas Injection	5.0(-4)	1.7(-4)	3.0
Total (instrumentation and electronic devices)	1.6(-1)	7.1(-2)	2.3
PIC Electronic Devices (separated from total value)	6.3(-2)	4.6(-2)	1.4

4. DISCUSSION

Instrument failure rates are mostly in the range of 2.0(-6) to 2.0(-5) while module failure rates are between 8.0(-7) and 8.0(-6). Using 2/3 logic would essentially replace the instrument failures with the electronic voting module failure. Down times would also be expected to decrease from 3 hours to perhaps 1.0 hours. Decreasing the unavailability due to instrument failures by a factor of for instance 5 yields a total PIC availability of 92 %. Better results could be obtained by reducing the number of measured parameters or by cross checking several related parameters. If cross checking is done using software, the hardware unavailability reported here need not be increased.

Reducing the contribution from the PIC electronic device hardware is more problematic since 2/3 logic at this level would produce a very complex and expensive network. In this case it may be necessary to increase the reliability of the modules themselves. The limited analysis of JET module failures performed for this study yielded failure rates in good agreement with or somewhat lower than U.S. commercial grade nuclear plant modules. Reference 3 indicates that using present generation military grade components can reduce module failure rates by an order of magnitude. However, the difference in cost was not clear.

Although the models used in the study are quite coarse they represent a first attempt to clearly identify those instrument-generated signals and subsequent signal paths which are vital for the pulse cycle. Failure rates, although not well defined, lie in a relatively narrow range so order of magnitude estimates are possible.

5. FURTHER STUDIES

The necessity of carrying out this type of study in an iterative way is stressed. Consequently a second phase of the PIC availability study is in progress and will be completed in 1989. The scope of work for this second study includes:

- Rearrangement of PIC 1 according to a general study formalism.
- Storage of the study using suitable IBM PC/2 software.
- Amendment of basic reliability information.
- Adjustment of the Control and Interlock System to a new NET architechture.

In order to finally arrive at a reliable fusion plant, this type of iterative study together with a fruitful exchange of information between the reliability analysts and the system experts is of great importance.

ACKNOWLEDGEMENT

We would like to thank Dr Rolf Bünde of the NET team for his continuous support and guidance and all the design and operation engineers at NET, JET, KFK Jülich who patiently explained some very complex systems to us.

REFERENCES

1. R. Bunde, Availability Assurance of NET, Strategy and Status of Work, NET Report 33, August 1984.
2. S. Dinsmore, J. Lorenzen, K. Pörn, H. Unneberg, Avialability Study of the NET Protection, Instrumentation and Control System, Studsvik Report NP-86/131.
3. C.W. Mayo et al., Improved Reliability for Analog Instrument and Control Systems, Vols. 1 and 2, ERRI NP-4483, May 1986.
4. IEEE Std 500-1984, Library of Congress Catalog No. 83-082816 (1984).
5. Joint European Torus, CODAS Division electrical module failure data base.

FUSION TECHNOLOGY 1988
A.M. Van Ingen, A. Nijsen-Vis, H.T. Klippel (editors)
Elsevier Science Publishers B.V., 1989

ASPECTS OF TECHNOLOGY IN THE TRITIUM INDUSTRY

P. GIROUX and P. BOUCQUEY

Commissariat à l'Energie Atomique - Centre d'Etudes de VALDUC
Boîte postale n° 14 - 21120 IS SUR TILLE (FRANCE)
Tel. (33) 80.35.13.05.

ABSTRACT

The importance of "technological" components employed in processes involving the use of tritium, as compared to that of other items specific to the process, tends in general to be underestimated. It manifests itself mainly in terms of costs (capital and maintenance) and of safety performance.
In order to optimize the cost/safety equation, therefore, very close attention has to be given to the selection and qualification of these components, with particular reference to : optimization of requirements ; optimization of dimensions ; standardization ; quality assurance ; reliability ; maintainability ; connections to control, instrumentation and data acquisition systems ; potential generation of waste (in normal use or on final dismantling).

1. DEFINITIONS

In the case of complex facilities involving the use of tritium or its compounds, such as :

- plants/laboratories producing, using or carrying out research on tritium, and
- waste reprocessing or packing plants,

the operation of the process is generally segmented and employs a number of quite distinct treatment *units* (for purification, storage, solid waste treatment, etc.).

Again, these units are, more often than not, physically distinguished by indispensable containment barriers (glove boxes, enclosures, rigid containment systems, etc.).

For the purposes of our discussion, the description *installation* refers :

- to anything to do with the process, inclusive of containment, connections to fluids and ventilation systems and between items of equipment (vacuum, tritium, detritiation),
- to electrical relay, automatic control and data acquisition racks, and
- to electronic and computer interfaces.

It *does not include* building infrastructures or attendant utilities and services such as ventilation, power supplies, remote alarms or fire protection.

Similarly, a *technological component* is defined* as any physical subassembly possessing the following characteristics :

- it is not specific to a particular unit ;
- it is employed on the majority of units ;
- it is produced in quantity (more than five specimens, sometimes in hundreds or thousands).

Typical examples are : Normetex pumps, flanges, etc.

* This definition can be extended, for instance, to include software used for remote alarms, remote signalling, remote monitoring, process automation, etc. In this case, the term *software component* could perhaps be employed.

The term *elementary technological component* will be confined to items or subassemblies that are not stripped down for maintenance purposes :

- either because they are of one-piece construction, e.g. a flange gasket,
- or because dismantling is impractical or would be unduly expensive (contaminated parts, welds, etc.), e.g. a pressure sensor.

The description *standardized technological assembly* will be used for a set of technological components assembled as per a standardized geometrical arrangement and clear-cut technical specifications.
An example is a gas transfer assembly.

In contrast to the above, the term *specific component* will be used for any physical assembly determined by the process served by a process unit, such as a purification reactor, melting furnace and the like. Specific components are normally distinguished by existing only in very small numbers and by their particular morphology.

Appendix I lists the most important technological assemblies and components.

2. RELATIVE IMPORTANCE OF TECHNOLOGICAL/SPECIFIC COMPONENTS EMPLOYED IN AN INSTALLATION

2.1. With respect to cost

INSTALLATION EMPLOYING TRITIUM OR COMPOUNDS OF TRITIUM		PROPORTION OF COMPONENTS technological %	specific %
NATURE OF COMPONENT			
Mechanical	by volume	85	15
	by weight	80	20
Electronic Computer	space taken up	90	10
CAPITAL COSTS			
Most items of equipment (typically)		85	15
In special cases :			
Complex process (incl. studies)		45	55
Glove box,transfer or storage (or similar)		100	0
MAINTENANCE COSTS			
Highly reliable processes (involving no consumption of reactants)		95	5
Processes employing components of limited life or consumable reactants such as uranium		75	25
DISMANTLING COSTS			
Disassembly		85	15
Tritiated waste reprocessing		80	20
Storage post-processing		90	10

2.2. With respect to safety

Certain risks of a specific nature attach to the majority of specific components, as determined by conditions of use (severe temperatures, high pressures, corrosion, material fatigue, etc.). Generally speaking, however, such risks will have been clearly identified and a comprehensive safety analysis will have been carried out (fault trees, multiple barriers, safety devices, technical acceptance procedures, etc.). Added to which, Safety Committees pay special attention to these items. This being so, *malfunctions affecting installations are rarely ascribable to specific components.*

Technological components are, in the main, subjected to conditions of service of no great severity as regards temperature and pressure, with the exception of rotating machinery, where wear and/or fatigue are comparatively substantial.

The basic difference arises from the use of technological components in far greater numbers than specific components. Even though, taken individually, components of each type meet standards of reliability that can be deemed sufficient, the probability of a malfunction increases rapidly with their number (e.g. on going from 100 to 1 000 components).

In the final analysis, experience shows that the majority (> 90 %) of physical failures endangering the safety of installations can be traced to technological components.

3. SELECTION AND QUALIFICATION OF TECHNOLOGICAL COMPONENTS

We have seen how important technological components are in terms of costs and safety performance. The selection and qualification of these components must, therefore, always take full account of a number of factors, as discussed below.

3.1. Factors with a bearing on costs

3.1.1. Optimization of requirements

The most effective optimization of the number of technological components to be used, based on a properly conducted value analysis, is an essential preliminary.

Any component (valve, sensor, etc.) included in an installation for which there is no demonstrable necessity represents a very considerable expense (more than ten times purchase cost) when allowance is made for fitting, electrical connections, computer links, augmenting the capacity of the remote control system, detritiation systems, maintenance and dismantling costs.

3.1.2. Optimization of sizing

On the same basis, one of the most important routes to savings on installation costs consists in optimizing the size, weight and bulk of each type of technological component, while not losing sight of the needs of maintenance.

3.1.3. Standardization

Standardization of technological components is important in order :

- to hold down capital costs, via improved purchasing policies and more cost-effective technical acceptance procedures ;
- to hold down operating costs, by trimming spare parts inventories and the costs of inhouse or contracted-out maintenance.

3.1.4. Quality assurance - Reliability - Maintainability

Enhanced reliability and proper attention to the question of maintenance in an environment geared to quality assurance are the two factors fundamental to reductions in operating costs.

A distinction also has to be drawn between :

- quality assurance as this relates to plant and equipment (soundness of materials, compliance with specifications, etc.) ;
- quality assurance as it relates to suppliers (qualification, compliance with laid-down procedures, etc.).

3.1.5. Liaison with control, instrumentation and data acquisition systems

The capital and operating costs of control, instrumentation and data acquisition systems are directly proportional to the size of the installation and hence to the number of technological components.

Again, these systems consist of components (microprocessor cards, input/output interfaces, relay circuitry, etc.) which have to be designed to meet the same criteria as mechanical technological components (optimization of requirements, quality assurance, maintainability, etc.).

There are, however, two special features to be allowed for :

- the sensitivity of these computerized systems to external interference is such that their reliability and protection must never be sacrificed to false economies ;
- the importance of the software to the proper functioning of the system is such that software quality has to be allowed for on a par with the quality of the mechanical, electrical, electronic and data processing types of technological components.

3.1.6. Technological components as a source of tritiated waste

We have already considered the costs associated with the dismantling of technological components, either during operations (replacement) or at the stage of final dismantling.

Clearly, any reduction in such costs involves optimizing the number and size of these components and enhancing their reliability.

3.2. Factors with a bearing on safety

Enhancing the safety of an installation mainly means enhancing the standards of safety of the technological components:

Keeping down the number of components Standardization Quality assurance Enhancing reliability	
↙	↘
Will reduce the probability of malfunction (loss of vacuum, explosion, corrosion, etc.) thereby limiting the radiological, electrical and mechanical hazards to which operators are exposed and the risk of emissions affecting the environment.	Will reduce the need for remedial action concerned with contaminated plant and equipment the major hazard to which maintenance personnel are exposed.

Keeping down the number of components Optimizing dimensions Reliability	
↙	↘
Will contribute to curtailing sources of tritiated waste (environmentally-friendly factors).	Will contribute to the improved effectiveness of systems for the removal of tritium from glove box atmospheres and hence limit the risks of emission to the environment.

Appendix 1
PRINCIPAL TECHNOLOGICAL COMPONENTS

MECHANICAL :

Connecting flanges	low/high pressure
Valves	electropneumatic/ manual/magnetic, for flow/pressure regulation, low/high pressure, process/vacuum lines, detritiation circuits.
Pumps	transfer, vacuum, turbomolecular, circulating, turbine, compressor.
Measurement sensors	temperature/pressure/ flow, conductivity (catharometer), hygrometer, ionization chamber, quadrupoles.

ELECTRICAL

Magnetic valve power packs
Voltage regulation
Control modules (pumps)

ELECTRONIC

Regulators (temperature/flow/pressure)
Acquisition modules (vacuum gauges and sensors)

COMPUTER SYSTEMS

Cards	microprocessor, memory extension, disk interface, asynchronous links.

Floppy/hard disk readers
Printers and recorders
VDU's and keyboards
Programmable automatic controllers.

FUSION TECHNOLOGY 1988
A.M. Van Ingen, A. Nijsen-Vis, H.T. Klippel (editors)
Elsevier Science Publishers B.V., 1989

FEEDBACK CONTROL SYSTEM DESIGN FOR THE PLASMA SHAPE IN ASDEX UPGRADE

W.Woyke, H.Bruhns*, P.McCarthy+, S.C.Cha*, K.Lackner*, U.Seidel*, R.Weiner*

Lehrstuhl für Steuerungs- und Regelungstechnik, Technische Universität München, P.O. Box 202420, D-8000 München, West Germany

Models for the feedback control system design are based on a two-dimensional equilibrium code. Analysis in the frequency domain renders the aggregation of dynamic behavior in relevant input-output-channels for feedback control. Bang-Bang control and a proper linear feedback control for the plasma position is introduced. Remaining currents of inner control coils for fast feedback control are distributed to vertical field coils on the slow time scale.

1. INTRODUCTION

ASDEX Upgrade is a Tokamak with a reactor-relevant poloidal divertor configuration now under construction at IPP. For simulation purposes and the inclusion of other aspects, i.e. plasma boundary physics, plasma wall interaction and the configuration control problems of a fusion reactor, the Tokamak has been designed with the poloidal field coils outside the toroidal coil and at a considerable distance from the plasma. Due to this fact and the Tokamak's strongly elongated plasma cross-section ($b/a = 1.6 - 1.9$), feedback control of the position and shape parameters of the plasma column is a major issue.

The modelling of position and shape parameters is based on the output of simulation with a two-dimensional MHD-Code. This metamodelling yields a lumped parameters model including dynamic behavior as functions of the applied voltages in active poloidal field coils. Single input-output channels in this multi-input multi-output model can be ordered to radial and vertical plasma position control. Their dynamic properties are approximated by *A*ctuator-*E*ffect-*M*odels (AEM) of low order and special structure. Comparison of simulations between these models and the complex Tokamak Simulation Code gives good results in accuracy of modelling of the dynamic behavior.

Two poloidal field coils, within the toroidal one, are used for the fast feedback control of radial and vertical plasma position. Based on AEM PID-controllers can be developed that nearly approximate the properties of a minimum time bang-bang control. On the next timescale the outer poloidal field coils support the inner control coils to compensate the steady state error. Engineering constraints and decoupling are criteria for distribution of control currents to vertical field coils to feedback some additional plasma shape parameters like elongation and triangularity.

2. ASDEX UPGRADE POLOIDAL FIELD COIL CONFIGURATION

ASDEX Upgrade is equipped with eleven distinct poloidal field coils (Figure 1). Within the vessel a galvanically crosswise coupled pair of stabilizing conductors diminishes the inherent vertical instability of the elongated plasma column. The most significant toroidal current modes of the vessel have a decay time of about $3 - 10\,ms$. Inside the toroidal field coils two inner control coils take care of fast plasma position control. In a later stage of experimentation this task will be assigned to control coils outside of the toroidal field coils. Three pairs of vertical field coils are placed to produce vertical field compensating hoopforce and to shape the plasma column.

* Max-Planck-Institut für Plasmaphysik, Garching.
+ attachment from University College, Cork, Ireland.

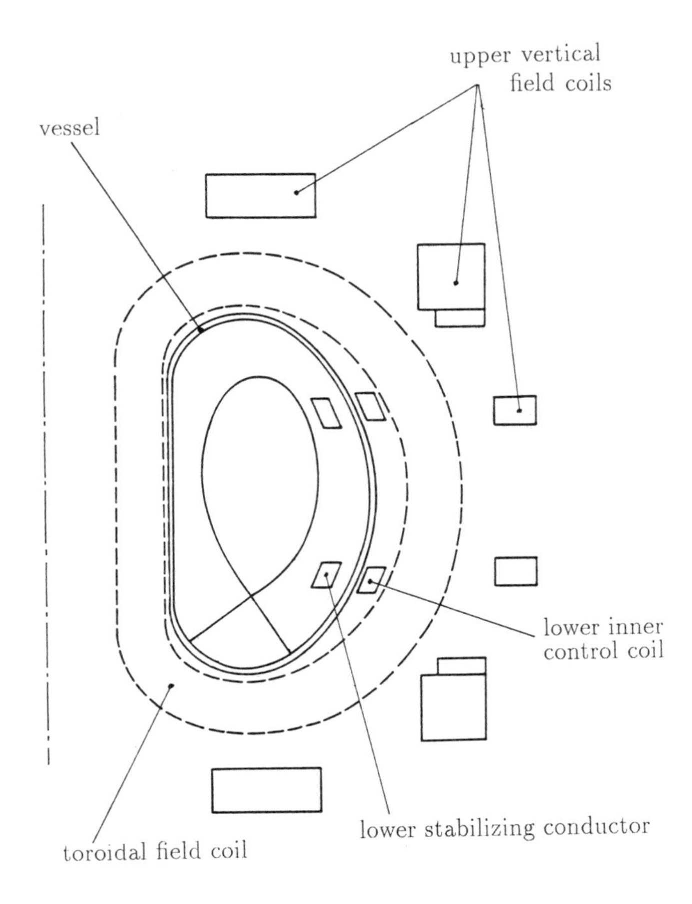

Figure 1: Poloidal field coil configuration of ASDEX Upgrade

3. MODELLING DYNAMIC BEHAVIOR

3.1. MHD - Codes.

The *T*okamak *S*imulation *C*ode[1] (TSC), developed by Princeton University (Plasma Physics Laboratory), slows down and damp Alfven waves by artificial enhancement of ion mass and viscosity. It calculates free-boundary evolutions of the plasma eqilibrium consistent with currents in external poloidal field coils and passive conductors. Thus external feedback control can be included in this code. It serves for validation via comparison of simulation experiments testing feedback control and lumped dynamic models based on zero-mass method described below[4]. It has to be checked that the results obtained with TSC are independent of the enhancement factor.

The free boundary *G*arching *E*quilibriums *C*ode[2] (GEC) computes ideal MHD Equilibria. In this application it is used to calculate shape parameters for a set of twelve input variables[5].

3.2. Metamodelling.

The plasma position and shape feedback control works on the resistive time scale applying voltages to poloidal field coils, here inner control coils and vertical field coils. The dynamic behavior is determined by the L/R-time of inductively coupled poloidal field coils and the interaction of the plasma column to the poloidal flux in the plasma region. While influence of shape swing, e.g. elongation variation, is negligibly small, interaction of radial and vertical plasma movements dominates some modes. This model operates with mutual inductancies between external currents and the plasma column. The parameters of this dynamic *L*umped *P*arameters *M*odell[3] (LPM) are computed by integration of spatial distributed parameters that are results of GEC simulation.

In addition GEC yields a database of shape parameters $\underline{g}$ related to currents $\underline{I}$ in poloidal field coils and parameters $\underline{T}$ nearly independent of the poloidal flux function, e.g. β_p and internal inductance l_i. Metamodelling is done by regression on this database to parametrize quadratic functions of shape parameters depending on external currents $\underline{I}$ and on $\underline{T}$.

$$\underline{g} = \underline{g}_o + (\mathbf{A}_1 + \mathbf{A}_2 \begin{pmatrix} \Delta\underline{I} \\ \Delta\underline{T} \end{pmatrix}) * \begin{pmatrix} \Delta\underline{I} \\ \Delta\underline{T} \end{pmatrix} \quad (1)$$

$\mathbf{A}_1$ and $\mathbf{A}_2$ are tensors containing the parameters of metamodelling. Δ denotes deviation to means I_o and T_o.

3.3. Analysis in the Frequency Domain.

Up-down symmetry of the poloidal field coil configuration is disturbed only by an asymmetric vertical arrengement of passive stabilizing conductors and by shaping the plasma into a Single Null configuration. In good approximation the sum (u_r) and the difference (u_z) of voltages applied to the control coils allow to command independently the sum (i_r) and difference (i_z) of the associated currents as well as radial (r) and vertical (z) plasma position. Prescribing u_r, respectively u_z, leads to single-input double-output *A*ctuator-*E*ffect-*M*odells (AEM) describing the dynamic behavior to be controlled by fast plasma position feedback.

These models contain the same modes as LPM. In the frequency domain, input-output-channels can be described by the residues r_i related to the modes λ_i,

$$z(s)/u_z(s) = \sum_i \frac{r_{zi}}{s - \lambda_i} \qquad (2)$$

$$i_z(s)/u_z(s) = \sum_i \frac{r_{ci}}{s - \lambda_i} \qquad (3)$$

where s is the argument of Laplace transformation. Normalization of residues yields a survey of the dominance of particular modes:

$$\tilde{r}_i = \frac{-r_i/\lambda_i}{\sum_j |r_j/\lambda_j|} \qquad (4)$$

Table 1 shows timeconstants of modes and normalized residues for the Single Null High Beta plasma configuration.

Table 1: Timeconstants and normalized residues in I-O-channels of fast plasma position control

No.	$-1/\lambda_i[ms]$	$\tilde{r}_i[\%]$			
		r/u_r	i_r/u_r	z/u_z	i_z/u_z
1	3,6	0,00	0,00	- 0,08	0,33
2	5,0	- 0,01	0,09	- 0,00	0,00
3	5,7	0,00	0,00	- 0,01	0,02
4	7,1	- 0,16	0,24	- 0,01	0,00
5	8,5	- 0,65	0,87	0,00	0,00
6	12,5	- 0,09	0,08	- 0,29	0,03
7	-296,0*)	- 0,01	0,01	- 54,55	5,15
8	312,0	0,06	0,04	- 42,35	93,97
9	490,0	56,96	90,44	- 0,18	0,05
10	942,0	0,00	0,00	- 0,71	0,25
11	1304,0	5,10	1,84	0,00	0,00
12	1643,0	2,12	0,66	- 0,03	0,00
13	2676,0	- 0,04	0,00	- 1,88	0,16
14	2889,0	- 1,16	0,03	0,22	0,03
15	3230,0	- 2,52	1,32	0,23	0,02
16	3940,0	- 0,10	0,00	0,68	0,00
17	7337,0	35,46	4,37	- 0,08	0,00

*) unstable mode

Six fast modes are associated with vessel currents. Remaining modes of flux dissipation in poloidal field coils and plasma movement contain an unstable mode (No. 7). Concerning the residues input-output-channels of voltages to currents are dominated by a single mode. The channels to associated positions are governed by an additional second mode.

3.4. Aggregation of AEM.

The marked dominance of these modes renders high accuracy in the aggregation of the dynamic behavior to an AEM with only three dynamic modes (Figure 2). Though eigenvalues are variable parameters of the reduction algorithm to improve approximation of gain and phase, they still can be assigned to vertical plasma displacement ($\lambda_z \sim \lambda_7$), flux dissipation of inner control coils ($\lambda_c \sim \lambda_8$) and vessel currents ($\lambda_v \sim \lambda_1$). K_i and K_z denote the quasi steady state gains in related I-O-channels. Similar results can be obtained for AEM of radial position control, even if the outer vertical field coils are used for control . I-O-channels for additional shape parameters do not show a difference in dominance between position and currents.

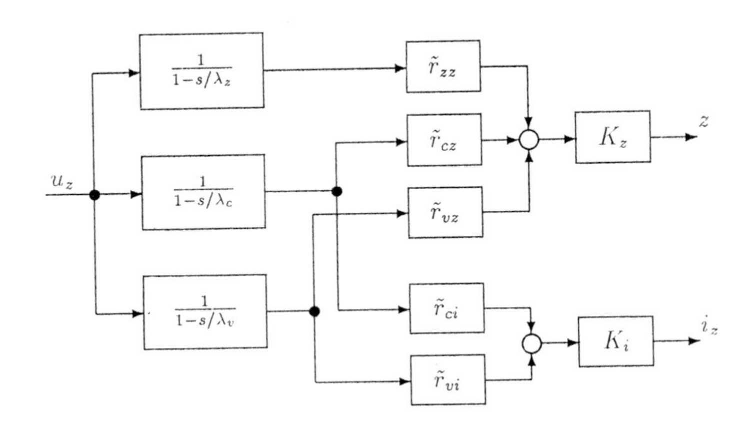

Figure 2: Structure of aggregated AEM

Table 2: Parameters of aggregated AEM

	$K_z= 19{,}5 * 10^{-3}$	$K_i= 3{,}1 * 10^3$
$-1/\lambda_z=-266{,}0ms$	$\tilde{r}_{zz}=-53{,}9\%$	–
$-1/\lambda_c= 292{,}0ms$	$\tilde{r}_{cz}=-45{,}9\%$	$\tilde{r}_{ci}=99{,}4\%$
$-1/\lambda_v= 5{,}5ms$	$\tilde{r}_{vz}= -0{,}2\%$	$\tilde{r}_{vi}= 0{,}6\%$

4. FEEDBACK CONTROL.

AEM and LPM models serve for parametrization and test of feedback control.

4.1. Fast Plasma Position Control.

Aggregated AEM renders the development of a Bang-Bang control to push back the plasma column to the nominal position in minimum time for a given perturbation. Thereby voltages of inner control coils are switched to the vertices of their limited amplitude. Line 1 of figure 3 shows Bang-Bang control of vertical position. There is a big overshot in negative direction of about 60% of the maximum deviation upwards to the nominal position. But vertical position as well as currents (line 1 of figure 4) return to the nominal value within 70 ms.

In comparison to this a PD-feedback control of vertical plasma position and additional proportional feedback of currents i_z to inner control coils applied to LPM yields good approximation of that minimum time control regarding timeconstants. Concerning vertical plasma position there results a better dynamic behavior, without any overshot (line 2 of figure 3). But the according currents i_z (line 2 of figure 4) are finite even after 100 ms. This residual value depends on nature and amount of perturbation.

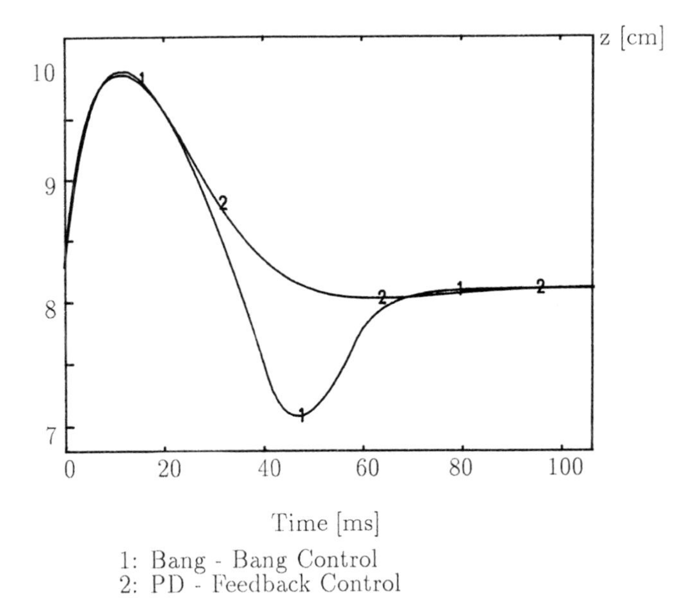

Figure 3: Controlled vertical displacement under rough perturbation.

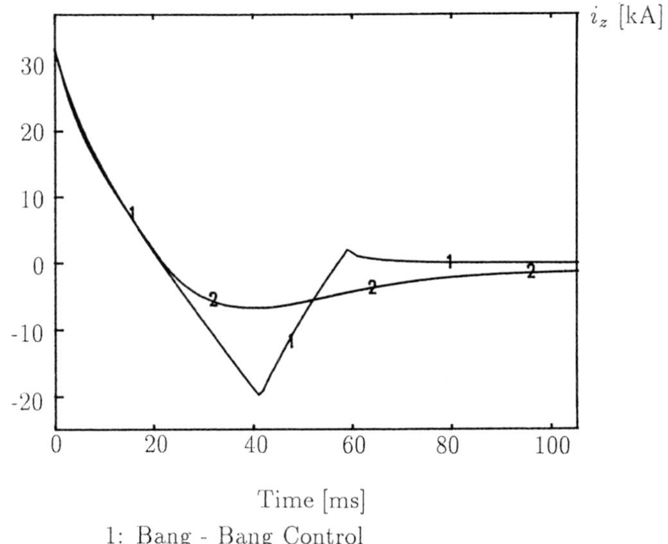

Figure 4: Currents i_z for fast plasma position control.

4.2. Load Distribution to Vertical Field Coils.

Simulation experiments show, that a command of vertical position over a range of 8 cm can be performed by the feedback control applied to the inner control coils even in presence of hard limitations on available voltage (figure 5). There remains a residual error of about 4 mm. But the more serious problem is the finite current amplitude, whose value exceeds a quarter of the maximum permissible current load of these coils (line 1 of figure 6). Therefore vertical field coils have to support the feedback of the plasma position on a slow time scale. The use of current feedback also for the vertical field coils, and the feedback of inner control coil currents to them, allows to reduce this residual current load on the average to zero (line 2 of figure 6). An additional result of this compensating feedback control is that the steady state error of the vertical displacement vanishes (line 2 of figure 5).

4.3. Plasma Shape Control

Beside the supporting control of radial and vertical position feedback control some shape parameters can be programmed by feedback control using the six vertical field coils, e.g. elongation, triangularity and the position of the lower saddle point. This is the inverse problem to metamodelling by regression over the database, because

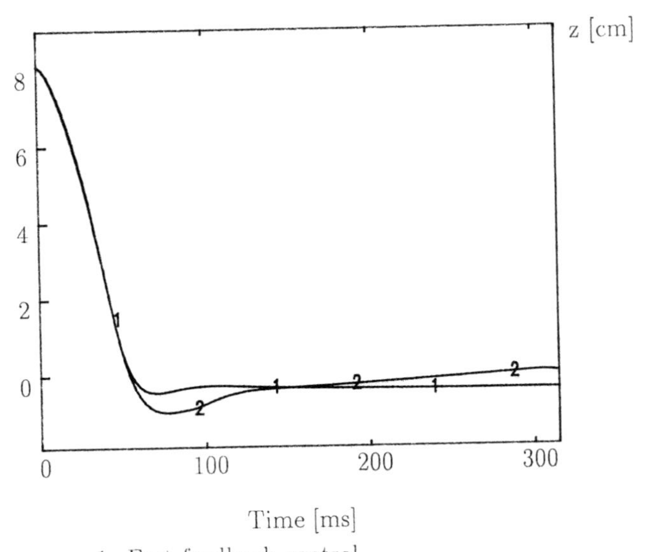

Figure 5: Command of vertical position using feedback control.

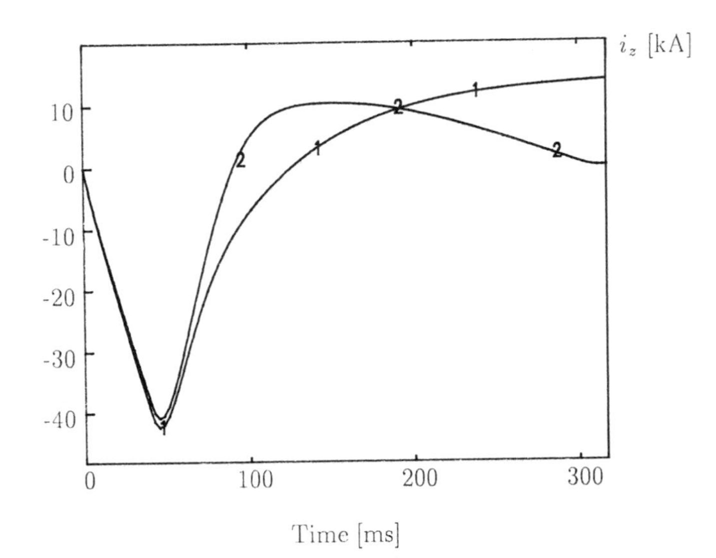

Figure 6: Currents i_z for command of vertical position.

currents of vertical field coils have to be a function of shape parameters[5]. But the lack of uniqueness enforces the inclusion of additional constraints such as the available voltage and current load of the vertical field coils. The dynamic behavior is determined by the current feedback applied to the vertical field coils.

5. CONCLUSIONS

Metamodelling and aggregation are methods to get proper models of dynamic behavior for the design of a feedback control system. Though they result in models similar to ones obtained with previous methods, the parametrization can be performed on a very accurate description of the complex geometry of ASDEX Upgrade. Therefore the feedback control system can be designed to accommodate the various distinct demands of dynamic performance.

REFERENCES

1. S.C. Jardin, Multiple Time-Scale Methods in Tokamak Magnetohydrodynamics, in: Multiple Time Scales, Computational Techniques, Vol. 3, eds. J.U. Brackbill and B.I. Cohen (New York 1985) pp. 185-232.

2. K. Lackner, Computation of Ideal MHD Equilibria, Computer Physics Communications 12 (1976) pp. 33-44.

3. U. Seidel, K. Lackner, G. Lappus, H. Preis and W. Woyke, Plasma Position Control in ASDEX Upgrade, in: Tokamak Start-up, Proc. of the 7th course of international school of fusion reactor technology, eds. H. Knoepfl (Plenum Publishing Corporation 1985) pp. 325-336.

4. W. Woyke, Simulationen des Plasmalageregelkreises mit dem Entwurfsmodell, Zwischenbericht zum F + E Vorhaben IPP 17182 (1988).

5. P.J. McCarthy, Optimal selection of plasma parameters and magnetic measurement locations for tokamak feedback control, this volume.

6. H. Bruhns et al., The control system of ASDEX Upgrade, this volume.

FUSION TECHNOLOGY 1988
A.M. Van Ingen, A. Nijsen-Vis, H.T. Klippel (editors)
Elsevier Science Publishers B.V., 1989

THE CONTROL SYSTEM FOR ASDEX UPGRADE

H. BRUHNS, S. CHA, P. HEIMANN, F. HERTWECK, A. JÜLICH, K. LACKNER,
P. McCARTHY, G. RAUPP, H. RICHTER, U. SEIDEL, M. TROPPMANN, R. WEINER
W. WOYKE*

Max-Planck-Institut für Plasmaphysik, D-8046 Garching EURATOM-Association.
*Technische Universität München, Federal Republic of Germany

The plasma position control in ASDEX-Upgrade will be performed by statistical correlation of equilibria and magnetic measurements. A control system capable to provide sufficient computing power for this task and to be expandable to a complete control of the experiment is described.

INTRODUCTION

ASDEX-Upgrade at the IPP Garching is a divertor tokamak with a noncircular plasma crossection aiming at the investigation of diverted plasmas in an embodiment similar to that of a tokamak reactor[1]. Hence the experiment is equipped with the main poloidal field coils located external to the toroidal magnet (TF-coils).

The main issue of the experimental program will be the plasma boundary and plasma wall physics. However, also the control of the plasma is a nontrivial problem. Due to the up-down asymmetry and the strong elongation (1.6 to 1.9) intended to allow operation at higher relative plasma pressure the plasma is highly sensitive to vertical instability. As it would be impossible to control this mode on the original time scale (about 1 ms) a passive 360° saddle coil is built inside the vacuum vessel which will enhance the vertical time constant to 100 - 250 ms. In addition, a pair of control coils are located between the TF-coils and the vacuum vessel in order to allow faster (25 - 50 ms) response than possible with the main poloidal field system (250 - 500 ms).

While most present experiments use methods based on flux extrapolation to determine the position of a boundary flux surface (and hence the plasma position) ASDEX-Upgrade will recover the plasma position and the plasma shape parameters by function parametrization i.e. by using predetermined statistically based correlations of computed equilibria and the corresponding fields and fluxes to be measured at probe locations close to the inside of the vacuum vessel[2,3]. The advantage of this method is that it is largely independent from the choice of control parameters, hence easily expandable to the control of additional features, and that it is little affected by up-down asymmetry and large flux differences between inner and outer probe locations. In addition, the probe positions can be arranged rather arbitrarily. (ASDEX-Upgrade will use probes almost equally spaced around the poloidal circumference according to technical feasibility.)

The plasma state determination with this method relies on the completeness and accuracy of the statistical basis of computed equilibria. For these, a static twodimensional code is used. There are, however, shortcomings due to lack of physical features in the equilibrium code. One example is that skin current profiles can occur (especially in fast dynamic situations) which are not considered by the equilibrium code.

Therefore, dynamic simulations of the plasma are performed with the Princeton Tokamak Simulation Code[4] covering the whole shot including start up and shut down. Although these computations have not been completed it is hoped to get sufficient information on features which are necessary to be observed in the data basis. Also, the TSC code shall serve as a model of the 'tokamak plant' for testing the control loop in pseudo real-time.

Another problem consist in the occurrence of plasma flows especially if auxiliary heating methods, e.g. neutral injection, are applied. We therefore envisage appending to the present databasis flow equilibria computed with NIVA[5] .

THE CONTROL COMPUTER SYSTEM

While conventional control systems use hard-wired (analog) techniques for the plasma position determination which can be very fast, a real time function parametrization - based method needs a high real-time digital computing power. Therefore a fully digital computer system was designed for the complete shot control. This includes master diagnostics, program set-up, shot program performance (feed forward), plasma position determination, feed back, systems control. Furthermore various features like coil-stresses, plasma density- and q-limits shall be computed in order to ensure operation within predetermined operational windows.

There are different requirements concerning the time scale of response. As most of the coil systems are slow, the corresponding control signals can be issued in 10 ms intervals except for special events. Only the inner control coils which regulate the plasma position need faster response. Here, an ultimate limitation is imposed by the thyristor units which can fire every .8 - 1.1 ms. The corresponding feed back computer (R2) will issue a command for new currents about every 1.0 or 2.0 ms.

Control computations for technical and (machine) safety purposes will be provided in 10 to 50 ms intervals with respect to the electrical and mechanical features. Most thermal properties are requested to be controlled in a 500 ms interval sequence.

VME-specifications were chosen for the basic computer units. These definitions are becoming a widely introduced standard. Independently, VME has already been chosen for all diagnostic computers of the experiment.

The control system is made up of five VME - workstations (Integrated Solutions/Racomp) which are connected by ETHERNET for program loading and communication which is not time critical. Each workstation is equipped with a MC 68020 CPU and various peripheral boards (fig. 1).

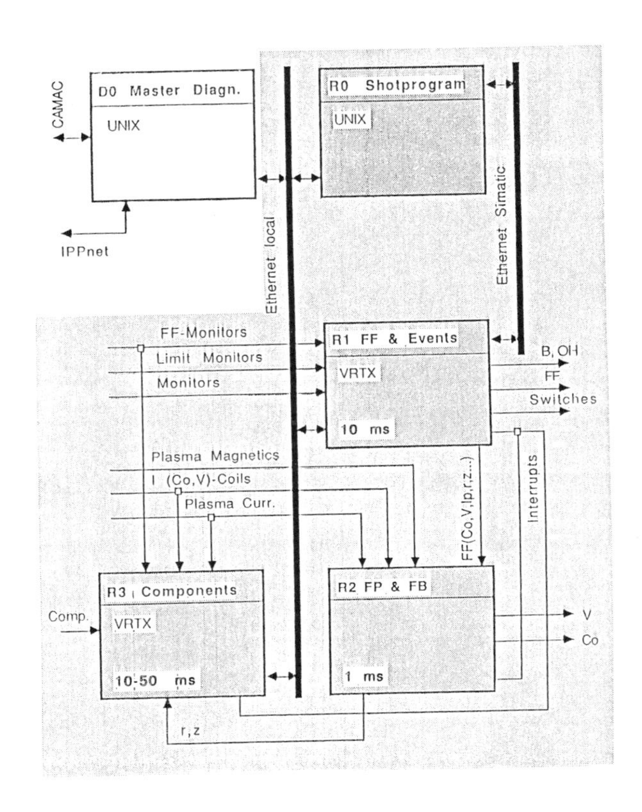

fig. 1

Besides this control system there are free programmable control computers for the various technical subsystems (coil power supplies, thyristor units, vacuum system, cooling system etc.) which are based on SIMATIC S5 units. These computers are connected to the 'fast' control system by a SIMATIC H1 network which is Ethernet compatible and which is shown in fig. 1 as well.

For off-line purposes, after-shot diagnostics and shot preparation, UNIX is an adequate operating system supplying all possibilities for the preparation of a comfortable user interface and is used for the units D0 and R0.

The computer D0 serves as a master diagnostics monitor. It is used by the program leader in order to get information from the last shot. This workstation has no real-time function.

The computer R0 handles the shot preparation. Here all communication with the shot programmer takes place, beginning with the authorization check in the morning until the final signal for shut down in the evening. All means for setting up the shot program are available on this computer, also manual operations are performed on this device like starting the next shot, abnormal break in the case of an external event which requires shot termination etc. However, from the systems point of view, R0 has no real time tasks. Fig. 2 shows the typical set up of a VME workstation.

There are three (groups of) real-time computers, R1, R2 and R3. All of these devices have to comply with requirements which cannot be handled by a single VME 68020 workstation. Therefore parallel computers or powerful input boards have been implemented into these units. The systems R1 and R3 are operated with Uniworks/VRTX, a real-time operating system which is prepared to follow the newly devised VME-Exec standard.

The 'feed forward' computer, R1, handles the task to provide all feed forward signals to the individual subsystems including to the plasma position computer R2. Presumably, the hardware will consist of one or two workstations, however, without the graphics facilities.

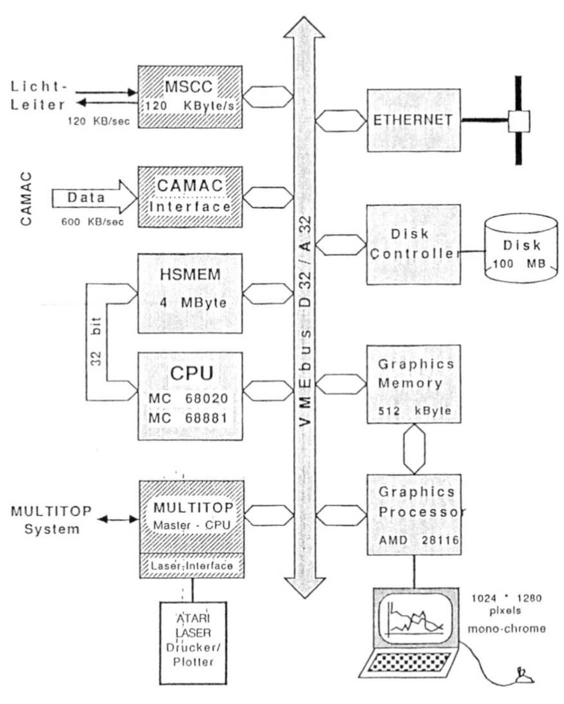

fig. 2

While, during most time of the shot duration, 10 ms intervals will suffice for feed forward, a 1 ms sequence is demanded during plasma start up and some exception handlings e.g. disruptions. Correspondingly there will be a inner control loop (which issues only a reduced set of commands) within the standard one. This is similar to the procedure at other machines[6].

The unit R2 in fig. 1 includes a MULTITOP/12 parallel computing unit (fig. 3). The system MULTITOP has been developed at the IPP by Richter and Hertweck[7] using INMOS T800 or T414 transputers. Apart from the usual features of transputers, to have four fully duplex I/O channels together with a CPU, memory and floating point unit (T800) merged into one chip, Multitop is equipped with a coupling network which allows to change the logical topology of the computer during task performance. Hence a great versatility and optimization of speed is achieved.

During the shot, a large amount of computations for safety purposes has to be performed. For this purpose a special (group) of computers, R3, has been defined. One, most important, task is to check whether the

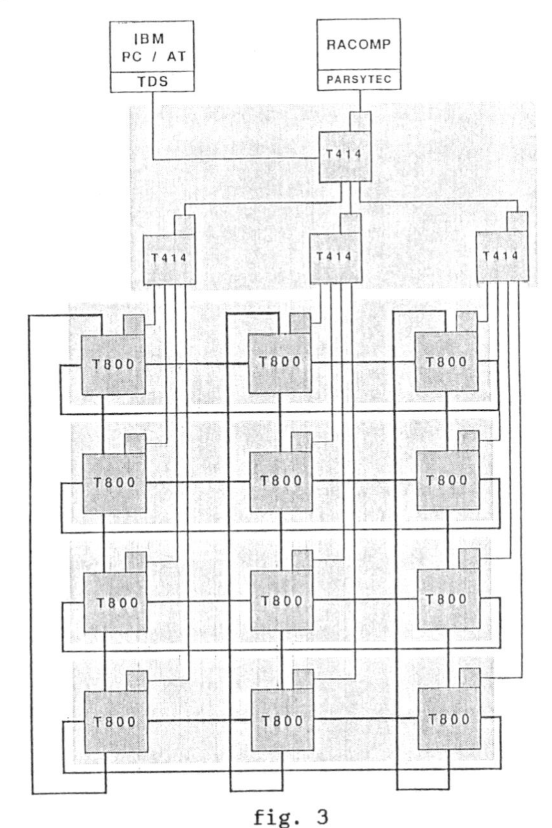

fig. 3

mechanical stresses exerted by the currents and fields lie within the permissible limits of the device. It is planned to perform the corresponding computations not only during the shot using the actual currents in the coil system but also before the shot, simultaneously with the shot program design (R0). Thus, a critical constellation of the different currents can be interactively detected and immediately corrected already when the shot programmer asks the shot program data to be stored.

Since the data input and output consumes a considerable fraction of the total time available per control, feed forward or feed back time step, special attention has been paid to optimize these procedures. Three different boards have been designed to handle the various tasks using transputer links for the I/O channels.

THE TIMER SYSTEM

A central role is given to the timer system, fig. 4. This system consists of a central timer which issues signals and local timers which receive these signals. Every technical or diagnostic system will be equipped with one or more local timers.

In conventional systems the only function of a timer system is to provide a few general time marks and possibly to synchronize the system. ASDEX-Upgrade will use a timing system which is fully digital and operates with 40 bit wide signals - after error handling 32 bit remain for effective use. Signals can be issued with 250

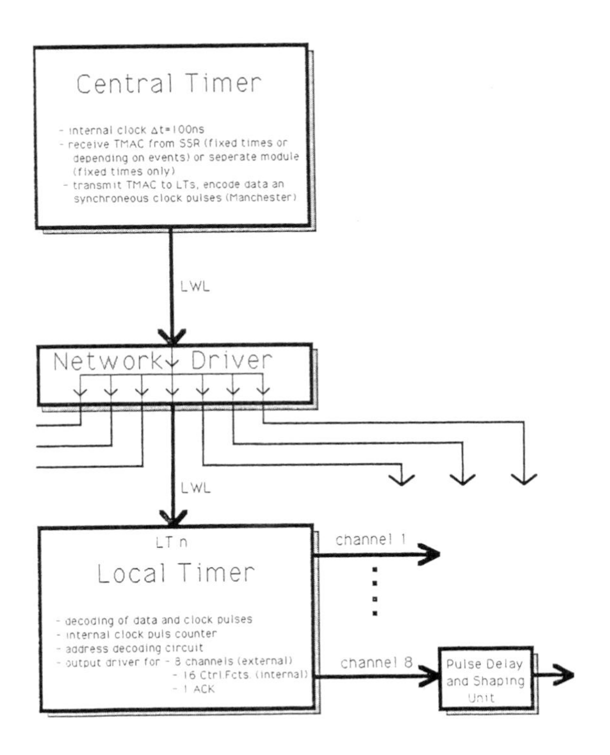

fig. 4

kHz. The system is designed so that either each individual local timer can be adressed or groups of various constitution. Both functions (flags) or amplitudes (8 bit) can be transmitted to connected systems so that a major part of the data transfer can be handled via this timer system. For the conventional tasks a subset of 256 time marks is available to indicate events like 'OH charged', 'Plasma achieved', 'Flat top begin', 'Flat top end', etc.

The timer system uses the serial link adapter protocol with additional 10 MHz Manchester coding. Hence it is compatible with all transputer system I/O as well as with other serial ports via Adapters. It is built largely from standard chips and designed fully modularly so that, at a later time, it will be easy to replace the protocol, the logical width of the signal, or other features, like decoding time marks, output channel selection, the shot-number, shot-time information, etc.

CONCLUSIONS

Compared to existing systems the (fast) control system for ASDEX-Upgrade has a number of features which will allow the implementation of a full centralized automatic control of the complete experiment. While in the initial operation period only a limited range of these features will be available, the systems design is prepared to embody the full range of possibilities without changes.

1. W. Köppendörfer et al. "ASDEX-Upgrade Project Proposal" IPP 1/217 Max-Planck-Institut für Plasmaphysik (1984)
2. H. Wind, "Function Parametrization", CERN 72-21, 1972, pp. 53-106; B.J. Braams, K. Lackner, "A Proposed Method for Fast Determination of Plasma Parameters, Report IPP 1/228, MPI für Plasmaphysik, 1984, B. J. Braams, W. Jilge, K. Lackner, Nucl. Fusion **26** (1986), 699
3. P. McCarthy et al., this conference
4. S.C. Jardin, W. Park Phys. Fluids **24** (1981) 679, S. C. Jardin, N. Pomphrey, J. DeLucia J. Comput. Phys. **66** (1986) 481
5. H.P. Zehrfeld, in: Proc. 13.th Europ. Conf. on Controlled Fusion and Plasma Heating, Schliersee (1986), Europhys. Conf. Abstr. **10C** Part 1, 57 (1986)
6. Y. Suzuki, I. Kondo, A. Ogata, T. Kumahara, T. Hatakeyama, K. Murai, D. Iba "System Design of Zenkei, the Central Control System for JT-60" Fusion Technology 1980 (CEC/Pergamon, Oxford 1981) p. 757
7. H. Richter, F. Hertweck, to be published; H. Richter, Ph.D. thesis, Technische Universität München, 1987

FUSION TECHNOLOGY 1988
A.M. Van Ingen, A. Nijsen-Vis, H.T. Klippel (editors)
Elsevier Science Publishers B.V., 1989

A DUAL SYSTEM FOR THE STABILISATION OF THE VERTICAL PLASMA POSITION OF THE JET EXPERIMENT

M Garribba, D Ciscato*, M Browne, S Dorling, P Noll

JET Joint Undertaking, Abingdon, Oxon OX14 3EA, UK
*Universita' di Padova, Padova, Italy

The vertical position of the JET plasma is unstable without feedback stabilisation. The growth rate can reach $200s^{-1}$ for strongly elongated plasmas ($b/a \simeq 2$). A failure of the stabilisation at large plasma current (7MA) can cause damage to the vessel due to large vertical forces. A dual system which continues the stabilisation even in case of faults in the measurement system or in the conditioning electronics is needed for high reliability. A significant simplification in the control was also achieved by avoiding magnetic signals integration.

1. INTRODUCTION

The JET vertical position control system comprises a Poloidal Radial Field Amplifier (PRFA) and a radial field coil represented by series connetcted subsections of PF coils P2 and P3 (fig 1). For feedback stabilisation, the rate of radial flux change is measured between symmetric locations $Z \simeq \pm 1.7$m, $R \simeq 3$m inside the vessel, using the flux propagation method [2]. A linear combination of magnetic signals from four internal pick-up coils and from to external differential loops (saddle coils) is used for this purpose. For control of the vertical position at a slower time scale, a PRFA current feedback loop is employed. Without feedback stabilisation the plasma position is unstable with a growth rate γ in the range 100 to 200 s^{-1}. It is governed by the decay time of eddy currents in the vacuum vessel $T_V \simeq 3$ms, and also to a minor extent by eddy currents in the mechanical shell. A short amplifier time constant $T_A < T_V$ is therefore essential.

The PRFA uses 50 Hz thyristor bridges, allowing 4 quadrant operation with voltage range ±4kV at full load current ±3kA. Two separately controlled units (range ±2kA) are connected in series with the radial field coil, employing 12 pulse operation corresponding to an average time lag of 1.67ms for small amplitude changes. The PRFA is configured as a voltage amplifier due to internal voltage feedback (fig 2). A rate limiter ensures safe operation with inductive load and at large positive or negative amplitude demands.

2. CONTROL SYSTEM

For the definition of suitable controller functions F_{CL}, F_{CI} (fig 3) the previous system analysis [1,3] on the basis of a simplified model, has been extended. The model comprises three circuit equations for currents in the radial field coil, the vessel and the mechanical shell and a vertical force balance equation for the plasma. Rigid vertical displacements δZ_p of a vertically elongated plasma with peaked current profile were assumed (no skin effect, no change of current). The amplifier transfer function is taken as $F_A \sim (1 + sT_A)^{-2}$, $T_A \simeq 2$ms.

The loop gain K_0 for voltage signals V_L (fig 3) is the dominant parameter characterising the stability region (fig 4). For a given degree of open loop instability (growth rate γ) a certain range of K_0 is possible. The upper limit gives a growing eigenmode at a frequency in the order of 100Hz, at the lower limit the frequency is in the order of 10Hz. The curve R_0 in fig 4 shows the maximum theoretical region, for pure voltage feedback in the stabilisation loop and absence of PRFA current feedback.

Previously, both non-integrated and integrated magnetic signals were used and proportional control was applied in the stabilisation loop, corresponding to a transfer function $F_{CL} = P_L(1 + 1/sT_L)$ in fig 3. The PRFA current was controlled by integral control $F_{CI} = 1/sT_I$, whereby the reference V_{OZ} is usually set to zero. A slightly reduced stabilisation range R_1 is obtained (fig 4).

The stabilisation system was simplified significantly by omission of all signal integrators (corresponding to $F_{CL} = P_L$). This required however the inclusion of some proportional feedback in the current loop ($F_{CI} = P_I + 1/sT_I$). The resulting stabilisation range R_2 (fig 4) is insignificantly different from R_1.

The stability region was tested experimentally at low plasma current, using a step of the current reference V_{OZ}. In pulse 14691 K_0 as increased 25% beyond the nominal value $K_0 = 1.0$. A damped oscillation ith f $\simeq$ 100Hz is visible (fig 5). In pulse 16972 K_0 was reduced to 0.8. The vertical position is almost unstable and a frequency f $\simeq$ 7Hz is seen (fig 6).

3. DUAL SYSTEM

Fig 7 shows the block diagram of the dual system. At power level a duplication is already implemented: two indipendent PRFA units are connected in series. If one unit trips it is put into freewheeling mode and the other one takes over. For high reliability this concept is extended to the measurement and control part. Two opposite octants are equipped to route the magnetic signals to the controller. The feedback signals are first individually filtered with a passive first order filter, which also adds a test signal. An isolation amplifier stage provides the necessary galvanic insulation between the torus and the control electronics. The six signals are then summed with appropriate weights to obtain the feedback signal DVZ. The signal of the active channel is compared with one passed through a passive dummy channel. A comparator decides wether one active channel is faulty and operates a switching network so as to take the PRFA input signal from the healthy channel only. With this configuration a complete independendence of action between the PPCC fault detection system and the PRFA fault detection system is achieved.

4. TEST SIGNAL INJECTION

The test signal injection is performed by signal transformers, which avoid any common mode voltage injection between two different channels. The test signal is summed with opposite polarities in all channels as shown in fig 8. When one injected signal is missing the faulty state is detected according to fig 8. The adopted scheme allows the detection of a cable disconnection. In the case of an open circuit the test signal is however not completely lost, but a contribution of ~24% of the original value is expected, due to the capacitive current in the cable.

5. FAULT DETECTION SYSTEM

Fig 9 shows the block diagram of the fault detection. An input high pass filter prevents transient saturations of the next gain stage. The signal A_3 coming from the active channel can be split in two terms:

$$A_3(t) = A_{30}(t) + A_{31} \sin \omega t$$

$A_{30}(t)$ is the feedback signal from the magnetis measurements, A_{31} is the injected test signal. A_d is a pure sine wave because it comes from the dummy channel. It matches the active channel phase within a certain error $\delta\varphi$. After multiplying the two signals a dc component M_{dc} can be retrieved:

$$M_{dc} = \frac{1}{2} A_{31} A_d \cos \delta\varphi$$

A fault occurring anywhere in the signal conditioning changes the amplitude of M_{dc} in both directions. A normal operation indow is defined together with the filter time constant for retrieving M_{dc}. Two opposite constraints are taken into account: a) the filter must provide a clean dc component; b) the reaction in case of fault must be within few milliseconds. The resulting compromise is such that the worst fault considered (saturation of a summing amplifier) does not imply the loss of the control.

An analysis of the robustness of the fault detection against plasma oscillations in the 1kHz range is also carried out. This is based on amplitudes and frequencies of oscillations observed on a certain range of JET shots. The result is that a first order low pass filter with 20Hz bandwidth and two thresholds at ±40% of the normal level can reject the disturbances and contain the response time of the fault detection in the range of 1.6 - 4.8ms. Table II shows the actions in case of faults occurring.

6. CONCLUSIONS

Since high reliability is required for JET 7MA operation, a dual stabilisation system including a protection system against faults in the electronics and in the measurements is proposed here. The intermediate result of this design is that a substantial simplification in the control system is achieved by omitting the integrators in the stabilising loop. The implementation of the dual system will also reduce the pick up of helical plasma modes, because the plasma position is measured at two opposite points in the torus. These modes have occasionally lead to abnormal stabilisation behaviour prior to a disruption.

REFERENCES

1. E Bertolini, P L Mondino, P Noll, Fusion Technology Vol 11, January 1987, pp 71-119

2. F Scheider, Proceedings 10th SOFT, Padova, September 4-8, 1978, p 1013

3. P Noll et al, Proceedings of 11th Symposium of Fusion Engineering, Austin, Texas, November 18-22, 1985, Vol 1, p33

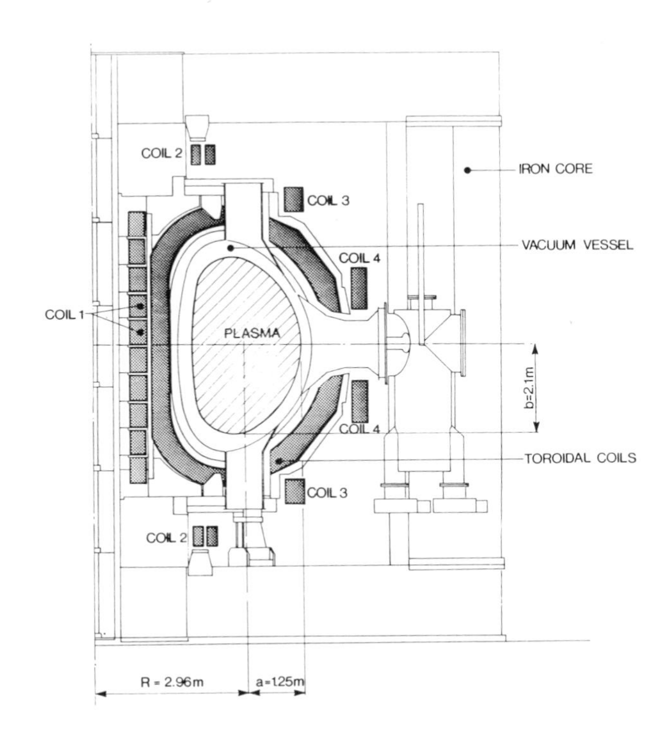

Fig 1: Layout of JET device

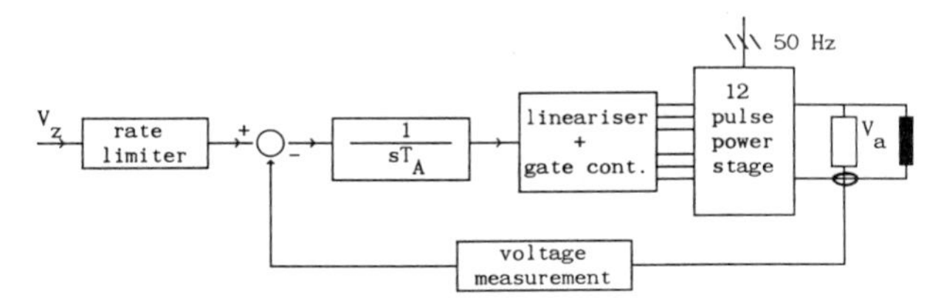

Fig 2: PRFA amplifier model

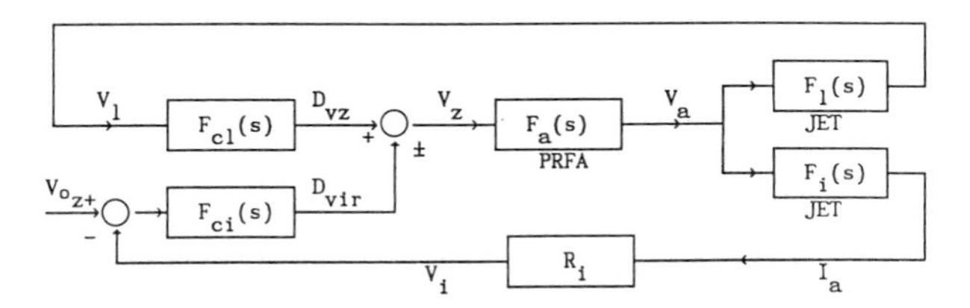

Fig 3: Block diagram of control system

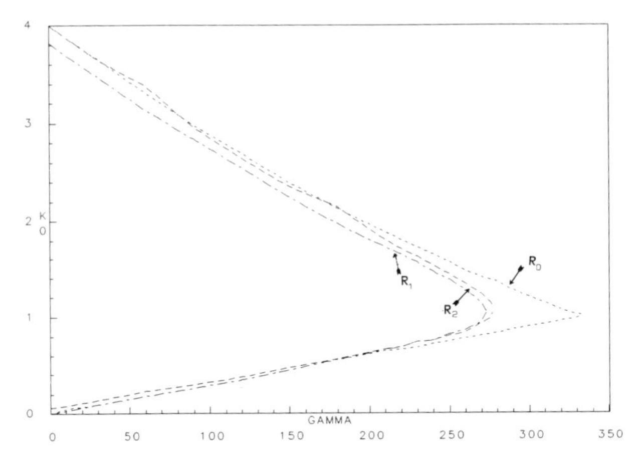

Fig 4: Stability region of the controller

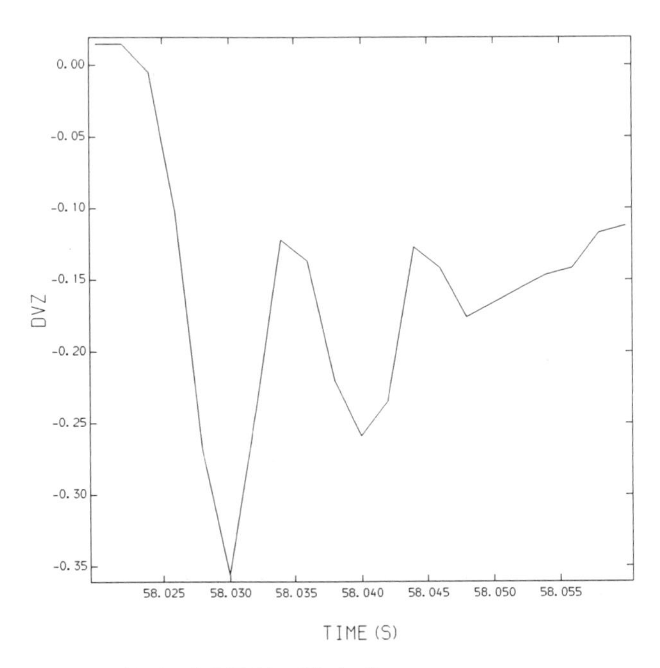

Fig 5: Shot # 14631, High frequency oscillations

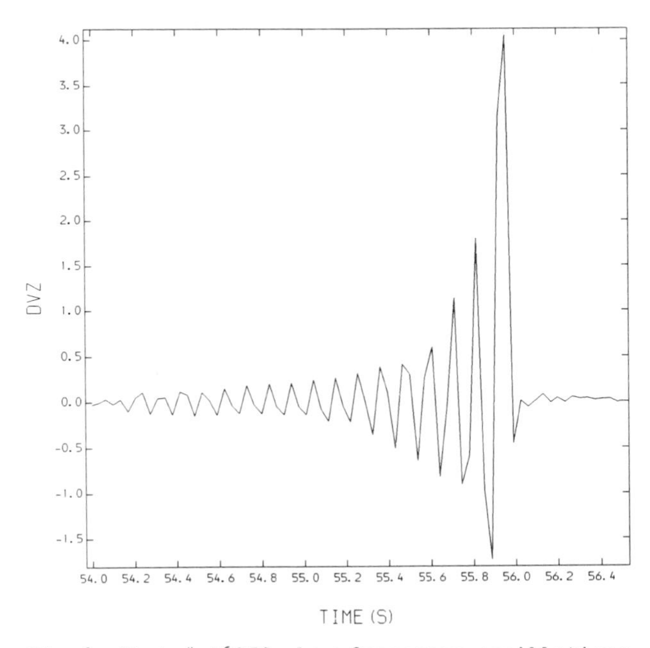

Fig 6: Shot # 16372, Low frequency oscillations

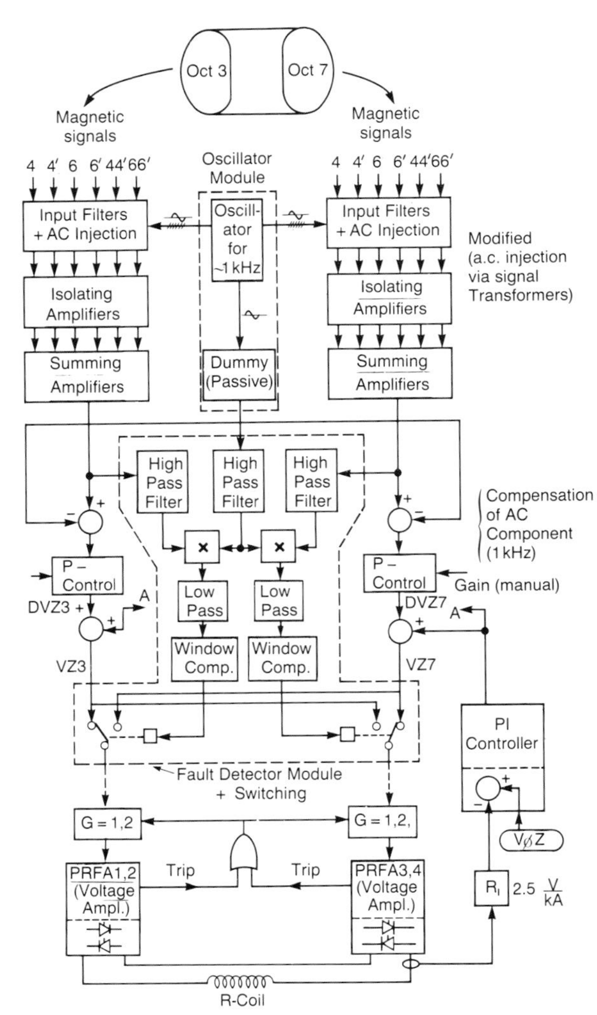

Fig 7: Block diagram of dual stabilisation system

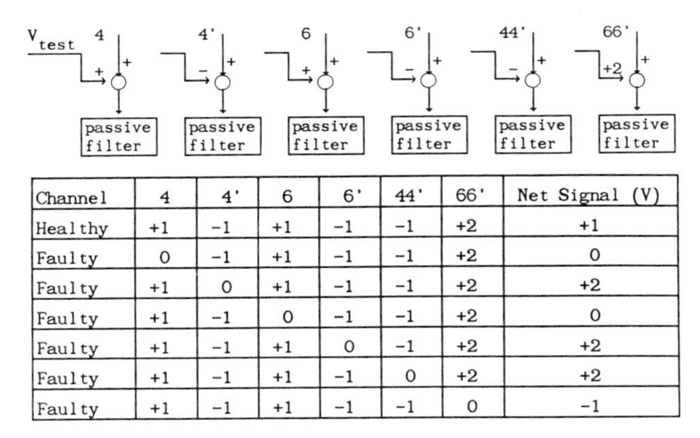

Channel	4	4'	6	6'	44'	66'	Net Signal (V)
Healthy	+1	-1	+1	-1	-1	+2	+1
Faulty	0	-1	+1	-1	-1	+2	0
Faulty	+1	0	+1	-1	-1	+2	+2
Faulty	+1	-1	0	-1	-1	+2	0
Faulty	+1	-1	+1	0	-1	+2	+2
Faulty	+1	-1	+1	-1	0	+2	+2
Faulty	+1	-1	+1	-1	-1	0	-1

Fig 8: Signal injection and working possible conditions

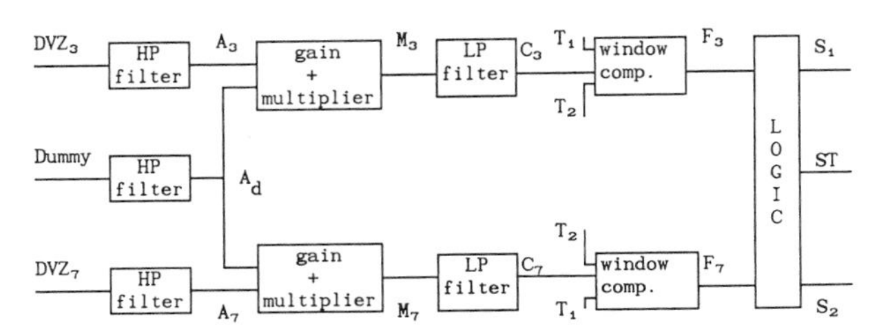

Fig 9: Block diagram of the fault detection

Table II: Actions in case of PPCC fault

ST = 0	Normal situation with V_{Z_3} driving PRFA A,B and V_{Z_7} PRFA C,D.
ST = 1	Channel 3 disconnected V_{Z_7} is driving both PRFA units.
	Possible faults:
	a] disconnection of one magnetic signal.
	b] saturation of any signal amplifier.
	c] no power supply to signal amplifiers.
	d] saturation of channel 3 of the fault detector.
ST = 2	Channel 7 disconnected V_{Z_3} is driving both PRFA units.
	Possible faults:
	same as in the case ST = 1 but in channel 7
ST = 3	Abnormal functioning with V_{Z_3} driving PRFA C,D and V_{Z_7} driving PRFA A,B.
	Possible faults:
	e] no test signal from the oscillator.
	f] channel 3 and 7 of the fault detector saturated.

FUSION TECHNOLOGY 1988
A.M. Van Ingen, A. Nijsen-Vis, H.T. Klippel (editors)

DISTRIBUTED PROCESS CONTROL SYSTEM FOR REMOTE CONTROL AND MONITORING OF THE TFTR TRITIUM SYSTEMS

G. SCHOBERT, N. ARNOLD, D. BASHORE, R. MIKA, G. OLIARO

Princeton Plasma Physics Laboratory, Princeton University, P.O. Box 451, Princeton, N.J. 08543 U.S.A.

This paper reviews the progress made in the application of a commercially available distributed process control system to support the requirements established for the Tritium REmote Control And Monitoring System (TRECAMS) of the Tokamak Fusion Test Reactor (TFTR). The system that will be discussed was purchased from Texas (TI) Instruments Automation Controls Division), previously marketed by Rexnord Automation. It consists of three, fully redundant, distributed process controllers interfaced to over 1800 analog and digital I/O points. The operator consoles located throughout the facility are supported by four Digital Equipment Corporation (DEC) PDP-11/73 computers. The PDP-11/73's and the three process controllers communicate over a fully redundant one megabaud fiber optic network. All system functionality is based on a set of completely integrated databases loaded to the process controllers and the PDP-11/73's.

1. INTRODUCTION

The goal of the Tokamak Fusion Test Reactor (TFTR)[1] is to attain "scientific breakeven" in which the fusion power produced by the plasma equals the power being applied to heat it. Due to the radioactive nature of tritium, one of the fuels required for these experiments, safe and efficient handling of tritium is an essential part of TFTR's program. The TRECAMS will provide remote control, continuous monitoring, automatic alarm responses, and control sequence execution of the tritium handling equipment at the TFTR. Critical hardwired interlocks for equipment protection and personnel safety are not a function of the TRECAMS. The preliminary design of the TRECAMS considered the process control requirements of the tritium handling equipment, operator interface requirements, high availability requirements, and currently available technology[2]. The result of the preliminary design effort was a "target system design" that included a generic system configuration and a detailed Performance Specification for the system components and support software.

The preliminary design for the TRECAMS determined that the process control requirements (permissives, control sequences, alarm responses, etc.) of the six tritium related subsystems could be supported by three redundant microprocessor-based controllers. These controllers would transfer status information and receive control requests from the Operator Interface System (OIS) via a redundant communication network. Five operator stations distributed around the facility were required as the human interface to the OIS. Three of these operator stations required multiple color graphic CRT screens to enable a tritium system operator to conveniently display sufficient status information of the subsystems.

2. VENDOR SELECTION

While surveying currently available products that could meet such requirements, it was recognized that either a Distributed Control System (DCS) or a high-end Programmable Logic Controller (PLC) based system would suffice. To induce the most competitive procurement possible, proposals from vendors of each type of system were solicited. The TRECAMS Performance Specification included in the Request For Proposal (RFP) was written as to allow each type of vendor to respond. The proposals were evaluated on their technical compliance to the specification (while also considering cost) to determine which approach best suited this application[2].

3. THE TEXAS INSTRUMENTS D/3 DISTRIBUTED CONTROL SYSTEM

The system selected by the Princeton Plasma Physics Laboratory (PPPL) via the competitive bid process was the D/3 Distributed Control System from Texas Instruments (TI). The system configuration that was supplied to meet PPPL's requirements is shown in Figure 1.

3.1. System architecture

Three D/3 Process Control Modules (PCM's) were provided as the three specified redundant microprocessor-based controllers. Each PCM was configured with the optional redundant Micro-Computer Unit (MCU) and a sufficient number of I/O Multiplexer Chassis (I/O Mux's) to accommodate the required amount of I/O modules for its subsystem interface (each I/O Mux can support 8 I/O modules, most I/O modules support 16 points).

The redundant MCU's are configured such that one MCU is active (the primary MCU) and the other one is a backup. A

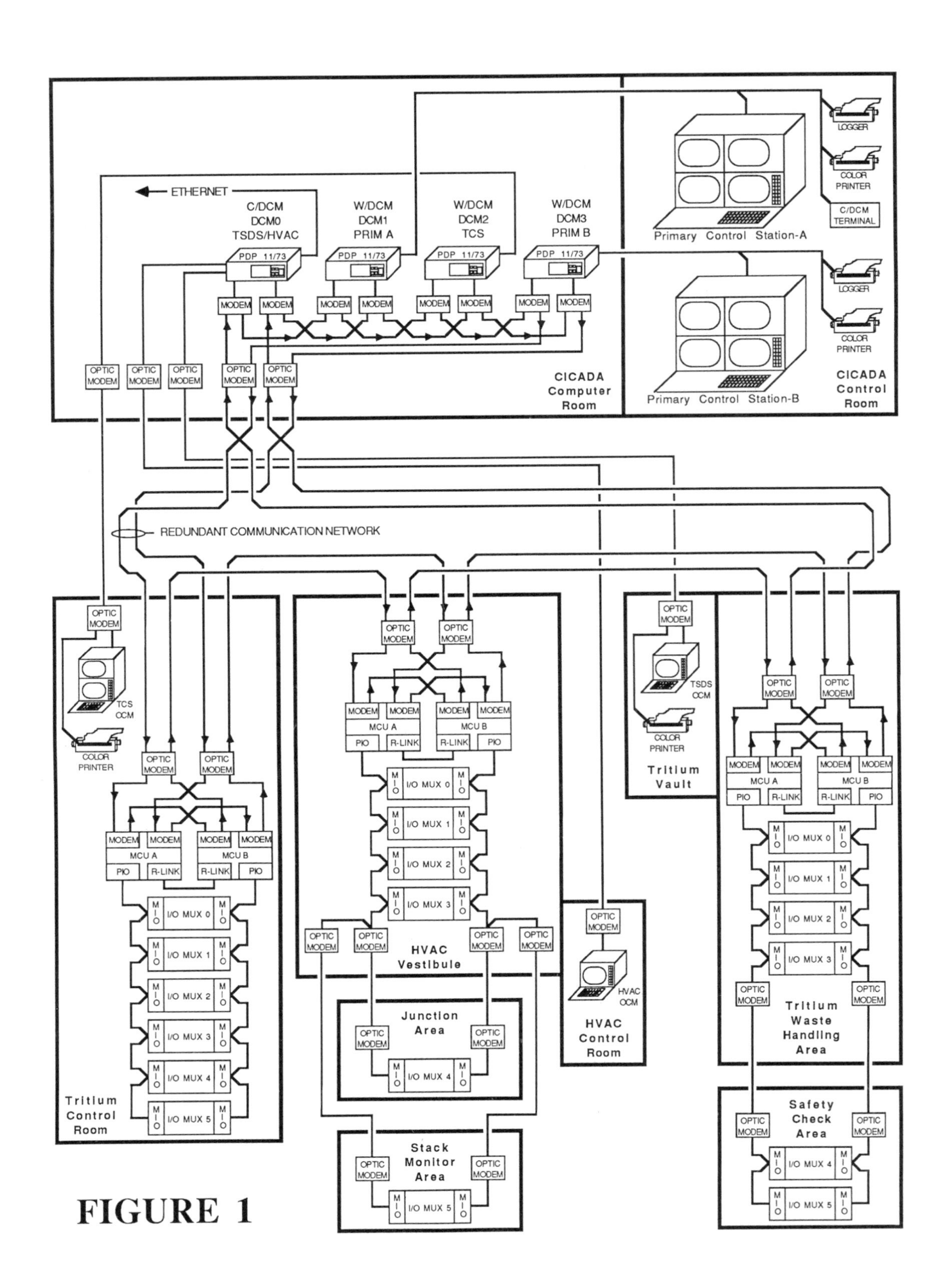
ETHERNET
C/DCM
DCM0
TSDS/HVAC
W/DCM
DCM1
PRIM A
W/DCM
DCM2
TCS
W/DCM
DCM3
PRIM B
PDP 11/73
MODEM
OPTIC MODEM
CICADA Computer Room
Primary Control Station-A
Primary Control Station-B
LOGGER
COLOR PRINTER
C/DCM TERMINAL
CICADA Control Room
REDUNDANT COMMUNICATION NETWORK
TCS CCM
MCU A
MCU B
PIO
R-LINK
I/O MUX 0
I/O MUX 1
I/O MUX 2
I/O MUX 3
I/O MUX 4
I/O MUX 5
Tritium Control Room
HVAC Vestibule
Junction Area
Stack Monitor Area
HVAC CCM
HVAC Control Room
TSDS CCM
Tritium Vault
Tritium Waste Handling Area
Safety Check Area

FIGURE 1

separate Utility Board continuously monitors the health of the MCU's and transfers control to the backup unit when necessary. Each MCU is a Multibus chassis that contains a processor board (Intel 8086 with 1MB memory), two network communication boards, an interface board to the I/O Mux's (PIO), and a communication card to the other MCU.

Due to potential electrical faults in certain experimental areas of TFTR, a fiber optic link for "remote I/O" was specified by PPPL to provide electrical isolation. Such a fiber optic link is a standard feature of the D/3 system when the distance between the MCU and the I/O Mux is of a significant distance (greater than 50 ft). In such a situation the parallel interface between the MCU and I/O Mux is serialized, transmitted optically, and then reconstructed into a parallel interface at the remote location.

Five D/3 Operator Console Modules (OCM's) were provided as the five control stations specified by PPPL. Each OCM consists of a 19" color graphic CRT screen, a 124 key keyboard arranged specifically for process control, a 64 key/indicator panel, and an alarm annunciator. An important PPPL requirement was the provision for multiple screens at some control stations to permit an operator to simultaneously view a number of the dynamic displays available (greater than 80). TI fulfilled this requirement by placing up to three additional screens at the specified OCM's for the operator to use at his discretion. A single key depression on the key/indicator panel provides keyboard access to each screen. This provision is the only non-standard feature supplied to PPPL. Its implementation was carefully evaluated by PPPL so as not to compromise the field proven, "off the shelf" nature of the system.

To support the OIS functions specified by PPPL, D/3 Display Control Modules (DCM's) were provided. A DCM consists of a DEC PDP-11/73 microcomputer with additional Q-bus cards installed for network communications and display generation. In addition to generating OCM displays and receiving operator input from the OCM keyboard(s), a DCM provides the OIS functions of alarm management, collection of current trend data, collection and storage of historical trend data, continuous monitoring of "D/3 System" status, and system security via password protection. A DCM also logs all alarms and operator commands to a disk file and printer.

One PDP-11/73 is designated in a D/3 system as the Configurator/Display Control Module (C/DCM). In addition to performing all standard DCM functions the Configurator is also used for application development and the online downloading of databases and control logic to the PCM's and DCM's as required by the RFP. To support the need for the TRECAMS to be linked to the PPPL VAX cluster, the C/DCM in the TRECAMS also contains an Ethernet card and DECNET software. This allows it to be a node on the facility's Ethernet network, providing easy transfer of historical trend and alarm data to the VAX cluster for long term archiving.

Since there is a D/3 system constraint that each DCM can only support four screens (except the C/DCM which can support only two), four DCM's were supplied (one being the C/DCM) in the TRECAMS to provide the required number of display screens.

As required in the RFP the four DCM's configured as the OIS and the six MCU's making up the three microprocessor-based controllers are all nodes on a redundant communication network. The topology of the network is a masterless token passing ring, implemented with a modified SDLC protocol (modified to eliminate the master). Data is transmitted at 1 Mb/sec using Frequency Shift Keying (FSK). Both coaxial and fiber optic transmission media are utilized in the installation. The fiber optic links are standard analog modems which optically transmit the serial FSK data.

3.2. Data flow

Figure 2 illustrates which system component performs each hierarchical control function discussed in Reference 2. Preprogrammed permissive checks and automatic sequences reside in

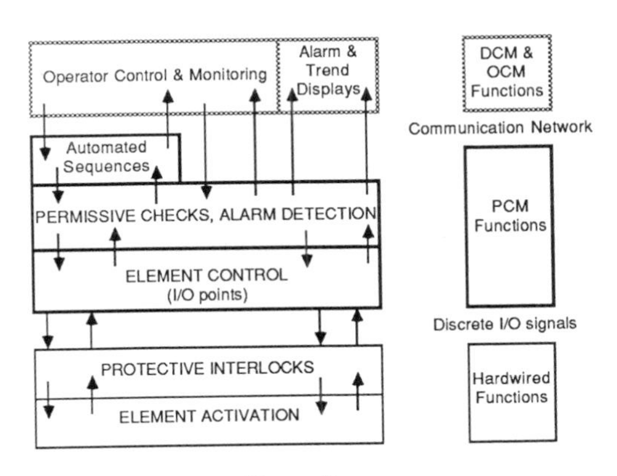

Figure 2.

the PCM and will execute without operator intervention. The operator can interact with an automatic sequence or with a process element directly. In both cases, the operator request is subject to the predefined permissive checks before actual activation is performed. In addition, certain process elements have hardwired protective interlocks that must be satisfied before a request from the TRECAMS is executed. The distribution of control functions allows the D/3 system to continue proper control of the tritium hardware despite network or OIS failures.

4. FEATURES OF THE TRECAMS HARDWARE

4.1. I/O

To provide the electrical interface to the tritium handling equipment both analog (4-20ma and thermocouple) and digital (contact and voltage) I/O were required. The D/3 System supports this need with a large variety of both digital and analog I/O types. Digital I/O can be of either voltage or contact varieties. Relay multiplexers channel analog signals to A/D converters that use a voltage to frequency technique to provide 13.5 bit accuracy. Each analog output is provided with its own dedicated D/A converter.

All analog and digital I/O are implemented with optical or relay isolation, separate ground planes and isolated DC to DC converters. These methods provide a high level of isolation (greater than 220 VRMS) and prevent noise and transients on the field wiring from contaminating the environment of the digital processing circuitry.

4.2. Reliability

Faultless handling of tritium is recognized by PPPL to be of paramount importance to the success of the TFTR project. Therefore, a major goal of the TRECAMS is to provide a dependable control system with extremely high availability. The system requirements, which call for continuous process control and sequence execution and sophisticated operator interfacing, such as multiple dynamic displays and alarm management, drive the need for computer and microprocessor-based process control and display capabilities. The resolution of these generally conflicting needs was addressed by applying three individual but complimentary philosophies in choosing the system that was supplied: redundancy, diversity and the utilization of field proven, "off the shelf" equipment and software.

At the I/O level the redundancy of critical control and monitoring points within the tritium handling equipment is carried through to the TRECAMS by having such redundant control or monitor points terminate on different I/O modules. This diversity permits failure of any I/O module without the loss of control or monitoring of a critical function. Should any I/O module fail, it can be replaced, while powered, without interrupting the function of any other I/O module (an important specification of the RFP).

From past experience at the TFTR facility one of the most likely events to bring a system down is the interruption of AC power to one or more of the system's components. This possibility was addressed by PPPL in the system installation and by TI in the design of the PCM's ordered by PPPL.

Each of the PCM (MCU and I/O Mux) power supplies contain batteries capable of maintaining complete system operation for approximately ten minutes when AC input power is removed. Additionally, the installation provides all PCM cabinets with two sources of AC power; one from an Uninterruptable Power System (UPS) and the other from the Utility Company power backed up by the TFTR standby generator (Standby power). These two sources are switched through a cabinet mounted transfer relay to the AC input of that cabinet. Normally the cabinet is fed from the UPS source which will provide protection from a power failure at the input of the UPS. Failure of the UPS or its output circuits will cause the transfer relay to switch to the Standby power until the UPS is restored.

Due to the rapidly changing magnetic fields inherent in the TFTR facility, high noise levels associated with AC power and grounding were considered by PPPL as a serious hazard to the reliable performance of the many microprocessors operating within the PCM's. Therefore, isolation transformers were installed by PPPL in the AC power feeds to the cabinets to permit the neutrals on the cabinet (secondary) side and the cabinet ground to be connected to a common grounding point. This approach eliminates electrical noise induced on the incoming neutral from being coupled to the cabinet and its electronics. As a further precaution, care was taken during the installation to provide each cabinet with only a single path to ground. This precludes ground loops from causing induced currents and noise within the cabinet structure itself. To completely eliminate ground loops and noise on the communication networks, the final system design provides all inter-area connections over fiber optic links.

The completely redundant communication network, required by PPPL between the process controllers and OIS, is provided by the optional redundant D/3 communication system supplied by TI. Each MCU and DCM is a node on two independent communication networks. Each network has its own dedicated modems, cables, communication cards and fiber optic links. Each transmitting node sends data on both networks while each receiving node selects one network from which to use the data. If a component failure causes several communication errors to be detected by a receiving node, it switches its selection to the other network. Because this selection is done on a node by node basis the full network redundancy is maintained at all other nodes during the failure.

As was mentioned, to support the necessary number of console screens, four DCM's were required in the system. To capitalize on this inherent redundancy a self imposed restriction has been followed by PPPL to have all four DCM's run identical configurations of the system software. With this approach the loss of any one DCM will only affect the OCM('s) that it was supporting. In the event a DCM fails, complete system functionality will still be provided at all remaining OCM's. In addition, the distribution of the AC power to the DCM's and OCM's was designed to preclude any single point of failure in the AC power system from affecting more than two of the four DCM's and one of the two Primary consoles.

4.3. Operator controls

Another area recognized as crucial to the faultless handling of tritium is the TRECAMS man-machine interface that permits the transformation of complex control requirements into manageable and understandable operations for the tritium system operators. The D/3 system supports PPPL's complex need for an informative and useful operating environment with a sophisticated complement of hardware and software.

As the operator's window to the TRECAMS, each operator console is provided with one to four 19 inch color screen(s). Where multiple screens are provided, keyboard access to each screen is selected by the operator. A collection of hierarchical displays, monitored on the color screen(s), are modified and controlled via a process oriented keyboard. The keyboard is divided into five keypad groups. Each group provides a specific function during operator interaction.

The SKID Panel is a software-supported 64 key/indicator panel located to the immediate right of each console's screen(s). The panel consists of a four wide by sixteen high matrix of LED backlighted touch-sensitive switches. The SKID Panel will be used to select keyboard access to screens at multiple screen consoles and to call up graphics, mimics, or other frequently used TRECAMS displays.

Two DEC LA100 Logging Printers document process alarms, system alarms, operator logins and operator-initiated process changes. An EPSON EX-800 color printer is provided at most consoles as the specified hard-copy unit for the screen displays. A "COPY" key on the OCM keyboard invokes this function.

5. FEATURES OF THE TRECAMS SOFTWARE

In the RFP, PPPL specified the need for a sophisticated database management package in which all information regarding each process point would be referenced by a single "tag name" and organized into one or more database(s). The process controllers were required to use this information to provide the local I/O processing and the OIS was required to use related information to generate process mimic displays, trend displays and annunciate alarms. The OIS was required to have a powerful graphics editor to support the generation of custom process mimic pages, a sophisticated alarm management system, and the capability to trend a large number of data points.

Another PPPL requirement was that all necessary logic and mathematical manipulation of the I/O and the execution of predefined sequences to change the operating modes of the tritium handling equipment would be performed by the process controllers. In addition, each process controller would autonomously support this requirement regardless of the condition of the OIS, another process controller, or the communication network.

These requirements are well met by TI's Continuous Control Package software, their Sequence And Batch Language (SABL), and their operator display software for alarm management, trending, faceplates (graphical representations of hardware instruments), and custom graphics generation.

5.1. The Continuous Control Package

A PCM is provided with complete instructions on how it should process each of its analog and digital I/O points via the Continuous Control Database loaded to its memory. The PCM continuously executes a set of programs that scan all inputs in a database defined sequence, processes them using database defined parameters, annunciates alarms, and controls outputs per database defined functions and logic. Because these programs are continuously executed at the PCM level, they provide the required autonomous support of the tritium subsystems.

The basis for organizing all processing within the Continuous Control Database is the External Point Name (EPN), an alphanumeric name of up to nine characters (analogous to the required "tag name") assigned to each data point. These points can be defined as individual inputs or outputs, groups of inputs or outputs, or calculated values. There are many parameters which can be specified for each EPN plus a large assortment of processing blocks can be applied to an EPN's raw input data.

Analog EPN's can be assigned alarm limits, scaling factors, engineering units, deadbands and many other parameters. Functions such as averaging, integration, calculation, mass flow and PID operations can be performed on any analog EPN's data. These "function blocks" can be linked together so that, for example, an average of a number of thermocouples could be calculated in an "average block" and then this value could be used in a "PID block" to control the power to a heater.

Digital EPN's may be assigned alarms when changing to either state and can access user defined "master devices". These "master devices" contain digital logic configured with AND gates, OR gates, Exclusive OR gates, Invertors, JK Flip Flops, One Shots, and Time Delays and can provide EPN specific control, status, and alarm outputs. EPN's that have different monitor and control point addresses but perform the same logic operations can access the same "master device" logic.

5.2. The Sequence And Batch Language (SABL)

To support the sequential control necessary to change the operating modes of the tritium handling equipment, the D/3 system provides SABL, a proprietary programming language. Using this language a PCM can be programmed to execute a sequence of functions based upon an operator's commands or upon the states of process inputs. These functions can include opening valves, turning pumps or heaters on and off, and verifying correct equipment status for each step of the sequence. During the execution of the sequence, SABL provides special status and alarm messages to the operator.

The execution of a SABL program occurs at the PCM level without the need for higher level support. This makes it an ideal approach to perform the required autonomous procedures such as system shutdown or the automatic processing of a tritium contaminated gas stream.

5.3. Alarm management

The D/3 alarm management system provides for the annunciation of both critical (P1) and non-critical (P2) alarms defined by the alarm states or limits of the EPN's plus a host of system level (P3) alarms to annunciate D/3 system failures. Each of these three types of alarms have there own dedicated set of Alarm Summary pages for display at the operator consoles. Critical alarms result in an audible signal and a change to red of associated graphics, non-critical alarms result in a change to yellow of associated graphics but no audible alarm. Both critical and non-critical alarms must be acknowledged by the operator before they are removed from their associated pages.

6. SYSTEM STATUS

PPPL has currently installed the system shown in Figure 1 and is in the process of making it fully operational. The relative ease and reduced costs of installation has emphasized another great advantage of using a commercial system. Also, the PPPL engineering effort was concentrated on the installation and support of the equipment rather than on the design of the equipment itself. Because the entire system was operational and underwent a Factory Acceptance Test and 48 hour controlled "burn-in" at TI's facility prior to shipment, it has been a straightforward effort to make the system operational again at PPPL. The application of a commercial system has saved large amounts of time and labor, corresponding to greatly reduced costs, over designing and building the TRECAMS in house, from component parts.

ACKNOWLEDGEMENTS

The authors thank the members of the TFTR DT Systems Division under R. Little; the Tritium Systems Branch under R.

Sissingh, including P. Chambers, R. Cherdack, C. Gentile, G. Levitsky and R. Rossmassler; the Engineering Department under J. Joyce; and the Computer Division under N. Sauthoff.

This work was supported by the U.S. Department of Energy under contract DE-AC02-76CH03073.

REFERENCES

1. Information Services, Information Bulletin NT-8 TFTR (Princeton University, Plasma Physics Laboratory, May 1986).

2. N. Arnold et al., Preliminary Design of the TFTR Tritium Remote Control and Monitoring System, in: Proceedings of the IEEE 12th Symposium on Fusion Engineering (California-USA, Monterey, October 1987) pp. 569-572.

FUSION TECHNOLOGY 1988
A.M. Van Ingen, A. Nijsen-Vis, H.T. Klippel (editors)

CONTROLLER DESIGN FOR PLASMA POSITION AND CURRENT CONTROL IN NET

G. Ambrosino[1], G. Celentano[1], E. Coccorese[2], F. Garofalo[1]

[1]University of Naples, Via Claudio 21, I-80125 Naples, Italy
[2]University of Reggio Calabria, Via Veneto 69, Reggio Calabria, Italy

A dynamic mathematical model of a plasma in a tokamak device is presented. The model, based on a circuital approach, is essentially addressed to the design of a control system for plasma positional stabilization and current regulation. It is shown how the model can be analysed with the methodologies of modern control theory.
The paper reports the first attempts in using this model for the design of the feedback system for the control of the plasma radial position and current in a next generation-type tokamak like NET. Preliminary simulations are carried out to analyse the interaction of the plasma radius and current with the external PF coils.

1. INTRODUCTION

As a difference to present day devices, next generation tokamaks shall have long burn phases and complex start-up procedures with significant changes of the plasma parameters: in addition the strong vertical elongation of the plasma shape introduces an inherent instability to be controlled on a fast time scale.

This makes the plasma position and current control one of the critical issues for the overall design of the machine, because of the constraints introduced on the reactor concept(e.g. maintenance scheme, passive structures, AC losses on superconductors, diagnostics, etc.).

The use of modern control theory methodologies is appealing in this respect, because of the inherent multivariable and time-varying nature of the system.

The aim of the paper is to present a generalized model of the PF system, well suited to the application of a wide class of control techniques.

First of all, a non-linear, lumped parameters model of the PF systems is presented. The approach utilized is the circuital one. Following this approach the plasma is considered as being made up of a finite number of free concentric loops of negligible mass acted on by the following forces:

1) forces due to electromagnetic interaction with external circuits;
2) forces due to eddy currents circulating in the passive structure;
3) forces due to externally applied magnetic fields;
4) additional forces of non-electromagnetic nature.

A linearized version of this model about a nominal trajectory is derived. It is shown to be particularly useful for the following purposes:

a) design of various control systems needed for plasma positional stabilization and current regulation;
b) evaluation of the power requirements of all control systems;
c) assessment of the closed loop system performance in terms of global parameters of the time response.

At this stage, the model has been used for the study of the radial position and plasma current control in NET. In this paper preliminary results are presented and commented.

2. THE MODEL

Let us consider an electromagnetic system made up with:

a) N_m floating circuits (representing the plasma) whose equilibrium is parametrized by N_q lagrangian coordinates q_i (in the following their electrical parameters will be identified by the subscript tilde);

b) N_f fixed circuits (representing the external coils and passive structure) (their electrical parameters will be identified by the subscript bar).

Both the fixed and the floating circuits are considered influenced by a magnetic field.

Besides the electromagnetic forces, the floating circuits are assumed influenced by forces of a different nature; these forces are assumed to be quadratically dependent on the currents flowing in the floating conductors.

The circuit equations for the overall system can be written in compact form as:

$$\frac{d}{dt}(\underline{L}\underline{i}) + \frac{d}{dt}(\underline{\underline{L}}\underset{\sim}{i}) + \frac{d}{dt}\underline{\Phi} + \underline{R}\underline{i} = \underline{De} \qquad (1)$$

$$\frac{d}{dt}(\underset{\sim}{L}\underset{\sim}{i}) + \frac{d}{dt}(\underset{\approx}{L}\underline{i}) + \frac{d}{dt}\underset{\sim}{\Phi} + \underset{\sim}{R}\underset{\sim}{i} = 0 \qquad (2)$$

where

L denote matrices of self and mutual inductances;

i are the vectors of currents;

Φ are the flux linkages due to external fields;

R are resistance matrices;

De are vectors of external voltages.

In accordance with the previous hypotheses, the equilibrium equation can be written as

$$\frac{\partial}{\partial q}\left(\frac{1}{2}\underset{\sim}{i}^T\underset{\sim}{L}\underset{\sim}{i}\right) + \frac{\partial}{\partial q}\left(\underset{\sim}{i}^T\underset{\approx}{L}\underline{i}\right) + B\underset{\sim}{i} + \frac{1}{2}\begin{pmatrix}\underset{\sim}{i}^T\Gamma_1\underset{\sim}{i}\\ \cdots\cdots \\ \underset{\sim}{i}^T\Gamma_{N_q}\underset{\sim}{i}\end{pmatrix} = 0 \qquad (3)$$

where

q is the vector of lagrangian coordinates;

B is the matrix whose generic element B_{ij} gives the force per unit of current flowing in the j-th floating circuit and acting on the system in the direction of the i-th lagrangian coordinate;

Γ_i is a diagonal matrix whose generic elements Γ_{ijj} represents the coefficient of proportionality between the square of the current in the j-th floating circuit and the force of non-electromagnetic nature acting on the system in the direction of the i-th lagrangian coordinate.

The electrical parameters of the floating circuits obviously depend upon the lagrangian coordinates q_i. Moreover, in order to utilize this circuital model to simulate the plasma, we assume that these parameters are affected by an additive time-varying term which accounts for plasma unmodelled phenomena, i.e.

$$\underset{\sim}{L} = \underset{\sim}{L}(q) + \Delta\underset{\sim}{L}(t)$$
$$\underset{\approx}{L} = \underset{\approx}{L}(q) + \Delta\underset{\approx}{L}(t)$$
$$\underset{\sim}{R} = \underset{\sim}{R}(q) + \Delta\underset{\sim}{R}(t)$$

As regards the parameters which take into account the influence of the external fields on the floating circuits, they depend on the lagrangian coordinates and on the spatial field configuration described by a vector function C(x,y,z,t), so that we can write

$$B = B(q,C)$$
$$\underset{\sim}{\Phi} = \underset{\sim}{\Phi}(q,C)$$
$$\underline{\Phi} = \underline{\Phi}(C)$$

In what follows we assume that the function C(x,y,z,t) describing the external fields is given by the sum of a nominal term $C_0(x,y,z,t)$ and of a time varying disturbance $\Delta C(x,y,z,t)$.

Since all the inductance terms that relate

to the floating circuits (i.e. the plasma) are in general functions of the lagrangian coordinates q_i, the model represented by eqs. (1), (2), (3) is intrinsically non-linear.

To get a model useful for the control systems design, eqs. (1)-(3) are linearized about a nominal equilibrium trajectory. Assuming

$$\underline{i} = \underline{i}_o(t) + \delta \underline{i} \qquad (4)$$
$$\underset{\sim}{i} = \underset{\sim}{i}_o(t) + \delta \underset{\sim}{i}$$
$$q = q_o(t) + \delta q$$

and substituting (4) into (1), (2), (3), using formal rules for mathemathical manipulation[1], the linearized model can be expressed in the standard linear state variable form

$$\dot{x} = Ax + Bu + Ed \qquad (5)$$
$$y = Cx + Du + Fd$$

where x is the state vector (i.e. the currents), u is the control vector (i.e. the voltages acting on the control coils), and y is the output vector (i.e. variables to be controlled), and d is the vector of external disturbances (i.e. forcing terms due to the variations of systems parameters with respect to their nominal values).

Matrices A,B,C,D,E,F, in general time-varying, depend on the electrical parameters, on their derivatives with respect to the time and to the lagrangian coordinates, and on the equilibrium trajectory.

The computer code (called NAPS-G) which has been developed on the basis of this model is subdivided into four independent sections, as follows:

Sect. 1 (preprocessor): the physical case to be studied is specified and the coefficients required in the linearized model are computed;

Sect. 2 (translator): the coefficients are arranged and manipulated in order to set up the system in the form of eqs. (5);

Sect. 3 (designer): the selected control technique is applied to determine the parameters of the controllers;

Sect. 4 (simulator): the simulation of the system subjected to a specified disturbance is performed.

The computation of the coefficients is not straightforward in the cases where shape parameters (e.g. elongation, triangularity, null point position, etc.) are involved; for this reason it is envisaged to interface the NAPS-G code with PROTEUS[2], an evolutive equilibrium code capable of providing the physical information required.

At the same time the coefficients are difficult to compute in the cases where a detailed description of the passive structures is required; for this reason the NAPS-G code is being interfaced with CARIDDI[3], A 3-D eddy current code capable of providing, after suitable post-treatment, the required inductance matrices.

3. APPLICATION TO NET

The above described model was applied to a typical NET flat-top configuration. As regards the vertical dynamics of the plasma, the code gave results in agreement with those obtained in previous works with a simulator of the sole vertical motions[4,5]. Here we report some preliminary results concerning the control of the plasma radial position and current.

For this application the plasma was considered as a single multi-filament circuit free to move only radially with no change in the shape and current profile (i.e. $N_m = N_q = 1$ in the model). A force of non-electromagnetic nature was assumed to act on the plasma in outwards radial direction, due to the kinetic pressure and having the well known expression $1/2\, \mu_o\, \beta_p\, I_p^2$.

The feedback circuit is made of two pairs of

up-down series connected PF external coils, namely the central solenoid (OH coil) for plasma current regulation, and the outermost coil (EF coil) for plasma major radius regulation. For simplicity only the vacuum vessel was considered as a passive structure, simulated by a set of axisymmetric conductors.

All the coefficients of the model were computed numerically, with the only exception being the radial derivative of the plasma self-inductance, for which it was assumed a simple analytic expression and the conservation of the toroidal flux.

The main data concerning the case under consideration are listed in Table I.

Table I

Plasma current (MA)	10.8
Plasma major radius (m)	5.18
Plasma minor radius (m)	1.32
Plasma elongation	2.1
Poloidal beta	1.4
Plasma self-inductance (μH)	10.4
Coil for plasma current regulation:	
R-coord. (m)	1.48
Z-coord. (m)	±1.25
Coil for radial position regulation:	
R-coord. (m)	9.65
Z-coord. (m)	±2.5
Vac. Vess. resistance (mΩ)	0.2

The objectives of the control system are to keep the plasma major radius and current close to their nominal values in the presence of a "soft disruption" here simulated by means of a 25% decrease of the poloidal beta with a time constant of 5 ms.

Although the coupling between the two active coils controlling plasma current and radial position would suggest the application of multivariable control design techniques, the following considerations enable, in a first approximation, the application of standard control methodology for single-input single-output (SISO) plants.

First of all the precision in the regulation of the plasma radius is much more critical than the one required for plasma current where it is possible to tolerate relatively large deviations from the nominal value. This allows to face the plasma current regulation problem by using the techniques of classical control for SISO plant, considering the influence of the EF coils as an additional time-varying disturbance.

Much more attention should be paid to the problem of plasma radius control in which a good rejection of external disturbances should be achieved. As the direct influence of the current in the OH coil on the plasma radius dynamics is negligible, so in this case the use of the classical approach for controller design is, at this first stage, justified.

These considerations are confirmed by numerical simulations carried out using a PI controller for the plasma radius control and a simple P controller for plasma current.

The tuning of the controller parameters has been made as to guarantee satisfactory stability margins and good performance in terms of disturbance rejection and response time, with a constraint of 100 MVA as maximum control power of the actuators.

The results of the simulation are shown in Figs. 1,2,3, which give a qualitatively acceptable picture of the expected system behaviour, under the assumpions made. For a more realistic controller design, further investigations are needed both on the procedures for the evaluation of the open-loop parameters and on the power supply concept and constraints.

4. CONCLUSIONS AND PERSPECTIVES

The computer code NAPS-G has been presented. The code is essentially addressed to the design of the control systems required in a new generation tokamak device, namely the plasma vertical stabilization system, the radial control, the current control and the shape control.

The code has been conceived so as to provide a multivariable model of plasma dynamics in the form required for the application of modern control methodologies. It also allows the evaluation of the performance of the control system in the presence of variations of the physical parameters of the plasma-external circuits system and its geometry.

A simple application to the plasma current and radial position control during the flat-top phase has been presented.

Work in progress includes the study of more sophisticated controllers for plasma position and current. As regards plasma shape control and the control during the start-up phase, more work will be needed because of difficulties in the evaluation of the physical parameters.

REFERENCES

1. Final report on NET contract No. 221/86-1/FU-I/NET, Annex B

2. R. Albanese et Al., Analysis of the plasma equilibrium evolution in the presence of circuits and massive conducting structures, this Conference.

3. R. Albanese et Al., 3-D Eddy Currents effects in the NET Vacuum Vessel structure, this Conference

4. E. Coccorese, F. Garofalo, Tokamak Start-up, E. Majorana series, Vol. 26, ed. H. Knoefel, Plenum Press, New York , 1986, p. 337

5. E. Salpietro et Al., Fusion Technology, 14, 1988, p.145

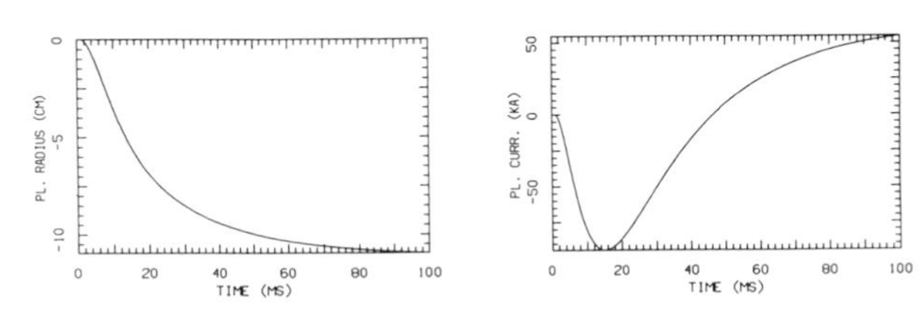

FIGURE 1
Time history of the open loop system.

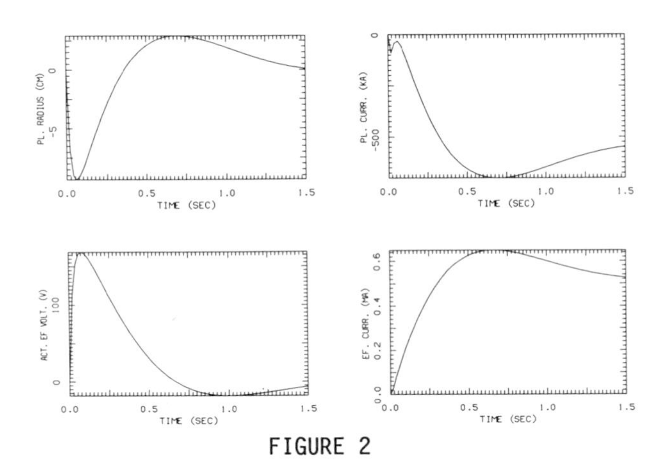

FIGURE 2
Time history of the closed loop system with only plasma major radius regulation.

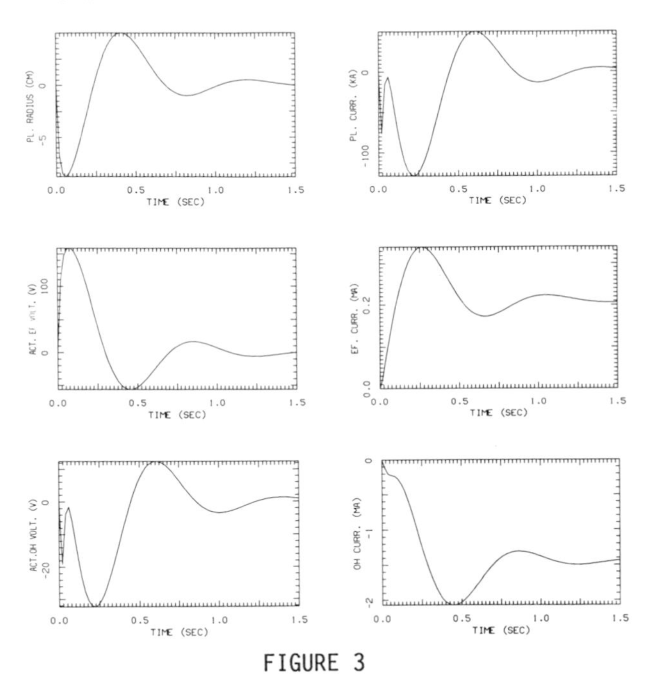

FIGURE 3
Time history of the closed loop system with both plasma current and major radius regulation.

FUSION TECHNOLOGY 1988
A.M. Van Ingen, A. Nijsen-Vis, H.T. Klippel (editors)

A REAL TIME DIGITAL COMPUTER FOR PLASMA SHAPE AND CURRENT CONTROL IN TORE SUPRA

J.M. BOTTEREAU, B. COUTURIER, B. GAGEY, C. LELOUP, D. MOULIN, P. OUVRIER-BUFFET, F. PARLANGE, P. RAZZAROLI, J. WEISSE

ASSOCIATION EURATOM-CEA SUR LA FUSION CONTROLEE
Département de recherches sur la Fusion Contrôlée
Centre d'Etudes Nucléaires de CADARACHE
F - 13108 SAINT PAUL LEZ DURANCE CEDEX - FRANCE

ABSTRACT

The poloidal field control method selected for TORE SUPRA : global approach, isoflux method, large number of measurements, requires a large amount of computation in real time. This is done on a microcomputer with a one millisecond response time.

1. INTRODUCTION

On TORE SUPRA all the plasma parameters are controlled by a single set of 9 coils and 9 power supplies : premagnetization, ohmic heating, vertical and radial field, plasma current and shape[1]. The 9 voltage references for the 9 generators are necessarily a combination of a lot of parameters as flux measurements, poloidal field measurements, plasma and coil currents. The combination has to be time dependant in order to provide plasma current and plasma positions variations. Figure 1 shows the scheme of the computation.

The equations to be solved are :

Error vector ΔX :

- $\Delta I_p = I_p - I_{ref}$ plasma current error

- $\Delta\Psi_i = |a_{ij}|\ |M_j|$ (i=1,7 ; j=1,30) plasma flux error (isoflux method)

- $\Delta I_{h,b}=(I_F-I_E)_{h,b}$ coil current error (current sharing in upper and lower largest coil) (E and F)

$|M_j|$ is a corrected measurement vector with 13 flux and 17 B_θ components.

$|a_{ij}|$ is a conversion matrix which depends on the required plasma position. See Référence 1 for $|a_{ij}|$ construction.

The PD control equation

The usual PD equation

$$V = |\alpha|\ \Delta X(t) + |\beta|\ \frac{d\ \Delta X\ (t)}{dt}$$ has been

transformed into :

$$V_i = \gamma_P(t)\ |\alpha_{ij}|\ \Delta X_j(t)+\gamma_D(t)\ |\beta_{ij}|\ \frac{d\Delta X_j(t)}{dt}$$

i = 1 to 9

j = 1 to 10

then :

$$V_i = |\alpha_{ij}|\ (F(t)\ \Delta X_j(t) + G(t)\ \Delta X_j\ (t-1))$$

by writing $\frac{d}{dt} = \frac{1}{\Delta t}\ (\Delta X(t) - \Delta X(t-1))$

and choosing $\beta_{ij} = \tau\ \alpha_{ij}$

In fact 3 different values of τ, of γ_P and γ_D have been retained corresponding to ΔI_p, $\Delta\Psi_i$, and $\Delta I_{h,b}$.

Output voltage

$U_i = V_i + W_i(t)$ where $W_i(t)$ are feedforward programmed voltages.

To design the RT computer we have to consider several objectives :

- computation of the equations,
- total cycle-time below 1 ms[1],
- connection with the central computer system of TORE-SUPRA,
- standardisation with other computers as data acquisition ones,
- input and output capabilities.

So a system based on an Intel 80386 microprocessor has been choosen.

The 1 ms target time could be met on the Intel 80386 microcomputer using the following tricks :

- use of low-level languages, PLM and Assembler for matrix multiplication,
- fixed-point arithmetic,
- no interpolation in real time ; all intermediate values are interpolated in floating point values, converted to integers and stored before the shot,
- derivative control matrix taken to be a linear function of proportionnal control matrix to reduce the number of operations.

2. SOFTWARE DESCRIPTION

2.1. The computer tasks can be divided in different phases :

- Reception of parameters and commands (2 mn before shot)

The main ones are plasma current reference, gains, programmed (feed forward) voltages, several flux conversion matrixes corresponding to different plasma positions, plasma position (matrix index) as a function of time. Except for the matrixes, these data are defined as 32 times amplitude points.

- Shot preparation phase (30 s before shot)

Initialization. Scaling and interpolation every 4 ms of commands and parameters.

- Shot phase to be detailed later.
- After shot phase Selection and transmission of stored datas as required by the Data Acquisition System.

2.2. Shot phase

During the shot phase, the computer fulfills the following tasks :

- data acquisition of signals,
- computation of flux at 13 points from the measurement of 13 saddle loops. Actually the absolute fluxes are not calculated but instead the flux differences with respect to a 14th point which acts as a reference flux. In such a way the common flux - mainly ohmic flux in the iron core - is subtracted from the measurements,
- addition of an offset value to the flux and B_θ values in order to take into account both electrical measurement offset and magnetic offset due to a small misalignement of the DC toroidal field,
- computation of 7 flux differences between the 7 points of the plasma surface and a reference point on this surface. This reference point was chosen as the internal point in the equatorial plane. It was for all discharges up to now, the contact point with the limiter, but this will not be necessarily the case in the future. These 7 flux differences and the flux at the reference point are obtained by a 7 x 30 matrix multiplication[1] in an assembler subroutine,
- computation of plasma current and plasma current error,
- computation of current differences in the E and F coils,

- comparison of all components of the error vector to limit values in order to prevent overflow in following calculation. Values in excess trigger an early soft ending of the discharge,

- multiplication of the present error and the previous step error (derivative term) by an adjustable time dependent general gain,

- multiplication of the error vector by a 9 x 10 gain matrix to obtain the feedback voltages. This matrix has many zeros so that the 9 voltages are obtained through 59 multiplications only,

- addition of time dependent programmed voltages,

- digital to analog conversion and output,

- storage of acquired datas, flux errors, surface flux, calculated voltages every 4 steps (4 ms).

With minor changes the same code works for the magnetic circuit prebiasing phase and the plasma phase although it proved easier for the prebiasing phase to control the power supplies through their constant current internal loops.

Fixed point arithmetic has the disadvantage of limited dynamic range. In order to prevent overflow, the manipulated quantities were shifted left or right which is equivalent to multiplications by powers of 2. Anyway, a compromise has to be made between resolution and maximum error quantities (overflow limit).

Error in	Resolution	Overflow limit	Value giving maximum voltage	Unit
Plasma current	1	573	150	kA
Flux	2	1000	250	mWb

A 16 mWb resolution instead of 2, has been used up to now, which corresponds to a deviation of plasma position of 6 mm for a 650 kA 65 cm plasma.

3. HARWARE DESCRIPTION

3.1. Connection with the TORE-SUPRA central computer system[2]

Figure n° 2 shows the scheme of the system. The preparation of the shot (parameters files programmation) and the data outputs are done at the NORSK DATA computer level. The Ethernet network is used to load the parameters file into the Feedback Loop Real Time Computer (FLRT Computer) just before the shot ; it is also used for the transmission of the stored datas after the shot.

3.2. The FLRT Computer itself

The computer is based around a INTEL multibus 1 system and uses ADAS and INTEL cards. Figure n° 3 shows the details of the computer architecture.

4. RESULTS

The poloidal field control was quite easy to operate and provided smooth plasma operation. Figure 4 shows the feedback reaction to an external perturbation. The feedforward voltage of D, E and F coil was abruptly changed during current plateau at 640 kA. 3 s after plasma initiation. Shown here as segmented line (linear interpolation) is lower F coil feedforward voltage. The measured voltage return in 30 ms to a value close to the original one. The small difference supports an error in flux corresponding to an 9 mm increase in plasma small radius.

This error is due to limited loop gain. Because of different perturbations in E and F coil voltages, while gains for plasma position are equal[1], the current difference start to increase which explains that F coil voltage slowly decreases during the following two seconds.

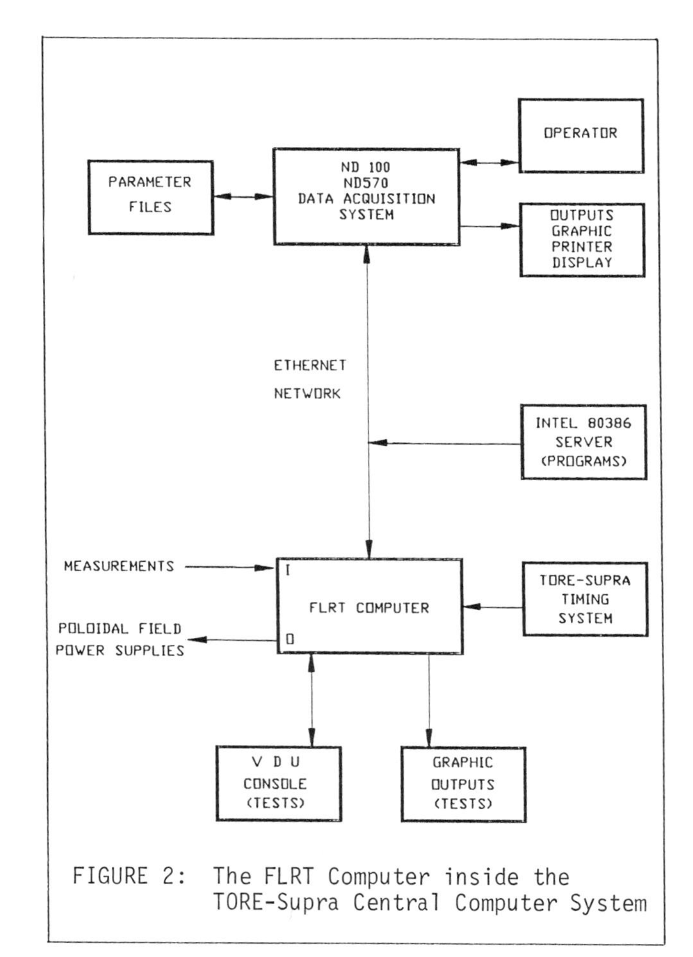

FIGURE 2: The FLRT Computer inside the TORE-Supra Central Computer System

REFERENCES

1 J.M. ANE and al.
A fully integrated field coil and power supply system for plasma boundary shape and position control in TORE SUPRA.
This conference.

2 B. GUILLERMINET and al.
The TORE SUPRA data acquisition system.
This conference.

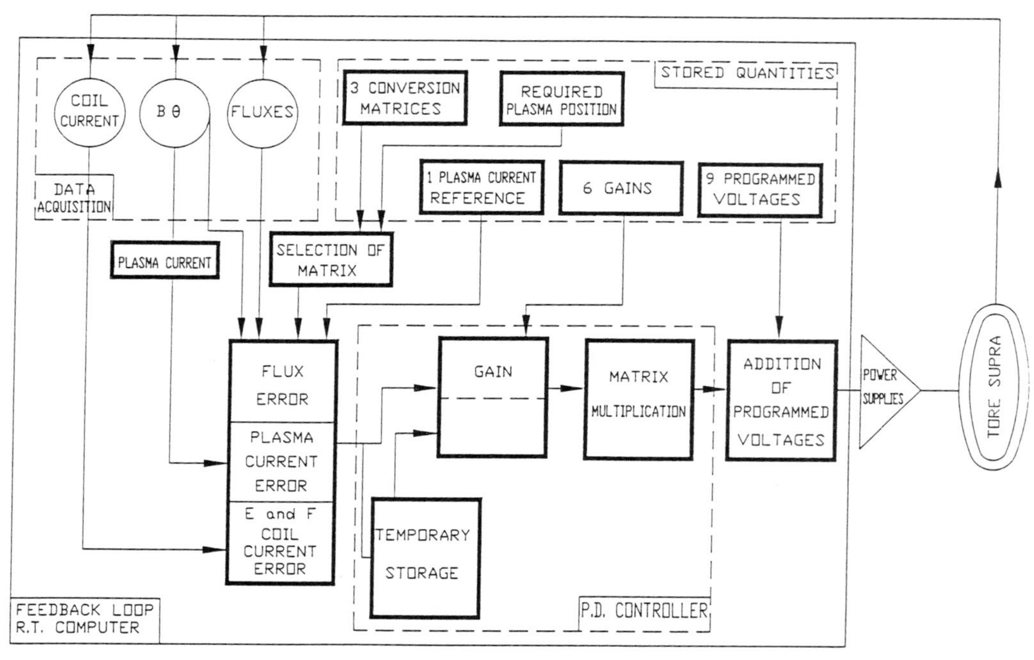

FIGURE 1: Scheme of the Real Time Computation

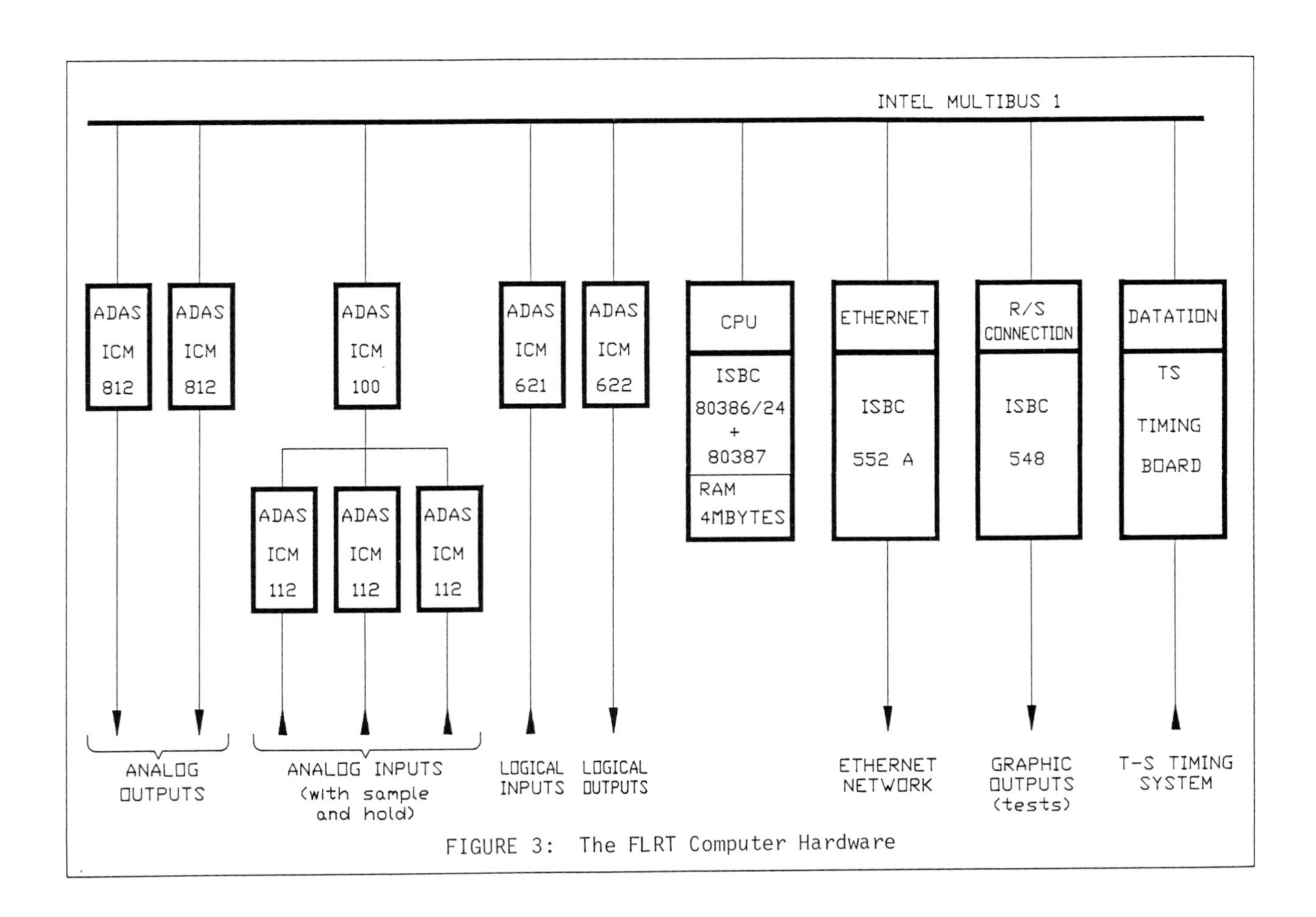

FIGURE 3: The FLRT Computer Hardware

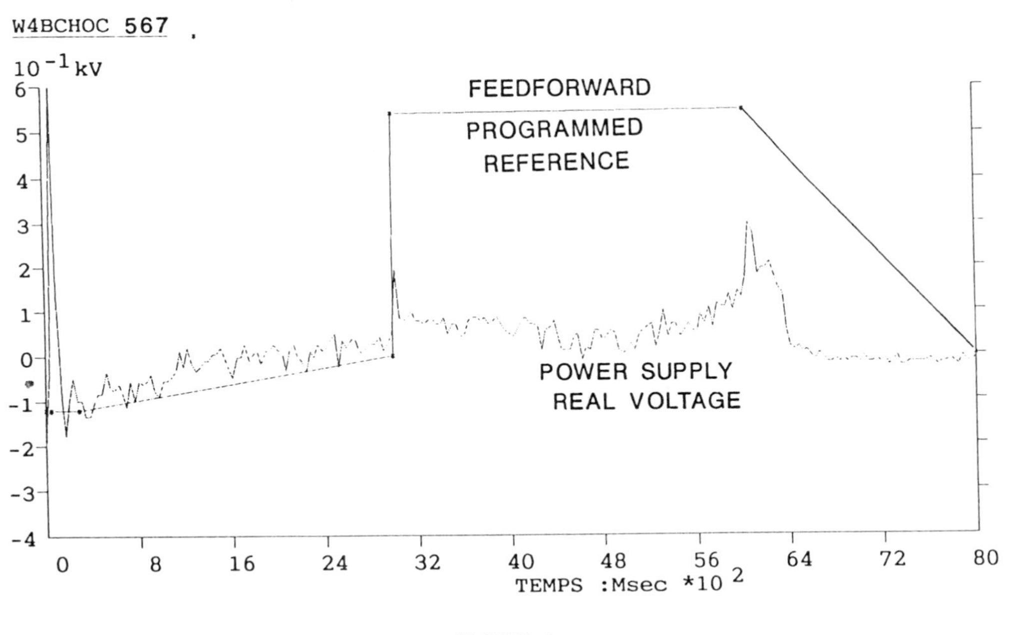

FIGURE 4

FUSION TECHNOLOGY 1988
A.M. Van Ingen, A. Nijsen-Vis, H.T. Klippel (editors)

THE TORE SUPRA DATA ACQUISITION SYSTEM

S. BALME - J. BRETON - B. COUTURIER - H. DEMARTHE
B. GUILLERMINET - B. GAGEY - F. HENNION - F. JUNIQUE
M. LELUYER - D. MOULIN - B. ROTHAN - N. UTZEL

Centre d'Etudes Nucléaires de Cadarache
Association EURATOM-CEA sur la fusion
Département de recherche sur la fusion contrôlée
13108 - SAINT PAUL LEZ DURANCE CEDEX FRANCE

1. ABSTRACT

The TORE SUPRA real-time acquisition system was designed to allow continuous data acquisition over long plasma pulse duration. The network is distributed in three levels of computers that realize simultaneously data acquisition, transport, processings and storage. This paper describes the hardware and software architectures.

2. INTRODUCTION

The TORE SUPRA data acquisition system was designed to provide continuous data taking and processing. TORE SUPRA started data-taking in April 1988, and during the last five months more than 500 plasma pulses, with a duration up to 9 s, have been processed and recorded. During this period, 3 up to 11 diagnostics have been connected to the data acquisition system giving an overall data size around 1 Mbytes per pulse. Works in progress are the automatic working mode, Camac data acquisition, time sharing data processing and display and performance increase ; nominal data rate is up to 1MBytes/s during short pulses and 200KBytes/s in continuous. Improvements under study are the connection of heterogeneous data acquisition computers and of graphic workstations, and the creation of a reduced shots database. Hardware computer architecture has been designed to allow modularity (easy way to add or remove a data acquisition computer or a front-end computer) and at the same time, to allow correlations analysis between signals coming from various experiments. Modularity is achieved by a large number of processors (usually, one for each diagnostic) related to a front-end computer via an Ethernet network. Correlations are done on a dedicated computer (Norsk Data ND570).

3. HARDWARE CONFIGURATION

Here, we remind the hardware architecture :
TORE SUPRA computer architecture scheme :

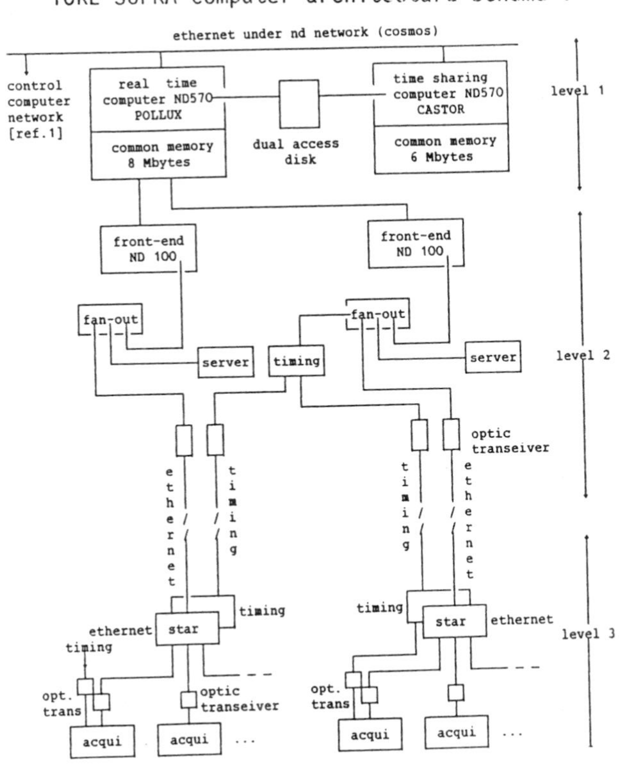

Data Acquisition components :

Data acquisition computers :

- 15 packaged microcomputers :
  readout boards in Multibus 1 format, INTEL 80386 microprocessors, 80387 floating-point processor, 2 up to 4Mbytes RAM, diskless,

Ethernet board, interface for parallel CAMAC without direct memory access and with simple interruptions handling (LAM).
- 1 micro-Vax, complete serial in CAMAC, Ethernet board (connection in progress).
- 1 ND110 (from Norsk Data) 3MBytes RAM, 75MBytes disk, complete parallel and serial CAMAC, Ethernet board, shared memory with the real-time ND570 computer (connection in progress).

Server unit :
- 2 INTEL 80386 + 80387,2Mb RAM, with a 65Mbytes disk, Ethernet board.

Front-end computers :
- 2ND110 (from Norsk Data) 3MBytes RAM, 75MBytes disk, Ethernet board, shared memory with the real-time ND570 computer.

Main computers :
- 1ND570 (from Norsk Data) used for real-time data collecting and processing (12MBytes RAM, 6250 bpi 125 ips magnetic tape).
- 1ND570 for time sharing processing and software developments (8MBytes RAM, 6250 bpi 125 ips).
- 1ND570 for time consuming analysis (6250 bpi 125 ips).
- 1ND510 used for data exchange with the control/command system / réf. 1 /.

Readout :
- Boards in MULTIBUS 1 format with flip-flop buffers (64 Kbytes) for analog and digital inputs.
- CAMAC modules.

Links :
- Ethernet with fiberoptic insulation are used between the acquisition units and the front-end computers. Optical star and transceivers are implemented.
- Fast shared memory (MPM5) with CPU synchronization (OCTOBUS) between the real-time and the front-end computers.
- Ethernet Cosmos between the main computers.
- Disk switch for fast data transfer between the real-time main computer and the time sharing computer.

Timing system :
- INTEL 80386 + 80387 with fiber-optic channels.

4. OVERALL WORKING SCHEME

Plasma pulse is mainly divided in three phases :

Phase 1 : OUT OF PULSE

In this phase, there is no data taking. Physicists are working in time sharing on the earlier recorded plasma pulses and doing some modifications of their diagnostic parameters. Physicists parameters are used to describe diagnostic acquisition and process, pilot parameters to describe timing system, acquisition strategies according to the events, and real-time displays. Each physicist defines the parameters tables used by his diagnostic with a specialized macro-language which is syntactically and semantically analysed.

Phase 2 : START OF PULSE

At this stage, the pilot decides to start a plasma pulse, the parameters are then read from the users files and stored in a memory structure. A message is sent to each data acquisition computer ; parameters from the memory structure are "compiled" and downloaded to the corresponding computers via the messages service. Then, an acknowledgment, which is the present data acquisition computer phase state, is sent by each data acquisition computer to the acquisition supervisor task, called CHORAL. Overall synchronization (computers and tasks) is done via the messages service.

Parameters compilation consists mainly, in a format translation depending on the computer type, and in a parameters ordering suitable to the receiver, according to the time sharing user description.

Phase 3 : DATA TAKING

Once we got all the acknowledgments, CHORAL sends a message to the timing system which is in charge of starting an accurate distributed clock on all the data acquisition crates. From now, the tasks synchronization is done by the data flow. At the low level, flip-flop buffers are used to get data continuously, which are then reduced, timed, formatted in frames and written in a multi-frames buffer. Frames are some amount of data, with a header made of a signal identifier, a frame number, the acquisition time, the period between each data, and an event identifier. Frames go to a front-end computer via an Ethernet network and then to the central computer through a shared memory. They are then available in a multi-frames buffer to all the real-time tasks, some of them are used for continuous monitoring, one is used for fast disk storage (up to 1.5 MBytes/s).

In the high level real-time computer, one task is in charge to modify some parameters in the memory structure, on request (message) of a data acquisition unit or of the control command system.

Once the pulse time is elapsed, the timing system sends an interrupt to each low level computer, which stops the data acquisition, waits for the end of the network sender task and sends a message to CHORAL. In the high level computer, we update the frames pointers in the parameters memory structure and record them in a file. In order to let long time plasma pulse analysis with the time-shared computer (CASTOR), we transmit the two files, parameters and frames, in a fast way using the disk switch : at present time, we use the internal Norsk network called Cosmos. The computers are then, all again, in the "phase 1" state.

5. MESSAGES SERVICE

This messages service is available on all the TORE SUPRA computers and is of prime importance for the data acquisition system because it assures the computers synchronization and parameters loading. Messages are sent without or with acknowledgment. In the last case, a time-out must be supplied by the message sender. Messages from a computer to an other (among the data acquisition messages service), are always transmitted to the real-time computer (POLLUX) ; hence there is only one routage table in the system.

Internally, user's messages are encapsulated with 16 bytes header defining the sender and the receiver. This header is the same than in the TORE SUPRA control command system, giving so the possibility to communicate directly from one task in the data acquisition system to an other in the control command system for instance. Sender, or receiver, are defined by two bytes : the computer number and the task number. Message length is up to 32760 bytes.

Between the high level computer (ND570) and the front-end computer (ND110), we use the shared memory for the data transmission and a Norsk Data specific bus (Octobus) for the task synchronization. Outside Norsk Data set, we use Ethernet at level 2 with a home-made protocol, based on HDLC/X25/BSC, in order to achieve high performance both for the messages service and the frames transport. We manage the packet loss, the duplicated packets, the flow rate and the connections. Packets collisions are in charge of the sublevel of Ethernet while we take into account the packets assembly/deassembly for messages larger than 1500 bytes. The front-end computers are mainly dedicated to this work in an efficient way ; each of them has to manage up to 18 low level computers which can send Ethernet packets in any order. To achieve this, we have written a library to handle messages queue and to manage a dynamic message area. At the present time, we have 256Kbytes for the dynamic messages area as for the dynamic frames area. On the Intel side, we have written a Fortran 77 messages service with messages queue handling, dynamic memory allocation, garbage collector,... in a way as portable as possible in order to easy the heterogeneous computer connection.

Between one low level user task and one high level user task, we achieve 70Kbytes/s for large messages (20000 bytes/message).

6. PARAMETERS EDITING AND STORAGE

Each diagnostic has some parameters files which are structured in indexed sequential mode. For a shot, one of them must be selected ; standard parameters are then checked and if not valid, the file is rejected. Standard parameters describe acquisition modules, signals features, acquisition rates and so on ; specifics parameters are anything else the physicist needs. Parameters files size is 5000 up to 50000 bytes. Pilot is a special diagnostic which settles on the timing system, the diagnostic acquisition rates according to the events, the signals he wants to display in real-time... . Physicists have the PREPA program for parameters handling : it is an interactive software which allows parameters manipulation as self-described entities ; these entities are often oriented (modules, signals...), but the general structure is opened. In addition of PREPA, there is an other program called PRE-COU, which is a graphic interactive program for parameters editing : reference curves which may be compared with real associated curves.

During the real-time acquisition (phase 3), data frames are got from the multi-frames buffers of the low and high level computers. In time sharing, they are got from the parameters and frames files : to assure the link between the two files, the parameters file contents as many tables as signals, filled, at the pulse end, with the frames pointers. A library provides time sharing access to parameters and frames ; an other one gives real-time access to the parameters only. Modules to access parameters are the same in real or shared time, apart from link-editing. It uses a special ND500 feature, called file segment, to get a fast and easy access. In order to allow other computers access, a parameters/frames network server is under development.

7. TIMING SYSTEM

A peculiar data acquisition unit is in charge of the timing system. During the phase 2, timing parameters are described and downloaded from the central computer : timing codes are 8 bits length Manchester II format (10Mb/s rate). At any event time, the corresponding code is sent to each data acquisition unit, which is interrupted.

A feedback network, also based on a 10Mb/s Manchester format, can collect external events, which are (or not, depending on parameters) re-emitted to the data acquisition units (for instance, some POLOIDAL's phases, disruptions, ...).

The timing codes can also be decoded by hardware devices (EUROPE and MULTIBUS I format) to provide pulses, triggers or crenels to drive some mechanisms. The accuracy is 1 μs. A code is used to reset counters on the timing board. These counters (32 bits, 1 MHz) are read "on-fly" by the acquisition units when it is necessary.

All the systems use the same basic clock (1 MHz), which is issued from Manchester, to provide the acquisition rate (from 1 μs to a few seconds) giving an accurate sampling clock.

A second network, based on the same principle but at 100 kHz rate, is used to synchronize the control/command devices, with a 1 ms timing accuracy.

8. DATA READOUT AND LOW LEVEL COMPUTERS

At the low level computers, we have one task for each acquisition board which is in charge to sort the data (depending on the timing strategy), formats them in frames and gives them to the data consumers tasks. In order to manage the multiple data streams and the tasks synchronization between the various producers and consumers, we use a data acquisition system originally designed at CERN. The data frames are stored in a circular buffer, and used in parallel by the consumers which are allowed to take e.g. all data, a fraction or only on request (respectively, case of storage, monitoring and display tasks). Low level buffering is needed to derandomize the data flow coming from the readout tasks and to allow the Ethernet sender task to wait when the network is busy for a small amount of time. Easy tasks configuration (consumer/producer may be added or removed) allows us to have the same software either in a stand alone mode (during tests period) or network connected.

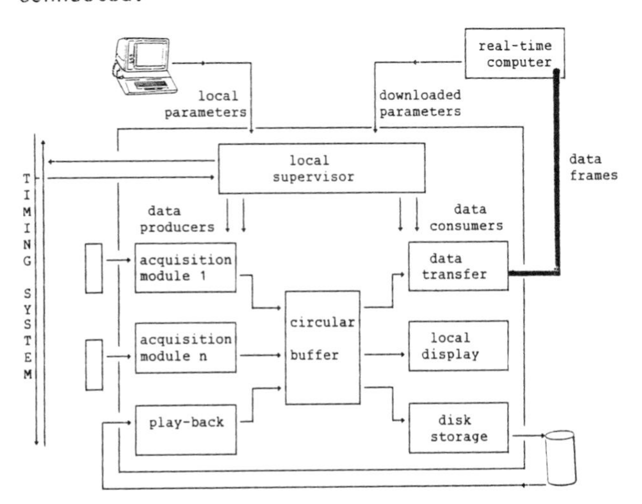

Low level data acquisition structure

The local supervisor

Out of pulse (phase 1), it provides some diagnostic controls and, at the end of a pulse, aborts all the data producers and consumers except display. At the start of a pulse (phase 2), it dispatches the downloaded (TS pulses) or local (tests) parameters to the producers and consumers, then provides some diagnostic controls and, at last, initializes the configured data producers and consumers, according to the working mode : autonomous or network connected. During a pulse (phase 3), it fills a table of timed events as soon as received from the timing system. Besides, during connected runnings, each phase change is reported to the real-time computer, via the messages service.

The data producers

Each acquisition task handles one continuous acquisition module with memory switching : an acquisition buffer is read as soon as filled. Then, the measurements are sorted and preprocessed according to the strategy of the current event at the acquisition time : each measurement is timed. A library provides different kinds of sorts : general or peculiar to a diagnostic (neutron flux computation,..). At last, the data frames are formatted according to the TS data structure and written in the circular buffer. Reading, sorting and formatting the measurements takes from 8 μs/byte up to 30 μs/byte depending mainly on the complexity of the sort.

The play-back task reads back the data frames stored during a previous pulse.

The data consumers

The data transfer task translates integer and real data in the ND format (4 μs/byte). Then it sends the translated data to a front-end computer via Ethernet with the same protocol as in the messages service. Frames are sent as soon as some amount of data is available. The data flow regulation is done by the front-end receiver task. The frames transport channel in the front-end computer is similar to the messages channel.

The disk storage task stores the data either on a local disk, or on the server disk.

On request, the display task plots the measurements of one or more signals on a local graphic terminal.

9. HIGH LEVEL COMPUTERS

One interactive program (CHORAL) assumes the overall synchronization of all the computers involved in the data acquisition system through the messages service. It sends commands to the low level local supervisors and receives acknowledgments. Errors are reported to it, giving to the pilot any choice to stop the data acquisition system or to go on. Connections with the TORE SUPRA control/command system will be achieved later. CHORAL has in charge to start/stop the parameters memory loading task (STRUCT), the parameters downloading task (TELE) and the real-time supervisor task (INIDAS). This last task schedules the frames producers and consumers which are managed around a circular buffer (1.5Mbytes actually but a typical size will be 3Mbytes).

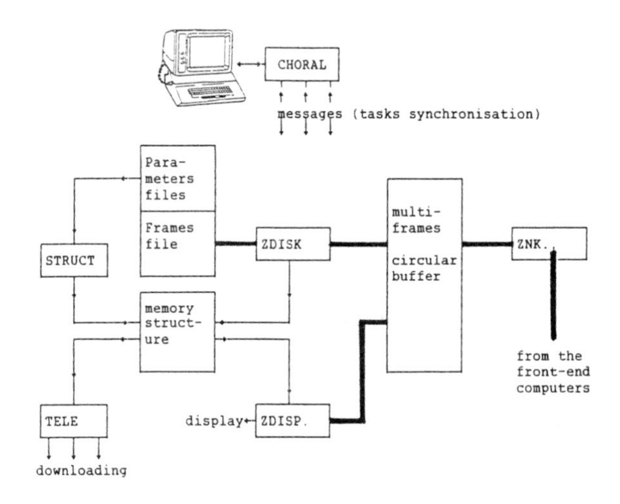

Real-time computer software architecture

STRUCT collects the parameters from the users files if some of them have been updated, and loads them in a memory structure during the phase 2 (start of pulse). This structure is a concatenation of all the diagnostic parameters files concerned by the shot, preceded by a general header allowing direct access to any parameters from any diagnostic.

TELE is the parameters "compiling" and downloading task which keeps informations about which diagnostic parameters files have been updated and how successful has been the previous load, avoiding unnecessary parameters transfer.

ZNK... take the frames packets from the memory shared with the front-end computers and put them in the circular buffer where they are available for the consumers. Actually, we have two producers for each front-end computer.

ZDISK is a special consumer which is not waked up for each frame (standard activation mode for the consumers) but for a large amount of data (64Kbytes), and writes them directly on disk. The data rate is up to 1.5Mbytes/s if the file has been previously created. ZDISK has to keep trace of the frames number for each signal and must update the frames pointers tables at the pulse end.

ZDISP... are some versatile display consumers. At each start of pulse, they read the requested parameters in the memory structure (up to 8 signals/page) and get the signals frames as soon as they are written by the producers. Hard copy on a laser printer is automatically done if asked, at the pulse end. Graphic library is μGD3, originally designed at CERN, for the real-time display (either in the data acquisition unit or in the central computer), and GKS for the time sharing visualisation.

Parameters and frames files are copied on a large disk using the COSMOS network, but soon, we will use a disk shared between the real-time and the time-sharing computer. Long time storage is actually under study.

10. CONCLUSION

The TORE SUPRA data acquisition system was ready at the end of February 1988. It illustrates a number of trends in data acquisition computer architecture, namely a high continuous data rate, on line data processing, data reduction and large amount of data storage. Improvements under study are mainly the performance increase at each level and a data base creation for the long time storage.

11. REFERENCES

/ 1 / "The TORE SUPRA control computer System" B. GAGEY and Al., proceedings of the XIVth Symposium on Fusion Technology, Avignon 1986.

FUSION TECHNOLOGY 1988
A.M. Van Ingen, A. Nijsen-Vis, H.T. Klippel (editors)

THE DATA ACQUISITION SYSTEM OF THE TOKAMAK DE VARENNES

J.M. LARSEN[#], D. BEAULIEU[*], P. de VILLERS[*] and E.S. ROBINS[*]

Centre canadien de fusion magnétique (CCFM)[+], 1804 Montée Ste-Julie, Varennes, QC, J0L 2P0, Canada

The data acquisition system of the Tokamak de Varennes is presently composed of two data acquisition and one data processing computers. The software system has two components: DIOS-V which handles the CAMAC hardware, and GALE-V which manages the data acquisition process from parameter entry to data processing, display and archiving. The system is being modified to run on a network of MicroVAXes.

1. INTRODUCTION

The operation of the Varennes Tokamak requires two different machine interfaces and, consequently, two separate systems for machine control, and for diagnostic data acquisition and processing.[1]

This paper concerns the data acquisition system. After reviewing the basic requirements, it presents the hardware configuration of the system and describes the software systems developed to support the experiment. Finally, some comments on further developments are presented.

2. BASIC REQUIREMENTS

2.1 Software configuration

Since the tokamak diagnostics are modified constantly, a data acquisition system capable of evolving without major reprogramming is required.

2.2. Large data handling capacity

The large quantity of collected data requires temporary storage in the data acquisition units, fast data retrieval and transfer software, large computing power for data processing and display.

2.3. Permanent archiving of the data

The archiving system must permit the examination of the data and the parameters of a given shot at any time.

2.4. Instrumentation standard

The tokamak environment requires many data acquisition units spread over a large area and electrically isolated from each other. A modular standardized system such as CAMAC is desirable to support changing requirements.

2.5. Event scheduling and central timing

A central timing facility is needed to control and synchronize the tokamak operations.

3. DESCRIPTION OF THE HARDWARE

The data acquisition system of the Tokamak de Varennes is composed of three computers, as shown in Figure 1. The computers are linked through an ETHERNET/DECNET communication link.

3.1 Data acquisition computers

The three computers have different duties. The PDP 11/44 and the MicroPDP 11/73 drive a

---

INRS-Energie, 1650 Montée Ste-Julie, Varennes, QC, J0L 2P0, Canada

* MPB Technologies, 1725 North Service Road, Trans-Canada Highway, Dorval, QC, H9P 1J1, Canada.

+ The CCFM is a joint venture of Hydro-Québec, Atomic Energy of Canada and the Institut national de la Recherche scientifique (INRS).

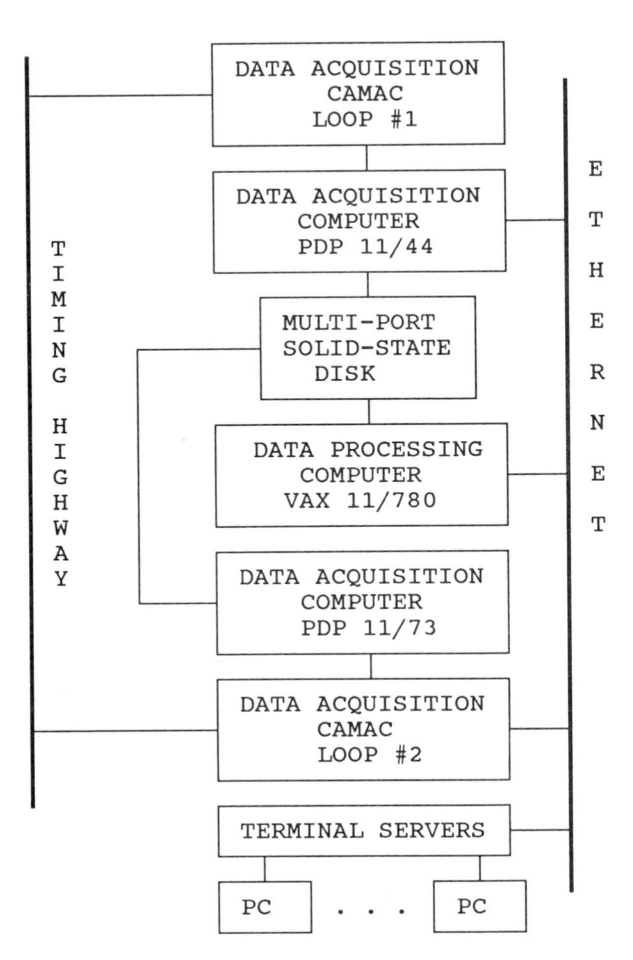

FIGURE 1
Hardware configuration

fibre optic, 5 megabyte per second, CAMAC serial highway using a 2050-2060 Kinetic System Serial Highway Driver and LeCroy 5211 U-Port adapters. They read the data from the CAMAC modules, and store them on the solid-state disk. A VAX 11/780 computer is used to process the data and display them on the physicists' terminals.

3.2 Data acquisition modules

Many CAMAC module types are in use, the most common being LeCroy 8212, 32 channel, 5 kHz digitizers, and DSP TRAK I systems, with 8 channel, 100 kHz digitizers. The digitizers are linked to CAMAC memories having 32 to 256 Kwords of memory. There are approximately 200 channels installed in the present phase of the experiment, with about 2 megabytes of CAMAC memory.

3.3 Central timing system

The central timing system run by the control computer provides facility-wide timing signals. It consists of CAMAC modules linked by an optical fibre highway carrying a 1 MHz biphase clock. There are two types of timing modules: encoders which accept hardware or software triggers to transmit codes on the timing highway, and decoders which monitor the highway, recognize the coded signals and use them to trigger the equipment. The decoders also contain five programmable counter timers providing a wide range of pulses, delays and clocks derived from the signals on the highway. These modules where designed for the Tokamak de Varennes by MPB Technologies Inc.

4. DESCRIPTION OF THE DATA ACQUISITION SOFTWARE

The data acquisition software is made up of two parts: DIOS (Dynamic I/O System), which manages the CAMAC loop on the PDPs, and GALE (Generalized Acquisition of Laboratory Experiments), which manages the data acquisition process on the PDPs and the VAX. Both systems were initially developed at the "Max Planck Institut für Plasma Physik", in Garching bei München, West Germany and were obtained through the TEXTOR project of the Kernforschungsanlage Jülich (KFA), West Germany, where they had been modified. The Varennes version of both systems has been extensively modified to support new equipment and local needs.

4.1. The DIOS-V system

The CAMAC software handles the transfer of large blocks of data in a multi-user environment. It also drives the CAMAC modules through simple functions at the user level. This requires a CAMAC I/O system tightly coupled with the PDP operating system, RSX-11M, which dictates its structure, its operation and its limitations.

DIOS considers the CAMAC system as a peripheral having one controller, the serial highway driver, and many units, the CAMAC modules. The CAMAC I/O requests are handled by RSX-11M as for other types of peripheral. Since the units have different characteristics, DIOS-V performs the work necessary to hide that fact from RSX-11M.

4.1.1. The DIOS DA: driver.

The only CAMAC device known by RSX-11M is the single unit DA device. DA:, its driver, receives some of the CAMAC I/O requests and handles all the LAMs from the CAMAC modules. Since the size of a RSX-11M device driver is limited, only the basic DIOS functions can be handled by DA:. The rest of the work is done in privileged tasks called ACPs (Ancillary Control Processors).

4.1.2. CAMAC module support.

The CAMAC modules form a constantly changing set of units to be linked to DA:. The loader ACP regulates access to DIOS and handles the request to use a CAMAC module. It creates the control blocks needed for I/O request processing and links them to the system tables. It then transfers control to the module specific driver ACP. Any further I/O request for the module is sent to the driver ACP, while system events (interrupts, cancel requests, time-out) are handled by DA:.

The module drivers have initialize and terminate functions to set up and reset the module and its database. Other functions, such as read and write functions, are included in the driver, as needed. These functions perform high level operations involving many interventions on a module, including interrupt service.

4.2. The GALE-V system

GALE-V is the interface between the physicists and the tokamak. Its components relate to various aspects of the data acquisition process: parameter entry, data collection, archiving, processing and display.

4.2.1. Basic assumptions.

GALE-V supports an experiment producing data during a SHOT of limited duration. A description of the equipment is used by the data acquisition software to perform its tasks. The content of this file is saved with the shot data, ensuring that the data files can be examined at a later time.

4.2.2. Components of GALE-V.

GALE-V comprises the following components:

- The configuration file contains the parameters describing the configuration and the equipment.
- The symbol file enables access to fields in the configuration file by name.
- The data file contains the data a shot, and a copy of the configuration file at that time.
- The DLG program is used to create and modify the configuration file for a diagnostic, and to adjust the equipment parameters for a shot.
- The ACQ program manages the data acquisition process according to the configuration file. It uses DIOS-V to perform CAMAC I/O requests.
- The data acquisition procedures manage the communication between the user interface (VAX) and the machine interface (PDPs).
- The symbolic access library contains routines used to access configuration file data from system and user programs.
- The data processing library contains routines to extract data from a data file.
- The GRF program enables the display of raw and processed data from a GALE-V data file.
- The GUQ program enables the user to query interactively a data file.

4.2.3. Experiment description: ideal model.

An experiment can be considered as a set of diagnostics connected to the computer through CAMAC modules. It can then be described by a binary tree of diagnostics, which are binary trees of CAMAC modules, as shown in Figure 2.

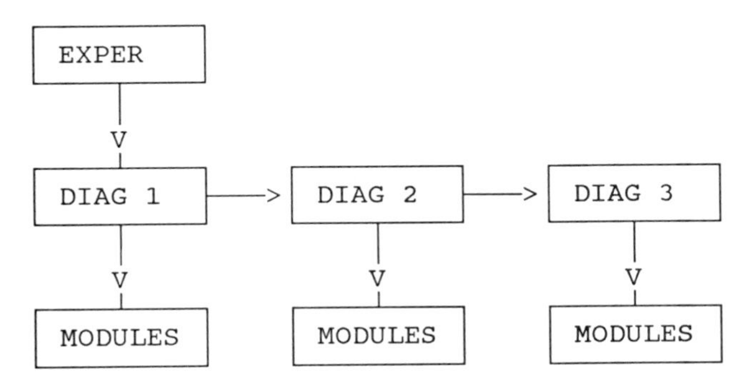

FIGURE 2
Binary tree representation of an experiment

The translation of this model for use by the computer is straightforward. Each item of the model - the experiment, the diagnostic and the module - is described by a block of data composed of a set of fields. Each field can be of different type (integer, real, ASCII, vector of the preceding types, etc). A symbol file contains the names, with the offset from the start of the block, the type and the length of each field.

The configuration file blocks have three sections:

- The link section is a management block of standard format.
- The constant parameter section contains the parameters used by DIOS-V and GALE-V. Its structure depends on the type of block.
- The variable parameter section is available to users to define new fields for use in programs or for data entered manually to be archived with the experimental data.

4.2.4. Experiment description: actual model.

The model described in Section 4.2.3 is essentially the original GALE model from Garching. It uses a single configuration file per experiment, and it handles the data acquisition process sequentially. The data generating capacity of the Tokamak de Varennes and the large disparity in data volumes from the various diagnostics require a multi-user approach to data collection and an asynchronous data acquisition. The remote operation of diagnostics requires more programming capabilities than the original ACQ program provided.

In order to parallelize data collection, each diagnostic has a configuration file and a data acquisition task. Each ACQ competes with the others for system resources so that a certain overlapping of operations occurs naturally. This has a marked effect for diagnostics which collect small quantities of data since they obtain their results relatively rapidly.

The present ACQ program supports asynchronous I/O requests to CAMAC modules. These are issued during the phase chosen for each module by the physicist. The time necessary to satisfy a request varies and can cover more than one phase. Other requests for different modules can be processed during that time.

Even if ACQ can handle many situations, it is restricted to the most common data acquisition and control needs: initialization of modules and collection of large blocks of data. If a module cannot be initialized in one I/O request, or if any kind of decision making is necessary, another program must be used. In order to synchronize this program with the shot scheduling, the concept of module is stretched to include program modules. These are started by ACQ as if they are CAMAC modules and they receive their data, included in the configuration file, through the inter-task communication facilities of RSX-11M. The same concept holds for data processing modules which are started on the VAX after the shot and which store processed data.

4.2.5. Graphics and data processing

In order to facilitate system wide access to data, two FORTRAN routines are available which provide a simple interface between user written programs and GALE-V data files. CAMAC channels are referred to using a signal name; hence all data is retrievable by specifying the signal name, shot number, and diagnostic code. Data ranges can also be specified. Parameters stored in the data file are also accessible by reference to the appropriate field. Limited signal processing during data recovery through the use of keywords and parameter file names can be included in the argument list of the interface routines.

The graphics package is based on a set of subroutines using the TEKTRONIX 4010 standard. Two and three-dimensional plots are supported. GRF uses these subroutines to produce a variety of 2D formats and 3D screen displays. GRF can interactively acquire and display data from any data file available on the system, and amongst other features can zoom in or out on one or a set of signals. GRF parameters for a diagnostic are contained in a graphics control file accessible via DLG. Using a combination of GRF and DLG different sets of signals and screen formats can be viewed relatively quickly after a shot. GRF performs filtering and sampling for large data sets which would otherwise take a long time to transmit and draw on a terminal.

5. FUTURE DEVELOPMENTS

The data acquisition system is presently being modified. The amount of data to be collected has quadrupled since the specification of the present system. The data acquisition, processing and display needs have been driven upwards by the availability of cheap and efficient hardware. They require sophisticated software and efficient processing which is outside the capabilities of older processors. Finally, the complete process for a diagnostic must run on a single computer to avoid data transmission delays.

The system is being redesigned as a distributed system based on a MicroVAX network. One MicroVAX II has been used for development and two 3600 MicroVAXes have been purchased to support diagnostics. The VMS CAMAC system driver from the Oak Ridge National Laboratory has been chosen and the data acquisition software is being rewritten for the VAX. A few of the most common CAMAC modules are supported and some diagnostics are being transferred to the new system.

In a later upgrade, the basic GALE approach will be updated using the ORACLE database management system distributed on the MicroVAXes. These computers will be joined into a Local Area Vax Cluster (LAVC) which will eventually include intelligent workstations to increase the processing and graphics capability of the system.

ACKNOWLEDGMENTS

The authors want to recognize the following members of the Tokamak de Varennes computer group. J.M. Guay has been involved in CAMAC software and hardware support and development. C. Bureau and B. Pronovost are involved in programming and operating the computer system.

REFERENCE

1. J.M. Larsen and C. Strong, Proc. of the Nuclear Power and Fusion Programs of the Canadian Nuclear Society, Canadian Engineering Centennial Conference, (1987), pp. 196-203

FUSION TECHNOLOGY 1988
A.M. Van Ingen, A. Nijsen-Vis, H.T. Klippel (editors)
Elsevier Science Publishers B.V., 1989

GAS INJECTION CONTROL SYSTEM FOR TOKAMAK T-10

V.A.GULJAEV, B.S.LEVKOV, V.G.KOVALENKO, G.E.NOTKIN, L.A.MATALIN

I.V.Kurchatov Institute of Atomic Energy, Moscow, USSR

The whole set of algorythms used on T-10 under low-q operating conditions and under ECRH are described in this paper. The complex multicomputer gas injection control system which regulates the gas puffing into the T-10 chamber is also described in accordance with above-mentioned algorythms.

1.INTRODUCTION

Magnetic fusion devices require accurate, fast-responding and versatile plasma parameters control systems. Horizontal position, Δ_H, plasma current, I_p, and toroidal field, B_z, control systems have already been present on T-10. It is necessary to control over the above-mentioned electromagnetic parameters to provide a satisfactory complex gas injection control (GIC). The GIC-system is needed for optimization of gas parameters in time in the T-10 chamber:

- at any break-down in the ionized gas;
- under Ohmic heating, when $q_{edge} \geqslant 2$;
- under ECRH.

The development of such a system is performed in the following way:

1. Optima algorithms are found for the GIC-system above-mentioned operating conditions.
2. The control systems are designed in accordance with optima algorithms.
3. These systems are included into the general T-10 control system circuit.
4. The results of experiments are analyzed and the requirements for the T-15 GIC-system are determined.

The first step is more complicated. It includes:

- analysis of static and dynamic dependences of plasma parameters on the rate of gas puffing into the T-10 vessel;
- option of some parameters necessary for the control performance:
- option of a block-diagram for the stable control system operation;
- option of a feed-back control system and the gas puffing rate calculations;
- multiparametric control system optimization, according to a certain criterion.

The optimum gas injection algorithm which provides an optimum gas breakdown and low q-operation with uncontrolled gas flow from the chamber walls and from the T-10 limiter is described by the authors in 1,2,3,4.

As known the radiation losses at the discharge start up essentially determine the discharge stability. Possible discharges can be (other conditions being constant:

A - with a disruption in case of a small amount of gas;

B - stable discharge;

2. $\bar{n}_e$ is controlled and $\tilde{B}_2$ is monitored so that $\tilde{B}_2 \leq \tilde{B}_{2\ max}$. This is the main algorithm, when $q \geq 3$. This control system has a very low dynamic error as a difference between $\bar{n}_e$ and $\bar{n}_{e\ ref.}$, when $\bar{n}_e < \bar{n}_{e\ limit}$. The automatic switching from one operating conditions to another ones is used when $\tilde{B}_2$ is monitored. When $\tilde{B}_2 \geq \tilde{B}_{2\ max}$, the gas puffing is switched off and $\bar{n}_e$ is not controlled. However, when $\tilde{B}_2 < \tilde{B}_{2\ max}$, the gas values are "on" and a difference between $\bar{n}_e$ and $\bar{n}_{e\ ref}$ tends to zero.
3. $\bar{n}_e$, $\tilde{B}_2$, PRL under plasma break-down, and Δ_H, I_p, B_z are controlled and monitored.
4. $\bar{n}_e$ is stabilized during ECRH. In this case, the product $\bar{n}_e \times \bar{T}_e$ rises. The gas puffing synchronized with leading edges of the ECR-pulses. Stabilization error is rather low.

All the control algorithms are needed for the T-10 experimental run.

An analytical and experimental identification for the control system design has been made.

The statical functions $\bar{n}_e=F_1(G)$ and $\tilde{B}_2=F_2(G)$, and $d\tilde{B}_2/dt=F_4(G,t)$ have been obtained, also $d\bar{n}_e/kt=F_3(G,t)$.

The proportional-integral-difference (PID) functions for the calculation of the system output (gas puffing) in the controller routines are used. It digital controllers are used for some requirements should be taken into account. First of all, the time for calculations of PID-functions and others, T_{cal}, should be $T_{cal} \ll \tau_i$, where τ_i is the time constant of controlled plasma processes. Moreover, the digital integration and differencetion error should be varied in a low range. For plasma processes controlled by gas puffing $\tau_i=(20\text{-}100).10^{-3}$sec, therefore $T_{cal} \leq 2.10^{-3}$sec and the digital words in arithmetic processors include 32 bits.

The control system inputs are connected with some diagnostics, electromagnetic probes for measuring the m/n=2/1 modes; the probes for radiation losses; and interferometer and the probe for measuring the pressure in the chamber before a discharge included. The pulsed valve developed at the Kurchatov Institute is used as a gas puffing valve. A special capacitor bank is used for diagnostic discharges. The interface set includes besides preamplifiers, a linear optically-insulated modeles (made according to the CAMAC-standard) used a galvanic decoupling in the channel of measurements and the normalizing amplifiers. Moreover, a specially-designed peak detector for the solution to a given problem was made. The peak detector was made in two modifications. The first, is an analogue-digital, the second one is microprocessor made in the CAMAK-standard.

At present, the hardware for the data aquisition, control, dialogue, TV-monitoring and for the archive is presented by the front-end intellegence controllers and by the CAMAC-crates. The controllers have microprocessors Intel 8080 and coprocessors AM 9511. The complex gas injection control system is a part of the general T-10 Data Acquisition System where the common archive center keeps the data from diagnostics and from control systems. Thus, the leading experimentalists

C - with a disruption in case of a large amount of gas.

The general view of the peak radiation losses (PRL) versus the initial pressure at the break-down is shown in Fig.1.

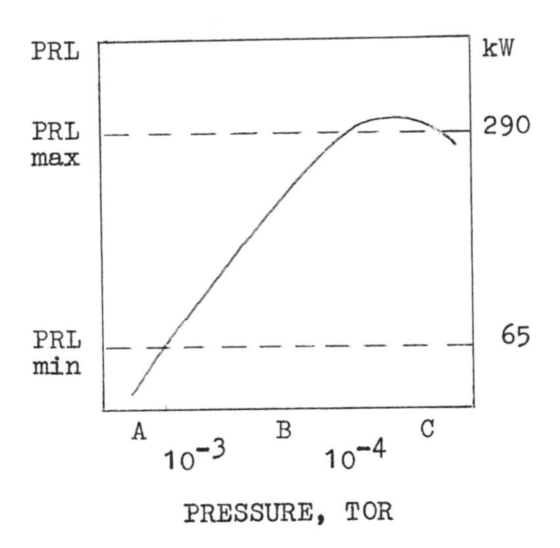

FIGURE 1

The PRL-time dependence in stable (B) and in unstable discharges (A and C) is shown in Fig.2.

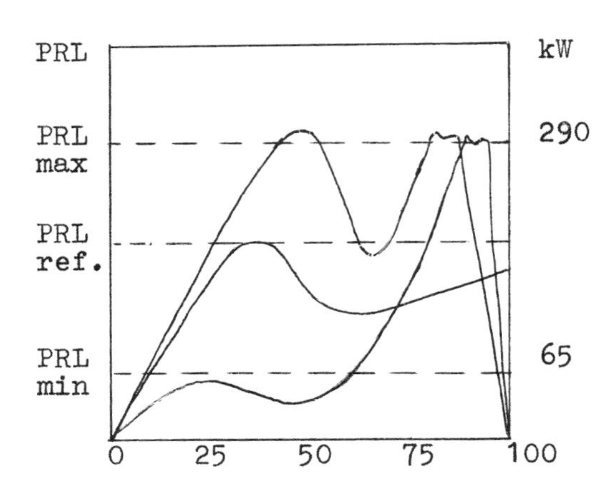

FIGURE 2

The peak power loss control seems to be impossible on T-10, when the gas puffing operation corresponds to the recycling coefficient $R \approx 1$. In this case, the initial pressure magnitude at a break-down is not only the function of gas puffing but also the function of the chamber wall state. The knowledge of a wall contribution to the break-down pressure gives an opportunity to compensate the wall effect by corresponding response of the gas puffing system operation.

A special diagnostic discharge is used on T-10 before the main one. A level of the wall saturation by gas is registered with the PRL-power signal magnitude and a preliminary gas puffing before the main discharge is corrected. The diagnostic discharge takes place 500 msec before the main one. This PRL-feed-forward control system compensates the main disturbance and provides the operation on the straight part of the PRL-pressure dependence from the first tokamak discharges.

However, the parameters characterizing the compensation circuit (such as a working gas pressure in the booster, the preliminary gas puffing rate and the gas puffing valve characteristics) are varied in different discharges providing an error in the pressure compensation during the break-downs. As a result a low PRL-dispersion and the PRL unsatisfactory evolution were observed on T-10. Therefore the modified gas puffing control system is being developed. It calculates the preliminary gas puffing by two signals. The first signal is a peak radiation loss from the preliminary diagnostic discharges. Another signal is a digital sum of the differences between a reference power levels - PRL_{ref} (this value should be found experimentally, see Fig.2), and the PRL from the preceeding main discharges.

Such a system is a discrete feedback control one and the time interval between the discharges which follow each other is a delay in this system. A part of the whole system in Z-transform is shown in Fig.3.

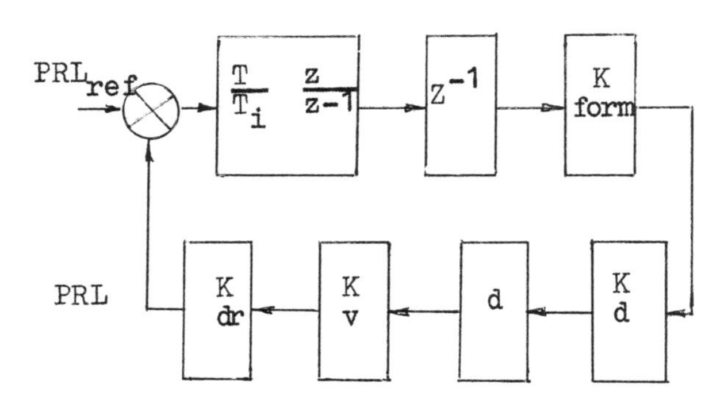

FIGURE 3

All the coefficients on this diagram are statical becouse all the tokamak parameters are unchanged under initial pressure in the T-10 chamber. Here: K_{form} is the ratio of the PRL-magnitude under break-down to the gas puffing pulse duration; K_{dr} is the driver conversion coefficient for the pulsed valve control; K_v is the pulsed gas valve conversion coefficient; d is the ratio between the initial plasma pressure under break-down and the PRL-magnitude; K_d is the data aquisition set conversion factor; $\frac{T}{T_i} x \frac{Z}{Z-1}$ is the discrete function of the integral controller in the Z-transform; Z^{-1} is the Z-transform delay for the designed system. As mentioned above in this system the PRL_{ref} signal should be found experimentally. The controller coefficient is a variable T_i=var. T_i is determined as a number of discharges which are used for calculations of the digital sum. The quality creterion is the Ψ-damping factor. When Ψ = 0.9, T_i=13 discharges.

The PRL_{ref} signal can also be calculated automatically by an adaptive system. In this case, the algorithm is:

$$\overline{PRL_{ref}} = \begin{cases} PRL_{max} & PRL \geqslant PRL_{max} \\ \frac{1}{i} x \sum_{k=1}^{i} PRL_k & PRL_{min} \leqslant PRL_k \leqslant PRL_{max} \\ PRL_{min} & PRL_k \leqslant PRL_{min} \end{cases}$$

i.e. only a difference $PRL_{min} \div PRL_{max}$ should be preset. "i" is the number of discharges. In a such adaptive system the PRL_k - magnitude undergoes an evolution from $\overline{PRL_{ref}}$ with a minimum dispersion. The preliminary GIC-system, together with the control over electromagnetic parameters such as Δ_H, I_p and B_z, makes identified start conditions and discharge stabilization.

The algorithm of the general GIC-system under break-down, Ohmic heating and ECRH includes the electron average density and the magnitude of the m/n= = 2/1 MHD-mode control and stabilization, $\overline{n}_e$ and $\widetilde{B}_2$. This algorithm include four parts:

1. Only $\widetilde{B}_2$ is controlled by two-functions: $\widetilde{B}_2 = k_o + k_1 e^{k_2 t}$ (where "t" is the time, and k_i=const) at the plasma current rise stage, and $\widetilde{B}_2$=const at the constant current stage. This algorithm is used for controlled discharges with $dI_p/dt \leqslant 1.7$MA/sec; B_z= = 1.5-3T; I_p = (100-560)kA; q= = 1.8÷2.9.

from T-10 can observe and discuss the quality of control system operation immediately after the discharges.

REFERENCES

1. G.E.Notkin. Fizika plasmy issue 1,v.11 1985 p. 62-67.

2. V.A.Guljaev, B.S.Levkov, G.E.Notkin, Plasma current density shaping at the initial stage of discharge on T-10. Proc. of 12-th Int. Conf. on Plasma Phys. and Contr. Nucl. Fusion Research, Budapest 1985, p.355-358.

3. V.A.Guljaev, V.G.Kovalenko, B.S. Levkov, G.E.Notkin. The control of plasma breakdown radiation losses on tokamak by preliminary discharge information. Proc. of 4-th All-Union Conf. on Engineering Problems in Fusion Reactors, Leningrad 1988, p. 208-209.

4. V.A.Guljaev, V.G.Kovalenko, B.S. Levkov, G.E.Notkin. The complex control system of the electron average plasma density $\bar{n}_e$ and of the level 2-nd mode MGD-activity $\tilde{B}_{2MGD}$ on installation T-10. Proc. of 4-th All-Union Conf. on Engineering Problems in Fusion Reactors, Leningrad 1988, p. 210-211.

FUSION TECHNOLOGY 1988
A.M. Van Ingen, A. Nijsen-Vis, H.T. Klippel (editors)
Elsevier Science Publishers B.V., 1989

TIMING AND CONTROL PROBLEMS IN PELLET INJECTION

Knud-V. WEISBERG

Association Euratom-Risø National Laboratory, DK-4000, Roskilde, Denmark

The incorporation of a pellet injector in a plasma experiment involves problems that may place restrictions on the design and performance of either the injector or the experiment itself. The solutions applied at some typical plasma experiments where we have installed injectors are described, and an outline is given of the control and timing systems being planned for RFX in Padova and FTU at Frascati.

Protecting the liner of the experiment against the salvo from a multishot injector if the plasma is missing is a fairly new problem. Such protection may demand special real-time signals from the instrumentation of the experiment.

1. INJECTOR AND EXPERIMENT, THE TIMING PROBLEM

Matching a number of pellet injectors to different types of plasma experiments (1,2,3,4,5) have disclosed some common problems. Most important, in general, are the mutual timing of fixed sequences and prediction of the time of interaction between the pellet and the experiment. The timing problems arise because the conditions at the firing moment may be so critical in both the injector and the experiment that neither is able to wait for a long period, if at all, for the other to be ready. One of them has to be able to accept the variations in the preparation time of the other, however.

2. PELLET PRODUCTION AND DELIVERY

The production and delivery cycle of a pellet injector may be divided into a number of stages as shown in Fig. 1. The stages may not all be present in a given injector application, but they are shown here to indicate where problems may be encountered.

When the injector receives a start signal it produces a pellet which is stored and eventually aged for some time at a suitably low temperature. A ready signal is transmitted to the experiment when the pellet preparation is finished, and the injector may now wait for an order to continue.

The arm signal and the arm period are needed only in some situations. In one of these the injector supplies or pulse circuits have to be reset to a start condition just before the firing. Usually no timing problems are encountered. Another situation, which usually occurs with timing problems, is one in which pre-heating of the pellet is necessary to obtain a low break-away pressure in order to get a low pellet velocity. It should then be included in the pellet delay as the pellet will be fired immediately at the end of the pre-heating interval. In this case the experiment must therefore be ready to accept the pellet when it is fired.

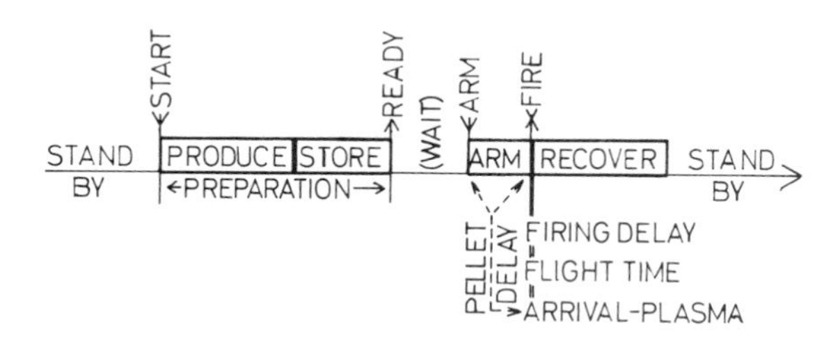

FIGURE 1
Stages in pellet production and delivery

The pre-heating is generally not needed for high speed pellets that may be fired at the storage temperature, and the injector will then be able to wait for the experiment to take place.

Apart from the arm period required for pre-heating, the pellet delay is made up by the firing delay and the flight time of the pellet. The firing delay is the time from the fire signal to the moment the pellet passes a detector just outside the barrel.

After the firing, some time is needed to allow the injector to return to the standby state.

The preparation time of an injector is typically 30 - 60 s, and the pellet delay for a high velocity pellet is 1 - 2 ms + flight time.

3.1. THE INJECTOR - EXPERIMENT INTERACTION

The moment of pellet - plasma interaction must be controllable and accurately known. The range of timing and control problems may be demonstrated by two groups of typical examples. The division is based on the ratio, PD/PP, between pellet delay and plasma period. The sequence of the signals/-operations is indicated by the numbers in the diagrams. The words "pulse preparation" cover the continous part of the preparation of the experiment before the "pulse", which includes the turn-on of B- and E-fields etc., and "pellet preparation" includes, for space reasons, also an eventual waiting time prior to the firing.

A pellet timer measuring the time of flight of the pellet between a set of optical detectors may be used just for velocity measurement or for producing trigger signals at appropriate times for the experiment or the diagnostics.

3.2. Long pellet delay / Short plasma period

3.2.1. The Puffatron

(Risø National Laboratory, ca 1972-76, pellet velocity 10 m/s.)

The Puffatron (2) was a rotating plasma experiment with a mechanical injector. The PD/PP ratio was around 0.1s/0.1 ms. The timing problems were small as it was possible to let the injector govern the run of the experiment as shown in Fig. 2.

The pellet preparation and pulse preparation of the experiment were initiated manually in the proper sequence. The pellet injector turned the B-field on and got a fire order in return.

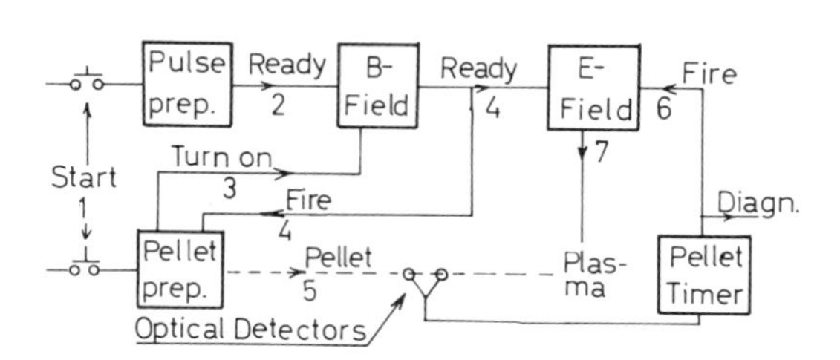

FIGURE 2
Puffatron time sequence

The very short build-up time, 4-5 µs, for the E-field and plasma and the very low pellet speed made it possible to design a simple trigger system based on a velocity measurement and a time of flight calculation.

The pellet velocity is measured by optical detectors and a timer which, after the necessary calculation, triggers the E-field the moment the pellet passes the plasma region.

3.2.2. ETA-BETA 2

(Padova, Italy. 1986-88, pellet velocity 100 m/s).

The PD/PP ratio at the reversed field pinch ETA-BETA 2 (5,6) is about 40 ms/2 ms, so here too the final trigger signals are produced by a pellet timer. And, as is seen from figure 3, these signals represent the only interaction between the injector and experiment.

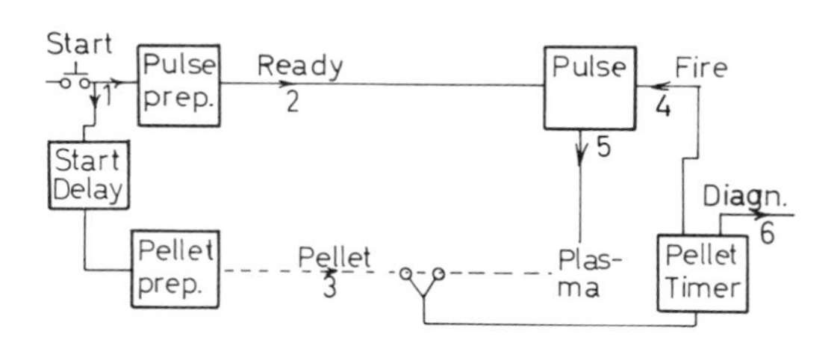

FIGURE 3
ETA-BETA 2 sequence

The injector is completely autonomous because a low pellet velocity, 100 m/s, makes it necessary to fire the pellet immediately after the extrusion to get good reproducibility. A problem is also present in the capacitor banks of the experiment which can be kept charged only for a few seconds to insure reliable operation. A start delay has therefore been inserted in front of the pellet preparation to place the firing of the pellet just after the end of the pulse preparation.

The real timing problem comes after firing the pellet, the pellet delay is long, some 40 ms, due to a long flight path, the plasma period is short, 2 ms, and the build-up period for the fields and plasma long, 5 ms. The pellet timer has therefore been made to produce two signals, one able to be set, independent of the pellet velocity, some 5 ms before the pellet reaches the plasma region, and the other, for the diagnostics, when it reaches the region.

3.3. Short pellet delay / Long plasma period

3.3.1. DANTE

(Risø National Laboratory, ca. 1977-86, pellet velocity 100 - 600 m/s, (several experimental injectors)).

The small tokamak DANTE was used in the early injector development at Risø (3, 4). The PD/PP ratio was typically 10 - 20 ms/ 50 - 100 ms and all the timing takes place before the firing as seen from Fig. 4.

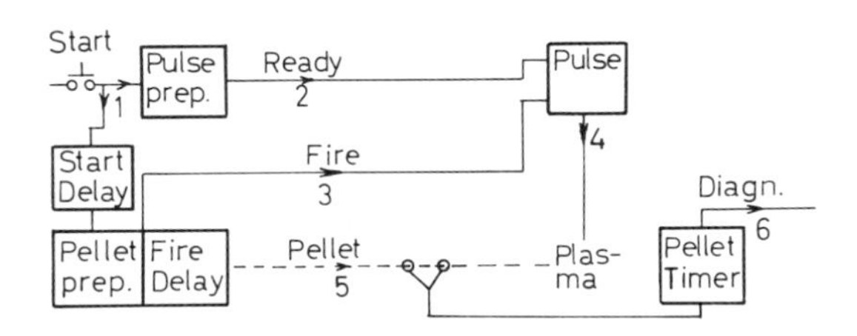

FIGURE 4
The DANTE sequence

A start delay before the pellet preparation allows the pulse preparation to finish just before the pellet is ready. The signal to indicate that the pellet is ready acts as a fire signal turning on both the plasma pulse and a fire delay in the injector. The firing of the pellet is retarded by the fire delay to allow the build-up of the plasma. The pellet timer is used only for timing the diagnostics.

3.3.2. TFR

(Fontenay-aux-Roses, France, 1983-85, pellet velocity 640 m/s).

The sequence at TFR (4) is shown in fig. 5. The PD/PP ratio was about 10 ms/ 0.1-0.6 s. The time sequence of TFR was rigid going straight from preparation to pulse without any pause. The pellet preparation time was much shorter than the pulse preparation time so a start delay was included to keep the waiting period of the high speed pellet reasonably short. The plasma is created about 0.6 s after the fire signal from TFR so a delay of 0.6 - 1.2 s is needed to retard the firing of the pellet.

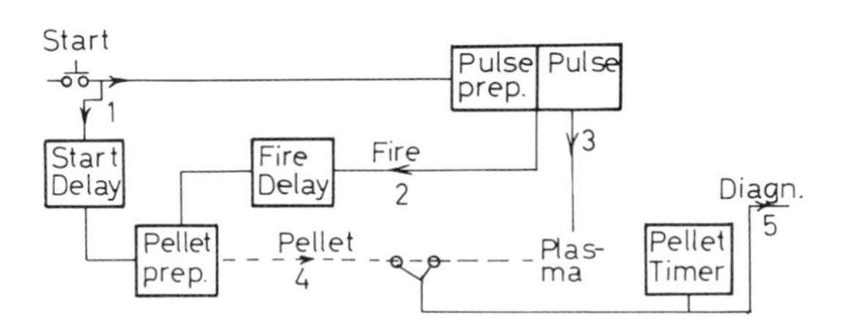

FIGURE 5
The TFR Sequence

A simple pellet timer is used only for measuring the velocity of the pellet. The variations in velocity were sufficiently small to allow the trigger signal for the diagnostics to be taken from one of the optical pellet detectors.

3.3.3. RFX and FTU.

(Padova and Frascati, Italy, development, expected velocity 1200 m/s).

The expected PD/PP ratio for the injectors that are under development for RFX and FTU (1) is 5 to 10 ms / 0.25 to 1 s or about the same as for TFR, but major changes in design and performance of the injectors compared to those used at the earlier experiments justify a brief description.

4. INJECTORS FOR RFX AND FTU

The injectors under develpoment for RFX and FTU differ radically from the injectors described above on three points:

1. The injectors are not the old extruder type (5) but the new simple pipe gun type (1).
2. Each injector will have 8 barrels to be fired in sequence.
3. High pellet velocity and high mass give the pellets sufficient impact energy to cause serious damage to graphite liners and some damage to metal liners if the plasma is absent during a salvo.

The switch from the extruder type to the pipe gun type simplifies both the mechanical and electrical parts of the injector, as all mechanical movements and the related interlocks disappear. The task of the micro-PLC controlling the pellet preparation is thus reduced to that of timing the single steps of the process. The reproducibility of the preparation time will therefore be markedly improved.

The introduction of the pipe gun also brings another improvement as it enables the temperature of the pellet during the formation and before the firing to be controlled more rigidly.

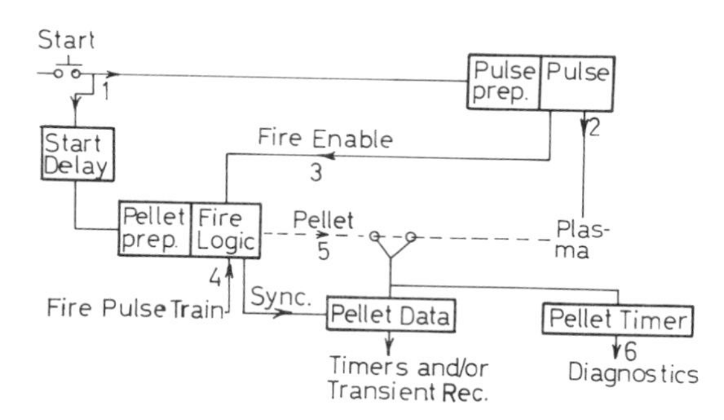

FIGURE 6
Provisional control and timing lay-out for RFX and FTU

The capability of the injector to produce 8 pellets with fair speed and mass in a single charge is a characteristic that is most important to the user; however, it also presents new problems on the control side. The problems occur both before and after the firing of a pellet; they may be tackled by the two blocks "Fire Logic" and "Pellet Data" introduced in Fig. 6.

The purpose of the fire logic is twofold: to enable the firing of the barrels in the right sequence by a train of pulses applied to one common input, and to provide an interlock to the plasma by the "fire enable" signal to protect the liner against pellets fired in the absence of a plasma. The instrumentation of the experiment will have to meet certain demands if the fire and fire enable signals are to be fully utilised.

The "fire enable" signal can be conditioned only by the presence of the plasma if a suitable, real time, signal is available from the experiment. The protection will be limited to preventing further firing. Pellets already under way cannot be stopped due to the short pellet delay.

The "Fire" signal supplied from the instrumentation may be either a simple equidistant pulse train or a pulse train programmed to give a wanted plasma development. The possibility of applying feedback between plasma density and pellet firing is limited by the signalprocessing delay and the pellet delay.

The "Pellet Data" box receives a synchronising signal from the fire logic when a pellet is fired and allows the recording of the time from the firing of each pellet to the passage of the first optical detector, the time of flight between detectors and, eventually, an analog signal from one of the detectors by associated timers or a transient recorder. The resolution of the analog signal is expected to be sufficient to make an evaluation of the integrity of the pellet possible.

The pellet timer is included only to provide signals telling the diagnostics when the pellets are at, e.g., the plasma center.

REFERENCES

1. H. Sørensen et al., A Multishot Pellet Injector Design, this conference.

2. L.W. Jørgensen, A.H. Sillesen and F. Øster, Ablation of Hydrogen Pellets in Hydrogen and Helium Plasmas, Plasma Physics Vol. 17 (1975) 453 to 461.

3. P.B. Jensen, V. Andersen, H. Bejder, M. Gadeberg and P. Nielsen, Ablation studies in Dante. Bull. Am. Phys. Soc. 26 (1981) 877.

4. H. Sørensen, P. Andersen, S.A. Andersen, A. Nordskov Nielsen, B. Sass, and K-V. Weisberg, On the injection of deuterium pellets, 12th Symposium on Fusion technology, Jülich, September 1982.

5. H. Sørensen, S.A. Andersen, A. Nordskov, B. Sass, P. Scarin, T. Visler and K-V. Weisberg, D_2 pellet injector for the reversed field pinch Eta Beta II in Padova, 14th Symposium on Fusion Technology, Avignon, September 1986

6. V. Antonini, M. Bagatin, E. Baseggio, M. Bassan, A. Buffa, L. Carraro, S. Costa, G. Flor, L. Giudicotti, P. Innocente, P. Martin, S. Martini, P.G. Noonan, S. Ortolani, R. Paccagnella, M.E. Puiatti, B. Sass, P. Scarin, H. Sørensen, M. Valisa, P. Villoresi, K-V. Weisberg and S. Zago. Recent results from the ETA-BETA II RFP experiment, Dubrovnik, 16 - 20 May 1988.

FUSION TECHNOLOGY 1988
A.M. Van Ingen, A. Nijsen-Vis, H.T. Klippel (editors)

CONTROL AND DATA-ACQUISITION SYSTEM FOR THE NEUTRAL BEAM PROBE DIAGNOSTIC ON TEXTOR

A.A.E. VAN BLOKLAND[1], M.K. KORTEN[2], F.D.A. DE WINTER[1], E.P. BARBIAN[1], H.W. VAN DER VEN[1].

[1]FOM Institute for Plasma Physics "Rijnhuizen", Association Euratom-FOM, P.O.Box 1207, 3430 BE, Nieuwegein, The Netherlands

[2]Institut für Plasmaphysik, Association Euratom-KFA, Kernforschungsanlage Jülich, P.O.Box 1913, 5170 Jülich, FRG

For the neutral beam probe diagnostic, to be applied for ion temperature measurements at TEXTOR, a control and data-acquisition system has been developed. The system is substantially autonomous and is controlled by a VAX Workstation and an IBM PC. For the required control and data-acquisition tasks CAMAC and the G-64 bus are used.
Communication with the autonomous vacuum system is performed through RS-232 lines connected to the VAX.

1. INTRODUCTION

A neutral beam probe diagnostic using Rutherford scattering (RUSC) has been developed for time- and space-resolved measurements of the ion temperature on the TEXTOR tokamak. The measuring method is based on elastic scattering of injected mono-energetic helium atoms by the plasma ions. The ion temperature is determined from the broadening of the energy distribution of the scattered particles[1].

The diagnostic consists of two main parts, firstly a beam probe which is produced by an ion-source with a neutralizing chamber, and secondly a Time-Of-Flight (TOF) analyser to determine the energy of the scattered particles[2].

The control and data-acquisition has to control and monitor the beam line parameters (beam current, acceleration and extraction voltage) and the power supplies of the analyser. Furthermore, it serves for determination of the flight times during the total TEXTOR discharge (4-5 s) to obtain the ion temperature with a time resolution of about 1 ms.This paper presents the control and data-acquisition system of the RUSC diagnostic.

2. BASIC SYSTEM CONSIDERATIONS

Hardware and software had to be selected for a high degree of compatibility with the TEXTOR environment. Moreover, the RUSC control and data-acquisition system has to satisfy the following requirements. It must

- be stand-alone for functional tests carried out at FOM Rijnhuizen preceeding the implementation at TEXTOR,
- fit into TEXTOR's control and data-acquisition environment during the subsequent operational stage in order to make use of the full resources of the TEXTOR host system and to contribute to the scientific data-base,
- have a separate and fail-safe control of the beam line, analyser and vacuum system parameters,
- allow for the previously described control and measurement requirements and processing of the data.

3. OVERVIEW OF THE SYSTEM

The RUSC control and data-acquisition system (Fig.1) is composed of:

- a VAX/GPX Workstation with a CAMAC process interface for data-acquisition and TEXTOR pulse-oriented control,
- (GESPAC) for beam line and analyser,
- vacuum components,
- an IBM PS/2 system to display the status of beam line and analyser parameters.

3.1 Computer systems

The system will be operated by a VAX Workstation and an IBM PC. They are equipped with the following components:

VAX/GPX Workstation

- VMS operating system
- 13 Mbyte main memory
- TK 50 magnetic tape drive
- 19" color graphics monitor with graphic processor VS-290.
- Q bus
- 159 Mbyte RD 54 fixed head disk storage
- Ethernet interface.

IBM PS/2 model 80

- MS DOS 3.3 operating system
- Turbo Pascal
- 30 Mbyte fixed head disk storage
- RS 232 port.

A DECserver 200 interfaced by a DELNI to Ethernet provides eight RS-232 connectors for a terminal, the matrix- and laserprinter and the 4 pumping control units. Another spare port is used for connection to the GESPAC equipment at a later phase. The DELNI is also used to link the VAX to the TEXTOR host system during shot operation.

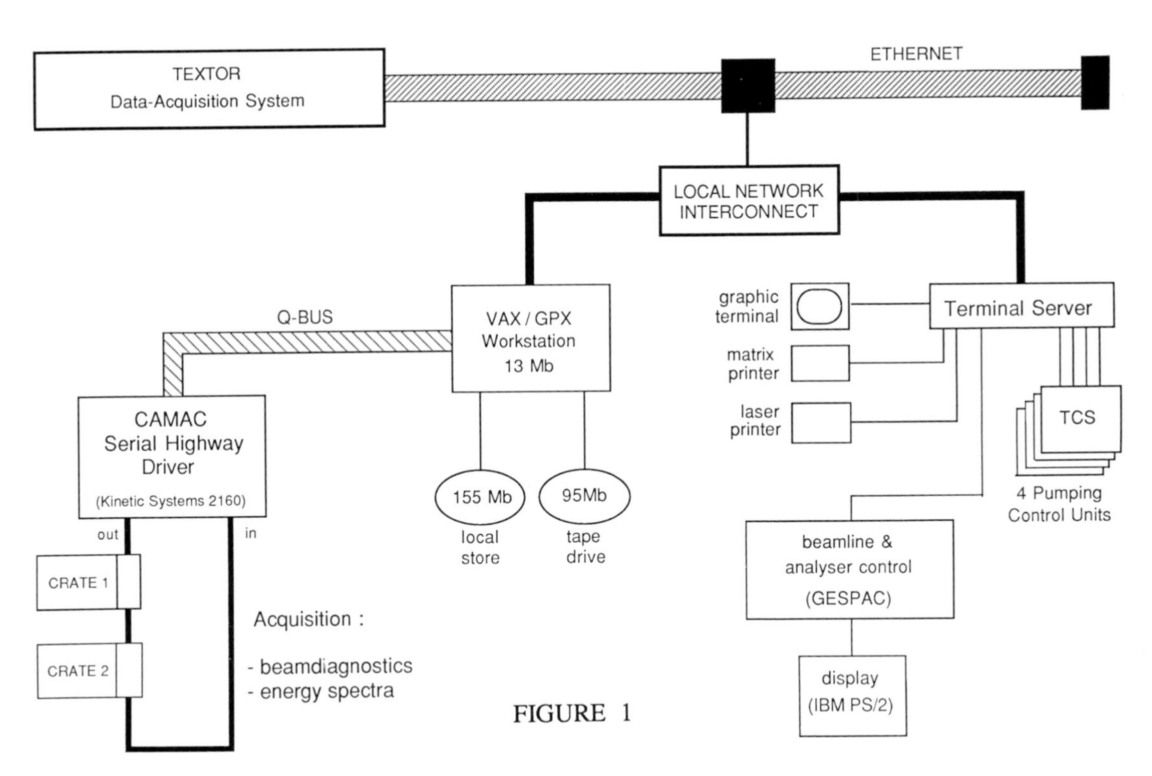

FIGURE 1

The RUSC control and data-acquisition system

3.2 The data-acquisition system

The measuring of the energy of the scattered helium particles is based on the Time-Of-Flight (TOF) principle. The TOF-analyser is shown schematically in Fig.2. The start detector is composed of a very thin carbon foil and two chevron channelplate detectors. When a helium particle passes the foil, electrons are released on both sides of this foil. These secondary electrons are accelerated by means of two high-transparent wire grids and deflected over 180° and finally detected by the chevron channelplate detectors. The stop signal is obtained when the particle itself is recorded by a channelplate at the end of the flight path. A triple coincidence measurement of both start signals and the stop signal enables the analyser to discriminate between real flight times and signals which are triggered by neutrons hitting the channelplates.The flight times are recorded by means of a Time-to-Digital Converter (TDC), LeCroy type 4204 (1 μs dead time), yielding a time resolution of 625 ps. The data are histogrammed in a memory (LeCroy type 3588, 16k 24 bit). The flight times recorded in 64 time intervals of 1 ms length form 64 spectra of 256 channels, filling the 16 k word memory.

Due to the requirement of collecting spectra at a 1 kHz rate during the total discharge, a large histogramming memory would be needed to buffer the data (4000 spectra, 2 Mchannels). As options we discussed a solution to use one large memory to store all data from one shot and alternatively the use of two smaller memories, which however have to be read out during the shot. We preferred the latter option because it minimizes the hardware equipment and appears to be less expensive than the first option. The chosen solution is based on the use of two memories which are filled in turn with data and read out during the shot. The switch control between these memories is performed by a memory switch module. As a consequence of this alternating read-out, a memory read-out time of 64 ms (0.5 Mbyte/s) is required.

CAMAC, which is used as a process I/O standard at TEXTOR was found to be suitable to fulfil the above mentioned demands. The LeCroy modules are available in CAMAC anyway. Shot-oriented control and timing are performed through a set of CAMAC modules as used at the TEXTOR Data-Acquisition System[3]. This allows for stand-alone operation as well to synchronize the control of the diagnostic with TEXTOR.

The CAMAC configuration is composed of a byte serial highway driver, Kinetic Systems type 2160, and two CAMAC crates which are controlled by KSC 3952 enhanced crate controllers and interfaced by U-port adapters KSC 3926 (fibre optic). The KSC type 2160 has been selected since it provides the enhanced block mode, which was found to be suitable to achieve the required data transfer velocity. A VMS driver for the CAMAC controller has been developed for the enhanced mode and tailored such that it fits into the Oak-Ridge software environment and raw-data file system used for data-acquisition[4,5].

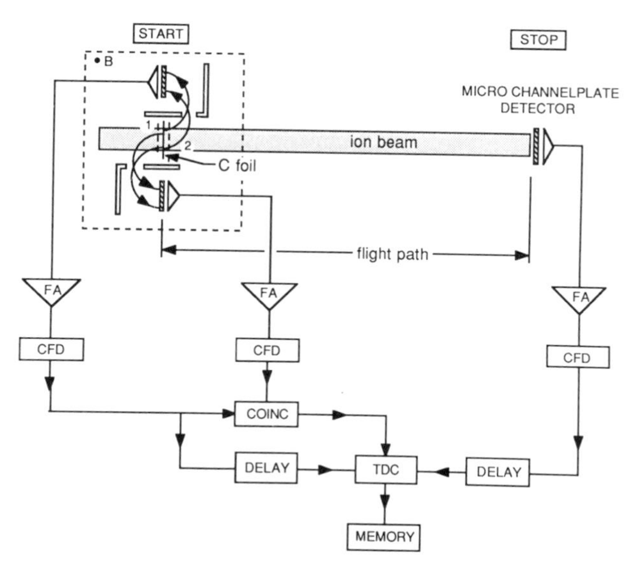

FIGURE 2

Schematic lay-out of the time-of-flight analyser

C foil	- carbon foil
B	- direction of magnetic field
1,2	- acceleration grids
FA	- fast amplifier
CFD	- constant fraction discriminator
COINC	- coincidence unit
TDC	- time to digital converter
MEMORY	- 16 k word histogramming memory

3.3 The control and monitoring system

The control equipment for the diagnostic will be well integrated into the total system. Beam line and analyser can be remotely controlled as the diagnostic is not accessible during shot operation of TEXTOR.

One part of the system is deduced from the GESPAC G-64 bus system which originates from CERN and is based on the Motorola 6809 microprocessor, which has been adapted for similar applications at FOM[6]. Here, it is employed for the conditioning of the ion-source, and the parameter setting of the TOF-analyser and neutralizing tube. The controls are executed through a joy-stick panel. The status of beam line and analyser will be life-displayed by means of an IBM PS/2 system linked with the G-64 system through RS-232. This IBM system has been selected, since the software for communication and graphical display was already available.

The vacuum system is entirely controlled by means of four independently operating pumping control units, Balzers/Pfeiffer type TCS 1000, which serve for automatic or manual control of the vacuum components. The control units are interfaced by RS-232 to the DECserver 200. The TCS 1000 units are controlled by a server software which has been developed at IPP/KFA[7]. A program to control and display the TCS 1000 system by an interactive graphical terminal input-output has been written which obtains its data by task-to-task communication from the TCS 1000 server.

4. STATUS OF THE SYSTEM

The components of the system have already been tested individually and are presently integrated. A 25 ms read-out time of the LS 3588 memory has been achieved, which is much faster than the required transfer time.

ACKNOWLEDGEMENTS

The authors wish to thank J. de Hoop, J. Kamp, F. Koenen, H. van Ramele and A. Putter from the FOM Electro-technical Department, and B. Becks and U. Schaufert, from the EDV Division IPP-KFA Jülich for substantial advice and assistance.

Part of this work was performed under the Euratom-FOM association agreement with financial support from NWO and Euratom.

REFERENCES

1. A.J.H. Donné, E.P. Barbian and H.W. van der Ven, J. Appl. Phys. **62** (1987) 3130.
2. A. Cosler et al., this conference A16.
3. M.K. Korten, 12th Proc. Symp. on Fusion Technology, Jülich, (1982) pp 1443-1448.
4. W.R. Wing, Data-Acquisition in Support of Physics, Basic and Advanced Diagnostic Techniques for Fusion Plasmas, P.E. Stott (1986), pp 55.
5. M.K. Korten, A KS 2160 VAX/VMS Driver, KFA/IPP Report, to be published .
6. J.P. Nijman et al., this conference K 12.
7. M.K. Korten and B. Becks, KFA/IPP Report, to be published.

FUSION TECHNOLOGY 1988
A.M. Van Ingen, A. Nijsen-Vis, H.T. Klippel (editors)

THE CONTROL AND MONITORING SYSTEM OF RTP

J.P. NIJMAN, W. KOOIJMAN, A.A.M. OOMENS, P.H.M. SMEETS, A.J.H. TIELEMANS

FOM Institute for Plasma Physics "Rijnhuizen", Association Euratom-FOM, P.O. Box 1207, 3430 BE, Nieuwegein, The Netherlands

The Rijnhuizen Tokamak Petula will use a computerized Control and Monitoring System to allow machine operation from a single multiscreen console (VAX Workstation). This paper presents an overview of both the hardware and the software structure and discusses the main system components and their interaction. It also gives the present status of implementation.

1. INTRODUCTION

The Rijnhuizen Tokamak Petula (RTP), which is being constructed from major components of the Petula tokamak from Grenoble, will be the main home based experimental device to support the Rijnhuizen fusion research programme on transport mechanisms in tokamak plasmas [1].

This medium sized tokamak is very well suited for the intended research programme. Not only does it fit the size of the laboratory, the available power supplies, diagnostics and ECRH equipment, it is also an ideal base for the education of students in plasma physics and for the development of new advanced diagnostic techniques.

Tokamaks are operated following a scenario that is often depicted in a state transition diagram such as shown in Figure 1.

Each state in the diagram represents a tokamak operation phase, in which a limited number of actions have to be performed. Apart from the short tokamak pulse itself, the actions are essentially not time critical and could in principle be performed by manual control.

It is therefore logical to split the control system into a (central) timing system that masters the time critical tokamak pulse sequencing and data acquisition, and a (computerized) system to assist the operator(s) with the (automatic) configuration of timers, power supplies, data recorders and additional equipment. Data transport from the data acquisition is more than an order greater than from the control system and is therefore is best handled by a separate system.

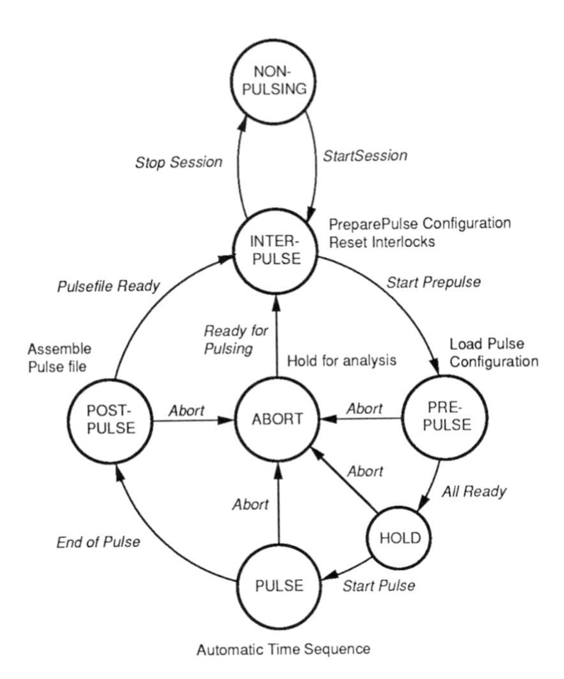

FIGURE 1
RTP (Supervisor) State Transition Diagram

Further it is assumed that each of the RTP units is self-governing and supporting its own local protection system and relies only on a central system for configuration and synchronization. This minimum amount of interaction with a central control system though, requires more local intelligence in the form of local feedback loops and pre-programmed local sequencing than in a centralized system, but this conforms nicely with the present tendency of equipping even the simplest instruments with (microprocessor) intelligence and to connect such units by means of a communication network.

At Rijnhuizen there is a need to easily reproduce previous tokamak pulses with a minimum amount of operator involvement. This asks for a sophisticated control system connected to a central database containing the pulse configuration library.

2. REQUIREMENTS

In addition to the forementioned guidelines we can estimate the hardware requirements for the RTP control and monitoring system, using the following list of signals.

1500	slow digital input/outputs (max. sampling frequency 10 Hz)
150	slow analog input/outputs (max. sampling frequency 10 Hz)
200	fast analog data acquisition channels (sampling frequency 10 kHz up to a few MHz, 4k samples/shot)
100	timing signals (resolution 1 μs)
50	fast interlock signals ('interrupts') (resolution 1 μs)
100	slow interlock signals (typically relais contacts)

To minimize the amount of interaction between the different units, the hardware and software of the control and monitoring system must be organized hierarchically into groups of functionally coherent units, thus forming largely autonomous functioning subsystems and local controllers. We estimate that there will be 7-10 subsystems and up to 50 local units.

An independent (central) interlock system is required to ensure the safety of personnel and machine, also in case of computer or network failure.

There should be no real-time control through the communication network. Data transport to and from the different units should, except for experimental data taken by the fast data acquisition channels, be limited to 4kbytes/pulse.

Apart from initial start-up of equipment in the morning, the whole machine should ultimately be controllable from a single multiscreen console, where response to operator action should never exceed 500 ms.
It is important to realize that, due to the available resources, this goal can only be reached in stages.

Because of the high level of interference, groups of signals should be galvanically isolated from each other and where possible transported via fibre-optic links. Low pass filters should be used where practicable.

3. HARDWARE STRUCTURE

The hardware structure that resulted from the guidelines and requirements mentioned above is shown in Figure 2.

Virtually all units have been connected to a main (Ethernet) backbone, and following "Rijnhuizen" tradition, the RTP computer (microVAX II) and the (CAMAC) data acquisition system are placed in a Faraday cage which also houses the subsystem controllers, and the central timing system and interrupt circuitry.
The main console, which is equipped with a multiscreen VAX 2000 Workstation, will be located in a control room outside the cage. This console also contains the relais logic (organized by subsystem) and (hardware) status panel of the central interlock circuit.

Initially there will be only one acquisition system for fast data, consisting of one or two CAMAC branches each with a maximum of 8 crates directly connected to the RTP VAX. At a later date this system will be extended with VME based data recording and processing units, to replace the existing analog plasma position control equipment and to allow extensive front-end processing for some of the diagnostics. These VME crates will be equipped with Ethernet interfaces.

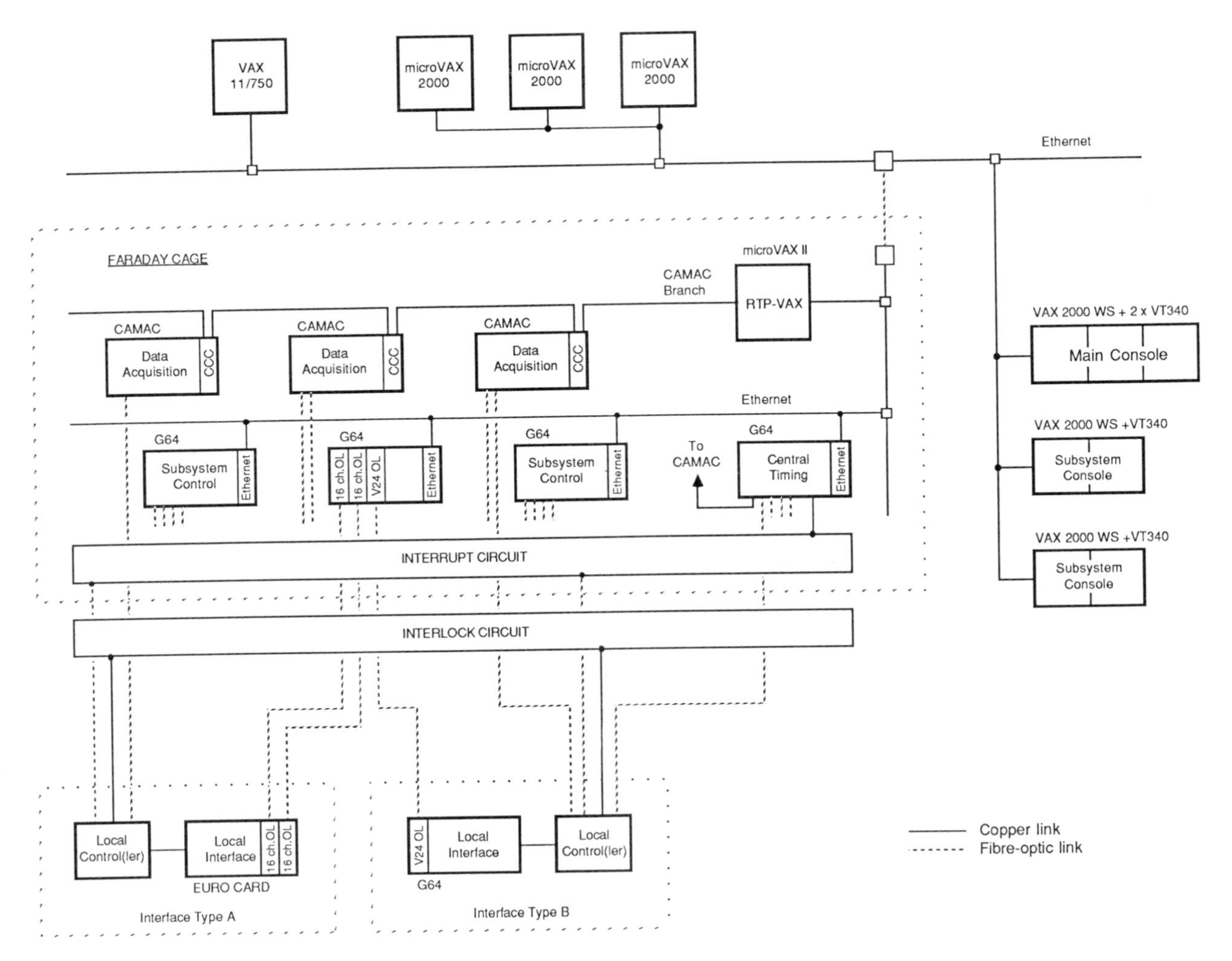

FIGURE 2. Overview RTP Control and Monitoring System

The subsystem controllers and the Central Timing System are built in 19" Eurocard racks with G64-bus backplanes, around a MC6809 8/16-bit microprocessor card and a MC68000 based intelligent Ethernet interface. The same type of controllers have been successfully used at CERN for slow control tasks and will also be used for their new DELPHI experiment.

Each controller can handle a maximum of 32 interface cards which will normally either be a 16-channel digital input or output, or a 2-channel 10-bits analog input or output card (both interface type A in Figure 2). The interface cards actually consist of two parts connected through a one-way (or optionally two-way) plastic fibre link. One of the cards is located in or near a local unit and contains circuitry for signal conditioning; the other card (in the controller) contains the G64-bus interface. These cards were developed at "Rijnhuizen" and have already successfully been used for the very noisy SPICA-II pinch experiment.

Intelligent local units and multiplexors may be linked to subsystems through a V24 optical link (marked type B in Figure 2).

4. SOFTWARE STRUCTURE

Although this is not clearly reflected in the hardware structure, the RTP control system has a strong software hierarchy. All control tasks are performed as close to the hardware as possible and are (within each subsystem) governed by a subsystem supervisor. This supervisor is a Finite State Machine (implemented in Pascal, like most of the control software) with a structure similar to that of Figure 1. Synchronization between the subsystem supervisors is performed by a central supervisor with the same structure, located in the RTP VAX.

Also other tasks with a supervisory nature, like collection of CAMAC data, composition of pulse files and the database for pulse configuration files, have been allocated to the RTP VAX. The actual storage of the pulse files though, and the computation and interpretation of the pulse data is done on a VAX 11/750 (also on the Ethernet network). For the operator interface and first data interpretation, which requires a lot of graphics processing, a VAX workstation (using a GKS library) has been chosen.

Communication between the different subsystems and the RTP VAX has been implemented using a Remote Procedure Call (RPC) system designed and in use at CERN in the same configuration [2]. The RPC system keeps the communication protocols hidden from the user, and is available both for Ethernet and for RS232 links between microVAXes and G64 micro systems. Communication between the RTP VAX, the consoles and the VAX 11/750 has been implemented using 'VAX clusters' for high speed data access and transport.

The G64 MC6809 microprocessor systems are programmed in OmegaSoft Pascal using the FLEX operating system. This system allows modular programming but does not support multi-tasking. In order to still guarantee proper 'task-switching' and limited response time, the main program loop has been structured around an event mechanism. This event mechanism may be compared with a software bus that handles the communication and synchronization between the different modules.

Each of the subsystem programs is based on the same structure of standard modules, forming the 'subsystem skeleton program', which is adapted and expanded with routines as required for the subsystem.
Figure 3 shows the modules of the skeleton program. The arrows indicate the interaction between the modules which is taken care of by the Event Manager.

A simple command interpreter has been added, to allow subsystem testing via an ordinary terminal, also when the network is not available.

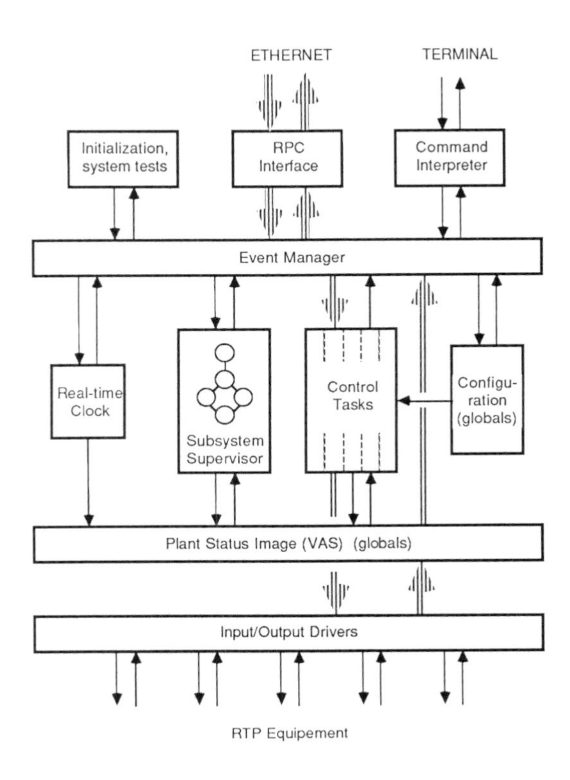

FIGURE 3
Blockdiagram Subsystem Skeleton

5. USER INTERFACE

Operation of the RTP machine will typically be coordinated from the main console in the RTP control room. The console is therefore equipped with a VAX 2000 Workstation and 19" colour screen, through which, by means of pull-down menus and 'mouse' action, control mimics and configuration 'forms' can be selected. The pull-down menus will be organized in such a way that, starting from the main menubar at the top of the screen, most of the available mimics can be selected with only one mouse action.

A mimic will typically display the actual status of (part of) a subsystem, and enables the operator (by further mouse action) to directly or indirectly (through 'dialog boxes') activate or configure pieces of equipment.

Adjacent to the main screen there will be two (non-interactive) VT340 colour screens, one normally showing global system information and pulse sequence (supervisor) status, and the other to be used to 'park' a copy of a mimic from the main screen.

The problem of finding or designing a suitable mimic editor or mimic programming language has been solved by using the standard AutoCAD drawing package[3] to edit the symbols and background pictures of the mimics, and to use a simple interpreter to translate the AutoCAD DXF (transfer) files into Pascal GKS source code to be used as library modules in the system.

6. PRESENT STATUS

All major computer components are in house and the microVAX II has been tested and used, clustered with the other "Rijnhuizen" computers, for program development during the last few months. The VAX 2000 Workstations have been tested and the console system software and the 'AutoCAD DXF interpreter' are almost finished, though work on actual control mimics has not yet been started.

The CAMAC equipment required for the first year of operation is available. A new CAMAC branch driver for the microVAX II is due for delivery in September 1988.

All the G64 equipment has been ordered and there is already a sufficient number of G64 cards available for the control of the 'base load'. New interface cards are expected to arrive early 1989. A large part of the subsystem skeleton program has been completed, but not tested with RPCs. Actual subsystem programming has not yet been started.

The design and construction of a new Central Timing System has been postponed, but the timing system from the former SPICA II experiment has been modified for use on RTP, and is currently under test.

ACKNOWLEDGEMENTS

This work was performed under the Euratom-FOM association agreement with financial support from NWO and Euratom.

REFERENCES

1. O.G. Kruijt et al., Engineering aspects of RTP, this conference.

2. T.J. Berners-Lee and A. Pastore, RPC User Manual, CERN DD/OC, 22 August 1986.

3. AutoCAD is a trademark of AutoDesk Ltd., South Bank Technopark, 90 London Road, LONDON SE1 6LN.

FUSION TECHNOLOGY 1988
A.M. Van Ingen, A. Nijsen-Vis, H.T. Klippel (editors)

SOME IMPLEMENTATION DETAILS OF THE MIXED HARDWARE CONTROL AND DATA ACQUISITION SYSTEM OF THE RFX EXPERIMENT

O.N.Hemming+, A. Luchetta, V. Schmidt++, S. Vitturi

Istituto Gas Ionizzati - Associazione EURATOM - ENEA - CNR
Corso Stati Uniti 4, - 35020 CAMIN (PADOVA), ITALY

The RFX control and data acquisition system utilizes a fibre-optic based local area network which connects a mix of programmable controllers, minicomputers and personal computers. The paper describes the implementation of the communication functions based on the uniform low-level protocols according to the ISO-OSI recommendations for levels 1 to 4. The paper also describes how the operation of the subsystems are synchronized to a state diagram which defines the system-wide operation of the experiment. Also given is a brief description of the operator interface via the subsystem control consoles.

1. INTRODUCTION

The Control, Monitoring and Data Acquisition System, SIGMA (Sistema di Gestione Monitoraggio ed Acquisizione Dati), is implemented as an integrated system, providing machine operation from one central machine control room. The structure of the SIGMA hardware and software has been designed to reflect the control and monitoring requirements of the functional subsystems of the RFX plant. Since presented at SOFT 86[1] the design has undergone the following modifications/ developments:

a) The Supervisor is now a programmable controller based system;

b) IBM PC compatible personal computers are now utilized for the subsystem consoles;

c) A scheduler task has been introduced to co-ordinate the subsystems (see Section 4);

d) Communication software has been chosen;

e) Fibre-optic star couplers have been introduced into the LAN implementation.

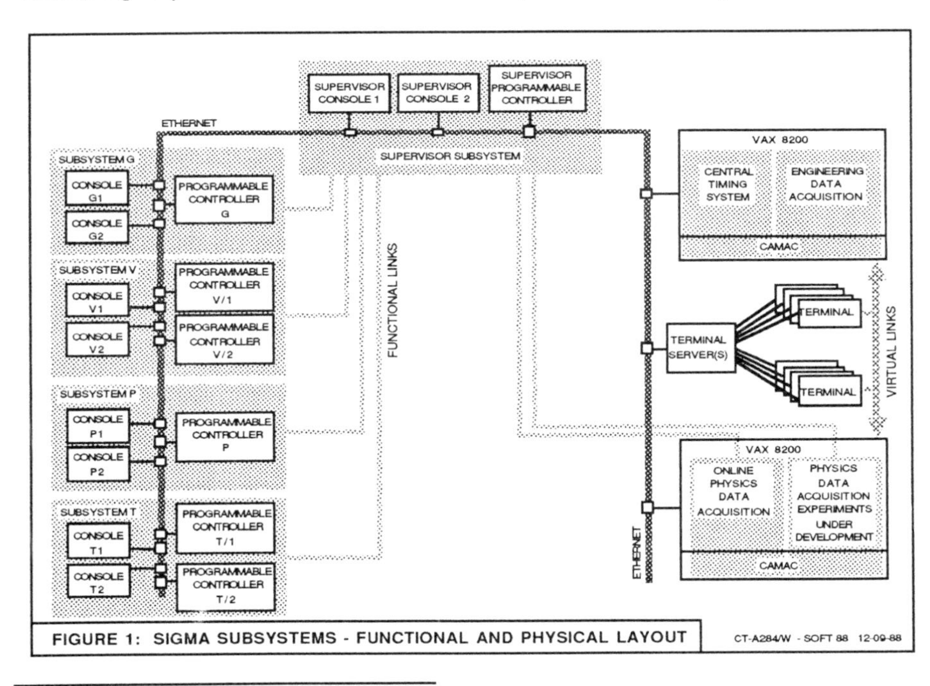

FIGURE 1: SIGMA SUBSYSTEMS - FUNCTIONAL AND PHYSICAL LAYOUT

2. STRUCTURE OF SIGMA

Figure 1 illustrates the functional structure of SIGMA.

There are five SIEMENS S5-150U[2] programmable controller (PLC) based components, comprising four machine subsystems: General Services (G), Vessel (V), Poloidal (P) and Toroidal (T). The Supervisor system is the fifth PLC component, and is responsible for

+ under contract from Systems Designers Spa, Milano, Italy
++ under contract from Hahn-Meitner-Institut Berlin GmbH, Berlin, FRG

the monitoring and control of all SIGMA subsystems. In particular it is responsible for the synchronization of the state changes of those subsystems.

A VAX 8200 is used for the On-line computer supporting the Engineering Data Acquisition (EDA) and the Central Timing System; the latter produces waveforms and all precision trigger signals for synchronized plant operation and data acquisition.

A second VAX 8200 (the Physics Data Acquisition, PDA, computer) supports (at least) two subsystems: the On-line PDA for established diagnostics, and the diagnostic development subsystem.

A VAX 8530 computer (not shown in Figure 1) is dedicated to the organization of the long term data archivation and retrieval, and to off-line data analysis.

3. THE SIGMA LOCAL AREA NETWORK

3.1. LAN Components

The SIGMA LAN is a fibre-optic implementation of IEEE 802.3, using active star couplers[3].

The SIEMENS S5 series of programmable controllers are interfaced to the network using plug-in communication processor boards (CP-535). The protocols used conform to those of the ISO reference model for Open Systems Interconnection (OSI)[4] up to level 4 (Transport level).

Taking into account the large distance between the actual locations of the programmable controllers and the consoles, plus the requirement to have mobile consoles, it was decided that consoles attached directly to the network would provide the most suitable solution. With their open programming capability and the availability of a suitable Ethernet interface card including an OSI Transport level driver (ISOLAN[5]), IBM PC AT type personal computers have been chosen for the consoles.

3.2. Communication Requirements

The communication has to provide data exchange between all LAN components with the following functionality:

data transfers:

- cyclic transfer of data from subsystem programmable controllers to their subsystem console or to the supervisor console, for status displays (mimics);
- downline loading of PLC operational parameters from an experiment database maintained on the On-line computer or subsystem console.
- the cyclic transfer of an alarm status block to the consoles from the subsystems.

messages:

- used by the scheduling system to transmit subsystem state information to/from the supervisor subsystem.
- for operator commands and setpoints from consoles.

3.3. Communication Implementation

Programmable Controllers

The CP-535 provides datagram (broadcast and multicast) and connection oriented functions of the ISO/OSI Transport level services (level 4). These are used directly, calling SEND and RECEIVE functions within the PLC programs, allowing the transfer of data buffers between partners. Above this level, the SIEMENS software implements functions to READ or WRITE data from/to a remote device. With these, the source or destination of the data is specified by the user, and no partner task is required on the remote device.

SEND/RECEIVE are used for the message system (using message tasks in each subsystem), and READ/WRITE for data transfers to the console and for downline loading of programmable controllers.

Computers

Communication links between VAX and SIEMENS PLCs are achieved using the DEC implementation of the OSI transport services, VOTS[6]. Using this it has been possible to successfully emulate a SIEMENS CP535 communication processor and perform the SEND, RECEIVE, READ and WRITE functions to a PLC from a VAX program. VOTS can be used concurrently with other Digital Architecture Network, DNA, products on the same Ethernet (DECnet and the remote terminal Local Area Transport, LAT, protocols)

DEC, with the co-operation of SIEMENS, have developed a product, VSH1[7], which offers the possibility to define symbolic names of PLC devices and data-items (memory locations) in those devices. Application tasks may then call the VSH1 routines SEND, RECEIVE, READ and WRITE specifying only those symbolic names. The VSH1 software also allows certain functions which emulate the networked programming devices for SIEMENS programmable controllers. The functions available allow start and stop of the controller and uplinedumping/downline-loading of the complete memory space.

Consoles

The ISOLAN Ethernet interface board used for the consoles is available with a driver (XPORT) providing the ISO/OSI Transport level services. Using this it is possible to create communication links with SIEMENS PLCs and emulate the functions of the communication processor.

The consoles have similar requirements for communication with the subsystems as have the VAX. As a product like VSH1 is not available, it was developed inhouse. The product OSH1, layered on top of the XPORT driver, provides SEND, RECEIVE, READ and WRITE functions using symbolic names for devices and data-items, similar to VSH1. With this it has been possible to interface a commercial plant supervisor package (see Section 5) to the SIEMENS PLCs via Ethernet.

4. SUBSYSTEM SYNCHRONIZATION

The RFX experiment will have various modes of operation (e.g. high power discharge, baking). The machine functions in a certain mode of operation for a number of experiment shots. Each experiment shot is divided into a number of states. Those states, and the transitions between them are described by a state machine. For the present a single state machine is used for the known modes of operation, Figure 2.

To co-ordinate and synchronize the transitions of subsystem tasks from one state to another, a scheduler program has been implemented to be installed on all SIGMA subsystems. Each task has the responsibility of informing the scheduler of its current status, using a set of predefined interface functions. In turn the scheduler maintains some global flags which inform the tasks of the current active state.

The scheduler of the Supervisor is called the master scheduler. The tasks that it monitors also include the schedulers of the other subsystems.

Normally all subsystems will operate together under the direction of the master scheduler. However, in certain modes of operation it may be necessary that a subsystem works as a standalone system. In such cases the subsystem scheduler can be set to operate in Standalone Mode and then acts as a master scheduler, but does not interact with the other subsystems.

The scheduler, as implemented, can deal with 32 tasks per subsystem, and with state machines having 32 states. It is driven by a mode of operation (MOP) table, which defines the state machine, and hence will allow in future for different state machines to be used. The MOP table has an entry for each state which defines the default next state and the allowed state transitions that may be requested by a task. Time-out periods (the maximum time for which a state should remain active before the next state transition occurs) and Wait-time periods (the minimum period for which the state must remain active) may also be set.

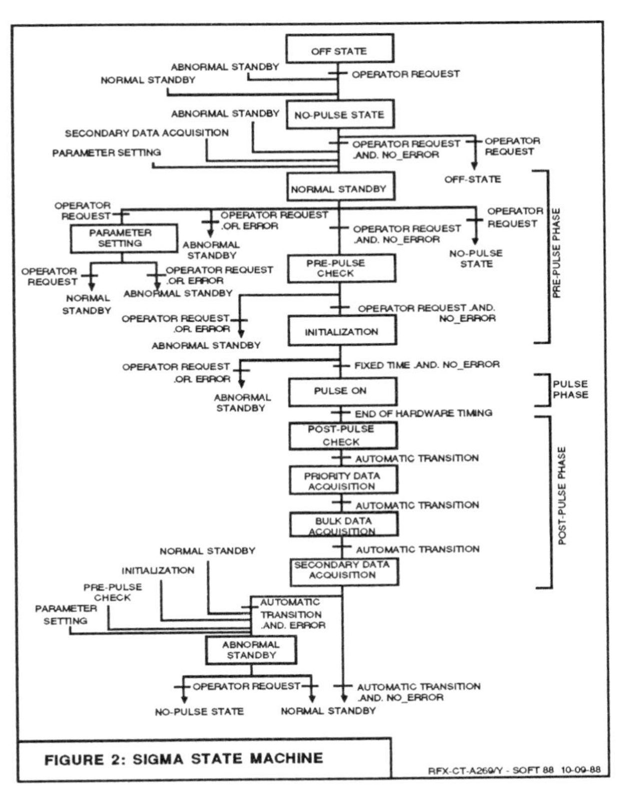

FIGURE 2: SIGMA STATE MACHINE

Task Status

For each state of an experiment shot, a task may have one of the following statuses:

State Not Started

- Indicates the task has not seen the state transition.

State Preparation

- The task is performing operations necessary to put itself (and hence any plant it is controlling) into the current state.

In State

- The task has completed all preparations necessary to

put itself in the current state.

Next State Enabled

- All tasks being monitored must indicate they are ready to proceed to the next state before the scheduler will grant the next state transition.

Ignored

- The task has indicated that the scheduler should not monitor the task during this state. Used where a state has no meaning for the task (e.g., data-acquisition states for a task performing no data-acquisition).

Once a state transition has been enabled, the scheduler begins monitoring the status of all the active tasks of the new state. The status of a subsystem is defined to be the status of the scheduler of that subsystem; the status of a subsystem scheduler is itself an 'AND' of the statuses of the tasks of the subsystem (i.e., when all active tasks have the status 'State Preparation' the scheduler status is 'State Preparation').

The master scheduler also derives its status as the 'AND' of the statuses of the tasks that it is monitoring, and hence those of the subsystem schedulers. The subsystem schedulers inform their change of status to the master scheduler via messages.

The status of the master scheduler is defined to be the status of SIGMA.

State Transitions

Each scheduler maintains on its subsystem a number of variables. One of these is the 'Subsystem State' of which each flag represents a state of the current mode.

When the master scheduler status becomes 'Next State Enabled', it determines what the next state should be (from the MOP table) and sets the relevant flag for that state in the Subsystem State variable. When a task sees a flag set in the Subsystem State, it may make the transition to the state which is represented by that flag; the scheduler ensures that only one flag is set. A 'Previous State' variable allows a task on transferring to a new state, to know from where it has come.

The master scheduler also sends this change of state information together with the MOP table entry of the new state to all subsystem schedulers, which then update their Subsystem State and other variables.

Request for State Transition

Any task may request a transition to a state which is not the default transition of the state machine. If granted, the next state transition will be to the requested state, unless it is overridden in the meantime by a request for transition to Abnormal Standby, which takes precedence over all transitions.

Apart from Abnormal Standby, a request will not be granted if another, different, transition request is outstanding, or if the transition is not a permitted one for the current state as defined in the MOP table.

A granted state-transition request is passed to the master scheduler, which then enables that transition rather than the one defined by the MOP table.

Abnormal Standby

The Abnormal Standby state may be entered if a tasks requests it, or if the scheduler detects a Time-out.

A request to go to Abnormal Standby will always be granted, and cancels any other requested transition. However, if the transition to Abnormal Standby is not permitted for the current state (i.e. if the MOP table has not been set to allow it), then the request will be delayed until the system is in a state where it is permitted; the requesting task may go to Abnormal Standby immediately.

5. SUBSYSTEM OPERATOR INTERFACE

The operator interface to the PLC based SIGMA subsystems is via the IBM PC compatible consoles running a commercial plant supervisor package, FactoryLink[8]. This provides pixel graphic mimics, and uses a memory resident database for digital, fixed-point, floating-point and text data-elements. It consists of a number of tasks running in parallel which exchange data via the database. The operator may enter data using the keyboard, or by using function keys or a mouse to set pre-determined values.

A task has been defined and is being implemented to interface FactoryLink to the Ethernet driver OSH1. Configuration tables define the data which is to be transferred to or from a subsystem PLC and any conversions required. Data can be read from the subsystem on either a continuous cyclic basis, or cyclically while only a certain mimic is displayed, or when triggered by the change of a database element. Data are written to the subsystem when triggered by the change of some specified database element.

The READ function of OSH1 is used for reading data, whilst data writing (setpoints or commands) uses the message handling functions of the communication system. This allows checks to be built into the message receiving task of the subsystem to ensure that a data location can be written to, and also allows (using special message codes) for single bit flags of the SIEMENS memory to be modified (the shortest data unit that can be written using the normal WRITE function is a byte).

The division between the functions of the console and the subsystem is defined roughly by the decision that the consoles should not be involved in any control or supervision of the plant. Consoles are used only for status displays and alarm logging and to accept operator input to be passed to the subsystem for validification. This has resulted in defining two special data types recognized by the console:

Composite Digital Status

A byte value in which each bit represents a particular state of a hardware or software component of the plant.

Composite Analogue Values

The analogue signal handling performed by the PLC subsystems includes raw value range checks, conversion to physical units and threshold limit checking for input signals, and vice-versa for output signals (setpoints). Tasks define, access and manage analogue signals using dedicated software functions written for the PLC subsystems; All values associated with an analogue signal (threshold limits etc.) are stored as a group in fixed memory blocks of the PLC; these are Composite Analogue Values.

Alarms of a subsystem are also managed by a system task; application tasks use standard functions to define alarms and to set or clear them. Alarms indicate conditions which need to be brought to the operators attention, and hence with each alarm is associated an alarm text which is displayed on the subsystem console. They are classified into Notification, Warning, Fault, and Central Fault types, with the latter two being the most serious and requiring operator acknowledgement. Central Faults are also displayed on the console of the Supervisor. The alarm manager of the subsystem uses a special memory area of the subsystem which is polled by the console.

6. STATUS (AUGUST '88)

At the beginning of August the project team transferred to the new RFX buildings. The computers are installed; two development PLC subsystems are working complete with consoles.

The VOTS, VSH1, and OSH1 software products are installed and functioning. A prototype version of the FactoryLink Ethernet interface to the SIEMENS programmable controllers is running; the final version is specified and under implementation. The message handling system has been specified and is under implementation .

Of the PLC based application software, the 380kV/20kV substation control and monitoring (a major part of subsystem G) has been implemented as a prototype in house. The remaining subsystems are in the final stages of specification.

The CAMAC based data acquisition system is the subject of a joint software development effort between the Istituto Gas Ionizzati and two other U.S. fusion laboratories[9].

REFERENCES

1. V. Schmidt et al, Structure of the RFX Control and Monitoring System, in: Proceedings of the 14th SOFT, Vol. 2, pp. 1445-1450.

2. SIEMENS Catalogue ST 54 and ST 54.1, S5-135U and S5-150U Programmable Controllers, 1986 Edition

3. Hirschmann Active Optical Data Network (LAN) on Ethernet base. Manufacturer: Richard Hirschmann Radiotechnisches Werk, P.O.B. 110, D-7300 Esslingen, West Germany.

4. 'Information Processing Systems' - Open Systems Interconnection, Basic Reference Model, ISO 7498, 1983

5. ISOLAN 4100 Series Controllers for IEEE 802.3 and ISO Transport Software. Manufacturer: BICC Data Networks, 1 Frogmore Lane, Hemel Hempstead, Herts HP3 9RJ, United Kingdom.

6. VOTS - VAX OSI Transport Service Version 1.2. DEC Networks & Communications Europe, P.O.Box 121, Reading, Berkshire RG2 0TU, United Kingdom.

7. VSH1 - VMS SIEMENS H1 Version 1.0. DEC ACT, Torino, Italy.

8. FactoryLink. Manufacturer: USDATA. 1551 Glenville Drive, P.O. Box 850058, Richardson, Texas 75085.

9. G. Flor, 'Design Proposal for the RFX Data Acquisition Software', internal paper CT-62 08/10/87.

FUSION TECHNOLOGY 1988
A.M. Van Ingen, A. Nijsen-Vis, H.T. Klippel (editors)
Elsevier Science Publishers B.V., 1989

USE OF FORTH IN INTEGRATED CONTROL SYSTEM DEVELOPMENT

Richard ENDSOR

Culham Laboratory, Abingdon, Oxfordshire, OX14 3DB, UK*

1. INTRODUCTION

FORTH is a structured computing system based on multi-layered symbolic operators. These operators essentially process data stored on a stack and use this stack directly for parameter passing. Parameter passing is untyped in the sense that there is no syntactic checking of stacked items when an operator is invoked.

At the heart of FORTH is a compiler/interpreter manipulating operator definitions in a single dictionary data structure. New operators are defined in terms of sequences of existing operators or by using assembler code for the underlying computer using FORTH defined assembler operators. There is no distinction between symbols defined in these different ways as each symbol appears in the dictionary followed by a "data structure" defining its behaviour. Execution of FORTH code by the interpreter uses the indirect, threaded code generated by succeeding layers of operator definitions to access the "primitive" operators written in FORTH assembler.

The FORTH system is an open system in that all system code is defined in the dictionary and can be invoked by subsequent additions to the dictionary. This includes the symbols which make up the compiler/interpreter. Usually a FORTH system, supplied with a computer, implements a standard set of basic arithmetic, logical and character operations and performs basic system operations such as controlling terminal input/output, disk operations, printing and task scheduling. The FORTH compiler/interpreter, together with its assembler, is also supplied, as is a screen based editor. A FORTH user application is then built by the addition and invocation of further symbolic operators. This distinction is somewhat blurred by the fact that the user application may include redefinition of any of the underlying FORTH symbols so modifying or extending the computer system.

The multi-layered, open structure of FORTH gives rise to a powerful form of structural abstraction in that an application can be expressed in a user defined abstract language which may be executed in different modes and configurations of underlying hardware. This arises from:-

(a) The ability to redefine FORTH compiler operators to form a compiler for an overlying, user defined language.

(b) Changes to underlying system operators to map onto different hardware configurations; to invoke either internal memory operations or external channel signalling.

These features of FORTH have been exploited in a particular control system project which will be briefly described with particular reference to the relevance of FORTH's structural abstraction.

2. RDS PROJECT

The COMPASS Requirements Definition System (RDS) is briefly described, from the user viewpoint, in an accompanying paper[2]. The aim of the system was to define a distributed control system such that individual sub-system

*Euratom/UKAEA Fusion Association

components and a central control system could be constructed by contractors. (The sub-system would be based on Programmable Logic Controllers and the Central System on a digital computer.) The operation of the control system was described; partly by an Input/Output Database listing all analogue and digital signals passing between the controlled plant and the sub-system and all signals passing between the sub-systems and the central control system. Operation of the system components was described by concurrent processes expressed as logical operations in a language known as the Requirements Definition Language (RDL). Timing of operations could be expressed by timer variables.

To check the logical completeness of the control system definition, the users could also construct RDL processes exhibiting the responses of the controlled plant and then run these with the processes representing the control system in a simulation mode. A separate language, known as a Display Definition Language (DDL), allowed mimic diagrams, exhibiting total system operation, to be displayed on attached screens. DDL also allowed control panel simulation by allowing attached mice to be used as pointers. A logging file enabled operation of the system to be recorded for later checking.

It was decided that FORTH was a suitable system in which to implement RDS. The implementation divided into three separate components. The Input/Output Database was expressed as FORTH variables and known as the Database Definition Language (DBDL). Systems operations were expressed in RDL and the system operation was displayed using DDL. As the DDL was only used for system checking during specification and was not part of the specification itself it was decided that DDL could be implemented as an addition to the underlying FORTH system and be expressed in a "basic" FORTH style (ie with input parameters being expressed before the operator in reverse Polish form). DBDL, which essentially consisted of a list of sub-system input/output signals of either analogue or digital form, could be expressed as "basic" FORTH variables. However, it was decided that RDL should express the operation of the sub-system in a way which would be immediately comprehensible to both the specifiers and implementers. RDL was, therefore, designed as a "standard" infix language with usual operator precedence which could be overriden by brackets. Assignment statements were allowed; also synchronisation statements using timer variables and structured conditional statements. The structured conditional statements could be nested and parameterless procedures could be defined and invoked. It was decided that the FORTH compiler/interpreter could be modified to act as a compiler for RDL.

3. IMPLEMENTATION OF A 'HIGH-LEVEL' COMPILER

The RDL language allowed expressions to be written in infix notation. These operated on logical or integer variables and constants. The definitions of standard FORTH operators (such as +, *, AND etc) were changed in two ways. Firstly, a precedence was attached to each operator allowing "normal" infix interpretation of expressions (such as A + B * C) and then the compile time action was altered. Instead of generating the indirect pointer to the operator coding at the current position in the dictionary (this would occur in the definition of the current RDL process), the indirect pointer, together with its precedence, would be stored on an 'alternate' stack (ie not the normal FORTH stack). At the same time, any indirect pointers to operators already stored on the alternate stack, of higher precedence, would be moved to the current position in the dictionary definition. At the termination of the expression, the alternate stack would be flushed and any remaining entries moved in precedence order to the current dictionary position. Bracket symbols were defined and caused the current precedence to be suitably incremented and decremented to enable correct evaluation of expressions such as (A + B) * C.

FORTH is structured such that redefinition of all existing symbols is straight forward. A new definition of the symbol will always be found, on future reference, in preference to an existing definition. The existing definition, however, still remains in the dictionary and may be referred to during the creation of the new definition. Extra actions can, therefore, always be added to any existing symbol. If, at some future stage, the new definition was removed from the dictionary, then FORTH would revert to using the previous definition.

Overall syntax checking for correct nesting of RDL process and conditional statement usage was performed by creating a state checking symbol operating on a single state variable. The various RDL keywords (such as "process", "operation", "continue-process", "if" etc) were defined, using the state checking symbol to check the existing state and then perform the desired transition if this were valid. Run-time behaviour relied on compiling indirect pointers to underlying FORTH operators, such as IF and THEN.

The RDL compiler using the FORTH compiler/interpreter was relatively succinct occupying about 120 lines of FORTH statements. It had, however, two weaknesses that could have been avoided if a separate RDL parser/code generator (which would have required more FORTH code) had been written. Firstly, the implementation required that all RDL elements were space separated when used to define user processes. This arose from the FORTH interpreter's symbol definition being based on space separation. Secondly, error conditions often produced terse responses. The least helpful was probably a question mark following what FORTH considered to be an undefined symbol. This could arise if the user of RDL had forgotten to define a variable in the accompanying DBDL before using it or failed to separate two elements by a space.

4. CONCURRENCY OF SIMULATION

Once user defined RDL processes had been compiled, these were run in simulation mode, with accompanying DDL screen displays, to ensure that correct system behaviour had been specified. Each process was linked into the FORTH "round-robin" scheduler as a background task. The scheduler worked on a linked list, each task running until it executed a PAUSE statement. This caused the state of the task to be stored and the following task in the list continued after its previous PAUSE. To ensure that execution of RDL processes appeared to be concurrent, PAUSE statements were compiler into code generated by RDL assignment statements. This ensured that a change of a variable in one process could be "immediately" detected and acted upon in other processes. This was especially useful as each process could have an exception condition associated with it. In a control system this could be, for example, an earthing failure. The compiler FORTH code checked the exception condition after each entry from a previous PAUSE. Thus, one RDL process detecting a failure could set a variable which would trigger the exception conditions in other processes and lead to a system "close-down".

The DDL language was interpreted for display on two associated screens. One screen was designated as a Control screen and the other a Status screen. Associated with each screen was a mouse which allowed a software driven pointer to be moved around on the screen. An interactive FORTH process ran for each screen; this being linked into the FORTH round-robin. DDL statements allowed the values of DBDL variables to be monitored and any changes to be displayed. Also areas of each screen could be designated as "Hit" areas and specified actions performed whenever the button of a mouse pointing to that area was pressed. These actions could simulate the System Controller pressing a "START" button in the simulated system.

5. RDS MEMORY ENHANCEMENT

The original RDS system was implemented on a DEC LSI-11 processor using 16 bit addressing to access the 64 Kbytes of main store. This meant that the basic FORTH system, the definitions for RDS and the user-specified control system simulation all had to fit into 56 Kbytes (the final 4 Kbytes being used for system input/output). This limitation proved to be a restriction and DEC had introduced, for other systems, a hardware memory mapping unit which allowed addressing up to 22 bits. Extra 64 Kbyte main store modules were also available allowing up to 256 Kbytes to be attached to the processor (using an 18 bit backplane). It was decided that RDS should be enhanced to support DBDL variables and RDL user processes running in extended memory with memory mapping.

Memory mapping was based on two sets of eight 16 bit registers, known as the Kernel and User registers. The FORTH system was modified to initialise these registers with seven Kernel registers pointing to the first 56 Kbytes of main store and the final Kernel register pointing to the last 4 Kbytes. FORTH system loading then continued in Kernel mode. The basic FORTH system and a modified RDS system were then loaded into the first 40 Kbytes of main store. The next 16 Kbytes of main store were reserved for loading a FORTH defined screen editor and for "Hit" and "Scan" tables for the Control and Status screens.

The modified RDS system contained renamed definitions of the FORTH dictionary handling routines which allowed access to a further directory to be stored in extended main store. When compiling the user defined control system, the dictionary routines would switch the processor from Kernel to User state and create user DBDL definitions in 56 Kbytes of extended memory. Data storage for variables associated with these definitions was assigned in a further 24 Kbytes of extended memory. Code generated by the RDL processes was mapped into 16 Kbytes of extended memory. Further areas of extended memory were assigned for run-time workspace for each RDL process. All these areas had previously been mapped into the DBDL variable and RDL process dictionary entries.

When the simulation was to be run the FORTH round-robin included the foreground tasks for the Control and Status screens and a single entry for a small, specially written extended memory scheduler. When entered this scheduler would either immediately exit or map three Kernel registers to run each RDL defined process in turn. Once every process had its turn the scheduler reset the Kernel registers to point to the initialised FORTH state and re-entered the FORTH round-robin. When an RDL process was running it accessed variable values by special fetch and store actions which switched the processor to User state, accessed the variable and then returned the processor to Kernel state.

It should be emphasised that as far as the users of RDS were concerned, the extended memory system was identical to the original RDS system and allowed RDS defined systems to be run without modification (of course the extension allowed far more extensive systems to be defined). The important point of the extension was the flexibility of the FORTH system and the relatively short period in which the modified system was produced by an experienced FORTH system programmer.

6. TEST SYSTEM MODE

The final area in which the RDS system was extended is of more general interest. RDS was used to create specifications for supply of a distributed control system based on Programmable Logic Controllers and a Central Computer. These were to be produced by an external contractor and there was a need for the specifiers to check that the delivered system met the specification. The LSI-11 system used for RDS had the Hytec modifications allowing access to a CAMAC crate. This was the specified communication medium for the control system; it was, therefore, decided to link the controlled plant models, written in RDL for the original simulation, to the delivered control system. The operation of the delivered system could then be checked using the Control or Status screens to monitor the operation. This method of testing allowed operation of the control system to be tested before the plant which it was to control had been delivered.

The required modification to RDS was to map specified DBDL variables to signals on Digital and Analogue channels in input/output modules accessible through the CAMAC crate. Some modification to the specification DBDL was required as the user had to intersperse a few statements giving the module addresses to which the following DBDL input/output variables would correspond (the specifications have already been drawn up with a clear definition of the grouping in which control signals would be wired on the control system). Code to initialise and perform input/output to CAMAC modules was written in about 100 lines of FORTH.

It remained for RDL and DDL to be modified to handle both CAMAC variables and memory variables in main store. One way would have been to test for variable type when accessing the variable and then perform different actions. FORTH offers instead a more structured solution. With the definition of each variable type it is possible to store references to separate store and fetch actions. Generic store and fetch operators may then be defined which are called whenever a DBDL variable is to be accessed. These generic operators then use the indirect pointers in the FORTH-generated data structures to access the actions needed for the current variable and return, if necessary, the required result on the stack (in the case of Analogue input/output appropriate scaling is also performed). Using this method a Test Mode RDS system could be constructed from the Simulation Mode coding with changes to about 40 lines of code.

7. CONCLUSIONS

FORTH offers an open, multi-layered structure on which flexible, abstract systems may be built. This has been demonstrated by RDS where an initial Requirements and Simulation system has been transformed into a tightly-coupled Validation and Acceptance system. The ability to base different stages of system development on an abstract, high-level specification is seen as an advance in system development techniques.

ACKNOWLEDGEMENTS

The FORTH implementation of RDS was performed mainly by Chris Stevens and Brian Mercer of Computer Solutions Ltd (UK) under contract to the UKAEA.

REFERENCES

1. L Brodie, Thinking FORTH, (Prentice-Hall 1986).

2. D C Edwards et al, COMPASS Electrical Systems Development and Commissioning, this volume.

FUSION TECHNOLOGY 1988
A.M. Van Ingen, A. Nijsen-Vis, H.T. Klippel (editors)

COMPUTER SYSTEM OF GAMMA 10

Akiyosi ITAKURA[1], Denji TSUBOUCHI[1]
Akihiro YAMAO[2], Hirofumi SHINOHARA[2]

(1) UNIVERSITY OF TSUKUBA, IBARAKI 305, JAPAN
(2) TOSHIBA CORPORATION, 1, TOSHIBACHO, FUCHU, TOKYO 183, JAPAN

Tandem mirror GAMMA 10, a large scale mirror equipment of Plasma Research Center, University of Tsukuba is one of the leading (mirror type) machines in the world. Three computers are used for controlling, monitoring, and data acquisition. Integrated operation of GAMMA 10 is performed by these three computers which are interconnected by computer communication links.
This report describes the computer system for GAMMA 10 system.

1.INTRODUCTION

GAMMA 10 is mainly composed of the vacuum vessel, the coils, AC power supply (a 250 MVA motor generator set (MG)), the DC power supply (rectifiers), the heating systems (NBI, ECH, ICRF etc.) and their control systems. They are separately located in several buildings. To enhance safety and operability, the control of the entire GAMMA 10 equipment needs to be centralized. A high performance computer system was introduced to meet these demands on the monitoring and control function as well as the data acquisition and processing function.

The computer system is composed of three computers (the GAMMA 10 control computer, the MG control computer and the plasma data acquisition computer). In this paper, the main functions and features of the computer system of GAMMA 10 are described.

2.CONTROL SYSTEM OF GAMMA 10

The control system of GAMMA 10 is schematically shown in Figure 1. It is organized in 3 hierarchical levels, supervisory level, subsystem level and local control level.

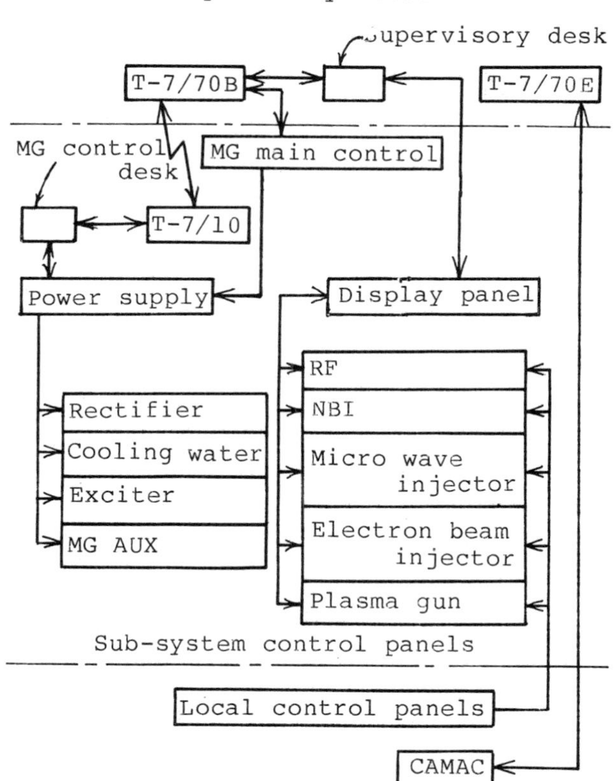

Fig.1 Control system configuration of GAMMA 10

The supervisory level is the GAMMA 10 control computer which provides the function of overall control and monitoring. Integrated operation of GAMMA 10 is executed by this computer. The supervisory desk has two CRTs and two keyboards for monitoring and parameter set up. Subsystem level is composed of subsystem control panels and MG control computer. Independent operation of each subsystem is realized by this level. Local control level is composed of local control panels which are dedicated to the equipment.

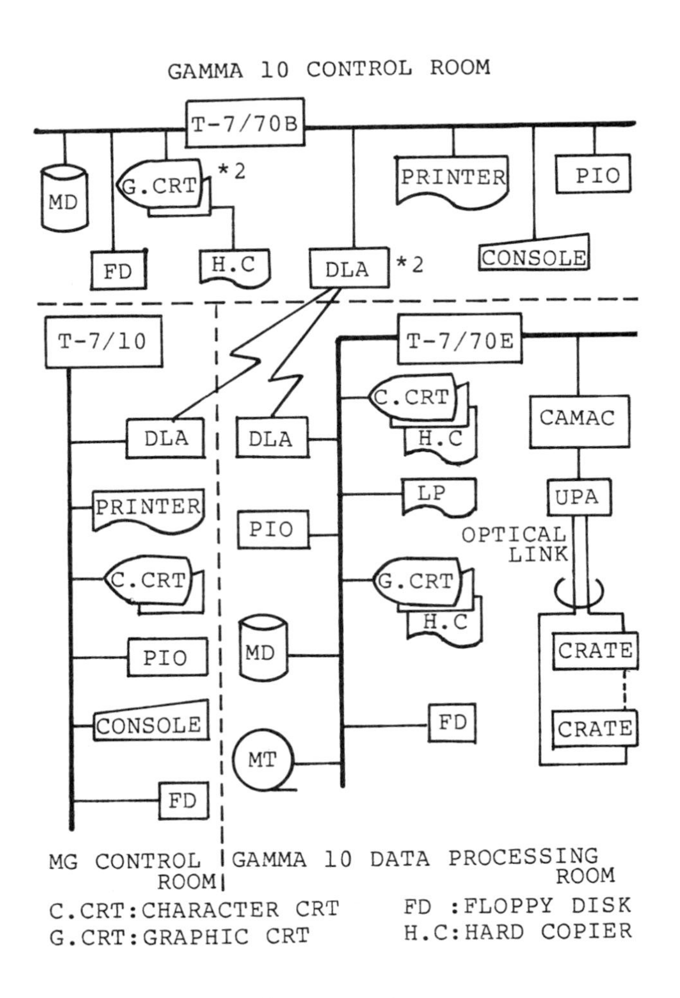

Fig.2 Hardware

3.COMPUTER HARDWARE CONFIGURATION

As shown in Fig.2, the computer system is composed of three computers, the GAMMA 10 control computer, the MG control computer and the plasma data acquisition computer, which are interconnected by computer communication links (device name DLA).

3.1 GAMMA 10 control computer

The computer is a 32-bit minicomputer of TOSBAC Series 7 Model 70B(T-7/70B).

The major peripheral devices are a process input/output device (PIO) that inputs/outputs process data, a semigraphic CRT, a hardcopier, a printer that prints out the records of GAMMA 10 operation.

3.2 MG control computer

The computer is a 16-bit minicomputer of TOSBAC T-7/10.

The major peripheral devices are a PIO that inputs the MG status and outputs the MG drive command, a character CRT, and a printer.

3.3 Plasma data acquisition computer

The computer is a 32-bit minicomputer of TOSBAC T-7/70E.

The major peripheral devices are a CAMAC interface that transfers the plasma diagnostic data from the CAMAC modules to the computer, a PIO that is used for the interface with the GAMMA 10 control console, a character CRT, a graphic CRT, a line printer (LP) and a magnetic tape device (MT) used for saving a large amount of plasma diagnostic data.

4.FUNCTIONS OF THE COMPUTER SYSTEM

Main functions of the computer system are as follows :

1. Overall control and monitoring of GAMMA 10.
2. Control of the MG set.
3. Plasma data acquisition.

4.1 Overall control and monitoring of GAMMA 10

The GAMMA 10 control computer is the main computer that provides this function. The details of this function are as follows :

(1) Monitoring - The computer system informs the operators of the status of GAMMA 10 equipment by schematic diagrams, graphs and messages displayed on the CRT, and it prints out equipment information. The data of the MG set are input periodically by the MG control computer and are sent to the GAMMA 10 control computer through computer communication link (DLA), while the other data is periodically input by the GAMMA 10 control computer through its PIO. The validity and the alarm status of the data are checked by the computer periodically. When an alarm status is detected, the message is displayed on the CRT and also typed out on the typewriter. About 120 analog inputs and 260 digital inputs are monitored by the computer.

(2) Parameter set up for the experiments - The operators can set the operational parameters of GAMMA 10 through the CRT keyboard. These parameters, when they are valid after the validity check by the computer, are set to the equipment before the shots. The parameters for the MG operation are sent to the MG control computer which controls the MG speed according to the cycle of the experiments.

(3) Execution of the operation sequence - The computer executes the sequence of the GAMMA 10 operation periodically according to the parameters. During the repetitive sequence of the experiments, the computer supervises and executes the overall sequential control and the MG speed control. Before every experiment the parameters are automatically set to the equipment and the MG is accelerated to the required speed. During the shot the operation data and experiment data are gathered by the computers, which are automatically saved in the magnetic disk after the experiment.

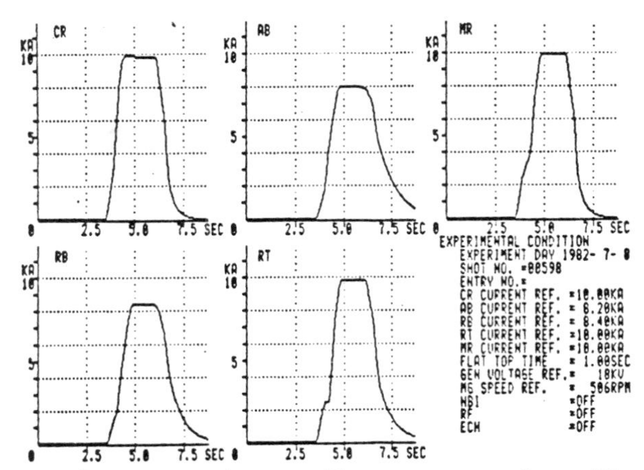

AB:Min.-B-anchor coil CR:Central coil
MR:Axisym.mirror coil RT:Racetrack coil
RB:Recircularizing coil

Fig.3 Coil current data

(4) Recording data of power supply - The data of the power supply such as the coil current, the coil voltage and the MG output voltage are acquired by the MG control computer and transmitted to GAMMA 10 control computer after the shots. These data are dislayed on the CRT (and printed on the hardcopier), and saved in the magnetic disk (MD). As for the data on the NBI and the RF equipment the data are acquired by the

GAMMA 10 control computer. The example of the coil currents are shown in Figure 3.(CRT hardcopy)

CRT based flexible man-machine communication assists the operator's work of above four main functions.

4.2 Control of the MG of GAMMA 10

The MG control computer provides the function of control and monitoring of the MG set. Power saving operation of the MG is realized by direct digital control(DDC) of the MG speed.

The MG speed is controlled by adjusting the secondary coil current of the induction motor, which is controlled by the position of the LRH(Liquid Rheostat). The MG is accelerated to 70% of its rated speed by the hardwired logic for controlling the LRH. When the MG speed reaches 70% of its rated value, the LRH control is switched to the computer. After that the LRH position is directly controlled by the MG control computer which calculates it from the current MG speed, the reference speed and the primary coil current of the induction motor at every 100 milliseconds. Figure 4 shows the typical sequence of the operation pattern. The experiments are executed over the 70% MG speed. Before the experiment the MG is accelerated to the required speed which is also calculated by the computer from the load patterns of the coil current. The time required for the acceleration from the current speed to the reference speed is also calculated by the computer. The computer adjusts the start timing of the acceleration in order to complete the acceleration just before the experiment start. This function realizes the saving energy. Otherwise it consumes more energy in order to sustain the MG at reference speed for a long time.

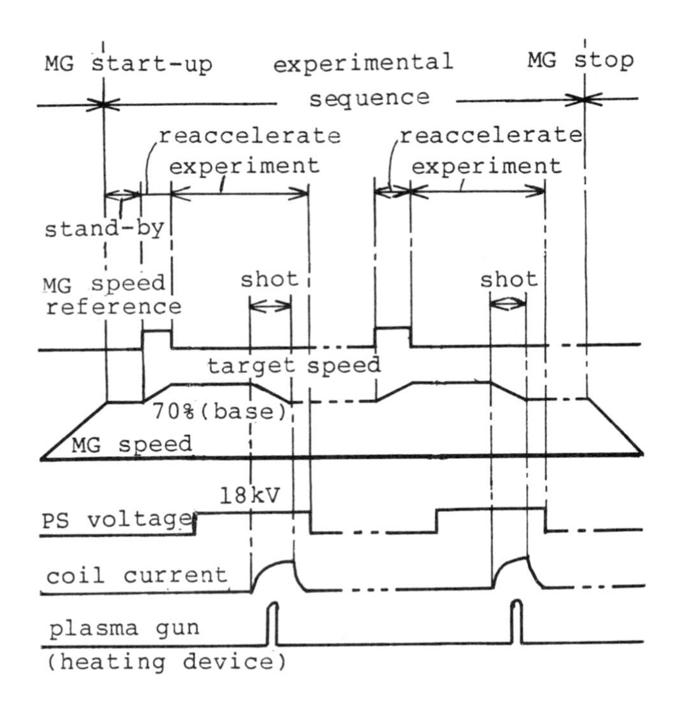

Fig.4 Operation Pattern

4.3 Plasma data acquisition

The function of the acquisition and storage of the plasma diagnostic data is provided by the plasma data acquisition computer. The computer also acquires the operation data of the GAMMA 10 heating device through DLA from GAMMA 10 control computer (200 microseconds data sampling). These data (800KB/shot) are stored in MT. The data acquisition devices are the CAMAC and the PIO. CAMAC modules are used for the plasma data acquisition because of its high speed data acquisition capability and the wide variety of modules. The optical (byte) serial CAMAC highway is applied because of the long distance between the computer and the crates, the flexibility , the expandability of the system and the noise immunity.

The computer provides the function of data gathering, data processing, data saving and data display. They are executed in accordance with the experiment cycle automatically without the aid of the operator. The timing signals required for the synchronous operation with the experiment are supplied by the master timing system. The CRT based flexible man-machine communication supports the operator's work of parameter setting and data display. The data gathering software is so designed as to be independent of the CAMAC system configuration. The parameters are easily changed from the keyboard even if the CAMAC system is re-configured.

5.MAIN FEATURES

The main features of the computer system are as follows :

(1) The computer provides the overall control function, while the hard-wired logic provides the protection logic (interlock), which realizes the flexibility of the operation while maintaining the reliability and the safety.

(2) The CAMAC optical type byte serial highway realizes the high speed data transmission and high noise resistance.

(3) Becuase of the automatic operation by the computer system, the operators can execute the experiments of GAMMA 10 very efficiently.

(4) The computer prevents the miss shots by the validity check of the experimental parameters before the shots, and it calculates required MG speed for the power saving operation of MG.

(5) The computer system is easily extended or changed according to the extention of GAMMA 10 equipment. GAMMA 10 has experienced some modification since its commissioning such as the installation of additional heating device and the additional diagnostics. The flexibility of the computer system has been proven through these experiences. Particularly, in case of change plasma diagnostics, it is not necessary to change the data acquisition programs or file layout.

6.CONCLUSION

The computer system is required to be highly reliable since a high power operation sequence is performed without interruption for about a week. During this operation sequence about 100 shots are executed per day. The GAMMA 10 has already experienced more than 55,000 shots since its first operation.

ACKNOWLEGDGEMENTS

We would like to thank all participants of GAMMA 10 in University of Tsukuba and all our colleagues in Toshiba corporation who helped us in the preparation of this paper.

REFERENCES

1.S.Miyoshi et al.: Tandem Mirror GAMMA 10 and its Contribution to Reactor Design, Proceedings of the third IAEA Technical Committee Meetings and Workshop on Fusion Reactor Design and Technology, Tokyo, 5-16 Oct.(1981)

2.T.Cho,M.Ichimura,M.inutake et al.: Stuies of Potential Formation and Transport in the Tandem Mirror GAMMA 10, in Plasma Phisics and Controlled Nuclear Fusion Research 1986(Proceedings of Eleventh International Conference, Kyoto, 1986) vol 2, IAEA, Vienna(1987),243.

FUSION TECHNOLOGY 1988
A.M. Van Ingen, A. Nijsen-Vis, H.T. Klippel (editors)
Elsevier Science Publishers B.V., 1989

RADIOLOGICAL PROBLEMS RELATED WITH THE PLASMA INDUCED EROSION OF THE FIRST WALL

L. GIANCARLI and H. DJERASSI

CEA - CEN Saclay, DPT/SPIN, Gif-sur-Yvette Cedex, FRANCE

The First Wall (FW) of a D-T tokamak is submitted to both plasma induced erosion and to neutron irradiation. The eroded atoms form particulates of diamater up to some microns and redeposit all over the FW with various levels of adhesion. Such a radioactive metallic dust (mainly SS in the case of NET) could represent a serious threat to the safety provisions of the plant. It has been found that, if continuous human acces into the NET building hall during downtime is required, less than one gram of SS-dust is allowed to escape from the machine.
It is also discussed the possibility of utilisation in the NET plasma chamber of standard or advanced techniques of decontamination and/or fixation of the radioactive dust in order to avoid or to simplify containment through physical barriers. Two dust fixing methods, carbonization and strippable coating, are discussed from the point of view of performances and of compatibility with the machine operations. Some decontamination methods, such as high pressure jet cleaning, electrochemical polishing, and aspiration, are also discussed.

1. INTRODUCTION

During the normal operations of NET it is expected that there will be erosion of the FW due to plasma-wall interactions which could result in the generation and accumulation of dust particulates on the plasma chamber internal surface. The physical parameters defining the dust, such as particle size distribution, chemical composition, weight, and activation level, depend on the machine design and operating conditions. In the present work the caracteristics of the dust expected to be formed in NET plasma chamber are given and the radiological consequence analysed. Starting from this analysis the second part of the paper concerns the discussion of possible methods of avoiding the spread of the radioactive dust into the building hall (Fig. 1) during maintenance operations[1].

2. FIRST WALL EROSION AND ACTIVATION

The NET-DN design has been considered[2]. The FW wall material is SS, the protective tiles are made of C, the divertor plates of W. We assumed an average neutron wall loading of 1 MW/m^2 and 0.8 year of continuos operations (end-of-life fluence).

In normal operations the FW erosion is largely due to sputtering. Typical erosion rates[2] are 0.2 mm/a for SS and 2.4 mm/a for graphite (1800°K). The behaviour of the eroded atoms is not known because no consistent redeposition theory or experimental data exist but it is expected that more than 99 % of them will redeposit in a stable form (INTOR assumption). The remaining part will form loosely attached particulates defined in this paper as 'dust'. Such a dust formation (metallic and graphite) has been found in JET, where it has recently been measured[3] about 7.5 g of metallic dust and 91 g of graphite dust (from the graphite tiles). The average aerodynamic diameter of the particulates was found to be about 1 μm, which means that they can be easily resuspended through mechanical action. If for NET one assumes similar dust characteristics, considering that an amount of about 3 kg is expected to be produced in the whole machine lifetime and that it will be activated, the problem of the possible spread into the hall has to be considered.

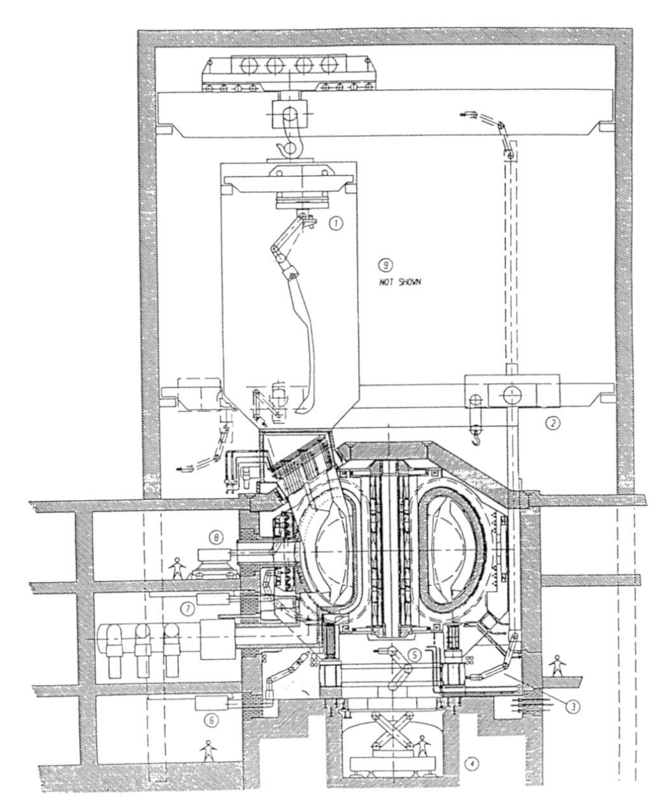

Fig. 1 - Example of NET builiding cross-section

Table 1 : Activity level in NET SS first wall (Bq/kg)

Isotope	Half-life	15 d operating time				0.8 a operating time			
		1 d cooling time		30 d cooling time		1 d cooling time		30 d cooling time	
Cr^{51}	27.8d	8.9+12	37.1%	4.3+12	43%	2.8+13	25.5%	1.4+13	18.4%
Co^{57}	270d	9.4+11	3.9%	8.7+11	8.7%	1.3+13	11.8%	1.2+13	15.8%
Mn^{54}	303d	6.6+11	2.8%	6.2+11	6.2%	9.5+12	8.6%	8.9+12	11.7%
Co^{58}	71.3d	3.2+12	13.3%	2.4+12	24%	1.3+13	11.8%	9.8+12	12.9%
Mo^{99}	67h	2.4+12	10%	-	-	2.4+12	2.2%	-	-
Co^{60}	5.26a	1.4+10	<0.1%	1.4+10	0.1%	2.6+11	0.2%	2.6+11	0.3%
Fe^{59}	45d	6.1+10	0.3%	3.9+10	0.4%	3.1+11	0.3%	2.0+11	0.3%
Fe^{55}	2.6a no γ	1.8+12	7.5%	1.7+12	17%	3.1+13	28.2%	3.0+13	39.5%
Co^{58m}	9.0h no γ	2.1+12	8.8%	-	-	2.1+12	1.9%	-	-
Tc^{99m}	6h no γ	2.3+12	9.6%	-	-	2.3+12	2.1%	-	-
V^{49}	330d no γ	5.6+10	0.2%	5.2+10	0.5%	8.1+11	0.7%	7.6+11	1.0%
Total		2.4+13		1.0+13		1.1+14		7.6+13	

The FW-steel activation level has been calculated[1] after 15 days of operations (approximatively the end of the physics phase, at present expected to be equivalent at ∿ 11d) and at the end of life. The results are given in Table 1 for two cooling times, 1d and 30d, which correspond to the relevant time internal for maintenance operations. It must be stressed that after 15d of operations the activation level is already more than 10 % of the end-of-life level.

3. HALL CONTAMINATION LIMITS FOR SS-DUST

The limits for the amount of dust allowed to spread in the building hall depend on the NET safety provisions. For instance, assuming that during downtime the dust is the only radiation source, we fixed the requirement of keeping the dose rate received by a person present in the hall lower than 25 μSv/h. Two cases must be considered : a) dose due to dust inhalation, b) dose due to the immersion in a dust cloud (no inhalation, only γ-rays effective). Calculations have been done in both cases for each radioisotope present in SS, assuming a hall volume of 50,000 m^3, homogeneous dust concentration and end-of-life activation level. The results are given in Table 2 where it is shown that less than 1g of SS-dust is allowed to spread in the hall in all cases. If the dust is assumed to be deposited on the floor only, the immersion limit could be increased by a factor 3. Calculations done for W-dust give similar results. One must stress that the immersion limit could be the effective one because inhalation effect could be reduced by the use of some reasonable protections (i.e. masks).

4. DUST SPREAD LIMITING METHODS

During maintenance operations the machine has to be opened and one or more blanket seg-

Table 2 : Amount of SS that can be realesead into the hall due to inhalation and immersion criteria for each isotope

Isotope	Half life	SS mass limit inhalation (g)	SS mass limit immersion (g)
Cr^{51}	27.8d	0.54	5.4
Co^{57}	270d	0.038	0.38
Mn^{54}	303d	0.053	0.53
Co^{58}	71.3d	0.039	0.39
Mo^{99}	67h	0.42	14.7
Co^{60}	5.27a	0.10	7.7
Fe^{59}	45d	0.81	14.5
Fe^{55}	2.6a no γ	0.048	100
Co^{58m}	9.0h no γ	24	NL
Tc^{99m}	6.0h no γ	44	17.6
V^{49}	330d no γ	19	NL
All isotopes		$9.3x10^{-3}$	0.13

NL : No Limit

ments have to be carried by a crane through the building hall. In such a situation the spread of the deposited dust is highly probable and methods for avoiding it must be found. Three general methods are in principle available :

a) Containment[4] through the so-called Tight-Intermediate Containment system (TIC) or a Closed Transfer Unit system (CTU) that use physical barriers for preventing the spread. These solutions, although always possible in principle, tend to complicate the maintenance operations.

b) Decontamination of the plasma-chamber internal surface before opening. This method is likely to be not sufficient when applied alone, although it would be the most satisfactory because it leads to a clean chamber (limited dust inventory).

c) Fixation of the activated dust on the wall for a time sufficiently long to perform all maintenance operations.

In the following sections some fixation and decontamination techniques will be discussed as a possible alternative to containment.

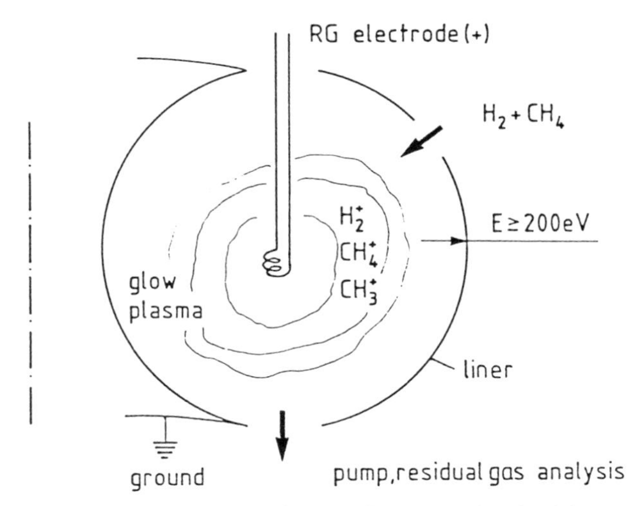

Fig. 2 - Schematic sketch of the carbonisation technique [5]

5. DUST FIXING TECHNIQUES

5.1. Strippable coating

Typical example is the Ionifixeau Meyer varnish, whose composition is 45 % acrylic emulsion + titanium oxyde, 55 % of water. It is a peelable varnish, allowing remote airless spray application, presently used in French nuclear industry. The main problems are the compatibility with further chamber operations, the low temperature of application (T wall < 70°C), the unknown behaviour if the wall temperature increases after application, and the evacuation of solid waste from the chamber.

5.2. Carbonization

This technique is already applied in tokamaks for start-up purposes[5]. Following the principle showed in Fig. 2, a carbon layer is deposited on the plasma chamber surface (thickness 0.03-0.1 μm) with a good degree of adhesion for most fusion-relevant materials. It can be immagined that, if a series of layers are produced up to a total thickness of about 1 μm, the dust particulates present on the surface remain trapped. Carbonization then could be used as a dust fixing technique. No serious problems related with the good working

Table 3 : Fraction of particulates extracted from the torus

Particulates diameter µm	Pumps throughput (m^3/h)						
	1,000	2,000	4,000	8,000	16,000	32,000	64,000
0.5	0.997	0.999	0.999	1.000	1.000	1.000	1.000
1	0.991	0.996	0.998	0.999	0.999	1.000	1.000
2	0.969	0.985	0.992	0.996	0.998	0.999	0.999
4	0.892	0.945	0.972	0.986	0.993	0.996	0.998
8	0.648	0.805	0.897	0.947	0.973	0.986	0.993
16	0.173	0.416	0.645	0.803	0.896	0.947	0.964
32	1.40-3	3.74-2	0.193	0.440	0.663	0.810	0.893

of the machine are expected. Nevertheless feasibility and efficiency are not proved yet; in particular, a) the technique has only be performed on polished surfaces, and b) a critical film thickness in relation with its stability (peeling) could appear. Experiments are then required before reaching any conclusion .

6. PLASMA CHAMBER DECONTAMINATION

Many techniques are available in the nuclear industry. A first basic selection can be done following some requirements and constraints strictly related with the application in NET. Two methods have been selected[1] as examples of agressive and weak techniques respectively.

6.1. In-situ electropolishing

The chamber wall could be used as anode and a remotely controlled perforated plate scanning the whole chamber surface could act as cathode. The consequent flow of electrolyte dissolves the surface material and the dust. The main problems are the important erosion produced in the existing application (∿ 0.05 mm of material removed) and the large amount of residual chemical in the chamber.

6.2. High-pressure jet cleaning

The jet fluid can be water, freon or gas. In the case of water (possible option of adding chemicals) good efficiency can be reached but there are the drawbacks of the large amount of liquid waste produced and of the oxygen residuals in the plasma chamber that have already given many problems in existing tokamaks. The use of Freon in closed loop appeared to be a good solution because of the limited amount of liquid waste coupled with a good obtainable efficiency but the limitation in temperature (boiling point 47°C) seems to be inacceptable. The last option is the use of inert gas (i.e. helium), in this case coupled with an aspirator. This technique needs some developments; feasibility and efficiency are at present unknown. A set of experiments should be defined in the near future.

7. AEROSOL ASPIRATION

It could be envisaged to extract the airborn particulates present in the torus (resuspended dust for instance) through direct aspiration by the existing vacuum pumps or additional ones. It is interesting to estimate the efficiency of such a procedure. A simple model of the torus (taken as a cylinder) has been considered and, neglecting chemicals effects, the pumping efficiency, as a function of the particle diamater and of the pumps throughtput, has been determined. The results are given in Table 3. It can be seen that, for diameters of 1 µm, more than 99 % of the particulates will be extracted, even for the lower throughput considered (1,000 m^3/h).

8. COMBINED TECHNIQUES

Considering the very limited amount of radioactive dust allowed in the NET building hall the required efficiency for decontamination, fixation or containment techniques will be anyway very high. The consequence is that, whichever technique is chosen, advanced technology and high standard system have to be developped, probably at high cost. It could be of interest to look at the possibility of using two (or more) combined techniques leading to a relaxation of the requirements to be satisfied by each of them. The increased complexity could be compensate by an easier feasibility. An optimization and a detailed analysis is required. Two possible examples are :

A) High-pressure gas decontamination plus carbonization. One could extract large-size perticulates through the helium jet system and then perform carbonization on a relatively cleaner surface with consequent easier application.

B) Carbonization plus containment. In fixing most particulates the required efficiency of the containment system is lowered, leading to a probably simplified maintenance procedure.

9. CONCLUSIONS

In this work it has been shown that the radioactive erosion dust likely to be present in the plasma chamber is a major safety concern for NET. For maintaining human access into the NET building hall one needs to avoid the spread of such a dust using high-efficiency techniques. This task is expecially difficult during maintenance operations because the plasma chamber has to remain open for several hours or even several days. Containment is the presently assumed solutions but considerably complicates the maintenance operations.

Some other methods, dust fixing and decontamination, have been discussed but, at present, none of them appears a suitable alternative. Nevertheless carbonization and high-pressure gas jet represent potential solutions provided some improvements and developments are made, possibly supported by a set of experiments to be defined in the short term.

The utilisation of one single technique could be found unreliable or unfeasible. Combined techniques, despite the apparently increased complexity, could be found more acceptable. A detailed analysis is required.

REFERENCES

1. L. Giancarli, 'Preliminary Study to Evaluate the Feasibility of Eliminating the Spread of Radioactive Contamination During the Maintenance of NET Internal Component', CEA-CEN Saclay, DPT/SPIN Fusion, Report SE.8/88.02 (1988).
2. NET Status Report - December 1985.
3. J. Charuau et al., 'First Experiment on Erosion Dust Measurement in a Tokamak', this volume.
4. M. Maupou et al., 'Comparison of Vertical versus Horizontal Maintenance Approach', CEA Report (1987), to be published.
5. J. Winter, 'Carbonization in Tokamaks', J. of Nucl. Mat. 145-147 (1987) 131.

FUSION TECHNOLOGY 1988
A.M. Van Ingen, A. Nijsen-Vis, H.T. Klippel (editors)

AIR INGRESS ACCIDENTS IN TOKAMAKS

A.V. JONES and P. ROCCO

Commission of the European Communities, Joint Research Centre, Ispra Establishment, 21020 Ispra (Va) - Italy

Accidents are considered consisting in a breach in the vacuum wall of a fusion reactor, the ingress of air into the torus and the mobilization and expulsion of activated eroded-sputtered material. Previous evaluations on NET, assuming bare plasma-facing components, have been reviewed and new estimates of the rate of material discharge and of the total fraction of discharged material are calculated, including resettlement processes. The case of graphite-tiled walls, without and with combustion, is also examined. Radiological evaluations are then performed to assess the hazard posed by the discharged material both within the NET building and outside it.

1. GENERAL

Plasma-wall interaction in tokamak reactors will result in the generation of particulates (erosion dust), the redeposition characteristics of which are not completely assessed at present. An accidental air ingress into the plasma chamber can mobilize this dust and expel it from the torus. In the case of walls tiled with reactive materials, such as graphite, combustion can also occur. Input data for the modelling of these accidents are taken from NET[1], but results are easily extrapolable to power reactors.

2. MODELS AND RESULTS FOR THE BARE WALL CASE

The authors have previously[2,3] considered the loss of vacuum accident in NET, pointing out the potential for radioactive contamination of the reactor hall and possibly of the external environment as a result of the suspension of erosion dust by the inrushing air and its subsequent transport out of the plasma chamber as the air within the torus is warmed by the hot first wall and expands. The configuration then considered was that of a bare steel first wall, and dust discharge rates and quantities were calculated for various postulated wall temperatures and for breach sizes ranging from that of a maintenance port for an articulated boom (1.2 m^2) down to a small hole of area 0.00785 m^2. It was also shown that only modest overpressures developed, so that no further breach need be expected.

At the time of publication[2] little was known about the granulometry of the erosion dust, and hence very conservatively it was assumed that there would be no resettlement within the reactor and no trapping of the dust in the sometimes tortuous paths between the first wall and the reactor hall. More recently, information on the granulometry of the dust in the torus of JET has become available[4], and it is now possible to estimate the potential for resettlement and trapping. In addition, the opportunity has been taken to revise certain details of the model employed in [2,3].

2.1 The model for air ingress and discharge

The PUFFER code[3], which calculates the inrush of air into the torus and its subsequent expulsion, assumes that the reactor hall is filled with air at atmospheric pressure, while the torus is initially under hard vacuum. It is further assumed that the breach is small compared with the dimensions of the torus, so that the internal pressure may be considered uniform. It is found that because of the large pressure difference the air flow is choked for much of the inrush process, so that details of the flow path do not greatly affect the flow rate. The inrushing air, which is cooled by the expansion, comes into contact with the first wall and divertor, which both during operation and during the first hours of a maintenance shutdown are at a temperature well above that in the reactor hall. An exact calculation of the warming of the air by the hot structures would require the solution of a problem of transient three-dimensional turbulent flow, which has not been deemed worthwhile in a safety evaluation. To a suitable approximation the mean heat transfer coefficient h is calculated as follows: if k is the thermal conductivity of the air and D the effective hydraulic diameter of the torus, h is given by the correlation[5]:

$$h = 0.023 \cdot \left(\frac{k}{D}\right) \cdot D^{0.8} \cdot \left(\frac{\rho \cdot v}{\mu}\right)^{0.8} \cdot Pr^{0.4} \qquad (1)$$

where ρ is the density of the air, μ its dynamic viscosity, Pr its Prandtl number and v a "typical" air flow velocity within the torus. For calculational purposes D was estimated as 4xvolume/surface area, where the volume and area values refer to the torus, excluding vacuum ducts and dust traps. Numerically $D = 2.7$ m. For v a value of 20% of the air velocity just downstream of the breach was assumed. Note that the h values used in [2,3] were too low, resulting in overlong discharge times and overestimates of the quantity of dust expelled.

2.2 Type of breach assumed and reactor state at rupture

References [2,3] calculated air ingress and expulsion for a wide range of breach sizes. This paper considers just one type of breach, the rupture of the double ceramic seals which provide the tightness for an ICRH launcher. There are four seals per launcher (inner diameter 185 mm, outer diameter 305 mm) and two cases are analysed: a) rupture of four seals (flow area 0.185 m^2); and b) rupture of one seal (flow area 0.0462 m^2). Two initial wall temperatures are considered, 473 K (maintenance) and 623 K (a maximum value during operation). It may be shown[9] that the wall temperature can be assumed constant during the ingress and discharge process. The air flow path for an ICRH launcher is tortuous, and its aerodynamic resistance must be taken into account in the discharge; a resistance coefficient of 4.3 is appropriate[9]. Results calculated with PUFFER are given in Table I, and Figure 1 shows the development of the mass of air in the torus and its mean temperature in Case 3 (rupture of four seals, mean wall temperature 623 K). The phases of air entry and expulsion are clearly marked. As explained in [2,3] the mass of air within the torus decays exponentially from M_{max} to its final value, with a half-life $t_{\frac{1}{2}}$ of 30 s or less.

Table I - Results for the dust expulsion problem - no combustion

Case	Hole area (m^2)	T_{wall} (K)	$\bar{v}_{in}$ (m/s)	t_{char} (s)	T_{char} (K)	M_{max} (kg)	$t_{1/2}$ (s)	% exp.	$\bar{v}_{out}$ (m/s)	ΔP_{max} (kPa)	t_P (s)	
1	0.0462	623	320	61	565	537	24	9	33	1.5	78.5	
2	0.0462	473	360	79	440	689	31	7	21	0.75	96	
3	0.185	623	325	19	449	675	28.5	28	32	1.75	28	
4	0.185	473	306	23.5	370	320	36	22	18.5	0.85	31.5	
5	0.185	623	280	19	524	828	15.5	16	29	1.6	26.5	$\gamma = 1.666$ (argon)
6	0.0462	973	314	40	859	353	15	12	53	2.8	54.5	
7	0.185	973	312	13.5	640	475	19	34	61	3.9	22	
8	0.0462	1823	320	21.5	1572	192	8	13	84	4.5	31.5	
9	0.185	1823	310	8	1127	272	10.5	38	114	7.8	14	

Cases 1-5: bare wall
Cases 6-9: graphite-coated wall
T_{wall}: wall temperature
$\bar{v}_{in}$: characteristic inrush velocity
t_{char}: time of which torus contains the maximum mass of gas
T_{char}: temperature of gas at t_{char}
M_{max} : maximum mass of gas in the torus
% exp: percentage of M_{max} expelled from the torus
$t_{\frac{1}{2}}$: characteristic time for the gas expulsion process
$\bar{v}_{out}$: characteristic outflow velocity
ΔP_{max}: maximum overpressure
t_p : time at which ΔP_{max} is attained

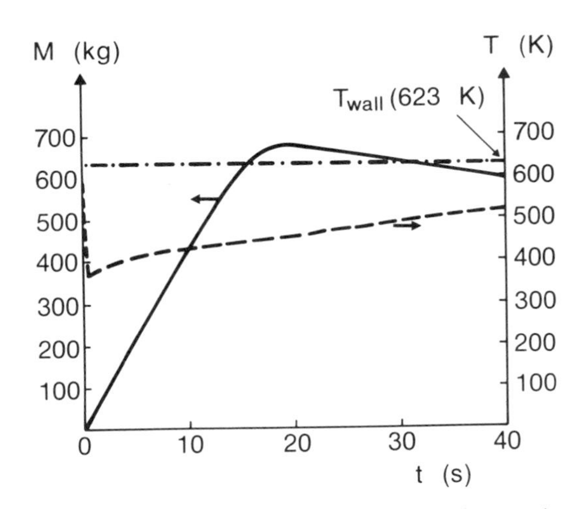

Fig. 1 Mass and temperature of gas in the torus. Bare wall case 3

These quantities are given in the table, which also indicates the percentage of the air mass expelled during the warming process; this is in the range 7-28% for the bare wall cases considered and up to 38% for the case of a coated wall without combustion (see Section 3). A special entry in Table I is Case 5, which refers to a reactor hall inerted with argon. The different material properties of argon result in a lower percentage of expelled gas and hence in reduced contamination.

2.3 Dust resuspension, settlement and trapping

A fraction of the dust sputtered from the first wall will be resuspended by the inrushing air and some of it may then settle out again before being discharged. Both processes depend on the granulometry. Table II gives the free fall speed in air and the time required to fall 1 m for steel spheres in air. For radii below 1 μm the free fall velocity is essentially zero, while for 1 mm particles and larger a strong upflow is necessary to prevent them settling out. Conversely, fine particles tend to cohere through van der Waals forces and so may be less readily resuspended.

Measurements in JET[4] indicate a resuspension factor (ratio of airborne dust to total dust) without air inrush of 10^{-5} to 10^{-4}, while the

Table II - Free fall speeds, v and time to fall 1 m, τ for steel spheres in air

a (μm)	v (m/s)	τ (s/m)
0.1	7.6×10^{-6}	1.32×10^{5}
1	7.6×10^{-4}	1316
10	7.6×10^{-2}	13.2
100	3.2	0.31
1000	13.3	0.075

dust particles have radii in the micron range, with a large variation. The granulometry may be similar in NET. It is expected that the resuspension factor will be larger however, because of the high velocity of the air inrush (see Table I for typical inrush and outflow velocities). Concerning settlement, since characteristic discharge phase half-lives are $\leq$ 36 s, it appears from Table II that sub-micron resuspended dust will not settle out before it is discharged. For larger particles the local velocity is important. Close to the breach the velocity will be of the order of v_{out}, which is sufficient to sustain particles up to 10 μm. Away from the breach velocities will be typical of natural convection (0.5-10 m/s) and particles above 10 μm would likely settle out. In summary, during the discharge phase a good fraction of the dust of radius $\geq$ 10 μm will settle out, but much of the remainder will pass through the breach in proportion to the percentage of gas expelled (as listed in Table I).

There is a probability that dust will be trapped along the flow path between the torus and the reactor hall. The small size of the particles and the relatively slow discharge velocities make purely inertial trapping inefficient. More effective may be sticking of particles to the duct walls. Sticking and redetachment depends on many parameters besides the granulometry, and must be evaluated through experimentation at velocities of the order of v_{out}. In the absence of reliable data we assume conservatively that all dust reaching the breach arrives in the reactor hall. In this case one may read off the percentage dust expelled from Table I, summarised in Figure 2. Figure 2 also gives the masses of dust expelled, assuming that 3 kg of dust are suspended within the torus. 2 kg corresponds to < 1% of the eroded material expected in NET for an integrated wall load of 1 MW·a/m^2 [6]. The masses are to be taken in conjunction with the radiological considerations of Section 4.

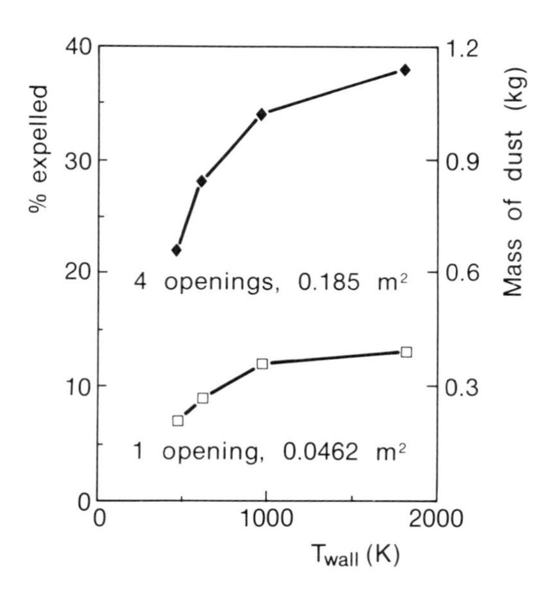

Fig. 2 Dust expelled vs wall temperature and break size

3. MODELS AND RESULTS FOR THE GRAPHITE-TILED WALL CASE

In the physics phase of NET it is proposed to cover the first wall and divertor with graphite tiles, 20 mm thick on the first wall, and 5 mm thick on the divertor. The loss of vacuum accident may be considered for this situation also. If combustion is assumed negligible (for instance, if the reactor hall is inerted), the model and conclusions are identical to those of Section 2. If the reactor hall contains air, however, the burning of the tiles may contribute decisively to the energy balance within the torus. The rate of burning depends on the tile surface temperature. Tiles may be mechanically attached (typical temperature 1823 K) or bonded (typical temperature 973 K). For cool tiles the fire risk is low[7], and we may assume that combustion is not significant. As in the bare wall case the wall temperature remains almost constant during the accident, and the model of Section 2 is valid. Cases 6 and 7 of Table I summarise calculations for the case of cool tiles. From the table and Figure 2 we see that the higher wall temperature results in a faster transient, a higher percentage expelled and a slightly higher excess pressure.

With the hotter mechanically attached tiles some burning of the graphite is inevitable. The theoretical treatment of this process can be kept simple, since graphite is always present in excess, and the oxygen in the incoming air is burnt to CO_2 (we assume) almost as soon as it enters. A maximum burn rate of 1.48×10^{-3} $kg\cdot m^{-2}\cdot s^{-1}$ of carbon is assumed in the calculation[8]. It can be shown[9] that the tile surface temperature may again be assumed constant. The apportionment of the heat of reaction between gas and tile is difficult to determine, and in the new extended PUFFER model which includes combustion (PUFFER-B), the user may specify a parameter DIS to control this feature (DIS = 1: all heat to gas; DIS = 0: all heat to tiles). PUFFER-B has been applied to the case of a hot graphite-tiled wall and ruptures of one or four seals as in Section 2, the main results being presented in Table III. DIS values of 1 and 0.5 have been specified to reveal the influence of this parameter. Figure 3 compares the evolutions of the gas mass and temperature for the case of the larger breach without burning (Case 9 of Table I) and with burning (Case D of Table III). Without burning the mass of gas within the torus peaks at t = 8 s and then falls off; the gas temperature is low at first and gradually rises to the wall temperature, and a slight excess pressure is developed. 38% of the gas mass is expelled. With burning the mass of gas peaks earlier and at a lower value. Burning continues to raise the temperature, and a peak of nearly 2350 K is attained when the oxygen in the torus is exhausted. There is a greater overpressure, with a maximum value of 17 kPa. Once the combustion process is complete the gas gradually cools to the wall temperature, so that there are two phases of air ingress in this case. Figure 4 compares case C (smaller breach) with case D (larger breach), specifying DIS = 0.5 in both cases. The evolution of gas mass and temperature has already been examined for case D. Case C merits discussion because of an interesting phenomenon. As may be seen from

Table III - Results for the dust expulsion problem with combustion

Case	Hole area (m^2)	T_{wall} (K)	$\bar{v}_{in}$ (m/s)	dis	t_{char} (s)	T_{char} (K)	M_{max} (kg)	T_{max} (K)	t_T (s)	ΔP_{max} (kPa)	t_P (s)	M_{CO_2} (kg)
A	0.0462	1823	320	1	34	1968	171	2188	15	0	—	—
B	0.185	1823	310	1	5.5	1597	203	2847	14	26	11	12.7
C	0.0462	1823	320	0.5	30	1844	184	1875	18	0	—	—
D	0.185	1823	310	0.5	6.5	1359	237	2349	19.5	17.6	12.5	12.5

T_{wall}: initial wall temperature
$\bar{v}_{in}$: characteristic inrush velocity
dis : fraction of heat of combustion deposited in gas phase
t_{char}: time at which the torus contains the maximum mass of gas
T_{char}: temperature of gas at t_{char}
M_{max} : maximum mass of gas in the torus
T_{max} : maximum gas temperature
t_T : time at which the maximum temperature is attained
ΔP_{max}: maximum overpressure
t_P : time at which ΔP_{max} is attained
M_{CO_2} : mass of CO_2 expelled

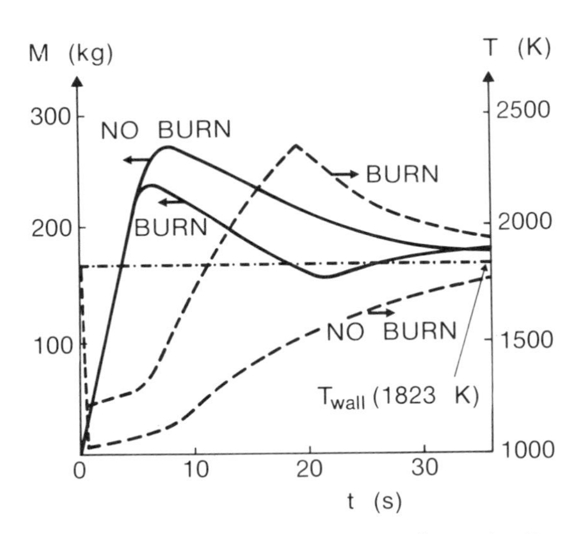

Fig. 3 Mass and temperature of gas in the torus. Comparison of graphite-coated wall cases 9 and D. (Effect of combustion)

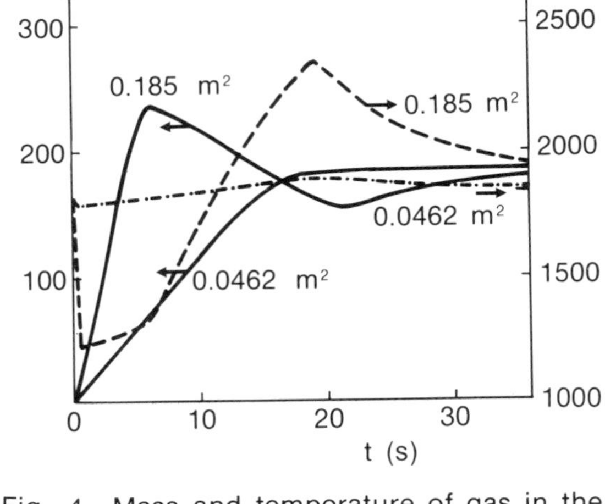

Fig. 4 Mass and temperature of gas in the torus. Comparison of graphite-coated wall cases C and D. (Effect of hole-size)

Figure 4 the combustion of the incoming air raises the gas temperature to a value close to the wall temperature immediately on entry. Consequently, no subsequent overpressure develops and there is no discharge phase. The same is found to occur in Case A (DIS = 1).

The calculations and figures provided are sufficient to demonstrate the dramatic difference made to the accident scenario by the burning of the graphite tiles and the potential both for overpressure and high temperatures and for virtual suppression of the discharge process, depending on the breach size and wall temperature. Note that this paper does not examine the possibility of a further breach or any stack effect; we merely remark that the energy which would be released by burning all the graphite in the torus is very large (∿ 500 GJ).

4. ACCIDENT CONSEQUENCES

The radiological impact is assessed at the end of the NET technological phase, i.e. for an average neutron wall load of 1 MW/m^2 and 7,000 hours of irradiation, spread over 7 years of operation. The accident condition examined for bare wall surfaces corresponds to Case 3 of Table I with 0.84 kg of erosion dust expelled (28% of the 3 kg suspended within the torus). The tiled wall case examined is Case D of Table III, with the expulsion of 12.5 kg of CO_2 and of all the tritium inventory present in the graphite.

The steel examined is U.K. nuclear grade AISI 316, while the graphite composition is that used in JET.

The volume of the reactor hall is assumed to be $3x10^5$ m^3 [10].

4.1 Contamination of the reactor hall

The neutron-induced radioactivity in the AISI 316 erosion dust is 0.345 TBq/g (9.32 Ci/g)[11]. 11 radioisotopes, namely: Ta-182, Sn-117m, Ni-57, Co-60, Co-58, Co-57, Fe-59, Fe-55, Mn-56, Mn-54, Cr-51, constitute about 80% of the total radioactivity and more than 95% of the radioactive hazard. The fine granulometry and the high outflow velocity will produce a large dust diffusion in the reactor hall. The average contamination in the ambient air will attain about $6x10^4$ DAC (Derived Air Concentration for Workers). In the case of a graphite-tiled wall, the graphite activation will be small, mainly due to S-35 and P-32 from impurities[12]. The average contamination will be 6 DAC.

The radioactive hazard in this case arises from the tritium permeated in the graphite tiles during NET operation. At the assumed graphite temperature of 1823 K this tritium inventory is evaluated as 1 PBq (3 g), whereas the inventory would be much higher at 973 K[13]. The release of 3 g of tritium in oxide form as a consequence of the fire would produce an average contamination of 4500 DAC.

4.2 Environmental effects

Two release conditions are examined, assuming that all the radioactive products expelled from the NET torus can reach the environment:

a) release from the reactor building, at a height of 20 m;

b) release from a 100 m stack.

a) The AISI 316 erosion dust emitted at 20 m in meteorological conditions F2 will produce a maximum acute dose (inhalation + 1 week groundshine) of less than 15 mSv (1.5 rem) between 800 and 1,200 m; it will be 8 mSv at 2000 m, 2 mSv at 5000 m. In the accident involving graphite-tiled walls, the radioactive impact will be essentially due to the tritium emission. 3 g of tritium will produce a maximum acute dose of 0.75 mSv if released as HTO, and about 0.4% this value in case of elemental tritium release[14].

b) The AISI 316 erosion dust emitted at a 100 m height will produce a maximum acute dose at 6000 m. It will be 3% of the maximum of case a), i.e. 0.45 mSv. The dose at 12,000 m will be reduced to 0.35 mSv. The doses due to 3 g of tritium as HTO will be 22 and 17 μSv at 6000 and 12,000 m, respectively.

The foregoing results require the following

comments.

Release condition a) requires two simultaneous breaks in the vacuum wall and in the containment building. Even in this severe and low probability accident condition, a large amount of erosion dust will settle inside the reactor hall. The corresponding environmental dose will be a fraction of that evaluated previously, 15 mSv. By comparison, the limit for early warning of evacuation is 50-100 mSv in most countries.

In condition b) the erosion dust release through the stack would be a small fraction of that evaluated (0.45 mSv), since:

- HEPA or sand filters will retain more than 99.5% and 99%, respectively of the particles with granulometry greater than 0.5 μm;
- stopping the ventilation system will retain most of the radioactivity within the building even if the filtering systems fail.

Regarding the tritium release, the emitted amount will be reduced by deposition inside the building and by the employment of clean-up systems.

5. CONCLUSIONS

With respect to previous evaluations, the revised values of the wall heat transfer and other improvements to PUFFER have significantly reduced the predicted fraction of dust released.

Data from JET concerning the dust granulometry suggest that little settling will occur within the torus. The potential for dust trapping along the expulsion duct requires experimental investigation.

The model has been extended to include burning of the graphite-tile coating. Combustion raises the gas temperature above that of the wall and can result either in non-expulsion of material for small break sizes or significant excess pressure (∿ 20 kPa) and gas expulsion for large break sizes. The possibility of further breaks and the potential for a stack effect have not been evaluated. They must be assessed in the future as they can cause the release of the total combustion heat of the graphite.

In the bare wall case, the model has now been significantly improved, so that the evaluated fraction of dust expelled will not be far from the real case. However, the margin of error on the dust carried out of the torus may be substantial due to the uncertainty concerning the dust production and mobilization.

In the case of graphite tiles, in the absence of stack effects, the evaluated CO_2 amount expelled from the torus is probably near to reality.

Whereas this type of accidents will produce modest environmental effects, the radioactive products expelled from the torus constitute a severe contamination hazard inside the building requiring extensive repair work. It should be noted that the amount of radioactivity released into the reactor hall can be greater by orders of magnitude than that caused by other severe accidents, for instance the tritium and corrosion product contamination consequent on a spillage of the primary coolant system. This means that the average failure probability of the vacuum wall must be reduced, by appropriate engineering design, to very low values, less than 10^{-4} a^{-1}. Emergency pumping systems, with dust filtering and collection should also be envisaged.

Extrapolating the results of this assessment from NET to power reactor conditions, the hazard of contamination will remain. In fact, the plasma-wall interaction and the consequent erosion dust production should be reduced due to expected improvements in the control of plasma-wall interaction. Conversely, the integrated wall load will be much higher than in NET with a correspondingly higher specific radioactivity.

ACKNOWLEDGEMENTS

The environmental doses due to the activation products and tritium have been assessed from data by G. Graziani, JRC-Ispra, and O. Edlund, Studsvik A.B.

REFERENCES

1. NET Status Report, December 1987, EUR-FU/XII-80/88-84.
2. A.V. Jones, P. Rocco, L. Deleanu and H. Djerassi, Safety analysis related to the possible release of activated erosion dust in fusion reactors, Proc. of the 14th SOFT, Pergamon Press (1986), Vol.2, pp.1329-1334.
3. A.V. Jones, Erosion dust contamination. A possible accident in NET. Calculation of the expulsion process, JRC-Ispra, TN No. I.05.B1.86.72 (1986).
4. J. Charuau and H. Djerassi, Plasma-wall interaction - JET experiment, draft final report, CEN-Saclay, December 1987.
5. J.M. Kay, An introduction to fluid mechanics and heat transfer, Cambridge University Press (1963).
6. M.F.A. Harrison, Culham Laboratory, Plasma interaction with the first wall in NET, communication at the NET meeting at Culham, June 19th, 1985.
7. F. Mazille, First wall and blanket analysis, CEA/IPSN report S - E 4.1.1.1/RE1/88.01, March 1988.
8. Groupe Français d'Etudes des Charbones, Les Charbones, Tome II, Masson & Cie (1965).
9. A.V. Jones, JRC-Ispra TN, in preparation.
10. G. Shaw and B. Stasko, Development of preliminary lay-out for the NET facility, CFFTP report 1-87001 (1987).
11. C. Ponti, JRC-Ispra, personal communication, 1988.
12. C. Ponti, Activation calculation for NET shielding blanket, JRC-Ispra TN, No. I.88.73 (1988).
13. J. Raeder, NET-Team Garching, Summary of fusion accident analyses performed in the E.C., ITER S & E Specialists meeting, Garching, 6-10 June 1988.
14. Environmental tritium behaviour - French experiment, final report, IPSN-DPT, CEN-Saclay (1988).

FUSION TECHNOLOGY 1988
A.M. Van Ingen, A. Nijsen-Vis, H.T. Klippel (editors)

ANALYSIS OF LOCA/LOFA RISKS FOR THE WATER COOLED FIRST WALL OF A NET TYPE FUSION REACTOR

F. MAZILLE

TECHNICATOME, Physics and Safety DPT, Rue A. Ampère BP 34, 13762 - LES MILLES CEDEX - FRANCE

H. DJERASSI

CEA - CEN SACLAY, IPSN/DPT/SPIN, 91191 - GIF SUR YVETTE - FRANCE

This paper presents a safety study of the first wall of NET (physical phase) related to coolant accidents. Both loss of coolant and loss of flow accidents are analysed. The consequences of a water leakage inside the vacuum chamber are quantified and the possible environnement impact is approached.

1. INTRODUCTION

In recent years, the investigation related to the safety and the environnemental impact of fusion reactors increased significantly. This paper presents an example of accident analysis which were performed by CEA and TECHNICATOME with the friendly and conclusive contribution of the NET TEAM safety group members. This presentation has been voluntarily restricted to the loss of the cooling function in the first wall during the NET physical phase.

2. CHOICE OF INITIATING EVENTS

A coherent accident analysis requires a rational choice of the initiating events. In a pre-design phase the use of a probabilistic approach is an appropriate method. In recent years, reliability studies were performed in order to estimate the failure rates of the NET first wall[1] and the availability of its cooling circuits[2]. Using the results of these two analyses, we have selected the following two initiating events and accident scenarii :

- The first event is a single pipe guillotine rupture inducing a water leakage inside the torus (LOCA). Assuming that the NET first wall would be built with classical quality assurance procedures of a nuclear power plant, the probability of such an event is estimated between 10^{-2} y^{-1} and 10^{-4} for NET III A by a WOHLER graph method[3].
- The second event is a loss of flow accident in the NET FW cooling circuit induced by an inadvertent valve closure. By analogy with classical PWR components (valves, pumps...), the probability of a total loss of flow in a FW segment is estimated just below 1 time per year (the failure rate of the corrective action will highly depend on the efficiency of the control and instrumentation).

3. CONSEQUENCES OF A LOSS OF COOLANT

A water leakage inside the torus of NET induces two kinds of consequences :

- the pressurization of the torus due to the double phase flow inlet,
- steam chemical reactions on the plasma facing components at high temperatures (FW protection tiles, divertor, NBI grids...).

3.1. Torus pressurization

The pressure transient induced by a LOCA depends highly on several parameters :

- the thermodynamical features (pressure, temperature) of the cooling circuit,
- the diameter of the broken pipe,
- the free volume of the vaccum chamber,
- the heat exchange between steam and internal structures,
- the heat exchange between internal structures and the outside.

Concerning the physical phase of NET, the pressure transient due to a guillotine rupture of one pipe (Ø = 13 mm, 10 b, 140°C) was calculated with the PWR code PAC, with initial conditions as described in fig. 1.

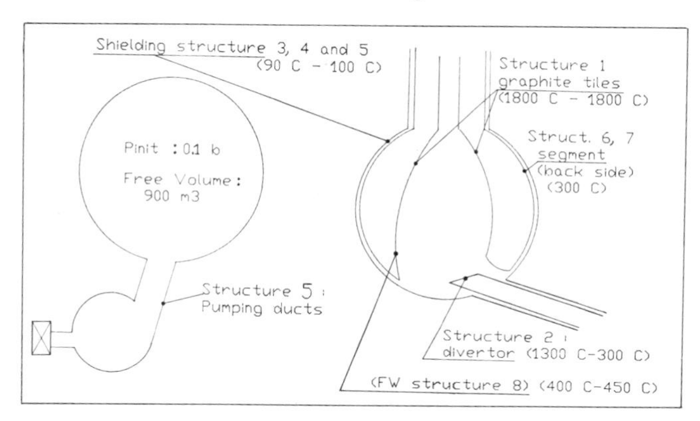

FIGURE 1
NET internal structures

As shown in fig. 2, the peak pressure has been estimated below 1.1 b (absolute pressure without any chemical reactions). If a high pressure, high temperature circuit (300°C, 100 b, Ø = 13 mm) is used for the first wall cooling, for instance during the technological phase of NET, a peak pressure between 2 b and 3 b is expected[3].

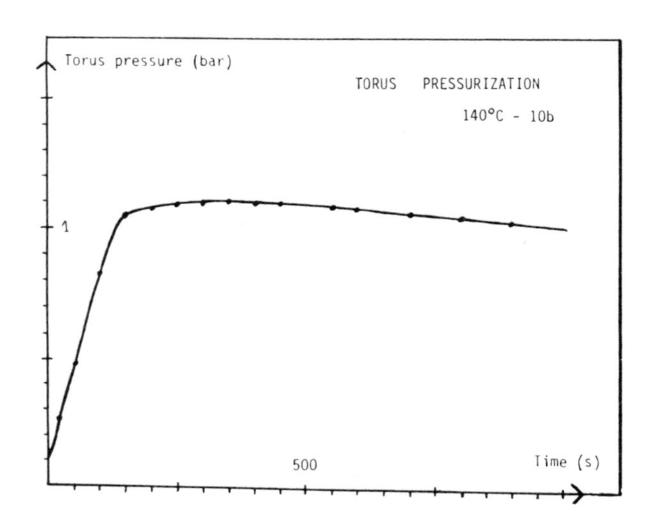

FIGURE 2
Torus pressurization (140°C - 10 b)

3.2. Chemical reactions

The consequences of chemical reactions on hot plasma facing structures are more difficult to quantify because of various uncertainties, related to the kinetics of chemical reactions, in particular for graphite.

Concerning the physical phase of NET, graphite is expected to assure a thermal protection of the first wall steel structure. As shown in fig. 3, graphite may be mechanically attached to the structure[4] (the tiles are then radiatively cooled) or directly attached to the structure (conductively cooled tiles).

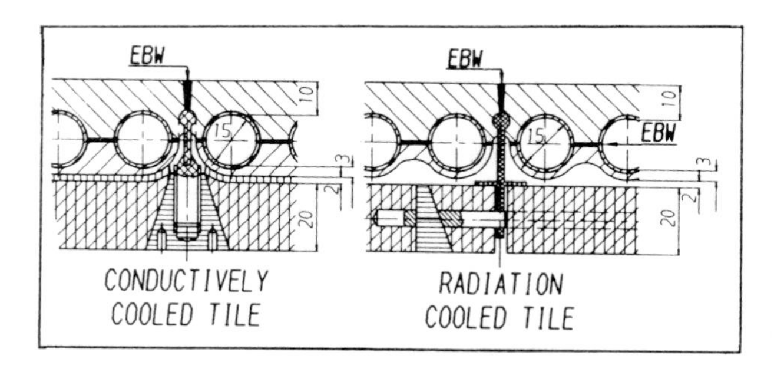

FIGURE 3
NET FW concepts

Considering, 400 m² of graphite protection, we have estimated the hydrogen production through $C + H_2O <===> CO + H_2$ reaction, by coupling a 1D thermic code with a $C + H_2O$ kinectics law based on large scale experiments performed in KFA JULICH[5].

For 400 m² of conductively cooled tiles (fig. 3), the hydrogen production is presented in fig. 4, assuming an initial tile temperature of 1500°C.

In spite of these severe assumptions, hydrogen production remains low because thermal conduction provides very efficient cooling.

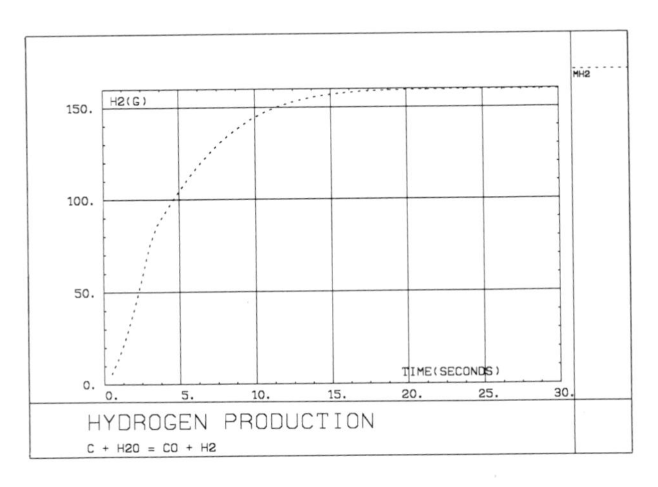

FIGURE 4
Conductively cooled tiles

For 400 m² of radiatively cooled tiles, the hydrogen production is presented in fig. 5, assuming an initial tile temperature of 1500°C. In this case, the hydrogen amount is significative, even though considered as an upper bound (inhibition effect at very high temperature is not taken into account).

In this case the pressure transient as presented in fig. 2 is modified by the produced gas (CO and H_2). As shown on fig. 6, the peak pressure is then 1.4 b.

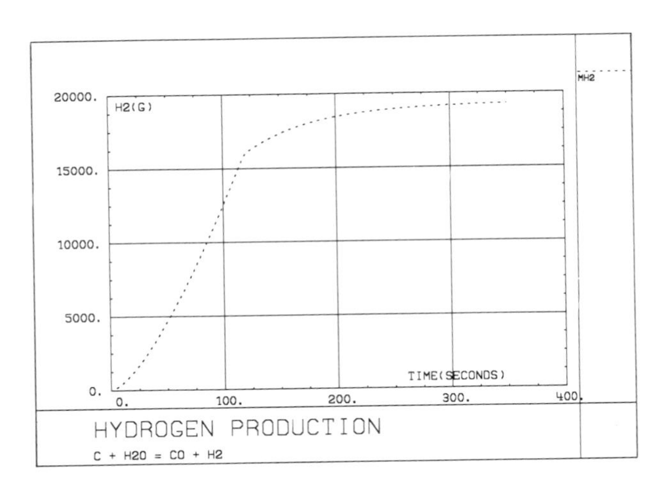

FIGURE 5
Radiatively cooled tiles

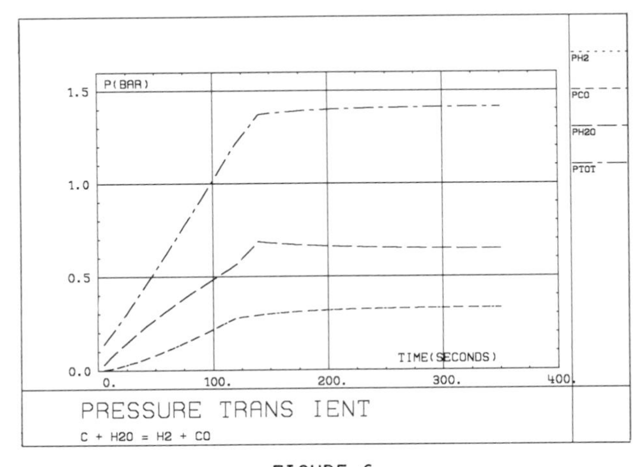

FIGURE 6
Pressure transient

4. CONSEQUENCES OF A LOSS OF FLOW ACCIDENT

A total loss of flow inside a FW segment induces overheating of the FW. In the first 20 s, coolant will be totally evaporated so that heat exchange would drastically drop[3]. Beyond this point, the FW will have a quasi adiabatic evolution. In fig. 7, we show the evolution of the FW maximum temperature with radiatively cooled graphite tiles, as soon as heat exchange coefficients can be considered as negligible. This calculation performed with the 2D code DELPHINE shows clearly that the FW is reaching its melting point within 25 s, for an heat flux of 0.4 MW m^{-2} (this result does not however integrate the 20 s necessary to dry out the FW cooling pipes).

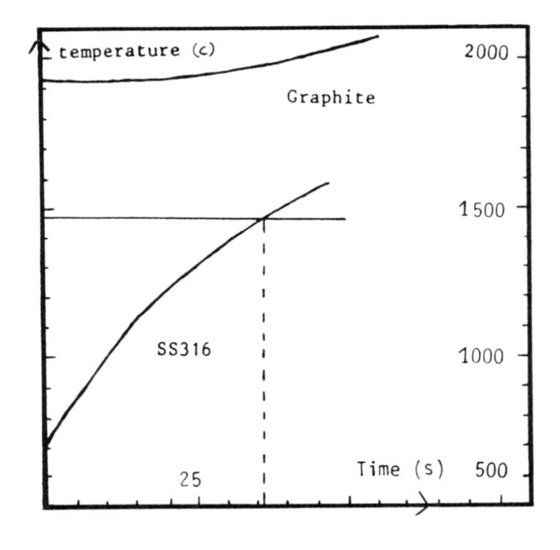

FIGURE 7
Temperature increase after a LOFA

Similar conclusions were drawn even considering a double loop and "half redundant" cooling circuit (2 x 50%)[6]. In this case, the delay time before the first appearance of dammages on the FW is larger.

An another consequence of the LOFA is the increase of the FW sublimation rate at such high temperatures. This evaporation produces impurities which, will be removed in part by the pumping system while the remaining part will contaminate the plasma. If we simply use a LANGMUIR law to modelize "in vacuo" sublimation process, disregarding the pumping effiency we may conclude that the sublimation rate not high enough during a LOFA to provoke a self plasma quench by its own contamination, before the FW is broken[3].

5. ENVIRONMENTAL IMPACT OF LOCA/LOFA

In this section, we present only orders of magnitude which have to be handled very carefully, in particular because of numerous uncertainties about the knowledge of the source terms. Anyway, the environmental impact due to a coolant accident in the FW will be significative only if all normal and emergency radioactive confinement facilities (cryostat, building, ventilation) are simultaneously failing, which is highly unlikely.

5.1. Tritium compounds

Tritium will be mobilized during a LOCA/LOFA from different parts of the reactor[7] :

- the primary coolant (if we assume a tritium concentration of about 10 Ci/g, the mobile inventory is then between 50 g and 200 g, depending on the cooling circuit volume),
- the plasma chamber evacuation (depending on the ability to isolate quickly the cryopumps from the torus chamber just after a LOCA, the mobile inventory may be estimated between few grams and 150 g),
- the fuel injection devices (upperbound 65 g),
- the plasma facing components (the mobile inventory is not yet determined with a sufficient accuracy).

The fraction of this mobile inventory which might be released effectively in the environment after the failure of all radioactive confinement means would be mostly in HTO form (concerning occupational safety much attention has to be paid to tritiated graphite fine dusts).

The inhalation dose at 1000 m from the release point was estimated around 4.8 10^{-4} Sv per gram of released HTO, under unfavourable weather conditions[8].

5.2. Metallic dust

Concerning metallic dust, the radiological source term is again very difficult to quantify mostly because of numerous uncertainties linked to erosion, sputtering and disruption phenomenons.

For metallic dust coming from plasma facing components, the estimation is not advanced and accurate enough but simple extrapolations from JET sampling analyses show that dust cannot be considered arbitrarily as a negligible contribution to the radioactive source term[9].

An important contribution to the source term may come from chemical reactions between H_2O and divertor coverage if high Z materials were used. For instance if 5 mm W5Re is used as divertor coverage for the technological phase of NET, we have estimated a possible release of 300 g (WO3), which means near 270 TBq (conservative estimation).

The last contribution is related to corrosion products inside the coolant. Calculations performed by CEA experts with the code PACTOLE[10], have estimated the metallic oxides release below 55 GBq, under PWR working conditions PH = 7, Li = 2 ppm) but without taking into account the resuspension of oxide deposits.

As a conclusion,if compared to tritium, the inhalation dose at 1000 m from the source point was estimated near 4.1 10^{-6} Sv per gram of released stainless steal (at 1.9 10^{11} Bq/g).

6. CONCLUSION

On the basis of this short and partial presentation, it may be concluded that coolant accidents have to be one of the major concern of the fusion safety experts in their future studies. In parallel another conclusive task will concern a better quantification of the "radiological source term" for fusion reactors.

REFERENCES

1. R. BUNDE and Al
Reliability assesment of first wall and blanket design for NET (1986) - BARRITZ - FRANCE, 5th international conference on reliability and maintainability

2. J. ROUILLARD and Al
Availability of nuclear plant components of NET - CEA report n° MC/100-284

3. F. MAZILLE
Development and evaluation of accident scenarii, CEA final report

4. G. VIEIDER
Progress in the development of a first wall for NET - int. symp. on fusion nuclear technology - TOKYO (1988)

5. R. MOORMANN
Restricted diffusion

6. H.T. KLIPPEL
Transient analysis of the water cooled eutectic LiPb blanket of NET - Soft conference - AVIGNON (1986)

7. P.J. DINNER
NET tritium systems design status (1987), NET internal communication

8. H. DJERASSI and Al
Environmental tritium behaviour french experiment, final report, ref. 85 07 R1

9. J. CHARUAU and Al
First JET experiment on erosion dust measurement in a TOKAMAK, this volume.

10. S. ANTHONI
Contamination du réacteur à fusion NET, SEN/ECC/86/236

FUSION TECHNOLOGY 1988
A.M. Van Ingen, A. Nijsen-Vis, H.T. Klippel (editors)

PASSIVE RESIDUAL POWER REMOVAL IN NET IN CASE OF TOTAL LOSS OF PUMPING SYSTEM

V. RENDA[1] and F. FENOGLIO[2]

[1]CEC, Joint Research Centre, Ispra Establishment, 21020 Ispra (Va) - Italy
[2]Politecnico di Torino, Dip. Energetica, Torino - Italy

- Safety and environmental impact will play an important role for the acceptability of Fusion Energy in the future.
- JRC-Ispra is engaged in research for the development of an inherent safe reactor design looking, as far as possible, for natural mechanisms and passive systems to prevent or control possible sources of hazard.
- Residual power of a like NET (Next European Torus) fusion reactor, essentially due to decay heat of activated materials, must be removed in case of total loss of pumping systems.
- The present work shows that residual power of NET can be removed, also for a loss of coolant flow in the whole reactor (case of electric black-out), by radiation between faced components, inertia in the pumping systems and thermosyphon effects.
- A proposal for the IFW circuit, allowing natural cooling, and the detailed design of major components, have been given.

1. INTRODUCTION

NET is conceived to show the technological feasibility of a fusion reactor, so that the geometry and the components (Figure 1) are like for a future demonstrative power plant.

Decay heat of activated plasma facing components was calculated with the data of Table 1 and the volumetric heat transients obtained for the main reactor materials are shown in Figure 2. Residual power is due both to decay heat and energy stored in the components and in the graphite.

2. THERMAL RADIATION REACTOR MODEL

Residual power transferred from the whole reactor to IFW (Inboard First Wall) circuit has been assessed (JRC program THESS) by a radial conduction-radiation model shown in Figure 3. The total power to be removed from the IFW is

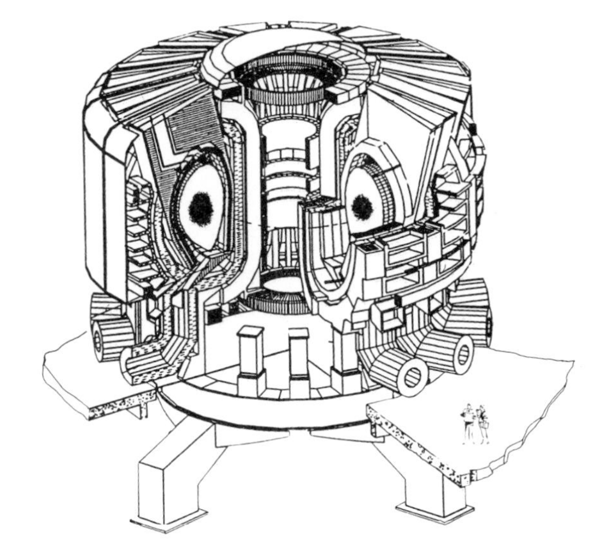

FIGURE 1
NET general configuration.

TABLE 1. Data for residual power evaluations

Neutron wall loading (MW/m^2)	1
Years of operation (a)	1
Availability (%)	100
Integrated wall loading (MW a/m^2)	1

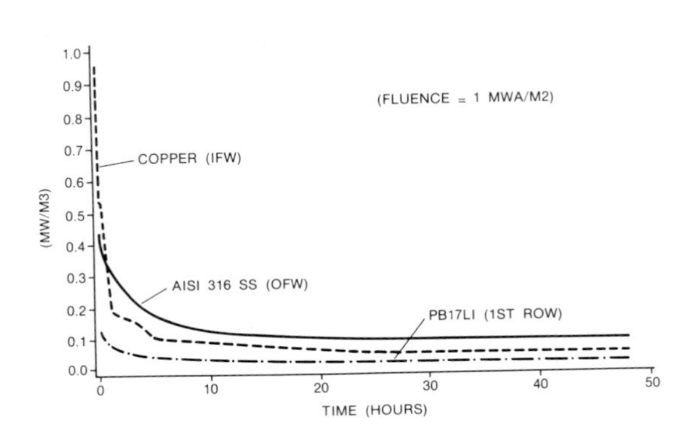

FIGURE 2
Residual power transients.

shown in Figure 4; in the first 100 s the transient from 205 MW (normal power) to 30 MW is governed by energy stored in the graphite, while in the following decay heat prevails.

3. INBOARD FIRST WALL COOLING CIRCUIT

An IFW cooling circuit, allowing thermosyphon effect, has been proposed and assessed for preliminary thermohydraulic analyses. The circuit arrangement is compatible with maintenance and reactor building. The main original components are the two great toroidal collectors

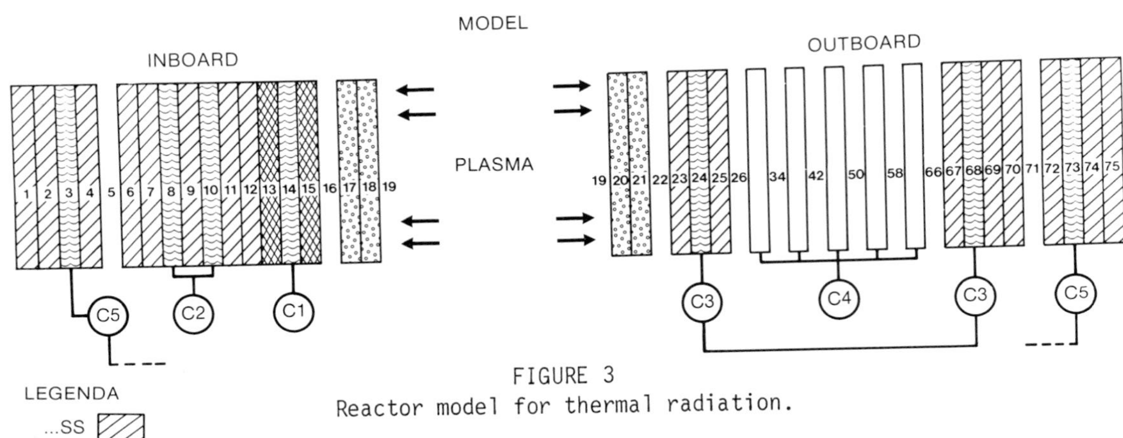

FIGURE 3
Reactor model for thermal radiation.

(cold and hot legs) placed at the top of the reactor. The other main components can be considered as classical ones.

4. THERMOHYDRAULIC ANALYSIS

4.1 Geometry and basic data

The proposed IFW/divertor system is composed by 48 panels, arranged in the 16 reactor segments (3 panels per segment). Each panel is made of 42 U-tubes embedded in a copper matrix protected against plasma interaction by a graphite armour. The function of pressure boundary (coolant vessel) is done by the U-tubes; for reasons of reliability AISI 316SS has been

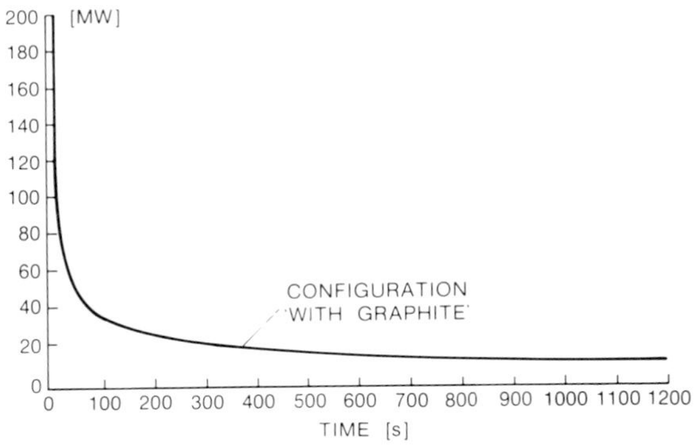

FIGURE 4
Thermal load to IFW after shutdown.

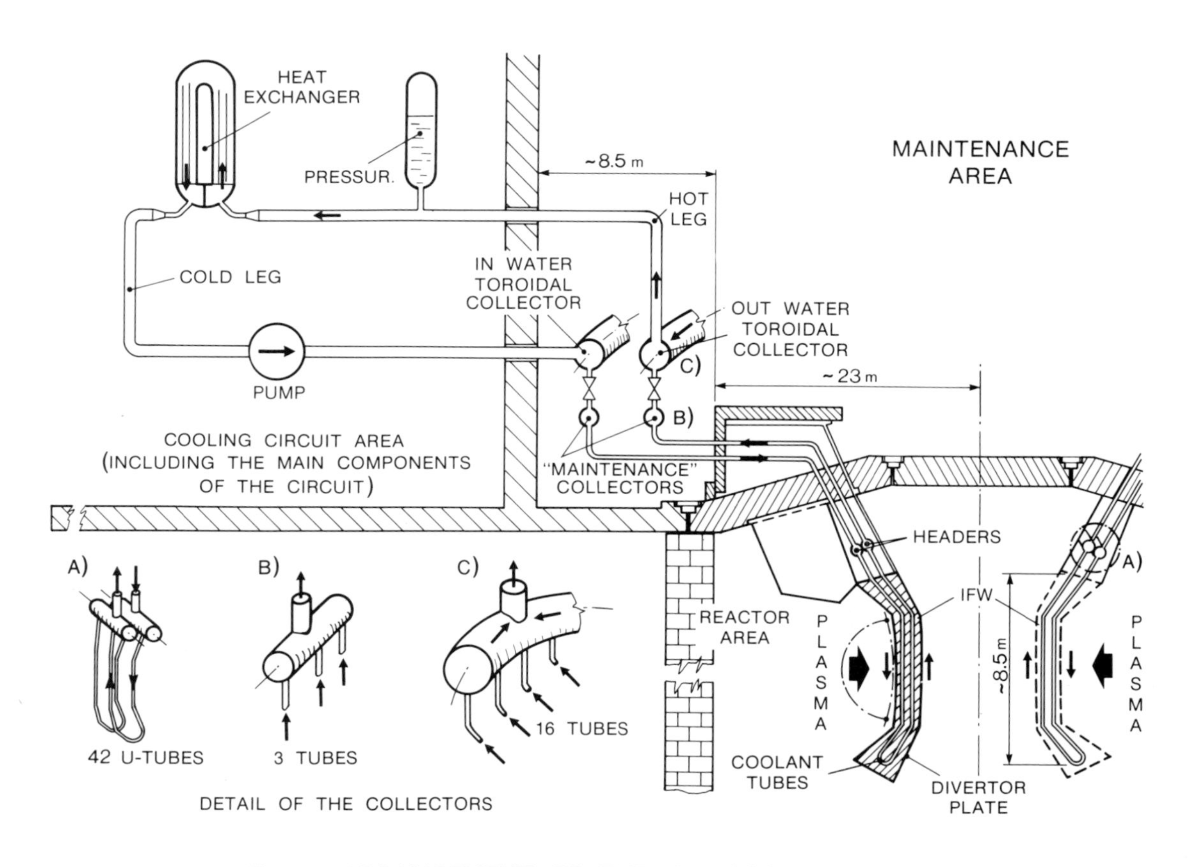

Fig. 5. ARRANGEMENT OF THE IFW COOLING CIRCUIT

chosen as material being well known for nuclear applications. An artistic view of the panel is shown in Figure 6 and the dimensions and shape of the cross section are shown in Figure 7. A total power of 205 MW must be removed from the IFW system. The basic thermohydraulic data are given in Table 2. Three different inertia for the pumping system have been considered ((100-200-500) kg·m^2) for the transient analysis to optimise the thermohydraulic behaviour in the first ∿ 100 seconds. The analysis was performed using the program CATHARE.

4.2 Analysis of the results

The power per tube generated in the IFW and transmitted to the cooler is shown in Figure 8; the influence of the pump inertia on the beginning of the transient (due to graphite) is evident.

The long-term transient (> 100 seconds) is essentially governed by the thermosyphon effect as can be seen in Figures 9 and 10. Water boiling can be avoided for a pump inertia of the order of 200 kg·m^2 or greater as shown in Figure 11; for an inertia of 500 kg·m^2 a minimum subcooling of 40°C is assured.

TABLE 2. Basic thermohydraulic data

Number of U-tubes per IFW panel	42	U-tube inlet temperature (°C)	50
Total U-tubes in the IFW	2016	U-tube outlet temperature (°C)	130
U-tube inner diameter (mm)	7	U-tube inlet saturation temp. (°C)	250
U-tube outer diameter (mm)	8	U-tube water velocity (m/s)	8
Minimum U-tube center distance (mm)	11	U-tube pressure drop (MPa)	1.7
Pressure at the pumphead (MPa)	4	U-tube mass flow (kg/s)	0.3

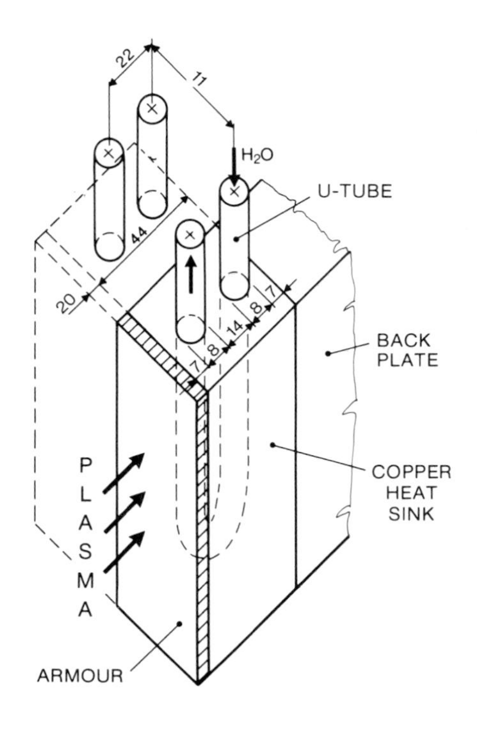

FIGURE 6
Detail of the IFW.

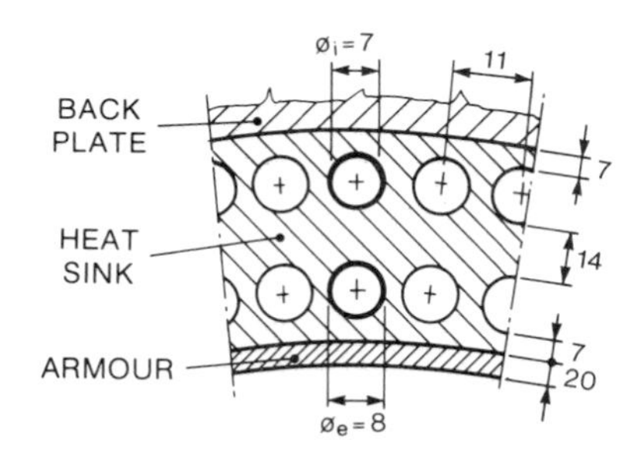

FIGURE 7
Cross section of the IFW.

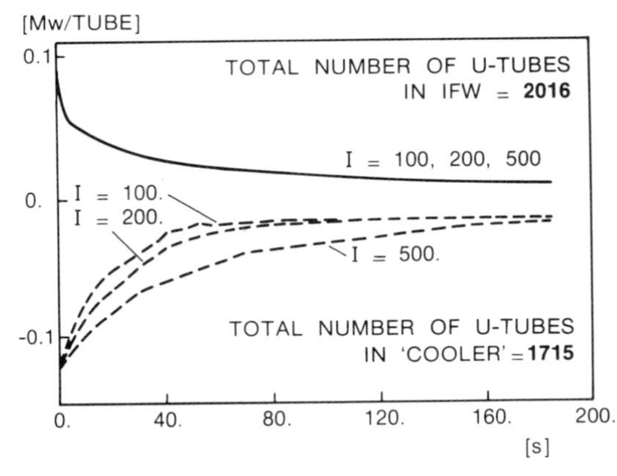

FIGURE 8
Heat power in IFW and cooler.

The pressure transient at the U-tube outlet (Figure 12) is small, and no mechanical consequences must be expected in the circuit. The liquid level in the pressurizer (Figure 13) shows that the component is correctly dimensioned also for long-term transients. The water and AISI 316SS temperatures along the U-tube for different times are shown in Figure 14 for an inertia of 200 kg·m^2

4.3 Interpretation of the results

The transient of the residual power to be removed by the IFW circuit is governed in the short term (< ∿ 100 s) by the transfer of the energy stored in the graphite to the coolant, while in the long term (> ∿ 100 s) decay heat becomes more and more important.

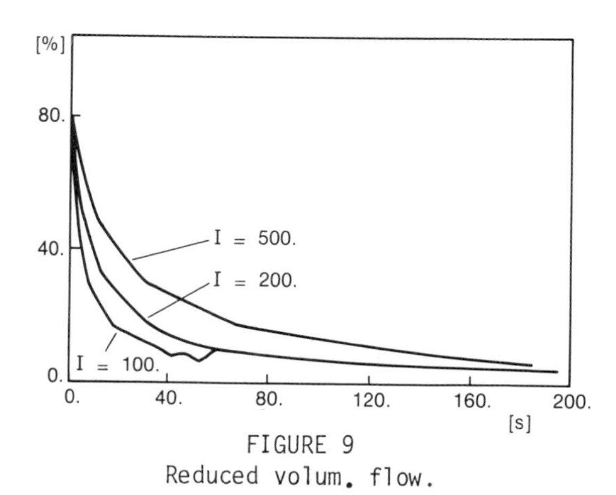

FIGURE 9
Reduced volum. flow.

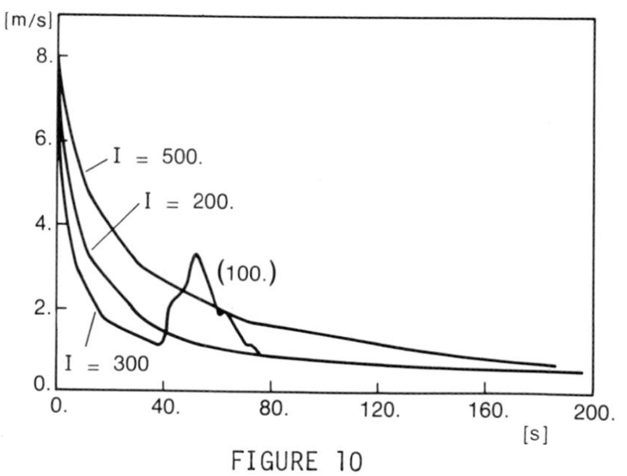

FIGURE 10
Liquid velocity at U-tube outlet.

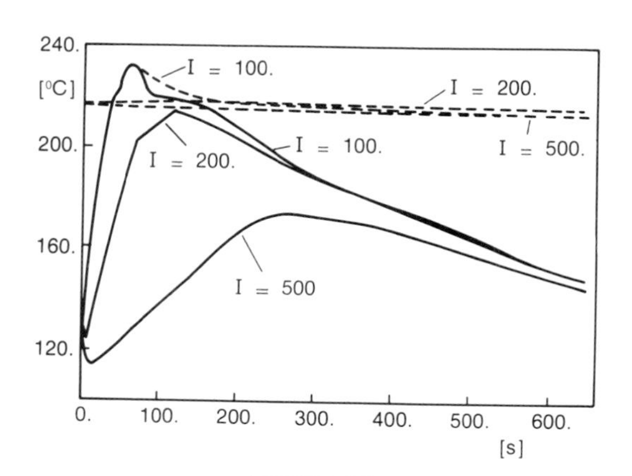

FIGURE 11
Liquid and saturation temperatures at U-tube outlet.

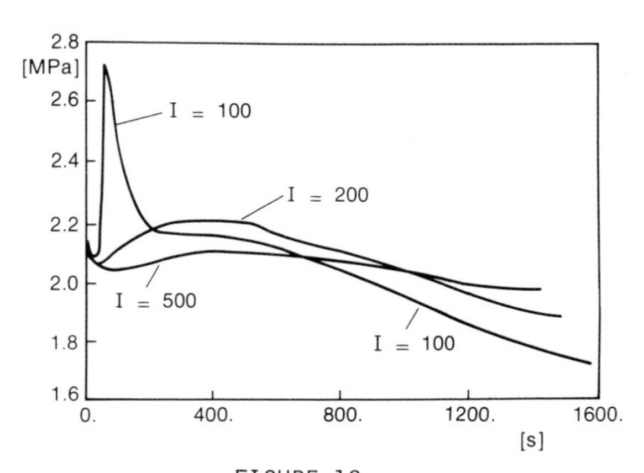

FIGURE 12
Pressure at U-tube outlet.

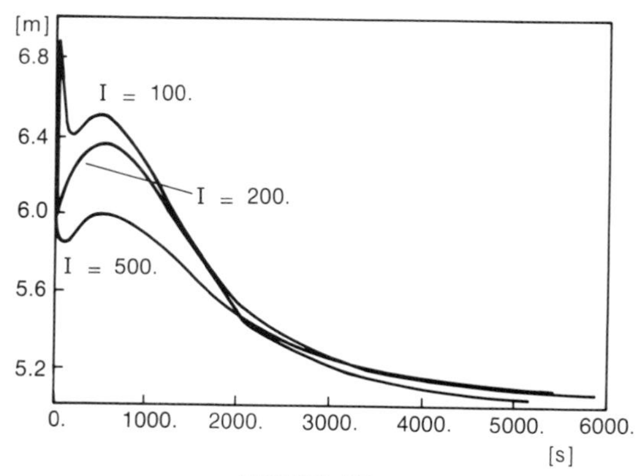

FIGURE 13
Liquid level in the pressurizer.

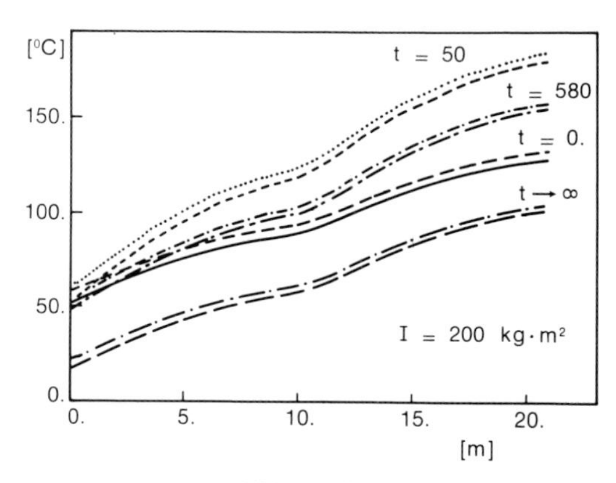

FIGURE 14
Temperature profile in the U-tube.

The transient can be managed in the short term driving the water velocity by the inertia of the pumping system and in the long term by the thermosyphon effect provided that the circuit is properly designed; the only new components, whose behaviour must be better checked, are the great toroidal collectors.

5. CONCLUSIONS

The JRC-Ispra effort to contribute to a safe fusion reactor design has been applied to the management of the hazards due to residual power in case of total loss of pumping systems.

The target to use only inherent/passive safety system (as far as possible) led to take into account radial conduction/radiation heat transfer and thermosyphon effect as natural heat removal mechanisms.

The analysis, based on NET geometry and data, has shown that the residual power of the whole reactor can be removed by the IFW cooling circuit if it is properly designed for natural cooling; moreover, water boiling can be avoided designing an adequate inertia for the pumping system.

From the point of view of the system analysis the accident investigated could be classified as UPSET condition for the circuit.

FUSION TECHNOLOGY 1988
A.M. Van Ingen, A. Nijsen-Vis, H.T. Klippel (editors)

APPLICATION OF THE SYSTEM STUDY METHODOLOGY TO THE SAFETY ANALYSIS OF THE F.W. AND BLANKET SYSTEMS OF NET

G.CAMBI(1),H.DJERASSI(2),T.PALMA(3),J.ROUILLARD(2),S.SARTO(3)G.ZAPPELLINI(4)

(1) Bologna Univ.-Nucl.Eng.Lab.-Via dei Colli 16 - I 40125 Bologna
(2) C.E.A.-I.P.S.N./DPT-Saclay,Gif sur Yvette,France
(3) E.N.E.A.- DISP - Via V.Brancati 48 - I 00144 Roma
(4) NIER Scrl - Via S.Stefano 16 - I 40125 Bologna

ABSTRACT

In the frame of the ECC/ENEA and ECC/CEA Association Contract for Fusion Technology Activities a system study methodological approach [1],[2],[3] is jointly applied by ENEA and CEA to the First Wall and Blanket Systems of the NET plant for the accidental scenario identification and risk evaluation.
A first qualitative and quantitative iteration of the analysis has been performed with reference to the NET Basic Machine [4],that is with a shielding blanket ,with addition of lithium salt in the primary coolant to achieve a significant production of tritium.
The main results of the work are described in this paper. More details are given in ref. [5],[6].

1.INTRODUCTION

A probabilistic safety approach is applied to the F.W. and Blanket Systems of the NET plant (Basic Machine) to evaluate the risk contribution from "Direct Failures" (see [3]) arising in these two systems.
The ENEA-CEA joint approach is based on the following main items:

- the methodology is a "systemic study":that allows to study a "system" as a totality of elements dynamically interacting;
- the methodology is based on a functional analysis : that allows to start from the conceptual design stage (i.e. when the missions of the "system" have been outlined);
- the analysis is of probabilistic type: the data for the probability evaluations have been taken from the existing reliability data banks, mainly based on fission plant data sources and adapted for fusion field. Data from fusion field are taken into account,when available;
- the methodology starts at the plant level and develops to the level of systems, subsystems and components when the details of the design are available;
- the methodology involves an iterative procedure.

The two analyzed systems are briefly described thereafter, in chapt.2.

2.FUNCTIONAL DESCRIPTION OF THE FIRST WALL AND BLANKET SYSTEMS OF THE NET BASIC MACHINE

Starting from the conceptual design outlined for NET [4], the following functional descriptions have been carried out for the two analyzed systems:

- Shielding Blanket

The system missions are the following:

. to provide neutron energy conversion into thermal energy;

. to provide a significant fraction of NET tritium needs;

. to provide the possibility oftesting some breeding test modules for DEMO (Demonstrative Fusion Reactor);

. to protect vacuum vessel and toroidal coils from excessive heat deposition and damage from neutrons and fluxes during the plant operations.

The principal process functions of the blanket system are :

1. conversion of both neutronic and γ flux energy in thermal energy;
2. transmission of that thermal energy and decay heat to main coolant;
3. transfer of thermal energy from coolant to the heat sink;
4. provision of a physical barrier against the radiation generated by plasma and by the First Wall activation.

The secondary process functions are:

5. provision of a fraction of the thermal energy required for the baking of the in-vessel components before the normal plasma operations;
6. provision of "driver blanket" sectors, with the aim at producing a fraction of the NET tritium requirement.
7. Furthermore the blanket must allow its drive, control and maintenance.

The Fig. 1 shows the functional flow sheet of the Blanket System (B.S.). According to its missions the B.S. is constituted by :

- a part inside the vacuum vessel (inboard and outboard blanket segment, see Fig. 2; this part is located in a region with high

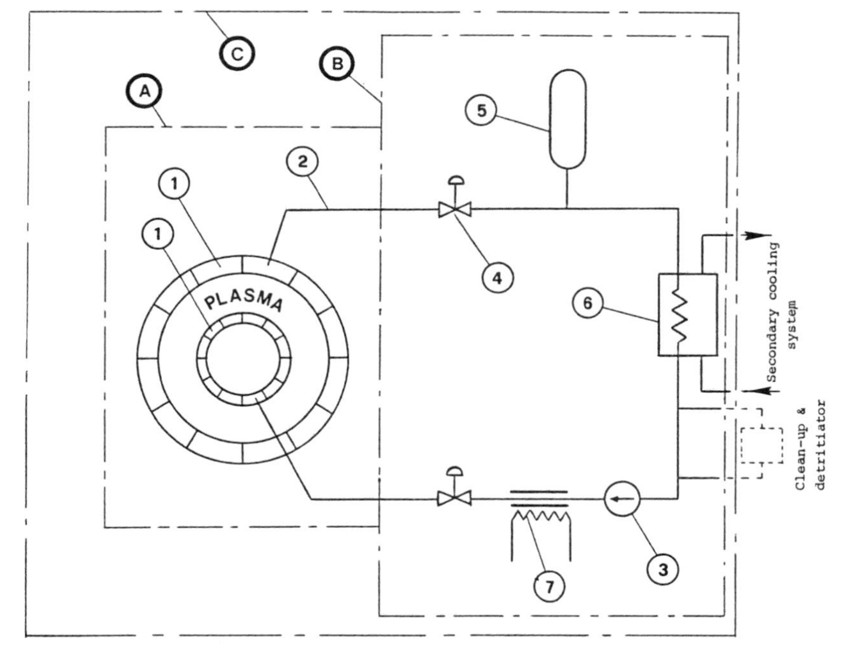

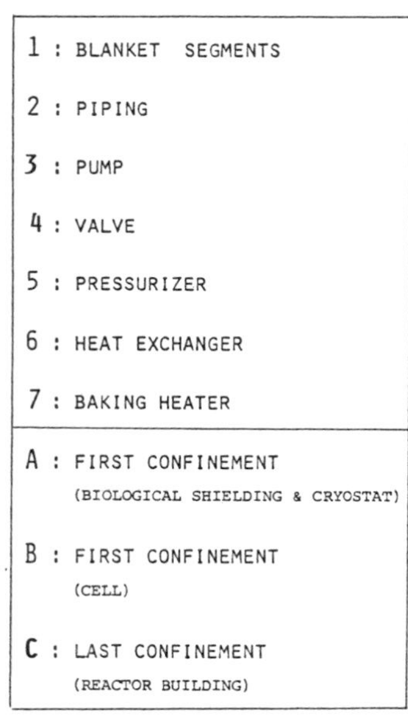

Fig 1

Blanket System: Functional flow - sheet

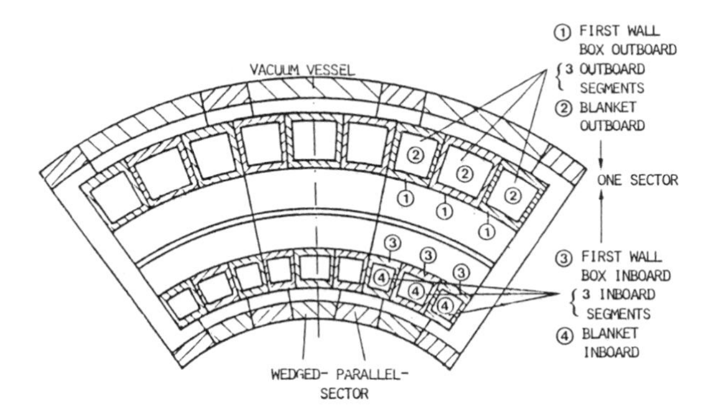

Fig. 2

NET F.W./Blanket segmentation

neutron flux, to generate thermal power by neutron collisions with coolant and structural materials (about 470 MW, see [7]) and to produce tritium (through the reaction Li + n ⟶ He + T). This part must guarantee also a contribution to the shielding of the Toroidal Field Coils.

- a part outside the vacuum vessel; that is constituted by one or more piping loops to transfer the thermal power and the tritium (produced in the blanket segments) to the heat sink and tothe tritium recovery, respectively.This part of the B.S. could be confined inside one or more cells. An additional heater may be necessary to warm-up the coolant before operation for the baking of the internal structures of the vacuum vessel.

Different alternative (<u>OPTIONS</u>) can be envisaged to accomplish the Functions needed for System Missions. They are mainly referred to the inboard and outboard blanket segmentation and to the conceptual design of the cooling loops (number of cooling loops and physical characteristics of the coolants).

For blanket segmentation only the option with 3 inboard and 3 outboard segment for each sector (16 blanket sectors in total) has been taken into account in this analysis.

For water cooling of the blanket segment the following options have been considered:

OPTION 1 (see ref.[7])

. one primary cooling loop dedicated to the inboard blanket segments, that allows to transfer about 120 MW to the heat sink;

. two independent cooling loops dedicated to the outboard blanket segments, that allow to transfer about 175 MW + 175 MW.

For this option improved passive safety by means of natural convention has been requested to be suitable to remove decay heat and to transfer it to an air cooler.

OPTION 2

. one 100% main capacity cooling loop and an additional emergency cooling system (active safety function,like the option 2 envisaged for the NETplasma facing component,see ref.[8]).

The general architecture of this system can be integrated (for the in-vessel part) in a box enclosing the First Wall panels; this arrangement has a containment function and it contributes to passive plasma stabilization,too.

- <u>First Wall</u>

As a consequence of the last provision, the Missions and the Functions of the F.W. are similar to those of the B.S. except the following: the blanket protection instead of the breeding function of the blanket itself.

The functional flow sheet of the F.W. is quite similar to that is shown in Fig. 1 for the B.S..

According to its missions the First Wall system (F.W.) is constituted by:

- a part inside the vacuum vessel (inboard and outboard F.W. segments, see Fig. 2); this part is a plasma-facing component and is located in a region with high thermal and neutron loads;

- a part outside the vacuum vessel; that is constituted by one or more coolant circuits to transfer the thermal power to the heat sink. As for B.S., this part could be confined inside one or more cells. An additional heater may be necessary for the baking.

Different alternatives (OPTIONS) can be envisaged to accomplish the Functions needed for System Missions. They are mainly referred to the choiche of the F.W. protective materials and to the conceptual design of the cooling loops (number of cooling loops and physical characteristics of the coolant).

For F.W. protection mechanically - attached tiles have been considered (conductivity cooled or radiation cooled tiles,see [9]).

For water cooling of the F.W. panels the following options have been considered:

OPTION 1

. two independent 50% cooling loops, that allow to transfer about 140 MW each to the heat sink. For this option improved passive safety by means of natural convention has been requested to be suitable to remove decay heat and to transfer it to an air cooler.

OPTION 2

. one 100% main capacity cooling loop and an additional emergency cooling system (active safety function, see ref.[8]).

3. SAFETY RELATED INITIATING EVENTS (see ref.[10])

From Functional Fault Trees (F.F.T.) and Functional Event Trees (F.E.T.) the following safety related Initiating Events (I.E.) have been identified as Direct Failures concerning the two analyzed systems:

h = 1 Small/Medium/Large Loss of Flow Accident (LOFA) outside vacuum vessel (as in B.S. and in F.W.)

h = 2 Small/Medium LOFA inside vacuum vessel (as in B.S. and in F.W.)

h = 3 Small/Medium/Large Loss of Coolant Accident (LOCA) outside vacuum vessel (as in B.S. and in F.W.)

h = 4 Small LOCA inside vacuum vessel (as in B.S. and in F.W.)

h = 5 Protective tiles failures.

4. RISK ASSESSMENT

The results carried out up to now are referred to the Blanket System, OPTION 1 (see chapt.2).

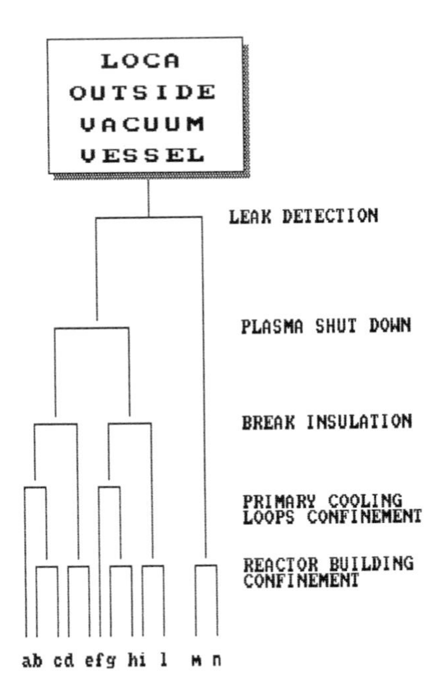

EVENTS

a,f : LOSS OF COOLANT IN THE PRIMARY COOLANT CELL; ROOM CONTAMINATION (TRITIUM AND ACTIVATED DUSTS); THE ROOM WITHSTAND THE PRESSURIZATION.

b,d,g,i,m : THE CELL (SECOND CONFINEMENT) FAILS; REACTOR BUILDING CONTAMINATION ; LAST CONFINEMENT EFFECTIVE.

c,e,h,l,n : EXTERNAL RELEASE (THROUGH STACK OR REACTOR BUILDING).

Fig.3

Event tree for the Initiating Event "LOCA in B.S. outside V.V."

The risk assessment has been performed mainly by means of the technique of event trees (see, as an example, Fig.3) and linked fault trees. The probability of the Initiating Events has been determined by fault trees or by Markov graphs.

The main assumptions that have been maid for consequence and probability evaluations are the following:

- water physical parameters (blanket coolant,see [4],pag.100): pressure ≅ 0.5 MPa ; Inlet temperature ≅ 60C; Outlet temperature <100 C water velocity < 3 m/s;
- total volume for each cooling loop [11] ≅ 80 m^3 ;
- Tritium content in aqueous salt blanket coolant = 3.7 x 10^{11} Bq/Kg (see ref.[12]);
- only the tritium releases (considered in HTO form) have been retained for consequence evaluations; no account for the activated corrosion products, up to now;
- doses to Maximum Exposed Individual (M.E.I.) have been evaluated assuming the following conditions for the releases (see ref.[13]):release height = 20 m ; release time = 1 h ; Pasquill class F ; wind speed = 2 m/s;
- different reliability data sources have been considered for probability evaluations.

5. RESULTS

The results of the analysis concern the risk evaluation associated to the B.S. for the Initiating Events h=3 and h=4.

These results are presented in Tab. 1 as probability of occurrence for categories of consequences. Two possibilities are presented:

A - the out of vessel part of the B.S. has a double confinement (cell + reactor building);

B - the out of vessel part of the B.S. has a single confinement (reactor building).

TABLE 1

Probability of occurence for categories of consequence.

1A: with double confinement
1B: without double confinement

CATEGORY OF CONSEQUENCES DOSES TO THE M.E.I. (mSv)	PROBABILITIES (y^{-1})
	OPTION 1A
1 - 5	9.6E-4
5 - 10	/
10 - 50	/
50 - 100	/
	OPTION 1B
1 - 5	9.6E-4
5 - 10	2.9E-6
10 - 50	/
50 - 100	/

The Fig. 4 shows the results versus targets comparisons. The Safety Design Criteria for Public have been taken from ref.[4],pag. 219.

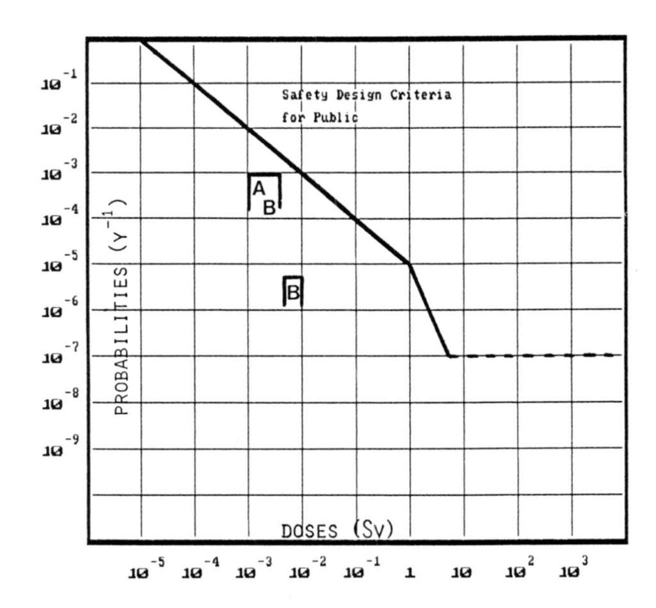

Fig. 4

Results versus target comparison for OPTION 1.
A = with double confinement
B = without double confinement

References

[1] S.Sarto, H.Djerassi, J.Rouillard, G.Zappellini,G.Cambi:First ENEA-CEA joint report on the general methodological approach for NET Overall Plant Accident Scenarios analysis.Doc. NIER T-RN-X-01215, 21.07.1987

[2] H.Djerassi, J.Rouillard, D.Leger, S.Sarto,G.Zappellini,G.Cambi:System Study Methodology Development and Potential Utilization for Fusion 1st I.S.F.N.T, 10-19 April 1988, Tokyo

[3] H.Djerassi, S.Sarto, J.Rouillard, G.Zappellini, G.Cambi: General safety approach for system studies and application to the NET exhausts and tritium systems.This conference

[4] American Nuclear Society: Fusion Technology, Volume 14, Number 1,July 1988

[5] T. Palma, G. Cambi: NET Blanket System. Overall Plant Accident Scenarios Identification and Risk Evaluation. Doc NIER - Draft progress report - September 1988

[6] T. Palma, G. Cambi: NET First Wall System. Overall Plant Accident Scenarios Identification and Risk Evaluation Doc NIER - Preliminary draft - May 1988

[7] J.Stich:Design and Availability of the Cooling System for the NET Fusion Device Eingereichte Diplomarbeit - Techn.Univ.Munchen - Betreuer R.Bunde,10.12.87

[8] G.Vieider: Cooling System Options for the NET Plasma Facing ComponentsNET/87/IN-018, 97-05-20 (me)

[9] G. Vieider et al.:Progress in the development of a First Wall for NET 1st I.S.F.N.T., 10-19 April 1988, Tokyo

[10]S.Sarto, G.Zappellini, G.Cambi: First Wall and 17Li-83Pb Blanket Systems Accident Analysis for NET 14th S.O.F.T. , 8-12 September 1986 , Avignon

[11]H.Th.Klippel: Transient Analysis of water-cooled eutectic LiPb blanket ECN-PB-88-4, February 1988, ECN-Petten

[12]M.Galley,P.Gierszewski,et al.: Preliminary Design of a Tritium Recovery System for an Aqueous Salt Blanket of NET Status Report #2, 24 November 1987

[13]W.Gulden:Environmental impact of Nuclear Fusion - Based on EUR/FU BRU/XII 828/86 Workshop on Fusion Safety and Environmental Impact, Rome 10/11 Nov. 1987

FUSION TECHNOLOGY 1988
A.M. Van Ingen, A. Nijsen-Vis, H.T. Klippel (editors)

GENERAL SAFETY APPROACH TO SYSTEM STUDY AND APPLICATION TO THE NET EXHAUST AND TRITIUM & FUEL HANDLING SYSTEMS

H. Djerassi* - S. Sarto** - G. Zappellini, G. Cambi*** - J. Rouillard****

* CEA-CEN SACLAY IPSN/DPT/SPIN - 91191 Gif-sur-Yvette FRANCE
** ENEA/DISP Via V. Branceti, 48. I 00144 - ROMA ITALY
*** NIER Santo Stephano 16. I 40124 - BOLOGNA ITALY
**** Consultant, 46 rue des Ecoles - Sucé - 44240 Sucé la Chapelle sur Erdre FRANCE

A summary of the General Safety approach to the System Study is presented, identifying the main objectives of this method, defining and quantifying the major risk entailed by a new machine like NET. The different phases of this method are described for the analysis of Direct Failures, Induced Failures, Complementary Cause Failures, with assessment of resultant risks, comments and safety recommendations.
Application of the method to the Tritium and Fuel Handling System (T.F.H.S) and Exhaust System (E.S.) is presented. Examples of quantification are given. The overall results for both systems are summarized, with comparisons with the safety criteria.

This new methodology, known as System Study, which was originally developped for new designs such as fusion plants, is being employed for optimization of design and construction costs; in this case, it is adapted to safety studies.

An exhaustive description of this method is given in System Study Methodology[1]. Presently this method is applied to the safety approach to the NET type machine, with a specific safety analysis of the Exhaust System (E.S) and Tritium and Fuel Handling System (T.F.H.S), forming part of the third phase of the general method.

2. APPLICATION TO A NET-TYPE MACHINE

System Safety Studies

The approach breaks down into five sections :

2.1. Direct failures - Section 1

A - Functional Analysis of the Overall Plant

This phase consists in identifying the NET Missions (To demonstrate the feasability of the DT reaction in a safe and reliable way), in defining the main process functions necessary to achieve these missions, and thus specifyting the systems necessary to fulfill between the systems and functions and the interfaces between the different systems must be defined and analyzed, as shown in the matrices of Figure 1.

Each dedicated system is analyzed with the same method, as for instance, the mission of the T.F.H.S. which is to provide tritium and deuterium to the plasma chamber, via the Fuel Injection System (F.I.S.) in conditions adequate for the DT. reaction. Therefore, the functions of T.F.H.S. and E.S. may be identified and defined, as shown in Figures 2 and 3 emphasizing the interactions between the systems and functions, and the functional interfaces between the sub-systems. The functional diagrams of T.F.H.S. and E.S. are shown in Figure 4.

B - Qualitative Event and Failure Trees

Failure Trees illustrate the linking of failures, and Event Trees define the functional relationship which exists between an initiating event (loss of a process function) and its consequences such as the environmental

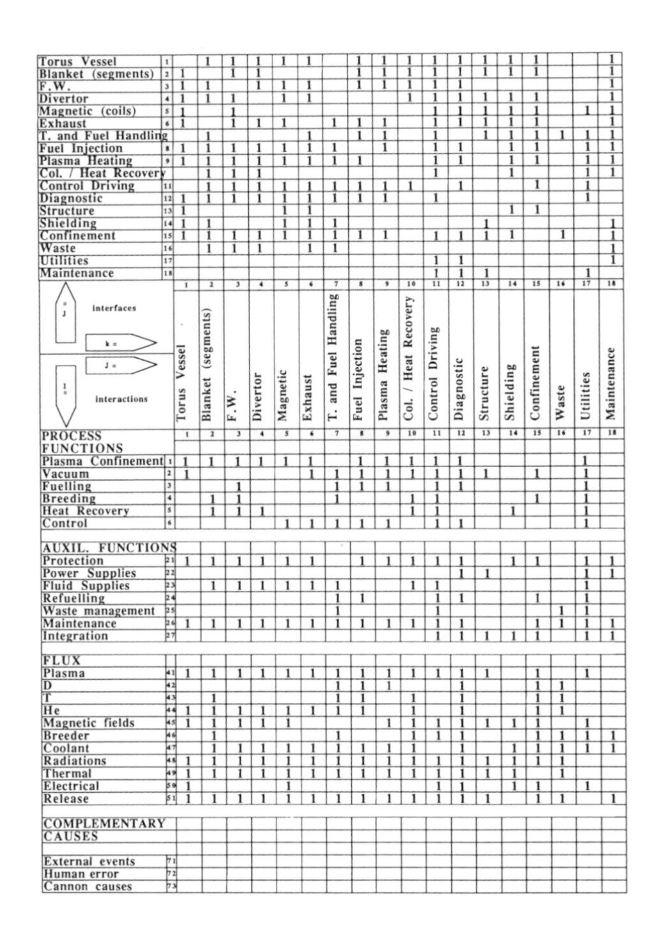

FIGURE 1 - General interface and interaction matrix of NET - type machine

impact, hazards induced on workers, and the damages to the machine itself, are worked out for calculating the occurrence probability of such an event. Examples of Fault Trees are shown in Figures 5 and 6

C - Risk Evaluation

This phase comprises :

- A failure mode analysis consisting in the evaluation of the failure probability using reliability data, or reliability allocation for components, units and/or sub-systems.
- An assessment of consequences by calculating an occurrence probability for each event, and characterizing the consequences for the radiological hazards and the other kinds of hazards.

The risks are then evaluated, as being the product of the consequences resulting from the initiating events by the probability of occurrence of the event. Subsequently, they are compared with the known safety criteria, for each kind of hazards such as environmental impact, detriments to the public, personnel and process (machine), and for each type of consequences (radiological, chemical, magnetic, thermal, etc.). Finally, preliminary safety recommendations are prepared to meet the prevailing safety regulations.

2.2. Induced Failures - Section 2

These failures are those induced by any failing system on the other ones. They are identified in the interaction and/or interface matrices as shown in Figure 1.

The overall risks of the studied system are calculated, this resulting in specific safety recommendations.

2.3. Complementary Cause of Failure - Section 3

This section accounts for other events such as external events (earthquakes, external explosion etc.) and human errors. In the design phase, these errors relate to incidence both on failure and common modes as usually meant in common mode assessments, and on internal common modes (Figure 6).

2.4. Assessment of Resultant Risk - Section 4

Risk inventory : totalization of the occurrence probabilities of each kind of possible consequences, and comparison with the safety criteria and, when necessary, with the safety regulatory rules.

2.5. Final Comments and Recommendations - Section 5

All the specific safety recommendations are synthetized. A sensitivity analysis is perfor-

FUNCTIONS	0	1	2	3	4	5	6	7	8	9	10	11	12	13	14	15	16	17	18	19	20	21	22
Fuelling	1	1			1					1	1				1						1	1	
Fuel Treatment	2		1	1		1	1	1	1	1	1		1		1						1	1	
Fuel Storage	3				1					1	1		1		1						1	1	
Tritium Recovery	4			1		1	1	1	1	1	1		1	1	1				1		1	1	
Control	5									1	1				1	1							
Tritium Inventory	6	1	1	1	1	1	1	1	1		1		1	1	1		1		1	1			
Protection	7									1		1	1	1	1	1					1		
Power Supplies	8									1					1								
Fluid Supplies	9									1	1			1		1					1		
Fuel Supplies	10				1					1	1		1				1	1			1		
Effluents Treatment	11									1	1		1	1		1			1				
Wastes	12									1	1		1	1		1				1			
Maintenance	13		1	1		1	1			1	1	1	1	1	1				1	1	1	1	1
Integration	14				1			1	1	1	1	1	1	1	1	1			1	1	1	1	1

INTERACTION: i = , j = ; INTERFACE: j = , k =		Fuelling (T2, DT, D2)	Purification	Isotopic Separation	Main Storage	Blanket Tritium Recovery	Coolant Tritium Recovery	Atmospheric Clean-up	Air Clean-up	Control Driving	Analysis Laboratory	Shielding	Confinement	Ventilation	Electrical Supplies	Vacuum, Neutral gas	Refuelling T2	Refuelling D2	Tritium Effluents	Tritium Waste	Maintenance Equipment	Structures	Facilities
		1	2	3	4	5	6	7	8	9	10	11	12	13	14	15	16	17	18	19	20	21	22
Fuelling (T2, DT, D2)	1				1					1	1		1		1							1	
Purification	2			1						1	1	1	1		1				1		1	1	1
Isotopic Separation	3		1		1	1	1			1	1		1		1				1		1	1	1
Main Storage	4	1		1						1	1				1		1	1	1		1	1	
Blanket Tritium Recovery	5			1			1			1	1	1	1		1				1		1	1	1
Coolant Tritium Recovery	6			1				1	1	1	1	1	1		1	1			1		1	1	1
Atmospheric Clean-up	7						1			1	1		1	1	1				1			1	
Air Clean-up	8						1			1	1			1	1				1			1	
Control Driving	9							1	1		1												1
Analysis Laboratory	10							1	1	1			1		1	1							1
Shielding	11																				1	1	
Confinement	12							1	1	1				1	1						1	1	
Ventilation	13							1	1				1		1	1							1
Electrical Supplies	14									1													1
Vacuum, Neutral gas	15									1													1
Refuelling T2	16				1			1	1	1	1		1		1				1		1	1	
Refuelling D2	17				1				1	1	1		1		1						1	1	
Tritium Effluents	18		1	1		1	1	1	1	1	1		1		1	1				1	1	1	
Tritium Wastes	19									1	1	1	1		1	1			1		1	1	
Maintenance Equipment	20	1	1	1	1	1	1				1	1	1		1	1	1			1		1	
Structures	21	1	1	1	1	1	1	1	1		1	1	1				1	1	1	1			
Facilities	22									1	1				1	1							

FIGURE 2 - T.F.H.S. Interface/interaction matrix

med to point out the major (sizing) accident scenarii and to emphasize the residual risks.

2.6. Conclusions on the methodology

This systematic approach allows to build up mathematical models for evaluating each accidental sequence, as regards the different systems, sub-systems, up to each component. These models must be assessed for one sub-system with subsequent extrapolation to all the other sub-systems.

3. APPLICATIONS TO T.F.H.S. AND E.S.

3.1. Direct Failure - Section 1

A - Functional Analysis as shown in Figure 4 presenting the general functional diagrams of the T.F.H.S and E.S., with their different sub-systems, such as storage, purification,

PROCESS FUNCTIONS										
Vacuum Pumping	1	1	1	1	1		1		1	
Circulation	2	1	1	1	1		1		1	
Filtering	3	1	1	1						1
Exhausted Gases Stor.	4	1	1	1						
Exhausted Gases Transf.	5	1	1	1	1				1	
Tightness	6	1	1	1				1		
Drive and Control	7				1				1	
AUXILIARY FUNCTIONS										
Protection	21	1	1	1						
Power Supply	22	1	1	1	1			1		1
Cooling	23	1	1	1		1	1		1	1
Waste Management	24	1	1	1						1
Monitoring	25	1	1	1	1	1	1	1	1	1
Maintenance	26	1	1	1	1			1	1	1
Integration	27	1	1	1		1	1			1
Supervision	28	1	1	1	1			1		1
FLUXES										
H	41	1								
T	42	1	1	1				1		1
D	43	1	1	1				1		1
He	44	1								
Impurities	45	1	1	1						1
Coolants	46	1	1	1		1	1		1	1
Thermal (heat)	47	1	1	1		1	1		1	1
Radiations	48	1	1	1		1	1	1		1
Electric Signals	49	1	1	1	1			1	1	1
Electric Power	50	1	1	1	1			1	1	1
Compressed Air	51	1	1	1						1
		1	2	3	4	5	6	7	8	9
INTERACTION: α, β; INTERFACE: j =, k =	SYSTEMS	Plasma Vacuum Pumping	N.B.I. Vacuum Pumping	F.I. Vacuum Pumping	Drive and Control	Supporting Structures	Shielding	Confinement	Utilities	Maintenance Equipment
SUB SYSTEMS		1	2	3	4	5	6	7	8	9
Plasma Vacuum Pumping	1				1	1	1	1	1	1
N.B.I. Vacuum Pumping	2				1	1	1	1	1	1
F.I. Vacuum Pumping	3				1	1	1	1	1	1
Drive and Control	4								1	
Supporting Structures	5	1	1	1	1		1	1		
Shielding	6	1	1	1		1		1		
Confinement	7	1	1	1	1	1	1			
Utilities	8					1				
Maintenance Equipment	9					1			1	

FIGURE 3 - Exhaust system Interface/interaction matrix

isotopic separation, etc. for the T.F.H.S., and plasma vacuum pumping sub-system, N.B.I. vacuum pumping sub-system etc. for the E.S.

For the T.F.H.S., the main functions concern :

- purification and isotopic separation of the exhaust gas,

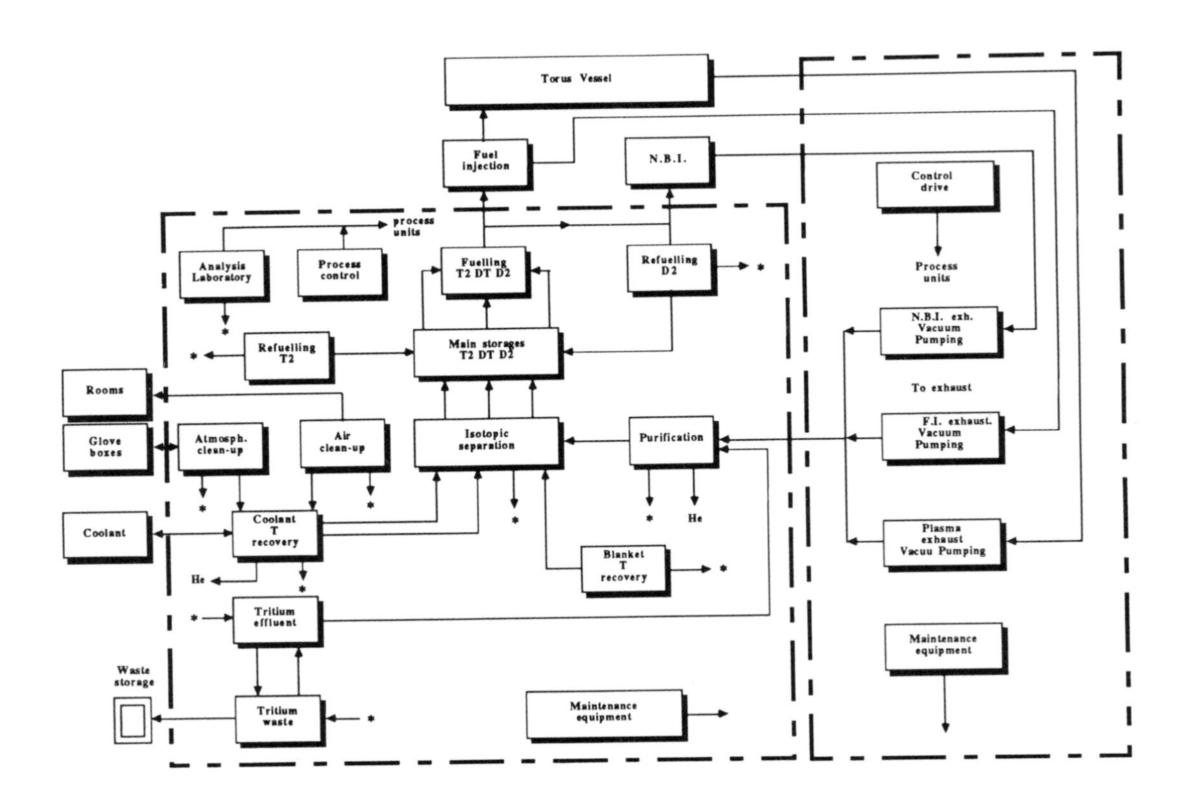

FIGURE 4 - Functional Diagrams

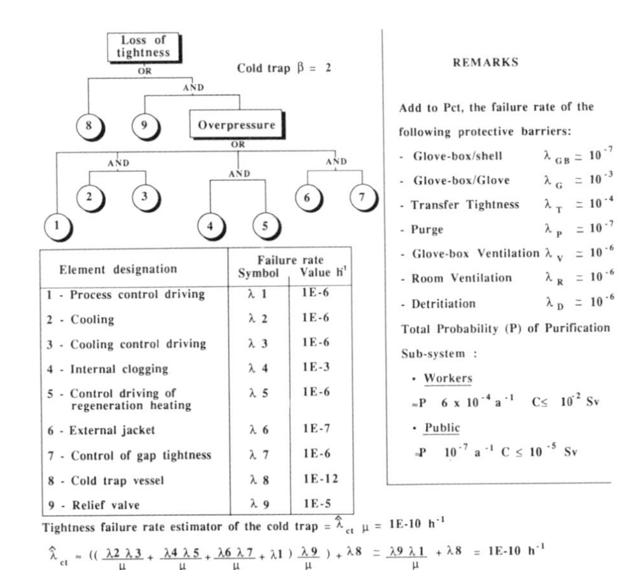

Element designation	Failure rate Symbol	Value h^{-1}
1 - Process control driving	$\lambda 1$	1E-6
2 - Cooling	$\lambda 2$	1E-6
3 - Cooling control driving	$\lambda 3$	1E-6
4 - Internal clogging	$\lambda 4$	1E-3
5 - Control driving of regeneration heating	$\lambda 5$	1E-6
6 - External jacket	$\lambda 6$	1E-7
7 - Control of gap tightness	$\lambda 7$	1E-6
8 - Cold trap vessel	$\lambda 8$	1E-12
9 - Relief valve	$\lambda 9$	1E-5

Tightness failure rate estimator of the cold trap = $\hat{\lambda}_{ct}$ μ = 1E-10 h^{-1}

$$\hat{\lambda}_{ct} \approx \left(\left(\frac{\lambda 2\,\lambda 3}{\mu} + \frac{\lambda 4\,\lambda 5}{\mu} + \frac{\lambda 6\,\lambda 7}{\mu} + \lambda 1\right)\frac{\lambda 9}{\mu}\right) + \lambda 8 \simeq \frac{\lambda 9\,\lambda 1}{\mu} + \lambda 8 \simeq 1E\text{-}10\ h^{-1}$$

REMARKS

Add to Pct, the failure rate of the following protective barriers:

- Glove-box/shell $\lambda_{GB} \simeq 10^{-7}$
- Glove-box/Glove $\lambda_{G} \simeq 10^{-3}$
- Transfer Tightness $\lambda_{T} \simeq 10^{-4}$
- Purge $\lambda_{P} \simeq 10^{-7}$
- Glove-box Ventilation $\lambda_{V} \simeq 10^{-6}$
- Room Ventilation $\lambda_{R} \simeq 10^{-6}$
- Detritiation $\lambda_{D} \simeq 10^{-6}$

Total Probability (P) of Purification Sub-system :

- Workers

  $\approx P$ $6 \times 10^{-4}\ a^{-1}$ $C \leq 10^{-2}$ Sv

- Public

  $\approx P$ $10^{-7}\ a^{-1}$ $C \leq 10^{-5}$ Sv

FIGURE 5 - T.F.H.S. Example fo Fault Tree

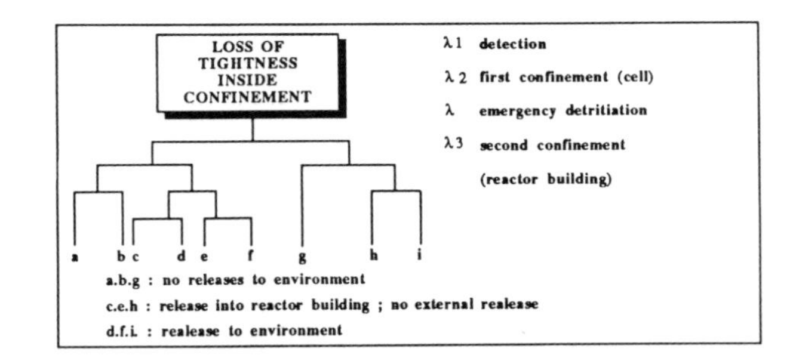

FIGURE 6 - E.S. Example of Fault Tree

- reduction of the concentration of impurities in the fuel down to levels allowable for refuelling,
- provision of means for interim storage of the fuel,
- recovery of tritium from the breeder, coolant, atmosphere of glove boxes, atmosphere of the rooms, effluents and waste.

For the E.S., the main functions cover the following :

- pumping out in the torus vacuum vessel prior to operation and during dwell time,
- removal of unburnt fuel,
- transfer of the purified exhaust gases to the T.F.H.S.

B - Functional Fault Trees (FFT) and Functional Event Trees (FET)

FFT shows the logical links between the failure of the units and/or components of the T.F.H.S. or E.S., which participates in a loss of process function, and entails safety-related consequences.

FET shows the functional relationship between the previous failure (loss of function) and the failure of other barriers (loss of protection). Examples for T.F.H.S. and E.S. are shown in Figures 5 and 6. (For each subsystem, the FFTs and FETs are worked out and evaluated, using reliability data available from data bases and improved for the specific conditions of the systems.

Using computer models, the overall risks induced by E.S. and T.F.H.S. are calculated as shown in Table 1, emphasizing the main results and the comparison of the overall risks with the safety criteria as illustrated in Figure 7. The resultant risks are generally below the criteria to be met for the chosen options.

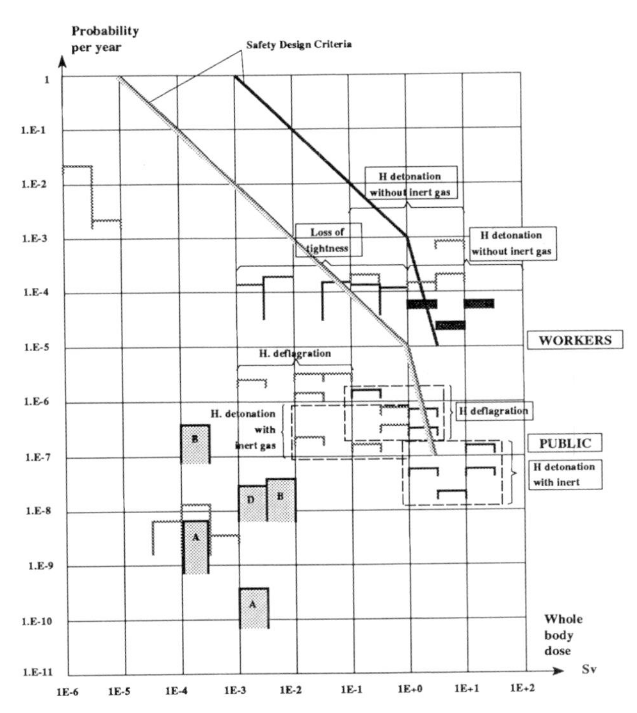

FIGURE 7 - Results of risk analysis for workers and public

4. CONCLUSIONS

This powerfull method, briefly described and illustrated, enables to analyze and quantify in a systematic way the risk entailed by each system, and thus by the plant.

The tasks referred to in the Sections 3, 4 and 5 need prior completion of the analysis of the 18 systems of NET Machine. Nevertheless, a few cardinal safety recommendations may be drawn at this step of the study such as :

- active protective barriers are necessary to allow safe operation of the systems (provision of passive barriers only, would be too stringent for users),
- controlled neutral atmosphere is necessary to avoid fire outbreaks and/or explosions,
- each tritium room must be isolated from the others to limit the probability of tritium leakage.

The possibility of modelizing safety analyses allows to follow the modifications in the design, to discuss with designers, to change the input data without undertaking new analyses.

T.F.H.S SUMMARY TABLE

- WORKERS

Sub-systems designation	Probabilities (a^{-1})	Consequences (Sv)	Risks workers (individual)
Fuelling T2	1.4 E-4	1 E-1 to 5 E-1	6.9 E-5
Fuelling DT	1.4 E-4	1 E-1 to 5 E-1	6.9 E-5
Purification	2.6 E-4	5 E-2 to 1 E-1	2.6 E-5
I.S.S.	1.2 E-4	5 E-1 to 1 E+0	1.2 E-4
Storage bed	9 E-5	1 E-3 to 5 E-3	4.5 E-7
Storage	1.2 E-4	5 E-2 to 1 E-1	1.2 E-5
Blanket Tritium Recovery	1.7 E-4	1 E-1 to 5 E-1	8.3 E-5
Coolant Tritium Recovery	1.7 E-4	1 E-1 to 5 E-1	8.5 E-5
Air Clean-up (room)	2.4 E-4	5 E-3 to 1 E-2	2.4 E-6
Analysis Laboratory	2.1 E-4	5 E-3 to 1 E-2	2.1 E-6
Refuelling T2	1.1 E-4	1 E-1 to 5 E-1	5.4 E-5
Tritium Effluents	3.1 E-4	1 E-3 to 5 E-3	1.5 E-6
Waste	3.4 E-4	5 E-3 to 1 E-2	3.4 E-6

- PUBLIC

System Overall Risks to Public	2.8 E-2	0 to 5E-6	1.4E-7
	5E-3	5E-6 to 1E-5	5E-8
	7.1E-9	5E-5 to 1E-4	7.1E-13
	1.1E-8	1E-4 to 5E-4	5.4E-12
	5E-9	5E-4 to 1E-3	5E-12

E.S. SUMMARY TABLE

a) With double process barrier in area A

Consequences (mSv)	probabilities (a^{-1})		
	OPTION 1 16 turbopumps	OPTION 2 16 cryopumps	OPTION 3 16 cryopumps
1to 5	2.2 E-9	3.9 E-9	5.1E-9
5 to 10	/	/	/
10 to 50	2.5 E-10	2.4 E-10	2.3 E-10
50 to 100	/	/	/

b) With double process barrier in area B

Consequences (mSv)	probabilities (a^{-1})		
	OPTION 1	OPTION 2	OPTION 3
1 to 5	7.8 E-9	9.3 E-9	1.3 E-8
5 to 10	/	/	/
10 to 50	9.4 E-10	1.1 E-10	9.1 E-10
50 to 100	/	/	/

c) With double process barrier with single confinement

Consequences (mSv)	probabilities (a^{-1})		
	OPTION 1	OPTION 2	OPTION 3
1 to 5	5.3 E-7	4 E-7	8.8 E-7
5 to 10	/	/	/
10 to 50	1.9 E-8	3.7 E-8	3.5 E-8
50 to 100	2.4 E-8	/	/

TABLE 1 - Summary table

REFERENCES

1. H. Djerassi, J. Rouillard, D. Leger, S. Sarto, G. Zappellini, G. Cambi, System study methodology development and potential utilization for fusion. I.S.F.N.T, Tokyo - 10-19 April 1988.
2. Net Status Report - 1985
3. Fusion Technology - July 1988 - Vol 14 - Next European Torus
4. CEA/ENEA - Overall Plant Accident Scenario. Reference report 1987-03
5. H. Djerassi, D. Léger, J. Rouillard, Tritium and Fuel Handling System CEA Report - August 1988 - S+E 5.4-01.
6. G. Cambi, S. Sarto, G. Zappellini Exhaust System - ENEA Report
7. G. Cambi et al. - System Study Application to the Safety Analysis of the Exhaust and the Tritium and Fuel Handling System of a fusion reactor, I.S.F.N.T., Tokyo - 10-19 April 1988.

FUSION TECHNOLOGY 1988
A.M. Van Ingen, A. Nijsen-Vis, H.T. Klippel (editors)

WASTE MANAGEMENT FOR NET

W. Gulden[1)], C. Ponti[2)], Ph. Guetat[3)], K. Broden[4)], G. Olsson[4)], and G. J. Butterworth[5)]

1) The NET Team, c/o IPP, Boltzmannstrasse 2, D-8046 Garching
2) Joint Research Centre, I-21027 Ispra/Varese
3) CEA, IPSN, Boite Postale 6, F-92260 Fontenay aux Roses
4) Studsvik Nuclear, S-611 82 Nykoeping
5) Culham Laboratory, GB-Abingdon OX14 3DB

Studies are under way to quantify and qualify radioactive wastes to be expected from NET (Next European Torus) and to identify a tentative strategy for its handling, conditioning and disposal.

Waste management and disposal strategies developed for fission plants can be applied to low and medium level fusion wastes, provided that tritium has been sufficiently removed and/or immobilized.

Handling and treatment of dismantled first wall and blanket segments (high level waste) will involve more complex procedures because of their volume, weight, afterheat and activation level. Assuming AISI-316 as structural material, an initial decay time in a short-term storage is needed before the spent components can be fragmented, compacted, detritiated and conditioned for intermediate and/or final storage.

A first evaluation indicates that the steel components in NET have a total mass of about 7000 tonnes. For decommissioning the corresponding waste volume will be about 2,000 m^3 including packaging. Eighty percent of this waste can be disposed of into a shallow geological repository and 20 percent in a deep geological repository. Each replacement of all first wall and blanket segments will produce additional waste of about 400 m^3 after packaging to be disposed of in a deep geological repository.

1. INTRODUCTION

To date, some internationally accepted criteria exist for characterisation of the radiological impact of waste disposal and waste disposal sites (e. g. dose limits for the most exposed individual of the public). Due to different country-specific waste management and disposal strategies, however, the derived limits expressed in contact dose rates, specific activities or total activities show considerable discrepancies. Some tendency exists to divide waste into the 3 categories HLW, MLW and LLW (high, medium and low level waste), but big differences in the classification are obvious from country to country.

No disposal sites for HLW are operational. However, apparently many countries plan to eventually dispose of HLW in deep geological repositories. Waste technologies for LLW and MLW exist in most countries but vary considerably. There are some shallow land burial (SLB) sites in Europe, but there is a trend toward requiring some sort of geological (deep or shallow) repositories for both MLW and LLW.

In a report on the present situation and prospects in the field of radioactive waste management in the European Community[1] LLW in general is characterized by dose rates not exceeding 2 mSv/h (0.2 rem/h), HLW is any waste that releases significant amounts of decay heat, which can be correlated to a specific activity of above 100 MBq/cm^3. IAEA[2] at present propose 2 mSv/h as the LLW upper limit and 20 mSv/h as the lower limit for HLW to characterize fission wastes at short term (some years).

Of the small number of the European sites available for SLB each has different acceptance criteria. The largest are the Drigg site in UK[3] and the Centre de Stockage de la Manche[4] (CSM) in France.

According to present UK regulations[5], waste should contain less than 12 MBq/kg (0.32 Ci/t) beta or gamma activity to be acceptable for SLB. France applies total activity limits for long lived isotopes and specific activity limits for short lived isotopes.

In the USA isotope-specific limits apply to the acceptability of waste for SLB according to the Code of Federal Regulations[6]. Some limits for class C wastes are relevant for steels used as structural material in fusion devices: 8 MBq/cm^3 (220 Ci/m^3) for Ni 59, 260 MBq/cm^3 (7000 Ci/m^3) for Ni 63 and 7400 Bq/cm^3 (0.2 Ci/m^3) for Nb 94. These limits lead to efforts in minimizing Mo and Nb in the structural materials to meet these requirements.

The contact dose rate at the surface of the waste is used by the IAEA to define three categories for handling and transport of beta/gamma solid wastes. Category 1 waste ($< 2 \times 10^{-3}$ Sv/h) can be handled and transported without special precautions. Category 2 waste (2×10^{-3} to 2×10^{-2} Sv/h) can be transported in simple concrete or lead shielded containers. Category 3 waste ($> 2 \times 10^{-2}$ Sv/h) needs special precautions for handling and transportation.

2. CHARACTERISATION OF NET STEEL WASTES

Quantification and qualification for final disposal of radioactive waste produced by NET will strongly depend on the site chosen. Indicative data to qualify the waste, however, have been derived by comparison with some criteria given in the previous section.

As soon as NET begins to operate, activation products will be formed. According to Guetat[7], after the first D-T pulse NET will have produced waste requiring special handling facilities. The first wall contact dose rate will be about 50 mSv/h after one minute of exposure and a decay time of one month, exceeding by far the IAEA category 1 limit of 2 mSv/h.

After about 25 full-power days of operation (the end of the physics phase corresponds to about 14 full-power days or a fluence of 0.03 MWa/m^2 for the first wall), NET first-wall and blanket structures will exceed the U.S. limits for shallow land burial (as defined in the U.S. Nuclear Regulatory Commission Regulation 10CFR61[6] and assuming the long-lived Nb 94 as the dominating isotope), even after a decay time of 100 years.

Figure 1 shows the contact dose rate versus time for the relevant components of NET after 1 full-power year of operation (corresponding to a fluence of 1 MWa/m^2 for the first wall). These values can be compared with the notional "hands-on limit" of 2.5×10^{-5} Sv/h which would allow the material to be worked on and machined in a controlled workshop or with the IAEA waste category 1 limit of 2×10^{-3} Sv/h (handling and transport without special precautions). First wall and blanket structures will not meet these limits within practicable cooling times. Back plates and inner shield, however, could reach the "hands-on limit" after about 150 years, the coils after 4 years.

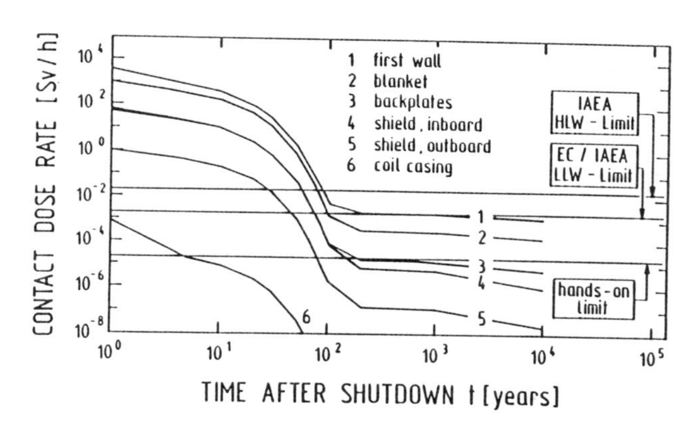

FIGURE 1
Contact dose rates for NET components after 1 full-power year of operation

Comparison with the limits derived from the EC report[1] for HLW and LLW shows that the first wall and blanket will be HLW for about 90 years.

3. WASTE MANAGEMENT STRATEGY FOR NET

LLW and MLW fusion operational wastes (wet and dry) contain tritium and activation products. They are similar to fission waste streams, having a somewhat higher volume. The waste management and disposal stategies developed for fission plants can be applied, provided that tritium has been adequately removed and/or immobilized.

Handling and treatment of dismantled first wall and blanket segments (high-level waste) will involve more complex procedures because of their volume, weight, tritium content and activation level.

A tentative strategy proposed for handling and disposal of wastes from NET is summarized in Figure 2. After a certain time in a short-term storage acting as a buffer or required for the decrease of activity and decay heat, some valuable materials with low specific activity may be separated. The remaining highly active materials will be broken up and sorted according to different materials categories, (e.g. steel, copper, tungsten, graphite, ceramics etc) and tritium contents.

All materials containing tritium above a certain concentration (typically 74 MBq/kg = 2 Ci/t) will have to undergo a detritiation process that will be specific to the chemical nature of the material. After the detritiation process, the waste has to be conditioned (which might include a tritium immobilisation treatment) for intermediate storage.

After the decay heat becomes negligible (depending on the composition of the materials involved, this lasts from a few years to many decades), the waste can be adequately conditioned for disposal in the corresponding final repositories.

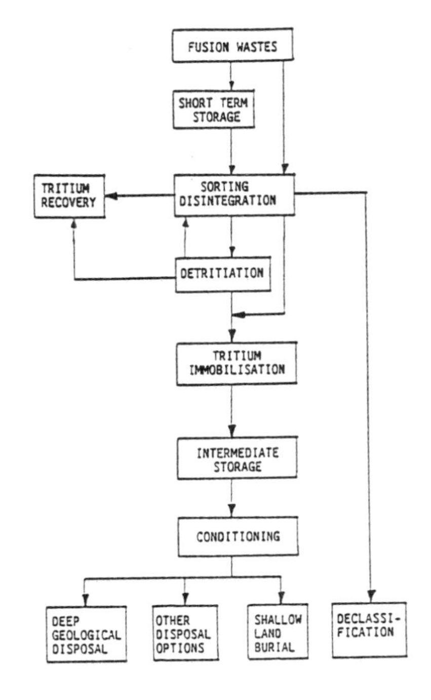

FIGURE 2

Tentative strategy for fusion waste management and disposal

Tritium recovery requirements are determined by both economics and regulatory limits for disposal. While typical values of 100 to 1000 Ci/t have been derived from fusion waste management studies for economic recovery of tritium from steel by melting[8], much lower limits exist in some national regulations for disposal (e.g. 74 MBq/kg = 2 Ci/t in France for the SLB site Centre de la Manche[4]). This item is less important, however, for tritiated highly activated waste which has to go to intermediate storage for decades and/or be disposed of in a deep geological repository.

Work is under way to experimentally determine actual detritiation factors which may be achieved by heating and melting. First experimental results[9] obtained by CEA indicate that for metallic components a level of about 1 to 30 Ci/t can be reached by melting.

No specific or additional facilities are needed for decommissioning of the highly activated NET components. Intermediate storage, hot cells and facilities have to be provided anyway to handle replaced in-vessel components at the end of the physics phase. Their decommissioning could simply be performed as another "replacement" at the end of the technology phase.

4. QUANTIFICATION OF NET WASTES ACCORDING TO A SPECIAL SCENARIO

Sweden is the only country to date having well defined specificatios for final disposal of LLW, MLW and HLW waste[10], with one repository for medium level waste already commissioned. Therefore the specifications of this country have been used to obtain at least some indicative values for the volumes that will have to be disposed of in final repositories.

LLW and MLW waste arising from the processing systems (i.e. fuel cycle and coolant purification systems), from decontamination and maintenance operations, and packed in steel drums and/or concrete blocks, could be disposed of in a shallow geological repository (SFR), located below sea bottom with 50 to 60 m of rock cover.

The highly active first wall and adjacent components would need about 1 year of cooling time before transport is allowed under present regulations. After a period of 10 to 30 years in an intermediate storage, final disposal of the adequately shielded packages would be possible in the deep geological repository (SFL), located at a depth of about 500 m in the bedrock.

Some estimates of volumes arising from replacement and decommissioning of the NET steel components and a tentative choice for NET waste repositories have been summarized in fusion waste management studies[9]. The total mass of about 7000 tonnes of the NET steel components correspond to a waste volume of about 2000 m^3 including packaging and assuming detritiation by melting. Of this waste, 80% can be disposed of into a shallow geological repository (of the SFR type) and 20% in a deep geological repository (of the SFL type). Each replacement of first wall, blanket structure, and back plate during the lifetime will produce additional waste of 400 m^3 after packaging, to be disposed of in a deep geological repository.

In summary, first wall and breeder blanket units will have to be disposed of in a deep geological repository of the SFL type, the back plates of the blanket segments and the inner shield could be disposed of in either a deep geological or a shallow geological repository of the SFR type, the outer shield in a shallow geological repository (or a shallow land burial site of the U.S. type), and the TF coils (steel and other materials) in a shallow land burial site.

5. CONCLUSIONS

The management and disposal strategies developed for fission waste can be applied for LLW and MLW fusion waste after sufficient tritium recovery/ removal and/or immobilization has been carried out.

Repositories originally designed for HLW and MLW fission waste appear suitable for the disposal of HLW and MLW wastes from NET.

After the first D-T pulse, NET will have produced waste requiring special handling facilities.

After ~25 full-power days of operation, NET first wall and blanket structures will pass the limit of shallow land burial (10CFR61).

At decommissioning, the steel components in NET will lead to a waste volume of ~2000 m^3.

Of the steel waste 80% can be disposed of into a shallow geological repository (Swedish SFR type), 20% needs a deep geological repository.

Each replacement of first wall, blanket structures, and back plates will produce ~400 m^3 of waste after packaging and will require disposal in a deep geological repository.

REFERENCES

1. Present situation and prospects in the field of radioactive waste management in the European Community. Documents of the Commission of the European Communities, COM (87) 312.

2. Standardisation of Radioactive Waste Categories, IAEA - Technical Reports Series No. 101 (1970).

3. Certificate of authorisation for the disposal of radioactive waste at Drigg. Issued jointly by the Ministry of Agriculture, Fisheries and Food (MAFF) and the Department of the Environment (DOE), United Kingdom (1971).

4. Prescriptions techniques relatives a l'exploitation du site de stockage de la Manche. Annexe II a la lettre SIN A 693/85 du 0.6.02.85. Service Central de Surete des Installations Nucleaires, Ministere de l'Industrie et de la Recherche, France (1985).

5. Radioactive Waste, Vol.I, first report from the Environment Committee of the House of Commons, Session 1985-86.

6. USNRC, 10 CFR Part 61: Licensing requirements for land disposal of radioactive waste. Final Rule. Federal Register (47 FR 57446), 27 December 1982.

7. P. Guetat. Fusion Reactor Wastes: Technical and Radiological Aspects for the Management of Wastes from NET and a Commerical Reactor. Presented at the IAEA Technical Committee Meeting on Fusion Reactor Safety, Culham, U.K., November 3-7, 1986.

8. K. Broden, A. Hultgren, G. Olsson, H. Djerassi, P. Giroux, J-L Rouyer. Fusion Waste Management - Safety and Environment Studies 1983-84 - European Fusion Technology Programme, EUR-FU/XII-361/85/35 (1985).

9. W. Gulden, C. Ponti, P. Guetat, D. Ochem, K. Broden, G. Olsson, G. J. Butterworth. Fusion Waste Management - Safety and Environment Studies 1985-1986. EUR-FU/XII-80/87/72 (1987).

10. SKBF/KBS Radioactive Waste Management Plan, Plan 82, Part 2, Facilities and Costs, SKBF/KBS Technical Report 82-09:2 (1982).

FUSION TECHNOLOGY 1988
A.M. Van Ingen, A. Nijsen-Vis, H.T. Klippel (editors)
Elsevier Science Publishers B.V., 1989

THERMAL SYSTEM ANALYSIS OF COMPACT IGNITION TOKAMAK*

C.B. Baxi

General Atomics, P.O. Box 85608, San Diego, CA 92138-5608, U.S.A.

This paper describes the thermal system analysis performed for the conceptual design of the Compact Ignition Tokamak (CIT). The CIT vacuum vessel and the first wall will be baked at a temperature of 350°C. The maximum fusion power during plasma operation will be 600 MW for 5 seconds at an interval of 1 hour. The coolant during bakeout and operation will be nitrogen or other suitable gas.

In order to analyze the above conditions, a thermal system analysis code (THESYS), previously used for design and analysis of the DIII–D tokamak, was modified and improved for application to CIT. THESYS is a 2-D transient thermal analysis code which models components of the first wall of the tokamak, such as limiters and divertors, the vacuum vessel wall, the insulation system, and the coolant loop. Analytical results for the DIII–D tokamak that were obtained with the THESYS code show good agreement with the observed results. The THESYS code was used for design and analysis of the CIT vacuum vessel system for bakeout and plasma operation. The results of the bakeout analysis indicate that a bakeout at 350°C is feasible by using hot nitrogen gas. A system analysis was also performed for the double-null plasma divertor operation and shows that 1 hour is sufficient to cool down the vessel between pulses. THESYS could be used for analysis of other devices, such as JET, JT-60, and ITER.

1. INTRODUCTION

The Compact Ignition Tokamak (CIT) has the objective of reaching ignition in order to address the scientific issues associated with ignited plasma.[1] Table 1 summarizes the selected CIT parameters.

The CIT vacuum vessel consists of solid formed Inconel plates weld together. Coolant passages are formed by welded channels on the outside of the vacuum vessel. The vessel and the ports are insulated with Q–Fiber® (98.5% pure silica fibers manufactured by the Manville Corporation) felt insulation. The insulation thickness will be 1.27 cm on the inner wall area and 3 cm over rest of the vessel and ports. Vessel heating will be achieved by a hot gas circulation in the cooling passages. The choice of coolant has not yet been finalized; however, for this study, calculations have been made with gaseous nitrogen (GN_2) as the reference coolant. A comparison of nitrogen, carbon dioxide, and helium coolants from thermal considerations is also presented.

It is planned to bake the CIT vacuum vessel and ports at a temperature of 350°C. It is also planned to maintain the vessel and the first wall above 330°C during operation. In order to analyze both of these scenarios, a computer code previously developed for thermal system (THESYS) design and analysis of DIII–D[2,3] was modified and upgraded.

TABLE 1
SELECTED CIT PARAMETERS

Major radius (m)	2.1
Minor radius (m)	0.65
Plasma elongation	2.0
Plasma current (MA)	11.0
Field on axis (T)	10.0
Plasma burn time (s)	5.0
Plasma heating power (MW)	Up to 40.0
Fusion power (MW)	600
Full power pulses	3000
Bakeout temperature (°C)	350
Minimum vessel temperature during operation (°C)	330
Coolant	GN_2
Flow rate (kg/s)	4
Cooling time (hr)	1

2. DESCRIPTION OF THE THESYS CODE

Figure 1 shows a cross section of the CIT vacuum vessel. THESYS analyzes the vacuum vessel system by dividing the vacuum vessel in a number of axisymmetric poloidal elements. Each poloidal element (Figure 2) consists of the first wall, the vessel wall, the coolant channel, and the insulation on the outside of the vessel.

* This work is sponsored by the United States Department of Energy under Contract No. DE–AC–84ER53158.

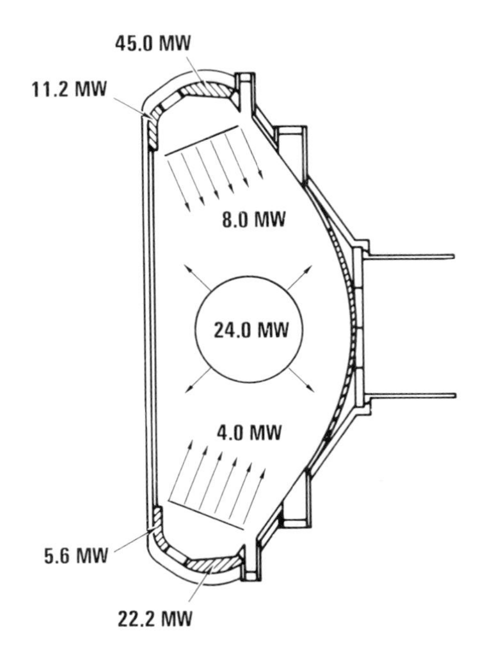

FIGURE 1
Cross section of the CIT vacuum vessel with heat flux deposition during double-null operation

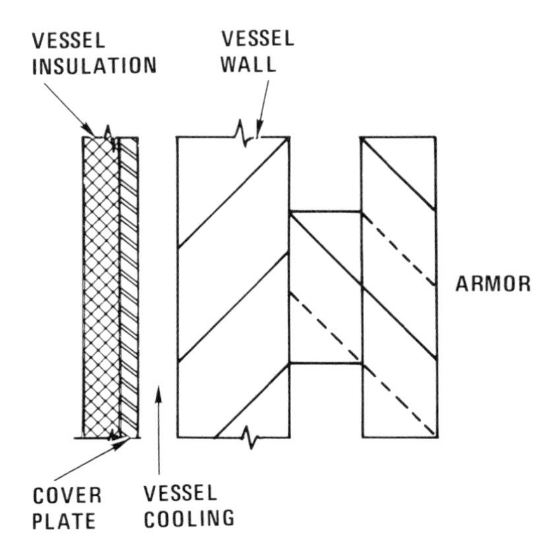

FIGURE 2
Details of an element in THESYS

Radiation view factors from each element to all other elements in the vessel are calculated by using a modified version of the code FACET.[4] The view factors are converted into exchange factors by Hottel's method.[5] Another auxiliary program is used to calculate the areas and hydraulic diameters of flow channels and to calculate the conductances between the vessel wall and the coolant channel cover plate.

The finite difference formulation in THESYS models the heat flux on the first wall and the nuclear heat generation in the first wall and the vessel wall. The model includes radiation heat transfer from the front of the first wall to all other elements, radiation, and conduction heat transfer from the first wall to vessel wall, convection from the coolant to the vessel wall and the coolant channel cover plate, conduction between the vessel wall and the cover plate, and heat loss through the insulation. Poloidal conduction in the vessel wall between adjacent elements is also taken into consideration. Thus, the code is flexible enough so that it can be used for thermal system analyses of any tokamak and can be used for analysis of either bakeout or plasma operation.

The results obtained from THESYS are useful to:

- Determine the heatup/cooldown times for bakeout.
- Study the effect of different coolants and their flow rates and inlet temperatures on heating and cooling.
- Determine the critical poloidal locations in the vessel for further detailed thermal and stress analysis.
- Provide boundary conditions for detailed analysis.
- Provide poloidal vessel temperature distribution as an input to structural analysis of the vessel.
- Determine heat loss from the vacuum vessel system.

3. APPLICATION OF THESYS TO DIII–D

In the DIII–D, the maximum energy removal is 60 MJ over a time period of 10 minutes. This can be easily achieved, since water cooling of the vessel and ports is used during plasma operation.

The bakeout of DIII–D at temperatures up to 400°C is accomplished by a combination of ohmic heating of the vessel wall and air circulation through the poloidal channels in the corrugated vessel walls. The THESYS code was used to analyze this condition. Results predicted by the code agreed quite well with measured values from the DIII–D vessel. Figure 3 shows such a comparison.

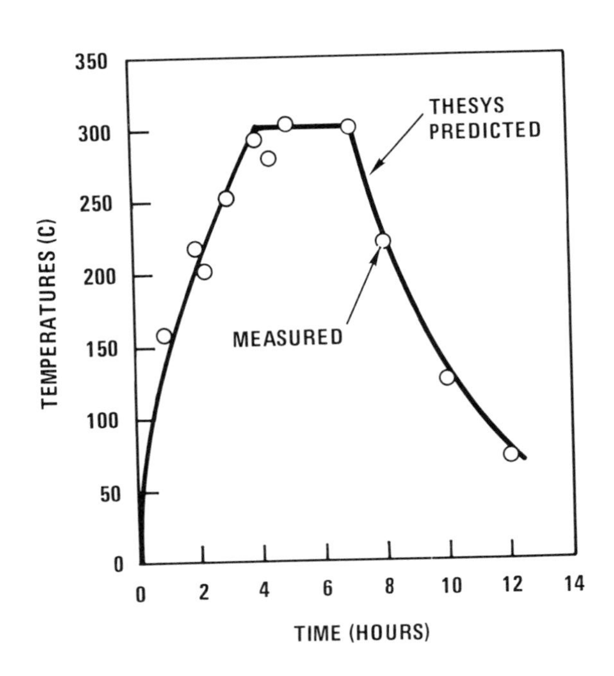

FIGURE 3
Comparison of THESYS predicted with measured temperature for DIII–D during bakeout

4. CIT THERMAL SYSTEM ANALYSIS

The THESYS code was used for two types of analysis. The first was the plasma operation during double-null divertor, and the second was for bakeout. At the writing of this paper, the choice of coolant for the CIT was not finalized. Hence, all the calculations were performed with gaseous nitrogen as the reference coolant. Various coolants are compared for thermal and other criteria, and a recommendation is made on a preferred coolant.

During plasma operation, it is desired to keep the minimum vessel temperature above 330°C. The distribution of energy deposition on the surface of the first wall for the double-null operation is shown in Figure 1. In addition, about 1000 MJ of energy is deposited in the first wall tiles and the vacuum vessel walls as volumetric generation due to neutrons during the plasma burn. In order to analyze these conditions, the vacuum vessel was divided into 12 poloidal elements. Gaseous nitrogen at a flow rate of 4 kg/s, with an inlet temperature of 330°C and 600 kPa average pressure was used to cool the vessel.

Figures 4 to 6 show the results obtained under these conditions. Figure 4 shows the bulk temperatures of first wall tile at the beginning and end of the burn. Since these are quasi-steady state results, at the end of one hour cooling time the tiles and vessel wall will cool down to initial conditions. The coolant temperature rise as a function of time is shown in Figure 5, indicating that the peak coolant temperature rise is almost twice the average. Figure 6 shows the poloidal vessel and coolant temperature distribution at 240 seconds after the beginning of the burn. A number of these results provide an input to the vessel structural analysis and the boundary conditions for detailed local analysis.

Since the minimum temperature of the vessel during operation will be close to the bakeout temperature, routine bakeout will consist of maintaining the hot GN_2 flow rate such that the energy input is equal to the heat loss. However, the vessel may have to be brought to bakeout temperature from cold conditions a few times. Figure 7 shows the result of heating up the vessel from −193°C to

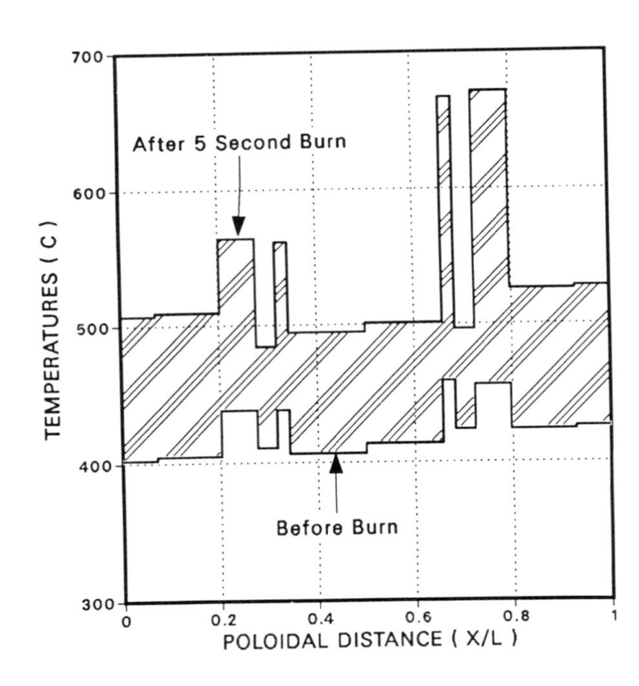

FIGURE 4
CIT first wall and divertor bulk temperatures at beginning and end of the pulse

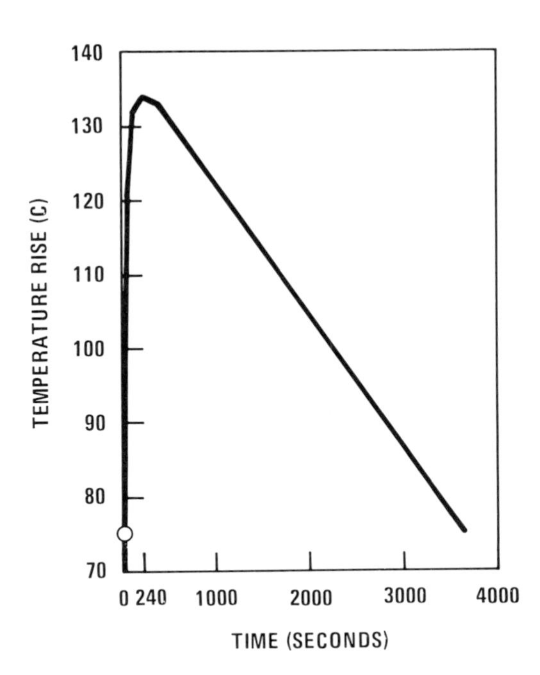

FIGURE 5
Difference between outlet and inlet coolant temperature for CIT during cooldown after the burn

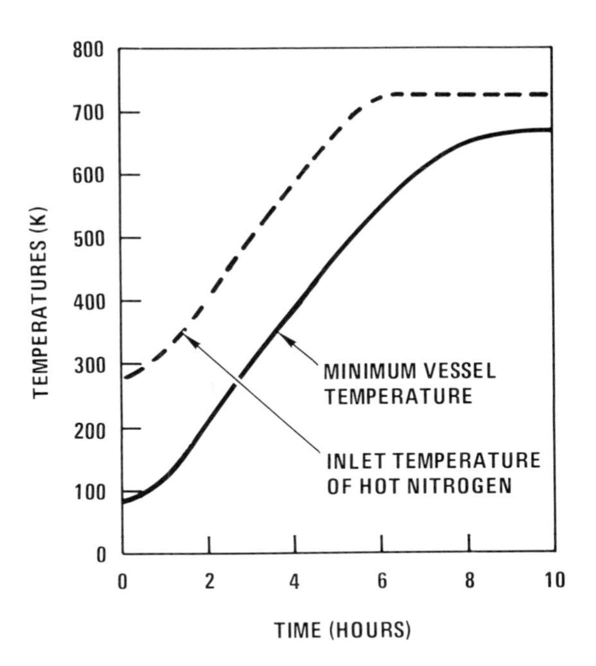

FIGURE 7
Heat-up of the CIT vessel from 80 K to 623 K (350°C) by hot nitrogen gas

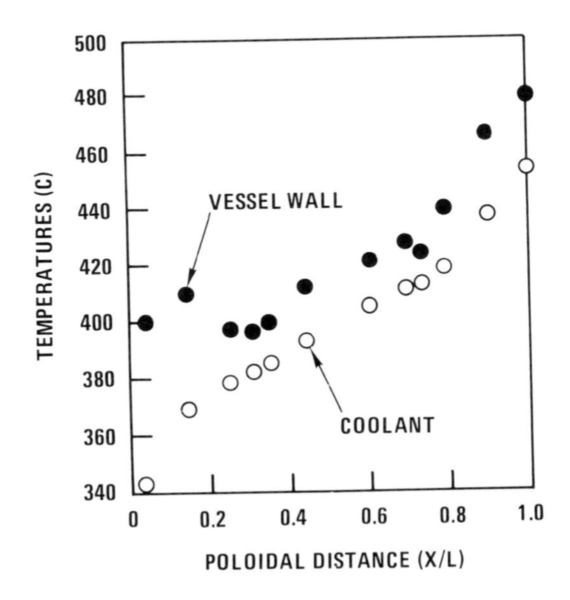

FIGURE 6
Poloidal distribution of vessel wall and coolant temperature at 240 seconds after the burn

350°C. For this analysis it was assumed that the inlet temperature of the heating gas is 200°C above the minimum vessel temperature, but not higher than 450°C. Additional parametric studies will be done to determine the optimum procedure to bring the vessel to bakeout temperature without creating excessive stresses in the vessel wall.

5. DISCUSSION

It has been demonstrated that the system analysis approach leads to useful information for designing the cooling system for the CIT. This analysis shows that a gaseous coolant such as nitrogen is thermally adequate to cool and heat the CIT. Two other easily available coolants were compared to nitrogen for thermal performance. Results in Table 2 show helium to be a superior coolant. Helium is also better from activation considerations due to production of C_{14} in nitrogen and carbon-dioxide. These advantages of helium must be viewed in light of higher cost of the coolant and higher capital cost for the helium loop.

TABLE 2
COMPARISON OF COOLANTS FOR CIT
FROM THERMAL CONSIDERATIONS

(Based on 1500 MJ energy removal rate, at an average coolant temperature rise of 100°C, and at a 600 kPa average pressure)

Coolant	Flow Rate (kg/s)	Pressure Drop (kPa)	Pumping Power (kW)	HTC (W/m^2–K)
Nitrogen	4	100	100	150
Helium	0.8	31	46	210
Carbon dioxide	4	64	40	127

6. CONCLUSIONS

Use of the THESYS code has resulted in the following thermal design conclusions for CIT:

- For the fusion energy of 600 MW for 5 seconds, CIT vacuum vessel system can be cooled by a gas coolant in 1 hour (Figure 4).
- During plasma operation, the minimum vessel temperature can be maintained above 330°C (Figure 4).
- Bakeout at a temperature of 350°C is feasible with gas heating alone (Figure 7).
- The flow rate required with a gaseous nitrogen coolant is about 4 kg/s for the vacuum vessel (Table 2). Additional flow will be required for ports.
- Helium is a superior coolant from thermal and activation considerations. The capital and operating costs will be higher with helium.
- Thermal stress, material properties, and tritium permeation need to be examined further to finalize high-temperature plasma operation.
- Due to its flexibility and adaptibility, THESYS can be applied to thermal system analyses of other tokamaks.

REFERENCES

1. J.A. Schmidt and H.P. Furth, Compact Ignition Tokamak Conceptual Design Report, A-860606-P-01, Princeton Plasma Physics Laboratory (June 1986).
2. J.L. Luxon and L.G. Davis, Fus. Technol. 8 (1985) 441.
3. C.B. Baxi, Design of Vacuum Vessel Heating System for DIII–D, in Proc. of the IEEE (1985), Cat. No. 85CH2251-7, Vol. 2, pp. 952–955.
4. A.B. Shapiro, FACET – A Radiation View Factor Computer Code for Axisymmetric, 1D Planar, and 3D Geometries with Shadowing, University of California, LLNL Report UCID-19887 (August 1983).
5. H.C. Hottel and A.F. Sarofim, Radiative Heat Transfer (McGraw Hill, New York, 1967).

FUSION TECHNOLOGY 1988
A.M. Van Ingen, A. Nijsen-Vis, H.T. Klippel (editors)

SYSTEM ANALYSIS OF TOKAMAK FUSION REACTOR PLANTS

T. Ida, M. Kasai, and I. Yanagisawa
Mitsubishi Atomic Power Industries, Inc., 4-1, Shibakouen 2-chome, Minato-ku, Tokyo 105, Japan

F. Matsuoka, Y. Imamura, and N. Asami
Mitsubishi Heavy Industries, Ltd., 4-1, Shibakouen 2-chome, Minato-ku, Tokyo 105, Japan

T. Takuma, K. Yamaji, T. Nanahara, S. Akita, A. Minato, and K. Nemoto
Central Research Institute of Electric Power Industry, 2-11-1, Iwato Kita, Komae-shi, Tokyo 201, Japan

In order to evaluate the economic potentiality of tokamak fusion reactor plants, system analyses of a pure fusion type and two kinds of fusion-fission hybrid types of tokamak reactors have been performed by using a systems analysis code, a revised version of NEW-TORSAC. Hybridization of fusion with fission leads to an approximately 40% reduction of cost of electricity (COE) of the tokamak fusion reactor. The Troyon coefficient and the normalized current drive efficiency are key plasma parameters useful to enhance the economic viability of the fusion reactor. Higher plasma elongation is more attractive in the economics of the reactor, although it brings about a severe restriction in the engineering design from a viewpoint of plasma position control. Improvement in the key plasma parameters is essential to enhance the economic potentiality of the tokamak fusion reactor enough to compete with alternative energy source systems such as fission and coal plants.

1. INTRODUCTION

The tokamak fusion reactor has been researched most actively and is thought as the first reactor to be commercialized. Although a research on plasma physics has progressed steadily, there remain many uncertainties to be solved. It seems that these uncertainties have strong impacts on the economics of a fusion reactor system. Accordingly, it is important to take into consideration the economics in a fusion reactor study. There are two major types of reactor systems: a pure fusion and a fusion-fission hybrid. The purpose of this paper is to evaluate the economic potentiality of these tokamak fusion reactors as a commercial reactor. Therefore, we analyze the effect of main plasma parameters on the economics of the tokamak fusion reactor. Furthermore, we compare the economics among the pure fusion type and two kinds of fusion-fission hybrid type.

2. METHOD AND MODEL

The economics of power plants can be evaluated using its cost of electricity (COE). Estimation of the COE of the fusion reactor requires a consistent reactor design according to its plasma condition and reactor specification. In our study, the estimation of the COE has been accomplished by using a systems analysis code, a revised version of NEW-TORSAC,[1] which has functions to design a tokamak reactor automatically and calculate the COE. Brief descriptions are provided below for the reactor design scheme and the economic model adopted in the systems analysis code. The reactor specification used in this study is also summarized in this section.

2.1. Reactor Design Scheme

A rough design of the reactor required for the economic analysis is performed through a series of a plasma design, a tokamak structure design, and a reactor plant equipment design. In the plasma design, plasma parameters are determined to achieve the specified reactor performance of the fusion power and the first wall loading. In the tokamak structure design, a fundamental configuration of the toroidal structure consisting of the first wall, blanket, shield, vacuum vessel, and toroidal field (TF) coil is determined to be consistent with a plasma configuration and given engineering constraints such as allowable toroidal ripple and allowable radiation damage. Further, poloidal field (PF) coils are located based on a plasma equilibrium calculation taking account of a maintenance space for the tokamak structure. In the reactor plant equipment design, an operation pattern of the PF coil currents is calculated according to the given operation scenario. Requirements are also estimated for the power supply system, the exhaust system, the cooling system, the cryogenic system, and the tritium system.

2.2. Economic Model

The US-DOE (Department of Energy) guideline[2] is known as a suitable model for a consistent, uniform, and complete economic evaluation of fusion reactors. This guideline is adopted basically in our economic analysis and some modifications are introduced to match it with the practical Japanese cost estimation scheme. For example, contingency allowance is not considered in this analysis.

Nine percent of the total capital cost is summed up as an interest during construction. The cost for escalation during construction is not considered because a constant money analysis model is adopted. An annual operating cost is assumed to be 2% of the total direct cost. Both tritium and plutonium (Pu) credits are considered as negative fuel costs. The COE, averaged during the depreciation period which is assumed to be sixteen years, is estimated using a discounting method with a discount rate of 6%. A fixed charge rate, utilized to determine the cost of capital component, is set to 12%.

2.3. Reactor Specification

Three types of the tokamak fusion reactor are selected to evaluate the economic potentiality of the tokamak fusion reactor plant. They are a pure fusion type (PF type), a fast-fission hybrid type (FFH type), and a fission-suppressed hybrid type (FSH type). The essential engineering parameters of each type are listed in Table 1. The net electric power is fixed to nearly 1000 MWe and the steady state operation is assumed to be realized by RF current drive. The blanket nuclear performance is derived from our previous studies[3,4,5] and is assumed to be independent of the plasma parameters and the reactor configuration. The FSH type in our study is not so prolific of Pu in comparison

Table 1 Engineering parameters of the three types of reactors

Reactor Type	PF	FFH	FSH
Fusion output (MW)	3500	668	1516
Energy multiplication factor(*1)	1.20	4.20(*2)	1.86
Tritium breeding ratio	1.15	1.10(*2)	1.05
Pu production rate (kg/cycle)	—	8.7	10.1
Thermal-conversion efficiency (%)	33.3	40.0	40.0
Capacity factor (%)	75	75	75
Blanket materials			
coolant	H_2O(*3)	M.S.(*4)	M.S.(*5)
breeder	Li_2O	Li_2O(*6)	M.S.(*5)
fissile	—	U_3Si	M.S.(*5)

(*1) Based on the fusion output.
(*2) Average value over one cycle (5 years and 4 months).
(*3) Pressurized water.
(*4) Molten salt of 47LiF-53BeF_2 with 40% enriched ^6Li.
(*5) Molten salt of 70LiF-20BeF_2-10UF_4 with 25% enriched ^6LI.
(*6) 40% enriched ^6Li.

with the FFH type because of the requirement that the tritium breeding ratio must be greater than unity. The thermal-conversion efficiencies are determined considering the properties of the coolant. The thicknesses of the blanket and shield are fixed a priori according to our previous design studies. The PF coils are prohibited to be located in the space for a horizontal access type of maintenance of the reactor core components.

3. RESULTS

3.1. Sensitivity to Main plasma parameters

Parametric surveys on plasma temperature, T, and safety factor, q, have been carried out to determine the reference values of these parameters. The safety factor is represented as

$$q = \frac{\pi a_p^2 B_t}{\mu_0 I_p R_p} \{ 1 + \kappa^2 (1 + 2 \gamma^2) \}$$

where R_p and a_p are the plasma major and minor radii (m), I_p is the plasma current (A), B_t is the toroidal field (T) at the plasma center, and κ and γ are the plasma elongation and triangularity. Furthermore, other parametric surveys have been also made on a coefficient of Troyon's toroidal beta limit scaling (Troyon coefficient)[6], g, plasma elongation, κ, and a normalized current drive efficiency[7], η, independently. The efficiency, represented as

$$\eta = (n_e/10^{20}) R_p I_p / P_{CD}$$

where n_e is the mean electron density and P_{CD} is the required current drive power (W), is a significant parameter because the steady state operation is assumed. The plasma parameters summarized in Table 2 are used in these surveys for the PF and the FFH type fusion reactors.

3.1.1. Plasma Temperature and Safety Factor

Figure 1 illustrates the dependences of R_p, a_p, I_p, and P_{CD} on T fixing q = 2.4 in the case of the FFH type reactor. The increase of T from

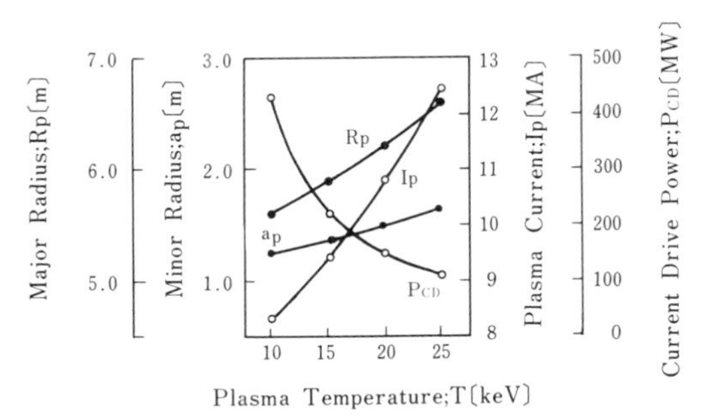

Figure 1 Dependences of R_p, a_p, I_p, and P_{CD} on T fixing q = 2.4 in the case of the FFH type reactor.

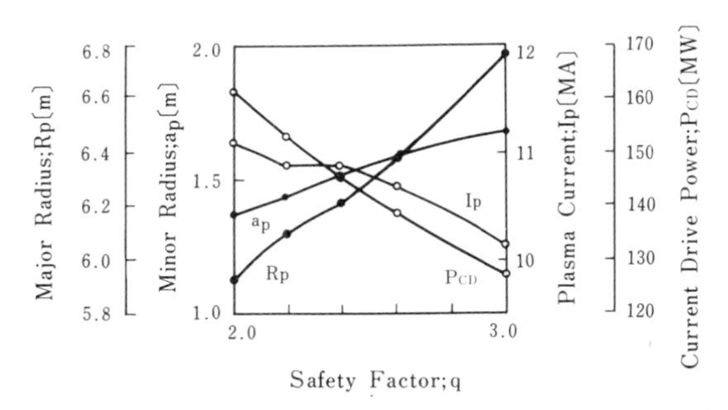

Figure 2 Dependences of R_p, a_p, I_p, and P_{CD} on q fixing T = 20 keV in the case of the FFH type reactor.

Table 2 Main plasma parameters

Parameter	Value	Reference
Plasma temperature, T (keV)	10/15/20/25	20(*1)
Safety factor, q	2.0/2.2/2.4/2.6/3.0	2.4(*1)
Troyon coefficient, g	3.5/7.0	3.5(*2)
Plasma elongation, κ(*3)	1.6/1.8/2.0	1.8(*2)
Plasma triangularity, γ(*3)	0.35	
Neutron wall load, P_W	maximum(*4)	
Current drive efficiency, η	0.35/1.05	0.35(*2)

(*1) Values are set to be references by these surveys.
(*2) Values are set to be references a priori.
(*3) Double null divertor configuration is assumed.
(*4) First wall loading is maximized under given plasma parameters.

10 keV to 25 keV brings about gradual increases of R_p, a_p, and I_p and rapid decrease of P_{CD}. The dependences of the same variables on q from 2.0 to 3.0 fixing T = 20 keV in the same type reactor are shown in Fig. 2. The effects of q on R_p, a_p, and P_{CD} are similar to those of T, although the effects of q on P_{CD} is weaker than that of T.

The effects of T and q on the COE are summarized in Figs. 3-(a) and (b) for the PF and the FFH type reactor, respectively. The minimum-COE point on T is found between 15 keV and 25 keV for each fixed safety factor and it shifts toward higher T as q is increased. This behavior results from the trade-off between the cost increasing with the increase of plasma size and the cost and inhouse power reduction with the decrease of P_{CD}. For T above 15 keV, as q is decreased, the COE is reduced. Since the plasma with lower q is apt to be disrupted, q is set to 2.4 as a reference value judging from current experimental results. The optimal plasma temperature is approximately 20 keV with q = 2.4 for both types of reactors. Accordingly, 20 keV is selected as a reference value of T in this study.

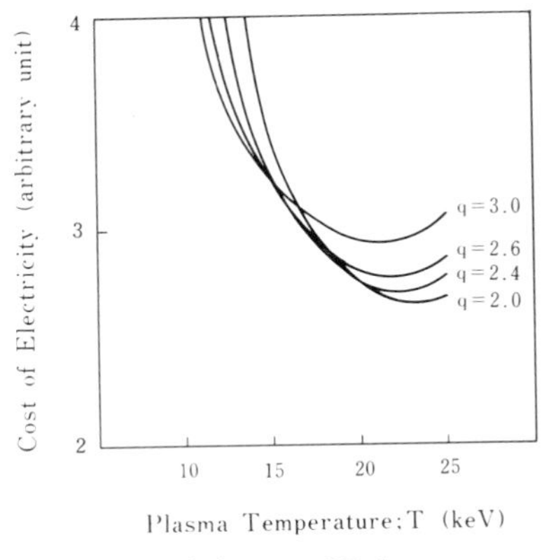

(a) the PF type

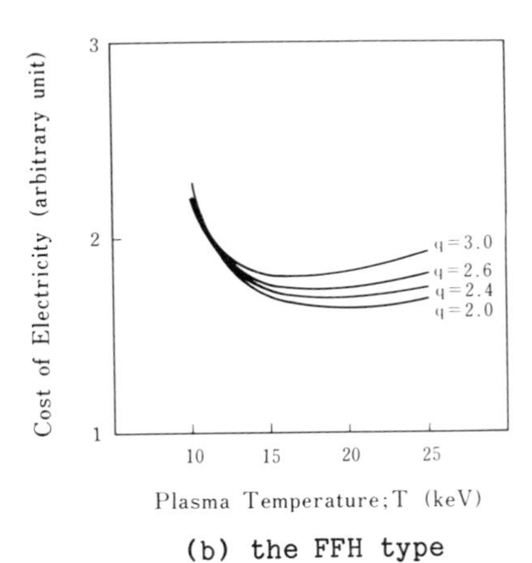

(b) the FFH type

Figure 3 Dependence of the COE on safety factor q as a function of plasma temperature T in the cases of (a) the PF and (b)the FFH types.

3.1.2. Troyon Coefficient

The dependence of the COE on the Troyon coefficient, g, is shown in Fig. 4. The decrease of fusion core size with the increase of g leads to the reduction of the total capital cost. An increase in this coefficient from 3.5 to 7.0 reduces the COE by 38% and 22% for the PF and the FFH type reactors, respectively. The stronger effect for the PF type comes from its larger size of the fusion core. The reference value for g is set to 3.5 taking into account current experimental studies.

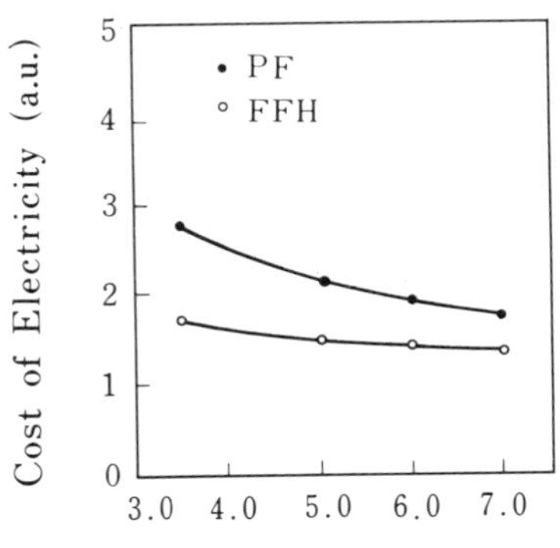

Figure 4 Dependence of the COE on the Troyon coefficient g in the cases of the PF and the FFH types of fusion reactors.

3.1.3. Plasma Elongation

The dependence of the COE on the plasma elongation, κ, is shown in Fig. 5. It is obvious that the increase of κ leads to the effective reduction of the COE. The reason of this reduction is that the increase of the toroidal beta limit with increasing κ reduces the reactor size and the capital cost. The instability of plasma vertical position will be enhanced in the case of highly elongated plasma. Accordingly, a plasma elongation of 1.8 is determined as the reference from the viewpoint of plasma position controllability.

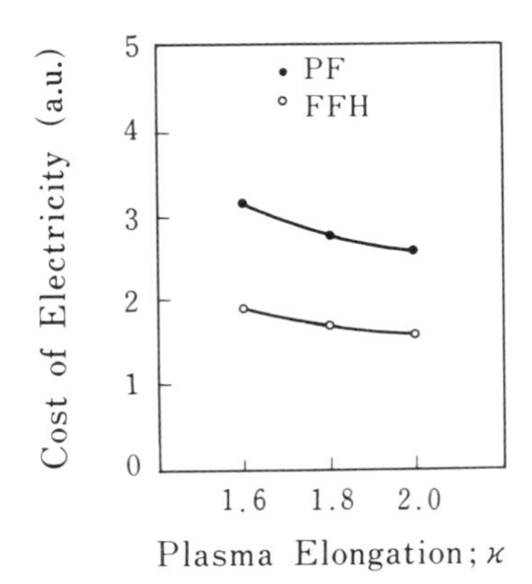

Figure 5 Dependence of the COE on the plasma elongation κ in the cases of the PF and the FFH types of fusion reactors.

3.1.4. Current Drive Efficiency

The reference value 0.35 for the normalized current drive efficiency is determined from the INTOR guideline.[8] Since this efficiency affects the required current drive power directly, the inhouse power and the capital cost for the RF device are reduced with an increasing efficiency. Improvement of three times in the efficiency reduces the COE by 32% and 14% for the PF and the FFH type, respectively.

3.2. Comparison among Reactor Types

Table 3 summarizes the main results of the economic analyses for the three types of reactors using the reference values of the plasma and engineering parameters. It is revealed that the COE of the FFH type is approximately 60% of that of the PF type. This difference comes mainly from the reduction of the total capital cost for the FFH type associated with the less-massive tokamak fusion structure, although higher thermal-conversion efficiency for the FFH type also emphasizes the difference. The COE of the FSH type is evaluated to be higher by approximately 27% than that of the FFH type. The effect of greater Pu benefit of the FSH type on the COE is canceled out due to the increase of the reactor size. The COE is ascending in the following order: the FFH, the FSH, and the PF types. It is indicated that the compactness of the reactor by hybridization of fusion with fission leads to the reduction of the COE.

4. DISCUSSION

The plasma temperature of 20 keV is identified as optimal from the viewpoint of the COE for the steady state tokamak fusion reactors with the reference values of plasma parameters. Strong impacts of the Troyon coefficient and the plasma elongation on the COE are derived from the dependence of the reactor size on these plasma parameters. The normalized current drive

Table 3 Main output of the economic analysis

Reactor Type	PF	FFH	FSH
Major radius, R_p (m)	8.19	6.20	7.26
Minor radius, a_p (m)	2.30	1.50	1.86
Plasma current, I_p (MA)	20.4	10.8	14.7
Neutron wall load (MW/m²)	2.47	0.94	1.48
Total power output (MWe)	1556	1179	1232
Net power output (MWe)	976	979	904
Total capital cost (G-yen)	1.28	0.78	0.97
Material weight (10^4 tonne)	3.38	1.95	2.92
Cost of electricity (yen/kWh)	27.65	16.93	21.52

efficiency is also a key parameter effective to the reduction of the COE because the inhouse power is very sensitive to this efficiency.

The economic evaluation for the tokamak fusion reactors shows that the COEs for the PF, the FFH, and the FSH types are approximately 2.5, 1.5, and 2.0 times respectively as much as those of competitive energy source systems such as fission and coal plants. Additional analysis using the Troyon coefficient of 7.0 and the normalized current drive of 1.05, which are twice and three times as much as the reference values respectively, results in the COEs of 12.72 yen/kWh, 12.01 yen/kWh, and 12.42 yen/kWh for the PF, the FFH, and the FSH types, respectively. The COEs computed in this study and their dependence on the Troyon coefficient are close to the values in the other estimation[9] also recently made by CRIEPI and others, although engineering parameters such as coolants are significantly different between the two analyses. Such improvements of plasma parameters make the COE of the fusion reactor close to those of the competitive systems. Furthermore, they reduce the difference in the COE among the three types of tokamak fusion reactors. Accordingly, innovative progress in plasma physics is required for the tokamak fusion reactors to become commercially viable.

Hybridization of fusion with fission is a promising option to enhance the economic potentiality of the fusion reactor. In our analysis, the FSH type reactor is not so attractive as the FFH type. However, the COE of the FSH (fissile fuel producing hybrid) is much lower in the reference 9 than in the present analysis. This difference is primarily due to the difference in the computation method, that is, a symbiotic system has been considered in the reference 9 including light water reactors supported. Such a system usually leads to the considerable reduction of the COE of the FSH type reactors. The viability of the FSH type would be also enhanced if the reactor could be designed to be more prolific of Pu and the price of fissile materials would rise.

5. CONCLUSION

The plasma temperature of 20 keV gives a minimum-COE of the steady state tokamak fusion reactor with the safety factor of 2.4 and the reference values of the other parameters. The hybridization of fusion with fission can be expected to bring about a large reduction in the COE of the fusion reactor, although it can not yet make the fusion reactor to be comparable to the alternative power systems. Such plasma parameters as the Troyon coefficient, the plasma elongation, and the normalized current drive efficiency have the strong impacts on the COE. Improvements in these plasma parameters could enhance the economic potentiality of the fusion reactor enough to compete with the alternatives.

REFERENCES

1. M. Kasai et al., JAERI-M 87-103, 1987.
2. S.C. Schulte et al., PNL-2648, 1978.
3. T. Tone et al., JAERI-M 83-031, 1983.
4. M. Inoue et al., Conceptual Design of Demonstration Molten Salt Hybrid Reactor (DMHR), Proc. 4th Int. Conf. on Emerging Nuclear Energy Systems, 1986, pp.70-74.
5. F. Matsuoka et al., Conceptual Design of a Tokamak Hybrid Power Reactor (THPR), Proc. 12th Symp. on Fusion Engrg. 1987, pp.943-946.
6. F. Troyon et al., Plas. Phys. and Contr. Fus. 26, 209 (1984).
7. N.J. Fisch, Phys. Rev. A. 24(6), 3245 ,1981.
8. INTOR Group, International Tokamak Reactor: Phase Two A Part 3, Vienna, IAEA ,1988.
9. M. Yamaoka et al., Parametric Study for Blanket Neutronics and Economics of Fusion-Fission Hybrid Reactor, Int. Symp. on Fusion Nucl. Technol. Tokyo, Japan, April 10-15, 1988.

FUSION TECHNOLOGY 1988
A.M. Van Ingen, A. Nijsen-Vis, H.T. Klippel (editors)
Elsevier Science Publishers B.V., 1989

ENGINEERING AND PHYSICS OF HIGH-POWER-DENSITY, COMPACT, REVERSED-FIELD-PINCH FUSION REACTORS: THE TITAN STUDY†

F. Najmabadi[1], R. W. Conn[1], R. A. Krakowski[2], K. R. Schultz[3], D. Steiner[4], and The TITAN Research Group

[1] University of California, Los Angeles, CA,
[3] General Atomics, San Diego, CA,
United States of America

[2] Los Alamos National Laboratory, Los Alamos, NM
[4] Rensselaer Polytechnic Institute, Troy, NY,

The technical feasibility and key developmental issues of compact, high-power-density Reversed-Field-Pinch (RFP) reactors are the primary results of the TITAN RFP reactor study. Two design approaches emerged, TITAN-I and TITAN-II, both of which are steady-state, DT-burning, ~ 1000 MWe power reactors. The TITAN designs are physically compact and have a high neutron wall loading of 18 MW/m^2. Detailed analyses indicate that: a) each design is technically feasible; b) attractive features of compact RFP reactors can be realized without sacrificing the safety and environmental potential of fusion; and c) major features of this particular embodiment of the RFP reactor are retained in a design window of neutron wall loading ranging from 10 to 20 MW/m^2. A major product of the TITAN study is the identification and quantification of major engineering and physics requirements for this class of RFP reactors. These finding are the focus of this paper.

1. INTRODUCTION

The TITAN Reversed-Field Pinch (RFP) fusion reactor study has been a multi-institutional research effort[1,2] to determine the technical feasibility and key developmental issues for an RFP fusion reactor operating at high power density. Other objectives of the TITAN study include: a) assessment of the potential economics (cost of electricity, COE), operational (maintenance and availability), safety and environmental features of high-mass-power density systems; b) determination of the design and operational window for this embodiment of the RFP reactor; and c) identification and evaluation of the plasma physics and engineering constraints and requirements for high mass-power-density RFP reactors. Mass power density (MPD) is defined as the ratio of the net electric power to the mass of the fusion power core (FPC), which includes the plasma chamber, first wall, blanket, shield, magnets, and related structure.

Parametric systems studies were used extensively to select minimum-cost design points and to evaluate the size and sensitivity of the avaliable design window; the parametric models were continually refined as better definitions and engineering designs for reactor subsystems became avaliable. These parametric studies point to attractive and compact RFP reactors operating with the neutron wall loading in the range of 10-20 MW/m^2 with a shallow minimum in COE at about 18 MW/m^2, as is shown in Figure 1. Even though operation at the lower end of the this range of wall loading (10-12 MW/m^2) is possible with a small cost penalty, the TITAN study adopted design points at the upper end of the design window, corresponding to the minimum-COE (18 MW/m^2), primarily to quantify and assess the technical feasibility and physics limits for the major subsystems of high-MPD RFP reactors.

In order to demonstrate the possibility of multiple engineering approaches to high-MPD reactors, two different detailed designs, TITAN-I and TITAN-II, were studied. Both conceptual reactor designs are based on the DT fuel cycle, have a net electric output of about 1000 MWe, and are compact. The plasma major radius is 3.9 m, plasma minor radius is 0.6 m and fusion power density is 80 MW/m^3. The TITAN designs have a high mass power density of

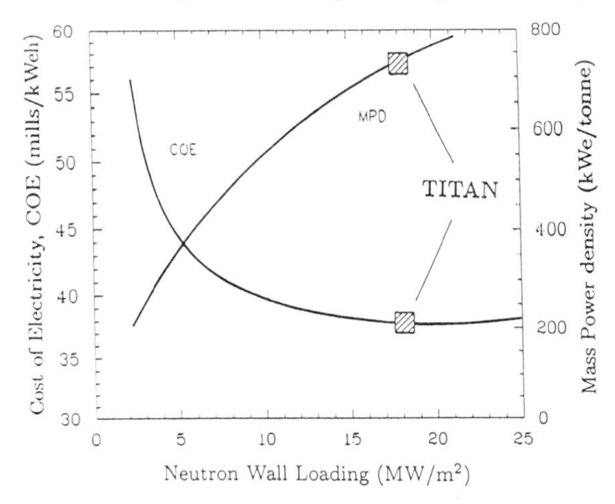

Figure 1. Variation of cost of electricity, COE, and mass power density, MPD, with neutron wall loading for high-power-density RFP reactors. The TITAN design point is also shown.

† This research was supported by the United States Department of Energy, Office of Fusion Energy.

TABLE I.
Major Characterisitics and Parameters of TITAN Reactors

a. PLASMA PARAMETERS

Major plasma radius (m)	3.9	Plasma current (MA)	17.8
Minor plasma radius (m)	0.6	Poloidal beta	0.22
Electron density ($10^{20}/m^3$)	9.3	Plasma temperature (keV)	10.
Poloidal field at first wall (T)	5.9	Pinch parameter, Θ	1.56
Toroidal field at first wall (T)	0.38	Reversal parameter, F	-0.1
Heating method	Ohmic	Current rise time (s)	10.
Current-drive method	OFCD	Current-drive efficiency (A/W to the power supply)	0.3
Impurity control System	3 toroidal divertors		
Fueling method	Pellet injection		

b. REACTOR PARAMETERS

	TITAN-I	TITAN-II
Neutron wall loading (MW/m^2)	18.	18.
Mass power density (kWe/tonne of FPC)	760.	800.
Cost of electricity (mills/kWeh)	40.	38.
Unit direct cost ($/kWe)	1531.	1543.
Net electric output (MWe)	970.	900.
Fusion power (MW)	2300.	2290.
Thermal power (MW)	2935.	2986.
Thermal cycle efficiency	0.44	0.35
Recirculating power fraction	0.25	0.14
Net plant efficiency	0.33	0.30

about 800 kWe/tonne of FPC, which is equivalent to the MPD of a fission pressurized-water reactor (PWR). By contrast, earlier studies of tokamak and tandem mirror reactors such as STARFIRE[3] and MARS[4], have MPD values around 50 kWe/tonne. More recent work suggests that tokamaks may achieve values up to 200 kWe/tonne[5,6]. It is the inherent physical characteristics of the RFP confinement concept[7] that permits the design of compact fusion reactors with such high values of mass power density. The major parameters of the TITAN designs are listed in Table I. Both designs are based on similar plasma parameters but have entirely different high-power-density fusion power core designs.

Opertional (maintenance and availability) and safety features have been incorporated into the TITAN designs from the beginning, with major goals being passive safety, simplicity, high availability and low cost. As a result, the TITAN designs would satisfy U. S. criteria for near-surface disposal of radioactive waste (Class C, 10CFR61) and achieve a high Level of Safety Assurance[8] with respect to FPC damage by decay after-heat and radioactivity release caused by accidents. Furthermore, the compact FPC designs of TITAN reduce each system to a few small and relatively low mass components, making toroidal segmentation of the FPC unnecessary as part of the normal maintenance procedure. A unique "single-piece" FPC maintenance procedure in which the first wall, blanket and divertor modules are replaced as a single unit is, therefore, utilized.

In the following sections, we briefly review the major features of TITAN designs and then concentrate on the physics and engineering requirements and constraints for achieving TITAN-class high-MPD reactors. Complete details of the TITAN-I and TITAN-II designs can be found in Ref. 1.

2. TITAN REACTOR DESIGNS

Because of the high plasma current density (10-15 MA/m^2), ohmic heating of the TITAN RFP plasma to ignition would be possible; no auxiliary heating system is required. The reactors operate at steady state, using oscillating-field current-drive (OFCD)[9,10] to maintain the 18 MA plasma current. The calculated efficiency of the OFCD system, including the effects of eddy currents induced in the FPC, is 0.3 A/W delivered to the power supply (0.8 A/W delivered to the plasma)[1].

The impurity control and particle exhaust system consists of three high-recycling, toroidal-field divertors. The TITAN plasma engineering designs take advantage of the beta-limited confinement observed in RFP experiments [7,11,12] to operate with a highly radiative core plasma, deliberately doped with a trace amount of high-Z, Xe impurities ($n_{Xe}/n_e \simeq 10^{-4}$, $Z_{eff} \simeq 1.7$). This intensely radiative mode of operation distributes the majority of plasma power (i. e., ohmic and alpha-particle power) uniformly on the first wall (4.5 MW/m^2). Simultaneously, the total charged-particle power to the divertor is small (< 5% of plasma power), and the peak heat load on the divertor target plates is reduced to less than $\simeq 9$ MW/m^2. The "open" magnetic geometry of the divertors together with the intensive radiative cooling leads to a high-recycling divertor with high density and low temperature near the divertor target. Negligible neutral-particle leakage from the divertor chamber to the core plasma, adequate particle exhaust rates, and very small first wall erosion rates are predicted[1].

The TITAN-I fusion power core is a self-cooled, liquid-lithium design using a vanadium-alloy (V-3Ti-1Si) structural material. The lithium coolant flows in the poloidal direction, and the blanket is configured as an integrated blanket coil (IBC)[13] which serves as both a tritium-breeding and energy-recovery blanket and as divertor and toroidal-field coils. The weak toroidal field at the edge of an RFP plasma makes liquid-metal cooling and the IBC attractive.

The fusion power core of TITAN-II is an aqueous "loop-in-pool" design with a dissolved $LiNO_3$ salt (5 *at.*% lithium) as the breeder. The structural material is 9C ferritic steel[14], a reduced-activation and high-strength alloy. The first-wall and blanket lobes are integrated and contain the pressurized coolant at 12 MPa. The blanket zone contains beryllium rods as neutron multiplier with 9C ferritic-steel cladding. The FPC and the entire primary loop of TITAN-II is submerged in a pool of low-temperature, low-pressure water. In the case of a major break in the primary-coolant pipes, the cold pool would absorb the thermal and afterheat energy from the hot loop. Calculations show that the pool would remain at a sufficiently low temperature to prevent the release of tritium and other radioactivity in the primary coolant system.

Operational (maintenance and availability), safety and environmental issues have been taken into account throughout both TITAN designs. The small size and mass of the TITAN designs permits "single-piece" maintenance procedures whereby the complete reactor torus is replaced as a single unit. The FPC is made of a few factory-fabricated pieces that are assembled on-site into a single torus, and tested to fully operational conditions before committment to nuclear service. This unique single-piece maintenance procedure is expected to increase the plant availability because: a) the period of down time resulting from scheduled and unscheduled FPC repairs is reduced; b) the FPC reliability is improved because of factory fabrication and integrated FPC pretesting on-site; c) no adverse effect is expected from the interaction of newly replaced materials operating in parallel to radiation-damaged materials; and d) it is possible to modify the FPC design continually to enhance the reactor performance as new technological developments emerge throughout the life of the plant.

Safety analyses have been performed in key areas: a) loss-of-flow (LOFA) and loss-of-coolant (LOCA) accidents; b) lithium fires for TITAN-I; c) radioactivity inventory; and d) site-boundary dose in case of major accidents. Both designs are configured to enhance their safety performance. Low-activation, low-after-heat alloys are used as the structural material throughout the FPC in order to minimize the peak temperature during LOFA and LOCA, while permitting near-surface disposal of waste under U. S. regulations. The safety analyses indicate that both TITAN designs would be passively safe, without reliance on any active safety systems.

3. PHYSICS/ENGINEERING REQUIREMENTS

A major objective of the TITAN study has been the identification and quantification of critical physics and engineering issues for high-MPD RFP reactors. In addition to achieving the necessary confinement and plasma beta, a high-MPD reactor plasma must operate a) with a highly radiative plasma core; b) with toroidal divertors; and c) at steady state. The foremost critical engineering subsystems are the in-vessel components that are subject to high surface heat flux.

3.1. Confinement

In most fusion reactor designs, for a given net-electric output and ignition condition, optimization of the neutron wall loading (corresponding to minimum COE) results in an "optimum" economic confinement time, which then must be compared with the physics predictions. The value of the global confinement time for TITAN design points is about 0.2 s ($\chi_E = 3r_p^2/16\tau_E = 0.34$ m^2/s) at a poloidal beta, β_θ, of 0.22.

The value chosen for β_θ is close to present-day experimental achievements. If this level of β_θ cannot be realized for a reactor-grade plasma, the plasma current in TITAN plasma would have to be increased. The main cost penalties would then be more expensive start-up equipement, equilibrium-field (EF) coils, and a larger recirculating power required by the current drive system. Parametric studies predict a 8% cost penalty if β_θ were limited to 0.15 and a 17% cost penalty for $\beta_\theta = 0.1$.

The available data on energy confinement time, τ_E, from experiments covering a limited range of both plasma current, I_ϕ, and minor radius, r_p, can be fitted to an expression of the form

$$\tau_E = C_\nu \, I_\phi^\nu \, r_p^2 \, f(\beta_\theta) \tag{1}$$

with $\nu \sim 1.0$. A similar form for τ_E with $\nu = 1.5$ can be derived from a simplified energy balance for an ohmic and non-radiating plasma, assuming constant β_θ and $I_\phi/\pi r_p^2 n$. Until detailed data on plasma profiles are avaliable, this empirical formula for τ_E is the best available. Figure 2 shows the variation of the Lawson parameter, $n\tau_E$ with the plasma current[15] for several RFP experiments[16–21]. The

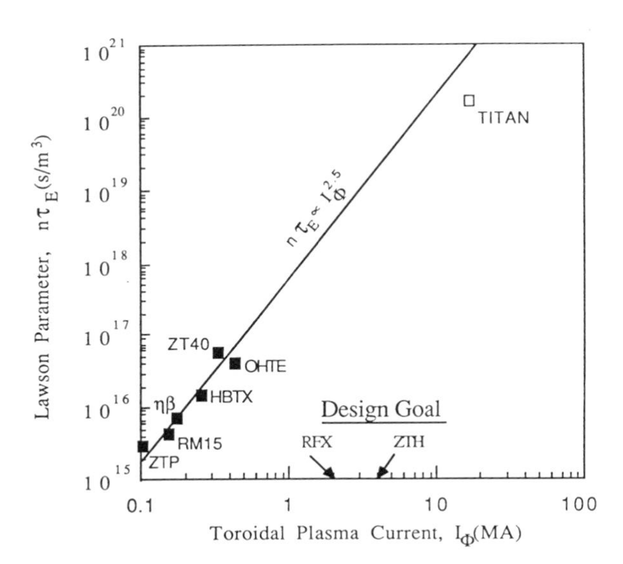

Figure 2. The Lawson parameter, $n\tau_E$, as a function of plasma current with data from several RFP experiments[15.-20] The TITAN design point and the expected operational points of the next generation RFP experiment, ZTH[12] and RFX[22] are also shown.

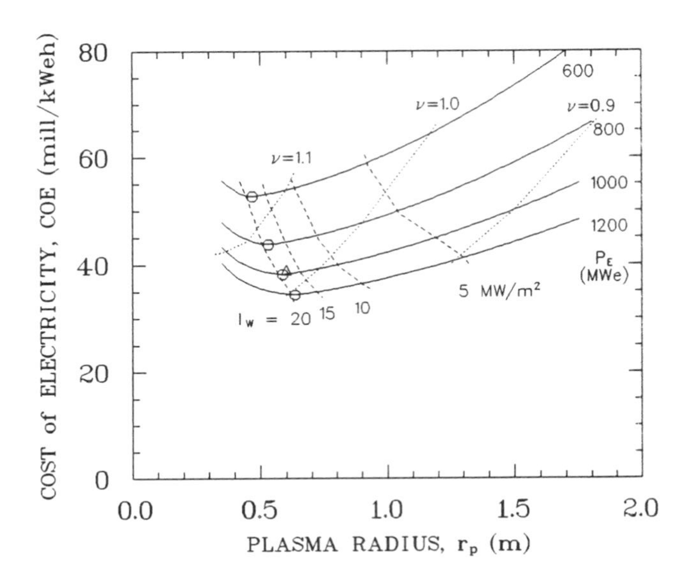

Figure 3. Variation of cost of electricity, COE, with the neutron wall loading for high-power density RFP reactors with a range of net output power. Contours of constant ν ($\tau_E \sim I_\phi^\nu$) are also shown.

TITAN minimum-COE design point is also indicated and shows the large degree of extrapolation from present-day achievements. The next generation of RFP experiments, ZTH[12] and RFX[22], are expected to extend the RFP data base closer to reactor-relevant regimes of operation (Fig. 2).

Given the uncertainty in achievable confinement time, a parametric survey of the impact of the confinement time on the reactor cost was performed. Figure 3 shows the COE for compact RFP reactors with different net electric output as a function of plasma size, r_p, or neutron wall loading, I_w. Also shown are dotted contours of the required confinement time, expressed in terms constant ν of Eq. 1. For TITAN-class reactors (1000 MWe and 10-20 MW/m^2 of neutron wall loading), roughly linear scaling of confinement with current is required. If this level of confinement cannot be achieved, the minimum-COE design points become inaccessible; the reactor design point would have to move to machines with larger plasma size and lower neutron wall loading resulting in an increase in cost.

The scaling of plasma pressure with current that is observed in RFP experiments suggests a beta-limited confinement, wherein, the level of intrinsic transport seems to adjust through changing MHD activity if other loss channels (e. g., radiation) become available[7,11,12]. A preliminary test of this hypothesis was made on ZT-40M by adding trace quantities of krypton impurity to the discharge[11,12]. The plasma beta remained relatively constant, and the global energy confinement remained relatively unaffected for radiation fractions of up to 90% of the total ohmic input power. This behavior should be contrasted with that from tokamaks, where increases in radiation decreases the global confinement time, ultimately leading to a plasma disruption. The TITAN designs take advantage of this experimental result and operate with an intensely radiating plasma, achieved by adding a trace of Xe impurities. As a result, 70% of the plasma power (i. e., ohmic and alpha-particle power) is radiated from the core plasma. An additional 25% of the plasma power is radiated from the edge-plasma that flows to the divertor, leaving only a small amount of charged-particle power incident on the divertor target plates. Operation with highly-radiative plasmas is central to achieving high-power-density RFP reactors, and this important issue should be resolved in the next generation of RFP experiments[12,22].

3.2. Impurity Control

The impurity control system of the compact reactors represent the most challenging and constraining subsystem of the design because of the large particle and heat fluxes incident on the in-vessel components. Both poloidal pump-limiters and toroidal divertors require that the majority of the plasma energy be radiated. Poloidal pump-limiters would suffer from the possibility of serious erosion of the limiter blades (and probably the first wall). The TITAN study, therefore, adopted three "open-geometry" toroidal-field divertors which together with the intensive radiation cooling of the core and edge-plasma leads to a high-recycling divertor with high density and low temperature near the divertor target and the first wall. Operation with a highly radiative plasma reduces the main functions of the divertor system to the maintenance of a dense edge plasma (to reduce sputtering and erosion of the first wall) and to particle exhaust.

The local ripple in the toroidal field caused by the divertor coils is large. Detailed magnetic design analysis of the divertor coils was performed to ensure that the size of the associated magnetic islands is a small fraction of the distance between the reversal surface and the divertor separatrix. The open geometry of the divertor system was exploited to expand the field lines near the divertor plate in order to reduce the charged-particle and heat fluxes. Detailed analysis of edge-plasma performance was made by a combination of core-plasma transport, edge-plasma transport, and Monte Carlo neutral-atom transport codes to develop a self-consistent picture of plasma behavior. These analyses were coupled with detailed thermal-hydraulic and stress calculations to find the optimum target shape and to complete the engineering design.

Since the divertor performance can strongly influence the overall design, an extensive parametric survey of the TITAN divertor system for a range of conditions was performed. First, increasing the number of divertor modules (i. e., increasing the total surface area of divertor target plates) does not substantially reduce the heat flux on the target plates. As the number of divertors is increased, the distance between the divertor modules (and the field-line connection length) is reduced, the edge-plasma parameters change, and a slightly warmer, less-radiative edge-plasma results. The radial decay lengths for both density and temperature become smaller. The peak heat flux on the divertor plates, therefore, is only slightly reduced at the cost of increased recirculating power needed to operate the additional divertor coils. Second, operation of the divertor system at the lower neutron wall loadings in the TITAN design window (10-12 MW/m^2) was investigated. With the same radiation fraction for the core plasma, the heat flux to the divertors would be reduced by about 20%. In principle, operation at the lower end of the design window would neither greatly ease the heat fluxes nor change the requirement for impurity injection and a highly-radiative plasma.

It appears that high-recycling toroidal divertors with open-geometry provide the most viable option for the impurity control system in compact RFPs. However, there are no experimental data either on the physics of toroidal-field divertors or the impact of the magnetic separatrix on RFP confinement. Physics and operational issues of toroidal-field divertors should receive greater emphasis in plans for the next generation of RFP experiments[12,22].

3.3. Current Drive

Parametric studies strongly suggest that compact RFPs must operate at steady state. Specifically, pulsed operation would lead to insurmountable energy-balance and engineering problems, given the start-up power requirements (maximum of 500 MWe for TITAN), the plasma stored energy ($\sim$ 2 GJ), very large heat fluxes to the in-vessel components, and rather short plasma L/R time (100-200 s). On the other hand, the requirements for the impurity control system, namely high density and low temperature operation with a highly-radiative plasma and increased Z_{eff}, makes the design of the current drive system more difficult. The TITAN reactors utilizes an oscillating-field current-drive (OFCD) system[9,10]. Helicity injection by OFCD relies on the natural relaxation process which maintains the RFP profiles[23], and is predicted to be efficient. In this scheme, the nonlinear coupling between toroidal and poloidal fluxes through the RFP natural relaxation process is used to rectify these oscillations, resulting in net plasma current. Even though the magnitude of oscillations in the plasma current and toroidal flux is small ($\delta I_\phi / I_\phi \simeq 0.02$, $\delta\Phi/\Phi \simeq 0.035$ for the TITAN design), the reactive power crossing the plasma surface (i. e., Poynting vectors) and, to a lesser extent, at the OFCD power supply terminals can be large, requiring a high gain ($Q \sim 100$) power supply and driver circuit. Also, the components of FPC should be carefully designed in order to minimize the amount of the eddy currents by introducing insulating gaps in the first wall and the blanket. These eddy current decrease the driver efficiency because larger magnetic fluxes must be produced by the driver coils for given fluxes crossing the plasma surface. In addition, eddy currents change the phase and shape of flux modulations and lead to a decrease in the overall driver efficiency.

Experimental results for OFCD are inconclusive, and testing of this concept in hotter (i. e., less resistive) plasmas must await the next generation of RFP experiments[12,22]. In order to quantify the technological features of the OFCD system, detailed plasma/circuit simulations of the OFCD cycle were performed. The predicted TITAN current-drive efficiency is 0.3 A/W delivered to the power supply for a high-gain driver circuit ($Q \simeq 100$). Parametric studies of the economic impact of the driver-circuit gain shows that the COE remains insensitive for $Q > 100$ and increases dramatically for $Q < 20$.

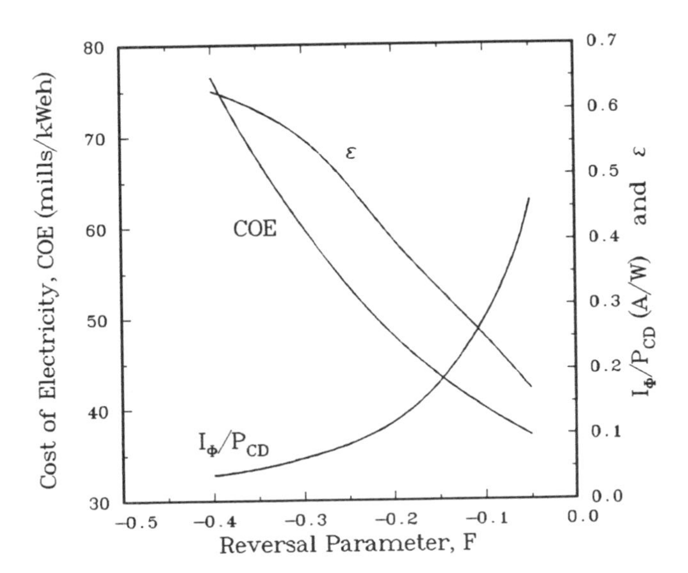

Figure 4. The variation of the cost of electricity, COE, the recirculating power fraction, ϵ, and the current drive efficiency, I_ϕ/P_{CD}, as a function of the reversal parameter, F.

The performance of the current-drive system strongly depends on plasma parameters. A particularly important variable is the reversal parameter, F. Deeper reversal (more negative F values) are desirable since the OFCD design for TITAN operates close to loss of reversal ($F \rightarrow 0$). Increasing the reversal, however, increases the reactive power in the toroidal-field circuit, already requiring the highest power in the OFCD system. The steady-state power consumption in the toroidal-field and divertor coils also increases for deeper levels of field reversal. Figure 4 shows the economic impact of deeper reversals. The increased costs are caused mainly by the sharp increase in the recirculating power for deeper reversals. It should be mentioned that the higher toroidal field at the first wall required for plasmas with deeper reversal would have adverse effects on the thermal-hydraulics performance of the liquid-metal-cooled TITAN-I design (e. g., increased pressure drops and higher pumping power); these effects are not included in Fig. 4.

3.4. First-Wall Design

Engineering design of the first wall for compact reactors is another critical issue. Generally, MHD effects precluded the use of liquid-metal coolants for high-heat-flux components (e. g., in tokamak designs). But the weak toroidal field at the edge of an RFP plasma is favorable for high-velocity liquid-metal cooling. In the TITAN-I design, the first wall and blanket consist of single-pass, poloidal flows aligned with the dominant poloidal field. Other major features include: a) separation of the first wall and blanket coolant circuits to allow a lower exit temperature for the first wall coolant; and b) use of MHD turbulent-flow heat transfer at the first wall made possible by the low magnetic interaction parameter (ratio of Nusselt number to the magnetic Reynolds number). The TITAN-I first wall thermal-hydraulic design can accommodate up to 5 MW/m^2 of heat flux with resonable MHD pressure drops (10 MPa in the first wall, 3 MPa in the blanket and 12 MPa in the divertor cooling circuit), a high thermal cycle efficiency (44%) and a modest pumping power (45 MWe)[1].

The MHD heat-transfer rate and associated pressure drops were calculated by extrapolating theoretical and empirical corelations to the regime of interest using dimensionless parameters such as the Nusselt number[24]. Confirmation of TITAN-I first wall design, however, requires experimental data, especially for high-velocity coolant in a relatively weak magnetic field, and computer simulations using multi-dimensional, liquid-metal MHD codes.

The parametric studies of the thermal-hydraulic designs for neutron wall loading in the range 10-20 MW/m^2 show that turbulent-flow heat transfer has to be utilized for the entire range of the design. Laminar-flow heat transfer would limit the maximum heat flux for TITAN-I configuration to about 2.5 MW/m^2, adequate only for the lower end of the design window. For lower neutron wall loadings, however, the first-wall coolant velocity can be somewhat reduced and/or the coolant exit temperature can be increased.

The TITAN-II first wall is cooled by an aqueous solution of $LiNO_3$ salt. Given the high heat flux on the first wall, sub-cooled flow boiling (SFB) heat transfer should be utilized. Semi-empirical correlations for SFB heat transfer for pure water[25] were modified to take into account the thermo-physical properties of TITAN-II $LiNO_3$ salt-solution coolant[1]. Any application of SFB heat transfer must ensure that the maximum heat flux is less than the critical heat flux by a certain safety factor. Semi-empirical corelations for critical heat flux for pure water exist[26] and are extrapolated to the $LiNO_3$ salt solution. The TITAN-II first-wall design can accommodate up to 5 MW/m^2 of heat flux, but experimental data on SFB heat transfer and critical heat fluxes for salt solutions are required to confirm these results. The use of SFB heat transfer is required even for operation of the TITAN-II design at lower neutron wall loadings (10 MW/m^2), albeit with a larger safety margin with respect to the critical-heat-flux.

3.5. Safety and Maintenance

The TITAN designs are configured to enhance both safety and operational (maintenance) characteristics. The reactors at the lower end of the TITAN design window (10-12 MW/m^2) may be too large and heavy for single-piece maintenance, although the MPD is still very high. The safety consequences of loss-of-flow and loss-of-coolant accidents, on the other hand, are directly related to the level of the decay-after-heat in the first wall and to the passive-cooling capability of the FPC design through radiation and conduction paths. The concentration of short-lived radioisotopes in the first wall is a linear function of the neutron wall loading and is not a function of the neutron fluence. As such, reactors with higher wall loading will have a higher level of decay after-heat for a given structural material composition and will require a more careful design to achieve a high Level of Safety Assurance[8].

4. CONCLUSIONS

The results from the TITAN study supports the technical feasibility, economic incentive and operational attractiveness of high-mass-power-density RFP fusion reactors. Parametric surveys of cost, physics and engineering performance of the reactor components show that the attributes of these systems are retained for neutron wall loadings ranging from 10 to 20 MW/m^2. Reactors operating in this "design window" are physically small, and the associated "compactness" results in improved economics. The relatively low cost of the fusion power core (<10% of the overall plant cost for TITAN) makes the economics of the reactor less sensitive to changes in the plasma performance or unit costs of FPC components. It should also lead to a more flexible and less-expensive development program for RFP reactors. A unique single-piece maintenance procedure can be utilized because of the relatively small size and mass of the FPC. The TITAN study also shows that with proper choice of material and FPC configuration, compact reactors can be made passively safe and their competitive economics can be realized without sacrificing the potentially attractive safety and environmental features of fusion.

Of course, the road towards compact RFP reactors contain major challenges and uncertainties, and many critical issues must be resolved. The next generation of reversed-field-pinch experiments, ZTH[12] and RFX[22], with hotter plasmas will substantially extend the RFP data base closer to reactor relevant regimes (Fig. 2). The key physics issues of high-MPD RFP reactors (particularly, operation with highly-radiative plasma, current drive and toroidal divertors) requires greater attention in the next generation of RFP experiments and designs.

REFERENCES

1. F. Najmabadi, R. W. Conn, *et al.*, "The TITAN Reversed-Field Pinch Fusion Reactor Study; The Final Report," UCLA-PPG-1200, Joint report of University of California Los Angeles, GA Technologies, Los Alamos National Laboratory and Rensselaer Polytechnic Institute (1988).
2. R. W. Conn, F. Najmabadi and the TITAN Research Group, *IEEE Proc. of 12th Symposium on Fusion Engineering, Monterey, California, 1987* **1** (1987) 503.
3. C. C. Baker, *et al.*, Argonne National Laboratory report ANL/FPP-80-1 (1980).
4. B. G. Logan, *et al.*, Lawrence Livermore National Laboratory report, UCRL-53480 (1984).
5. D. Ehst, *et al.*, Argonne national Laboratory report ANL/FPP/86-1 (1987).
6. R. A. Krakowski and J. G. Delene, *J. Fusion Energy* **7** (1988) 49.
7. H. A. Bodin, R. A. Krakowski, and O. Ortolani, *Fusion Technology* **10** (1986) 307.
8. J. P. Holdren, *et al.*, *Fusion Technology* **13** (1988) 7. Also, J. P. Holdren, *et al.*, Lawrence Livermore National Laboratory report, UCRL-53766 (1987).
9. M. K. Bevir and J. W. Gray, Proc. of RFP Theory Workshop, Los Alamos, NM, U.S.A. (1980), Los Alamos National Laboratory report LA-8944-C (1982) 176.
10. K. F. Schoenberg, *et al.*, Los Alamos National Laboratory report LA-UR-88-122 (1988). Also, R. A. Scardovelli, *et al.*, Los Alamos National Laboratory report LA-UR-2802 (1988).
11. M. M. Pickrell, *et al.*, "Evidence for a Poloidal Beta Limit on ZT-40M," *Bull. Am. Phys. Soc.* **29** (1984) 1403.
12. P. Thullen and K. Schoenberg (Eds.), Los Alamos National Laboratory Report LA-UR-84-2602 (1984) 26-28.
13. D. Steiner, *et al.*, *Fusion Technology* **7** (1985) 66.
14. D. S. Gelles, *et al.*, UCLA report, UCLA-PPG-1049, (1987).
15. J. N. DiMarco, Private Communications (1988).
16. T. Shimada, *et al.*, *IAEA Proc. of 11th Int. Conf. on Plasma Phys. and Controlled Nucl. Fusion Research, Kyoto, 1986* **2** (1987) 453. Also, Y. Hirano, *et al.*, *IAEA Proc. of 10th Int. Conf. on Plasma Phys. and Controlled Nucl. Fusion Research, London, 1984* **2** (1985) 475.
17. B. Alper and S. Matini, Istituto Gas Ionizzati report IGI 85/04 (1985). Also, B. Alper, *et al.*, Proc. of 12th European Conf. on Controlled Fusion and Plasma Phys., Budapest, 1985, *European Phys. Soc.* **1** (1985) 578.
18. T. Tamano, *et al.*, *IAEA Proc. of 10th Int. Conf. on Plasma Phys. and Controlled Nucl. Fusion Research, London, 1984* **2** (1985) 431.
19. D. A. Baker, *et al.*, *IAEA Proc. of 10th Int. Conf. on Plasma Phys. and Controlled Nucl. Fusion Research, London, 1984* **2** (1985) 439. Also, P. G. Weber, *et al.*, Proc. of 12th European Conf. on Controlled Fusion and Plasma Phys., Budapest, 1985, *European Phys. Soc.* **1** (1985) 570.
20. M. M. Pickrell, *et al.*, *Bull. Am. Phys. Soc.* (1988).
21. H. A. Bodin, *Plasma Phys. and Controlled Fusion* **29** (1987) 1297.
22. G. Malesani and G. Rostagni, *Proc. of 14th Symposium on Fusion Technology, Avignon* (1986) 173.
23. J. B. Taylor, *Rev. Mod. Phys.* **58** (1986) 741. Also, J. B. Taylor, *Phys. Rev. Lett.* **33** (1974) 1139.
24. D. S. Kovner, *et al.*, *Magnitnaya Gidrodinamika* **7** (1966).
25. W. H. Jens and P. A. Lottes, Argonne National Laboratory (USAEC) report, ANL-6675 (1962).
26. R. D. Boyd, *et al.*, Sandia National Laboratory report SAND-84-0159 (1985).

FUSION TECHNOLOGY 1988
A.M. Van Ingen, A. Nijsen-Vis, H.T. Klippel (editors)

NAVIGATOR (FER) --- Negative-Ion-Grounded Advanced Tokamak Reactor ---

S. Yamamoto, Y. Ohara, M. Azumi, N. Fujisawa, T. Horie, H. Iida, S. Nishio, Y. Seki, Y. Shimomura, M. Sugihara, S. Tanaka, K. Tachikawa, K. Tani, Y. Okumura, K. Shibanuma, S. Matsuda and FER Team

Japan Atomic Energy Research Institute, Naka Fusion Research Establishment, Ibaraki, Japan. 311-01

The NAVIGATOR concept is based on the negative-ion-grounded 500 keV/20 MW neutral beam injection system (NBI system), which has been proposed and studied at JAERI. The NAVIGATOR concept contains two categories; one is the NAVIGATOR machine as a tokamak reactor, and the other is the NAVIGATOR philosophy as a guiding principle in fusion research. The NAVIGATOR machine implies an NBI heated and full inductive ramped-up reactor. The NAVIGATOR concept should be applied in a phased approach to and beyond the operating goal for the FER (Fusion Experimental Reactor, the next generation tokamak machine in Japan). The mission of the FER is to realize self-ignition and a long controlled burn of about 800 seconds and to develop and test fusion technologies, including the tritium fuel cycle, superconducting magnet, remote maintenance and breeding blanket test modules.

The NAVIGATOR concept is composed of three major elements, that is, reliable operation scenarios, reliable maintenability and sufficient flexibility of the reactor. The NAVIGATOR concept well supports the ideas of phased operation and phased construction of the FER, which will result in the reduction of technological risk.

1. INTRODUCTION

Recently, the scientific feasibility of the fusion reactor has been successfully demonstrated by the JT-60. The result is encouraging in promoting the realization of the fusion reactor. Next, the engineering feasibility of nuclear fusion as a power reactor should be clarified as soon as possible by the next generation tokamak machine. In order to make the design of the next generation machine attractive and cost effective, sufficient innovative ideas should be adopted. At the same time, the design should be based on the present databases. In order to reconcile the difference between innovative ideas and realistic ideas, the NAVIGATOR concept has been proposed at JAERI[1]. The NAVIGATOR concept is a guiding principle in fusion research and should be applied in a phased approach to and beyond the operating goal for the FER.

2. The NAVIGATOR concept

The NAVIGATOR concept is composed of three major elements, that is, reliable operation scenarios, reliable maintenability and sufficient flexibility of the reactor. In this paper, the NAVIGATOR machine is shown to satisfy the above three elements.

3. The NAVIGATOR machine

NBI, LHRF, ICRF and ECRF are candidates for the heating facility to realize a self-ignited plasma condition. H-mode confinement is obtained in all the NBI heated tokamak with a separatrix magnetic surface. The overall efficiency of the negative-ion-grounded NBI system is expected to become comparable with that of ICRF and much better than those of other RF facilities. This value can be improved further by employing a plasma neutralizer and/or an energy convertor system. The progress in the R&D of the negative ion source is remarkable[2]. The negative-ion-grounded NBI system has many functions of heating, current ramp-up, transformer recharge, steady state and current profile control[1,3]. However, in order to support the reliable operation scenarios, at first, only the heating function of the NBI system is demanded in the NAVIGATOR machine. That is, the NAVIGATOR machine implies an NBI heated and full-inductive ramped-up reactor.

4. Negative-Ion-Grounded Advanced Tokamak Reactor.

4.1. Reliable maintenability

The top and side views of a negative-ion-grounded advanced tokamak reactor are shown in Fig. 1. Two beam line systems are enough for realizing a self-ignition plasma in the H-mode confinement case. Three beam line systems are prepared for the FER, considering flexibility in operation scenarios and redundancy. The NBI system consists of a beam line, a 500 keV power supply system, an evacuating system including a tritium handling system, a cooling water circulating system and auxiliary sub-systems[4]. The beam line consists of an intense negative ion source, a beam profile controller, a long gas neutralizer, an ion beam dump, a neutron shutter, a drift duct including an injection port of the torus and a shine-through armor plate against the neutral beams into the torus. In both the ion source room and the beam dump room, cryopanels are mounted. The neutralizer and the ion source room are shielded magnetically, and the beam line is covered with neutron shields. The D^- ions are produced by a volume production method and extracted by DC acceleration.

The most important feature of the NBI system is characterized by a long, slender neutralizer, which is 0.3 meter wide and 24 meters long. The required beam divergence is 0.3 deg or 5 mrad. Such a low beam divergence can be realized by using a volume D^- ion source. The long, slender neutralizer concept results in a considerable reduction in required pumping speed and in neutron damage of the ion sources. The total pumping speed is 850 m^3/s for 20 MW injection. From the evaluation of the backstream of tritium molecules from the torus into the ion source room, the gas handling system is selected. The backstream into the ion source room is very low. Thus, the NBI gas from ion source is recirculated until the buildup of tritium becomes a problem. A purification system is used for impurities in the deuterium gas recirculating system. After a number of uses, the NBI gas from the ion source is transferred to the main reactor fuel recycle system or detritiation system. The neutron flux above 1 MeV is estimated to be 10^7 $n/cm^2 \cdot s$ at the position of the ion source insulator. The database on which to judge the performance limits of the NBI components, especially for organic materials, is poor. Concerning mechanical strength, the radiation exposure limit under gamma-ray irradiation for epoxy may be supposed to be 1 MGy (10^8rad) or less. Then, under neutron irradiation, assuming that the irradiation limit value for mechanical life time is 10^{16} n/cm^2, it may be expected to have a life time (operation time) of the magnitude of 10^9 seconds. But the limit value for electrical resistivity has a tendency to be reduced by a few orders of magnitude. Thus, it is desirable that the ion source insulator has enough distance from the beam axis so that it will have a moderate life time. For an operation time of a magnitude of 10^7 seconds (for several years), it is expected that it will be necessary to exchange the ion source insulator only a few times. The thickness of the neutron shield for the NBI system has been estimated to satisfy a dose value outside the shield of less than 2.5 mrem/hour one day after shutdown following a 2-year full power operation.

Another feature is that the poloidal magnetic field from the tokamak is used to deflect the unneutralized negative ions and reionized positive ions toward the ion dump near the injection port. Therefore, no bending magnet, which takes much space, is used in this design. The space for the ion dump room is designed to be narrow in order to make the reactor maintenance easier. In the FER case, the bore of the TF coils, set by the ripple condition of 0.75% enables a part of the first wall (removable shield) to be withdrawn by a radial straight-line motion through the access port, as shown in Fig. 2. The divertor is also segmented into 12 sectors toroidally and withdrawn by a single straight-line radial motion. The divertor segment is withdrawn without interference with the NBI system. The movable shield with the drift duct is also withdrawn, after taking off only a part of the neutron shutter, as shown in Fig. 1. Each shine-through armor plate is installed on the same movable shield for another NBI injection port, which keeps the number of ports employed by such NBI systems minimum. The shine-through armor plate is extractable while keeping vacuum condition of the reactor vacuum chamber, as shown in Fig. 1.

The choice of the first wall material is one of the key factor in the design of the next machine. Recent experimental results show that a low-Z material is necessary for the production of a high performance plasma. However, for example, the erosion width of the graphite divertor plate of the FER is estimated to become several meters during the reactor life. In order to overcome such a large erosion and offer reliable maintenability, in-situ and vacuum repair of the divertor surface is adopted. If these design concepts are adopted, the NBI reactor will become more reliable, have much longer lifetime and will be more easily maintained.

4.2. Phased operation, phased construction and sufficient flexibility

In order to realize a disruption free and self-ignited controlled long burn plasma, phased operation and phased construction are attached great importance in the design of the FER. The phased operation consists of three phases; the first is a cold phase with hydrogen and/or deuterium discharges, in which a person can enter into the vacuum vessel of the FER, the second is a semi-hot phase with deuterium and tritium discharges in order to realize very short self-ignited plasma, in which a person can enter into the vacuum vessel after many hydrogen and/or deuterium discharges and the last is a hot phase with deuterium and tritium discharges in order to make a mission-oriented long self-ignited plasma.

In the cold phase of the NAVIGATOR machine, a simple vacuum vessel without a neutron shield, as shown in Fig. 3, will be used. The vacuum chamber may be easily exchanged without interference with the heating facility, if the negative-ion-grounded NBI system is used. The vacuum of the NBI system is maintained during the exchange of the vacuum vessel, and full power injection is possible just after the exchange. Thus, the NAVIGATOR machine involves the present large tokamak machine in itself and has sufficient flexibility. An NBI heated plasma offers a good and reliable target plasma for the RF. When well-accustomed diagnostics and simple RF launchers without a neutron shield, developed in JT-60 or JT-60 up-grade, are set up on the flexible reactor, the plasma in the cold-phase and antenna-plasma coupling will be able to be well examined. In the NAVIGATOR machine, the reference plasma configuration is a single null divertor configuration with a medium elongation of around 1.7 and triangularity of around 0.2. The vertical positional instability is expected to be well stabilized. In the medium elongation case, the poloidal field power supply for horizontal access is possible to be smaller than that for vertical access. During the cold phase, plasma parameters around the reference value will be examined, using a high elongation plasma of up to 2.0, high triangularity up to 0.3, a larger plasma size and higher plasma current. Through these examinations, disruption free plasma is expected in the hot phase. The NAVIGATOR cold and semi-hot phase flexible

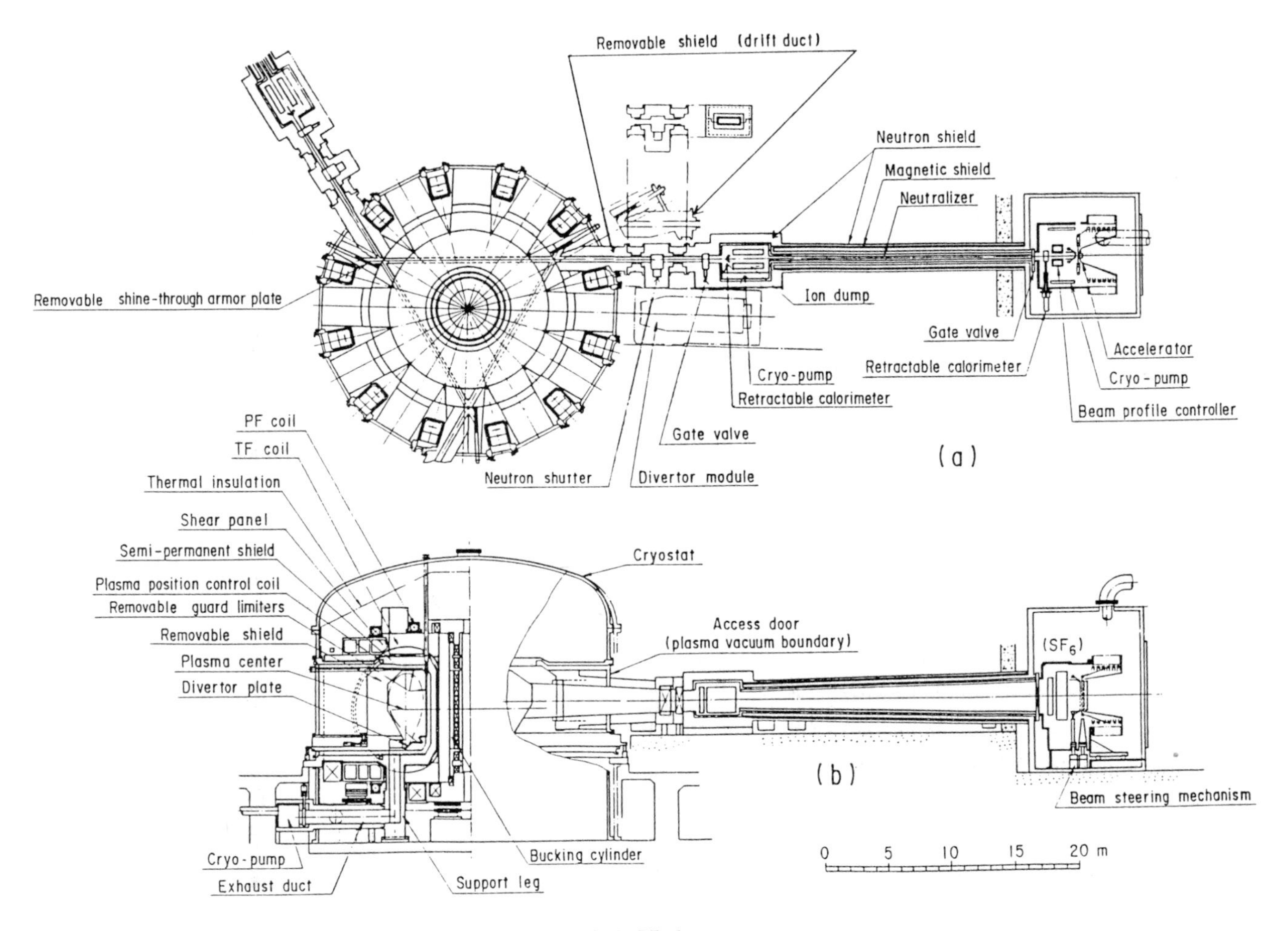

FIGURE 1
Top and side view of the Negative-Ion-Grounded Advanced Tokamak Reactor

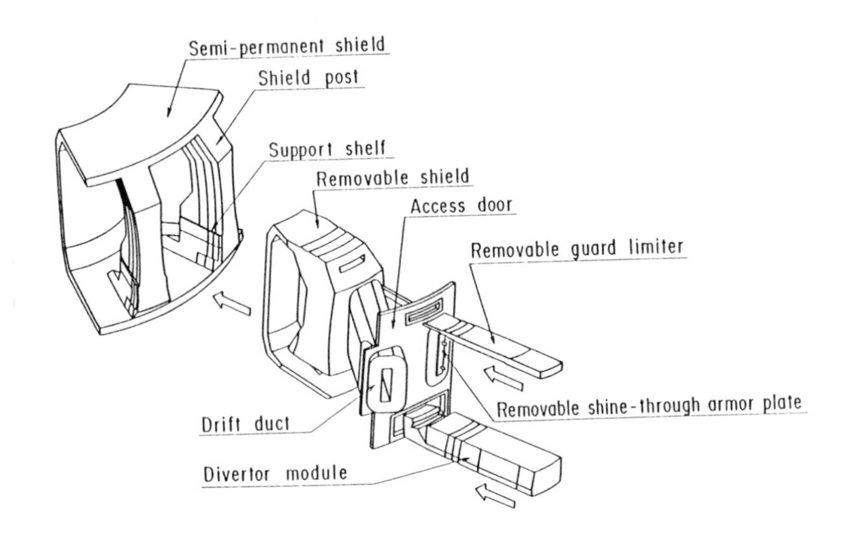

FIGURE 2
Torus segmentation and structural concept of injection port, including the access door

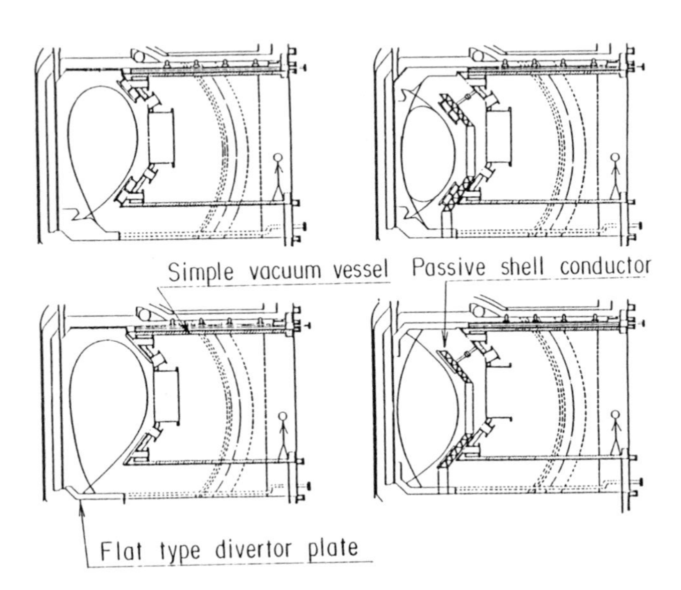

FIGURE 3
Simple vacuum vessel without neutron shield during the cold phase. During the cold phase, various types of plasma configurations are obtained by exchanging the position of the passive shell conductor and the configuration of the divertor plate, including a flat type divertor plate.

reactor is used for the NAVIGATOR hot phase mission oriented reactor in order to make the plasma well-accustomed, except for thermal instability and to confirm the specification for the reactor components. We can obtain information concerning the slowing down process and the ripple-loss mechanism, for high energy hydrogen and deuterium particles, in the cold phase. For example, by using the simultaneous injection of 500 keV NBI and ICRF (second harmonic ICRF heating or minority heating), information on alpha particles in the semi-hot phase and estimation of the plasma behavior of the hot phase may be obtained by using both the databases for the cold and semi-hot phases and simulation codes. Therefore, it may not be necessary any longer that the mechanical configuration of the NAVIGATOR machine be changed, after going into the hot operation. Even if the ignition condition were not realized, the NAVIGATOR concept will satisfy the technological mission of the next generation tokamak machines and will supply databases for the DEMO reactor. Well-considered phased operation and phased construction will result in the reduction of technological risk and initial cost and in the enhancement of cost benefit.

5. CONCLUSIONS

The NAVIGATOR concept has been proposed in this paper. The NAVIGATOR concept is composed of three major elements, that is, reliable operation scenarios, reliable maintenability and sufficient flexibility of the reactor. The negative-ion-grounded advanced tokamak reactor based on a 500 keV/20 MW neutral beam injection system (Now, we have just started to design a 1 MeV/25 MW neutral beam injection system in order to increase the current drive efficiency) satisfies the above three components and well supports the ideas of phased operation and phased construction of the FER.

ACKNOWLEDGEMENTS

Thanks are due to S. Shimamoto, Y. Tanaka, S. Tamura, T. Iijima, M. Tanaka, M. Yoshikawa, K. Tomabechi, Y. Iso and S. Mori for support and encouragement.

References

1. S. Yamamoto et al., JAERI-M 88-086.

2. Y. Okumura et al., Proc. 4th Int. Symp. on the Production and Neutralization of Negative Ion and Beams, Brookhaven, 1986.

3. S. Yamamoto et al., in Plasma Physics and Controlled Nuclear Fusion Research (Proc. 11th Int. Conf. Kyoto, 1986), Vol. 3, IAEA, Vienna (1987) 267.

4. Y. Ohara et al., IAEA Technical Committee Meeting on Negative Ion Beam Heating, Culham Lab., July (1987).

FUSION TECHNOLOGY 1988
A.M. Van Ingen, A. Nijsen-Vis, H.T. Klippel (editors)

SCALING LAWS AND DESIGN CRITERIA FOR A PRESS - SUPPORTED COMPACT TOKAMAK

R.Albanese [1],E. Coccorese [2],F. Esposito [3],L. De Menna [4],R. Martone [1],G. Miano [4],G. Rubinacci [1],A. Sestero [5].

[1] Istituto di Ingegneria Elettronica, Università di Salerno, Italy.[2] Istituto di Ingegneria Civile ed Energetica, Università di Reggio Calabria, Italy.[3] Dipartimento di Fisica, Università di Napoli, Italy.[4] Dipartimento Elettrico per l'Energia, Università di Napoli, Italy.[5] Centro Ricerche Energia,ENEA, Frascati, Italy.

The paper collects preliminary engineering results of a study on Omitron like press-supported high field tokamak experiments. Attention is called on a particular invariance property of the equations describing the behavior of the system of conductors in the machines, such invariance leading to the recognition that families of homologous devices can be defined, the systems of conductors in any given family behaving all in a self-similar way. The present investigation shows that tokamak devices with a magnetic field of 17.5 Tesla in a configuration with a plasma aspect ratio of 2.5 can apparently be built, compatibly with all the mechanical and electrotechnical constraints.

1. INTRODUCTORY REMARKS

The fusion community is presently whitenessing a surge of interest for the high field tokamaks - the latter being held capable of producing high grade plasmas (possibly burning plasmas) with a moderate drain on financial resources. To really capitalize on the advantages of the high field, however, it would appear to be desirable to go beyond the magnetic field value of, say, 12 Tesla (on the plasma axis), which is the presumable limit of conventional toroidal magnet technology. Thus new ideas would seem to be called for. One such new concept has been previously introduced, which relies on an active support structure in the form of "press" - see the OMITRON proposal [1,2,3]. Such press concept has been subsequently called upon in the design of certain advanced machines (CIT, latest versions of IGNITOR), even though their target magnetic field values are possibly still within the limits of conventional magnet technology. The maximum performance capabilities of the press concept are however thought to be, in principle, considerably more rewarding[1,2,3]. Among other things, the apparent capability of bringing about plasma ignition purely by ohmic heating has the foremost appeal [4]. The latter goal is in fact what is being targeted by the recent IGNITEX proposal (which indeed appears to follow quite closely a number of prescriptions from the original OMITRON concept). Finally with on eye to future more exotic fusion applications - it is also appropriate here to record the circumstance that the burning of such advanced fusion fuels as D-^{3}He and catalyzed D-D has been deemed to be a non totally implausible goal for press-supported devices [4,5].

The key engineering novelty in the OMITRON magnet design is the application of a compensative mechanical load in active, controlled form - a solution which appears particularly well suited to meet the design needs of compact toroidal magnets for high-field fusion applications. We recall that in mechanical engineering compensative loads are usually brought about by way of positional constraints introduced in the mechanical structure at the moment of assembly (pre-loading procedures), with hence the consequence that the magnitude of the pre-load is determined once and for all from the beginning . Therefore, whenever the primary mechanical load (which one aims at compensating with the help of the pre-load) is a significantly variable function of time (as is, indeed, in the case of the magnets), the above circumstance already entails a less than optimal state of affairs. The situation becomes particularly unfortunate if, moreover, some parts of the mechanical structure are subjected to strong heating (as it is also the case for compact high-field magnets): indeed, the need to allow for a free expansion of the heated material (to the end of avoiding excessive thermal stresses) is clearly at odds with the need to physically constrain it for the purpose of applying the desired pre-load. As a further inconvenience, finally, it is to be born in mind that a mechanical load applied for very long times does actually induce some permanent deformation even if the produced stress level stays always below the yield value ("creep" phenomenon): this leads to a gradual deterioration of the pre-load initially built in, which in some cases - as, e.g., in the case of the construction of the heavily pre-loaded magnets which are here of concern - may turn out to be unacceptable.

As a way off the above difficulties, in the OMITRON proposal the suggestion was made of applying the required compensative load as a controlled function of time by way of a press. It was also further suggested to implement the latter in the form of an electromagnetic press, the latter having the advantage that an automatic, optimal programming of the compensative load with time can be simply obtained by placing the press coils in series with the main magnet coils. Finally, the prescription was given of filling up the configuration doughnut-hole completely with either conducting or structural material - which is an obvious stride toward robustness (in an intuitive way), but also (in a less intuitive way) allows for the reaching, locally, of conditions of quasi-isotropy of the stress tensor - the latter circumstance being quite favorable from the mechanical point of view.

In order to fully implement the last suggestion, one can hardly avoid the choice of placing the transformer coils inside the magnet toroidal bore. Such a solution is usually avoided in conventional tokamaks, due to the entailed greater difficulties in the assembly procedures: and indeed new schemes for assembling the magnet-transformer system have had to be figured out for the OMITRON proposal. It is however to be emphasized out that under almost any other respect the placing of the transformer inside the magnet toroidal bore is in fact advantageous - as in particular much more magnetic flux is thereby made available under the same constraints on stresses and heating in the transformer windings.

As next topic in this compact review of issues, we note that whenever in a project of engineering one has to draw perilously near the yield point of materials, one would of course like to find comfort in the practical testing of scaled-down models of the devices under concern. Now, it turns out, luckily - as pointed out already within the original OMITRON proposal - that all phenomena (stationary as well as transient) which are important in the behavior of magnets are in fact reproduced very precisely (in a self-similar fashion) in scaled-down models, provided only the time is scaled as the square of the linear dimensions. Due to the practical importance of such a statement, we have deemed appropriate to report a formal proof of it in Section 3 of the present contribution.

The advantages of operating with high-field tokamaks and the nature of the associated physical scenarios have been repeatedly discussed in the last decade - notably, above all, by B. Coppi and the other people who have contributed to the IGNITOR program. A compact account of such arguments (without aiming at completeness) will be presented, for the benefit of the readers, in Section 2 of the present contribution.

Finally as a mean to give the reader a tangible feeling of how large a design improvement is attempted in the OMITRON proposal, we draw attention to the fact that a reasonable mechanical factor of merit (MFM) for toroidal magnets - if any such one exists -, can be chosen to be the ratio of the maximum magnetic field pressure existing in the toroidal bore to the maximum Von Mises stress existing in the conductor. A glance at Table 1.1 shows that while the MFM's of a number of other important high field devices lie between the values 0.40 and 0.70, the OMITRON MFM exhibits a net leap forward to about the value 1.00.

TABLE 1.1 Values of MFM for various high field devices

	B_ϕ(axis) (Tesla)	B_ϕ(max) (Tesla)	σ_{VM}(max) (MPa)	MFM
FT	10.0	15.7	240	0.41
CIT	10.4	17.8	300	0.42
IGNITOR-L	12.4	20.4	300	0.55
ALCATOR -C	14.0	21.0	300	0.59
IGNITEX	20.0	30.0	540	0.67
OMITRON	17.5	35.0	490	1.00

2.REQUIREMENTS FOR PURELY OHMIC IGNITION

Let us briefly review the requirements for achieving ignition purely by ohmic heating in a tokamak experiment. To this end, we consider the energy balance of the plasma, keeping only such terms in the latter (for the sake of simplicity) which play crucial roles when approaching and first entering ignition conditions:

$$\frac{d(3nT)}{dt}=P_{OH}+P_{\alpha}-P_{BS}-P_{D} \qquad (2.1)$$

where t is the time, n is a suitably averaged value of the plasma density, T is a suitably averaged value of the plasma temperature, and P_{OH}, P_{α}, P_{BS}, P_D are suitably averaged power density terms corresponding respectively to the ohmic input, α-particle energy input, Bremsstrahlung loss, and diffusion loss. Most notably, the synchrotron radiation loss has been neglected in eq.(2.1), as such loss becomes important only at higher temperatures, well inside the ignited regime (at least for the high density approach to ignition with which we are here concerned).

For an ohmically heated device, the most critical time along the path to ignition is the crossing of the range of temperatures around the value T_o (4.3 keV) for which $P_\alpha = P_{BS}$: in fact in this range the term P_α cannot yet unleash its full effect, being substantially compensated by the P_{BS} term, while the ohmic heating term P_{OH} has already been considerably weakened due to the temperature increase. Indeed, in a successful ignition approach the curve T(t) of the temperature evolution with respect to time is expected to exhibit an inflection point (point of minimum of the first derivative with respect to time) when crossing some temperature value close to T_o - the rate of temperature increase first getting duller and duller due to the weakening of the ohmic heating, and then quickening up again, as the α-particle energy deposition becomes important. In the following, for the sake of simplicity, we shall actually draw attention to the specific case whereby the inflection point occurs exactly at the value T_o, the latter case being anyway well representative also of other nearby paths to ignition. In order to proceed, hence, it is convenient to spell out the specific temperature dependences that the various terms on the right hand side of eq.(2.1) exhibit in the vicinity of $T=T_o$ - assuming for simplicity that during the crossing of the critical range of temperatures around $T=T_o$ one maintains constant values of plasma density and plasma current density:

$$3n\frac{dT}{dt} = P^o_{OH}\,(T/T_o)^{-3/2} + P^o_\alpha\,(T/T_o)^4 - P^o_{BS}\,(T/T_o)^{1/2} + - P^o_D\,T/T_o \qquad (2.2)$$

where P^o_{OH}, P^o_α, P^o_{BS}, P^o_D are the values that the power densities P_{OH}, P_α, P_{BS}, P_D respectively assume for $T=T_o$. Note that the exponent "4" in the α-particle term is just an approximation (reasonably acceptable for the range of temperatures of concern), while the exponent "1" in the diffusive loss term results from the assumption of the so-called "Neo-Alcator" scaling for the associated energy confinement time (as appropriate for ohmically-heated tokamak plasmas). From eq.(2.2) one can also compute, recursively, d^2T/dt^2 and d^3T/dt^3 and from the requirement that dT/dt and d^2T/dt^2 be approximately zero for $T=T_o$ one obtains the approximate constraints :

$$P^o_{OH} = P^o_D, \qquad P^o_1 = \frac{5}{7} P^o_2, \qquad (2.3)$$

where we have introduced P^o_1 and P^o_2 defined as

$$P^o_1 = P^o_\alpha = P^o_{BS}, \quad P^o_2 = P^o_{OH} = P^o_D \qquad (2.4)$$

Note, finally, that by exploiting the results specified by eqs.(2.3) into the computed expression of d^3T/dt^3 one obtains that for $T=T_o$ the condition $d^3T/dt^3 > 0$ is verified - which was of course required in order that the curve T(t) could cross over its tangent at inflection in the appropriate direction .

We must now spell out the meaning, in physical terms, of the two constraints that have been obtained. We recall that:

$$P^o_{OH} = \eta(T_o)J^2 = \eta(T_o)\left(\frac{2B_P}{\mu_o\, a}\right)^2 = \eta(T_o)\left(\frac{2B_T}{\mu_o qR}\right)^2 \qquad (2.5)$$

$$P^o_D = \frac{3nT_o}{\tau} = \frac{3nT_o}{C_{NA} naR^2} = \frac{3T_o}{C_{NA} aR^2}, \qquad (2.6)$$

where η(T) is the plasma resistivity (which is a function of the temperature only), J is the average plasma current density, B_P is the poloidal magnetic field, B_T is the toroidal magnetic field, a is the plasma minor radius, R is the configuration major radius, μ_o is the vacuum magnetic permeability, q is the so-called "safety factor" at the plasma border, τ is the energy confinement time due to the diffusive losses, and C_{NA} is the constant multiplicative factor entering into the expression of the "Neo-Alcator" scaling, the latter being expressed by:

$$\tau = C_{NA}\; n\; a\; R^2, \qquad (2.7)$$

By using eqs.(2.4) and (2.5) and (2.6) into eqs.(2.3) one may rearrange the two constraints in a different form:

$$\frac{B_T^2\, a}{q^2} = \frac{3}{4} \cdot \frac{n^2 T_o}{C_{NA}\eta(T_o)}, \qquad (2.8)$$

$$A^2\, n^2\, a^3 = \frac{3T_o}{C_{NA}\, C_{BS}}, \qquad (2.9)$$

where the right hand sides are constant expressions. Here A is the aspect ratio R/a and C_{BS} is defined by the relation $P^o_{BS} = C_{BS}\, n^2$.

The meaning of eqs.(2.8) and (2.9) is as follows. Taking, for the safety factor q, the lowest value compatible with the MHD stability requirements of the discharge, eq.(2.8) yields the value that the "factor of merit" aB_T^2 is required to have in order for the considered marginal path to ignition to be attainable. Then, for given values of the minor radius and of the aspect ratio, eq.(2.9) fixes the value that the plasma density n must acquire at the crossing of the temperature value T_o. Of course, the greater is the margin that one has in aB_T^2 over the latter's threshold value, the more comfortably wide are the available paths to ignition. Note, in particular, that the plasma density - which for marginal values of aB_T^2 is quite sharply prescribed by eq.(2.9) - for larger values of aB_T^2 can take up values in a substantially wider band.

Note also that, for a family of machines all having the same factor of merit aB_T^2 and the same aspect ratio, from eq.(2.9) one obtains the following scaling of the plasma density n with the magnetic field:

$$n \propto B_T^3 , \tag{2.10}$$

Now, for large values of B_T the latter requirement does clearly become incompatible with the well known MHD limit on the ratio of plasma pressure to magnetic pressure, which requires n to be less that some value that scales as B_T^2. It turns out, however, that there is no such incompatibility for the range of values of B_T which can be possibly achieved with the help of the solution developed in the present paper.

We draw finally attention to some considerations of machine cost. In a crude approximation, the latter is often estimated to be proportional to the device's magnetic energy content, or:

$$\text{Cost} \propto a^2 R B_T^2 = a^3 \; A B_T^2 , \tag{2.11}$$

whence for a family of devices all having the same value of aB_T^2 one obtains:

$$\text{Cost} \propto A a^2 \propto A/B_T^4 , \tag{2.12}$$

We see from the latter relation that is does not pay to go to large values of the aspect ratio, because this increases the cost without bringing any benefits to the plasma physics (at least with reference to the selection of plasma physics features which have been taken into account in the present paper) . But the most important information that eq.(2.12) conveys to us is the scaling of the machine cost with the magnetic field. In order for the reader to fully appreciate what the latter result means, in Table 2.1 we have listed the cost function - calculated from eq.(2.12) assuming a fixed value of A - of a number of possible ohmic ignition experiments at various fields (the reference normalization figure being the cost function value 100 for CIT ignition experiment).

While the figures of Table 2.1 must not be taken too strictly at their face value because of the uncertainties and approximations that have been involved in their derivation they are undoubtedly indicative of a well definite trend, and offer plenty of matter for thought - in particular, providing certainly enough of a justification for pursuing purely ohmic ignition in compact high field tokamak experiments.

TABLE 2.1 Some possible ohmic ignition experiments

Type of experiment	B_T(Tesla)	Cost function
OMITRON-like field	17.5	13
IGNITOR-like field	12.4	50
CIT-like field	10.4	100
NET-like field	5.5	1200

3. SCALING LAWS

We consider different models of a tokamak device having similar geometry i.e., having the same shape but different overall dimensions. A unique characteristic length L may be used to identify each model. Points correspondent in the linear transformation between two models are said to be 'homologous'. As we are dealing with dynamic phenomena, we introduce also a linear time transformation and the concept of 'homologous times'; the characteristic time identifying each model in the time scale will be T.

Now we consider the distribution of a given property $\pi(r,t)$ and we compare, on this respect, two models, namely the 'prototype' and the 'model', by comparing the values of π and π' in homologous points and times. The two functions are said to be similar provided that the ratio π'/π is a constant. In the set of models having similar distributions π each model can be identified by a constant Π, that, for example, can be taken as the maximum value of $\pi(r,t)$.

The first step in the dimensional analysis of a problem is to decide which variables enter the problem. In particular let us examine the stresses $\sigma(r, t)$ due to the electromagnetic force density and to the temperature distribution in the conductors of the toroidal magnet and of the poloidal system; in this case we may easily state that the stress parameter Σ characterizing each model within a set of similar ones, must functionally depend on a characteristic magnetic field B_o - let us say the toroidal field on the axis -, a force F acting on the boundary, the Young's modules E, the Poisson's ratio ν, the thermal expansion coefficient β, the vacuum permeability μ_o and the maximum temperature jump in the conductors Θ. This can be formally expressed by a relation of the following type:

$$\Sigma = g_1(B_o , F, L , E , \nu , \mu_o , \Theta , \beta) , \qquad (3.1)$$

A dimensional analysis of the involved variables and parameters leads to identifying the following adimensional products:

$$\nu; \qquad B_o^2/(\Sigma \mu_o); \qquad \beta \Theta \; ; \quad F/(E L^2), \quad (3.2)$$

that by virtue of the Buckingham's[6] theorem, must be considered the elements of a complete set of adimensional products. It follows that relationship (3.1) must be of the form:

$$\Sigma = (B_o^2/ \mu_o)f_1(F/(E L^2) , \nu \; , \beta \Theta ,) . \qquad (3.3)$$

From equation (3.3) we see that for a set of models, made of the same materials, if B_o and the external pressure are fixed and the temperature jump is the same in homologous points, the distribution of the stresses $\sigma(r,t)$ must be identical in all the models.

If we now consider the thermal diffusion process that produces the temperature distribution, the following relation can be easily understood:

$$\Theta = g_2 (L , k , \rho C , \eta , J , T), \qquad (3.4)$$

where k is the coefficient of thermal conductivity , ρC the heat capacity per unit volume,η the resistivity and J the maximum current density.

By inspection a complete set of adimensional products it is easily found:

$$\rho C\, L^2 /k\, T \quad ; \qquad \eta J^2\, T /\Theta \rho C. \qquad (3.5)$$

The functional dependence of Θ must then be given by a relationship of the form:

$$f_2(\rho C\, L^2 /k\, T, \; \eta J^2\, T /\Theta \rho C) = 0. \qquad (3.6)$$

Finally we have to analyze the current density distribution j(r,t) due to the magnetic field diffusion . Neglecting the $\partial D/\partial t$ term in Maxwell's equations, we may state that the constant J characterizing a model in the set having similar distribution of j(r,t), must functionally depend on the characteristic length L, the characteristic time T, the permeability μ_o , the resistivity η and the magnetic field B_o:

$$J = g_3 (\mu_o , \eta, L , T , B_o), \qquad (3.7)$$

A dimensional analysis again leads to identifying two adimensional products:

$$\mu_o L^2 /\eta T \quad , \quad B_o/\mu_o J\, L \quad , \qquad (3.8)$$

so that the following relationship must hold:

$$J = (B_o/\mu_o\, L)\, f_3 (\mu_o L^2 /\eta T). \qquad (3.9)$$

In conclusion from equations (3.3), (3.6) and (3.9) it is evident that all the models in a set having the same magnetic field distribution and times scaling with the square of the length scaling factor k_L will have the same the temperature and stress distributions.It is important to emphasize that not only the stationary phenomena but also all transients associated with the time evolution of the current pulse, and with the deposition and transport of heat can be faithfully reproduced in all the models of a set.

Boundary conditions and auxiliary conditions may of course vary from case to case, and their sharing in the above invariance property must therefore be verified separately for each application: but all sensible choices of boundary and auxiliary conditions are in fact expected to be also compatible with the stated invariance.

These results can be very well described in the $(L^2,1/T)$ plane, or better in the plane $(R_o^2, 1/T)$, having chosen as characteristic length the major radius R_o of the magnet. In this plane, for a given B_o, the branches of the hyperboles R_o^2/T = const. in fact do describe geometrical configurations for which the distributions of the stresses and of the temperatures are the same. Along these curves the current densities decrease as R_o increases keeping constant the product J R_o, so that the same temperature increase is obtained by decreasing the time scale T.

On a certain curve the value of Σ may reach the maximum allowable value; all the configurations to the left of this curve will be then forbidden. In a similar way on a certain curve the value of Θ may reach its maximum allowable value. If this curve is to the right of the previous one, it will be the new feasibility boundary.

We cannot leave this topic without a comment on the behavior of the insulating layers in the winding - a most notable item which lies, in principle, outside the scope of the previous considerations of invariance. For composite materials such as glass-epoxy insulators, indeed, the quality of performance depends sensitively on the absolute thickness of the insulating layer - something which therefore one would prefer to have the same both in the original apparatus and in the model (rather than scaling it with the device dimensions). Now, it is lucky that this goes along nicely also with respect to the electrical requirements on the insulator: indeed, from the scaling of the electric field given before, one has that the difference of electric potential between corresponding points (e.g., across corresponding layers of insulator) is the same in both the original apparatus and in the model, and hence the electric field in the insulator is the same if the insulator thickness is the same. Let it be noted, in any case, that whichever choice one makes of the thickness of the insulator, there are no consequences on the overall behavior of the rest of the apparatus, since the volume taken up by the insulating layers is always a small fraction of the volume of the conductors.

4.NUMERICAL COMPUTATION : STRESS ANALYSIS AND DESIGNE OF THE POLOIDAL SYSTEM

The magnetic field and the heat diffusion in the magnet has been studied by using the finite element code[7,8]. The code computes the current density, the temperature and the body force in each element of the mesh during the time evolution of the total current of the magnet, assumed to be given. In particular in our computations a linearly rising current such to give a final temperature not greater than 370 K, has been assumed.

We have considered two different conducting materials: copper and GLIDCop AL/15 (copper with diffused Al_2O_3). In comparison with the copper the GLIDCop has an higher resistivity (3.2$\mu\Omega$ cm at room temperature and 0.7 $\mu\Omega$ cm at the liquid nitrogen temperature). The maximum allowable rise times, taking into account the practical limitations on the temperature increase, have been computed as 2.5 sec and 1.5 sec respectively for the two materials. In spite of the shorter rising time, the temperature distribution produced by the diffusion is more favorable in the GLIDCop that in the copper. If we consider that the mechanical strength of this material is roughly two times that of the copper - yield strength of 58 Kg/mm^2 against 30 Kg/mm^2 - we conclude that better performances can be expected with the GLIDCop built toroidal magnets.

In a high field toroidal magnet one of the most critical problem is the electrical insulation between each coil and against the steel supporting structure. To assure optimal working conditions the mechanical contact between the conductors and the supporting structure has to be capable of transmitting stresses of considerable intensity through the insulation layer. The epoxy glass-reinforced resins, if in a thin layer, are certainly able of transmitting strong normal stresses but not strong shear stresses. Shear stresses are not significative on the contact surface between coils, but can be considerable on the contact surface conductor-steel. To solve this problem a "sandwich" structure of the insulation can be adopted: the insulating layer is enclosed in a thin sandwich that has a lubricated contact with the steel supporting structure. By means of this technique the two surfaces can slide with respect to each other and no shear stress has to be transmitted. From the design point of view this solution requires a smooth contact surface between conductors and steel casing, which has to be a smooth surface without corners.

The choice that has given the best results is shown in Fig.4.1. The contact surface is obtained from two segments of ellipse matched at the highest point with a common horizontal tangent. In the same figure the supporting structure and the foot of the press- in a single body - are also shown.

As in the original Omitron proposal the press is envisaged to be electromagnetic, and driven by the same current fed into the toroidal magnet, so that the stresses in the whole structure rise as the current rises, reaching their maximum values when the current attains its maximum. For this reason we have computed the stresses distribution at the time of the maximum current. The calculations have been made by means of the finite elements code SAP IV.

The SAP IV code does not consent the sliding of two adjacent surfaces. However, as this is a crucial point in our analysis, we have circumvented such a difficulty by adopting the following procedure: in the first place, all the supporting structure and the conductors are separately analyzed and a SAP run is carried out for the conductors alone under the action of the body electromagnetic forces and of the thermal stresses, but without any boundary constrain on the contact surface. In this way we determine

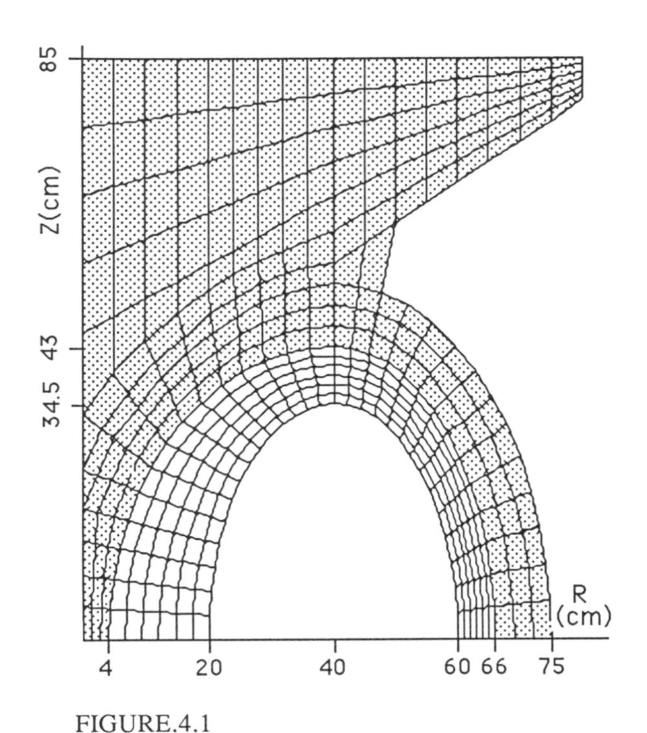

FIGURE.4.1
Meridian cross section of the toroidal magnet.

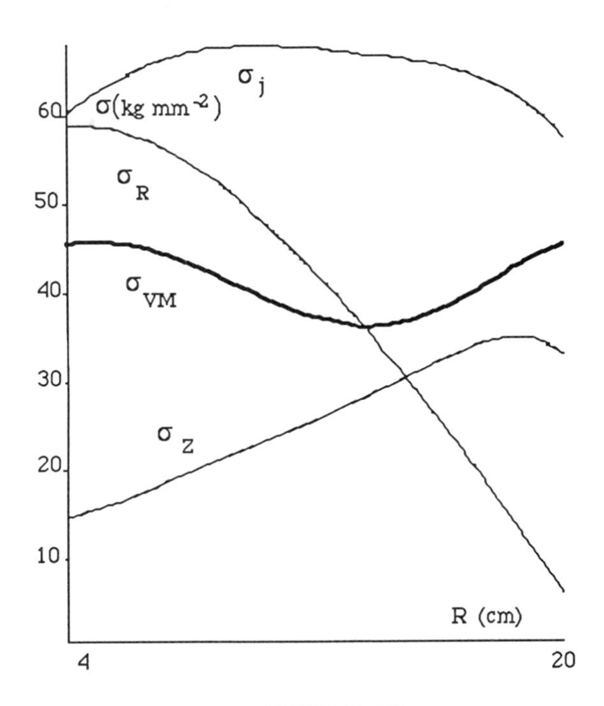

FIGURE 4.2
Stresses in the equatorial cross section of the conductor at the time of maximum magnetic field.

the free displacements of all nodes with respect to their equilibrium positions. Then, we perform as many SAP runs as there are nodes along the contact line between the conductor and the structure, applying each time a unitary force normal to the surface in a single node. As can be easily understood, because of the linearity, on the basis of the displacements of the nodes computed in this way, and of the ones computed with the actual force distribution, but without constrains, a set of coefficients can be constructed such to give, by means of a linear combination, the actual displacement under the action of an arbitrary force distribution on the contact surface.

The described procedure has been implemented in a computer code. The results show that the relative shear displacements are typically of the order of 10^{-2} cm, and thus actually negligible, while the normal relative displacements are less than 10^{-4}, the latter confirming the validity of the algorithm.

In the poloidal section shown in Fig.4.1 the stresses in the magnet reach their maximum value on the equatorial plane. By varying the dimensions of the foot of the press and the total vertical applied force, we have looked for an optimal condition.

The dimensions of the foot of the press determine the way in which the vertical force is transmitted to the magnet conductors: if the press radius is small the pressure is transmitted essentially toward the central column, while if the radius is larger the pressure is primarily absorbed by the external part of the magnet. The optimal radius value has been found to be, in our design, just slightly larger than the major radius of the magnet. The pressure of the press must produce a suitable compression along Z in an area near the axis of symmetry in the equational plane: indeed in this area σ_R and σ_ϕ are negative (compression) while, without the press, σ_Z would be positive, causing a large value of the equivalent Von Mises stress.

The total vertical force of the press must be computed in such a way as to produce a large Z- compression on the equatorial plane for small values of R, without producing too much compression for large values of R in those regions where σ_R is considerably lower. The optimal value has been computed to be 19000 metric tons, equivalent to a pressure, on the upper press surface, of 10 Kg/mm^2.

In Fig.4.2 we have shown the behavior of the stresses in the equatorial plane along the radius. The maximum value of the Von Mises stress is 49 Kg/mm^2, which when compared to the yield strength of the CLIDCop AL/15 gives a safety factor of 1.2. In particular we note that only σ_ϕ attains a value larger than 58 Kg/mm^2. It would be possible, in principle, to reduce this stress component by artificially producing the expansion of the central steel column by means of the thermal effect of an electrical current.

In the conceptual design of OMITRON the equilibrium and transformer windings are located in the magnet toroidal bore. In particular the novelty is represented by the transformer inside the magnet toroidal bore: such solution is usually avoided in tokamaks, because of the complexity of the assembling, but it is in fact an advantage from the electromagnetic point of view because much more magnetic flux is made available under the same constraints on stresses and heating in the transformer windings.

Because of the limited space between the toroidal magnet and the plasma column, among all the possible poloidal field configuration able to provide equilibrium and ohmic heating, we have chosen the one that minimizes the total ampere turns and the required volume.

The design of the poloidal field systems has been effectively carried out by means of a linear programming method which evaluates the transformer and the equilibrium field coils[8,9]. The current density profile has been assumed to be parabolic. In Table 4.1 and in Fig. 4.1 the results of the computations and the behavior of the field lines are shown for two cases: in the case a) the plasma current is 5 MA and the elongation 1.5, while in the case b) the plasma current is 6.3 MA and the elongation 2. The figures show the map of the field in the cross section and the current distribution can be inferred by the dimensions of the conductors (the current density being the same everywhere).

TAB 4.1	case a	case b
Major radius (m)	.40	.40
Minor radius (m)	.16	.145
Elongation ratio	1.5	2.0
Plasma current (MA)	5.0	6.3
Resistive volt-secs (Wb)	.74	.95
q at the plasma border	2.2	2.3
Plasma inductance (μH)	.62	.59
Max current density (kA/cm^2)	30	30
Total ampere - turns (MA)	3.36	3.95
Flux swing (Wb)	2.97	3.77

In conclusion, the stress analysis that we have carried out shows that an OMITRON like magnet can be designed in the frame of the available technologies and materials, with safety margins that can be accepted for this kind of device.

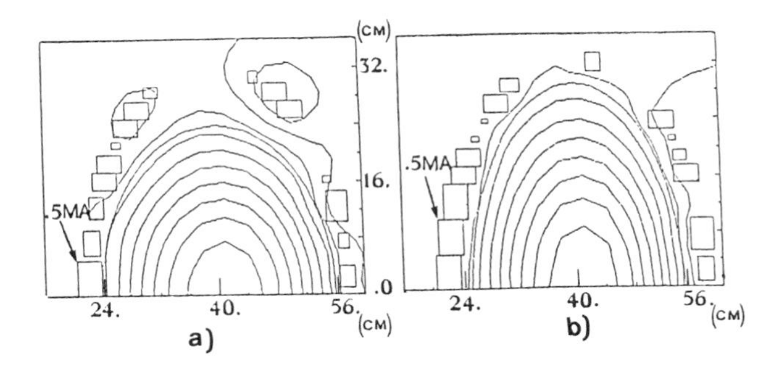

FIGURE 4.1
The figure show the map of the field in the cross section. The current distribution can be inferred by the dimension of the conductors, the current density being the same everywhere

REFERENCES

1 U.S. Patent no. 4,475,096 of October 2nd, 1984, priority based in Italian Patent Application no.48482 - A/81 of October 14th, 1981.

2 A. Sestero, Proc. of the 12th Symposium on Fusion Technology, Jülich, September 1982, pp. 909 - 912.

3 A. Sestero, Comments on Plasma Physics and Controlled Fusion 8, 31 (1983).

4 G. Vlad, Nuovo Cimento 84B, 141 (1984).

5 A. Sestero and G. Vlad, Atomkernenergie - Kerntechnick 44, 23 (1984).

6 H.L. Langhaar, Dimensional Analysis and Theory of Models, John Wiley & Sons, Inc. N.Y.

7 L.De Menna, G. Rubinacci and F. Esposito, 8th Symp. on Engineering Problems of Fusion Research, S. Francisco (1979)

8 R. Albanese et all., Press-Supported High Field Tokamaks, Department of Electrical Engineering, University of Naples, Report 2.2/88,(1988).

9 S. Bobbio, E.Coccorese and R. Martone, Fusion Technology, vol 2, p 999, Pergamon Press (1980).

FUSION TECHNOLOGY 1988
A.M. Van Ingen, A. Nijsen-Vis, H.T. Klippel (editors)
Elsevier Science Publishers B.V., 1989

THE EXTRAP FUSION REACTOR CONCEPT*

J.E. ENINGER and B. LEHNERT

Royal Institute of Technology, S-100 44 Stockholm, Sweden

ABSTRACT A study has recently been initiated to assess the fusion reactor potential of the Extrap high-beta toroidal z-pinch concept. A reactor model is defined that fullfills certain economic and operational criteria that are characteristic of compact toroidal systems, including moderately large electric power output, high power density, high first wall loading, and simple construction. This model is applied to Extrap, and a 1000 MW_e reference reactor having a first wall neutron loading of 10 MW/m^2 is outlined. The minor plasma radius is 1.5 m, the major radius 4.5 m and the pinch current 10 MA. A 0.7 m thick blanket/reflector/shield is chosen to achieve sufficient breeding of tritium, good energy multiplication, and shielding of normal copper coils.

1. INTRODUCTION

The Extrap concept[1] for magnetic plasma confinement is based on the stabilization of a high-beta z-pinch by an applied poloidal field generated by currents in a set of external conductors. This applied field combines with the field of the pinch current to generate a separatrix with x-points, and makes the plasma cross-section non-circular, as shown in Fig. 1. The Extrap confinement system can be either linear or toroidal, and experiments have been performed in both geometries.

Linear experiments[2] first showed that the global stability of a linear z-pinch was improved when the discharge was generated along the axis of an octupole magnetic field. Although the configuration is unstable according to ideal MHD theory, the 0.2 m long, 2 x 10^7 A/m^2 discharges exhibited Bennett equilibrium and were sustained for over 50 μs which was much longer than the 0.5 μs radial Alfvén transit time.

Subsequent experiments with the Extrap T1 toroidal device[3] (plasma major radius 0.45 m,

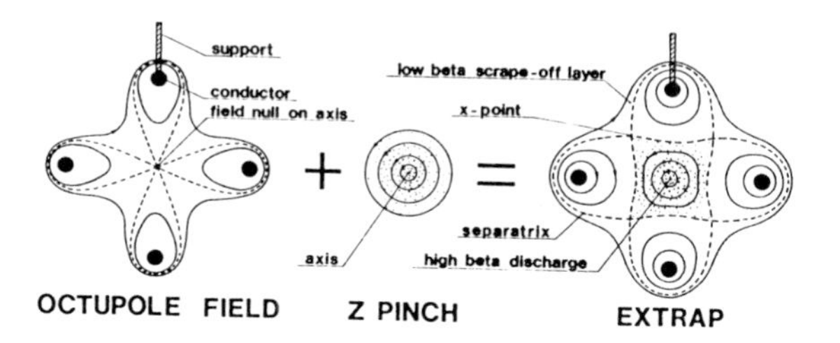

FIGURE 1
An Extrap z-pinch configuration results when a z-pinch is generated along the axis of a vacuum octupole magnetic field produced by currents in four conductors.

minor radius 0.04 m) have shown that stable, high-beta (∿ 50%) equilibria can be generated with a small safety factor q (∿ 0.1).

Measured electron temperatures in both linear and toroidal experiments indicate a scaling of 1-2 eV/kA for the pinch currents up to 30 kA used, with a ratio of pinch current to line

*Work supported by the European Communities under an association contract between Euratom and Sweden.

density in the range of 0.5 - 1 x 10^{-14} Am.

These results form an interesting base for a fusion reactor, and motivates continued investigation of the Extrap concept. In addition to the further scaling of temperature with current, an important physics question is how the stability of Extrap depends on the ratio Θ_i of plasma minor radius to ion Larmor radius. Kinetic LLR-effects might explain the observed stability in the present experiments ($\Theta_i \sim 3$-8), but these effects would probably be less important under reactor conditions ($\Theta_i \sim 30$-50).

2. REACTOR DESIGN CRITERIA

In order to make the fusion reactor a competitive alternative for producing electrical power[4], it is necessary to substantially increase the power density and mass utilization of the fusion power core (FPC), including vacuum chamber, first wall, blanket, shield and coil system over conventional designs, thereby reducing the capital cost of the plant and the cost of the generated electricity. A reasonable target[4,5,6] should be to make the fusion FPC comparable in size to the reactor core of LWR fission power plants, having a thermal power density of $\sim$10 MW_{th}/m^3 and a mass power density of 0.5-1 MW_e/ton. This means an improvement of more than a factor of ten over conventional tokamak reactor designs[7]. The main features of a compact reactor design are:

- The power density of a reactor FPC is proportional to $\beta^2 B^4$. A high-beta ($\beta \geq 0.2$) system such as Extrap can therefore achieve high power density at low magnetic field and low stress at the coils. The low external field may be produced by normally conducting copper coils of simple design and low weight at an acceptable fraction (10-20%) of recirculating power.
- A high power density in the reactor core implies a high neutron flux level at the first wall. A neutron wall loading of 5-20 MW/m^2 is characteristic of compact reactors[6]. The neutron fluence limit of the first wall at these flux levels is about 15 $MWyrs/m^2$.
- The heat load at the first wall can be substantially reduced by effective impurity control. Here the separatrix and the x-points of the Extrap magnetic field geometry suggest the possibility of naturally achieving an efficient divertor. Also, a cold gas mantle can help control impurities and plasma profiles in an Extrap reactor.
- The simplest and most compact magnet coil system is usually obtained with normal conductors rather than superconductors which require extra shielding and are more expensive to build and to operate. In order to get good utilization of the applied field, and to minimize ohmic losses, it is important to locate the coils as close to the plasma boundary as possible.
- The blanket of a compact reactor should be as thin and low in mass as possible[6,8]. The tritium breeding in Li is enhanced by a neutron multiplier (Pb, Be). The blanket energy multiplication can be increased by a high-temperature reflector/shield.
- Steady-state operation[4,6] is an important requirement of a magnetic fusion reactor. Two possibilities exist to achieve steady-state current-drive in Extrap[9], one is by a diffusion driven "bootstrap-like" pinch current (see Section 5), the other is by a high-frequency current-drive.
- The net electric power output should be relatively low, 500-1000 MW_e to satisfy operational and environmental requirements[4].
- To attain high availability, a compact high power density reactor must be of simple design,

with easy maintenance and scheduled replacement of modular units[6].

3. EXTRAP REFERENCE REACTOR

Based on the design criteria outlined above, a reactor model for the Extrap concept has been defined. The characteristic parameters of this reference reactor are listed in Table 1 (and deduced in Section 4). A schematic drawing of the reactor core cross-section is shown in Figure 2.

TABLE 1
Extrap reference reactor design parameters

Parameter	Symbol	Value
Electrical power (net)	P_e	1000 MW_e
Neutron first-wall loading	π_w	10 MW/m^2
Major plasma radius	R	4.5 m
Minor plasma radius	$\bar{a}$	1.5 m
Plasma current	J_ϕ	10 MA
Octupole ring current (each)	J_r	10 MA
Magnetic field at plasma edge	B_a	1.33 T
Maximum field at coil	B_r	5 T
Coil ohmic dissipation (20K)*	P_c	81 MW
Toroidal field coils etc.	P_x	50 MW
Recirculated power fraction	ε	12%
Fusion power	P_f	2750 MW
Thermal power	P_t	3250 MW
Blanket/shield thickness	b	0.70 m
Coil weight	M_c	670 tons
Blanket/shield weight	M_b	1840 tons
Larmor radius parameter	Θ_i	~40

* Includes refrigerator power

First, in order to compare the Extrap reactor with other reactor concepts[5], a net electric power output of 1000 MW_e is chosen. Second, a neutron wall loading of 10 MW/m^2 is selected, which is in the middle of the range used for other compact designs[6]. Third, the toroidal aspect ratio (= major plasma radius/minor plasma radius) is fixed at 3.0. These selections lead to a minor plasma radius of 1.5 m and a major plasma radius of 4.5 m.

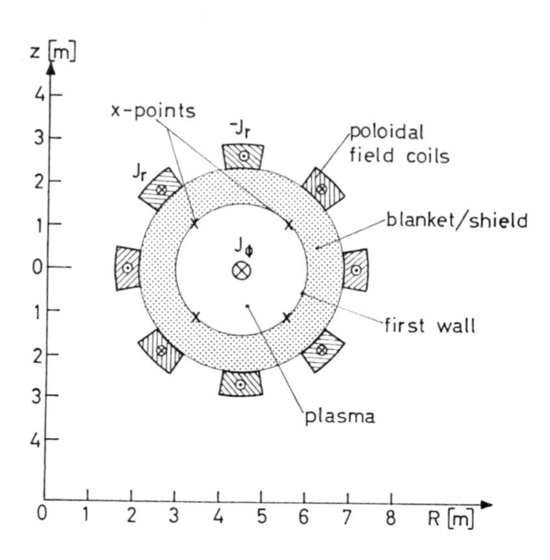

FIGURE 2
Schematic cross-section of Extrap reference reactor

A blanket/reflector/shield total thickness of 0.7 m is needed to achieve a) sufficient breeding of tritium, b) good energy multiplication, and c) shielding of normal copper coils for long (30 year) lifetime[6].

A conservative approach is taken in determining the required octupole field coil current. The plasma boundary is assumed to coincide with the magnetic separatrix, and therefore the four x-points of the field are placed just inside the first wall of the reactor. This also permits full poloidal field divertor operation.

The octupole field coils system is based on refrigerated, normally conducting copper coils[10]. In principle, this coil system can generate the fields necessary for both start-up and long-term plasma confinement. Thus, ohmic heating and equilibrium fields are generated by programming of the current distribution in the eight separate coils.

A weak toroidal field is included to facilitate start-up of the discharge. The temperature is ramped up at low density and low beta, with subsequent increase of the density and of beta. Steady state operation is achieved by a method of diffusion driven "bootstrap-like" pinch current generated by pellet injection and alpha particle heating (see section 5).

4. REACTOR PARAMETER SCALING

4.1. Reactor power output

The power produced by DT fusion reactions is

$$P_f = n_D n_T \langle\sigma v\rangle_{DT}\, Q_{DT}\, f_p V_p \tag{1}$$

where V_p is plasma volume, f_p reacting fraction of volume, n_D density of deuterium, n_T density of tritium, $\langle\sigma v\rangle_{DT}$ DT reaction rate, and Q_{DT} energy produced in each reaction.

The pressure balance of the pinch is given by the Bennett relation

$$J_\phi^2 = C \cdot \bar{a}^2 n_i k\, T \tag{2}$$

where I is pinch current, $\bar{a}$ plasma radius, n_i ion density, T plasma temperature on pinch axis, and C profile dependent constant.

Combining Eq. (1) and (2) yields

$$\bar{a}P_f = \frac{\pi^2 \langle\sigma v\rangle_{DT} Q_{DT} f_R f_p}{2(CkT)^2} \cdot J_\phi^4 \tag{3}$$

where it is assumed that $n_D = n_T = 1/2\, n_i$ and where the toroidal aspect ratio f_R has been introduced. The value of $Q_{DT} = 2.82 \times 10^{-12}$ Joule (17.6 MeV). The quantity $\langle\sigma v\rangle_{DT}/T^2$ varies only slowly with the temperature and has a maximum value of $1.22 \times 10^{-24} m^3/s$ (keV) at T ≅ 30 keV. For a parabolic current profile, the value of $C = 24\pi^2/5\mu_0$. Taking $f_p = 0.4$ and $f_R = 3$, the fusion power then becomes

$$P_f = 4.12\; 10^{-19} J_\phi^4/\bar{a} \tag{4}$$

The net electrical power can be expressed as

$$P_e = \eta_t(1-\varepsilon)(0.75 f_b + 0.25)P_f$$

which then, using (4) and taking $\eta_t = 0.35$, $\varepsilon = 0.12$ gives

$$P_e[MW] = 0.158\, \frac{(J_\phi[MA])^4}{\bar{a}[m]} \tag{5}$$

The pinch current meeded to produce 1000 MW_e power in a reactor with $\bar{a} = 1.5$ m minor radius is then 10 MA.

4.2. Coil power dissipation

The ohmic power dissipation of a poloidal field coil system having n toroidal coils with an average radius of $\bar{a}f_R$, and a cross-sectional area of A_c, is

$$P_c = n\, \frac{\rho \cdot 2\pi a_p f_R f_K}{A_c f_c}\, J_r^2 \tag{6}$$

where ρ is the resistivity of the conductors occupying a fraction f_c of the total coil area. The coil current is J_r ampereturns in each conductor. Furthermore, the effective dissipation is reduced by a factor f_K by refrigeration compared to room temperature operation. Taking $f_K = 0.07$ which corresponds to copper at 20 K and allows for refrigeration losses[10], $\rho = \rho_{cu} = 1.8 \times 10^{-8}$ m, $f_c = 0.7$ allowing for insulation and cooling channels, $f_R = 3$, $n = 8$, $\bar{a} = 1.5$ m, $A_c = 0.5\ m^2$, and $J_r = 10$ MA gives $P_c = 81$ MW.

4.3. Location of x-points

The radial location a_x of the x-points of 2n conductors with their centers uniformly distributed around a circle of radius a_r can be expressed as[11]

$$\frac{a_x}{a_r} = \left[\sqrt{1+\left(n\frac{J_r}{J_\phi}\right)^2} - n\frac{J_r}{J_\phi}\right]^{1/n} \tag{7}$$

where J_ϕ is the plasma current, and J_r the current in each conductor, with currents in adjacent conductors going in opposite directions.

The function a_x/a_r is shown in Figure 3 for n = 3 (curve I) and n = 4(curve II). Two other cases are also shown in the Figure, viz. n = 4 with coparallel currents (curve III) and n = 4 with antiparallel currents (curve IV).

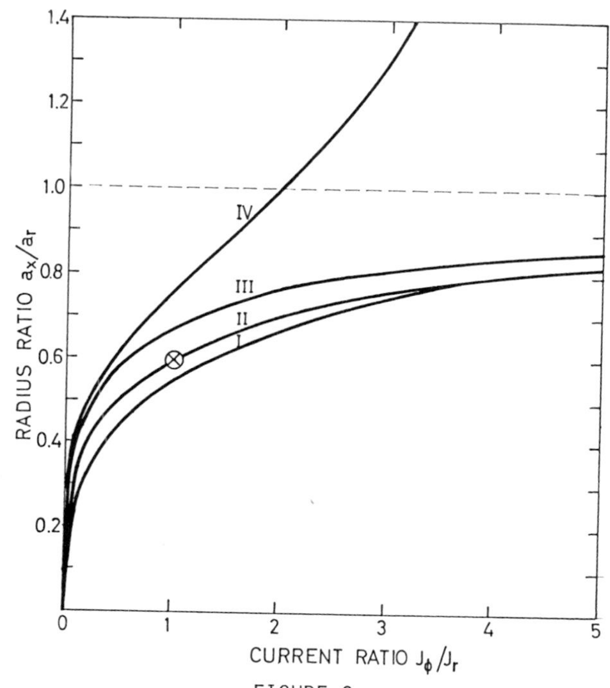

FIGURE 3
Location of x-points for different conductor configurations. The x-point radius a_x is normalized to the radius of current centers a_r. ⊗ indicates reference reactor design point.

The latter two configurations have n conductor currents at a_r and return currents at infinity, and the x-point locations is given by

$$\frac{a_x}{a_r} = \left[\frac{1}{n\cdot\frac{J_r}{J_\phi} \pm 1}\right]^{1/n} \tag{8}$$

where n = 4 and the plus sign is used for J_r and J_ϕ being coparallel, and the minus sign for antiparallel currents.

5. STEADY STATE OPERATION OF THE EXTRAP REACTOR

There are in principle two ways in which steady state operation of the Extrap reactor appears to be possible, namely by a diffusion-driven "bootstrap-like" pinch current, or by a high-frequency current-drive. The latter may have to be investigated in more detail as a future option but will not be considered in this context. Recent results on the former[9] are summarized as follows.

The mechanisms of a diffusion-driven current density in a toroidal configuration is based on the electric field induced by the velocity of plasma diffusion across the confining magnetic field. The driving force is provided by the plasma pressure gradient, which is sustained by volume sources of fuel and heat. The sources of fuel can be due either to pellet injection or to neutral gas penetrating from a cold-mantle into the plasma. The sources of heat can be provided by thermonuclear reactions, or by auxiliary heating. The realization of volume distributions of matter and heat sinks introduces certain constraints on the steady-state plasma density, temperature and current profiles.

The analysis and associated kinetic considerations indicate that steady-state operation should be possible for certain classes of plasma profiles, without running into singularity problems at the pinch axis[12]. Such operation leads to higher axial currents in a z-pinch without an axial magnetic field component than in a tokamak-like case under similar plasma conditions. Moreover, diffusion-driven toroidal pinch currents which allow for alpha particle containment might be realized in Extrap. This appears to be the case for reactor relevant parameter values, and for acceptable pressure and current profiles, as dictated by stability.

6. DESIGN OPTIONS AND IMPROVEMENTS

The reference reactor outlined in Section 3 is one possible reactor scenario based on Extrap concept. It is not intended to be an optimum solution, merely one plausible design to illustrate the general feasibility of the concept. Other design options will be discussed briefly in the following. These options represent trade-offs either in the engineering or in the physics, or between engineering and physics.

6.1. Engineering trade-offs

An example of engineering trade-offs is to consider other positions for the poloidal magnetic field coils. A cryogenic coil system was chosen for the reference design, with a 0.7 m thick blanket/shield between the plasma and the coils. One other option is to use a blanket of minimum thickness, 0.2 m (just sufficient for tritium breeding at a ratio ∿ 1.1). For a 1.5 m plasma radius, it is then possible to locate the current center of the coils at 1.9 m and have the x-points inside the first wall for a ring current of $J_r = J_\phi/4$ (see Fig. 3). In this location, the coils would have to be run at ≧ 400^0 C in order to achieve efficient (η_t = 35%) recovery of the thermal power deposited in the coils in addition to the heating due to ohmic dissipation. The net coil dissipation in this case is calculated to be 230 MW or 19% of the gross electrical power output of the reactor.

Another option is to use superconducting coils. Additional shielding would be needed, where the thickness and composition depends on the radiation limit of the superconductor used. Conventional low-temperature materials such as NbTi require 0.3-0.5 m of shielding in addition to the 0.7 m used in the reference design. Using the optimistic value of 1.0 m total thickness for the blanket/shielded structure leads to $J_r = 2J_\phi$ (see Fig. 3) if the objective is to keep the x-points inside the first wall. The feasibility of a 8 x 20 MA superconducting coil system would have to be further investigated. The rapid development presently seen in high-temperature superconductors is clearly of interest to this application.

6.2. Relaxed physics constraints

In all cases discussed so far, it is assumed that the x-points of the magnetic field have to be located inside the plasma chamber. The engineering constraints dictated by the breeding and shielding requirements then relate the minimum coil current to the plasma current. Thus, a ratio of J_ϕ/J_r ranging between 0.5 and 4 results from these considerations, with the reference reactor case at J_ϕ/J_r= 1. Recent experimental results[3] using four conductors show good plasma stability for J_ϕ/J_r of about 2. From a stability point of view, then, reactor designs using eight conductors at $J_\phi/J_r < 4$ are conservative and imply overdesign of the coil system.

Experiments also show that the plasma does not generally extend all the way out to the separatrix boundary. This suggests that one could perhaps locate the x-points at some distance outside the first wall and still maintain good plasma confinement (although the divertor question then must be reassessed). A significant reduction of the coil current and the associated power dissipation could be obtained if this proves feasible. In addition to the Larmor radius parameter Θ_i, the plasma stability conditions can be expressed in terms of the magnetic field ratio at the plasma boundary, $M = B_a/B_r$, where B_a and B_r are the fields generated by the plasma current and coil current, respectively. From the experimentally determined value of M, one can therefore determine the coil current needed to generate B_r at a given radius.

The reference design has M = 1 and uses coils refrigerated to 20K to obtain low dissipation. The same dissipation can be achieved at room temperature for M = 4. This illustrates a key trade-off between physics and engineering design which should be explored further for improving and optimizing the reactor performance of Extrap.

6.3. Other magnet configurations

As shown in Fig. 3, the most favorable octupole coil configuration uses eight conductors (curve II) which doubles the field strength and locates the x-points at a smaller minor radius compared to a four conductor system (curves III and IV).

Going to a hexapole configuration (curve I) accentuates this advantage further. However, the physics of the hexapole Extrap has not yet been explored, and to now base a reactor concept on this would be too uncertain. It should be kept in mind as a possibility for achieving improved reactor performance.

Poloidal fields can be generated by magnets using pole pieces of magnetic material, and this might be a design option for an Extrap reactor. Advantages over external ring conductor systems would be in lower coil losses and higher engineering beta due to reduction of leakage fields. A design constraint on this kind of magnet is the saturation limit of candidate materials. It would probably be necessary to run the magnets at elevated temperature in order to get them as close as possible to the plasma. The current-carrying coils could still be well shielded since their location now becomes less critical. Finally, the pole pieces would cause a distortion of the magnetic field pattern which must be taken into account in the equilibrium and stability analysis of the plasma.

7. CONCLUSIONS

The high beta value achieved under stable toroidal equilibrium conditions makes the Extrap confinement concept interesting for a fusion reactor. The present study shows that a compact reactor with moderate power output should be feasible using normally conducting, refrigerated copper coils. Startup is accomplished primarily by ohmic heating, and steady state operation is achieved by a method of diffusion driven pinch current which is sustained by pellet injection and alpha particle heating. Design tradeoffs include changing the number of poloidal field coils, their location, and the ratio of coil current to plasma current. Among constraints on the parameter space are plasma stability limits and ohmic dissipation in the coils.

REFERENCES

1. Lehnert, B., Physica Scripta 10, 139(1974).
2. Drake, J.R., Hellsten, T., Landberg, R., Lehnert B., and Wilner, B., Controlled Nuclear Fusion Research 1980 (Proc. 8th Int. Conf. Brussels, 1980) vol. II, IAEA, Vienna (1981) 717.
3. Drake, J.R., Brunsell, P., Brzozowski, J., Eninger, J.E., Hedin, E.R., Karlsson, P., Lehnert, B., Jin Li, Scheffel, J., Sätherblom, H.E., Tennfors, E., and Wilner, B. "Experimental Studies of a High-Beta, Noncircular Cross-section, Toroidal Pinch", 12th Int. Conf. on Plasma Physics and Controlled Nuclear Fusion Research, Nice, France, October 1988, Paper IAEA-CN-50/C-5-19.
4. Carruthers, R., in Unconventional Approaches to Fusion (Ed. by Brunelli, B. and Leotta, G.G.), Plenum Press, New York and London 1982, pp. 3,39.
5. Hagenson, R.L., et al., "The Reversed-Field Pinch: A Compact Approach to Fusion Power", Nuclear Fusion, Suppl., 3(1985)373.
6. The Titan Reversed-Field Pinch Fusion Reactor Study, Scoping Phase Report, UCLA-PPG-1100, Jan. 1987.
7. STARFIRE "A Commercial Tokamak Fusion Power Plant Study", Argonne National Laboratory Report, ANL/FPP-80-1, Sept. 1980.
8. Böhme, G., et al "Studies of a Modular Advanced Stellarator Reactor ASRA 6C, Max-Planck-Institut für Plasmaphysik, IPP 2/285, May 1987.
9. Lehnert, B., "Diffusion-Driven Steady States of the Z-pinch", Fusion Technology, 13 (1987).
10. Rose, D.J. and Clark, M.Jr., Plasmas and Controlled Fusion, MIT Press and Wiley and Sons, Inc., New York and London 1961, p.304.
11. Bonnevier, B., "External Magnetic Field Configurations for Extrap", Royal Institute of Technology, Stockholm, TRITA-PFU-82-10.
12. Braginskii, S.I., "Investigation of the Axial Region of a Plasma Beam", Plasma Physics and the Problem of Controlled Thermonuclear Reactions, vol. I, p. 279, Leontovich, M.A., Ed. Pergamon Press, Oxford (1961).

FUSION TECHNOLOGY 1988
A.M. Van Ingen, A. Nijsen-Vis, H.T. Klippel (editors)

QUICK REPLACEMENT TECHNOLOGY USING SMA DRIVER FOR HIGH LOAD CORE ELEMENTS OF FUSION REACTOR

Masahiro NISHIKAWA, Masamichi KAWAI, Eizaburo TACHIBANA, Seiichi GOTO, Saburo TODA*, Makoto OKAMOTO**, Hiromasa IIDA***, Taiji HOSHIYA+, Masamichi KONDO++, Yoshio SAWADA++, Shigeru TADO+++, Kazuhiro YOSHIKAWA++++ and Kenji WATANABE

Osaka University, Faculty of Engineering, 2-1 Yamada-oka, Suita, Osaka, Japan
*Tohoku University, Faculty of Engineering, **Tokyo Institute of Technology, ***JAERI, Naka, +JAERI, Oarai, ++Toshiba Co., +++Mitsubishi Electric Co., ++++Fujikin International, Inc.

The quick replacement has been accomplished by using the shape memory alloy (SMA) coupling in the conceptual design of a cassette compact toroid reactor (CCTR). The SMA joint can be used for connecting or disconnecting the coupling by simply controlling the SMA temperature. And further by using SMA driving element, the compact large gate valve has been newly devised. This large gate valve will enable the in-situ quick handling without releasing vacuum condition, so that the baking of the vacuum boundary for every replacements become unnecessary except the initial baking i.e. vacuum- vacuum replacement.

In this report, the compact large gate valve using SMA driver has been designed in more detail and the fundamental behavior of SMA driver supporting this design has been studied. And this technology can be used for the present experimental tokamak machines and the fusion test reactor designs.

1. INTRODUCTION

Conceptual designs of fusion reactor are on the newer trend toward the compact reactor that is smaller and cheaper than the other competitive energy source. In this reactor, the neutron wall loading is designed to be about 16 MW/m^2, which value is very high in comparison with that of about 3.6 MW/m^2 in the conventional low density reactor design[1]. The first wall under so high neutron loading suffers from various damages and must be replaced frequently at short duration in the whole life of the fusion power plant. It becomes very important to develop the quick handling technology of replacements and maintenances to release from the damages, supported by newly devised mechanical and structural methods on the system engineering side.

The quick replacement has been accomplished by using the shape memory alloy (SMA) coupling in the conceptual design of a cassette compact toroid reactor (CCTR)[2]. The SMA joint can be used for connecting or disconnecting the coupling by simply controlling the SMA temperature. And further by using SMA driving element, the compact large gate valve has been developed. This large gate valve will enable the in-situ quick handling without releasing vacuum condition, so that the baking of the vacuum boundary for every replacements become unnecessary except the initial baking, i.e. vacuum-vacuum replacement. The use of the large compact gate valve will save the baking time taken by 50% of the time required for the replacements of elements near the plasma core. This quick replacement technology (QRT) can apply extensively to the conventional tokamak design as well as other alternative high power density fusion reactor.

In this report, the compact large gate valve using SMA driver has been designed in more detail and the fundamental behavior of SMA driver supporting this design has been studied on the scaling of the driving force for a large SMA driver, and the neutron irradiation effects on the performance of SMA.

2. QUICK REPLACEMENT TECHNOLOGY (QRT) BY USING SMA

The scheme of replacement of reactor core

parts should be better to be quick and simple procedure, and QRT is one of significant methods to release from the problem of high neutron wall loading. Here the coupling system and the compact large gate valve are newly devised.

2.1. Coupling system by using SMA joints

The SMA coupling performances is illustrated in Fig.1. Over the austenite transformation temperature(Af), the coupler is connected between the faced flanges with vacuum tight. Under the martensite transformation temperature(Mf), this coupler is expanded a little clearance by itself but enough to be disconnected easily without friction. Using these functional SMA coupling controlled only by SMA temperature, the replacement and maintenance can be operated rapidly and simply without the extremely intelligent robot system.

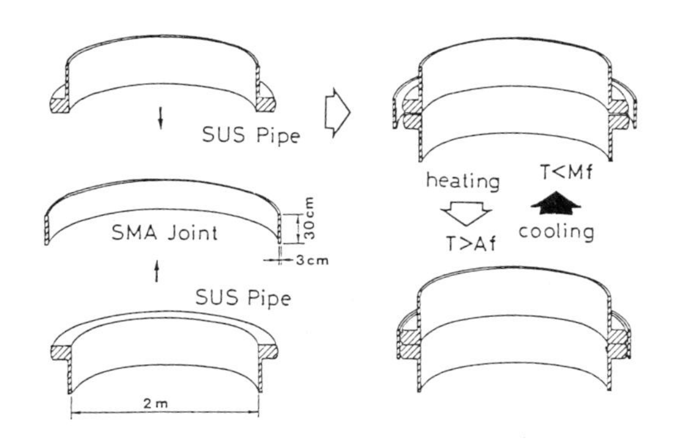

FIGURE 1
Illustration of direct contact seal method

Process of disassembly of the reactor core parts of CCTR is shown in Fig.2. At first, the upper divertors are disconnected from the coaxial gun body by controlling the temperature of the SMA coupler below the martensite transformation point. Then the divertors are removed. As repeating the same operation, the gun body is hung up and removed to the reservoir room. Next, the plasma container is also disconnected from the lower divertor and removed. Finally the SMA coupler is taken away after one service. To use these systems is found to save the time for replacement and maintenance and also to save the working space in comparison with the conventional bolt maneuvering system.

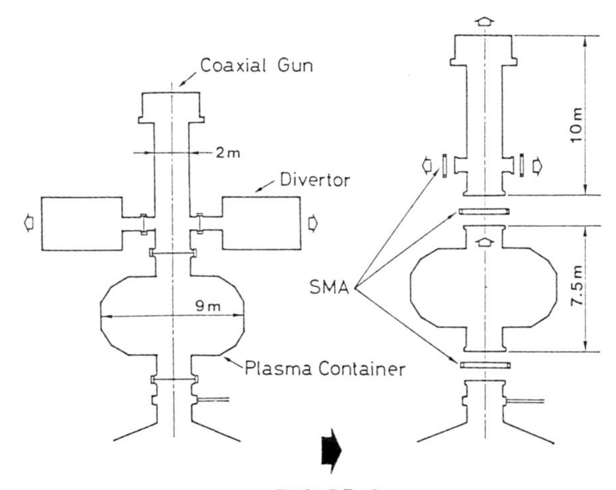

FIGURE 2
Disassembly of fusion core parts in CCTR

The direct contact seal method is easiest one for replacement in all of sealing method. The shape of contacting part on the flange side is designed to be like a convex due to increment of tightening force. The tightening force of this joint is calculated by the modified NIKE2D[3] to achieve the pressure of about 150 MPa, whose value is equivalent pressure applied on the metallic O-ring of the vacuum seal. In the conventional vacuum sealing, the metal O ring has been used by inserting between flanges. In this case, the tightening force is also calculated by the same code[4].

2.2 Compact gate valve

On disassembly, it is important to bake out the components in the vacuum vessel for prevention against the pollution by tritium. On assembly, it is also important to bake the inner surface for degas and to evacuate for cleaning. This baking time is reported to be taken by 50% of the time required for the first wall and blanket replacement in INTOR study[5]. QRT may be limited by this baking process in the conventional replacement. By using SMA driving element, the compact large gate valve can be newly devised. This large gate valve will enable the in-situ handling without breaking vacuum, so that the baking of the vacuum boundary for every replacement become unnecessary except the initial baking.

The conceptual design of the compact gate valve is shown in Fig.3. The vacuum boundary

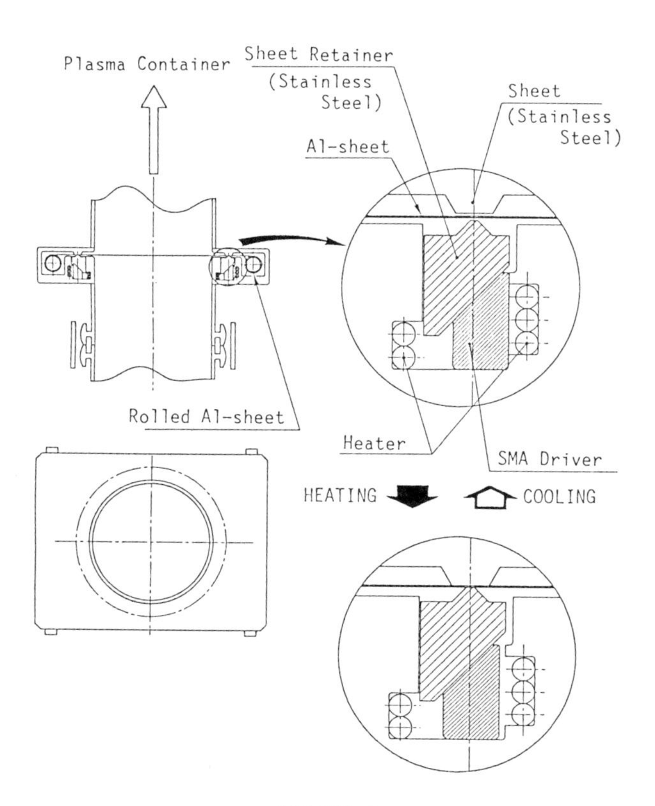

FIGURE 3
Conceptual design of compact circular gate valve using SMA driver

is set by the aluminum sheet rolled in the cylinder equipped in the both sides. When the cylindrical SMA driver is heated, the SMA driver shrinks inward by itself to make the cylindrical stainless steel retainer move upward and finally to fasten the aluminum sheet between the retainer and the valve seat. When the SMA driver is cooled, the driver expands reversely by itself to be loose.(see the detail view of valve seat in Fig.3.) If the gate valve is opened, the aluminum sheet is loose by cooling the SMA driver and rolled up till the aperture part is appeared. And by heating the SMA driver, the sheet with the aperture is fasten by the retainer again. Thus, a SMA ring driver can fasten the Al sheet homogeneously as mentioned above.

On the other hand, the rectangular flanges are used in many parts of tokamak machine. The rectangular compact gate valve is also developed as shown in Fig.4. Figure 4 shows an example that the gate valve is used in the tokamak divertor parts which need frequent replacements for various high loadings. The fastening method of Al sheet between the retainer and the seat is the same as the circular one. The retainer can be driven by some localized SMA drivers. We consider two driving methods: one is the method by the SMA bolt and the other is the method by the SMA ring driver. The performance of the SMA ring driver is the same principal as one of the the circular gate valve. The SMA ring driver may be fitted to fine control of the retainer.

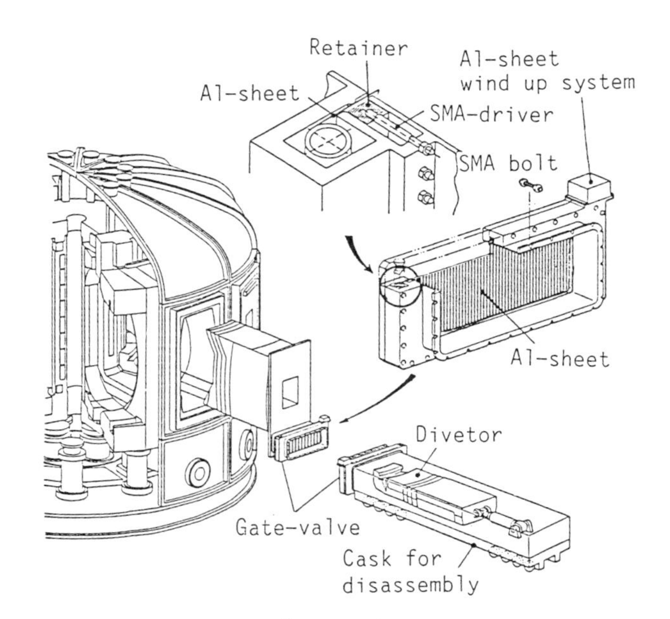

Figure 4
Application of rectangular compact gate valve to tokamak divertor replacement

2.3. Merit of QRT by using SMA

The time required for replacement of the first wall and blanket is estimated on the reference of INTOR study[5] as shown in Table 1.

By QRT using SMA devices such as SMA coupler, SMA bolt and so on, the time required for replacement is reduced at the ratio of 73% when the baking process is included in every replacement. If the baking process is excluded in replacement process, the time for the replacement by QRT can be reduced at the ratio of 42%

Table 1. Effect of baking process on replacement

		including baking process	excluding baking process
(A)	Conventional design(INTOR)	570 hours (23.8 days)	266 hours (11.1 days)
(B)	New design with SMA driver	417 hours (17.4 days)	113 hours (4.7 days)
	(B)/(A)	73%	42%

in comparison with the time for the replacement without QRT. And further, the ratio becomes about 20% as compared with the conventional replacement including the baking process without QRT. Thus, the compact gate valve may realize the replacement excluding the baking process i.e. vacuum-vacuum replacement and support QRT.

3. FUNDAMENTAL BEHAVIOR OF SMA DRIVER

A reversible behavior of SMA driver is obtained by the training effect i.e. appearing a reversible performance with repeat operations between expansion by a external force under the martensite transformation temperature (Ms) and shrinkage by the shape memory effect above the inverse martensite transformation temperature (As).[6]

3.1 Tightening force

The tightening force can be measured by a shrinkage strain of the stainless steel pipe (SUS 304) inserting to the cylindrical SMA driver[7]. Figure 5 shows the temperature dependence of the tightening pressure in the case where the stainless steel tube inserted to the inside of SMA driver. Increasing the temperature of the SMA driver, the diameter of the

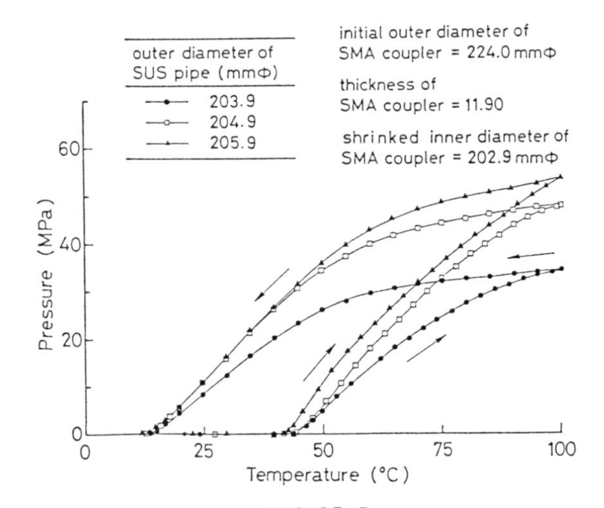

FIGURE 5
Tightening pressure under various restrained strain

driver begins to be contracted above As point and it will contact with the SUS pipe at about 45 C. The tightening pressure increases with the temperature above 45 C after the SMA driver contacts with the SUS pipe. At 100 C, the tightening pressure becomes from 45 MPa to 55 MPa in order of interferences between them. Decreasing the temperature from 100 C to 45 C, the pressure is still imposed to the SUS pipe, keeping the constant value, followed by decreasing. At about 15 C where the pressure falls into zero, it will be found that the SMA cylindrical driver will detach from the SUS pipe. When the temperature rises, the tightening pressure P_u is approximately obtained as

$$P_u = (1-2\varepsilon)\,\sigma\,(T,\varepsilon,A_s)\ \ln\frac{R_o+\Delta R_o}{R_o} \quad ,(1)$$

$$\sigma = \frac{1}{\alpha_1}\left[\ 1-\mathrm{EXP}\left(-\alpha_1 Q_1\left[1+\frac{\eta}{2}(T+A_s)\right](T-A_s)\right)\right],$$

where σ, T, ε, R_o and ΔR_o represent the recovery stress, the temperature, the strain, the ring driver radius without stress and the thickness of the ring, respectively. And the constants for the temperature variation, α_1, Q_1 and η can be estimated from experimental results.

The equation(1) is found to be useful for the scaling formula of a large ring driver.

3.2. Neutron irradiation effect of TiNi SMA

Specimens are prepared to perform tensile tests, hardness tests and annealing tests. Specimens are neutron irradiated at JMTR (Japan Materials Testing Reactor) in JAERI. Irradiation conditions are as follows:

Fast neutron fluences (>1MeV): $7.7\times10^{23}/m^2$
Thermal neutron fluences (<0.683eV) : $2.9\times10^{24}/m^2$
Irradiation temperature : 323 K
Irradiation period of time : 1.9 Ms (22 d)

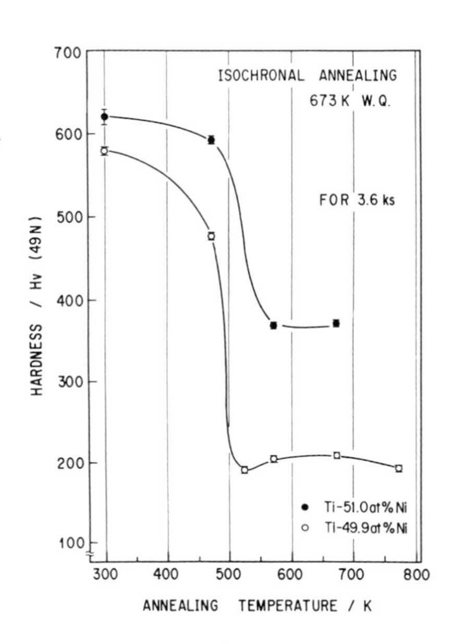

FIGURE 6
Vickers hardness versus annealing temperature curves for irradiated Ti-Ni alloys (Ti-49.9at%Ni and Ti-51.0at%Ni) which are aged at 673 K for 3.6 ks, after 3.6 ks of isochronal annealing at temperature between 298 and 773 K.

The effect of post-irradiation annealing (isochronal annealing for 3.6 ks) on the hardness of TiNi is shown in Fig.6. As the annealing temperature is increased gradually, it is confirmed that irradiated alloys come to a completely restored state after holding at temperatures above 523 K. These hardening properties are entirely recovered with annealing treatments, peculiar to the irradiated TiNi SMA. On the stress-strain curves, the irradiated alloys are restored to the initial condition after post-annealing above a threshold temperature of 523 K.[8]

The SMA of TiNi may be sensitive to the build-up of irradiation damage and the local irradiation modification of the ordered B2 structure(crystal structure of CsCl type) to be in a highly strained state, however, is observed to recover from the neutron-irradiated effects by annealing above a threshold temperature over 523 K.[8] These effects may give us a solution for the key issues how to use SMA in the neutron irradiation field.

4. SUMMARY

The compact large gate valve using SMA driver has been designed for vacuum-vacuum replacement, which enable to advance QRT.

And the fundamental behavior of SMA driver supporting this design has been studied on the scaling of the driving force for a large SMA driver, and the neutron irradiation effects on the performance of SMA.

REFERENCES

1. C.C.Baker et al.,"STARFIRE",ANL/FPP-80-1, ANL(Sep.1980).

2. M. Nishikawa et al., Fusion Technology, 9 (1986) 101.

3. E. Tachibana et al., Computational Mechanics, IV113-118, Springer-Verlag (1986).

4. M. Nishikawa et al., Fusion Engineering and Design 5 (1988) 401.

5. INTOR Workshop Phase IIA Part3(1988) under publication.

6. J. Perkins, Materials with Shape Memory, Jabtonna, Poland (1981).

7. M. Nishikawa et al., Proc. of ICOMAT (1986), 1077-1082, The Japan Institute of Metals.

8. T. Hoshiya et al., Proc. of ICOMAT (1986), 685-690, The Japan Institute of Metals.

FUSION TECHNOLOGY 1988
A.M. Van Ingen, A. Nijsen-Vis, H.T. Klippel (editors)
Elsevier Science Publishers B.V., 1989

THE SPHEROMAK FUSION MACHINE AS PROPULSION IN SPACEFLIGHT

Hans BERGMAN

Capitol Radio Engineering Institute, U.S.A.+

Modification of the practical lightweight spheromak produces a FUSION-ROCKET for interplanetary spacetravel.

1. THE S-1 SPHEROMAK FUSION MACHINE

The S-1 Spheromak device, the first large-scale test of the spheromak magnetic confinement concept, began operation at the Princeton Plasma Physics Laboratory early in 1983. As in a tokamak a spheromak uses toroidal and poloidal magnetic fields to confine a dough-nut-shaped plasma. In a spheromak, however, the currents generating the toroidal field flow in the plasma itself, eliminating the large external toroidal field coils required on tokamaks.

Higher current densities minimize the need for supplemental plasma heating apparatus, thereby simplifying the task of reaching ignition temperatures. The spheromak's high engineering beta value (the ratio of the plasma pressure to the magnetic field pressure, measured at the coils rather than within the plasma) allows maintenance of a higher plasma energy with less stress on the coils themselves in comparison with tokamaks.

The spheromak consists of 3 main sections: the vacuum vessel, 6 outer equilibrium field (EF) coils, and a flux core, which contains a 6-turn poloidal field (PF) coil, a 90-turn toroidal field (TF) coil and one additional EF coil. Recently, addition of a loose-fitting figure-8 stabilization system has improved plasma stability and parameters.

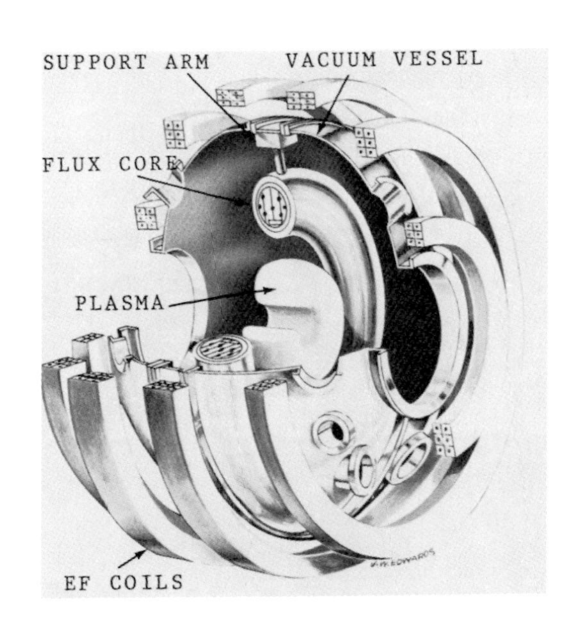

The EF coils are powered by 2 motor-generator sets, which provide a steady-state EF field over the duration of a plasma discharge by supplying currents up to 22,300 amperes for up to 5 seconds.

The PF and TF systems are powered by separate capacitor banks; power is

+Waddenzeestraat 200, 1784VH, Den Helder, The Netherlands.

switched from the banks to the coils by ignitron tubes.

A 1/8-inch-thick aluminum shell covers the coil and fiberglass combination. This shell helps smooth the magnetic fields generated by the PF and TF coils, thereby reducing field errors and plasma-core interaction. The shell also helps stabilize the plasma during formation. The flux core is covered by a metallic "liner", which functions as a clean vacuum vessel wall. Since the liner must allow the magnetic fluxes from inside the core to be transferred to the plasma, Inconel, a very resistive nickel-based alloy, is fabricated into a thin (0.010-inch thick) cover.

Creating the flux core liner began when 0.125-inch thick Inconel sheets were shaped into 2 "half-toroids" by explosion forming (an improved method over the previously used spinning formation process). The thickness was reduced by chemical milling, a process whereby the liners were dipped into an acid bath. The acid removed material from areas not masked off, thus achieving a final product of uniform thick-

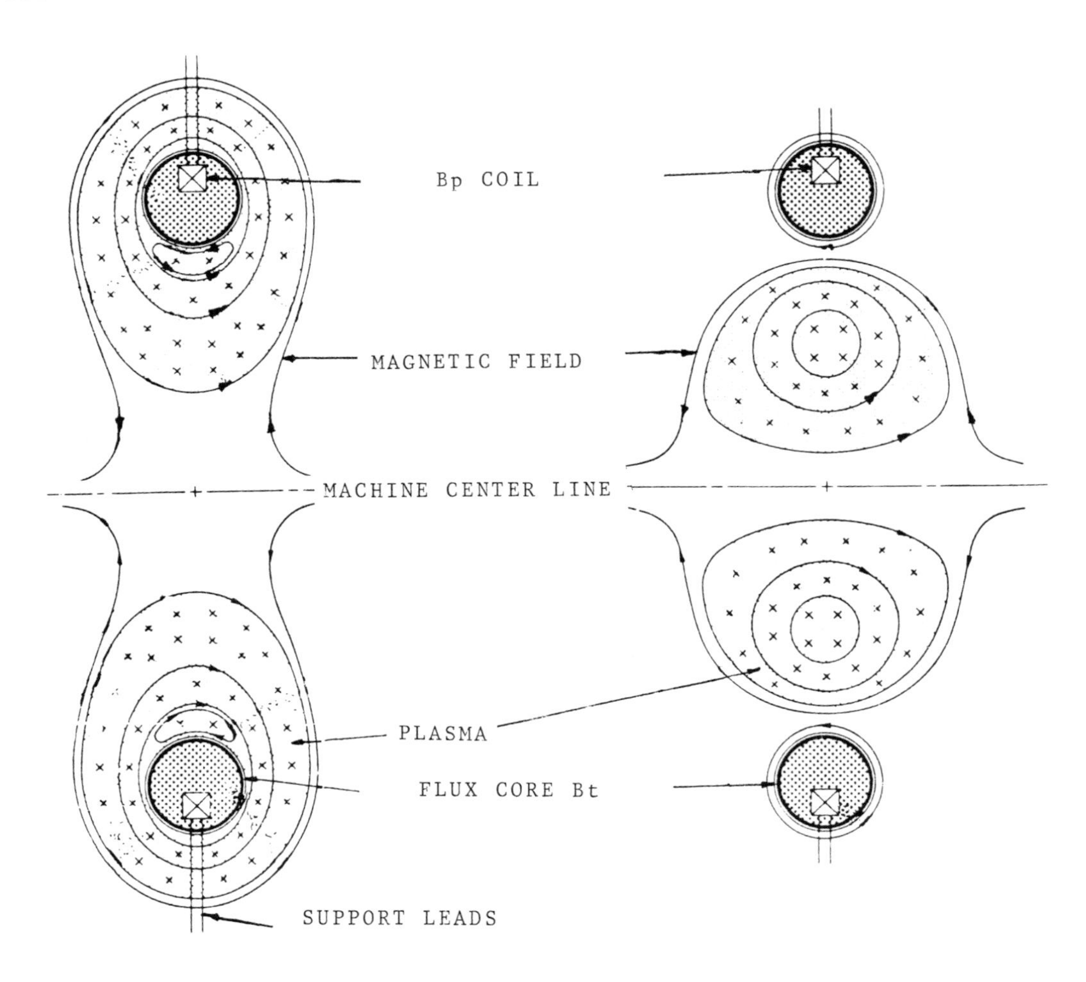

Initial plasma formation and final plasma configuration.

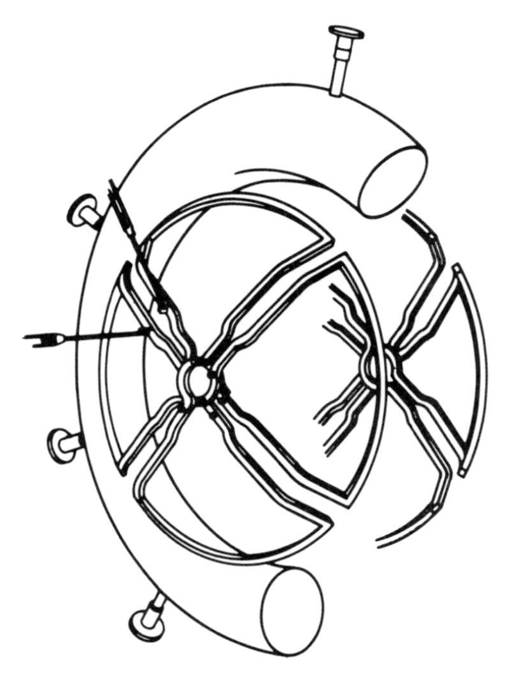

Passive figure-8 coil stabilization system and S-1 flux core

ness. The process has recently been used for the fabrication of the Space Shuttle's liquid fuel tanks.

2. MAKING AN S-1 SPHEROMAK PLASMA

In the S-1 Spheromak, plasmas are formed by using the flux core's coils to induce plasma currents.

At the start of an experiment, a neutral gas surrounds the flux core. The outer EF coils are switched on, generating a magnetic field that is strongest at the outer edges of the core.

The PF coils are then energized as is the TF coil about 75 micro-seconds later. This creates a high voltage near the surface of the flux core, initiating the ionization that creates the plasma. The current in the PF and TF windings is changed, inducing poloidal and toroidal currents in the plasma. Simultaneously, the equilibrium fields from the EF coils push the ionized gas to the center of the machine. Here the plasma's self generated fields, aided by those of the external coils, keep it confined.

The S-1 device reached in 1986 the milestones established when the S-1 project was proposed in 1979: attainment of hot (100 eV) plasmas with stable lifetimes of 1 msec or more. The plasma has a temperature of approximately 1 million degrees Kelvin.

The S-1 device forms spheromak plasmas with major radii ranging from 45 to 60 cm, and minor radii of 25 to 50 cm. Toroidal plasma currents of up to 350 kA have been achieved at moderate powerlevels. Measured peak electron temperatures range from 40 to 110 eV.

3. FUTURE EXPERIMENTAL PROGRAM

Preliminary experiments indicate betas from 5 to 50% are possible.

Another major emphasis of S-1 research will be the inductive sustainment of spheromaks. Sustainment is any method of actively driving plasma currents to extend the plasma lifetime, maintain a steady-state discharge, or increase plasma currents.

One scheme for sustainment is the use of a poloidal-flux transformer along the major axis. Toroidal plasma currents of 1MA are expected to be achieved with this system. Preliminary experimental evidence from a prototype device and preliminary theoritical evidence from numerical simulations demonstrate the successful use of a poloidal-flux transformer for sustaining a spheromak by increasing and prolonging not only the toroidal plasma current but also the toroidal magnetic flux in the plasma.

A further upgrade of the spheromak system would be aimed at achieving 0.5 to 1.0 keV plasmas. This would involve moving the spheromak plasma away from the plane of the flux core by use of a magnetic field, followed by compression by a factor of approximately 2.5. In a power reactor, this would allow a non-reacting plasma to be formed, moved away from the flux core, then ignited; the core would thus be protected from neutron exposure.

4. THE PALENKOMAK FUSION ROCKET

In the modified spheromak only the right halve of the external EF coils is used to serve as a "magnetic catchnet" for the plasma ball. The left halve of the EF coils is replaced by a flux core with a small diameter. A miniature flux core is placed inside the flux core. This PF coil with TF windings moves the plasma directly 30 degrees into the "magnetic catchnet" of 3 EF coils. A small stabilizing PF coil with opposite windings pushes the plasma towards the plasma limiter. The 2 opposite ellips

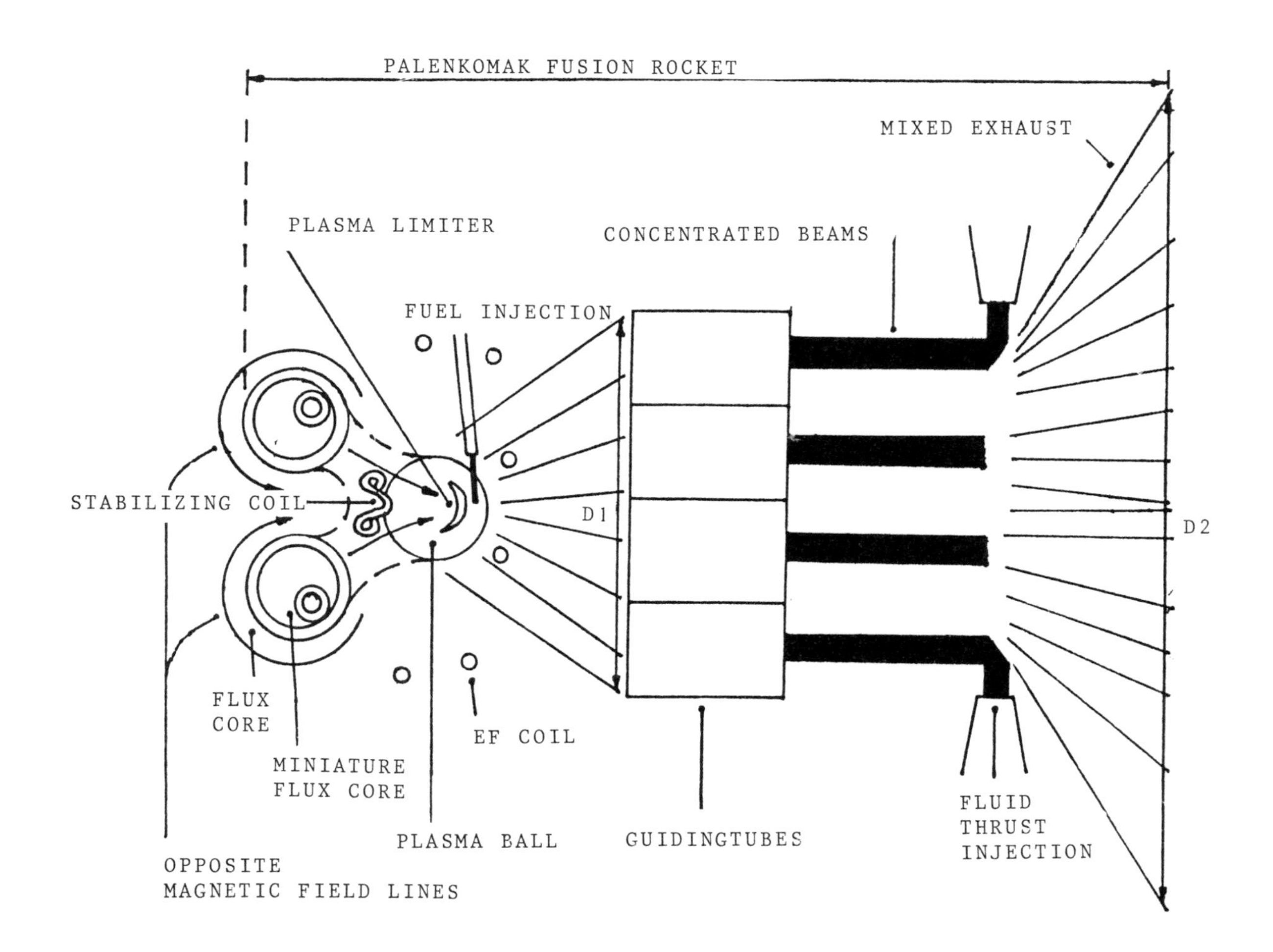

The Palenkomak fusion rocket for interplanetary spacetravel.

shaped plasma currents come together in the plasma ball and confine the plasma ball as a bi-polar complimentary unity. The number of revolutions of the opposite ellips shaped plasma currents determine the final stability by creating opposite magnetic moments.

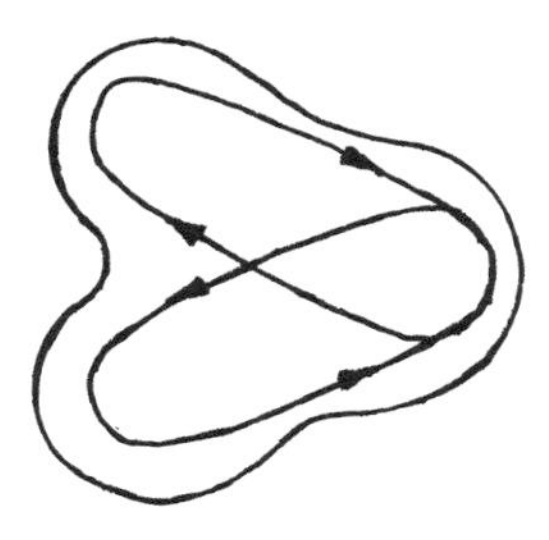

Opposite ellips shaped electron currents confine the plasma ball.

In the balanced plasma ball a fuel injection takes place emitting a plasma flux of particles creating an outlet with a diameter of D1. The particles from the plasma are concentrated in 14 guidingtubes creating 14 beams of particles. A fluid injection of thrust material takes place in the 14 beams creating a mixture of plasma- and thrust particles to increase the final thrust and rocketspeed. This "mixed exhaust" has a diameter D2, which is 2.5 times larger than the first outlet diameter D1. After the "expansion treatment" the "mixed exhaust" will be shot into space to provide the thrust for the fusion rocket. This happens according to the third law of Newton: action = reaction.

5. PALENKOMAK PARAMETERS

The escape velocity for earth is 11.2 km/s.

The fusion rocket has a plasma temperature of 100 million degrees Kelvin, a low fuel consumption with a high specific impuls Isp = 100,000 seconds and a high exhaust speed of thrust particles. The fusion rocket can work for a long time, because of the low fuel consumption. A low fuel consumption however is not favourable for the thrust, so that a thrust injection in the second outlet is necessary. This increase of thrust material results in a higher thrust and rocket speed.

If we have the power available P, the mass flow rate through engine $\dot{m}$ and the exhaust speed c, then

$$P = \frac{\dot{m}}{2} c^2 \quad \text{and thrust } T$$

$$T = \dot{m}c \text{ . From this}$$

$$P = \frac{1}{2\dot{m}} T^2 \text{, or } T = \sqrt{2\dot{m}P}$$

We see that T will go up if we increase $\dot{m}$, at constant power. Increasing T is of advantage, because time requirement to accumulate some speed increment Δv decreases.

A chemical rocket has a gas temperature of 3573 degrees Kelvin in the combustion chamber (Space Shuttle thruster), a high fuel consumption with a low specific impuls Isp = 450 seconds and a low exhaust speed. The disadvantage of a chemical rocket is the high fuel consumption or low specific impuls, so that the payload is always very low compared to the lift-off weight. A chemical rocket travels in 9 months to the important planet Mars.

The Palenkomak fusion rocket can travel in the future to Mars in a few months, involving the machine losses.

FUSION TECHNOLOGY 1988
A.M. Van Ingen, A. Nijsen-Vis, H.T. Klippel (editors)

APPROACHING IGNITION IN VIEW OF SOME OF THE MOST RECENT TRENDS IN THE GLOBAL ENERGY CONFINEMENT SCALING

R. Giannella, M. Roccella

Associazione EURATOM-ENEA sulla Fusione, Centro Ricerche Energia Frascati,
C.P. 65 - 00044 Frascati, Rome, Italy

A numerical code has been developed to attempt an analysis according to different scaling laws and operating scenarios of several recently proposed tokamaks. In the used plasma model, profile consistency between current density and temperature have been assumed taking into account neoclassical conductivity. Given the scaling law and the scenario, the code scans all the configuration space looking for the "machines" capable of reaching ignition, according to some simple technological constraints. The results for the most conservative situation considered are shown.

1. INTRODUCTION

The trends shown by the most recent systematic analyses of experimental data lead to quite uncertain estimates for the expected performances of a tokamak fusion reactor. Due to the large extrapolation required by the scaling laws suggested, several proposals have been put forward for intermediate steps (Ignitor, CIT, Ignitex, JIT, JET-Upgrade). These machines should access, with reasonable certainty and wide enough margin, the alpha heating domain, in order to allow a correct sizing of plants whose objectives, are comparable to those of NET or ITER. Using a numerical code that analyzes the plasma stationary states at different additional power values, an analysis of some proposals has been carried out considering the different confinement scaling laws and operation scenarios. We also looked for tokamak configurations capable of ignition consistently with some of the most commonly accepted technological constraints for a fusion reactor. At this end, for any given scaling law and plasma scenario, the code scans the configuration space characterizing each "machine" according to its potential performance. As the correct scaling to be used is questionable, we have looked for configurations capable of ignition in the most conservative situation, the so-called Goldston L-mode.

2. PLASMA MODEL

A very good coupling between ions and electrons is required for thermonuclear plasmas. In such conditions, the approximation $T_e = T_i = T$ holds.

The other two usual operational limits related to the line average density and to the total β are set by the relations

$$M = \overline{n}\ (R/B) < \min(1,\ 1.5\ k/q_{cyl}) \qquad (1)$$

$$\langle\beta\rangle < 0.03\, I/a\,B$$

where $q_{cyl} = 5\ k\ a^2\ B/(RI)$, k is the elongation of the plasma cross section, I the plasma current, B the toroidal magnetic field, R is the torus major radius and a the plasma "horizontal minor radius" (the units used are MA, Tesla, m, keV, MW and $\bar{n}$, the electron linear density, is normalized to $10^{20}/m^3$) M is the Murakami parameter.

In order to describe also plasmas with noncircular cross sections, the volume V enclosed by the generic flux surface has been chosen as radial-like coordinate[1]. The "radial" profile of the electron density n will be described by the following expression

$$n = n_0 (1 - x)^{\nu}$$

where $x = V/V_p$ and V_p is the plasma volume.

For the current density j, the following profile was assumed

$$j(x) = j_0 \qquad \text{for } x < x_0$$
$$j(x) = j_0 \left(\frac{1 - x}{1 - x_0}\right)^{\lambda} \qquad \text{for } x > x_0 \tag{2}$$

where the constant j_0 is related to the central value of the safety factor q_0. The profile for the temperature T is found using the consistency relation

$$\sigma_{nc\|}(x, n, T(x)) \approx \frac{j(x)}{j_0} \tag{3}$$

for $x > x_0$ and assuming $T = T_0 = T(x_0)$ for $x < x_0$. We assume moreover that all the plasma parameters such as k, q, etc., are also constant for $x < x_0$. For $\sigma_{nc\|}$, an analytical approximation has been used[2]. We choose the values of λ and x_0 consistently with the conditions that the temperature profile is not hollow in the center and with the prefixed ratio $q(a)/q(x_0)$. The temperature profiles obtained in this way are not far from the experimental data for both high and low collisionality and for different values of the safety factor at the plasma edge (comparisons between the profiles obtained and data from JET[3] and FT[4] are given in Figs. 1 and 2).

3. ENERGY BALANCE

If U is the kinetic energy of the plasma, the balance equation reads

$$\frac{dU}{dt} = P_{oh} + P_{ad} + P_{\alpha} - P_{rad} - P_1 \tag{4}$$

where P_{oh}, P_{ad}, P_{α} are the ohmic, auxiliary and alpha input power, P_{rad} and P_1 are the radiative and convection/conduction power losses. We conservatively assume that only $P_1 = U/\tau_E$ is accounted for by the confinement time. P_{rad} is assumed due to both cyclotron radiation[5] and to

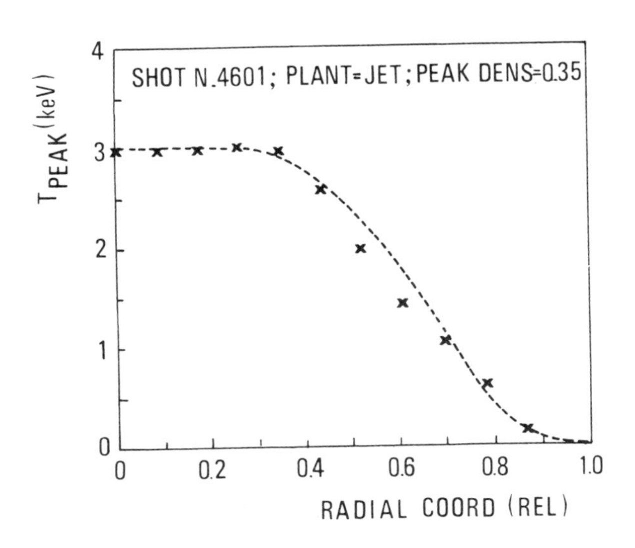

FIGURE 1
Comparison between computed and measured temperature profiles at JET

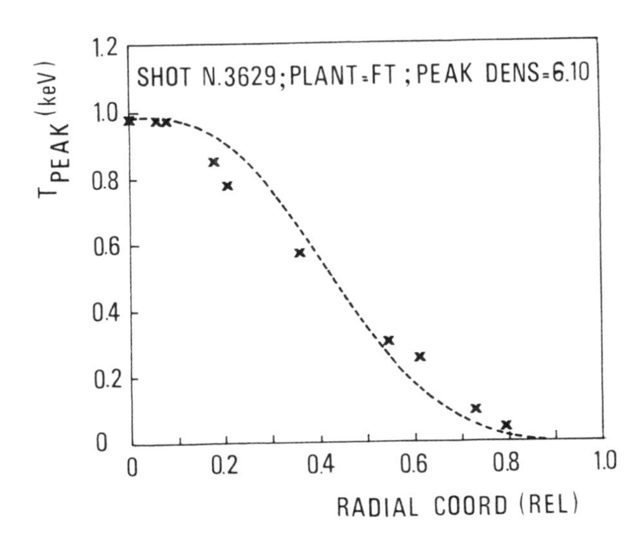

FIGURE 2
Comparison between computed and measured temperature profile at FT

bremsstrahlung losses.

The global energy confinement time of Eq. (4) was assumed, according to the Goldston's suggestion, to be given by

$$\tau_E = (\tau_{oh}^{-2} + \tau_{aux}^{-2})^{-1/2}$$

where for the "ohmic" confinement time the neo-alcator expression[6]

$$\tau_{oh} = 0.103\, R^{2.04}\, a^{1.04}\, q_{cyl}^{1/2}\, \bar{n}$$

was chosen, and for τ_{aux} both the scaling given by Goldston[6],

$$\tau_{aux} = 0.037\, k^{1/2}\, R^{1.75}\, a^{-0.37}\, I\, P_1^{-1/2}$$

and later by Kaye and Goldston[7]

$$\tau_{aux} = 0.055\, k^{0.28}\, R^{1.65}\, a^{-0.49}\, I^{1.24}\, B^{-0.09}\, n^{0.26}\, P_1^{-0.58}$$

were used. For each scaling, possible improved confinement conditions (H-mode) were also taken into account by simply doubling the numeric coefficient of the corresponding scaling law. The peaking parameter of the density profile: ν=1, and ν=2.5 was used to simulate gas-puffing and pellet injection respectively.

4. ESTIMATED PERFORMANCE OF SOME PROPOSALS FOR THE NEXT TOKAMAKS

A stationary solution of Eq. 4 in the (T_o, n_o) plane in terms of P_{ad} is shown in Fig. 3, for a given machine and plasma scenario. The short-dashed curves correspond to negative values of $P_{ad,}$ where self heating occurs. The dot-dashed contours correspond to equilibria for which auxiliary heating is needed. The full lines separating the two families of curves represent the ohmic equilibrium curve (OEC). The operational limits, as well as some Q=const curves, where Q $\simeq$ 5 $P\alpha/(P_{oh}+P_{ad})$ are also shown in the figures. We will assume in the following that ignition is reached when a high Q (e.g., > 10) stationary equilibrium condition, attained within the operational limits, can be ohmically sustained.

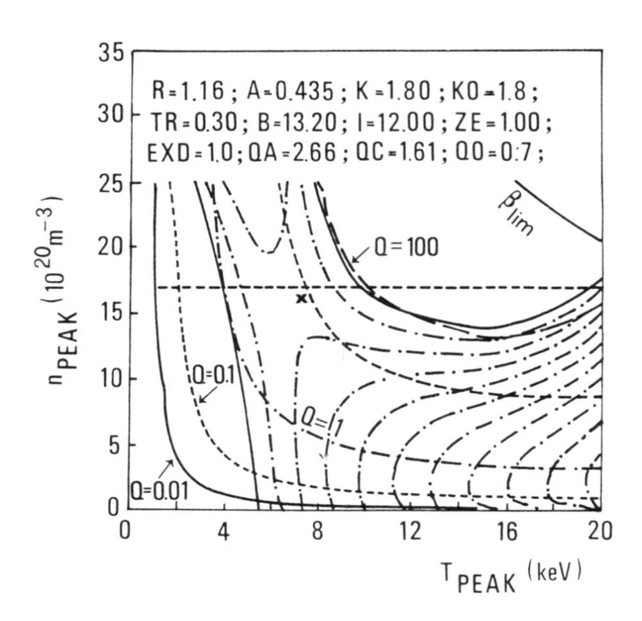

FIGURE 3
Equilibrium countours at constant auxiliary power for Ignitor[8] assuming Kaye-Goldston L-mode scaling and q_o = 0.7. The density limit corresponding to M = 1 (dashed line) and beta limit (full upper line) are also presented in the figure

For each machine, a variety of operation scenarios is conceivable, although the effectiveness or the feasibility of some operation procedures are still uncertain. We have considered a number of combinations of such possibilities, including peaked density profile (such as can be obtained by pellet injection), and stable peaked temperature profiles with $q_0 < 1$ and H-modes for both Goldston and Kaye-Goldston scalings. An example of such an analysis is given in Table I where the best performances compatible with the operational limits are presented for NET.

5. PARAMETER SPACE ANALYSIS

We looked for machine configurations which allow ignition even in the most restrictive hypothesis. Some technological constraints have been imposed on the problems, varying according to whether one plans to build an experimental machine, with or without superconducting

Table I: QMA, QMI maximum or minimum values of Q achievable in stable conditions along the OEC courves; P_{ON}, P_A, P_{RA} ohmic alpha and radiation powers in these conditions; $\tau = \tau_E(s)$; $Tn\tau = Tn\tau_E 10^{20} m^{-3}$ keV; SCEN = Plasma scenario: STA = (ν = 1 q = 1 L-mode Goldston (G) or Keye-Goldston (kG) scaling; H = H-mode; P = = Pellet injection (ν=2.5); Q = low central q(q = 0.7).

LOW Q OHMIC BRANCH							PTH	M	HIGH Q OEC (IGN. BRANCH)					
SCEN	QMA	PHO	PA	PRA	τ	$n\tau T$			QMI	PHO	PA	PRA	τ	$n\tau T$
STA.G	.14	13.	0.4	2.9	2.6	9.2								
H...G	.22	12.	0.5	3.8	3.6	15.1								
P...G	.22	12.	0.5	3.0	2.6	13.5								
HP..G	.33	12.	0.8	4.5	3.8	24.0	11	.83	389	1.1	83	14	2.4	82.6
Q...G	.19	12.	0.5	2.3	2.4	9.2								
HQ..G	.32	13.	0.8	4.2	3.8	18.7								
PQ..G	.29	.13.	0.8	3.9	2.7	16.7								
HPQ.G	.45	12.	1.1	4.4	3.6	25.4	9	.79	303	1.2	74	12	2.6	82.9
STAKG	.17	3.13.	0.4	3.3	3.0	11.5								
H..KG	.24	112.	0.6	4.2	4.0	17.4	13	.96	461	1.1	99	17	2.6	71.8
P..KG	.27	2.12.	0.7	3.8	3.1	18.0								
HP.KG	.36	12.	0.9	4.8	3.1	27.0	9	.78	344	1.1	74	13	2.6	82.9
Q..KG	.26	14.	0.7	5.0	3.4	17.0								
HQ.KG	.35	12.	0.9	4.4	4.0	20.1	11	.91	328	1.2	81	14	2.7	74.1
PQ.KG	.35	13.	0.9	3.9	3.0	19.8								
HPQKG	.49	12.	1.2	4.5	3.8	27.1	7	.75	327	1.2	67	11	2.6	73.1

NET extended version: R=5.4; A=1.4; K=2.2;δ=0.4; B=5; I=14.8; ZE=2; ZI=7Fα=1 QPSI=2.5

magnet, or a real breeding reactor intended to operate at full power for several years. In this case the tokamak should necessarily be designed with superconducting magnet coils and enough room should be allowed between the plasma and the coils for adequate shielding. The latter problem, can be described by a simple model. The scaling vs the tokamak geometry of the maximum tolerable magnetic field B_m at the magnet surface and the distance s needed to house the magnet shielding and the blanket can, practically, be neglected. For all the geometries we assumed B_m=12 T and s=1.3 m9. Together with these constraints, the following conditions (considered with "reasonable safety" in the range of the actual possibilities) were assumed: q_ψ=2.8 at the plasma edge and q_ψ=1 at the plasma axis, Z_{eff}=2, Z_i=7, ν=1, the fraction of alpha energy absorbed by the plasma f_α=1 and the energy confinement given by Goldston L-mode scaling. Each run of the code scans the (a,R) plane for a given value of the plasma current I. The results of the analysis are shown in Fig. 4.

6. CONCLUSIONS

From the analysis made on several of the most recent proposals (CIT, IGNITOR, IGNITEX, JET, NET, JET-Upgrade) it can be concluded that:

a) never an additional power greater than 15 MW seems required to reach ignition, whenever it is possible;

```
I=15 MA                ELONGATION X 2
A0=          0         1         2         3         4
             01234567890123456789012345678901234567890123456789
R0= 2.50      YY
R0= 3.00      YYYYYYY
R0= 3.50      YYYYYYY00000
R0= 4.00      YYYYYY00000000000
R0= 4.50      YYYYYY0000000000000000                       a)
R0= 5.00      YYYYYYY000000000000000000
R0= 5.50      YYYYYYY700000000000000000000
R0= 6.00      YYYYYYY77000000000000000000000
R0= 6.50      YYYYYYY87000000000000000000000000
R0= 7.00      YYYYYYY8760000000000000000000000000
R0= 7.50      YYYYYYY876600000000000000000000000XXXX
R0= 8.00      YYYYYYYY7766000000000000000000000XXXXXXX
R0= 8.50      YYYYYYYY776650000000000000000000XXXXXXXXXXX
R0= 9.00      YYYYYYYY876650000000000000000000XXXXXXXXXXXXX

I=30 MA                ELONGATION X 2
A0=          0         1         2         3         4
             01234567890123456789012345678901234567890123456789
R0= 2.50      YY
R0= 3.00      YYYYYYY
R0= 3.50      YYYYYYYYYYYY
R0= 4.00      YYYYYYYYYYY000000
R0= 4.50      YYYYYYYYYY000000000000                       b)
R0= 5.00      YYYYYYYYYY870000000000000
R0= 5.50      YYYYYYYYYY877600000000000000
R0= 6.00      YYYYYYYYYY87766000000000000000
R0= 6.50      YYYYYYYYYY8877665000000000000000
R0= 7.00      YYYYYYYYYY88776665000000000000000000
R0= 7.50      YYYYYYYYYYY777666555000000000000000000
R0= 8.00      YYYYYYYYYYY87766655500000000000000000000
R0= 8.50      YYYYYYYYYYY8877666555000000000000000000000
R0= 9.00      YYYYYYYYYYY88776665554000000000000000000000000

I=50 MA                ELONGATION X 2
A0=          0         1         2         3         4
             01234567890123456789012345678901234567890123456789
R0= 2.50      YY
R0= 3.00      YYYYYYY
R0= 3.50      YYYYYYYYYYYY
R0= 4.00      YYYYYYYYYYYYYYYYY
R0= 4.50      YYYYYYYYYYYYYYYY000000                       c)
R0= 5.00      YYYYYYYYYYYYY887700000000
R0= 5.50      YYYYYYYYYYYYY8877766000000000
R0= 6.00      YYYYYYYYYYYYY88777666000000000
R0= 6.50      YYYYYYYYYYYYY8Y87766665550000000
R0= 7.00      YYYYYYYYYYYYY8Y8777666655500000000
R0= 7.50      YYYYYYYYYYYYY8Y877766665555400000000
R0= 8.00      YYYYYYYYYYYYY8YY8776666555555440000000000
R0= 8.50      YYYYYYYYYYYYYY8Y877766665555544400000000000
R0= 9.00      YYYYYYYYYYYYYY8Y87776666555555444400000000000
```

FIGURE 4
Parametric study using Goldston L-mode at 15 MA (a), 30 MA (b) and 50 MA (c) plasma current. The value of elongation (times 2) approximated to the nearest integer is indicated at the intersection of the row of the major axis (R_O) with the column of the minor axis (A_O). The variation of these parameters is made by steps of 0.5 and 0.1 m respectively. The symbols X, Y or 0 are used respectively depending on wether the elongation calculated consistently with our constraints results >4, < 1 or that ignition doesn't occur.

b) in some cases where the ignition is not reachable, still stable burn situation which can provide high values of Q (say greater than 30), can be attained with the aid of auxiliary heating;

c) from Fig. 4 it appears that if elongation should be maintained below 2.2 (the NET reference value), plasma currents over 30 MA are required for tokamaks whose major radius is greater than 9 m, or greater than 50 MA if radius is maintained below 7.5 m. On the other hand, if current is to be kept below 15 MA and the major radius below 7.5 m, the elongation should be about 3.

ACKNOWLEDGMENTS

Particular thanks are due to R. Andreani, F. De Marco, M. Gasparotto and L. Pieroni for careful revision of the manuscript, useful discussions and for their continuous interest and encouragement.

REFERENCES

1. B. Coppi, M.I.T. Report PTP 84/4 and Plasma Phys., 11 (1985) 83.
2. S.P. Hirshmann, R.J. Hawryluk, B. Birge, Nucl. Fusion, 17 (1977) 611.
3. D.V. Barlett, et al., Nucl. Fusion, 28 (1988) 73.
4. L. Pieroni, private communication.
5. B.A. Trubnikov, Rev. Plasma Phys., 7, (1979) 345.
6. R.J. Goldston, Plasma Phys. and Contr. Nucl. Fusion, 26, (1984) 87.
7. S.M. Kaye, R.J. Goldston, Nucl. Fusion, 25, (1985) 65.
8. B. Coppi, M.I.T. Report PTP 87/4, March 1987.
9. NET Status Report December 1987 - Eur-FU/XII-80/88/84

AUTHOR INDEX